(Continued)

BIOLOGY

A GUIDE TO THE NATURAL WORLD

ACADEMIC ADVISORS

Gary Anderson
University of California, Davis

Tania Baker
Massachusetts Institute of Technology

David Berrigan
University of Washington

Leon W. Browder
University of Calgary

Anthony Ives
University of Wisconsin

Michelle Murphy
University of Notre Dame

Anu Singh-Cundy
Western Washington University

Erica Lynn Suchman
Colorado State University

Sara Via
University of Maryland

Nicholas Wade
The New York Times

John Whitmarsh
University of Illinois

BIOLOGY

A GUIDE TO THE NATURAL WORLD

Second Edition

David Krogh
Berkeley, California

Prentice Hall
Upper Saddle River, New Jersey 07458

Library of Congress Cataloging-in-Publication Data

Krogh, David.
  Biology: a guide to the natural world / David Krogh—[2nd ed.].
    p. cm
  ISBN 0-13-090726-X—ISBN 0-13-092178-5 (pbk.)
  1. Biology. I. Title.
  QH308.2 .K76 2002
  570—dc21

00-066580
CIP

Executive Editor: *Gary Carlson*
Editor-in-Chief, Life and Geosciences: *Sheri L. Snavely*
Project Manager: *Karen Horton*
Art Development Editor: *Kim Quillin*
Text Development Editor: *Annie Reid*
MediaLabs Development Editor: *Peggy Brickman*
Editorial Assistant: *Lisa Tarabokjia*
Vice President of Production and Manufacturing: *David W. Riccardi*
Executive Managing Editor: *Kathleen Schiaparelli*
Senior Production Editor: *Nicole M. Bush*
Production Support: *Susan Fisher, Ed Thomas*
Assistant Managing Editor, Science Media: *Elizabeth Wright, Nicole M. Bush*
Executive Marketing Manager, Biology and Geosciences: *Jennifer Welchans*
Marketing Manager: *Shari Meffert*
Marketing Assistant: *Anke Braun*
Manufacturing Manager: *Trudy Pisciotti*
Assistant Manufacturing Manager: *Michael Bell*
Director of Creative Services: *Paul Belfanti*

Manager of Electronic Composition and Digital Content: *Jim Sullivan*
Electronic Composition/Production Specialist: *Donna Marie Paukovits*
Managing Editor, Audio/Visual Assets: *Grace Hazeldine*
A/V Editor: *Adam Velthaus*
Art Support: *Julita Nazario, Shannon Sims*
Director of Design: *Carole Anson*
Art Director: *Jonathan Boylan*
Cover Designer: *Luke Daigle*
Interior Designers: *Lynn Stiles, Jonathan Boylan*
Media Developer: *Mike Guidry/Lightcone Interactive*
Media Editor: *Andrew Stull*
Media Production Editors: *Anthony Maffia, Nicole M. Bush*
Photo Research Administrator: *Melinda Reo*
Photo Researcher: *Diane Austin*
Copy Editor: *Chris Thillen*
Image Permission Coordinator: *Tony Arabia*
Art Studio/Illustrator: *Imagineering*

Cover Photo Credits: *Front, from left*: Butterfly Cethosia Biblis (J. Y. Grospas/Peter Arnold, Inc); Na'v flower (Middleton/Liittschwager); Great egret (SuperStock, Inc.). *Spine*: Zebras at Etosha National Park, Namibia (Wolfgang Kaehler Photography). *Rear, from left*: Pine forest in Landes, France (SuperStock, Inc.); Mushrooms in tree trunk, Amazon rain forest (Wolfgang Kaehler Photography); Wood fern Thelyteridaceae in Swamp Sanctuary, Florida (Fritz Polking/Peter Arnold, Inc.)

© 2002, 2000 by Prentice-Hall, Inc.
Upper Saddle River, New Jersey 07458

Printed in the United States of America

10 9 8 7 6 5 4 3

ISBN 0-13-090726-X (case)/0-13-092178-5 (paper)/
0-13-093685-5 (Instructor's Edition)

Pearson Education LTD., *London*
Pearson Education Australia PTY, Limited, *Sydney*
Pearson Education Singapore, Pte. Ltd
Pearson Education North Asia Ltd, *Hong Kong*
Pearson Education Canada, Ltd., *Toronto*
Pearson Educación de Mexico, S.A. de C.V.
Pearson Education—Japan, *Tokyo*
Pearson Education Malaysia, Pte. Ltd.

For my friends Jerry and Teresa
Far away, but always in my heart

Essays

Forty-four essays appear in the book, most of them having an applied slant. They deal with such topics as acid rain, fad diets, DNA fingerprinting, osteoporosis and young women, and the nature of human sexuality.

 In the How Did We Learn? essays, students can come into understand the inventiveness and the plain hard work that generally are prerequisites to scientific discovery.

MediaLabs

There are fifteen MediaLabs throughout the book. The topics were carefully chosen not only for student interest but also because they highlight issues that students may come across in their daily lives. Each MediaLab takes the reader on a journey of discovery through CD-ROM activities and web investigations.

Brief Contents

Contents

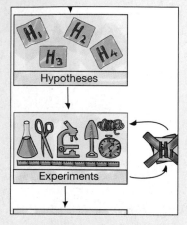

Scientific Method. **8**

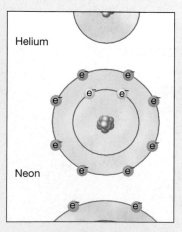

Helium

Neon

Electron configuration. **24**

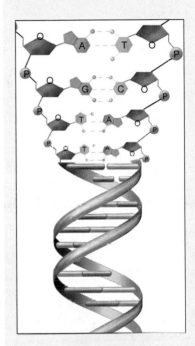

DNA's structure.   **61**

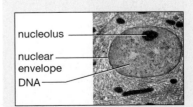

nucleolus
nuclear envelope
DNA

The cell's nucleus.   **76**

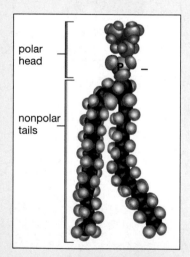

Dual-natured molecule.   **98**

Contents

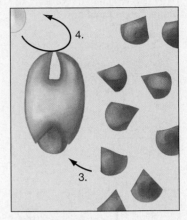
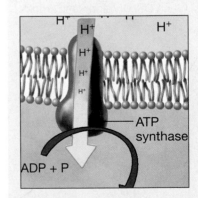

Contents

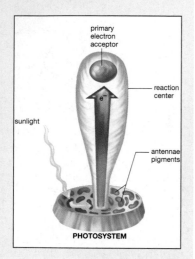

Photosystem. **156**

Unit 3 How Life Goes On: Genetics

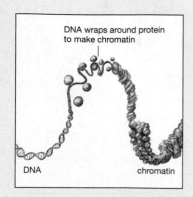

Chromatin. **177**

Contents

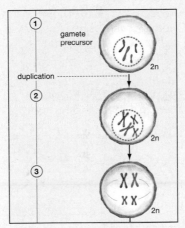

Steps in meiosis. **193**

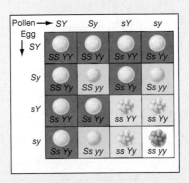

Punnett square. **217**

Essays
Testing for Genetic
Trouble **238**
How Did We Learn? Thomas
Hunt Morgan: Using Fruit Flies
to Look More Deeply into
Genetics **244**

MediaLab
Do We Know Too Much?
Human Genetic Testing **252**

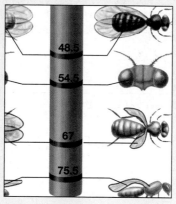

The wild and the mutant flies. **246**

Essay
How Did We Learn? Getting
Clear about What Genes Do:
Beadle and Tatum **264**

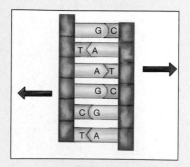

Normal DNA. **261**

Contents

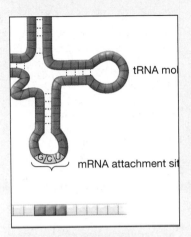

Transfer RNA structure. **277**

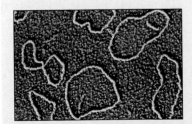

Plasmids. **295**

HIGHLIGHTS

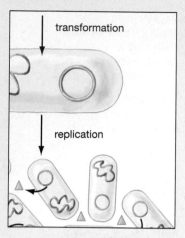

Genetic engineering. **296**

Unit 4   Life's Organizing Principle: Evolution and the Diversity of Life

Essay
Can Darwinian Theory Make Us
Healthier?   **331**

Charles Darwin.   **323**

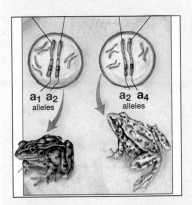

The genetic basis of evolution.   **341**

En route to speciation?   **360**

Essay
Physical Forces that Have Shaped Evolution: Climate, Extraterrestrial Objects, and Continental Drift **380**

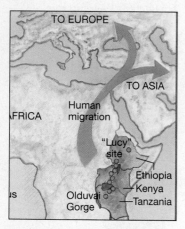

Africa: Cradle of the humans. **398**

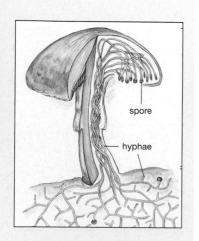

Structure of a fungus. **421**

Contents

Lichens growing on a boulder. **423**

Cheetah. **458**

Contents

Unit 5 A Bounty That Feeds Us All: Plants

Essays
What Is Plant Food? **477**
Keeping Cut Flowers
Fresh **483**
Ripening Fruit Is a Gas **485**

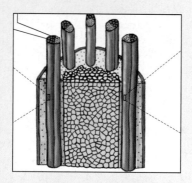

Fluid-transport structure. **483**

Essays
A Tree's History Can Be Seen in
Its Wood **507**
The Syrup for Your Pancakes
Comes from Xylem **510**

MediaLab
Why Do We Need Plants Any-
way? The Importance of Plant
Diversity **524**

Unit 6 What Makes the Organism Tick? Animal Anatomy and Physiology

HIGHLIGHTS

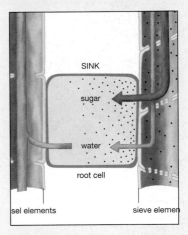

Food transport. **512**

Essay
Doing Something about
Osteoporosis While You Are
Young **540**

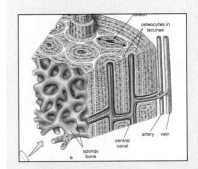

Bone features. **539**

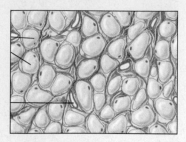

Adipose tissue. **531**

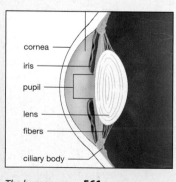

The human eye. **561**

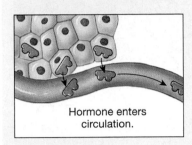

Endocrine cells releasing hormone.   **566**

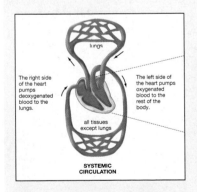

Circulation's big picture.   **595**

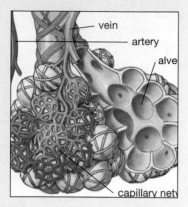
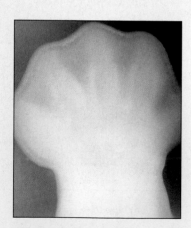

Shaping by removing. **628**

Contents

Unit 7 The Living World as a Whole: Ecology and Behavior

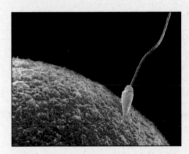

Near the moment of conception. **647**

Keystone species: Pisaster ochraceus. **674**

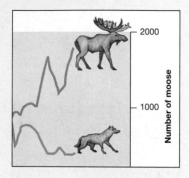

Predator and prey populations.   **679**

The High Plains Aquifer.   **699**

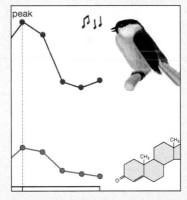

Preface

David Krogh has been writing about science for 20 years in newspapers, magazines, books, and for educational institutions. He is the author of *Smoking: The Artificial Passion*, an account of the pharmacological and cultural motivations behind the use of tobacco, which was nominated for the *Los Angeles Times* Book Prize in Science and Technology. David has written on physics and on technology issues, but his primary interest has been in biology. He has written on the possible effect methane may be having on global warming; on early research into the role that growth factors may play in neural regeneration following injury; on the synthesis of naturally occurring neurotoxins and their possible use in heart disease; on the use of imported drugs to treat cancer; and on the relationship between alcohol and mood states in women. He has a particular interest in the history of biology and in the relationship between biological research and modern American culture. He holds bachelor's degrees in both journalism and history from the University of Missouri. In another facet of his writing career, he is the director of communications for the Academic Senate of the University of California.

From the Author

Book titles may be the first thing any reader sees in a book, but they're often the last thing an author ponders. Not so with *Biology: A Guide to the Natural World*. The title arrived fairly early on, courtesy of the muse, and then stuck because it so aptly expresses what I think is special about this book.

Flip through these pages, and you'll see all the elements that students and teachers look for in any modern introductory textbook—rich, full-color art, an extensive study apparatus, and a full complement of digital learning tools. When you leaf slowly through the book and start to read a little of it, however, I think that something a little more subtle starts coming through. This second quality has to do with a sense of connection with students. The sensibility that I hope is apparent in *A Guide to the Natural World* is that there's a wonderful living world to be explored; that we who produced this book would like nothing better than to show this world to students; and that we want to take them on an instructive walk through this world, rather than a difficult march.

All the members of the team who produced both the first, and now the second edition of *A Guide to the Natural World* worked with this idea in mind. We felt that we were taking students on a journey through the living world and that, rather like tour guides, we needed to be mindful of where students were at any given point. Would they remember this term from earlier in the chapter? Had we created enough of a bridge between one subject and the next? The idea was never to leave students with the feeling that they were wandering alone through terrain that lacked signposts. Rather, we aimed to give them the sense that they had a companion—this book—that would guide them through the subject of biology. *A Guide to the Natural World*, then, really is intended as a kind of guide, with its audience being students who are taking biology but not majoring in it.

Biology is complex, however, and if students are to understand it at anything beyond the most superficial level, details are necessary. It won't do to make what one faculty member called "magical leaps" over the difficult parts of complex subjects. Our goal was to make the difficult comprehensible, not to make it disappear altogether. Thus, the reader will find in this book fairly detailed accounts of such subjects as cellular respiration, photosynthesis, immune-system function, and plant reproduction. It was in covering such topics that our concern for student comprehension was put to its greatest test. We like the way we handled these subjects and other key topics, however, and we hope readers will feel the same way.

What's New in the Second Edition?

Much has changed in the *Guide* from the first edition to the second. Here's a brief listing of the subject matter that is new in the second edition.

- Increased coverage of the diversity of the living world, including a new chapter on animal diversity

- A new chapter on animal behavior
- Increased coverage of human evolution
- Coverage of many of the new developments in biotechnology: stem-cell research, the possibility of human cloning and xenotransplantation, the results of the sequencing of the human genome, and the controversy surrounding genetically modified foods
- Expanded coverage of the issue of global warming
- Updated or new information on such issues as Mad Cow disease, acid rain, and fad diets

Some detail on these additions probably is in order. Anyone who writes a textbook has to carry out a balancing act between putting in too much and putting in too little. Following publication of the first edition, faculty convinced us that we had erred on the side of too little in connection with two topics: the diversity of life and animal behavior. Therefore, with this edition, readers will see expanded coverage of both topics. Where once we covered diversity in a single chapter, we now cover it in two, the second of which is devoted to animals. The diversity coverage has also been rearranged, so that faculty who want to review plants without going into the details of their anatomy and physiology can do so with the help of the book's first diversity chapter. Meanwhile, animal behavior got its own chapter in the second edition. Students seem to find this a fascinating subject, and their author did too, after diving into it. The *Guide's* diversity coverage begins with Chapter 20, while its animal behavior Chapter is 31.

Faculty and students also wanted more coverage of human evolution in the book, and to that end, we have substantially expanded our coverage of this subject. The long, last module of Chapter 19 is given over to it. I'm happy to say that we are as up-to-date as a textbook can be on this fast-moving field. Faculty who wanted to see coverage of the senses will find, in Chapter 25, a long section on vision as an example of our sensory capabilities.

Apart from expanding into new areas, the second edition of the *Guide* also needed to take account of new developments in biology.

There has been plenty to take account of. As one who has followed perhaps a score of research areas for several years now, I can attest that there is no grass growing under the feet of biologists. The sequencing of the human genome has brought with it a tidal wave of new findings—new fields of biology, even. (It would be interesting to pinpoint the first published use of such terms as bioinformatics.) As a result, this book's biotechnology coverage, in Chapter 15, has changed greatly. It wasn't just the sequencing of the human genome that necessitated this change, however. Reproductive cloning has raised the possibility of human cloning and xenotransplantation. Meanwhile, the fight over genetically modified foods has greatly intensified in the past couple of years. Readers will find expanded coverage of all these issues in Chapter 15. Another fast-emerging and controversial field in biology is that of stem-cell research. This topic seemed a natural fit with the book's Chapter 27, which covers development.

With each passing month since the first edition was published, biology seems to have figured ever more prominently in other societal issues as well. Accordingly, the second edition of the *Guide* has retained and updated its coverage of such subjects as DNA fingerprinting, cancer, and acid rain, while adding new essays on such subjects as Mad Cow disease (Chapter 20), fad diets (Chapter 3), and human sexuality (Chapter 31). Global warming has emerged in the past two years as perhaps the planet's single most worrisome environmental issue. Readers will find updated and expanded coverage of it in Chapter 30.

Coverage of the Process of Discovery
One of the priorities for the second edition was to continue to impart to students a sense of *how* research results are arrived at in biology. Most of the book's chapters weave information on the process of discovery into explanations of what has been discovered. See, for example, Chapter 13 on Watson, Crick, and the DNA molecule; or Chapter 31 on proximate and ultimate causes in animal behavior. The first edition of the book also had a series of stand-alone "How Did We

Preface

Learn?" essays, and these have been updated and expanded for the second edition. (See the box on animal navigation in Chapter 31.) We also noted that, while faculty and students like these essays, they didn't like them interrupting the flow of a chapter's main text. Thus, "How Did We Learn?" boxes now appear at the end of chapters, rather than in the middle of them.

Electronic Media and the Second Edition

One of the most exciting features of the second edition concerns not what the book covers, but enhancements in its coverage that have been made possible by electronic media. Students and faculty have come to expect sophisticated media components in textbooks, but with the second edition of *A Guide to the Natural World*, I think we will exceed their expectations.

The book's media offerings for students can be conceptualized as falling into two categories. First, there are the CD-ROM Tutorials—well-named because collectively they function as a kind of book-length tutor. Each of them leads students through a series of related biological concepts with the help of the specialized teaching tool of animation. If, upon reading Chapter 14 on genetic transcription and translation, a student isn't able to visualize how transfer RNA and messenger RNA work together at ribosomes, he or she can turn to the chapter's CD-ROM Tutorial and see this process laid out, step by step, with all the kinetics presented in animations. This story, of manufacturing proteins, is a CD-ROM "learning module" for Chapter 14—one of four contained in that chapter's CD-ROM Tutorial. Each module walks students through a key chapter concept; each contains an interactive activity or exercise; and each ends with its own summary and mini-quiz.

All the CD-ROM Tutorials were developed by Mike Guidry and his colleagues at Light-Cone Interactive. Mike's team produced a tutorial for every chapter in the book, each one identified in the text with an icon like the one shown at left.

Of course, students can turn to tutorial animations simply to make a given book illustration come to life; but they can also use the tutorials as just that—as learning sessions that employ interactive, step-by-step progressions. The proof here is in the pudding; take a look at some of the tutorials, and I think you'll agree they are a strong addition to the book.

Apart from the CD-ROM Tutorials, the *Guide* has, in its second edition, an expanded roster of the MediaLabs that proved so popular in the first edition. Produced by Peggy Brickman of the University of Georgia, these MediaLabs are aimed at making plain the linkage between biological concepts and real-world issues, and at fostering critical thinking about this linkage. A given lab starts by having students review, through a CD-ROM Tutorial, certain key concepts in a chapter. Then students are asked to investigate real-world issues connected to these concepts by going to suggested websites. (The cell cycle, covered in Chapter 9, may be intimately involved in the initiation of cancer, but what environmental factors are most important in getting cancer going? A *Scientific American* Web page tells the tale.) Having done this digging, students are then asked to communicate what they have learned by writing brief essays on questions that are put to them. The book now has 15 MediaLabs, each integrated with the content of a specific chapter. Each MediaLab begins within the book itself (at the end of selected chapters), but then broadens out to the CD-ROM and the wide world of the Internet.

Many more digital tools are available to students in this second edition of the *Guide*. The book's website http://www.prenhall.com/ krogh, developed by Prentice Hall's Andrew Stull, provides a host of resources. Students looking at any chapter at the website can click on a "Destinations" hyperlink and be presented with a rich roster of chapter-specific Internet links. Self-quizzes for each chapter also are posted on the website, with quiz questions divided into "basic" and "challenge" sets. (To make things easy for students, the CD-ROM Tutorials contain links to both the Companion Website and the MediaLabs.) Beyond this, there is a set of audio files that can

Tutorial 14.3.4
Protein Translation

be launched from the website. These are National Public Radio Biocast programs that have been integrated by their author, Bruce Hofkin, into each chapter in the book. Upon launching the *Biocasts* for Chapter 10, for example, a student can listen to a short program on a new technology that helps parents choose the gender of their child. This technology is connected to a basic concept covered in Chapter 10, sex determination in meiosis. Hofkin then brings the basic and applied science together in questions he poses at the end of the program.

All of these digital resources (and more) are available to students in the second edition of the *Guide*, but faculty have additional resources at their disposal. The Instructor's CD-ROM contains all of the key animations in the student CD-ROM; these are in turn part of a bank of images, known as the Media-Portfolio, containing every illustration and most of the photos in the book. The Media-Portfolio makes all the figures available in several formats, including PowerPoint slides that can be mixed and matched as desired, with figure parts, labels, and captions that can be edited. In addition, the *Instructor's Guide* and test-item file are embedded as a Word document in the CD-ROM, so that faculty can cut and paste what they need. Beyond these things, all the traditional media, such as transparencies, are available to faculty.

Notable Features in *A Guide to the Natural World*

Design and Illustrations

As in the first edition of the book, each chapter in the second edition is divided into numbered modules (1.1, 1.2, and so forth), so that instructors can easily assign selected parts of a given chapter. The chapter sections are listed at the start of each chapter, and end-of-chapter summaries are indexed by section. On the first page of each chapter is a visual "filmstrip" that offers an intriguing preview of what's to come.

Flip through the pages of the *Guide,* and you'll note another useful design element right away: Text almost always occupies the top left of a page, with illustrations at the bottom. As a result, text continued from one page to the next is almost never broken up by a photo or illustration. Students reading text will not have their concentration broken by graphics when they turn to new pages.

Regarding the book's illustrations, I think *A Guide to the Natural World* is first-rate for reasons of both process and personnel. The process was that illustrations were once again constructed chapter by chapter in a collaboration between myself and artist Kim Quillin. Kim and I now have to communicate through electronic files, whisking them from one coast to the other, whereas in the first edition we communicated at a Berkeley Starbuck's. (Kim moved back to her native coastal Maryland after finishing her Ph.D. in biomechanics at UC Berkeley.) But our method of working has remained the same: We revise chapters at an early stage, based on the illustrations that Kim comes up with, thus ensuring a tight integration between text and illustrations. Put another way, the figures in the book aren't just adjuncts to the text. Rather, figures and text have shaped each other in a back-and-forth process.

The *Guide to the Natural World* Team

Given all the names I've mentioned so far, it may go without saying that production of this book has been a team effort. It is my good fortune to have been given great teams for both editions of *A Guide to the Natural World*. So large is an effort such as this that there are many people I've never met who have put in long hours on the book. I've noted Kim Quillin and her role in the book's art program. Annie Reid served ably as the book's developmental editor—the person who looked over everything Kim and I came up with and said whether it worked, after which she put the revised product together in a package that could be made into a book. Chris Thillen copyedited the manuscript, patiently making sure that the English language was used correctly. Nicole Bush has been a fine production editor, bringing together pieces of art and blocks of text into the nicely laid out final product you see before you. Peggy Brickman not only developed the MediaLabs but also contributed greatly to the CD-ROM Tutorials produced by Mike Guidry and his coworkers. Thanks needs to go out in advance to Jennifer Welchans and Shari Meffert, who are just beginning to get the word out about the new edition of the *Guide*. Finally, we had great support at the top from Prentice Hall Editors Gary Carlson and Sheri Snavely, who managed the project on its largest scale.

Apart from these team members, more than two-hundred faculty have now carefully critiqued every word and image you see in *A Guide to the Natural World*. (Is any written work more carefully reviewed than a textbook? Peer-reviewed scientific papers are the only other contenders that come to mind.) The names of reviewing faculty can be found beginning on page xxxiv. Of these faculty, I need to make special note of the team of academic advisors who have provided advice not only on the details of the book, but on its overall structure and coverage. These advisors are listed across from the title page.

Finally, my thanks to all the faculty who used the first edition of *A Guide to the Natural World* in their courses and then let us know how it worked. Some of these faculty were reviewers, but some were instructors who sent in comments by e-mail or by old-fashioned letter just because they thought their feedback might be helpful. If they said the book needed some tweaking, we listened—the result being what you see in front of you. The main message from these faculty, however, was gratifying indeed. From them, we learned that we had done what we intended to do with the first edition of *A Guide to the Natural World*: We had created a book that their students could understand. Moreover, they said, we did this not by leaving out the hard parts, but by thinking carefully about how all the parts should be presented. Here's hoping that the second edition works as well.

David Krogh
Berkeley, California

The Book Team

Kim Quillin received her B.A. in biology at Oberlin College and her Ph.D. in integrative biology from the University of California, Berkeley. Her teaching experience ranges from elementary school science to undergraduate biology at both Oberlin College and UC, Berkeley. She has studied birds in the Smithsonian Museum of Natural History, howler monkey social behavior in Costa Rica, and restoration ecology of aquatic plants in Ohio. Kim has studied art for over two decades. Her formal art training ranges from the Maryland Summer Center for the Arts to college courses and professional workshops. This book is evidence of her dedicated efforts toward the effective visual communication of biological principles.

Marguerite (Peggy) Brickman received her B.A. from Columbia (College) University and her Ph.D. in genetics from the University of California, Berkeley. In teaching non-science majors, she focuses on making the material both relevant and entertaining. As an Assistant Professor in the Botany Department at the University of Georgia, she has been ranked number one by student evaluations in the Division of Biology for the past three semesters and is the recipient of an Excellence in Undergraduate Education teaching award. One of the reasons her classes are so popular is her ability to integrate media into her lectures. Peggy teaches her students to view electronic information critically, so that after they finish the class they can approach these same media with intelligence and savvy. As MediaLab editor for the book, Peggy has designed the labs to encourage students to think through the material logically and critically.

We express sincere gratitude to the expert reviewers who worked closely with the author in reviewing final pages to ensure the scientific accuracy of the text and art.

Anthony Ives, *University of Wisconsin, Madison*
Leslie Roldan, *Massachusetts Institute of Technology*
Anu Singh-Cundy, *Western Washington University*
Ellen Smith, *Arizona State University West*
Erica Suchman, *Colorado State University*
Christine Tachibana, *University of Washington*
Sara Via, *University of Maryland*
John Whitmarsh, *University of Illinois*

The end-of-chapter questions were carefully crafted by a team of dedicated instructors, and we wish to acknowledge their contribution.

Ed Bartholomew, *Maui Community College*
David Berrigan, *University of Washington*
Gail E. Gasparich, *Towson University*
Carol A. Hurney, *Virginia Commonwealth University*

Kate Lajtha, *University of Oregon*
Michelle Murphy, *University of Notre Dame*
Rhoda E. Perozzi, *Virginia Commonwealth University*
Heidi Rottschafer, *University of Notre Dame*
Anu Singh-Cundy, *Western Washington University*
Sara Via, *University of Maryland*

Media Reviewers

Robert S. Boyd, *Auburn University, University of Georgia*
Carolyn Glaubensklee, *University of Southern Colorado*
Gregory J. Podgorski, *Utah State University*
David A. Rintoul, *Kansas State University*
Ron Ruppert, *Cuesta College*
Brian Sailer, *Sam Houston State University*
Rebekah J. Thomas, *Saint Leo University*
Jennifer M. Warner, *University of North Carolina, Charlotte*
Jamie Welling, *South Suburban College*

(continued on the next page)

Second Edition Reviewers

John Alcock, *Arizona State University*
Sylvester Allred, *Northern Arizona University*
Gary Anderson, *University of California, Davis*
Marjay A. Anderson, *Howard University*
Jessica Baack, *Montgomery College*
Tania Baker, *Massachusetts Institute of Technology*
Peter Bednekoff, *Eastern Michigan University*
David Berrigan, *University of Washington*
Andrew Blaustein, *Oregon State University*
Robert S. Boyd, *Auburn University*
Leon W. Browder, *University of Calgary*
Warren Burggren, *University of North Texas*
David Byres, *Florida Central Community College, South Campus*
Van D. Christman, *Ricks College*
Deborah C. Clark, *Middle Tennessee State University*
Patricia Cox, *University of Tennessee, Knoxville*
Garry Davies, *University of Alaska, Anchorage*
Paula Dedmon, *Gaston College*
Miriam del Campo, *Miami Dade Community College*
Llewellyn Densmore, *Texas Technical University*
Jean Dickey, *Clemson University*
Christopher Dobson, *Front Range Community College*
Deborah Dodson, *Vincennes University*
Richard H. Falk, *University of California, Davis*
Christine M. Foreman, *University of Toledo*
Carl S. Frankel, *Pennsylvania State University*
Lawrence Friedman, *University of Missouri, St. Louis*
Matt Geisler, *University of California, Riverside*
Carolyn Glaubensklee, *University of Southern Colorado*
Judith Goodenough, *University of Massachusetts, Amherst*
G. A. Griffith, *South Suburban College*
Edward Hale, *Ball State University*
Kelly Hamilton, *Shoreline Community College*
Steve Heard, *University of Iowa*
Eva Horne, *Kansas State University*
Michael Hudecki, *State University of New York, Buffalo*
Michael Hudspeth, *Northern Illinois University*
Catherine J. Hurlbut, *Florida Community College*

Anthony Ives, *University of Wisconsin, Green Bay*
Kevin M. Kelly, *California State University, Long Beach*
Jeanette J. Kiem, *Guilford Technical Community College*
Jocelyn Krebs, *University of Alaska, Anchorage*
Kate Lajtha, *Oregon State University*
Paul Lurquin, *Washington State University*
James Manser, *Harvey Mudd College*
Paul Mason, *Butte Community College*
Lee H. Mitchell, *Mount Hood Community College*
Janice Moore, *Colorado State University*
Jorge A. Moreno, *University of Colorado*
Michael D. Morgan, *University of Wisconsin, Green Bay*
David Mork, *Saint Cloud State University*
Michelle Murphy, *University of Notre Dame*
Courtney Murren, *University of Tennessee, Knoxville*
Harry Nickla, *Creighton University*
Maya Patel, *Ithaca College*
Carolyn Peters, *Spoon River College*
Holly C. Pinkart, *Central Washington University*
Barbara Pleasants, *Iowa State University*
Gregory J. Podgorski, *Utah State University*
Lynn Polasek, *Los Angeles Valley College*
F. Harvey Pough, *Arizona State University West*
Regina Rector, *William Rainey Harper College*
Dennis Richardson, *Quinnipiac University*
David A. Rintoul, *Kansas State University*
Laurel Roberts, *University of Pittsburgh*
Rodney A. Rogers, *Drake University*
Leslie Ann Roldan, *Massachusetts Institute of Technology*
Ron Ruppert, *Cuesta College*
Julie Schroer, *Bismarck State College*
Anu Singh-Cundy, *Western Washington University*
Peter Slater, *University of St. Andrews, UK*
Nancy G. Solomon, *Miami University*
Allan R. Stevens, *Snow College*
Erica Lynn Suchman, *Colorado State University*
Christine Tachibana, *University of Washington*
Rebekah J. Thomas, *Saint Leo University*
Todd T Tracy., *Colorado State University*
Joseph W. Vanable, Jr., *Purdue University*
Sara Via, *University of Maryland*

Tanya Vickers, *University of Utah*
Janet Vigna, *Southwest State University*
Allan Hayes Vogel, *Chemeketa Community College*
Nicholas Wade, *The New York Times*
Jyoti R. Wagle, *Houston Community College*
Timothy S. Wakefield , *John Brown University*
Charles Walcott, *Cornell University*
Gene Walton, *Tallahassee Community College*
Jennifer M. Warner, *University of North Carolina*
Jamie Welling, *South Suburban College*
John Whitmarsh, *University of Illinois*
Susan Whittemore, *Keene State University*
Mark A. Woelfe, *Vanderbilt University*
Lorne Wolfe, *Georgia Southern University*

First Edition Reviewers

Dawn Adams, *Baylor University*
David L. Alles, *Western Washington University*
Gary Anderson, *University of California, Davis*
Michael F. Antolin, *Colorado State University*
Kerri Armstrong, *Community College of Philadelphia*
Mary Ashley, *University of Illinois, Chicago*
Kemuel Badger, *Ball State University*
Michael C. Bell, *Richland College*
William J. Bell, *University of Kansas*
David Berrigan, *University of Washington*
Lois A. Bichler, *Stephens College*
A. W. Blackler, *Cornell University*
Robert S. Boyd, *Auburn University*
Bonnie L. Brenner, *Wilbur Wright College (City College of Chicago)*
Mimi Bres, *Prince George's Community College*
Peggy Brickman, *University of Georgia*
Leon Browder, *University of Calgary*
Arthur L. Buikema, *Virginia Polytechnic Institute and State University (Allegheny College)*
Steven K. Burian, *Southern Connecticut State University*
Janis K. Bush, *University of Texas, San Antonio*
Linda Butler, *University of Texas, Austin*
W. Barkley Butler, *Indiana University of Pennsylvania*
William S. Cohen, *University of Kentucky*

Tricia Cooley, *Laredo Community College*

Karen A. Conzelman, *Glendale Community College*

Patricia B. Cox, *University of Tennessee*

John Crane, *Washington State University*

Brent DeMars, *Lakeland Community College*

Jean DeSaix, *University of North Carolina, Chapel Hill*

Matthew M. Douglas, *Grand Rapids Community College (University of Kansas)*

Lee C. Drickamer, *Southern Illinois University*

Charles Duggins, Jr., *University of South Carolina*

Susan A. Dunford, *University of Cincinnati*

Ron Edwards, *University of Florida*

Douglas J. Eernisse, *California State University, Fullerton*

Jamin Eisenbach, *Eastern Michigan University*

George Ellmore, *Tufts University*

Patrick E. Elvander, *University of California, Santa Cruz*

Michael Emsley, *George Mason University*

David W. Essar, *Winona State University*

Michael Farabee, *Estrella Mountain Community College*

Rita Farrar, *Louisiana State University*

John Philip Fawley, *Westminster College*

Eugene J. Fenster, *Longview Community College*

John L. Frola, *University of Akron*

Larry Fulton, *American River College*

Gail E. Gasparich, *Towson University*

Claudette Giscombe, *University of Southern Indiana*

Jack M. Goldberg, *University of California, Davis*

Glenn A. Gorelick, *Citrus College*

Melvin H. Green, *University of California, San Diego*

Gail Hall, *Trinity College*

Linnea S. Hall, *California State University, Sacramento*

Madeline Hall, *Cleveland State University*

Steven C. Harris, *Clarion University*

Walter Hewitson, *Bridgewater State College*

Jane Aloi Horlings, *Saddleback College*

Eva Horne, *Kansas State University*

Terry L. Hufford, *The George Washington University*

Carol A. Hurney, *James Madison University*

Andrea Huvard, *California Lutheran University*

Martin Ikkanda, *Los Angeles Pierce College*

Rose M. Isgrigg, *Ohio University*

Anthony Ives, *University of Wisconsin, Madison*

Tom Jurik, *Iowa State University*

Anne Keddy-Hector, *Austin Community College*

Kathleen Keeler, *University of Nebraska*

Nancy Keene, *Pellissippi State Technical Community College*

Kevin M. Kelley, *California State University, Long Beach*

Tom Knoedler, *Ohio State University, Lima Campus*

Don E. Krane, *Wright State University*

Erika Ann Lawson, *Columbia College*

Mike Lawson, *Missouri Southern State College*

Ann Lumsden, *Florida State University*

Michael M. Martin, *University of Michigan, Ann Arbor*

Michel Masson, *Santa Barbara City College*

Mary Colleen McNamara, *Albuquerque T-VI A Community College*

Scott M. Moody, *Ohio University*

Joseph Moore, *California State University, Northridge*

Deborah A. Morris, *Portland State University*

Allison Morrison-Shetlar, *Georgia Southern University*

Richard Mortensen, *Albion College*

Michelle Murphy, *University of Notre Dame*

Royden Nakamura, *California Polytechnic State University*

Jane Noble-Harvey, *University of Delaware*

Marcy P. Osgood, *University of Michigan*

Andrea Ostrofsky, *University of Maine*

Patricia A. Peroni, *Davidson College*

Rhoda E. Perozzi, *Virginia Commonwealth University*

John S. Peters, *College of Charleston*

Kim M. Peterson, *University of Alaska, Anchorage*

Raleigh K. Pettegrew, *Denison University*

Gary W. Pettibone, *State University of New York, College at Buffalo*

Barbara Pleasants, *Iowa State University*

John M. Pleasants, *Iowa State University*

Don Pribor, *University of Toledo*

Louis Primavera, *Hawaii Pacific University*

Paul Ramp, *Pellissippi State and Technical Community College*

Sonia J. Ringstrom, *Loyola University*

Leslie Ann Roldan, *Massachusetts Institute of Technology*

Heidi Rottschafer, *University of Notre Dame*

John Rueter, *Portland State University*

Nancy Sanders, *Northeast Missouri State University*

Gary Sarinsky, *City University of New York, Kingsborough Community College*

Edna Seaman, *University of Massachusetts, Boston*

Ralph W. Seelke, *University of Wisconsin, Superior*

Prem P. Sehgal, *East Carolina University*

C. Thomas Settlemire, *Bowdoin College*

Robert Shetlar, *Georgia Southern University*

Mark A. Shotwell, *University of Slippery Rock*

Linda Simpson, *University of North Carolina, Charlotte*

Anu Singh-Cundy, *Western Washington University*

Ellen Smith, *Arizona State University West*

Philip J. Snider, *University of Houston*

Frederick W. Spiegel, *University of Arkansas*

Kathleen M. Steinert, *Bellevue Community College*

Donald P. Streubel, *Idaho State University*

Erica Suchman, *Colorado State University*

Gerald Summers, *University of Missouri, Columbia*

Joanne Tornow, *University of Southern Mississippi*

Robin W. Tyser, *University of Wisconsin, LaCrosse*

Sara Via, *University of Maryland*

Dennis Vrba, *North Iowa Area Community College*

Nicholas Wade, *The New York Times*

John H. Wahlert, *Baruch College, The City University of New York*

Timothy S. Wakefield, *Auburn University*

Sarah Ward, *Colorado State University*

R. Barry Welch, *San Antonio College*

John Whitmarsh, *University of Illinois at Urbana-Champaign*

Sandra Winicur, *Indiana University, South Bend*

William Wischusen, *Louisiana State University*

Deborah Wisti-Peterson, *University of Washington*

Rachel Witcher, *University of Central Florida*

Wade B. Worthen, *Furman University*

Robert Yost, *Indiana University Purdue University Indianapolis*

1

Science as a Way of Learning
A Guide to the Natural World

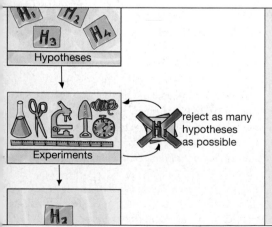

How scientists think about the world—
the scientific method.
(Section 1.3, page 8)

Pasteur's Experiments:

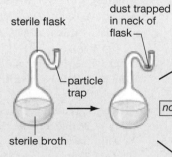
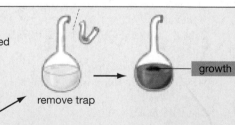

A famous experiment by a famous scientist.
(Section 1.3, page 9)

Science has great impact on our lives now and stands to have greater impact on them in the future. Science is both a body of knowledge and a means of acquiring knowledge. Biology, a branch of science, is the study of life.

About 20 years ago, a friend of mine showed me a magazine article which asserted that science and technology were becoming such important parts of our society that people would either have to jump on an approaching technological train or get run over by it. At the time, it just didn't look that way to me. Why, I wondered, would members of my generation have to know any more about science than members of my parents' generation did?

Looking back on this, it's easy to see that the magazine article had it about right. Indeed, the beginning of the 1980s can be seen as about the time when science and technology began knocking on the average person's door with great frequency. The fundamental breakthrough that brought about modern electronics, including the computer, was the invention of the transistor at Bell Labs in 1947. Average American workers then *heard* about computers for 30 years, but by the mid 1980s, they were *using* computers right on their desktops and having other electronic innovations, such as VCRs and CD players, thrust at them from every angle (**see Figure 1.1**). The fundamental breakthrough that brought about the biotechnology industry was the description of the DNA molecule in 1953 by James Watson

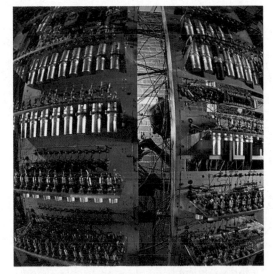
a

b

Figure 1.1
Then and Now

a A technician enters data into the world's first programmable computer, run initially in 1948. Called "Baby," the computer was more than 2 meters (6 feet) tall and almost 5 meters (15.5 feet) wide, but had a total memory of only 128 bytes—less than one-ten-thousandth the amount of data that can be put on a common floppy disk today.

b A businesswoman using a laptop computer participates in a video conference while viewing colleagues in a window on her computer screen. A small video camera sits on top of her computer.

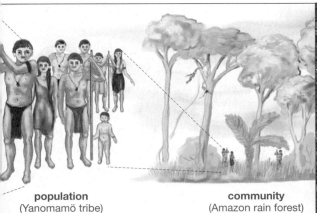

population
(Yanomamö tribe)

community
(Amazon rain forest)

From atom to rain forest, the hierarchy of life.
(Section 1.4, page 12)

Nature's finery—it's evolutionary.
(Section 1.5, page 14)

You don't know what this insect is, but it scares you. Why?
(Section 1.5, page 15)

and Francis Crick. People heard more about genes in subsequent years, but only in 1978 did we get the first genetically engineered medicine (human insulin).

What once was a trickle in both these fields became a stream in the 1980s and a torrent in the 1990s. Indeed, in the 1990s it would have been quite a challenge to *avoid* the technological changes going on all around. (You might decline a computer at home, but try evading one at work.) Strictly biological innovations didn't have this kind of presence for the average person, but it was not hard to see signs of a coming wave. Consider in vitro fertilization ("test-tube babies"), DNA testing for paternity, the wide popularity of new antidepressant drugs such as Prozac, the controversy over genetically modified foods, and genetic testing for such diseases as cystic fibrosis. Then, there was the cloning of Dolly the sheep from a single cell of an adult sheep. Since Dolly, scientists have gone on to clone cows, mice, monkeys, and goats and are working hard to see if cloned pigs might be used to produce human "spare parts," such as hearts and livers. (**see Figure 1.2**). Beyond these things, scarcely a week goes by without an environmental issue being in the news, with scientists being turned to for guidance on how to deal with it.

Figure 1.2
A Step Beyond Dolly
Dolly the sheep was merely the first step in animal cloning. In March 2000, the same British pharmaceutical company that cloned Dolly announced that it had cloned these five healthy piglets from a single cell of an adult sow. Pig-cloning research has a practical goal: the growing of organs, in pigs, that can be transplanted to human beings. Tens of thousands of Americans need organ transplants but cannot get them because of a shortage of human organs. Pig organs are about the same size as human organs and carry out the same functions. Pig cloning is aimed at producing organs that, once transplanted, will not be attacked by the human immune system.

If we ask, then, whether science and technology are becoming ever more important in our society, the answer is plain. To fully participate in the workforce, to make everyday decisions, and to make informed choices at the ballot box, the average person must now be more technologically and scientifically literate than at any time in the past, and this trend only stands to accelerate in coming years.

Given this, it seems worthwhile to take a look at the relationship between science and society. What practical effect does science have on us? What do average citizens think of science and scientists? These questions are examined at the beginning of this chapter. This will be followed by a review of the nature of science and of biology.

1.1 How Does Science Impact the Everyday World?

How is science in general—and biology in particular—likely to be relevant to a person's life in contemporary society?

A Look at the News

To get an idea, let's look at some of the biology-related news that came to Americans through one magazine (*Time*) during one randomly selected period (six months in 2000). More than 60 stories connected to biology appeared in *Time* during this period (**see Figure 1.3**). Six are noted here.

Vitamin Overdose, said the headline in an April 24 *Time* story that reported on the increasingly popular American practice of taking "megadoses" of vitamin C and vitamin E in the hope of living longer, healthier lives. A national Institute of Medicine study threw cold water on the idea, however, saying there is insufficient evidence to conclude that large doses of vitamins can protect people from chronic diseases. Worse, the study said, *really* large doses of these supplements may actually cause health problems. People take large doses of these vitamins, the *Time* story noted, because some studies have raised the possibility that they may serve as "antioxidants," soaking up the "free radicals" that seem to play roles in aging and in such afflictions as heart disease and cancer. But what is a free radical? For that matter, what is an oxidant? *Time* didn't have the space to let its readers know, but readers of this *textbook* who are interested can find out about free radicals in the box "Free Radicals" on **page 29**, while readers who want to know more about oxidation can turn to Chapter 7, on **page 130**.

The Science of Dissent was one article among several in a special "Life on the Mississippi" section *Time* ran in its July 10 issue. The article reported on a public high school teacher in Faribault, Minnesota, who has gone to court to be allowed to teach what his critics say is creationism—the idea that the living world is so diverse and complex it could not have been shaped by the forces of nature alone. The teacher claims, meanwhile, that all he wants to do is teach *evolution* with "an honest look at the difficulties and inconsistencies in the theory." Debates over the teaching of evolution and creationism have raged for decades in America and show no signs of tapering off. In Chapter 16, beginning on **page 320**, you can find an account of why the vast majority of scientists are convinced that all of Earth's life-forms did indeed develop through the process of evolution.

The Big Meltdown was the title *Time* gave to a September 4 story on the effect global warming is having in Earth's far north region, the Arctic. Sea ice in the Arctic is now 40 percent thinner and covers 6 percent less area than in 1980. *Time*'s story provided vivid evidence of the down-to-Earth effects of global warming. In Alaska, for example, much of the "permafrost" is no longer permanently frozen. The result is "power lines tilted at crazy angles and houses sinking up to their window sashes as the ground liquifies." Scientific evidence clearly indicates that Earth is warming, and a scientific consensus appears to be emerging that a human activity—the burning of fossil fuels—is at least partly to blame. But how could human activity cause the Earth to warm? You can find an explanation in "The Worrisome Issue of Global Warming" on **page 708** of Chapter 30.

Battle Pending, *Time* said in an April 17 article that noted what a big business human genes have become. Old-line pharmaceutical companies and newer biotech start-ups are in a feverish race to file patents on newly discovered genes, hoping to cash in should knowledge about the genes lead to products that can be used in medicine, agriculture, manufacturing,

or even home lawn care. (Imagine grass that almost never has to be cut because it stays short thanks to genetic programming.) The problem, as the *Time* article pointed out, is that firms are filing patents on genes that merely look like they might be important, while having little idea of what this importance may be. Readers who want to know what a gene is can read Chapter 9, beginning on **page 170**.

Little Hope, Less Help, said a July 24 *Time* article on AIDS in sub-Saharan Africa, which is expected to suffer 23 million AIDS deaths by 2005. By the end of the decade, life expectancy in Botswana, Zimbabwe, and South Africa will plummet to 30; without AIDS, it would have been about 70. AIDS is devastating because it attacks the human body's own defenses—its immune system. How does the immune system work and how can AIDS destroy it? For an account, see "The Immune System: Defending the Body from Invaders" on **page 572**.

Grains of Hope, the cover story in *Time*'s July 31 issue, concerned a new chapter in the debate over genetically modified foods. Until recently, such foods have primarily benefited food *producers*, such as farmers and seed suppliers, by making crops more resistant to pests or to herbicides (the chemicals used to control weeds). Now, however, European researchers have developed a genetically modified rice that contains its own beta carotene, meaning that this rice, unlike the natural variety, can be

Figure 1.3
Science in the News
The importance of science to everyday life is reflected in the large number of news stories that focus on science. The text contains a summary of six stories concerning biology that appeared in *Time* magazine during a six-month period in 2000. Pictured is a cover story *Time* ran in July on the sequencing of the human genome.

a source of vitamin A. This is no small matter, as at least a million children die each year because they are weakened by vitamin A deficiency. Yet genetically modified foods have numerous critics who charge that they are "Frankenfoods," whose production stands to harm the environment and possibly human beings as well. The newly developed variety of rice certainly is genetically modified, as it contains genes from two other living things (daffodils and a species of bacteria). But how can genes from one species be spliced into another? More generally, what is the promise—and the peril—of biotechnology? To find out, see Chapter 15, beginning on **page 292**.

1.2 What Does the Public Think, and Know, about Science?

Science issues may be thrust upon the public today, but what do Americans think about science, and how "scientifically literate" are they? A wealth of information on these topics was contained in a report titled *Science and Engineering Indicators—1998*. This document was one in a series of reports on the state of science and engineering in America produced by the National Science Foundation (NSF). In a section of the 1998 report that dealt with the American public and its relation to science, two things stood out. One is that Americans are positive about science in many ways; the other is that their *knowledge* of science is uneven—surprisingly high in some areas and surprisingly low in others.

Public Attitudes toward Science

On American attitudes toward science, surveys used in the NSF report revealed that Americans were more interested in science in the late 1990s than at any time in the previous 20 years. They have more confidence in American scientific leadership than they do in the leadership of Congress, corporations, the press, organized religion, even the U.S. Supreme Court. (Only leaders in medicine ranked higher in the public's esteem.) And they believe science has brought much to them already, and that it stands to bring a lot more. More than 85 percent of all Americans believe that the world is a better place because of science.

Public Knowledge of Science

Given Americans' attitudes toward science, it's somewhat surprising to learn what they *know* about science. If you look at **Figure 1.4**, you can see the proportion of a representative sample of adult Americans who could correctly answer some basic scientific questions. Better than 75 percent understand that light travels faster than sound, but only about 20 percent understand what DNA is. Almost 80 percent of Americans know that the continents are moving about the face of the Earth, but less than half know that it takes a year for the Earth to go around the Sun. (The alternatives were that it made the journey in a day or a month.) Meanwhile, another question revealed that more than a quarter of all Americans think the Sun goes around the Earth. With respect to how science works, 27 percent of those questioned were classified as having at least a minimal understanding of scientific inquiry, which is to say they understood that science involves formulating hypotheses and testing them with experiments, a topic you'll learn more about later in this chapter.

Figure 1.4
What Do Americans Know about Science?
Some results published by the National Science Foundation.

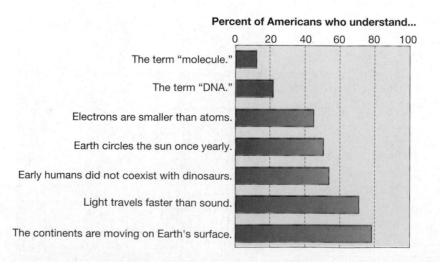

Percent of Americans who understand...

The term "molecule."

The term "DNA."

Electrons are smaller than atoms.

Earth circles the sun once yearly.

Early humans did not coexist with dinosaurs.

Light travels faster than sound.

The continents are moving on Earth's surface.

Science Education Makes for Informed Citizens

Connected to the issue of scientific literacy is the question of science education. According to the NSF study, there is a high correlation between how much American adults know about general topics in science and how many science courses they have taken in school. In short, science education is retained in a significant way in that it makes for better-informed citizens.

1.3   What Is Science?

Having looked a little at the ways science affects our everyday world, let's now review something about science in general and about biology as one of its disciplines. The point here is to give you some sense of the underpinnings of science and biology—to review something about the how and why of them before getting to what they have revealed. This discussion starts with the big picture and progressively gets more specific.

Science as a Body of Knowledge

Science is in one sense a process—a *way* of learning. In this respect, it is an activity carried out under certain loosely agreed-to rules, which you'll get to shortly. **Science** is also a body of knowledge, however. It is a collection of unified insights about nature, the evidence for which is an array of facts. The unified insights of science are commonly referred to as *theories*.

It's unfortunate but true that *theory* means one thing in everyday speech and something almost completely different in scientific communication. In everyday speech, a theory can be little more than a hunch. It is an unproven idea that may or may not have any evidence to support it. In science, meanwhile, a **theory** is a general set of principles, supported by evidence, that explains some aspect of nature. There is, for example, a Big Bang theory of the universe. It is a general set of principles that explains how our universe came to be and how it developed. Among its principles are that a cataclysmic explosion occurred 10–15 billion years ago; and that, after it, matter first developed in the form of gases that then coalesced into the stars we can see all around us. There are numerous facts supporting these principles, such as the current size of the universe and its average temperature.

As you might imagine, with any theory this grand some *pieces* of it are in dispute; some facts don't fit with the theory, and scientists disagree about how to *interpret* this piece of information or that. On the whole, though, these general insights have withstood the questioning of critics, and together they stand as a scientific theory.

The Importance of Theories

Far from being a hunch, a scientific theory actually is a much more valued entity than is a scientific fact, for the theory has an *explanatory* power, while a fact is generally an isolated piece of information. That the universe is at least 10 billion years old is a wonderfully interesting fact, but it explains very little in comparison with the Big Bang theory. Facts are important; theories could not be supported or refuted without them. But science is first and foremost in the theory-building business, not the fact-finding business.

Science as a Process: Arriving at Scientific Insights

So how does a body of facts and theories come about? What is the process of scientific investigation, in other words? When **science** is viewed as a process, it could be defined as a means of coming to understand the natural world through the testing of hypotheses. This process generally is referred to as the **scientific method**. The starting state for scientific inquiry is always *observation:* A piece of the natural world is observed to work in a certain way. Then follows the *question*, which broadly speaking is one of three types: a "what" question, a "why" question, or a "how" question. Biologists have asked, for example, What are genes made of? Why does the number of species decrease as we move from the equator to the poles? How does the brain make sense of visual images?

Formulating Hypotheses, Performing Experiments

Following the formulation of the question, various hypotheses are proposed that might answer it. A **hypothesis** is a tentative, testable explanation for an observed phenomenon. In almost any scientific question, several hypotheses are proposed to account for the same observation. Which one is correct? Most frequently in science, the answer is provided by a series of *experiments*, meaning controlled tests of the question at hand (**see Figure 1.5**). It may go without saying that scientists don't regard all hypotheses as being equally worthy of undergoing experimental test. By the time scientists arrive at the experimental stage, they usually have an idea of which is the most promising hypothesis among the contenders, and they then proceed to put that hypothesis to the test. Let's see how this worked in an example from history.

The Test of Experiment: Pasteur and Spontaneous Generation

Does life regularly arise from anything *but* life, or can it be created "spontaneously," through the coming together of basic chemicals? The latter idea had a wide acceptance from the time of the ancient Romans forward, and as late as the nineteenth century it was championed by a number of the leading *scientists* of the day. So how could the issue be decided? The famous French chemist and medical researcher Louis Pasteur formulated a hypothesis to address this question (**see Figure 1.6**). He believed that many purported examples of life arising spontaneously were simply instances of airborne microscopic

organisms landing on a suitable substance and then multiplying in such profusion that they could be seen. Life came from life, in other words, not from spontaneous generation. But how could this be demonstrated? In 1860, Pasteur sterilized a meat broth in glass flasks by heating it, while at the same time heating the glass *necks* of the flasks, after which he bent the necks into a "swan" or S-shape. The ends of the flasks remained open to the air, but inside the flasks there was not a sign of life. Why? The broth remained sterile because microbe-bearing dust particles got trapped in the bend of the flask's neck. If Pasteur broke the neck off before the bend, however, the flask soon had a riot of bacterial life growing within it. In another test, Pasteur tilted the flask so that the broth *touched* the bend in the neck, a change that likewise got the microbes growing.

Elements in Pasteur's Experiments

Now, note what was at work here. Pasteur had a preconceived notion of what the truth was, and designed experiments to test his hypothesis. Critically, he performed the same set of steps several times in the experiments, keeping all the elements the same each time—except for one. The nutrient broth was the same in each test; it was heated the same amount of time and in the same kind of flask. What *changed* each time was one critical **variable**, meaning an adjustable condition in an experiment. In this case, the variable was either the shape of the flask neck, or the tilt of the flask. Given the fact that all other elements of the experiments were kept the same, the experiments had rigorous controls: All conditions were held constant over several trials, except for a single variable. A **control** can be defined as a comparative condition in an experiment. Pasteur's finding that no life grew in the bent-necked flask is interesting, but tells us very little by itself. We only learn something by comparing this finding to the result in the control condition: that life did grow when the flask neck was straight.

Note also that the idea of spontaneous generation was not banished with this one set of experiments—nor should it have been. Pasteur's experiments provided one of the *facts* mentioned earlier, in this case the fact that flasks of liquid will remain sterile under certain conditions. The idea that life arises only from life is, however, one of the scientific *theories* noted earlier, meaning that it requires the accumulation of many facts pointing in the same direction.

Figure 1.5
Scientific Method
The scientific method enables us to answer questions by testing hypotheses.

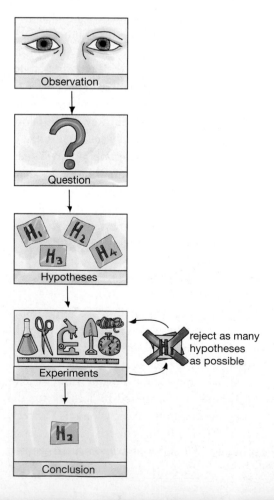

Observation

Question

Hypotheses

Experiments

reject as many hypotheses as possible

H_2

Conclusion

Other Kinds of Support for Hypotheses

Some scientific questions are difficult or impossible to test purely through experiment. For example, there currently is a controversy over whether birds are the direct descendants of dinosaurs. What kind of experiment could be run to test this hypothesis? Certain modern-day evidence is available to us—the DNA of living birds, for example—but examining DNA does not amount to an experiment. Instead it is observation, which is another valid way to test a hypothesis. Evidence from the past can also be observed, of course, which in this case means the observation of dinosaur and bird fossils. Indeed, fossils have been the key evidence in convincing most experts that birds are the descendants of dinosaurs.

Joining experiment and observation, statistics is a tool used frequently in science, as you can see in "Lung-cancer, Smoking, and Statistics in Science" on page 10.

From Hypothesis to Theory

When does an idea move from hypothesis to theory? One of the ironies of the orderly undertaking called science is that there's nothing orderly about the change from hypothesis to theory. No scientific supreme court exists to make a decision. Scientists aren't polled for their views on such questions, and even if they were, at what point would we say something had been "proven"? When more than 50 percent of the experts in the field assent to it? When there are no dissenters left?

Tutorial 1.1.1
The Scientific Method (Pasteur's Experiments)

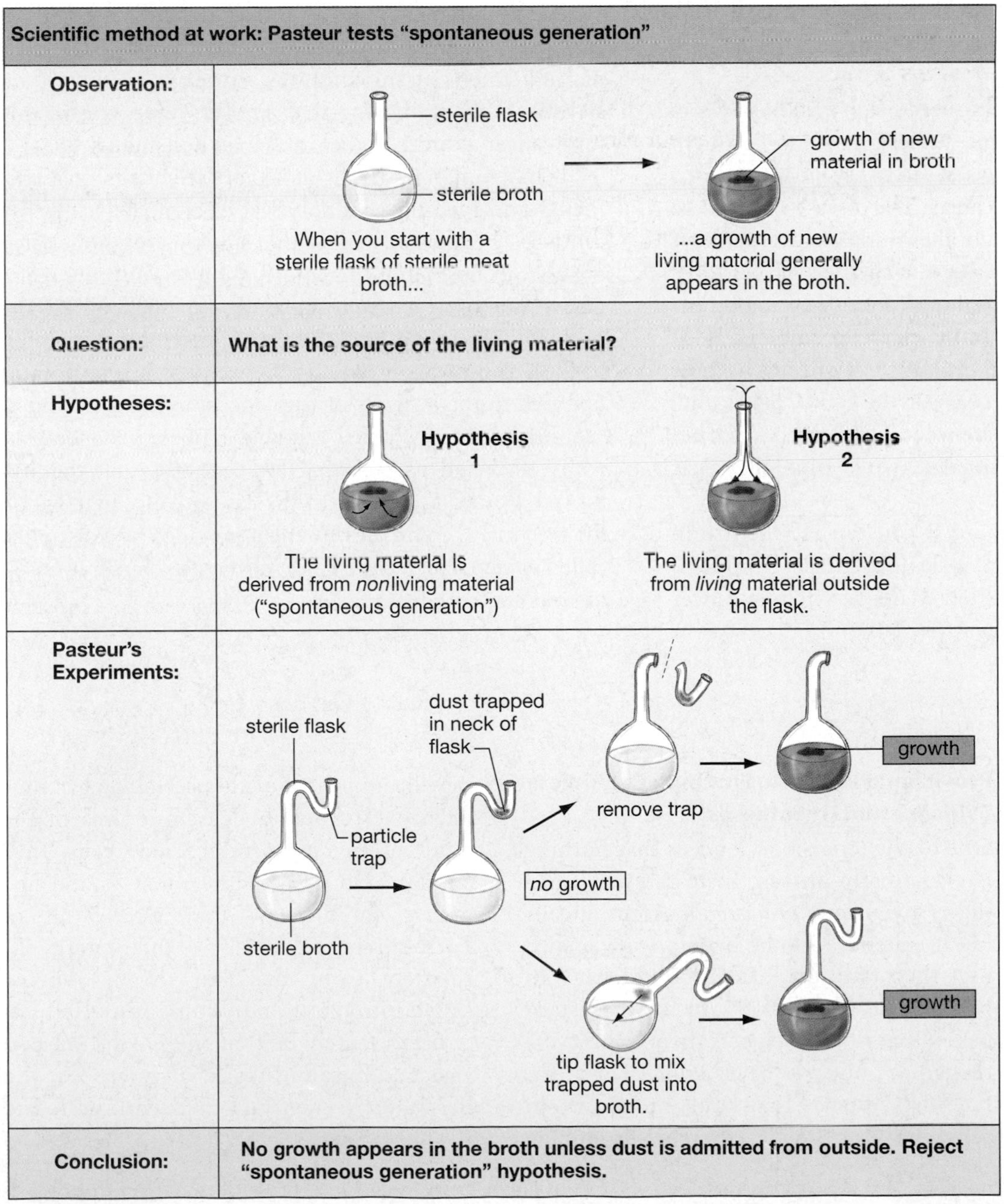

Figure 1.6
Pasteur's Experiments
Pasteur's spontaneous generation experiments and the scientific method. Nineteenth-century observation made clear that life would appear in a medium, such as broth, that had been sterilized. But what was the source of this life? One hypothesis was that it arose through "spontaneous generation," meaning it formed from the simple chemicals in the broth. Conversely, Pasteur hypothesized that it originated from airborne microorganisms. He was able to design an experiment that offered evidence for this hypothesis. The device he used was an S-shaped flask, which enabled air to enter the flask freely while trapping all particles (including invisible microorganisms) in a bend in the neck.

Scientific method at work: Pasteur tests "spontaneous generation"	
Observation:	When you start with a sterile flask of sterile meat broth... ...a growth of new living material generally appears in the broth. (sterile flask, sterile broth → growth of new material in broth)
Question:	What is the source of the living material?
Hypotheses:	**Hypothesis 1** The living material is derived from *non*living material ("spontaneous generation") **Hypothesis 2** The living material is derived from *living* material outside the flask.
Pasteur's Experiments:	sterile flask, particle trap, sterile broth → dust trapped in neck of flask → no growth; remove trap → growth; tip flask to mix trapped dust into broth. → growth
Conclusion:	No growth appears in the broth unless dust is admitted from outside. Reject "spontaneous generation" hypothesis.

Lung Cancer, Smoking, and Statistics in Science

Valuable as they are, experimental and observational tests often are not enough to provide answers to scientific questions. In countless instances, scientists employ an additional tool in coming to comprehend reality—a mathematical tool—as you'll see in the following example.

The evidence that cigarette smoking causes lung cancer (and heart disease and emphysema and on and on) has been around for so long that most people have no idea why smoking was looked into as a health hazard in the first place. You might think that scientists were suspicious of tobacco decades ago and thus began experimenting with it in the laboratory, but this wasn't the case. Instead, the trail that led to tobacco as a health hazard started with a mystery about disease.

When the lung-cancer pioneer Alton Ochsner was in medical school in 1919, his surgery professor brought both the junior and senior classes in to see an autopsy of a man who had died of lung cancer. The disease was then so rare that the professor thought the young medical students might never see another case during their professional lifetimes. Prior to the 1920s, lung cancer was among the rarest forms of cancer, because cigarette smoking itself was rare before the twentieth century. It did not become the dominant form of tobacco use in the United States until the 1920s. This made a difference in lung-cancer rates because cigarette smoke is inhaled, while pipe and cigar smoke generally are not.

If you look at essay **Figure 1**, you can see the rise in lung-cancer mortality in U.S. males and females from 1930 forward. Note that women show a later rise in lung-cancer deaths; this is because women started smoking en masse later. (Also note that in the 1990s, lung-cancer rates finally began to level off—or drop in the case of men. This was a direct result of a decline in smoking that began in the 1970s.)

Given the lung-cancer trends that were apparent in males by the 1930s forward, the task before scientists was to explain the alarming increase in this disease. What could the cause of this scourge be, the medical detectives wondered? The effects of men being gassed in World War I? Increased road tar? Pollution from power plants? Through the 1940s, cigarette smoking was only one suspect among many.

Laboratory experiment eventually would play a part in fingering tobacco as the lung-cancer culprit, but the original indictment of smoking was written in numbers—in statistical tables showing that smokers were contracting lung cancer at much higher rates than nonsmokers.

It has sometimes been said that "science is measurement," and the phrase is a marvel of compact truth. For centuries, people had an idea that smoking might be causing serious harm, but this information fell into the realm of guessing or of *anecdote,* meaning personal stories. The problem with anecdote is that there is no measurement in it; there is no way of judging the validity of one story as opposed to the next. Related to anecdote is the notion of "common sense," which is valuable in many instances, but which also had us believing for centuries that the Sun moved around the Earth. In the case of smoking, it took the extremely careful measurement provided by a discipline called *epidemiology*—the study of disease distributions— to separate truth from fiction.

Provisional Assent to Findings: Legitimate Evidence and Hypotheses

One of the tenets of science is that nothing is ever finally proven. Instead, every finding is given only *provisional* assent, meaning it is believed to be true for now, pending the addition of new evidence. In practice, some theories are so well established that no one expects them to be overturned . . . but you never know! For years, scientists "knew" that adult human brain cells did not divide, thereby producing new brain cells. But this certainty has now been overturned; at least one area of the adult brain *does* produce new cells. In a similar vein, one of the most established beliefs in biology—that nearly all of Earth's living things derive their energy ultimately from the Sun—is now being called into question by some respected scientists. (It may be that the Earth houses a vast underground world of microbes who derive their energy from chemicals found in soil and rocks.)

Probability in Science

Note that "measurement" in this instance was a matter of calculating *probability*, which is often the case in science. Epidemiologists found a linkage between smoking and lung cancer, in the sense that those who smoked were more likely to get the disease. But having seen this, scientists then had to ask: Could this result be a matter of pure chance? A person tossing a coin might get heads five times in a row, and it might be written off to chance. But would it be the same if the person came up with heads *seventy* times in a row? No; at that point there would be justification for assuming that some force other than chance was in operation (such as a rigged coin). When the epidemiologists looked at their statistical tables and saw so many more smokers than nonsmokers getting lung cancer, they had to ask whether this result fell into the realm of seven heads in a row, or seventy. Even in the earliest studies they concluded that more than chance was at work in the results. After many studies, they concluded that smoking was *causing* lung-cancer. But how did they judge what was probable and what was not in an issue as complicated as this one? The researchers relied on techniques developed in the branch of mathematics called *statistics*.

The importance of probability and statistics to science can hardly be overstated. These tools are used frequently in nearly every scientific discipline. Imagine that 10 experimental plots of land are being compared, five with fertilizer added to them, the other five without. The plots with the added fertilizer end up with more growth but fewer kinds of plants. Could the differences between the two kinds of plots be a matter of chance? Here, as in so many other tests, scientists would use the tools of statistics to get at the truth.

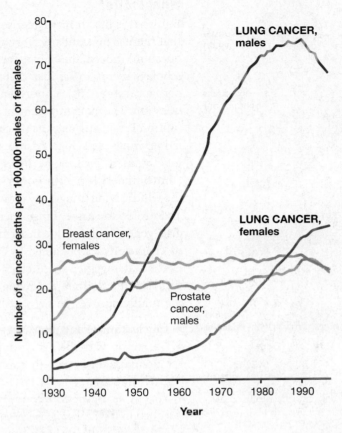

Figure 1
Rise in Lung-Cancer Mortality in U.S. Males and Females from 1930 Forward

So in the final analysis, scientists must always be open to the possibility that what we "know" is wrong. Indeed, this is one of the fundamental principles of science. Here it is, stated another way, along with some other important scientific principles regarding hypotheses and evidence:

- Every assertion regarding the natural world is subject to challenge and revision.

- Results obtained in experiments must be *reproducible*. Different investigators must be able to obtain the same results from the same sets of procedures and materials.
- Any scientific hypothesis or claim must be *falsifiable*, meaning open to negation through means of scientific inquiry. The assertion that "UFOs are visiting the Earth" does not rise to the level of a scientific claim, because there is no way to prove that this is *not* so.

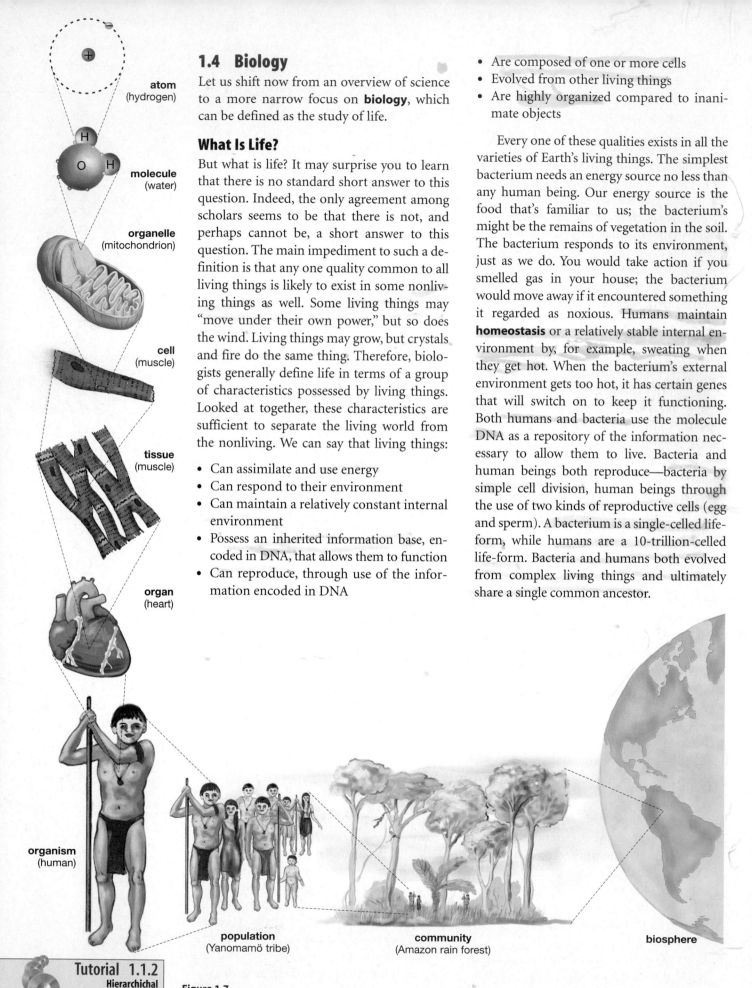

1.4 Biology

Let us shift now from an overview of science to a more narrow focus on **biology**, which can be defined as the study of life.

What Is Life?

But what is life? It may surprise you to learn that there is no standard short answer to this question. Indeed, the only agreement among scholars seems to be that there is not, and perhaps cannot be, a short answer to this question. The main impediment to such a definition is that any one quality common to all living things is likely to exist in some nonliving things as well. Some living things may "move under their own power," but so does the wind. Living things may grow, but crystals and fire do the same thing. Therefore, biologists generally define life in terms of a group of characteristics possessed by living things. Looked at together, these characteristics are sufficient to separate the living world from the nonliving. We can say that living things:

- Can assimilate and use energy
- Can respond to their environment
- Can maintain a relatively constant internal environment
- Possess an inherited information base, encoded in DNA, that allows them to function
- Can reproduce, through use of the information encoded in DNA

- Are composed of one or more cells
- Evolved from other living things
- Are highly organized compared to inanimate objects

Every one of these qualities exists in all the varieties of Earth's living things. The simplest bacterium needs an energy source no less than any human being. Our energy source is the food that's familiar to us; the bacterium's might be the remains of vegetation in the soil. The bacterium responds to its environment, just as we do. You would take action if you smelled gas in your house; the bacterium would move away if it encountered something it regarded as noxious. Humans maintain **homeostasis** or a relatively stable internal environment by, for example, sweating when they get hot. When the bacterium's external environment gets too hot, it has certain genes that will switch on to keep it functioning. Both humans and bacteria use the molecule DNA as a repository of the information necessary to allow them to live. Bacteria and human beings both reproduce—bacteria by simple cell division, human beings through the use of two kinds of reproductive cells (egg and sperm). A bacterium is a single-celled lifeform, while humans are a 10-trillion-celled life-form. Bacteria and humans both evolved from complex living things and ultimately share a single common ancestor.

atom
(hydrogen)

molecule
(water)

organelle
(mitochondrion)

cell
(muscle)

tissue
(muscle)

organ
(heart)

organism
(human)

population
(Yanomamö tribe)

community
(Amazon rain forest)

biosphere

Tutorial 1.1.2
Hierarchichal
Organization of Life

Figure 1.7
Levels of Organization in Living Things

There are some exceptions to these "universals." For example, the overwhelming majority of honeybees and ants are sterile females; they can't reproduce, but no one would doubt that they're alive. In the main, however, if something is living, it has all these qualities.

Life Is Highly Organized, in a Hierarchical Manner

One item on the list of qualities requires a little more explanation. It is that living things are highly organized compared to inanimate matter. More specifically, they are organized in a "hierarchical" manner, meaning one in which lower levels of organization are progressively integrated to make up higher levels. The main levels in this hierarchy could be compared to the organization of a business. In a corporation, there may be individuals making up an office, several offices making up a department, several departments making up a division, and so forth. In life, there is one set of organized "building blocks" making up another (**see Figure 1.7**).

Actually life is not just "highly" organized. Nothing else comes *close* to it in organizational complexity. The Sun is a large thing, but it is an uncomplicated thing compared to even the simplest organism. Consider that you have about 10 trillion cells in your body and that, with some exceptions, each of these cells has in it a complement of DNA that is made up of chemical building blocks. How many building blocks? Three billion of them. Now, you probably know that most cells divide regularly, one cell becoming two, the two becoming four, and so on. Each time this happens, each of the 3 billion DNA building blocks must be faithfully *copied*, so that both of the cells resulting from cell division will have their own complete copy of DNA. And this is just the copying of the molecule, before anything is actually done with it. Complex indeed. Let's see what life's levels of organization are.

Levels of Organization in Living Things

The building blocks of matter, called *atoms*, lie at the base of life's organizational structure. (See Chapter 2 for an account of them.) Atoms come together to form *molecules*, meaning entities consisting of a defined number of atoms in a defined spatial relationship to one another. A molecule of water is one atom of oxygen bonded to two atoms of hydrogen, with these atoms *arranged* in a very precise way. Molecules in turn form what are called *organelles*, meaning "tiny organs" in a cell. Each of your cells has, for example, hundreds of organelles in it, called *mitochondria*, that transform the energy from food into an energy form your body can use. Such an organelle is not just a collection of molecules that exist close to one another. It is a highly organized structure, as you can tell just from looking at the rendering of it in Figure 1.7.

At the next step up the organizational chain are entities that are actually *living*, as opposed to entities that are components of life. *Cells* are units that can do all of the things listed earlier: assimilate energy, reproduce, react to their environment, and so forth. Indeed, most experts would agree that cells are the *only* place that life exists. You may say: But isn't there a lot of material in between my cells? The answer is yes; it's mostly water with a good number of other molecules in the mix. But if all the cells were removed from this watery milieu, there would be nothing resembling life left in it.

The next step up is to a *tissue*, meaning a collection of cells that serve a common function. Your body contains collections of muscle cells that serve the same function (contraction). Each concentration of these cells constitutes muscle tissue. Several *kinds* of tissues can come together to form a functioning unit known as an *organ*. Your heart, for example, is a collection of nerve tissue and muscle tissue, among other types. An assemblage of cells, tissues, and organs can then form a multicelled *organism*. (Of course, back down at the cell level, a one-celled bacterium is also an organism; it's just not one with organs and so forth.)

From here on out, life's levels of organization all involve *many* organisms. Members of a single type of living thing (a species), living together in a defined area, make up what is known as a *population*. When you look at *all* the kinds of living things in a given area, you are looking at a *community*. Finally, all the communities of the Earth—and the physical environment with which they interact—make up the *biosphere*.

1.5 Special Qualities of Biology

Almost all scientific disciplines can trace their origins to the ancient Greeks, and biology is no exception. In the work of such Greeks as Hippocrates and Galen, we can find the origins of modern medical science. In the work of Aristotle and others, we can find the origins of "natural history," which led to what we think of today as mainstream biology and the larger category of the **life sciences**, which includes not only biology, but medicine, forestry, and the like.

Despite these ancient origins, biology is, in a sense, a much younger science, than, say, physics, which is one of the **physical sciences**, meaning the natural sciences not concerned with life. Western Europe's revolution in the physical sciences probably can be dated from the sixteenth century, when Nicholas Copernicus published his work *On the Revolution of Heavenly Spheres,* which demonstrated that Earth moves around the Sun. Meanwhile, biology did not come into its own as a science until the *nineteenth* century.

Prior to the 1800s, biology was almost purely *descriptive,* meaning that the "naturalists" that we would today call biologists largely confined themselves to describing living things—what kinds there were, where they lived, what features they had, and so forth. Beginning in about the 1820s, however, biologists began to formulate biological *theories* as that term was defined earlier. They began to postulate that all life exists within cells, that life comes only from life, that life is passed on through small packets of information that we now call genes, and so forth. To put this another way, biologists in the nineteenth century began describing the *rules* of the living world, whereas before they were largely describing *forms* in the living world.

This change moved biology closer to the same scientific footing as physics. But biology was then, and remains now, a very different kind of science from any of the physical sciences, with physics being a clear case in point. One reason for this difference is that the constituent parts of physics are very uniform and far fewer in number than is the case in biology. Physics deals with only 92 stable elements, such as hydrogen and gold, and to a first approximation, if you've seen one electron, you've seen them all.

Meanwhile, in biology, if you've seen one species you've seen just that—one species. Each species is at least marginally different from another, and many are greatly dissimilar. Moreover, there are thought to be at least 10 million species on Earth. And each of these species has all the organizational levels of elements in physics *and more.* (They not only have electrons and atoms, they have organelles, cells, tissues, and so on.) Biology is concerned with the rules that govern all species, and you've seen that there are some biological "universals." But when cancer researchers are looking for the principles that underlie cell division, they are likely to be looking at only one of two main kinds of cells; when ecologists are looking at what causes dry grassland to turn into desert, their findings are likely to have little relevance to the rain forest. Put simply, the living world is tremendously diverse compared to the nonliving world, and such diversity means that universal rules in biology are likely to be few

Figure 1.8
Evolution Has Shaped the Living World

a A peacock displaying his plumage

b A poison dart frog in Colombia

c New Caledonia pine trees towering above palm trees in the New Caledonia islands, east of Australia

a

b

c

and far between. Biology is concerned with the *particular* to a far greater degree than is the case in the physical sciences. Note also that "universals" in biology may not apply beyond Earth; we don't know if life even exists anywhere else, much less what its rules are. Meanwhile, the rules of physics truly are universal in that they are equally applicable on Earth or in the farthest reaches of the cosmos.

Biology's Chief Unifying Principle

Almost all biologists would agree that the most important thread that runs through biology is **evolution**, meaning the gradual modification of populations of living things over time, with this modification sometimes resulting in the development of new species. Evolution is central to biology, because every living thing has been *shaped* by evolution. (There are no exceptions to this universal.) Given this, the explanatory power of evolution is immense. Why do peacocks have their finery, or frogs their coloration, or trees their height (**see Figure 1.8**)? All these things stand as wonders of nature's diversity, but with knowledge of evolution they are wonders of diversity that *make sense*. For example, why do so many unrelated stinging insects look alike? Evolutionary principles suggest they *evolved* to look alike because of the general protection this provides from predators. Think of yourself for a moment as a bee predator. Having once gotten stung, would you annoy *any* roundish insect that had a black-and-yellow-striped coloration? You probably learned your lesson about this in connection with one species, but *many* species of insects are now protected from you simply by virtue of the coloration they share with the others (**see Figure 1.9**). Thus, there were reproductive benefits to individuals who, through genetic chance, happened to get a slightly more striped coloration: They left more offspring, because they were bothered

less by predators. Over time, entire populations moved in this direction. They evolved, in other words.

The means by which living things can evolve is a topic this book takes up beginning in Chapter 16. Suffice it to say for now that a consideration of evolution is never far from most biological observations. So strong is evolution's explanatory power that, in uncovering something new about, say, a sequence of DNA or the life cycle of a given organism, one of the first things a biologist will ask is: Why would evolution shape things in this way?

The Organization of This Book

This book has something in common with the levels of organization you looked at earlier, in that it too goes from constituent parts to the larger whole. It begins with atoms, moves on to the biological molecules that atoms make up, and then goes to cells. The end of the book covers the highest levels of biological organization, which is to say natural communities and Earth's biosphere. In between, however, are tours of such facets of life as energy, DNA-encoded information, and reproduction. Even here, however, you'll be moving in a general way from the small to the large, because much of the first part of the book is given over to **molecular biology**—meaning the study of individual molecules (such as DNA) as they affect living things. Then you'll move into evolution, which touches on **organismal biology**, meaning the study of whole organisms. Next is the **physiology** or physical functioning of plants and animals, which largely concerns tissues and organs. Finally there is **ecology**, which is the study of the interactions of organisms with each other and with their physical environment. And so, let's begin to look at biology—as a body of knowledge and a way of learning.

a

b

**Figure 1.9
Similar Enough to Yield a Benefit**

a The golden northern bumblebee

b The Sandhills hornet

These are two of the many stinging insects that have the black-and-yellow-striped coloration that offers protection from predators.

Chapter Review

Summary

1.1 How Does Science Impact the Everyday World?

- Science is playing an increasingly important role in the everyday lives of Americans, as evidenced by weekly news regarding such issues as genetically modified food, disease, and the biotech industry.

1.2 What Does the Public Think, and Know, about Science?

- Americans are interested in science, have a great deal of confidence in American scientific leadership, and overwhelmingly believe that the world is a better place because of science.

- Americans have an uneven knowledge about science. Almost 80 percent of adult Americans know that the continents are moving about the face of the Earth, for example, but more than a quarter think the Sun goes around the Earth.

1.3 What Is Science?

- Science is a body of knowledge, a collection of unified insights about nature, the evidence for which is an array of facts.

- The unified insights of science are known as theories. A theory is a general set of principles, supported by evidence, that explains some aspect of nature.

- Science can also be defined as a way of learning: a process of coming to understand the natural world through the testing of hypotheses.

- Science works through the scientific method, in which an observation leads to the formulation of a question about the natural world. Then comes a hypothesis—an explanation that has not been proven to be true. The hypothesis may be tested through observation, through a series of experiments, or by statistical means.
  TUTORIAL 1.1.1: The Scientific Method (Pasteur's Experiments)

- Every assertion regarding the natural world is subject to challenge and revision. Results obtained in experiments must be reproducible. Any scientific hypothesis or claim must be falsifiable, meaning open to negation through means of scientific inquiry.

1.4 Biology

- Biology is the study of life. Life is defined by a group of characteristics possessed by living things. Living things can assimilate energy, respond to their environment, maintain a relatively constant internal environment, and possess an inherited information base, encoded in DNA, that allows them to function. Living things can also reproduce, are composed of one or more cells, are evolved from other living things, and are highly organized compared to inanimate objects.

- Life is organized in a hierarchical manner, running in increasing complexity from atoms to molecules and then in sequence to organelles, cells, tissues, organs, organisms, populations, communities, and the biosphere.
  TUTORIAL 1.1.2: Hierarchichal Organization of Life

1.5 Special Qualities of Biology

- Until the early nineteenth century, biology was largely a descriptive science, meaning it largely catalogued and described the Earth's living things. Beginning about the 1820s, however, life science researchers began to formulate biological theories, such as that life comes only from life and exists only within cells.

- Biology's subject matter—the living world—is notable for its diversity.

- Biology's chief unifying principle is evolution, which can be defined as the gradual modification of populations of living things over time, with this modification sometimes resulting in the development of new species.

Key Terms

biology 12	organismal biology 15
control 8	physical science 14
ecology 15	physiology 15
evolution 15	science 7
homeostasis 12	scientific method 7
hypothesis 7	theory 7
life science 14	variable 8
molecular biology 15	

Understanding the Basics

Multiple-Choice Questions

1. Which of the following statements best describes the nature of a scientific hypothesis?
   a. A hypothesis is an idea that is widely accepted as a description of objective reality by a majority of scientists.
   b. A hypothesis must stand alone, and not be based on prior knowledge.
   c. A scientific hypothesis must be testable through experimentation, observation, or mathematical demonstration.
   d. Experiments can be designed that will prove the validity of a hypothesis.
   e. A hypothesis when accepted becomes a scientific law.

2. Those who wish to berate a scientific theory sometimes say, "that's only a theory." The use of the word *theory* for a biological concept means that
   a. There is absolute certainty about the validity of the concept.
   b. Most scientists would agree that there is a preponderance of evidence in support of the concept.
   c. The concept is in doubt among most scientists.
   d. The concept is no more than a hypothesis.
   e. The concept has no basis in fact.

3. It may be argued that an automobile constitutes living matter because the burning of gasoline represents metabolism. Also, because a car picks up speed when the accelerator is depressed, one might claim it is responding to stimuli. Which of these reasons would you give for definitively concluding that a car is a nonliving entity?
   a. It does not store or use energy.
   b. It does not reproduce by transmitting genetic information through DNA.
   c. Carbon is not a major component of its chemical makeup.
   d. It exhibits cellular organization.
   e. It breaks down sometimes.

4. Evolution is a central, unifying theme in biology because
   a. Evolution is a falsifiable hypothesis.
   b. Humans have evolved from ancestors we share with present-day monkeys.
   c. Evolution has occurred in the past, even though it no longer operates today.
   d. The enormously diverse forms of life on Earth have all been shaped by evolution.
   e. Almost all biologists believe in it.

5. Biologists generally define life in terms of a group of characteristics possessed by living things. Which of the following is not a characteristic of living things?
   a. All living things possess an inherited information base, encoded in DNA, that allows them to function.
   b. All living things can respond to their environment.
   c. All living things can maintain a relatively constant internal environment.
   d. All living things evolved from other living things.
   e. All living things are composed of two or more cells.

Brief Review

1. What is science? In what ways is science similar to, and different from, belief systems such as religious faith?

2. What is a controlled experiment? Why is it important to keep all variables but one constant in a scientific experiment?

3. How did Louis Pasteur cast doubt on the idea of spontaneous generation?

4. Describe the defining features of life as we know it on the Earth.

5. Living systems can be described at various hierarchical levels. List as many levels of biological organization as you can think of, from the microscopic to the largest levels imaginable.

Applying Your Knowledge

1. Would you agree that it is valuable for a nation to have a citizenry that is reasonably well versed in science? Give reasons for your answer. Would you say this need has become especially urgent in the last two decades? If so, why?

2. Is it harder to prove a hypothesis than to disprove it? Imagine you wanted to establish that cheetahs are the fastest land animals, and assume you have the ability to clock any animal moving at its top speed. Now, what would it take to disprove the idea that cheetahs are the fastest land animals? What would it take to prove that cheetahs are the fastest land mammals, meaning no other land mammal could run faster than they?

3. If you were sent on an interplanetary mission to investigate the presence of life on Mars, what would you look for? Would you explore the land and the atmosphere? Imagine you discover an entity you suspect is a living being. Realizing that life elsewhere in the universe may not be organized by the same rules as on Earth, which of the features of life on Earth, if any, would you insist that the entity display before you would declare it living?

2

The Fundamental Building Blocks
Chemistry and Life

Pure gold. It's elementary.
(Section 2.1, page 22)

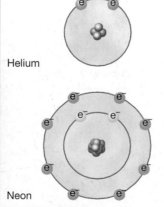

Helium

Neon

Filled shells mean stability.
(Section 2.2, page 24)

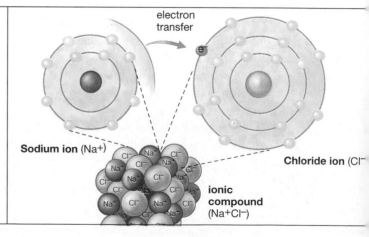

electron
transfer

Sodium ion (Na+)

Chloride ion (Cl⁻

**ionic
compound**
(Na+Cl⁻)

You put it on popcorn.
(Section 2.2, page 28)

Life is carried on through chains of chemical reactions.

Cities are made of buildings and buildings are made of bricks, bricks are made of earth, earth is made of . . . ? To answer this question, in this chapter we will look at what the material world is made of. Biology is our subject, but to fully understand it, you need to learn a little about what underlies biology. You need to learn a little about what *biology* is made of, in a sense. And to do this, you need to understand some of the basics in two other fields: chemistry and physics.

How are these disciplines relevant to biology? Well, consider chemistry. The average person probably is aware that living things are made up of individual units called cells. But beyond this bit of knowledge, reality fades and a kind of fantasy takes over. In it, the cells that populate people or plants or birds carry on their activities under the direction of their own low-level consciousness. A cell *decides* to move, it *decides* to divide, and so on. Not so. By the time our story is finished, many chapters from now, it will be clear to you that the cells that make up living things have no such consciousness, but rather do what they do as the result of a chain of chemical reactions. Repulsion and bonding, latching on and reforming, depositing and breaking down—all these things take place without a scrap of consciousness underlying them. What makes people and plants and birds function at this level is *chemistry.*

What is chemistry concerned with? Look around you. Do you see a table, light from a lamp, a patch of night or daytime sky? Everything that exists can be viewed as falling into one of two categories: matter or energy. You will learn something about energy in this chapter, but we are most concerned here with *matter and its transformations,* which is the subject of chemistry. Matter can be defined as anything that takes up space and has **mass**. This latter term is a measure of the *quantity* of matter in any given object. How much space does an object occupy—how much "volume," to put it another way—and how *dense* is the matter within that space? These are the things that define mass. For our purposes, we may think of mass as being equivalent to weight, though in physics they make a distinction between these two things.

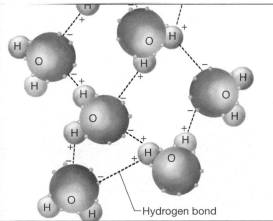

The ties that bond water molecules together.
(Section 2.2, page 30)

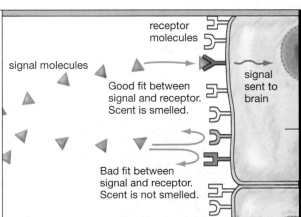

Do I smell bread in the oven?
(Section 2.3, page 31)

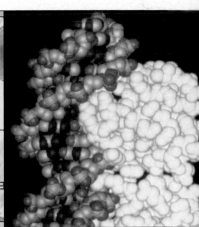

Many molecules are large and complex.
(Section 2.3, page 31)

2.1 The Nature of Matter: The Atom

Beholding matter all around us, it is natural to ask, what is its nature? A child sees a grain of sand, pounds it with a rock, sees the smaller bits that result and wonders: What is this stuff like at the end of these divisions? Not surprisingly, adults too have wondered about this question—for centuries. About 2,400 years ago, the Greek philosopher Plato accepted the notion that all matter is made up of four primary substances: earth, air, fire, and water. A near-contemporary of his, Democritus, believed that these substances were in turn made up of smaller units that were both invisible and in*div*isible—they could not be broken down further. He called these units atoms (**see Figure 2.1**).

Well, at least one cheer for Democritus, because he had it partly right. Centuries of painstaking work lying between his time and ours has confirmed that matter is indeed composed of tiny pieces of matter, which we still call atoms. But these atoms are not indivisible, as Democritus thought. Rather, they are themselves composed of constituent parts. A superficial account of *all* the parts scientists have discovered to date would go on for pages and still be incomplete. Physicists are continually slamming together parts of atoms with ever-greater force in an effort to determine what *else* there may be at the heart of matter. (This is what the machines called "atom smashers" do. The physicists who run them could be compared to people who, in trying to find out what parts a watch has, throw it on the ground and record the way its various mechanisms fly out upon impact; **see Figure 2.2**).

Interesting stuff, but it is purely the business of physics, with little relation to biology. We are not concerned here with what's at the very end of these divisions. We do care a good deal, however, about what's *nearly* at the end of them.

Protons, Neutrons, and Electrons

For our purposes there are three important constituent parts of an atom: **protons**, **neutrons**, and **electrons**. These three parts exist in a spatial arrangement that is uniform in all matter. Protons and neutrons are packed tightly together in a core (the atom's **nucleus**), and electrons move around this core some distance away (**see Figure 2.3**). The one variation on this theme is the substance hydrogen, the lightest of all the kinds of matter we will run into. Hydrogen has no neutrons, but rather only one proton in its nucleus and one electron in motion around it.

These three "subatomic" particles have mind-bending sizes and proportions. As P. W. Atkins has pointed out, an atom is so small that 100 million carbon atoms would lie end to end in a line of carbon about this long: —————————— (3 centimeters). Things are just as disorienting when we consider the size of the atom as a whole, relative to the nucleus. The whole atom, with electrons at its edge, is 100,000 times bigger than the nucleus. If you were to draw a model of an atom *to scale* and began by sketching a

Figure 2.1
The Building Blocks of Life
Viewing this idealized feather at different levels of magnification, we eventually arrive at the building block of all matter, the atom. The atom selected here, from among a multitude that make up the feather, is a single hydrogen atom, composed of one proton and one electron.

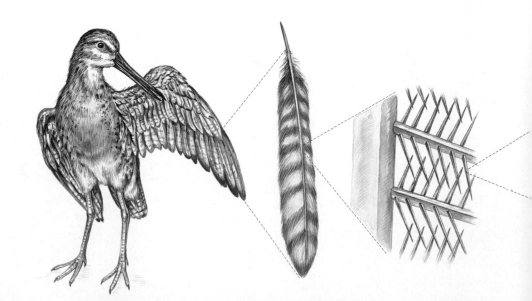

nucleus of, say, half an inch, you'd have to draw some of its electrons more than three-quarters of a mile away.

Although the nucleus accounts for very little of the *space* an atom takes up, it accounts for almost all of the *mass* an atom has. So negligible are electrons in this regard, in fact, that all of the mass (or weight) of an atom is considered to reside with the nucleus' protons and neutrons.

The components of atoms have another quality that interests us: **electrical charge**. Protons are positively charged and electrons are negatively charged. Meanwhile, neutrons—as their name implies—have no charge, but are electrically neutral. Because all these particles do not exist separately, but *combine* to form an atom, as a whole the atom may be electrically neutral as well. The negative charge of the electrons balances out the positive charge of the protons. Why? Because in this state the *number* of protons an atom has is exactly equal to the number of electrons it has (though we'll see a different, "ionic" state later in this chapter). In contrast, the number of *neutrons* an atom has can vary in relation to the other two particles.

With this picture of atoms in mind, we can begin to answer the question that has been handed down to us through history: What is matter? We certainly have a commonsense answer to this question. Matter is any substance that exists in our everyday experience. For example, the iron that goes into cars is matter. But what is it that differentiates this iron from, say, gold? The answer is that an iron atom has 26 protons in its nucleus, while a gold atom has 79.

Figure 2.2
Getting to the Heart of Matter
Understanding what the tiny objects called atoms are made of requires the use of extremely large particle accelerators or "atom smashers," such as this one at Fermilab in Batavia, Illinois. Subatomic particles are accelerated around Fermilab's magnetic ring, some 6.3 kilometers or 3.8 miles in circumference, and then slammed into each other or into a fixed target. How big is this operation? The detectors that record the collisions weigh 5,000 tons apiece and are three stories high.

Tutorial 2.1.1
Structure of the Atom, Elements, Isotopes

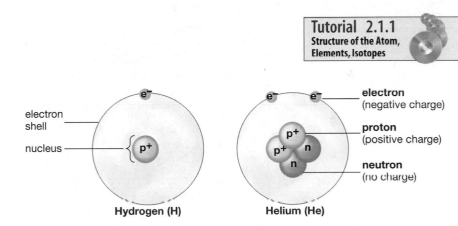

Figure 2.3
Representations of Atoms
One conceptualization of two separate atoms, hydrogen and helium. The model is not drawn to scale; if it were, the electrons would be perhaps a third of a mile away from the nuclei. The model also is simplified, giving the appearance that electrons exist in track-like orbits around an atom's nucleus. In fact, electrons spend time in volumes of space that have several different shapes.

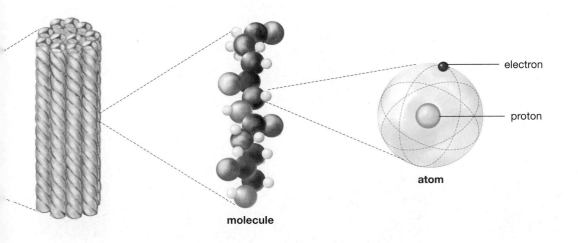

Fundamental Forms of Matter: The Element

Gold is an **element**—a substance that is "pure" because it cannot be reduced to any simpler set of component substances through chemical processes. And the thing that defines each element is the number of protons it has in its nucleus. A solid-gold bar, then, represents a huge collection of identical atoms, each of which has 79 protons in its nucleus (**Figure 2.4**). In making gold jewelry,

an artist may combine gold with another metal such as silver or copper to form an alloy that is stronger than pure gold, but the gold atoms are still present, all of them retaining their 79-proton nuclei.

Given what you've just read about protons, neutrons, and electrons, you may wonder why gold—or any other element—cannot be reduced to any "simpler set of component substances." Aren't protons and neutrons components of atoms? Yes, but they are not component *substances*, because they cannot exist by themselves as matter. Rather, protons and neutrons must *combine* with each other to make up atoms.

Assigning Numbers to the Elements

In the same way that buildings can be defined by a location, and thus have a street number assigned to them, these elements, which are defined by protons in their nuclei, have an **atomic number** assigned to them. We have observed that hydrogen has but one proton in its nucleus, and it turns out that scientists have constructed the atomic numbering system so that it goes from smallest number of protons to largest. Thus, hydrogen has the atomic number 1. The next element, helium, has two protons, so it is assigned the atomic number 2. Continuing on this scale all the way up through the elements found in nature, we would end with uranium, which has an atomic number of 92.

Figure 2.4
Pure Gold
Gold is an element because it cannot be reduced to any simpler set of substances through chemical means. Each gold bar and nugget is made up of a vast collection of identical atoms—those with 79 protons in their nuclei.

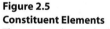

Figure 2.5
Constituent Elements
The major chemical elements found in Earth's crust (including the oceans and the atmosphere) and in the human body.

Earth's crust

other 8%

oxygen 50%

silicon 26%

aluminum 8%
calcium 3%
iron 5%

Human body

other 7%

hydrogen 10%

oxygen 65%

carbon 18%

Given this view of the nature of matter, we are now in a position to answer the question posed at the beginning of the chapter: What is a handful of earth—or anything else—made of? The answer is one or more elements. If you look at **Figure 2.5**, you can see the most important elements that go into making up both the earth's crust and human beings.

Isotopes

All this seems like a nice, tidy way to identify elements—one element, one atomic number, based on number of protons—except that we're leaving out something. Recall that atoms also have neutrons in their nuclei, that these neutrons add weight to the atom, and that the number of neutrons can vary independently of the number of protons. What this means is that in thinking about an element in terms of its weight, we have to take neutrons into account. Furthermore, because the number of neutrons in an element's nucleus may vary, we can have various *forms* of elements, called **isotopes**. Most people have heard of one example of an isotope, whether or not they recognize it as such. The element carbon has six protons, giving it an atomic number of 6. In its most common form, it also has six neutrons. However, a relatively small amount of carbon exists in a form that has *eight* neutrons. Well, the element is still carbon, and in this form the number of its protons and neutrons equals 14, so the *isotope* is carbon-14, which is used in determining the ages of fossils and geologic samples.

Most elements have several isotopes. Hydrogen, for example, which usually has one proton and one electron, also exists in two other forms: deuterium, which has the proton, electron, and one neutron; and tritium, which has one proton, one electron, and two neutrons (**see Figure 2.6**). **Figure 2.7** shows you how isotopes are used in medicine.

The Importance of Electrons

In our account so far of the subatomic trio, we have had much to say about protons and neutrons, but little to say about electrons. This was necessary because we needed to go over the nature of matter, but in a sense you can regard what has been set forth to this point as so much stage-setting, because what's most important in biology is the way

elements *combine* with other elements. And in this combining, it is the outermost electrons that play a critical role. Just as you come into contact with the world through what lies at your surface—your eyes, your ears, your hands—so atoms link up with one another through what lies at their outer edges. The interior of an atom is very quiet in a sense, while the atom's outer electrons exist in a world that can be one of continual forming and breaking of alliances.

2.2   Matter Is Transformed through Chemical Bonding

The process of chemical combination and rearrangement is called **chemical bonding**, and for us it represents the heart of the story in

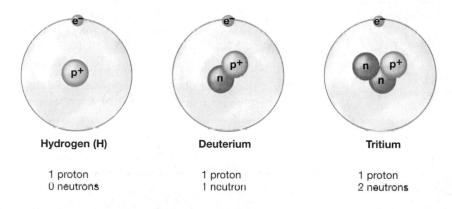

Hydrogen (H)

1 proton
0 neutrons

Deuterium

1 proton
1 neutron

Tritium

1 proton
2 neutrons

**Figure 2.6
Same Element, Different Forms**
Pictured are three isotopes of hydrogen. Like all isotopes, they differ in the number of neutrons they have.

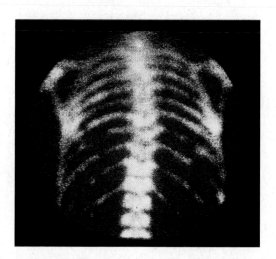

**Figure 2.7
Cancer Diagnosis with Isotopes**
The yellow spots on this patient's spine and right scapula (shoulder blade) represent sites of cancerous cell growth. Doctors were able to pinpoint the cancer by injecting the patient with a radioactive isotope. Cancerous bone concentrates the isotope more strongly than normal bone, which results in the isotope showing up as brighter "hot spots" in cancerous tissue.

chemistry. When the outermost electrons of two atoms come into contact, it becomes possible for these electrons to reshuffle themselves in a way that allows the atoms to become attached to one another. This can take place in two ways: One atom can *give up* one or more electrons to another, or one atom can *share* one or more electrons with another atom. Giving up electrons is called ionic bonding; sharing electrons is called covalent bonding. A third type of bond, which we'll get to shortly, also is important for our purposes: the hydrogen bond.

Energy Always Seeks Its Lowest State

Atoms that undertake bonding with one another do so because they are in a more *stable* state after the bonding than before it. A frequently used phrase is helpful in understanding this kind of stability: Energy always seeks its lowest state. Imagine a boulder perched precariously on a hill. A mere shove might send it rolling toward its lower energy state—at the bottom of the hill. It would not then roll *up* the hill, either spontaneously or with a light shove, because it is now existing in a lower energy state than it did before—one that is clearly

more stable than its former precarious perch. When we turn to electrons, the energy is not gravitational, but electrical. Atoms bond with one another to the extent that doing so moves them to a lower, more stable energy state.

But what determines the *likelihood* of any given atom taking part in this bonding? An analogy might be helpful here. At one time, certain fancy social dances used what were known as *dance cards* to make sure the young men and women present actually danced. For the first dance, you'd agree in advance with so-and-so to dance together (and write that down on your card); for the second dance, you'd agree with someone else, and so on. Now the important point was, *everyone wanted to have a full dance card.* Otherwise you risked standing around by yourself on the sidelines feeling foolish. But, having filled up your card, you were no longer free to pair up with anyone else.

Seeking a Full Outer Shell: Covalent Bonding

What an atom "seeks" to fill is not a dance card, but its outer shell. It does so because, with a filled outer shell, it is in a lower energy state. Having achieved this, however, it is far less likely to "pair up" with any other atom.

What are these "outer shells"? As it happens, electrons reside in certain well-defined "energy levels" outside the nuclei of atoms. The number of these energy levels varies depending on the element in question. Here we need only note the practical effect of these levels on bonding: *Two* electrons are required to fill the first energy level (or shell) of any given atom, and *eight* are required to fill all the levels thereafter. If you look at the electron configurations pictured in **Figure 2.8**, you can see that two elements—hydrogen and helium—have so few electrons in orbit around them that they have nothing *but* a first energy level, while the other elements pictured have two or three energy levels. This means that hydrogen and helium require only two electrons in orbit around their nuclei to have filled outer shells, but that all other elements require eight electrons to have this kind of complete outer electron complement—to have full dance cards, we might say.

How Chemical Bonding Works in One Instance: Water

To see how chemical bonding works in connection with this concept of filled outer shells, take a look at the bonding that

Figure 2.8
Electron Configurations in Some Representative Elements

The concentric rings represent energy levels or "shells" of the elements, and the dots on the rings represent electrons. Hydrogen has but a single shell and a single electron within it, while carbon has two shells with a total of six electrons in them. Helium, neon, and argon have filled outer shells and are thus unreactive. Hydrogen, carbon, and sodium do not have filled outer shells and are thus reactive—they readily combine with other elements.

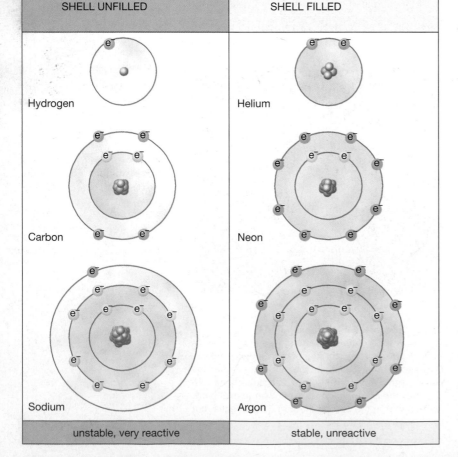

OUTERMOST ELECTRON SHELL UNFILLED	OUTERMOST ELECTRON SHELL FILLED
Hydrogen	Helium
Carbon	Neon
Sodium	Argon
unstable, very reactive	stable, unreactive

occurs with the constituent parts of one of the most simple (and important) substances on Earth: water. In so doing, you'll see one of the kinds of bonding we talked about—covalent bonding.

The familiar chemical symbol for water is H_2O. This means that two atoms of hydrogen (H) have combined with one atom of oxygen (O) to form water. (See "Notating Chemistry" on page 26 for an explanation of symbols in chemistry.) Recall that hydrogen has but one electron running around in its single energy level. Also recall, however, that this first level is not completed until it has *two* electrons in it. Hydrogen could fill this shell in any number of ways. It might, for example, come into contact with another hydrogen atom. These two atoms can then form a **covalent bond** by sharing a *pair* of electrons with each other—one electron from each atom. These electrons can now be found orbiting the nuclei of both atoms. As a result, both have two electrons in their outer energy levels, which makes them filled.

Our hydrogen atom might also, however, come into contact with an oxygen atom, which has eight electrons. Looking at Figure 2.9, you can see what this means: Two electrons fill oxygen's first energy level, which leaves six left over for its second. But remember that the second shell of any atom is not completed until it holds *eight* electrons. Thus oxygen, like hydrogen, would welcome a partner. Only it needs *two* electrons to fill its outer shell, which means that two atoms of hydrogen would do. Once again, pairs of electrons are shared. The oxygen atom and first hydrogen atom donate one electron each for the first pair; and the oxygen and second hydrogen atom each donate one electron for the second pair. The result? H_2O: Two atoms of hydrogen and one atom of oxygen, covalently bonded together and all of them "satisfied" to be in that condition. (Occasionally in nature, covalent bonding will take place in a way that leaves one atom with an unpaired electron, a potentially harmful phenomenon you can read about in "Free Radicals" on page 29.)

Matter Is Not Gained or Lost in Chemical Reactions

Note that when this pairing up of electrons happens, no matter has been gained or lost. We started with two atoms of hydrogen and

one atom of oxygen, and we finish that way. The difference is that these atoms are now bonded. This points up an important principle, known as the **law of conservation of mass**, which states that matter is neither created nor destroyed in a chemical reaction.

What Is a Molecule?

When two or more atoms combine in this kind of covalent reaction, the result is a **molecule**: a compound of a defined number of atoms in a defined spatial relationship. Here, one atom of oxygen has combined with two atoms of hydrogen to create *one water molecule.* (What we commonly think of as water, then, is an enormous collection of these individual water molecules.) A molecule need not be made of two different elements, however. Two hydrogen atoms can covalently bond to form one *hydrogen* molecule. On the other side of the coin, a molecule could contain many different elements bonded together. Consider sucrose, or regular table sugar, which is $C_{12}H_{22}O_{11}$ (12 carbon atoms bonded to 22 hydrogen atoms and 11 oxygen atoms).

Tutorial 2.2.1
Covalent Bonding

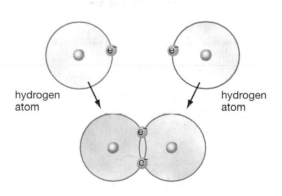

hydrogen
atom

hydrogen
atom

a   hydrogen molecule

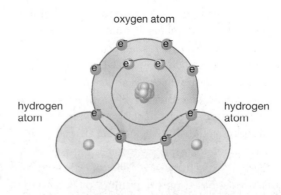

oxygen atom

hydrogen
atom

hydrogen
atom

b   water molecule

Figure 2.9
Covalent Bonding
A covalent bond is formed when two atoms share one or more pairs of electrons.

a Two atoms of hydrogen have come together, and each shares its lone electron with the other. This gives both atoms a filled outer shell—and stability.

b Two hydrogen atoms have linked with one oxygen atom; in this case, two pairs of electrons are shared, one pair between each of the hydrogen atoms and the oxygen atom.

Notating Chemistry

One of the "languages of science" is the system of symbols that chemistry uses. Somewhat forbidding at first viewing, this system soon comes to serve its intended purpose of conveying a lot of information very quickly.

Our starting place is that each chemical element has its own symbol, so that hydrogen becomes H, carbon C, and platinum Pt.

When we begin to combine these elements into molecules, it is necessary to specify how *many* atoms of each element are part of the molecule. If we have two atoms of oxygen together—which is the way oxygen is usually packaged in our atmosphere—we have the molecule O_2. Three molecules of O_2 is written as $3O_2$. This kind of notation is known as a **molecular formula**. It is very helpful in stipulating the makeup of molecules, from the simple, such as oxygen, to the complex, such as chlorophyll, which is notated $C_{55}H_{72}MgN_4O_5$.

To see a molecular formula is to learn a lot about what atoms are in a molecule, but nothing about the way the atoms are *arranged* in relation to one another. (Look at chlorophyll's formula. Is there a line of 55 carbon atoms followed by 72 hydrogen atoms? From the molecular formula, how could you tell?) To convey this ordering information,

chemists and biologists use what are known as **structural formulas**—two-dimensional representations of a given molecule. Methane (CH_4) is a very simple molecule composed, as the molecular formula shows, of one atom of carbon and four of hydrogen. In a structural formula, these constituent parts are conceptualized like this:

$$\begin{array}{c} H \\ | \\ H-C-H \\ | \\ H \end{array}$$

methane

Note that there is a single line between each hydrogen atom and the central carbon atom. This has not been done just because a single line is easy to draw. The bond between any of the hydrogen atoms and carbon is a **single bond**: Each line represents *one* pair of electrons being shared. Thus:

$$\begin{array}{c} H \\ H:C:H \\ H \end{array} \text{---- electron pairs}$$

There can also be **double bonds** and **triple bonds**. When carbon dioxide forms, there are two oxygen atoms. Each shares two pairs of electrons with a lone carbon

Reactive and Unreactive Elements

The elements considered so far all welcome bonding partners, because all of them have incomplete outer shells. This is not true of all elements, however. There is, for example, the helium atom, which has two electrons. It thus *comes equipped*, we might say, with a filled outer shell. As such, it is extremely stable—it is unreactive with other elements. It is so unreactive that it is part of a family of elements that at one time were known as the inert gases, because it was thought that these elements never combined with anything. At the opposite end of the spectrum are elements that are extremely reactive. Look again at the representation of the sodium molecule in Figure 2.8. It has 11 electrons, two in the first shell and eight in the second, which leaves but one electron in the third shell—a very unstable state. Between the extremes of sodium and helium are elements with a range of outer (or *valence*) electrons. Thus, there is a spectrum of stability in the chemical elements, based on the number of outer-shell electrons each element has—from 1 to 8.

Polar and Nonpolar Bonding

Not all covalent bonds are created alike. When two hydrogen atoms come together, the result is a hydrogen molecule (H_2). Now, in the hydrogen molecule, the electrons are shared *equally*. That is, the two electrons the hydrogen atoms are sharing are equally attracted to each hydrogen atom. This is not the case, however, with the water molecule.

Look at the representation of the water molecule in **Figure 2.10a**. As it turns out, the oxygen atom has a greater power to attract electrons to itself than do the hydrogen atoms. The term for measuring this kind of pull is **electronegativity**. Because the oxygen atom has more electronegativity than do the hydrogen atoms, it tends to pull the shared electrons away from the hydrogen and toward itself. When this happens, the molecule takes on a **polarity** or a difference in electrical charge at one end as opposed to the other. Because electrons are negatively charged, and because they can be found closer to the oxygen nucleus, the oxygen end of the molecule

atom. Here's how we would notate the double bond in a carbon dioxide (CO_2) molecule:

$$O = C = O$$

carbon dioxide

Though structural (or "skeletal") formulas can tell us a good deal about the ordering of atoms in a molecule, they tell us very little about the *three-dimensional* arrangements of atoms. For this, we rely on two other kinds of representations, the **ball-and-stick model** and the **space-filling model**. An ammonia molecule (NH_3) is pictured next in both forms, with the molecular and structural formulas added to show the progression:

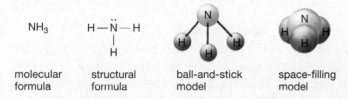

molecular formula · structural formula · ball-and-stick model · space-filling model

Note that the ball-and-stick model gives us a better idea of molecular angles of the atoms, but that the space-filling model gives us a better idea of the relative size of these atoms and how one actually hugs the other.

Finally, it is useful to have a way to notate the "before" and "after" stages of a chemical reaction. This is done by employing a simple arrow, as when carbon reacts with hydrogen to form methane: $C + 4H \rightarrow CH_4$. The carbon and hydrogen atoms on the left are **reactants**, the arrow means "yields," and the methane molecule on the right is the **product** of the reaction. Here's a graphic representation of what is happening:

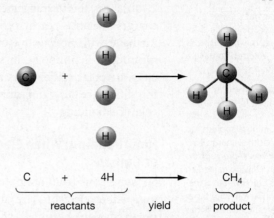

C + 4H ⟶ CH₄
reactants · yield · product

An important thing to keep in mind about notation goes back to the law of conservation of mass (page 25): Matter is neither created nor destroyed in a chemical reaction. It follows that reactions such as the one above must be *balanced*: We must have the same number of atoms when the reaction is finished (on the right) as when it started (on the left). Thus, we could not have $C_2 + 4H \rightarrow CH_4$. This is an unbalanced reaction. We started out with two atoms of carbon on the left, but somehow ended up with one atom of carbon on the right. Nature doesn't play that way.

becomes slightly negatively charged, while the hydrogen regions become slightly positively charged. We still have a covalent bond, but it is a specific type: a **polar covalent bond**. Conversely, with the hydrogen molecule—where electrons are being shared equally—we have a **nonpolar covalent bond**.

To grasp the importance of this, consider the water molecule, with its positive and negative regions. What's going to happen when it comes into contact with *other* polar molecules? The oppositely charged parts of the molecules will attract and the similarly charged parts will repel. It's like having a bar magnet and trying to bring its positive end into contact with the positive end of another magnet: Left on its own, the second magnet just flips around, so that positive is now linked to negative. In the same way, molecules flip around in relation to their polarity.

It is possible for atoms with different electronegativity to link together and still have the resulting molecule be nonpolar. Water is polar because the atom with more electronegativity

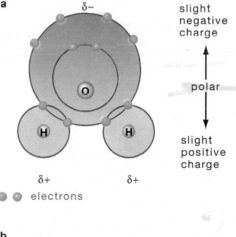

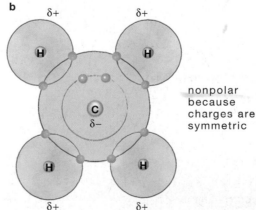

a — δ− / slight negative charge / polar / slight positive charge / δ+ / δ+ / electrons

b — δ+ / δ+ / δ+ / δ+ / δ− / nonpolar because charges are symmetric

Figure 2.10
Polar and Nonpolar Covalent Bonding

a In the water molecule, the oxygen atom exerts a greater attraction on the shared electrons than do the hydrogen atoms. Thus the electrons are shifted toward the oxygen atom, giving the oxygen atom a partial negative charge (because electrons are negatively charged) and the hydrogen atoms a partial positive charge. ("Partial" here is indicated by the Greek delta symbol δ.) The molecule as a whole is polar, meaning it has a difference in charge at one end, as opposed to the other.

b In the methane molecule, the carbon atom is more electronegative than the hydrogen atoms, but the methane molecule as a whole is nonpolar because its hydrogen atoms are arranged symmetrically around the central carbon atom, meaning the partial charges that exist balance each other out. Thus, methane has no difference in charge at one end, as opposed to the other.

Figure 2.11
Ionic Bonding

a Sodium has but a single electron in its outer shell, while chlorine has seven, meaning it lacks only a single electron to have a completed outer shell.

b When these two atoms come together, sodium loses its third-shell electron to chlorine, in the process becoming a sodium ion with a net positive charge (because it now has more protons than electrons). Having gained an electron, the chlorine atom becomes a chloride ion, with a net negative charge (because it has more electrons than protons).

c The sodium and chloride ions are now attracted to each other because they are oppositely charged.

d The result of this "electrostatic" attraction, involving many sodium and chloride ions, is a sodium chloride crystal (NaCl), better known as table salt.

(the oxygen) lies to one *side* of the two hydrogen atoms. Meanwhile, in a molecule such as methane, four hydrogen atoms are arranged in a symmetrical way around a central, and more electronegative, carbon atom (**see Figure 2.10b**). In this arrangement the differing charges balance each other out, leaving methane with no positive or negative end—meaning it is nonpolar. In sum, some molecules are polar while others are nonpolar, and this difference has significant consequences for chemical bonding.

Ionic Bonding: When Electrons Are Lost or Gained

So we've gone from nonpolar covalent bonding, where electrons are shared equally, to polar covalent bonding, where electrons are pulled to one side of the resulting molecule. What if we carried this just one step further and had instances in which the electronegativity differences between two atoms were so extreme that electrons were pulled *off* of one atom altogether, only to latch on to the atom that was attracting them? This is what happens in our second type of bonding, **ionic**

bonding. The classic illustration of this type of bonding involves the sodium we looked at earlier and the element chlorine. Recall that sodium has 11 electrons, meaning that there's a lone electron flying around in its third electron shell. Chlorine, meanwhile, has 17 electrons, meaning it has 7 electrons in the third shell. Remember that 8 is a magic number for outer-shell stability. Sodium could get to this number by *losing* one electron, while chlorine could get to it by *gaining* one electron. That's just how this encounter goes: Sodium does in fact lose its one electron, chlorine gains it, and both parties become stable in the process (**see Figure 2.11**).

What Is an Ion?

But this story has a postscript. Having lost an electron (with its negative charge), sodium (Na) then takes on an overall *positive* charge. Having gained an electron, chlorine (Cl) takes on a negative charge. Each is then said to be an **ion**: a charged atom; or, to put it another way, an atom whose number of electrons differs from its number of protons. We denote the ionized forms of these atoms like this: Na^+, Cl^-. Were an atom to gain or lose more electrons than this, we would put a number in front of the charge sign. For example, to show that the magnesium atom has lost two electrons and thus become a positively charged magnesium ion, we would write Mg^{2+}.

Note that we now have two ions, Na^+ and Cl^-, with differing charges in proximity to one another. They are thus attracted to one another through an *electrostatic attraction*, and have an ionic bond between them. This hardly ever happens with just *two* atoms, of course. Many billions of atoms are bonded together in this way, up, down, and sideways from each other. This whole collection is, likewise, called an ion; or, if two or more elements are mixed together this way, an **ionic compound**. The particular ionic compound just described actually is very familiar. Sodium and chlorine combine to create sodium chloride, which is better known as table salt. The notation should properly be written Na^+Cl^-, but it is usually denoted as just plain NaCl.

Is it apparent how an ionic compound differs from a molecule? In an ionic compound, there is no fixed number of atoms linked up in a defined spatial relationship, as in H_2O. Rather, an undefined number of charged atoms are bonded together, as in NaCl.

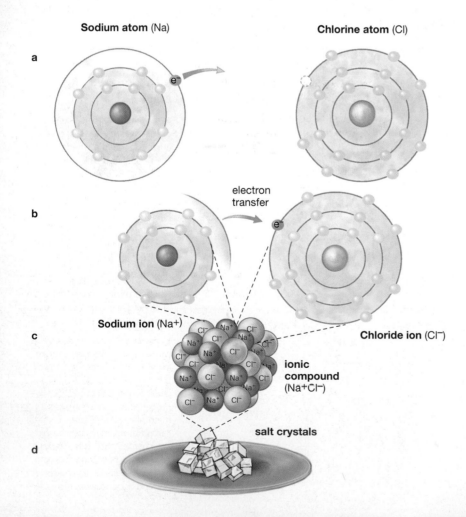

Sodium atom (Na)   **Chlorine atom (Cl)**

a

electron transfer

b

Sodium ion (Na+)   **Chloride ion (Cl⁻)**

c   **ionic compound (Na+Cl⁻)**

salt crystals

d

Free Radicals

Their name makes them sound like a group of sixties activists set loose after years in jail, but **free radicals** aren't people at all; they are atoms or molecules. They *have* been set loose, however—to damage human bodies in illnesses that may range from cancer to coronary heart disease.

The way the public normally hears about these culprits is through the recommended means of limiting their harm: a diet rich in vitamins C, E, and the substance beta carotene. Looking at free radicals from another angle, however, we can see how they represent a damaging exception to the rules of chemical bonding.

You know that atoms "seek" to have a full outer energy shell, which in most cases means eight electrons. In covalent bonding, atoms achieve this state by sharing *pairs* of electrons with one another, one electron of each pair coming from each of the atoms involved. Occasionally in nature, however, atoms come together to create a molecule in which one of the component atoms has an *unpaired electron* in its outer shell. Nitrogen can come together with oxygen, for example, to form nitric oxide (NO). Recall that oxygen has six outer electrons (and thus needs two for stability), while nitrogen has five outer electrons (and thus needs three). When oxygen and nitrogen hook up, they can share only two electrons with one another before oxygen's outer shell is filled. This, however, leaves nitrogen with an unpaired electron.

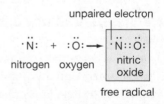

unpaired electron

nitrogen   oxygen   nitric oxide

free radical

Unstable molecules like this usually exist only briefly, as intermediate molecules in chemical reactions. And in people, that's just the problem. Human beings are among the many species that use a terrific amount of oxygen to extract energy from food. In this process, oxygen is constantly picking up electrons. Through the many steps in metabolism, oxygen may come together with other substances to create a type of free radical called reactive oxygen. Though any one reactive oxygen molecule is short-lived, the damage comes

by way of a destructive chain reaction: Seeking partners, one free radical begets more, which beget more.

Where's the harm? Well, for one thing, free radicals may irritate or scar artery walls, which invites artery-clogging fatty deposits around the damage. They also may have a

Free radicals may irritate or scar artery walls, which invites artery-clogging fatty deposits around the damage.

mutation-causing or "mutagenic" effect on human DNA, which can be a factor leading to cancer. Some primary sites of free-radical generation and damage are the "powerhouses" in cells—structures called mitochondria—which are primary sites at which energy is transferred from food. Indeed, a growing body of evidence supports a long-standing theory of human aging, which holds that many of the things we associate with getting older—memory loss, hearing impairment—can be traced to the cumulative effects of free radicals damaging DNA in the mitochondria, thus diminishing the body's energy supply.

Free radicals are the natural product of metabolism in human beings; they are the price we pay for being alive. However, they can be created in us in *greater* numbers in accordance with our behavior. Some of the usual suspects seem to be involved here—cigarette smoking, alcohol consumption, and sunlight exposure. Radiation provides a good example of how free radicals can be produced in us. Medical workers guard against getting excessive exposure to x-rays, because, like other forms of "ionizing" radiation, x-rays cause water molecules in living tissue to break down in such a way that they yield free radicals.

Against this production of free radicals, however, nature has also provided its own set of free-radical scavengers, among them the aforementioned beta carotene and vitamins C and E. We can control these, in the sense of making sure we have plenty of them in our diets. We can buy them in pill form, of course, but the jury is still out on the effect of very large doses of these substances. The best bet is to eat the right kinds of foods—meaning a lot of citrus fruits, whole grains, and vegetables of the green leafy, orange, and yellow variety.

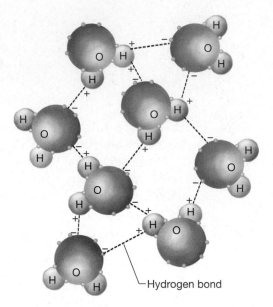

Tutorial 2.2.6

Hydrogen Bonding

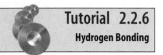

Figure 2.12
Hydrogen Bonding
The hydrogen bond, in this case between water molecules, is indicated by the dotted line. It exists because of the attraction between hydrogen atoms, with their partial positive charge, and the unshared electrons of the oxygen atom, with their partial negative charge.

To take a step back for a second, recall that bonding runs a gamut from the nonpolar covalent bonding (where electrons are shared equally) to the slightly charged polar covalent bonding (where they are shared somewhat unequally) to the charged ionic bonding (where electrons are gained or lost altogether). It's important to recognize that there is a *spectrum of polarity*, and that within it, some bonds are almost completely ionic (as with sodium chloride) while others are completely nonpolar (as with the hydrogen molecule).

A Third Form of Bonding: Hydrogen Bonding

We need to look at one more variant on bonding, called hydrogen bonding. Recall that in any water molecule, the stronger electronegativity of the oxygen atom pulls the electrons *shared* with the hydrogen atoms toward the oxygen nucleus, giving the oxygen end of the molecule a partial negative charge and the hydrogen end of the molecule a partial positive charge. So what happens when you place several water molecules together? A positive hydrogen atom of one molecule is weakly attracted to the negative, *unshared* electrons of its oxygen neighbor. Thus is created the **hydrogen bond**, which links an already covalently bonded hydrogen atom with an electronegative atom (in this case with oxygen; **see Figure 2.12**). Hydrogen bonding is a linkage that, for our purposes, nearly always pairs hydrogen with either oxygen or nitrogen. These relatively weak bonds are important in linking the atoms of a single molecule to one another, but they are just as important in creating bonds *between* molecules, as in the example. The hydrogen bond, indicated by a dotted line, exists in many of the molecules of life —in DNA, proteins, and elsewhere.

2.3 Some Qualities of Chemical Compounds

Molecules Have a Three-Dimensional Shape

We now need to make more explicit what has been noted only by implication in our diagrams of water molecules: that molecules and ionic compounds have a three-dimensional shape. It is useful to depict them as two-dimensional chains and rings and such, but in real life a molecule is as three-dimensional as a sculpture. A fair number of shapes are possible, even in simpler molecules. Atoms may be lined up in a row, or in triangles or pyramid shapes. As an example, look at another representation of the water and methane molecules (**Figure 2.13**). You can see that in water there is a very definite spatial configuration: Its hydrogen atoms are splayed out from its oxygen atom at an angle of 104.5°.

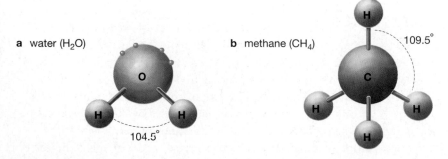

Figure 2.13
Three-Dimensional Representations of Molecules

a In the case of water there is an angle of 104.5° between hydrogen atoms.

b Methane is a molecule with an angle of 109.5° between hydrogen atoms.

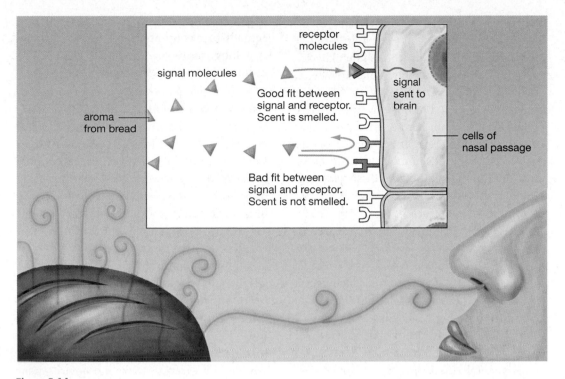

Figure 2.14
The Importance of Shape
Gas molecules wafting off from bread (the triangles) bind with specific receptors on the surface of the cell, thus acting as signaling molecules that set a cellular process in motion. For this binding to take place, the gas and nasal receptor molecules must fit together; this fit is governed by the shape of each molecule.

Tutorial 2.3.1
Molecular Shape
and Biology

Molecular Shape Is Very Important in Biology

Why does molecular shape matter? It is critical in enabling biological molecules to carry out the activities they do. This is so because molecular shape determines the capacity of molecules to latch onto or "bind" with one another. When, for example, you smell the aroma of fresh-baked bread, gas molecules wafting off the bread bind with receptor molecules in your nasal passages, thus sending a message to the brain about the presence of bread. It is the precise shape of the gas molecules and nasal receptor molecules that allows them to bind with one another. Look at **Figure 2.14** to see how this works. If you look at **Figure 2.15**, you can get an idea of how large some biological molecules are, relative to the simple molecules considered so far.

Having learned a little about atoms and molecules, we now need to review a few final concepts that will aid us in understanding chemistry on a slightly larger scale.

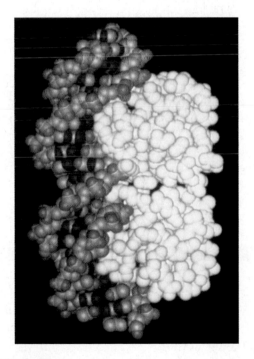

Figure 2.15
Complex Binding
A computer model of some real-life molecular binding. In this case a protein (the yellow atoms) binds to a length of DNA. (The protein is a "repressor" that turns off the activity of this section of DNA.) This space-filling model provides some idea of the enormous number of atoms and the complicated shapes that make up some of the molecules employed by living things.

Solutes, Solvents, and Solutions

Take a glass of water and pour a little salt in it. When you stir that up, the salt quickly disappears. It has not actually gone anywhere, of course. It has simply mixed with the water. Now, if it has mixed uniformly, so that there are no lumps of salt here or there, you have created a **solution**: a homogeneous mixture of two or more kinds of molecules, atoms, or ions. The salt is what's being dissolved here, so it is the **solute**. The water is doing the dissolving, so it is the **solvent** (see **Figure 2.16**).

Molecules vary greatly in the degree to which they are *soluble*—are able to be dissolved—in different solvents. This is because, for something to act as a solvent, it must be able to form chemical bonds with the solute. The more bonding a solvent can do with a solute, the greater capacity that solvent will have to break down the solute.

A general rule is that like dissolves like. Substances that are nonpolar dissolve best in nonpolar solvents; substances that are polar dissolve best in polar solvents. Salt and water are both polar, and the one dissolves in the other. Conversely, the ingredient in soap that actually breaks up nonpolar greases and dirt is a long, nonpolar molecular chain composed of hydrogen and carbon atoms.

On to Some Detail Regarding Water

In biology, you will often see casual references to a molecule being *soluble* in a certain way. And you will see this most commonly mentioned in two ways: as something being fat-soluble or water-soluble. Does it dissolve in fat or does it dissolve in water? Both things are important to life, but the importance of water as a solvent cannot be overstated. Water accounts for about 75–85 percent of a cell's weight, and most cells are surrounded by it. It is not difficult to see why water should be so prominent in this way. Life began in water, much life lives in it today, and our own ancestors made the transition from sea to land essentially by carrying their wet environment with them—inside themselves. So important to life is this simple, common molecule that we end with it here, only to take it up again in our next chapter in a more detailed way.

Figure 2.16
A Solute Dissolved by a Solvent Makes a Solution
When we pour a small amount of table salt (an ionic compound composed of sodium and chloride ions) into water, the salt crystals dissolve into the water.

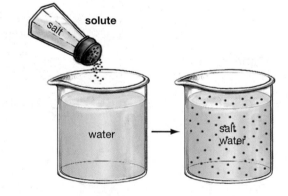

Chapter Review

Summary

2.1 The Nature of Matter: The Atom

- The fundamental unit of matter is the atom. The three most important constituent parts of an atom are protons, neutrons, and electrons. Protons and neutrons exist in the atom's nucleus, while electrons move around the nucleus, at some distance from it. Atoms have an electrical charge. Protons are positively charged, electrons are negatively charged, but neutrons carry no charge.

- An element is any substance that cannot be reduced to any simpler set of constituent substances through chemical means. Each element is defined by the number of protons in its nucleus.

- The number of neutrons in an atom can vary independently of the number of protons. Thus a single element can exist in various forms, called isotopes, depending on the number of neutrons it possesses.
TUTORIAL 2.1.1: Structure of the Atom, Elements, Isotopes

2.2 Matter Is Transformed through Chemical Bonding

- Atoms can link to one another in the process of chemical bonding. Among the forms this bonding can take are covalent bonding, in which atoms share one or more electrons, and ionic bonding, in which atoms lose and accept electrons from each other.

- Chemical bonding comes about as atoms "seek" their lowest energy state. An atom achieves this state when it has a filled outer electron shell. Hydrogen and helium require two electrons in orbit around their nuclei to have filled outer shells, while all other elements require eight electrons to have filled outer shells.

- A molecule is a compound of a defined number of atoms in a defined spatial relationship. For example, two hydrogen atoms can link with one oxygen atom to form one water molecule.

- Atoms of different elements differ in their power to attract electrons. The term for measuring this power is electronegativity. Through electronegativity, a molecule can take on a polarity, meaning a difference in electrical charge at one end compared to the other. Covalent chemical bonds can be polar or nonpolar. A polar covalent bond exists when shared electrons are not being shared equally among atoms in a molecule, due to electronegativity differences.
  TUTORIAL 2.2.1: Covalent Bonding

- Two atoms will undergo a process of ionization when the electronegativity differences between them are great enough that one atom loses one or more electrons to the other. This process creates ions, meaning atoms whose number of electrons differs from their number of protons. The charge differences that result from ionization can produce an electrostatic attraction between ions. This attraction is an ionic bond. When atoms of two or more elements bond together ionically, the result is an ionic compound.
  TUTORIAL 2.2.4: Ionic Bonding

- Hydrogen bonding links a covalently bonded hydrogen atom with an electronegative atom. In water, a hydrogen atom of one water molecule will form a hydrogen bond with an unshared oxygen electron of a neighboring water molecule.
  TUTORIAL 2.2.6: Hydrogen Bonding

2.3 Some Qualities of Chemical Compounds

- Three-dimensional molecular shape is important in biology because this shape determines the capacity molecules have to bind with one another.
  TUTORIAL 2.3.1: Molecular Shape and Biology

- A solution is a homogeneous mixture of two or more kinds of molecules, atoms, or ions. The compound being dissolved in solution is the solute; the compound doing the dissolving is the solvent.

- A general rule in chemistry is that like dissolves like: Substances that are nonpolar dissolve best in nonpolar solvents, while substances that are polar dissolve best in polar solvents.

Key Terms

atomic number 22	molecular formula 26
ball-and-stick model 27	molecule 25
chemical bonding 23	neutron 20
covalent bond 25	nonpolar covalent bond 27
double bond 26	nucleus 20
electrical charge 21	polar covalent bond 27
electron 20	polarity 26
electronegativity 26	product 27
element 22	proton 20
free radical 29	reactant 27
hydrogen bond 30	single bond 26
ion 28	solute 32
ionic bonding 28	solution 32
ionic compound 28	solvent 32
isotope 23	space-filling model 27
law of conservation of mass 25	structural formula 26
mass 19	triple bond 26

Understanding the Basics

Multiple-Choice Questions

1. Carbon is an element with an atomic number of 6. Based on this information, which of the following statements is true? (More than one may be true.)
   a. Carbon can be broken down into simpler component substances.
   b. Carbon cannot be broken down into simpler component substances.
   c. Each carbon atom will always have 6 neutrons.
   d. Each carbon atom will always have 6 protons.
   e. Protons + electrons = 6.

2. Suppose that you are reviewing for a test, and a fellow student says that the equation for photosynthesis is $6CO_2 + H_2O \rightarrow C_6H_{12}O_6 + 6O_2$. How would you reply?
   a. They are right.
   b. The way this equation is written violates the law of the conservation of matter.
   c. The CO_2 is held together by ionic bonds.
   d. A hydrogen bond holds the two hydrogens to the oxygen in the water molecule.
   e. There are 12 carbons in sugar.

3. Neon used to be called an inert gas. Thus it
   a. easily forms perfect covalent bonds
   b. easily forms ionic bonds
   c. has a filled outer shell
   d. is polar
   e. all of these

4. Oxygen and hydrogen differ in their electronegativity. Thus
   a. They share electrons, but unequally.
   b. Sometimes oxygen takes electrons completely away from hydrogen.
   c. They share electrons equally.
   d. Hydrogen is attracted to oxygen, but does not bond with it.
   e. They have the same number of protons.

5. A molecule that does not have a net electrical charge at one end as opposed to the other is:
   a. an isotope
   b. a polar molecule
   c. a reactant
   d. a nonpolar molecule
   e. a solvent

6. You add sugar to your coffee, and the sugar dissolves. Thus the coffee is the _____ and the sugar is the _____.
   a. solute . . . solvent
   b. solvent . . . solute

   c. polar covalent bond . . . nonpolar covalent bond
   d. nonpolar covalent bond . . . polar covalent bond
   e. ionic bond . . . hydrogen bond

7. The two strands of a DNA molecule are held together because large numbers of hydrogens that are covalently bonded to oxygen or nitrogen in one strand are weakly attracted to oxygens or nitrogens in the opposite strand. Therefore the two strands of DNA are held together by
   a. polar covalent bonds
   b. nonpolar covalent bonds
   c. ionic bonds
   d. inert bonds
   e. hydrogen bonds

8. While baking cookies for a friend, you're having a hard time keeping your roommate from eating them because the smell is driving her wild. She is able to smell the cookies because the molecules wafting from them
   a. have a shape that allows them to bind to receptors in her nose
   b. form permanent attractions to nerves in her nose by ionic bonding

c. form temporary attractions to nerves in her nose by covalent bonding

d. dissolve in her blood and travel to her brain

e. travel along her nerves—causing changes as they go—until they reach the brain, where they are recognized

Brief Review

1. As with most elements, carbon comes in several forms, one of which is carbon-14. What are these forms called, and how does one differ from the other?

2. Draw a line and label one end "complete + or − charge" and the other end "no charge" to indicate the charges on the molecules or ions after bonding has occurred. Along the line, indicate where polar covalent bonds, nonpolar covalent bonds, and ionic bonds should be placed.

3. Why are free radicals so dangerous when they are produced in our bodies?

4. Compare the size of an atom with the size of its nucleus. Where are the electrons? In light of this, what occupies most of an atom?

5. Why are atoms unlikely to react when they have their outer shell filled with electrons?

Applying Your Knowledge

1. In the Middle Ages alchemists labored to turn common materials such as iron into precious metals such as gold. If you could journey back in time, how could you convince an alchemist that iron cannot be changed into gold? Remember that nothing was really known about atoms and elements in the Middle Ages.

2. Go to the Internet and find three references to free radicals. Are any of these related to selling a product designed to decrease the number of free radicals? Based on what you have learned in this chapter, do you think that taking these products is a good idea?

3. The next time you go to a large warehouse or discount store, look around. List as many reasons as you can to explain why it is important that molecules have a three-dimensional shape. How would sensations differ if we lived in a two-dimensional world?

3

Water, pH, and Biological Molecules

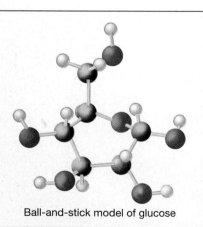

Ball-and-stick model of glucose

The most important energy
source for our bodies.
(Section 3.3, page 47)

Energy used:
Glucose can serve as
an immediate source
of energy.

glucose

Energy stored:
Or, glucose can be
made into glycogen and
stored in the liver and
muscle for later use.

glycogen

Energy from plants—use it now or
store it for later on.
(Section 3.4, page 53)

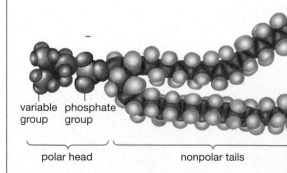

variable   phosphate
group   group

polar head   nonpolar tails

A dual-natured molecule with head and tails.
(Section 3.4, page 56)

The qualities of water have shaped life. The degree to which solutions are acidic or basic—the pH they have—has strong effects on living things. Four kinds of biological molecules form the living world.

Imagine that it's raining outside as you read this, the rhythm of individual drops combining to create a sound that is like no other. It's a sound that most people find comforting. And isn't that the way with other sounds water makes when it's in motion? With ocean waves or a running stream? It may be that these things are soothing purely *as sounds*—something about their tone, or their steady pace. But isn't it possible that we are calmed by the sounds of rain or streams because they are the sounds *of water*? Rain has fallen on leaves for millennia, sending our ancestors—and now us—a timeless message: Water's here; you and your family will have enough to drink; food will be abundant; all is well.

3.1 The Importance of Water to Life

Human societies tend to come together precisely where water exists, of course. In places where it's plentiful, water seems less like a substance than an environment: People drink it, cook in it, bathe in it, wash wastes away in it, harness it for power, swim in it. Some 71 percent of Earth's surface is ocean water, and human bodies are about 66 percent water by weight, so that if we have, say, a 128-pound person, about 85 pounds of him or her will be water. If Earth amounts to a watery environment flecked by the landmasses we call continents, the human body amounts to a watery mass with significant proportions of other materials immersed in it.

Water Is a Major Player in Many of Life's Processes

This preponderance of water in living things gives rise to an important point. Recall that you learned, in Chapter 2, what a solution is. Well, a specific kind of solution is one in which water is the solvent. This is an **aqueous solution**. Thus, when biologists talk about a given process taking place "in solution," what they mean, unless otherwise noted, is that it is occurring in *aqueous* solution.

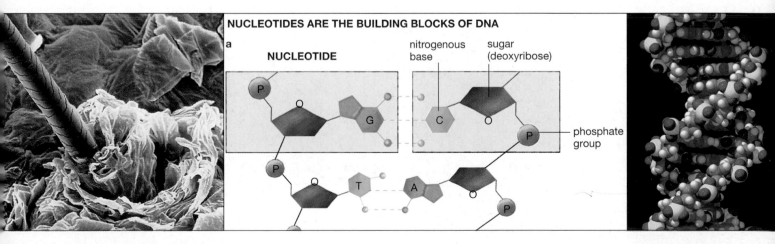

A protein found over almost all of your body.
(Section 3.4, page 59)

NUCLEOTIDES ARE THE BUILDING BLOCKS OF DNA

a
NUCLEOTIDE

nitrogenous base

sugar (deoxyribose)

phosphate group

The building blocks of DNA.
(Section 3.4, page 61)

A portrait of the most famous molecule in all of biology.
(Section 3.4, page 61)

Because water is the medium in which life's processes take place, it is of course a major player in many of life's chemical reactions. One role that water plays is to break down the bonds of compounds that have been placed into it. (These compounds are the *solutes* to water's *solvent*.) This role is reviewed in Chapter 2 in connection with table salt, or sodium chloride.

Look at **Figure 3.1** for a more detailed view of water's solvent power. Attracted by the polar nature of the water molecule, the sodium and chloride ions separate from a crystal—and from each other. Each ion is then surrounded by several water molecules (Figure 3.1b). These units keep the sodium and chloride ions from getting back together. In other words, they keep the ions evenly dispersed throughout the water, which is what makes this a solution (Figure 3.1c). Water works as a solvent here because the ionic compound sodium chloride carries an electrical charge. What *generally* makes water work as a solvent, however, is its ability to form hydrogen bonds with other molecules.

Water's Structure Gives It Many Unusual Properties

When we note water's ability to act as a solvent, we're actually not giving water its due. Water is not just *a* solvent: Over the range of substances, nothing can match it as a solvent. It can dissolve more compounds in greater amounts than can any other liquid.

Ice Floats Because It Is Less Dense than Water

But solvency power is merely the beginning of water's abilities. It is a multi-talented performer. And it achieves this status because it is . . . odd. Compared to other molecules, water is like some zany eccentric whose powers stem precisely from its eccentricity. Consider the fact that ice floats on water. This is so because the solid form of H_2O is less dense than the liquid form—a strange reversal of nature's normal pattern. Things work this way with H_2O because, when water molecules slow their motion in cooling, they are able to form the maximum number of hydrogen bonds with each other. The result is that water molecules are spaced *farther apart* when frozen. Thus, ice is less dense than water. This may seem like some minor, quirky quality but it actually has the effect of making possible life as we know it. Ice on the surface of water acts to insulate the water beneath it from the freezing surface temperatures and wind above, creating a warmer environment for organisms such as fish (**see Figure 3.2**). If ice *sank*, on the other hand, the entire body of water would freeze solid at colder latitudes, creating an environment in which few living things could survive for long.

Water Has a Great Capacity to Absorb and Store Heat

Water serves as an insulator not only when it is frozen, but also when it is liquid or gas. This is

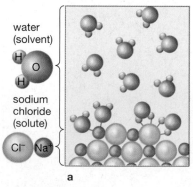
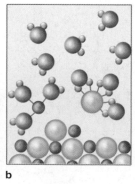
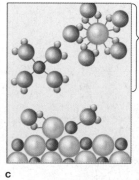

water
(solvent)

H
O
H

sodium
chloride
(solute)

Cl⁻ Na⁺

a

b

c

sodium and
chloride ions
dissolved
in water

Tutorial 3.1.1
**Chemical Properties
of Water**

Figure 3.1
Water's Power as a Solvent

a Both sodium ions (Na^+) and chloride ions (Cl^-), the component parts of this crystal of table salt, carry an electric charge. Thus, both are attracted to the charged regions of the water molecule—sodium to the negative charge of water's oxygen atom, and chloride to the positive charge of water's hydrogen atoms.

b Pulled from the crystal, and separated from each other by this attraction, sodium and chloride ions become surrounded by water molecules.

c This process of separating sodium and chloride ions repeats until both ions are evenly dispersed, making this an aqueous solution.

so because water has what is known as a high **specific heat**: Compared to other substances, it takes a relatively large amount of energy to raise the temperature of water. Put a gram container of drinking alcohol (ethanol) side by side with one of water, heat them both, and it will take almost twice as much energy to raise the temperature of the water one degree Celsius as it will the alcohol. Having absorbed this much heat, however, water then has the capacity to *release* it when the environment around it is colder than the water itself. The result? Water acts as a great heat buffer for Earth. The oceans absorb tremendous amounts of radiant energy from the Sun, only to release this heat when the temperature of the air above the ocean gets colder. Without this buffering, temperature on Earth would be less stable. People who have spent a day and a night in a desert can attest to this effect. The searing heat of the desert day radiates off the desert floor; but at night, with little water vapor in the air to capture this heat, the desert cools off dramatically. In the same way, our *internal* temperature can remain much more stable because the water that makes up so much of us is first able to absorb and then to release great amounts of heat. The sweat that we throw off in exercise has considerable cooling power because each drop of perspiration carries with it a great deal of heat.

Water derives these powers from its chemical structure, in particular from the hydrogen bonding described earlier. Weak and shifting though individual hydrogen bonds may be, collectively they have a great strength. Heat is the motion of molecules. To get molecules moving, though, chemical bonds must first be broken. Because water has a formidable set of hydrogen bonds, it takes a lot of energy to break them and get its molecules moving. This very same set of bonds gives water molecules another notable characteristic: cohesion, meaning a tendency to stay together. With this quality, water can be drawn (by evaporation) in one continuous column from a plant's roots all the way up to its leaves and out into the air as water vapor.

Water's Cohesion Gives It Surface Tension

Water's cohesion imparts a special quality, called **surface tension**, in places where water meets air. Water molecules *below* the surface are equally attracted in all directions to other water molecules. *At* the surface, however, water molecules have no such attraction to the air above them—they are pulled down and to the side, but not up. This causes the "beading" that water droplets do on surfaces. More important for our purposes, surface water molecules pack together more closely

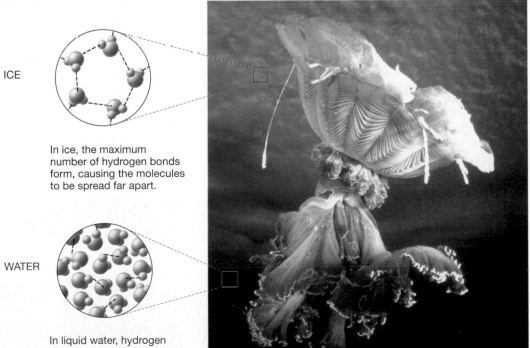

ICE

In ice, the maximum number of hydrogen bonds form, causing the molecules to be spread far apart.

WATER

In liquid water, hydrogen bonds constantly break and reform, enabling a more dense spacing than in ice.

Figure 3.2
Under the Ice
Water is unusual in that its solid form (ice) Is less dense than its liquid form. The fact that ice floats makes life possible in cold-weather aquatic environments. Here a giant jellyfish floats under the ice in the waters off Antarctica.

than do interior molecules, allowing all kinds of small animals to move across the surface of water rather than sinking into it. Note the familiar water strider in **Figure 3.3**.

Having looked at the qualities of water, you may be tempted to think how *uncanny* it is that water has all these qualities that are conducive to life. But there actually is nothing surprising about this. Life on Earth began in water; thus water has fundamentally conditioned life as we know it. Being surprised about water's life-enhancing qualities is like being surprised that a stage is a good place to put on a play.

What Water Cannot Do

With its great complexity, life requires molecules that *can't* be dissolved by water. Such is the case with nonpolar covalent molecules. Compounds made of hydrogen and carbon (**hydrocarbons**) are good examples of nonpolar molecules. Recall that Chapter 2 presented methane (CH_4) as an example of a nonpolar molecule. Petroleum products are more complex hydrocarbons than is methane, and you can see a vivid demonstration of their water-insolubility in oil spills (**see Figure 3.4**). Oil doesn't dissolve in water because the oil carries no electrical charge that water can bond with; thus water has no way to separate one oil molecule from another.

Two Important Terms: Hydrophobic and Hydrophilic

The ability of molecules to form bonds with water has a couple of important names attached to it. Compounds that will interact with water—such as the sodium chloride considered earlier—are known as **hydrophilic** ("water-loving"), while those that do not interact with water are known as **hydrophobic** ("water-fearing"). Both terms are misleading, in that no substance has any known emotional relationship with water. *Hydrophobic* is particularly off the mark, because water does not repel hydrophobic molecules; rather, in bonding together, water molecules form circles around concentrations of hydrophobic molecules, as if they had lassoed them.

The importance of hydrophobic molecules can be illustrated in part by the common milk carton. Why is the milk carton important? Because it can keep milk separate from everything else. We living organisms need some kind of "carton" that can separate the world outside of us from ourselves. Likewise, organisms have great use *within* themselves for compartments that can be sealed off to one degree or another. If water broke down every molecule of life it came in contact with, then it would break down all these divisions of living systems.

Molecules do not have to be completely hydrophilic or hydrophobic. Indeed, as you'll see, a number of important molecules have both hydrophilic and hydrophobic portions.

3.2 Acids and Bases Are Important to Life

When considering the question of aqueous solutions, an important concept is that of acids, bases, and the pH scale used to measure their levels.

We've all had experience with acids and bases, whether we've called them by these names or not. Acidic substances tend to be a little more familiar: lemon juice, vinegar, tomatoes. Substances that are strongly acidic have a well-deserved reputation for being dangerous: The unadorned term *acid* is often

**Figure 3.3
Walking on Water**

a Water's high surface tension enables small animals, such as this pond skater, to walk on it.

b This same surface tension also causes water to bead up on waxy or oily surfaces, such as this goose feather.

used to mean something that can sear human flesh. It might seem to follow that bases are benign, but ammonia is a strong base, as are many oven cleaners. The safe zone for living tissue in general lies with substances that are neither strongly acidic nor strongly basic.

Science has developed a way of measuring the degree to which something is acidic or basic—the pH scale. So widespread is pH usage that it pops up from time to time in television advertising ("It's pH-balanced!").

Acids Yield Hydrogen Ions in Solution; Bases Accept Them

The "H" in pH stands for hydrogen, while the "p" can helpfully be thought of as standing for power. Thus we get "hydrogen power," which describes what lies at the root of pH. An **acid** is any substance that *yields hydrogen ions* when put in solution. A **base** is any substance that *accepts* hydrogen ions in solution.

How might this yielding or accepting come about? Recall first that an ion is a charged atom, and that atoms become charged through the gain or loss of one or more electrons. Because electrons carry a negative charge, the loss of an electron leaves an atom with a net *positive* charge. Also recall that the hydrogen atom amounts to one central proton and one electron that circles around it. A hydrogen ion, then, is a lone proton that has lost its electron, thus yielding a positively charged ion whose symbol is H^+.

Now, suppose you put an acid—hydrochloric acid (HCl)—into some water. What happens is that HCl *dissociates* or breaks apart into its ionic components, H^+ and Cl^-. The HCl has therefore yielded a hydrogen ion (H^+). Now, with a greater concentration of hydrogen ions in it, the water is more acidic than it was (**see Figure 3.5**).

What about bases? There is a compound called sodium hydroxide (NaOH)—better known as lye—that, when poured into water, dissociates into Na^+ (sodium) and OH^- (hydroxide) ions. The place to look here is the OH^- ions. Negatively charged as they are, they would readily bond with positively charged H^+ ions. In other words, they would *accept* H^+ ions in solution, which is the definition of a base. Put another way, were OH^- ions to be put into a solution with H^+ ions in it, they would accept the H^+ ions, making the solution more basic—or, to look at it another

Figure 3.4
Oil and Water Do Not Mix
When there is an oil spill in the ocean, the oil stays concentrated even as it spreads out, because oil and water do not form chemical bonds with each other. Here trawlers are using a boom to clean up after an oil spill in Great Britain.

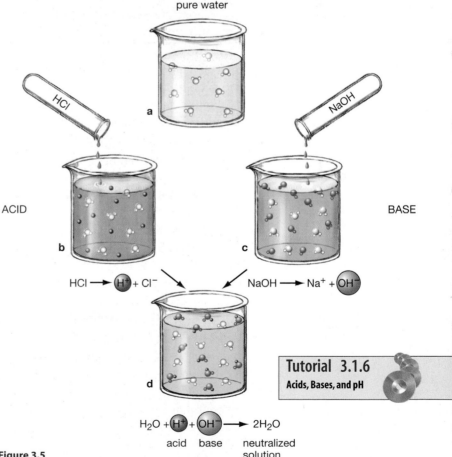

Tutorial 3.1.6
Acids, Bases, and pH

Figure 3.5
Hydrogen Ions and pH

a Pure water is a "neutral" substance in terms of its pH levels.

b Hydrochloric acid (HCl), poured into the water, dissociates into H^+ and Cl^- ions. With a higher concentration of H^+ ions in it, the water moves toward the acidic end of the pH scale.

c An equal concentration of sodium hydroxide, poured into water, dissociates into Na^+ and OH^- ions, moving the water toward the basic end of the scale.

d When the acid and base solutions are poured together, the OH^- ions from **c** accept the H^+ ions from **b**, forming more water and bringing the solution back to a neutral pH.

way, *less acidic*. Thus, acids and bases are something like a teeter-totter: When one goes up, the other comes down. As you might have guessed, in the right proportions they can balance each other out perfectly. Look at Figure 3.5 to see how this would play out with the solutions you've looked at so far. Mixing them, the collection of H^+ and OH^- ions now comes together in water, and for each pair of ions that interacts, the result is:

$$H^+ + OH^- \rightarrow H_2O$$

Water, which is neutral on the pH scale. The acid and the base have perfectly balanced one another out. The OH^- ion, generally referred to as the **hydroxide ion**, is important because compounds that yield them in quantity are strongly basic and can be used to move solutions from the acidic toward the basic.

Many Common Substances Can Be Ranked According to How Acidic or Basic They Are

Look at **Figure 3.6** to get an idea of how acidic or basic some common substances are. Following from the notion of what pH amounts to, it's clear that battery acid, for example, is strongly acidic because, when it dissociates in solution, it yields a large number of hydrogen ions. As you move on to lemon juice and then tomatoes, however, you run into weaker acids, meaning substances that yield fewer H^+ ions in solution. By the time you get to seawater, you've arrived at substances that *accept* hydrogen ions.

The pH Scale Allows Us to Quantify How Acidic or Basic Compounds Are

The net effect of all this yielding and accepting of hydrogen ions is the *concentration* of H^+ ions in solution, and it is through this that the notion of pH is *quantified*—has numbers attached to it—in the form of the **pH scale**. Look at Figure 3.6 again, this time in connection with the numbers that mark it off. You can see that zero on the scale is the most acidic while 14 is the most basic. It's important to note that the pH scale is *logarithmic*: A substance with a pH of 9 is *10* times as basic as a substance with a pH of 8 and *100* times as basic as a substance with a pH of 7.

Some Terms Used When Dealing with pH

Here are some notes on pH terminology:

- As a solution becomes more basic, its pH *rises*. Thus the higher the pH, the more basic the solution; and the lower the pH, the more acidic the solution. Oven cleaner is said to have a high pH, while lemon juice has a low pH.
- Given that, in ionization, the lone hydrogen proton separates from its electron, the hydrogen ion amounts to nothing *but* a proton. It is also correct, therefore, to say that an acid is something that yields *protons* in solution, while a base is something that accepts protons. This is how you will often hear hydrogen ions talked about in biology.
- A solution that is basic is also referred to as an **alkaline** solution.

Why Does pH Matter?

So, why do we care about pH? The brief answer is: because living things are sensitive to its levels in many ways. There is, for example, a class of biological molecules called enzymes that you'll be reading about later in this chapter. Enzymes are chemical tools that must retain a very specific shape to function. However, if you put an enzyme in a solution whose pH is, say, too acidic, the enzyme loses it shape. Why? Because the charged nature of the acidic solution starts breaking down the enzyme's chemical bonds. (Remember, lots of positively charged protons are floating around in an acidic solution.)

Beyond this, consider that scientists discovered early in 2000 that the breath of asthmatics who are having an asthma attack is a hundred times more acidic than is usually the case. There are two lessons in this. One is simply that pH imbalances can have harmful consequences: It is the excess acidity in the lungs of these asthmatics that is partly responsible for closing down their airways. The other lesson has to do with pH and life in general. An asthma attack is an "autoimmune disorder" that comes about when the body mistakenly believes it is being invaded by microorganisms. And how has evolution shaped our bodies to respond to such an invasion? By lowering the pH enormously at the site of the attack, because living things in general (including

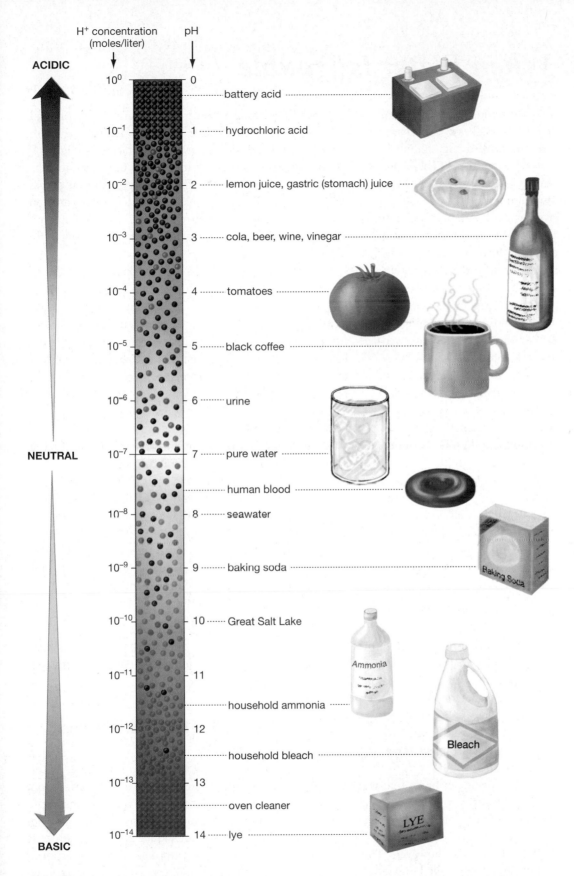

Figure 3.6
Common Substances and the pH Scale
Chemists use units called moles per liter to measure the concentration of substances in solution. The pH scale, derived from this framework, measures the concentration of hydrogen ions per liter of solution. The most acidic substances on the scale have the greatest concentration of hydrogen ions, while the most basic (or alkaline) substances have the least concentration of hydrogen ions. The scale is logarithmic, so that wine, for example, is 10 times as acidic as tomatoes, and 100 times as acidic as black coffee.

Acid Rain: When Water Is Trouble

If the concept of pH seems remote from the real world, consider this: Hundreds of lakes in the eastern and midwestern United States are dying, and their condition has everything to do with pH. They have reached this state not by natural processes, however, but by civilization's impact on the environment.

This situation has been brought about by a phenomenon commonly called acid rain, but more properly known as acid deposition, since acid *snow* and a dry form of acid discharge are problems as well. In acid deposition, you can see an application of pH principles, only played out this time on an enormous scale. You've learned something about acids and bases and the importance of *internal* pH to living things. It makes sense, then, that when rainwater itself has a skewed pH, there is going to be widespread trouble. And this is indeed what has happened, not only in the United States, but in Canada, in Eastern Europe and England, in China—in short, worldwide, though industrialized nations are the most affected. In the 1980s, residents of the Scottish Highlands had the dubious distinction of being able to experience *black snow*, which was extremely acidic. In the pristine Adirondacks region of upstate New York some 200 lakes were, by the early 1990s, incapable of supporting fish. Scandinavian residents at that time might well have envied New York residents, however, because some 16,000 lakes in Norway and Sweden by then had no fish in them.

Acid precipitation comes primarily from two sources: the sulfur dioxide (SO_2) emissions of coal- and oil-burning factories (in particular power plants), and the nitric oxide (NO) and nitrogen dioxide (NO_2) emissions that come mostly from cars, with an assist from factories (**Figure 1**). When these compounds rise into the air, they come together with a culprit you've seen before: an oxygen radical, specifically the hydroxyl radical (OH). In so doing, they are converted to sulfuric acid and nitric acid. *These* compounds then combine with the rainwater in clouds. The result of this is acidic rain or snow.

Acid deposition can do its damage directly or indirectly. By simply making a lake more acidic, it begins to interfere with the metabolic processes of the animals who live there. Acid rain also has the effect, however, of leaching metals from the soil it passes through. Aluminum brought into lakes in this way damages the gills of fish, destroying their ability to absorb oxygen from the water. This same effect takes place in connection with land vegetation, where leached metals can kill plant roots outright or keep them from absorbing nutrients.

Acid deposition will have different effects on different environments, and here again you can see the basics of pH at work. Soils are not neutral in their pH; should acid rain fall on soil that is significantly basic (or "alkaline"), the buffering process you've read about will take place: Basic compounds in the soil will combine with the water and accept hydrogen ions from the rain, thus resulting in water that is less acidic. Likewise, aquatic environments may contain basic ions that can neutralize incoming water.

most microorganisms) just cannot tolerate extremes of pH. Life generally thrives at a pH that hovers around neutral: From about 6–8 is the usual range, with the pH of the cell being about 7. So sensitive are living things to pH that many have developed elaborate **buffering systems** to keep pH hovering at optimal levels. Buffers are generally weak acids or bases that work to neutralize any sudden infusion of acid or base by accepting or donating hydrogen ions.

Different parts of living things have different pH needs, however. The digestive structures in cells, called lysosomes, require a mildly acidic environment to get their work done. But human stomachs require a *very* acidic environment in order to function properly: Gastric juices in the stomach have a pH of between 1.6 and 1.8. pH matters on larger scales as well; acid rain has become a great problem for whole environments. (See "Acid Rain," above.)

3.3 Carbon Is a Central Element in Life

You're crossing a bridge of sorts here. To this point, the discussion has largely been about water and hydrogen and various ions. From time to time, mention was made of the element *carbon* (**see Figure 3.7**), although only as one substance among many. But carbon is not that at all.

Carbon as a Starting Ingredient

Life is based on carbon, or if you will, on carbon compounds immersed in water. You could think of carbon in life in the same way as you do flour

The extent to which such environments possess basic ions is known as their *acid-neutralizing capacity*. Given what you've read about pH, it might be apparent that this capacity is not limitless but instead can dwindle as more acidic compounds are added. Indeed, soils can suffer from a *long-term* depletion of basic compounds, and this has happened in some eastern U.S. forests. A 1990 amendment to the federal Clean Air Act resulted in a reduction in U.S. sulfur emissions, but left some nitrogen emissions unchanged through the middle of the decade. This meant there was an overall decline in acid *deposition* in the Adirondacks in the period 1992–1999. Despite this, however, many Adirondack lakes continued to get more acidic through end of the 1990s. Why would lakes get more acidic if acid deposition has been reduced? Because the acid-neutralizing capacity of the lakes and the soil around them has been so depleted by years of acid rain.

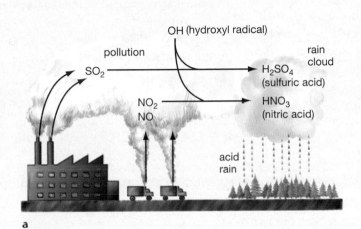

Figure 1
How Acid Rain Forms

(a) Sulfur dioxide (SO_2) from coal- and oil-burning power plants rises into the air along with nitric oxide (NO) and nitrogen dioxide (NO_2), which come mostly from cars. These compounds combine with hydroxyl radicals (OH) in the atmosphere to produce sulfuric acid (H_2SO_4) and nitric acid (HNO_3), which combine with atmospheric water, making it acidic.

(b) The effects of acid rain can be seen in pine trees in the Adirondacks region of New York State.

in baking. How many times does a baking recipe begin: "Start with two cups of flour." With the molecules of life, the recipe often begins: "Start with a carbon ring," or "start with a chain of carbon atoms." But just as cinnamon or fruit will be folded into flour, thus imparting special qualities to the basic ingredient, so will other elements join carbon: A little addition of this here, a little of that on the end, and voilà! A complex molecule that serves a very specific purpose.

Carbon's Importance Stems from Its Bonding Capacity

How does carbon come by its great powers? Linkage. Recall from Chapter 2 that carbon has four outer-shell, or "valence," electrons, but that all elements (except hydrogen and helium) need *eight* outer-shell electrons for maximum stability. This means that carbon

Figure 3.7
Pure Carbon, Mostly Carbon
A diamond is pure carbon and the hardest natural material known. Surrounding the diamond is a lump of coal, which has a high carbon content because it is made up mostly of the remains of carbon-rich, ancient vegetation.

Tutorial 3.2.1
Chemistry of Carbon

needs to link up with four more electrons to achieve stability. Moreover, the bonds that carbon creates are covalent—it is *sharing* electrons with other atoms—which means its linkages are more stable than those formed by elements that bond ionically. Given the natural forces that can buffet life about, such as ultraviolet radiation and heat, this stability is an important quality in first getting life going and then keeping it going. So important is carbon that it forms one whole branch of chemistry, **organic chemistry**, meaning the chemistry of compounds that include carbon.

How does carbon link up with itself or other atoms? You've already had a look at a model of a very simple carbon compound, methane (CH_4). Look at it again:

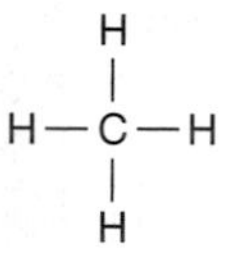

Note that the only elements present are hydrogen and carbon atoms. Now observe a slightly more complex hydrocarbon, the familiar gas propane (C_3H_8):

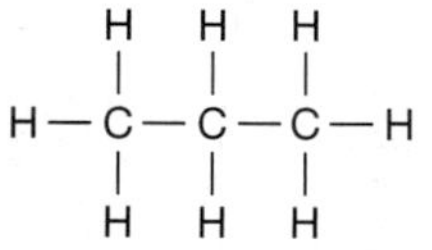

You can see that instead of being surrounded by four hydrogen atoms, the carbons here link up with other carbons, as well as with the hydrogens. This process of extension keeps going with still more complex hydrocarbons. For obvious reasons, the preceding configuration is known as a *straight-chain* carbon molecule. But carbon has more tricks up its sleeve than simple straight-line extensions. The next hydrocarbon up the line is butane (C_4H_{10}). It can be just another straight-chain extension, looking like this:

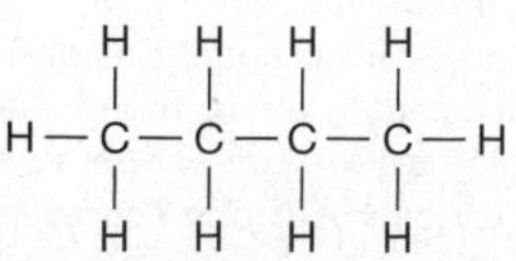

But it can also look like this:

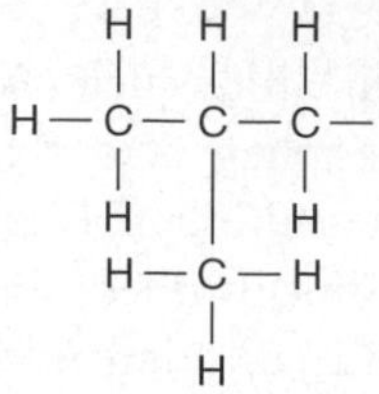

These two forms of butane are known as **isomers**—molecules that are the same in their chemical formulas, but differ in the spatial arrangement of their elements.

Not content to branch only in this way, however, carbon can also form rings. Here is the structure of benzene (C_6H_6):

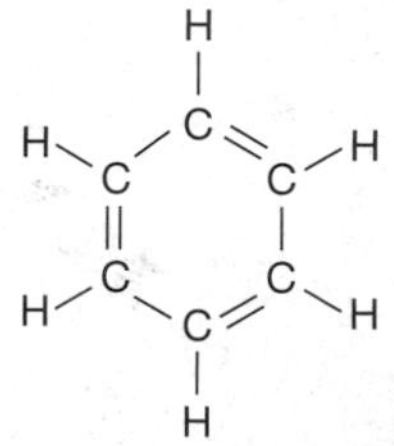

Note the three sets of double lines in the molecule. Recall that this means there are double *bonds* between these atoms; the atoms involved are sharing *two* pairs of electrons.

It just so happens that all the carbon molecules introduced so far are made of nothing but carbon and hydrogen. But this is the exception rather than the rule. Here are two representations of a molecule mentioned already—glucose, better known as blood sugar:

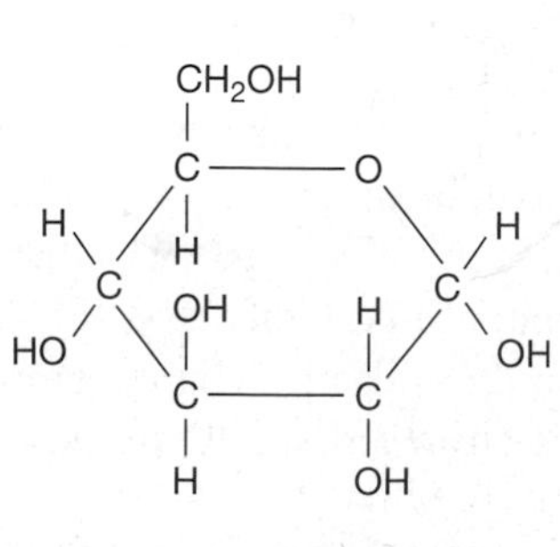

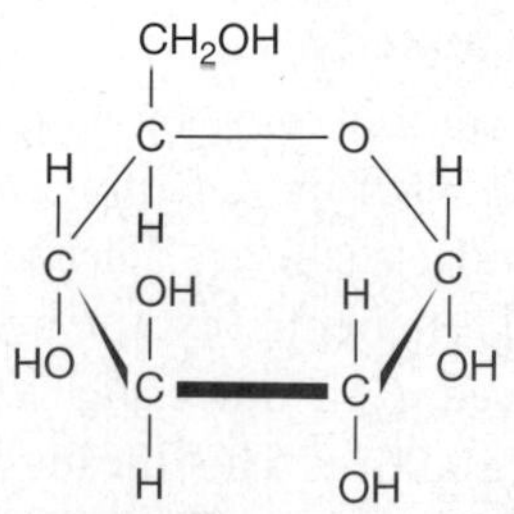

So here are some added oxygen atoms. Notice how the second model, below the first, with its heavy line down at the bottom, looks a little different than the benzene ring? Such a model is meant to give you a slightly more realistic picture of the actual arrangement of atoms when a carbon ring is present. Think of the ring as lying at an angle to the paper, with the heavy line closer to you than the line in the back; the H and OH groups then lie above and below the plane of the ring, as shown. You may also notice that when things get this complicated, space starts getting a little tight for writing in the letters that stand for the various atoms that are a part of the molecule. Because most rings are formed predominately of carbon, it is common to dispense with the Cs when a structural formula is presented. Here is a third representation of the glucose model, written in this stripped-down form:

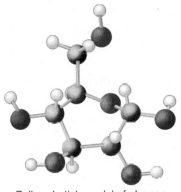

You can see that carbon is *assumed* to exist at each bond juncture in the ring; if some other element occupies one of these points, it is explicitly noted. Often the solitary Hs that are attached to carbons are left out as well.

Such models as these, as you'll recall from the last chapter, don't show the actual three-dimensional form of a molecule. To do so, ball-and-stick or space-filling models will do. Shown in **Figure 3.8** are examples of these models.

3.4 The Molecules of Life: Carbohydrates, Lipids, Proteins, and Nucleic Acids

With this review of carbon structures under your belt, you're ready to begin looking at some of the classes of carbon-based molecules—the molecules of living things. You'll explore four groupings of these organic compounds: carbohydrates, lipids, proteins, and nucleic acids.

The Building-Blocks Model of Organic Molecules

As you get into this section, it will be a great help to keep in mind that *complex* organic molecules often are made from *simpler* molecules. Many of the molecules you'll be reading about have a building-blocks quality to them: Take a simple sugar, or monosaccharide, such as glucose, put it together with another monosaccharide (fructose), and you have a larger *disaccharide* called sucrose (better known as table sugar). Put *many* monosaccharide units together and you have a polysaccharide, such as starch. The large units are called **polymers**, while the smaller building blocks of polymers are called **monomers**. Look at **Table 3.1** for examples of both.

Ball-and-stick model of glucose   Space-filling model of glucose

Figure 3.8
Ball-and-Stick and Space-Filling Models of Glucose
The simple sugar glucose is the single most important energy source for our bodies. Because it has several OH groups, it is highly hydrophilic and thus readily breaks down in water. Note how the carbon atoms (in black) form the core of the molecule, with oxygen (in red) and hydrogen (in white) at the periphery.

Table 3.1	
Monomers, Polymers	
If the Monomer is...	**The Polymer is...**
A monosaccharide (for example, glucose, fructose)	A polysaccharide (for example, starch, glycogen, cellulose)
An amino acid (for example, arginine, leucine)	A polypeptide or protein [for example, A- and B-chains of insulin (polypeptides) and insulin itself (protein)]
A nucleotide (sugar, phosphate, base in combination)	A nucleic acid (for example, DNA, RNAs)

Carbohydrates: From Simple Sugars to Cellulose

Happily, for purposes of memory, the elements in the first molecules we'll look at, carbohydrates, are all hinted at in the name: **Carbohydrates** must contain carbon, oxygen, and hydrogen. In many instances, they will contain nothing *but* carbon, oxygen, and hydrogen. Furthermore, they usually contain exactly twice as much hydrogen as oxygen. For example, the carbohydrate glucose ($C_6H_{12}O_6$) contains 12 atoms of hydrogen and 6 of oxygen. Most people think of carbohydrates purely in terms of foods such as breads and pasta (**see Figure 3.9**); but as you'll see, carbohydrates have more roles than this in nature.

The building blocks of the carbohydrates are the **monosaccharides** mentioned earlier, which also are known as **simple sugars**. You've already seen several views of one of these molecules—glucose. Glucose has a use in and of itself: Much of the food we eat is broken down into it, at which point it becomes our most important energy source. Glucose can also bond with other monosaccharides, however, to form more complex carbohydrates. Let's take a look at how this happens. If you look at **Figure 3.10**, you can see an example of two glucose molecules bonding to create the disaccharide called maltose.

Several things are worth noting here. In maltose, as you can see, the link that joins the two glucose monomers is a single oxygen atom linked to carbons of each of the glucose units. To get this, two atoms of hydrogen and one atom of oxygen are split off (on the left side of the equation) from the original glucose molecules. What becomes of these three atoms? Look at the far right-hand side of the reaction and see the $+ H_2O$. The product of this reaction is a molecule of maltose and a molecule of water.

Now note the arrows in the middle of the reaction. You've been used to seeing a single, rightward-pointing arrow, but here the arrows go both ways. What this means is that this is a **reversible reaction**; it can go both ways. When maltose is placed in water, it can be split apart to yield two *glucose* molecules.

Kinds of Simple Carbohydrates

You have thus far seen examples of one of the most simple carbohydrates, the monosaccharide glucose, and one slightly more complex carbohydrate, a disaccharide called maltose. There are many kinds of mono- and disaccharides, however. Among the monosaccharides are, for example, fructose and deoxyribose. Among disaccharides, there are sucrose and lactose. At the risk of pointing out the obvious, note that all these sugars have -*ose* at the end of their name: If it's an -*ose*, it's a sugar (**see Figure 3.11**).

Complex Carbohydrates Are Made of Chains of Simple Carbohydrates

You've seen carbohydrates built up from monosaccharides to disaccharides. If you go another couple of bumps up, you get to the **polysaccharides**. The *poly* here means "many" sugars, and the term is apt. In the polysaccharide molecule cellulose, for example, there may be 10,000 glucose units linked

Figure 3.9
Carbohydrates in Foods
Breads, cereals, and pasta make up a significant proportion of our diets. These foods are all rich in carbohydrates, one of the four main types of biological molecules.

up with one another. The basic unit here is the six-carbon monosaccharide glucose, $C_6H_{12}O_6$, from which chains of glucose units are built up. The complexity of these molecules gives them another name: complex carbohydrates. Four different types of complex carbohydrates interest us: starch, glycogen, cellulose, and chitin (**see Figure 3.12** on page 50).

Starch is familiar to most people, because it comes to us in the form of food. Potatoes, rice, carrots, corn; all these are starchy foods. Note that all of them come from plants. A closer look reveals that *in plants*, these starches serve as the main form of carbohydrate *storage*, sometimes as seeds (rice and wheat grains), or sometimes as roots (carrots or beets). (**See Figure 3.12a.**)

Glycogen does for animals what starch does for plants: It serves as the primary form in which carbohydrates are stored. For this reason, glycogen is sometimes called animal starch. The starches or sugars we eat are broken down, eventually into glucose, at which point some of the glucose may be used immediately. Some may not be needed right away, however, in which case it is moved into the muscle cells and liver to be stored as the more complex carbohydrate glycogen. The "carbohydrate loading" that swimmers and runners do before races (by eating lots of pasta, for example) could just as easily be called glycogen loading (**see Figure 3.12b**).

Cellulose is contained in the walls of the cells of many organisms. Despite this innocuous-sounding function, cellulose is important because it makes up so much of the natural world. It is easily the most abundant carbohydrate on Earth: Trees, cotton, leaves, and grasses are largely made of it. When cellulose is enmeshed with a hardening compound called *lignin*, the result is a set of cell walls that can hold up giant redwood trees. Because cellulose is so dense and rigid, it is not surprising that human beings and other mammals cannot digest it. This statement may make you wonder about grass-eating *cows*, but in fact cows have cellulose digested for them by special bacteria that reside in their digestive tract. Cellulose is important to humans in that it is our major source of insoluble fiber, which helps move foods through the digestive tract. Because cellulose exists in the cell walls of plants, we can get our fiber from such foods as whole grains and fresh fruits. (**See Figure 3.12c.**)

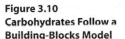

Tutorial 3.3.2
Molecules of Life:
Carbohydrates

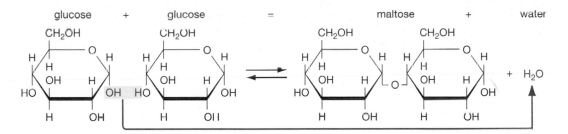

Figure 3.10
Carbohydrates Follow a Building-Blocks Model
In this example, two units of the monosaccharide (or simple sugar) glucose link to form the disaccharide maltose. In addition to maltose, the reaction yields water. The double arrows indicate that this reaction is reversible; a single maltose molecule can yield two glucose molecules.

Figure 3.11
Sugars Come in Many Forms
Sucrose or table sugar comes to us from sugar cane or sugar beets, and glucose is found in corn syrup. Fructose comes to us in sweet fruits and in high-fructose corn syrup, which often is used to sweeten soft drinks.

Chitin is a carbohydrate that forms the external skeleton or "exoskeleton" of the category of animals known as arthropods, a grouping so large it includes all insects, spiders, and "crustaceans," such as crabs. In all these animals, chitin plays a "structural" role similar to that of cellulose in plants: It gives shape and strength to the structure of the organism (**see Figure 3.12d**).

Lipids: Oils, Fats, Hormones, and the Outer Lining of Cells

The second class of biological molecules, **lipids**, turns out to be made of the same elements as carbohydrates—carbon, hydrogen, and oxygen. But lipids have much more hydrogen, relative to oxygen, than do the carbohydrates. We're all familiar with some lipids; they exist as fats, as oils, as cholesterol, and as hormones such as testosterone and estrogen. Unlike the other biological molecules you'll be studying, a lipid is not a polymer composed of component-part monomers; no single structural unit is common to all lipids. The one characteristic that pure lipids share is that they do not readily break down in water. Remember an earlier discussion about the need life has to create internal containers? Well, lipids are able to serve this function

Figure 3.12
Four Examples of Complex Carbohydrates
All complex carbohydrates are composed of chains of glucose, but they differ in the details of their chemical structure.

a Starch serves as a form of carbohydrate storage in many plants. Here starch granules can be seen within the cells of a slice of raw potato.

b Glycogen serves as a form of carbohydrate storage, as is the case in this photo of glycogen globules in the liver.

c Cellulose, running as fibers through cell walls, provides structural support for plants. The photo is of sets of cellulose fibers running at right angles to one another in the cell wall of marine algae.

d Chitin provides structural support for some animals. The outer "skin" or cuticle of insects is composed mostly of chitin. The photo shows the exoskeleton of a tick.

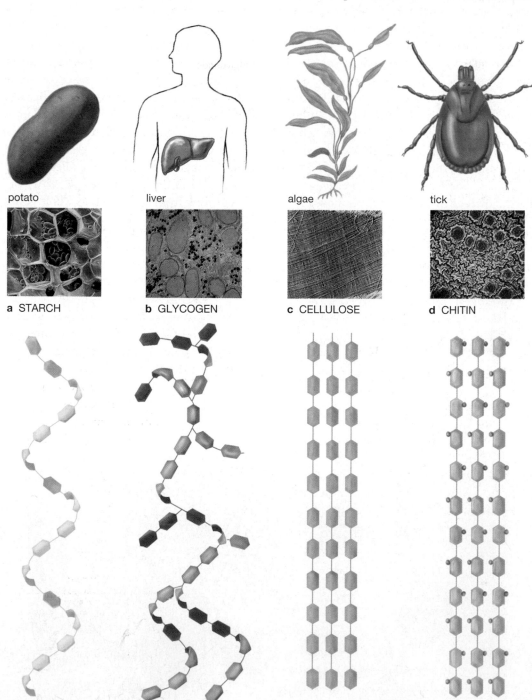

potato   liver   algae   tick

a STARCH   **b** GLYCOGEN   **c** CELLULOSE   **d** CHITIN

O H H H H H H H H H H H H H H H H H H
‖ | | | | | | | | | | | | | | | | | |
HO—C—C—C—C—C—C—C—C—C—C—C—C—C—C—C—C—C—C—H
| | | | | | | | | | | | | | | | |
H H H H H H H H H H H H H H H H H

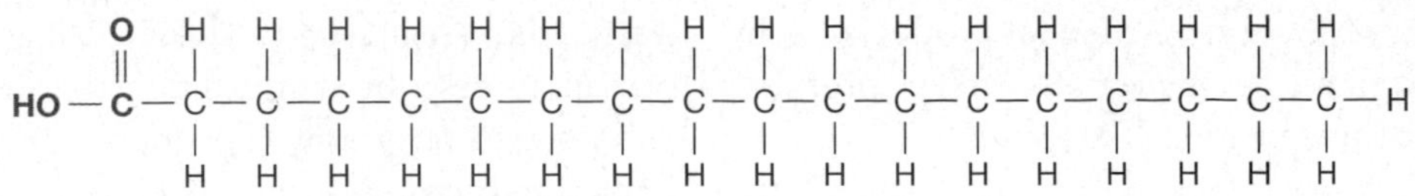

Figure 3.13
Structural Formula for Stearic Acid

glycerol + 3 fatty acids = triglyceride + water

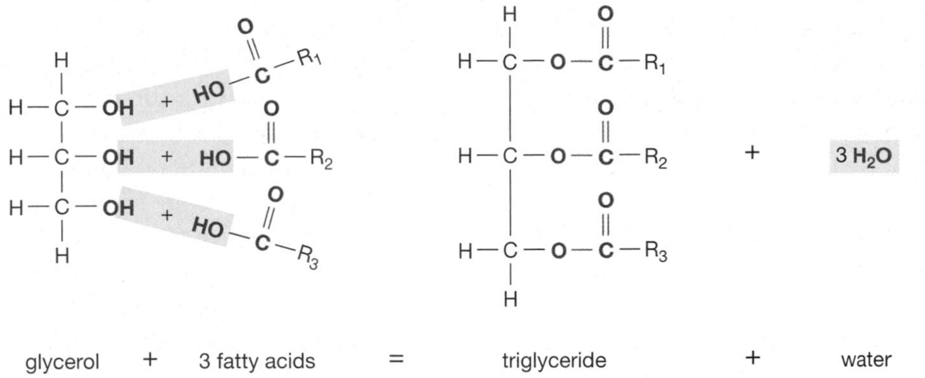

Figure 3.14
Formation of a Triglyceride

because they are relatively insoluable in water. In addition, lipids have considerable powers to store energy and to provide insulation.

One Class of Lipids Is the Glycerides

Glycerides, the most common kind of lipid, can be thought of as a molecule in two parts. The first is a "head" composed of a particular kind of alcohol, usually **glycerol**. Attached to the glycerol head is the second part of the molecule, **fatty acids**, each consisting of a long chain of carbon and hydrogen atoms (see Figure 3.15.).

The structural formula for stearic acid, one of the fatty acids found in animal fat, is shown in **Figure 3.13**. This monotonous chain of hydrogen and carbon is, as you might have guessed, the hydrocarbon portion of the fatty acid molecule. Fatty acids can have from 4 to 22 carbons linked up like this. On the left side, you can see that the chain terminates with a COOH chemical group (indicated in red), which is properly known as a *carboxyl group*.

Now look at the other part of the fat molecule, the alcohol known as glycerol:

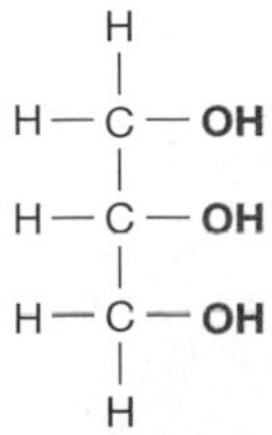

the carboxyl part of the fatty acid chain is what makes a glyceride. But you can see that the glycerol has *three* OH groups on the right. Thus there is, you might say, docking space on glycerol for three fatty acids, and in the synthesis of many glycerides, this linkage takes place: Three fatty acids link up with glycerol to form a **triglyceride**, which is an important form of lipid. **Figure 3.14** shows how it works schematically.

The Rs on the right end of the COOH group stand for whatever the hydrocarbon chain is of that particular fatty acid. To get a better idea of the shape of a completed triglyceride, look at the space-filling model in **Figure 3.15**. This particular triglyceride, called tristearin, has three stearic fatty acids stemming like tines on a fork from the glycerol. But this is only one possibility among many, and is exceptional, rather than usual, in that all three fatty acids are the same. Among the dozens of fatty acids that exist, several *different* kinds of fatty acids generally will hook up in glycerol's three "slots" to

H
|
H—C—OH
|
H—C—OH
|
H—C—OH
|
H

The hallmark of alcohols is that they have an OH group, which you can see here. Bringing the OH portion of glycerol together with

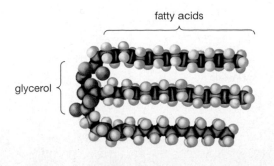

Figure 3.15
The Triglyceride Tristearin
This lipid molecule is composed of three stearic fatty acids, stemming rightward from the glycerol OH "heads." Tristearin is found both in beef fat and in the cocoa butter that helps make up chocolate.

form a triglyceride. Actual fat products, such as butter, are composed of different proportions of these various fatty acids.

We've seen three fatty acids linking with glycerol to form *triglycerides*, but monoglycerides (one fatty acid joined to glycerol) and diglycerides (two fatty acids joined with glycerol) can be formed as well. Triglycerides are the most important of the glycerides, however, because they constitute about 90 percent of the lipid weight in foods. In short, the substances we call fats usually are composed mostly of triglycerides.

Saturated and Unsaturated Fatty Acids: A Linkage with Solids and Liquids and with Health

Look now at **Figure 3.16**, which shows three different fatty acids: palmitic, oleic, and linoleic. On first glance, there is seemingly not much difference among the three. One thing that does jump out at you is the "kinks" in the oleic and linoleic acids compared to the straight-line palmitic acid. Upon closer inspection, you can see that the palmitic acid has an unbroken line of single bonds linking its carbon atoms. Meanwhile, the oleic acid has one double bond between the carbons in its chain and the linoleic acid has two. Furthermore, note that these double bonds exist precisely where the kinks appear in these molecules.

Fatty Acids: From Solid to Liquid What these variations describe are the differences between a **saturated fatty acid** (no double bonds), a **monounsaturated fatty acid** (one double bond), and a **polyunsaturated fatty acid** (two or more double bonds). What a saturated fatty acid is saturated *with* is hydrogen atoms. If a given fatty acid has a carbon double bond, it can take on more hydrogen atoms; if it doesn't have such a bond, it's saturated.

This may not sound like much of a difference, but it has several important consequences. First, at room temperature, as you move from saturated to unsaturated, you also move in general from fats in their solid form to fats in their liquid form, which we call **oils**. You can turn naturally occurring oils *into* fats by saturating them—by moving hydrogen into the oils in a process known as hydrogenation. How solid do you want a lipid to be? Just decide, and saturate a given portion of its double bonds. This is one step in creating margarine from vegetable oils. Why does a linkage exist between the degree of saturation and a solid or liquid form? This has to do with the kinks in these molecules—or the lack of them. Being all of one straight-line form, the saturated fatty acids pack together tightly, like so many boards in a lumber yard. The unsaturated fatty acids lack this uniformity, however,

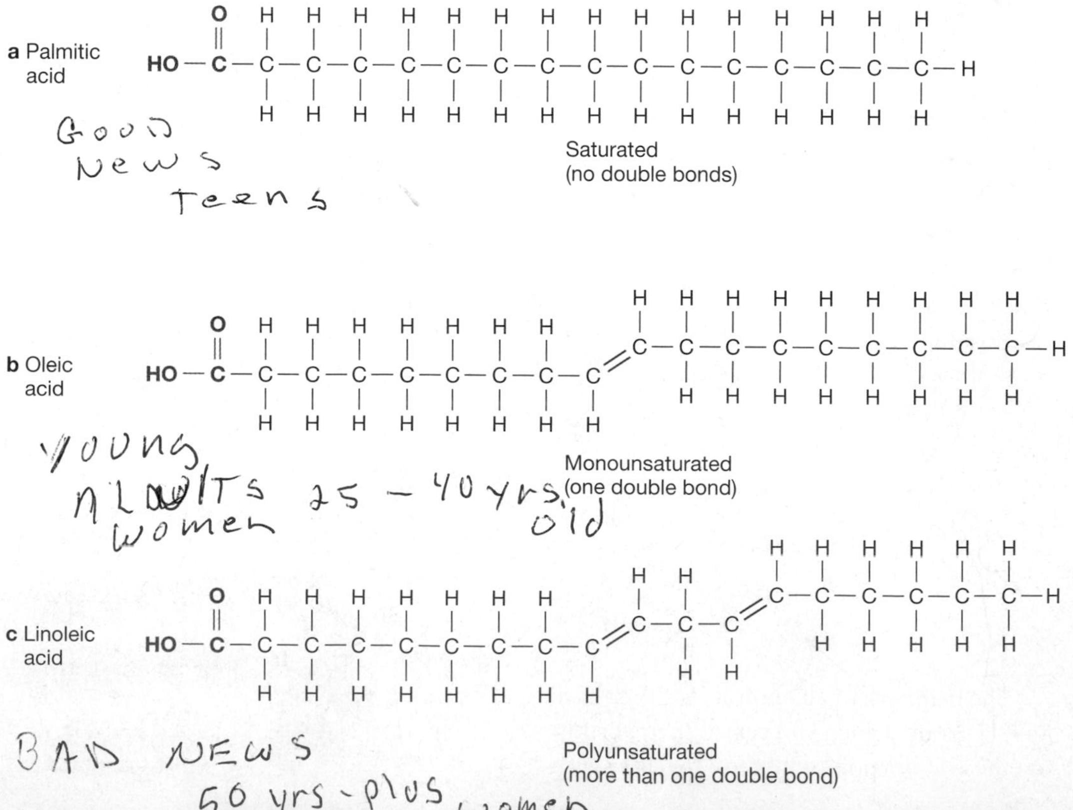

Figure 3.16
Saturated and Unsaturated Fatty Acids
The degree to which fatty acid hydrocarbon chains are "saturated" with hydrogen atoms has consequences for both the form these lipids take and for human health.

a The hydrocarbon "tail" in palmitic acid is formed by an unbroken line of carbons, each with a single bond to the next.

b In oleic acid, a double bond exists at one point between two carbon atoms. An additional hydrogen atom could link to each of these carbon atoms instead, which would make this a saturated fatty acid—saturated with hydrogen atoms. As things stand, this is a monounsaturated fatty acid.

c The carbons in linoleic acid have double bonds in two locations, making this a polyunsaturated fatty acid. The "kinks" imparted by double carbon-bonds make unsaturated and polyunsaturated fatty acids more likely to be liquid oils, rather than solid fats, at room temperature.

a Palmitic acid — Saturated (no double bonds)

b Oleic acid — Monounsaturated (one double bond)

c Linoleic acid — Polyunsaturated (more than one double bond)

and thus stick out at varying angles to one another, a relative disorder that generally makes them liquid at room temperature.

Saturated Fatty Acids and Health The distinction between saturated and unsaturated fatty acids also has another important consequence, this one related to human health. Think of it this way: To the degree that *fatty acids* are saturated, the *fats* they make up will be saturated—and saturated fats have been linked with heart disease. They are, in fact, more important than dietary cholesterol in determining how much circulating or "blood" cholesterol a person will have. This is a complicated linkage, but its practical effect is that public health groups have advised people to limit their intake of saturated fats, which in their natural form are found primarily in animal products (meat, high-fat dairy products) and tropical oils (coconut and cocoa butter, which is used in making chocolate). Fats and oils aren't tempting to human beings because nature had

it in for us, however. These lipids are a terrific source of energy, they're a great way to store energy, and they act as valuable insulators in cold weather. (Think of seals and polar bears, with their thick layers of fat.)

Energy Use and Energy Storage via Lipids and Carbohydrates

Lipids have something in common with carbohydrates in connection with energy *storage* on the one hand and energy *use* on the other. To be stored, lipids must be in their triglyceride form; to be used to provide energy, they must first be broken down into glycerol and fatty acids. Carbohydrates, meanwhile, are stored as the complex carbohydrate glycogen; to be used for energy expenditure, they must be broken down to simple carbohydrates, often glucose. This process of alternately building up molecules for energy storage, or breaking them down for energy expenditure, is a major task of living things. If you look at **Figure 3.17**, you

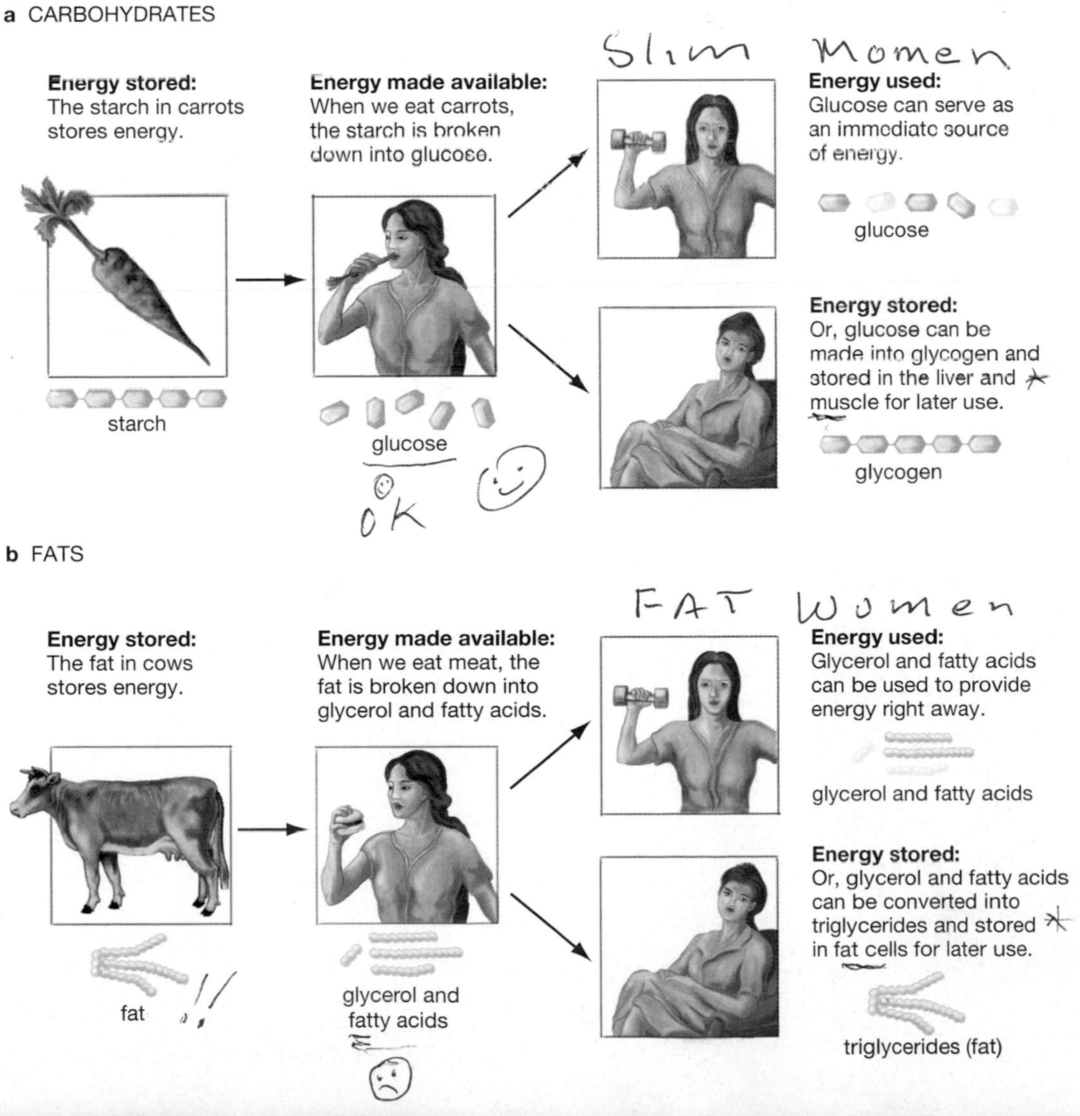

a CARBOHYDRATES

Energy stored:
The starch in carrots stores energy.

starch

Energy made available:
When we eat carrots, the starch is broken down into glucose.

glucose

Energy used:
Glucose can serve as an immediate source of energy.

glucose

Energy stored:
Or, glucose can be made into glycogen and stored in the liver and muscle for later use.

glycogen

b FATS

Energy stored:
The fat in cows stores energy.

fat

Energy made available:
When we eat meat, the fat is broken down into glycerol and fatty acids.

glycerol and fatty acids

Energy used:
Glycerol and fatty acids can be used to provide energy right away.

glycerol and fatty acids

Energy stored:
Or, glycerol and fatty acids can be converted into triglycerides and stored in fat cells for later use.

triglycerides (fat)

Figure 3.17
Storage and Use of Carbohydrates and Lipids
Carbohydrates and lipids generally are stored in one form, but used in another.

a In plants, carbohydrates commonly are stored as starch, a complex carbohydrate that is a major component of such vegetables as carrots. Human beings can consume such a starch, after which it may be broken down into the smaller, simple carbohydrate glucose and used immediately—for exercise or other purposes—or stored in the liver and muscles as the larger, more complex carbohydrate glycogen.

b Cows use their fat for energy storage and insulation. The fat portion of the meat that humans consume may be used for energy, but only after being broken down into its glycerol and fatty acid components. Conversely, humans may convert the components back into triglycerides to be stored in fat cells, providing energy reserves for later.

Dietary Decisions: Should You Cut Carbohydrates?

Can people stay slimmer by carefully controlling what *kinds* of foods they eat, as opposed to watching how *much* food they eat? In the late 1990s, a theme emerged in the $50 billion-a-year American dieting industry, which was that the way to lose weight is by eating less of a very specific kind of food: carbohydrates. Robert Atkins' *New Diet Revolution, Sugar Busters!*, *The Zone*, and other diet books differed in details; but all carried a similar message: that people who want to lose weight need to restrict carbohydrates, relative to other kinds of foods. In some diet books, carbohydrates were even alleged to addict people or lessen the body's ability to burn up food.

But is it true? Do the low-carbohydrate diets work? Do they allow people to take weight off, keep it off, and remain healthy? The answer is that no one knows—not even the authors of the diets. The true test of a diet is to get a group of similar people together and then randomly assign some of them to go on the diet while keeping others off it. Only with the latter "control group" in place can anything be learned about the diet's effectiveness. None of the diets touted in the books noted above has ever been put to this test. Thankfully, this may change. The U.S. Department of Agriculture announced, in May 2000, that it was embarking on a systematic study of the popular diets.

Until the USDA's results are in, the only way to judge the low-carbohydrate diets is to ask whether they make sense based on what scientists know about nutrition and metabolism. Here, the news is not encouraging. The consensus among researchers seems to be that there is no reason to believe that, compared to other foods, carbohydrates are addictive, leave a person feeling hungrier, or cause additional fat to be stored in the body. To the extent that the low-carbohydrate diets work their effect probably comes from a one-two series of events that has nothing to do with the claims that are made for them.

The first, short-term effect of the diets stems from our body's ability to either use carbohydrates immediately or store them away as glycogen. But every time we store away a gram of carbohydrate as glycogen, we store away three grams of *water* with it. (Fats, conversely, are stored without water.) Most researchers believe that the rapid loss of a few pounds experienced by so many low-carbohydrate dieters simply represents water loss. Moreover, this loss is temporary; it will be wiped out should the dieter eat even a few servings of potatoes or rice.

A different process operates with later-stage weight losses. As it happens, the body needs carbohydrates to break down *fats* completely, and the body burns fat in the absence of dietary carbohydrates. If carbohydrate reduction is extreme enough, a class of acids (called ketone bodies) will accumulate in bodily fluids. The result is a condition, called ketosis, that can bring about nausea, thus depressing appetite. In addition, the dieter's food choices are now limited because carbohydrates form a large portion of the foods people like to eat. The result? Low-carbohydrate dieters start eating less—cutting calories, in other words. But this is weight loss the old-fashioned way!

All this might seem harmless enough, but remember that low-carb diets substitute *fats* (along with protein) for carbohydrates. Fats are, of course, what doctors want people to eat less of to keep a healthy heart. Moreover, it is carbohydrates—beans, grains, vegetables, and fruits—that overwhelmingly provide fiber and the cancer-fighting compounds known as phytochemicals.

In sum, based on established principles of nutrition, there seem to be good reasons to avoid low-carbohydrate diets. Of course, the USDA diet studies could prove the experts wrong, but in the meantime each of us has a bet to make regarding what we eat.

Figure 1
Recent Diet Books

can see how carbohydrates and lipids are alternately stored and used in living things. (In "Dietary Decisions" on page 54 you can read about the current spate of books that make claims about losing weight by reducing the proportion of carbohydrates we eat.)

A Second Class of Lipids Is the Steroids

The class of molecules known as **steroids** represents a smaller kind of lipid, different from the glycerides. What all steroids have in common is a linked set of four carbon rings. What separates one steroid from another are the various side chains that can be attached to this four-ring structure (**see Figure 3.18**). When you see how different steroids are structurally from the triglycerides, you can understand why the monomers-to-polymers framework doesn't apply to lipids.

Among the most well-known steroids are cholesterol and two of the steroid hormones, testosterone and estrogen. Like fats in general, **cholesterol** has a bad reputation; but also like fats in general, it serves good purposes too. Compounds that come from cholesterol help break down fats in the first place. Cholesterol is part of the outer membrane of many cells, and it acts as a precursor for many steroids, among them the steroid hormones testosterone (a principal "male" hormone), and estrogen (a principal "female" hormone). The term *steroids* by itself undoubtedly rings a bell, because the phrase "on steroids" has come to mean artificially bulked-up or supercharged. In this common usage, steroids refers to manufactured drugs that are close chemical cousins of the muscle-building "male" steroid hormones (which actually exist in smaller amounts in females as well; see Figure 3.19).

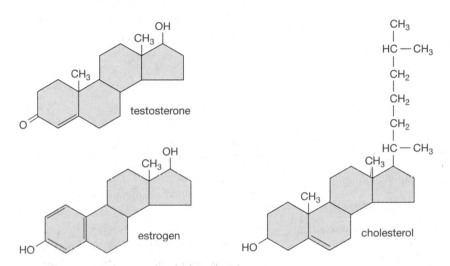

a What all steroids have in common is a four-ring carbon unit:

b What makes individual steroids unique are the side chains attached to the rings:

testosterone

estrogen

cholesterol

Figure 3.18
Structure of Steroids

a The basic unit of steroids, four interlocked carbon rings.

b Types of steroids, each differentiated from the other by the side chains that extend from the four-ring skeleton. Testosterone is a principal "male" hormone, while estrogen is a principal "female" hormone. Both of these steroid hormones actually are found in both men and women, though in differing amounts. Cholesterol is also a prevalent and important steroid in both men and women. Although cholesterol has a bad reputation, it has several important functions—for example, breaking down fats.

Figure 3.19
Steroids, in Uses Natural and Unnatural

a Some steroid hormones such as estrogen and testosterone are important in natural processes, such as reproduction.

b Pharmaceutical steroids can be used in potentially harmful ways, for example, to build muscle mass. Pictured is a Chinese swimmer who, at a 1998 competition, tested positive for a banned substance that helps flush evidence of steroids from the body.

A Third Class of Lipids Is the Phospholipids

The final class of lipids, **phospholipids**, has something of the same makeup as triglycerides, in that a phospholipid has a glycerol head that has fatty acids attached to it. But where triglycerides have three fatty acids stemming from the glycerol, phospholipids only two (**see Figure 3.20**). Linking up with glycerol's third OH group is a **phosphate group**, meaning a phosphorus atom surrounded by four oxygen atoms. This change is extremely important because it gives phospholipids a dual nature: The fatty acid tails, being hydrocarbons, are hydrophobic; but the phosphate head is hydro*philic* because it is *charged*.

In Figure 3.20, you can see the effect of this in solution. Imagine a phospholipid as a marker buoy in deep water. No matter how you push the hydrocarbon tail around, it's going to end up waving free *out* of the water, while the head is going to be submerged in it. You will learn more about these molecules later, when you get to cells; for now, just note that the material on the periphery of cells—the outer membrane of a cell—is largely made of phospholipids. Living things need the kind of partitions described earlier, and these partitions are composed to a significant extent of phospholipids.

Proteins

Living things must accomplish a great number of tasks just to get through a day, and the diverse biological molecules you've been looking at allow this to happen. You've seen carbohydrates do some things, and you've seen lipids do some more. But in the range of tasks that molecules accomplish, proteins reign supreme. Witness the fact that almost every chemical reaction that takes place in living things is hastened—or, in practical terms, *enabled*—by a particular kind of protein called an enzyme. These molecules function in nature like some vast group of tools, each one taking on a specific chemical task. Accordingly, an animal cell might contain up to 4,000 different types of enzymes.

We might marvel at proteins solely because of what enzymes can do, but the amazing thing is that enzymes are only *one class* of proteins. Proteins also form the scaffolding, or structure, of a good deal of tissue; they're active in transporting molecules from one site to another; they allow muscles to contract and cells to move; some hormones are made from them. If you factor out water, they account for about half the weight of the average cell. In short, it's hard to overestimate the importance of these molecules. **Table 3.2** lists some of the different kinds of proteins.

Proteins Are Made from Chains of Amino Acids

Proteins are prime examples of the building-block type of molecule described earlier. The monomers in this case are called amino acids. String an arbitrary number of them together in a chain—some say 10, some say 30—and you have a **polypeptide**; when the polypeptide *folds up* in a specific three-dimensional manner, you have a **protein**. As a practical matter, proteins are likely to be made of hundreds of amino acids strung together and folded up.

Figure 3.21a gives the fundamental structural unit for amino acids, followed by a couple of examples. You can see carbon at the center of this unit, an amino group off to its left, and a carboxyl group to its right. (The name *amino acid* comes from the fact that there is an *amino* group and, in the carboxyl group, an *acid* group.) What differentiates one amino

Figure 3.20
A Dual-Natured Molecule

a Phospholipids are composed of two long fatty-acid "tails" attached to a "head" containing a phosphate group (which carries a negative charge) and another variable group (which often carries a charge).

b Because the head is polarized, it can bond with water and thus will remain submerged in it; the tails, on the other hand, have no such bonding capability.

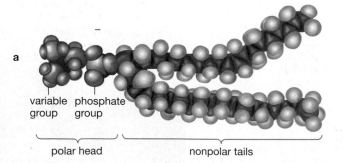

a

variable group / phosphate group

polar head   nonpolar tails

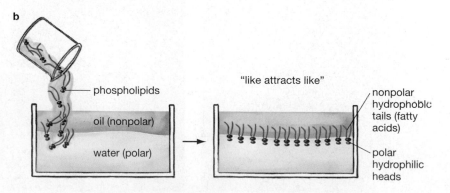

b

phospholipids

oil (nonpolar)

water (polar)

"like attracts like"

nonpolar hydrophobic tails (fatty acids)

polar hydrophilic heads

**Table 3.2
Types of Proteins**

Type	Role	Examples
Enzymes	Quicken chemical reactions	Sucrase: Positions sucrose (table sugar) in such a way that it can be broken down into component parts of glucose and fructose
Hormones	Chemical messengers	Growth hormone: Stimulates growth of bones
Transport	Move other molecules	Hemoglobin: Transports oxygen through blood
Contractile	Movement	Myosin and actin: Allow muscles to contract
Protective	Healing; defense against invader	Fibrinogen: Stops bleeding Antibodies: Kill bacterial invaders
Structural	Mechanical support	Keratin: Hair Collagen: Cartilage
Storage	Stores nutrients	Ovalbumin: Egg white, used as nutrient for embryos
Toxins	Defense, predation	Bacterial diphtheria toxin
Communication	Cell signaling	Glycoprotein: Receptors on cell surface

acid from another is the group of atoms that occupies the R or "side-chain" position. In **Figure 3.21b** you can see examples of actual amino acids, tyrosine and glutamine, with their different occupants of the R position.

A Group of Only 20 Amino Acids Is the Basis for All Proteins in Living Things

Although only two examples are shown here, it is a group of *20* amino acids in total that is the basis for all the proteins that occur in living organisms. The thousands of proteins that exist can be made from a mere 20 amino acids, because these amino acids can be strung together in different *order*. Substitute an alanine here for a glutamine there, and you've got a different protein. In this, amino acids commonly are compared to letters of the alphabet. In English, substituting one letter can take us from *bat* to *hat*. In the natural world, 20 amino acids can be put together in different order to create a multitude of proteins, each with a different function.

The stringing together of amino acids happens in a regular way: The carboxyl

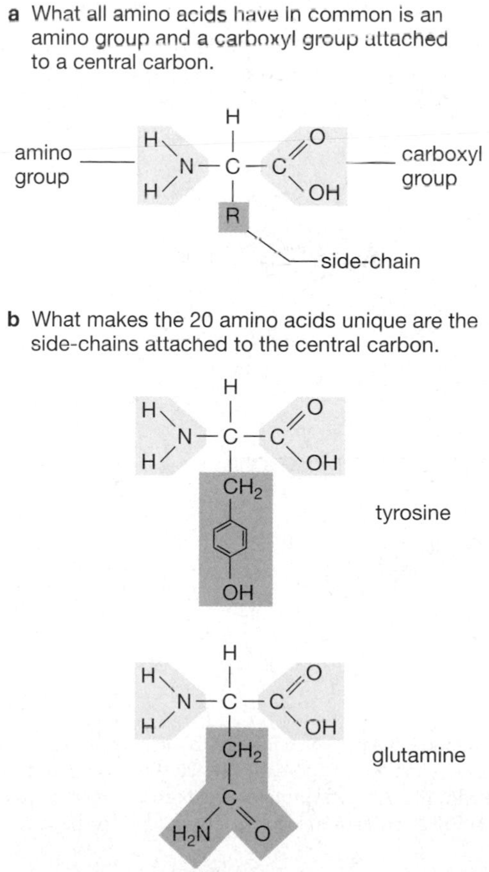

a What all amino acids have in common is an amino group and a carboxyl group attached to a central carbon.

b What makes the 20 amino acids unique are the side-chains attached to the central carbon.

tyrosine

glutamine

**Figure 3.21
Structure of Amino Acids**

a The structural element that is common to all amino acids is an amino group and a carboxyl group (on the left and right), linked by a central carbon with a hydrogen attached to it. What makes one amino acid different from another is the side-chain of atoms that occupies the R position.

b Two examples of actual amino acids; the differing occupants of the R position give us the amino acids tyrosine and glutamine.

group of one amino acid joins to the amino group of another, with the loss of a water molecule resulting. Look at **Figure 3.22** to see how three amino acids come together.

Shape Is Critical to the Functioning of All Proteins

Now, recall that a protein is a chain of amino acids that has become *folded up* in a specific way. As the amino acids are being strung together in sequence, all the kinds of chemical forces discussed earlier begin to work on the chain; as a result, it begins to twist and turn and fold into a unique three-dimensional shape. And it turns out that in the functioning of proteins, this shape, or **protein conformation**, is utterly crucial. Here's an example of why. The hormone insulin, which is a protein, is released from a person's pancreas; it moves through the bloodstream and latches onto muscle cells; its presence then allows blood sugar to get into the muscles and provide energy. Now, how does insulin "latch onto" muscle cells? By *linking* itself with a molecule, called an insulin receptor, that lies on the surface of the muscle cell. But protein and receptor must be precisely shaped for such linkage or "binding" to occur.

This binding of a protein hormone to surface receptor sounds important enough; but recall that in addition to being hormones, proteins are the chemical enablers called enzymes, and they're transport molecules, and so forth. In *all* these functions, shape is critical. The American architect Frank Lloyd Wright had a famous dictum about his designs: "Form follows function." With proteins, we can turn this around and say that function follows form (**see Figure 3.23**).

There Are Four Levels of Protein Structure

So, what forms do proteins take? Well, the answer to this depends on what vantage point you adopt in looking at them. There are four levels of structure in proteins that determine their final shape. You can see all four levels in **Figure 3.24**. The first of these levels, the **primary structure** of a protein, is simply its sequence of amino acids. Everything about the final shape of a protein is dictated by this sequence. Electrochemical bonding and repulsion forces act on this structure, and the result is the folded-up protein.

As it turns out, when these forces begin to operate on the amino acid sequence, a couple of common shapes begin to emerge in the **secondary structure** of proteins. The first of these common shapes, called the **alpha helix**, has a shape much like a corkscrew. The other common secondary structure in proteins, called the **beta pleated sheet**, is shaped something like the folds of an accordion. Proteins can be made almost entirely of alpha helices. This is the case with hair, nails, horns, and the like. Likewise, proteins can be made entirely of beta pleated sheets. The most familiar example of this is silk, in which the beta sheets lie pancake-style on one another. Often, however, alpha helices and beta pleated sheets form what we might think of as design motifs within a larger protein structure; they periodically give way to the less regular segments called *random coils*. The larger-scale twists and turns that the protein takes form its **tertiary structure**. The way in which *two or more* polypeptide chains come together to form a protein results in that protein's **quaternary structure**.

Tutorial 3.3.11
Molecules of Life:
Proteins

The linkage of several amino acids...

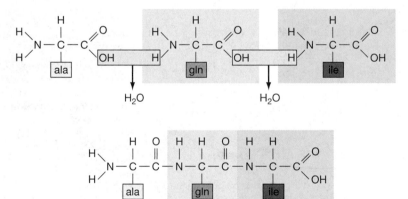

...produces a polypeptide chain like this:

Figure 3.22
Beginnings of a Protein
Amino acids join together to form polypeptide chains, which fold up to become proteins. The linking of amino acids yields water as a byproduct. In this figure, alanine (ala) first joins with glutamine (gln), which then is linked to isoleucine (ile). A typical protein would consist of hundreds of amino acids linked up sequentially. (A list of all 20 primary amino acids can be found on page 271.)

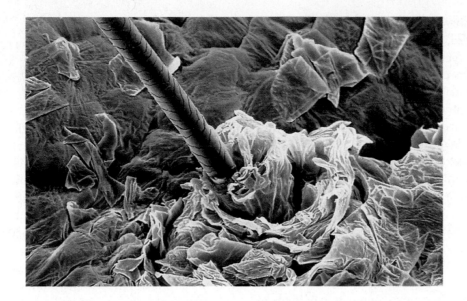

Figure 3.23
Made of Protein
A human hair, shown here emerging from a follicle in the skin, is composed mostly of the protein keratin.

FOUR LEVELS OF STRUCTURE IN PROTEINS

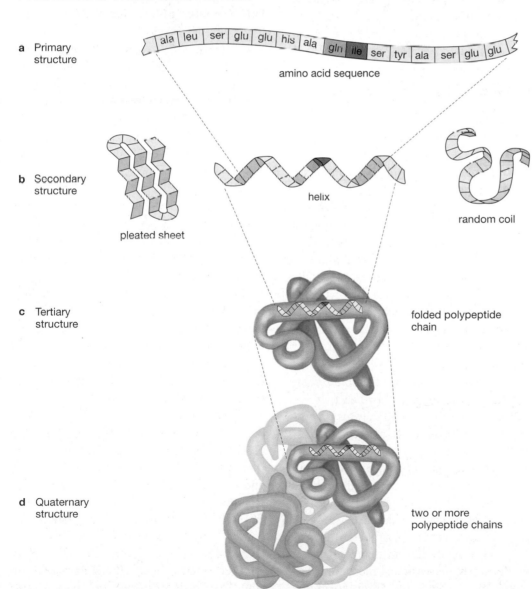

a Primary structure

| ala | leu | ser | glu | glu | his | ala | gln | ile | ser | tyr | ala | ser | glu | glu |

amino acid sequence

b Secondary structure

pleated sheet

helix

random coil

c Tertiary structure

folded polypeptide chain

d Quaternary structure

two or more polypeptide chains

Figure 3.24
Four Levels of Structure in Proteins

a The primary structure of any protein is simply its sequence of amino acids. This sequence determines everything else about the protein's final shape.

b Structural motifs, such as the corkscrew-like alpha helix, beta pleated sheet, and the less organized "random coils" are parts of many polypeptide chains, forming their secondary structure.

c These motifs may persist through a set of larger-scale turns that make up the tertiary structure of the molecule.

d Several polypeptide chains may be linked together in a given protein, in this case hemoglobin, with their configuration forming its quaternary structure.

Proteins Can Come Undone

As noted, proteins have to maintain a precise conformation in order to function. However, proteins can *lose* their shape, and thus their functionality. You saw an example of this earlier in connection with pH and enzymes. In the wrong pH environment, an enzyme can *un*fold, thus losing its ability to hasten a chemical process. Alcohol works as a disinfectant on skin because it **denatures** or alters the shape of the proteins of bacteria.

Lipoproteins and Glycoproteins

Some molecules in living things are hybrids, or combinations of the various types of molecules you've been looking at. **Lipoproteins**, as their name implies, are a combination of lipids and proteins. Active in transporting fats throughout the body, lipoproteins amount to a capsule of protein surrounding a globule of fat.

Two kinds of lipoproteins have managed to enter public consciousness despite the handicap of having long names. These are high-density lipoproteins and low-density lipoproteins, also known as HDLs and LDLs. What makes them more or less dense is the ratio of protein to lipid in them; protein is more dense than lipid, so a high-density lipoprotein is one that contains a relatively greater amount of protein. LDLs have acquired a reputation as the villains of lipoproteins, because they carry cholesterol *to* outlying tissues including the coronary arteries of the heart, where this cholesterol may come to reside, thickening eventually into "plaques" that can block coronary arteries and bring about a heart attack. HDLs, meanwhile, are regarded as the cavalry of lipoproteins; they carry cholesterol *away* from outlying cells to the liver. A high proportion of HDLs in relation to cholesterol is predictive of keeping a healthy heart.

Glycoproteins are combinations of proteins and *carbohydrates*. Remember the discussion about insulin traveling through the bloodstream and latching onto a *receptor* on the surface of a muscle cell? Well, such receptors are usually glycoproteins, meaning mostly protein with a side-chain made of carbohydrate. A profusion of these receptors sits on cell surfaces like so many antennae, ready for a partner with just the right shape to come by and latch on. Some hormones themselves are glycoproteins, along with many other proteins released from cells.

Nucleotides and Nucleic Acids

Nucleotides are the last major type of biological molecule we'll study. They are important molecules in and of themselves and as building blocks for a couple of very important molecules.

As molecules in their own right, some nucleotides, called adenosine phosphates, serve as chemical energy carriers. When you study energy within cells, you'll find many references to a molecule called adenosine triphosphate, or ATP. As money is to shopping, so ATP is to getting things done in living organisms. In this text the most detailed discussion of nucleotides, however, comes in connection with their role as building blocks in two very large and important molecules, which are briefly considered next.

DNA Provides Information for the Structure of Proteins

You've learned that proteins perform a large number of biological functions, and that one class of proteins, the enzymes, may be represented with up to 4,000 types in a single animal cell. If you had a factory that turned out four thousand different kinds of tools, you would obviously need some direction on how each of these tools was to be manufactured: This part of the tool goes here first, and then that goes there, and so on. There is a molecule that in essence provides this kind of information for the construction of proteins. It is **DNA**, or **deoxyribonucleic acid**, and it is comprised of a huge number of nucleotides. In human beings, about 3 billion nucleotides are strung together to form our main molecule of DNA, and one copy of this molecule exists in each of our cells.

The information contained in the DNA molecule is much like the information contained in a cookbook, only what the DNA "recipes" are calling for are precisely ordered chains of amino acids, the building blocks of proteins. "Start with an alanine, then add a cysteine, then a tyrosine . . . " and so on for hundreds of steps, the DNA-encoded instructions will say, after which—through a series of steps—a protein becomes synthesized and then gets busy on some task. A player in this series of steps is another nucleic acid, **ribonucleic acid** or **RNA**, which is involved in ferrying the DNA-encoded instructions to the sites in the cell where proteins are put together.

The Structural Unit of DNA Is the Nucleotide

Look at **Figure 3.25a** and you'll see the structure of the building blocks that make up DNA. Each nucleic acid **nucleotide** is a molecule in three parts: A *phosphate* group, a *sugar* (deoxyribose), and a nitrogen-containing *base*. One nucleotide then attaches to another, forming a chain. *Two* of these chains then link together—as if a ladder, split down the middle, were coming together—forming the most famous molecule in all of biology, the DNA double helix.

Tutorial 3.3.15
Molecules of Life:
Nucleic Acids

NUCLEOTIDES ARE THE BUILDING BLOCKS OF DNA

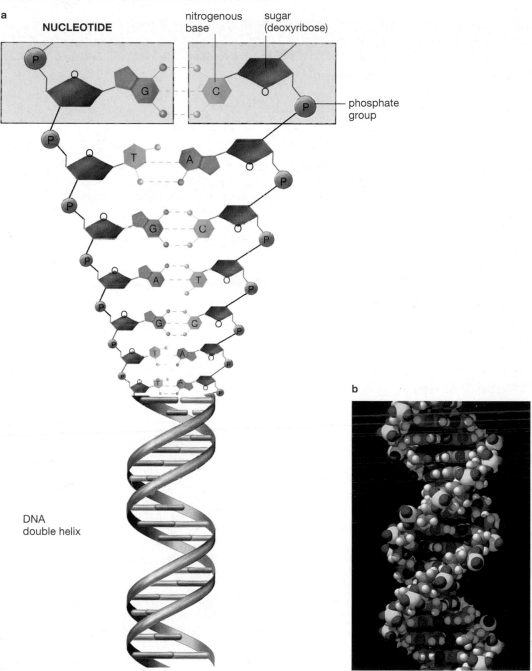

Figure 3.25
Nucleotides Are the Building Blocks of DNA

a The basic unit of the DNA molecule is the nucleotide, a molecule in three parts: a sugar, (deoxyribose), a phosphate group, and a nitrogen-containing base. A given nucleotide might contain any of four bases: Adenine (A), Guanine (G), Cytosine (C), or Thymine (T). The sugar and phosphate components of the nucleotides link up to form the outer "rails" of the DNA molecule, while the bases point toward the molecule's interior. Two chains of nucleotides are linked, via hydrogen bonds, to form DNA's double helix. Note that the hydrogen bond is between one base and another.

b A computer-generated space-filling model of DNA.

On to Cells

Before getting to the story of this elegant DNA molecule, you need to learn a little bit about the territory in which it does its work. What you'll be looking at is a profusion of jostling, roiling, ceaselessly working chemical factories that make up all living things. These factories are called cells. Look at **Table 3.3**, and you'll find a summary of all the types of molecules that were reviewed in this chapter.

Table 3.3
Summary Table of Biological Molecules

Type of Molecule	Subgroups	Examples and Roles
Carbohydrates	Monosaccharides	Glucose: Energy source
	Disaccharides	Sucrose: Energy source
	Polysaccharides	Glycogen: Storage form of glucose Starch: Carbohydrate storage in plants; used by animals in nutrition Cellulose: Plant cell walls, structure; fiber in animal digestion Chitin: External skeleton of arthropods
Lipids	Triglycerides 3 Fatty acids and glycerol	Fats, Oils (butter, corn oil): Food, energy, storage, insulation
	Fatty acids Components of Triglycerides	Stearic Acid: Food, energy sources
	Steroids Four-ring structure	Cholesterol: Fat digestion, hormone precursor, cell membrane component
	Phospholipids Polar head, nonpolar tails	Cell membrane structure
Proteins	Enzymes Chemically active	Sucrase: Breaks down sugar
	Structural	Keratin: Hair
	Lipoproteins Protein-lipid molecule	HDLs, LDLs: Transport of lipids
	Glycoproteins Protein-sugar molecule	Cell surface receptors
Nucleotides	Adenosine phosphates	Adenosine triphosphate (ATP): Energy transfer
	Nucleic acids Sugar, phosphate group, base	DNA, RNAs: Contain information for and facilitate synthesis of proteins

Chapter Review

Summary

3.1 The Importance of Water to Life

- Water has several qualities that have strongly affected life on Earth. It is a powerful solvent, being able to dissolve more compounds in greater amounts than any other liquid. Because water's solid form (ice) is less dense than its liquid form, bodies of water in colder climates do not freeze solid in winter, which allows life to flourish under the ice.

- Water has a great capacity to absorb and retain heat. Because of this, the oceans act as heat buffers for the Earth, thus stabilizing Earth's temperature. Water has a high degree of cohesion, which allows water to be drawn up through plants, via evaporation, in one continuous column, from roots through leaves.

- Some compounds do not interact with water. Hydrocarbons such as petroleum are examples of such hydrophobic compounds. Water cannot break down hydrophobic compounds, which is why oil and water don't mix. Compounds that do interact with water are polar or carry an electric charge and are called hydrophilic compounds.
  TUTORIAL 3.1.1: Chemical Properties of Water

3.2 Acids and Bases Are Important to Life

- An acid is any substance that yields hydrogen ions when put in solution. A base is any substance that accepts hydrogen ions in solution. A base added to an acidic solution makes that solution less acidic, while an acid added to a basic solution makes that solution less basic.
  TUTORIAL 3.1.6: Acids, Bases, and pH

- The concentration of hydrogen ions that a given solution has determines how basic or acidic that solution is, as measured on the pH scale. This scale runs from 0 to 14, with 0 being most acidic, 14 being most basic, and 7 being neutral. The pH scale is logarithmic; a substance with a pH of 9 is 10 times as basic as a substance with a pH of 8. Living things function best in a near-neutral pH, though some systems in living things have different pH requirements.

3.3 Carbon Is a Central Element in Life

- Carbon is a central element to life, because most biological molecules are built on a carbon framework. Carbon plays this central role because its outer shell has only four of the eight electrons necessary for maximum stability. Carbon atoms are thus able to form stable, covalent bonds with a wide variety of atoms, including other carbon atoms. The complexity of living things is facilitated by carbon's linkage capacity.
  TUTORIAL 3.2.1: Chemistry of Carbon

3.4 The Molecules of Life: Carbohydrates, Lipids, Proteins, and Nucleic Acids

- Carbohydrates are formed from the building blocks or monomers of simple sugars, such as glucose. These can be linked to form the larger carbohydrate polymers such as starch, glycogen, cellulose, and chitin. Carbohydrates generally are stored in polymer form in living things (glycogen in animals, for example) but broken down into monomers to be used (glucose in animals).
  TUTORIAL 3.3.2: Molecules of Life: Carbohydrates

- There are several different varieties of lipids. Among the most important are the triglycerides, composed of a glyceride and three fatty acids. Most of the fats we consume are triglycerides. Another important variety of lipids is the steroids, which include cholesterol, and such hormones as testosterone and estrogen.
  TUTORIAL 3.3.5: Molecules of Life: Lipids

- Proteins are a diverse group of biological molecules composed of the monomers of amino acids. Important groups of proteins include enzymes, which hasten chemical reactions, and structural proteins, which make up such structures as hair. The primary structure of a protein is its amino acid sequence; this sequence determines how a protein folds up. The activities of proteins are determined by their final, folded shapes.
  TUTORIAL 3.3.11: Molecules of Life: Proteins

- Nucleic acids are polymers composed of nucleotides. The nucleic acid DNA is composed of nucleotides that contain a phosphate group, a sugar (deoxyribose), and one of four nitrogen-containing bases. DNA is a repository of genetic information. The sequence of its bases encodes the information for the production of the huge array of proteins produced by living things.
  TUTORIAL 3.3.15: Molecules of Life: Nucleic Acids

Key Terms

acid 41	denatured 60
alkaline 42	deoxyribonucleic acid (DNA) 60
alpha helix 58	fatty acid 51
aqueous solution 37	glyceride 51
base 41	glycerol 51
beta pleated sheet 58	glycogen 49
buffering system 44	glycoprotein 60
carbohydrate 48	hydrocarbon 40
cellulose 49	hydrophilic 40
chitin 50	hydrophobic 40
cholesterol 55	hydroxide ion 42

Understanding the Basics

Multiple-Choice Questions

1. Near an ocean or other large body of water, air temperatures do not vary as much with the seasons as they do in the middle of a continent. This tendency of water to resist changes in temperature is the result of water's
   a. high density
   b. low density
   c. being a good solvent
   d. low specific heat
   e. high specific heat

2. A frog survives a freezing-cold winter on the bottom of a pond because
   a. Ice, which floats on water, insulates the water beneath it.
   b. The cells of a frog's body cannot freeze.
   c. Water has a low specific heat.
   d. The surface tension of water protects the frog's body.
   e. All of these are factors.

3. Janine has dry skin, so she uses body oil every morning. The oil seals in some of the water on her skin, so that it doesn't get as dry. This is possible because oils:
   a. are hydrophilic
   b. are rare in nature
   c. have a high specific heat
   d. are more dense than water
   e. are hydrophobic

4. Some plants live in bogs in which the pH is about 2. Thus these plants are able to survive in a(n) _____ external environment.
   a. basic
   b. buffered
   c. acidic
   d. neutral
   e. alkaline

5. When you eat starch such as spaghetti, an enzyme in your mouth breaks it down to maltose. Eventually, the maltose enters your small intestine, where it is broken down to glucose, which you can absorb into your bloodstream. The starch is a _____, the maltose is a _____, and the glucose is a(n) _____.
   a. protein . . . dipeptide . . . amino acid
   b. monosaccharide . . . disaccharide . . . polysaccharide
   c. triglyceride . . . fatty acid . . . glycerol
   d. amino acid . . . dipeptide . . . protein
   e. polysaccharide . . . disaccharide . . . monosaccharide

6. Which of these is *not* an actual difference between saturated and unsaturated fats?
   a. Saturated fats are more likely to be solid at room temperature; unsaturated fats are more likely to be liquid.
   b. Saturated fats are a type of cholesterol or steroid, whereas unsaturated fats are a triglyceride.
   c. Saturated fats have fatty acids that pack closely together, owing to their straight-line construction. Unsaturated fats have fatty acids arranged at varying angles to one another, because of their double bonds.
   d. Saturated fats are composed of fatty acids that have no double bonds in their chemical structure. Unsaturated fats are composed of fatty acids that have one or more double bonds in their chemical structure.
   e. All of the above are differences between saturated and unsaturated fats.

7. The myoglobin protein, which carries oxygen in muscle cells, has only the first three levels of protein structure. In other words, it lacks a quaternary level. From this you can conclude that myoglobin
   a. is made of nucleic acids
   b. is made of only one polypeptide chain
   c. lacks hydrogen bonds
   d. is not helical or pleated
   e. is a fiber

8. You received your genetic information from your parents in the form of DNA. This DNA carried the instructions for making
   a. carbohydrates such as glycogen
   b. fatty acids
   c. phospholipids for making the membranes of your cells
   d. proteins
   e. all of the above

9. John is lactose intolerant. The *-ose* ending indicates that John cannot digest a certain
   a. sugar
   b. polysaccharide
   c. protein
   d. steroid
   e. enzyme

Brief Review

1. Describe two ways that water serves as a heat buffer.

2. Both low-density lipoproteins (LDLs) and high-density lipoproteins (HDLs) are involved with carrying fats through the bloodstream. If your LDL count is unusually high, should you be concerned? What if your HDL count is high? Why are they different?

3. List as many functions of proteins as possible. Why are proteins able to do so many different types of jobs? How does this affect the world we live in?

4. How does cohesion allow water to move through plants?

5. What effects does hydrogen bonding have on the properties of water?

Applying Your Knowledge

1. Would you buy a pH-balanced shampoo, or one that is not? Explain your answer.

2. Suppose that you are designing clothing to be worn at a space station on a very cold planet. Should you consider designing clothing that has a layer of water between two layers of plastic? Why or why not? What else might you need to consider?

3. Many species of beans produce a poisonous group of glycoproteins called lectins, which cause red blood cells to clump together (agglutinate) and cease to function. (Fortunately, cooking destroys most of them.) Using what you learned about the structure and function of glycoproteins, suggest how they might do this. What benefit do you think plants might get just by causing red blood cells to clump together?

MediaLab

You Are What You Eat: Food and the Molecules of Life

Not only do we obtain all of the matter that makes up our body from the food and water we take in, but the quality of the foods we consume is critical to our overall health. To learn more, enter the *CD-ROM Tutorial* and learn to analyze food labels so you can interpret the energy sources and classes of macromolecules found in your diet. Using the *Web Investigation*, you can determine your specific daily Calorie requirements, and recommended dietary allowances (RDAs) for macromolecules, minerals, and vitamins. In the *Communicate Your Results* section, you will develop strategies for designing a healthy diet.

This *MediaLab* can be found in Chapter 3 on your CD-ROM (Tutorial 3.3.17) and Companion Website (http://www.prenhall.com/krogh3).

CD-ROM TUTORIAL

Obtaining enough Calories to perform our daily tasks is not hard for most Americans (although there is hunger in America). Indeed, many of us struggle with the problem of consuming too many Calories. Diets high in lipids (fats and oils) contribute to many health problems, including heart disease and cancer. How can you tell if your diet is healthy? In this *CD-ROM Tutorial*, you'll learn how to evaluate your diet using food labels.

Activity

1. *First you will determine the Calories (energy content) in fats, carbohydrates, and protein molecules, and learn to find Calorie content on a nutrition label.*

2. *Then you will use this knowledge in calculating serving sizes and determining what percentage of Calories come from fats, carbohydrates, and proteins so that you can compare your results with the dietary levels suggested for a healthy diet.*

After learning to read a nutrition label, you'll be ready to tailor your diet to your specific energy requirements in the *Web Investigation*.

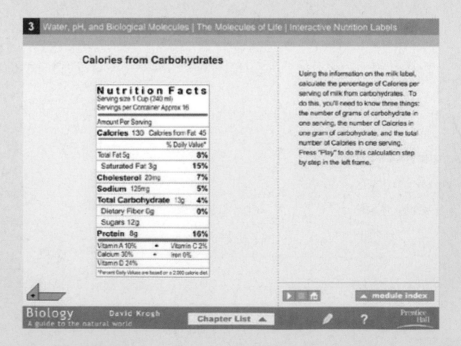

WEB INVESTIGATION

Investigation 1

Estimated time for completion = 5 minutes

Now that you have learned to read a food label and understand what you are eating, wouldn't it be nice to have a personal nutritionist determine the ideal diet for you? Select the Keyword **HEALTHY BODY CALCULATOR** on your CD or Website for this *MediaLab*. Then enter your personal description and submit it. You'll receive an analysis of your health status, including whether you are at a healthy weight and body fat percentage. You'll also receive a personalized daily guide for your food consumption.

Investigation 2

Estimated time for completion = 15 minutes, after keeping a 24-hour diet diary

Dieters often consider Calories when analyzing their diet, but forget to include healthy amounts of all food groups. To compare your recommended diet (from *Web Investigation 1*) with your actual diet, keep a 24-hour record of what you eat. When you are ready to analyze that diet, select the Keyword **DIET ANALYSIS TOOL** on your CD or Website for this *MediaLab*, and enter your personal demographic and diet information. The site will calculate the total Calories consumed (for fats, proteins, and carbohydrates) and compare the result to your specific requirements.

Investigation 3

Time for completion = 5 minutes

The United States has one of the fattest populations on Earth. Over 20 percent of U.S. adults can be classified as clinically obese (due largely to our inability to resist a barrage of good-tasting, inexpensive, high-fat foods. By adding low-calorie fat substitutes, food manufacturers have responded to public pressure for foods with reduced fat content. How can you recognize a fat substitute, and are they good for you? Clip an ingredients label from a package of reduced or low-fat food, and make a list of any fat substitutes. To help you identify them, select the Keyword **FAT SUBSTITUTES** on your CD or Website for this *MediaLab*.

Now that you have analyzed your own diet, discuss with your peers what you have learned about creating a healthy diet.

COMMUNICATE YOUR RESULTS

Exercise 1

Estimated time for completion = 25 minutes

In the United States, many states now impose taxes on tobacco to discourage smoking and to help offset the health costs associated with smoking. However, heart disease, which has been linked to a high-fat diet, is the country's number one killer. In addition, more cancer cases in the United States are related to diet than to smoking. Compose a 250-word argument supporting or refuting the following plan: Taxing high-fat foods would be a great way to reduce the incidence of heart disease and help offset the cost of its treatment. The taxes could also be used to educate the public to the dangers of a high-fat diet, in the same way that taxes on cigarettes are used for antismoking ads.

Exercise 2

Estimated time for completion = 15 minutes

Compare the diet analysis that you did in *Web Investigation 2* with the diet of another student to determine ways you might help each other create a more healthy diet. How are your diets related? What nutrients or food groups are you eating too much of, or too little of? Revise your day's diet to match RDA guidelines, especially with less than 30 percent of total Calories from fat, and post it on the textbook Website for other students to view.

Exercise 3

Estimated time for completion = 5 minutes

Make a list of all the fat substitutes that you found on your ingredient label from *Web Investigation 3*. Indicate what type of fat substitute you found (like the synthetic fat Olestra). Can you determine how many Calories you would save by eating this reduced-fat product instead of the high-fat one?

4

Life's Home
The Cell

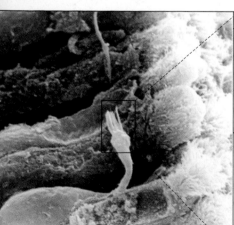

Are those muffins I smell?
(Section 4.1, page 70)

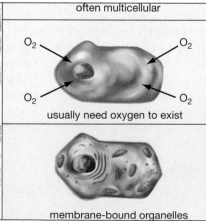

often multicellular

usually need oxygen to exist

membrane-bound organelles

Some cells need oxygen.
(Section 4.2, page 71)

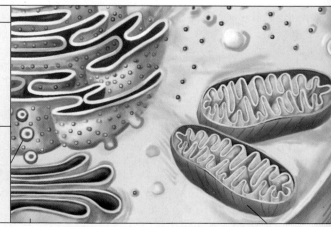

Inside the cell.
(Section 4.3, page 74)

All life exists within cells. These tiny entities can be compared to factories whose products maintain life.

The United States Senate has 100 members in it, and all of them do occasionally gather together to consider given issues. But if you want to know how the Senate actually gets something *done*, you must look to the specialized working units of the Senate known as committees. Things work this way because the business of the nation is simply too complex for every senator to deal with every issue. Each senator must instead specialize in certain issues.

Specialization goes hand in hand with complexity, and this is as true of the natural world as it is of the political. So complex are the tasks that birds and trees and people carry out that if you want to know how any of these creatures actually gets something *done*, you must look again to specialized units, in this case the units known as cells.

4.1 Cells Are the Working Units of Life

Muffins are in the oven and you are in the living room. Gas molecules from the baking muffins waft into the living room, and some of them happen to make their way to your nose, there to travel a short distance to your upper nasal cavity and land on a set of ceaselessly moving hair-like projections called cilia. These actually are the extensions of some specialized nerve cells called olfactory receptors (**see Figure 4.1**, p. 70). If enough muffin molecules bind with enough of the cilia, an impulse is passed along (through other nerve cells) to trigger not only the *sensation* of smell, but the *association* of this smell with the memory of muffins past. How do we know muffins are in the oven? Through cells. How do we move our hands or read this page? Through cells. Life's working units are cells, and in our amazingly complex natural world, there is a great specialization in them.

Cells Bring Unity and Continuity to Life

And yet there is unity. Every form of life either is a cell, or is composed of cells. The one possible exception to this is viruses, but even they must commandeer the machinery of cells in order to reproduce. There is unity, too, in the way cells come about: Every cell comes *from* a cell. Human beings are incapable of producing cells from scratch in the laboratory, and so far as we can tell, nature has fashioned cells from simple molecules

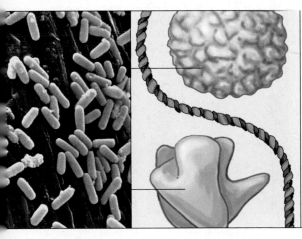

Hidden life on a pin. (Essay, page 72)

It's a small world. (Essay, page 73)

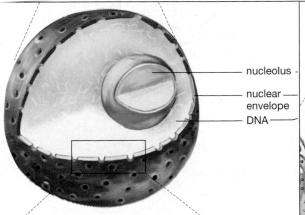

nucleolus

nuclear envelope

DNA

The center of attention. (Section 4.4, page 76)

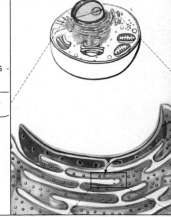

Where proteins shape up. (Section 4.4, page 77)

only once—back when life on Earth got started. The fact that all cells come from cells means that each cell in your body is a link in a cellular chain that stretches back more than 3.5 billion years.

4.2 All Cells Are Either Prokaryotic or Eukaryotic

So, what are these tiny, working units we call cells? You've seen that cells can specialize and come in a variety of forms. It follows from this that there is no such thing as a "typical" cell. There are, however, certain *categories* of cells that are important, and the two most important of these are **prokaryotic cells** and **eukaryotic cells**. Every cell that exists is one or the other, and this simple either-or quality extends to the organisms that fall into these camps. All prokaryotic cells either are bacteria or another microscopic form of life known as archaea. Setting bacteria and archaea aside, *all other cells* are eukaryotic. This means all the cells in plants, in animals, in fungi, and in another grouping called protists that you'll be introduced to in a later chapter. (For a look at the person who first beheld the micro-world of cells, see "First Sightings: Anton van Leeuwenhoek," on page 92.)

Prokaryotic and Eukaryotic Differences

The name *eukaryote* comes from the Greek *eu*, meaning "true," and *karyon*, meaning "nucleus," while *prokaryote* means "before nucleus." These terms describe the most critical distinction between the two cell types. Eukaryotes have a nucleus, bound within a thin membrane, that contains almost all their DNA. While the DNA of prokaryotes is localized (in a "nucleoid" region), it is not bound within a membrane (**see Figure 4.2**).

This distinction, though, only begins to describe the differences between these two cell types. However appealing it may be to think of two single-celled creatures—one a prokaryote, the other a eukaryote—as "alike," the distance between them as life-forms is immense. Human beings and chimpanzees are nearly identical in comparison. Eukaryotes tend to be much larger than prokaryotes; indeed, thousands of bacteria could easily fit into an average eukaryotic cell (see "The Size of Cells" on page 72). Eukaryotes are quite often multicelled organisms, while prokaryotes are single-celled. And most eukaryotes are aerobic—they need oxygen to exist—whereas many prokaryotes can get along with or without oxygen while others actually are poisoned by it.

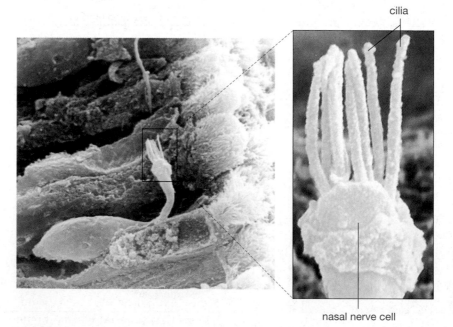

Figure 4.1
Cells Can Specialize
In more complex organisms, different cells carry out different functions. In the picture at left below, you can see one type of cell, a nerve cell, in this case located in the lining of the nose and surrounded by gray accessory cells. A closer look at one of these nerve cells, in the picture at right, shows a number of hair-like extensions, called cilia, protruding from it. When we smell muffins in the oven, gas molecules that waft off the muffins bind with the cilia.

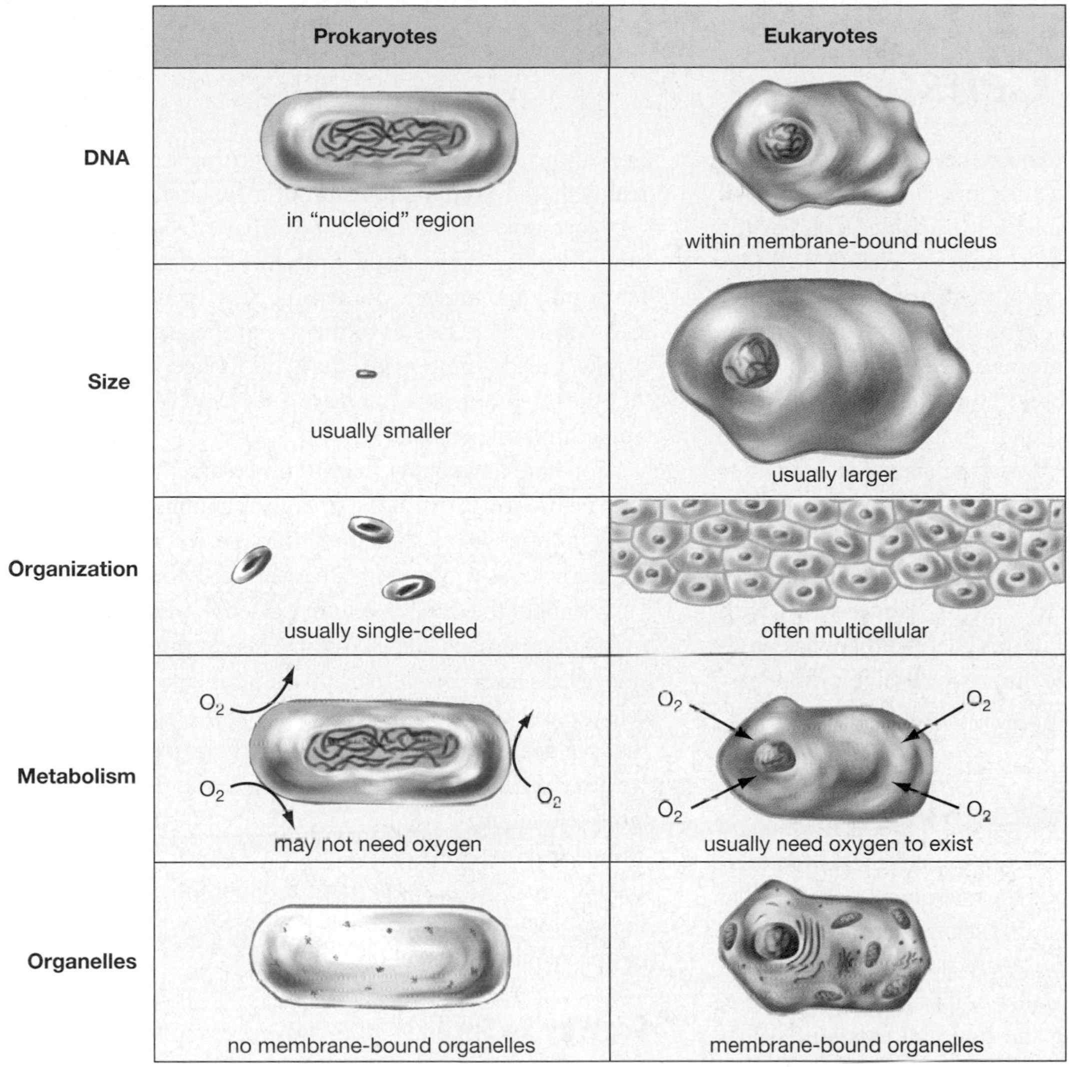

	Prokaryotes	Eukaryotes
DNA	in "nucleoid" region	within membrane-bound nucleus
Size	usually smaller	usually larger
Organization	usually single-celled	often multicellular
Metabolism	O_2 O_2 O_2 may not need oxygen	O_2 O_2 O_2 O_2 usually need oxygen to exist
Organelles	no membrane-bound organelles	membrane-bound organelles

Tutorial 4.1.1
**Prokaryotic and
Eukaryotic Cells**

Figure 4.2
**Prokaryote and Eukaryote
Cells Compared**
A prokaryote cell is a self-contained organism, since the prokaryotes—bacteria and archaea are essentially single-celled. Eukaryote organisms, which may be single- or multi-celled, include plants, animals, fungi, and protists.

Compartmentalization in Eukaryotic Cells

Perhaps the most notable distinction between prokaryotes and eukaryotes, though, is that eukaryotes are *compartmentalized* to a far greater degree than are prokaryotes. The nucleus in eukaryotic cells is just one of their **organelles**, or "tiny organs"—internal compartments that are absent almost altogether in prokaryotes. For example, eukaryotes employ organelles called mitochondria that transform energy from food, while prokaryotes have no such structures.

What's the importance of this compartmentalization? Recall the observation made at the beginning of this chapter: Specialization and complexity go hand in hand. Eukaryotic cells are specialized *internally*, and this characteristic underlies their ability to specialize *as cells*. They have organelles, a much larger

DNA "blueprint," and more proteins to go with it. The creatures made from eukaryotic cells can hear and see and fly. Prokaryotes, meanwhile, exist in a soundless, sightless, flightless world in which one cell equals one living thing. This is not to say that prokaryotes are uniform, nor that they are unsuccessful. On the contrary, they were the first to exist on Earth, and they undoubtedly will be the last to go, if it comes to that. They are diverse, and extremely successful, if success is defined as living in a lot of places in huge numbers. As Lynn Margulis and Karlene Schwartz have observed, more bacteria are living in your mouth right now than the number of people who have ever existed. But prokaryotic cells are *limited* compared to the eukaryotic variety, and so are not the focus of this chapter. For the range of biological processes to be studied here, eukaryotes are the place to look.

The Size of Cells

For several chapters, you've been going over atoms and ions and molecules and such—things small enough to be invisible to the unaided eye. For the most part, you've had to imagine what these things look like, simply because most of them are so small that we either have no pictures of them at all (as with electrons) or few clear pictures (as with atoms). If you flip through the pages of this chapter and those to come, however, you begin to see a fair number of actual photographs. They don't have the same quality as summer vacation snapshots, but they are recognizable as pictures, or more properly as **micrographs** (pictures taken with the aid of a microscope). Micrographs enable us to see surprising things; for example, hundreds of bacteria on the tip of a pin (**Figure 1**). So, with the cells that are introduced in this chapter, there has obviously been a bump up in size into a world that is more easily visible with the help of various kinds of technology.

How Small Are They?

In taking stock of the micro-world, two units of measure are particularly valuable. The smaller of them is the **nanometer**, abbreviated as nm; the larger is the **micrometer**, abbreviated μm. (The unfamiliar-looking first letter there is the Greek symbol for a small *m*. Scientists are not trying to be purposely obscure in using it; another unit of the metric system, the millimeter, lays claim to the mm abbreviation.) What these stand for is a billionth of a meter (nm) and a millionth of a meter (μm). A meter equals about 39.6 inches, or just over a yard, which gives you some starting sense of physical reality in understanding the rest of these sizes.

Now look at **Figure 2** to see what size various objects are. Atoms are down at the bottom of the scale, at about a tenth of a nanometer. Something less than a tenfold increase from there gets us to the size of the protein building-blocks, called amino acids, that you looked at last chapter. Another tenfold-plus increase and you've reached the upper limit on proteins.

You have to go better than 10 times larger than this, however, before you arrive at the size of something that is actually *living*, as opposed to something that is a component part of life. Because, as you've seen, life means cells, this means we're talking about the smallest cells in existence, which are bacteria measuring perhaps 200–300 nm. This extreme on the small side of cells has a counterpart on the large side with the single cells we call chicken or ostrich eggs and certain nerve cells that can stretch out to a meter in length. In general, however, we just cross into the micrometer range with the smaller bacterial cells: about 1–10 μm. The cell size for most plants and animals falls in a range that is 10 times larger than this, about 10–100 μm. At 10 μm, perhaps a billion cells could fit into the tip of your finger. And how about your whole body? One estimate is about 10 trillion.

So, Why So Small?

Having learned how small cells are, you might then well ask: *Why* are they so small? As noted, cells are small chemical

 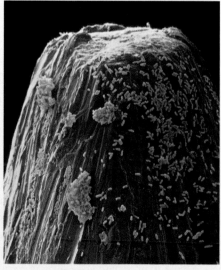 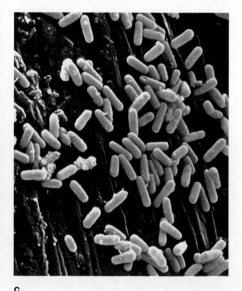

a   b   c

Figure 1
Hidden Life
Microscope enlargements of the tip of this pin show an abundance of life—in this case bacteria—thriving on an object that we normally think of as being devoid of living organisms. (a: ×85; b: ×425; c: ×2100)

factories, and just like any factory, they are constantly shipping things in and out. The size-limiting factor for cells is having enough surface area to export and import all that they need.

This constraint comes about because of a fundamental mathematical principle: As the surface area of an object increases, its *volume* increases even more. Say you have a cube, 1 inch long on each of its sides. Its surface area is 6 square inches (length × width × number of sides), while its volume is one cubic inch (length × width × height). Now say you increase the side dimension to 8 inches. The surface area goes from 6 to 384, but the *volume* goes from 1 to 512. Where, at a one-inch dimension, there were *six* square inches of surface area for every cubic inch of volume, now there are only *three-quarters* of an inch of surface for every cubic inch of volume. Beyond a certain volume, then, a cell simply would not have enough surface area to import and export all the materials it needs to. This effectively sets an upper limit on how big cells can be and explains why most of the cells in an elephant are no bigger than those in an ant, though the elephant does have more cells than the ant. **Figure 3** shows you some size comparisons in the micro-world.

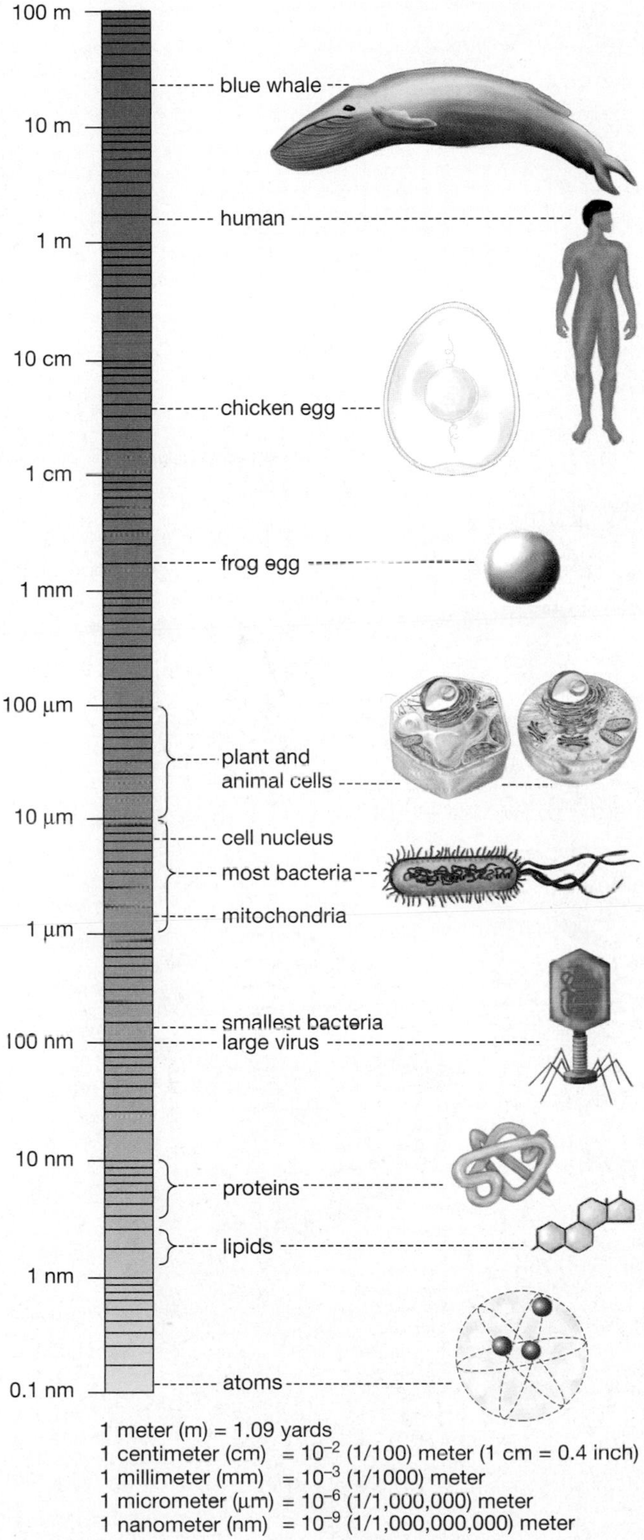

![Figure 2 scale illustration showing sizes of objects]

100 m	blue whale
10 m	
1 m	human
10 cm	chicken egg
1 cm	
1 mm	frog egg
100 μm	plant and animal cells
10 μm	cell nucleus / most bacteria
1 μm	mitochondria
100 nm	smallest bacteria / large virus
10 nm	proteins
1 nm	lipids
0.1 nm	atoms

1 meter (m) = 1.09 yards
1 centimeter (cm) = 10^{-2} (1/100) meter (1 cm = 0.4 inch)
1 millimeter (mm) = 10^{-3} (1/1000) meter
1 micrometer (μm) = 10^{-6} (1/1,000,000) meter
1 nanometer (nm) = 10^{-9} (1/1,000,000,000) meter

Figure 2
Little and Big
The sizes of some selected objects in the natural world.

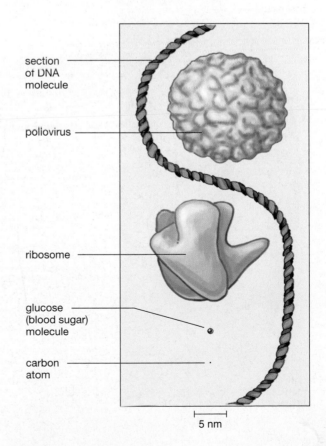

section of DNA molecule

poliovirus

ribosome

glucose (blood sugar) molecule

carbon atom

5 nm

Figure 3
Small Is a Relative Thing
The sizes and shapes of five natural-world entities, each magnified a million times.

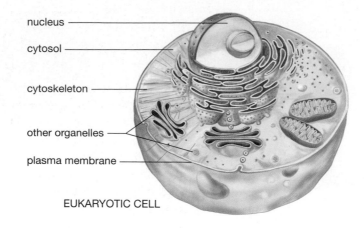

Figure 4.3
Eukaryotic Cell
All eukaryotic cells possess a nucleus, other membrane-bound organelles, jelly-like cytosol, a cytoskeleton, and an outer plasma membrane.

nucleus

cytosol

cytoskeleton

other organelles

plasma membrane

EUKARYOTIC CELL

Tutorial 4.1.2
The Animal Cell

Figure 4.4
The Animal Cell

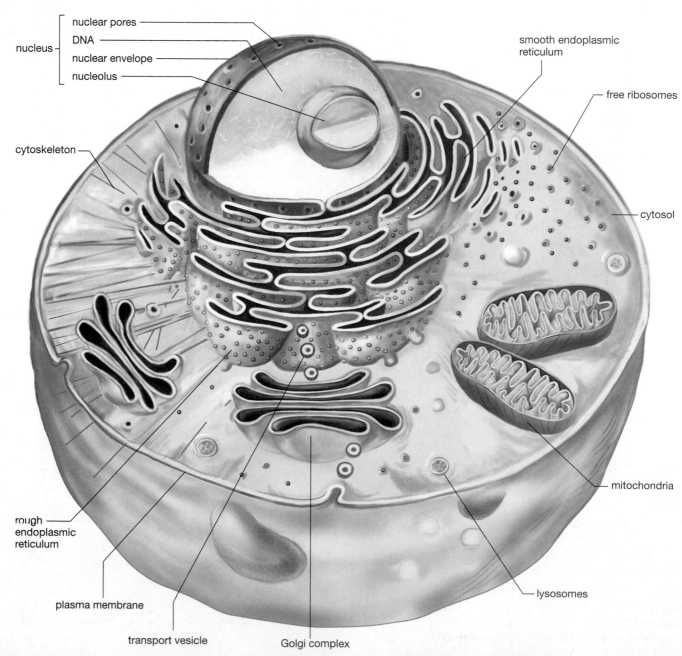

nuclear pores

DNA

nuclear envelope

nucleolus

nucleus

cytoskeleton

smooth endoplasmic reticulum

free ribosomes

cytosol

mitochondria

lysosomes

rough endoplasmic reticulum

plasma membrane

transport vesicle

Golgi complex

4.3 The Eukaryotic Cell

Eukaryotic cells are made of many different components, but in overview these can be broken down into five parts (**see Figure 4.3**).

- The cell's **nucleus**, a membrane-lined compartment that serves as the cell's information center.
- Its other **organelles**, which lie *outside* the cell nucleus.
- The **cytosol**, a protein-rich, jelly-like fluid in which the cell's organelles outside the nucleus are immersed.
- The **cytoskeleton**, a kind of internal scaffolding that has several different kinds of units; some of these can be likened to tent poles, others to monorails.
- The outer boundary of the cell, the **plasma membrane**.

In any discussion of a cell, you are likely to hear the term **cytoplasm**, which simply means the *region* inside the plasma membrane but outside the nucleus. The cytoplasm is different from the cyto*sol*. If you removed all the structures of the cytoplasm—meaning the organelles and the cytoskeleton—what would be left is the cytosol, which is mostly water. This does not mean that the cytosol is simply a passive medium for the other structures. But it is not an organized structure in the way the organelles are. Almost all the organelles you'll see are encased in their own membranes, just as the whole cell is encased in its plasma membrane. In the balance of this chapter, you'll explore all of these components except for the plasma membrane, which is so special it gets its own chapter. You've been hearing the term *membrane* a lot, and will be hearing it a good deal more in the pages to come. What are membranes? Until you get the formal definition next chapter, think of biological membranes as the flexible, chemically active linings of cell compartments.

The Animal Cell

Scientists sometimes make a convenient division of eukaryotic cells into two types: animal cells and plant cells. These cell types have more similarities than they do differences, but they are different *enough* that it will be helpful to look at them separately. We'll start by examining animal cells and then look at how plant cells differ from them.

Insofar as we can characterize that elusive creature, the "typical" animal cell (see Figure 4.4), it is roughly spherical, probably about 25 micrometers (μm) in diameter, surrounded by, and linked to, cells of similar type, and immersed in water.

4.4 A Tour of the Animal Cell: Along the Protein Production Path

You are now going to take an extended tour of the animal cell; in this tour, it will often be helpful to think of a cell as a kind of living factory. Much of the first part of your trip will be spent tracing the way a cell puts together a protein for "export" outside of itself.

Figure 4.5 shows the path you'll be taking, from nucleus to the outer edge of the cell. Don't be bothered by the unfamiliar terms in Figure 4.5, because they'll all be explained in the text. The important thing is to have some sense of the path that protein production takes.

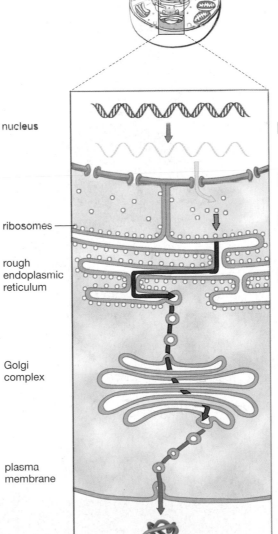

**Figure 4.5
Path of Protein Production in Cells**

nucleus

ribosomes

rough endoplasmic reticulum

Golgi complex

plasma membrane

1. Instructions from DNA are copied onto mRNA.

2. mRNA moves to ribosome.

3. Ribosome moves to endoplasmic reticulum and "reads" mRNA instructions.

4. Amino acid chain growing from ribosome is dropped inside endoplasmic reticulum membrane. Chain folds into protein.

5. Protein moves to Golgi complex for additional processing and for sorting.

6. Protein moves to plasma membrane for export.

Beginning in the Control Center: The Nucleus

As noted in Chapter 3, proteins are critical working molecules in living things, and DNA contains the information for producing these proteins. DNA is like a cookbook whose chemical building blocks in effect say, "Now give me some of this, now some of this, then some of this," the final result being that each DNA "recipe"—each gene on the DNA molecule—gives the complete specifications for a protein. (See Chapter 3, p. 60.) In the eukaryotic cell, DNA is largely confined within the cell **nucleus**, the outer boundary of which is a concentric, double membrane called the **nuclear envelope** (see Figure 4.6).

There comes a point in the life of most cells when they divide, one cell becoming two. Because (with a few exceptions) all cells must possess the set of instructions that are contained in DNA, it follows that when a cell divides, its original complement of DNA must *duplicate*, so that both cells that result from the cell division can have their own DNA. The nucleus, then, is not just the site where DNA exists; it is the site where new DNA is put together, or "synthesized," for this duplication.

Messenger RNA

At the end of Chapter 3, you saw that the process of protein synthesis requires that DNA's instructions first get copied onto another long-chain molecule, RNA. This step is akin to having a cassette tape (of DNA) and then making a copy of it (onto RNA). Our RNA "tape" then moves out of the nucleus to continue the process of protein synthesis. As it turns out, RNA comes in several forms. The one that the DNA instructions are copied onto is called **messenger RNA** (mRNA). Given mRNA's function, it must have, of course, a way to *get out* of the nucleus; its exit points turn out to be thousands of **nuclear pores** that stud the surface of the nuclear envelope. Materials can go the other way through the nuclear pores as well: Proteins, energy molecules, and other "raw

Figure 4.6
The Cell's Nucleus
The DNA of eukaryotic cells is sequestered inside a compartment, the nucleus, which is lined by a double membrane known as the nuclear envelope. Compounds pass into and out of the nucleus through a series of microscopic channels called nuclear pores. The prominent spherical structure within the nucleus is the nucleolus, an area that specializes in the production of ribosomal RNA—the material that helps make up ribosomes. Protein production is dependent upon the information encoded in DNA's sequence of chemical building blocks. This information is copied onto a length of messenger RNA (mRNA), which then exits from the nucleus through a nuclear pore. (Micrograph: ×4400)

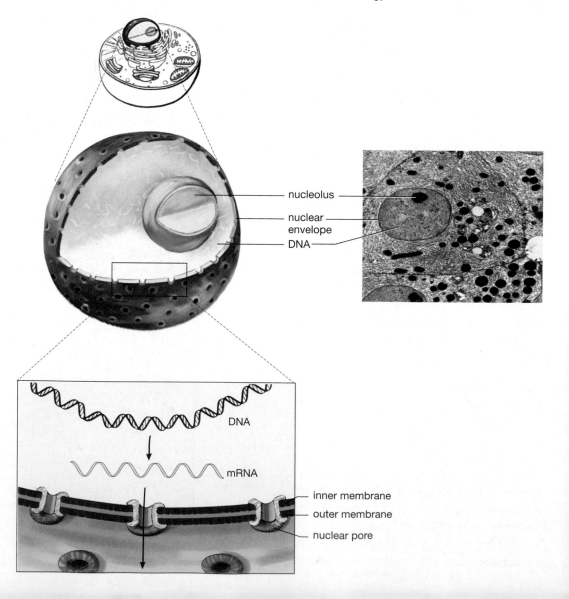

nucleolus

nuclear envelope

DNA

DNA

mRNA

inner membrane

outer membrane

nuclear pore

materials" pass from the cytoplasm into the nucleus by way of them.

mRNA Moves Out of the Nucleus

Imagine shrinking down in size so that the nucleus seems about as big as a house, with you standing outside it. What you would see in protein synthesis is lengths of mRNA, rapidly moving out through nuclear pores and dropping off into the cytoplasm. Several varieties of RNA actually come out like this, but for now, let's just follow the trail of the mRNA. You're about ready to leave the nucleus to continue your cell tour, but before you do, note two things. First, DNA contains information for making proteins, and the mRNA coming out of the nuclear pores amounts to a means of disseminating this information. Thus, it's not hard to conceptualize the nucleus as a control center for the cell. Beyond this, in looking at Figure 4.6, you've probably noticed that there is a rather imposing structure *within* the nucleus, called the **nucleolus**. For now, just *hold that thought*; the mRNA tapes have come out of the nuclear pores and are making a short trip to another part of the cell.

Ribosomes

Small structures called **ribosomes** are the destination for our mRNA tapes. Ribosomes are commonly described as the "workbenches" of protein synthesis, which is a fine metaphor. But following the notion of mRNA as a cassette tape, you might look at a ribosome as a kind of *playback head* on a cassette deck. What does such a head do in an actual deck? As a tape is run through it, it reads signals that have been laid down on it (as magnetic bits) and turns these into sound. Likewise, each ribosome acts as a site that an mRNA tape runs through, only the information on this tape results in the production of an *object* that grows from the ribosome: A chain of amino acids that folds up into the molecule we call a protein (**see Figure 4.7**).

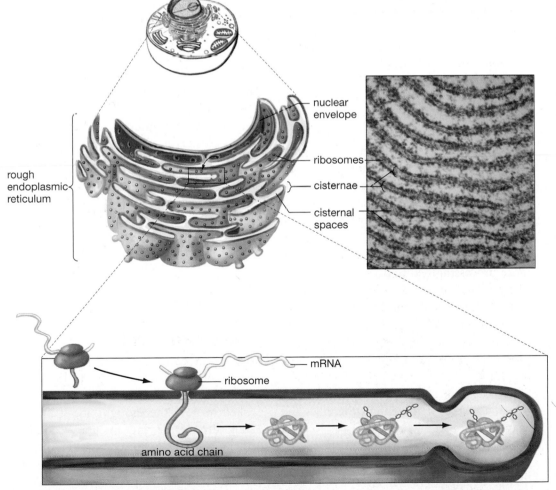

Figure 4.7
Where Many Proteins Take Shape: The Rough Endoplasmic Reticulum (RER)
Messenger RNA "tapes" move out from the nucleus to ribosomes that are free-standing in the cytoplasm. There the tapes are "read" by the ribosomes. The result is an amino acid (or polypeptide) chain that begins to grow from one part of the ribosome. Many of these chains contain chemical sequences that prompt the ribosome first to stop chain production and then to migrate to the rough endoplasmic reticulum. The ribosome then attaches to the outside face of the rough ER. When chain elongation resumes, polypeptide chains such as this one drop into the cisternal space of the rough ER. There they will fold up into their protein shape and undergo processing (for example, with a side-chain being added). Each protein is then encased in a membrane-lined vesicle, ready for transport to the Golgi complex. (Micrograph: ×90,500)

nuclear envelope
rough endoplasmic reticulum
ribosomes
cisternae
cisternal spaces

mRNA
ribosome
amino acid chain

1. mRNA docks on ribosome. Amino acid chain production begins.

2. Ribosome docks on ER. Amino acid chain moves into cisternal space as it is completed.

3. Amino acid chain folds up making a protein.

4. Side chains added to protein.

5. Vesicle formed to house protein while in transport.

You may remember that the kind of protein we are tracking will eventually be exported out of the cell altogether. The synthesis of this kind of protein actually stops when only a very short sequence of the amino acid (or "polypeptide") chain has exited from the ribosome. Why? "Export" polypeptide chains need to be processed within other structures in the cell before they can become fully functional proteins. The first step in this process is for the ribosome, and its associated cargo, to migrate a short distance in the cell and then attach to another cell structure.

The Rough Endoplasmic Reticulum

If you look again at Figure 4.4, you can see that, though the nucleus cuts a roughly spherical figure out of the cell, there is, in essence, a folded-up continuation of the nuclear envelope on one side. This mass of membrane has a name that is a mouthful: the **rough endoplasmic reticulum**. It is rough because it is studded with ribosomes; it is endo*plasmic* because it lies within (endo) the cyto*plasm*; and it is a reticulum because it is a network, which is what *reticulum* means in Latin. Understandably, it is nearly always referred to as the rough ER or RER.

Our ribosome, bound up with its mRNA and polypeptide chains, will migrate to the rough ER and dock on its outside face, thus joining a multitude of other ribosomes that have done the same thing. As noted, the polypeptide chain that is being output from the ribosome is, in essence, an *unfinished* protein that needs to go through more processing before it can be exported. The first step in this processing leads only to the other side of the ER wall the ribosome is embedded in. As the ribosome goes on with its work, the polypeptide chain it is producing drops into *chambers* inside the rough ER.

If you look at Figure 4.7, you can see that the whole of the rough ER takes the shape of a set of flattened sacs (called **cisternae**). The membrane that the rough ER is composed of forms the periphery of these sacs. Inside are the **cisternal spaces** of the RER. (These spaces are sometimes called the RER's *lumen*.) As polypeptide chains enter the cisternal spaces, they first fold up into their protein shapes, as you saw in Chapter 3. Beyond this, most proteins that are exported from cells have sugar side-chains added to them here. Quality control of the production line is in operation in the RER as well. Polypeptide chains that have faulty sequences are detected in the RER cisternal space and ejected out of the protein production line altogether, after which they are degraded into their component parts. Other proteins will pass the quality control tests, however, and will then move out of the RER for more processing.

Several Locations for Ribosomes

All of the mRNA that comes out of the nuclear pores goes to ribosomes, but only some of these ribosomes end up migrating to the rough ER. A multitude of ribosomes will remain **free ribosomes**, which is to say free-standing, in the cytosol. What makes the difference? Remember how, in the ribosome we looked at, only a small stretch of the polypeptide chain it was producing emerged before the ribosome first halted its work and then migrated? That small stretch of the chain contained a chemical signal that said, in effect, "RER processing needed." Many polypeptide chains, however, contain no such signal. In general, RER-bound ribosomes produce proteins that will ultimately reside in the cell's membranes or that will be exported out of the cell altogether (the **secretory proteins**). Meanwhile, most of the proteins that will be used within the cell's cytoplasm or nucleus are made within free ribosomes.

A Pause for the Nucleolus

Before you continue on the path of protein processing, think back a bit to the discussion about the nucleus, when you were asked to hold the thought about the large structure *inside* the nucleus, called the nucleolus. This is the point where its story can be told, because now you know what ribosomes are.

It turns out that *ribosomes* are mostly made of RNA. (They are made of a mixture of proteins and *ribo*nucleic acid.) But so great is the cell's need for ribosomes that a special section of the nucleus is devoted to their synthesis. This is the nucleolus. The *type* of RNA that's part of the ribosomes is, fittingly enough, called ribosomal RNA or rRNA, and it's one of the multiple varieties of RNA mentioned before. Ribosomes are brought to an unfinished state within the nucleolus, after which they pass through the nuclear pores and into the cytoplasm; when put together in final form there, they begin receiving mRNA tapes. They are the one variety of organelle that is not surrounded by a membrane. (They are also the one variety of organelle that prokaryotic cells have.) The cell's traffic in ribosomes is considerable; with millions of them in existence, lots are going to be wearing out all the time. Perhaps a thousand need to be replaced every minute.

Elegant Transportation: Transport Vesicles

The proteins that have been processed within the rough ER need to move out of it and to their "downstream" destinations before being exported. But how do proteins move from one location to another within the cell? Recall that the rough ER and the nuclear membrane amount to one long, convoluted membrane. And, as just noted, all the organelles in the cell except ribosomes are "membrane-bound," meaning they have membranes at their periphery. Each of these membranes has its own chemical structure, but collectively they have an amazing ability to work together: A piece of one membrane can *bud off,* as the term goes, carrying inside it some of our proteins-in-process. Moving through the cytosol, this tiny sphere of membrane can then *fuse* with another membrane-bound organelle, releasing its protein cargo in the process. This network of organelle membranes and their budding and fusing spheres is known as the **endomembrane system**. The spheres that

move within it, carrying proteins and other molecules, are called **transport vesicles**.

This system gives cells a remarkable capability. One minute a piece of membrane may be an integral part of, say, the rough ER; the next it is separating off as a spheroid and moving through the cytosol, carrying proteins within. It is this system that makes it possible for our proteins-in-process to move out of the rough ER. Note, though, that many different *kinds* of proteins are being processed at any one time in the rough ER cisternae. It is as if the cellular factory has a lot of different assembly lines working at once. Most of the proteins under construction are, however, initially bound for the same place—the Golgi complex.

Downstream from the Rough ER: The Golgi Complex

Once a transport vesicle, bearing proteins, has budded off from the rough ER, it then moves through the cytosol to fuse with the membrane of another organelle, one first noticed by Italian biologist Camillo Golgi at the beginning of the twentieth century. The **Golgi complex** further processes proteins. Some side-chains of sugar may be trimmed here, or phosphate groups may be added. But the Golgi complex does something else as well. Recall that, on the one hand, some proteins in this production line are bound for export outside the cell, while other proteins will end up being used within various membranes in the cell. It follows that proteins have to be *sorted and shipped* appropriately, and the Golgi does just this, acting as a kind of distribution center. Chemical "tags" that are part of the proteins often allow for this routing. Remember how, in the rough ER, carbohydrate side-chains might be attached to a newly formed protein? Oftentimes it is these side-chains that serve as the routing tags; other times a section of the protein's amino acid sequence will serve this function.

The Golgi is similar to the ER in that it amounts to a series of cisternae, or connected membranous sacs with internal spaces (**see Figure 4.8**). Proteins arrive at the Golgi housed in transport vesicles that fuse with the Golgi "face" nearest the RER, at which point the vesicles release their protein cargo into the Golgi cisternal sacs for processing. How the Golgi then carries out its work is a matter of some debate. Some evidence indicates that the successive cisternal sacs of the Golgi are relatively fixed entities that first receive and then release vesicles bearing proteins. Other evidence indicates that the sacs themselves physically move forward, toward the outer Golgi face, changing in composition and activity as they do, in accordance with new enzymes they receive. Whatever the case, once processed, proteins of the sort we are following eventually bud off from the outside face of the Golgi, now housed in their final transport vesicles.

From the Golgi to the Surface

For secretory proteins, the journey that began with the transcription of DNA is almost over.

Once a vesicle buds off from the Golgi, all that remains is for it to make its way through the cytosol to the plasma membrane at the outer reaches of the cell. There, the vesicle fuses with the plasma membrane and the protein is ejected into the extracellular world. This last step, called **exocytosis**, is a process you'll be looking at next chapter. With it, one finished product of the cellular factory has rolled out the door.

4.5 Outside the Protein Production Path: Other Cell Structures

A functioning cell engages in more activities than the protein synthesis and shipment just reviewed.

The Smooth Endoplasmic Reticulum

If you look back to Figure 4.4, you can see that there actually are *two* kinds of endoplasmic reticuli. The part of the ER membrane, farther out from the nucleus, that has no ribosomes is

Figure 4.8
Processing and Routing:
The Golgi Complex
Transport vesicles from the rough endoplasmic reticulum (RER) move to the Golgi complex, where they unload their protein contents by fusing with the Golgi membrane. Under one current hypothesis, the protein is then passed, within other vesicles, through the layers of disk-shaped Golgi cisternae, where editing of the protein may occur. Conversely, it may be that the layers themselves physically move forward with their protein cargo inside. At the part of the Golgi furthest from the rough ER, the proteins are sorted, packed in vesicles, and shipped to sites mostly in cell membranes or outside the cell altogether. The vesicles in the micrograph are the pink and purple spheres.

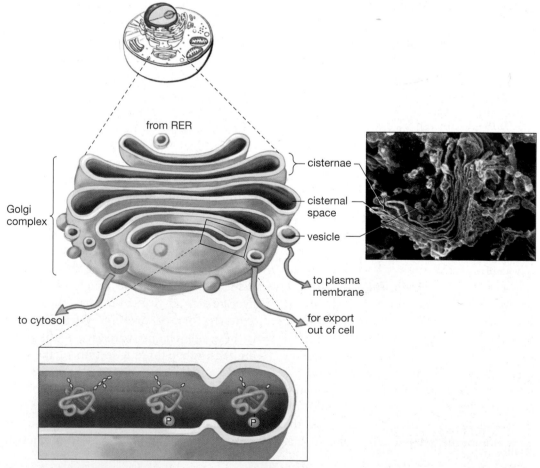

from RER

Golgi complex

cisternae

cisternal space

vesicle

to plasma membrane

to cytosol

for export out of cell

1. Side chains are edited (sugars may be trimmed, phosphate groups added).

2. Vesicle formed for protein transport.

called the **smooth endoplasmic reticulum**, or smooth ER. It's "smooth" because it is not peppered with ribosomes, and this very quality means it is not a site of protein synthesis. In general, the smooth ER is the site of the synthesis of various kinds of lipids, and a site at which potentially harmful substances are detoxified. The tasks the smooth ER undertakes, however, will vary in accordance with cell type. The lipids we normally think of as "fats" are synthesized and stored in the smooth ER of liver and fat cells, while the "steroid" lipids reviewed last chapter—testosterone and estrogen—are synthesized in the smooth ER of the ovaries and testes. The detoxification of potentially harmful substances, such as alcohol, takes place largely in the smooth ER of liver cells.

Tiny Acid Vats: Lysosomes and Cellular Recycling

Any factory must be able to get rid of some old materials, while recycling others. A factory also needs new materials, brought in from the outside, that probably will have to undergo some processing before being used. A single organelle in the animal cell aids in doing all these things, and it is called a **lysosome**. Several hundred of these membrane-bound organelles may exist in any given cell. You could think of them as scaled-off acid vats that take in large molecules, break them down, and then return the resulting smaller molecules to the cytosol. What they cannot return, they retain inside themselves or expel outside the cell. They carry out this work not only on molecules entering the cell from the outside (say, vesicles filled with food) but on materials that exist inside the cell—on worn-out organelle parts, for example (**see Figure 4.9**).

A given lysosome may be filled with scores of different enzymes that can break larger molecules into their component parts—an enzymatic array that allows each lysosome to break down whatever comes its way. A lysosome gets ahold of its macromolecule prey through the endomembrane system. A lysosome will fuse with the membrane surrounding a worn-out organelle part; proceeding to engulf it, it then goes to work breaking the organelle down. The small molecules that result then pass freely out of the lysosome and into the cytosol for reuse elsewhere. Thus there is recycling at the cellular level. Cells carry out this kind of self-renewal

at an amazing rate. Christian de Duve, who with his colleagues discovered lysosomes in the 1950s, has noted the effect of this activity on human brain cells. In an elderly person, he notes, such cells

have been there for decades. Yet most of their mitochondria, ribosomes, membranes, and other organelles are less than a month old. Over the years, the cells have destroyed and remade most of their constituent molecules from hundreds to thousands of times, some even more than 100,000 times.

So why do we grow old? Alas, we don't have any definitive answers to that question yet. One clue, however, is the fact that there are certain things even lysosomes cannot digest, and over time these substances begin to accumulate. In multicellular organisms, the

**Figure 4.9
Cellular Recycling:
Lysosomes**
Lysosomes are membrane-bound organelles that contain potent enzymes capable of digesting large molecules. When a lysosome fuses with a worn-out organelle, its enzymes break the organelle down so that the small resulting molecules can be returned to the cytosol and used elsewhere—a form of cellular recycling. If a lysosome cannot digest a given material, it may expel it outside the cell, though, in multicelled organisms, lysosomes generally will hold on to such materials, so as not to harm structures outside the cell. Lysosomes also digest small particles, such as food, that come from outside the cell.

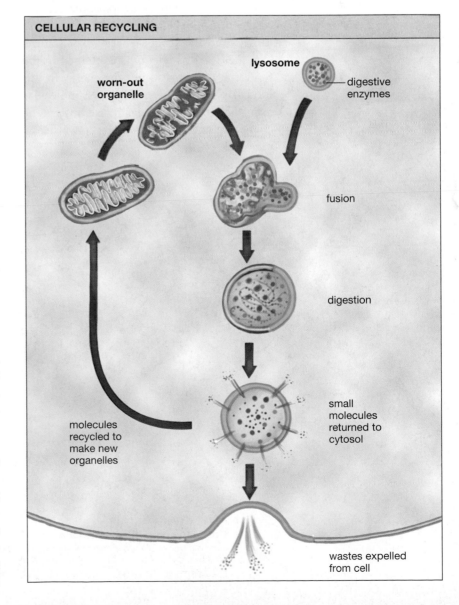

CELLULAR RECYCLING

lysosome
worn-out
organelle
digestive
enzymes
fusion
digestion
small
molecules
returned to
cytosol
molecules
recycled to
make new
organelles
wastes expelled
from cell

lysosome's response is to hold on to this material, rather than releasing it outside the cell, since dumping it could harm extracellular structures. The result of this response, however, is a kind of permanent cellular bloat or constipation as greater numbers of lysosomes cease to function in digestion and become mere holding tanks for indigestible material. Comic as this may seem, there is a serious side to it. The lysosomes of some people are deficient in certain enzymes; one manifestation of this can be Tay-Sachs disease, in which lysosomes are unable to digest great amounts of material and thus swell enormously, killing the cells that house them. The result is mental retardation and death at an early age.

Extracting Energy from Food: Mitochondria

Just as there is no such thing as a free lunch, there is no such thing as free ribosomal action, or protein export, or lysosome activity. There is a price to be paid for all these things, and it is called energy expenditure. The fuel for this energy is contained in the food that cells ingest. But the energy in this food has to be converted into a molecular *form* that the cell can easily use, just as the energy in, say, coal needs to be converted by a power plant into a form that home appliances can easily use—electricity. In eukaryotic cells, the place to look for most of this conversion is inside a group of organelles called **mitochondria**.

Figure 4.10 shows what a mitochondrion has in the way of structure: A continuous outer membrane enclosing an inner membrane that has a series of convolutions in it. The effect of these infoldings is to give mitochondria a larger internal surface area for carrying out their energy transformation activities.

Few details about mitochondria are included here, because much of Chapter 7 is devoted to them. Suffice it to say that to carry out their work, mitochondria need not only food, but *oxygen*. (Ever wonder why you need to breathe?) The *products* of mitochondrial activity, meanwhile, are carbon dioxide, water, and the energy-currency molecule adenosine triphosphate, or ATP, reviewed briefly last chapter.

Another thing to note about mitochondria is that they are probably the descendants of resident aliens. It is likely that they are the end products of bacterial cells that invaded eukaryotic cells more than a billion years ago, only to end up living within them (see "The Stranger within," on page 83).

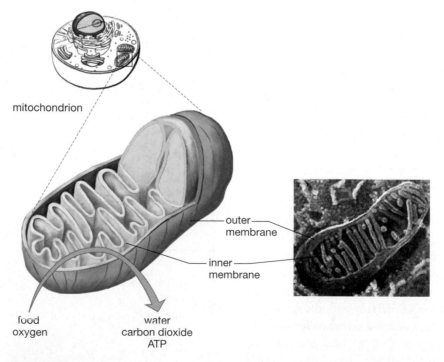

mitochondrion

outer membrane

inner membrane

food
oxygen

water
carbon dioxide
ATP

Figure 4.10
Energy Transformers: Mitochondria
Just as a power plant converts the energy contained in coal into useful electrical energy, mitochondria convert the energy contained in food into a useful molecular form of energy for the cell—ATP. Cells can contain anywhere from a few mitochondria to a few thousand.

The Stranger within:
Lynn Margulis and Endosymbiosis

If pressed to think about it, most people would probably agree that there's something slightly creepy about being made up of cells. Jostling and slithering as they are, dividing, working and dying within us without so much as asking permission, they give us the feeling that the unitary *self* that we so cherish actually amounts to an unruly collection of creatures within.

How sobering it is, then, to realize that we may be composite beings in more ways than one. Each of our *cells* may have within it the vestiges of yet other living things: the descendants of bacteria that long ago invaded our ancestors' cells, only to take up residence there. In plant and animal cells, the mitochondria that serve as cellular "powerhouses" appear to be the descendants of bacteria. Plant and algae cells may have a different set of bacterial descendants in them: the plastids, which include the chloroplasts that carry out photosynthesis.

The idea that these structures are descended from free-standing bacteria is called the endosymbiotic theory—*endo* for "within" the cell, and *symbiotic* for "symbiosis," meaning a situation in which two organisms not of the same species live in close association. Endosymbiosis is not a new idea; it was proposed as far back as the nineteenth century. It has, however, made quite a journey in recent years. Regarded as a crackpot notion as late as the 1960s, it came to be given some grudging credibility over time. By the mid-1980s it had made the transition from interesting hypothesis to generally accepted theory.

This is a kind of heroic journey, of course; and fittingly, one person is primarily responsible for it. She is Lynn Margulis, a professor of botany at the University of Massachusetts, Amherst. Resurrecting endosymbiosis in her 1965 Ph.D. thesis, Margulis went on to be its champion, at first taking it through derision and skepticism, and then seeing it through to vindication.

What makes us think that endosymbiosis really happened? First of all, mitochondria and plastids—which, remember, are *organelles* within eukaryotic cells—have many of the characteristics of free-standing *cells*, specifically free-standing bacterial *cells*. They have their own ribosomes and their own DNA, both of a bacterial type, and they reproduce through division, like bacterial cells, partly under their own genetic control. All of this makes mitochondria look suspiciously like once-independent cells that now find themselves living inside other cells. Moreover, sequencing of the chemical building blocks of mitochondrial DNA has allowed researchers to trace the evolutionary origins of these organelles to a single, ancient bacterial species. Fittingly, this species is a close relative of a group of modern bacteria that make their living as parasites inside other cells.

Margulis believes that, for mitochondria, the transition to endosymbiosis was completed about 1.4 billion years ago; prior to this, they existed as an aggressive form of oxygen-using bacteria. Seizing an opportunity, they invaded a form of bacteria that was fairly intolerant of oxygen. In more recent versions of this theory, the bacteria invaded eukaryotic cells. The arrangement the two species eventually came to was simple: The host provided food to the invader; and the oxygen-using invader, now domesticated, allowed the host to survive in an oxygenated world.

Endosymbiosis has been challenged recently by scientists who have suggested that mitochondria came about by means of bacterial cells *evolving into* eukaryotic cells, rather than invading them. Hopefully, time will tell where the truth lies.

4.6 The Cytoskeleton: Internal Scaffolding

When you see a set of eukaryotic cells in action, the word that comes to mind is hyperactive. Cells are not passive entities. Jostling, narrowing, expanding, moving about, capturing objects and bringing them in, expelling other objects: This is life as a bubbling cauldron of activity. It was once thought that cells did all these things with pretty much the equipment you've looked at so far, meaning that the Golgi complex, the ribosomes, and so forth were thought to be floating in an undifferentiated soup that was the cytosol. But advanced microscopy showed that there is, in fact, a tangled forest within the cytoplasm: Protein strands that give cells their shape, that anchor the structures you've been reading about, that act as monorails for particles moving within, and that allow the cells themselves to move. Taken as a whole, these protein strands are called the **cytoskeleton**.

Some of the cytoskeleton's parts are permanent and relatively static, but many are moving and some are assembled or disassembled very rapidly. The cytoskeleton usually is divided into three component parts. Ordered by size, going from smallest to largest in diameter, they are **microfilaments**, **intermediate filaments**, and **microtubules** (**see Figure 4.11**). Let's take a look at some of the characteristics of each.

Microfilaments

The most slender of our cytoskeletal fibers, microfilaments are made of the protein *actin*, which, by itself, exists as a support or "structural" filament in almost all eukaryotic cells. Actin microfilaments can also help cells move or capture prey, essentially by growing very rapidly at one end—in the direction of the movement or extension—while decomposing rapidly at the other end. Actin works with another protein, myosin, to bring about muscle contraction in muscle cells. You can see a vivid example of microfilament-aided extension in **Figure 4.12**.

Intermediate Filaments

These in-between-sized proteins are the most permanent of the cytoskeletal elements, perhaps coming closest to our everyday notion of what a skeleton is like. They stabilize the positions of the nucleus and organelles within the cell.

Microtubules

These largest of the cytoskeletal elements also play a structural role in the cell; in fact, theirs seems to be the preeminent structural role in the sense of determining the shape of the cell. But they take on several other tasks in addition. They serve, for example, as the monorails discussed earlier. Recall that protein-laden vesicles move from one organelle to another in the cell. These spheres are moving along the "rails" of microtubules, while sitting atop the "engine" of one of the so-called motor proteins. (**see Figure 4.13**).

Figure 4.11
Structure and Movement: The Cytoskeleton
Three types of fibers form the inner scaffolding or cytoskeleton of the cell: microfilaments, intermediate filaments, and microtubules.

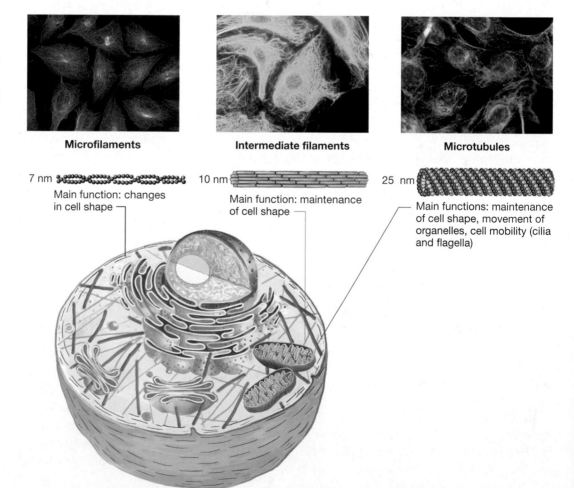

Microfilaments

7 nm — Main function: changes in cell shape

Intermediate filaments

10 nm — Main function: maintenance of cell shape

Microtubules

25 nm — Main functions: maintenance of cell shape, movement of organelles, cell mobility (cilia and flagella)

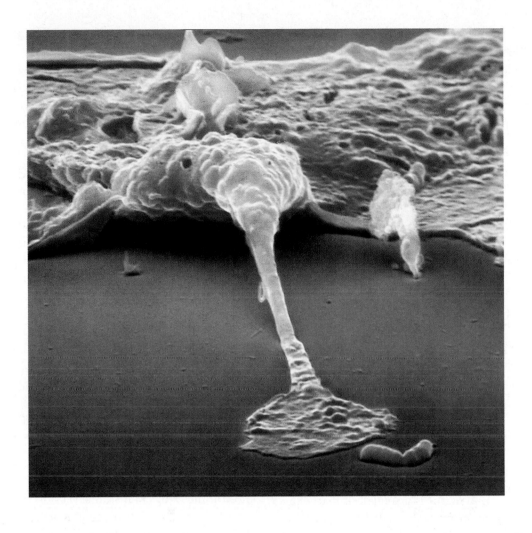

Figure 4.12
Microfilaments in Action
Certain cells can move or capture prey by sending out extensions of themselves called pseudopodia ("false feet"). It is the rapid construction of actin microfilaments—in the direction of the extension—that makes this possible. Here a type of blood borne guard cell called a macrophage is about to use a pseudopodium it is constructing to capture a green bacterium.

a transport monorails

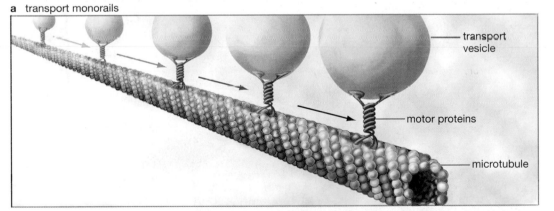

transport
vesicle

motor proteins

microtubule

b cilia

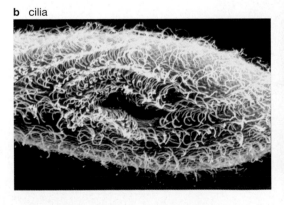

c flagellum

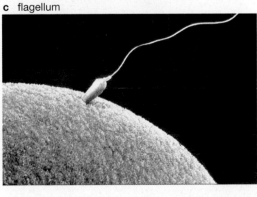

Figure 4.13
Several Functions for Microtubules

a They are the "rails" on which vesicles move through the cell, carried along by "motor proteins."

b They exist outside the cell in the form of cilia, which are profuse collections of hair-like projections that beat rapidly, forming currents that can propel a cell or move material around it.

c Microtubules are also found outside the cells in the form of flagella. The flagellum on this sperm cell is enabling it to seek entry into an egg.

Cell Extensions Made of Microtubules: Cilia and Flagella Microtubules also form the underlying structure for two kinds of cell extensions, **cilia** and **flagella** (see Figure 4.13). Cilia exist as a profusion of hair-like growths extending from cells. Their function is simple: Move back and forth very rapidly, perhaps 10 to 40 times per second. The purpose of this movement can be either to propel a cell or to move material *around* a cell. Cilia are extremely common among single-celled organisms and in some of the cells of simple animals (sponges, jellyfish). You saw an example of cilia in humans earlier in connection with our sense of smell. Our lungs also are lined with cilia, whose job it is to sweep the lungs clean of whatever foreign matter has been inhaled. In this work, like most cilia, lung cilia all beat at once in the same direction, acting like rowers in a crew.

Cilia grow from eukaryotic cells in great profusion, but it is a different story with flagella. Only a few generally will sprout from a given cell—indeed, there often is just a single tail-like flagellum. The function of flagella is cell movement. Only one kind of animal cell is flagellated, and it scarcely needs an introduction: A sperm is a single cell that whips its flagellum mightily in order to get to an unfertilized egg.

In Summary: Structures in the Animal Cell

In your tour of the cell so far, you've pictured a cell as a factory, one that synthesizes proteins in a "production line" that starts with DNA in the nucleus and then goes to the ribosomes (via mRNA), to the rough ER, to the Golgi complex, and finally to the protein's destination (the plasma membrane, export, and so on). You've also seen that cells have other structures such as lysosomes for digestion and recycling, mitochondria for energy transformation, the smooth endoplasmic reticulum for detoxification and lipid synthesis, and the cytoskeleton for structure and movement. If you look at **Figure 4.14**, you can see, in metaphorical form, a "map" of these component parts within the cell. **Table 4.1** (on page 87 in the discussion of plant cells) lists cellular elements found in plants and animals, as well as some elements found only in plant cells.

4.7 The Plant Cell

As noted earlier, plant and animal cells have more similarities than they do differences. Among these similarities is that plant cells do most of the things that animal cells do (produce proteins, transform energy, and so on.) A quick look at **Figure 4.15** will confirm for

Figure 4.14
The Cell as a Factory
This comparison may help you to remember some roles of the different parts of the cell.

CONTROL CENTER
(nucleus)

STRUCTURE
(cytoskeleton)

ASSEMBLY LINE
(endoplasmic
reticulum)

WORKBENCHES
(ribosomes)

DISTRIBUTION
CENTER
(Golgi complex)

POWERHOUSES
(mitochondria)

CLEANING
CREW
(lysosomes)

SECURITY GATE
(cell membrane)

Table 4.1
Structures in Plant and Animal Cells

Name	Location	Function
Cytoskeleton	Cytoplasm	Maintains cell shape, facilitates cell movement and movement of materials within cell
Cytosol	Cytoplasm	Protein-rich fluid in which organelles and cytoskeleton are immersed
Golgi complex	Cytoplasm	Processing, sorting of proteins
Lysosomes (in animal cells only)	Cytoplasm	Digestion of imported materials and cell's own used materials
Mitochondria	Cytoplasm	Transform energy from food
Nucleolus	Nucleus	Synthesis of ribosomal RNA
Nucleus	Inside nuclear envelope	Site of most of cell's DNA
Ribosomes	Rough ER Free-standing in cytoplasm	Sites of protein synthesis
Rough endoplasmic reticulum	Cytoplasm	Protein processing
Smooth endoplasmic reticulum	Cytoplasm	Lipid synthesis, storage; detoxification of harmful substances
Vesicles	Cytoplasm	Transport of proteins and other cellular materials
Cell walls (in plant cells only)	Outside plasma membrane	Limit water uptake; maintain cell membrane shape, protect from outside influences
Central vacuole (in plant cells only)	Cytoplasm	Cell metabolism, pH balance, digestion, water maintenance
Plastids (in plant cells only)	Cytoplasm	Nutrient storage, pigmentation, photosynthesis (chloroplasts)

Tutorial 4.2.3
Structures in Plant and Animal Cells

Figure 4.15
Common Structures in Animal and Plant Cells

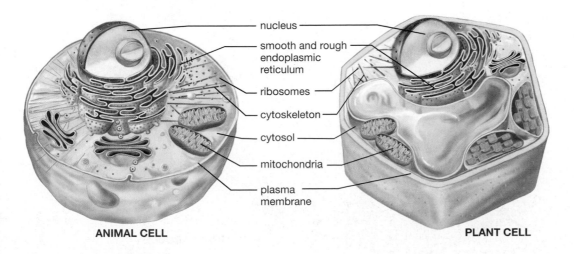

nucleus

smooth and rough endoplasmic reticulum

ribosomes

cytoskeleton

cytosol

mitochondria

plasma membrane

ANIMAL CELL

PLANT CELL

Tutorial 4.2.2
The Plant Cell

Figure 4.16
The Plant Cell
The cell wall, central vacuole, and chloroplasts do not exist in animal cells, but the other components are common to both plant and animal cells.

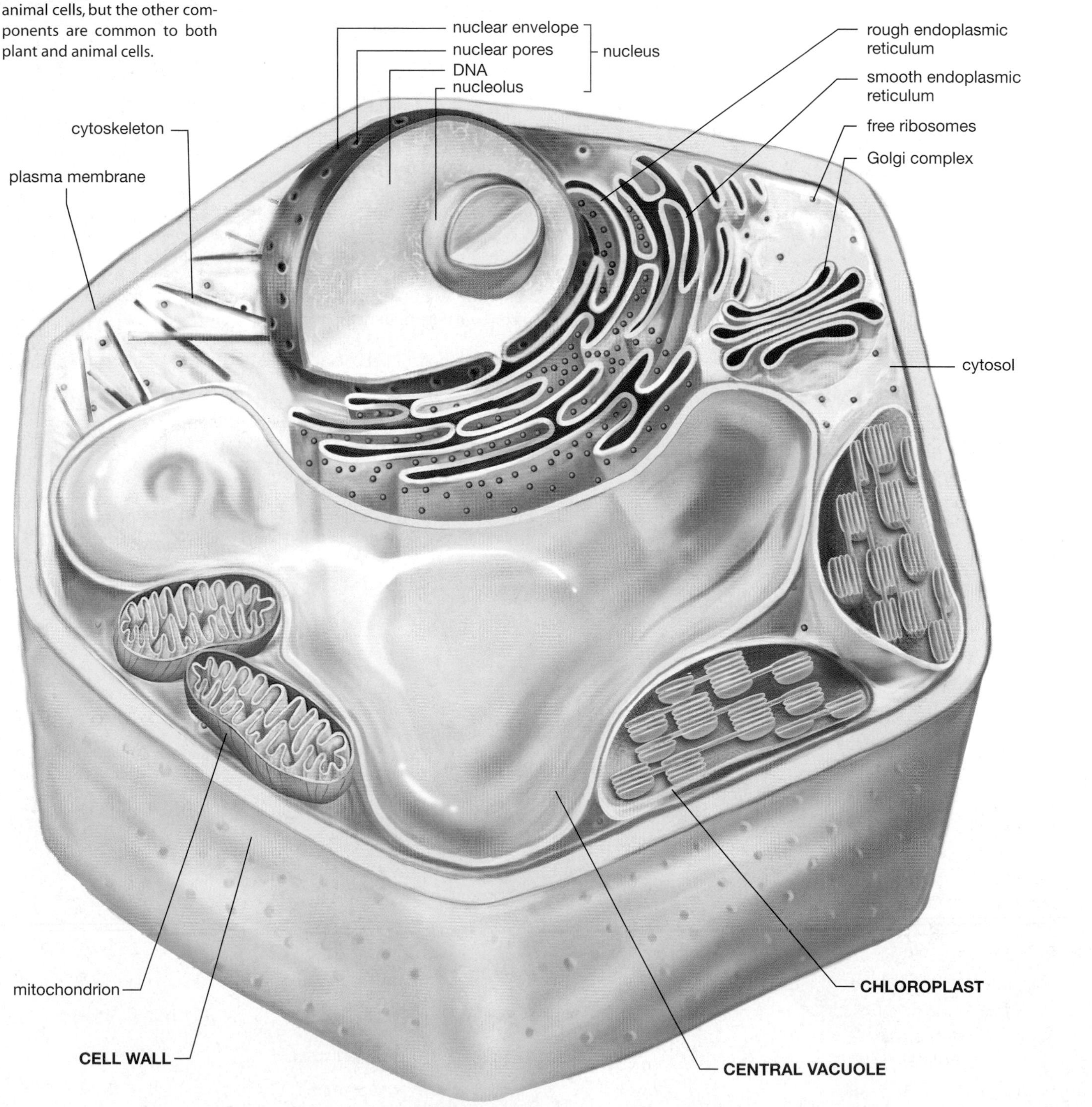

you how structurally similar plant and animal cells are as well. As you see, a plant cell has a nucleus (with a nucleolus), the smooth and rough ERs, a cytoskeleton—most of the things you've just gone over in animal cells. Indeed, there is only one structure present in the animal cells you've looked at that plant cells don't have: the lysosome. What jumps out at you when you look at plant cells is not what they lack compared to animal cells, but what they *have* that animal cells do not. As you can see in **Figure 4.16**, these additions are

- A thick cell wall
- A large structure called a central vacuole
- Structures called plastids, one important variety of which is the chloroplast

The Cell Wall

Plant cells have an outside layer that makes their plasma membrane, just inside it, look

nuclear envelope
nuclear pores   nucleus
DNA
nucleolus

rough endoplasmic reticulum

smooth endoplasmic reticulum

free ribosomes

Golgi complex

cytoskeleton

plasma membrane

cytosol

mitochondrion

CELL WALL

CHLOROPLAST

CENTRAL VACUOLE

rather thin and frail by comparison. This is because it is thin and frail by comparison; the plasma membrane of a plant cell may be 0.01 µm thick, while the combined units of a cell wall may stretch to 7 µm or more. Cell walls are nearly always present in plant cells, whereas animal cells don't have them. You should note, however, that many organisms that are neither plant nor animal—bacteria, protists, and fungi—also have cell walls.

What do cell walls do? They provide plant cells with structural strength, put a limit on their absorption of water (as you'll see next chapter), and generally protect plants from harmful outside influences. So why don't *animal* cells have cell walls? Because cell walls make for a rather rigid, inflexible organism—like plants, which are stationary. Animals, meanwhile, are mobile and thus must remain flexible.

Cell walls in plants can come in several forms, but all such forms will be composed chiefly of a molecule you were introduced to last chapter: cellulose, a polysaccharide that is embedded within cell walls in the way reinforcing bars run through concrete. In some cell walls, cellulose is joined by a compound called lignin, which imparts considerable structural strength. You can see a vivid demonstration of this in the material we know as wood, which is largely made of cell walls (**see Figure 4.17a**).

Cell walls can serve different functions over the life of an organism. Generally, they are the site of a good deal of metabolic activity, and thus should not be considered mere barriers. On the other hand, the outer portion of the barrier we call tree bark consists mostly of the cell walls of dead cells (**see Figure 4.17b**).

a

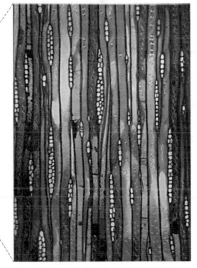

b

Figure 4.17
Great Strength from Small Things

a Redwoods such as this one can reach enormous heights because of the strength of their wood, which is largely made of cell walls.

b Cell walls play important roles in living cells, but they also are valuable as the strong, remaining components of dead cells; here they help make up this redwood bark. (×240)

The Central Vacuole

In looking at the diagram of the plant cell, you'll see one structure, called a **central vacuole**, that is so prominent it appears to be a kind of organelle continent surrounded by a mere moat of cytosol. In a mature plant cell, one or two central vacuoles may comprise 90 percent of cell volume.

Although animal cells can have vacuoles, the imposing central vacuole in plants is different. For a start, it is composed mostly of water, which demonstrates just how watery plant cells are. A typical animal cell may be 70 to 85 percent water, but for plant cells the water proportion is likely to be 90 to 98 percent.

The watery milieu of the central vacuole contains hundreds of other substances. Many of these are nutrients, others are waste products. There are also hydrogen ions, pumped in to keep the cell's cytoplasm at a near-neutral pH. Given these materials, it may be obvious that the central vacuole does a lot of things: It stores nutrients, it is involved in metabolism, and it retains waste products or degrades them with digestive enzymes, like the lysosomes in animal cells. There is even an aesthetic side to the central vacuole: Many red and blue flowers owe their colors to the pigments it contains.

Plastids

Plastids are a diverse group of organelles that are found only in plants and algae. Some gather and store nutrients for plant cells; others are pigment-containing organelles that give us, for example, the red color of the tomato skin. The best known of the plastids, however, are the **chloroplasts**.

Photosynthesis: The Chloroplasts

The pigment chlorophyll, which is contained in chloroplasts, is the central reason the living world is largely a green world. Color is the least of the reasons we should be grateful to chloroplast-containing plants, however, because we're indebted to them for almost all the food we eat and much of the oxygen we breathe. Chloroplasts are the sites of *photosynthesis*, in which sunlight is captured and used to produce a carbohydrate—a sugar that functions as the plant's food—from nothing more than carbon dioxide, water, and a few minerals (**see Figure 4.18**). This may not seem like such a big deal until you try to name a food you eat that is not a plant or does not itself eat plants. A byproduct of photosynthesis is oxygen, whose significance to us can scarcely be overstated. From this, it may be obvious that plants could get along just fine without people, but the reverse is not true. A plantless world would soon be a personless world.

4.8 Cell Communication: Why Cells Need Not Be Islands

Most of what you've seen so far has made the cell seem like an isolated entity, but this is not the case. Single-celled organisms can exist as separate entities, along with certain plant or animal cells (red blood cells, for example), but most plant and animal cells are linked together in organized collections referred to as **tissues**. Not surprisingly, these assemblages of cells—be they plant or animal—have the ability to communicate with one another.

Communication among Plant Cells

Having noted the thickness of something like the cell walls in plants, you might wonder how one plant cell could interact with another. Communication between plant cells takes place quite readily, however, through a series of tiny

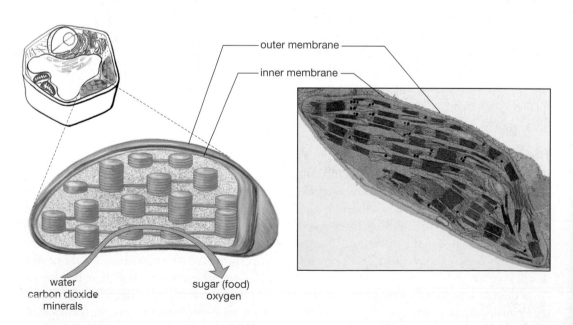

outer membrane
inner membrane

water
carbon dioxide
minerals

sugar (food)
oxygen

Figure 4.18
Food Source for the World
Chloroplasts, the tiny organelles that exist in plant cells, are sites of photosynthesis—the process that provides food for most of the living world. Using the starting materials of water, carbon dioxide, and a few minerals, plants use energy from sunlight to produce their own food. A double membrane lines the chloroplasts. (Micrograph: ×13,000)

channels in the cell wall called **plasmodesmata** (singular, plasmodesma). The structure of these channels is such that the cytoplasm of one plant cell is continuous with that of another—so much so that the cytoplasm of an entire plant can be properly looked at as one continuous whole (**see Figure 4.19a**). The structure of plasmodesmata is more complex than that of a simple opening, but the basic idea is of a channel-like linkage between two plant cells.

Communication among Animal Cells

There are no plasmodesmata in animal cells, but there are three other kinds of **cell junctions**, or linkages, one of which serves to facilitate cell communication. It is called a **gap junction**, and it consists of clusters of protein structures that shoot through the plasma membrane of a cell from one side to the other. When these structures line up in adjacent cells, the result is a channel for passage of small molecules and electrical signals (see Figure 4.19b). The animal gap junctions and the plant plasmodesmata are very different kinds of channels. Plasmodesmata can be thought of as permanent channels between plant cells, whereas gap junctions open only as necessary.

On to the Periphery

Having looked at what is inside the cell, you've arrived at the cell's periphery, the plasma membrane, which is where you'll be staying for awhile. It may at first seem strange to devote a whole chapter to an outer boundary. How much attention would you pay, after all, to a factory's wall as opposed to its contents? The answer: A lot, if that wall could move, continually renew itself, and let some things in while keeping others out. Such is the case with the plasma membrane, a slender lining that manages to make one of the most fundamental distinctions on Earth: Inside, life goes on; outside it does not.

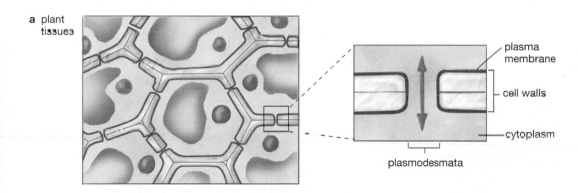

a plant tissues

plasma membrane

cell walls

cytoplasm

plasmodesmata

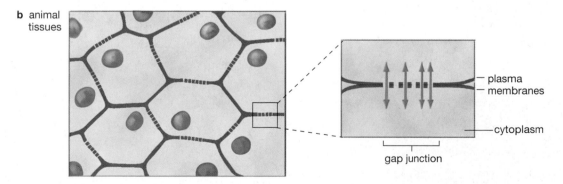

b animal tissues

plasma membranes

cytoplasm

gap junction

Figure 4.19
Cell Communication

a Plasmodesmata In plants, a series of tiny pores between plant cells, the plasmodesmata, allow for the movement of materials among cells. Thanks to the plasmodesmata channels, the cytoplasm of one cell is continuous with the cytoplasm of the next; the plant as a whole can be thought of as having a single complement of continuous cytoplasm.

b Gap junctions In animals, protein assemblies come into alignment with one another, forming communication channels between cells. A cluster of many such assemblies—perhaps several hundred—is called a gap junction.

First Sightings: Anton van Leeuwenhoek

To the list of explorers that includes Columbus and Balboa we could, in a sense, add the name Anton van Leeuwenhoek. This unassuming Dutchman was the great early voyager into another world, the micro-world.

It was not until the seventeenth century that human beings realized that things as small as cells existed. Leeuwenhoek began to report in the 1670s on what he saw with the aid of a device that had been invented at the end of the 1500s—the microscope (**Figure 1**). (Note, though, that Leeuwenhoek actually was using devices that we would refer to today as magnifying glasses.) One of Leeuwenhoek's contemporaries, Englishman Robert Hooke, coined the term *cell* after viewing a slice of cork under a simple microscope, but Hooke's purpose was to reveal the detailed structure of familiar, small objects, such as the flea. Leeuwenhoek, by contrast, revealed the *existence* of creatures unimagined until his time. Moreover, he carried out this work in the most extraordinary fashion: Laboring alone in the small town of Delft with palm-sized magnifiers he himself had created, looking at anything that struck his fancy. (And many things struck his fancy; he once looked at exploding gunpowder under a microscope, nearly blinding himself in the process.) For 50 years, while working as shopkeeper and minor city official, this untrained amateur of boundless curiosity examined the micro-world and reported on it in letters he posted to the Royal Society in London.

Who could believe what he uncovered? How was it possible that there was a buzzing, blooming universe of "animalcules" (little animals) whose existence had been completely unsuspected? Prior to his work, no one thought that any creature smaller than a worm could exist within the human body. Yet, examining scrapings from his own mouth, Leeuwenhoek tells us:

> I saw, with as great a wonderment as ever before, an inconceivably great number of little animalcules, and in so unbelievably small a quantity of the foresaid stuff, that those who didn't see it with their own eyes could scarce credit it.

The animalcules that Leeuwenhoek beheld over his career were single-celled organisms that today are known as bacteria and protists. It would be two hundred years before Leeuwenhoek's findings were fully integrated into a modern theory of cells. Yet the man from Delft had shown that only a small portion of the world's living things are visible things.

a

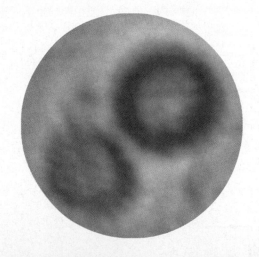

b

Figure 1

a What Leeuwenhoek Could See
Biologist Brian Ford used an actual 300-year-old Leeuwenhoek microscope to capture this image of spiny spores from a truffle. (×600)

b Leeuwenhoek's Microscopes Were Hand-Held and Paddle-Shaped
Leeuwenhoek revealing the micro-world to Queen Catherine of England, wife of King Charles II (1630–1685).

Chapter Review

Summary

4.1 Cells Are the Working Units of Life

- With the possible exception of viruses, every form of life on Earth either is a cell or is composed of cells. All cells come into existence through the activity of other cells.

4.2 All Cells Are Either Prokaryotic or Eukaryotic

- All cells can be classified as prokaryotic or eukaryotic. All plants, animals, fungi, and protists either are single eukaryotic cells or are composed of eukaryotic cells. Prokaryotic cells are either bacteria or archaea. Eukaryotic cells have most of their DNA contained in a membrane-lined nucleus, whereas prokaryotic cells do not have a nucleus. Eukaryotic cells also have more specialized structures, called organelles, than do prokaryotic cells. All prokaryotes are single-celled, whereas many eukaryotes are multicelled.
TUTORIAL 4.1.1: Prokaryotic and Eukaryotic Cells

4.3 The Eukaryotic Cell

- There are five principal components to the eukaryotic animal cell: the nucleus, other organelles, the cytosol, the cytoskeleton, and the plasma membrane. Organelles are "tiny organs" within the cell that carry out specialized functions, such as energy transfer and material recycling. The cytosol is the fluid in which these organelles are immersed. (The cytoplasm is the region of the cell inside the plasma membrane but outside the nucleus.) The cytoskeleton is composed of several groups of proteins that give the cell support and facilitate transportation of cellular elements. The plasma membrane is the chemically active outer boundary of the animal cell. Plant cells have a plasma membrane that has, outside of it, a cell wall.
TUTORIAL 4.2.1: The Animal Cell

4.4 A Tour of the Animal Cell: Along the Protein Production Path

- Information for the construction of proteins is contained in the DNA located in the cell nucleus. This information is copied onto an informational "tape" of messenger RNA that departs the cell nucleus through nuclear pores and goes to the sites of protein synthesis, structures called ribosomes. Many ribosomes first migrate to, and then embed in, a series of membrane sacs called the rough endoplasmic reticulum. Polypeptide (or amino acid) chains that will become proteins are dropped from ribosomes into the internal spaces of the rough endoplasmic reticulum, where these chains fold up and undergo editing. Some ribosomes are not embedded in rough endoplasmic reticulum but instead remain free-standing in the cytosol.

- Materials move from one structure to another in the cell via the endomembrane system, in which a piece of membrane, with proteins or other materials inside, can bud off from one organelle, move through the cell, and then fuse with another membrane-lined structure. Membrane-lined structures that carry cellular materials are called transport vesicles.

- Transport vesicles move proteins that are being produced from the rough endoplasmic reticulum to the Golgi complex, where the proteins are processed further and marked for shipment to appropriate cellular locations.

4.5 Outside the Protein Production Path: Other Cell Structures

- The smooth endoplasmic reticulum is a network of membranes that functions to synthesize lipids and to detoxify potentially harmful substances. Lysosomes are organelles that break down material coming into the cell (such as food) or worn-out cellular structures, returning the components of these structures to the cytoplasm for further use. Lysosomes will hold onto, or expel outside the cell, material that cannot be recycled inside the cell. Mitochondria function to extract energy from food and to transform this energy into a chemical form the cell can use.

4.6 The Cytoskeleton: Internal Scaffolding

- Cells have within them a web of protein strands, called a cytoskeleton, that provide the cell with structure, facilitate the movement of materials inside the cell, facilitate cell movement, and help cells capture prey.

- There are three principal types of cytoskeleton elements. Ordered by size, going from smallest to largest in diameter, they are microfilaments, intermediate filaments, and microtubules. Microfilaments are made of the protein actin. They help the cell move and capture prey. Actin works with the protein myosin to bring about muscle contraction. Intermediate filaments provide support and structure to the cell. Microtubules play a structural role in cells and facilitate the movement of materials inside the cell by serving as transport "rails."
TUTORIAL 4.2.3: Structures in Plant and Animal Cells

- Cilia and flagella are extensions of cells composed of microtubules. Cilia extend from cells in great numbers, serving to move the cell or to move material around the cell. By contrast, one—or at most a few—flagella extend from cells that have them. The function of flagella is cell movement.

4.7 The Plant Cell

- Plant cells have almost all the structures found in animal cells. Plant cells also have a cell wall, a large central vacuole, and organelles called plastids, one variety of which is the chloroplasts that are the sites of photosynthesis. The cell wall gives the plant structural strength and helps regulate the intake and retention of water.

  TUTORIAL 4.2.2: The Plant Cell

4.8 Cell Communication: Why Cells Need Not Be Islands

- Cells are able to communicate with each other through special structures: channels called plasmodesmata in plants and channels called gap junctions in animal cells.

Key Terms

cell junction 91	micrograph 72
central vacuole 89	micrometer 72
chloroplast 90	microtubule 84
cilia 86	mitochondria 82
cisternae 78	nanometer 72
cisternal space 78	nuclear envelope 76
cytoplasm 75	nuclear pore 76
cytoskeleton 75	nucleolus 77
cytosol 75	nucleus 75
endomembrane system 79	organelle 71
eukaryotic cell 70	plasma membrane 75
exocytosis 80	plasmodesmata 91
flagella 86	prokaryotic cell 70
free ribosome 78	ribosome 77
gap junction 91	rough endoplasmic reticulum 78
Golgi complex 79	secretory protein 78
intermediate filament 84	smooth endoplasmic reticulum 81
lysosome 81	tissue 90
messenger RNA (mRNA) 76	transport vesicle 79
microfilament 84	

Understanding the Basics

Multiple-Choice Questions

1. Jerome has strep throat, a bacterial infection. The cause of the infection is
   a. the growth of a virus
   b. the presence of archaea
   c. eukaryotic cells dividing in his throat
   d. organelles that take control of his organs—in this case, his throat
   e. prokaryotic cells

2. Where would you expect to find a cytoskeleton?
   a. primarily inside the nucleus
   b. as the internal structure of a mitochondrion
   c. holding the organelles together in bacterial cells
   d. throughout the cytosol
   e. as an outer coat on an insect

3. Suppose Dr. Hyde found a cell that had many mitochondria, a nucleus, a cell wall made of cellulose, and an endoplasmic reticulum, as well as many other parts. He might assume that he has found
   a. a plant cell
   b. an animal cell
   c. a bacterial cell
   d. one of these three types, but he will not know which type without further investigation
   e. either a plant cell or an animal cell, but not a bacterial cell

4. Cells in the pancreas manufacture large amounts of protein. Which of these would you expect to find a large amount or number of in pancreatic cells?
   a. DNA
   b. rough endoplasmic reticulum
   c. smooth endoplasmic reticulum
   d. chloroplasts
   e. plasma membrane

5. Suppose that a mutation caused a child to be unable to form intermediate filaments. Which of these problems would the child be most likely to develop?
   a. the inability to get energy out of food
   b. an inability to manufacture proteins
   c. a tendency for the nucleus and organelles in the cell to drift around inside the cell
   d. an inability of the skeletal muscles to contract
   e. an inability to move proteins from one part of the cell to another

6. Heart muscle cells have a number of gap junctions connecting them to the adjoining cells. From this you can conclude that heart muscle cells
   a. exchange nuclei very frequently
   b. have plasmodesmata
   c. move vacuoles from cell to cell
   d. communicate easily
   e. lack the ability to divide

7. Which is the correct ranking of these "small" things, from smallest to largest? (Use typical sizes.)
   a. animal cells<atoms<bacteria<proteins<amino acids
   b. atoms<amino acids<proteins<bacteria<animal cells
   c. atoms<proteins<amino acids<animal cells<bacteria
   d. bacteria<atoms<amino acids<proteins<animal cells
   e. bacteria<animal cells<atoms<amino acids<proteins

8. Cells need large amounts of ribosomal RNA to make proteins. The ribosomal RNA is made in a specialized structure known as _____, which is found in _____.
   a. a chloroplast . . . the cytosol
   b. a ribosome . . . the cytosol
   c. the endoplasmic reticulum . . . the nucleus
   d. the nuclear envelope . . . the nucleus
   e. the nucleolus . . . the nucleus

Brief Review

1. Suppose your entire body were just one gigantic cell with one central nucleus and lots of organelles outside to perform the various functions. What problems can you envision with this system?

2. What are some of the differences between eukaryotic cells and prokaryotic cells?

3. Insulin, the hormone that controls sugar levels in the body, is a protein that is secreted from special cells in the pancreas. Trace the path from a piece of DNA in the nucleus that carries the code for insulin to the release of the hormone in the bloodstream.

4. What are cilia and flagella? What are some of the roles they play in the human body?

5. Consider what plants would be like if they had no plasmodesmata. Explain why this would be a problem.

6. What is a gap junction, and what is its function?

Applying Your Knowledge

1. What would life on Earth be like if there were no plants?

2. In what ways is the cytoskeleton like bones?

5

Life's Border
The Plasma Membrane

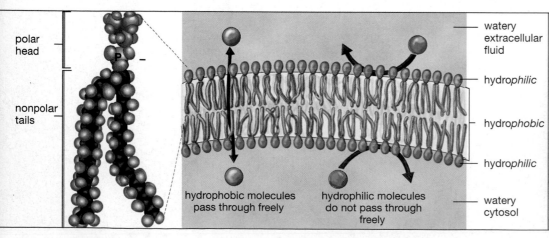

polar head

nonpolar tails

watery extracellular fluid

hydro*philic*

hydro*phobic*

hydro*philic*

watery cytosol

hydrophobic molecules pass through freely

hydrophilic molecules do not pass through freely

An essential molecule.
(Section 5.3, page 98)

A cellular sandwich.
(Section 5.3, page 98)

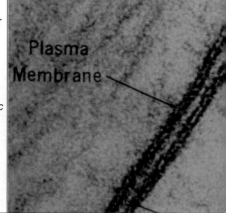

Plasma Membrane

Portrait of the plasma membrane
(Essay, page 110)

The outer lining of cells is in a sense the outer border of life. In its roles as protector, gatekeeper, and message carrier, this lining is indispensable to life.

As a professional football quarterback, Boomer Esiason led the Cincinnati Bengals to the Super Bowl in 1989. His toughest challenge, however, has come not on the football field, but at home, where he and his wife have struggled to care for their son Gunnar, who was born in 1991 with the disease cystic fibrosis.

Gunnar's treatment has included the heartbreaking ritual of "percussive" therapy, which is used on young victims of the disease. They lie at an angle, head toward the ground, while their parents pound gently on their back, chest, and sides in an effort to dislodge the mucus that accumulates in their lungs. The lungs of healthy children are lined by mucus too, but in a layer that is thin and wet enough to be regularly swept away by the hair-like cilia that extend into lung passages like so many tiny brooms. In cystic fibrosis patients, the mucous layer is thicker and drier and becomes a site for repeated bacterial infections. In time, these infections can result in the lung passages being destroyed. The average life expectancy for a cystic fibrosis victim is 29 years.

5.1 The Importance of Activity at the Cell's Periphery

The difference between a healthy child and Gunnar Esiason comes down to this: Because of a faulty gene, the substance chloride cannot be transported in sufficient quantity from the inside of Gunnar's cells to the outside of them. Healthy individuals have a protein that acts as a channel for chloride—a kind of passageway that spans the cell's outer membrane. In contrast, the cells of cystic fibrosis patients have either defective chloride channels or none at all. The result is a lack of chloride outside their cells, which causes the mucus buildup.

As noted last chapter, life goes on only inside cells. The fact that the cell's outer, or "plasma," membrane is out on life's edge may prompt the thought that it's *merely* at the edge, as if the real action is taking place deep within the cell. But consider, as in cystic fibrosis, the effect of having a *defective* plasma membrane. The focus in this chapter is to learn more about the nature of this important border.

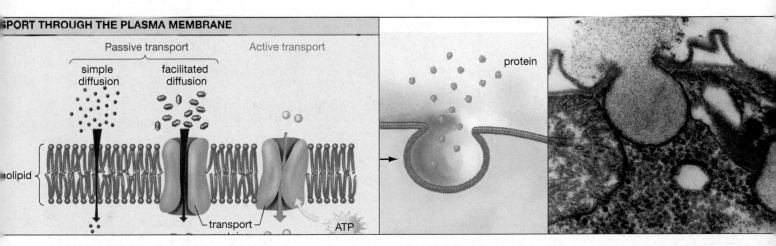

Journey through the plasma membrane.
(Section 5.4, page 104)

Moving out of the cell.
(Section 5.6, page 106)

Moving out of the cell.
(Section 5.6, page 106)

5.2 Why Do We Need the Plasma Membrane?

The plasma membrane does perform one seemingly pedestrian task, but it is absolutely critical: *keeping the goods concentrated*. For life to get going on Earth, there had to be some way that biological molecules such as proteins and fats could be collected in certain concentrations, and the plasma membrane helped accomplish this. Not surprisingly, this task is related to a second function, which is keeping harmful materials *out*.

Cells are not isolated units, however. They need to have many substances moving in and out of them. This kind of controlled passage can be counted as the third of the major activities of the plasma membrane.

In a fourth function, the plasma membrane is critical in what is known as cell signaling. Almost all cells are constantly communicating with one another by means of chemical signals. Such communication has the ability to change the activity of the cells. (A signal might, for example, increase a cell's production of a protein.) The initial points of contact for cell signaling often are molecules that lie like so many antennae on the surface of the plasma membrane.

Figure 5.1
Dual-Natured Lining
The essential building block of the cell's plasma membrane is the phospholipid molecule, which has both a hydrophilic "head" that bonds with water, and hydrophobic "tails" that do not. Two layers of phospholipids sandwich together to form the plasma membrane. The phospholipids' hydrophobic tails form the interior of the membrane, while their hydrophilic heads jut out toward the watery environments that exist both inside and outside of the cell. The bilayer forms a barrier to all but the smallest hydrophilic molecules, but hydrophobic molecules can pass through fairly freely.

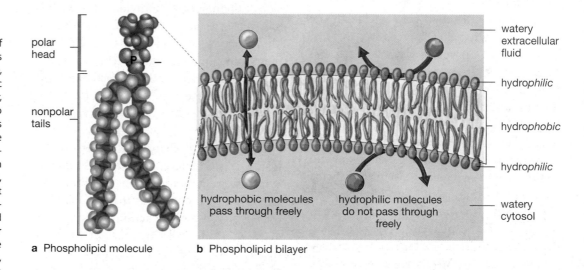

polar head

nonpolar tails

a Phospholipid molecule

hydrophobic molecules pass through freely

hydrophilic molecules do not pass through freely

b Phospholipid bilayer

watery extracellular fluid

hydro*philic*

hydro*phobic*

hydro*philic*

watery cytosol

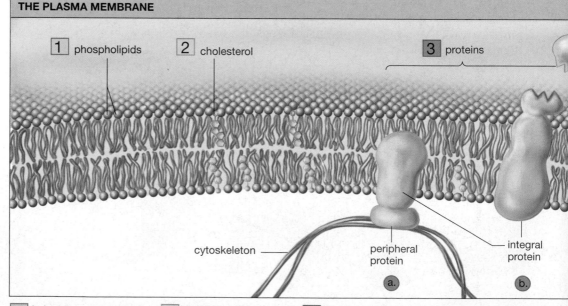

THE PLASMA MEMBRANE

1 phospholipids   2 cholesterol   3 proteins

cytoskeleton

peripheral protein

integral protein

a.

b.

1 A double or "bilayer" of **phospholipid molecules**, with their hydrophilic "heads" facing outward, toward the watery environment that lies both inside and outside the cell, and their hydrophobic "tails" pointing inward, toward each other.

2 **Cholesterol** molecules that act as a patching substance and that help the cell maintain an optimal level of fluidity.

3 **Proteins**, which are integral, meaning bound to the hydrophobic interior of the membrane, or peripheral, meaning not bound in this way. Membrane proteins serve four main functions:

a. **Structural support**, often when attached to parts of the cell's scaffolding, or "cytoskeleton."

b. **Recognition**. Binding sites on some proteins can serve to identify the cell to other cells, such as those of the immune system.

Tutorial 5.1.1
The Plasma Membrane

Figure 5.2
The Plasma Membrane

5.3 Four Components of the Plasma Membrane

What manner of material is the plasma membrane? First, it is very much like the membranes described in Chapter 4. Much of what follows about the plasma membrane could be said of all the membranes that are part of the endomembrane system—the membranes of the rough endoplasmic reticulum or the Golgi complex, for example. It's necessary to take a more detailed look at membrane structure here, though. To do that, you'll need to reacquaint yourself with a molecule you've met before.

First Component: The Phospholipid Bilayer

Recall the discussion in Chapter 3 of phospholipids—molecules that have two long fatty-acid chains and a phosphate-bearing group (**see Figure 5.1**). The phosphate group is hydrophilic, or water-seeking, and the fatty-acid chains are hydrophobic, or water-evading. The result is a molecule that, once placed in water, has its phosphate "head" pointing in a different direction from its fatty-acid "tails." It behaves, in fact, like a buoy does when bobbing in deep water. No matter how you push the hydrophobic tails,

they end up waving free out of the water, while the hydrophilic head stays submerged.

The form the plasma membrane takes is of two *layers* of phospholipids sandwiched together, the tails of each layer pointing inward, toward each other, and the heads pointing outward. This makes sense because what *lies* outward in both directions is a watery environment—attractive to phospholipid heads, but not to the tails. (Remember that the interior, or cytoplasm, of the cell is mostly water. This is usually the case with the extracellular fluid as well.) This basic structure of the plasma membrane is known as the **phospholipid bilayer**, which you can see in the context of the other membrane components in **Figure 5.2.**

This phospholipid composition makes the plasma membrane a very fluid structure. The two layers of phospholipid tails mingle with each other in the membrane and are able to slide past one another, giving the membrane its loose nature. Recall, however, that these are two *hydrophobic* layers of tails. This gives the plasma membrane one of its most important characteristics: The only substances that can pass through it with any degree of freedom are *other* hydrophobic substances or very small hydrophilic molecules, such as water.

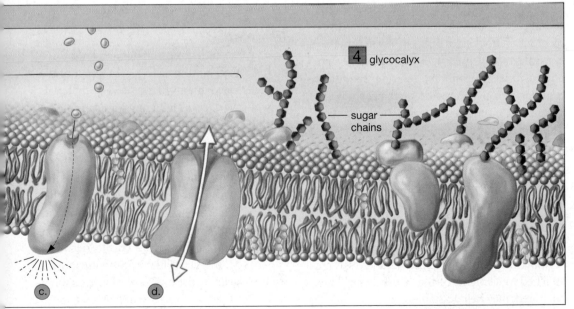

4 glycocalyx

sugar chains

c.

d.

c. Communication. Receptor proteins, protruding out from the plasma membrane, can be the point of contact for signals sent to the cell via traveling molecules, such as hormones.

d. Transport. Proteins can serve as channels through which materials can pass in and out of the cell.

4 The **glycocalyx.** Sugar chains that attach to communication or recognition proteins, serving as their binding sites. The glycocalyx can also lubricate cells and act as an adhesion layer for them.

This stems from the rule of thumb about solvents in Chapter 3: like dissolves like. The bilayer is a hydrophobic substance—a *lipid* substance. Thus, various steroid hormones (which, remember, are lipids) gain fairly easy entry, as do fatty acids, because these substances will dissolve in the lipid bilayer.

If the bilayer allows some things to pass through, however, it keeps many others out. Hydro*philic* substances—ions, polar molecules—do none of the dissolving that lipids do in the membrane and are thus not allowed in or out of the cell without help. Water, which is polar, gets to pass through the bilayer only because it is such a small molecule.

Second Component: Cholesterol

Molecules of the lipid material cholesterol nestle between phospholipid molecules throughout the plasma membrane, performing two functions. First, they act as a kind of patching substance on the bilayer, keeping some small molecules from getting through. Second, they help keep the membrane at an optimum level of fluidity.

Third Component: Proteins

Embedded within, or lying on, the membrane lipids is a third major group of membrane components: proteins, of which there are two major types, integral and peripheral. As usual, the names tell a good part of the story. **Integral proteins** may span the entire membrane, popping out on both sides of it, or they may merely extend *partway* into it. Either way, they are bound to its hydrophobic interior. **Peripheral proteins** lie on either side of the membrane, but are not bound to the hydrophobic interior. They are usually attached to integral proteins at the surface of the membrane.

Taken together, peripheral and integral proteins have a diverse range of functions. Here is a selected set of them.

Structural Support

Peripheral proteins that lie on the interior or "cytoplasmic" side of the membrane often are attached to elements of the cytoskeleton. Thus anchored, they help to stabilize various parts of the cell and play a part in giving animal cells their characteristic shape. (Why not plant cells? Because their shape is determined by the cell wall.)

Recognition

Much like military sentries, some cells need to know the answer to a critical question: Who goes there, friend or foe? The sentries are immune system cells, moving through the blood system and interacting with proteins that extend from the cell surface. The specific shape of **binding sites** on these proteins can convey the message: Self here, pass on by. Conversely, a cell with a *foreign* set of binding sites on its proteins (generally an invader) elicits an immune system attack.

Communication

Cells communicate with one another in various ways. Signals can be sent from one cell to a neighboring cell, and communication over longer distances is possible through the chemical messengers known as hormones. The nervous system provides a means of fast communication throughout the body. In all these instances, signals are likely to be channeled through **receptor proteins** (often referred to simply as receptors). Many of these are integral proteins that protrude from the plasma membrane to the space outside the cell.

Each receptor has on it a binding site, shaped so that it can latch onto only a single communication molecule, or perhaps a group of closely related molecules. The usual outcome of such binding is a change in the cell's activity. For example, when a hormone binds with a receptor on the surface of a cell, the result can be a cascade of cellular reactions *inside* the cell that culminate in a protein being produced.

Transport

You have seen that most materials cannot simply pass through the cell's plasma membrane. Yet cells need many of these very materials. How do they get in? Different kinds of integral proteins take on a transport task, in ways that will be reviewed shortly.

Recall that life depends on the ability to make fine distinctions—to get more of this, but less of that; to respond to this, but not to that. Proteins greatly enhance the cell's ability to interact in very *particular* ways with its environment. Some proteins say, in effect, "self here." Others say, "muscle cell here." Still others stand ready to facilitate the transport of a given ion—and only that ion—across the plasma membrane. Proteins have been likened before to a group of specialized tools; the proteins of the plasma membrane provide more cases in point.

Fourth Component: The Glycocalyx

If you look at Figure 5.2 and focus on the extracellular face of the plasma membrane, you can see, protruding from proteins and phospholipids alike, a number of small, branched extensions. These are carbohydrate or sugar chains that serve as the actual binding sites for many of the proteins mentioned earlier. Sugar chains also serve to lubricate cells, or to keep cells in place by acting as a sticky adhesion layer. Add to these some peripheral proteins, and you get an outer coat for the cell that goes by the name of **glycocalyx**, meaning "sugar coat."

The Fluid-Mosaic Membrane Model

With this review of component parts under your belt, you're ready for a definition of the **plasma membrane**. It is a membrane, forming the outer boundary of many cells, composed of a phospholipid bilayer that is interspersed with proteins and cholesterol molecules and coated on its exterior face with carbohydrate chains associated with proteins and lipids.

Taking a step back, the general message here is that the plasma membrane is a loose, lipid structure peppered with proteins and coated with sugars. A better image, however, might be of a *sea* of lipids that has proteins floating on it, because as it turns out, the plasma membrane is fluid enough that most of its elements are able to move sideways or "laterally" through it fairly freely. Indeed, membrane proteins frequently are compared to icebergs drifting through an ocean. (The proteins are numerous enough, however, that it's tempting to

liken the cell surface to the Arctic Ocean at about the time the ice pack starts to break up.) When we overlay this quality of movement on the basic membrane structure just reviewed, the result is the modern view of the plasma membrane, the **fluid-mosaic model**: A mosaic of proteins moving within the fluid that is the phospholipid bilayer. For details on how this model was arrived at, see "How Did We Learn?" on page 109.

5.4 Moving Materials In and Out: Diffusions and Gradients

As we observed earlier, one of the main roles of the plasma membrane is to let in whatever's needed and keep out whatever's not. Having learned something about the membrane, you will now see how it carries out this task of passage and blockage.

It is necessary to begin, however, not with the cell, but with the more general notion of how it is that substances go from places where they are more concentrated to places where they are less concentrated. Anyone who has put some food dye in water has a sense of how this works: A few drops of, say, red dye will tumble into the liquid, start dispersing, and eventually the result is a solution that is uniformly just a shade more reddish than plain water (**see Figure 5.3**). The question that will be addressed here is why this takes place.

a　　　　　　　b　　　　　　　c

water
molecules

dye
molecules

Figure 5.3
From Concentrated to Dispersed
Diffusion is the movement of molecules or ions from areas of their greater concentration to areas of their lesser concentration. In this sequence of photos and diagrams, a few drops of red dye, added to a beaker of water, are at first heavily concentrated in one area but then begin to diffuse, eventually becoming evenly distributed throughout the solution.

Random Movement and Even Distribution

In the real world, several forces are at work in the distribution of dye in a liquid. But for purposes of explanation, here you'll look at just one of them. All molecules or ions are constantly in motion, and this motion is random. (The *degree* to which molecules are in motion defines their temperature; absolute zero is the point at which all molecular motion has ceased.) Liquid molecules are able to slide *past* one another, which increases their ability to move in comparison with solids. So, with random molecular motion and a liquid substance, the dye molecules are now in motion.

But how do these molecules come to be *evenly* distributed through the water? A rigorous explanation of this would require a tour of an area of science you'll be looking at in Chapter 6, thermodynamics. But to get a sense of why this distribution should happen, let us take another tack. Imagine the dye in water, only this time the process has been videotaped.

Tutorial 5.1.2
Diffusion through Membranes

Figure 5.4
Osmosis in Action

a An aqueous solution divided by a semipermeable membrane has a solute—in this case, salt—poured into its right chamber.

b As a result, though water continues to flow in both directions through the membrane, there is a net movement of water toward the side with the greater concentration of solutes in it.

c Why does this occur? Water molecules that are bonded to the sodium (Na⁺) and chloride (Cl⁻) ions that make up salt are not free to pass through the membrane to the left chamber of the container.

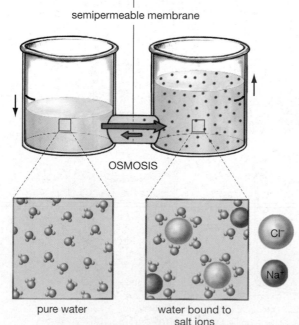

a
solute
solvent
semipermeable membrane
b
OSMOSIS
c
pure water
water bound to salt ions
Cl⁻
Na⁺

The dye drops in and the tape rolls for just one tick of the molecular clock—just long enough for each molecule of the dye to make one of its motions. Then freeze-frame the action. Now, on the *second* tick of the clock, what is the probability that each molecule will move *back* to its starting-state position, versus the probability that it will move somewhere else? The odds, of course, greatly favor the latter possibility. There are simply more places for these molecules to go than places they have been. This process continues until you get an even distribution of the dye over time.

Gradients and Diffusion

The dispersal of dye in liquid is an example of **diffusion**, meaning the movement of molecules or ions from a region of their higher concentration to a region of their lower concentration. This concept implies the notion of a *gradient*. Just as a road might have a grade in which it goes from higher to lower elevation, so there is a **concentration gradient** in solutions. This gradient is defined as the difference between the highest and lowest concentration of a solute within a given medium. As with bicycles coasting down a grade, the natural tendency for any solute is to move *down* its concentration gradient, from higher concentration to lower. Bikes and solutes can move *up* a grade or gradient, but there is a price to be paid for this—the expenditure of energy.

Diffusion through Membranes

So far, you have looked at molecules diffusing in an undivided container. But the subject here is divisions—those provided by the plasma membrane. Let us consider, therefore, what happens to a solution in a container that is divided by a membrane. If that membrane is *permeable* to both water and the solute—that is, if both water and the solute can freely pass through it—and the solute lies only on one side of the membrane, then the predictable happens: The solute moves down its concentration gradient, diffusing right through the membrane, eventually becoming evenly distributed on both sides of it.

Now let's imagine a *semi*permeable membrane, one that water can freely move through but that solutes cannot. **Figure 5.4** shows you what happens if more solute is put on the right side of the membrane than the left. Water flows through the membrane both ways, but *more* water flows into the right chamber, which has a greater concentration of solutes in it. The result is that the solution on the right side rises to a higher level than the one on the left.

This seems strange on first viewing, as if gravity were taking a vacation on the right side of the container. But what has been demonstrated here is **osmosis**: the net movement of water across a semipermeable membrane from an area of lower solute concentration to an area of higher solute concentration. Why should this occur? In the case of a solute like salt, water molecules will surround and bond with the sodium and chloride ions that salt separates into in solution. Because these solutes are not free to pass through the membrane, the water molecules bound to them will likewise remain confined to the right side. This means that more "free" water will exist on the left side, and the result is a net movement of water into the right side.

The Plasma Membrane as a Semipermeable Membrane

So what does movement of water have to do with the cell's outer lining? The cell's phospholipid bilayer is itself a semipermeable membrane. It is somewhat permeable to water and lipid substances but not permeable to larger charged substances. Thus osmosis can take place across the plasma membrane. As an

extreme example of this effect, remember the lament of the sailor stranded in the salt sea: "Water, water, everywhere/Nor any drop to drink." If that sailor went out of his head and gulped down lots of seawater, what he would get is an enormous concentration of sodium chloride ions in his extracellular fluid—*outside* the plasma membrane. This is no different in principle than dumping more solutes into the right side of the container. The result would be water flowing *out* of his cells, cell dehydration, and in extreme cases, death. (What actually kills people in these cases is a shrinkage of brain cells.)

Osmosis is not an inherently negative force, however. It is the primary means by which plants *get* water, and it is a player in all sorts of routine metabolic processes. Harm comes from it only in cases of extreme solute imbalance.

Osmosis and Plant Cells

It is in connection with the possibilities of solute imbalances that plants have a great advantage over animals—their cell walls. Remember from Chapter 4 that one function of the cell wall is to regulate water uptake. Well, now we can see why this is so valuable. Animal cells, which do not have a wall, can expand until they break when water comes in (**see Figure 5.5**). Plant cells, conversely, will

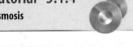

Tutorial 5.1.4
Osmosis

Figure 5.5
Osmosis in Cells

a When solutes (such as salt) exist in greater concentration outside the cell than inside, water moves out of the cell by osmosis and the cell shrinks. Here, the fluid surrounding the cell is hypertonic to the cell's cytoplasm.

b When solute concentrations inside and outside the cell are balanced, there is a balance of water movement into and out of the cell. Here, the fluid surrounding the cell is isotonic to the cell's cytoplasm.

c When solutes exist in greater concentration inside the cell than outside, water moves into the cell by osmosis. This influx may cause animal cells to burst, but plant cells are reinforced with cell walls and thus remain turgid—generally a healthy state for them. Here, the fluid surrounding the cell is hypotonic to the cell's cytoplasm.

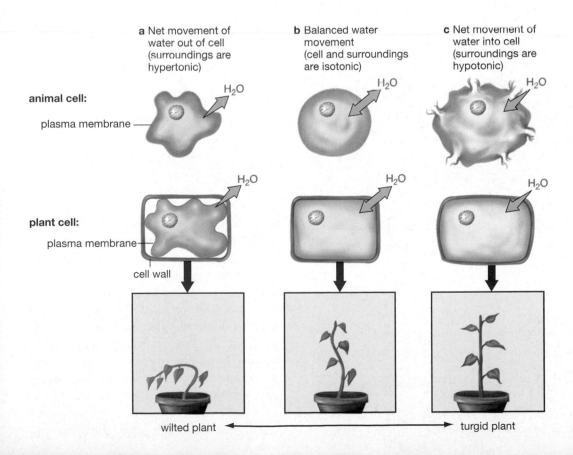

a Net movement of water out of cell (surroundings are hypertonic)

b Balanced water movement (cell and surroundings are isotonic)

c Net movement of water into cell (surroundings are hypotonic)

animal cell:
plasma membrane
H_2O

plant cell:
plasma membrane
cell wall
H_2O

wilted plant ⟵   ⟶ turgid plant

expand only until their membranes push up against the cell wall with some force, setting up a pressure, or turgor, that keeps more water from coming in. Such tight quarters may sound uncomfortable, but this actually is an optimal condition for plants. A nice, crisp celery stick is one that has achieved this kind of *turgid* state, while a droopy stick that has lost this quality has cells that are *flaccid;* flowers in the latter condition are *wilted.*

Osmosis and Cell Environments

Is a given cell likely to lose water to its surroundings, gain water from them, or have a balanced flow back and forth with them? Any of these things is possible, depending on what the solute concentration is outside the cell as opposed to inside. Three terms are helpful in describing the various conditions that can exist. A fluid that has a higher concentration of solutes than another is said to be a **hypertonic solution**. If a cell's surroundings are hypertonic to the cell's cytoplasm, water will flow out of the cell. Two solutions that have equal concentrations of solutes are said to be **isotonic**. If one of these solutions is the cell's

cytoplasm and the other the fluid surrounding the cell, fluid flow will be balanced between cell and surroundings. Finally, a fluid that has a lower concentration of solutes than another is a **hypotonic solution**. If a cell's surroundings are hypotonic to the cell's cytoplasm, water will flow into the cell. If these terms seem confusing, just keep in mind that what is hyper- or hypo- (high or low) is the *solute* concentration. Figure 5.5 gives examples of what can happen to cells in all three types of environments.

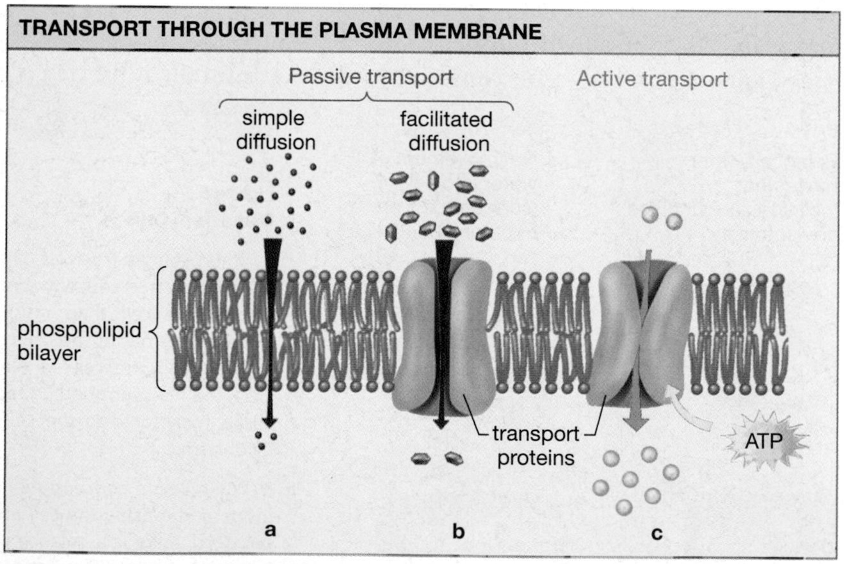

5.5 How Do Materials Get In and Out of the Cell?

This excursion into the land of diffusion and osmosis has prepared us to start looking at the ways materials actually move into and out of the cell, across the plasma membrane. The big picture here is that some molecules are able to cross with no more assistance than is provided by diffusion or osmosis. Other molecules require these forces and special protein channels to move through, and still others require channels and the expenditure of energy to get across. Any time energy is required for movement across the plasma membrane, **active transport** has occurred. When such energy does not need to be expended, **passive transport** is in operation (**see Figure 5.6**).

Passive Transport

Our review of transport begins with two varieties of passive transport.

Simple Diffusion

Water molecules and such gases as oxygen and carbon dioxide need only move down their concentration gradients to pass into or out of the cell. Having been delivered by blood capillaries to an area just outside the cell, oxygen exists there in greater concentration than it does inside the cell. Moving down its concentration gradient, it dissolves into the plasma membrane, travels through it, and emerges on the other side. This is an example of **simple diffusion**, meaning diffusion that does not require a special protein channel. Molecules that are larger than oxygen, but that are fat soluble, such as the steroid hormones mentioned before, also move into the cell in this manner. Carbon dioxide, which is formed *in* the cell as a result of cellular respiration, has a net movement *out* of the cell through simple diffusion.

TRANSPORT THROUGH THE PLASMA MEMBRANE

Passive transport — simple diffusion, facilitated diffusion. Active transport.

phospholipid bilayer. transport proteins. ATP. a b c

Tutorial 5.2.1
Passive Transport

Figure 5.6
Transport through the Plasma Membrane
Passive transport requires no expenditure of energy, but active transport does.

a In simple diffusion, materials move down their concentration gradient through the phospholipid bilayer.

b In facilitated diffusion, the passage of materials is aided both by a concentration gradient and by a transport protein.

c In active transport, molecules again move through a transport protein, but now energy must be expended to move them against their concentration gradient.

Facilitated Diffusion: Help from Proteins

Water can traverse the lipid bilayer through simple diffusion, moving through *despite* the bonding it does. It can pass through in this way (rather slowly) primarily because it's such a small molecule. Yet with water itself—and certainly with polar molecules larger than water—something else begins to be required for adequate amounts to pass through the plasma membrane. What's required are special channels—passageways formed by some of the integral proteins described earlier. With this, the method of passage is no longer *simple* diffusion; it is **facilitated diffusion**, meaning the passage of materials through the plasma membrane, aided by a concentration gradient and a transport protein.

Here you can begin to see the protein specificity referred to earlier. The proteins that serve as channels through the plasma membrane are called **transport proteins**, and each acts as a conduit for only one substance, or at most a small group of associated substances. The process of transport begins with the kind of binding explained earlier. A circulating molecule of glucose, for example, latches onto the binding site of a glucose transport protein. This binding causes the protein to change its shape, thus allowing the glucose molecule to pass through. This protein is providing a channel, a hydrophilic passageway through a hydrophobic environment. In addition to glucose, other hydrophilic molecules, such as amino acids, move through the plasma membrane in this way.

Be aware that this form of transport does not require the expenditure of energy, because it has a concentration gradient working in its favor. Glucose exists in greater concentration outside the cell than in it. It is moving *down* its concentration gradient, then, by passing into the cell. So, this type of facilitated diffusion has something in common with simple diffusion. *Both* processes are driven by concentration gradients, meaning that neither requires metabolic energy. Thus, both are specific examples of passive transport.

Though the glucose in our example moved only one way in facilitated diffusion (into the cell), in general transport proteins are a channel for movement *either* way through the membrane. All that's required is a concentration gradient and a binding of the material with the transport protein.

Active Transport

If passive transport—either in simple or facilitated diffusion—were the only means of membrane passage available, cells would be totally dependent on concentration gradients. If, for example, molecules of a given amino acid come to exist in the same concentration on both sides of the plasma membrane, these amino acid travelers will continue to be transported, but they will be flowing out of the cell at the same rate they are flowing in.

For cells, the problem is that some solutes are *needed* in greater concentration, say, inside the cell as opposed to outside. Yet passive transport constantly works to equalize solute concentrations.

Moving Compounds against Their Concentration Gradient

The cell's solution to moving solutes against their concentration gradients is *pumps*. The cell does this by expending energy to move substances across the plasma membrane. Many kinds of pumps are in operation in active transport, but each is specific for one or perhaps two substances. The energy source for such transport often is ATP (adenosine triphosphate), the "energy currency" molecule introduced in Chapter 3, though the cell also harnesses the electrochemical power of charged ions to move substances across the membrane. One of the most important and best studied of the active transport mechanisms is known as the sodium-potassium pump.

The Sodium-Potassium Pump

The sodium-potassium pump allows the cell to maintain an environment of high potassium (K^+) *inside* the cell, and high sodium (Na^+) *outside* the cell. Imagine that we could take a snapshot of a cell at a given moment in time that would allow us to see these ion concentrations. With lots of Na^+ ions outside the cell, the result is predictable: A slow "leakage" of them into the cell as they follow their concentration gradient down and move in through transport proteins. K^+ ions, meanwhile, are moving out.

But our fantasy snapshot would also reveal something else. An abundance of protruding proteins that are busy pumping these molecules the *other way*, acting like so many bilge pumps moving seawater that has leaked into a ship back out into the ocean. If

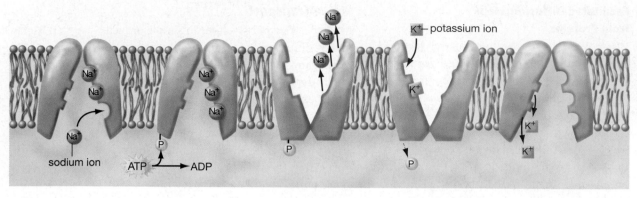

extracellular
fluid

phospholipid
bilayer

cytosol

sodium ion

ATP → ADP

K⁺—potassium ion

1. Three sodium ions (Na⁺) located within the cell's cytoplasm bind with a transport protein.

2. The energy molecule ATP gives up an energetic phosphate group to the transport protein.

3. This binding causes the protein to open its channel to the extracellular fluid; to lose its Na⁺ binding sites, thus releasing the ions into the fluid; and to create binding sites for potassium ions (K⁺).

4. Two K⁺ move into the protein's K⁺ binding sites, which brings about the release of the phosphate group.

5. This loss returns the protein to its original shape, releasing the K⁺ into the cytoplasm and readying the protein for binding with another set of Na⁺.

Figure 5.7
Active Transport in the Sodium- Potassium Pump

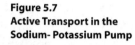
Tutorial 5.2.2
Active Transport

Tutorial 5.3.1
Exocytosis

Figure 5.8
Movement Out of the Cell

a In exocytosis, a transport vesicle—perhaps loaded with proteins or waste products—moves to the plasma membrane and fuses with it. This section of the membrane then opens, and the contents of the former vesicle are released to the extracellular fluid.

b Micrograph of material being expelled from the cell through exocytosis.

you look at **Figure 5.7**, you can see how the sodium-potassium pump works. The point here is not to count the movement of so many molecules in and out, but just to observe the process. One critical aspect of this is the shape-changing noted before with transport proteins. This is how a substance can be isolated within a cell one moment, but then free to move into the extracellular environment the next.

5.6 Getting the Big Stuff In and Out

Pumps and channels, diffusion and osmosis—these mechanisms move substances in and out of the cell, but as it turns out they only move relatively *small* substances.

Remember an earlier discussion about the immune system cells that check to see if a given cell is friend or foe? Well, if an immune sentry finds a foe, it may have to ingest this *whole cell*. As you might guess, a cell cannot do this by employing little channels or pumps. You learned a bit in Chapter 4 about what does happen in these cases; now you'll be looking at it in somewhat greater detail.

The methods in question here are endocytosis, which brings materials into the cell; and exocytosis, which sends them out. What these mechanisms have in common is their use of vesicles, the membrane-lined enclosures that alternately bud off from membranes or fuse with them.

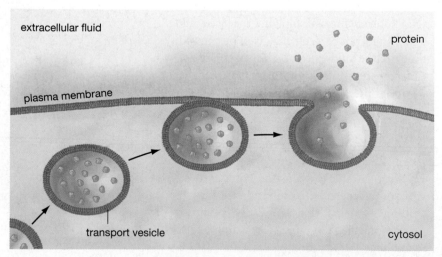

extracellular fluid

protein

plasma membrane

transport vesicle

cytosol

a

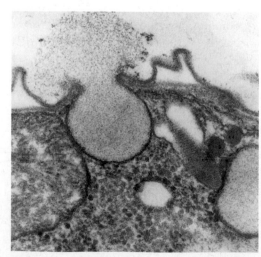

b

Movement Out: Exocytosis

Exocytosis is defined as the movement of materials out of the cell through a fusion of vesicles with the plasma membrane. As **Figure 5.8** shows, exocytosis involves a transport vesicle making its way to the plasma membrane and fusing with it, whereupon the vesicle's contents are released into the extracellular fluid. You observed in Chapter 4 that cells use exocytosis when they are exporting proteins. For single-celled creatures, waste products may be released into the extracellular fluid through this process.

Movement In: Endocytosis

Endocytosis is the movement of relatively large materials into the cell by infolding of the plasma membrane, and it can take several forms: pinocytosis, receptor-mediated endocytosis, and phagocytosis.

Pinocytosis

Pinocytosis means "cell drinking," and if you look at **Figure 5.9**, you can see why this is fairly accurate. The cell invaginates, creating a kind of harbor on its exterior. Whatever material happens to be enclosed in the

Tutorial 5.3.2
Endocytosis

Figure 5.9
Three Ways to Get Relatively Large Materials into the Cell

a In pinocytosis, the plasma membrane invaginates to create a kind of harbor. The harbor then encloses completely, pinches off as a vesicle, and moves into the cell's cytoplasm, carrying with it whatever material was enclosed.

b In receptor-mediated endocytosis, many receptors bind to molecules. Then, while holding on to the molecules, the receptors migrate laterally through the cell membrane, arriving at a depression called a coated pit. The coated pit pinches off, delivering its receptor-held molecules into the cytoplasm.

c In phagocytosis, food particles—or perhaps whole organisms (such as bacteria)—are taken in by means of "false feet" or pseudopodia that surround the material. Pseudopodia then fuse together, forming a vesicle that moves into the cell's interior with its catch enclosed.

a Pinocytosis
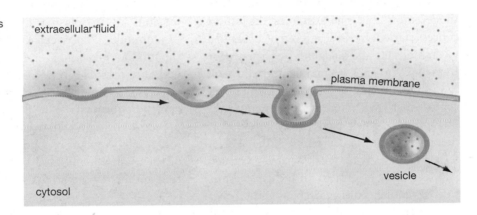

b Receptor mediated endocytosis
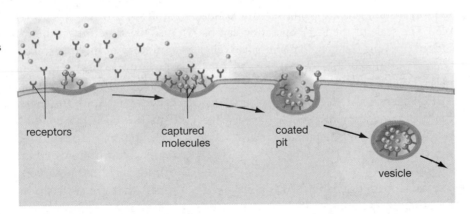

c Phagocytosis
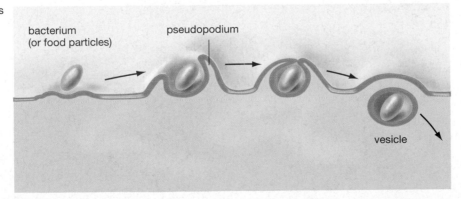

harbor when it pinches off to become a vesicle is brought into the cell. What is brought in, of course, is mostly water, with some solutes in it. This process is not, however, a random act; cells are prompted to begin pinocytosis because of the binding of specific substances to receptors. **Pinocytosis** can be defined as a form of endocytosis that brings into the cell a small volume of extracellular fluid and the materials suspended in it.

Receptor-Mediated Endocytosis (RME)

As its name implies, the second form of endocytosis, **receptor-mediated endocytosis** (RME), depends on receptors as well. But here the receptors' role is not just to get the import process started. In RME, a receptor binds to a given molecule and *holds onto it.* The receptor then makes a quick lateral migration through the cell membrane to join other identical receptors that have bound with the same kinds of molecules. Their place of congregation is literally a pit; it is a depression in the cell, referred to as a *coated pit.* (What it is coated with is a layer of protein on the cytoplasmic side.) Eventually, the pit deepens and pinches off, creating the familiar vesicle moving into the cell. RME is very important in getting nutrients and other substances into cells. (It is the way insulin gets into your cells, for example.) The scale of the RME operation can be judged from the fact that, in mammalian cells, coated pits might take up 20 percent of a plasma membrane's surface area. RME is also a clear example of the kind of icebergs-on-the-ocean migration that proteins are capable of within the plasma membrane.

Phagocytosis

The third form of endocytosis is a means of of bringing even larger materials into the cell. It is phagocytosis (literally, "cell eating"). This is the mechanism mentioned earlier, by which a human immune system cell might ingest a whole bacterium (see Figure 4.12 on page 85). This is also the way that many one-celled creatures eat, so it's probably apparent that relatively large materials are being ingested here. How large? Vesicles of perhaps 1–2 µm, as opposed to pinocytosis and RME, where the vesicles might be a tenth this size. Not all the materials brought into the cell by phagocytosis are whole cells; parts of cells and large nutrients are fair game as well.

As you can see in Figure 5.9, phagocytosis begins when the cell sends out extensions of its plasma membrane called pseudopodia ("false feet"). These surround the food and fuse their ends together. What was once outside is now inside, encased in a vesicle and moving toward the cell's interior. **Phagocytosis** can be defined as a process of bringing relatively large materials into a cell by means of wrapping extensions of the plasma membrane around the materials and fusing the extensions together.

On to Energy

Your tour of the cell and its plasma membrane is now complete. What's coming up next is something called bioenergetics. In our story of biology so far, there have been continuing vague references to proteins that are called enzymes and to a molecule called ATP and to energy expenditures. This has been rather like a description of an elephant that avoids the subject of its trunk. We will now embark on the subject of bioenergetics, however, which describes forces that are so basic they condition all of biology.

The Fluid-Mosaic Model of the Plasma Membrane

HOW DID WE LEARN?

The richly detailed illustrations found in science textbooks do a great job of presenting information, but they also can be a little misleading. Given their clarity, such illustrations might leave students with the impression that many of the discoveries that scientists make would be apparent to anyone who simply looked in the right place with the right microscope.

Figure 1a presents a good example of the kind of artist's rendering of the plasma membrane you've been seeing in this chapter, while **Figure 1b** presents an example of the kind of pictures (or "micrographs") we have of it.

Our view of the cell membrane was constructed piece by piece in an intellectual chain that stretched out for decades.

You'll probably agree that no one could have discerned the detail in 1a just by looking at 1b. How, then, do we know that the plasma membrane actually is structured as you see it in 1a? How do we know that it is largely lipid, and that proteins move about on the surface of it like so many icebergs? As it turns out, our view of the cell membrane was constructed piece by piece in an intellectual chain that stretched out for decades.

The idea that the surface of the cell is largely made of lipid material was set forth as early as 1855 by investigator Carl Nägeli. In 1899, Charles E. Overton came to the same conclusion by working with some cellular extensions of plant roots. What Overton found was that lipid-soluble substances moved quite easily into these "root hairs," while water-soluble substances did not. Remember now the adage about like dissolving like: Lipid substances would dissolve within *lipid materials*. Overton's conclusion was that the outside lining of the root cells was made of just such materials. From our vantage point—knowing about the phospholipid bilayer that makes up much of the membrane—we can see that he was right.

In 1925, two Dutch scientists, E. G. Gorter and F. Grendel, decided to take red blood cells and measure their lipid content. They knew from the work of another scientist that a given volume of phospholipids would cover a certain *area* when arrayed in a layer. What they found, however, was that the lipids they extracted from the blood cells spread out into an area *twice* as large as would be expected from this benchmark. This doubling, they concluded, came about because the cell membrane was doubled; it existed as the *bi*layer you've been reading about.

Following this advance, however, it was apparent that the plasma membrane had to be composed of more than just two layers of lipids pressed together. For one thing, it was clear that closely related types of hydrophilic molecules passed through the membrane at different rates. This was difficult to reconcile with a uniform phospholipid membrane, which would have let all such substances pass through at nearly the same rate. Given such evidence, James Danielli, writing with Hugh Davson, proposed a view of the plasma membrane that ended up having a long run in science. With modifications over time, the Davson-Danielli model, as it was called, wore the mantle of being "generally accepted" by scientists from the time it was set forth in 1935 until the early 1970s. It proposed that a *layer of proteins* coated each side of the phospholipid bilayer.

By the 1950s, electron microscopes had become refined enough that they yielded a "railroad track" picture of the plasma membrane: two dark bands separated by a space in

Figure 1
How Much Can a Picture Tell Us?
(**a**) An artist's rendering of the plasma membrane, and
(**b**) a micrograph of it.

a

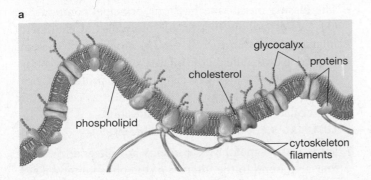

glycocalyx
proteins
cholesterol
phospholipid
cytoskeleton filaments

b

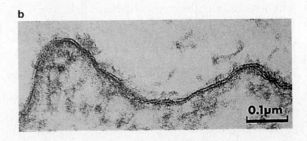

0.1µm

(continued)

The Fluid-Mosaic Model of the Plasma Membrane

(continued)

the middle (the view you see in **Figure 2**). Seduced, perhaps, by the comforts of what was "known" to be true, scientists took such pictures to be visual proof of Davson-Danielli. The two dark lines were protein bands, it was thought, and the space in the middle was the phospholipid bilayer. We now know, however, that the "tracks" actually were electrons that bound with both membrane proteins and the charged portions of the phospholipid bilayer, thus creating the impression of a continuous track. In reality, there were no protein bands.

Through these years, it was also becoming apparent that there wasn't just one membrane in the cell, there were many: membranes encasing the mitochondria, membranes making up the Golgi complex, and so forth. In 1960, J. David Robertson set forth a proposal that saw *all* the cell's membranes as existing within the framework of Davson-Danielli. By this time the venerable model had been modified to include the notion of some proteins *spanning* the bilayer, from cytoplasm to outside face. This, then, was the generally accepted view of cell membranes in the early 1960s: Davson-Danielli-Robertson, which posited a phospholipid bilayer coated on each side with a protein layer, and shot through here and there with other proteins.

Into this situation came a chemist, then at Yale, named Jonathan Singer. In 1962, Singer got to thinking about cell membranes and realized that Davson-Danielli-Robertson

had a real problem: It would require hydro*phobic* parts of the supposed protein coat to be in contact with *water*. This was like assuming you could set one marble on top of another and expect it to stay there. By 1964, Singer was convinced not only that Davson-Danielli-Robertson had some shortcomings, but that he had a model that improved on it.

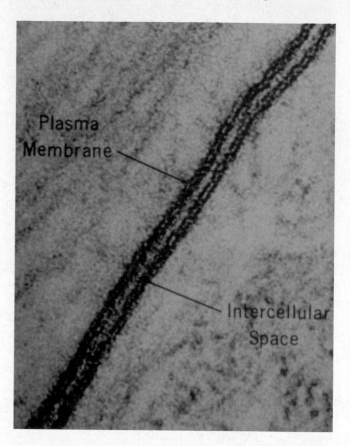

Figure 2
A Closeup of the Plasma Membrane
Electron micrograph of two adjoining cells, each of which has a plasma membrane with a "railroad track" appearance. The intercellular space is the space between the two cells. (×470,000)

Chapter Review

Summary

5.1 The Importance of Activity at the Cell's Periphery

- The plasma membrane, a structure that forms the outer boundary of many cells, plays an important role in many cellular processes.

5.2 The Plasma Membrane's Functions in Overview

- The cell's plasma membrane functions to (1) concentrate the chemical materials necessary to sustain life; (2) serve as a barrier, keeping harmful compounds out of the cell; (3) allow the controlled movement of compounds into and out of the cell; and (4) play a part in cell signaling.

5.3 Four Components of the Plasma Membrane

- The plasma membrane has four principal components: (1) A phospholipid bilayer, (2) molecules of cholesterol interspersed within the bilayer, (3) proteins that are embedded in or that lie on the bilayer, and (4) short carbohydrate chains and their associated surface proteins, collectively called the glycocalyx, that function as cell binding sites and in cell lubrication and adhesion.
TUTORIAL 5.1.1: The Plasma Membrane

- Some plasma membrane proteins are integral, meaning they are bound to the interior of the phospolipid bilayer. Others are peripheral, meaning they lie on either side of the membrane but are not bound to its interior.

It's one thing, however, to believe you're right, based on general chemical principles, and quite another to prove it.

"What happened was that I wasn't going to propose a model without some kind of experiments to go with it," he said in a telephone interview from his office at the University of California, San Diego. Working with a post-doctoral student of his, John Lenard, Singer published a scientific paper in 1966 that made the basic break with Davson-Danielli-Robertson. The view that emerged from this paper was that the membrane had a series of *individual* proteins that either were embedded in the bilayer or lay on it. By 1971, Singer was able to set forth a case for his model in great detail in a book chapter. The problem was that nobody paid much attention to it.

"This was a very thorough thing and it took up a lot of space," he says, "but ultimately I came to realize that nobody reads books. However, at that time, at least, if you wrote a review in *Science* [magazine] it would come to people's attention." So he set out to write just such a review. In the interval between the book chapter and the magazine article, however, one last piece of the membrane puzzle fell into place.

Recall that the modern view of the cell membrane is the fluid-mosaic model. It is *mosaic* because it posits discrete proteins within the phospholipid bilayer; but it is *fluid* because these proteins (and the phospholipids themselves) can *move* laterally through the membrane. In 1971, Singer had the mosaic part; by 1972, he had the fluid as well.

"I came to Dr. Singer's lab in 1967," says Garth Nicolson, who was a graduate student with Singer in those days and now heads the Tumor Biology Department at the M. D. Anderson Cancer Center, University of Texas, in Houston.

"I became very interested in the polarity of membranes . . . what I found when doing this work was that the distribution of components [on the surface of the cell] seemed to change dynamically." What Nicolson saw, in other words, was that the constituent parts of the membrane were shifting around on it.

Singer and Nicolson are agreed that a paper by L. D. Frye and Michael Edidin made Singer a believer in the rapid movement of materials across the cell surface. Their results, Singer says, "triggered the release of our inhibitions about proposing this kind of thing, which was, in a sense, the natural consequence of our model."

The upshot was a paper called "The Fluid Mosaic Model of the Structure of Cell Membranes," which appeared in *Science* magazine in February of 1972, with Singer and Nicolson as authors. It turned out to have great impact. Read any biology textbook today, and the account of the cell's plasma membrane is essentially the view set forth in this paper. The illustrations contained in textbooks likewise all look like a drawing in the *Science* article. Such was its effect that the fluid-mosaic model is just as likely to be referred to as the "Singer-Nicolson" model, even though, as Garth Nicolson is the first to say, "this is really [Singer's] model . . . he should get the credit for the membrane as far as I'm concerned."

Does this mean that the fluid-mosaic model has at last given us an accurate picture of the membrane? We might wonder about this because, after all, Davson-Danielli was generally accepted for more than 30 years. Singer believes, however, that fluid-mosaic does describe reality. "It's not going to change," he says. "Things rattle around for awhile, but then you get settled in on the right picture."

- Membrane proteins serve four primary functions: (1) structural support; (2) cell recognition; (3) communication, by serving as receptors on the periphery of the membrane; and (4) transport, by providing channels for the movement of compounds into or out of the cell.

- The plasma membrane is conceptualized within the fluid-mosaic model: A fluid medium of the phospholipid bilayer that has within it and on it a mosaic of proteins that can move laterally with relative ease.

5.4 Moving Materials In and Out: Diffusions and Gradients

- Diffusion is the movement of molecules or ions from a region of their higher concentration to a region of their lower concentration. A concentration gradient defines the difference between the highest and lowest concentrations of a solute within a given medium. Compounds can move from higher to lower concentrations—meaning down their concentration gradients—through diffusion. Energy must be expended, however, to move compounds up their concentration gradients—from regions of their lower concentration to regions of their higher concentration.

- A semipermeable membrane is one that allows some compounds to pass through freely while blocking the passage of other compounds. The plasma membrane is a semipermeable membrane.
TUTORIAL 5.1.2: Diffusion through Membranes

- Osmosis is the net movement of water across a semipermeable membrane from an area of lower solute concentration to an area of higher solute concentration. Osmosis is a major force in living things, since it is responsible for much of the movement of fluids into and out of cells.
TUTORIAL 5.1.4: Osmosis

5.5 How Do Materials Get In and Out of the Cell?

- Some compounds are able to cross the plasma membrane strictly through diffusion and osmosis; others require these forces and special protein channels; still others require protein channels and the expenditure of cellular energy.

- Any time energy is required for movement across the plasma membrane, active transport has occurred. When such movement occurs without energy expenditure, passive transport has occurred.
  TUTORIALS: 5.2.1 Passive Transport; 5.2.2 ActiveTransport

- Simple diffusion is diffusion that does not require a protein channel. Facilitated diffusion is diffusion that requires a protein channel. In facilitated diffusion, transport proteins function as channels for hydrophilic substances—substances that because of their size and electrical charge cannot diffuse through the plasma membrane.

- Cells employ chemical pumps to move compounds across the plasma membrane against their concentration gradients.

5.6 Getting the Big Stuff In and Out

- Larger materials are brought into the cell through endocytosis and moved out through exocytosis. Both mechanisms employ vesicles: the membrane-lined enclosures that alternately bud off from membranes or fuse with them.
  TUTORIAL 5.3.1: Exocytosis

- There are three principal forms of endocytosis: (1) pinocytosis, in which the plasma membrane invaginates, creating an enclosure that pinches off to become a vesicle that moves into the cell; (2) receptor-mediated endocytosis, in which cell-surface receptors bind with materials to be brought into the cell and then migrate laterally through the cell membrane, congregating in a location where vesicle-budding will bring them into the cell; and (3) phagocytosis, in which certain cells engulf whole cells, fragments of them, or other organic materials.
  TUTORIAL 5.3.2: Endocytosis

Key Terms

active transport 104	**isotonic solution** 104
binding site 100	**osmosis** 103
concentration gradient 102	**passive transport** 104
diffusion 102	**peripheral protein** 100
endocytosis 107	**phagocytosis** 108
exocytosis 107	**phospholipid bilayer** 99
facilitated diffusion 105	**pinocytosis** 108
fluid-mosaic model 101	**plasma membrane** 101
glycocalyx 101	**receptor-mediated endocytosis (RME)** 108
hypertonic solution 104	**receptor protein** 100
hypotonic solution 104	**simple diffusion** 104
integral protein 100	**transport protein** 105

Understanding the Basics

Multiple-Choice Questions

1. If the plasma membrane lacked a glycocalyx, the cell would have difficulty
   a. keeping its fluid state
   b. transporting materials across the membrane
   c. recognizing proteins on the outside of the membrane
   d. binding to the cytoskeleton on the inside of the membrane
   e. keeping small particles from passing easily across the membrane

2. You eat spaghetti for dinner and digest the starch in it to glucose. The glucose is now in high concentration in your small intestine, where food particles are absorbed into the bloodstream. If the glucose molecules move into your bloodstream by diffusion, they have moved from the high concentration in the intestine to the low concentration in the bloodstream.
   a. True.
   b. False, because particles cannot move across membranes by diffusion.
   c. False, because diffusion moves materials from low concentration to high concentration.
   d. False, because molecules are too large to move across membranes.
   e. False, because energy is required for diffusion.

3. In osmosis, water moves
   a. away from dissolved materials
   b. toward dissolved materials
   c. wherever energy pulls it
   d. independently of dissolved materials
   e. any of the above could be true in different situations

4. Tyrone gave Lakeisha a rose for the anniversary of their first date. Now the rose has wilted, because
   a. The cells have burst.
   b. The cell walls have broken down.
   c. The plasma membrane has lost the proteins that keep water in.
   d. The phospholipid bilayer has broken down.
   e. The plant has insufficient water available to it for the cells' water to push optimally against the cell wall.

5. The protein that carries out the sodium-potassium pump is an *integral* protein. The word *integral* tells you that this protein
   a. functions on only one side of the membrane
   b. is connected to a cholesterol molecule
   c. is essential to the cell
   d. is connected to the hydrophobic interior of the plasma membrane
   e. is connected to the glycocalyx

6. You get a cut on your finger and some bacteria enter. Your immune system cells kill off the invaders by eating them. This involves

a. pinocytosis
b. phagocytosis
c. receptor-mediated endocytosis
d. exocytosis
e. active transport

7. According to the fluid-mosaic model of membrane structure, the membrane is made up of a
   a. nearly continuous base of proteins with phospholipids stuck in it
   b. a bilayer of phospholipds with proteins stuck in it that have fairly free lateral movement
   c. bilayer of protein with layers of phospholipids on the outside and inside
   d. bilayer of phospholipid with layers of proteins on the outside and inside
   e. bilayer of phospholipids sandwiched between two bilayers of proteins

8. The sodium-potassium pump uses _____ to move sodium ions _____ the cell and potassium ions _____ the cell.
   a. lipids . . . into . . . out of
   b. proteins . . . into . . . out of
   c. energy . . . out of . . . into
   d. cholesterol . . . out of . . . into
   e. binding . . . into . . . out of

9. Endocytosis and exocytosis are similar in that both involve the use of
   a. vesicles that carry materials into and out of the cell
   b. the Golgi apparatus to package materials
   c. diffusion to carry food and wastes across the plasma membrane
   d. starch to supply large amounts of energy
   e. rough ER to manufacture proteins

10. Osmosis refers to the movement of _____ a concentration gradient.
    a. energy down
    b. dissolved materials against
    c. water along
    d. dissolved materials along
    e. water against

Brief Review

1. Mentally picture a taste-bud cell on your tongue. Assume that this particular taste bud responds to sweet tastes. Explain how the surface of the cell—its plasma membrane—might allow it to respond to sugar, but not to salt or other tastes.

2. You often hear how dangerous it is for athletes to take steroids to "bulk up" and how many problems steroids cause in various parts of the body. Why do these drugs affect all parts of the body—not just the muscles?

3. Explain why the phospholipid "heads" of the plasma membrane are always pointed toward the cytosol and extracellular fluid, whereas the "tails" are always oriented toward the middle of the membrane.

4. Most cells in your body do not respond to most hormones. For example, a hormone known as *erythropoietin* causes the bone marrow to make more red blood cells, but has little effect elsewhere in the body. Explain how the plasma membrane could control this effect.

5. As you saw in this chapter, all cellular membranes are essentially the same. However, the specifics given in the chapter relate primarily to the plasma membrane on the outside of the cell. What differences do you suppose might be true of membranes in the interior of the cell?

6. Place a number of marbles on one end of a cafeteria tray. Shake the tray gently a few times, keeping it level. What happens to the marbles? Start over and shake faster the next time. What happens now? Explain how this is like diffusion. What is shaking the tray similar to in the process of diffusion?

7. The cells that line your stomach make it very acidic by pumping large numbers of H$^+$ ions from the inside of the cells into the stomach. They have to do this *against* a concentration gradient. What process do these cells use to do this pumping, and what do they need to carry it out?

Applying Your Knowledge

1. In the introduction to this chapter, you read about what happens in cystic fibrosis when a single ion (chloride) cannot be transported to the surface of the cell. Think about what would happen if any of the three major components of the cell membrane (phospholipid bilayer, glycocalyx, cholesterol) could not be made. What would happen to the cell? What general conclusion can you then reach about the necessity of all parts of the cell membrane?

2. Because materials move from high concentration to low concentration by diffusion, and they move the same way by facilitated diffusion, why do cells need to make proteins to carry out facilitated diffusion? Isn't this just a waste of energy?

3. In 1998 scientists determined the structure of a channel protein that allows potassium ions to move across the plasma membrane. This channel works with the sodium-potassium pump to enable nerve cells to respond to changes in the environment. If you were a scientist, how might you use this information in understanding how cells communicate?

4. Recently, researchers discovered the structure of the protein that allows about 70 percent of all cold viruses to bind to the plasma membrane of a human cell, and enter to cause a cold. What benefit would there be to knowing the shape of that protein? How might you use this information, if you were working for a pharmaceutical company?

MediaLab

Balancing Your Imports and Exports: Membrane Transport

In this chapter, you read that without the efficient, rapid movement of substances across your membranes, you would not be able to move, absorb nutrients, or even see this page. The *CD-ROM Tutorial* illustrates the different types of membrane transport involved in all these processes. In the following *Web Investigation*, you will examine the importance of membranes in organisms ranging from bacteria to humans, so that in the *Communicate Your Results* section you can analyze how membrane transport has been manipulated to treat wounds, cystic fibrosis, or drug addiction.

This *MediaLab* can be found in Chapter 5 on your CD-ROM (Tutorial 5.2.1) and Companion Website (http://www.prenhall.com/krogh3).

CD-ROM TUTORIAL

Cells cannot live without exchanging substances with the environment. There are four different ways of transporting substances across membranes; each is illustrated through graphic animations in this *CD-ROM Tutorial*.

Activity

1. First, you will discover how small molecules, including water, may be able to move across the lipid portion of the membrane using **diffusion** in a process called **passive transport**.

2. You will then compare this to the process of **active transport**, where the energy of ATP must be used to pump molecules across protein channels in the membrane. You can then try your hand at matching which mechanism would be used to transport specific molecules.

3. Finally, you will learn how other substances—critical in nerve transmission—are moved in bulk within membrane "bubbles," a process called **endocytosis** or **exocytosis**.

After completing this activity, you should be familiar enough with these transport mechanisms to begin the *Web Investigation* into how they work together in cells.

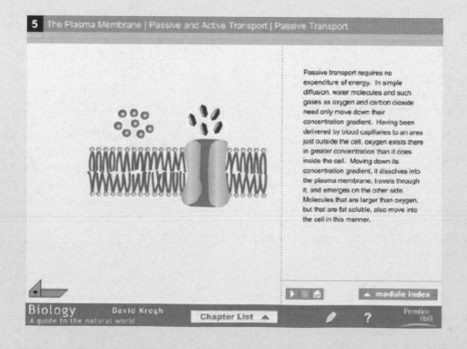

WEB INVESTIGATION

Investigation 1

Estimated time for completion = 5 minutes

Cigarettes, chocolate, and cocaine: Have you ever wondered how they make us feel so good? Our brain contains millions of neurons that help us respond to the environment, coordinating sight, smell, hearing, touch, and even pleasure. All these neurons must communicate with each other using membrane transport. Most addictive drugs are able to elevate the natural neuron communication that occurs after pleasant sensations like a hug, or a bite of candy. Select the Keyword **ADDICTION** on your CD or Website to read more about nerve-cell communication. Once you have read about nerve cells, click on the next link, "drugs," at the bottom of the page to view how cocaine acts. After reading the article, select the Keyword **NERVES** to view an animation of one nerve cell communicating with another.

Investigation 2

Estimated time for completion = 5 minutes

Why do we wash our hands with soap, or cleanse a wound with alcohol or iodine? They are all antiseptics, substances that kill cells. What is it in soap or alcohol that would harm bacteria but not our own cells? Select the Keyword **ANTISEPTICS** on your CD or Website and read Table 2 in the linked article to find out how these antiseptics harm bacteria. Antiseptics are commonly found in many different household products, including cleaning supplies and mouthwashes. Select the Keyword **MOUTHWASH** to read an article listing the active ingredients in mouthwash. Then, clip a label from an antiseptic product and use these two websites to determine its active ingredients.

Investigation 3

Estimated time for completion = 10 minutes

At the opening of this chapter you read about cystic fibrosis, a debilitating disease caused by a genetic disorder. The disease is caused by a defect in the gene that produces the protein CFTR, which forms channels in the cell membranes involved in the active transport of chloride ions. Because this disease greatly reduces life expectancy, you would expect few of these defective genes to be passed on to future generations. Yet, cystic fibrosis has been around for centuries. Could there possibly be an up side to having a defective gene like this? Select the Keyword **ADVANTAGE**, and read about evolution and the selective advantage of the cystic fibrosis gene.

COMMUNICATE YOUR RESULTS

Exercise 1

Estimated time for completion = 10 minutes

In *Web Investigation 1*, you saw how membrane transport operates to conduct nerve impulses. In 50 words, identify the types of membrane transport involved in passing nervous impulses, and describe their roles. These nerve impulses may be sending messages of pain (or pleasure in the case of addictive drugs) to your central nervous system. How does cocaine cause pleasurable sensations? Can you propose a way to interrupt neuron communications such as pain signals? Does this knowledge help to explain how painkillers and other drugs like cocaine or morphine work?

Exercise 2

Estimated time for completion = 5 minutes

What's the best way to cleanse your mouth, hands, or wounds of harmful bacteria? In *Web Investigation 2*, you read a list of antiseptics and saw that many of them kill bacteria by disrupting cell membranes. Create a list of all the active ingredients from the labels of your common household antiseptics that act by disrupting cell membranes.

Exercise 3

Estimated time for completion = 5 minutes

The article in *Web Investigation 3* explores the idea that if detrimental alleles like cystic fibrosis remain in the population, they must have some selective advantage. In 50 words, describe the selective advantage provided by the cystic fibrosis allele. Propose a hypothesis to explain this advantage, given what you know about the cause of cystic fibrosis.

6

Life's Mainspring
An Introduction to Energy

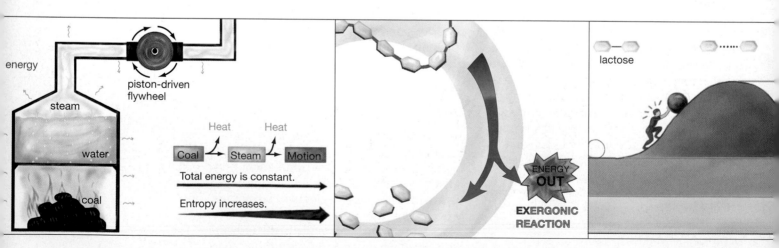

Freeing energy from chemical bonds.
(Section 6.2, page 119)

A downhill reaction releases energy.
(Section 6.3, page 121)

There must be an easier way—and there is.
(Section 6.6, page 125)

Energy flows through life in a never-ending pattern of acquisition, storage, and use.

My young daughter Tessa has a teddy bear with a music box in it that lets us listen to "Brahms's Lullaby." All that's required is that I wind up the key that attaches to the music box. After a few turns of the wrist, the melody cycles through several times, then slows for a last few bars, then stops. This process can be repeated as many times as I wind the music box. However—no surprise here—the music box never seems to wind itself. As we listen, a mechanism in the music box keeps the melody coming out at an even tempo. Even the slowing at the end seems just right for a lullaby, since the audience is a sleepy child.

6.1 Energy Is Central to Life

The forces that determine how Tessa's teddy bear works also turn out to condition life, since, in order to function, both the music box and life must be supplied with the same thing: energy. Just as the music from the teddy bear depends on energy from an "outside hand" (mine) to continue, so life on Earth depends on energy from an outside source: the Sun, whose daily showering of energetic rays lets the green world of plants and algae flourish. These organisms then pass their bounty along to animals in the form of food to eat and oxygen to breathe.

Beyond this, the Sun and the music box have another thing in common; their energy always runs downhill, from more concentrated to less. The spring in the music box spontaneously unwinds, but it does not rewind itself. The Sun sends out light and heat, but this energy does not spontaneously come together to form another sun. The music from the music box eventually must come to a stop, and even the Sun will not shine forever. With all its grandeur, its energy is winding down, its brilliance dispersing. The difference is that once the Sun's showering of energy comes to a stop, nothing will restart it.

While the Sun's gift lasts, however, living things can transform its energy and use it to do things like sprout leaves and swim through oceans. Such complex activities make it imperative, though, that living things be able to *control* the energy they capture. The spring within the music box unwinds at a measured rate, spinning out the melody at just the right pace; similarly, living things have elaborate mechanisms in place that allow them to make the most out of the energy they receive.

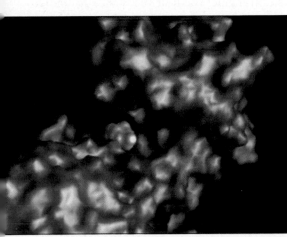
Shaping up to fit in grooves and pockets. (Section 6.6, page 125)

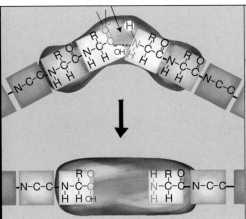
Enzymes at work. (Section 6.6, page 126)

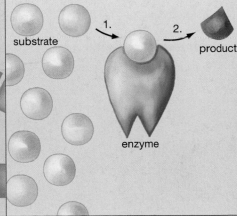
Regulating an enzyme. (Section 6.7, page 127)

Principles such as this are important enough to biology that this chapter and the two that follow are devoted to the subject of energy and life. In this chapter you'll learn some basics about energy, including information on how organisms employ molecules called enzymes to lower the amount of energy needed to get chemical reactions going. Then, in Chapter 7, you'll see how living things "harvest" energy, meaning how they extract energy from food. Finally, in Chapter 8, you'll look at the ways that plants capture and transform the Sun's energy in the process called photosynthesis.

6.2 What Is Energy?

Energy is commonly defined as the capacity to do work, but this definition merely passes the question mark along, to the word *work*. Given this, here's one possible definition of energy: The capacity to bring about movement against an opposing force. Speaking more generally yet, some observers have defined energy as the capacity to bring about change. The thing that makes energy so tricky as a concept is that although we can measure it with great precision, and we can experience its effects, we cannot grasp it or see it. We can see the water that drives a waterwheel, but who has seen the energy the water contains? (**See Figure 6.1.**)

The Forms of Energy

Energy can be thought of as coming in different forms, many of them familiar. Mechanical energy is captured in the wound-up spring of the musical teddy bear. Chemical

Figure 6.1
Where's the Energy?
Part of the water moving over these falls is channeled to a water wheel, which turns as a result. Though the fall of the water is plainly visible, the energy released is not.

energy is the energy held in the chemical bonds of a lump of coal. One other way to conceptualize energy is as potential and kinetic. **Potential energy** is stored energy: The rock perched precariously at the top of the hill; the charged ions kept on one side of a cell membrane. Conversely, **kinetic energy** is energy in motion, as with the rock tumbling down the hill or water driving a waterwheel.

The Study of Energy: Thermodynamics

Given how important energy is to life, it makes sense that there is a branch of biology, called *bioenergetics*, that links biology with a scientific discipline known as **thermodynamics**, the study of energy. Though thermodynamics is a mathematical and rather abstract discipline, research into it was prompted originally by a very down-to-earth goal—the desire to build a better steam engine. British inventors had harnessed steam power in the eighteenth century, and this discovery had the awesome impact of sparking the Industrial Revolution. The word *revolution* is fitting here because, before the advent of steam power, people used the stored energy in, say, wood or coal solely to produce *heat*. The steam engine was a device that allowed us to use heat to perform *work* on a scale that was previously impossible. Heat could be channeled, it was discovered, to drive an engine, which in turn could power any number of industrial processes.

With this development, new questions confronted scientists: How exactly did heat drive a steam engine? Was heat a substance that moved from one place to another? Was there a finite quantity of energy, such that it could all be used up? Research carried out in England, Germany, and France during the nineteenth century yielded the answers to these questions, ultimately resulting in the development of some insights known as the laws of thermodynamics.

The First Law of Thermodynamics: The Transformation of Energy

The **first law of thermodynamics** states that energy is never created or destroyed, but is only transformed. The Sun's energy is not used up by green plants; rather, some of it is *converted* by the plants into chemical form. Plants use the Sun's energy to put together carbohydrates, whose *chemical bonds* contain some of the energy that previously existed in the Sun's rays.

A critical point here is the qualification that *some of* the Sun's energy takes on a chemical form in plants. As it turns out, a plant cannot convert *all* of the solar energy it receives into carbohydrates. Likewise, a steam engine actually converts only a fraction of the energy contained in coal into the movement of a piston. What happens to the energy that is not stored in carbohydrates or transformed into motion? It can't be destroyed; that's clear from the first law. What happens is that it is converted into *heat*. For every energy "transaction" that takes place, at least some of the original energy is converted into heat. This is true for every transaction in every energy system: cells burning food, cars burning gasoline, electric utilities burning coal, and your brain decoding this sentence.

The way this concept is usually phrased, however, is that some of the original energy is *lost* to heat. At first glance, this may seem like an unfair knock on heat, which after all has intrinsic value. Indeed, you may wonder about how this concept applies to the work of a steam engine. The piston is *driven* by heat that is produced by burning coal. Why, then, is energy "lost" to heat?

Well, consider what might be called the before-and-after of energy transfer. In the "before," a lump of coal is a very ordered object, containing carbon atoms that exist in a precise spatial relationship to one another. Thus bonded, these atoms are not free to move. However, bringing a flame to the lump of coal causes the coal's carbon atoms to react with oxygen. Chemical bonds are broken through this process, and the energy stored in these bonds is released as heat.

Now cut to the "after" of the energy transformation. Heat is the *random* motion of molecules. In the case of the steam engine, there are molecules in random motion— better known as hot air—inside and outside the apparatus. In comparison to the "before" of the transaction, it's clear that the energy contained in the *ordered, concentrated* chemical bonds of coal has been transformed into the *disordered, dispersed* energy of heat.

The Second Law of Thermodynamics: The Natural Tendency toward Disorder

And this gets us to the critical thing. Energy transformations will spontaneously run *only* from greater order to lesser order. Some of the heat produced in a steam engine is useful; it's driving a piston. Some of it is not, however—it's simply dissipated. But *all* of it

amounts to a form of energy that is more dispersed than coal and that will quickly disperse further. Once this happens, there is no chance that it will spontaneously convert into anything *but* heat. We expect to see a lump of coal burn to ashes, and to feel the hot air that results disperse. But we do not expect to see the ashes reform into a lump of coal, nor to feel the hot air concentrate itself again.

In yielding energy, matter goes from a more-ordered state to a less-ordered state. The scientific principle that speaks to this is the **second law of thermodynamics**: Energy transfer always results in a greater amount of disorder in the universe. In connection with this law there is a term you will hear—*entropy*, which is a measure of the amount of disorder in a system. The greater the entropy, the greater the disorder. **Figure 6.2** summarizes the thermodynamic principles of the first and second laws of thermodynamics.

The Consequences of Thermodynamics

All of this may seem like just so much—one might say—*hot air*. But few things so profoundly condition the universe we live in as these thermodynamic rules. Indeed, the relentless increase in entropy that the second law entails may dictate nothing less than the fate of the universe.

Allow yourself, for a second, to think of the universe as consisting solely of the Sun, the Earth, and the space between them. The Sun is a concentration of hydrogen that is undergoing a nuclear reaction, thus releasing light and heat.

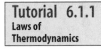

Tutorial 6.1.1
Laws of
Thermodynamics

Figure 6.2
The Transformations of Energy
In a steam engine, energy locked up in the chemical bonds of coal is transformed into heat energy and mechanical energy. There is no loss of energy in this process, but energy is transformed from a more-ordered, concentrated form (the chemical bonds of coal) to a less-ordered, more dispersed form (heat). Thus, the amount of disorder—or entropy —increases in the transaction.

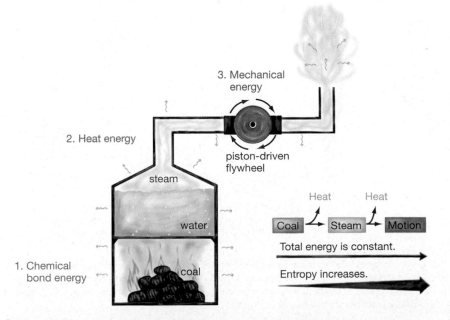

Table 6.1
Relative Efficiencies of Several Energy Systems

The efficiency of several energy systems, as measured by the proportion of energy they receive relative to what they then make available to perform work. In measuring the efficiency of the car engine, the question is, how much of the energy contained in the chemical bonds of gasoline is converted by the car into the kinetic energy of wheel movement? In each system, most of the energy not available for work is lost to heat.

System	Efficiency
Conventional power plant	34%
Human respiration of glucose	37%
Automobile engine	25%

Living things on Earth manage to capture buckets of this energy, primarily through the activities of plants. But every time one of these buckets is poured into another—every time light energy from the Sun is transformed into carbohydrate energy in plants, for example—there is some spillage into heat. And once heat is generated, there is no spontaneous way it can make its way back into the orderly form of sunlight or carbohydrates. With every firing of a car's cylinder or contraction of a muscle, then, some amount of energy is transformed into heat.

Against this, life as we know it is made possible by ordered concentrations of energy. The Sun sustains life, but dispersed heat does not. One day, our Sun will have wound down like the spring in a music box, bringing to an end the energy source for life on Earth. One possible scenario for the universe as a whole is that *all* its stars eventually will darken, leaving a cold universe with little potential for life—indeed, with little activity of any sort. The laws of thermodynamics are powerful indeed.

6.3 How Is Energy Used by Living Things?

A sprouting plant is constantly building itself up; it is making larger, more complex molecules (proteins, polysaccharides) from smaller, simpler ones (amino acids, simple sugars). This growing degree of organization stands in contrast to the spontaneous course of things, which is breakdown and disorder. In the context of the universe as a whole, living things are contributors to its entropy. Any given biological activity—building a protein, say—generates heat and thus increases the disorder of the universe. But living things can bring about *local* increases in order (in themselves). It takes energy to do this, however, and on Earth the source of this energy ultimately is the Sun. Living things can be quite efficient at converting energy from outside sources, as you can see by looking at **Table 6.1.**

Kinds of Work for Living Things

In considering *how* it is that living things use energy, there are really only a few kinds of work that they carry out. The most obvious of these is **mechanical work**, exemplified by the contraction of a muscle. There also is **transport work**; for example, moving sodium ions against their concentration gradient, as you saw in Chapter 5. Finally, there is **synthetic work**, which can be seen in the buildup of complex molecules (proteins) from simpler ones (amino acids).

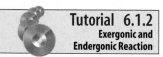

Tutorial 6.1.2
Exergonic and Endergonic Reaction

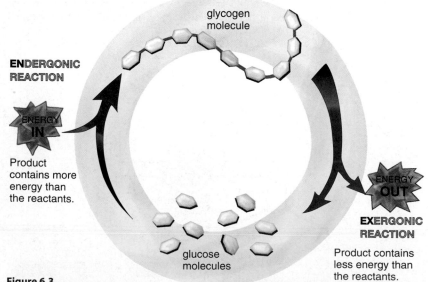

ENDERGONIC REACTION

ENERGY IN

Product contains more energy than the reactants.

glycogen molecule

ENERGY OUT

EXERGONIC REACTION

Product contains less energy than the reactants.

glucose molecules

Figure 6.3
Energy Stored and Released
It takes energy to build up more complex molecules (in this case glycogen) from simpler molecules (in this case glucose). Such a buildup is thus an endergonic or uphill reaction. Conversely, energy is released in reactions in which more complex molecules are broken down into simpler ones. Such reactions are downhill or exergonic.

Up and Down the Great Energy Hill

Synthetic work leads to the idea of an energy cycle in living things. A starchy carbohydrate is a more complex thing than the simple sugars it is made of. It is a more *ordered* arrangement of atoms than is a group of individual simple sugars. As such, the second law of thermodynamics tells us, it should *take* energy to make such a substance from simple sugars, and this is indeed the case.

Going in the opposite direction, the breakdown of a starchy carbohydrate into simple sugars is an action favored by the second law, because it brings about *less* order. Such a process therefore should not take energy—no more than it takes energy for a boulder to roll down a hill. Indeed, this process should *release* energy, and this too is the case. This understanding leads to the concept of a great uphill and downhill cycle of energy in living things.

Downhill (Exergonic) and Uphill (Endergonic) Reactions

The breakdown of a carbohydrate is a specific example of a general kind of energy reaction. Such a breakdown is a form of **exergonic** energy exchange. An exergonic (meaning "energy out") chemical reaction is one that runs downhill, releasing energy; it is a reaction in which the starting set of molecules (the reactants) contain more energy than the final set of molecules (the products).

Conversely, an **endergonic** ("energy in") reaction is one that runs uphill; the products of such a reaction contain more energy than the reactants. Energy, in other words, is stored away in such a reaction. This is what happens when, for example, simple glucose molecules are brought together to form glycogen, which is a storage form of carbohydrates (**see Figure 6.3**).

Coupled Reactions

Now, as it happens, the two kinds of reactions are constantly *linked* in living things. Why should this be so? If you want to get something uphill—into a higher energy state—you have to power that process. And that means *using* some energy. (Think of getting a bicycle uphill.) Thus, endergonic reactions require exergonic reactions. When the two are brought together, it is known as a **coupled reaction**. You actually observed some of these reactions in Chapter 5, though they weren't labeled as such then. Remember how the sodium-potassium pump moved sodium ions

out of the cell against their concentration gradient? (See pages 102–106.) For the sodium ions, this was an uphill reaction: They finished the reaction in a higher energy state than they started, because they were moved up their electrical and concentration gradients.

And how was this movement powered? Through a downhill reaction. All you observed in Chapter 5 was that an ATP molecule furnished the energy that powered the sodium pump. You have probably guessed by now, however, that what this really meant was the breaking down of a complex molecule into a simpler one, thus yielding the pump energy. The molecule being broken down was one that deserves a close consideration—ATP.

6.4   The Energy Molecule: ATP

Adenosine triphosphate. If you look at **Figure 6.4**, you can see that ATP is a nitrogen-containing molecule with three "phosphate groups"

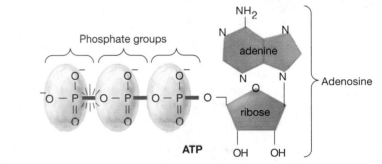

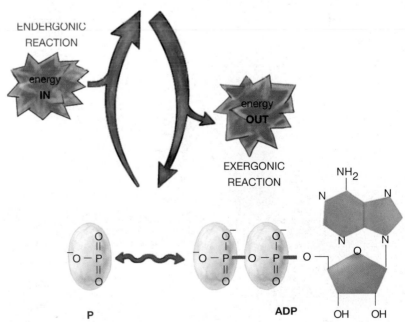

Figure 6.4
Energy Release from Breakdown of ATP
ATP (adenosine triphosphate) is life's most important energy transfer molecule. It stores energy in the form of chemical bonds (shown in red) between its phosphate groups. When the bond between the second and outermost phosphate group is broken, the outermost phosphate separates from ATP and energy is released. This separation transforms ATP into adenosine diphosphate or ADP, which then goes on to pick up another phosphate group, becoming ATP again.

attached to it. Each group amounts to one phosphate atom and three oxygen atoms. Notice that each of the phosphate groups is negatively charged. This is the key to understanding why ATP serves as such an effective energy transfer molecule. Remember that like charges repel each other, and in this case three phosphate groups are doing just that. Energy was required to put these phosphate groups together as they are, in this relatively unstable state. This linkage represents a move *up* the energy hill.

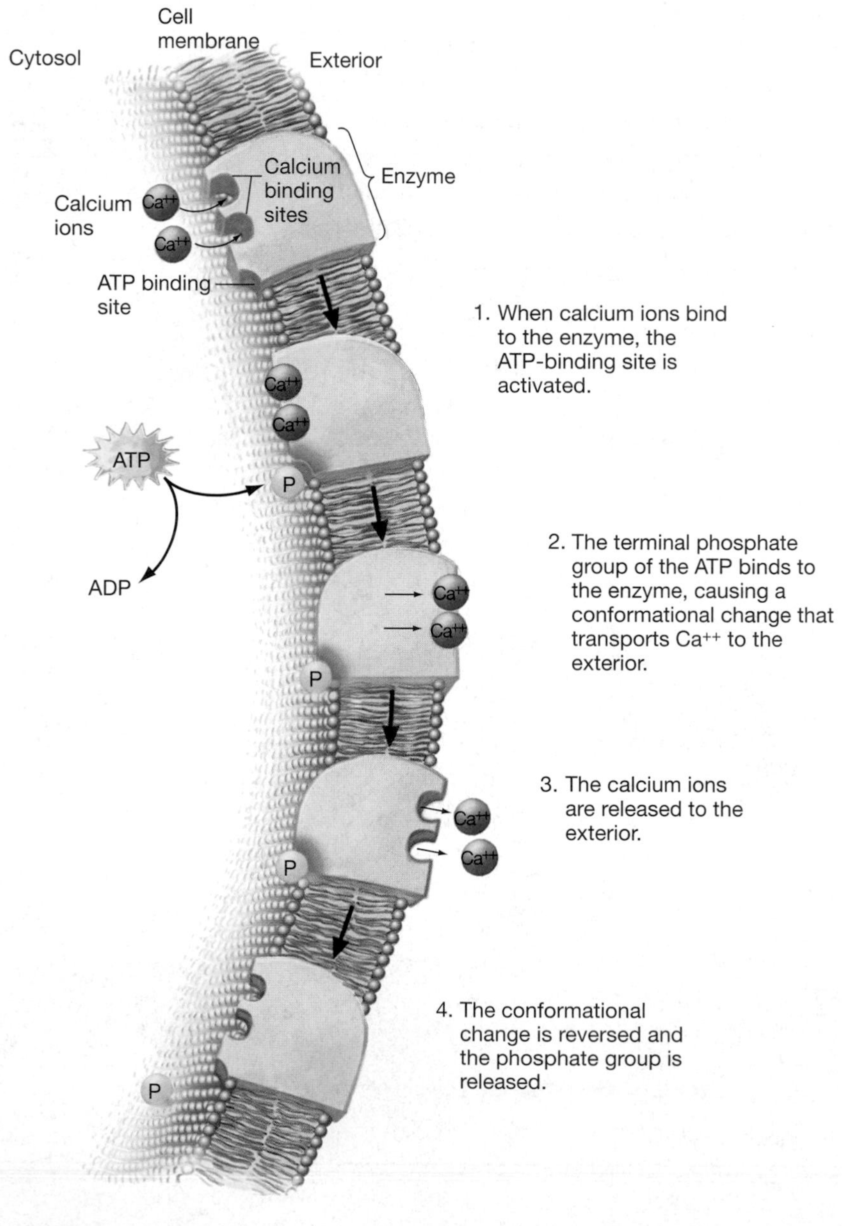

Cytosol

Cell membrane

Exterior

Calcium binding sites

Enzyme

Calcium ions Ca^{++}

Ca^{++}

ATP binding site

Ca^{++}

Ca^{++}

ATP

P

ADP

Ca^{++}

Ca^{++}

P

P

Ca^{++}

Ca^{++}

P

P

1. When calcium ions bind to the enzyme, the ATP-binding site is activated.

2. The terminal phosphate group of the ATP binds to the enzyme, causing a conformational change that transports Ca^{++} to the exterior.

3. The calcium ions are released to the exterior.

4. The conformational change is reversed and the phosphate group is released.

Figure 6.5
How ATP Functions
By transferring a phosphate group to the enzyme shown in this figure, ATP causes the enzyme to change shape in a manner that transports calcium ions across the cell membrane.

With ATP, it is as though someone squeezed a jack-in-the-box back into the box and shut the lid. It should be no surprise that when the lid is opened, a good deal of energy will be released.

How Does ATP Function?

Now, how is the energy from ATP used? As a starting point, consider one type of action ATP molecules will carry out within, say, a muscle cell (**see Figure 6.5**). Under the right conditions, ATP will bind to a type of protein, called an enzyme, that spans the membrane of this cell. The ATP molecule is then split, with the outermost of its phosphate groups breaking off from the rest of the molecule. This is a downhill reaction: It took energy to get the third of ATP's phosphate groups onto it (in an uphill push), and this stored energy is now being released. The subsequent attachment of this phosphate group to the protein causes the protein to change its shape. This shape change happens to drive the transport of calcium ions across the protein. For our purposes, however, what's important is that something that would not have happened on its own took place because of the energy provided by the ATP molecule.

The ATP/ADP Cycle

There is another side to this ATP-driven reaction, however. Once the outermost phosphate group has split off, ATP has become a molecule containing *two* phosphate groups, adenosine *di*phosphate or ADP. It is now free to have a third phosphate group added to it again (in an uphill reaction). This is just what happens: ADP returns to being ATP, which is capable of providing energy for yet another reaction. This shuttling from ATP to ADP and back takes place constantly in cells.

ATP as Money

In the economy of the cell, ATP really is like money. Imagine working at a typical job, in which you're called upon to do dozens of different things. You do this and you do that, but in the end what you get for all this activity is *money*. You can then take this cash and pay the rent, buy some CDs, or go bowling. Now think of animals such as ourselves and their breakdown of food. In the end, what they get for all their complex efforts in this regard is ATP. It is the final outcome of the energy-harvesting process. And, like money, ATP then can be *used* to do any number of things, such as transport calcium ions across a cell membrane. The big difference between ATP and money is that money can be stored for long periods (in wallets, banks, and so on.) while ATP cannot. It is a good energy transfer molecule, but not a good energy storage molecule.

Between Food and ATP

The essence of our story so far is that, in animals, energy harvesting starts with the ingestion of food and ends with the synthesis and expenditure of ATP. This is, however, such a bare-bones description that it is rather like a synopsis of *Cinderella* that goes: There was a young girl who was mistreated by her stepmother; then she married a prince. What's missing is everything that happened in between to make things come out this way. ATP, as noted, is produced through an *uphill* process: Energy must be expended in order for its three phosphate groups to be bonded together. How is it that people go from eating something to getting the ATP they need? It is almost time to answer this question—it will be in Chapter 7—but first, you need to know something about a group of proteins that living things use to control energy.

6.5 Efficient Energy Use in Living Things: Enzymes

Lactose—better known as milk sugar—is composed of two simple sugars, glucose and galactose, that are linked together by a single atom of oxygen. It follows from what you've seen so far that lactose will *split* into its glucose and galactose parts without any energy being required. Indeed, some energy will be released, because chemical bonds have been broken and two smaller molecules have been produced. Such a splitting is therefore a downhill reaction.

If we actually took a small amount of lactose, however, and put it in some plain water, it would certainly be hours—it might be days—before *any* of the lactose molecules would break down into glucose and galactose. How can this be? Most of us drink milk and it doesn't seem to pile up in us.

Hastening Reactions

The secret is that when lactose is metabolized in living things, it isn't being split up in plain water. Something else is present that speeds up this process immensely, hastening it perhaps a billionfold. That something is an *enzyme*. This particular enzyme is called lactase, but it is merely one of more than 5,000 enzymes known to exist. Each of these compounds is working on some chemical process, but not all are involved in splitting molecules. Some enzymes combine molecules, some rearrange them. This work goes on mostly inside of cells, but enzymes also function elsewhere: in the bloodstream, the digestive tract, and the extracellular fluid, for example.

Given their diverse work sites, it may be apparent that enzymes are involved in a lot of different activities; but this doesn't begin to give enzymes the recognition they deserve. Enzymes facilitate nearly every chemical process that takes place in living things. No organism could survive without them. Technically, these compounds are only *accelerating* chemical reactions that would happen anyway, as with lactase splitting lactose. But in practical terms, they are *enabling* these reactions, because no living thing could wait days or months for the milk sugar it ingests to be broken down, or for hormones to be put together, or for bleeding to stop.

To get an idea of the importance of enzymes, consider the millions of Americans who are known as lactose intolerant because, after childhood, their bodies reduced their production of lactase. When these people consume milk, the lactose in it *does* sit in their intestines until it is partially digested not by them, but by the bacteria that all humans have in their digestive tract. The end result can be gas from the bacterial digestion and bloating from a buildup of fluids in the intestine.

Specific Tasks and Metabolic Pathways

From the sheer number of enzymes that exist, you can get a sense of how specifically they are matched to given tasks. Although some enzymes can work on *groups* of similar substances, in the main, a specific enzyme facilitates a specific reaction. This means that a particular enzyme works on one or perhaps two molecules and no others. Thus, lactase breaks down lactose—and *only* lactose—and the products of its activity are always glucose and galactose.

Certain activities in living things, such as breast-milk production, are much more complex than the breakdown of lactose. They are multistep processes, *each step of which* requires its own enzyme. Most large-scale activities in living things work this way: leaf growth, digestion, hormonal balance. These processes are carried out through a sequential set of enzymatically controlled steps. Collectively, these steps are known as a **metabolic pathway**. In such a process, each enzyme does a particular job and then leaves the succeeding task to the next enzyme in the pathway (**see Figure 6.6**).

The substance that is being worked on by enzymes is known as a **substrate**. In our example, lactose is the substrate for lactase. The *-ase* ending you see in *lactase* is common to most enzymes. As you can see, the substrate is identified in the enzyme's name. The sum of all the chemical reactions that a cell or larger organism carries out is known as its **metabolism**. To put things simply, enzymes are active in all facets of the metabolism of all living things.

6.6 Lowering the Activation Barrier through Enzymes

All this information about enzymes is well and good, you may be saying at this point, but what does it have to do with energy? The answer is that in carrying out their tasks, enzymes are in the business of lowering the amount of energy needed to get chemical reactions going. And this means that the reactions can get going faster.

An analogy may be helpful here. Consider the now-familiar rock perched at the top of a hill. Situated in this way, it has a good deal of potential energy. If it rolled down the hill, it would release this potential energy as kinetic energy. But here is the critical question: What would be required to get this rock going? It is *perched* at the top of the hill because it is *stable* at the top of the hill. To get it going would require *additional* energy in the form of, say, a push. This additional energy is referred to as the activation energy of this energy transaction. In living things, **activation energy** can be defined as the energy required to initiate a chemical reaction.

Now consider the lactose molecule. It is "perched" like the rock, in a sense, because it lies "uphill" from the glucose and galactose it can break into. But it is also stable like the rock. Left alone—sitting in a milk carton—it *stays* lactose. This kind of stability is typical of most molecules in our everyday experience.

But the body needs the glucose and galactose that come from lactose breakdown, and it can't wait days to get these molecules. How is it going to be supplied with them as needed? The elegant solution is enzymes. They lower the activation energy needed to carry out chemical reactions. If you look at **Figure 6.7**, you can see how this works. Imagine that you had to push a boulder up the hill in (a) to get it to roll down the other side. Now imagine that you only had to push a boulder up the hill in (b) to get the same result. In which situation would you be able to get your desired result *faster*?

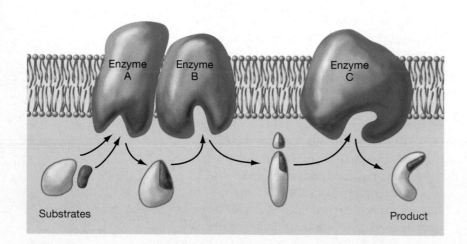

Substrates

Product

Figure 6.6
Metabolic Pathway: Sequence of Enzyme Action
Most processes in living organisms are carried out through a metabolic pathway—a sequential set of enzymatically controlled reactions in which the product of one reaction serves as the substrate for the next. Enzymes perform specific tasks on specific substrates. In this example, enzyme A combines two substrates, enzyme B removes part of its substrate, and enzyme C changes the shape of its substrate.

How Do Enzymes Work?

How do enzymes carry out this task of accelerating chemical reactions? The short answer is that they bind to their substrates, and in so doing make these substances more vulnerable to chemical alteration. The amazing thing is that they do this without being permanently altered themselves. Enzymes are **catalysts**—substances that retain their original chemical composition while bringing about a change in a substrate. At the end of its cleaving of the lactose molecule, lactase has exactly the same chemical structure as it did before. It is thus free to pick up another lactose molecule and split *it*. All this takes place in a flash: In one second, a given enzyme can carry out hundreds or even thousands of chemical transformations.

With only a handful of exceptions, all known enzymes are proteins. They are generally globular or ball-like proteins whose shape includes a kind of pocket into which the substrate fits. If you look at **Figure 6.8**, you can see a space-filling model of one enzyme, called hexokinase. You can also see, buried there in the middle, its substrate, glucose—better known as blood sugar.

As with all proteins, enzymes are made up of amino acids, but only a few of the hundreds of amino acids in an enzyme are typically involved in actually binding with the substrate. Five or six would be common. The place in the enzyme where the action occurs is called the **active site**. There, in the pocket of the enzyme, is where the substrate is both bound and transformed. In some cases, the participants at the active site include, along with amino acids, one or more accessory molecules. One group of these molecules is known as **coenzymes**. If you've ever wondered what vitamins do, participation in enzyme binding is a big part of it. Once we have ingested them, many vitamins are transformed into coenzymes that allow enzymes to function.

A Case in Point: Chymotrypsin

The details of how enzymes carry out their work are complex, and they vary according to enzyme. Let's consider in a very general way, however, the activity of a much-studied enzyme called chymotrypsin.

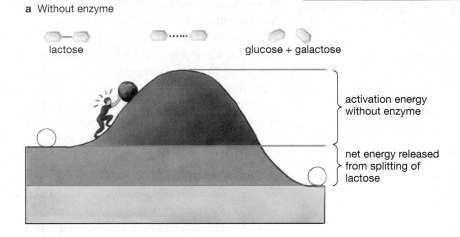

a Without enzyme

lactose   glucose + galactose

activation energy without enzyme

net energy released from splitting of lactose

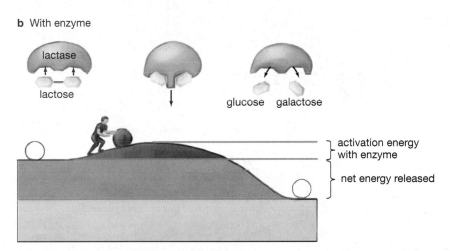

b With enzyme

lactase
lactose

glucose   galactose

activation energy with enzyme

net energy released

Figure 6.7
Enzymes Accelerate Chemical Reactions
How is lactose split into glucose and galactose? Without an enzyme, the amount of energy necessary to activate this reaction is high. In the presence of the enzyme lactase, however, a low activation energy is sufficient to get the process started. The energy released from the splitting of lactose is the same in both cases.

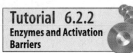

Tutorial 6.2.2
Enzymes and Activation Barriers

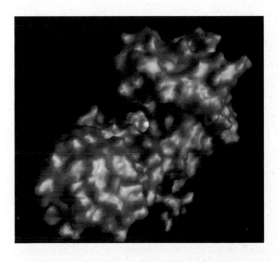

Figure 6.8
Shape Is Important in Enzymes
Substrates generally fit into small grooves or pockets in enzymes. The location at which the enzyme binds the substrate is known as the enzyme's active site. Pictured is a computer-generated model of a glucose molecule (in yellow) binding to the active site of an enzyme called hexokinase (in blue).

Chymotrypsin is delivered from the human pancreas to the small intestine, where it works with water to break down proteins we have ingested. Its function is to clip protein chains in between their building-block amino acids. It does this by breaking the single bond that binds one amino acid to the next (**see Figure 6.9**). After having bound to part of a protein chain, chymotrypsin then interacts with it to create a transition-state molecule. In effect, it distorts the shape of the protein—and holds it in this new shape briefly—in such a way that the protein becomes vulnerable to bonding with ionized water molecules. This allows carbon and nitrogen atoms to latch onto new partners, and

the protein chain is clipped. Chymotrypsin then returns to its original form and proceeds to a new reaction.

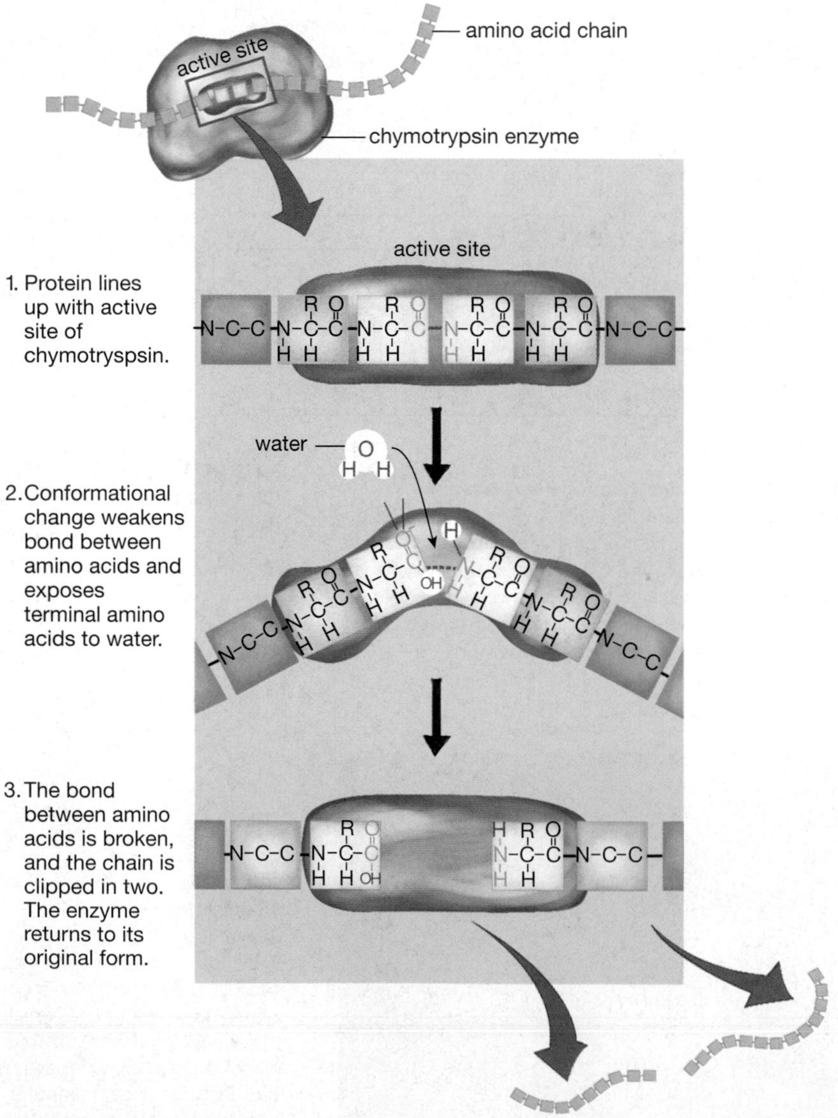

6.7 Regulating Enzymatic Activity

The activity of enzymes is regulated in several ways. The question here is, what factors influence the amount of "product" an enzyme turns out? One of these factors, quite logically, is the amount of substrate in the enzyme's vicinity. If there's no substrate to work on, no product can be turned out.

The work of enzymes can, however, be reduced in several other ways as well. For example, some molecules that are not the normal substrate for an enzyme can still bind to the enzyme at its active site anyway. An enzyme that is "occupied" in this way is not free to bind with its normal substrate and thus is not able to carry out its usual catalytic activity.

Allosteric Regulation of Enzymes

Enzyme activity can also be lowered by binding that occurs at a location different from that of the active site. This binding brings into play the concept of *feedback* in a system.

Most of us have experienced the phenomenon known as **negative feedback** in connection with a common home thermostat. Falling temperature causes the thermostat to kick on the furnace. The *product* of this activity is hot air. When this air has warmed the room enough, the thermostat senses the heat and shuts the furnace down. Meanwhile, the product of an *enzyme's* work is an altered substrate. This chemical product, like heat feeding back to a thermostat, can diminish the activity of that enzyme.

How does this work in real life? Earlier, you saw a model of an enzyme called hexokinase, which works on the substrate glucose. The product of this reaction is something called glucose 6-phosphate. This product is to the enzyme as hot air is to a thermostat: It binds to hexokinase at a site other than the active site, and in so doing changes hexokinase's shape. This change in shape renders hexokinase unable to bind with its substrate, glucose, meaning hexokinase's work is temporarily shut down. If you look at **Figure 6.10**, you can see how this process works schematically.

Figure 6.9
How the Enzyme Chymotrypsin Works
In this example, the enzyme chymotrypsin is facilitating the breakdown of a protein by changing the protein's shape. In the absence of chymotrypsin, this process would take a billion times longer.

This process is known as **allosteric** regulation (*allo* means "other" in Greek, and *steric* means "shape"). The importance of such regulation is that enzymes are not simply fated to turn out product as long as there is substrate to work on. If a cell has too much of a product, allosteric control can cut down on it; too little and its concentration can be increased. The product of a reaction can slow down the reaction itself; isn't that a neat trick?

On to Harvesting Energy from Food

So, to review for a second, you've seen that for animals, energy can come packaged in food—and that the harvesting of this energy begins with food being broken down. The end result of energy harvesting is that the energy currency molecule, ATP, is produced and then used to power various processes. The breakdown of food and the synthesis of ATP occur in individual steps that are catalyzed by enzymes.

In Chapter 26, you'll look at what might be called the upstream part of this process—how we go from eating a potato, say, through digesting it and thus breaking it down into its constituent parts—in this case simple sugars. But for now, you'll be looking at the downstream part of this process: How does the body extract energy from food and use this energy to put together ATP? That is the subject of Chapter 7.

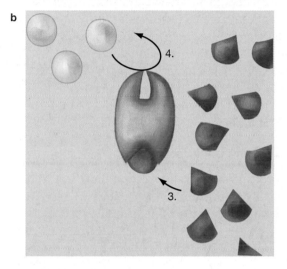

1. Substrate binds to enzyme.

2. Enzyme transforms substrate to product.

3. Product binds to a *different* site on the enzyme, causing the enzyme to change shape.

4. The new shape of the enzyme prevents it from binding to any more substrate.

Figure 6.10
When the Product Slows the Reaction
The frequency of chemical reactions can be controlled by negative feedback.

a In step 1, the substrate (yellow circles) binds to the enzyme. This transforms the substrate, in step 2, into the product (the orange wedges).

b The product now binds to a different site on the enzyme (step 3), causing a conformational change in the enzyme that prevents it from binding with any more substrate (step 4).

Tutorial 6.2.3
Regulating Enzymatic Activity

Chapter Review

Summary

6.1 Energy Is Central to Life

- Living things require a source of energy. The Sun is the ultimate source of energy for most living things on Earth.

6.2 What Is Energy?

- Energy is the capacity to bring about movement against an opposing force.

- Two fundamental principles of energy are the first law of thermodynamics, which states that energy is never created or destroyed, but is only transformed; and the second law of thermodynamics, which states that energy

transfer will always result in a greater amount of disorder in the universe.
TUTORIAL 6.1.1: Laws of Thermodynamics

- In every energy transaction, some energy will be lost to the disordered form of heat. Entropy is a measure of the amount of disorder in a system; the greater the entropy, the greater the disorder.

- Living things can bring about local increases in order (in themselves), through their metabolic processes, but it takes energy to do this. For most living things, the source of this energy ultimately is the Sun. Plants are able to grow by harvesting solar energy, and animals feed on plants.

6.3 How Is Energy Used by Living Things?

- Energy is stored away in endergonic (uphill) reactions, in which the products contain more energy than the reactants. Conversely, energy is released in exergonic (downhill) reactions, in which the reactants contain more energy than the products. The linkage of simple sugars to form a complex carbohydrate is an endergonic reaction, which requires an input of energy. The breakdown of a complex carbohydrate into simple sugars is an exergonic reaction, which releases energy.
  TUTORIAL 6.1.2: Exergonic and Endergonic Reaction

6.4 The Energy Currency Molecule: ATP

- Adenosine triphosphate (ATP) is the most important energy transfer molecule in living things. It is the final product of the energy-harvesting process, in which energy is extracted from food. ATP drives reactions by donating the third of its three phosphate groups to those reactions, in the process becoming the two-phosphate molecule adenosine diphosphate (ADP). The energy supplied by food powers the process by which a third phosphate group is attached to ADP, making it ATP once again.
  TUTORIAL 6.1.3: The ATP Molecule

6.5 Efficient Energy Use in Living Things: Enzymes

- Enzymes are a class of proteins that greatly increase the rate of chemical reactions in an organism. Nearly every chemical process that takes place in living things is facilitated by an enzyme.

- The sum of all the chemical reactions that a cell or larger living thing carries out is its metabolism. Many activities in living things are controlled by metabolic pathways in which a series of interrelated steps is undertaken, each one of them facilitated by an enzyme.

6.6 Lowering the Activation Barrier through Enzymes

- Enzymes lower activation energy, meaning the energy required to initiate a chemical reaction. Enzymes are catalysts: They retain their original chemical composition while bringing about a change in the molecules they bind with, called their substrates.
  TUTORIAL 6.2.2: Enzymes and Activation Barriers

6.7 Regulating Enzymatic Activity

- Enzyme activity can be controlled in several ways. One of these is allosteric regulation, in which the product of an enzyme-controlled reaction binds with the enzyme, thus reducing its activity.
  TUTORIAL 6.2.3: Regulating Enzymatic Activity

Key Terms

Understanding the Basics

Multiple-Choice Questions

1. According to the first law of thermodynamics, energy
   a. is never lost or gained, but is only transformed
   b. always requires an ultimate source such as the Sun
   c. can never be gained, but can be lost
   d. can never really be harnessed
   e. can never be transformed

2. Each time there is a chemical reaction, some energy is exchanged. According to the second law of thermodynamics, with each exchange
   a. Some energy is lost, but other energy is created.
   b. Energy cannot be exchanged in chemical reactions, but must come from the Sun every time.
   c. A system goes from a more-ordered to a less-ordered state, and some energy becomes unavailable because it is lost as heat.
   d. Energy is gained for future use.
   e. Some energy is permanently and completely destroyed.

3. You are running a marathon and need lots of energy to keep your muscles contracting. ATP is broken down to ADP to supply the energy. Thus, the breaking down of ATP to ADP is
   a. caused by a coenzyme
   b. caused by a vitamin
   c. endergonic
   d. exergonic
   e. a, b, and c are all correct

4. ATP stores energy in the form of
   a. mechanical energy
   b. heat
   c. complex carbohydrates
   d. bond energy
   e. amino acids

5. People who are lactose intolerant lack a compound called lactase in their digestive tract. You know that this compound, lactase, is probably a(n)
   a. hormone, because it causes a change in the person
   b. coenzyme, because it helps the enzyme, lactose, do its job

c. vitamin, because it is essential to health

d. substrate, because it is involved with food

e. enzyme, because it ends in -ase.

6. Glycolysis, a process that is discussed in the next chapter, is a metabolic pathway. From this we know that glycolysis

a. consists of a series of steps

b. uses a number of enzymes

c. most likely is a large-scale activity

d. a and c only

e. a, b, and c

7. If you accidentally took a poison that prevented any ATP from being made, you would

a. die

b. have to use ADP for your direct source of energy

c. have to use lipids for your direct source of energy

d. have to use glucose for your direct source of energy

e. have to use enzymes for your direct source of energy

8. Which of the following will lower the energy of activation of a reaction in a cell?

a. lowering the temperature

b. lowering the pressure

c. using an enzyme

d. changing the amount of the reactants

e. supplying ATP

Brief Review

1. A plant leaf is more ordered and complex than are carbon dioxide and water. In photosynthesis plants take small, relatively disordered carbon dioxide and water molecules and make complex, ordered sugars out of them. How is it possible for this to happen without violating the second law of thermodynamics?

2. The text notes that in the sodium-potassium pump, sodium ions are moved out of the cell against their concentration gradient and that ATP plays a part in this. Explain how this is a coupled reaction.

3. Where does the energy come from that is stored and released by ATP? In what way is ATP an energy currency molecule?

4. The average person thinks of enzymes only as substances in their digestive tract that break down food. How would you explain the role of enzymes to your little sister, who has heard about digestion in grade school biology?

5. What is meant by allosteric regulation of an enzyme?

Applying Your Knowledge

1. In light of the laws of thermodynamics, explain why you get so much hotter when you run than when you are sitting still.

2. A box of macaroni, which is made almost entirely of starch, sits on a shelf. If you cook the macaroni and eat it, the starch starts to break down to sugars in the few seconds that the food is in your mouth. What helps to break down the starch? Would the box of macaroni eventually convert to sugar if you left it on the shelf?

7

Vital Harvest
Deriving Energy from Food

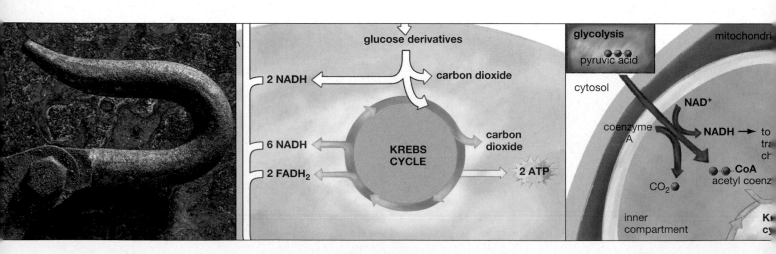

One effect of oxidation: rust.
(Section 7.1, page 133)

One stage in energy harvesting.
(Section 7.3, page 136)

A transition step
in respiration.
(Section 7.5, page 142)

To remain alive, living things must acquire energy and then put that energy into a form they can use.

In May of 1996, the world's attention was riveted on news coming out of Nepal, in Southeast Asia. Disaster had struck several groups of mountain climbers who had been attempting to scale the world's highest peak, Mt. Everest. A vicious storm had taken the members of two Everest expeditions by surprise as they were moving down from Everest's summit, toward the safety of lower altitudes. Within 36 hours, five climbers had lost their lives and another was so badly frostbitten that one of his hands had to be amputated.

One of the expeditions was led by a highly respected guide from New Zealand, Rob Hall. Courageously remaining near the summit in an effort to bring down one of his clients (who had collapsed), Hall ended up spending a night in a howling storm on Everest without a tent. When the Sun rose the next day, the factors working against him included cold and wind, which he might have encountered in lots of inhospitable places on Earth. Also bedeviling him, however, was something that human beings rarely encounter: a lack of oxygen. At his altitude—sitting in the snow at 8,700 meters (or about 28,500 feet)—the oxygen he took in with each breath was little more than a third of the amount he would have inhaled with each breath at sea level. Technology might have come to his aid, as he had two oxygen bottles with him, but his intake valve for them had become clogged with ice.

Down at Everest's lower altitudes, other climbers—some of them Hall's friends—had to endure the agony of talking to him, by two-way radio, while not being able to reach him physically. All of them knew that a lack of oxygen was draining not only his muscles but his mind. Here is Hall in one of his radio transmissions asking about another guide on the ill-fated expedition: "Harold was with me last night, but he doesn't seem to be with me now. Was Harold with me? Can you tell me that?" In the end, the mountain without mercy claimed Rob Hall. He never got up from the place where he spent his night on Everest.

Exhausted Near the Everest Summit
Two members of the ill-fated expeditions that climbed Mount Everest in May 1996 approach the mountain's summit. At this altitude, each breath brings with it only about one-third the amount of oxygen available at sea level.

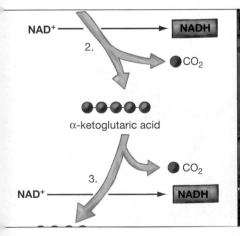

A closer look at
energy harvesting.
(Section 7.5, page 143)

Fruit of the vine.
(Essay, page 139)

Traveling electrons.
(Section 7.6, page 144)

The tragedy of the 1996 Everest expeditions drives home a point and raises a paradox: No one needs to be reminded that we need to breathe in order to live, and most people are aware that oxygen is the most important thing that comes in with breath. That said, of the next 100 people you meet, how many could tell you what oxygen is *doing* to sustain life? Put another way, why do we need to breathe?

The short answer is that breathing and oxygen are in the energy transfer business. They are part of a system that allows us to extract, from food, energy that is then used to put together the "energy currency" molecule, ATP. As you'll recall from Chapter 6, the body uses ATP (adenosine triphosphate) to power activities that range from muscle contraction to thinking to cell repair. Living things need large amounts of ATP to live, and organisms such as ourselves use oxygen to produce much of our ATP. If we don't get enough oxygen to keep energy transfer moving, our bodies and minds start failing.

Oxygen is not required for all the "harvesting" we do of the energy contained in food. When you lift a box, the short burst of energy that is required comes largely from a different sort of energy harvesting, which you'll learn more about shortly. But even here, the essential *product* of energy extraction is ATP. Keep that in mind and you won't get lost in the byways of energy transfer. How does the body produce enough ATP to allow us to read a book or climb a mountain? In this chapter, you'll find out.

7.1 Energizing ATP: Adding a Phosphate Group to ADP

Given the importance of ATP to this story, let us begin with a brief review of how this molecule works. Recall from Chapter 6 that each ATP molecule has three phosphate groups attached to it, and that ATP powers a given reaction by losing the outermost of these phosphate groups. In this process ATP is transformed into a molecule with *two* phosphate groups, adenosine diphosphate or ADP. To return to its more "energized" ATP state, it must have a third phosphate group attached again. This is, however, a trip *up* the energy hill, because the product (ATP) contains more energy than the reactant (ADP). Thus, getting that third phosphate group onto ADP requires energy. It is like preparing a spring-loaded mousetrap for action. It takes energy to pull a mousetrap bar back, and the same is true of putting a third phosphate group onto ATP (**see Figure 7.1**). Where does the energy come from? For animals such as ourselves, it comes from food; energy that is extracted from food powers the phosphate group up the energy hill—and literally onto ADP.

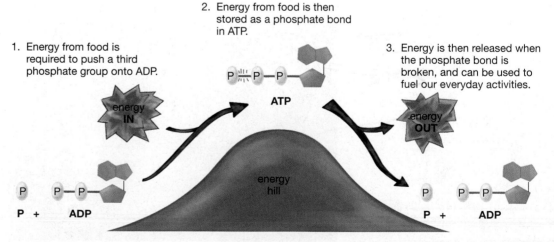

1. Energy from food is required to push a third phosphate group onto ADP.

2. Energy from food is then stored as a phosphate bond in ATP.

3. Energy is then released when the phosphate bond is broken, and can be used to fuel our everyday activities.

energy IN

ATP

energy OUT

energy hill

P + ADP

P + ADP

Figure 7.1
Storing and Releasing Energy
Adenosine triphosphate (ATP) is the most important energy-releasing molecule in our bodies. The energy it contains is used to power everything from muscle contraction to thinking.

Tutorial 7.1.2
Oxidation-reduction Reactions

7.2   Electrons Fall Down the Energy Hill to Drive the Uphill Production of ATP

In tracking the extraction of energy from food, we will use as an example one particular molecule, glucose, to see how energy is harvested from it. Though the details here are complex, the essential story is simple. *Electrons* derived from glucose, which is high in energy, will be running downhill; they will be channeled off, a few at a time, and their downhill drop will power the *uphill* push needed to attach a phosphate group onto ADP. The glucose-derived electrons will be transferred to several intermediate molecules in their downhill journey, but the *final* molecule that will receive them at the bottom of the energy hill is oxygen. We need to breathe because we need oxygen to serve as this final electron acceptor.

As you are following this process, you may well wonder why so many steps are required to transfer energy from food to ATP. Why isn't there just *one* step in energy transfer—food to ATP? It turns out that the gradual transfer of energy, through many steps, allows the body to make the most of the energy it receives. Think of it this way. You could have water drop 1,000 feet straight off a cliff onto a waterwheel below. In this case, the wheel would simply be pulverized by the water crashing onto it. In the process, the potential energy that had been contained in the water at the top of the cliff would be transformed almost entirely into useless heat. Conversely, you could have a stream that ran briskly downhill for 1,000 vertical feet with waterwheels spaced periodically to channel the energy off a little at a time, thereby conserving the energy for work. The steps you are going to read about amount to this kind of controlled channeling off of energy—the energy that's contained in food.

The Great Energy Conveyors: Redox Reactions

The basis for electron transfers down the energy hill is straightforward: Some substances more strongly attract electrons than do others. A substance that loses one or more electrons to another is said to have been **oxidized**. We hear this word all the time—when paint on an outdoor surface has become dulled, for example, or when metal rusts (**see Figure 7.2**). Meanwhile, the substance that *gains* electrons in this reaction is said to have undergone **reduction**. (This seems about as logical as saying a country lost a war by winning. One way to think about it is that, because electrons carry a negative charge, any substance that gains electrons has had a reduction in its *positive charge*.)

In cells, oxidation and reduction never occur independently. If one substance is oxidized, another must be reduced. It's like a teeter-totter; if one side goes down, the other must come up. The combined operation is known as a reduction-oxidation reaction, or simply a **redox reaction**. Critically, the substance being oxidized in a redox reaction has its electrons traveling energetically *downhill*.

Many Molecules Can Oxidize Other Molecules

The term *oxidation* might give you the idea that *oxygen* must be involved in any redox reaction, but this is not the case. Any compound that serves to accept electrons from another is a so-called oxidizing agent. In living things a large number of molecules are involved in energy transfer, and each has a certain tendency to gain or lose electrons relative to the others.

Figure 7.2
An Effect of Oxidation
Rust has formed on a hook atop a corroded steel sheet.

This is how electrons can be passed down the energy hill: The starting "energetic" molecule of glucose is oxidized by another molecule, which in turn is oxidized by the *next* molecule down the hill. The whole thing might be thought of as a kind of downhill electron bucket brigade. Molecules that serve to shuttle electrons down the energy hill are known as **electron carriers**. The thing that makes their role a little complicated is that many of the electrons they accept are bound up originally in hydrogen atoms. If you can remember back to Chapter 3, a hydrogen atom amounts to one proton and one electron. In transferring a hydrogen atom, then, a molecule is transferring a single electron (bound to a proton), which means a redox reaction has taken place.

Redox through Intermediates: NAD

The most important electron carrier is a molecule known as **nicotinamide adenine dinucleotide**, or **NAD**, which can helpfully be thought of as a city cab. It can exist in two states: loaded with passengers or empty. And like a cab, it can switch very easily between those two states. The passengers that NAD picks up and drops off are electrons.

The "empty" state that NAD comes in is ionic: NAD^+. Remembering back to the discussion about ions, you can see that NAD^+ is positively charged, meaning that it has lost an electron. In a redox reaction, what NAD^+ does is pick up one hydrogen atom (an electron and a proton) and one solo electron

(from a second hydrogen atom). The isolated electron that NAD^+ picks up turns it from positively charged to neutral ($NAD^+ \rightarrow$ NAD); the whole hydrogen atom takes it from NAD to NADH. Keeping an eye on redox reactions here, NAD^+ has become NADH by oxidizing a substance—by accepting electrons from it.

So much for half of NAD's role: picking up passengers. *Now*, as NADH, it is loaded with these passengers. It can proceed down the energy hill to donate them to molecules that have a greater potential to *accept* electrons than it does. Having dropped its passengers off with such a molecule, it returns to being the empty NAD^+ and is ready for another pickup (**see Figure 7.3**). Through this process, NADH transfers energy from one molecule to another.

How Does NAD Do Its Job?

The molecule that NAD^+ is oxidizing is the starting glucose molecule (actually derivatives of it). To carry out its oxidizing role, NAD^+ obviously needs to be brought together with the glucose derivatives. What have you been looking at that brings substances together in this way? Enzymes! One of the things you'll be seeing, therefore, is NAD^+ and glucose derivatives being brought together by enzymes, at which point NAD^+ accepts electrons from these derivatives, later to hand them off to another molecule. Electron transfer through intermediate molecules such as NAD^+ provides the energy for

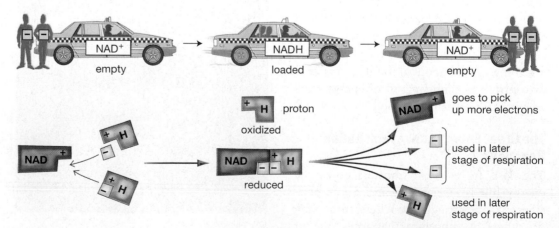

Figure 7.3
The Electron Carrier NAD$^+$
In its unloaded form (NAD$^+$) and its loaded form (NADH), this molecule is a critical player in energy transfer, picking up energetic electrons from food and transferring them to later stages of respiration.

1. NAD$^+$ within a cell, along with two hydrogen atoms that are part of the food that is supplying energy for the body.

2. NAD$^+$ is reduced to NAD by accepting an electron from a hydrogen atom. It also picks up another hydrogen atom to become NADH.

3. NADH carries the electrons to a later stage of respiration then drops them off, becoming oxidized to its original form, NAD$^+$.

most of the ATP produced. Thus, when looking at the diagrams in this chapter that outline respiration, you'll see:

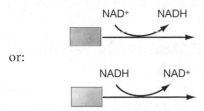

or:

What these drawings indicate is the electron carrier shifting between its empty state (NAD$^+$) and its loaded state (NADH), or vice versa.

7.3 The Three Stages of Cellular Respiration: Glycolysis, the Krebs Cycle, and the Electron Transport Chain

All the energy players are finally in place: glucose, redox reactions, electron carriers, enzymes, and the rest. In terms of a molecular formula, the big picture on respiration looks like this:

$$C_6H_{12}O_6 + 6O_2 + ADP \rightarrow 6CO_2 + 6H_2O + ATP$$

The starting molecule on the left ($C_6H_{12}O_6$) is glucose, which is storing chemical energy. You can also see on the left the oxygen ($6O_2$) that is needed as the final electron acceptor, and ADP. The *products* of this reaction, to the right of the arrow, are carbon dioxide ($6CO_2$) and water ($6H_2O$) as by-products, and energy in the form of ATP. How *much* ATP does the breakdown of one molecule of glucose yield? A maximum of about 36 molecules of ATP. Though many separate steps are involved in actually getting energy from glucose to ATP, respiration can be divided into three *sets* of steps. These are known as *glycolysis*, the *Krebs cycle*, and the *electron transport chain*.

Glycolysis Is the First Stage in Energy Harvesting

The division among these respiratory steps is, in one sense, a division of evolutionary history. For most eukaryotic creatures—ourselves, birds, trees—the first of these sets of steps, glycolysis, could be likened to a nineteenth-century steam engine working side by side with a state-of-the art electric turbine. Glycolysis does produce ATP, but

its net production is only two ATP molecules per glucose molecule, which is pretty small in comparison to the 36 ATP molecules likely to come from all three sets of steps. Human beings make use of the energy yielded in glycolysis, but we could not function for 3 minutes if it were our *sole* source of energy. The real import of glycolysis for us is that it's critical in supplying energy to us in certain situations—in a short burst of activity, for example—and it is a necessary precursor to the other stages of respiration. Meanwhile, for certain bacteria and one-celled eukaryotes, glycolysis can be the sole source of energy transfer. Glycolysis does not use oxygen directly. Many of the organisms relying on it as their sole means of energy transfer live in oxygenless environments.

The Krebs Cycle and the Electron Transport Chain Were Later in Evolving and Are More Efficient

In this division of who uses what, you can probably begin to guess where glycolysis fits in. Scientists believe that it is a more ancient form of energy harvesting than the other two sets of steps. This is why it exists in all organisms. However, at some time in the evolutionary past, processes we now call the Krebs cycle and the electron transport chain were added. The electron transport chain requires one critical thing that glycolysis doesn't: the ability to use oxygen. Because the products of glycolysis feed into the Krebs cycle and the oxygen-using electron transport chain, the entire three-stage process is referred to as oxygen-dependent or *aerobic* energy transfer. But, as noted, glycolysis can take place in the absence of oxygen.

The separation between glycolysis and the other two steps is also a physical separation, in that glycolysis takes place in the area of the cell outside the nucleus—the cytoplasm—while, for eukaryotes, the Krebs cycle and the electron transport chain take place within cell structures noted earlier—the mitochondria—that are immersed in the cytosol.

An Overview of the Three Stages

In overview, energy harvesting through these three phases goes like this: We get some ATP yield in glycolysis and the Krebs cycle, but for eukaryotes such as ourselves the *main* function of these phases is the transfer of electrons to the electron carriers such as NAD⁺. Bearing their electron cargo, these carriers then move to the electron transport chain, where they are themselves oxidized (losing electrons). The movement of these electrons through the chain then yields about 32 of the approximately 36 molecules of ATP that are netted per molecule of glucose. If you look at **Figure 7.4**, you can see two representations of the three stages of energy extraction. Taken as a whole, this three-stage harvesting of energy is known as **cellular respiration**. Now let's see how the three phases of this process work in detail.

7.4 First Stage of Respiration: Glycolysis

Glycolysis, the first stage of aerobic energy harvesting, means "sugar splitting," which is appropriate because our starting molecule is the simple sugar glucose. Briefly, here is what

a

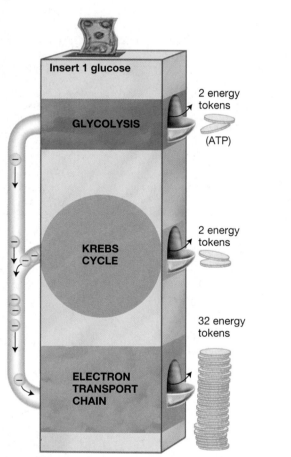

b

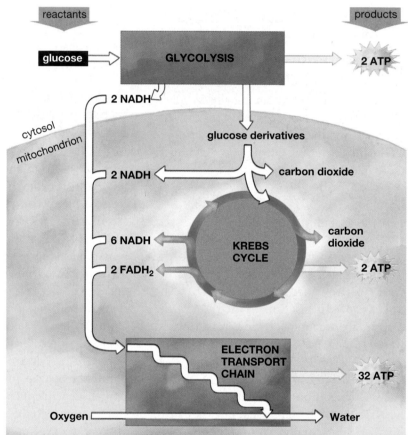

36 ATP maximum
per glucose molecule

Figure 7.4
Overview of Energy Harvesting

a In metaphorical terms

Just as the video games in some arcades can use only tokens (rather than money) to make them function, so our bodies can use only ATP (rather than food) as a direct source of energy. The energy contained in food—glucose in this example—is transferred to ATP in three major steps: glycolysis, the Krebs cycle, and the electron transport chain. Though glycolysis and the Krebs cycle contribute only small amounts of ATP directly, they also contribute electrons (on the left of the token machine) that help bring about the large yield of ATP in the electron transport chain. Our energy-transfer mechanisms are not quite as efficient as the arcade machine makes them appear. At each stage of the conversion process, some of the original energy contained in the glucose is lost to heat.

b In schematic terms

As with the arcade machine, the starting point in this example is a single molecule of glucose, which again yields ATP in three major sets of steps: glycolysis, the Krebs cycle, and the electron transport chain (ETC). These steps can yield a maximum of about 36 molecules of ATP: 2 in gylcolysis, 2 in the Krebs cycle, and 32 in the ETC. As noted, however, glycolysis and the Krebs cycle also yield electrons that move to the ETC, aiding in its ATP production. These electrons get to the ETC via the electron carriers NADH and FADH₂, shown on the left. Oxygen is consumed in energy harvesting, while water and carbon dioxide are produced in it. Glycolysis takes place in the cytoplasm of the cell, but the Krebs cycle and ETC take place in the cellular organelles called mitochondria.

---

happens during glycolysis. First, this one molecule of glucose has to be prepared, in a sense, for energy release. ATP actually has to be *used*, rather than synthesized, in two of the first steps of glycolysis, so that the relatively stable glucose can be put into the form of a less-stable sugar. This sugar is then split in half. The two molecules formed have three carbons each (whereas the starting glucose has six). Once this split is accomplished, the steps of glycolysis take place in *duplicate*: It is like taking one long piece of cloth, dividing it in two, and then proceeding to do the same things to both of the resulting pieces. Look now at **Figure 7.5** to see how glycolysis proceeds.

SUMMARY OF GLYCOLYSIS

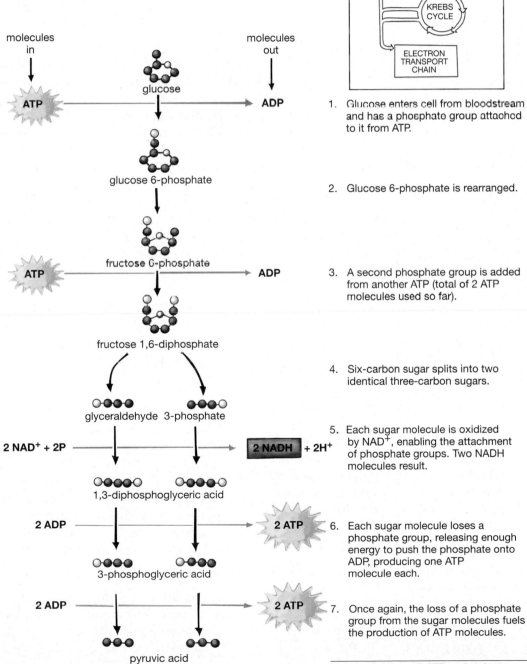

Figure 7.5
Summary of Glycolysis
In glycolysis, the single glucose molecule is transformed in a series of steps into two molecules of a substance called pyruvic acid. These two molecules then move on to the next stage of cellular respiration (the Krebs cycle). Meanwhile, glycolysis also produces two molecules of electron-carrying NADH, which move directly to the electron transport chain. Although two molecules of ATP are used up in the earlier stages of glycolysis, four more are produced in the later stages, for a net production of two ATP molecules per glucose molecule. The carbon atoms are represented by red circles, and the phosphate groups are represented by yellow circles.

1. Glucose enters cell from bloodstream and has a phosphate group attached to it from ATP.

2. Glucose 6-phosphate is rearranged.

3. A second phosphate group is added from another ATP (total of 2 ATP molecules used so far).

4. Six-carbon sugar splits into two identical three-carbon sugars.

5. Each sugar molecule is oxidized by NAD$^+$, enabling the attachment of phosphate groups. Two NADH molecules result.

6. Each sugar molecule loses a phosphate group, releasing enough energy to push the phosphate onto ADP, producing one ATP molecule each.

7. Once again, the loss of a phosphate group from the sugar molecules fuels the production of ATP molecules.

Net production: **2 ATP** molecules produced for each 1 glucose molecule used

When Energy Harvesting Ends at Glycolysis, Beer Can Be the Result

For many bacteria—and even sometimes for people,—energy harvesting ends with the set of steps called glycolysis, rather than proceeding on through the Krebs cycle and electron transport chain. When this happens, organisms need a way to recycle their energy transfer molecules in a way that doesn't depend on oxygen. (The NADH they produced in glycolysis needs to lose its added electrons and become NAD$^+$ again, so it can be reused in glycolysis.) The solution to this dilemma is a kind of electron dumping, the *products* of which turn out to be some of the most familiar substances in the world.

Fermentation in Yeast

Yeast is a good example of how this works, because it is a busy single-celled fungus. Yeasts can live by glycolysis alone, or they can go through aerobic respiration. Say, then, that yeasts are working away on sugar, going through glycolysis, but doing so in an oxygenless environment. Recall that the final "substrate" product of glycolysis is two molecules of a substance called pyruvic acid. In yeasts, the pyruvic acid they end up with is converted to a molecule called acetaldehyde, and *it* takes on the electrons

from NADH, meaning the recycling problem is now solved. Having taken on NADH's electrons, though, acetaldehyde now is converted to something very familiar: ethanol, better known as drinking alcohol.

For yeasts, alcohol is simply a by-product of the fermentation they carry out.

Human beings *put* yeasts in environments in which this will happen, of course, because this is how we make wine and beer. Imagine yeast cells inside a dark, airless wine cask, working away on burgundy grape juice, harvesting energy from the grape-juice sugars, but in so doing turning out pyruvic acid, which is turned into alcohol. Things are going along fine for these yeast cells, until the alcohol content of the wine reaches a level (about 14 percent) at which they can no longer survive in it. This whole process is known as **alcoholic fermentation**, the process by which yeasts produce alcohol as a by-product of glycolysis they perform in an oxygenless environment.

Here are the steps of glycolysis in detail.

The Steps of Glycolysis in Human Beings

1. Delivered by the bloodstream, glucose enters the cell and immediately has a phosphate group from ATP attached to it. This process, called phosphorylation, is facilitated by hexokinase, an enzyme introduced in Chapter 6. Because the phosphate is attached to the sixth carbon of glucose, it now goes under the name glucose 6-phosphate. Note that one molecule of ATP has been *used* in this step. The ATP ledger thus reads: minus-one.

2. Glucose 6-phosphate is rearranged to become a molecule called fructose 6-phosphate.

3. Fructose 6-phosphate has a second phosphate group added, so that it becomes a molecule called fructose 1,6-diphosphate. Another molecule of ATP has been used in getting this arrangement,

however, which makes the ATP ledger now read: minus-two. But fructose 1,6-diphosphate now has phosphate groups on each end, which is important because it is about to be split.

4. Fructose 1,6-diphosphate becomes two molecules of glyceraldehyde 3-phosphate. We started with a six-carbon sugar, glucose. We now have two separate three-carbon sugars, each with phosphate groups attached. From here on out, glycolysis begins to proceed in duplicate. What happens to one of the glyceraldehyde molecules happens to the other.

5. At last, some energy from glucose. An enzyme brings together the glyceraldehyde 3-phosphate molecule, NAD$^+$, and a phosphate group. The outcome of this is that NAD$^+$ oxidizes glyceraldehyde 3-phosphate. Now NAD$^+$ in its new form, NADH, moves down the energy hill, occupied with one new hydrogen atom

For yeasts, alcohol is simply a waste by-product of the glycolysis they utilize in *anaerobic* energy conversion, meaning conversion in an oxygenless environment. The same is true of the carbon dioxide that also results from their conversion of pyruvic acid. Here again, humans have found great use for the refuse of nature. Put the right yeast into dough, and its continuing fermentation produces the CO_2 that causes bread to rise and become "light," because of the air holes now in it. (And the alcohol? It evaporates.)

Fermentation in Animals

Alcoholic fermentation is the solution that fungi (and occasionally plants) employ to sustain glycolysis in the absence of oxygen. But animals take a different tack, because in them the product of glycolysis, pyruvic acid, accepts the electrons from NADH. In this process, pyruvic acid is turned into another substance: **lactic acid**. Thus, the kind of fermentation that animals (and certain bacteria) carry out is called **lactate fermentation**. Have you ever experienced the muscle "burn" that comes with, for example, climbing several flights of stairs? If so, you have experienced a buildup of lactic acid in your muscles. Human beings cannot function in oxygenless environments, of course,

but we do rely on glycolysis and lactate fermentation when we need a quick burst of energy. In such a situation, our relatively slow oxygen-using system—glycolysis, the Krebs cycle, and electron transport chain—just doesn't have time to respond.

Yeast Inside, Performing Fermentation
This man is inspecting a gauge on a huge metal tank of fermenting wine. The alcohol in the wine is a by-product of the glycolysis carried out by yeast in an oxygenless environment.

(a proton plus an electron) and one new solo electron. The oxidation it has carried out is energetic enough that it allows the phosphate group to become attached to the main molecule. Now called 1,3-diphosphoglyceric acid, it has two phosphates attached to it. Because everything in this step is happening in duplicate, this yields two NADH molecules.

6. ATP at last. Energy-enriched as it was, 1,3-diphosphoglyceric acid loses one of its phosphate groups downhill, thus becoming 3-phosphoglyceric acid. The fall is energetic enough to push this phosphate onto an ADP molecule and make it ATP. Because this is happening in duplicate, two molecules of ATP are produced. Recall that two ATP were used in steps 1 and 3, however, so that the ATP ledger now stands at zero.

7. Through a couple of reactions, 3-phosphoglyceric acid becomes a molecule called phosphoenolpyruvic acid, which generates more ATP when it transfers its phosphate group to ADP. As this takes place in duplicate, two more ATP molecules are created, which leaves the ATP ledger, at the end of glycolysis, at plus-two. The reaction turns phosphoenolpyruvic acid into pyruvic acid, the final form of the original glucose molecule in glycolysis.

Before continuing on to the other stages of energy harvesting, you may wish to read more about what might be called the consequences of glycolysis. For some organisms, this is the end of the harvesting line, and even oxygen-using organisms such as ourselves rely on it more heavily for energy transfer in certain situations. You can read about this in "When Energy Harvesting Ends at Glycolysis," above. Glycolysis fits into a special place in providing energy for human exercise; you can read about this in "Energy and Exercise" on page 140.

Energy and Exercise

One minute we're sleeping, the next we're looking for our running shoes, and 10 minutes after that we're bounding down the road trying to cover three miles at a faster pace than yesterday. Coming to the steep hill at the end of our run, we go all out—charging up to the top of it, slowing down only after we start going downhill, breathless and spent. How can the human body cope with a range of energy demands that runs from sleeping to sprinting? Its secret is having a kind of energy trio at its service, with each member of the group specializing in delivering energy in particular situations.

The essence of any kind of exercise is the contraction of muscles, and the only energy molecule that can power the contraction of skeletal muscle is ATP. You have seen that the human body uses a three-stage process to produce ATP: glycolysis, the Krebs cycle, and the electron transport chain. As a whole, this system is dependent on oxygen, but glycolysis is not directly dependent on it. For brief periods of time we can produce a good deal of the ATP we need through a glycolysis that is de-coupled from the Krebs cycle and the electron transport chain, though there is a price to be paid for this, which is the buildup of lactic acid in our muscles (see "When Energy Harvesting Ends at Glycolysis" on page 138). Oxygen-independent glycolysis can thus be thought of as the first member of the body's energy trio, while *aerobic* or oxygen-using energy transfer can be thought of as the second member. Our cells also have a capacity to *store* small amounts of ATP, however, and they can also stockpile another molecule, phosphocreatine (PCr), that acts as a reservoir of phosphate groups that can be used to produce ATP. This stored ATP/PCr is the third member of the energy ensemble. Let's now see how all three players would work together in the course of a bike ride, starting out at a fairly brisk pace.

With the first turns of the pedals, bike riders bump up their body's energy demands tremendously. In their cycling, they could easily burn up 10 times the calories they would while in a resting state. (As a point of comparison, in doing housework a person typically burns up about three times the calories he or she does while at rest.)

So, how do muscle cells accommodate such large increases in their need for ATP? In the first few seconds, the small reservoirs of ATP/PCr take the lead role, with a proportionately smaller contribution from glycolysis and a smaller contribution yet from aerobic metabolism. These differences reflect the time it takes for each of these systems to get going. You've seen how many steps there are in glycolysis, and how many *more* steps there are in the Krebs cycle and electron transport chain. By comparison, the output of ATP from stored ATP/PCr is instantaneous. Remember, though, that the ATP/PCr reservoir is small. A person who had to depend solely on it for a full-ahead sprint would have only enough energy to last about six seconds. On a longer bike ride, glycolysis and ATP/PCr are contributing roughly equal amounts of ATP within 10 seconds. Within 30 seconds, glycolysis has greatly eclipsed PCr as an energy provider.

At about this same 30-second mark, the third player, aerobic respiration, is supplying only about 30 percent of the ATP, but its contribution is growing fast. In **Figure 1**,

7.5 Second Stage of Respiration: The Krebs Cycle

Glycolysis has accomplished three valuable things in energy harvesting: It yielded two ATP molecules, it yielded two energized molecules of NADH, and it resulted in two molecules of pyruvic acid. These pyruvic acid molecules are the derivatives of the original glucose molecule. Now *they* are the molecules that will be oxidized—that will have electrons removed from them—in the next stage of energy harvesting. They serve this purpose because glycolysis by itself could not wring all the energy out of the starting glucose molecule. (After all, it yielded only two of the 36 molecules of ATP that eventually are harvested.)

After glycolysis, the second stage of respiration is called the **Krebs cycle**. This stage is named for the German and English biochemist Hans Krebs, who in the 1930s used

Duration of maximal exercise								
Seconds			Minutes					
10	30	60	2	4	10	30	60	120
Percent anaerobic 90	80	70	50	35	15	5	2	1
Percent aerobic 10	20	30	50	65	85	95	98	99

Figure 1
Different Contributions over Time
Relative contributions of anaerobic and aerobic respiration to exercise during the duration of a workout. "Anaerobic" here means a combination of glycolysis and stored ATP/PCr release. At the 1-minute mark, aerobic respiration is supplying only 30 percent of the body's energy needs; at the 10-minute mark, it is supplying 85 percent. (Adapted from Astrand, P. O., and Rodahl, K. *Textbook of Work Physiology*. New York: McGraw Hill Book Company, © 1977.)

you can see how the aerobic contribution to ATP yield grows over the course of a long, strenuous workout. At the start, oxygen-dependent respiration is supplying less than 10 percent of our ATP, but 10 minutes into the workout it is delivering 85 percent.

You might think that this would be the final division of ATP contributions among the three players, but the amounts they supply can vary over time depending on the intensity of the activity. **Figure 2** shows how this would work in an actual professional-level bicycle race. Going along at a steady state in "pack" riding, you can see the overriding contribution of aerobic metabolism. In the final sprint, however, stores of ATP/PCr—replenished

during the pack riding—are brought out to take the leading role, with the contribution of glycolysis increasing as well. In stretches of difficult hill climbing, meanwhile, glycolysis predominates.

Of course, lots of sports have no aerobic component to them at all. The energy for a sport like football is supplied almost entirely by stored ATP/PCr. Even in a competition as long as a 400-meter sprint—about once around a regulation track—about 70 percent of the energy comes from a combination of ATP/PCr and glycolysis.

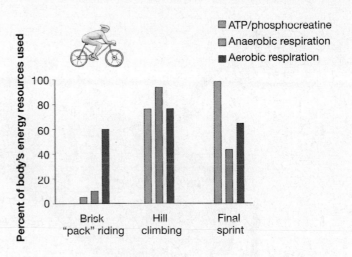

Figure 2
Different Activities, Different Energy Stores
Energy harvesting processes used at different stages in a bicycle race. (Adapted from Kearney, J. T. "Training the Olympic Athletes." *Scientific American*, June 1996, p. 54.)

the flight muscles of pigeons to find out how aerobic respiration works. Because the first product of the Krebs cycle is citric acid, the cycle is sometimes referred to as the **citric acid cycle**.

The ATP yield in this cycle is once again a paltry two molecules. The big harvest of ATP comes in the *final* stage of respiration, the electron transport chain. The Krebs cycle serves, however, to set the stage for this

harvest by supplying energy-bearing electron carriers, as you will see.

Site of Action Moves from the Cytoplasm to the Mitochondria

Glycolysis yielded its two molecules of pyruvic acid in the cytoplasm, but the site of energy harvesting quickly shifts, with the pyruvic acid, to the new site of energy transfer, the mitochondria. These organelles

Figure 7.6
Energy Transfer in the Mitochondria
Mitochondria are organelles, or "tiny organs," that exist within cells. They are the location for the second and third sets of steps in cellular respiration, the Krebs cycle and the electron transport chain. Following a transitional step (see Figure 7.7), the products of glycolysis—the downstream products of the original glucose molecule—pass into the inner compartment of a mitochondrion, where the Krebs cycle takes place. Electrons derived from the Krebs cycle then migrate, via electron carriers, from the Krebs cycle site into the highly folded inner membrane of the mitochondrion, where the bulk of ATP is produced in the electron transport chain.

are where all the action takes place from here on out, so now would be a good time to look at **Figure 7.6** and examine their structure. You can see that mitochondria have both an inner and an outer membrane. The Krebs cycle reactions take place to the interior of the inner membrane—in an area known as the inner compartment—while the reactions of the electron transport chain take place *within* the inner membrane.

Between Glycolysis and the Krebs Cycle, an Intermediate Step

There actually is a transition step in respiration between glycolysis and the Krebs cycle. In it, the three-carbon pyruvic acid molecule combines with a substance called coenzyme A, thus forming acetyl coenzyme A (acetyl CoA). One outcome of this reaction is that acetyl CoA

enters the Krebs cycle. There are, however, two other products of this reaction. One is a carbon dioxide molecule that, like others you'll be seeing, eventually diffuses out of the cell and into the bloodstream. This is one source of the CO_2 that we exhale when we breathe. The other product is one more molecule of NADH, which finds its way to the electron transport chain. This reaction takes place twice for each glucose molecule (**see Figure 7.7**).

Into the Krebs Cycle: Why Is It a Cycle?

Acetyl CoA now enters the Krebs cycle. The reason this is a *cycle* becomes clear when you look at **Figure 7.8**. You can see that, in the first step of the cycle, acetyl CoA combines with a substance called oxaloacetic acid to produce citric acid. If you look over at about 11 o'clock on the circle, however, you can see that the *last* step in the

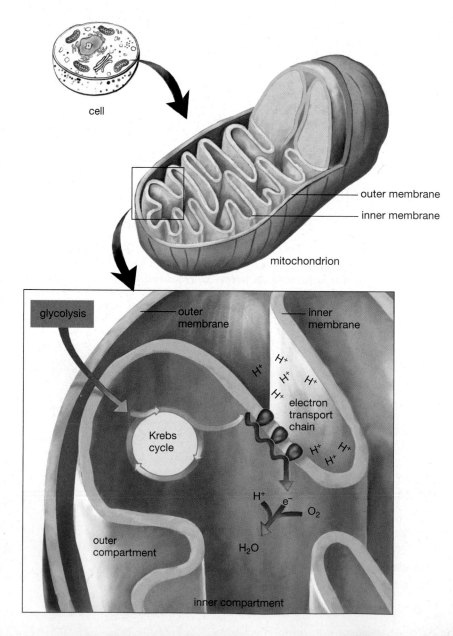

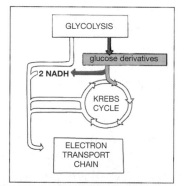

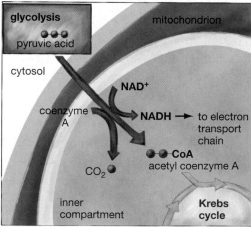

Figure 7.7
Transition between Glycolysis and the Krebs Cycle
The pyruvic acid product of glycolysis does not enter directly into the Krebs cycle. Rather, it must first be transformed into acetyl coenzyme A. The consequences of this reaction are the production of CO_2, which dissolves into the bloodstream, and the production of an NADH molecule, which continues onto the electron transport chain. Because one molecule of glucose produces two molecules of pyruvic acid, two molecules of NADH are produced per glucose molecule in this transitional step.

cycle is the *synthesis* of oxaloacetic acid. Thus, a substance that's necessary for this chain of events to take place is itself a product of the chain.

Stroll around the circle now to see how the Krebs cycle functions in a little more detail. In essence, what's happening is that, as the entering acetyl CoA molecule is transformed into these various molecules, it is being oxidized by electron carriers, with the resulting electrons moving on to the electron transport chain. ATP is also derived, and CO_2 is a product. The only major player that's unfamiliar here is the FAD-$FADH_2$ that can be seen taking part in a redox reaction at about 8 o'clock. It is simply another electron carrier, similar to NAD^+/NADH.

In counting up the Krebs cycle's yield of both ATP and electron carriers (NADHs and so on), recall that *two turns* around this cycle result from the original glucose molecule (because glucose provided two molecules of the starting pyruvic acid). This means a total yield of 6 NADH, 2 $FADH_2$, and 2 ATP from Krebs for each glucose molecule.

Here are the steps of the Krebs cycle in detail.

The Steps of the Krebs Cycle

1. Acetyl CoA combines with the four-carbon oxaloacetic acid. The CoA fragment separates from this compound; the six-carbon citric acid that results is an energetic substrate that will now be oxidized.

2. A citric acid derivative is oxidized in the $NAD^+ \rightarrow$ NADH reaction, with NADH going to the ETC. An intermediate molecule then loses a CO_2 molecule. Citric acid is now alpha-ketoglutaric acid.

SUMMARY OF THE KREBS CYCLE

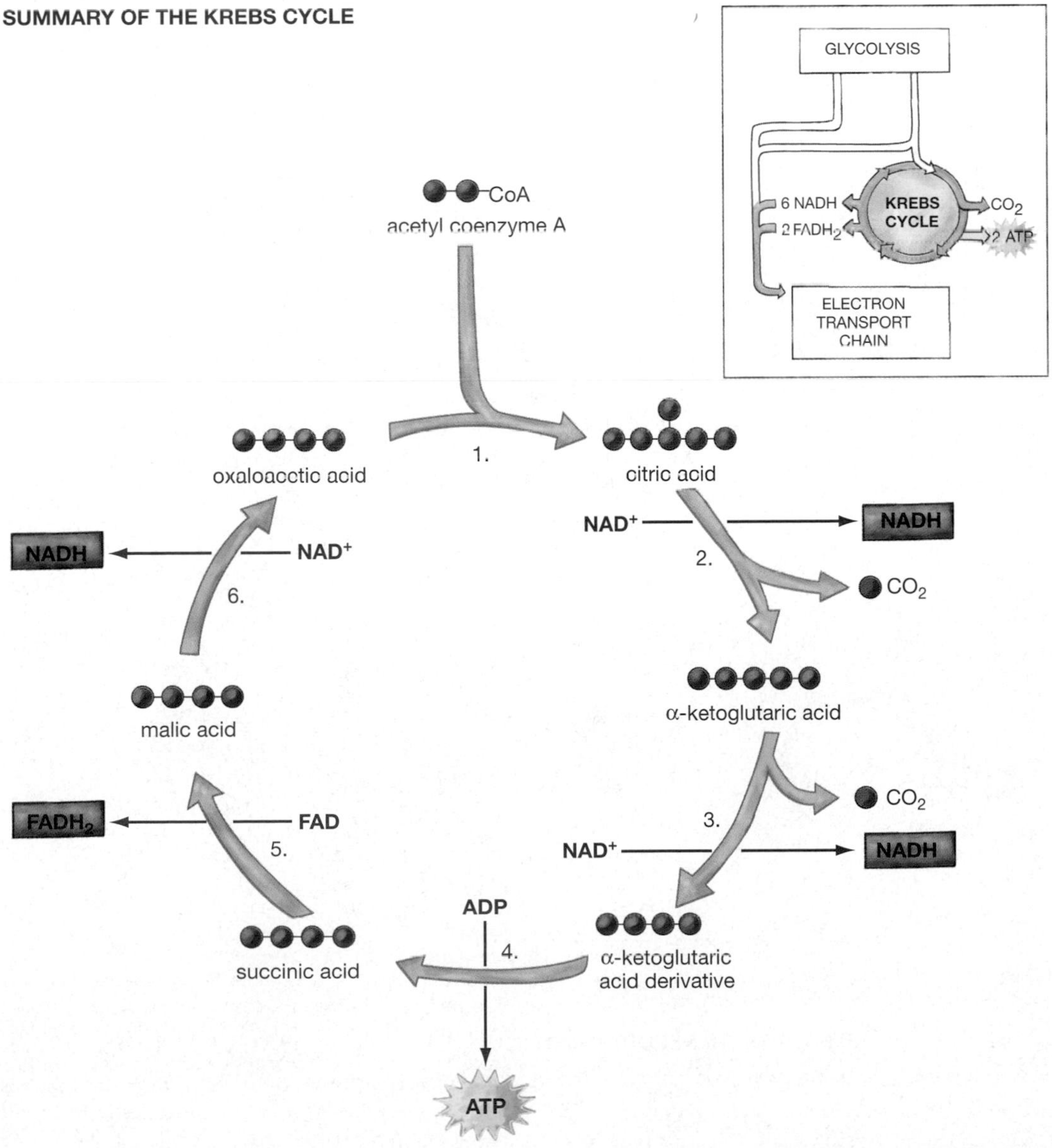

Tutorial 7.2.4
Krebs Cycle

Figure 7.8
Summary of the Krebs Cycle
The Krebs cycle is the major source of electrons that are transported to the electron transport chain by the electron carriers NADH and $FADH_2$. For each molecule of glucose, two molecules of acetyl coenzyme A enter the Krebs cycle. Through a series of reactions, a total of 6 NADH, 2 $FADH_2$, and 2 ATP are produced per glucose molecule. (From counting the number of NADH and $FADH_2$ around the cycle, it would appear that only 3 NADH and 1 $FADH_2$ are produced, but remember that one molecule of glucose results in two "trips" around the cycle, as two molecules of acetyl coenzyme A will enter the Krebs cycle for every molecule of glucose that is metabolized.)

3. Alpha-ketoglutaric acid loses a CO_2 molecule, and the resulting four-carbon molecule is oxidized by NAD^+.

4. A derivative of alpha-ketoglutaric acid is split, releasing enough energy to attach a phosphate group to ADP. ATP is thus produced. Alpha-ketoglutaric acid has become succinic acid.

5. Succinic acid is oxidized by FAD, losing two hydrogen atoms (complete with electrons) to it. FAD is thus reduced to $FADH_2$ and moves to the ETC. In a series of steps, succinic acid is transformed into malic acid.

6. Malic acid is oxidized by NAD^+. This oxidation transforms malic acid into oxaloacetic acid, which is now ready to enter into another turn of the cycle.

7.6 Third Stage of Respiration: The Electron Transport Chain

Having moved through the Krebs Cycle, you've reached the main event in cellular respiration: The production of 32 more ATP molecules through the work of the **electron transport chain (ETC)**, the third and most productive stage of aerobic energy harvesting. Glycolysis took place in the cytoplasm, and the Krebs cycle took place in the inner compartment of the mitochondria. Now, however, the action shifts to the mitochondrial inner membrane, the site of the electron transport chain (**see Figure 7.9**). The "links" in this chain are a series of molecules. You've been looking for some time now at how NADH and $FADH_2$ carry off the electrons derived from glycolysis and the Krebs cycle. This is the destination of these electron carriers and their cargo.

Tutorial 7.2.7
Electron Transport Chain

Figure 7.9
The Electron Transport Chain (ETC)
The movement of electrons through the ETC powers the process that provides the bulk of the ATP yield in respiration. The electrons carried by NADH and $FADH_2$ are released into the ETC and transported along its chain of molecules. The movement of electrons along the chain releases enough energy to power the pumping of hydrogen ions (H^+) across the membrane into the outer compartment of the mitochondrion. It is the subsequent energetic "fall" of the H^+ ions back into the inner compartment that drives the synthesis of ATP molecules by the enzyme ATP synthase.

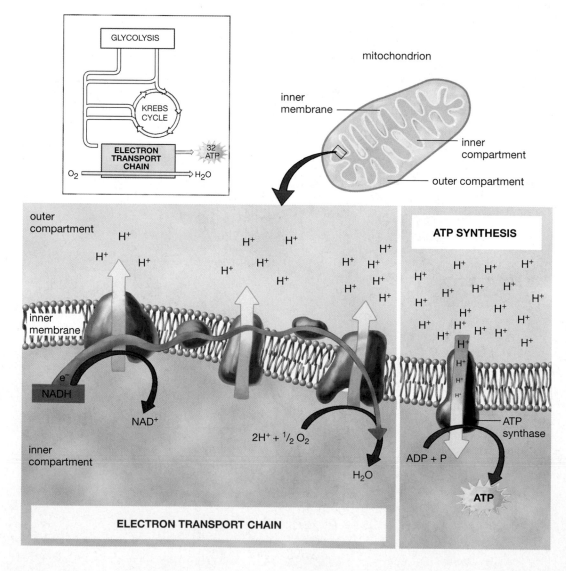

Upon reaching the mitochondrial inner membrane, the electron carriers donate electrons and hydrogen ions to the ETC. Once again, this is a trip down the energy hill, only *NADH* is now the molecule whose electrons exist at a higher energy level. Thus, when NADH runs into the appropriate enzyme in the ETC, NADH is oxidized, donating its electrons and proton to the ETC. This process then is simply repeated down the whole ETC, each carrier donating electrons to the next electron carrier in line. A carrier is reduced by receiving these electrons; donating them to the next carrier, it then returns to an oxidized state. Each carrier thus alternates between oxidized and reduced states.

Visualizing the ETC

There are several ways to conceptualize the ETC. In Figure 7.9, you can see that one way is to view the ETC as three large enzyme complexes with two smaller mobile molecules that link them. When NADH arrives at the inner membrane, it bumps into the ETC's first carrier, which accepts electrons from NADH; this carrier then donates these electrons to the next carrier, and so on down the line to the last electron acceptor, which is oxygen.

Where's the ATP?

At this point you may be tapping your foot, waiting for this promised harvest of ATP to appear. So far, all that's taken place is a transfer of electrons in the ETC. Well, note what happens in the first ETC enzyme complex. The movement of electrons through it releases enough energy to power the movement of *hydrogen ions* (H$^+$ ions) through the complex, pushing them from the inner compartment into the outer compartment. (The fall of the electrons causes the enzyme complex to change its shape in a way that facilitates the passage of H$^+$ ions through it and into the outer compartment.) By the time the electrons have completed their movement through the ETC, this pumping of H$^+$ ions will occur with two other complexes. Critically, these H$^+$ ions are being pumped *against* their concentration and electrical gradients (which were reviewed in Chapter 5). Put another way, the ions are being pumped *up* the energy hill with energy supplied by the *downhill* fall of electrons through the ETC.

Here's where the ATP is produced at last. The H$^+$ ions that have been pumped into the outer compartment move *back down* their concentration and energy gradients, into the inner compartment, through a special enzyme called **ATP synthase**. This remarkable enzyme is driven by the H$^+$ ions flowing through it, which cause part of the enzyme to rotate (as fast as 100 revolutions per second). This spinning drives the synthesis of ATP from the constituent parts of ADP and phosphate.

This mechanism—the pumping of hydrogen ions powering the synthesis of ATP—was first proposed in 1961 by British biochemist Peter Mitchell, who won the Nobel Prize in 1978 for his contribution to bioenergetics. Lest it pass by too quickly, *this* is the essence of aerobic respiration. The downhill drop of electrons derived from food has powered the uphill synthesis of the molecule that powers the vast majority of our activities.

Bountiful Harvest: ATP Accounting

In the ETC, each NADH molecule will be responsible for the production of three molecules of ATP. And, of course, every molecule of NADH that comes to the chain will result in this amount of ATP. And how many are coming? Ten: two produced in glycolysis, two from pyruvic acid conversion, and six from the Krebs cycle. The FADH$_2$ carrier that came from the Krebs cycle joins the ETC a little later than does NADH and as such produces only about two ATP molecules, compared to NADH's three. Given this, here's the final count on our energy harvest.

Stage	NADH	FADH$_2$	ATP Yield
Glycolysis	2		2
Pyruvic acid conversion	2		
Krebs cycle	6	2	2
Electron transport chain	10 ⟶		30
		2 ⟶	4
Loss, due to active transport:			−2
Total ATP:			36

What's apparent from looking at this is that the ETC is a mighty energizer. Think about how efficient this process is compared to glycolysis (34 ATP compared to 2). Imagine one car that got 34 miles to the gallon compared to another that got 2. The 36 total ATP yield actually is the amount that can be derived under optimal conditions. (The "loss, due to active transport" in the table refers to the expenditure of ATP required to move the NADH produced in glycolysis into the mitochondria.)

Finally, Oxygen Is Reduced, Producing Water

The story of the breakdown of glucose—the saga of *one molecule's* transformation—is nearly complete. The only unfinished business lies at the end of the ETC. As noted earlier, oxygen is the final acceptor of the working electrons. In the mitochondrial inner compartment, oxygen accepts two electrons and two H^+ ions (from the inner compartment). The result is H_2O: water. That's why the formula for respiration is written as:

$$ADP + C_6H_{12}O_6 + 6O_2 \rightarrow ATP + 6CO_2 + 6H_2O$$

ADP plus glucose plus oxygen yields ATP plus carbon dioxide plus water: That's the short story of aerobic respiration.

7.7 Other Foods, Other Respiratory Pathways

By looking at aerobic respiration as it applies to glucose, which was used as the starting molecule, you've been able to go through the whole respiratory chain. But it goes without saying that there are many kinds of nutrients besides glucose—for example, fats, proteins, and other sugars. All these can provide energy by being oxidized within the chain of reactions you've just gone over. However, none of them proceeds through the chain in exactly the same way as glucose. In addition, the body has more uses for foods than breaking them down to produce ATP. At any given moment, an organism may need to go the other way: toward the *buildup* of proteins from amino acids or the synthesis of the fats we call triglycerides. Organisms are able to handle this variability because nutrients and their derivatives can be *channeled* through pathways in different directions in the respiratory chain in accordance with cell needs.

The alternation of the breakdown of food molecules and the buildup of food storage molecules is one of the central functions that many living things carry out.

Alternate Respiratory Pathways: Fats as an Example

To give one example of how this channeling of nutrients works, consider the possible fates of triglycerides. As you learned in Chapter 4, triglycerides are fats made of a three-carbon glycerol "head" and three long hydrocarbon chains that look like tines on a fork. Imagine that a given cell needs energy from fat. The question is, how do *triglycerides* go through aerobic respiration, as opposed to glucose?

The first thing that happens is that a triglyceride molecule is split by enzymes into its constituent parts of fatty acids and glycerol. Glycerol is then converted into glyceraldehyde phosphate. This formidable name may not ring a bell, but if you look back to step 4 in Figure 7.5, you'll see that it is one of the *downstream* derivatives of glucose in glycolysis. To put this another way, glycerol does not first get converted to glucose and then march through *all* of the steps of glycolysis. Instead, it joins the glycolytic pathway several steps down from glucose and is then converted to pyruvic acid. Then, as pyruvic acid, it goes through the Krebs cycle and ETC, producing ATP. And the fatty acids? They are converted to acetyl CoA, which you may remember is the substrate that enters the Krebs cycle, yielding energy by becoming oxidized there.

What Happens When Less Energy Is Needed?

So much for what happens when an organism needs a good deal of energy from fats. But what happens to the triglyceride components when this need is reduced? Glycerol, which had been converted to glyceraldehyde phosphate and begun marching through glycolysis, can be converted to *glucose* in a process that, in its main steps, is glycolysis in reverse. The resulting glucose molecules can then be used to do any number of things. They might, for example, be put together to form the storage form of glucose, glycogen.

You may be getting a sense that the metabolism is rather like a traffic circle: Molecules can enter at different points and leave at different points for varying destinations, depending on the needs of the organism. In Figure 7.10, you can see a summary of the way foods are broken down through various respiratory pathways.

On to Photosynthesis

This long walk through cellular respiration has illustrated how living things harvest energy from food. Recall, however, that this chapter began with the observation that we ultimately have one source to thank for this food: the Sun's energy, trapped by plants in the chemical bonds of carbohydrates. This energy conversion takes place through a process that has a beautiful symmetry with the respiration you have just looked at. That process is called photosynthesis.

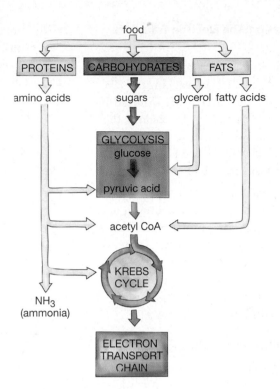

7.7   Other Foods, Other Respiratory Pathways

Tutorial 7.2.8
Many Respiratory Pathways

Figure 7.10
Many Respiratory Pathways
Glucose is not the only starting material for cellular respiration. Other carbohydrates, proteins, and fats can also be used as fuel for cellular respiration. These reactants enter the process at different stages.

Chapter Review

Summary

7.1   Energizing ATP: Adding a Phosphate Group to ADP

- The molecule adenosine triphosphate (ATP) supplies the energy for nearly all the activities of living things. For ATP to be produced, a third phosphate group must be added to adenosine diphosphate (ADP), a process that requires energy.
  TUTORIAL 7.1.2: Oxidation-reduction Reactions

7.2   Electrons Fall Down the Energy Hill to Drive the Uphill Production of ATP

- In animals, the energetic fall of electrons derived from food powers the process by which the third phosphate group is attached to ADP, making it ATP.

- Electron transfer in the production of ATP works through redox reactions, meaning reactions in which one substance loses electrons to another substance. Electrons are carried between one part of the energy harvesting process and another by electron carriers, the most important of which is nicotinamide adenine dinucleotide or NAD^+.

7.3   The Three Stages of Cellular Respiration: Glycolysis, the Krebs Cycle, and the Electron Transport Chain

- In more complex organisms, the harvesting of energy from food takes place in three principal stages, glycolysis, the Krebs cycle, and the electron transport chain (ETC).

- Some organisms rely solely on glycolysis for energy harvesting. For most organisms, glycolysis is a primary process of energy extraction only in certain situations—when quick bursts of energy are required, for example—but it is a necessary precursor to the other two stages of energy harvesting. Together, the three stages of energy harvesting are known as cellular respiration.

- Glycolysis takes place in the cell's cytoplasm, while the Krebs cycle and the electron transport chain (ETC) take place in the cellular organelles called mitochondria. Glycolysis yields two net molecules of ATP per molecule of glucose, as does the Krebs cycle. The net yield in the electron transport chain is a maximum of about 32 ATP molecules per molecule of glucose. In addition to ATP, glycolysis and the Krebs cycle both yield electrons that are carried to the ETC via electron carriers.

7.4   First Stage of Respiration: Glycolysis

- The net energy yield of glycolysis is two molecules of NADH and two molecules of ATP per molecule of glucose.
  TUTORIAL 7.2.3: Glycolysis

7.5   Second Stage of Respiration: The Krebs Cycle

- The net energy yield of the Krebs cycle is six molecules of NADH, two molecules of $FADH_2$, and two molecules of ATP per molecule of glucose.
  TUTORIAL 7.2.4: Krebs Cycle

7.6 Third Stage of Respiration: The Electron Transport Chain

- The ETC is a series of molecules that are located in the mitochondrial inner membrane. Upon reaching the ETC, the electron carriers NADH and FADH$_2$ are oxidized by molecules in the chain. Each carrier in the chain is then reduced by accepting electrons from the carrier that came before it. The last electron acceptor in the ETC is oxygen.
  TUTORIAL 7.2.7: Electron Transport Chain

- The movement of electrons through the ETC releases enough energy to power the movement of hydrogen ions (H$^+$ ions) through the three ETC protein complexes, moving them from the mitochondria's inner compartment to its outer compartment. The movement of these ions down their concentration and energy gradients, back into the inner compartment through an enzyme called ATP synthase, drives the synthesis of ATP from ADP.

7.7 Other Foods, Other Respiratory Pathways

- Nutrients and their derivatives can be channeled through different pathways in cellular metabolism, in accordance with the needs of the organism. At any given moment, a cell may need to work more on synthesizing organic molecules than on breaking them down.
  TUTORIAL 7.2.8: Many Respiratory Pathways

Key Terms

alcoholic fermentation 138	Krebs cycle 140
ATP synthase 145	lactate fermentation 139
cellular respiration 136	lactic acid 139
citric acid cycle 141	nicotinamide adenine dinucleotide (NAD) 134
electron transport chain (ETC) 144	oxidized 133
glycolysis 136	redox reaction 133
intermediate electron carrier 134	reduction 133

Understanding the Basics

Multiple-Choice Questions

1. You need energy to think, to keep your heart beating, to play a sport, and to study this book. This energy is directly supplied by _____, which is produced in the process of respiration.
   a. enzymes
   b. ATP
   c. NAD$^+$
   d. vitamins
   e. proteins

2. Which of these statements best explains why most of the energy you capture from food comes from cellular respiration's three stages—glycolysis, the Krebs cycle, and the electron transport chain—rather than from glycolysis alone.
   a. Cellular respiration is easier to carry out.
   b. Glycolysis requires more enzymes and thus more energy.
   c. Most cells lack the ability to make the enzymes needed to carry out glycolysis.
   d. Glycolysis produces less ATP than does cellular respiration.
   e. Glycolysis does not require NAD$^+$.

3. Imagine a prokaryotic (bacterial) cell in an environment that did not contain oxygen but that did contain food. In such a situation, the cell could not
   a. extract energy from food
   b. perform glycolysis
   c. utilize the Krebs cycle
   d. utilize the electron transport chain
   e. utilize either the Krebs cycle or the electron transport chain

4. What happens when a cell in your body momentarily has an abundance of ATP?
   a. It stops oxidizing food and starts building up storage food molecules from smaller food molecules.
   b. It continues to produce ATP at the same rate so that it is ready for the next emergency.
   c. It starts harvesting energy solely through glycolysis.
   d. It forms glucose out of extra ATP.
   e. It dies.

5. At most, how many molecules of ATP can be produced per glucose molecule in cellular respiration?
   a. 2
   b. 8
   c. 24
   d. 36
   e. 75

6. Where does the Krebs cycle occur in this model of a mitochondrion?

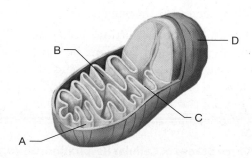

   a. A
   b. B
   c. C
   d. D
   e. none of these

7. When NADH passes its electrons to the ETC, it is
   a. electrolyzed
   b. polarized
   c. reduced
   d. oxidized
   e. deformed

8. In the first step of glycolysis, glucose enters the cell and immediately has a phosphate group from ATP attached to it. This process is called
   a. phosphorylation
   b. oxidation
   c. photosynthesis
   d. electron transport
   e. the citric acid cycle

Brief Review

1. Picture yourself on a trampoline. Use the fact that you have to jump in order to start bouncing to explain how ATP can be formed as a result of electrons from glucose falling down an energy hill.

2. Explain why you should expect to have more mitochondria in your muscle cells than in your skin cells.

3. Since ATP is used for our cellular needs, why don't we just eat ATP?

4. A few desert animals are able to get by without ever drinking water. They use the water in their food to supply some of their need for water. They also use the water they produce in respiration. Where does that water come from?

5. FAD is derived from the B-vitamin, riboflavin. What is the function of FAD? What do you think happens if you fail to get enough riboflavin?

Applying Your Knowledge

1. When you eat foods that contain small amounts of the B-vitamin niacin, the vitamin is converted into NAD^+ in your body. Using what you know about the role of NAD^+, predict what will happen to people whose diets are deficient in niacin.

2. Why do you think that we use the same word—*respiration*—for breathing as we do for breaking down food to extract its energy?

3. Enzymes are now used for a host of processes outside of cells, including such things as cleaning clothes and hands. In the late 1990s researchers found an enzyme in spinach that will break down explosives such as TNT. In light of what you know about enzymes, how could this work? Think of other environmental problems that might be solved by using enzymes.

4. In Madeleine L'Engle's children's novel *A Wrinkle in Time*, the mitochondria in one of the characters start to die. Describe what would happen to people who lost their mitochondria, and explain why it would happen.

MediaLab

Dietary Fad or Miracle Drug? Using Science to Understand Metabolism

You've heard the claims: "Burns fat while you sleep!" "Energy fuel capsules!" and even "Magic weight-loss pills!" Companies entice us to try some new product by exploiting our ignorance about the actual process of cellular metabolism—how cells extract energy from food. This *MediaLab* covers the basic concepts needed to understand how cells really use the energy from food. To learn more, enter the *CD-ROM Tutorial* for an animated review of metabolism. Once you have completed the *CD-ROM Tutorial*, try the *Web Investigation* to uncover how metabolism may be misrepresented in diet claims on the Internet. Finally, *Communicate Your Results* will help you practice the skills you'll need to critically appraise the material you read.

This *MediaLab* can be found in Chapter 7 on your CD-ROM (Tutorial 7.1.2) and Companion Website (http://www.prenhall.com/krogh3).

CD-ROM TUTORIAL

Americans are presented with such an abundance of food that they tend to equate the word *diet* with limiting the consumption of Calories, not as a necessary source of energy and nutrients. However, if you can see how your body uses the molecules in food, then you can begin to see why you need to consume certain foods. This *CD-ROM Tutorial* is designed to review metabolism.

Activity

1. First, you will examine the role of the cellular energy currency, ATP, and the energy conveyors, Redox reactions.

2. Then you will investigate why we eat, by following the series of steps needed to convert the energy from food (carbohydrates, proteins, and fats) into ATP.

3. Finally, you can play an interactive metabolism game where you are given a list of starting materials, a cell, and asked to make ATP.

After reviewing the concepts of metabolism in these processes, you'll be ready to test your knowledge in the next section by uncovering unfounded, impossible diet and exercise claims on the Internet.

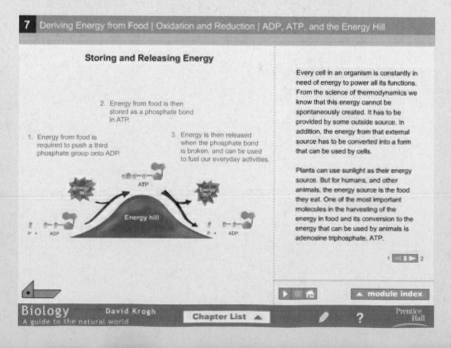

WEB INVESTIGATION

Investigation 1
Estimated time for completion = 10 minutes

How do we know this diet stuff won't live up to its promises? There is an excellent Web page devoted to disseminating information on how to recognize fraudulent weight-loss products and programs. Select the Keyword **WEIGHT-LOSS FRAUD** on your CD or Website for this *MediaLab*. After you've read the criteria, perform a search of the Internet to locate examples of weight-loss Websites that make one of the promises described in the fraud article, like magic weight loss without dieting or exercise.

Investigation 2
Estimated time for completion = 5 minutes

Chromium picolinate is a common weight-loss supplement found in diet pills like Metabolife.

What is it? Does it work? How can you tell? Select the Keyword **CHROMIUM SUPPLE-MENTS 1** or **2** on your CD or Website for this *MediaLab* to read two contradictory articles about chromium. Which one do you believe?

Investigation 3
Estimated time for completion = 5 minutes

Over 50 percent of adult Americans purchase dietary supplements to improve their health, but who tests the accuracy of the health claims made by supplement manufacturers? The responsibility for checking the safety of products and the truthfulness of label claims lies in the hands of consumers and manufacturers, not in an impartial government agency like the FDA. Select the Keyword **FDA GUIDE TO SUPPLEMENTS** on your CD or Website for this *MediaLab* to read the FDA's guide to dietary supplements.

COMMUNICATE YOUR RESULTS

Determine the integrity of the websites from the *Web Investigation* section by applying rigorous questions in the following examples.

Exercise 1
Estimated time for completion = 5 minutes

How can you determine what to believe on the Internet, when websites include scientific terminology and definitions that may not be based on tested, controlled experiments? Consumers can be embarrassingly gullible when they believe incredible diet claims. Make a difference by posting information about one of these too-good-to-be-true sites (a message board is provided at the textbook's Companion Website). Describe in two or three sentences why you think the product falls under the examples listed in the diet fraud article.

Exercise 2
Estimated time for completion = 20 minutes

Many diet-supplement companies promote the idea that human health problems can be remedied through ingesting supplements. How would a scientist determine the credibility of these claims? Here is a list of guidelines:

- *Do the websites refer to studies published in reputable science journals with rigorous standards, or do they refer to testimonials and stories?*
- *Do the author's credentials lend credibility, or are they unrelated?*

- *Does the author stand to benefit financially by the sales of the product?*

Compose a 250-word argument describing the claims made at the two websites listed in Web Investigation 2, and using the scientific criteria just listed, explain which one is most credible.

Exercise 3
Estimated time for completion = 15 minutes

After reading the FDA report on supplements in *Web Investigation 3*, compose a letter to the FDA expressing whether you think the current system provides adequate protection to Internet consumers. Include your opinion on whether supplements should be subjected to the same rigorous clinical studies as drugs to determine their effectiveness, safety, possible interactions with other substances, and appropriate dosages. Should the FDA take more responsibility for removing potentially unsafe dietary supplements? Should the FDA take a greater role in enforcing prosecution of companies making false or erroneous health claims for its supplements?

Applying rigorous questions like these helps prepare you to critically analyze future material that you may find on the Internet. Internet material is not always filtered or checked for accuracy, so you'll need the skills practiced in this section to assess it on your own.

8

The Green World's Gift
Photosynthesis

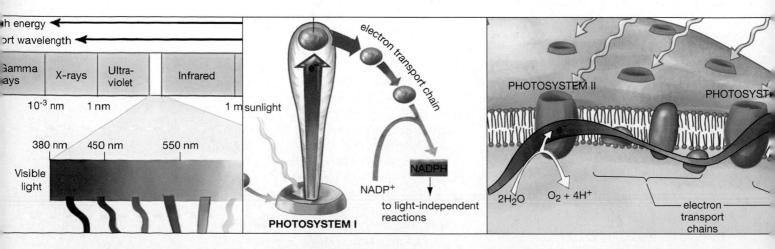

The electromagnetic spectrum.
(Section 8.2, page 154)

Capturing the Sun's energy.
(Section 8.2, page 157)

Photosynthesis in action.
(Section 8.3, page 159)

Plants use the Sun's energy to make their own food — and ours.

On a summer's evening, out by a pond as the Sun is going down, nature tunes up her symphony. The crickets bring their chirping to the fore, while frogs chime in with their own countermelody. Silently, something else is changing as the last light fades: The green world is shutting down. Microscopic pores on the leaves of bushes and trees are closing up, ceasing to be openings for the carbon dioxide that flows in and the water vapor that flows out during the day. The green world is alive at night, but the activities it carries out are not so much different from those a human listener might be undertaking while relaxing at water's edge. Both are building up tissue, storing away some foodstuffs while breaking down others, getting rid of the by-products that result from all these processes. The green world has a special talent, but sunlight must be falling for it to be put into practice.

8.1 Photosynthesis and Energy

The green world's special talent is called photosynthesis, and the wonder of it can perhaps best be appreciated by imagining what life would be like if we humans possessed it. The bone and muscle within us—and the energy to make them work—come from the foods we eat: carbohydrates and fats and proteins. Imagine, though, being able to flourish simply by spending time in the Sun and having access to three things: water, small amounts of minerals,

and carbon dioxide. In this condition we would not *eat* carbohydrates in the usual forms (bread, sugar, potatoes). Rather, we would transform the simple gas carbon dioxide *into* carbohydrates using nothing more than water, minerals, and sunlight. Carbohydrates produced in this way are as useful as any food; they can be stored away as starches, moved from one part of the body to another as simple sugars, or broken down to provide ATP for any energy needs. This, then, is what plants do: They *make their own food.* They use the energy provided by the Sun to take a simple gas, carbon dioxide, join it to a carbohydrate (a sugar) and then *energize* that sugar, thus transforming it into food. In this activity, they are joined by a variety of other photosynthesizing organisms (**see Figure 8.1**).

**Figure 8.1
Three Types of Photosynthesizers**
Photosynthesis is carried out not only by familiar plants, such as these sunflowers **(a)**, but by algae, such as this giant kelp **(b)**, and by some bacteria, such as these cyanobacteria **(c)**. (magnified ×1025)

a

b

c

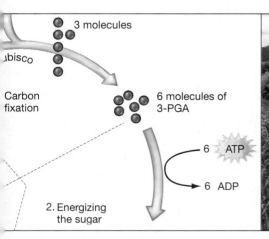

3 molecules

Rubisco

Carbon fixation

6 molecules of 3-PGA

6 ATP

6 ADP

2. Energizing the sugar

An important cycle begins.
(Section 8.5, page 160)

Warm-weather photosynthesis.
(Section 8.7, page 162)

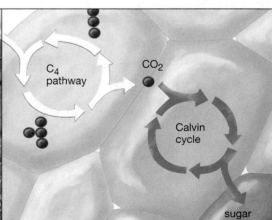

C_4 pathway

CO_2

Calvin cycle

sugar

A different path for photosynthesis.
(Section 8.7, page 162)

From Plants, a Great Bounty for Animals

The process of photosynthesis is not only good for plants; it's good for other living things, including animals, since most of the living world ultimately depends upon photosynthesis for food. Try naming something you eat that isn't a plant or doesn't itself eat plants to survive, and you will come up with a very short list indeed.

Food, however, is only half the bounty that plants provide. A by-product of photosynthesis—a kind of castoff from the work of plants—is oxygen. As part of photosynthesis, plants break water molecules apart. In doing so, they *use* electrons and protons from H_2O, but they *leave behind* oxygen molecules (O_2). This is the oxygen that exists in our atmosphere; it is the oxygen that we breathe in every minute of every day, or suffer the mortal consequences. Meanwhile, one of the starting ingredients for photosynthesis is the carbon dioxide we breathe *out* with each breath, thus adding a small amount to the atmospheric store that exists. In the living world, one organism's by-product is another's life-sustaining substance.

With this, you can step back and view the largest picture of them all with respect to energy flow in living things. It is a great pathway in which energy comes from the Sun and then, in photosynthesis, is stored in plants in the complex molecules we call carbohydrates. Thus stored—in such forms as grass leaves or wheat grains—these carbohydrates can be broken down and used, either by the plants themselves or by the animals that eat them. One by-product of the breakdown of food is carbon dioxide, which keeps photosynthesis going. Meanwhile, one by-product of photosynthesis is oxygen, which keeps aerobic respiration going.

Up and Down the Energy Hill Again

Looked at another way, these two processes are simply trips up and down the energy hill that became so familiar in Chapter 7. In respiration, the trip is down the hill—from more stored energy (in food) to less, as the food is being broken down. Photosynthesis is a trip back up the energy hill. Here electrons are being removed from water, boosted to a more energetic state by the power of sunlight, and then brought together with a sugar and carbon dioxide, resulting in an *energy-rich* sugar—food in the form of a carbohydrate. The story of how this is accomplished is the subject of this chapter.

8.2 The Components of Photosynthesis

Photosynthesis can be defined as the process by which certain groups of organisms capture energy from sunlight and convert it into chemical energy, with this energy initially being stored in a carbohydrate. Think of it as starting with the absorption of sunlight by leaves. *Absorption* here means just what it sounds like: Light is taken in by the leaves. Actually, leaves capture a *portion* of the light that falls on them. The sunlight that makes it to the Earth's surface is composed of a spectrum of energetic rays (measured by their "wavelengths") that range from very short, ultraviolet rays, through visible light rays, to the longer and less-energetic infrared rays. (In **Figure 8.2** you can see that these wavelengths are, in turn, part of a larger range of electromagnetic radiation.)

What Kind of Light Drives Photosynthesis?

Photosynthesis is driven by part of the *visible light* spectrum—mainly by blue and red light of certain wavelengths. Tune a laser to emit blue light of a certain wavelength, shine this beam on a plant, and the plant will absorb a great deal of this light. On the other hand, tune the laser to emit *green* light, and the plant will scatter most of this light. This is why plants are green: They strongly scatter the green portion of the visible light spectrum.

Tutorial 8.1.2
The Electromagnetic Spectrum

Figure 8.2
The Electromagnetic Spectrum

Sunlight is composed of rays of many different wavelengths, but we can see only those in the visible light range (represented by the color spectrum). When the light of the sun shines on a green leaf, it is primarily the Sun's green rays that are scattered, while frequencies in other ranges are absorbed; frequencies in the red and blue ranges drive photosynthesis most strongly.

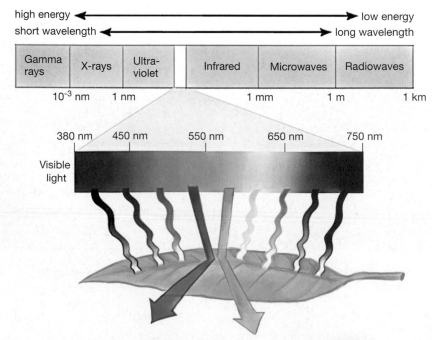

Where in the Plant Does Photosynthesis Occur?

So, where is this light absorption occurring within plants? To answer this question, you need to know something about the playing fields of photosynthesis, leaves.

The broad, flat leaves that are so common in nature have in essence a two-part structure: A *blade* (which we usually think of as the leaf itself) and a *petiole*, more commonly referred to as the leaf stem. If you look at the idealized leaf in **Figure 8.3**, you can see that the blade can be likened to a kind of cellular sandwich, with one layer of outer (or epidermal) cells at the top, one layer at the bottom, and layers of mesophyll cells in between.

The leaf's epidermal cells are dotted with numerous microscopic pores, called **stomata**. These are the openings mentioned at the beginning of the chapter that let carbon dioxide in and water vapor out. There is no shortage of stomata for this work; a given square centimeter of plant leaf may contain from a thousand to a hundred thousand of them.

Photosynthesis Central: The Chloroplasts

A journey to the actual site of photosynthesis would take you to the leaf's layers of mesophyll

Figure 8.3
Site of Photosynthesis

a

1. **Leaf.** The primary sites of photosynthesis in plants, leaves have a two-part structure: a petiole (or stem) and a blade (normally thought of as the leaf).

2. **Leaf section.** In cross section, leaves have a sandwich-like structure, with epidermal layers at top and bottom and mesophyll cells in between. Most photosynthesis is performed within mesophyll cells. Leaf epidermis is pocked with a large number of microscopic openings, called stomata, that allow carbon dioxide to pass in and water vapor to pass out.

3. **Mesophyll cell.** A single mesophyll cell within a leaf contains all the component parts of plant cells in general, including the organelles—called chloroplasts—that are the actual sites of photosynthesis.

4. **Chloroplast.** Each chloroplast has an outer membrane at its periphery; then an inner membrane; then a liquid material, called the stroma, that has immersed within it a network of membranes, the thylakoids. These thylakoids periodically stack on one another to create structures called grana.

5. **Grana.** Looking at thylakoids in a granum, each thylakoid is composed not only of a thylakoid membrane, but of a watery interior space called the thylakoid compartment. All the steps of photosynthesis take place either in the thylakoid membrane, or in the stroma.

b Top, a micrograph of a leaf cross section, with layers of mesophyll sandwiched between upper and lower epidermis layers. Bottom, a micrograph of a chloroplast. The groups of thin lines are the stacks of grana. The large white body is a starch body, produced through photosynthesis.

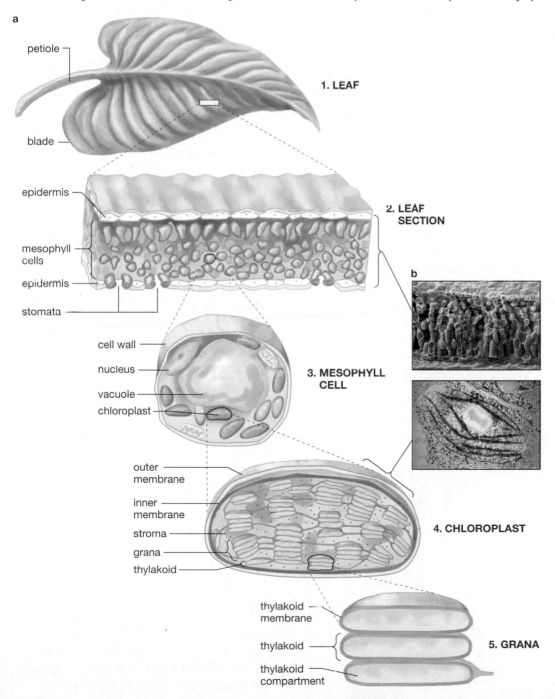

a

petiole

blade

1. LEAF

epidermis

mesophyll cells

epidermis

stomata

2. LEAF SECTION

b

cell wall

nucleus

vacuole

chloroplast

3. MESOPHYLL CELL

outer membrane

inner membrane

stroma

grana

thylakoid

4. CHLOROPLAST

thylakoid membrane

thylakoid

thylakoid compartment

5. GRANA

cells, then inside the cells themselves, then into a group of organelles within the cells—the **chloroplasts**. Anywhere from one to several hundred of these oblong structures might exist within a given leaf cell, each capable of carrying out photosynthesis on its own.

Traveling into the *chloroplasts*, you'd come to the molecules, called pigments, that actually absorb sunlight. Anything that strongly absorbs certain wavelengths of sunlight is called a pigment. Chloroplasts contain a primary pigment, embedded in their membranes, called **chlorophyll *a***, which is aided by several substances known as **accessory pigments**. Working together, these pigments perform the task of absorbing the light wavelengths mentioned earlier.

Figure 8.3 shows you the detailed structure of a chloroplast. As has so often been the case with cell structures, what chloroplasts present are membranes within membranes. The chloroplast has outer and inner membranes at its periphery. Moving inside them, to the interior of the chloroplast, you see a network of membranes, called **thylakoids**. At frequent intervals, these thylakoids stack on top of one another like so many pancakes, creating structures called **grana**. Thylakoids are immersed in the liquid material of the chloroplast, the **stroma**, and have an interior fluid space called the **thylakoid compartment.** All the steps of photosynthesis take place in the thylakoid membrane, or the stroma of the chloroplast.

There Are Two Essential Stages in Photosynthesis

Keeping these structures and concepts in mind, you can now begin tracing the steps of photosynthesis. It is helpful to divide photosynthesis into two main stages. In the first stage (the *photo* of photosynthesis), the power of sunlight will do two things: strip water of electrons and then boost these electrons to a higher energy level. Thus boosted, the electrons move from one electron carrier to another, finally attaching to a mobile electron carrier ($NADP^+$) that will carry them to the second set of reactions. In this second stage (the *synthesis* of photosynthesis), the electrons and carbon dioxide are brought together with a sugar. The attachment of the CO_2 and electrons to this sugar produces a high-energy sugar.

The processes in the first stage of photosynthesis obviously depend on sunlight, but those in the second stage do not (at least not directly). Given this division, scientists have named the first set of steps the **light-dependent reactions**, and the second set the **light-independent reactions**. You will be following these reactions as they occur in time, meaning from the collection of sunlight to the synthesis of carbohydrates.

The Working Unit of Photosynthesis Is Called a Photosystem

Absorption of sunlight takes place within *aggregations* of a few hundred pigment molecules. The majority of molecules in these units—most of the chlorophyll *a* and all of the accessory pigments—serve only as "antennae" that absorb energy from the Sun and pass it on. At the center of the antennae system, however, is a molecular complex known as the **reaction center**, which contains a pair of special chlorophyll *a* molecules that transform solar energy into chemical energy.

You could think of photosynthesis as beginning when sunlight is absorbed by any of the hundreds of antennae pigment molecules in a photosystem unit. The absorbed energy is then passed on to the pair of chlorophyll *a* molecules in a reaction center. As a result, electrons from this pair are "moved" in a couple of ways, one physical and one metaphorical. Physically, these electrons jump to another electron carrier within the reaction center, a "primary electron acceptor." You can get a sense of how the reaction center and antennae pigments fit together to form a **photosystem** by looking at **Figure 8.4**.

Figure 8.4
The Working Units of Photosynthesis
Photosystems are multipart units that bring together electrons derived from water and energy derived from the Sun. Hundreds of antennae pigments absorb sunlight and transfer solar energy to the photosystem's reaction center. This energy gives a boost to an electron within the reaction center in two ways: It physically moves the electron to the center's primary electron acceptor, and it moves the electron up the energy hill to a more energetic state.

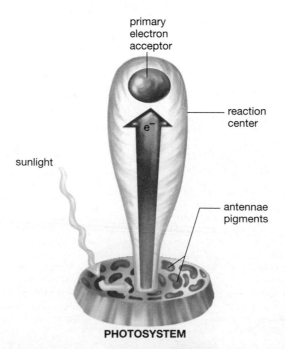

primary electron acceptor

reaction center

e⁻

sunlight

antennae pigments

PHOTOSYSTEM

You also could think of "distance traveled" in photosynthesis in a metaphorical way, however: as a movement up and down an energy hill. It is the energy from sunlight that is doing the pumping up the energy hill in photosynthesis.

Energy Transfer in Photosynthesis Works through Redox Reactions

How can electrons move up and down an energy hill? Recall from Chapter 7 that some substances have a tendency to *lose* electrons to other substances. Any time one substance loses (or "donates") electrons to another, it is said to have been oxidized. Meanwhile, the substance that gained electrons is said to have been reduced. These two reactions always happen together; if one substance is oxidized, another must be reduced. This kind of reduction-oxidation reaction has a shorter name: a *redox reaction.*

Many of the photosynthetic steps you will be seeing are redox reactions. In **Figure 8.5**, you can see where they lead in terms of movement up and down the energy hill. The electrons are pumped up a couple of formidable energy gradients (in photosystems II and I) only to come partway back down them, as they "seek" their lowest energy state, releasing energy as they fall—just as a boulder releases energy by rolling downhill. These are redox reactions, with each molecule in the chain undergoing successive oxidation and reduction reactions.

8.3 Stage 1: The Steps of the Light-Dependent Reactions

With these components and processes in mind, let's trace the steps of the light-dependent reactions (see Figure 8.5). The first step is that solar energy, collected by photosystem II's antennae molecules, arrives at the reaction center. This energy then gives a boost to an electron in the reaction center in the two ways noted before. The electron *physically* moves to another part of the reaction center protein complex, the primary electron acceptor. But it is also pumped up the energy hill. With this activity, the reaction center chlorophyll *a* has *lost* an electron. That loss leaves an energy "hole" in this chlorophyll,

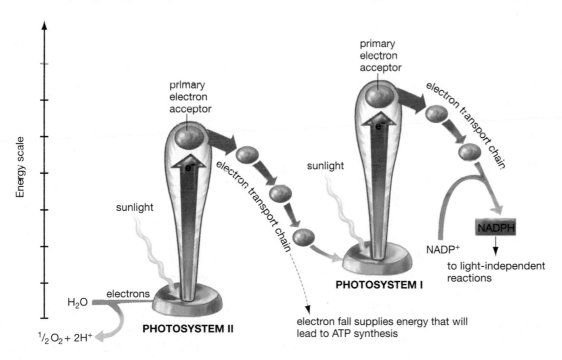

Figure 8.5
Collecting Solar Energy, Boosting Electrons
The solar energy gathered by photosystems II and I is used to energize electrons that exist within the photosystems' reaction centers. Energetically, the electrons are boosted up, and then move down, two energy hills. Physically, they first move to primary electron acceptors and then down electron transport chains until they are at last taken up by $NADP^+$ to form NADPH. The NADPH molecules then transfer the electrons to the light-independent reactions, where they are used to make sugars. When an electron is boosted in photosystem II, the resulting energy imbalance drives the process by which water molecules are split. Electrons derived from the water then move to the reaction center of the photosystem, where they are the next to be boosted by the Sun's energy. The energetic fall of electrons down the electron transport chain between photosystems II and I provides the energy for the synthesis of ATP, which is used to power the light-independent reactions.

making it an oxidizing agent. With the energy provided by this imbalance, a special enzyme in the reaction center splits water molecules that lie within the thylakoid compartment. These water molecules are now being oxidized, which means they are *losing* electrons. The electrons travel to the reaction center, where they will be the next electrons in line for an energy boost.

A Chain of Redox Reactions and Another Boost from the Sun

By following Figure 8.5 in a general way, you can see what happens next to electrons that are boosted in photosystem II. After arriving at the primary electron acceptor, they fall back down the energy hill as they are transferred through a series of electron transport molecules, each one oxidizing its predecessor. At the bottom of this hill they arrive at photosystem I, which includes a slightly different kind of reaction center. This center is also receiving solar energy, and uses it to boost electrons to a higher energy state. From this energetic state, electrons are transferred down the energy hill—until they are received by the electron carrier $NADP^+$, which is very much like the electron carrier NAD^+ reviewed in Chapter 7. In accepting electrons, $NADP^+$ becomes reduced to NADPH, an electron carrier that ferries the electrons into the second stage of photosynthesis: the light-independent reactions, which will yield the high-energy sugar that is the essential product of photosynthesis.

The Physical Movement of Electrons in the Light-Dependent Reactions

In this movement up and down the energy "hill," the electrons have moved physically through the chloroplast. As **Figure 8.6** shows, they started out in the water of the thylakoid compartment, moved into and then through the thylakoid membrane—handed off at each step by the various electron transport molecules—and then finally ended up in the stroma, attached to NADPH.

8.4 What Makes the Light-Dependent Reactions So Important?

The steps just reviewed are the essence of the light-dependent reactions. Lest you miss the forest for the trees, let's be clear about two momentous things that have taken place within these steps.

The Liberation of Oxygen from Water

The first notable event takes place near the start, when water is split. This reaction provides the traveling electrons whose path you've just followed. It also provides something else: the oxygen that today accounts for more than 20 percent of the Earth's atmosphere. *Here* is where we get the substance that is the breath of life itself for most species, all of it contributed by living things. Given its importance, it is no small irony that oxygen is, in another sense, a kind of green-world refuse—leftovers from the process by which plants strip water of the electrons they need to carry out photosynthesis.

The Transformation of Solar Energy to Chemical Energy

The second major feat that takes place during light-dependent reactions is the transformation of solar energy to chemical energy. When the energy of sunlight moves to a chlorophyll molecule in a reaction center, it boosts a chlorophyll electron from what is known as a ground state to an excited state. Thus far, this has been described as a movement up a metaphorical energy hill. In physical terms, however, this electron is moving farther out from the nucleus of an atom. To appreciate what's special about photosynthesis, it's necessary to note what the *common* fates are for such excited electrons. One fate is for the electrons to drop back down to the original (and more stable) state, in the process releasing as *heat* the energy they have absorbed. This is why black objects get hot on sunny days. Another possible fate is for falling electrons to release part of their energy as light in the process known as fluorescence.

In photosynthesis, however, the energized electrons are *transferred to a different molecule*—the initial electron acceptors of photosystems II and I. They don't fall back to their ground state, releasing relatively useless heat or light in the process. They are passed on in a redox reaction. This is the bridge to life as we know it. Without this step, the energy of the Sun merely makes the Earth warmer. With it, the green world grows profusely; it captures some of the Sun's energy and with it builds trunks and grains and leaves. By one reckoning, photosynthesis produces up to 155 billion tons of material each year. This is an amazing amount—about 25 tons for every person on Earth. Electrons are taken from water, boosted to a higher energy state by the Sun, and transferred to other molecules. The result is the food that sustains nearly all living things.

It's also worth noting that a primary function of the fall of electrons between photosystems II and I is a release of energy that is used for the production of ATP. This ATP is used to power the reactions that are coming up in the second stage of photosynthesis. Were we to walk through this process of ATP production, you would see great similarities between it and the ATP production process reviewed last chapter. In photosynthesis, however, the energy that powers ATP production comes from the Sun (not from food). It is the energetic fall of electrons

between photosystems II and I that releases the energy necessary to produce ATP. But these electrons are able to fall energetically only because they were *boosted* energetically by the Sun's power, in photosystem II.

8.5   Stage 2 of Photosynthesis: The Light-Independent Reactions

Now, back to the central story. You are ready to see how carbon dioxide and high-energy electrons are incorporated into carbohydrates in the light-*independent* reactions. To begin, let's take stock of where the first set of reactions left off. The short answer is: in the stroma. That's where NADP$^+$ has become NADPH by accepting the energized electrons pouring out from photosystem I. There's also ATP in the stroma, freshly made from the light-dependent reactions.

Tutorial 8.2.4
**Light-Dependent
Reactions**

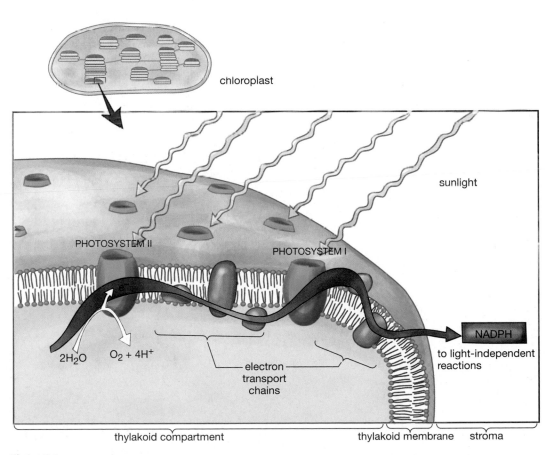

Figure 8.6
Light-Dependent Reactions
The light-dependent reactions take place in the thylakoid membranes within the chloroplasts. Electrons are donated by water molecules located in the thylakoid compartments. Powered by the Sun's energy, these electrons are passed along the electron transport chain embedded in the thylakoid membrane and end up stored in NADPH in the stroma. An additional product of the splitting of water molecules is oxygen atoms, which quickly combine into the O$_2$ form. This is the atmospheric oxygen that we breathe in.

Energized Sugar Comes from a Cycle of Reactions: The Calvin Cycle

Even with the right ingredients, however, it takes more than one step to go from carbon dioxide and a sugar to a sugar that is energetic enough to serve as food. It takes a set of steps that make up a cycle, commonly known as the Calvin cycle or the C_3 cycle. The highlights of this cycle are set forth in **Figure 8.7**. (The Calvin cycle is named after the chemist Melvin Calvin, whose work you can read about in "Plants Make Their Own Food, But How?" on page 164.)

The *synthesis* of photosynthesis is, in its first steps, a process of **fixation**—of a gas being incorporated into an organic molecule. Specifically, carbon dioxide is being fixed into our starting sugar, known as RuBP.

Here are the steps in the **Calvin cycle**:

1. **Bringing together carbon dioxide and sugar.** The process begins with three molecules of CO_2 (coming in through the leaf stomata) being joined to three molecules of RuBP, which has five carbon atoms. Given that one carbon of CO_2 is being added to a five-carbon molecule (and that it is being

SUMMARY OF THE CALVIN CYCLE

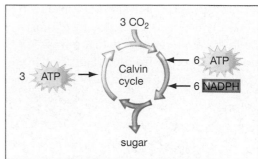

Tutorial 8.2.5
Light-Independent Reactions

Figure 8.7
Calvin Cycle

Overview: CO_2, ATP, and electrons (contained in hydrogen atoms) from NADPH are the input into the Calvin cycle, while a sugar (G3P) is its output.

1. **Carbon fixation.** The enzyme rubisco brings together three molecules of CO_2 with three molecules of the five-carbon sugar RuBP; the three resulting six-carbon molecules are immediately split into six three-carbon molecules named 3-PGA (3-phosphoglyceric acid).

2. **Energizing the sugar.** In two separate reactions, (a) Six ATP molecules react with six 3-PGA, in each case transferring a phosphate onto the 3-PGA. (b) The six 3-PGA derivatives oxidize (gain electrons from) six NADPH molecules; in so doing, they are transformed into the energy-rich sugar G3P (glyceraldehyde 3-phosphate).

3. **Exit of product.** One molecule of G3P exits as the output of the Calvin cycle. This molecule, the product of photosynthesis, can be used for energy or transformed into materials that make up the plant.

4. **Regeneration of RuBP.** In several reactions, five molecules of G3P are transformed into three molecules of RuBP.

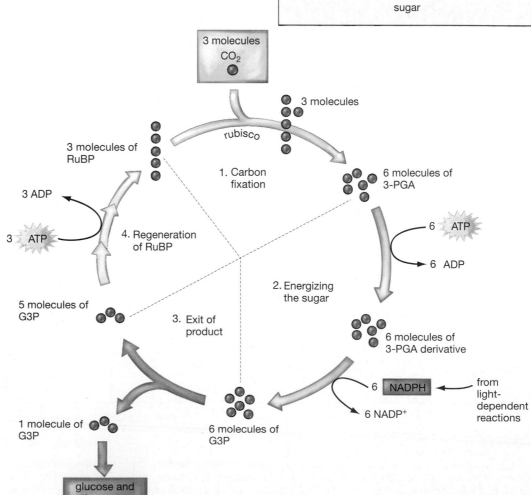

done three times), the result, of course, should be three *six-carbon* molecules. This is the case, but only for a vanishingly brief time. Each six-carbon molecule is so unstable that it is split apart immediately into two *three-carbon* molecules of a substance called 3-phosphoglyceric acid or 3-PGA. The three carbons here give the C_3 cycle its name.

A key player in photosynthesis is the enzyme that brings together CO_2 and RuBP. This massive protein goes under several aliases, but the one used here is **rubisco**. You need to be clear about what it is doing: bringing together in its active site CO_2 and the sugar RuBP. Here is where the small, low-energy gas molecule CO_2 is joined to the ranks of organic molecules, specifically sugars.

2. **Energizing the sugar.** Carbon dioxide has now been added to a sugar; what remains is for this sugar to be energized. These things take place in the next two steps. First, ATP reacts with the molecules of 3-PGA, putting a phosphate group onto each of them. Then NADPH donates a pair of electrons to each of the resulting molecules, yielding six molecules of a substance called glyceraldehyde 3-phosphate or G3P.

3. **Our food has arrived.** There should now be a little fanfare, because with this step the food has arrived. Note that G3P was produced by receiving energetic electrons from NADPH—the electrons that came originally from water and that were boosted up the energy hill by sunlight. G3P thus has more potential energy than did its predecessor in the Calvin cycle. It is precisely this position —a greater distance up the energy hill—that makes *food*.

G3P is thus the product of photosynthesis. It is, however, a product that is analogous to steel coming out of a factory: It can be turned into many things. Put two molecules of G3P together and you get the more familiar 6-carbon sugar, glucose. In turn, many molecules of glucose can come together to form the large storage molecules known as starches, familiar to us in such forms as potatoes or wheat grains. Beyond this, sugar is used to make proteins, which are then used as structural components of the plant or as enzymes. Given all this, you may ask, what is the *ultimate* product of photosynthesis? The answer is *the whole plant*.

4. **Regeneration of RuBP.** As you can see from Figure 8.7, the C_3 cycle still has some distance to go after the synthesis of G3P. Critically, only *one* of the six molecules of G3P will exit as product, while the other five will continue through the cycle to be used in the regeneration of RuBP. This is a process that, as you can see, is powered by the expenditure of more ATP. **Figure 8.8** summarizes the process of photosynthesis.

8.6 Photorespiration: Undercutting Photosynthesis

You have now gone through the basic process of photosynthesis. It's necessary, however, to say something about what might be called a *glitch* in this process, and then note a mechanism that plants have evolved to deal with it.

The glitch is called **photorespiration**, and it concerns a deviation from the process you just went through. Recall that the Calvin cycle begins with the important enzyme, rubisco, binding with carbon dioxide and then combining it with a five-carbon sugar. As it turns out, however, rubisco frequently binds with *oxygen* instead of carbon dioxide. When this happens, no carbohydrate is produced, which means no plant growth results from the reaction.

The amazing thing is how often photorespiration takes place. In C_3 photosynthesis, rubisco may fix up to one molecule of O_2 for every three of CO_2 it incorporates. This is no trivial matter, because many food crops use the C_3 cycle; in these instances, photorespiration is undercutting human food production.

Figure 8.8
Summary of Photosynthesis in the Chloroplasts of Plant Cells
In the light-dependent reactions, solar energy is converted to chemical energy in the thylakoids, and the chemical energy is stored temporarily in the form of ATP and NADPH. Water is required for this reaction, and oxygen is a by-product. The stored chemical energy is in turn used in the light-independent reactions (the Calvin cycle), taking place in the stroma, in which a high-energy sugar is made from carbon dioxide and the sugar RuBP. The sugar can be used for food, or may become part of the plant's structure.

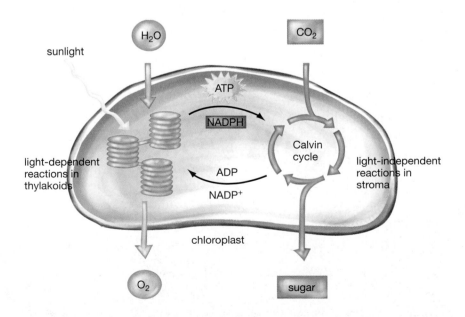

Figure 8.9
Warm-Weather Photosynthesis
Plants such as this sugarcane, shown here in Hawaii, utilize C_4 photosynthesis.

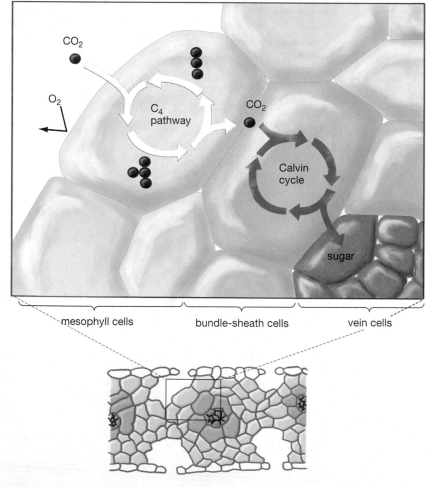

mesophyll cells   bundle-sheath cells   vein cells

Tutorial 8.4.2
C_4 **Photosynthesis**

Figure 8.10
The C_4 Pathway
The C_4 pathway is an adaptation of some warm-climate plants to the problem of photorespiration. Photorespiration occurs when the C_3 enzyme rubisco binds to oxygen rather than carbon dioxide (thereby stopping the production of sugar). C_4 plants contain a different enzyme in their mesophyll cells that binds to carbon dioxide, but not to oxygen. Then the carbon dioxide is escorted into special bundle-sheath cells, where rubisco can bind to it in the Calvin cycle reviewed earlier, meaning the production of sugar can proceed.

8.7 A Different Kind of Photosynthesis: The C_4 Pathway

Photorespiration is especially likely to take place in warm environments, because heat prompts the stomata on leaves to close in order to preserve water. The effect of closed stomata, however, is not only that water is kept *in*, but that CO_2 is kept *out*. With a deficit of CO_2, rubisco tends to bind more frequently with oxygen, prompting photorespiration— and plants that simply do not grow as much.

Evolution produced an adaptation that allows some warm-climate plant species to lessen this problem. It is a mechanism that utilizes a special carbon-fixing "front-end" and then a carbon-dioxide shuttle into the Calvin cycle. The so-called C_4 plants that make use of this mechanism (**see Figure 8.9**) don't initially use rubisco to fix carbon dioxide. Rather, they employ a different enzyme that does not bind with oxygen. The immediate product of this reaction is a four-carbon molecule (hence the C_4 cycle's name). The derivative of that molecule moves to a special group of cells and *releases* the CO_2, which will be used in the Calvin cycle. These special cells are wrapped around leaf veins and are hence named the **bundle-sheath cells** (**see Figure 8.10**). What the C_4 system provides—even at times when little CO_2 may be coming in through the stomata—is a relatively high concentration of CO_2 where it's needed: in the cells where the Calvin cycle takes place.

The C_4 Pathway Is Not Always Advantageous

The C_4 pathway is known to be employed in about 1,000 species of plants, most notably in some grasses, and in corn, sugarcane, and sorghum. But there are vastly more C_3 than C_4 plants. You might wonder why evolution hasn't weeded out C_3 plants in favor of the C_4 variety, since the latter seem to be so much more efficient at photosynthesis. The answer is that C_4 fixation does not confer an across-the-board advantage. It has a cost, which is the expenditure of ATP in shuttling carbon dioxide to the bundle-sheath cells. What this means is that the warm-weather advantage of C_4 fixation doesn't travel well; there are few C_4 species in cold-weather climates.

8.8 Another Photosynthetic Variation: CAM Plants

When plants live in climates that are not just warm but *dry*, a large part of their survival comes down to retaining water. Photosynthesis, however, works against water retention, because as you have seen, when CO_2 can pass in, water vapor can pass out.

The plants we call succulents (and some other plants) have a solution for this: Close the stomata during the day and open them at night. The plants then carry out a C_4-like metabolism at night, but only up to a point. They fix carbon dioxide into an initial four-carbon molecule and then stand pat. The CO_2 stays "banked" in them, awaiting the energy of the next day's Sun, which will power the production of the ATP they need to carry out the range of photosynthetic operations (**see Figure 8.11**).

This process, called **CAM metabolism**, gets the job done for cactus, pineapple, orchid, and mint family plants, among others. As you've no doubt guessed, CAM is an acronym; it stands for *crassulacean acid metabolism*. The succulent plant family *Crassulaceae* was the first group of plants in which CAM metabolism was discovered. The three methods of photosynthesis—C_3, C_4, and CAM—are summarized in **Figure 8.12**.

Closing Thoughts on Photosynthesis and Energy

Readers who have stayed the course on this account of photosynthesis now know at least one thing very well about it: How complicated it is, with its many components and long metabolic pathways. It's also probably clear by this point why scientists continue to study photosynthesis in such detail. It is the foundation of plant growth, and upon plant growth hinges nothing less than the survival of all animals—including human beings. Without an understanding of this linkage, it's easy to see plants as a set of mute fixtures whose main contribution to human life is aesthetic. But with this knowledge, you can begin to see the central position that plants occupy in the interconnected web of life.

Over the last three chapters, you have learned about the endless back-and-forth of oxygen, carbon dioxide, and energy in the living world. *Cycle* has been a recurring word in this long discussion, because the only one-way trip you've encountered has been the relentless "spillage" of energy from the Sun down into heat. Looked at in a cynical way, Earth and its inhabitants constitute a kind of leaky holding tank for energy that comes from the Sun. Looked at another way, however, the living

world has been able, through photosynthesis, to take the Sun's energy and build a remarkable edifice with it. Think of the forms and sheer mass of living material on Earth. One of the most amazing things about this structure is that we humans get to be both a part of it and witnesses to it.

Tutorial 8.4.3
CAM Photosynthesis

Figure 8.11
Dry-Weather Photosynthesis
Plants such as this saguaro cactus in Arizona utilize CAM photosynthesis, thereby preserving precious water.

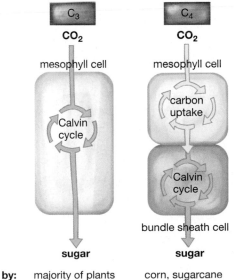
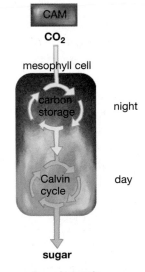

	C_3	C_4	CAM
	CO_2	CO_2	CO_2
	mesophyll cell Calvin cycle	mesophyll cell carbon uptake Calvin cycle bundle sheath cell	mesophyll cell carbon storage (night) Calvin cycle (day)
	sugar	sugar	sugar
Used by:	majority of plants	corn, sugarcane (warm environments)	cactus, pineapple, orchid (dry environments)
Benefits:	efficient use of ATP	less photorespiration	less water loss
Problems:	photorespiration	uses up more ATP	uses up more ATP; hard to "bank" enough CO_2

Figure 8.12
Three Modes of Photosynthesis

Plants Make Their Own Food, But How?

Science constantly is confronted with what are known as "black box" problems. In these problems, researchers know what goes into a process and they know what comes out of it, but what happens *in between* is a mystery. At the conclusion of World War II, plant science had a doozy of a black-box problem. It was clear that plants were starting with carbon dioxide from the air and ending with high-energy carbohydrates, but no one knew what came in between. What were the *intermediate* substances that carbon dioxide was being made part of on the route to food? Getting inside this black box turned out to require a three-part combination: insightful researchers, new scientific techniques, and old-fashioned hard work.

One of the peacetime benefits of World War II's atomic weapons research was that scientists could produce quantities of substances known as *radioactive isotopes*. As you saw in Chapter 2, isotopes result from variations in the number of neutrons an element has in its nucleus. Radioactivity, meanwhile, is the release of radiant energy from atoms undergoing rapid decay. The element carbon is everywhere in nature; and as it turns out, one of its isotopes, carbon-14, is radioactive. Carbon-14 occurs naturally, but the ability to *manufacture* this substance opened a host of possibilities for research in the mid-1940s.

Stepping up to the plate in 1946 to take advantage of these possibilities was a University of California, Berkeley, chemist named Melvin Calvin. Then just 35 years old, Calvin was encouraged by the physicist Ernest Lawrence of Berkeley to apply radiocarbon techniques to organic chemistry research. With carbon-14, Calvin realized, he could *tag* carbon dioxide to find out what it becomes part of during photosynthesis, much as a zoologist might tag a bear with a signaling device to find out where it goes during winter. The radiation that carbon-14 emits might be thought of as a signal whose constant message is: "Here I am."

The steps that Calvin and his colleagues followed were simple in conception but very difficult in execution. They first took some photosynthetic algae and put them in a flask that had light shining on it. Like any photosynthetic organism, these algae would, of course, be taking in carbon dioxide to carry out their work. In this case, however, the carbon dioxide in question contained carbon-14. The trick was to allow the algae to perform a little photosynthesis using this $^{14}CO_2$. Then the algae were plunged into boiling alcohol, which killed them. By the time the plunge was taken, the carbon dioxide would have made it to *some point* in its transformation to carbohydrate; it would have

been incorporated into one molecule in the series of intermediate molecules involved. Stopping photosynthesis at differing points—two seconds after it started, five seconds, and so on—meant stopping the progress of carbon dioxide as it marched through its changes.

All of this would have been for naught, however, without a way to figure out where the carbon was when the process was stopped. This is where the carbon-14 could do its job, but it needed to be joined to another technique that was also newly developed at that time: two-dimensional paper chromatography. If you look at **Figure 1a**, you can see how this technique works. Put a spot of ink on some filter paper and then stick one end of the paper in a solvent—water, for example. The water moves through the paper, carrying different components of the ink along with it at different rates. You can tell something about what the ink is made of by measuring how far its components move through the paper. You can tell still more by doing this in two dimensions: Let the solvent move through the spot one way, then set the paper down at a right angle in a *second* solvent and let it move through in another direction (**Figure 1b**). This is what Calvin and his colleagues did, using as their "ink" the algae that had been killed following photosynthesis.

The paper chromatography left the algae cells drawn out in two directions by the solvents. But where was the original carbon in this material? Because it was radioactively *tagged* carbon, it was sending out its signal saying, "I'm here now." But only a special "receiver" could pick up this signal: photographic film (**Figure 1c**). Setting x-ray film tightly on top of the chromatography paper for long periods—usually about two weeks—caused the radiation from the carbon-14 to expose the film. Spots would show up saying, in effect, "carbon-14 is now here, in this concentration."

There is a wide gap, however, between seeing a black blotch on a piece of film and *identifying* that blotch as a specific substance. The difficulty the researchers encountered can be measured in time—10 years in this case. That's how long it took these workers to complete their investigation. During that time, Calvin and his principal colleagues—postdoctoral researcher Andrew Benson and graduate student James Bassham—worked extraordinarily long hours in deciphering what the downstream products of photosynthesis were and the order in which they were produced.

This long operation was a great success, however. Calvin's team dismantled the black box of carbon fixation, and the information they uncovered can be *used* to aid in such things as increased food production. The set of steps these workers elucidated has variously been referred to as the C_3 cycle, Calvin-Benson cycle, or the Calvin cycle. In 1961,

Melvin Calvin received the ultimate in congratulations for a piece of scientific work when he was awarded the Nobel prize in chemistry for shining a light on photosynthesis.

Figure 1

Tutorial 8.3.1
The Calvin Experiments

a paper chromatography in first dimension

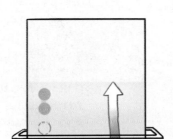
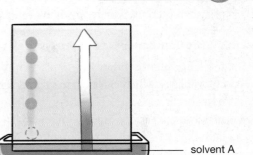

Material of interest (algae for Calvin) is put on paper in solution form.

Solvent A moves through paper, separating out components of original material. (Components, shown here as green dots, would probably not be visible in experiments.)

solvent A

b paper chromatography in second dimension

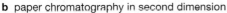
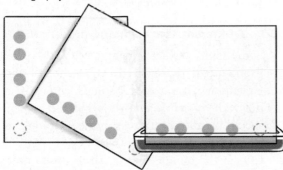
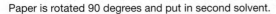
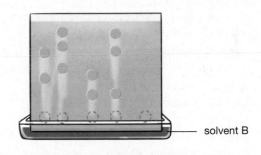

solvent B

Paper is rotated 90 degrees and put in second solvent.

Component compounds are separated in second dimension. Some compounds contain radioactively labeled carbon.

c autoradiography

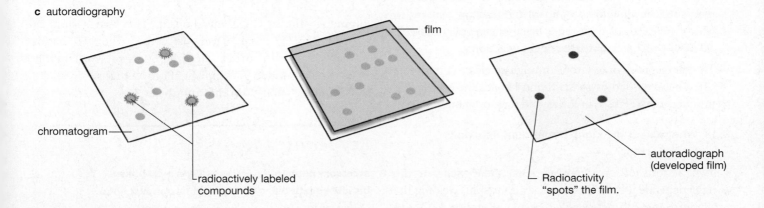

film

chromatogram

radioactively labeled compounds

Radioactivity "spots" the film.

autoradiograph (developed film)

Chapter Review

Summary

8.1 Photosynthesis and Energy

- Photosynthesis is the process by which certain organisms produce their own carbohydrates, using carbon dioxide, water, minerals, and energy captured from the Sun's rays.

- Photosynthesis has made possible life as we know it on Earth, because the organic material produced in photosynthesis (sugar) is the source of food for most of Earth's living things. Photosynthesis also is responsible for the atmospheric oxygen used by many living things in respiration.

8.2 The Components of Photosynthesis

- In plants, photosynthesis takes place in the organelles called chloroplasts, which can exist in great abundance in the mesophyll cells of leaves. The energy for photosynthesis comes mostly from various blue and red wavelengths of visible sunlight that are absorbed by pigments in the chloroplasts. Plant leaves contain microscopic pores called stomata that can open and close, letting carbon dioxide in and water vapor out.
  TUTORIAL 8.1.2: The Electromagnetic Spectrum

- There are two primary stages to photosynthesis. In the first stage, electrons derived from water are energetically boosted by the power of sunlight. The electrons physically move in this process from special chlorophyll molecules through a series of electron carriers, ending up as part of the electron carrier NADPH, which carries the electrons to the second stage of photosynthesis. In this second stage, the electrons are brought together with carbon dioxide and a sugar to produce a high-energy sugar.
  TUTORIAL 8.2.1: Overview of Photosynthesis

8.3 Stage 1: The Steps of the Light-Dependent Reactions

- In its first stage, photosynthesis works through a pair of complexes, photosystems II and I. These photosystems are composed of antennae chlorophyll and accessory molecules that absorb and transmit solar energy; reaction centers that accept both this energy and electrons derived from water; and primary electron receptors (a part of the reaction centers), to which the electrons move after being energetically boosted.
  TUTORIAL 8.2.4: Light-Dependent Reactions

- The energetic fall of electrons through an electron transport chain between photosystems II and I yields the energy that produces the ATP used in the second stage of photosynthesis.

8.4 What Makes the Light-Dependent Reactions So Important?

- Two actions of great consequence take place in the light-dependent reactions. First, water is split, yielding the electrons that ultimately play a part in providing food, and liberating oxygen from water as well. This is the oxygen that organisms such as ourselves breathe in. Second, the electrons that are given an energy boost in the light-dependent reactions are transferred to a different molecule, the initial electron acceptor. This is one of the keys to the plant's ability to make its own food.

8.5 Stage 2 of Photosynthesis: The Light-Independent Reactions

- In the second stage of photosynthesis, through the process of the Calvin cycle, carbon dioxide from the atmosphere is brought together with a sugar, RuBP, by the enzyme rubisco. The resulting compound is energized with the addition of phosphate groups and electrons supplied by the first stage of photosynthesis. The result is the high-energy sugar G3P, which is the product of photosynthesis.
  TUTORIALS: 8.2.5 Light-Independent Reactions;
  8.3.1 The Calvin Experiments

- G3P can be used for energy or for plant growth. Everything in the plant ultimately is derived from this sugar.

8.6 Photorespiration: Undercutting Photosynthesis

- In C_3 plants, the enzyme rubisco frequently binds with oxygen, rather than carbon dioxide, thus undercutting productivity in photosynthesis.

8.7 A Different Kind of Photosynthesis: The C_4 Pathway

- Some plants in warm climates, where this problem is most pronounced, have evolved a C_4 method of photosynthesis. C_4 employs an enzyme that binds with carbon dioxide but not with oxygen. The carbon dioxide is then shuttled to special bundle-sheath cells in the plant and released, after which it moves into the Calvin cycle. With high levels of CO_2 in the bundle-sheath cells, rubisco binds with CO_2 (and not oxygen), thus greatly reducing photorespiration.
  TUTORIAL 8.4.2: C_4 Photosynthesis

8.8 Another Photosynthetic Variation: CAM Plants

- Dry-weather plants such as cacti employ another form of photosynthesis, CAM metabolism. The plant's stomata open only at night, letting in carbon dioxide that is "banked" until sunrise, when the Sun's rays will supply the energy needed to complete photosynthesis.
  TUTORIAL 8.4.3: CAM Photosynthesis

Key Terms

accessory pigment 156
bundle-sheath cell 162
Calvin cycle 160

CAM metabolism 163
chlorophyll *a* 156
chloroplast 156

Understanding the Basics

Multiple-Choice Questions

1. Most of your house plants are green (if they are healthy) because they
   a. absorb green light
   b. take the energy out of white light and produce green light
   c. contain molecules that manufacture green light in the process of photosynthesis
   d. use green light in photosynthesis
   e. scatter green light

2. A college student asked his freshman biology teacher about the leaves on his plant, which were turning white. She told him he should
   a. be concerned because this means that the plant pigments are disappearing so the plant cannot photosynthesize
   b. be concerned because his plant was producing so much starch in photosynthesis that it would be unable to store any more
   c. not be concerned because plants use many colors of light for photosynthesis
   d. not be concerned because many parts (such as the roots) of plants are white, so plants photosynthesize just fine with white leaves
   e. be concerned because white pigments damage plants

3. A tulip tree turns yellow in the fall, because the accessory pigments show up when the chlorophyll has faded. These accessory pigments
   a. are formed to take over the function of chlorophyll in the fall
   b. are produced to protect the leaves in the fall
   c. are always present to help absorb light for photosynthesis
   d. work only when chlorophyll fails
   e. are found just outside the chloroplast to pass light into it

4. Suppose you become a biologist. On a bright summer day, while studying the rate of photosynthesis in a desert succulent, you discover that its stomata are closed—yet it is producing ATP. This plant must be carrying out
   a. photorespiration
   b. the C_3 cycle
   c. the C_4 pathway
   d. CAM metabolism
   e. CO_2 uptake

5. Most plant photosynthesis occurs in the _____ of leaves.
   a. upper epidermal cells
   b. lower epidermal cells
   c. guard cells
   d. stomata
   e. mesophyll cells

6. A leaf on a tree is photosynthesizing rapidly, capturing light energy through light-dependent reactions. Why do the cells of the tree also need light-independent reactions?
   a. to use up the excess energy
   b. to cause photorespiration to occur
   c. to make sugars so that the tree can use the energy it captures
   d. to release the energy to the atmosphere
   e. to form the woody portion of the tree

7. Oxygen is released to the atmosphere in the light-dependent reactions of photosynthesis when water is split to supply electrons to
   a. photosystem I
   b. photosystem II
   c. Calvin cycle
   d. C_4 pathway
   e. CAM

Brief Review

1. Because plants harness energy in photosynthesis, do they need cellular respiration as well? Why or why not?

2. Sunlight reaching Earth is composed of a spectrum of energetic rays. Describe the rays that make up this spectrum. Which portion of the spectrum drives photosynthesis?

3. How is a photosystem put together? What are its parts? Explain why these all need to work together for the photosystem to function.

4. Picture a world without photosynthesis. What would it be like? What would happen to plants, animals, fungi, and so on?

5. Which would you expect to grow more rapidly—a C_3 plant, a C_4 plant, or a CAM plant? Be careful; there may be more than one answer to this question. Explain your answer.

6. List the four steps in the Calvin cycle, and explain why they are necessary. Why should you assume that all these steps are necessary? Do plants or animals typically waste energy? Why or why not?

Applying Your Knowledge

1. Rubisco has been called the most important protein on Earth. What arguments can you make to support that hypothesis? Do you have any that would go against it?

2. One of the most dramatic environmental changes today is the increase in carbon-dioxide levels in the atmosphere. We usually hear about it in connection with "global warming," but that's not the only issue. What effect would increased carbon dioxide have on plants? If more trees are planted, would this stand to have any effect on atmospheric CO_2 levels?

MediaLab
Capturing Sunlight to Make Food: Photosynthesis

You've probably heard songs in which the singers attribute their life, the act of eating and breathing, to their romantic interest. Love is a powerful emotion, but you'd think at least one songwriter would give homage to the event that really does make our lives possible—the process of photosynthesis. Perhaps it wouldn't sell. The *CD-ROM Tutorial* reviews photosynthesis, so that in the following *Web Investigation,* you will be prepared to gather information on topics closely associated with it. In the *Communicate Your Results* section, you can interpret for yourself how important photosynthesis is in our lives.

This *MediaLab* can be found in Chapter 8 on your CD-ROM (Tutorial 8.2.1) and Companion Website (http://www.prenhall.com/krogh3).

CD-ROM TUTORIAL

The overall process of photosynthesis is often summarized in the following reaction:

water + carbon dioxide + light energy →
glucose + oxygen

As with most biological processes, photosynthesis does not occur in one step, but in a series of reactions. In this *CD-ROM Tutorial*, you'll review the three phases of photosynthesis, focusing on reactants and products.

Activity

1. First, you will learn about how light energy is absorbed by pigment molecules in plants.

2. Next, you will discover how light energy is converted to a stored chemical form for transport and later storage.

3. Finally, you will investigate the storage of energy in glucose, and test your skills in a simulation of photosynthesis.

After this review, you'll be prepared to apply, extend, and experience some of the consequences of the photosynthetic process in the following *Web Investigation* section.

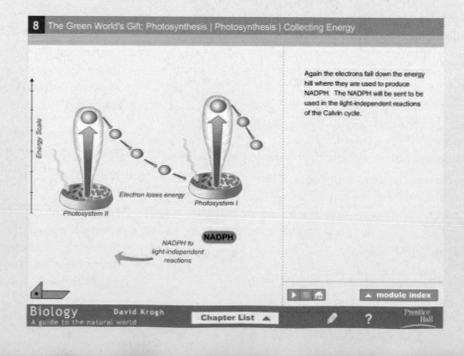

Investigation 1

Estimated time for completion = 10 minutes

If photosynthesis is to occur at all, the plant must be able to intercept light. The better a plant is in intercepting the light, the greater opportunities the plant has to produce food. Select the Keyword **LEAVES** on your CD or Website for this *MediaLab* to read an article describing how plants maximize the amount of light they can intercept. At the bottom of that page is a forward link to a discussion of leaf position in a canopy. Review this page. Why have so many plants developed broad, flattened leaves? What is heliotropism, and how does it make plants more efficient? What are sleep movements? Describe the model for leaf orientation in the upper, middle, and lower levels of a leaf canopy.

Investigation 2

Estimated time for completion = 10 minutes

Plants absorb light energy using special molecules called photosynthetic pigments (predominantly chlorophylls). These pigments absorb most of the visible light except for wavelengths in the green part of the spectrum. Consequently, plants appear green. But what happens in the fall when leaves change colors? Why are there so many different colors? And why are the colors more intense some years? Select the Keyword **COLORS** on your CD or Website for this *MediaLab* to access a site that will help you answer these questions.

Investigation 3

Estimated time for completion = 15 minutes

You've probably heard of global warming and the greenhouse effect, in which certain gases cause heat to be trapped in the Earth's atmosphere in the same way heat is trapped in a greenhouse. How hot will it be in your community in 100 years? Select the Keyword **HOTSPOT** on your CD or Website for this *MediaLab* to review predicted temperatures for the year 2100. One of the major greenhouse gases is carbon dioxide, which has been added to the atmosphere by the burning of fossil fuels. Of course, plants in photosynthesis trap CO_2 out of the atmosphere. Could plants help with global warming? Select the Keyword **CO_2** to view changing CO_2 levels over 25 years in Barrow, Alaska; Mauna Loa, Hawaii; Samoa, and the South Pole. What might account for the yearly fluctuations in CO_2 levels? Where fluctuations are the greatest, is there a seasonal difference in plant activity?

COMMUNICATE YOUR RESULTS

Exercise 1

Estimated time for completion = 15 minutes

Using the knowledge from the articles in *Web Investigation 1*, design a plant that can effectively intercept light while keeping the amount of plant tissues to a minimum. Here are some characteristics of this hypothetical plant to keep in mind: The plant grows in a seasonal environment, water conservation is important but not critical, and the plant typically has ten leaves located at five levels on the stem. Draw the leaves for this plant. Also draw a view of the plant from the side and a view of the plant from the top. Describe the decisions you made while creating this plant.

Exercise 2

Estimated time for completion = 15 minutes

If it's fall and the leaves are changing, make a collection of leaves from different species in your area. Otherwise, find pictures of fall leaves and clip or copy them to create a collection. Place these leaves on a poster or in a bound collection. Identify the species of plant as well as possible using available references. For each species, using the leaf's color as a guide, identify the dominant pigment present in the leaf. Is there a pattern of leaf color and plant type? Describe any patterns you observe.

Exercise 3

Estimated time for completion = 15 minutes

In *Web Investigation 3*, you saw an annual fluctuation in CO_2 levels. One possible explanation for the yearly cycle is that increased plant growth decreases atmospheric CO_2 levels during the growing season. How would you test this hypothesis? Prepare a short presentation or poster explaining the initial observation and hypothesis; and your test, observations, and conclusions. Include your conclusions on how photosynthesis could be used to fight global warming, and how deforestation would contribute to the problem.

9

Introduction to Genetics; Mitosis and Cytokinesis

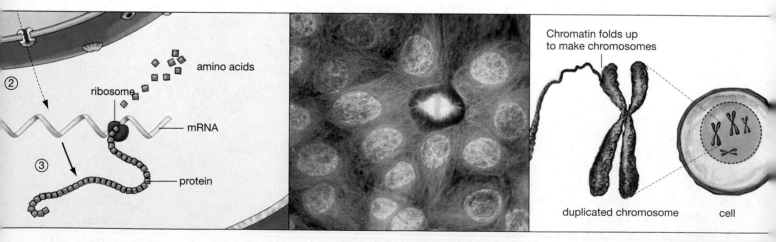

amino acids

ribosome

② mRNA

③ protein

Protein synthesis in action.
(Section 9.1, page 173)

What does a cancer cell look like?
(Essay, page 175)

Chromatin folds up
to make chromosomes

duplicated chromosome   cell

A closer look at
the chromosome.
(Section 9.3, page 177)

Life is heavily dependent on information that is stored in DNA. This information is duplicated and then apportioned every time a cell divides.

Everyone knows how one generation of living things gives rise to another, at least in a general sense. But the details of how this works are not apparent just by looking at Earth's creatures. A brief sexual encounter takes place, and the next thing we know there are new kittens or chicks or babies. Only the encounter and the newborns are obvious; what comes in between is mysterious.

The problem, of course, is that most of the component parts of reproduction are hidden away and, in any event, so small that the process is completely invisible to the naked eye. Given this, it's no wonder that for centuries human beings wondered what it was that one generation passed on so that the next generation could be formed.

A similar kind of puzzlement existed over a related question: What is the mechanism that controls the development of a living thing as it goes from being a microscopic cell to being a human baby in full cry? For that matter, how is it that the adult body is able to build muscle or to repair a minor wound?

What, in short, is the mechanism that controls day-to-day physical functioning?

9.1 An Introduction to Genetics

The Key to Reproduction, Development, and Maintenance Is DNA

It took roughly a century of hard-won scientific advances for these questions to be answered. In the end, it turned out that a single substance is central to the reproduction of living things, their development from single cells, and their day-to-day functioning. This substance is deoxyribonucleic acid, a long, vanishingly thin molecule known everywhere today by its acronym, DNA.

DNA contains what is known as an organism's **genome**: the complete collection of that organism's genetic information, strung together in functional units called genes that lie along DNA's famous double helix.

How is it that this one molecule can undertake all the tasks just described? To the question about what it is that humans *pass on*

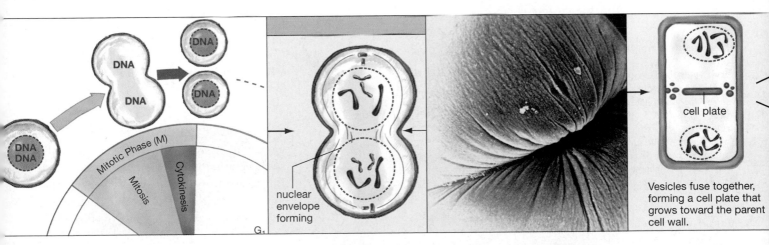

How does mitosis work?
(Section 9.3, page 179)

The mechanics of cell division.
(Section 9.4, page 180)

A frog egg undergoing cell division.
(Section 9.4, page 182)

Plant cell division.
(Section 9.5, page 182)

cell plate

Vesicles fuse together, forming a cell plate that grows toward the parent cell wall.

nuclear envelope forming

Mitotic Phase (M)

Mitosis

Cytokinesis

DNA

DNA

DNA

DNA

DNA

DNA

DNA

G₁

in reproduction, the answer is: half of a father's genome and half of a mother's genome, both these halves coming together in a fertilized egg to produce a whole, new genome (**see Figure 9.1**).

To the other questions about what controls the development and everyday functioning of living things, the answer is: the instructions provided by the genes in the genome.

DNA Contains Instructions for Protein Production

To grasp how genes can control the development and functioning of living things, note that there are logically two parts to such control. If genes are *giving* instructions, something needs to carry them out. As it turns out, the workers here are **proteins**; in tremendous diversity and number, they are the working products of the information contained in genes.

Readers who went through Chapter 3 (on biological molecules) and Chapter 6 (on energy) should have a good grasp already of what proteins are capable of. As we have seen, there are proteins that form cartilage and hair; there are communication proteins that help form the receptors on cells; there are hormonal proteins and transport proteins. Then there are **enzymes**, the chemically active proteins that speed up—or in practical terms, *enable*—chemical reactions in living things. Without these protein molecules, almost nothing could get done in the body; with them, a dazzling set of chemical reactions takes place every second in every living thing.

How Do Genes Direct the Production of Proteins?

What most genes do, then, is contain the information for the production of proteins. In Chapter 14, you'll get the details of how protein production comes about, but for now let's just look at the essentials of the story. Imagine you are a telegraph operator in the Old West, and your "key" starts tapping frantically. You leap forward in your chair, listening for the sound of short and long bursts on the key, and you begin writing down the letters these sounds stand for. Here's the message you get:

·—·· ——— —·—· —·—   —   ····   ·   ··· ·— ··—· ·

which means:

L o c k   t h e   s a f e

You walk over and do just that. There is thus a code here, Morse code, that has resulted in some action. In DNA there is likewise a code. But instead of a code specified in short or long sounds, it is a code specified in chemical substances. There are four of these substances, and they lie along the double helix. They are the "bases" adenine, thymine, guanine, and cytosine, abbreviated A, T, G, C.

A particular sequence of these bases— usually thousands of letters long—contains information that can be acted upon, just as the Morse code did. Only the action here is

Figure 9.1
Passing on Genetic Information
Half of each person's genome comes from the father, and half from the mother.

Figure 9.2
Information-Bearing DNA Molecule
The DNA molecule is composed partly of two handrails made of sugar and phosphate. The genetic information in the molecule is contained in the sequence of "bases" along one strand of the double helix. For example, the bases in the highlighted section of DNA are in the order CTGA

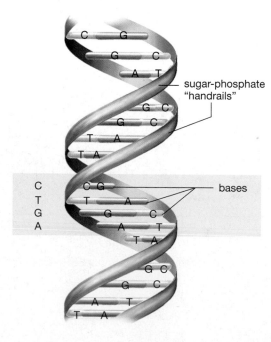

sugar-phosphate "handrails"

bases

the production of a particular protein. One series of A's, T's, C's, and G's contains the information for the production of one protein, but a *different* sequence of A's, T's, G's, and C's specifies a different protein. These separate sequences of bases are separate *genes*.

You saw in Chapter 3 that proteins are composed of building blocks called amino acids. The base sequence that makes up a gene, then, is like a message that says, "Give me this amino acid, now this one, then this one, . . ." and so on for hundreds of steps until a protein is created that folds up and gets busy on some task. This task can be one small part of *forming* a human being—as in embryonic development—or of *maintaining* one, as in an adult. One generation gives rise to another by passing on this entire complement of information.

The Architecture of DNA

It is the architecture of the DNA molecule that makes this protein production process possible. Look at the general scheme of it pictured in **Figure 9.2**. As you can see, DNA looks something like an open, spiral staircase, the handrails of which are a repeating series of sugar (deoxyribose) and phosphate molecules. Meanwhile, the messages you've just been reading about are contained in the steps of the staircase, in the order in which the bases A, T, C, and G are placed along the helix. As you can see, these bases extend inward from both of the DNA rails. The order in which any of the bases occurs *along the rails* is extremely varied. This order is what "encodes" the information for the production of proteins.

The Path of Protein Synthesis

Look now at **Figure 9.3** and see the path that is followed in protein synthesis. The DNA you have been reading about is contained in the nucleus of the cell. The first step is that a stretch of it unwinds there, and its message—the order of a string of bases—is copied onto a molecule called messenger RNA (mRNA). This length of mRNA, which could be thought of as an information tape being dubbed off a "master" DNA tape, then exits from the cell nucleus. Its destination is a molecular workbench in the cell's cytoplasm, a structure called a ribosome. It is here that both message (the mRNA tape) and raw materials (amino acids) come together to make

the product (a protein). The mRNA tape is "read" within the ribosome, and as this happens a growing chain of amino acids is linked together in the ribosome in the order called for by the mRNA. When the chain is finished and folded up, a protein has come into existence.

Genetics as Information Management

The story you have been reading about is that of **genetics**, meaning the study of heredity. In reflecting on this story, the first thing to note is that it concerns the storage, duplication, and transfer of *information*. True, it is information encoded in *chemical form*, but that shouldn't bother us. Information comes in lots of forms; there are the familiar words on a page, but there are also bar graphs, smoke signals, or a baby's smile.

To grasp the fact of genetics as molecular information, think back to Chapter 6, where you looked at a particular protein, an enzyme called lactase that helps break up milk sugar (lactose) into its components of glucose and galactose. In line with what you've seen so far, there is a gene that prompts the production of lactase. This gene's message is

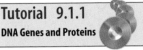

Tutorial 9.1.1
DNA Genes and Proteins

Figure 9.3
The Path of Protein Synthesis
(1) The information contained in a length of DNA is transcribed onto the "tape" of a length of messenger RNA (mRNA). (2) The mRNA then exits the cell's nucleus and goes to a structure in the cell's cytoplasm called a ribosome. (3) There, the mRNA tape is "read," and a string of amino acids is put together in the order specified by the tape. The result is a protein.

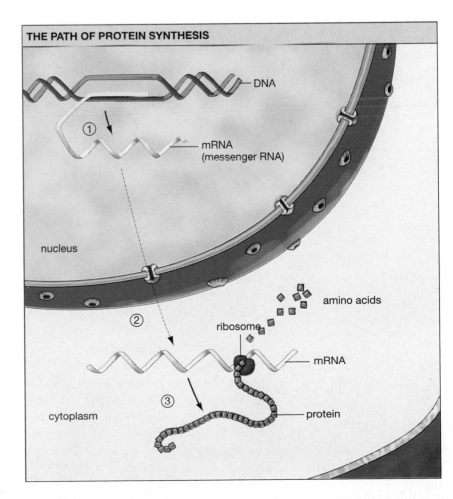

copied onto mRNA, which then migrates to a ribosome in the cytoplasm. Lactase is put together there, amino acid by amino acid, and, once synthesized, it folds up and gets busy clipping lactose molecules.

Now, consider the role of the gene in this process. It didn't do the lactose clipping; it didn't even migrate to the cytoplasm to be involved in the synthesis of lactase. In this case, its role was more like that of a . . . cookbook recipe. It specified this amino acid, then that amino acid, and eventually lactase was produced.

From One Gene to a Collection

Now, in considering not just one gene, but rather the entire *collection* of genes in a living thing—its genome, in other words—it is easy to see that we are dealing with a vast *library* of information. The size of such a collection can vary, but to give you some idea, the human genome includes at least 26,000 genes and may include as many as 40,000. We generally think that information comes to us after birth and from outside the body, but in fact we are born with a volume of information that has been amassed and edited over 3.5 billion years of evolution.

The amazing thing about such collections is that there is not just one copy of them in living things; there may be trillions. Most cells within an organism contain a complete copy of that organism's genome. A given type of cell puts to use (or "expresses") only some parts of this genome, while another type of cell utilizes other parts; but this is merely a way of saying that different genes are active in different cells. This is what distinguishes a liver cell, say, from a bone cell. These cells do different things and, accordingly, need to synthesize different proteins. But *both* kinds of cells have within them the whole genomic library; they merely put different parts of it to use.

Cells duplicate, of course. One cell divides to become two, the two divide to become four, and so on. Given this, if each cell in the body is to have its own copy of the genome, then each time a cell duplicates, the genome that lies within that cell must duplicate as well. This is just what happens, in a process that you'll be going over shortly. Before beginning, however, let's take a step back and see how genetics as a whole is approached in this book.

The Path of Study in Genetics

From what you've seen so far, it is probably apparent that genetics is important in biology. Many biologists would go further and argue that genetics lies at the *heart* of biology, because it is so central to what distinguishes the living world from the nonliving. The ability to reproduce is a distinguishing characteristic of life, and reproduction is the business of genetics. Beyond that, living things are notable for the degree of organization they possess. Genetics is, to a considerable extent, the means by which life's unimaginable complexity is managed. It is thus a key to the high level of organization we see in life.

Given genetics' importance, in this book you'll be looking at it from several angles. First, you will study DNA mostly as it comes packaged in units called chromosomes. Then you will learn how genetics functions in whole organisms, namely Gregor Mendel's famous pea plants. Only later will you return to DNA proper—to its replication and its protein-coding function. Finally, you will see how basic knowledge about genetics is being applied in the brave new world of genetic engineering. Let's begin the tour now, by seeing how life goes from one cell to many.

9.2 An Introduction to Cell Division

How does a baby grow, or a plant develop, or a wound heal? Always through cell division. As noted in Chapter 4, with the possible exception of viruses, life exists only inside cells. Further, we know that cells come only from other cells. And the *way* cells come from cells is by dividing. To understand the continuity in life, then, we need to have some understanding of how cells divide.

The sheer numbers involved in this process are mind-boggling. In your body, as many as four million cell divisions are taking place each second. It is likely that every one of these divisions serves some purpose in maintaining your body, yet there are instances in which cells divide excessively. How does the disease we call cancer take hold? Always through the *unrestrained* division of cells. As you can see in "When Cell Division Runs Amok" cancer is intimately related to basic cellular processes.

When Cell Division Runs Amok: Cancer

In the main text, there is a review of the cell cycle, meaning the repeating pattern of growth, genetic duplication, and division that most cells go through. But what controls this cycle? This topic is of compelling interest because the *unrestrained* division of cells goes by another name: cancer. As it turns out, much of the cancer research that goes on today is research on the control of the cell cycle.

What causes unrestrained cell growth? To use a common but apt metaphor, cells can come to this out-of-control state in one of two ways: They can get their accelerator stuck, or they can have their brakes fail. The control mechanisms that *induce* cell division can become hyperactive, or the mechanisms that *suppress* cell division can fail to perform. You may have heard a couple of terms used to describe the genetic components of such processes. There are normal genes that induce cell division, but that when mutated can cause cancer; these are the stuck-accelerator genes, called oncogenes. Then there are genes that normally suppress cell division, but that, in a mutated state can cause cancer by acting like failed brakes. These are tumor suppressor genes.

Investigations in the 1980s and 1990s revealed that the master controllers of cell division are chemical complexes composed of two types of proteins: kinases and cyclins, which associate to make cyclin-dependent kinases or CDKs. The details by which CDKs govern cell division are complex and need not concern us here. Suffice it to say that these compounds work together in a linked chain of protein activity. CDKs brought into production prompt some cell-division task—such as the breakdown of the nuclear envelope—after which they prompt the *next* step in cell division.

CDKs might appear to be the domain of basic cell biologists, while tumor suppressor genes might appear to be the domain of cancer researchers. But the work of these different types of researchers is increasingly coming together. In fact, cell and cancer scientists seem at times to be like two sets of miners who are burrowing in from opposite sides of a mountain, only to meet unexpectedly in the middle.

Consider, on the cancer research side, Utah scientist Mark Skolnick and his colleague Lisa Cannon-Albright. Some years ago, the two were looking at data from a group of families that had a high incidence of melanoma, which is an often deadly form of skin cancer. By 1992, the researchers found that these families had a genetic susceptibility to melanoma. Skolnick and researcher Alexander Kamb then looked at melanoma cells and saw that they seemed to be missing one small part of a chromosome. This suggested that the culprit was a *gene* that was missing. When present and functioning normally, such a gene might *restrain* the growth of cells. It would be, in other words, one of the sets of brakes mentioned earlier—a

tumor suppressor gene. This is exactly what Kamb and Skolnick believed they were on to. Accordingly, they set out to find it and describe it.

Meanwhile, digging in from the cell biology side of the mountain was one of the nation's experts on the cell cycle. In 1993 David Beach, at Cold Spring Harbor laboratory in Long Island, identified a gene that he called p16. As it turned out, p16 coded for a protein that inhibits cell growth. It does this by binding to, and thus suppressing, none other than a cyclin-dependent kinase (specifically one called cdk4). When genes are found and characterized, they are put into a kind of gene directory, which Beach did with p16.

Then, in early 1994, Kamb and Skolnick, looking for the missing gene, zeroed in on what they believed might be the culprit. Reading its sequence of T's, C's, G's and A's, they then looked up this sequence in the gene directory to see if it had been characterized before. It had, as a matter of fact—by Beach! Though Beach could not be sure at the time that it was a tumor suppressor gene, it turned out that the gene that caused *melanoma* by its dysfunction was none other than the p16 *growth-inhibition* gene that Beach had found. A malfunctioning p16 gene, then, is like a set of failed brakes—one that does not produce a protein that restrains cell division.

Researchers have also found the flip side of this: a gene, which codes for cyclin, that seems at times to simply "stay on." Because cyclins are required to make CDKs become active, a cyclin that was overproduced might well cause *excess* CDK activation and prompt excessive cell division. Here, in other words, is a stuck accelerator. It's worth noting that in general, for cancers to get started, cells have to experience both a stuck accelerator and failed brakes.

As may be apparent, the process of cancer initiation is very much bound up with the process of cell-cycle regulation. Only time will tell what gains in the fight against cancer will come from this knowledge.

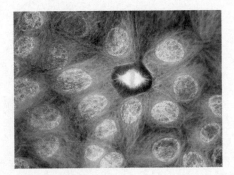

Figure 1
Harmful Division
Pictured are skin cancer cells, taken from a tumor and growing in a laboratory. The red and yellow cell in the center is undergoing cell division. The DNA of the cells is stained blue. (×300)

But why should cells divide at all? That is, why don't they just stay as they are? Cells tend to grow in their day-to-day functioning but, as noted in Chapter 4, they can't grow very *much* or soon they will not have enough surface area to carry out their essential import and export activities. (See "The Size of Cells" on page 72.) Also, cells die and need to be replaced.

Cell division may be a somewhat misleading term if it is taken to mean a simple separation of cellular material, akin to a candy bar being split between two friends. Indeed, a splitting does occur in cell division, but certain parts of the cell must *duplicate* before this happens. Then the duplicated material is parceled out with fine precision, half of it going to one "daughter" cell and half to the other.

What's being duplicated and divided is DNA. Because a cell's full complement of DNA contains such critical information, it would not do for a cell to be left with 50 or 75 percent of this information; rather, it needs the whole thing—no more and no less.

With this in mind, here is the big picture on cell division. First, there is a duplication of DNA; then there is the movement of two precisely matched quantities of it to opposite sides of the "parent" cell; finally there is the splitting of the parent cell into two daughter cells. The duplication of DNA is known as replication, the apportioning of it into two quantities is known as mitosis, and the splitting of the cellular material is known as cytokinesis (**see Figure 9.4**). The goal for the remainder of this chapter is to learn a little bit about replication and a good deal more about mitosis and cytokinesis.

The Replication of DNA

So how does the first part of cell division work—the DNA replication? If you look at **Figure 9.5**, you can see a simplified representation of the DNA molecule shown in Figure 9.2. Notice that the strands of the double helix have started to unwind, which is an initial step in DNA replication (Figure 9.5, step 1). Each of the two resulting single strands then serves as a template or pattern upon which a new strand is created (Figure 9.5, step 2). Earlier, DNA structure was compared to a spiral staircase. Now think of that staircase as splitting right down the middle of its steps (where the A and T or the G and C bases meet). Then a new *half*-staircase is created that bonds with one of the old DNA strands; a second half-staircase is likewise created that bonds with the *second* original strand. (These new strands are the green ones in Figure 9.5, step 2.) How does this happen? Free-floating DNA bases will bond, one base at a time, with bases on the original strand. This process continues down the line on the first DNA strand until a second strand is created. This replication process, which doubles the cell's DNA content, takes place in advance of the cell dividing.

In Chapter 13, you will metaphorically walk along the rungs of the double helix to grasp the details of this doubling of DNA strands. Right now, however, you'll be considering DNA on a larger scale, as it comes *packaged*.

Figure 9.4
Overview of Cell Division

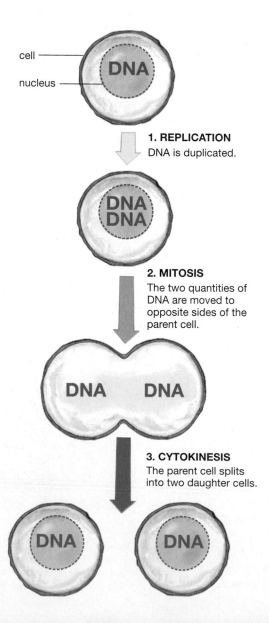

cell

nucleus

DNA

1. REPLICATION
DNA is duplicated.

DNA
DNA

2. MITOSIS
The two quantities of DNA are moved to opposite sides of the parent cell.

DNA   DNA

3. CYTOKINESIS
The parent cell splits into two daughter cells.

DNA   DNA

9.3 DNA Is Packaged in Chromosomes

You have thus far conceptualized each cell's DNA content as spiraling out in *one* long double helix; but it does not, in fact, take that form. Rather, the DNA in each cell comes divided up and packaged into units called **chromosomes** (from the Greek for "colored bodies," a name stemming from the dye these structures absorbed when they were first seen under a microscope). Different organisms have different numbers of chromosomes; human cells have 46, for example, while onion cells have 16. Chromosomes amount to DNA "packages" in several senses. They consist not only of DNA itself, but of a protein material around which DNA is wrapped. The resulting combination of DNA and protein is known as **chromatin**. Thus a number of chromosomes, composed of chromatin, exist in a cell's nucleus, detached but in close physical proximity to one another, and the *collection* of chromosomes makes up nearly the entire complement of a cell's DNA.

If you look at **Figure 9.6a**, you can see how to think about the double helix being packaged into chromosomes. **Figure 9.6b** shows you how to relate the "staircase-splitting" DNA replication just described to what this means at the

1. Original DNA
   molecule unwinds.

2. New DNA strands are synthesized
   from the two original strands.

**Figure 9.5
DNA Replication**

1. A DNA molecule unwinds.

2. Each of the single strands of the original molecule (in red) serves as a template or pattern for the creation of a second DNA strand (in green). Bases on the red strand pair, one base at a time, with free-floating bases until an entire second strand is created. With this, the quantity of DNA has effectively doubled.

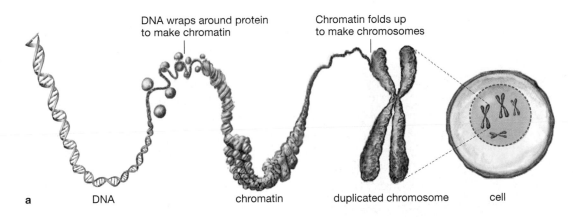

DNA wraps around protein
to make chromatin

Chromatin folds up
to make chromosomes

a   DNA   chromatin   duplicated chromosome   cell

Figure 9.6

a DNA Is Packaged in Units Called Chromosomes
The DNA molecule is bound up with proteins, with the resulting combination known as chromatin. This chromatin then makes up structures called chromosomes.

b DNA Replication at Two Levels
When DNA replicates, the result is two copies of the original DNA molecule. At the chromosomal level, the result of this replication is a single chromosome in duplicated state. The chromosome is composed of two sister chromatids—the two copies of the original DNA molecule. An unduplicated chromosome doesn't actually have the well-defined shape of the chromosome in the figure. It's shown this way merely for purposes of comparison with the duplicated chromosome.

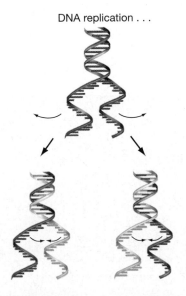

DNA replication . . .

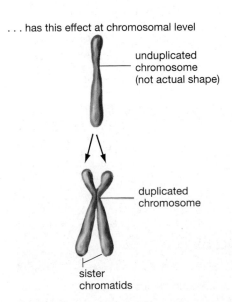

. . . has this effect at chromosomal level

unduplicated
chromosome
(not actual shape)

duplicated
chromosome

sister
chromatids

b

chromosomal level. When we say that DNA is replicating, this is another way of saying that the chromosomes that DNA helps make up are *duplicating*. Once that has happened, the result is chromosomes in "duplicated state," which means individual chromosomes, each of which is made up of two sister chromatids. Each chromatid is one of the newly replicated DNA double helices, as you can see in Figure 9.6b. More formally, a **chromatid** is one of the two identical strands of chromatin that make up a chromosome in its duplicated state. (It probably goes without saying that, despite the name, sister chromatids are neither siblings nor females.)

Matched Pairs of Chromosomes

Chromosomes are detached from one another, but that does not mean they are completely different from one another. In eukaryotes such as ourselves, in fact, chromosomes come in *pairs* that are close, but not exact matches. The 46 chromosomes we have come to us as 23 chromosomes from *each parent*. Critically, with one exception that you'll soon get to, these are 23 matched pairs of chromosomes, each chromosome from the mother matching with one from the father.

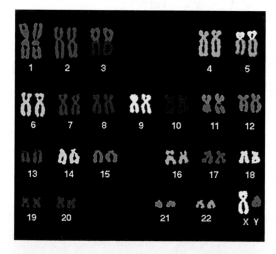

Defining a "Matched Pair"

What is a "matched pair" of chromosomes? We have a chromosome 1 that we inherit from our mother, and it contains a set of genes very similar to those that lie on the chromosome 1 we inherit from our *father*. The same is true for the maternal and paternal copies of chromosome 2, chromosome 3, and so on. Each of us, then, has 23 pairs of **homologous chromosomes**, "homologous" here meaning the same in size and function.

However, homologous chromosomes are not *exactly* alike. Any two of them will contain genes for the same kinds of protein products. If a given paternal chromosome has a gene that codes for hair color, it is a safe bet that the matching *ma*ternal chromosome will have a gene that codes for hair color. But there can be variations on these genes. A gene on the *pa*ternal chromosome may help code for *red* hair, while the gene on the matching maternal chromosome may help code for blonde hair. Nevertheless, both chromosomes have genes that code for hair color, as well as for thousands of other traits, and as such, are said to be homologous.

X and Y Chromosomes

The one exception to the matched-pairs rule in human chromosomes is the so-called sex chromosomes of males. Human females have the 23 pairs of homologous chromosomes mentioned before; 22 of these are *autosomes*, or nonsex chromosomes, and one is a homologous pair of *X* chromosomes that, when present, means a growing embryo will be a female. Males, on the other hand, have the 22 pairs of homologous autosomes and one X chromosome; but then they also have one *Y* chromosome, which has a gene on it that confers the *male* sex on an embryo.

This whole scheme is laid out for you in **Figure 9.7**. There you can see a **karyotype**, or pictorial arrangement of a full set of human chromosomes. There are 46 chromosomes in

Figure 9.7
A Karyotype Displays a Full Set of Chromosomes

One member of each chromosome pair comes from the individual's father and the other member from the mother. Each paired set of chromosomes is said to be "homologous," meaning the same in size and function. (The two chromosomes over the number 1 are a homologous pair, the two over number 2, and so forth.) Homologous chromosomes are not exactly alike, however; the genes on them may differ somewhat, meaning the effects they produce will differ. Because this karyotype set is from a human male, there are 22 pairs of homologous chromosomes and then one X and one Y chromosome (which are not homologous). All the chromosomes are in the duplicated state.

Figure 9.8
DNA Can Be Arranged in Two Ways

a Replication occurs when the chromosomes DNA comes packaged in are in a relatively formless state.

b Mitosis occurs after DNA condenses into the easily discernable shapes of duplicated chromosomes. The micrograph is of color-enhanced duplicated human chromosomes.

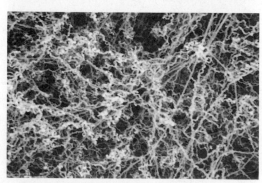

a DNA in uncondensed form

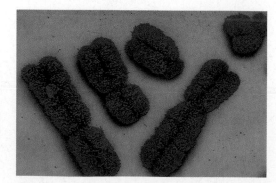

b DNA condensed into chromosomes

all, 22 of them matched pairs. In this case, one pair has one X chromosome and one Y chromosome, meaning this is a male.

The chromosomes in Figure 9.7 are in the duplicated state noted earlier, each of them being composed of two sister chromatids. For the sake of clarity, it would be nice to see a picture of some *un*duplicated chromosomes, but we have no pictures that are the counterpart to Figure 9.7. This is so because chromosomes take on an easily discernable shape only after they duplicate, prior to cell division. As cell division approaches, the DNA-protein complex changes its form; it tightens up, condensing mightily to produce well-defined chromosomes (**Figure 9.8**). After cell division is complete, the cell's chromosomes return to their relatively formless state.

Why this change in form? This condensing before cell division has the same effect as you taking all your scattered belongings and packing them into boxes just before moving from one apartment to another. Remember that DNA is packing up to leave as well, in this case for life in a successor cell. Were it not to tighten into its duplicated chromosomal form, its elongated fibers would get tangled up in the move.

Chromosome Duplication as a Part of Cell Division

With the chromosomes duplicated, you are now ready to look at the process by which they will split up. Before following this path, though, recognize that this separation occurs within the larger process that we are following: the division of the cell as a whole. As you saw in Chapter 5, there are many more things inside a cell than its complement of chromosomes. There are mitochondria, lysosomes, ribosomes, and so forth, and the cytosol in which they are immersed. Together, these things make up the cell's cytoplasm—the material *outside* the nucleus, as opposed to chromosomes, which lie *inside* the nucleus. With cell division, about half of the cytoplasmic material goes to one daughter cell and half to the other. It is helpful, then, to conceptualize cell division as having two separable components: **mitosis**, which is the division of the cell's chromosomes, and **cytokinesis**, which is the division of its cytoplasm.

The Cell Cycle

If you look at **Figure 9.9**, you can see how these two processes are linked over time. The figure illustrates the **cell cycle**: the repeating pattern of growth, genetic duplication, and

finally division that holds true for many kinds of cells. There are two main phases in this cycle. First there is **interphase**, in which the cell simultaneously carries out its work, grows, and—in preparation for division—duplicates its chromosomes. Second, there is **mitotic phase** (or M phase), in which the duplicated chromosomes separate properly and then the cell as a whole splits in two.

These two larger phases are in turn subdivided into smaller phases. In interphase, first there is G_1, standing for "gap-one." What goes on here are normal cell operations and cell growth. The cell then enters the *S* or "synthesis" phase, which is the synthesis of *DNA*, resulting in the duplication of the chromosomes. When this ends, interphase's G_2 period begins, during which there is more cytoplasmic growth and a preparation for cell division. Because DNA synthesis is stopped in G_2 as well as G_1, the word *gap* in G_1 and G_2 can be thought of as a gap in DNA synthesis.

The length of the cell cycle varies greatly from one type of cell to another. In a typical animal cell, the total cell cycle length is about 24 hours. Within this cycle, mitotic phase takes up only about 30 minutes. Within interphase, an animal cell spends roughly 12 hours in G_1, 6 hours in S, and 6 hours in G_2.

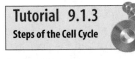

Tutorial 9.1.3
Steps of the Cell Cycle

Figure 9.9
The Cell Cycle
The three main stages of cell division—DNA replication, mitosis, and cytokinesis—can be seen here in the context of a complete cell cycle. This cycle traditionally is divided into two main phases, interphase and mitotic (or "M") phase, which are in turn divided into other phases.

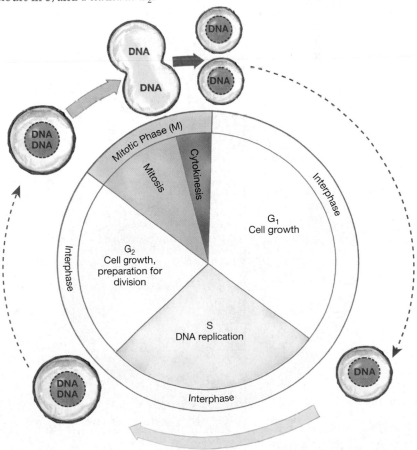

9.4 Mitosis and Cytokinesis

It is the second overarching phase of the cell cycle, mitotic phase, that is the focus for much of the rest of this chapter. Now the concern is not with how a cell carries out its general functions, but with how it *divides*. Mitotic phase includes both mitosis—in which the cell's newly duplicated chromosomes condense, align themselves, and then separate—and cytokinesis, in which a cell's cytoplasm divides and one cell becomes two. Starting with mitosis, you'll look at its four phases in order (**see Figure 9.10**). We will first consider mitosis and cytokinesis as they occur in animal cells.

The Phases of Mitosis

Prophase

The beginning of mitosis marks the end of the cell's interphase. The starting state is that the cell's DNA, which replicated during the S phase, has just begun to pack itself into well-defined chromosomes. When we can finally *see* such chromosomes with a microscope, we can say that mitosis has begun, with prophase. Now the packing job continues; before it is done, the DNA will have coiled and

condensed into one eight-thousandth of its S-phase length. When this is over, there are 46 well-defined chromosomes in the nucleus in duplicated state (meaning they are composed of 92 chromatids). Meanwhile, the nuclear envelope—the double membrane surrounding the nucleus—begins to break up.

While this is going on, big changes are taking place outside the nucleus. Recall from Chapter 4 the protein structures called **microtubules** that are part of the cell's cytoskeleton. In some of their roles, microtubules can be compared to tent poles that can shorten or lengthen as need be. Throughout mitosis, microtubules stretch the cell as a whole, and physically move the cell's chromosomes around.

In an interphase cell there exists, just outside the nucleus, a **centrosome**: a cellular structure that acts as an organizing center for the assembly of microtubules. This centrosome duplicates, so that now there are two microtubule organizing centers (see Figure 9.10). Now, in prophase, these two centrosomes start to move apart. Where are they going? To the poles. From here on out, it is convenient to think of mitosis and cytokinesis in terms of a global metaphor: The centrosomes

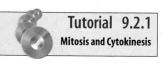

Tutorial 9.2.1
Mitosis and Cytokinesis

Figure 9.10
Mitosis and Cytokinesis
The micrographs at the bottom of the figure are pictures of mitosis and cytokinesis in a whitefish embryo.

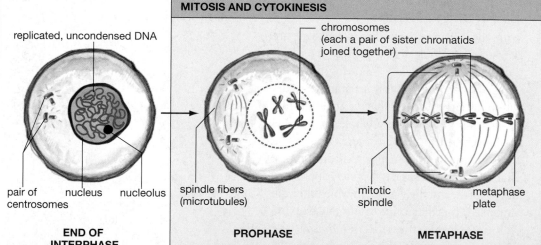

MITOSIS AND CYTOKINESIS

replicated, uncondensed DNA

chromosomes
(each a pair of sister chromatids
joined together)

pair of   nucleus   nucleolus
centrosomes

spindle fibers
(microtubules)

mitotic   metaphase
spindle   plate

**END OF
INTERPHASE**

DNA has already duplicated back in the S phase. Centrosome has doubled.

PROPHASE

Mitosis begins. The chromosomes take shape as the DNA condenses. The nuclear envelope begins to break down. The two centrosomes begin to move toward the cellular poles, sprouting microtubules as they go.

METAPHASE

Linkage and alignment. Some microtubules of the mitotic spindle form a cage around the cell's former nucleus while others attach to the sister chromatids and align them at the metaphase plate. Each chromatid now faces the pole opposite that of its sister chromatid.

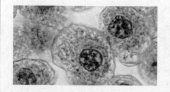

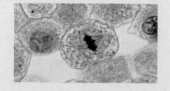

migrate to the cellular poles, while the chromosomes first align, and then separate, along a cellular equator called the metaphase plate. Unlike microtubules or the centrosome, the **metaphase plate** is not a structure, but instead is a plane located midway between the poles of a dividing cell. As the centrosomes move apart, they begin sprouting microtubules in all directions. By the time they have reached their respective poles, they will have several types of microtubules stemming from them. One variety of them forms a football-shaped cage around the nuclear material, while a second variety attaches to the chromosomes themselves. Taken together, the microtubules active in cell division are known as the **mitotic spindle**.

Metaphase

By the time metaphase begins, the nuclear envelope has disappeared completely, and the microtubules that were growing toward the chromosomes now *attach* to them. With this, what had been random becomes orderly. Through a lively back-and-forth movement, the chromosomes eventually are *aligned* at the equator: Each pair of chromatids is now moved to the metaphase plate by action of the microtubules. Just as important, each chromatid now faces the pole *opposite* that of its sister chromatid, and each chromatid is *attached* to its respective pole by perhaps 30 microtubules.

Anaphase

At last the genetic material divides. As you may have guessed, this is a parting of sisters. It is the sister chromatids that are pulled apart, each now becoming a full-fledged chromosome. All 46 chromatid pairs divide at the same time, and each member of a chromatid pair is pulled toward its respective pole by motor proteins that act like a train engine running on the tracks of the microtubules, which shorten as this process continues.

Telophase

Telophase represents a return to things as they were before mitosis started. The newly independent chromosomes, having arrived at their respective poles, now unwind and lose their clearly defined shape. New nuclear membranes are forming. When this work is complete, there are two finished daughter nuclei lying in one elongating cell. Even as this is going on, though, something else is taking place that will result in this one cell becoming two.

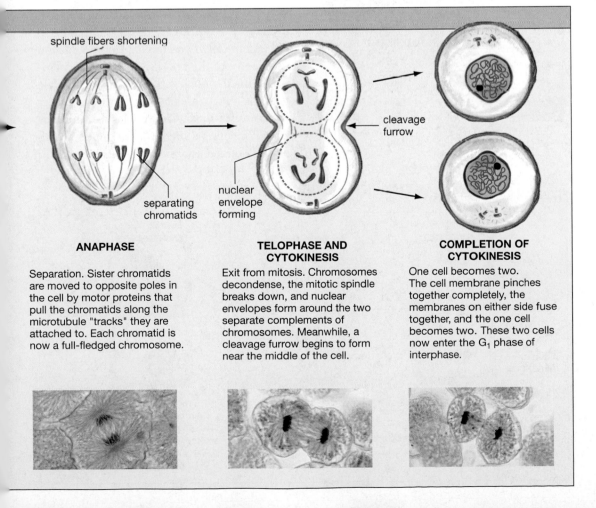

spindle fibers shortening

separating chromatids

nuclear envelope forming

cleavage furrow

ANAPHASE

Separation. Sister chromatids are moved to opposite poles in the cell by motor proteins that pull the chromatids along the microtubule "tracks" they are attached to. Each chromatid is now a full-fledged chromosome.

TELOPHASE AND CYTOKINESIS

Exit from mitosis. Chromosomes decondense, the mitotic spindle breaks down, and nuclear envelopes form around the two separate complements of chromosomes. Meanwhile, a cleavage furrow begins to form near the middle of the cell.

COMPLETION OF CYTOKINESIS

One cell becomes two. The cell membrane pinches together completely, the membranes on either side fuse together, and the one cell becomes two. These two cells now enter the G_1 phase of interphase.

Cytokinesis

Cytokinesis actually began back in anaphase and is well under way by the time of telophase. It works through the tightening of a cellular waistband that is composed of two sets of protein filaments working together. These materials form a *contractile ring* that narrows along the cellular equator (**see Figure 9.11**). A *cleavage furrow*, or indentation of the cell's surface, results from the ring's contraction; consequently, the fibers in the mitotic spindle are pushed closer and closer together, eventually forming one thick pole that is destined to break. The dividing cell now assumes an hourglass shape; as

the contractile ring continues to pinch in, one cell becomes two by means of something you looked at in Chapter 4: membrane fusion. The membranes on each half of the hourglass circle toward each other and then fuse. With this, the two cells become separate. Mitosis and cytokinesis are over, and the two daughter cells slip back into the relative quiet of interphase.

9.5 Variations in Cell Division

Now, we'll look at variations in cell division in certain different types of cells, including plant cells and prokaryotes.

Plant Cells

For the most part, plant cells carry out the steps of mitosis just as animal cells do. In the splitting of cytokinesis, however, plant cells must deal with something that animal cells don't have to: the cell wall noted in Chapter 4. The way animal cells carry out cytokinesis—pinching the plasma membrane inward via the contractile ring—wouldn't work with plant cells, because their cell wall lies *outside* the plasma membrane.

The plant cell's solution to cytokinesis is to grow a new cell wall and plasma membrane that run roughly down the middle of the parent cell. If you look at **Figure 9.12**, you can see how this works. A series of membrane-lined vesicles begins to accumulate near the metaphase plate; inside these vesicles is a complex sugar. The vesicles then fuse together, forming a single, flat **cell plate**, which can be defined as a membranous sac that, in

Figure 9.11
Cytokinesis in Animals
Cytokinesis in animal cells begins with an indentation of the cell surface, a cleavage furrow, shown here in a dividing frog egg. (×85)

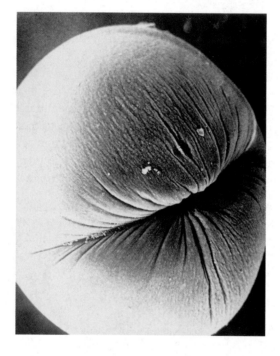

Tutorial 9.2.6
Cytokinesis in Plants

Figure 9.12
Cytokinesis in Plants
The cell walls of plants are too rigid to form cleavage furrows, as animal cells do when dividing. Therefore, plant cells use a different strategy in cytokinesis. They build a new cell wall down the middle of the parent cell to separate the two daughter cells from the inside out.

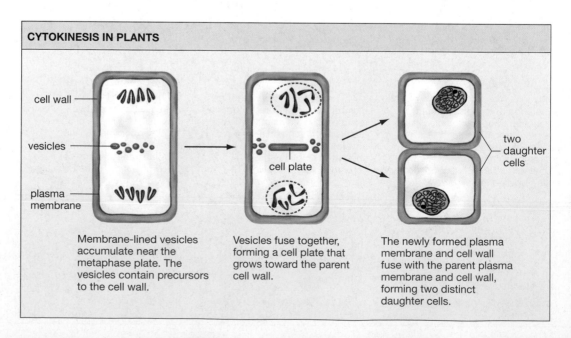

CYTOKINESIS IN PLANTS

cell wall

vesicles

plasma membrane

cell plate

two daughter cells

Membrane-lined vesicles accumulate near the metaphase plate. The vesicles contain precursors to the cell wall.

Vesicles fuse together, forming a cell plate that grows toward the parent cell wall.

The newly formed plasma membrane and cell wall fuse with the parent plasma membrane and cell wall, forming two distinct daughter cells.

BINARY FISSION IN BACTERIA

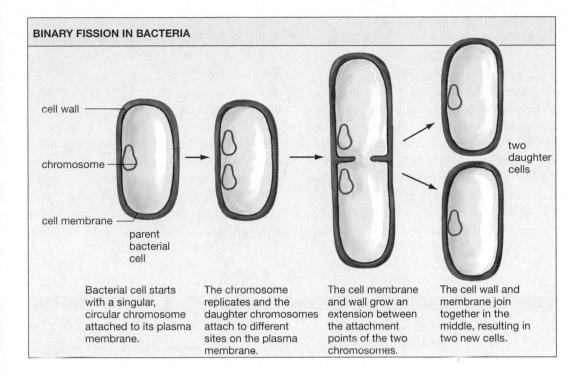

cell wall

chromosome

cell membrane

parent
bacterial
cell

two
daughter
cells

Bacterial cell starts
with a singular,
circular chromosome
attached to its plasma
membrane.

The chromosome
replicates and the
daughter chromosomes
attach to different
sites on the plasma
membrane.

The cell membrane
and wall grow an
extension between
the attachment
points of the two
chromosomes.

The cell wall and
membrane join
together in the
middle, resulting in
two new cells.

Tutorial 9.3.1
**Binary Fission
in Bacteria**

Figure 9.13
Binary Fission in Bacteria
The key to bacterial cell division
is that the bacterial chromosome
attaches to the cell's plasma
membrane. Both daughter chromosomes attach to the plasma
membrane, but in different locations. The cell then produces an
outgrowth of both plasma membrane and cell wall—called a
septum—that grows in from
each side of the cell, between the
two daughter chromosomes.
When the septum runs com-
pletely across the cell, it divides
the cell in two, leaving one chromosome in each daughter cell.

cell division, is the precursor to the plant
cell wall. The cell plate then grows toward
the parent cell wall. Its membrane material
fuses with the original plasma membrane,
while the cell-wall material within is released to fuse with the original cell wall.
The result is two distinct daughter cells,
complete with cell walls.

Prokaryotes

It's also important to recognize that some cells
have fundamentally different ways of replicating. The cells of prokaryotes—the bacteria
and archaea you looked at in Chapter 4—are a
case in point.

If you look at **Figure 9.13**, you can see
how cell division works among the prokaryotes. Bacteria are single-celled and have a
single, circular chromosome. Bacterial cells
do not, however, have their chromosome sequestered inside a nucleus, as with eukaryotic
cells, for the simple reason that they *have* no
nucleus. Instead, their lone chromosome is
attached to their plasma membrane.

A bacterial cell begins its division by duplicating its chromosome, utilizing the same
DNA unwinding and replication described
earlier. As the daughter chromosomes are
completed, however, they attach to *different
sites* on the plasma membrane. The cell then
begins developing an outgrowth of its
plasma membrane and cell wall in between
these attachment sites. This outgrowth,

called a septum, begins growing from opposite sides of the cell. When the two septum
extensions join in the middle, they divide the
one cell into two (**see Figure 9.14**). This
process is obviously very different from the
eukaryotic mitosis and cytokinesis you
looked at. As such, prokaryotic cell division
goes under an entirely different name: **binary
fission**. The simplicity of binary fission
makes for a short cell cycle. Remember how
the animal cells we looked at took about 24
hours to go through a cycle? In optimal
conditions, bacteria can complete their cycle
in 20 minutes.

Figure 9.14
**A Micrograph of an *E. coli*
Bacterium Completing Cell
Division**
These bacteria exist in great
quantity in the human digestive
tract, generally harmlessly. It is
certain rare varieties of *E. coli*,
generally found in food, that
have caused outbreaks of sickness and even death in recent
years. (×30,000)

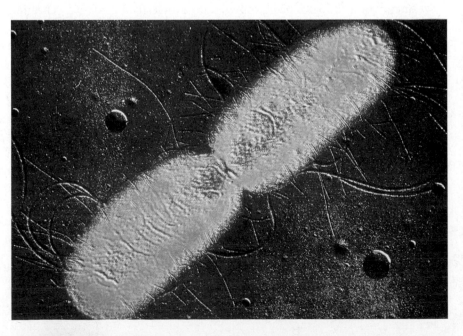

Variations in the Frequency of Cell Division

There are also variations in cell division among different types of eukaryotic cells. Not all cells divide throughout their existence. Most human brain cells are formed in the first three months of embryonic existence and then live for decades, with relatively few of them ever dividing again. Much the same is true of the leaf cells in plants; they divide only when the leaf is very small, then grow for a time, and then function at this mature size for as long as the leaf lives. At the other extreme, "stem" cells located in human bone marrow never *stop* dividing as they produce blood cells.

On to Meiosis

The cell division reviewed in this chapter concerns "somatic" cells—the kind that form blood, bone, muscle, nerves, and many other sorts of tissue. Given a distribution this wide, you might well ask: What kind of cells are *not* somatic? The answer is only one kind—the kind that forms the basis for each succeeding generation of sexually reproducing living things.

Chapter Review

Summary

9.1 An Introduction to Genetics

- DNA is an information-bearing molecule that plays a critical role in the reproduction, development, and everyday functioning of living things. DNA contains the information for the production of proteins, which carry out an array of tasks in living things.

- The information in DNA is encoded in chemical units called genes. Each gene is composed of a series of chemical bases, with the order of these bases determining what protein the gene codes for. There are four DNA bases: adenine (A), thymine (T), guanine (G), and cytosine (C). A typical gene is composed of thousands of bases sequenced in a given order.

- Protein synthesis begins with the information in a sequence of DNA bases being copied onto a molecule called messenger RNA (mRNA). This molecule moves out of the nucleus to a structure in the cell's cytoplasm called a ribosome. There, the mRNA "tape" is "read," and amino acids brought to the ribosome are put together in the sequence called for in the mRNA tape. The result is a chain of amino acids that folds into a protein.
  TUTORIAL 9.1.1: DNA Genes and Proteins

- Most of the cells in an organism contain a complete copy of that organism's genome, meaning its collection of genetic information. Before cells divide, their genome must first be copied and the resulting copies apportioned evenly into what will become two daughter cells.

9.2 An Introduction to Cell Division

- Cell division includes the copying of DNA (known as replication), the apportioning of it into two quantities (mitosis), and the splitting of cellular material into two (cytokinesis).

9.3 DNA Is Packaged in Chromosomes

- DNA comes packaged in units called chromosomes. These chromosomes are composed of DNA and its associated proteins—a combined chemical complex called chromatin. Chromosomes exist in an unduplicated state until such time as DNA replicates, prior to cell division. DNA replication results in chromosomes that are in duplicated state, meaning one chromosome composed of two identical sister chromatids.
  TUTORIAL 9.1.2: Chromosomes and Duplication

- Chromosomes in human beings (and many other species) come in matched pairs, with one member of each pair inherited from the mother, and the other member of each pair inherited from the father. Such homologous chromosomes have closely matched sets of genes on them, though many of these genes are not identical. A given paternal chromosome may have genes that code, for example, for slightly different hair or skin color than the counterpart genes on the homologous maternal chromosome. Human beings have 46 chromosomes—22 matched pairs and either a matched pair of X chromosomes (in females) or an X and a Y chromosome (in males).

- Cell division fits into the larger framework of the cell cycle, meaning a repeating pattern of growth, genetic replication, and cell division. The cell cycle has two main phases. One is interphase, in which the cell carries out its work, grows, and duplicates its chromosomes in preparation for division. The second is mitotic phase, in which the duplicated chromosomes separate and the cell splits in two.
  TUTORIAL 9.1.3: Steps of the Cell Cycle

9.4 Mitosis and Cytokinesis

- There are four stages in mitosis: prophase, metaphase, anaphase, and telophase. The essence of the process is that duplicated chromosomes line up along a midpoint plane of the parent cell (the metaphase plate) with the sister chromatids that make up each duplicated chromosome on opposite sides of the plate. The sister chromatids are then moved apart, to opposite poles of the parent cell. Once cell division is complete, sister chromatids will reside in separate daughter cells, with each sister chromatid now functioning as a full-fledged chromosome.

TUTORIAL 9.2.1: Mitosis and Cytokinesis

- Cytokinesis in animal cells works through a ring of protein filaments that tightens at the middle of a dividing cell. Membranes on the portions of the cell being pinched together then fuse, resulting in two daughter cells.

9.5 Variations in Cell Division

- Because of their cell walls, plant cells must carry out cytokinesis differently from animal cells. The plant's solution is to grow new cell walls and plasma membranes near the metaphase plate, thus dividing the parent cell into two daughter cells. Prokaryotes such as bacteria employ a process called binary fission: They double their single chromosome, with the two resulting chromosomes attaching to different sites on the plasma membrane. Then an outgrowth of plasma membrane and cell wall, called a septum, begins growing from opposite sides of the cell. When the two septum extensions join in the middle, they divide the one cell into two.

TUTORIALS: 9.2.6 Cytokinesis in Plants; 9.3.1 Binary Fission in Bacteria

Key Terms

binary fission 183	genome 171
cell cycle 179	homologous chromosomes 178
cell plate 182	Interphase 179
centrosome 180	karyotype 178
chromatid 178	metaphase plate 181
chromatin 177	microtubule 180
chromosome 177	mitosis 179
cytokinesis 179	mitotic phase 179
enzyme 172	mitotic spindle 181
genetics 173	protein 172

Understanding the Basics

Multiple-Choice Questions

1. A protein that accelerates chemical reactions in the body is
   a. an enzyme
   b. a histone
   c. a gene
   d. DNA
   e. glucose

2. The four bases used in the DNA code are
   a. adenine, thymine, guanine, and cytosol
   b. adenine, thyroxine, glucose, and cytosine
   c. adrenaline, thymine, glucosamine, and uracil
   d. adenine, thymine, guanine, and cytosine
   e. none of the above

3. The "handrails" of the DNA structural staircase are composed of
   a. adenine, paired with thymine
   b. fatty acids and sugars
   c. phosphate groups and nitrogen-containing bases (ATCG)
   d. deoxyribose and phosphate groups
   e. sugars and proteins

4. The human genome contains
   a. 22 chromosomes
   b. 44 chromosomes
   c. as many as 40,000 genes
   d. no more than 27,000 genes
   e. about 1,000,000 genes

5. DNA is primarily
   a. a molecule that protects cells from invaders
   b. a molecule that contains information
   c. a molecule that helps transfer energy
   d. a "workbench" on which proteins are put together
   e. a form of protein

6. Prophase has begun when
   a. We can see well-defined chromosomes.
   b. A cleavage furrow has formed.
   c. Chromosomes have aligned along the metaphase plate.
   d. Chromosomes have duplicated.
   e. Sister chromatids have separated.

7. A _____ is one set of protein filaments that helps bring about cell division during cytokinesis.
   a. dividing fiber
   b. metaphase plate
   c. spindle
   d. contractile ring
   e. centrosome

8. The length of the cell cycle in animals is about
   a. 16 hours
   b. 24 hours
   c. 30 minutes
   d. 2 hours
   e. 12 hours

9. By far, the longest phase of the cell cycle is
   a. prophase
   b. interphase
   c. metaphase
   d. cytokinesis
   e. telophase

10. A pictorial arrangement of a full set of chromosomes, organized by homologous pairs, is called a(n)
    a. chromosome grid
    b. genome
    c. genetic "picture"
    d. autosome alignment
    e. karyotype

11. If a cell is not dividing, but instead is in the process of normal operation and growth, what phase of the cell cycle would it be in?
    a. reproductive phase
    b. synthesis (S) phase
    c. gap-one (G₁) phase
    d. gap-two (G₂) phase
    e. either c or d would be correct

Brief Review

1. In what way does DNA help form and then maintain our bodies?

2. The drawing shows a cell in a phase of mitosis. What phase is shown? How many chromatids are present in this stage? What is the total number of chromosomes each daughter cell will have?

3. How do mitosis and cytokinesis differ?

4. Prokaryote cell division is called _____ _____.

5. Name the five phases of the cell cycle and give the importance of each, describing the major events occurring in each phase.

6. What additional cell structure must plant cells deal with when they are undergoing cytokinesis (as compared to animal cells)? Given this difference, how do plant cells carry out cytokinesis?

7. In what way are the 23 pairs of human chromosomes "matched" chromosomes?

8. How does the prokaryote chromosome structure differ from the eukaryote chromosome structure?

9. The protein synthesis pathway—DNA → mRNA → protein—is a very important concept in biology.
   a. What are the information-bearing "building blocks" of DNA?
   b. Where is DNA located in the cell?
   c. What does the "m" in mRNA stand for?
   d. Using this knowledge, what is the function of mRNA, and what is its cellular pathway?
   e. What are the "building blocks" of proteins?
   f. Where does protein formation occur?

Applying Your Knowledge

1. If a human cell fails to undergo cytokinesis, how many chromosomes will be in the final cell?

2. People who are born with the condition called Down's syndrome generally have three copies of a certain chromosome (chromosome 21), rather than the usual two copies. What can you deduce from this about the information contained in chromosomes?

3. Say the total number of chromosomes in a hypothetical cell is six. Draw the cell in each of the four phases of mitosis. How many chromosomes and chromatids are present in the first phase?

4. Why is it accurate to think of each human being as the owner of a library of ancient information?

5. If each cell contains the entire genome of an organism, why wouldn't a liver cell do exactly the same things as a muscle cell?

MediaLab

Too Much Division: The Cell Cycle and Cancer

This *MediaLab* is designed to stimulate interest in cell division by allowing you to examine the role of the cell cycle in cancer. To learn more, enter the *CD-ROM Tutorial* for an animated review of cell division. Once you have completed this section, you will be guided through a series of *Web Investigations*, where you can utilize information on the Internet to examine your own personal risk of contracting cancer, taking into account such factors as inheritance and exposure to environmental carcinogens. Finally, in *Communicate Your Results*, you can evaluate the relevance of these factors.

This *MediaLab* can be found in Chapter 9 on your CD-ROM (Tutorial 9.1.3) and Companion Website (http://www.prenhall.com/krogh3).

CD-ROM TUTORIAL

One cell dividing to create two is crucial for growth, for wound repair, and for creating the next generation of organisms. But if this process is not properly controlled, rogue cancer cells can be produced. You will be reviewing the correct steps in cell division in several interactive exercises as well as observing the consequences of uncontrolled cell division in cancer cells.

Activity

1. *First, you will observe the key to reproduction, the duplication of the genetic information in our DNA.*

2. *You will then review the cycle in which our body duplicates the DNA and other cellular components of existing cells to refresh and replace cells when we grow, or when a wound heals.*

3. *Then you will compare this to the events of uncontrolled cell division in cancer cells.*

4. *Finally, you can test your knowledge in an interactive exercise where you direct the events of the cell division itself.*

After reviewing the concepts of cell division, you can evaluate your fears and misconceptions about cancer in the *Web Investigations* and apply your knowledge of cell division to achieve a better understanding of cancer.

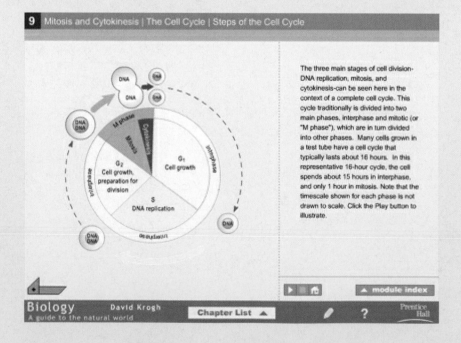

WEB INVESTIGATION

Investigation 1

Estimated time for completion = 15 minutes

Cells can become cancerous (accelerated and unregulated growth) because they have a mutation in one of the genes that normally controls the rate at which a cell travels through the cell cycle. Scientists believe that inheriting a defective gene may predispose a person toward developing some specific cancers. What proportion of cancers are inherited? Actually, a very small number. In families with these defective genes, the same type of cancer is seen in two or more close relatives, with an earlier-than-usual onset. Select the Keyword **INHERITED CANCERS** on your CD or Website for this *MediaLab* to view a list of inherited cancers. Then, select Keyword **PERSONAL HEALTH FORM** to complete a personal health history form, which will likely relieve you of your concern about inheriting these cancers.

Investigation 2

Estimated time for completion = 5 minutes

Daily media warnings expound on evidence that chemicals, drugs, and diet can increase our chances of acquiring cancer. But how can you determine which risk factors are serious, and which may be rare or insignificant? Select the Keyword **CANCER RISK FACTORS** on your CD or Website for this *MediaLab* to read an article comparing the relative risk factors in contracting cancer from natural and human-made compounds.

Investigation 3

Estimated time for completion = 5 minutes

Americans worry that cancer rates are rising because we are increasingly exposed to chemicals or other factors that increase cancer (carcinogens). Select the Keyword **NATIONAL CANCER INSTITUTE** on your CD or Website for this *MediaLab* to view research on suspected carcinogens, and read one of the articles posted at this site.

COMMUNICATE YOUR RESULTS

How well can scientists predict who will get cancer, when both heredity and the environment are such important factors? In the following exercises, you'll evaluate your own hereditary and environmental cancer risk.

Exercise 1

Estimated time for completion = 5 minutes

Now that you have filled out your personal health history form from *Web Investigation 1*, prepare a 50-word document indicating what type of cancer is found in your family. If it has an earlier-than-normal onset and is not found in two or more close relatives, it is probably not inherited.

Exercise 2

Estimated time for completion = 30 minutes

How can you evaluate the significance of news stories on such substances as chlorinated water or artificial sweeteners as potential cancer causes? After completing *Web Investigation 2*, compose a 250-word essay describing which statement you think is most correct:

1. *Synthetic chemicals manufactured by humans are responsible for more cancers (are more carcinogenic) than are natural compounds like sunlight and natural foods.*

2. *Chemicals found in natural foods and sunlight are more carcinogenic than are synthetic chemicals.*

3. *Both naturally occurring and synthetic chemicals are equally responsible for causing cancer.*

Exercise 3

Estimated time for completion = 15 minutes

Web Investigation 3 introduced you to cancer risk factors evaluated by the National Cancer Institute. Form a group of your peers, and have each member present one of these factors (for example, artificial sweeteners) to the group. In your presentation, be sure to inform the group whether they should be concerned about getting cancer from this factor. Also tell them about what tests could be done to relieve this concern.

10

Preparing for Sexual Reproduction
Meiosis

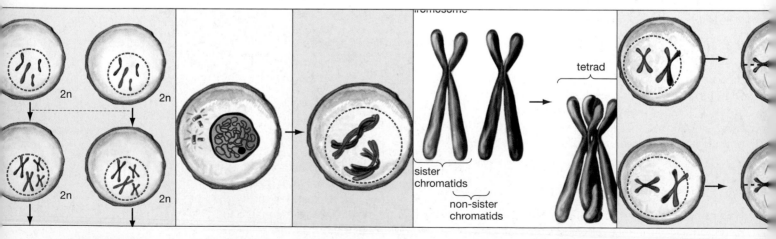

Meiosis and
Mitosis compared.
(Section 10.2, page 193)

Chromosomes link as they condense.
(Section 10.2, page 194)

Swapping sections.
(Section 10.2, page 194)

Sister chromatids
line up.
(Section 10.2,
page 194)

One special kind of cell division yields the sex cells that give rise to succeeding generations.

A real-life couple—let's call them Jack Fennington and Jill Kent—combine their last names when they get married and thus become Jack and Jill Fennington-Kent. Now, what would happen if the Fennington-Kent's children were to continue in this tradition? Their daughter Susie might marry, say, Ralph Reeson-Dodd, which would make her Susie Fennington-Kent-Reeson-Dodd. If *her* daughter, Carol, were to keep this up, she might be fated to become Carol Fennington-Kent-Reeson-Dodd-Smith-Lee-Minderbinder-Green, and so on.

This growing chain of names illustrates a kind of problem that living things have managed to address in carrying out reproduction. "Reproduction" in this context does not mean the mitotic cell division you looked at in Chapter 9. Reproduction here means *sexual* reproduction, in which specialized reproductive cells come together to produce offspring. The cells described in Chapter 9—the ones that undergo mitosis—are known as *somatic* cells, which in animals means all the cells in the organism except for one type. What type is this? Sex cells; cells called *gametes*. These are reproductive cells, known more commonly as eggs and sperm.

Eggs and sperm are not commonly thought of as cells, but that's just what they are, complete with plasma membranes, nuclei, chromosomes, and all the rest (though sperm eventually lose much of this material). Eggs and sperm are *special* cells, however, in that they are not destined to divide like somatic cells and thus make more eggs and sperm. Rather, they will *fuse*—sperm fertilizing egg—to make a zygote, which grows into a whole organism. It is sperm and egg that take part in sexual reproduction and that thus bear some relation to the Fennington-Kent problem. Here's how it plays out in humans. Human somatic cells, as you have seen, have 23 pairs of chromosomes or 46 chromosomes in all. If an egg and sperm each brought 46 chromosomes to their *union*, the result would be a zygote with 92 chromosomes. The *next* generation down would presumably have 184 chromosomes, the next after that 368, and so on. This would be about as functional as having Fennington-Kent-Reeson-Dodd-Smith-Lee-Minderbinder-Green as a last name.

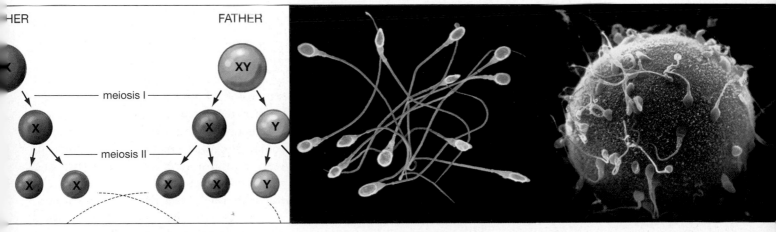

MOTHER | FATHER

meiosis I

meiosis II

On the way to a boy or a girl.
(Section 10.3, page 199)

About 250 million of these
are made every day.
(Section 10.4, page 201)

End of two journeys.
(Section 10.4, page 201)

10.1 An Overview of Meiosis

How do chromosomes avoid the problem of doubling with each generation?

A Chromosome Reduction before Union of Egg and Sperm

In sexual reproduction, chromosome union is preceded by chromosome *reduction*. The reduction comes in the cells that give rise to sperm and egg. When these cells divide, the result is sperm or egg cells that have only *half* the usual, somatic number of chromosomes. Human sperm or egg, in other words, have only 23 chromosomes in them. (Not the 23 *pairs* of chromosomes that somatic cells have, mind you; 23 chromosomes.) Each 23-chromosome sperm can then unite with a 23-chromosome egg to produce a *46-chromosome* zygote that develops into a new human being. In each generation, then, there is first a halving of chromosome number (when egg and sperm cells are produced), followed by a coming together of these two halves (when sperm and egg unite). This basic pattern holds true for sexual reproduction in all eukaryotes.

Some Helpful Terms

Now here are a few terms that will be helpful. The kind of cell division that results in the halving of chromosome number is called *meiosis*, in contrast with *mitosis*, which you looked at in Chapter 9. In mitosis, there is no halving of chromosome number; just a duplication of chromosomes and then a division of these chromosomes into each of two daughter cells.

In animals, cells that go through meiosis—thus getting the half-number of chromosomes—are always reproductive cells, the term for which is **gametes**, as we've seen. Such reduced-chromosome cells are said to be in the haploid state, the term *haploid* literally meaning "single vessel." When egg and sperm unite, however, it marks a return to the *diploid*, or "double vessel" state of cellular existence, meaning 46 chromosomes in human beings. (More formally, **haploid** means possessing a single set of chromosomes while **diploid** means possessing two sets of chromosomes.) **Meiosis** can thus be defined as a process in which a single diploid cell divides to produce haploid reproductive cells. Diploid cells are also sometimes referred to as **2n** cells (the "2" here standing for a doubled number of chromosomes), while haploid cells are said to be **1n**. Thus, eggs and sperm are 1n, while the diploid cells that give rise to them are 2n.

In what follows, we will be looking at meiosis as it occurs in human beings; toward the end of the chapter we'll go over some variations on reproduction in other kinds of organisms.

10.2 The Steps in Meiosis

How does the chromosomal halving take place? Let's go over the process of meiosis and see. **Figure 10.1** shows meiosis, in a stripped-down form, as compared to the mitosis described in Chapter 9. As you can see, meiosis includes one chromosome duplication followed by two cellular divisions. This is the key to how the chromosome numbers work out. In mitosis, there was one chromosome duplication and one subsequent division of the doubled chromosomes (into two cells). In meiosis there is one duplication followed by *two* divisions (into four cells). The first of the meiotic divisions is a separation not of chroma*tids*, as in mitosis, but of homologous chromo*somes*. The second meiotic division is then the separation of the chromatids that make up these homologous chromosomes.

But what is this term *homologous*? You may remember from Chapter 9 that homologous chromosomes are chromosomes that are the same in size and function. We have a chromosome 1 that we inherit from our mother, and it is homologous to the chromosome 1 we inherit from our father. In what sense? Both chromosomes have sets of genes on them that are similar, though not identical, in function. Likewise we inherit a chromosome 2 from our mother that is homologous with chromosome 2 from our father, and so forth for 23 pairs. The lone exception to this pairs rule is that the X and Y chromosomes that males have are not homologous with one another. As you will see, meiosis will separate homologous chromosomes from one another.

Let's walk through the steps of meiosis now. These details will reveal a process that not only divides up genetic material, but

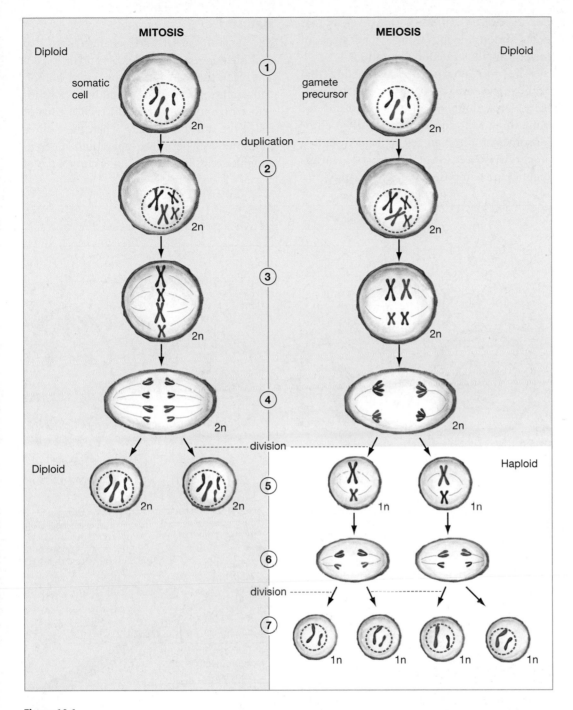

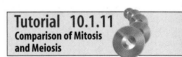

Tutorial 10.1.11
Comparison of Mitosis and Meiosis

Figure 10.1
Meiosis Compared to Mitosis

1. Both mitosis and meiosis begin with diploid cells, meaning cells that contain paired sets of chromosomes. The two members of each pair are homologous, meaning the same in size and function. Two sets of homologous chromosomes are shown in both the mitosis and meiosis figures. The larger chromosome pairs in each cell represent one homologous pair, while the smaller chromosome pairs represent the other homologous pair. One member of each homologous pair (in red) comes from the mother of the person whose cell is undergoing meiosis, while the other member of the pair (in blue) comes from the father of this person.

2. In both mitosis and meiosis, the chromosomes duplicate. Each chromosome is now composed of two sister chromatids.

3. In mitosis, the chromosomes line up on the metaphase plate, one sister chromatid on each side of the plate. In meiosis, meanwhile, homologous chromosomes—not sister chromatids—line up on opposite sides of the metaphase plate.

4. In mitosis, the sister chromatids separate. In meiosis, the homologous pairs of chromosomes separate.

5. In mitosis, cell division takes place, and each of the sister chromatids from step 4 is now a full-fledged chromosome. Mitosis is finished. In meiosis, in the first of two cell divisions, one member of each homologous pair has gone to one cell, the other member to the other cell. Because each of these cells now has only a single set of chromosomes, each is in the haploid state. Next, these single chromosomes line up on the metaphase plate, with their sister chromatids on opposite sides of the plate.

6. The sister chromatids of each chromosome then separate.

7. The cells divide again, yielding four haploid cells.

provides for a great deal of the diversity that exists among living things. The steps of meiosis are illustrated in **Figure 10.2**.

The process starts with the doubling (that is, the replication) of DNA you saw in Chapter 9. DNA doubling in meiosis means the same thing as it did in mitosis: All 46 chromosomes in a human cell replicate, yielding 46 chromosomes in duplicated state (meaning they are composed of 92 chromatids).

Meiosis I

The steps of meiosis are separated into two multistep stages, called meiosis I and meiosis II. The big picture here is that in meiosis I, homologous chromosomes are positioned close together, on opposite sides of the cellular equator called the metaphase plate. Then these chromosomes will be pulled apart into different daughter cells. In meiosis II, the chroma*tids* of

Tutorial 10.1.1

Steps in Meiosis

Figure 10.2

a The Steps of Meiosis

b Recombination

c Independent Assortment

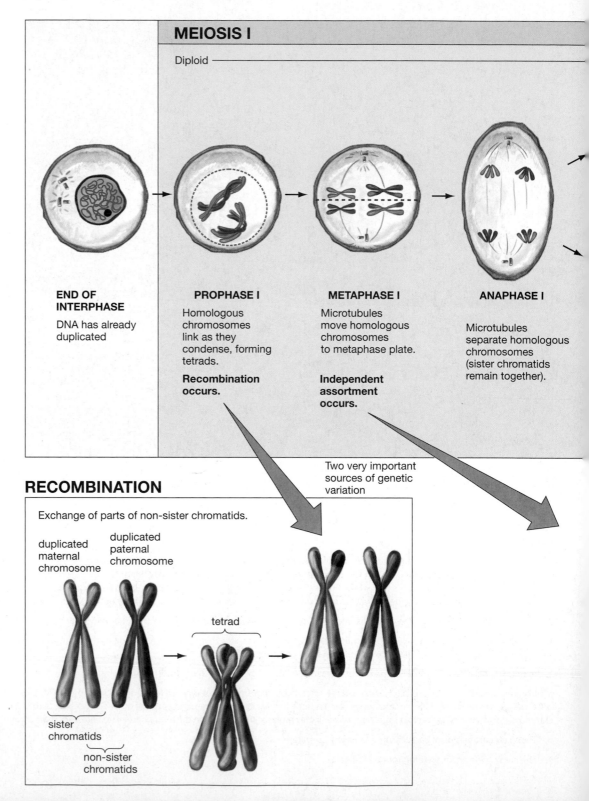

MEIOSIS I

Diploid

END OF INTERPHASE

DNA has already duplicated

PROPHASE I

Homologous chromosomes link as they condense, forming tetrads.

Recombination occurs.

METAPHASE I

Microtubules move homologous chromosomes to metaphase plate.

Independent assortment occurs.

ANAPHASE I

Microtubules separate homologous chromosomes (sister chromatids remain together).

Two very important sources of genetic variation

RECOMBINATION

Exchange of parts of non-sister chromatids.

duplicated maternal chromosome

duplicated paternal chromosome

tetrad

sister chromatids

non-sister chromatids

these now-separated chromosomes will, in turn, separate into different daughter cells.

Prophase I

Meiosis I begins, in prophase I, with the same appearance of 46 identifiable chromosomes that you observed in mitosis. The first big difference between meiosis and mitosis also appears in prophase I. It is that, in meiosis, *homologous chromosomes pair up*. In mitosis,

46 individual chromosomes condensed, but did not pair up in any way. In meiosis, however, each pair of homologous chromosomes link up as they condense. Maternal chromosome 5, for example, intertwines with *paternal* chromosome 5, maternal chromosome 6 with paternal 6, and so on. When two homologous chromosomes have paired up in this way, they are called a **tetrad** (see the "Recombination" box in Figure 10.2).

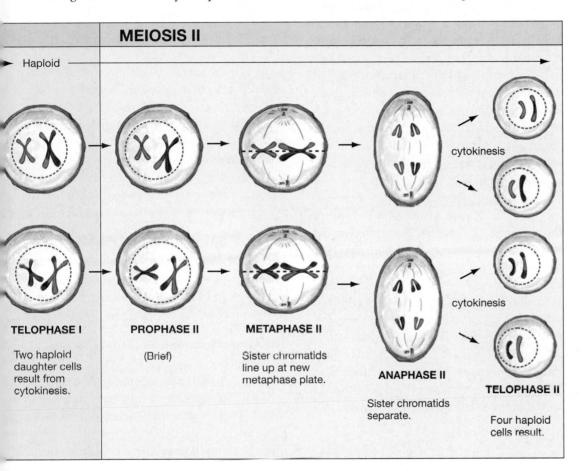

MEIOSIS II

Haploid

TELOPHASE I

Two haploid daughter cells result from cytokinesis.

PROPHASE II

(Brief)

METAPHASE II

Sister chromatids line up at new metaphase plate.

ANAPHASE II

Sister chromatids separate.

cytokinesis

cytokinesis

TELOPHASE II

Four haploid cells result.

INDEPENDENT ASSORTMENT

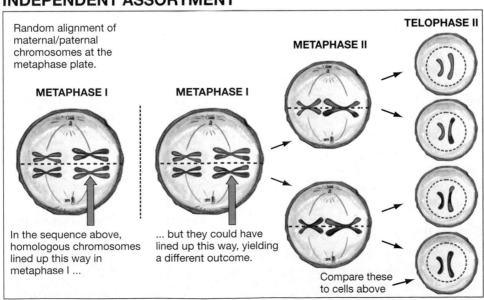

Random alignment of maternal/paternal chromosomes at the metaphase plate.

METAPHASE I

In the sequence above, homologous chromosomes lined up this way in metaphase I ...

METAPHASE I

... but they could have lined up this way, yielding a different outcome.

METAPHASE II

TELOPHASE II

Compare these to cells above

A critical bit of part-swapping then takes place between the non-sister chromatids of the paired chromosomes. This process is called *recombination* (or *crossing over*), and you'll be looking at it in detail later. Once this recombination has finished, the homologous chromosomes begin to unwind from one another, though they remain overlapped.

Metaphase I

Still existing as tetrads, the homologous chromosomes are moved by microtubules to the metaphase plate; in this step of meiosis, the maternal member of a given pair lies on one side of the plate and the paternal member on the other. A critical point about this, however, is that the alignment adopted by any one pair of chromosomes bears no relation to the alignment adopted by any other pair. It may be, for example, that paternal chromosome 5 will line up on what we might call side A of the metaphase plate; if so, then maternal chromosome 5 ends up on side B. Shift to chromosome 6, however, and things could just as easily be reversed: The *maternal* chromosome might end up on side A, and the paternal on side B. It is thus a throw of the dice as to which side of the plate a given chromosome lines up on. More important, it is this chance event that determines which *cell* a chromosome joins, because each *side* of the plate becomes a separate *cell* once cell division is finished. As you will see, this randomness is critical in the shuffling of genetic material.

Anaphase I

In Anaphase I, the paired, homologous chromosomes now begin to move away from each other, toward their respective poles, via the same microtubule spindle that operated in mitosis. Though homologous chromosomes have now separated, the sister chromatids that make them up are still together.

Telophase I

With chromosome movement toward the poles completed, the original cell now undergoes cytokinesis, dividing into two completely separate daughter cells. With this, there are two haploid cells, whereas in the beginning there was one diploid cell.

Meiosis II

There is little pause for interphase between meiosis I and meiosis II; little time for the daughter cells to regroup and carry on regular metabolic activities.

Prophase II and Metaphase II

What happens instead is another division—or set of divisions, because there are now two cells. Each of these cells now has 23 duplicated chromosomes in it. After a very brief prophase, these chromosomes line up at a new metaphase plate, with sister chromatids on opposite sides of each plate.

Anaphase II and Telophase II

What happens next is the same thing that happened in mitosis. The *sister chromatids* now separate, moving toward opposite poles, and with this assuming the role of full-fledged chromosomes. Once this is completed, cytokinesis occurs. Where once there were two cells, there are now four. The *difference* between this process and mitosis is that each of these cells has 23 chromosomes in it instead of 46.

10.3 What Is the Significance of Meiosis?

You have now examined the mechanics of meiosis (Figure 10.2). But what are the effects of this process?

The Chromosome Duplication Problem Is Solved

First of all, the "Fennington-Kent" problem noted at the beginning of the chapter has been solved. There will be no 92-chromosome zygotes, because egg and sperm will each bring only 23 chromosomes to their meeting, thus yielding a combined 46 chromosomes in human somatic cells. Just as notable, however, are the possibilities for diversity that meiosis brings about.

Meiosis Ensures Genetic Diversity in Two Ways

Meiosis provides genetic diversity and shuffles the genetic deck in two ways.

Diversity through Recombination

Recall that, back in prophase I of meiosis, the homologous chromosome pairs intertwined for a time and then engaged in something called **recombination** (or **crossing over**). In this, there is an exchange of *parts* of chromosomes prior to their lining up at the metaphase plate.

Look at the "Recombination" box in Figure 10.2 to see how this important process works. In the actual process, there is a physical breaking of non-sister chromatids and then a "reunion" of the separated sections onto new chromatid partners. With this, what had been a "maternal" chromosome now has a portion of the paternal chromosome within it, and vice versa.

The linkage between recombination and genetic diversity probably is apparent; the ability of reciprocal lengths of DNA to be exchanged between chromosomes provides a means by which the genetic deck can be reshuffled prior to the formation of gametes. This is so because of what you saw in Chapter 9 about homologous chromosomes: While the genetic information on any two will be similar, it will not be identical. A gene on one chromosome may code for red hair color, but the gene on its homologous chromosome may code for blonde hair color. In recombination, such genetic variants are being swapped between homologous chromosomes.

Diversity through the Independent Assortment of Chromosomes

Another way that meiosis ensures diversity is a crucial event that took place in metaphase I (right after recombination) with the random alignment or **independent assortment** of homologous chromosomes at the metaphase plate. (See the "Independent Assortment" box in Figure 10.2). Recall that the alignment of any one pair of chromosomes along the plate bears no relation to the alignment of any other pair. If you looked at, say, chromosome 5, the paternal member of the pair might line up on "side A" of the metaphase plate, meaning maternal chromosome 5 will be on side B. Shift to chromosome 6, however, and things could just as easily be reversed. If we wonder how traits from our mother and father can get "mixed up" in children we have, independent assortment joins recombination in providing an answer.

Independent assortment has to do with the possibilities that follow from random alignment at the metaphase plate. If human beings had only two chromosomes in each cell that went through meiosis—one from the father, one from the mother—there would be only two possible "states" of line-up at the plate: "maternal on side A" of the plate or "maternal on side B" (with the paternal homologue on the opposite side). The actual number of possibilities, however, is this number of states

(2) *raised to the power* of the number of chromosome pairs. Thus, if three chromosome pairs existed, the number of possible outcomes would go to 2^3, or 8. Now get your calculator ready. With the 23 pairs of chromosomes human beings have, the possibilities go to 2^{23}, or about 8 million. General Motors may be able to mix and match engines and upholstery to come up with scores of different models of a given car, but in any given human meiosis, chromosome pairs can line up along the metaphase plate in 8 million different ways. This means any given egg or sperm will be one of about 8 million possible "models." To get a basic idea of how this works out, look at the "Independent Assortment" box in Figure 10.2 to see what it would mean, regarding the final genetic makeup of eggs and sperm, to have a variation on how the homologous chromosomes lined up at the plate.

What's the importance of this? Remember that there are *genes* on these chromosomes, and that the genes on any given paternal chromosome may differ somewhat from those on its matched maternal chromosome. Which chromosome the offspring gets thus makes a difference—in eye color, height, or any of thousands of other characteristics. Given this, each of us is very much affected by the way paternal and maternal chromosomes happened to line up in two very special meioses: those that produced the egg cell and the sperm cell that gave rise to us.

When we add independent assortment to recombination, it's easy to see why separate children from the same parents can come out looking so different—from each other as well as from the parents themselves (**see Figure 10.3**). Overall, meiosis generates

Figure 10.3
Ensuring Variety
The shuffling of genetic material that occurs during meiosis is the primary reason that children look different from their parents, and from each other.

The Revealing Y Chromosome

A debate that existed in the United States for nearly 200 years was apparently settled in 1998 when genetic tests revealed a high probability that Thomas Jefferson, the third president of the United States, had fathered at least one son by a slave he owned named Sally Hemings. Historians had argued the case for years on the basis of fragmentary evidence, but it was genetic testing of the modern-day descendents of Hemings and Jefferson's paternal uncle that all but ended the debate.

What scientists examined in the case was the descendants' DNA—in every case, DNA extracted from their Y chromosomes. Why just the Y chromosomes? These chromosomes turn out to be something of an oddity in the human chromosome collection, and it was this very quality that made them useful in the Jefferson-Hemings investigation.

The Y chromosome is unusual because it is the only human chromosome that exists in one sex but not the other—in men, but not in women. Females, as we've seen, have two X chromosomes, but males have one X chromosome and one Y. The Y is also unusual in that it is a pipsqueak of a chromosome, having in it perhaps 20 genes in all, compared to the thousands of genes that typically exist in other chromosomes. Finally, the Y is different in that it does almost no part-swapping in meiosis. As explained in the main text, in the early stages of meiosis, homologous chromosomes pair up and exchange reciprocal parts in the process called *recombination* (or *crossing over*). The two X chromosomes that a woman has will come together in meiosis and do just this. The Y chromosome a man has, however, engages in only the slightest bit of recombination with his X chromosome. (Nature has kept the X and Y separate in this way in order to keep the sexes separate. Imagine that through crossing over, the single gene on the Y chromosome that initiates male development ended up on the female X chromosome. The whole male-female distinction would begin to break down.)

The Y's abstention from recombination means, however, that the DNA "book" that it contains is not having pages inserted in it, and removed from it, every time a new generation is produced. (Think what recombination is doing: removing one part of a chromosome and replacing it with a part of a chromosome from genetic *diversity*, while its counterpart, mitosis, does not. Mitosis makes genetically exact copies of cells; it *retains* the qualities that cells have from one generation to the next. Meiosis mixes genetic elements each time it produces reproductive cells and thus brings about genetic variation in succeeding generations.

Meiosis and Sex Outcome

Let us think now about what meiosis means in terms of passing on *sex* to offspring. In humans there is one exception to the rule that chromosomes come in homologous pairs. Human females do indeed have 23 pairs of matched chromosomes, including 22 pairs of "autosomes" and one matched pair of **sex chromosomes**, meaning the chromosomes that determine what sex an individual will be. The sex chromosomes in females are called X chromosomes, and each female possesses two of them. In males, conversely, there are 22 autosomes, one X sex chromosome, and then one Y sex chromosome. It is this Y chromosome that confers the male sex.

In meiosis I in a female, the female's two X chromosomes, being homologous, line up together at the metaphase plate. Then these chromosomes separate, each of them going to different cells (**see Figure 10.4**). In males, the non-homologous X and Y chromosomes line up as if they were homologues. Then, one resulting cell gets an X chromosome while the other gets the Y chromosome.

Sex determination is then simple. Each of the eggs produced by a female bears a single X chromosome. If this egg is fertilized by a *sperm* bearing an X chromosome, the resulting child will be a female. If, however, the egg is fertilized by a sperm bearing a Y chromosome, the child will be a male.

Because females don't have Y chromosomes to pass on, it follows that the Y chromosome that any male has had to come from his

another person.) In historical genetic testing, as was done in connection with Jefferson and Hemings, this absence of recombination is critical. With the Y chromosome, the genetic record—the sequence of A's, T's, C's, and G's—is not getting scrambled in every generation.

The Y chromosome's unusual qualities allowed it to play a role in a historical detective story.

The scientists who did the Jefferson and Hemings tests looked for similar repeating sequences of Y-chromosome DNA in 19 descendants of the Hemings and Jefferson families—all of them males, of course—and found such closely matching sequences that there is little doubt these families are related. The only remaining question is whether it could have been one of Jefferson's relatives—rather than Jefferson himself—who fathered Hemings' child.

Jefferson and Hemings
In the movie *Jefferson in Paris*, the actor Nick Nolte portrayed Thomas Jefferson while Thandie Newton portrayed Jefferson's slave, Sally Hemings. DNA tests on Jefferson's descendants have provided strong evidence that Jefferson and Hemings had a child together.

father. Meanwhile, the single X chromosome that any male has had to come from his mother. Females, conversely, carry one X chromosome from their mother and one from their father. (For details on how Y chromosomes can be used in genetic investigations, see "The Revealing Y Chromosome," above.)

Tutorial 10.2.1
Sex Outcome in Human Reproduction

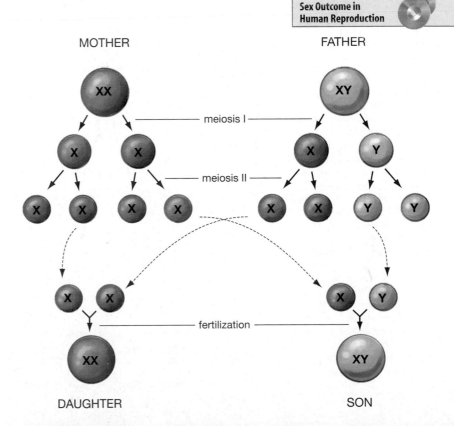

Figure 10.4
Sex Outcome in Human Reproduction
Early in meiosis I, the mother's two X chromosomes line up at the metaphase plate (XX). Meanwhile, in the father's meiosis, his X and Y chromosomes line up (XY). The X-X and X-Y pairs then separate into different cells. In meiosis II, the chromatids that made up the duplicated chromosomes separate, yielding individual eggs and sperm. In fertilization, should a Y-bearing sperm from the male reach the egg first, the child will be a boy; should the male's X-bearing sperm reach the egg first, the child will be a girl.

10.4 Gamete Formation in Humans

You've seen that, through meiosis, a single diploid cell gives rise to four haploid gametes. But what are the diploid cells that go through meiosis? As you might expect, there is a whole sequence of actions involved in the process by which egg and sperm are produced—a process known as *gamete formation*. The starting female cells are known as **oogonia**, while the male cells are **spermatogonia**. These are diploid cells that give rise to two other sets of diploid cells, the **primary oocytes** and **primary spermatocytes**. *These* then are the cells that undergo meiosis, yielding haploid sperm and egg cells.

Sperm Formation

Spermatogonia exist in the male testes and are capable either of giving rise to primary spermatocytes or of generating more spermatogonia. It is this latter, self-generating ability that qualifies spermatogonia as "stem"

cells and that allows males to *keep* producing sperm throughout their lives, beginning at puberty. These sperm-producing "factories," in other words, generate not only sperm but more factories as well (**see Figure 10.5a**).

The primary spermatocytes that come from spermatogonia go through meiosis I and thus produce the haploid **secondary spermatocytes**. Meiosis II then ensues, and the result is four haploid **spermatids**. These are not yet the familiar, tadpole-like sperm; those come only after spermatids develop further, a process that takes about 3 weeks. In this process, a given sperm cell must fully develop its tail-like flagellum. As it turns out, these "tails"—the only such structure in the human body—are composed of none other than the microtubules you have been seeing so much of in cell division. Moving the flagellum in a whip-like motion, microtubules serve to propel the sperm on their journey (**see Figure 10.6**). Sperm also need to "travel light,"

Figure 10.5
Sperm and Egg Formation in Humans

a In sperm formation (spermatogenesis), diploid cells called spermatogonia produce primary spermatocytes. The primary spermatocytes are the diploid cells that go through meiosis I, yielding haploid secondary spermatocytes. These spermatocytes then go through meiosis II, yielding four haploid spermatids that will develop into mature sperm cells.

b In egg formation (oogenesis), cells called oogonia, produced before the birth of the female, develop into primary oocytes. These diploid cells will remain in meiosis I until they mature in the female ovary, beginning at puberty. (Only one oocyte per month, on average, will continue through this maturation process.) Oocytes that mature will enter meiosis II, but their development will remain arrested there until they are fertilized by sperm. An unequal meiotic division of cellular material leads to the production of three polar bodies from the original oocyte and one well-endowed egg. The egg can go on to be fertilized, but the polar bodies will be degraded.

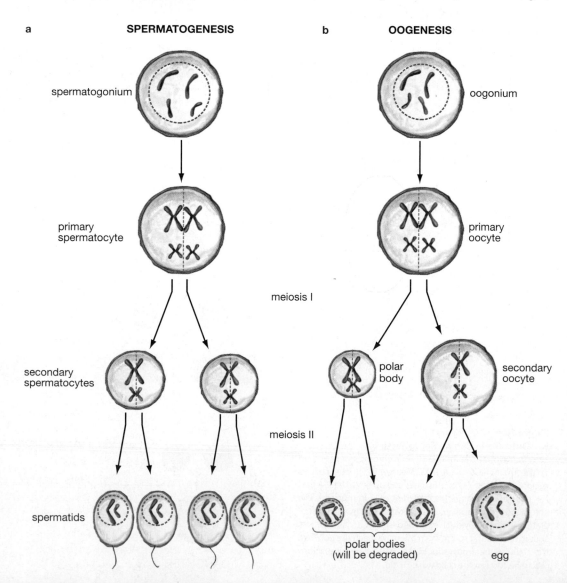

a   SPERMATOGENESIS

spermatogonium

primary spermatocyte

secondary spermatocytes

meiosis I

meiosis II

spermatids

b   OOGENESIS

oogonium

primary oocyte

polar body

secondary oocyte

polar bodies (will be degraded)

egg

to which end developing sperm get rid of nearly all the cellular organelles reviewed back in Chapter 4: lysosomes, the Golgi complex, and so forth. A mature sperm thus amounts to little more than a haploid set of DNA-containing chromosomes in front, a collection of "engines" (mitochondria) in the middle, and a propeller (the flagellum) in the back. About 250 million sperm are made each day, and about that number will be released with each ejaculation. Sperm that are not released will age and eventually are destroyed by the body's immune system.

Egg Formation

The need of male sperm to become stripped-down DNA packages stands in contrast to the needs of female ova or eggs, which must have rich resources of nutrients and building materials in order to sustain life once they have been fertilized. This difference in function is reflected in size: The volume of an egg is 200,000 times the volume of a sperm (**see Figure 10.7** to get an idea of this difference). Accordingly, eggs are created in a process that maximizes cell content, at least for the fraction of eggs that become viable.

Egg formation begins with the cells called the *oogonia* of the female. These may sound like the counterpart to the male spermatogonia, but actually they are very different. Whereas some spermatogonia are self-generating stem cells, oogonia are produced in large numbers only up to the seventh month of embryonic life, and none are created after birth. Of the millions produced, most will die while the female is still an embryo; but some survive to become primary oocytes and thus, as diploid cells, enter meiosis I (**see Figure 10.5b**).

Amazingly enough, meiosis I is where these oocytes *stay*—for years. After puberty, an average of only one oocyte per month is selected to complete maturation in the ovary, and thus enters into meiosis II. Even this meiosis is arrested, however, until this oocyte is fertilized by a sperm. It is the union with a sperm that prompts the completion of meiosis II. Thus, an oocyte that came into being when a female was an embryo might remain suspended in the cell cycle for years or even decades. Even then, it is not *all* primary oocytes that move through the cycle, but only a selected few; out of the millions initially created, only a few hundred will mature to ovulation during a woman's lifetime.

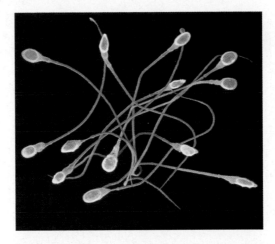

Figure 10.6
Mobile Cells, Bearing Chromosomes
Human sperm, colored to show detail.

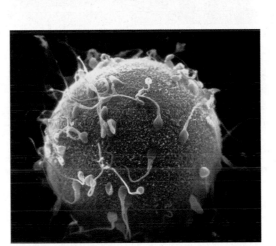

Figure 10.7
Big Difference in Size
A human egg surrounded by much smaller human sperm.

One Egg, Several Polar Bodies

The oocytes that do mature go through the rest of meiosis I and then meiosis II, as described earlier. However, in this progression—from one diploid oocyte to four haploid cells—almost all the cytoplasm will be shunted to *one* of the four cells, the better to build up its stores of nutrients and other cytoplasmic material. This happens in two rounds: meiosis I and meiosis II. When cytokinesis comes in meiosis I, one daughter cell gets almost all the cytoplasm while the second daughter cell gets almost none. This same thing happens in meiosis II. The result? One richly endowed haploid ovum and two, or perhaps three, very small haploid cells called **polar bodies** (Figure 10.5b). (Only two polar bodies and the ovum will be the end result if the polar body produced in meiosis I does not continue through meiosis II, which often happens.) The egg has a chance at taking part in producing offspring, but the polar bodies are bound for oblivion rather than a shot at immortality. They will eventually degrade into their constituent substances.

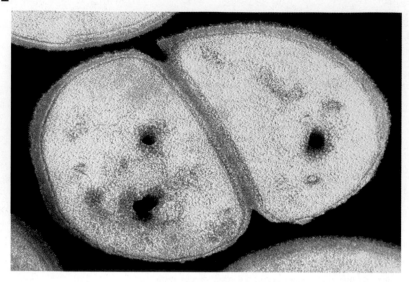

Figure 10.8
Reproduction in Bacteria
There is no fusion of sperm and eggs in bacteria, just a division of one cell into two. Pictured is a *Staphylococcus aureus* bacterium undergoing such cell division, referred to as binary fission.

How do certain oocytes get "selected" to mature out of prophase I? This occurs in connection with an intricate interplay that involves their surrounding tissue, hormones, and the happenstance of timing. Of the relatively few oocytes that do mature, in general only *one* will fully develop and be expelled from the ovary for a journey through the uterine tubes, where it can be fertilized by sperm. This journey happens only once every 28 days on average, a number that may ring a bell; the process of oocyte maturation is directly linked to the process of menstruation. You'll be going over all of this in Chapter 28.

Figure 10.9
Regenerating the Whole from a Part
A sea star off the coast of Papua New Guinea regenerating a body and other arms from a piece of one arm and part of its central disk.

10.5 Life Cycles: Humans and Other Organisms

The process of the formation of the human gametes called egg and sperm is followed, as you have seen, by the union of these gametes, which creates a fertilized egg. This egg (or zygote) then develops into a whole human being through the mitotic division of cells. But even as mitosis goes on, another group of egg and sperm is being readied in the new human being through the process of meiosis. This process amounts to a **life cycle**: the repeating series of steps that occur in the reproduction of an organism. You have looked at the human life cycle, but it's important to recognize that it amounts to only one among many.

Not All Reproduction Is Sexual

Human reproduction is sexual reproduction: the union of two reproductive cells to create a new organism. But in many other species there is **asexual reproduction**. Bacteria are single-celled organisms that possess a single, circular chromosome. (This latter fact makes them haploid by definition.) You saw in Chapter 9 that bacteria divide through the process of binary fission, one bacterium becoming two in the course of its cell cycle (**Figure 10.8**). As it turns out, that's it for bacteria as far as reproduction goes. For them, in other words, the cell cycle *is* the life cycle. There is no true diploid state or fusion of two cells; just an exact replication of DNA and a splitting of cytoplasm. There is asexual reproduction among eukaryotes as well; many single-celled eukaryotes reproduce through simple cell division, though there are lots of variations on this theme.

Asexual reproduction is also seen as we move through the branching evolutionary tree. Anyone who has ever seen a "cutting" taken from a plant knows that a snipped-off stem or branch will, when properly planted, sprout its own roots and grow into an independent plant, complete with the ability to take part in sexual reproduction. This process is called **vegetative reproduction**, and it has a counterpart in the animal world. Cut the arm and part of a central disk from a sea star, for example, and, through **regeneration**, this severed limb will grow into a whole, multi-armed sea star (see **Figure 10.9**).

Note that all these forms of asexual reproduction produce offspring that are genetically identical to their parents. There is no mixing and matching of chromosomes from different parents in these asexual processes, just a cloning of one genetically identical individual from its parent.

Variations in Sexual Reproduction

Within the framework of sexual reproduction, as you can imagine, there are many variations. For example, separate gametes need not come from separate organisms. Some animals, such as tapeworms, are hermaphrodites, meaning they have both male and female reproductive parts. In some cases, hermaphrodites fertilize themselves, but generally they are fertilized by another member of their species. Most plants contain both male and female parts, and some plants "self-fertilize"—their sperm can fertilize their eggs—as with a pea plant you will be looking at in Chapter 11.

On to Patterns of Inheritance

The tour of cell division over the last couple of chapters has yielded not only a look at how one cell becomes two, but at how chromosomes operate within this process. As you have seen, it is the DNA in these chromosomes that is the bearer of the "qualities" or "traits" that are passed on from one generation to the next. Given the paired nature of chromosomes and their precisely ordered activity during meiosis, it stands to reason that there would be some predictability or pattern to the passing on of traits. This is the case, as it turns out. And the person who first recognized this pattern was a monk who worked in nineteenth-century Europe in an obscurity he did not deserve.

Chapter Review

Summary

10.1 An Overview of Meiosis

- In human beings, nearly all cells have paired sets of chromosomes, meaning these cells are diploid. Meiosis is the process by which a single diploid cell divides to produce four haploid cells—cells that contain a single set of chromosomes.

- The haploid cells produced through meiosis are called gametes. Female gametes are eggs; male gametes are sperm. They are the reproductive cells of human beings and many other organisms.

- When the haploid sperm and haploid egg fuse, a diploid fertilized egg (or zygote) is produced, setting into development a new generation of organism.

10.2 The Steps in Meiosis

- In meiosis, there is one round of chromosome replication, followed by two rounds of cell division. There are two primary stages to meiosis, meiosis I and meiosis II. In meiosis I, chromosome duplication is followed by a pairing of homologous chromosomes with one another during which time they exchange reciprocal sections of themselves. These chromosome pairs are then separated, in the first round of cell division, into separate daughter cells. In meiosis II, the chromatids of the duplicated chromosomes are separated, in the second round of cell division, into four daughter cells.
TUTORIALS: 10.1.1 Steps in Meiosis; 10.1.11 Comparison of Mitosis and Meiosis

10.3 What Is the Significance of Meiosis?

- Meiosis generates diversity by ensuring that the gametes it gives rise to will differ from one another genetically. In this, it is unlike regular cell division, or mitosis, which produces daughter cells that are exact genetic copies of parent cells.

- Meiosis generates genetic diversity in two ways. First, in prophase I of meiosis, homologous chromosomes pair with each other and, in the process called recombination or crossing over, exchange reciprocal chromosomal segments with one another. Second, in metaphase I of meiosis, there is a random alignment or independent assortment of maternal and paternal chromosomes on either side of the metaphase plate. This chance alignment determines which daughter cell each chromosome will end up in.

- Human females have 23 matched pairs of chromosomes—22 autosomes and two X chromosomes. Human males have 22 autosomes, one X chromosome, and one Y chromosome. Each egg that a female produces has a single X chromosome in it. Each sperm that a male produces has either an X or a Y chromosome within it. If a sperm with a Y chromosome fertilizes an egg, the offspring will be male. If a sperm with an X chromosome fertilizes the egg, the offspring will be female.
TUTORIAL 10.2.1: Sex Outcome in Human Reproduction

10.4 Gamete Formation in Humans

- The human male diploid cells called primary spermatocytes go through meiosis, producing, by the end of meiosis II, spermatids that develop into mature sperm. Human egg formation begins with oogonia cells, all of which are produced prior to the birth of the female. These give rise to primary oocytes, whose development is arrested in meiosis I until they are released in ovulation and fertilized.

10.5 Life Cycles: Humans and Other Organisms

- Not all reproduction is sexual reproduction. Bacteria reproduce through binary fission, meaning a chromosomal replication followed by cell division. Plants can engage in vegetative reproduction, and other organisms, such as worms and sea stars, can carry out regeneration.

Key Terms

1n 192	**polar body** 201
2n 192	**primary oocyte** 200
asexual reproduction 202	**primary spermatocyte** 200
crossing over 196	**recombination** 196
diploid 192	**regeneration** 202
gamete 192	**secondary spermatocyte** 200
haploid 192	**sex chromosome** 198
independent assortment 197	**spermatid** 200
life cycle 202	**spermatogonium** 200
meiosis 192	**tetrad** 195
oogonium 200	**vegetative reproduction** 202

Understanding the Basics

Multiple-Choice Questions

1. In reference to chromosomes, meiosis involves

   a. no duplications, 1 reduction
   b. 1 duplication, 2 reductions
   c. 2 duplications, 1 reductions
   d. 2 duplications, 2 reductions
   e. none of the above

2. Imagine, at prophase of meiosis I, that a diploid (2n) cell that is precursor to the gamete has eight chromosomes. How many chromatids are in the cell?

   a. 16
   b. 8
   c. 4
   d. 2
   e. twice that of mitosis prophase

3. Independent assortment occurs during

   a. prophase I
   b. anaphase II
   c. metaphase I
   d. metaphase II
   e. both c and d

4. Which of the following statements is *not* true regarding egg formation in humans?

   a. Oogonia are not produced until the eleventh or twelfth year.
   b. At the completion of division, one egg and two or three polar bodies exist.
   c. In both meiosis I and II, the metaphase plate is shifted to one side of the cell.
   d. Oocyte maturation occurs in approximately a 28-day cycle.
   e. Each egg contains 23 chromosomes.

5. What are the two sources of genetic diversity in meiosis?

   a. binary fission and regeneration
   b. recombination and independent assortment
   c. oogenesis and recombination
   d. spermatogenesis and oogenesis
   e. vegetative reproduction and regeneration

6. An egg has approximately

   a. 200,000 times the volume of a sperm
   b. 100,000 times the volume of a sperm
   c. 50,000 times the volume of a sperm
   d. 25,000 times the volume of a sperm
   e. 0.5 times the volume of a sperm

7. The cytoplasmic content of sperm is very limited compared to that of an egg. The major ingredients of sperm cell are

   a. a nucleus and Golgi complex
   b. ribosomes and mRNA
   c. a haploid set of chromosomes and endoplasmic reticulum
   d. mitochondria and ribosomes
   e. a haploid set of chromosomes and mitochondria

8. In prophase I of meiosis,

   a. Maternal chromosome 1 joins with maternal chromosome 2.
   b. Maternal chromosomes exchange genetic material with other maternal chromosomes.
   c. Paternal chromosomes exchange genetic material with other paternal chromosomes.
   d. Tetrads form, joining homologous chromosomes (for example, maternal chromosome 1 with paternal chromosome 1).
   e. Both c and d are correct.

9. At the completion of meiosis I, each human cell contains

   a. 23 chromosomes, 46 chromatids
   b. 46 chromosomes, 46 chromatids
   c. 46 chromosomes, 92 chromatids
   d. 23 chromosomes, 23 chromatids
   e. none of the above

10. In sexual reproduction, there is first a(n) _____ of the normal number of chromosomes in somatic cells, followed by a _____ of male and female reproductive cells.

    a. multiplication, division
    b. doubling, decrease
    c. maintenance, halving
    d. tripling, splitting
    e. halving, union

Brief Review

1. Does chromosome reduction take place during meiosis I or II?

2. Is asexual reproduction confined to the plant world?

3. Why are men able to keep producing sperm throughout their lifetime, from puberty forward?

4. How does meiotic cytokinesis in the production of eggs differ from mitotic cytokinesis?

5. Define and distinguish between somatic cells and gametes.

6. A diploid organism has four chromosomes. Name the phases of mitosis or meiosis shown in the following figures.

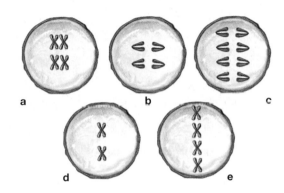

7. In what ways does meiosis ensure genetic diversity in offspring?

8. Which anaphase of meiosis is most like anaphase in mitosis? Why?

9. Why is an egg so big relative to a sperm?

10. Define the following terms:
    a. gamete
    b. recombination (or crossing over)
    c. asexual reproduction
    d. haploid

Applying Your Knowledge

1. Why would crossing over not be of significant value during prophase II of meiosis?

2. Ultimately, is it the paternal or maternal gamete that delivers the sex-determining gene? Why?

3. Does asexual reproduction or sexual reproduction allow for the greatest genetic variation? Why?

4. Draw one progeny cell in the following phases and indicate the number of chromosomes and chromatids present.

	number of	number of
drawing of cell	chromosomes	chromatids
$2n = 4$ during metaphase of mitosis		
$2n = 6$ during anaphase of meiosis II		
$2n = 4$ following telophase of meiosis I		

11

The First Geneticist
Mendel and His Discoveries

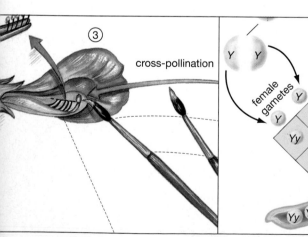

Cross-pollination.
(Section 11.2, page 209)

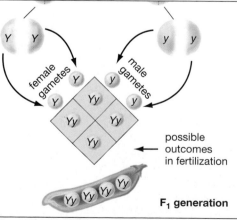

Green and yellow peas.
(Section 11.4, page 212)

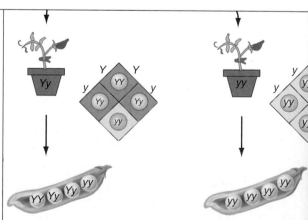

More pea experiments.
(Section 11.4, page 213)

A straightforward set of rules governs the way living things pass on many of their traits. Gregor Mendel was the first person to understand what those rules are.

In the Czech Republic, in the oldest part of the city of Brno, stands the former Monastery of St. Thomas. Now a research center, St. Thomas has a small, fenced-in garden that sits just outside its historic living quarters. At one end of the garden stands an ivy-covered tablet, inscribed with a few simple words: "Prelate Gregor Mendel made his experiments for his law here." Visitors to the garden can cast their eyes upward and see the rooms that Gregor Johann Mendel lived in while carrying out his work in the years between 1856 and 1863 (**Figure 11.1**). Mendel labored during this time on a species of common peas, *Pisum sativum*, using tweezers and an artist's paintbrush as his tools. Like many scientific discoverers, Mendel looked at ordinary objects; and as with many scientists, his work was tedious and exacting. Nevertheless, the distance from what Mendel did to what Mendel learned is the distance from the ordinary to the historic.

Coming to adulthood in the mid-nineteenth century, Mendel knew nothing of the elements of genetics we have reviewed so far. Chromosomes and the genes that lie along them had not even been discovered when he was working with his peas; little knowledge existed of meiosis or mitosis, to say nothing of DNA. And as it happened, Mendel himself played no direct role in discovering any of these things. Yet this unassuming monk, the son of eastern European peasant farmers, generally is accorded the title of the father of genetics. What contribution earned him this honor?

**Figure 11.1
Austrian Monk and Naturalist Gregor Mendel**

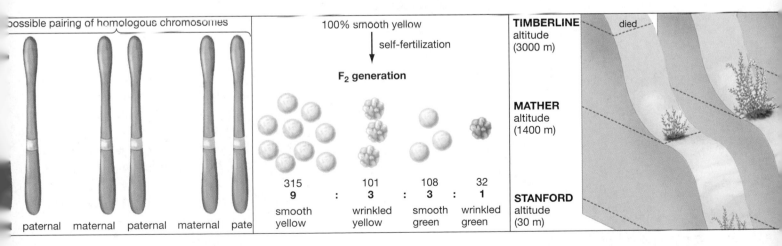

possible pairing of homologous chromosomes

paternal   maternal   paternal   maternal   pate

Alleles on chromosomes.
(Section 11.4, page 216)

100% smooth yellow

self-fertilization

F₂ generation

315	101	108	32
9	: 3	: 3	: 1
smooth yellow	wrinkled yellow	smooth green	wrinkled green

From a smooth yellow pea to four phenotypes.
(Section 11.5, page 217)

TIMBERLINE altitude (3000 m)   died

MATHER altitude (1400 m)

STANFORD altitude (30 m)

Genes and environment.
(Section 11.10, page 222)

11.1 Mendel and the Black Box

Mendel's achievement was to comprehend what was going on inside what might be called the "black box" of genetics without ever being able to look inside that box himself. Science is filled with so-called black-box problems, in which researchers know what goes *into* a given process and what comes *out*. It is what is going on in between—in the black box—that is a mystery. In the case of genetics, what lies inside the black box is DNA and chromosomes and meiosis, and so forth—all the component *parts* of genetics, in other words, and the way they work together.

Because of the timing of his birth, Mendel had no knowledge of any of these things. Until Mendel, however, nobody had looked carefully at even the starting and ending points of this black-box problem: at what went in and what came out. Mendel's original pea plants represented the "input" side to the black box of genetics, while the offspring he got from breeding these plants represented the output. By looking carefully at generations of parents and offspring—at both sides of the box, in a sense—Mendel was able to infer something about what had to be going on within. As you will see, his main inferences were correct: (1) That the basic units of genetics are material elements that come in pairs; (2) that these elements (today called *genes*) can retain their character through many generations; and (3) that gene pairs *separate* during the formation of gametes.

These insights may sound familiar, because all of them were approached from another direction in Chapters 9 and 10. Here are the lessons you went over then:

- Genes are material elements—lengths of DNA.
- In human beings, genes come in pairs, residing in pairs of chromosomes.
- Chromosomes make copies of themselves, thus giving the genes that lie along them the ability to be passed on intact through generations.
- In meiosis, homologous chromosomes line up next to each other and then *separate*, with each member of a pair ending up in a different egg or sperm cell.

By observing generations of pea plants and applying mathematics to his observations, Mendel inferred that something like this had to be happening in reproduction. Because Mendel was the first to perceive a set of principles that govern inheritance, we date our knowledge of genetics from him.

11.2 The Experimental Subjects: *Pisum sativum*

Beginning work in his monastery garden after a period of study in Vienna, Mendel managed to pick, in *Pisum sativum*, a nearly perfect species on which to carry out his experiments. If you look at **Figure 11.2**, you can see something of the anatomy and life cycle of this garden pea. Note that what we think of as peas in

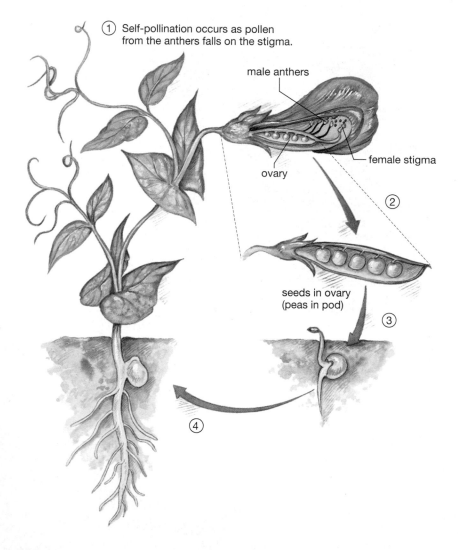

1. Self-pollination occurs as pollen from the anthers falls on the stigma.

male anthers

female stigma

ovary

2

seeds in ovary
(peas in pod)

3

4

Figure 11.2
Anatomy and Life-Cycle of the Pea Plant

1. Sperm-bearing pollen, which exists in the plant's anthers, lands on the plant's stigma.

2. This results in the fertilization of the plant's eggs, which are housed in the plant's ovary. These eggs will develop into seeds in the ovary (peas in a pod), which represent a new plant generation. Each seed is fertilized separately.

3. After being planted, each seed has the potential to grow into a separate plant.

4. The seedlings develop into mature seed plants, capable of producing their own offspring.

a pod are seeds in this plant's ovary. Each of these seeds begins as an unfertilized egg, just as a human baby begins as a maternal egg that is unfertilized. Sperm-bearing pollen, landing on the plant's stigma, then set in motion the fertilization of the eggs.

Importantly, *Pisum* plants can *self*-pollinate. The anthers of a given flower release pollen grains that land on that flower's stigma. Each seed that develops from the resulting fertilizations can then be planted in the ground and give rise to a new generation of plant. The way to think of the seeds in a pod is as multiple offspring from separate fertilizations—one pollen grain fertilizes this seed, another pollen grain fertilizes another. This is why, as you'll see, seeds that are in the same pod can have different characteristics.

Though the pea plants can self-pollinate, Mendel could also **cross-pollinate** the plants at will—he could have one plant pollinate another—by going to work with his tweezers and paintbrushes, as shown in **Figure 11.3**.

In directing pollination, Mendel could control for certain attributes in his pea plants— qualities now referred to as **characters**. If you look at **Table 11.1** on page 210, you can see that these included seven qualities in all—such things as stem length, seed color, and seed shape. Note that each of these characters comes in two varieties. There are *yellow* or *green* seeds, for example, and purple or white flowers. Such character variations are known as **traits**. Mendel referred to each of these variations as being either a "dominant" or a "recessive" trait. For now, you can think of a recessive trait as

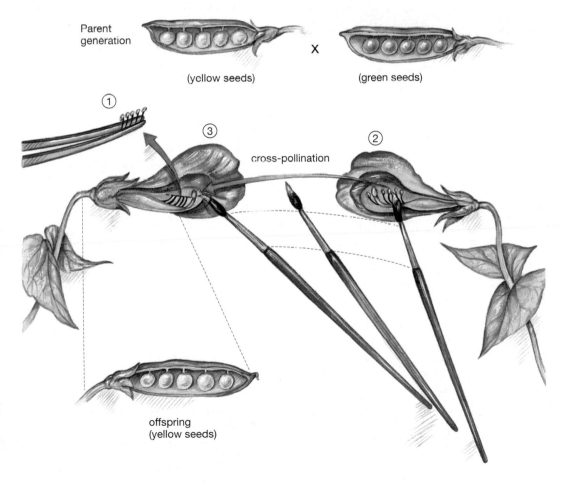

Figure 11.3
How to Cross-Pollinate Pea Plants

1. Before fertilization occurs, peel back the closed petals of a pea plant (in this case, one that came from a line that yielded yellow peas). Then pull out the pollen-bearing stamens with tweezers so that self-fertilization is no longer possible.

2. Next, gather pollen from another plant by dabbing its anthers with a paintbrush.

3. Finally, rub these pollen grains onto the stigma of the first plant. The results of the cross-pollination can be observed when the fertilized eggs mature into seeds in the ovary, meaning peas in a pod. The resulting seeds are yellow in this case, because yellow is dominant over green.

one that tends to remain hidden in certain generations of pea plants. A dominant trait, meanwhile, can be thought of as a trait that tends to appear in those same generations. You'll get a formal definition of dominant and recessive later.

Phenotype and Genotype

Now let's put these traits in another context. Taken together, traits represent what are called **phenotypes**. Broadly speaking, a phenotype is a physiological feature, bodily characteristic, or behavior of an organism. In the context of Mendel, phenotype means the pea plant's visible physical characteristics. Purple flowers are one phenotype, white flowers another; yellow seeds are one phenotype, green seeds another. Meanwhile, any phenotype is in significant part determined by an organism's underlying **genotype**, meaning its genetic makeup. You could think of phenotype as the observable outcome of genotype.

11.3 Starting the Experiments: Yellow and Green Peas

Mendel began by making sure that his starting plants "bred true" for the phenotypes under study. That is, he assured himself that, for example, all of his purple-flowered plants would produce nothing but generations of purple offspring.

Parental, F₁, and F₂ Generations

The **parental generation** in such experiments is often referred to in shorthand as the **P** generation. Meanwhile, the offspring of the parental generation are known as the **first filial generation**, the word *filial* indicating a son or a daughter. The shorthand for first filial is **F₁**. The F_x form can be used for any succeeding generation. You will be seeing a lot of the second filial, or F_2 generation, for example.

So how did Mendel proceed with his plants? Let's start with just one example. Mendel took, for his P generation, some plants from a line that bred true for yellow seeds and other plants from a line that bred true for green seeds. Then he took pollen from one variety of the plants and fertilized the other variety, as seen in Figure 11.3. The plants fertilized in this way then produced their own seeds—every one of which was yellow. (Remember that although these seeds were developing within the pods of the parental generation, they represented the new generation—the F_1 generation.) Getting all yellow seeds was an interesting result in itself. Because each seed in a pea-plant pod is fertilized separately, it *can* be the case that a given pod will contain both yellow and green seeds. (Other pea characters can be variable within a pod as well, as you can see in **Figure 11.4**.) Yet all the seeds in this generation were yellow.

Having viewed these results, Mendel then planted his F_1 seeds and let plants grow from them. These plants were then allowed to *self*-pollinate. What he got from those plants, in the F_2 generation, was 6,022 yellow seeds and 2,001 green seeds.

The Power of Counting

In counting the seeds, Mendel was taking a giant step forward. Why? Experimenters before Mendel had gotten the kind of results he had.

Table 11.1
Pea-Plant Characters Studied by Mendel

Character Studied	Dominant Trait		Recessive Trait	
Seed Shape	Smooth		Wrinkled	
Seed Color	Yellow		Green	
Pod Shape	Inflated		Wrinkled	
Pod Color	Green		Yellow	
Flower Color	Purple		White	
Flower Position	On stem		At tip	
Stem Length	Tall		Dwarf	

What they did not do, however, was undertake a careful *counting* of their results and then analyze the results in terms of proportions.

Looking at the specific results, two things stand out. First, green seeds disappeared in F$_1$, but came back in F$_2$. Recall that there was not a green seed to be found in the F$_1$ generation pods, but in F$_2$, there are 2,001 of them. Second, green seeds came back in F$_2$ as a *specific proportion* of the seeds as a whole. The F$_2$ generation isn't divided 50/50, half-yellow and half-green; instead, there are roughly 3 yellow seeds for every 1 green seed—or to put it another way, a 3:1 ratio of yellow to green seeds. As it turned out, Mendel got the same result with each of the seven characters he was studying, as you can see in **Table 11.2**. Note that the 3:1 ratio is a proportion of dominant to recessive traits—more yellow seeds than green, more purple flowers than white. It was *solely* the dominant traits that appeared in the F$_1$ generation, but in the F$_2$ generation the dominants are simply appearing in greater *proportion* than the recessives.

Interpreting the F$_1$ and F$_2$ Results

What did Mendel learn from these results?

No "Blending" in Inheritance

For one thing, he saw that inheritance for his peas was not a matter of the "blending" of their characteristics. For example, the flower colors purple and white were retained as just that. No intermediate phenotypes—say, light blue flowers—resulted from the cross in any generation. This finding ran contrary to the notion, popular in Mendel's time, that two given traits would blend into a homogenous third entity, as coffee and milk will blend together.

Dominant and Recessive Elements Come in Pairs

Beyond this finding, it was apparent that, dominance notwithstanding, plants could retain the *potential* for recessive phenotypes, even though those phenotypes might not appear in a given generation. Mendel's F$_1$ plants had no green seeds, but by F$_2$ the green seeds were back. It was reasonable to assume, therefore, that the yellow-seed F$_1$ plants retained a green-seed *element*, which got expressed only

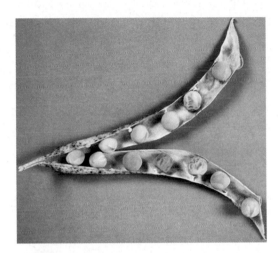

Figure 11.4
Variation within a Pea Pod
Since each garden pea is fertilized separately, individual peas within a pod can have different character traits. Note that some of these peas have a smooth texture, while others are wrinkled.

Table 11.2 Ratios of Dominant to Recessive in Mendel's Plants		
Dominant Trait	**Recessive Trait**	**Ratio of Dominant to Recessive in F$_2$ Generation**
Tall stem	Dwarf stem	2.84:1 (787 tall plants, 277 dwarfs)
Smooth seed	Wrinkled seed	2.96:1 (5,474 smooth, 1,850 wrinkled)
Yellow seed	Green seed	3.01:1 (6,022 yellow, 2,001 green)
Inflated pod	Wrinkled pod	2.95:1 (882 inflated, 299 wrinkled)
Green pod	Yellow pod	2.82:1 (428 green, 152 yellow)
Purple flower	White flower	3.14:1 (705 purple, 224 white)
Flower on stem	Flower at tip	3.14:1 (651 along stem, 207 at tip)
	Average ratio, all traits:	3:1

Tutorial 11.1.1
Mendel F₁ Crosses
(Punnett Square)

Figure 11.5

a Mendel's F₁ Crosses

1. Mendel started out by cross-breeding plants that for generations had yielded either all yellow seeds or all green seeds. In the example pictured, female gametes are being provided by a plant that has the dominant, yellow alleles (*YY*); while the male gametes are being provided by a plant that has the recessive, green alleles (*yy*).

2. The cells of the pea plants that give rise to gametes start to go through meiosis.

3. The two alleles for pea color, which lie on separate homologous chromosomes, separate in meiosis, yielding gametes that each bear a single allele for seed color. In the female, each gamete bears a *Y* allele; in the male, each bears a *y* allele.

4. The Punnett square shows the possible combinations that can result when the male and female gametes come together in the moment of fertilization. (If you have trouble reading the Punnett square, see Figure 11.5b.) The single possible outcome in this fertilization is a mixed genotype, *Yy*.

5. Because *Y* (yellow) is dominant over *y* (green), the result is that all the offspring in the F₁ generation are yellow, because they all contain a *Y* allele.

b How to Read a Punnett Square

This Punnett square concerns pea-plant flower color, which is either purple (dominant) or white (recessive). Two of the four possible combinations are shown.

1. A *p* gamete from the male combines with a *p* gamete from the female to produce an offspring of *pp* genotype (and white color).

2. A *p* gamete from the male combines with a *P* gamete from the female to produce an offspring of *Pp* genotype (and purple color). What are the other two possible genetic combinations in this crossing, and what are their phenotypic outcomes?

in the F₂ generation. Finally, because his phenotypes came in pairs, it was reasonable for Mendel to hypothesize that the elements likewise came in pairs.

If you think Mendel's *pairs of elements* sound suspiciously like the "pairs of genes" on homologous chromosomes we've seen before, you are right. However, it's time to start referring to these "matched pairs of genes" in scientific terminology. It is more accurate to think of matched pairs of genes as alternative forms of a single gene. The proper name for an alternative form of a gene is an **allele**. Thus the pea plant had a single gene for seed color that came in two alleles—one of which coded for yellow seeds, the other of which coded for green seeds. These alleles resided on separate homologous chromosomes.

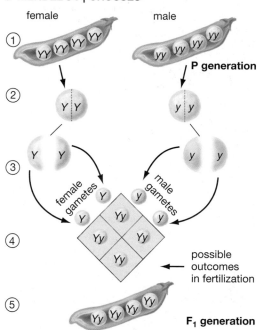

a MENDEL'S F₁ CROSSES

female   male

① P generation

②

③ female gametes / male gametes

④ possible outcomes in fertilization

⑤ F₁ generation

b HOW TO READ A PUNNETT SQUARE

female gametes

male gametes

① ②

11.4 Another Generation for Mendel

Having seen recessive phenotypes reappear in F₂, Mendel decided to extend his breeding. In his next experiment, he took a set of F₂ generation seeds—all of them yellow—planted them and let the plants they grew into self-pollinate. Then he looked at the color of the F₃ seeds contained in the pods of these plants. What he found was that, of 519 plants grown, 166 yielded only yellow seeds, while the remaining 353 plants had a mixture of yellow and green seeds in the pods, in the familiar 3:1 ratio. But when he similarly took F₂ green seeds, planted them, and let them self-pollinate, these plants produced nothing but green seeds.

What could have caused these results? The answer was the underlying elements Mendel had hit upon, meaning the alleles for the color gene. When he planted his yellow seeds, some of them yielded all-yellow offspring, but others yielded mixed (green and yellow) offspring. He reasoned that there must be *two kinds* of yellow plants: Those that are "pure" yellow (the 166 that yielded solely yellow seeds) and those that were "mixed" yellow (the 353 that produced both yellow and green seeds). Pure yellow had *nothing but* yellow alleles within, and would produce nothing but yellow seeds when self-pollinated. Meanwhile, the mixed yellow would have one yellow and one *green* allele within, and thus would produce pods that contained seeds of both colors. Finally, the green seeds had nothing but green alleles and thus produced nothing but green seeds.

Mendel's Generations in Pictures

Let's make these points clearer by reviewing all three generations of experiments schematically.

The F₁ Generation

It will be helpful to introduce a convention here. Let us refer, as Mendel did, to dominant and recessive alleles by uppercase and lowercase letters, with uppercase used for dominant types and lowercase used for recessive types. Thus a "pure" yellow-seeded plant, having *two* yellow alleles, would be symbolized as *YY*. Meanwhile, pure green-seeded plants are *yy*, while mixed seeds are *Yy*. Let us say, just as an example, that female gametes are being supplied by a plant that is

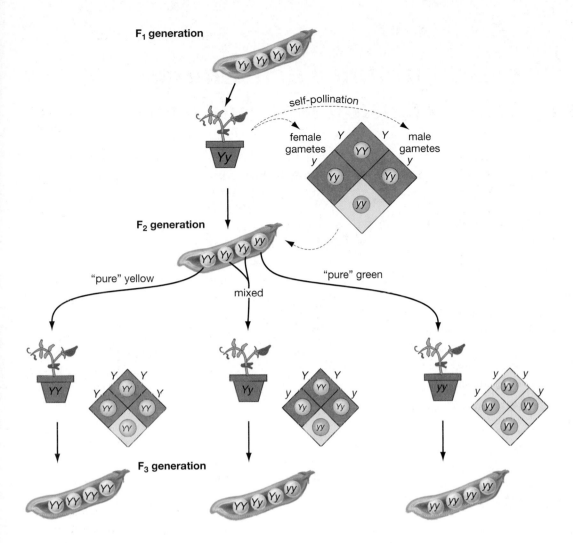

Figure 11.6
From the F₁ to the F₃ Generation

F₁ to F₂: The starting point is the F₁ generation, a set of seeds that all have the *Yy* genotype. These seeds are planted and the plants go through meiosis, yielding the gametes shown in the Punnett square. When these gametes come together in self-fertilization, the possibilities include *YY* and *yy* combinations, as well as the *Yy* combination seen in the F₁ generation. The existence of *yy* individuals is the reason green seeds reappear in the F₂ generation. Because *Y* is dominant, the green phenotype could not appear in seeds that had even a single *Y* allele. **F₂ to F₃:** With three starting genotypes (*YY, Yy, yy*) the F₂ generation yields plants that have these three genotypes, though there are more F₂ plants of "mixed" genotype than of either "pure" genotype.

Tutorial 11.1.3
The F₂ and F₃ Generations

YY, while male gametes are coming from a plant that is *yy*. In meiosis, as you've seen, homologous chromosomes separate. This happens in the pea plants, meaning that each of these plants will contribute *one member* of its gene pair to its respective gametes—the *YY* female contributing a *Y* gamete, and the *yy* male contributing a *y* gamete. When these gametes fuse in the moment of fertilization, the result is a *Yy* hybrid in the F₁ generation. If you look at **Figure 11.5**, you will be introduced to a time-honored way to represent such outcomes, the Punnett square, and see Mendel's F₁ results symbolized as well.

Note that, because *Y* is dominant, all the seeds in the F₁ offspring pods will have a yellow

phenotype, even though every one of these seeds contains a mixed *genotype*. (Every one is *Yy*.) It takes only one *Y* allele for a seed to be yellow; for a seed to be green, then, it must have two *yy* alleles.

The F₂ Generation
Next come the F₁ crosses that yielded the F₂ generation. The starting point here is the F₁ seeds, which are all of the mixed *Yy* type, as you can see in **Figure 11.6**. Meiosis then occurs in the gamete precursors, which results in a separation of these alleles: Half the gametes now contain *Y* alleles and the other half *y*. You can see from the figure why this cross can now give us back the green-seed phenotype, on

Proportions and Their Causes:
The Rules of Multiplication and Addition

To be a good songwriter, a person needs to be both a good musical composer and a good lyricist. Of all the people who decide to take up songwriting, say that 1 in 10 is a good composer and that 1 in 10 is a good lyricist. Now, what is the probability that a person who takes up songwriting will be *both* a good composer and a good lyricist? The answer to this question turns out to have relevance to all kinds of questions in our world, including the results that Gregor Mendel got.

When Mendel examined the F_2 pods that contained the F_3 seeds, he found that there were many more pods that had mixed seed colors within (yellow and green) than pods whose seeds were either all-yellow or all-green. Here's the F_1 meiosis that led to this outcome:

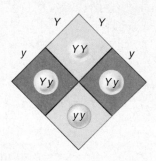

The Punnett square provides an intuitive visual sense of why things turned out the way they did. The F_1 meiosis produced twice as many mixed F_2 seeds (Yy) as either type of "pure" seeds (YY or yy). It's even more helpful, however, to understand a principle that underlies this outcome. A simple coin-tossing example can illustrate what is at work. Suppose that you are going to throw two coins in the air—a nickel and a dime. What is the likelihood that *both* coins will come up heads? There is only one way to get this result:

1. Nickel = heads; dime = heads

Now consider tossing the two coins and specifying a *mixed* result. There are *two* ways that this condition can be satisfied:

1. Nickel = heads; dime = tails
2. Dime = heads; nickel = tails

The seeds are in the same situation. What is the likelihood of getting a "pure" yellow seed? As you can see by looking again at the Punnett square, there is only one way to get this: female Y, male Y. Now consider specifying a mixed result. There are two ways to get this: female y, male Y; and female Y, male y.

average in a 1:3 proportion with yellow seeds. Note that there are twice as many mixed-genotype seeds (Yy) as either variety of pure seed (YY or yy). Important mathematical principles underlie this outcome. You can read more about them in "Proportions and Their Causes," above.

The F₃ Generation

Finally, in Figure 11.6, you can see why Mendel got the results he did when he went from F_2 to F_3. Recall that when Mendel planted and self-fertilized his *yellow* F_2 seeds, he got 519 plants, of which 166 produced nothing but yellow seeds. Now you can see why: These were not only phenotypically yellow; they were "pure" yellow in terms of genotype (YY). Mendel also got 353 plants that produced both yellow and green seeds. This is

because the seeds these plants grew from were of mixed genotype (Yy); in reproduction, their offspring would be of both pure (YY, yy) and mixed (Yy) genotype. Finally, when Mendel planted and self-fertilized *green* F_2 seeds, all the plants that resulted from these seeds bred true for green seeds, because all of them began as yy. **Figure 11.7** illustrates how the three genotypes discussed here yield only the two phenotypes of yellow or green.

The Law of Segregation

For inheritance to work this way, Mendel saw something we noted earlier: That, though plant cells may contain *two* copies (alleles) of a gene relating to a given character, these copies must *separate* in gamete formation. How else could two Yy parents ever give rise to yy (or YY) progeny, unless the Yy elements

215

The Rule of Multiplication

In order to state a principle that can generalize to any situation, let's look at this concept in a more formal way. What are the odds of tossing up one coin and getting

> *What are the odds of tossing up two coins and having both come up heads?*

heads? Fifty-fifty, right? This means a 50 percent chance, which equals 0.5. Now what are the odds of tossing up two coins and having both come up heads? A principle called the **rule of multiplication** comes into play here. It states that the probability of any *two* events happening is the product of their respective probabilities. In this case, the probability for each head coming up is 0.5, so the equation is 0.5 × 0.5 = 0.25, or a 25 percent probability of two heads.

The Rule of Addition

Now what is the probability of getting one coin that comes up heads and one that comes up tails—when either of them can be heads or tails? When an outcome can occur in two or more *different* ways, as is the case here, the probability of this happening is the *sum* of the respective probabilities. This principle is known as the **rule of addition**. The probability of nickel = heads, dime = tails is the same 0.25 probability you saw in the rule of multiplication. But the probability of nickel = tails, dime = heads is also 0.25. The probability that *either* of these outcomes will take place is thus 0.25 + 0.25 = 0.50, or 50 percent.

All this gives us a way to figure the songwriter probability—or any set of independent probabilities. Remember the stipulation that 1 in 10 aspiring songwriters is a good composer and 1 in 10 is a good lyricist. Thus, under the rule of multiplication, 0.10 × 0.10 = 0.01, meaning 1 in 100 will be good at both composing and lyric writing.

could first separate (in the parents) and then recombine in different ways (in the offspring)? Thus did Mendel derive his insight, sometimes called Mendel's First Law or the **Law of Segregation**: Organisms have two genetic elements (alleles) that separate in gamete formation. As noted, the physical basis for this law, which Mendel knew nothing of, is the separation of homologous chromosomes during meiosis.

Homozygous and Heterozygous Conditions

It's time to add a couple more terms to the concepts you have been considering. There is scientific terminology for the genotypically "pure" and "mixed" organisms noted earlier. An organism that has two identical alleles of a gene for a given character is said to be **homozygous** for that character (as with *YY* or *yy*). An organism that has differing alleles for a character is said to be **heterozygous** for that character (as with *Yy*). You often see homozygous used in combination with the terms dominant and recessive. For example, a *yy* plant is a **homozygous recessive**, while a

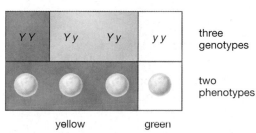

Figure 11.7
Three Genotypes, Two Phenotypes
The two alleles for seed color (*Y* = yellow and *y* = green) can result in three genotypes (*YY*, *Yy*, *yy*), but these can yield only two phenotypes (yellow and green).

YY is a **homozygous dominant** (**see Figure 11.8**). Knowledge of these terms puts you in a position to understand formal definitions of *dominant* and *recessive*. **Dominant** means: expressed in the heterozygous condition. In a heterozygous pea plant (Yy), the yellow allele (Y) is expressed, meaning it is dominant over the green allele (y). **Recessive** means: not expressed in the heterozygous condition. The green allele (y) is recessive because it is not expressed when it exists heterozygously with the yellow allele (Yy).

11.5 Crosses Involving Two Characters

Thus far, the subject has been how pea plants come to differ in *one* of their characters, seed color. When people breed plants for a single difference such as this, looking to see how the offspring will come out, the procedure is known as a **monohybrid cross**. Mendel, however, went on to ask: What happens if you breed plants for *two* characters? What happens, in other words, if you undertake what is known as **dihybrid cross**?

Crosses for Seed Color *and* Seed Shape

One dihybrid cross that Mendel performed involved, for its first character, the yellow and green seeds you've become so familiar with. In addition, this cross involved a second character of these seeds: their shape, which can be smooth or wrinkled (as you saw in Figure 11.4). It's clear that yellow color is dominant to green and, as Table 11.1 shows, smooth seed shape is dominant to wrinkled. We have been denoting the alleles for seed color as Y for yellow and y for green. In the same fashion, let us now denote seed-shape alleles as S for smooth and s for wrinkled.

You can see, in **Figure 11.9a**, what the phenotypic outcomes were for these crosses in the P and F_1 generations. In the F_1 generation, all the seeds were smooth and yellow. When the F_1s were self-pollinated, however, the F_2 phenotypes that resulted came out as: 315 smooth yellow seeds; 108 smooth green seeds; 101 wrinkled yellow seeds; and 32 wrinkled green seeds. These numbers work out to a 9:3:3:1 ratio, meaning 9 parts smooth yellow; 3 parts smooth green; 3 parts wrinkled yellow; and 1 part wrinkled green.

A Hidden, Underlying Ratio

For Mendel, these imposing figures represented another opportunity to perceive something about how the black box of genetics operated. He realized early on that this ratio might be a composite, hiding an underlying reality for each of the *single* characters he was studying. Think for a second, as Mendel did, of only one of these characters, the *color* of the seeds. Here, the result is:

315 (smooth) yellow seeds	108 (smooth) green seeds
101 (wrinkled) yellow	32 (wrinkled) green
416 yellow seeds	140 green seeds

This equals a yellow:green ratio of about 3:1. In other words, the familiar 3:1 ratio in the F_2 generation still holds. It held as well when Mendel looked only at seed *shape*. This suggested something to Mendel that turned out to be another of his major insights: That *characters*—in this case, seed shape and seed color—are transmitted *independently* of one another. When these two characters were being crossed at the same time, the results for each were the same as when they were crossed as single characters in the earlier experiments. Thus, one character's transmission did not appear to affect the other's.

Understanding the 9:3:3:1 Ratio

Still, what's the basis of the formidable 9:3:3:1 ratio that Mendel got in his dihybrid cross, with its mixture of wrinkled yellows, smooth greens, and so forth? This is simply a more

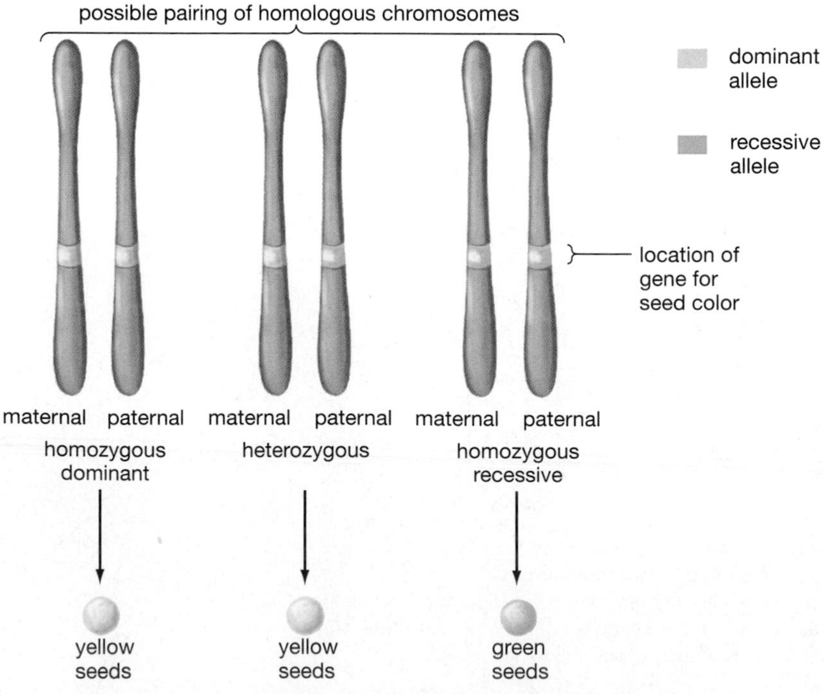

Figure 11.8
Chromosomes and Phenotypes
The figure shows how alleles on chromosomes yielded the pea-color phenotypes that Mendel observed.

possible pairing of homologous chromosomes

dominant allele

recessive allele

location of gene for seed color

maternal	paternal	maternal	paternal	maternal	paternal
homozygous dominant		heterozygous		homozygous recessive	

yellow seeds

yellow seeds

green seeds

complex example of the kind of outcomes you saw earlier with the aid of the Punnett square. If you look at the square in **Figure 11.9b** and start adding up internal squares *by phenotype*, Mendel's 9:3:3:1 ratio starts making sense. You can see, for example, where the 1 in the ratio comes from: There is only 1 part wrinkled green seeds, because it is only the *sy* (pollen) and *sy* (egg) combination that yields this phenotype. Likewise, you will get three parts wrinkled yellow seeds by adding up the three possible male and female gamete combinations that could bring this about.

The Law of Independent Assortment

This outcome could only come about, however, if Mendel's fundamental insight about dihybrid crosses was correct: That characters were being transmitted independently of one another. If one character had *affected* another's transmission, these ratios would have

been very different. This insight of Mendel's is now known as Mendel's Second Law, or the **Law of Independent Assortment**. It states that during gamete formation, gene pairs assort independently of one another.

Independent Assortment and Chromosomes

Now recall that the underlying physical basis for this law was set forth in Chapter 10: In meiosis, pairs of homologous chromosomes *assort independently* from one another at the metaphase plate. It may be, for example, that paternal chromosome 5 will line up on side A of the metaphase plate; if so, then maternal chromosome 5 ends up on side B. For chromosome 6, however, things could just as easily be reversed: The *maternal* chromosome might end up on side A, and the paternal on side B (see Figure 10.2c, page 195.) The genes for Mendel's seed color exist on the plant's

Tutorial 11.1.6
**Phenotype Ratios
in a Dihybrid Cross**

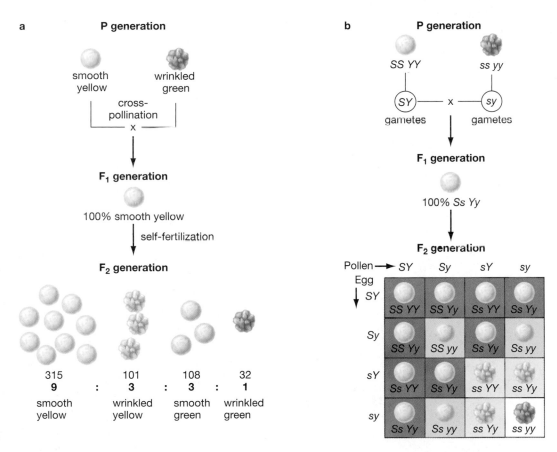

Figure 11.9
Phenotype Ratios in a Dihybrid Cross

a In one of his dihybrid crosses, Mendel cross-bred plants that had smooth yellow seeds with those that had green wrinkled seeds. The result was a generation of plants that all had smooth yellow seeds. When these plants self-fertilized, the result was an F_2 generation that had the phenotypes shown in a 9:3:3:1 ratio.

b The Punnett square demonstrates why Mendel got the 9:3:3:1 phenotypic ratio in his dihybrid cross. Nine combinations yield smooth yellow seeds, 3 yield smooth green seeds, 3 yield wrinkled yellow seeds, while only 1 results in a wrinkled green seed.

chromosome 1, while the genes for seed shape exist on its chromosome 7. Because these are separate chromosomes, they assort independently at the metaphase plate, meaning they are *passed on* independently to future generations.

11.6 Reception of Mendel's Ideas

When Mendel finished his experiments, he delivered two lectures on them to a local scientific society in 1865. It would be nice to report that the society members' jaws dropped open once Mendel let them in on how inheritance worked in the living world, but no such thing happened. His findings were ultimately published in a scientific journal that was distributed in Germany, Austria, the United States, and England. Mendel even took it upon himself to get 40 reprints of his paper, thereafter sending some out to various scientists in an effort to spark some exchange on his findings. All to no avail. His work sank nearly without a trace, finally to be rediscovered in 1900, 16 years after his death. Today, the consensus within the scientific community is that nobody cared about Mendel's findings in his own time simply because nobody grasped their significance (see "Why So Unrecognized?" on the next page).

This poor early reception notwithstanding, Mendel's insights have stood the test of time. For certain kinds of phenotypes, his rules of inheritance are extremely reliable. To this day, biologists will begin an observation by noting that "Mendelian rules tell us that . . ." or "Mendelian inheritance operates here." Even allowing that he had intellectual predecessors, the fact is that Mendel did not just *add* to the discipline of genetics; he founded it.

11.7 Incomplete Dominance

Mendel knew as well as anyone that the rules he discovered did not apply in all instances of inheritance. In his peas, crossing white- and purple-flowered plants resulted in an F_1 generation in which all flowers were purple. In some other species, however, the results are different. Crossing a true-breeding red-flowered *snapdragon* with a true-breeding white-flowered snapdragon produces neither red- nor white-flowered F_1 snapdragons, but *pink* snapdragons (**see Figure 11.10**). This might seem to indicate that inheritance can indeed work through the *blending* of genetic traits—a notion that Mendel's work had quashed. Blending is not taking place, however. When breeding is continued through the F_2 generation, red and white flowers come back, along with the pink, in a familiar ratio: 1 part red, 2 parts pink, 1 part white. This 1:2:1 ratio demonstrates that red and white alleles have not irretrievably blended into pink, but rather have come together in F_1, only to separate out again in F_2. Still, how can the pink F_1 snapdragons be explained?

Genes Code for Proteins

First, think about what genes do: They contain information regarding the production of proteins. In the case of colors, such proteins can bring about the formation of pigments. This is

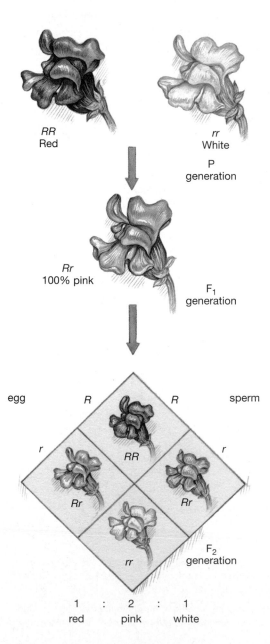

Tutorial 11.2.1
Incomplete Dominance

Figure 11.10
How Red and White Yield Pink: Incomplete Dominance in Snapdragons
A single red allele (*R*) yields only enough pigment to produce a flower that is pink—the only phenotype in the F_1 generation. In the F_2 generation, however, red alleles combine (*RR*) to produce the red-flower phenotype.

Why So Unrecognized?

"If fame belonged to me, I could not escape her—if she did not, the longest day would pass me on the chase," wrote Emily Dickinson in 1862, in seeming ambivalence about becoming a published poet. Dickinson's words came in a letter to Thomas Higginson, who had recently advised her against seeking publication of her verses on grounds that they were not yet strong enough. Higginson was the second, and as it turned out, last person to disparage Dickinson's efforts to publish her poetry. After his advice was offered, Dickinson kept her own company as a writer; by the time she died, only a close circle around her had seen her work.

Half a world away, Dickinson had a counterpart of sorts in her contemporary, Gregor Mendel (born eight years before her, and deceased two years before). Like Dickinson, Mendel had something special to offer the world. But it was an offer made to a world that was not prepared to listen. Mendel died in obscurity in 1884, his work appreciated by, so far as we can tell, no one but himself. Indeed, 34 years were to pass between publication of his research results and recognition of their significance. Yet from the time his work was rediscovered in 1900, Mendel has been treated with great respect, such that today he is generally recognized as the founder of genetics.

How, then, could Mendel's work have gone unrecognized in his own time? A couple of easy but unsatisfying answers are that he labored in a scientific backwater (in the hinterland of the Austrian empire) and that his results were published in an obscure scientific journal.

Mendel appears to have been a victim of his own originality.

Even granting these points, a good number of scientists still had access to Mendel's report—if not by reading it in the journal that published it, then by receiving one of the reprints of it that Mendel himself mailed out. If fame didn't follow from such exposure, why not simple recognition by at least one person?

The answer appears to be fairly straightforward: Mendel wasn't appreciated in his lifetime by anyone because he wasn't comprehended in his lifetime by anyone. Indeed, intense exposure to his ideas seemed to make no difference in this regard. Consider that he carried on a seven-year correspondence regarding his work with a famous botanist, Carl Nägeli, during which time Nägeli, like everyone else, seemed to miss the significance of Mendel's experiments entirely. Assuming that nineteenth-century scientists were as bright as those in the twentieth century, how could this have happened?

First, Mendel himself did not make much of the principles of heredity he uncovered. In his long paper, he punctuates a great deal of detail about his experiments with some understated formulations of general principles. In itself, this is not surprising. Scientists often let data "speak for themselves," resting assured that their colleagues will not miss a major point, however subtly it may be presented. Most scientists can do this, however, because they are confident that their colleagues will be *looking* for major points in their papers. In Mendel's case, nobody was looking for points about how heredity worked because scarcely anybody but Mendel had a concept of pure heredity. As contemporary geneticists Daniel L. Hartl and Vitezslav Orel have observed, in the minds of nineteenth-century scientists, heredity was merely a part of development—the process by which a fertilized egg becomes a fully formed organism. Mendel, meanwhile, thought of heredity as operating under its own independent set of rules, after which he went on to say what those rules were. Mendel thus delivered an answer to a question that nobody had asked: How does heredity work?

Other barriers existed to understanding as well. Mendel used mathematics to derive his insights on genetics, an approach that was novel in the biology of his day. Beyond this, in 1859 Charles Darwin's *On the Origin of Species* had been published, an event that sent scientists scurrying to look for "continuous," or subtle variations between organisms—a phenomenon that they thought might shed some light on evolution. Conversely, Mendel's paper, coming seven years later, talked about pairs of seemingly unchangeable genetic elements that gave rise to such either/or features as smooth or wrinkled seeds. It was difficult to fit such elements into a scheme of gradual evolutionary transformation.

Looking over all these factors, Mendel appears to have been, in essence, a victim of his own originality. (This again puts him in the company of Emily Dickinson, whose language was too idiosyncratic for her early readers.) There is a saying that it doesn't pay to be more than 10 minutes ahead of your time. Gregor Mendel had the misfortune of being about 34 years ahead of his.

the case with the snapdragons; they have a gene for color, and one of the alleles of this gene brings about the production of red pigment. Meanwhile, the second allele of this gene is nonfunctional—it brings about the production of no pigment at all. Two of the red alleles, then, yield a red color, while one red allele produces only enough pigment to yield a pink color. Neither the red nor the white allele, therefore, is completely dominant; each is thus said to be **incompletely dominant**.

11.8 Lessons from Blood Types: Codominance

The notion of genes as protein-producing entities can help us steer through another variation on Mendelian inheritance, this one having a great importance for variety in the living world. When people speak of blood "types," what they mean is types of proteins that cover the surface of red blood cells. These surface proteins come in many different varieties, the two most important of which are designated A and B.

Blood types are completely under genetic control, with a single gene that lies on chromosome 9 determining what blood type a person will have—A, B, AB, or O. There are two copies of this gene in each individual, because there are two copies of chromosome 9 in each individual—one inherited from the mother, the other inherited from the father. Thus, there exists the possibility for *alleles*, or gene variants in each individual.

Now, a single person could have two alleles that code for type A protein, or a person could have both alleles coding for type B molecules. In the first case a person would have type A blood, and in the second type B blood. However, a person could have one allele that codes for the type A molecule *and* one that codes for type B, in which case that person would have type AB blood. Finally, a person could have two alleles that are inactive—that code for neither surface protein—and would thus have type O blood.

Getting Both Types of Surface Proteins

What's new in this case is, first, another aspect of the idea of dominance in alleles. In the blood types, *both* the A and B alleles produce proteins (for the cell surface molecules). People who are heterozygous for these alleles thus do not get a phenotype that, like pink flowers, lies halfway between A and B. They get *both*

the A and B proteins on the surface of their blood cells. Alleles that have this kind of independent effect are said to be **codominant**.

How to Think of Dominance

All this gets us to an important point: how to conceptualize dominance. In common speech the word *dominance* is used all the time, to mean "Bringing another under submission by means of superior power." Alleles, however, are not locked in struggle with one another. Alleles contain information for the production of proteins. Such proteins (or lack of them) can create dominant phenotypes (smooth peas), intermediate phenotypes (pink snapdragons), or codominant phenotypes (AB blood). Moreover, remember the lesson from Mendel that alleles are passed on *intact* through generations. The pea plant's recessive green allele did not disappear when brought together with a dominant yellow allele. It simply was not expressed. It continued to be passed on to subsequent generations, and *was* expressed when brought together with another green allele. This has relevance for traits more familiar to us. People seem to have an idea that human blonde hair, to give one example, eventually will cease to exist because the genes for it are recessive to dark hair. The alleles for blonde hair are, however, not disappearing. To the extent that alleles for it come together in reproduction, there will always be blonde-haired people.

11.9 Multiple Alleles and Polygenic Inheritance

ABO blood types have one more lesson for us. In Mendel's seed shapes, there were two phenotypic variants: smooth and wrinkled. In blood types, there were three: A, B, and O. As you know, no one person can have more than *two* alleles for a given gene. As a result, a person may be, say A and B, but he or she cannot be A, B, and O. It is thus only in a *population* of humans that the full range of ABO alleles can be found. And therein lies the lesson. In a population, alleles can come not just in two, but in many variants.

When three or more alleles of the same gene exist in a population, they are known as **multiple alleles.** Such alleles are important in and of themselves, for they can bring diversity to traits that are governed by single genes (such as human blood types). Their more general significance comes, however, in connection with traits that are controlled not by one gene, but by many. Recall that seed texture in Mendel's peas was governed by a single gene (which had smooth and wrinkled allelic variants). This situation is, however, the *exception* rather than the rule in the living world. Most traits are governed by many genes. The human traits of height, weight, eye color, and skin color are each controlled by several genes. In the wider living world, the color of a wheat grain, the length of an ear of corn, and the amount of milk a cow gives are all controlled by many genes acting together. The term for such genetic influence is **polygenic inheritance**, meaning the inheritance of a genetic character that is determined by the interaction of multiple genes, with each gene having a small additive effect on the character.

When we have many genes contributing some small increment to a trait, the result is what you see in **Figure 11.11**. Human beings don't come in two heights, or three or four; they display what is known as a "continuous variation" in height, each person being just barely taller or shorter than the next.

Continuous variation also holds true for human skin color. Human beings don't really have "black" or "white" or "red" skin. Instead they have skin that comes in a *range* of hues, in which one color shades imperceptibly into another. Likewise, trees of a given species come in a range of heights, and the beaks of birds come in a range of lengths. Polygenic inheritance creates this fine-grained diversity.

Now, note something else about Figure 11.11a. Most of the students fall in the middle range of heights in the group. Put another way, there are fewer tall or short students than there are students of medium height. This height distribution means that the group as a whole takes on a kind of shape—a bell shape. This is, in fact, the famous **bell curve**, meaning a distribution of values that is symmetrical, and largest around the average. Most biological traits manifest in this way. Look, for example, at the beak depths in a group of Darwin's finches in Figure 11.11b. Graphs like this just confirm what we know intuitively—that traits in living things cluster around what is average, rather than what is extreme.

a

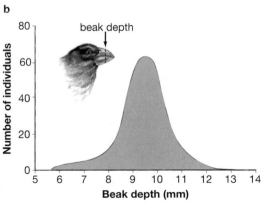

Figure 11.11
Continuous Variation and the Bell Curve

a Mendel's pea seeds may have been green or yellow, but traits such as human height do not have this either/or quality. Human heights exist in a range, with no fixed increments between heights of individuals. Such "continuous variation" is the result of polygenic inheritance, in which each of several genes contributes a small additive effect to a character. The Brigham Young students in the picture have been arranged by height to show how continuous variation works in this one human trait. Note that the group as a whole takes on the shape of a bell. The students' heights are distributed in a pattern that creates a bell curve.

b A bell curve can also be seen in this graph, which plots the average beak depth of a population of Darwin's finches from the Galpagos Islands. Note how most of the finches had beak depths close to the average depth of the population as a whole.

In polygenic inheritance, with its many genes and alleles, gene interactions are so complex that predictions about phenotype are a matter of *probability*, not certainty. Human height is largely under genetic control, and there are means of trying to predict the height of children based on the height of their parents. (The starting point generally is to add together the height of the parents and divide by two.) All that such predictions can do, however, is specify that if parents are of a given height, then there is a certain probability that their children will fall into a given range of heights. Think how different this is from the situation with Mendel's peas. There, you could confidently predict that a given cross would yield, say, all green peas.

Scientists work awfully hard at trying to determine the probabilities for certain polygenic outcomes—those related to disease. In carrying out this work, they are essentially doing what Mendel did: observing traits in a parental generation and then observing traits in the parents' offspring. Such work is the basis for the warnings we sometimes hear about how much a person's risk for a particular cancer goes up if that person has a relative who has contracted the disease. With polygenic illnesses such as cancer, however, it is difficult to separate the genetic and environmental factors. Smoking is responsible for about 90 percent of all lung cancer cases, but that still leaves some nonsmokers contracting lung cancer, and

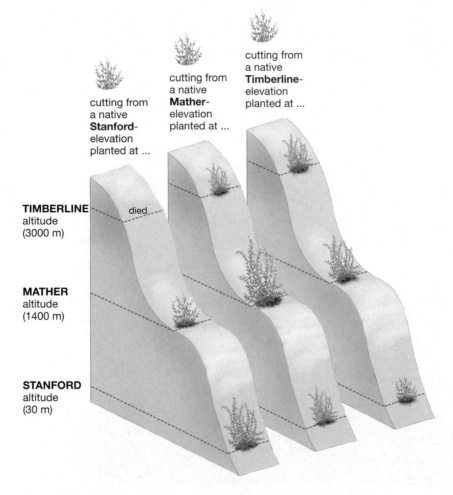

Figure 11.12
Environment and Genes Interact
The *Potentilla glandulosa* plant grew in nature at the three elevations shown. Cuttings were then taken from the plants and grown in experimental plots at all three elevations. For example, the Stanford *Potentilla*, which grew naturally at 30 meters (100 feet), was planted at 1,400 meters (4,600 feet) and 3,000 meters (10,000 feet) as well as at its native elevation. The size of each plant in the figure shows how well the cuttings grew at the various elevations. The results illustrate how environment and genes interact to produce the physical traits or phenotypes of organisms. You can see, for example, that the Stanford and Mather plants grew best at their native elevations, while the Timberline plant did not.

only a minority of smokers contracting it. This raises the question of the genetic *susceptibility* to a disease like lung cancer in relation to a powerful environmental influence such as smoking. As you'll now see, genes and environment nearly always work together to create physical outcomes in living things.

11.10 Genes and Environment

If you look at **Figure 11.12**, you will see a species of plant that is native to California, *Potentilla glandulosa*. Going into an experiment with this plant, researchers knew that varieties of it grew in the wild at three elevations: the Stanford plant at 30 meters (100 feet), the Mather plant at 1,400 meters (4,600 feet), and the Timberline plant at 3,000 meters (10,000 feet). Cuttings from each of these plants were then replanted in experimental plots, both at the native elevation and the other two elevations. Thus genetically identical versions of the Timberline plant, for example, were grown at 30 meters and 1,400 meters (as well as at its native 3,000 meters). You can see for yourself what happened: The Timberline cutting actually grew larger at 1,400 meters, rather than its native 3,000 meters; but it grew least well at 30 meters.

Now, the same genes existed in the Timberline plant no matter where it was grown, because all the plants were taken from cuttings. Yet this plant did much better in one location as opposed to another. The message here is one of genetic limitation: Genotype specifies only so much about phenotype. Protein products that bring about a relatively tall plant at 3,000 meters may bring about a relatively short one at 30 meters.

We can put a little finer point on this. Look at the *Potentillas* again. Is it fair to say that any of the three varieties contains a "tall" gene? Grown at its native elevation, the Mather is the tallest plant overall. But how can the idea of a "tall" Mather gene be squared with the plant's short stature at either the Timberline or Stanford elevations? It can't. The plant's stature depended partly on environment. A light bulb may come with a wattage rating that holds true whether the bulb is burning in Denver or Death Valley, but genes do not possess such unvarying strengths or potentials. Imagine a light bulb

whose brightness *changes* incrementally with each locale it operates in, and you begin to get the idea about genes.

11.11 One Gene, Several Effects: Pleiotropy

You've seen that many genes can work together to produce one effect, such as human height. It is also possible, however, for a single gene to have many effects. This is frequently the case, actually, for the simple reason that the processes of living things are so interrelated.

The phenomenon of one gene having many effects is called **pleiotropy**. Many examples of this phenomenon exist; one of the more interesting, perhaps, is a debilitation known as *fragile-X syndrome*, which is the most common cause of inherited mental retardation. The *X* in its name refers to the X chromosome. *Fragile* comes from the break on the long arm of the X chromosome that can be seen in people who suffer from the syndrome. Fragile-X is now thought to be caused by a defect in a single gene, dubbed *FMR-1* by the team of researchers who discovered it.

The upshot of this defect is that fragile-X victims lack a functional *FMR-1* gene, which is to say their *FMR-1* does not appear to prompt production of any protein. One clear result of this is mental retardation; people with fragile-X commonly have an IQ of about 40–70 (as opposed to the norm of 100). For our purposes, fragile-X also has effects that seemingly stand far afield from mental retardation. Those with fragile-X often have an abnormally long face; large, protuberant ears; and (when male) large testicles. This broad range of effects, seemingly stemming from the inactivation of a single gene, contains a message: Genes normally work together in an interrelated web, rather than functioning under a one-gene → one-phenotype model.

On to the Chromosome

This completes the survey of Mendel's work and the variations on his ideas of inheritance. As you have seen, Mendel's work lay dormant for some decades in the nineteenth century. What sent scientists scurrying back to his findings were experiments on those tiny entities you have looked at in detail before: chromosomes, the subject of the next chapter.

Chapter Review

Summary

11.1 Mendel and the Black Box

- Gregor Mendel was the first person to comprehend some of the most basic principles of genetics. He reached these understandings in the mid-nineteenth century, working in what is now the Czech Republic and using as his experimental subjects a species of garden pea, *Pisum sativum*.

11.2 The Experimental Subjects: *Pisum sativum*

- Mendel looked at seven characters in his plants—such attributes as seed color and texture. In his plants, each of these characters came in two varieties or traits, one of them dominant, the other recessive. His experiments involved breeding pea plants, starting with plants that had a given set of traits and then observing which of those traits showed up in succeeding generations.

- The physical functioning, bodily characteristics, or actions of an organism are its phenotype. In Mendel's plants, purple flowers were one phenotype and white flowers were another. Phenotypes in any organism are in significant part determined by that organism's genotype, meaning its genetic makeup. Mendel realized that the phenotypes in his plants were being controlled by what we would today call their genotypes.

11.3 Starting the Experiments: Yellow and Green Peas

- Mendel realized that it was possible for organisms to have identical phenotypes—for all his pea plants to have yellow seeds, for example—and yet to have differing underlying genotypes.

- One of Mendel's central insights was that the basic units of genetics are material elements that, in his pea plants, came in pairs. These elements, today called genes, come in alternative forms called alleles. One member of an allele pair resides on one chromosome, while the other allele resides on a second chromosome that is homologous to the first.

- Another of Mendel's insights was that genes retain their character through many generations, rather than being "blended" together. Genes that coded for yellow pea color, for example, were retained in their existing form over many generations.

11.4 Another Generation for Mendel

- A third insight of Mendel's was that alleles separate prior to the formation of gametes. Though Mendel did not know it, the physical basis for this is that the alleles he was observing resided on homologous chromosomes, which always separate in meiosis. An organism that has two identical alleles of a gene for a given character is said to be homozygous for that character. An organism that has differing alleles for a character is said to be heterozygous for that character. Dominant means: expressed in the heterozygous condition. Recessive means: not expressed in the heterozygous condition.
TUTORIALS: 11.1.1 Mendel F_1 Crosses (Punnett Square); 11.1.3 The F_2 and F_3 Generations; 11.1.4 Rules of Multiplication and Addition

11.5 Crosses Involving Two Characters

- Mendel observed that the genes for the different characters he studied were passed on independently of one another. This was so because the genes for these characters resided on separate, non-homologous chromosomes. The physical basis for what he found is the independent assortment of chromosomes during meiosis.
TUTORIAL 11.1.6: Phenotype Ratios in a Dihybrid Cross

11.6 Reception of Mendel's Ideas

- Gregor Mendel published his work, but the significance of it was never recognized in his lifetime. It was only rediscovered 16 years after his death, in 1900.

11.7 Incomplete Dominance

- Not all inheritance works through the principles Mendel perceived in his peas. Incomplete dominance operates when neither allele for a given gene is completely dominant, with the result that heterozygous genotypes can yield an intermediate phenotype (such as pink snapdragons).
TUTORIAL 11.2.1: Incomplete Dominance

11.8 Lessons from Blood Types: Codominance

- In some instances, differing alleles of the same gene will have independent effects, rather than one allele being dominant over the other. Such is the case with the genes that code for the human blood proteins, A and B. Neither of the alleles for these proteins is dominant over the other. Rather, a person can have a genotype that produces either or both of the proteins, or neither of them.

11.9 Multiple Alleles and Polygenic Inheritance

- Human beings and many other species can have no more than two alleles for a given gene, each allele residing on a separate, homologous chromosome. Many allelic variants of a gene can, however, exist in a population. Most traits in living things are governed not by one gene but by many, with these genes often having several allelic variants.

The term for such genetic influence is polygenic inheritance, meaning the inheritance of a genetic character that is determined by the interaction of multiple genes, with each gene having a small additive effect on the character.

- Polygenic inheritance produces continuous variation in phenotypes, meaning there are no fixed increments of difference between individuals. Human skin, for example, comes in a range of colors in which one color shades imperceptibly into the next.

- The traits produced in polygenic inheritance tend to manifest in bell-curve distributions, in which most individuals display near-average trait values, rather than extreme trait values. Gene interactions and gene-environment interactions are so complex in polygenic inheritance that predictions about phenotypes are a matter of probability, not certainty.

11.10 Genes and Environment

- The effects of genes can vary greatly in accordance with the environment in which the genes are expressed. An organism's genotype and environment interact to produce that organism's phenotype.

11.11 One Gene, Several Effects: Pleiotropy

- Genes work in an interrelated fashion, such that a single gene is likely to have multiple effects. Pleiotropy is a phenomenon in which one gene has many effects.

Key Terms

Understanding the Basics

Multiple-Choice Questions

1. The first filial generation, or F₁, results when
   a. two P generation organisms are crossed
   b. the progeny of a P generation is crossed
   c. a P generation is crossed with an F₂ organism
   d. an F₃ organism is crossed with an F₂ organism
   e. both a and d

2. When Mendel crossed pure-bred yellow peas with pure-bred green peas, the F₁ generation contained
   a. all yellow peas
   b. all green peas
   c. green and yellow peas
   d. green, wrinkled peas
   e. both b and d

3. A very important characteristic of the garden peas used by Mendel was that
   a. Each trait examined had more than two different varieties.
   b. All yellow peas had the same genotype.
   c. The F₁ generation always resulted in a 1:1 phenotypic ratio.
   d. All crosses resulted in a 1:3 genotypic ratio.
   e. The plants had the ability to self-fertilize.

4. If a cross is made between a pure-breeding green, round (yyRR) plant, and a pure-breeding yellow, wrinkled (YYrr) plant, what is the result?
   a. all green, wrinkled peas
   b. all yellow, wrinkled peas
   c. all yellow, round peas
   d. yellow wrinkled; green wrinkled
   e. none of the above

5. If plants in a cross have the following genotypes (YYRr × yyRr), the correct term for the type of cross is:
   a. recessive cross
   b. dominant cross
   c. monohybrid cross
   d. dihybrid cross
   e. codominant cross

6. Mendel's first law states:
   a. During the formation of gametes, gene pairs demonstrate independent assortment.
   b. During the formation of gametes, genetic elements (alleles) segregate from each other.
   c. The probability of any two events happening is the product of their respective probabilities.
   d. If an organism is pure-breeding for any trait, it will display incomplete dominance.
   e. Mendel did not write any laws regarding genetics.

7. A trait whose phenotype in the heterozygous condition is an intermediate of the two homozygous conditions (that is, heterozygous pink flowers, homozygous red and white flowers) demonstrates:
   a. pleiotropy
   b. incomplete dominance
   c. intermediate dominance
   d. codominance
   e. multiple alleles

8. Mendel's second law states that
   a. No organism can have more than two alleles for a given character.
   b. No two organisms can ever be genetically identical.
   c. Organisms will tend to be heterozygous for most traits, rather than homozygous.
   d. The way one gene pair assorts during gamete formation will affect the way other gene pairs assort.
   e. Gene pairs assort independently of one another during gamete formation.

9. Which of the following is true about polygenic inheritance?
   a. It is the exception rather than the rule in living things.
   b. Its outcomes are always predictable.
   c. It is never seen in human beings.
   d. It produces phenotypes that grade smoothly into one another.
   e. It produces only homozygous individuals.

10. Which of the following is true regarding the effects of genes and environment on phenotypic variation?
    a. Gene expression will always be greatest in the native environment.
    b. Gene expression is not affected by a change in environment.
    c. Phenotypes are based only on environmental influences.
    d. Environment may easily affect genotype, not phenotype.
    e. The phenotypes of genetically identical organisms may vary in different environments.

Brief Review

1. What are the differences between:
   a. phenotype and genotype?
   b. dominant and recessive?
   c. codominance and incomplete dominance?

2. True or False: When Mendel conducted his genetic research, neither DNA nor chromosomes had been discovered.

3. What genetic principle is demonstrated in the fragile-X syndrome? Explain how this principle differs from the relationship between genes and physical outcomes that Mendel observed in his pea plants.

4. In a case of Mendelian genetics where one gene affects one trait, how many different genotypes may represent a dominant phenotype (for example, yellow pea color)?

5. If a heterozygous yellow pea plant is crossed with a homozygous green pea plant, what percentage of the progeny peas will be yellow?

6. What term is now given to Mendel's "element of inheritance"?

7. What two laws are credited to Mendel and his research?

8. What Mendelian crosses would result in the following phenotypic ratios?
   a. 1:1
   b. 3:1
   c. 1:1:1:1
   d. 9:3:3:1

9. a. You would love to have some yellow supersweet corn for your garden, but all you can order are yellow starchy varieties, or white supersweet. Your gardening friends tell you that white is the only supersweet variety, but you decide to try a little breeding experiment to prove them wrong. You cross the yellow not-so-sweet to the white supersweet, and the ears are all yellow and not-so-sweet. You persist for another year, by planting this F_1 generation of seeds and letting them self-pollinate in an open field. The ears of the plants are a mixture of kernels, yellow and white, sweet and not-so-sweet. How would you explain this to your friends?

   b. Imagine that you counted 1000 kernels, assuming that you are dealing with two alleles of two different independently assorting genes: Y for yellow (y, white) and Su for starchy (su for supersweet). How many would be:

   Yellow and starchy
   White and starchy
   Yellow and supersweet
   White and supersweet

   c. One of your friends is convinced and asks for some seeds of the yellow supersweet to grow. You warn him that even though all the plants he grows will be supersweet, some will still give the mixture of white and yellow kernels. What percent of the plants he grows can be expected to give all yellow supersweet ears?

10. Horse breeders need to make sense of the genotypes of their mares and stallions to select for the most desirable traits, like coat color. Look at the following crosses and the resulting progeny colors, and decide what kind of variation on Mendellian inheritance must be at work. Second, indicate the genotypes of the parents (homozygous dominant, heterozygous, or homozygous recessive) for the Chestnut gene, referred to as C.

| Parents | | Progeny |
Female	Male	Colors
White	White	all White
Chestnut	Chestnut	all Chestnut
White	Chestnut	all Palomino
Palomino	Palomino	1/4 White, 1/4 Chestnut, 1/2 Palomino
Palomino	Chestnut	1/2 Chestnut, 1/2 Palomino
White	Palomino	1/2 White, 1/2 Palomino

Applying Your Knowledge

1. What is meant by the term *allele*? Why can no more than two alleles of any gene exist simultaneously in a person?

2. A researcher is studying the traits of a recently identified plant species. She is particularly interested in two characteristics: leaf shape (round or oval), and flower color (red or blue). After crossing two true-breeding plant lines, she self-pollinates the F_1 generation plants, and observes the following results in the F_2 generation:

   Round leaves, blue flowers: 110 plants
   Round leaves, red flowers: 29 plants
   Oval leaves, blue flowers: 317 plants
   Oval leaves, red flowers: 105 plants

   Using these data, which two traits are dominant? What were the genotypes of the parents?

3. A purple-flowering plant (*PP*) is crossed to a white-flowering plant (*pp*).
   a. Using a Punnett square, diagram the cross. What are the genotypic and phenotypic outcomes?
   b. Two F_1 individuals from the previous cross are crossed to each other. Draw a new Punnett square, diagramming the new outcome. What is the ratio of purple flowers to white flowers? How many genotypes are there? In what frequencies do they occur?

4. A pure breeding, purple-flowered, yellow pea plant (*PPYY*) is crossed to a white-flowered, green pea plant (*ppyy*). The F_1 generation is allowed to self-fertilize. What are the phenotypic and genotypic ratios of the F_2 plants?

5. Using what you have learned in this chapter, and the following genetic cross, predict the phenotypic expression of any F_1 mouse. (Assume capital abbreviations represent dominant phenotypes.)

   G = gray, g = brown
   L = long hair, l = short hair
   C = curly tail, c = straight tail
   W = whiskers, w = no whiskers
   E = big ears, e = small ears
   a. Cross: *GGLLCCwwee* × *ggllCCwwEE* What is the phenotype of the F_1 generation?
   b. If the F_1 generation is crossed, what is the probability that the offspring will have the following genotype: *GgLlCCwwee*?

6. ABO blood types: In many paternity cases the first test involves blood typing, since blood type is an easily detected genetic trait.
   a. In such a paternity case, a man denies fathering a child of blood type O. He is blood type AB, while the mother is blood type O. Based on this test, could he have fathered the type O child? Why or why not?
   b. If the man were type A, and the mother type O, would the conclusion be any different? Why or why not?

7. Stands of aspen trees often are a series of genetically identical individuals, with each succeeding tree growing from the severed shoot of another tree. Using what you've learned of genetics in this chapter, would you expect one aspen tree in a stand to differ greatly from another in its phenotype? Would you expect each to look exactly like the next in terms of phenotype?

8. In the years just before Mendel was formulating his laws of genetics, the Englishman Charles Darwin published a book on the subject of evolution. Darwin's hypothesis regarding the way new organisms evolved was that a given organism with a trait that favored its survival—say a slightly longer beak in a bird—would live to have more offspring than other such birds. Thus this trait would be passed on to more birds in a succeeding generation than would the trait of a shorter beak. Such changes would then build up over time until one species evolved from another. Do you see anything in the genetic rules uncovered by Mendel that tends to support Darwin's ideas about evolution?

MediaLab

Where Did I Get This Nose? Understanding Mendelian Genetics

Can you roll your tongue? Is there a cleft in your chin? What is the color of your eyes? These and countless other characteristics that make you a unique individual were determined by the traits you inherited from your parents. In this *MediaLab*, you'll firmly establish your understanding of the rules governing the inheritance of simple genetic traits. In the *CD-ROM Tutorial*, you will work through these rules using some of Gregor Mendel's original genetic studies of peas. In the *Web Investigation*, you can practice these rules of probability with your own family, and make predictions about the inheritance of these traits in the *Communicate Your Results* section.

This *MediaLab* can be found in Chapter 11 on your CD-ROM (Tutorial 11.2.1) and Companion Website (http://www.prenhall.com/krogh3).

CD-ROM TUTORIAL

It is obvious to most people that physical traits are inherited in families. But the rules that govern them were a mystery until a nineteenth-century Augustinian friar, Gregor Mendel, discovered the pattern behind inheritance of traits, by conducting breeding experiments on garden peas. With this *CD-ROM Tutorial*, you can review Mendel's experiments in order to lay the foundation for applying these rules to human genetics.

Activity

1. *First, you will examine how different versions of a trait (like green versus yellow seed color) are passed from parents to offspring, and how rules of probability can explain the outcome.*

2. *Then you will learn to use a Punnett square, a chart that shows all possible combinations of alleles in the offspring of any parents, allowing you to predict the chances of inheriting one or more different traits.*

3. *Finally, you can try setting up your own crosses and make predictions about the outcomes.*

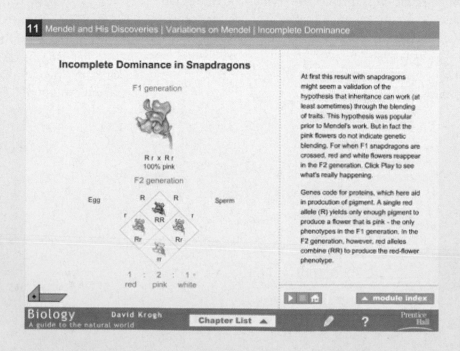

WEB INVESTIGATION

Investigation 1

Estimated time for completion = 10 minutes

Most complex human genetic traits are not as simple to understand as the pea traits that Mendel studied. For example, researchers have found that children of two obese parents have an 80 percent chance of being obese. Can we conclude that it is due to their genes, or is it due to their environment? Maybe people are more likely to overeat and not exercise if their parents do. Select the Keyword **OBESITY** on your CD or Website to read an article on the evidence for a genetic cause for obesity.

Investigation 2

Estimated time for completion = 10 minutes

Before the introduction of DNA testing, ABO blood type was commonly used as evidence in paternity tests. Select the Keyword **BLOOD TYPE** on your CD or Website and read a tutorial about the inheritance of blood type. Then, try your hand at answering some questions on blood type by clicking next on the Keyword, **ABO QUESTIONS**. Then select the blood type questions on the genetics quiz.

Investigation 3

Estimated time for completion = 10 minutes

Understanding the inheritance of Mendelian traits is just the beginning of understanding how genes help shape our identity. Our appearance, health, and even behavior is strongly influenced by our genes. Currently underway is a multinational project, called the Human Genome Project, which is intended to uncover all of the genes located on all the human chromosomes. Select the Keyword **GENOME** on your CD or Website to view a current map of all the human genes that have been located thus far. Select a chromosome and read as much information as you can about a human disease-causing gene that has been located there. What is the name of the gene, and what disease does it cause? Is there a genetic test available for the gene? Has identifying the gene helped create any new therapies?

Now that you have practiced some of Mendel's laws in the preceding investigations, apply and extend your abilities by addressing some genetics problems.

COMMUNICATE YOUR RESULTS

Exercise 1

Estimated time for completion = 10 minutes

To what extent is our physical appearance, behavior, and even personality shaped by our genes, as opposed to our environment? Having read the evidence given in the article in *Web Investigation 1* for a genetic basis for obesity, what is your conclusion about how our weight is determined? Be prepared to discuss this with several of your peers by answering the following questions: In your mind, what was the most convincing point in the article? How did the researchers control for environmental differences verses genetic differences? Can you really assume that two individuals, even siblings, are ever raised with the same environment? How many participants would need to be included in this type of study to make it statistically significant?

Exercise 2

Estimated time for completion = 5 minutes

You just learned how to use ABO blood type to determine parentage. Now find out your own blood type and the blood type of your parents, and try to determine your family's genotypes. If you don't know your blood type, you can find it out easily when you donate blood at your local blood drive. Create a problem like the ones from the **ABO QUESTIONS** in *Web Investigation 2*, and then generate a Punnett square showing all possible blood-type genotypes for the children of your family. What is your phenotype? Genotype? If your parents had more children, what is the probability that they could donate blood for you?

Exercise 3

Estimated time for completion = 5 minutes

Prepare a 50-word description of the gene you investigated in *Web Investigation 3* to inform your classmates about the current state of human genetics research.

12

Chromosomes and Inheritance

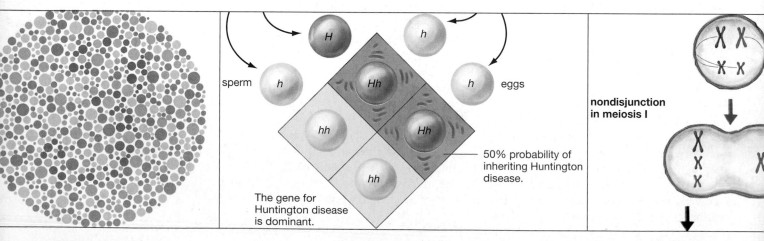

What number do you see?
(Section 12.1, page 232)

Inheriting Huntington disease.
(Section 12.2, page 235)

sperm

eggs

50% probability of inheriting Huntington disease.

The gene for Huntington disease is dominant.

nondisjunction in meiosis I

A mistake in meiosis.
(Section 12.4, page 237)

Chromosomes are critical players in the process by which living things pass on traits. Some of our worst diseases result from chromosomes that have failed to function properly.

The common condition known as Down syndrome is almost always caused by the inheritance of an extra chromosome. But how can a parent who does not have Down syndrome pass on a harmful extra chromosome to a son or daughter? The gene that causes sickle-cell anemia seems also to offer some protection against malaria. How can it have both effects? The chromosomal makeup of women offers them a kind of protection against being hemophiliacs. Why should this be so?

From these questions, it may be apparent that chromosomes and the genes on them are centrally involved in issues of human health. You saw last chapter that chromosomal activity lies at the root of several of the genetic principles that Gregor Mendel uncovered. Mendel's "pairs of elements," which separate in reproduction, had as their physical basis the pairs of homologous chromosomes—one inherited from the male, one from the female—that first pair up and then separate from one another in meiosis. The differing traits that Mendel investigated so thoroughly, such as yellow or green peas in a pod, had as their physical basis the pairs of

alleles or differing forms of a gene that lie on homologous chromosomes.

But there are aspects of chromosomal functioning that Mendel's work scarcely touched on. For example, in Chapter 10 you saw that a critical step in meiosis is the **recombination** or **crossing over** that occurs early in it. In this process, homologous chromosomes first intertwine and then swap pieces of themselves.

Then there is the issue of sex chromosomes. As noted in Chapter 10, human females have 22 pairs of autosomes and then two sex chromosomes—their X chromosomes, which are the chromosomes that make them female. Meanwhile, human males have 22 pairs of autosomes, one X chromosome, and one Y chromosome, the latter of which confers the male sex. But apart from conferring sex, what part do these special chromosomes play in heredity?

Finally, what happens when chromosomes fail to properly carry out their separation or recombination? The short answer is that such mistakes are responsible for a number of the diseases that affect human beings, such as Down syndrome.

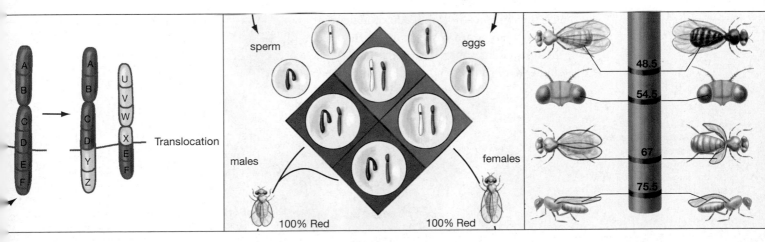

Translocation.
(Section 12.5, page 243)

Only the males had white eyes.
(Essay, page 245)

Linking genes to chromosomes.
(Essay, page 246)

These issues frame the three broad subjects covered in this chapter: (1) what special role sex chromosomes play in human heredity, (2) what human conditions can result from chromosomes that don't function properly, and (3) how the work that Mendel began was extended by later researchers. We'll start with sex and heredity.

12.1 X-Linked Inheritance in Humans

Minor cuts are simply an irritation to most of us, but for people suffering from the condition known as hemophilia, such cuts can be life-threatening. Hemophilia is a failure of blood to clot properly. A group of proteins interact to make blood clot, but about 80 percent of hemophiliacs lack a functioning version of just one of these proteins (called Factor VIII). It is *genes* that contain the information for the production of proteins, of course, so at root hemophilia is a genetic disease. In this respect, it turns out to have something in common with two other afflictions: a disease called Duchenne muscular dystrophy, and a far less serious condition, red-green color blindness.

What do all these disorders have in common? They are all known as X-linked conditions, and they all claim more male victims than female. Let's see why.

X Chromosome: Male Vulnerability, Female Protection

It is the chromosomal makeup of males that makes them more susceptible to these afflictions. Remember that men have but a single X chromosome, while women have two X chromosomes. As it happens, genes for blood clotting, color vision, and Duchenne muscular dystrophy *lie* on the X chromosome. Thus, a male will be red-green color blind if only one of his chromosomes—his lone X chromosome—carries a nonfunctional allele for it. A female, meanwhile, can carry one nonfunctional allele for color vision, but be protected by a "good" allele along her *second* X chromosome.

How Color Vision Works

How does such protection work? Think, once again, of genes as entities that code for proteins. In the case of color vision, within the eye there are molecules called pigments that absorb different colors of light. In humans, proteins make these pigments. A gene that codes for blue pigment lies on chromosome 7, while the genes for both red and green pigments lie very close to one another on the X chromosome. A red-green color-blind person, then, is one in whom these X-located genes fail to code for the proper proteins. The result, in the most severe cases, is a person who cannot distinguish red from green (**Figure 12.1**).

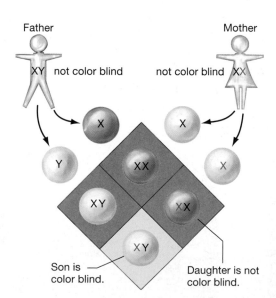

Son is color blind.

Daughter is not color blind.

Figure 12.2
The Value of Having Two X Chromosomes
The mother in the figure is not herself color blind but has one nonfunctional set of red-green alleles, which she passes on to one son and one daughter. The daughter is not color blind because she has inherited, from her father, a second X chromosome—one that has functional red-green alleles. Meanwhile the son is color blind because his only X chromosome is the flawed one he inherited from his mother. (The other son, shown in the square's leftmost cell, is not color blind because he has inherited his mother's functional alleles.)

Figure 12.1
Typical Test for Red-Green Color Blindness
If you do not see a number inside the large circle, you may have this recessive trait.

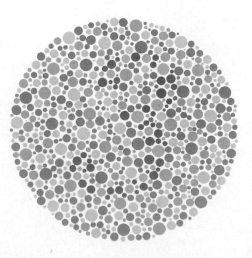

Color Blindness as a Recessive Condition

All it takes, however, is one set of functioning red and green alleles for a person *not* to be color blind. Red-green color blindness is thus a **recessive condition**: A genetic condition that will not exist in the presence of functional alleles. Given this, think about the interesting way that color blindness is passed along. Say there is a mother who is not color blind herself, but who is heterozygous for the trait. That is, she has functional color alleles on one of her X chromosomes but nonfunctional alleles on her other X chromosome. Should her son happen to inherit this second X chromosome, he will be color blind, because the X chromosome he got from his mother is his *only* X chromosome (**Figure 12.2**). Meanwhile, a daughter who inherited this chromosome would likely be protected by her second X chromosome—the one she got from her father. Not surprisingly, then,

more males are color blind than females. About 8 percent of the male population has some degree of color blindness, while for females the figure is about half a percent. Female color blindness comes about only when a daughter inherits a dysfunctional X-chromosome allele from her mother *and* one from her father as well.

12.2 Autosomal Genetic Disorders

X and Y chromosomes are, of course, only two of the chromosomes in the human collection, and any chromosome can have a malfunctioning gene on it. The chromosomes other than the X and Y chromosomes are called *autosomes*, as you've seen. A recessive dysfunction related to an autosome is thus known as an **autosomal recessive disorder.** (For a list of all the disorders you'll be looking at and more, see **Table 12.1**.)

Table 12.1
Selected Examples of Human Genetic Disorders

Type	Name of Condition	Effects
X-linked recessive disorders	Hemophilia	Faulty blood clotting
	Duchenne muscular dystrophy	Wasting of muscles
	Red-green color blindness	Inability to distinguish red from green
Autosomal recessive disorders	Albinism	No pigmentation in skin
	Sickle-cell anemia	Decreased oxygen to brain and muscles
	Cystic fibrosis	Impaired lung function, lung infections
	Phenylketonuria	Mental retardation
	Tay-Sachs disease	Nervous system degeneration in infants
	Werner syndrome	Premature aging
Autosomal dominant disorders	Polydactyly	Extra fingers or toes
	Campodactyly	Inability to straighten little finger
	Huntington disease	Brain tissue degeneration
Aberrations in chromosome number	Down syndrome	Mental retardation, shortened life span
	Turner syndrome	Sterility, short stature
	Kleinfelter syndrome	Dysfunctional testicles, feminized features
Aberrations in chromosome structure	Cri-du-chat syndrome	Mental retardation, malformed larynx
	Fragile-X syndrome	Mental retardation, facial deformities

Sickle-Cell Anemia

A well-known example of an autosomal recessive disorder is sickle-cell anemia, which affects populations derived from several areas on the globe, including Africa. In the United States it is, of course, most widely known as a disease affecting African Americans. The "sickle" in the name comes from the curved shape that is taken on by the red blood cells of its victims. Red blood cells carry oxygen to all parts of the body; in their normal shape they look a little like doughnuts with incomplete holes. When they take on a sickle shape, however (**Figure 12.3**), red cells clog up capillaries, thus resulting in decreased oxygen supplies to brain and muscle. The average life-expectancy for men in the United States with the condition is 42 years; for U.S. women, it is 48 years.

The question is, what causes a red blood cell to take on this lethal, sickled form? There is a protein, called *hemoglobin*, that carries the oxygen within red blood cells. The vast majority of people in the world have one form of hemoglobin, called *hemoglobin A*; but sickle-cell anemia sufferers have another form of this protein, *hemoglobin S*, which coalesces into crystals that distort the cell.

The pattern of inheritance for sickle-cell anemia is very simple. A person with one allele producing hemoglobin S is not affected by the condition to any serious extent, but is a **carrier** for it, meaning a person who does not suffer from a recessive genetic debilitation, but who carries genes for it that can be passed along to offspring. (The mother in Figure 12.2 is a carrier for color blindness.) Because sickle-cell anemia is an autosomal disorder, alleles for it will lie on *two* chromosomes in both female and male offspring. Thus, both mother and father must be at least heterozygous for sickle-cell anemia in order for an offspring to inherit two hemoglobin S alleles and suffer from the disease. The laws governing inheritance of this disorder therefore are simply those of Mendel's monohybrid cross. You can look at the Punnett square in **Figure 12.4a** to see how this works out.

Malaria Protection: Hemoglobin S as a Useful Protein

Sickle-cell anemia offers another lesson in connection with the notion of hemoglobin S as a "faulty" protein. It is in fact quite a functional protein in certain circumstances; namely, when it appears heterozygously (with hemoglobin A) in people who live in regions where malaria is common. The most severe form of malaria is caused by a single-celled parasite that is transmitted into human beings by the *Anopheles* mosquito. Traveling through the bloodstream, these parasites invade red blood cells, in the process destroying them. For reasons we still don't understand, having hemoglobin S in one's system makes red blood cells resistant to invasion. Thus, it is likely that hemoglobin S became as widespread as it is because of its value in resisting malaria. The price of this resistance, however, was that some offspring would not be heterozygous for hemoglobin S, but would be homozygous for it, meaning they would have sickle-cell anemia.

Dominant Disorders

Though sickle-cell anemia is an autosomal disorder and red-green color blindness a sex-linked disorder, they are both recessive conditions: A person with even a single properly functioning allele will not suffer from them. However, there are also **dominant disorders:** genetic conditions in which a single faulty allele can cause damage, even when a second, functional allele exists. This leads to the concept of an **autosomal dominant disorder,** simply meaning a dominant genetic disorder

Figure 12.3
Healthy Cells, Sickled Cells

a Normal red blood cells.

b Sickled red blood cells indicative of sickle-cell anemia.

a

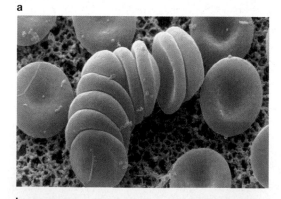

b

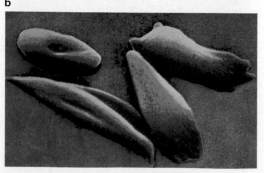

a

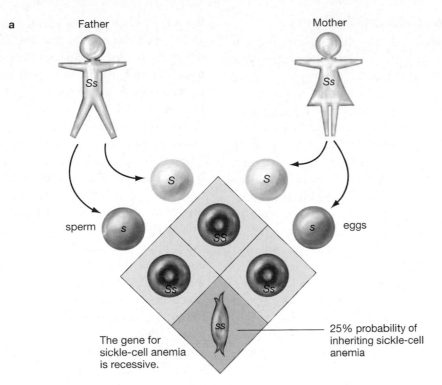

Father

Mother

Ss

Ss

S

S

sperm *s*

s eggs

SS

Ss

Ss

ss

The gene for
sickle-cell anemia
is recessive.

25% probability of
inheriting sickle-cell
anemia

b

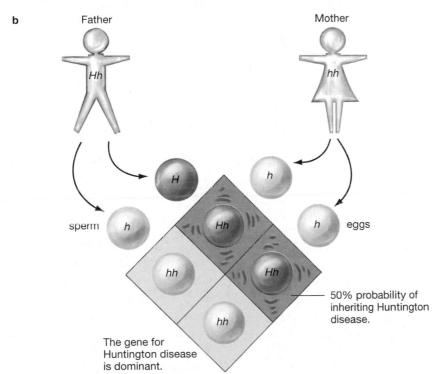

Father

Mother

Hh

hh

H

h

sperm *h*

h eggs

Hh

hh

Hh

hh

The gene for
Huntington disease
is dominant.

50% probability of
inheriting Huntington
disease.

Tutorial 12.2.1
Sickle-Cell Anemia and
Huntington Disease

Figure 12.4
Transmission of Recessive
and Dominant Disorders

a Sickle-cell anemia is a recessive autosomal disorder; both the mother and father must carry at least one allele for the trait in order for a son or daughter to be a sickle-cell victim. When both parents have one sickle-cell allele, there is a 25 percent chance that any given offspring will inherit the condition.

b Conversely, in Huntington disease, if only a single parent has a Huntington allele there is a 50 percent chance that a son or daughter will inherit the condition.

caused by a faulty allele that lies on an autosomal chromosome. There is, for example, an autosomal dominant disorder called Huntington disease. Affecting about 30,000 Americans, Huntington disease results in both mental impairment and uncontrollable spastic movements called *chorea*. Perhaps its most famous victim was the American folksinger Woody Guthrie. Like most Huntington sufferers, Guthrie did not begin to show symptoms of the disease until well into adulthood—*after*

he had children, who then may have had the disease passed along to them. You can look at the Punnett square in **Figure 12.4b** to see how the inheritance pattern of an autosomal dominant disease, such as Huntington, differs from that of an autosomal recessive illness. Because a parent need only pass on a single Huntington allele for a son or daughter to suffer from the condition, the chances of any given offspring getting the disease from a single affected parent are one out of two.

Tracking Traits with Pedigrees

Confronted with a medical condition that is running through a family, scientists sometimes find it helpful to construct a medical **pedigree**, meaning a familial history. Normally set forth as diagrams, medical pedigrees do more than give a family history of a disease. They can be used to ascertain whether a condition is dominant or recessive, which can help establish probabilities for *future* inheritance of the condition—something that can be very helpful for couples thinking of having a child.

If you look at **Figure 12.5**, you can see a simple pedigree for albinism, a lack of skin pigmentation, which is known to be an autosomal recessive condition. In the figure, you can see some of the standard symbols used in pedigrees. A circle is used for a female and a square for a male. Parents are indicated by a horizontal line connecting a male and a female, while a vertical line between the parents leads to a lower row that denotes the parents' offspring. This second row—a horizontal line of siblings—has an order to it: oldest child on the left, youngest on the right. A circle or square that is filled in indicates a family member who has the condition (in this case albinism), while a symbol that is half filled in indicates a person known to be heterozygous for a recessive condition. One of the strengths of a pedigree is that it can sometimes tell researchers which persons are heterozygous carriers of a recessive condition. Remember that a carrier does not display *symptoms* of the condition. So how can a pedigree reveal this?

Take a look at Figure 12.5. The only thing that would be apparent from looking at any of the people in the pedigree is that two of them—a female in generation II and a male in generation III—had the condition of albinism.

But a little knowledge of Mendelian genetics also allows some other deductions. If the condition had been dominant, then it would have manifested itself in at least one of the parents on the left in generation I. Because this was not the case, it is fair to deduce that this is a recessive condition. The fact that it is recessive, however, means that *both* parents in generation I had to be carriers for the allele—both had to be heterozygous for the condition, which is why their symbols can be half shaded in. Things are less clear with the parents on the right in generation I. One of their sons did not manifest the condition himself, but went on to have a son who did. From this, we know that the son in generation II had to have been heterozygous for the condition. But from which parent did this son get his albinism allele? Either, or both, of his parents in generation I could have been heterozygous for the condition and yet between them, only have passed along a single albinism allele.

This mixture of certainty and uncertainty leads to the genotype labeling you see in the figure, with *A* representing the dominant "normal" allele and *a* representing the recessive albinism allele. We know, for example, that both parents on the left in generation I had to be *Aa*, but all we can say for sure about the parents on the right in the same generation is that each of them had to have at least one *A* allele.

12.3 Aberrations in Chromosomal Sets: Polyploidy

All of the maladies you've looked at so far have resulted from dysfunctional *genes* that exist among a standard array of chromosomes. In humans this array is our 22 pairs of autosomes and either an XX combination (in females) or an XY (in males). Meanwhile, Mendel's peas had 7 pairs of chromosomes, while the *Drosophila melanogaster* fly has 4 pairs. Whatever the *number* of chromosomes, note the similarity in all these species: Their chromosomes come in pairs, or to put it another way, they all have two *sets* of chromosomes (meaning they are **diploid**).

In many situations, however, organisms don't end up with the two sets of chromosomes that are standard for their species. The condition in which one or more entire sets of chromosomes have been added to the genome of a diploid organism is called **polyploidy**.

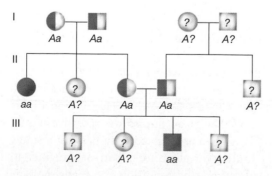

Figure 12.5
A Hypothetical Pedigree for Albinism through Three Generations

Valuable in Plants, a Disaster in Humans

Polyploidy is a calamity for human beings, but in plants it can result in perfectly robust organisms—so much so that plant polyploidy has come about countless times through evolution. You will look more closely at its effects during the review of evolution in Chapter 18 (see pages 362–363). For now, simply note that changes in chromosome set number are not necessarily a bad thing in all species. For human beings, however, polyploidy is such an unmitigated disaster that perhaps only 1 percent of human embryos with the condition will survive to birth—and none of these babies lives long. Concerns about chromosome number in living persons, then, center not on addition or deletion of whole sets of chromosomes, but rather on the gain or loss of *individual* chromosomes.

12.4 Incorrect Chromosome Number: Aneuploidy

The condition called **aneuploidy** is one in which an organism has either more or fewer chromosomes than normally exist in a full set for that organism. So common is aneuploidy in humans that at least 5 percent of all human pregnancies are thought to be affected by it. The vast majority of aneuploid embryos will spontaneously abort, but it is possible for embryos with the wrong number of chromosomes to become fetuses and then babies. It is possible, in other words, for human beings to end up with the wrong number of chromosomes in their diploid set, meaning the wrong number of chromosomes in nearly every kind of cell in their bodies.

A Common Cause of Aneuploidy: Nondisjunction

A common cause of human aneuploidy is a phenomenon known as **nondisjunction**, which simply means a failure of homologous chromosomes or sister chromatids to separate during meiosis. The result is that some daughter cells can end up with one chromosome too many, while others can end up with one chromosome too few.

Looking at **Figure 12.6**, you can see that nondisjunction can occur either in meiosis I

Tutorial 12.2.5
Aneuploidy and Nondisjunction

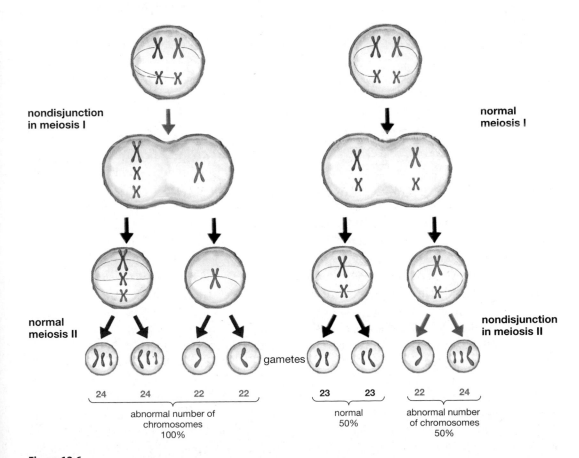

Figure 12.6
A Mistake in Meiosis Brings about an Abnormal Chromosome Count
Nondisjunction can occur either in meiosis I or meiosis II, when either chromosomes or chromatids fail to separate properly. When this occurs in meiosis I, 100 percent of the resulting gametes will be abnormal; when it takes place in meiosis II, only 50 percent will be abnormal.

Testing for Genetic Trouble

"Is the baby OK?" In a life full of questions, it's possible that none carries more weight than this one. In most cases the answer will be reassuring, but in some it will be devastating. Small wonder, then, that scientists constantly are trying to perfect ways to know in advance whether a newborn will be healthy.

You have seen that a large number of debilitating human conditions have genetic causes. Further, you know that nearly every cell in a person's body contains a complete copy of that person's genome, or set of genes. Therefore, to check on someone's genetic well-being, all that is necessary is to have access to a small collection of that person's cells. And such cells exist, of course, in fetuses as well as in newborn babies and adults. Thus, it's not hard to see what the procedure would be for prenatal genetic testing: Gather fetal cells and examine their DNA and chromosomes.

Given that such testing is now commonplace, it's worth pointing out that the first general genetic screening method, **amniocentesis**, became available only in the late 1960s. Indeed, it was only in 1956 that scientists finally arrived at the correct *number* of human chromosomes. This is obviously a piece of information that had to be in place before a technique like amniocentesis could perform one of its main tasks: checking for an *improper* number of chromosomes in a developing fetus.

If you look at **Figures 1** and **2**, (see Figure 2 on page 241), you can see how amniocentesis is carried out. A physician, using an ultrasound image of mother and fetus to "steer," inserts a needle through the mother's abdomen and into the amniotic fluid in which the fetus is suspended. The physician then draws up a small amount of amniotic fluid, which contains epidermal (skin) cells that the fetus has sloughed off as it has developed. Some tests are conducted right away on the fluid itself; other tests require isolated fetal cells, which are needed in relatively large quantity. To get the latter, technicians put the amniotic fluid into test tubes that are then centrifuged, which separates the cells from the fluid around them. The cells are then cultured, or grown to large numbers in the laboratory, after which their genetic material is examined. DNA itself can be tested for abnormalities, while the number of chromosomes

or in meiosis II. In meiosis I, two homologous chromosomes, on the opposite side of the metaphase plate, can be pulled to the *same* side of a dividing cell, producing daughter cells with imbalanced numbers of chromosomes. Conversely, nondisjunction can take place in meiosis II, by means of sister chromatids going to the same daughter cell after failing to "disjoin" (hence the cumbersome term *nondisjunction*). Such actions then produce an egg or sperm that has 24 or 22 chromosomes instead of the standard haploid number of 23; after union with a normal egg or sperm, this results in a zygote that has either 47 or 45 chromosomes, instead of the standard 46.

Down Syndrome

It is the gain of an autosome that brings about the most commonly noticed form of aneuploidy in human beings. **Down syndrome** is, in some 95 percent of cases, a condition in which a person has *three* copies of chromosome 21, rather than the standard two. The usual cause is nondisjunction in egg formation: Female eggs develop in such a way that they contain one too many copies of chromosome 21. However, about 10 percent of the nondisjunctions that lead to Down syndrome take place in the development of sperm. Down syndrome is seen in about 0.1 percent of all live births. It results in an array of effects: smallish, oval heads, IQs that are well below normal, infertility in males, short stature and reduced life span in both sexes.

Since the late 1960s, it has been possible to test developing fetuses for Down syndrome. Today, several methods are commonly employed, as you can see in "Testing for Genetic Trouble," above. It is well known that when women pass the age of about 35, their risk of giving birth to a Down syndrome child increases dramatically. It seems

the fetus has can be determined by creation of the **karyotype**, or picture of the set of fetal chromosomes.

You may wonder about the ultrasound technique that the physician uses to guide the needle in amniocentesis. Isn't this technique also used to warn about trouble with a growing fetus? Yes, but the technique's high-frequency sound waves simply create an image of the fetus and its surroundings. This image can confirm that the fetus is alive, that it is of a certain age, and that it is properly placed in the uterus. But it does nothing to give physicans access to the genetic material of the fetus.

An alternative to amniocentesis is **chorionic villus sampling** (CVS), which was introduced in the early 1980s. In it, cells derived from the embryo are suctioned from the "villi," or extensions of the chorionic membrane that surrounds the placenta. In CVS, a needle can be inserted across the mother's abdomen, as in amniocentesis. More often, however, a suctioning tube is guided through the mother's vagina and cervix to the chorionic membrane, as you can see in Figure 2 on page 241.

(continued)

Figure 1
An Image Produced with Sound
Through the technique of ultrasound, it's possible to get an image of a fetus while it is still in the mother's uterus. In the procedure, high-frequency sound waves are emitted from a hand-held device that generally is moved over the mother's abdomen by a physician. A computer interprets the echoes from the sound waves, thereby yielding an image of the fetus. The healthy fetus pictured here is about four months old.

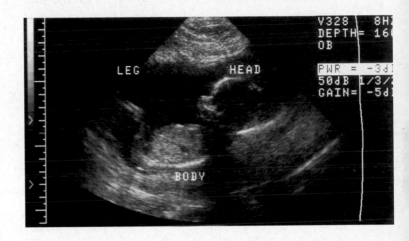

not as well known that even at maternal age 40, the odds of conceiving such a child are less than 1 in 100 (**Figure 12.7**). Scientists are not certain why the mother's age should figure so prominently in Down syndrome. One hypothesis is that older mothers are more

a

b

Mother's Age	Chances of Giving Birth to a Child with Down Syndrome
20	1 in 1925
25	1 in 1205
30	1 in 885
35	1 in 365
40	1 in 110
45	1 in 32

Figure 12.7
Down Syndrome: Increasing Risk with Age

a An adolescent girl with Down syndrome works with her teacher.

b The risk of giving birth to a Down syndrome child increases dramatically past maternal age 35.

Testing for Genetic Trouble (continued)

Though amniocentesis and CVS are powerful diagnostic tools, they are not everything we might hope for in a prenatal test. For one thing, there is the question of timing. Traditional amniocentesis is done 14 to 16 weeks into a pregnancy. Once the procedure is completed, it then takes an additional 10 to 14 days to culture the cells for DNA testing and karyotyping. As a result, more than four months of a nine-month pregnancy may elapse before parents can learn about the health of their unborn. The major advantage of CVS over amniocentesis is that it is faster. It usually is performed between 9 and 12 weeks after conception, and many of its results may be available within a few days.

> ## CVS and amniocentesis are invasive procedures: They involve inserting medical instruments into the delicate surroundings of the fetus.

Despite this advantage CVS is, along with amniocentesis, an invasive procedure: It involves inserting medical instruments into the delicate surroundings of the fetus. In the United States, the risk of fetal loss through CVS is about 1 percent of procedures performed, while the figure for amniocentesis is about half that. Beyond this, both amniocentesis and CVS are expensive, largely because they are labor-intensive.

Given these limitations, amniocentesis and CVS are not routine screening procedures. In general, they are performed only when some genetic risk factors exist in a pregnancy, the most common of these being a mother who is past the age of 35. Yet babies with genetic afflictions are most often born to couples who were thought to be at low risk. Testing for Down syndrome only among older women reveals fewer than a quarter of the cases that exist simply because most babies are born to younger women. What's clearly needed is a genetic test that *can* become routine—one that is inexpensive, quick, noninvasive, and that can be performed early in the pregnancy.

For some years now, scientists have been trying to perfect a technique that addresses all these issues.

The technique is experimental enough that it doesn't yet have a short name (like amniocentesis). Rather, it is referred to by the unwieldy title of *fetal cells derived from maternal circulation*. Its starting point is the now-established fact that fetal cells are present in small numbers in the *mother*'s bloodstream fairly early in pregnancy. This has raised the possibility of drawing a few ounces of blood from a mother's arm, isolating fetal cells from the blood, and then testing the genetic material in the cells for defects.

The central question about the technique is whether it will ever be able to reliably yield a few desired fetal cells from among an ocean of maternal cells in a cost-effective way. In 6 fluid ounces of blood, there may be 100 billion maternal red blood cells but only 100 to 1,000 fetal red blood cells. Further, physicians need to be confident that the fetal cells they do detect are those of the fetus currently in the mother's uterus, rather than cells from an earlier pregnancy. Nevertheless, it may be that analyzing fetal cells from maternal circulation will be the thing that transforms genetic testing from a specialized procedure into one that is routine.

likely than younger mothers to carry a Down syndrome fetus to term rather than spontaneously aborting it. We also know that as a woman ages, her developing eggs are more likely to make "mistakes" in meiosis, such mistakes being, as you've seen, the usual root cause of Down syndrome.

Abnormal Numbers of Sex Chromosomes

Living persons can also have numerous kinds of *sex* chromosome additions or deletions, usually resulting in deformities or debilitations. An example is Turner syndrome, which produces people who are phenotypically female, but who have only one X chromosome. Such females, then, have only 45 chromosomes, rather than the usual 46. Their state is sometimes referred to as XO, the "O" signifying the missing X chromosome. The absence of a second sex chromosome causes a range of afflictions. Females with Turner syndrome have ovaries that don't develop properly (which causes sterility), they are generally short and mentally retarded, and they often have brown spots (called *nevi*) over their bodies.

Turner syndrome results from a loss of a chromosome, but all manner of sex chromosome *additions* can also take place. There are, for example, XXY men, who while

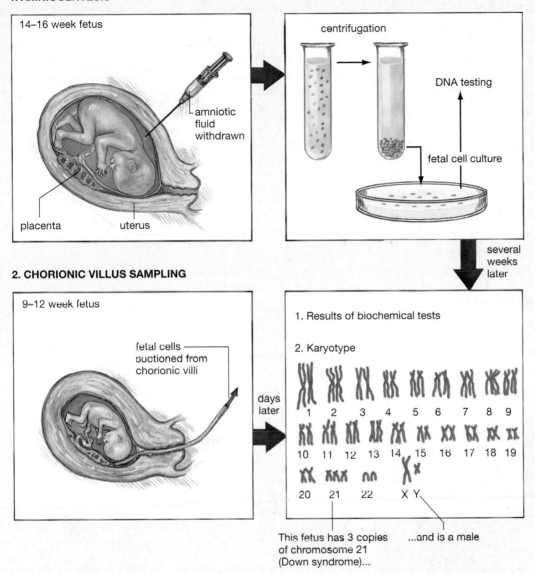

1. AMNIOCENTESIS

14–16 week fetus

amniotic fluid withdrawn

placenta   uterus

centrifugation

DNA testing

fetal cell culture

several weeks later

2. CHORIONIC VILLUS SAMPLING

9–12 week fetus

fetal cells suctioned from chorionic villi

days later

1. Results of biochemical tests

2. Karyotype

1 2 3 4 5 6 7 8 9

10 11 12 13 14 15 16 17 18 19

20 21 22 X Y

This fetus has 3 copies of chromosome 21 (Down syndrome)...

...and is a male

Figure 2
Two Means of Fetal Genetic Testing and Their Results
Amniocentesis and chorionic villus sampling compared.

phenotypically male in most respects, tend to have a number of feminine features: some breast development, a more feminine figure, and lack of facial hair. When coupled with other characteristics (such as tall stature and dysfunctional testicles), the result is a condition called Klinefelter syndrome.

12.5 Structural Aberrations in Chromosomes

Aberrations can occur *within* a given chromosome, sometimes resulting from interactions *between* chromosomes. A frequent cause of such change is that pieces of chromosomes can break off from the main chromosomal body. Such a chromosomal fragment may then be lost to further genetic activity, or it may rejoin a chromosome—either the one it came from or another—often to harmful effect. These structural aberrations may come about spontaneously, but they may also be caused by exposure to such factors as radiation, viruses, and chemicals.

Deletions

A chromosomal **deletion** occurs when a chromosome fragment breaks off and then does not rejoin any chromosome. Such an event might take place during meiosis in a parent whose own somatic cell chromosomes were perfectly normal. In such an instance, a healthy parent could pass along a complement of 23 chromosomes, one chromosome of which would be missing a segment. If the deleted piece were large enough, the zygote would likely not survive. You can see how critical even a portion of a chromosome is by looking at children who suffer from a rare condition, called *cri-du-chat* ("cry of the cat") syndrome, that results from the deletion of the far end of the "short arm" of chromosome 5. If you look at **Figure 12.8**, you can see a **karyotype**—a visual image of a chromosome set—that shows you this deletion. Children born with the cri-du-chat condition exhibit a host of maladies, among them mental retardation and an improperly constructed larynx that early in life produces sounds akin to those of a cat.

Inversions and Translocations

When a chromosome fragment rejoins the chromosome it came from, it may do so with its orientation "flipped," so that the fragment's chemical sequence is out of order—an inversion (see Figure 12.9). A **translocation** can occur when two chromosomes that are *not* homologous exchange pieces, leaving both with improper gene sequences, which can have phenotypic effects.

Duplications

There is, of course, a perfectly normal process of chromosomal part swapping—recombination. But consider what might happen if two homologous chromosomes exchanged *unequal* pieces of themselves in crossing over. One would lose genetic material (effectively a deletion) while the other would gain it. Because these are homologous (meaning paired) chromosomes, the fragment added to the latter chromosome would *duplicate* some of the material it already has. Duplications can take place in several different ways, and they are not always harmful. Nevertheless, some of our most devastating diseases are caused by genetic duplications. The Huntington disease noted earlier in this chapter has as its root cause an abnormal number of "repeats" of given genetic sequences.

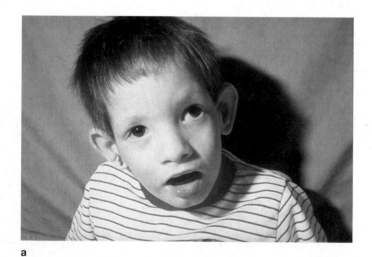

a

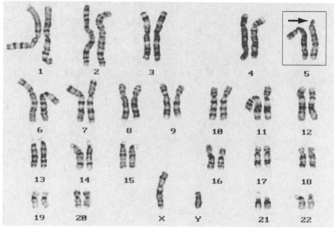

b

Figure 12.8
Cri-du-Chat Syndrome

a A five-year-old boy affected by cri-du-chat syndrome. Note the small head and low-set ears.

b Karyotype of a person with cri-du-chat syndrome. Note, in chromosome 5, the abnormality in the homologous chromosome on the right (see arrow). It lacks a section that is present in the left chromosome.

How Did We Learn?

For scientists to understand the chromosomal afflictions you've looked at in this chapter, they had to understand the basic linkage between chromosomes and inheritance. A scientist from Columbia University was the single most important figure in deepening our knowledge of this subject. You can read about his work in "Thomas Hunt Morgan" beginning on page 244.

On to DNA

For several chapters, you've been looking at genetics from the point of view of the genetic packages called *chromosomes*. Important as chromosomes are, it's been apparent that real genetic control lies with the units that help make them up: the DNA sequences we call *genes*. As you've seen, it is genes that make pea plant flowers purple or white—and that give a person Huntington disease or not. It's time to explore just what these genes are and how they operate.

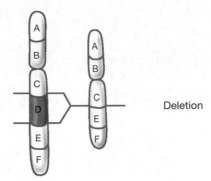

Deletion

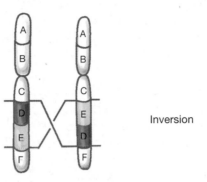

Inversion

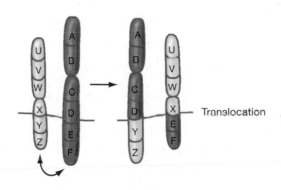

Translocation

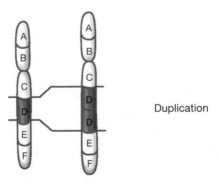

Duplication

Tutorial 12.2.6
Chromosomal
Structural Changes

Figure 12.9
Chromosomal Structural
Changes

Thomas Hunt Morgan: Using Fruit Flies to Look More Deeply into Genetics

Gregor Mendel used pea plants in his work, but generations of researchers after him have employed a common fruit fly, *Drosophila melanogaster*, in trying to unravel the secrets of genetics. One of the earliest *Drosophila* workers was a Kentuckian named Thomas Hunt Morgan, who greatly deepened our understanding of heredity with work he began in 1908 in his lab at Columbia University.

The tiny *Drosophila* fly has a lot going for it as an experimental subject, including a short generation time. Whereas Mendel got one generation of pea plants a year, Morgan got one generation of *Drosophila* every 12 days or so. In practical terms, this meant that Morgan's group had to wait at most a month to see the results of their experiments.

Lessons from a Mutation: White-Eyed Flies

Morgan's path to discovery began with a chance event. One day he looked into one of the empty milk bottles in which he kept his flies and saw something strange: Among numerous normal or "wild-type" *Drosophila* with red eyes was a single male with white eyes (**see Figure 1**). Morgan correctly assumed that the variation he saw had come about because of a spontaneous genetic change or "mutation" in the fly. He then crossed his white-eyed male with females that he knew bred true for red eyes, and got all red-eyed flies in the F₁ generation. No surprise there; it simply looked as though red eyes were the same type of

dominant trait that Mendel had seen in, for example, his yellow peas. Indeed, when Morgan went on to breed the F₁ generation—those that had mixed red and white alleles—he got a 3:1 red-to-white ratio in the F₂ generation. But the strange thing was that every white-eyed fly was a *male*. Why should that be?

The importance of Morgan's insight was that he had linked a particular trait to a particular chromosome, which was a first.

As it turns out, sex chromosomes exist in *Drosophila* as they do in human beings: Females have two X chromosomes, while males have one X and one Y. Doing further experiments on these flies, Morgan eventually decided that the gene for eye color had to lie on a *particular* chromosome: the X chromosome. Why? Any fly that was white-eyed had to have only white-eyed alleles (because white was recessive). Normally, of course, this would mean *two* white-eyed alleles, but there was an exception: Males could be white-eyed if they had but a single white allele—because they have only a *single X chromosome*. You can look at **Figure 2** to see how this plays out schematically. The importance of Morgan's insight was not that he had grasped something about eye color in the fly. It was that he had linked a particular trait to a particular chromosome, which was a first.

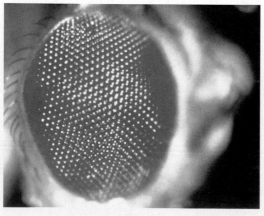
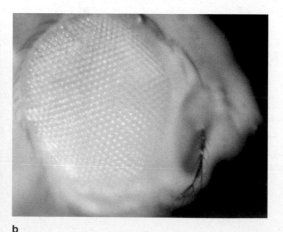

Figure 1

a The compound eye of a red-eyed *Drosophila*

b The compound eye of a white-eyed *Drosophila*

a

b

Tutorial 12.1.1
Why All Morgan's White-Eyed F$_2$s Were Males

When Traits Travel Together: Linkage in Genetic Transmission

Continuing with *Drosophila* breeding work, Morgan and his colleagues soon were finding all kinds of mutations in the flies. There was, for example, a mutation for "miniature" wings, as opposed to full ones. This trait followed the rules described earlier for the white-eyed mutation and accordingly was deemed to be an X-linked characteristic. Not surprisingly, the white-eyed and miniature-winged mutations tended to be transmitted *together* when both of these mutations were bred for, because the genes for both of them lay on the same chromosome. Morgan and his colleagues eventually were to find, however, not just one but several clusters of traits that tended to be transmitted together, thus giving rise to the concept of linked transmission in genetics.

Linked Genes Can Separate: Recombination

The question was, though, did genes on the same chromosome *always* travel together? Surprisingly, the answer was no. For example, miniature wings were usually passed on together with white eyes, but not always. Sometimes one trait would appear in an offspring but the other would not. Since genes for these traits were on one chromosome, how did they get separated from one another?

A paper published in Europe in 1909 gave Morgan a clue. Viewing cells under a microscope, F. A. Janssens had observed chromosomes entwining with each other during an early phase of meiosis. Given this and his own data, Morgan made a creative leap. He suggested that chromosomes can swap parts with each other during their meiotic intertwining. Sound at all familiar? Remember, in the Chapter 10 review of meiosis, the discussion of the phenomenon known as **recombination** or **crossing over**? Here, with Morgan, is the insight that led to this now-established tenet of science. To see how crossing over makes a difference in inheritance, look at Figure 10.2b on page 194.

(continued)

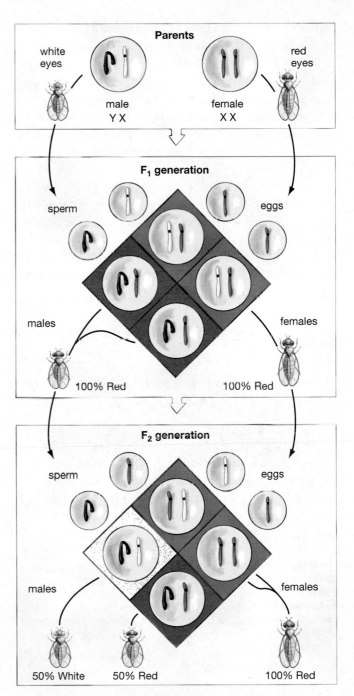

Figure 2
Why All Morgan's White-Eyed F$_2$s Were Males
The F$_2$ female flies that inherited one white-eyed X-chromosome allele also inherited a second red-eyed X allele, making them red-eyed. However, the males inheriting the white-eyed allele had no second X chromosome. Instead, they inherited a Y chromosome—which has no genes on it for eye color—leaving them white-eyed.

Thomas Hunt Morgan: Using Fruit Flies to Look More Deeply into Genetics *(continued)*

Making the First Chromosomal Maps

Morgan's group went on to correctly infer that the closer two genes lay on a chromosome, the less likely they were to separate during recombination. (Imagine having a stick with two lines on it; the closer the two lines are, the less likely they would be to separate should the stick be broken.) By tracking the *frequency* with which genes separated, the researchers were able to begin making the first maps of chromosomes. You can see one of these maps in **Figure 3**, showing the sequence of some of the genes on *Drosophila*'s chromosome II. The distance between each gene, measured in "map units," does not show the *physical* distance between the genes along the chromosome, but rather their relative distance from one another, as indicated by the frequency of their separation in meiosis. Different kinds of gene mapping remain a staple of genetics to this day.

> *Morgan's group went on to correctly infer that the closer two genes lay on a chromosome, the less likely they were to separate during recombination.*

In their work, Morgan and his colleagues managed to confirm the essence of Mendel's theories of inheritance, even as they ventured into areas that lay beyond Mendel's principles, such as gene linkage and crossing over. Appropriately enough, Morgan's efforts won him the Nobel prize in 1934.

Figure 3
Partial Map of Chromosome II in *Drosophila*
The "map units" mark off not physical distance but rather relative separation of the genes, calibrated by the frequency with which they separated from one another in meiosis.

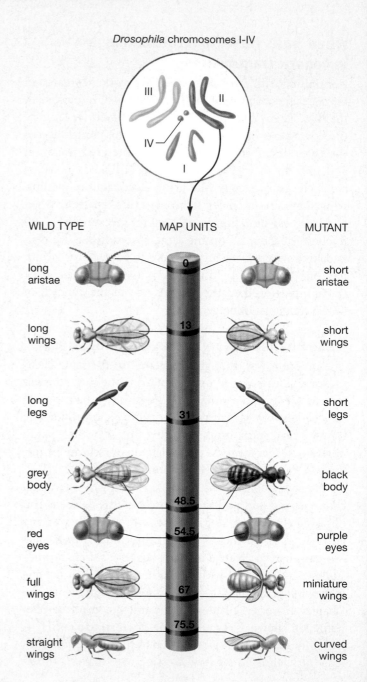

Drosophila chromosomes I-IV

WILD TYPE   MAP UNITS   MUTANT

long aristae — 0 — short aristae

long wings — 13 — short wings

long legs — 31 — short legs

grey body — 48.5 — black body

red eyes — 54.5 — purple eyes

full wings — 67 — miniature wings

— 75.5 —

straight wings — curved wings

Chapter Review ▮▮▮▮▮▮▮▮▮▮▮▮▮▮▮▮▮▮▮▮▮▮▮▮▮▮▮

Summary

12.1 X-Linked Inheritance in Humans

- Certain human conditions, such as hemophilia and Duchenne muscular dystrophy, are called **X-linked conditions**, because they stem from nonfunctional genes located on the X chromosome. Men are more likely than women to suffer from these conditions because men have only a single X chromosome. A woman with, for example, a nonfunctional blood-clotting allele on one of her X chromosomes often will be protected from hemophilia by a functional allele that lies on her second X chromosome.

- Hemophilia and Duchenne muscular dystrophy are examples of recessive genetic conditions, meaning conditions that will not exist in the presence of functional alleles.

- Given the nature of recessive genetic conditions, persons who do not suffer from the condition themselves may yet be able to pass the condition to their offspring. Such persons, referred to as **carriers** of the condition, are heterozygous for it—the alleles they have for the trait differ, one of them being functional, the other being nonfunctional.

12.2 Autosomal Genetic Disorders

- Sickle-cell anemia is an example of a recessive autosomal disorder. It is autosomal because the genetic defect that brings it about involves neither the X nor Y chromosome. It is recessive because persons must be homozygous for the sickle-cell allele in order to suffer from the condition—they must have two alleles that code for the same sickle-cell hemoglobin protein. This protein offers some protection against malaria, however. Therefore, persons who are heterozygous for hemoglobin proteins do not suffer from sickle-cell anemia and yet are afforded some protection against malaria.
  TUTORIAL 12.2.1: Sickle-Cell Anemia and Huntington Disease

- Some genetic disorders are referred to as **dominant disorders**, meaning those in which a single allele can bring about the condition, regardless of whether a person also has a normal allele.

12.3 Aberrations in Chromosomal Sets: Polyploidy

- Human beings and many other species have diploid or paired sets of chromosomes. In human beings, this means 46 chromosomes in all: 22 pairs of autosomes and either an XX chromosome pair (for females) or an XY pair (for males). The state of having more than two sets of chromosomes is called **polyploidy**. Many plants are polyploid, but human beings cannot tolerate the condition.

12.4 Incorrect Chromosome Number: Aneuploidy

- Aneuploidy is a condition in which an organism has either more or fewer chromosomes than normally exist in a full set for that organism. One example of aneuploidy in human beings is Down syndrome, whose victims generally have a third copy of chromosome 21.

- A common cause of aneuploidy is nondisjunction, in which homologous chromosomes or sister chromatids fail to separate correctly in meiosis, leading to eggs or sperm that have one too many or one too few chromosomes.
  TUTORIAL 12.2.5: Aneuploidy and Nondisjunction

12.5 Structural Aberrations in Chromosomes

- Aberrations can occur within chromosomes, including deletions, inversions and translocations, and sequence duplications.
  TUTORIALS: 12.2.6 Chromosomal Structural Changes; 12.1.1 Why All Morgan's White-Eyed F$_2$s Were Males

Key Terms

allele 231	dominant disorder 234
amniocentesis 238	Down syndrome 238
aneuploidy 237	inversion 242
autosomal dominant disorder 234	karyotype 242
autosomal recessive disorder 233	nondisjunction 237
carrier 234	pedigree 236
chorionic villus sampling (CVS) 239	polyploidy 236
crossing over 231	recessive condition 233
deletion 241	recombination 231
diploid 236	translocation 242

Understanding the Basics

Multiple-Choice Questions

1. Examples of sex-linked diseases include
   a. sickle-cell anemia, red-green color blindness, and polyploidy
   b. Duchenne muscular dystrophy, Down syndrome, and red-green color blindness
   c. hemophilia, red-green color blindness, and Duchenne muscular dystrophy
   d. Huntington disease, hemophilia, and Klinefelter syndrome
   e. Down syndrome, cri-du-chat, and hemophilia

2. Sickle-cell anemia patients have a different form of _____, which leads to _____.
   a. hemoglobin, increased muscle activity
   b. a sex chromosome, sterility
   c. pigment, color blindness
   d. skin cell, nevi
   e. hemoglobin, decreased oxygen transport

3. Individuals most protected against malaria are those who
   a. are heterozygous for sickle-cell anemia (one hemoglobin A, one hemoglobin S)
   b. are homozygous for sickle-cell anemia (two hemoglobin S)
   c. are hemophiliacs
   d. are homozygous for hemoglobin A
   e. both a and c

4. Inheritance of a dominant autosomal disorder differs from inheritance of an autosomal recessive disorder in that:
   a. A dominant disorder may be passed on only if both parents are affected.
   b. A dominant disorder is evident only if the offspring is homozygous for the allele.
   c. A dominant disorder is more often seen in females.
   d. A dominant disorder may be passed on even if only one parent is affected.
   e. both b and d

5. An individual having 44 autosomes and one X chromosome would be classified as:
   a. polyploid
   b. aneuploid
   c. having Klinefelter syndrome
   d. having Turner syndrome
   e. both b and d

6. A human embryo with 69 chromosomes would
   a. die in the womb or shortly after birth
   b. be considered polyploid
   c. have fewer problems than a plant with the same condition
   d. both a and b
   e. all of the above

7. Which of the following statements is *not* true regarding Down syndrome?
   a. Affected individuals usually have three copies of chromosome 21.
   b. The condition is often caused by nondisjunction in egg formation.
   c. A woman over 35 increases her chance of having a Down syndrome child to 1 out of every 40 births.
   d. Approximately 0.1 percent of all live births are children with Down syndrome.
   e. The physical characteristics associated with the disorder include mental retardation, male infertility, short stature, and reduced life span.

8. Inherited nondisjunction can occur in
   a. meiosis I or mitosis
   b. mitosis or meiosis II
   c. only meiosis II
   d. only meiosis I
   e. either meiosis I or II

9. A defect in recombination may lead to which of the following?
   a. polyploidy
   b. duplication of chromosome region
   c. translocation of chromosome region
   d. b and c
   e. a and b

10. Which of the following diseases is not evident until well into adulthood?
    a. Huntington disease
    b. Down syndrome
    c. cri-du-chat
    d. Turner syndrome
    e. none of the above

Brief Review

1. For the diseases listed below, indicate the mode of inheritance, using the following list: autosomal recessive, X-linked, chromosomal deletion, aneuploidy, autosomal dominant, polyploidy.
   a. Huntington disease
   b. cri-du-chat
   c. Turner syndrome
   d. Down syndrome
   e. red-green color blindness
   f. sickle-cell anemia
   g. Klinefelter syndrome
   h. hemophilia
   i. Duchenne muscular dystrophy

2. In sex-linked diseases, which gender is more frequently affected—male or female? Why?

3. Why can people with red-green color blindness still see other colors, such as blue?

4. Compare and contrast the four chromosome structural aberrations discussed in this chapter: deletions, inversions, translocations, and duplications.

5. Given the following situations, answer accordingly:
   a. A human cell with 47 chromosomes has probably undergone _____.
   b. A cell duplicates DNA, but fails to undergo cytokinesis. This would lead to _____.

6. Why would a person with Klinefelter syndrome exhibit both male and female features?

7. Certain genotypes may be advantageous in specific situations. Give an example of a heterozygous genotype that is

advantageous over either homozygous genotype. (Hint: especially in mosquito-infested areas.) Why is this the case?

Applying Your Knowledge

1. A red-green color-blind man marries a woman who is a carrier for red-green color blindness. (a) What percentage of their sons could be expected to be color blind? (b) What percentage of their daughters?

2. A couple has a son with Duchenne muscular dystrophy. It is known that external environment and exposure to harsh chemicals may lead to birth defects. The child's father works at a local chemical plant, and suspecting it played a part in the condition of his son, sues the company for damages. You are called in to testify as an expert in genetics. How would you interpret the data? Does the man have a valid case? Why or why not?

3. Genetics techniques are making it more feasible for couples to have healthy children. Some tests are currently in use for specific selection during in vitro fertilization, such as disease screening and sex preferences. When do you think these techniques are justified? (Sex determination for sex-linked disease prevention vs. choosing sex for "family balancing" only.) Where do we draw the line between cure and convenience?

4. A couple comes to you for genetic counseling, worried that if they have a child, it will inherit sickle-cell anemia. The mother has sickle-cell anemia, but the father is known not even to be a carrier for it. (That is, he has two alleles that code for hemoglobin A, rather than the sickle-cell-causing hemoglobin S.) Their question to you: What are the odds that any of our children would inherit sickle-cell anemia?

Genetics Problems

Note: In several of the problems that follow, you probably will need to construct Punnett squares to get an answer.

1. **a.** For an X-linked recessive condition such as red-green color blindness, what is the chance of having an affected son under the following two scenarios?

   Scenario I: The mother is a carrier and the father is normal.

   Scenario II: The father is affected and the mother has no family history of the disease.

   b. Using X for the dominant color-producing allele on the X chromosome, x for the recessive (color-blind) allele, and Y for the male chromosome, what are the genotypes of the parents in the two scenarios?

   c. In scenario I, if the parents have a daughter, what are the chances that she will be a carrier (that is, heterozygous for the condition)? What are the chances of a daughter being a carrier in scenario II?

2. Coat color in Labrador retrievers is governed by two genes: B/b and E/e.

   B = black pigment
   b = brown pigment
   E = allows pigment to be deposited
   e = does not allow pigment to be deposited

   The B allele is dominant to the b, and E is dominant to e. The ee genotype specifies a golden lab regardless of what B/b allele combination it carries. This is because the black or brown pigment cannot be deposited in a proper manner. If a brown lab is mated to a brown lab, and they have some golden puppies:

   a. What are the genotypes of the parents?

   b. What is the chance that these parents will be able to generate black lab puppies? Brown labs? Golden labs?

   c. In a litter of 12 puppies, how many theoretically will be of each color based on your answer in (b)?

3. Using the following pedigrees, can you ascertain whether the condition is dominant, recessive, on the X chromosome, or an autosome? (For these questions, no indication has been made for heterozygotes.) Be sure to indicate which individual or mating brought you to your conclusion.

   a. deafness

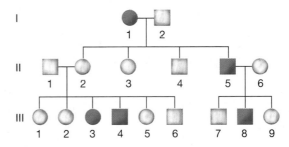

   b. muscle atrophy

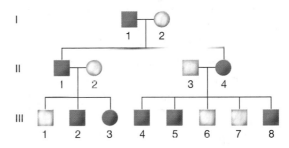

   c. ichthyosis (extremely dry skin)

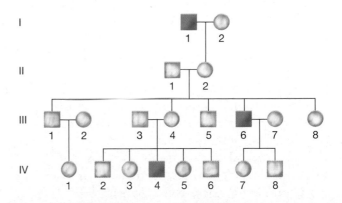

4. A couple, Jan and Barry Smith, see a genetic counselor because they are concerned about the possibility that their unborn son will have Marfan syndrome, a connective tissue disorder that can lead to serious problems with blood circulation. Both of their families have a history of Marfan syndrome—Jan's brother, and Barry's mother, father, and sister are affected. Jan and Barry are unaffected. No other information is known about the family history. Assume that Marfan syndrome is controlled by a single gene on an autosome, and answer the following questions.

   a. Draw a pedigree that includes all individuals mentioned above. Label Jan and Barry Smith with stars.
   b. Is this disease dominant or recessive? Why?
   c. What is the genotype of each of the following individuals? (Give all possibilities.) Barry's mom, Barry's sister, Barry, Jan. Use *A* to signify the dominant allele for the condition and *a* to signify the recessive allele.
   d. What are the chances that Barry and Jan will have a child with Marfan syndrome?

5. a. A young married couple is concerned about their child inheriting a rare disease called mucopolysaccharidosis from the wife's family. There are two forms of the disease: one called type IH (Hurler's syndrome) caused by an autosomal recessive mutation, and the other called type IIA (Hunter's syndrome) caused by an X-linked recessive mutation. Both can lead to a deficiency in an enzyme needed to cut up certain sugars, which eventually build up and lead to progressive mental retardation, disability, and often death before the age of 15. There is no history of this disease in the husband's family, so they assume he is not a carrier. If the wife were a carrier for this mutation, what would be the chances of her son suffering from the condition? Would it matter if she were a carrier for Hunter's or Hurler's syndrome?

   b. The wife informs you that the only member of her father's family to have the disease was her father's sister, but both her father's parents and his brother were normal. Construct a pedigree of the wife's family to help explain whether you believe she should be concerned about Hunter's or Hurler's syndrome. Be sure to indicate which relative was instrumental in ruling out one of the syndromes.

6. A mother with blood type A and a father with blood type B have a child that is blood type O. Write the genotypes of all three individuals using the following alleles: A, B, and O.

7. A man in his fifties is showing signs of Huntington disease, which has an autosomal dominant pattern of inheritance. Two of his daughters, both in their thirties, are worried that they will get the disease, and that they have already passed the gene to their children. The daughters get tested for the disease-causing allele. One daughter, Debbie, is free of the allele. The other daughter, June, carries one copy of the allele.

   a. What is the genotype of each daughter? (*H* and *h* are the alleles.)

b. What is the chance that each daughter will get Huntington disease?

c. What is the chance that each daughter has passed the gene to her children?

8. The following couples plan to marry and want to know the chance of having a child with the following diseases. You are a genetic counselor who needs to answer their questions.

a. two carrier parents for cystic fibrosis (autosomal recessive)

b. one parent with Marfan syndrome and the other parent with no family history for the syndrome (autosomal dominant)

c. an unaffected father and carrier mother for hemophilia (X-linked recessive)

d. two parents with achrondroplasia (an autosomal dominant bone disorder that can result in dwarfism)

9. **a.** What are the possible gametes from the following persons:

Person I: *Aa*

Person II: *aa*

Person III: *aaBb*

Person IV: *AaBb*

b. What are the genotypes and frequencies of the offspring produced if you cross person I with person II and person III with person IV?

10. Determine the most likely mode of inheritance for the following pedigrees. What is the genotype of the affected individuals in each case?

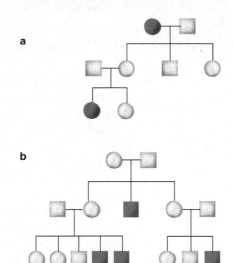

11. A curious genetics student wants to reproduce Mendel's pea-plant results. She crosses a true-breeding tall plant (*TT*) to a short plant (*tt*). She self-crosses the F_1 plants and obtains 200 F_2 plants. How many of these will be tall, short, or intermediate in height? (Assume height is controlled by a single gene, and that tall is dominant to short.)

MediaLab

Do We Know Too Much? Human Genetic Testing

Collum is an 8-month-old baby with an incurable neurodegenerative disorder. In a few short weeks, he changes from a playful, robust infant to being unable to lift his head, to move, and eventually even to breathe. When reading heart-breaking parental accounts about battling these kinds of rare conditions, you may ask why it happens, what medical science is doing for a cure, and if it could happen to your child. Doctors have recently discovered the cause of many of these illnesses; unfortunately, they are genetic mutations passed from parent to child during conception.

Today there are few treatment options for repairing mutations. With this discouraging news, the only solace may be that we can now predict the chance of a future child or grandchild suffering the same fate. Given what you know about human genetics from reading Chapter 12, can you predict your child's chance of inheriting some disorder? In the following *CD-ROM Tutorial*, you will review the three patterns of inheritance of genetic mutations, so that in the *Web Investigation* you can analyze your skills as a human geneticist.

This *MediaLab* can be found in Chapter 12 on your CD-ROM (Tutorial 12.2.1) and Companion Website (http://www.prenhall.com/krogh3).

CD-ROM TUTORIAL

Human genetics uses Mendel's rules about the probability of inheriting a specific genetic characteristic and extends them to predicting inheritance of human diseases. This *CD-ROM Tutorial* will lead you through the three main patterns of inheritance seen in human genetics.

Activity

1. *First, you will learn how to create a Punnett square to predict the chance of a child inheriting a recessive or dominant condition caused by a mutation on one of human chromosomes 1–22, which are called autosomes.*

2. *Then, you will compare this prediction with the special inheritance pattern seen for mutations on the X chromosome, and test your ability to discern the difference by practicing with some real pedigrees.*

3. *Finally, you will examine one of the most serious genetic disorders, chromosomal mutations, and see how they occur.*

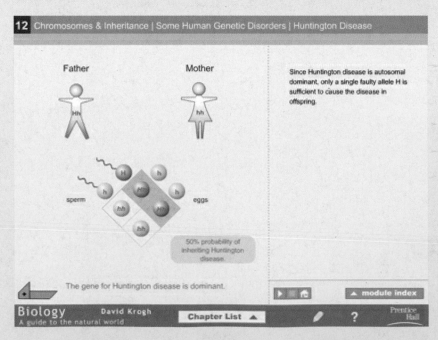

WEB INVESTIGATION

Investigation 1

Estimated time for completion = 10 minutes

Genetic counselors are health professionals who specialize in disseminating information on over 4,000 different inherited human diseases. They take detailed family histories, review basic genetics, and counsel patients on the likelihood of inheriting one of thousands of disorders, while dealing with the emotional and sometimes moral issues involved. How do you know if you should see a genetic counselor? Would you understand the basic genetics they explained to you? Click on the Keyword **COUNSELOR** on your CD or Website to read an article describing gene testing.

Investigation 2

Estimated time for completion = 5 minutes

In the future portrayed in the 1997 movie *Gattaca*, prenatal genetic screening helps couples chose the perfect child, free from mutations that would predispose heart disease, low IQ, or even myopia. Click on the Keyword **GATTACA** on your CD or Website to visit the future and try your hand at designing your offspring. What would you select for your child's height, weight, health problems, and even IQ if you could?

Investigation 3

Estimated time for completion = 5 minutes

How close is science getting to science fiction? In *Web Investigation 2*, you tried to design the perfect child. How many of the traits like Alzheimer's, drug addiction, arthritis, even height that you listed as undesirable in your family can now be tested for? Click on the Keyword **GENETESTS** on your CD or Website. When you get there, click on "View all Diseases" to investigate one of the hundreds of genetic mutations for which tests are currently available. Pick one of the genetic diseases that runs in your family, and investigate the state of research into treatments and predictive gene testing.

Now that you have investigated the current science of genetic testing, comment on what you learned and how it has influenced your feelings on some of the emotional issues involved.

COMMUNICATE YOUR RESULTS

Exercise 1

Estimated time for completion = 15 minutes

If someone could test your DNA and tell you what diseases you would likely suffer from in 25 years, would you want to know the results? Using the article from *Web Investigation 1*, compose a 250-word paper supporting or refuting your personal decision to know the results of your own gene test. Use examples, including the specific disease you would be tested for, the actual likelihood of contracting the disease if you have a positive test, how this knowledge could help you avoid dying from the disease, and how you think having this knowledge would affect you and your family.

Exercise 2

Estimated time for completion = 5 minutes

Is it even possible to design the perfect child using your genetic background? If there are undesirable traits in your family, you may need to add beneficial genes from outside your gene pool. The checklist from the *Gattaca* website uses questions similar to those a genetics counselor might ask you about your family health history. List all the "undesirable" inherited traits that run in your family.

Exercise 3

Estimated time for completion = 5 minutes

In *Web Investigation 3*, you looked at one of the hundreds of human diseases with a genetic component, such as familial breast cancer, that may run in your family and can currently be tested for. Prepare a 250-word report for your own family members about the current state of knowledge on this disease and the tests available. Describe the disease, and indicate how much of the severity of symptoms is due to environmental influences like diet and how much to genetic mutations. If a DNA test is available, indicate whether you would like to know the results of the test, and if the test can accurately predict who will get the disease.

13

DNA Structure and Replication

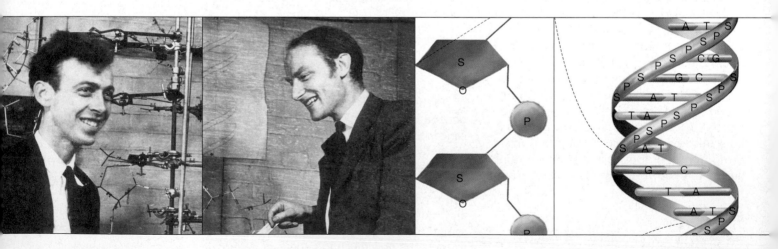

James Watson.
(Section 13.2, page 256)

Francis Crick.
(Section 13.2, page 256)

The double helix.
(Section 13.3, page 258)

DNA serves as a storehouse of information for living things. Scientists could not understand how DNA could play this role until they understood how all of DNA's component parts fit together.

"In research the front line is almost always in a fog," wrote Francis Crick in his 1988 memoir, *What Mad Pursuit.* Crick has firsthand knowledge of this. In 1953, with James Watson, he discovered the structure of DNA, an effort that brought him to the front line of research, which is where he has remained ever since. Crick's book makes clear what a tremendous difference there is between learning something about nature and discovering something about it. In learning, dozens of voices stand ready to instruct; there are books, lectures, videos, CDs—a world of well-organized information. The person who seeks to discover something about nature, meanwhile, is confronted with silence. Nature goes on: Cells divide, chromosomes condense, birds migrate. But *why* do these things operate the way they do? All the researcher has as a guide are the things themselves, working away. The trick is to devise some test (called an *experiment*) that can make nature's routine operations yield up information. In this process of teasing

out truth, knowledge is gained in very small increments. Scientists are like the first people down a darkened maze of tunnels; they must make their way exceedingly slowly, feeling each square inch of wallspace as they go, after which they can only leave lights on *behind* them (in the form of their scientific papers, books, and so forth).

13.1 What Do Genes Do, and What Are They Made of?

To get a feel for the process of discovery, consider the state of genetics early in the twentieth century. By 1920, it had been demonstrated beyond any doubt that genetic information resided on chromosomes. Within a decade or so, by looking at some abnormally large fly chromosomes, scientists could observe in great detail such chromosomal processes as recombination. By viewing the banding patterns on these chromosomes, they could even identify the rough location of some genes.

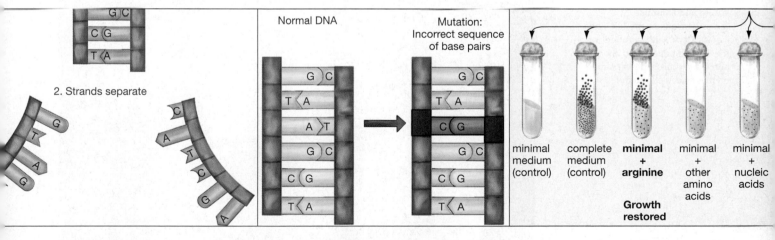

DNA replication begins.
(Section 13.3, page 259)

One kind of mutation.
(Section 13.4, page 261)

A famous experiment.
(Essay, page 265)

Yet what was a gene? What was the physical nature of this unit of heredity, and how did it work? No one knew. Because genes are much smaller than the chromosomes on which they reside, there was no hope of simply viewing one under a microscope. In observing chromosomes, scientists were only "looking" at genes in the way that any of us is "looking" at lunar rocks by glancing up at the Moon. Through the 1920s and 1930s, then, genes could be described only in vague, functional terms, as in: A gene is an entity that lies along a chromosome and brings about a phenotypic trait in an organism (for example, the green or yellow peas of Mendel's experiments).

Things were to clear up somewhat in the ensuing years. The seminal achievement in genetics in the late 1930s and early 1940s was a more concrete description of what genes do: They bring about the production of proteins. The central achievement in genetics from the mid-1940s to the early 1950s was strong evidence indicating what genes are composed of: deoxyribonucleic acid, or DNA for short.

DNA Structure and the Rise of Molecular Biology

By the early 1950s, a key question became *how* DNA carried out its genetic function. To find out, scientists had to piece together its exact chemical structure. This investigation turned out to be a watershed event in biology. Just as opening a mechanical watch and observing how its parts fit together would allow a person to understand how the watch works, so deciphering the structure of DNA allowed biologists to understand how genetics works at its most fundamental level. How could a section of DNA specify a protein? How could genetic information be copied? The structure of the DNA molecule suggested answers to these questions, as you'll see in this chapter.

Apart from what this investigation uncovered, the inquiry itself stood as a symbol of a new era in biology. When Gregor Mendel did his experiments with peas, he was looking at whole organisms. When T. H. Morgan was doing his work with *Drosophila* flies, he was interested in whole chromosomes, which he knew to be composed of several types of molecules. In the early 1950s, however, the search turned to the constituent parts of a single molecule. What atoms of oxygen, hydrogen, and carbon were in DNA, and how were they arranged? Biological research of this sort has grown ever more important in the decades since the 1950s. It is today known as **molecular biology**: The investigation of life at the level of the constituent parts of its individual molecules.

Figure 13.1
Young and Famous
James Watson, on the left, and Francis Crick, with a model of the DNA double helix, shortly after they published their paper on the molecule's structure.

13.2 Watson and Crick: The Double Helix

James Watson and Francis Crick may not be instantly recognized scientific names in the way that, say, Albert Einstein or Louis Pasteur are, but there's a certain public awareness that these two researchers did something important in connection with DNA (**Figure 13.1**). What they did was present to the world, in 1953, the structure of DNA: Atom-by-atom, bond-by-bond, this is how DNA fits together, they said in unveiling DNA's now-famous configuration, the double helix.

The two were a seemingly unlikely pair to make an epochal scientific discovery. Watson, an American, was a 23-year-old who was scarcely more than a year out of graduate school, and Crick, an Englishman, was a 35-year-old just then working on his doctorate when the two met in the fall of 1951 at Cambridge University in England. Ending up together by coincidence at the same laboratory, they realized in short order their mutual interest in the structure of DNA and, to the neglect of projects they were supposed to be working on, they began several rounds of model building and brainstorming that resulted in their breakthrough.

Watson and Crick were greatly aided in their investigation by the work of others. Though the DNA molecule was too small to be seen by even the most powerful microscopes of the time, something about its structure could be inferred from a technique called *x-ray crystallography*. In this process, a purified form of a molecule (a crystal) is "grown" to a considerable mass, and pieces of it are bombarded with x-rays. The way these rays scatter upon impact then reveals something about the structure of the molecule.

If you look at **Figure 13.2**, you can see the results of some x-ray crystallography. As you can imagine, it takes a highly trained observer to be able to perceive anything about the structure of a molecule from such an image. Fortunately for Watson and Crick, such a person was working just up the road from them. She was Rosalind Franklin, a researcher at King's College in London and one of the handful of individuals then skilled in performing x-ray crystallography on DNA. She and her colleague, Maurice Wilkins, were themselves working on the structure of DNA at the time, as were other researchers in America. Thus did Watson, at least, regard the search for DNA structure to be a race between several teams, a fact that concentrated his efforts wonderfully. In 1962 Watson, Crick, and Wilkins were awarded the Nobel Prize in Medicine or Physiology for their work on DNA. Rosalind Franklin died in 1958 at the age of 37. Nobel Prizes are not awarded posthumously; it's unknown what would have happened had she lived.

Let's start now to look at DNA's structure, which will put you in a position to appreciate the achievement of Watson, Crick, and their fellow researchers.

a

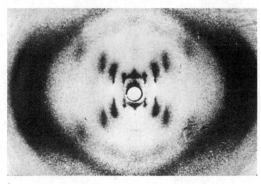

b

Figure 13.2

a DNA Investigator
Rosalind Franklin, whose work in x-ray crystallography was important in revealing the structure of the DNA molecule.

b Imaging DNA
One of Franklin's x-ray crystallography images of DNA. The "cross" formed of dark spots indicated that the molecule had a helical structure.

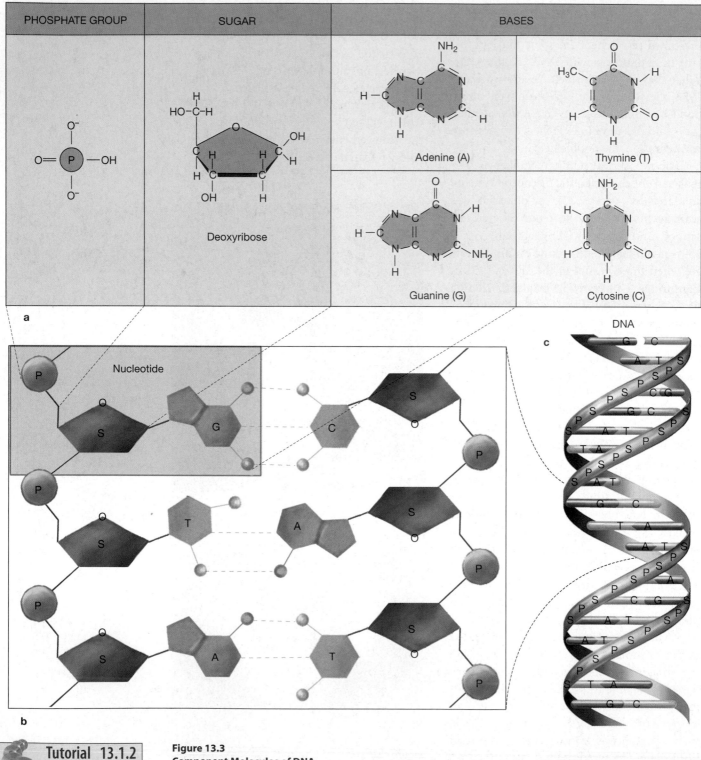

Tutorial 13.1.2
Structure of DNA

Figure 13.3
Component Molecules of DNA

a The DNA molecule is composed of three types of component molecules: phosphate groups, the sugar deoxyribose, and the bases adenine, thymine, guanine, and cytosine (A, T, G, and C).

b The building blocks of DNA are units called nucleotides, each of which is composed of one phosphate group, one sugar, and one of the four bases—in this example, G. Across the strands of the helix, G always pairs with C, and T with A.

c Sugar and phosphate link together to form the "handrails" of the double-helix "staircase," while the bases form the "stairsteps," each base extending across the helix to link with a complementary base extending from the other side.

13.3 The Components of DNA and Their Arrangement

If you look at **Figure 13.3a**, you can see the component parts of DNA. First, there is a phosphate group, and then a sugar called *deoxyribose*. By looking at Figure 13.3b and focusing on the red-colored molecules on the left, you can see how sugar and phosphate fit together to form a kind of chain, with its links going: sugar-phosphate-sugar-phosphate (symbolized in the figure by S and P). If you look now at Figure 13.3c and think of the entire DNA double helix as a staircase, the two "handrails" on it are formed by these sugar-phosphate chains.

There is then a third component to DNA, the "steps" lying between its handrails. These steps are composed of DNA's "bases": adenine, guanine, thymine, or cytosine—A, G, T, and C for short. If you look at Figure 13.3b, you can see how these bases are linked to the DNA handrails.

As the figure shows, each of the bases amounts to a *half*-stair-step, if you will. Each extends inward from one handrail of the double helix and is then joined to a complementary base extending inward from the *other* handrail of the DNA molecule. (The bases are linked via hydrogen bonds, symbolized by the blue dotted lines in Figure 13.3b.)

The figure shows some examples of an A base being paired with a T base, and one of a C base being paired to a G base. This turns out to be one of the fundamental rules about DNA structure. If you viewed a billion DNA *base pairs*, as they're called, you would find the same thing: A always pairing with T, and G always pairing with C, across the helix.

The Structure of DNA Gives Away the Secret of Replication

It was DNA's very structure that suggested the answer to one of the great questions of genetics: How is genetic information passed on? To put this another way, how does a cell make a copy of its own DNA? As you've seen, a full complement of our DNA is contained in nearly every cell in our body. Yet cells divide, and the *daughter* cells contain exactly the same DNA complement as the parent cell. This means that genetic information is passed on by means of *DNA being copied*, with one copy of this molecule ending up in each daughter cell. But how?

The structure Watson and Crick discovered suggested a way. We've observed that A must always pair with T, and G with C, across the two strands of the double helix. This rule, Watson and Crick saw, meant that each single strand of DNA could serve as a *template* for the synthesis of a new single strand (**see Figure 13.4**). Each A on an old strand would specify the place for a T on the new, each G on the old a place for C on the new, and so forth. All that was required was for the two old strands to separate—splitting the stair steps right down the middle—and for new strands to be synthesized that were complementary to the old. As it turned out, this was exactly how things worked, as you'll see shortly.

Tutorial 13.1.3
Replication of DNA

Figure 13.4
DNA Replication
The result of DNA replication is two identical strands of DNA, whereas the process began with one.

1. DNA to be replicated

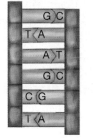

2. Strands separate

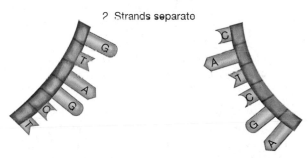

3. Each strand now serves as a template for the synthesis of a separate DNA molecule as free nucleotides base-pair with complementary nucleotides on the existing strands.

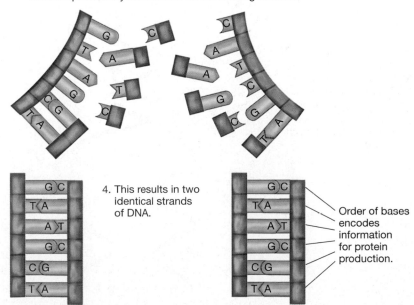

4. This results in two identical strands of DNA.

Order of bases encodes information for protein production.

The Structure of DNA Gives Away the Secret of Protein Production

The second thing the structure suggested was a partial answer to another great question of genetics: How the molecule of heredity could be versatile enough to specify the dazzling array of proteins that all living things produce. The handrails of DNA's double helix are monotonous—phosphate, sugar, phosphate, sugar. But the bases can be laid out along these handrails in an extremely varied manner. A has to pair with T and G with C *across* the helix, but *along* the handrails, the bases can come in any order. Look at the top drawing in Figure 13.4 and see the order this hypothetical group of bases comes in; starting from the bottom right, A-G-C-T-A-C. Strung together by the thousands, these bases can specify a particular protein, just as in Morse code a series of short and long clicks can specify a particular word.

The Building Blocks of DNA Replication

You have already looked briefly at the basic process by which DNA is copied or "replicated," but we now need to review some of the details of this process. The basic steps here are straightforward. The essential building block of DNA is the unit called the **nucleotide** which can be seen in the box at the upper left of Figure 13.3b. It contains one sugar and one phosphate molecule, joined to one of the four bases (in this case G). In replication, as you can see in Figure 13.4, the double helix unwinds. The nucleotides on

each of its single strands are paired with free-floating nucleotides that line up in a new, complementary strand. Because *both* strands of the original double helix are being paired with new strands, the end product is two double helices, where before there was one.

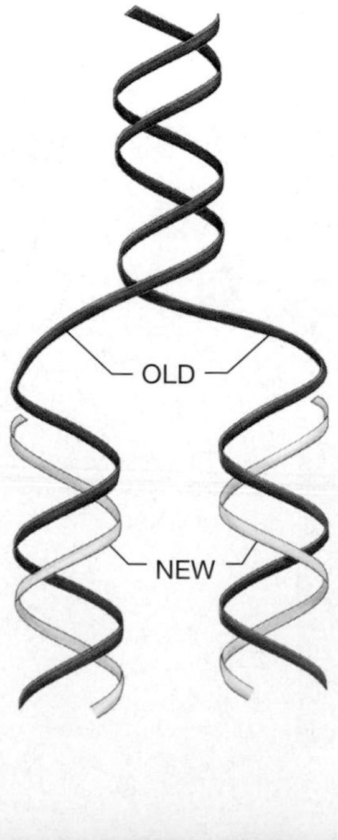

DNA Replication: Something Old, Something New

It's worth drawing a little finer point on one aspect of this process: Each resulting double helix is a combination of the old and the new. Each represents one "parental" strand of DNA and one newly synthesized complementary strand. This is conceptually important because it is how life builds upon itself. You can see this illustrated in **Figure 13.5**.

Eventually, the two double helices are fully formed. At that point they part company, during mitosis, with each double helix moving into a separate cell and thereafter serving as an independently functioning segment of DNA. (At the chromosomal level, the newly replicated helices of DNA are the sister chromatids you've seen so much of in the last few chapters.)

However simple this process may sound in overview, its details are enormously complicated. As you might imagine, such a process could not proceed without enzymes to catalyze it. To name just two groups of them, there are enzymes called *helicases* that unwind the double helix, separating its two strands to make the bases on them available for base-pairing. Another group of enzymes, collectively known as **DNA polymerases**, then moves along the double helix, joining together nucleotides as they are added—one by one—to the new, complementary strands of DNA.

Editing Out Mistakes

The base pairing that goes on in replication happens hundreds, then thousands, then millions of times in a given stretch of DNA: Free-standing A's line up with complementary T's, while C's align with G's. Given the numbers of base pairs involved, the amazing thing is how few mismatches there are by the time the process is completed. The error rate in DNA replication—the rate at which the *wrong* bases have been brought together— might be only a few in every billion bases by

Figure 13.5
How Life Builds on Itself
Each newly synthesized DNA molecule is a combination of the old and the new. An existing DNA molecule unwinds, and each of the resulting strands (the old) serves as a template for a complementary strand that will be formed through base pairing (the new).

the end of replication. Yet *during* replication, such a mistake might be made once in every 100,000 bases. Obviously, to start with one number of mistakes but to end up with far fewer, the cell's genetic machinery has to be capable of correcting its errors.

This happens partly through the services of the versatile DNA polymerases, which are able to perform a kind of DNA editing: They remove a mismatched nucleotide and replace it with a proper one. Interestingly enough, this happens through a kind of backspacing. Normally the DNA polymerases are moving along a DNA chain, linking together recently arrived nucleotides. When an error is detected, however, they stop, move "backward," remove the incorrect nucleotide, put in the correct one, and then move forward again.

13.4 Mutations: Another Name for a Permanent Change in DNA Structure

With this consideration of alterations in DNA structure, you have arrived at an important concept: that of a **mutation**, which can be defined as a permanent alteration of a DNA base sequence. Such alterations come about because the cell's various DNA error-correcting mechanisms are not foolproof; they do not correct all the mistakes that occur.

DNA's makeup can be altered in many ways. A slight change in the chemical form of a base might, for example, cause a G to link up across the helix with a T (instead of with its normal partner, C). Conversely, a sequence of base pairs *along* the handrails might be inserted in error. (A C-G base pair might be put in, for example, where an A-T normally resides. **See Figure 13.6** for these examples.) Permanent mistakes like these are called **point mutations**, meaning a mutation at a single location in the genome. Such mutations stand in contrast to the kind of whole-chromosome aberrations presented in Chapter 12, though these also qualify as mutations under some definitions.

In what way are mutations "permanent"? Think of how DNA replication works. Before any cell can divide, it must first make a copy of its complement of DNA. Should this DNA contain an uncorrected mistake, it too will be copied again and again with each succeeding cell division. Most mutations have no noticeable effect on an organism. But the concept of mutation is quite rightly fearful to us, because in relatively rare instances, mutations can have disastrous effects.

An Example of Mutations Passed Along in a Line of Cells: Cancer

A cancerous growth is a line of cells that has undergone a special kind of mutation—one that causes the affected cells to proliferate wildly. Once these cells start multiplying, they can become *invasive*—they can move into normal tissues and destroy their ability to function. If the cancerous cells invade tissues in a vital organ, such as the pancreas or liver, the result can be death. A group of invasive cells is known as an invasive *tumor*. In some cases, cells from the original or "primary" tumor will break away from it and *metastasize*: They will move through the bloodstream or lymphatic system and begin multiplying as secondary tumors at sites distant from the primary tumor.

But how does this process get going in the first place? Any mutation, including one that contributes to cancer, can come about because a cell's DNA replicating machinery has made a random, spontaneous error that goes uncorrected. Though such errors are made regularly in all organisms, most of them play no role in causing cancer. Mutation *rates* can be affected, however, by so-called environmental influences. Cigarette smoke is one such influence, and ultraviolet light is another. Any environmental influence

Tutorial 13.3.1
Nature of Mutations

Figure 13.6
The Nature of Mutations
Shown are two examples of mistakes that can be made during DNA replication. Errors such as these that go uncorrected are known as "point mutations."

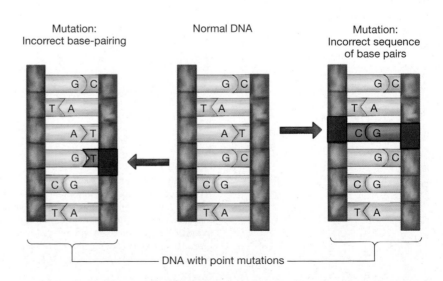

Mutation: Incorrect base-pairing   Normal DNA   Mutation: Incorrect sequence of base pairs

DNA with point mutations

that can contribute to the onset of cancer is a **carcinogen**. We normally think of carcinogens as being chemicals, and indeed there are hundreds of known chemical carcinogens, such as asbestos. But note that ultraviolet light is not a chemical; it is a form of radiation, and yet it is a carcinogen—one that is active in instigating various forms of skin cancer. Likewise, viruses can be carcinogens, as with the human papilloma virus that can contribute to cervical cancer.

Whatever form a carcinogen may take, in most instances its ultimate effect is to help bring about one or more of the mutations lying at the root of cancer. (This is not always the case; a substance can be a carcinogen without causing mutations.) The reason cancers aren't more common is that *multiple* mutations are required to produce cancerous cells. In colon cancer, for example, from four to seven mutations are required for a "malignant" or metastasizing tumor to start to grow. This is one reason that cancer rates generally rise with age; it takes time for all of the necessary mutation events to occur (**see Figure 13.7**). In addition, mutations often are the result of years of environmental exposure—to cigarette smoke, to sunlight, or to other carcinogens.

Heritable Mutations: Those That Occur in Germ-Line Cells

Most mutations come about in the body's **somatic cells**, which is to say cells that do not become eggs or sperm. Conversely, some mutations arise in **germ-line cells**, meaning the cells that do become eggs or sperm. The important point here is that germ-line cell mutations are *heritable*—they can be passed on from one generation to the next. Many generations ago, for example, the DNA in some germ-line cells underwent mutations that resulted in the conditions we now call sickle-cell anemia and Huntington disease. These conditions can be passed on from parents to children in the ways reviewed in Chapter 12.

Think of the difference between these mutations and the cancer-causing mutations reviewed earlier. In, say, pancreatic cancer, a line of cells in the pancreas undergoes a series of mutations that causes them to multiply wildly. However harmful this line of cells may be, it is separate from the

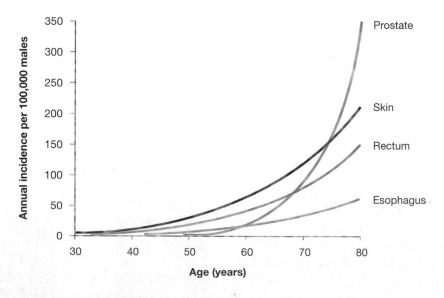

Figure 13.7
Cancer Incidence and Age
A cell becomes cancerous only after its DNA has undergone multiple mutations. It takes time for such a series of mutations to occur. Environmental influences, such as cigarette smoke, may take years to bring about mutations. Beyond these things, the body's mutation-correcting processes become less vigorous with advancing age. The result is what you see in the graph: The incidence for most forms of cancer rises dramatically with age.
Adapted from B. Vogelstein and K. Kinzler, *Trends. Genet.* 9 (1993):101.

line of cells that gives rise to eggs or sperm. Thus, pancreatic cancer cannot be passed on from one generation to the next. What can be passed on are genes that *predispose* a person to a disease like pancreatic cancer. Often a person inherits one mutated allele for a given cell-growth process, but also inherits a second "good" allele. It is a mutation to this second allele, now in a somatic cell, that gets the malignant process going. Up to 10 percent of cancers are thought to be influenced by genetic predisposition.

As observed earlier, DNA replication has a very low error rate. After learning about the kinds of effects mutations can have on organisms, you can understand why: Life as we know it could not exist with a *high* error rate. The cause of sickle-cell anemia turns out to be the substitution of a *single* T for an A in the gene that codes for protein hemoglobin in red blood cells. (This substitution produces a form of hemoglobin that causes blood cells to assume the harmful sickle shape, as you can see in Figure 12.3 on page 234). Given this, you can see why genetic error rates have to be low, at least for some DNA sequences. How could a human being or a bird or a tree survive a high error rate in the DNA that codes for proteins?

The Value of Mistakes: Evolutionary Adaptation

We should also not be surprised that some genetic mistakes go uncorrected, for the simple reason that mutations are a critical part of evolutionary adaptation in organisms. Germ-line mutations are the only means by which completely new genetic information can be added to a species' genome, in the form of new alleles (meaning variant forms of a gene). Organisms can combine *existing* alleles, in myriad ways. Think of meiosis, with its chromosomal part swapping (recombination or crossing over) and reshuffling of chromosomes (independent assortment). Valuable as these processes are, no amount of genetic recombination could have produced, for example, the eyes that some living things possess. To go from no eyes to eyes, there had to have been some mutations along the line—some accidental reorderings of DNA sequences such that entirely new proteins were produced.

To put these mistakes into perspective, most mutations do nothing to an organism, and when they do have some effect, it is generally harmful. Yet mutations are occasionally useful. In the big picture, they are *vital* to living things, given their struggle to get along in environments that are constantly changing. If environments change, species need to change too, in order to survive. And the only way major changes can come about is through mutations. You'll be looking at this topic again, but for now, isn't it interesting to ponder the fact that the living world adapts partly through its mistakes?

How Did We Learn?

It may be clear to us today that genes are information-bearing units and that the information they contain is used to produce proteins. But 60 years ago, things were not so clear. Indeed, scientists had to work hard to attain this insight. To find out more about their efforts, see "Getting Clear about What Genes Do" on page 264.

On to How Genetic Information Is Put to Use

Back in Chapter 9, you saw that since antiquity, human beings have speculated about how one generation of living things can give rise to another—about *how* it is that life goes on. The questions posed by the Greeks, by scientists in the Renaissance, by Mendel and Morgan and others were finally answered in the 1950s and 1960s, when scientists came to understand DNA replication at the molecular level. The something-old, something-new quality to this replication—parental DNA strands serving as templates for new strands—was the detailed answer to the ancient question of how the qualities of living things can be passed down through generations.

Splendid as this function is, it is only one of the two great tasks carried out by our genetic machinery. You have just reviewed the process by which genetic information is replicated; now let's look at the process by which this information is used. How can a stretch of DNA bring about the production of a protein? You'll see in the chapter coming up.

Getting Clear about What Genes Do: Beadle and Tatum

In the 1930s, biologists were confused not only about what genes were made of (DNA) but also about what it was that genes did. We now know that a large part of what they do is contain the information for the production of proteins, including the reaction-hastening proteins called *enzymes*. But it took work over a period of years by Stanford University researchers George Beadle and Edward Tatum to make this clear. Their experimental subject was a red fungus, the bread mold *Neurospora*.

Following the Growth of a Fungus

Research scientists are always looking for ways to control the confusing conditions of nature in order to tease out the rules it plays by. Beadle and Tatum's control mechanism was, first, to *induce* genetic mutations in *Neurospora* spores by bombarding them with x-rays. These mutated spores were then transferred one at a time into test tubes, where the second element of control came in. Natural or "wild type" *Neurospora* can manufacture most of what they need to live, given just a few basic biological molecules in their surroundings. But could the descendants of the *mutated* group of *Neurospora* do this? Could they be put in a so-called minimal test tube medium—a liquid environment that lacked the biological molecules the fungus normally has access to—and still multiply into a thriving colony? The answer for many of the mutant strains was no. The mutations these organisms had undergone had rendered

them unable to produce some substance that they needed to thrive. By being selective about *adding* substances to the minimal media, Beadle and Tatum could tell what the missing substance was. (If they added the right thing, the colony would grow, as you can see in **Figure 1**.) For some of the mutant strains, the missing substance turned out to be the amino acid arginine.

We now know that arginine is produced in one of the metabolic pathways reviewed in Chapter 6. In such a pathway, a precursor compound is modified several times in a

> *In the 1930s, biologists were confused not only about what genes were made of but also about what genes did.*

chain, with a different enzyme facilitating each step of the operation. By selectively supplying different strains of mutant spores with arginine's *precursor* substances, Beadle and Tatum could discern the step at which the arginine metabolic pathway had been blocked. One strain required only precursor A to produce arginine, for example, while another strain required precursors A and B. In the tradition of T. H. Morgan, the researchers linked these blockages to genes on three separate *Neurospora* chromosomes. The summary result then was clear. A given *Neurospora* strain had become a mutant by having

Chapter Review

Summary

13.1 What Do Genes Do, and What Are They Made of?

- James Watson and Francis Crick discovered the chemical structure of DNA in 1953. This event ushered in a new era in biology because it allowed researchers to understand some of the most fundamental processes in genetics.

- In trying to decipher the structure of DNA, Watson and Crick were performing work in molecular biology. This is the investigation of life at the level of the constituent parts of its individual molecules. Molecular biology has grown greatly in importance since the 1950s.

13.2 Watson and Crick: The Double Helix

- Watson and Crick met in the early 1950s at Cambridge University in England and set about to decipher the structure of DNA. Their research was aided by the work of others, including Rosalind Franklin, who was using x-ray crystallography to learn about DNA's structure.

13.3 The Components of DNA and Their Arrangement

- The DNA molecule is composed of building blocks called nucleotides, each of which consists of one sugar (deoxyribose), one phosphate group, and one of four bases: adenine, guanine, thymine, or cytosine (A, G, T, or C). The sugar and

its genes altered through radiation. As a result, it no longer carried out the transformation of, say, precursor A to precursor B. It took an *enzyme* to facilitate this transformation. What genes were doing, then, was bringing about the production of enzymes.

The One-Gene, One-Enzyme Hypothesis

Eventually, a famous label was applied to these experimental results: The *one-gene, one-enzyme hypothesis*. What each gene does is call up production of one enzyme, which undertakes a particular metabolic task. Though this notion is now dated, the essential insight was not only correct, it was critical. With it, scientists at last understood what genes were doing. For this work, which they completed in 1941, Beadle and Tatum shared in the Nobel prize in 1958.

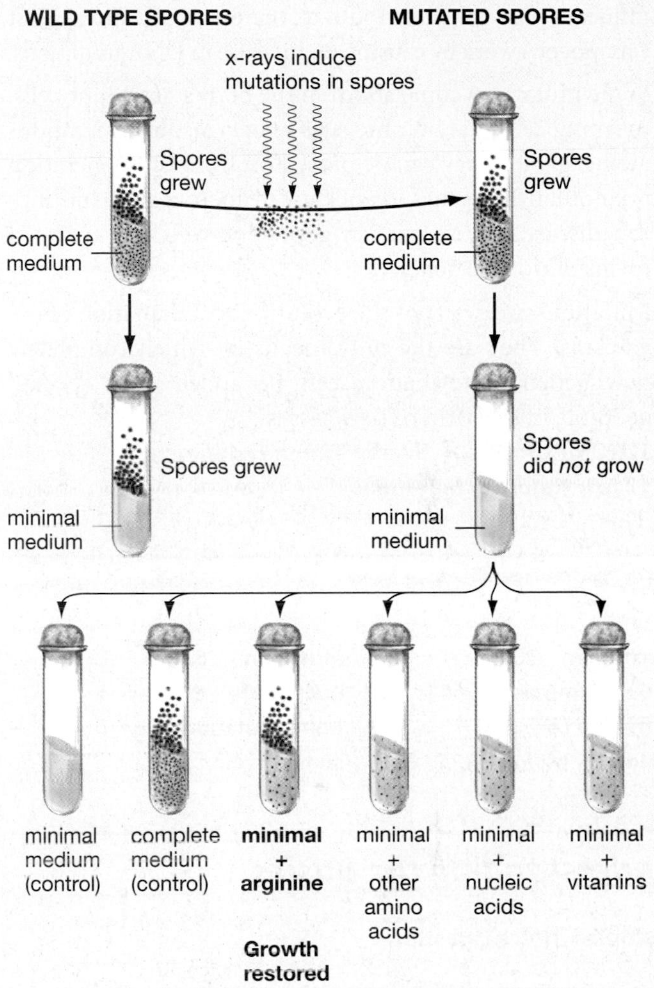

Figure 1
Beadle and Tatum Experiment with Mutated Spores
Using spores of the fungus *Neurospora*, Beadle and Tatum found that some mutated versions of the spores could not grow in a medium that lacked a basic set of organic molecules. By selectively adding molecules to this "minimal medium," they demonstrated that the mutation was preventing the spores from producing certain compounds, among them the amino acid arginine. By tracing the metabolic pathway leading to arginine, Beadle and Tatum realized that the *Neurospora* genes that had been made dysfunctional were coding for enzymes that facilitated the steps leading from the precursors of arginine to arginine itself.

Tutorial 13.2.1
Beadle-Tatum
Experiment

phosphate groups are linked together in a chain that forms the "handrails" of the DNA double helix. Bases then extend inward from the handrails, with base pairs joined to each other in the middle by hydrogen bonds. In this base pairing, A always pairs with T, while G always pairs with C.
TUTORIAL 13.1.2: Structure of DNA

- DNA is copied by means of each strand of DNA serving as a template for the synthesis of a new, complementary strand. The DNA double helix first divides down the middle. Each A on an original strand then specifies a place for a T in a new strand, each G specifies a place for a C on the new strand, and so forth.
TUTORIAL 13.1.3: Replication of DNA

- Each double helix produced in replication is a combination of one parental strand of DNA and one newly synthesized complementary strand. This is how life builds upon itself. A group of enzymes known as DNA polymerases are central to

DNA replication; they move along the double helix, bonding together new nucleotides in complementary DNA strands.

- DNA can encode the information for the huge number of proteins utilized by living things because the sequence of bases along DNA's handrails can be laid out in an extremely varied manner. A collection of bases in one order encodes the information for one protein, while a different sequence of basis encodes the information for a different protein.

13.4 Mutations: Another Name for a Permanent Change in DNA Structure

- The error rate in DNA replication is very low, partly because DNA polymerases are able to correct mistakes in base pairs. When such mistakes are made and then not corrected, the result is a mutation, meaning a permanent alteration in a cell's DNA base sequences.

- Most mutations have no effect on an organism; when they do have an effect, it is generally negative. Cancers result from a line of cells that have undergone types of mutations that cause them to proliferate wildly. Mutations occur regularly in all organisms as random, spontaneous events. The mutation rate of an organism can, however, be affected by environmental influences. A carcinogen is an environmental influence that can contribute to the onset of cancer. Most carcinogens work by causing an increase in DNA mutations.

- Some mutations come about in the body's germ-line cells, meaning cells that become eggs or sperm. Such mutations are heritable: They can be passed on from one generation to another. The genes for sickle-cell anemia and Huntington disease, which exist in germ-line cells, are mutated forms of normal genes.

- Mutations are a part of the evolutionary adaptation of organisms. They are the only means by which completely new genetic information can be added to a species' genome, in the form of new alleles.

  TUTORIALS: 13.2.1 Beadle-Tatum Experiment; 13.3.1 Nature of Mutations

Key Terms

carcinogen 262	**mutation** 261
DNA polymerase 260	**nucleotide** 260
germ-line cell 262	**point mutation** 261
molecular biology 256	**somatic cell** 262

Understanding the Basics

Multiple-Choice Questions

1. The structure of DNA was determined in _____ by Watson and Crick.
   a. 1943
   b. 1953
   c. 1921
   d. 1973
   e. 1962

2. The type of biological study that was exemplified through the discovery of DNA structure is called
   a. detailed biology
   b. component biology
   c. microscopic biology
   d. elemental biology
   e. molecular biology

3. The components of the DNA handrails are _____ and _____.
   a. nucleotides, phosphates
   b. sugars, nucleotides
   c. sugars, bases
   d. sugars, phosphates
   e. none of the above

4. Which of the following combinations is an example of a nucleotide?
   a. sugar + phosphate + adenine
   b. sugar + phosphate + guanine
   c. sugar + phosphate + thymine
   d. sugar + phosphate + cytosine
   e. all of the above

5. Which of the following DNA sequences contains a mutation?
   a. TCAA
      AGTT
   b. CAGC
      GTCG
   c. GATA
      CTCT
   d. AATT
      TTAA
   e. GACG
      CTGC

6. In DNA replication, the parent strand of DNA is called the _____ strand, while the daughter strand is called the _____ strand.
   a. template, complementary
   b. pattern, mirror
   c. template, opposite
   d. complementary, template
   e. opposite, mirror

7. Though the DNA structure could not be detected by a microscope, Rosalind Franklin aided in its discovery by using
   a. electron microscopy
   b. x-ray crystallography
   c. mass spectrometry
   d. high-pressure liquid chromatography
   e. a combination of b and c

8. DNA polymerases work to
   a. unwind the double helix during replication
   b. cause mutations in DNA
   c. join together nucleotides in a growing, complementary DNA strand
   d. correct errors in DNA base sequences during replication
   e. both c and d

9. Which of the following processes or substances is believed to have contributed the most to evolutionary adaptation?
   a. "normal" recombination or crossing over
   b. DNA polymerase
   c. cytokinesis
   d. mutation
   e. chromosome condensation

10. How are the bases of the DNA double helix linked together?
    a. phosphate bonds
    b. hydrogen bonds
    c. carbon–carbon bonds
    d. sugar bonds
    e. basic bonds

Brief Review

1. How does the phrase "something old, something new" describe the method of DNA replication employed by the cell?

2. Give one example from this chapter of a human disease that is caused by a heritable mutation.

3. True or False: DNA polymerases work only in a forward direction.

4. True or False: All mutations are detrimental.

5. Why can such a simple set of materials (bases A, T, C, and G) encode for every protein of our body?

6. Give an example of a point mutation that produces an allele that results in a disease.

Applying Your Knowledge

1. Given the following DNA sequence, determine the complementary strand that would be added in replication:

   ATTGCATGATAGCC

2. Why does a germline mutation carry greater potential significance than a somatic mutation?

3. A gene's nucleotide bases are known to be composed of 15 percent guanine. What percentages of each of the other three bases are contained in the same gene?

4. Why is it important that DNA replication have a low error rate?

5. How do we know that DNA replicates only once in every cell cycle, and not many times?

6. Would you expect cancer to arise more often in types of cells that divide frequently (such as skin cells) or in types of cells that divide rarely or not at all (such as nerve cells)? Explain your reasoning.

14

How Proteins Are Made
Genetic Transcription, Translation, and Regulation

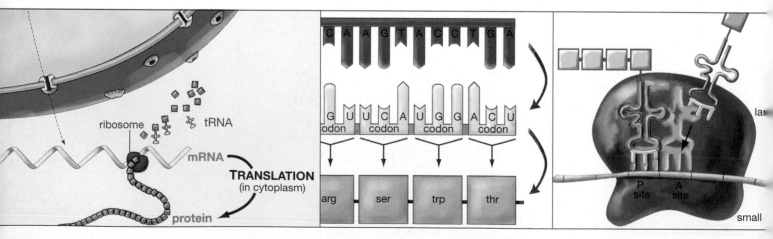

Transferring information.
(Section 14.2, page 270)

The triplet code.
(Section 14.2, page 270)

A ribosome.
(Section 14.4, page 276)

Like workers using a set of blueprints to construct a house, cells use the information contained in DNA to construct proteins. Each cell can finely tune its production of proteins in accordance with its needs.

The scientific advance that was reviewed in Chapter 13, the discovery of the structure of DNA in 1953, bears some comparison to the birth of Napoleon Bonaparte: Both events had momentous consequences, but not for a while. Only slowly did the discoveries come that, in time, were seen as hinging on James Watson and Francis Crick's work. So significant were these advances, however, that the years between 1953 and 1966 are now regarded by some scholars as a kind of golden age of genetics; a time when basic findings about this field came fast and furious, such that by the mid-1960s scientists had at last grasped the essentials of how the genetic machinery works.

These discoveries were made in connection with both of the two great tasks of genetics: On the one hand, the copying or "replication" of DNA that you reviewed in Chapter 13; on the other, the means by which DNA's genetic information brings about the production of proteins. The first task has to do with how genetic information is preserved; the second with how genetic information is utilized. It is the latter task that is the subject of this chapter.

14.1 The Structure of Proteins

Because you'll be dealing extensively with proteins here, now may be a good time to review some basic information about them. As you saw in Chapter 3, proteins fit into the "building blocks" model of biological molecules. The blocks in this case are amino acids. String a number of these together and you have a **polypeptide** chain, which then folds up in a specific three-dimensional manner, resulting in a protein. Proteins are likely to be made of hundreds of amino acids strung together and folded up. It is the three-dimensional shape of proteins that gives them their working ability—generally speaking, the ability to bind with other molecules in very specific ways. Many proteins are composed of several linked polypeptide chains.

Synthesizing Many Proteins from 20 Amino Acids

Though there are tens of thousands of different proteins, all of them are put together from a starting set of a mere 20 amino acids. If you wonder how such diversity can proceed from such simplicity, think of the English language,

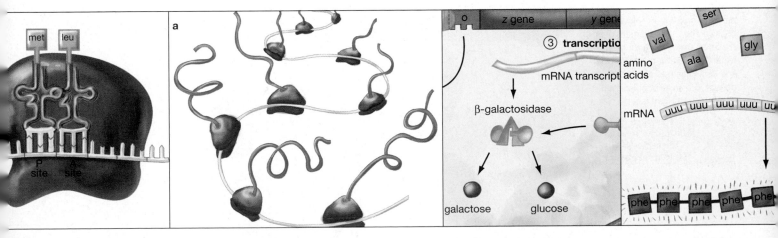

One step in a chain.
(Section 14.4, page 276)

Mass production.
(Section 14.4, page 276)

Genetic regulation in action.
(Section 14.5, page 281)

Cracking the code.
(Essay, page 285)

which has thousands of words, but only 26 letters in its alphabet. It is the *order* in which letters occur that determines whether, for example, "cat" or "act" is spelled out; just so, it is the order of amino acids that determines what protein is synthesized.

If you look at **Figure 14.1a**, you can see the chemical structure of two free-standing amino acids, glycine (gly) and isoleucine (ile). If you look at **Figure 14.1b**, you can see that these two amino acids are, in fact, the first two that occur in one of the two polypeptide chains that make up the unusually small protein insulin. **Figure 14.1c** then shows a three-dimensional representation of insulin—the "folded up" form this protein assumes before taking on its function of moving blood sugar into cells. A list of all 20 primary amino acids and their three-letter abbreviations can be found in **Table 14.1**.

For our purposes, the question is: How do the chains of amino acids that make up a protein come into being? How do gly, ile, val, and so forth come to be strung together in a specific order to create this protein? You know from what you've studied so far that genes "order up" proteins, and that genes amount to a series of chemical "bases"—the A's, T's, C's, and G's along a DNA strand. How, then, does *this* series of DNA bases come to specify *that* series of amino acids?

14.2 Protein Synthesis in Overview: Transcription and Translation

In overview, this process can be described fairly simply, as those of you who have read Chapter 9 know. The DNA just referred to is contained in the nucleus of the cell. The first

Tutorial 14.1.1
Amino Acids

Figure 14.1
The Structure of Proteins

a The building blocks of proteins are amino acids such as glycine and isoleucine, which differ only in their side-chain composition (white background).

b These amino acids are strung together to form polypeptide chains. Pictured is one of the two polypeptide chains that make up the unusually small protein, insulin.

c Polypeptide chains function as proteins only when folded into their proper three-dimensional shape, as shown here for insulin. Note the position of the glycine and isoleucine amino acids in one of the insulin polypeptide chains (colored light green).

Tutorial 14.1.2
Higher Level (3-D)
Structure of Proteins

a

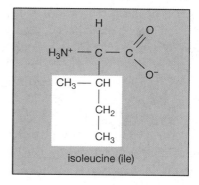

glycine (gly)

isoleucine (ile)

b

H₃N⁺– | gly | ile | val | glu | gln | cys | cys | ala | ser | val | cys | ser | leu | tyr | gln | leu | glu | asn | tyr | cys | asn |–C

c

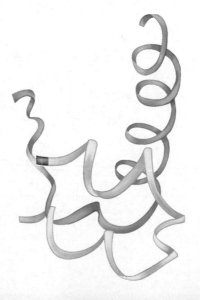

step is that a stretch of it unwinds there, and its message—the order of a string of its bases along one of its strands—is copied onto a molecule called messenger RNA (mRNA). This length of mRNA, which could be thought of as an information tape being dubbed off a "master" DNA tape, then exits from the cell nucleus (**Figure 14.2**).

The destination of this mRNA is a molecular workbench in the cell's cytoplasm, a structure called a *ribosome*. More formally, a **ribosome** is an organelle ("tiny organ"), located in the cell's cytoplasm, that is the site of protein synthesis. It is at the ribosome that both message (the mRNA tape) and raw materials (amino acids) come together to make the product (a protein). As the mRNA tape is "read" within the ribosome, something grows

from it: a chain of amino acids that have been linked together in the ribosome in the order specified by the mRNA. When the chain is finished and folded up, a protein has come into existence. And how do amino acids get to the ribosomes? They are brought there by a *second* type of RNA, transfer RNA (tRNA).

As may be apparent, protein synthesis divides neatly into two sets of steps. The first set is called **transcription**: the process by which the genetic information encoded in DNA is copied onto messenger RNA. The second set is called **translation**: the process by which information encoded in messenger RNA is used to assemble a protein at a ribosome. Let's look in more detail now at both of these processes, starting with transcription.

Table 14.1 Amino Acids	
Amino Acid	**Abbreviation**
alanine	ala
arginine	arg
asparagine	asn
aspartic acid	asp
cysteine	cys
glutamine	gln
glutamic acid	glu
glycine	gly
histidine	his
isoleucine	ile
leucine	leu
lysine	lys
methionine	met
phenylalanine	phe
proline	pro
serine	ser
threonine	thr
tryptophan	trp
tyrosine	tyr
valine	val

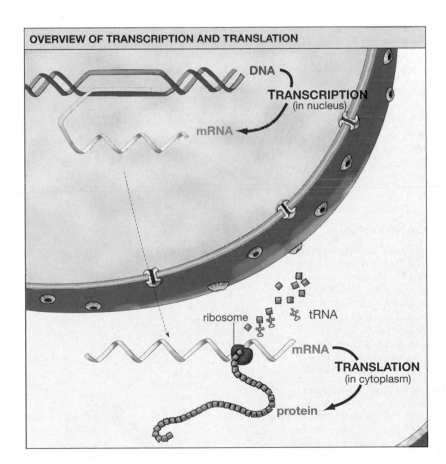

Figure 14.2
The Two Major Stages of Protein Synthesis
In transcription, a section of DNA unwinds and nucleotides on it form base pairs with nucleotides of messenger RNA, creating an mRNA "tape." This segment of mRNA then leaves the cell nucleus, bound for a ribosome in the cell's cytoplasm, where translation takes place. Joining the mRNA tape at the ribosome are amino acids, brought there by transfer RNA molecules. The length of messenger RNA is then "read" within the ribosome. The result? A chain of amino acids is linked together in the order specified by the mRNA tape. When the chain is finished and folded up, a protein has come into existence.

The First Stage of Protein Synthesis: Transcription

From what's been reviewed so far, you can see that a key player in transcription (and translation) is RNA, whose full name is ribonucleic acid.

An Important Player: RNA

If you look at **Figure 14.3**, you can see just how similar RNA is to DNA. For one thing, RNA has the sugar and phosphate "handrail" components you saw in DNA last chapter. Then, in both molecules, this two-part structure is joined to a third element, a base.

There are, however, differences between DNA and RNA. For one, RNA is usually single-stranded, whereas DNA is structured in two strands that form its famous double helix. Beyond this, recall that any given DNA nucleotide has one of four bases: adenine, guanine, cytosine or thymine (A, G, C, or T). RNA utilizes the first three of these, but then substitutes uracil (U) for the thymine (T) found in DNA.

Passing on the Message: Base Pairing Again

Given the chemical similarity between DNA and RNA, it's not hard to see how DNA's genetic message can be passed on to messenger RNA: Base pairing is at work again. Recall from Chapter 13 how DNA is replicated. The double helix is unwound, after which bases (the A's, T's, C's, and G's) along the now-single DNA strands pair up with *complementary* DNA bases from among free-floating nucleotides. Every T on a single DNA strand pairs with a free-floating A, and every C links up with a G, thus yielding a complementary DNA *chain*. Because of RNA's similarity to DNA, the bases RNA has can *also* form base pairs with DNA. The twist to this RNA-DNA base pairing is that each *A* on a DNA strand links up with a *U* on the RNA strand, instead of the T that would be A's partner in DNA-to-DNA base pairing. Thus, complementary base pairing is the means by which a DNA message is transferred to RNA. The length of mRNA that results from this process of transcription is called a **transcript** (see **Figure 14.4**). With all this in mind, here's a more formal definition of **messenger RNA (MRNA)**: a type of RNA that encodes, and carries, from the cell's nucleus to the cell's ribosomes, information for the synthesis of proteins.

A Triplet Code

Thus far, you have been looking at what might be called the flow of genetic information (from DNA to mRNA to ribosomes). A separate issue in protein synthesis, however, has to do with the linkage between DNA on the one hand and amino acids on the other. Think of this as scientists did in the 1950s. Allowing that DNA bases are coding for amino acids, a

Figure 14.3
RNA and DNA Compared

a Each building-block nucleotide of both DNA and RNA is composed of a phosphate group, a sugar—ribose in RNA, deoxyribose in DNA—and one of four bases. RNA and DNA both utilize the bases adenine, guanine, and cytosine (A, G, and C), but RNA utilizes the base uracil (U) instead of the thymine (T) that DNA utilizes.

b Both RNA and DNA amount to linked chains of these nucleotides. The "handrails" of both RNA and DNA are composed of the sugar molecule of one nucleotide linked to the phosphate molecule of the next nucleotide (thus the S-P-S-P labeling). The "stair steps" stemming from the handrails are formed by the bases of the nucleotides, as with the G and C bases extending inward at the top of the DNA strand. While DNA is double-stranded, RNA generally is single-stranded.

RNA nucleotide

a

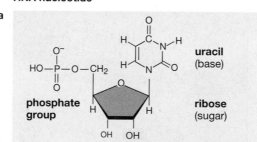

uracil (base)
phosphate group
ribose (sugar)

DNA nucleotide

thymine (base)
phosphate group
deoxyribose (sugar)

b

RNA strand

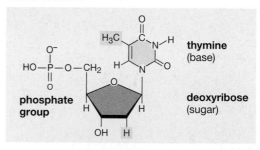

Sugar-phosphate handrail

Bases:
cytosine (C)
guanine (G)
adenine (A)
uracil (U)

DNA strand

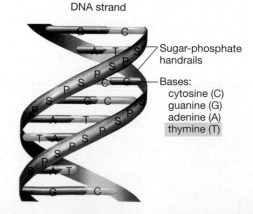

Sugar-phosphate handrails

Bases:
cytosine (C)
guanine (G)
adenine (A)
thymine (T)

1. A region of the DNA double helix is unwound.

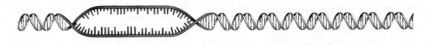

2. RNA nucleotides begin assembling on the DNA template.

Figure 14.4
Transcription Works through Base Pairing
Thanks to their chemical similarity, DNA and RNA can engage in base pairing, and this base pairing is how messenger RNA transcripts are synthesized. (1) A region of the DNA double helix is unwound. (2) Individual RNA nucleotides begin base pairing with complementary DNA nucleotides. (As you'll see later, an enzyme called RNA polymerase both unwinds the DNA and adds each RNA nucleotide to the growing mRNA chain.) (3) As transcription proceeds, the completed portion of the mRNA chain separates from the DNA strand. The region of DNA that has been transcribed is wound back up, even as DNA "downstream" from the transcription site is unwound, thus making more DNA nucleotides available for base pairing. (4) Transcription is ended when the finished mRNA transcript is cleaved from the DNA strand. The process is completed when the DNA is rewound into its original form.

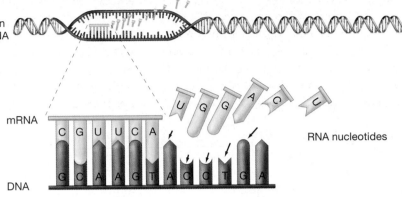

3. The completed portion of the mRNA transcript separates from the DNA. Meanwhile, more of the untranscribed region of the DNA is unwound.

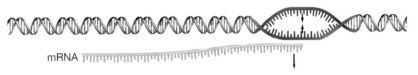

4. Transcription is completed. The mRNA transcript is released from the DNA, and the DNA is rewound into its original form.

Figure 14.5
Triplet Code
Each triplet of DNA bases codes for a triplet of mRNA bases (a codon), but it takes a complete codon to code for a single amino acid.

primary question that arose then was: how *many* DNA bases does it take to code for each amino acid in a protein? It seemed clear that the answer was not one. Because there are only four DNA bases (A, T, C, and G), if each coded for its own amino acid, only four amino acids could have been incorporated into proteins, as opposed to the 20 that actually are. Nor could the answer be two, because the number of possible amino acids this number could yield up was only 16. Thinking that nature probably would work as economically as possible, the South African-born biochemist Sydney Brenner suggested that DNA worked in a **triplet code**, meaning that each *three* DNA bases specified a single amino acid.

Brenner turned out to be right. As **Figure 14.5** shows, each three bases in a DNA sequence pairs with three mRNA bases, but each group of three mRNA bases then codes for a *single* amino acid. Each coding triplet of mRNA bases is known, appropriately enough, as a **codon**.

The Second Stage in Protein Synthesis: Translation

The completion of transcription in protein synthesis leaves an mRNA transcript arriving at the ribosomes—but with no protein synthesized from it yet. A protein does come together, however, in the second stage in synthesis noted

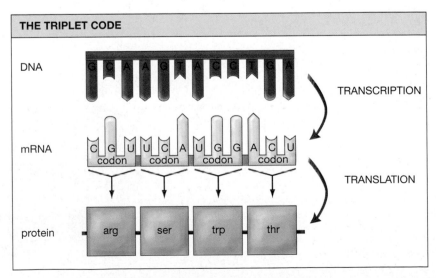

earlier, translation. Ribosomes are the molecular workbenches at which two ends of the protein-synthesizing process come together. On the one hand, the genetic message comes to the ribosomes in the form of the messenger RNA tape. On the other, amino acids must be brought to the ribosome to be strung together in the order specified by the tape. As noted, the molecule that brings the amino acids to ribosomes is a second form of RNA, transfer RNA (or tRNA).

Aptly Named: Transfer RNA

If you look now at **Figure 14.6**, you can see why this molecule is aptly placed within the translation phase of protein synthesis. When, in common discourse, we think of a *translator*, we think of someone who can communicate in *two* languages. Transfer RNA effectively does this. One end of each tRNA molecule links with a specific amino acid, which it finds floating free in the cytoplasm. Then, transferring this amino acid to the ribosome, this tRNA molecule employs its *opposite* end to form base pairs with *nucleic* acids—with a codon on the mRNA tape that is being "read" inside the ribosome. Thus tRNA is the translator between the molecular languages of amino acids and nucleic acids. **Transfer RNA** can be defined as a form of RNA that, in protein synthesis, bonds with amino acids, transfers them to ribosomes, and then bonds with messenger RNA.

Though there's many a detail to go, with this you've arrived at the essence of protein synthesis. Each tRNA carries a specific amino acid to a ribosome. There, three bases in the tRNA molecule form base pairs with three bases in the mRNA transcript (a codon). As a *succession* of mRNA codons is read in the ribosome, a corresponding succession of amino acids is brought to the ribosome. In this process, these amino acids are linked together end to end to form a growing polypeptide chain—in the order specified by the mRNA transcript. And, of course, this transcript was in turn specified originally by the DNA sequence back in the cell's nucleus. Thus you can see the truth of a saying James Watson once coined regarding protein synthesis: "DNA makes RNA makes protein."

14.3 The Importance of the Genetic Code

Very shortly, you'll look at this process of protein synthesis in greater detail. Before that, however,

Figure 14.6
Bridging Molecule
Transfer RNA (tRNA) molecules link up with amino acids on the one hand and mRNA codons on the other, thus forming a chemical link between the two kinds of molecules in protein synthesis. They also transfer amino acids to ribosomes, as shown in the steps of the figure.

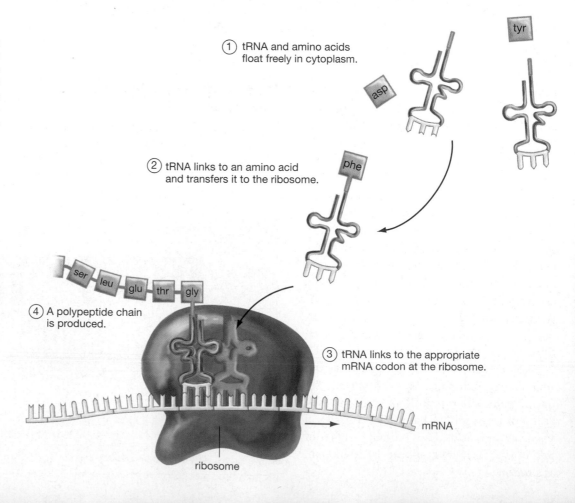

① tRNA and amino acids float freely in cytoplasm.

② tRNA links to an amino acid and transfers it to the ribosome.

④ A polypeptide chain is produced.

③ tRNA links to the appropriate mRNA codon at the ribosome.

mRNA

ribosome

it will be helpful to learn a little more about something only briefly touched on—the genetic code.

Cracking the Code

Earlier, it was noted that the genetic code is a triplet code, each three messenger RNA bases specifying a single amino acid. Scientists knew this by the 1960s. But the question that then confronted them was: If you have a given three bases, *which* amino acid do they specify? In Morse code, the sounds .__. (short-long-short) code for the letter P. But if we know that a given RNA codon has the base sequence UCC, what *amino acid* does this code for? Today we know that the answer is serine (ser), but at one time this was not clear at all. What scientists had to do was decipher the **genetic code**, which can be defined as the inventory of linkages between nucleotide triplets and the amino acids they code for. If you look at "Cracking the Genetic Code" on page 285, you can get an idea of how scientists solved this mystery. **Figure 14.7** shows the genetic code in its entirety. Now, a couple of notes about its nature and importance.

A Redundant Code

One of the notable things about the genetic code is that not every mRNA triplet in it codes for an amino acid. Note, in Figure 14.7, that three mRNA codons specify "stop" codes that bring an end to the synthesis of a polypeptide chain. One triplet (AUG) specifies a special amino acid, methionine (met), which serves as the starting amino acid for most polypeptide chains. Note also that the genetic code is redundant. The amino acid phenylalanine (phe), for example, is coded for not only by the UUU triplet, but by UUC as well. Indeed, almost all the amino acids are coded for by more than one mRNA codon. Leucine (leu) is coded for by no fewer than six.

The Genetic Code and Life's Unity

You have read a good deal in this book about the diversity and unity of life. The genetic code is a sterling example of the unity. With only a few exceptions, the genetic code is universal in all living things. This means that the base triplet CAC, for example, codes for the amino acid histidine, whether this coding is going on in a bacterium or in a human being.

Why is this important? First, it is evidence that all life on Earth is derived from a single ancestor. How else can we explain all of life's diverse organisms sharing this very specific code? Such a complex molecular linkage—this triplet equals that amino acid—is extremely unlikely to have evolved *more* than once. What seems likely is that it came to be employed in an ancient common ancestor, after which it was passed on to all of this organism's descendants—every creature that has subsequently lived on Earth. Note, then, that this code has been passed on from one generation to the next, and from one evolving species to the next, for billions of years. We humans share in an informational linkage stretching back billions of years—and running through all the contemporary living world.

Apart from this, the universality of the genetic code has a very practical consequence. It means that genes from one organism can function in another. This has both good and bad consequences for human beings. First the bad: Viruses that cause diseases ranging from colds to AIDS can "hijack" the human cellular machinery for their own purposes, precisely because their genes function in human cells. (The human cellular machinery will put together proteins whether these proteins are called for by human or viral DNA.) Now the good news: Using "biotechnology" processes you'll be learning about in Chapter 15, human beings can today use viruses and bacteria to manufacture all kinds of products, including medicines such as human insulin and human growth hormone. In these cases, human genes are being put to work inside microorganisms.

First Base	Second Base: U		C		A		G		Third Base
U	UUU	phe	UCU	ser	UAU	tyr	UGU	cys	U
	UUC	phe	UCC	ser	UAC	tyr	UGC	cys	C
	UUA	leu	UCA	ser	UAA	**Stop**	UGA	**Stop**	A
	UUG	leu	UCG	ser	UAG	**Stop**	UGG	trp	G
C	CUU	leu	CCU	pro	CAU	his	CGU	arg	U
	CUC	leu	CCC	pro	CAC	his	CGC	arg	C
	CUA	leu	CCA	pro	CAA	gln	CGA	arg	A
	CUG	leu	CCG	pro	CAG	gln	CGG	arg	G
A	AUU	ile	ACU	thr	AAU	asn	AGU	ser	U
	AUC	ile	ACC	thr	AAC	asn	AGC	ser	C
	AUA	ile	ACA	thr	AAA	lys	AGA	arg	A
	AUG	met (start)	ACG	thr	AAG	lys	AGG	arg	G
G	GUU	val	GCU	ala	GAU	asp	GGU	gly	U
	GUC	val	GCC	ala	GAC	asp	GGC	gly	C
	GUA	val	GCA	ala	GAA	glu	GGA	gly	A
	GUG	val	GCG	ala	GAG	glu	GGG	gly	G

Figure 14.7
The Genetic Code Dictionary
If we know what a given mRNA codon is, how can we find out what amino acid it codes for? This dictionary of the genetic code offers a way. In Figure 14.5, you saw that the codon CGU coded for the amino acid arginine (arg). Looking that up here, C is the first base (go to the C row along the "first base" line), G is the second base (go to the G column under the "second base" line) and U is the third (go to the codon parallel with the U in the "third base" line).

Figure 14.8
Editing Out Genetic Material
Special enzymes remove the noncoding regions of mRNA (introns) and splice the coding regions (exons) back together to create an edited transcript.

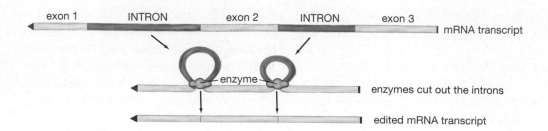

In other cases, genes from microorganisms are being inserted in, for example, plants to yield hardier, healthier strains of crops, such as rice and soybeans. This transferability of genes from one species to another has, as its basis, the universality of the genetic code.

14.4 A Closer Look at Protein Synthesis

Thus far, you've reviewed in outline the means by which DNA codes for proteins. Now this process is described in somewhat greater detail.

Protein Synthesis Begins: Transcription

Gene transcription begins with a section of the DNA double helix unwinding, after which one of the now-single DNA strands serves as a template for creation of a messenger RNA strand (see Figure 14.4). It will come as no surprise to learn that an enzyme is critically involved in this process. **RNA polymerase**, as this enzyme is known, undertakes both of these initial tasks: unwinding the DNA sequence and stringing together a chain of RNA nucleotides that is complementary to it. This is the messenger RNA "tape" that eventually will move to the ribosome.

Messenger RNA Processing

It turns out that, in eukaryotes such as ourselves, the mRNA tape is not functional from the moment it is synthesized. Instead, it must undergo some editing before it is a finished product. What form does this mRNA editing take? After some initial steps, the most important part of it is an operation that bears comparison to film editing. When directors put movies together, they first assemble footage of all the scenes they have shot, sequenced in their proper order. Then they go to work with "scissors and paste." In looking at the work, a director may decide that a given scene is useless in its entirety, in which case it is snipped out. Then the film segment that comes before this scene and the film segment that comes after it are *spliced together*.

Introns and Exons

This is roughly what happens with eukaryotic mRNA. As you can see in **Figure 14.8**, special molecular splicing complexes cut out sections of the mRNA transcript. Then the remaining portions of the mRNA tape are spliced together. The portions of the transcript that are cut out—portions called introns—do not code for amino acids. Eventually, the transcript will contain no introns at all. What remains are exons, which are

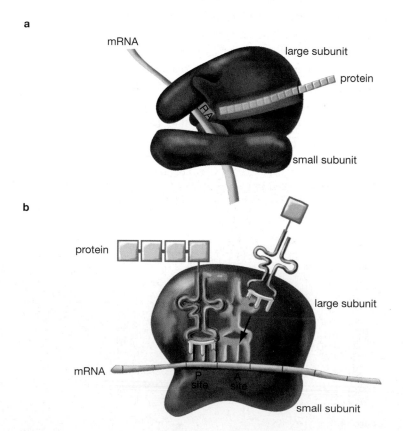

Figure 14.9
The Structure of Ribosomes

a Ribosomes are composed of two subunits that come together during translation.

b A simplified cross section of the ribosome illustrates the "P" and "A" sites where tRNA molecules bind during translation.

portions of the mRNA transcript that *do* code for amino acids. The terms *intron* and *exon* can also be used to refer to DNA sequences. This makes sense, because each coding DNA sequence with introns in it naturally leads to a corresponding mRNA sequence with complementary introns. Thus an **intron** can be defined as a segment of DNA, or the messenger RNA transcript complementary to it, that does not encode information for sequencing amino acids. An **exon**, meanwhile, can be defined as a segment of DNA, or the messenger RNA transcript complementary to it, that does encode information for sequencing amino acids. The intron and exon names are derived from their function—or lack of it. Introns are *in*tervening sequences of mRNA or DNA, while exons are *ex*pressed as amino acids. The obvious question that follows from this is: If introns don't code for anything, what are they doing in DNA in the first place? As you can see in "Making Sense of 'Junk' DNA" on page 280, clues are steadily emerging about the function of these sequences.

Translation in Detail

With editing concluded, the finished mRNA transcript moves out into the cell's cytoplasm, where it is destined to link up with a ribosome for the second phase of protein synthesis. As you've seen, another key player must also make its way to these ribosomal sites for this work to be completed: transfer RNA molecules, each one of which comes with an amino acid in tow. Ribosomes act in one sense as a "reading head" in a tape player, through which the mRNA tape will pass.

The Structure of Ribosomes

If you look at **Figure 14.9**, you can begin to get an idea of the structure of ribosomes. Note that they are composed of two "subunits"—one larger than the other—both made of a mixture of proteins and yet another type of RNA, **ribosomal RNA** (or **rRNA**). The two subunits generally float apart from another in the cytoplasm until prompted to come together by the process of translation. You can see that when the subunits have been joined, there exist two binding sites (which it is convenient to think of as slots), one of them a "P" site, the other an "A" site. You'll be looking at their roles shortly. **Table 14.2** reviews three types of RNA that are active in protein synthesis.

The Structure of Transfer RNA

Figure 14.10 provides a more detailed look at the transfer RNA molecule. The role of tRNA as a "translator" between the molecular languages of nucleic acids (RNA) and amino acids was noted earlier. Figure 14.10 gives you an idea of how this one molecule can serve as this kind of bridge. One end of the molecule

Table 14.2 Types of RNA		
Type of RNA	**Functions in**	**Function**
Messenger RNA (mRNA)	Nucleus, migrates to ribosomes in cytoplasm	Carries DNA sequence information to ribosomes
Transfer RNA (tRNA)	Cytoplasm	Provides linkage between mRNA and amino acids; transfers amino acids to ribosomes
Ribosomal RNA (rRNA)	Cytoplasm	Structural component of ribosomes

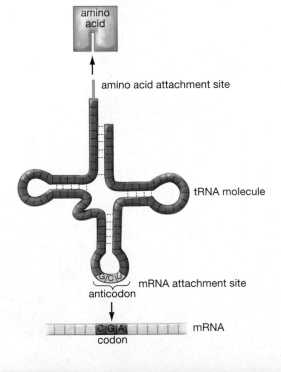

amino acid

amino acid attachment site

tRNA molecule

G/C/U
anticodon

mRNA attachment site

C G A
codon

mRNA

**Figure 14.10
The Two-Dimensional
Structure of Transfer RNA**
One end of the tRNA molecule binds to a specific amino acid, while the other end binds to a specific mRNA codon.

Tutorial 14.3.4
Protein Translation

Figure 14.11
The Steps of Translation

1. A messenger RNA transcript binds to the small subunit of a ribosome as the first transfer RNA is arriving. The mRNA codon AUG is the "start" sequence for most polypeptide chains. The tRNA, with its methionine (met) amino acid attached, then binds to this AUG codon.

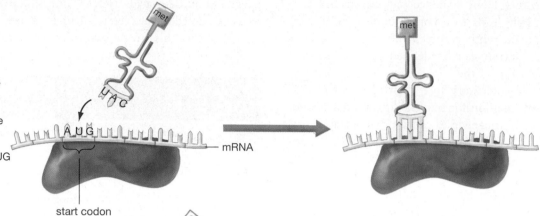

2. The large ribosomal subunit joins the ribosome, as a second tRNA arrives, bearing a leucine (leu) amino acid. The second tRNA binds to the mRNA chain, within the ribosome's A site.

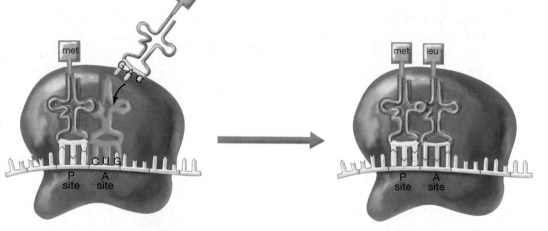

3. A bond is formed between the newly arrived leu amino acid, and the met amino acid, thus forming a polypeptide chain. The tRNA that had been in the P site leaves.

4. The ribosome in effect moves down the mRNA chain, one codon to the right, thus switching the second tRNA (bearing the polypeptide chain) into the P site and bringing a new codon into the A site. The A site is then available for a new tRNA, and the process of elongation continues.

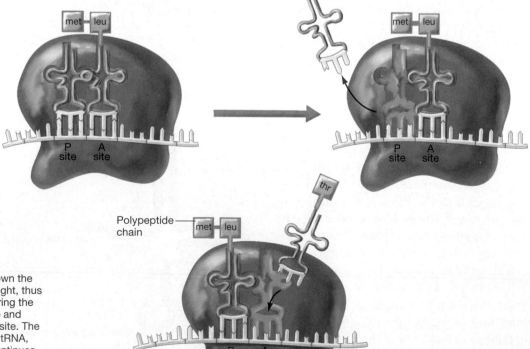

has three bases on it, called an **anticodon**, that can form a base pair with the appropriate codon in an mRNA tape. At the *opposite* end of the tRNA molecule (the 12 o'clock position in the figure) there is an attachment site for an *amino acid*. Each tRNA molecule is specific for a particular amino acid, as well as a particular codon, so that the correct amino acid sequence will result from a given mRNA sequence.

The Steps of Translation

Now translation can begin. An mRNA transcript has left the nucleus and is ready to begin binding with a ribosome. Meanwhile, nearby tRNA molecules have bound with their appropriate amino acids. For purposes of simplicity, we'll follow the process as it occurs in prokaryotes.

mRNA Binds to Ribosome, First tRNA Arrives   In this first step, the mRNA tape arrives at the ribosome and binds to the ribosome's small subunit (see **Figure 14.11**). The mRNA codon AUG is the usual "start" codon for a polypeptide chain. Next, a tRNA molecule with the appropriate anticodon sequence (UAC) binds to this AUG codon. This tRNA arrives bearing its appropriate amino acid, which is methionine (met). Following this, the large ribosomal subunit becomes part of the ribosome, providing the ribosome's A and P binding sites.

Polypeptide Chain Is Elongated   Now the polypeptide chain will have more amino acids added to it. As you can see, this chain elongation begins with a second incoming tRNA molecule binding to an mRNA codon in the A site. Because it happens to be a CUG codon, a tRNA with a GAC anticodon binds to it. This tRNA comes bearing the amino acid leucine (leu).

The met amino acid attached to the tRNA in the P site now bonds with the leu amino acid attached to the tRNA in the A site. In this process, the bond is broken between met and its original tRNA. A polypeptide chain is now attached to the tRNA in the A site, and, with this, the original tRNA is released from the P site. Now a kind of molecular musical chairs ensues: The ribosome can be thought of as moving down the mRNA line, three base pairs to the right. This action has two effects: The tRNA that had been in the A is relocated to the P site; and a new mRNA codon moves into the now-vacated A site. The rest of elongation amounts to a repetition of this process: A new incoming tRNA binds with the new codon in the A site, the growing polypeptide chain bonds with that amino acid, and so on.

Termination of the Growing Chain   Recall that when you looked at the genetic code, there were three codons that acted as stop signals for polypeptide synthesis. Any time one of these "termination" codons moves into the ribosome's A site, it doesn't bind with an incoming tRNA, but instead brings about a severing of the linkage between the P-site tRNA and the polypeptide chain. Indeed, the whole translation apparatus comes apart at this point, with the polypeptide chain being released to fold up and be processed as a protein. Translation has been completed.

Speed of the Process; Movement through Several Ribosomes   How fast does this process go? Approximately five amino acids are added to a growing polypeptide chain every second. This figure is somewhat misleading, however, in that mRNA tapes often are "read" not by one ribosome, but by many. As you can see in **Figure 14.12**, several ribosomes—perhaps scores of them—might move, in effect, over a given mRNA transcript, with identical polypeptide chains then growing out of each ribosome.

Tutorial 14.3.7
Polyribosomes

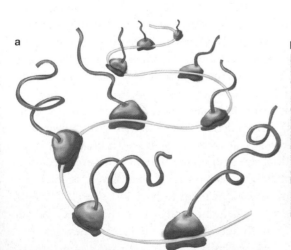

a

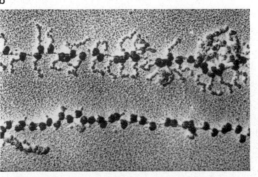

b

Figure 14.12
Mass Production

a An mRNA transcript can be translated by many ribosomes at once, resulting in the production of many copies of the same protein.

b A micrograph of this process in operation. The figure shows two mRNA strands with ribosomes spaced along their length. In the upper strand, translation is under way and polypeptides can be seen emerging from the ribosomes.

Making Sense of "Junk" DNA

Imagine that you pick up a book, a novel, that begins: "It was the best of times, it was the the pzknlku ljkh uiop nk; lkj bhji; lkjhjoiugh qpmzxd kjhb olkjb worst of times." Reading on, you see that things go on like this for hundreds of pages, right to the end. Baffled? So were scientists when, in 1977, they discovered that eukaryotes such as ourselves have genetic "books" that are filled with passages of such seeming nonsense.

You've already seen that the "sense" side to an organism's "genome," or store of genetic material, is the information contained in that organism's DNA. Such information comes in the form of A's, T's, C's, and G's that code for proteins. If a genome were nothing *but* sense, one "coding" triplet of these bases would follow another in succession, with only some housekeeping sequences in between them. What scientists began perceiving in the late 1970s, however, is that it is the *coding* sections of the genome that are few and far between. For many years following the 1970s revelations, estimates were that only 3 to 5 percent of the human genome coded for protein. Then, with the unveiling of the human genome sequence in early 2001, the estimates dropped again: At most, a mere 1.5 percent of the human genome seems to code for protein. To gain some perspective on this, consider that, while the DNA in any of our cells would stretch to about 6 feet in length if uncoiled, only about 1 inch of this length is DNA that codes for proteins.

So, what is all this noncoding DNA? Some portions are the housekeeping or "regulatory" sequences of DNA noted in the text, while other portions—at the tips of chromosomes—play a part in keeping chromosomes intact. Some DNA, however, has no known function at all and thus, years ago, was given a name that has stuck: junk DNA.

Such "junk" can be thought of as existing in two locations. First, there are the **introns** mentioned earlier, which are noncoding nucleotide strings *within* gene sequences. Introns bear some comparison to the nonsense strings in the first paragraph, in that they interrupt related strings of bases that *do* code for a given protein. It would not be unusual for a gene to contain half a dozen introns, with these intervening sections comprising most of the bases in the sequence.

Junk DNA can also lie, however, *in between* segments that code for genes. Some of this junk can be stupendously monotonous. For example, there are some 30,000 to 50,000 short segments of DNA in the human genome that consist of nothing more than a couple of bases repeated over and over, as in: CACACACACA . . .

This latter sequence, a molecular broken record, would certainly seem to fit the definition of junk, if by that we mean something so useless that it might as well be thrown away. And yet, research in recent years has cast doubt on whether any sequence in the genome is insignificant in this way. It is true that many sequences do not code for protein, but that does not mean that they serve no purpose at all.

Consider, for example, a sequence in the genomes of primates that was long thought to be the essence of junk. Known as the Alu repeat, this sequence is about 280 base pairs long, which is not particularly extensive in the context of the 3-billion-base-pair human genome, until you consider that about 1 million *copies* of Alu are believed to be dispersed throughout the human genome. Thus, this sequence alone makes up nearly 10 percent of our genome—far more than all the coding regions put together.

The Alu repeat is thought to be derived from a gene in the primate genome that was transformed into a DNA sequence capable of making copies of itself that could be inserted elsewhere in the genome. Under this view, the Alu sequence was a

Some DNA has no known function at all and thus, years ago, was given a name that has stuck: junk DNA.

kind of squatter capable of multiplication. It did no particular harm to its "host" genome, but did no good for it either; it simply made more and more copies of itself, which continued to be inserted into the primate genome. Thus, the Alu was thought of as falling into a particular category of junk DNA, *selfish DNA*, meaning DNA whose only function is to replicate itself.

But is this true? Human Genome Project researchers found that older Alu segments—those that came into the primate genome earlier in evolutionary history—tend to congregate near the gene-rich areas of the human genome. It's hard to explain why a sequence that is junk would link up with genes over evolutionary time unless that sequence was working with the genes in some fashion—perhaps helping to regulate them. Beyond this observation, researcher Carl Schmid of the University of California, Davis, has found that in mice, a close genetic relative of the Alu repeat is activated when mice become physiologically stressed by, for example, being made to drink lots of alcohol. The impression scientists are getting is that, while the Alu repeat may have arrived originally as a genetic squatter, it has long since conscripted into duty by the primate genome.

The seeming usefulness of something as repetitive and long-winded as the Alu repeat calls into question the whole notion of "junk" DNA. One view is that there is no junk DNA; there is only human ignorance of the functions that various DNA segments serve. Another view, however, is that there really are vast stretches of human and other genomes that serve no function at all—not even the function of making more copies of themselves, as many sequences of junk DNA seem to have lost this ability. With the entire human genome sequence nearly in hand, we may be on the road to finding out where the truth lies.

14.5   Genetic Regulation

At this point, you have a basic understanding of how genes code for proteins, of how they make copies of themselves, and so forth. Even with this, however, some significant questions need to be addressed. If you think about it, in the entire discussion of genetics over the past few chapters, you have seen no reason for a given gene to turn on or off—to begin transcription or to end it. Scientists know that this does happen; that the cells in our pancreas, for example, turn out the protein insulin *intermittently*—in response to what we've eaten—rather than ceaselessly transcribing the gene for insulin. So, what are the start and stop signals for this protein-synthesizing machinery? For that matter, why is it that only *pancreatic* cells turn out insulin when almost all other cells in the body possess copies of the insulin gene? To come to an understanding of these things, you need to learn something about how protein synthesis is regulated.

DNA is the Cookbook, not the Cook

We have thus far thought of the "genome" or genetic inventory of a living thing as a kind of cookbook, containing recipes (DNA sequences) that lead to products (proteins) that undertake a wide variety of tasks (hastening chemical reactions, serving as structural components of tissue, and so on). It is tempting to go from these ideas to thinking that DNA's the boss, at least for the metabolic processes lying outside our conscious control. After all, it's giving "orders" for the body to put together these all-important proteins, isn't it?

Well, sticking with the metaphor, a cookbook does not give orders; it merely contains information, which an *agent* (called a cook) then carries out. It becomes clear that DNA is not the cook when you consider that by itself, DNA can't synthesize anything. Think back to the "transcription" in which a DNA sequence is copied onto messenger RNA. The DNA double helix does not unwind itself, nor does it bring mRNA nucleotides into alignment with its own bases, nor does it snip out introns in the mRNA, nor does it effect any of the other steps you looked at. Instead, all these tasks are carried out by enzymes, which are *proteins*. (Recall that the protein complex known as RNA polymerase is central in carrying out the steps of transcription.) DNA is indeed the cookbook, but it is powerless to prepare its own recipes. A critical insight is that, through chemical bonding, information comes *to* the double helix as well as from it, in a process known as genetic regulation.

A Model System in Genetic Regulation: The Operon

Genetic regulation can occur at any step in the protein synthesis process. Here we'll focus on a single example of genetic regulation, one that takes place in the transcription stage of protein synthesis. You'll look at what's known as a "model" regulatory system in a model organism. The system is called the *operon*, and the organism is the *Escherichia coli* bacterium. *E. coli* has long been a model organism because it is simple, it can easily be experimented upon, and it reproduces in as little as 20 minutes. At the end of this account, you'll see that the very proteins DNA codes for can feed back onto it, regulating its activity by binding with it.

Jacques Monod and François Jacob's Experiments

The sterling qualities of *E. coli* as a research subject made it an ideal choice for two French researchers, Jacques Monod and François Jacob, who in the 1950s wanted to learn how genes turn on and off in living creatures. By then it was clear that in most cases, genes didn't simply *stay* on. The activity of at least some genes had to be *inducible*, meaning triggered by conditions in the organism's environment. One means of inducement seemed clear; it had been known for many years that bacteria would synthesize certain enzyme proteins only if the substance these enzymes worked on (their "substrate") was present in the bacterial cell. Accordingly, Monod and Jacob focused on a group of enzymes that allow *E. coli* first to obtain the sugar lactose and then to break it down into two simple sugars. This latter step is a clipping operation in which lactose (commonly known as milk sugar) is broken into its component simple sugars, galactose and glucose.

Tutorial 14.4.1
The Lac Operon

Three genes are involved in lactose metabolism in *E. coli*, two of which are important for our purposes. As you can see in **Figure 14.13a**, there is a *y* gene that codes for a so-called *permease* enzyme, which transports lactose into the cell. Then there is a *z* gene, which codes for the enzyme beta-galactosidase (or β-galactosidase), which does the lactose clipping mentioned before. (A third, *a*, gene produces a related enzyme; but this need not concern us, other than to note why it sits next to the other two.)

Activity Induced by Lactose

The genes involved in lactose metabolism were known to be inducible, and the *inducer*—the substance that prompted their activity—was known to be lactose. Put simply, in the presence of lactose, these genes turned on. Yet what was the control mechanism here? The experiments of Jacob and Monod eventually led them to propose that *E. coli*'s lactose metabolism is governed by an elegant multipart genetic system, which they dubbed the *operon*.

Figure 14.13
Genetic Regulation in Action

a Structure of the *lac* operon system. The system consists of (left to right) an *i* gene that codes for a repressor protein (the two green crosses); RNA polymerase, shown bound to the DNA strand's promotor region (P); a region of the DNA strand called the operator (O); two genes and their protein products: The *z* gene that codes for the lactose-clipping beta-galactosidase enzyme; and the *y* gene, which codes for a permease enzyme, which transports lactose into the cell. (The *a* gene, while part of the system, codes for a protein not relevant to the explanation.)

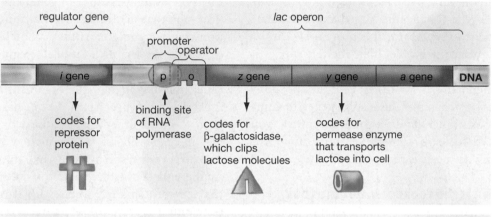

b In the absence of lactose, the repressor protein binds to the operator and inhibits the transcription of lactose-processing enzymes.

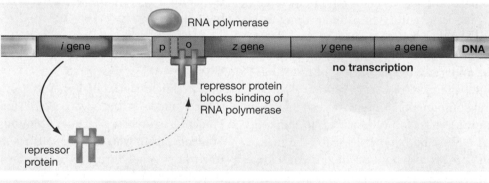

c When lactose is present, the repressor is inhibited from binding with the operator, thus allowing transcription and hence production of the lactose-processing enzymes.

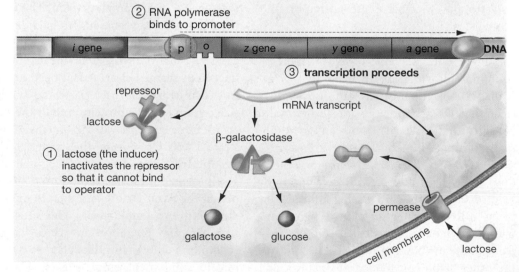

In the lac operon, the *z*, *y*, and *a* genes are transcribed as a unit onto a single messenger RNA transcript. "Upstream" from the *z* gene is a segment of DNA known as the promoter sequence. Any **promotor sequence** is the site on a segment of DNA at which transcription of a gene begins. More specifically, it is the binding site for the RNA polymerase—the enzyme that brings about transcription by going down the line, pairing mRNA nucleotides with their DNA counterparts. Note, however, that *in between* the promoter site and the *z* gene sequence is a region that Jacob and Monod called the *operator*. This sequence of DNA can exist in one of two states. In one, it is bound to a protein called a *repressor*, which effectively *blocks* the binding of RNA polymerase to the promoter, thus shutting down transcription. In the other state, it is unbound—free of the repressor—in which case transcription can proceed. Because the repressor is a protein, it is itself the product of a gene (the *i* gene) that lies just a little farther yet upstream. (You can see it in Figure 14.13a, on the left.)

Now, the critical question is: What causes the repressor either to bind with the operator (thus halting transcription) or to stay clear of it (thus allowing transcription)? The answer is the absence or presence of lactose. If you look at Figure 14.13b, you can see how this plays out. The critical factor is that the repressor protein has *two* binding sites: One of them binds to the operator, as already noted, but the other binds to *lactose*. When this lactose binding takes place, a repressor that is bound to an operator changes shape; like a clothespin that's been pinched, it opens up and is released from the operator site. The result? RNA polymerase is no longer blocked, meaning it can bind to the promoter site and begin transcription of the *z*, *y*, and *a* genes. They produce their enzymes, and lactose metabolism proceeds apace. When all the lactose has been broken down, however, the repressor molecules return to their original shape, meaning the shape that can bind to the operator, thereby shutting down gene transcription.

Thus did Jacob and Monod elucidate a detailed mechanism of gene regulation. Within this concept, they confirmed the existence of genetic sequences whose sole functions are *regulatory*. Think of it: The *i* gene sequence codes for a protein, but it is a protein whose sole function is to allow DNA transcription to go forward, or to turn it off. Meanwhile, the operator DNA segment doesn't code for a protein at all; instead it exists as a binding site for this repressor protein. Such regulatory sequences stand in distinction to the DNA segments that code for end-product proteins—in this series, the *z*, *y*, and *a* genes. As a piece of scientific research, Jacob and Monod's work on the operon was so stellar that, with André Lwoff, they were awarded the Nobel prize in 1965. For our purposes, the lac operon exemplifies the general concept of gene regulation in transcription. The proteins that are produced through DNA's instructions *feed back* on the DNA molecule itself, thus helping to control the production of other proteins.

14.6 The Magnitude of the Metabolic Operation

Before leaving the subject of protein synthesis, let's now consider the metabolic operation that is managed under genetic regulation and then something about the gene itself.

The Number of Proteins Utilized

The human genome is thought to code for somewhere between 50,000 and 100,000 proteins. Any one human cell, however, might produce as few as 5,000 of these proteins or perhaps as many as 20,000. Five thousand proteins seems, in any event, to be the minimum number found in all eukaryotic cells. A quantity this large may seem surprising, but think of what a cell must do. To take one example, immature red blood cells called erythroblasts are unique in their production of the protein hemoglobin; but like most cells, erythroblasts must be generalists as well as specialists. Like almost all cells, they need proteins for ribosomes, membranes, a cytoskeleton—the machinery, in short, that makes cells work—and they must respond to their environment as well. These factors mean that they must contain a huge number of proteins. Some of these proteins are produced almost continuously, while others are "inducible," as with the lactose-clipping β-galactosidase in *E. coli*.

The Size of the Genome

Not surprisingly, to code for all these proteins, eukaryotic cells must also have an immense genome. As noted, the human genome is about 3 billion base pairs long. To make this large number a little less abstract, consider the following exercise. If we took the base sequence of the human genome and simply arrayed the single-letter symbols for the bases on a printed page, going like this:

AATCCGTTTGGAGAAACGGCCCTATT
GGCAGCAAGGCTCTCGGGTCGTCAACG
CGTATTAAACATATTTCAAGGCTCTA . . .

it would take about 1,000 telephone books, each of them 1,000 pages long, merely to record it all. We're talking, then, about an unbroken series of these base symbols that goes on for a million pages. That is one measure of the size of the human genome. (And remember, each base must be faithfully copied every time a cell divides!) You can get an idea of the size of several different genomes if you look at **Table 14.3**.

14.7   What Is a Gene?

Finally, given the material presented in this chapter, you are ready for a more accurate definition of the most basic unit in genetics, the gene. Thus far a gene has been viewed as a length of DNA that codes for a protein. And as you've just seen, some of these proteins can be regulatory, while others are not. But recall from the section on translation that the "workbenches" of genetic translation, the ribosomes, are made partly of RNA. Where does this ribosomal RNA come from? Just like messenger RNA, ribosomal RNA must be coded for by DNA. A segment of DNA unwinds and then forms base pairs with RNA nucleotides to produce an RNA sequence. Only this is a *ribosomal* RNA sequence that doesn't code for anything; it simply migrates to the cytoplasm to become part of a ribosome. So great is the cell's need for ribosomes that there is an entire section of the nucleus (called the nucleolus) whose DNA is constantly being transcribed for rRNA. Likewise, there are sections of DNA that code for transfer RNA, which also migrates to the cytoplasm to be involved in translation. Note that in the case of both transfer RNA and ribosomal RNA, DNA is being transcribed. But the ultimate *product* of this transcription is not a protein; it's RNA in two forms. With all this in mind, here is a more accurate definition of a gene. A **gene** is a segment of DNA that brings about the transcription of a segment of RNA. Note that under this definition, a gene includes both the DNA regulatory sequences and the DNA segments that actually form base pairs with RNA.

How Did We Learn?

You saw earlier that in the 1960s, scientists were puzzling over the nature of the genetic code. If you had a given DNA or mRNA triplet, what amino acid did it code for? To find out how scientists solved this mystery, see "Cracking the Genetic Code" on page 285.

On to Biotechnology

You have seen in this chapter that DNA is only one of the molecules governing protein synthesis. But saying that something is simply one part of a chain is not saying it is not powerful. DNA may "only" be the cookbook, but think of the profound effect this single function stands to have. Following the metaphor, think how different a pie would be if an editing error in a recipe resulted in the insertion of the word *cinnamon* where *sugar* should be (and if our cook had to insert the cinnamon). Now imagine the change for the better if you could *change the recipe* to include sugar instead of cinnamon. The ability to manipulate DNA's instructions lies at the heart of what is commonly called biotechnology. This exciting, promising area of biology is the subject of Chapter 15.

**Table 14.3
Sizes of Different Genomes**

Different genetic sets. The genome sizes of a virus and six organisms, measured in number of base pairs.

Virus/Organism	Genome Size in Base Pairs
T2 virus	200,000
E. coli bacterium	4.6 million
Yeast	12 million
Fruit fly	180 million
Chicken	1.2 billion
Human	3 billion
Pea	5 billion

Cracking the Genetic Code

Children often are amused by giving common words special meanings. "If I say, 'flower,' that means to stand up," one will say to another. With this code in mind, the second child dutifully pops right up when the magic word is spoken. In the early 1960s, scientists were like observers in such a game. This one, however, involved genetics, and the whole point was to decipher a genetic code that was being employed. "If I say to you, 'the codon is UUA,'" the riddle went, "what amino acid gets added to a protein chain?"

At the time, this was a vexing question indeed. One of the difficulties in answering it was that researchers knew that a working cell is like a molecular Times Square: It is jammed with the traffic of many proteins working away or undergoing synthesis. Marshall Nirenberg and Heinrich Matthaei, scientists at the National Institutes of Health in the early 1960s, knew they had to work *outside* of this great complexity in order to follow the path of a *single* RNA triplet and see what amino acid it led to. Thus they utilized what is known as a "cell-free" system, meaning they watched amino acid translation play out in test tubes rather than cells.

You've seen that, within cells, stretches of messenger RNA are synthesized by means of free-standing RNA nucleotides forming base pairs with lengths of DNA. Working outside of cells, Nirenberg and Matthaei synthesized an *artificial* mRNA chain by using a special enzyme that can string together RNA nucleotides. Critically, it can string together *whatever* nucleotide chain researchers wish, depending on what bases the researchers start with.

> ## Running through the list, the researchers hit pay dirt with the amino acid phenylalanine.

For example, putting this enzyme together with the basic building blocks of RNA, but no base except uracil (U), scientists got an *mRNA chain* that had no bases but uracil in it (UUUU . . .).

So then the question became: What amino acid did the base triplet UUU code for? To find out, Nirenberg and Matthaei took their mRNA and put it into a test tube containing ribosomes, all 20 amino acids, and other molecules. Thus, this mix had in it all the molecular machinery necessary for protein translation, but it had only *one* mRNA tape to work with—the UUU mRNA transcript the researchers had made. It was therefore reasonable to

expect that this system would turn out polypeptide chains containing only one amino acid—whichever amino acid was coded for by the UUU triplet.

The trick was to look at all the amino acids in the mix by means of radioactively labeling *one* of them (out of the 20) with each experiment (see **Figure 1**). Such labeling effectively puts a kind of tag on a given amino acid. Other techniques allowed the researchers to tell whether such an amino acid had remained free-standing or had been incorporated into a polypeptide chain. Running through the list, the researchers hit pay dirt with the amino acid phenylalanine. It clearly was part of a polypeptide chain, and the only way it could have been incorporated in this way was through the activity of mRNA. Because the only mRNA triplet in the system was UUU, this meant that UUU coded for phenylalanine.

Several other techniques had to be worked out in the ensuing years (by several researchers) to decipher the entire genetic code. All of these techniques, however, relied on the cell-free systems and synthetic mRNAs that Nirenberg and Matthaei employed. By 1968, the entire code had been cracked. In that same year, Marshall Nirenberg shared the Nobel prize with researchers Robert Holley and Har Gobind Khorana for undertaking this critical piece of molecular deciphering.

Tutorial 14.2.4
Cracking The Code

Figure 1
Cracking the Code
In their work, Nirenberg and Matthaei radioactively labeled, in each experiment, one amino acid out of the total group of 20. Other techniques allowed them to tell whether such an amino acid remained free-standing or had been incorporated into a polypeptide chain. When they labeled phenylalanine, the researchers found that it had been incorporated into a chain. Because the only mRNA triplet in their system was UUU, this told them that UUU had to code for phenylalanine.

Chapter Review

Summary

14.1 The Structure of Proteins

- Proteins are composed of building blocks called amino acids. A string of amino acids is called a polypeptide chain. Once such a chain has folded into its working three-dimensional shape, it is a protein. Though there are tens of thousands of different proteins, all of them are put together from a starting set of 20 amino acids. It is the order in which the amino acids are linked in a polypeptide chain that determines which protein will be produced.
  TUTORIAL 14.1.1: Amino Acids

14.2 Protein Synthesis in Overview: Transcription and Translation

- There are two principal stages in protein synthesis. The first stage is transcription, in which the information encoded in DNA is copied onto a length of messenger RNA, which then moves from the cell nucleus to structures in the cytoplasm called ribosomes. The second stage is translation, in which a polypeptide chain is produced in accordance with the instructions encoded in the mRNA sequence.

- The information in DNA is transferred to messenger RNA through complementary base pairing. The length of mRNA that results from this base pairing is called a transcript.

- Each three coding bases of DNA pairs with three mRNA bases, but each group of three mRNA bases then codes for a single amino acid. Each triplet of mRNA bases that codes for an amino acid is called a codon.
  TUTORIAL 14.1.2: Higher Level (3-D) Structure of Proteins

- Transfer RNA serves as a bridging molecule, because it can bond with both mRNA and amino acids. A given tRNA molecule bonds with a specific amino acid in the cell's cytoplasm and then transfers that amino acid to a ribosome in which an mRNA transcript is being "read." When the tRNA arrives, another portion of it, called an anticodon, bonds with the appropriate codon in the messenger RNA chain.
  TUTORIAL 14.3.1: Protein Transcription

- Specific mRNA nucleotide triplets code for specific amino acids or for "start" and "stop" sequences in protein synthesis. The entire inventory of these linkages is known as the genetic code.
  TUTORIAL 14.2.2: The Triplet Genetic Code

14.3 The Importance of the Genetic Code

- The near-universality of the genetic code in living things is evidence for a single, common ancestor for all life that now exists on Earth. A common genetic code is the reason viruses

can utilize the genetic machinery of other living things, but it is also the basis for today's biotechnology industry.
  TUTORIAL 14.2.4: Cracking The Code

14.4 A Closer Look at Protein Synthesis

- Many of the sequences of DNA and messenger RNA in eukaryotes do not code for proteins. These intervening sequences, or introns, are spliced out of mRNA transcripts before the transcripts leave the cell nucleus. The protein-coding portions of the genome are called exons. Other sequences of DNA are purely regulatory, serving only to affect transcription. Still other sequences have no known function.
  TUTORIALS: 14.3.4 Protein Translation; 14.3.7 Polyribosomes

14.5 Genetic Regulation

- Protein synthesis can be facilitated or impeded at many points. This process is called genetic regulation. The operon system reviewed in the text regulated mRNA transcription. In such regulation, the proteins produced through DNA's instructions feed back on the DNA molecule itself, thus helping to control the production of other proteins.
  TUTORIAL 14.4.1: The Lac Operon

14.6 The Magnitude of the Metabolic Operation

- The human genome is thought to code for somewhere between 50,000 and 100,000 proteins. Any one human cell, however, might produce as few as 5,000 of these proteins or perhaps as many as 20,000. With the special tasks that eukaryotic cells take on, most of them must contain a huge number of proteins. Not surprisingly, to code for all these proteins, eukaryotic cells must also have an immense genome.

14.7 What is a Gene?

- A gene can be defined as a segment of DNA that brings about the transcription of a segment of RNA.

Key Terms

anticodon 277

codon 273

exon 277

gene 284

genetic code 275

intron 275

messenger RNA (mRNA) 272

polypeptide 269

promoter sequence 283

ribosome 271

ribosomal RNA 277

(rRNA) 277

RNA polymerase 276

transcript 272

transcription 271

transfer RNA (tRNA) 274

translation 271

triplet code 273

Understanding the Basics

Multiple-Choice Questions

1. The working ability of a protein is found in its
   a. polypeptide sequence
   b. ability to encode DNA polymerase
   c. two-dimensional structure
   d. three-dimensional structure
   e. both b and c

2. How does DNA differ from RNA?
   a. DNA uses the bases A, T, C, G; RNA uses the bases A, U, C, G.
   b. DNA is a double-stranded molecule; RNA is usually single-stranded.
   c. DNA is a nucleic acid; RNA is a protein.
   d. a and b
   e. all of the above

3. The "t" of tRNA stands for:
   a. tripartite
   b. teleporting
   c. transfer
   d. tracking
   e. transcribing

4. What mRNA sequence signals the start of a sequence to be translated?
   a. ATG
   b. UGA
   c. AUG
   d. UAG
   e. UAA

5. An intron is:
   a. the noncoding portion of a length of DNA or messenger RNA
   b. the structure in the cytoplasm that is the site of protein synthesis
   c. the coding portion of a length of DNA or messenger RNA
   d. retained in the messenger RNA transcript after that transcript has been edited
   e. both a and d

6. Which of the following statements is true regarding ribosomes?
   a. They are composed of two separate units of rRNA combined with protein.
   b. They are used to translate the mRNA sequence into a protein composed of amino acids.
   c. They have two different sites for tRNA binding: P, where tRNA exits the ribosome and A, where new tRNA attaches to the ribosome.
   d. a and b only
   e. all of the above

7. Why can transfer RNA be referred to as a "bridging" molecule?
   a. It forms a chemical bridge between DNA and messenger RNA.
   b. It creates the linkage between the base pairs of the double helix.
   c. It links species through evolution.
   d. It links ribosomes together.
   e. It forms chemical bonds with both amino acids and messenger RNA.

8. The minimum number of proteins found in all eukaryotic cells is believed to be
   a. 100,000
   b. 50,000
   c. 20,000
   d. 5,000
   e. 2,000

9. Feedback was employed in the example of genetic regulation in the text. What was feeding back onto what?
   a. Proteins produced through the information encoded in DNA went on to regulate the transcription of that same DNA.
   b. Ribosomes produced through the information encoded in DNA went on to serve as the sites of protein synthesis.
   c. Messenger RNA strands synthesized as a complement to DNA went on to reduce DNA transcription.
   d. Transfer RNA molecules limited translation at the ribosomes.
   e. Transfer RNA molecules sped up translation in mitochondria.

10. The genetic code is the same in:
    a. all organisms except bacteria
    b. nearly all organisms
    c. both organisms and crystals
    d. all organisms except plants
    e. all organisms except humans

Brief Review

1. What two great tasks are carried out by our genetic machinery?

2. What are the three types of RNA discussed in this chapter? What do their abbreviations stand for? What are their functions?

3. What is the difference between transcription and translation?

4. What is one detrimental effect of the universal nature of the genetic code?

5. What is the difference between introns and exons?

6. Why is it inaccurate to conceptualize DNA as the sole controller of a cell's production of proteins?

7. How many DNA bases does it take to code for an RNA codon? How many amino acids does an RNA codon code for?

8. In what sense does each human being share in an ancient informational linkage?

Applying Your Knowledge

1. Why does the activity of at least some genes need to be inducible?

2. Given the following sequence of DNA, what would the complementary mRNA sequence be?

   TACCCGTATACGATCATGGTCAAGTCGTAC

3. Given the following sequence of mRNA, what would the resulting amino acid sequence be?

   AUGAAACGGGGACCAAUGGAUAACUAA

4. In a hypothetical situation, you have identified a new species that uses six bases, instead of three, in its genetic code. It has also been discovered that the proteins this species forms are made up of 220 amino acids. To ensure that each amino

acid has a separate and distinct encoding sequence, how many bases need to be included in each codon?

5. In Chapter 13, point mutations were introduced as sources of error in DNA replication. Shown below are five mRNA sequences, including one "normal" sequence and four different point mutations. What effect will each of these point mutations have on the amino acid sequence? What does this indicate about the significance of the location and type of point mutation that occurs? (Remember that an mRNA initiates an amino acid sequence with a start codon and ends the sequence with a stop codon.)

Normal mRNA: AUG AAA CAU GCA CUA AUG UAA CCU
Point mutation 1: AUG UAA CAU GCA CUA AUG UAA CCU
Point mutation 2: AUG AAA CAU CCA CUA AUG UAA CCU
Point mutation 3: AUG AAA CAU GCA CUA AUG AAA CCU
Point mutation 4: AUG AAG CAU GCA CUA AUG UAA CCU

6. You hear that scientists have discovered an animal-like organism living at the bottom of the ocean. It not only looks different from anything we've seen but also employs a different genetic code. (It uses DNA and the standard 20 amino acids, but in a completely different linkage to one another.) What are some hypotheses that could account for this?

MediaLab

Can We Stop the Cycle? DNA to RNA to Protein

We take for granted our control over bacteria and viruses, confident that we can stop them dead with a simple antibiotic or antiviral medication. But, did you know that these medications work to control the same critical steps of transcription, translation, and protein manufacture that occur in our cells? In the *CD-ROM Tutorial* of this *MediaLab*, you'll review the cellular activities needed to manufacture proteins, so that in the *Web Investigation*, you can analyze the subtle differences between our cells and their invaders. These differences can be exploited in our fight against the enemies within us, as you'll see in the *Communicate Your Results* section.

This *MediaLab* can be found in Chapter 14 on your CD-ROM (Tutorial 14.3.4) and Companion Website (http://www.prenhall.com/krogh3).

CD-ROM TUTORIAL

Every single one of the thousands of different proteins made by a cell is created in the same way, and the steps are similar in all cells. This *CD-ROM Tutorial* leads you step by step through the protein production pathway.

Activity

1. *First, you will review the two major stages in protein synthesis: (a) copying the information of DNA into RNA so that it can exit the nucleus, and (b) the actual manufacture of protein on ribosomes in the cytoplasm.*

2. *Next, you can try your hand at directing these processes, by gathering all the ingredients needed to transcribe DNA and translate the resulting RNA message into protein.*

3. *Finally, you'll observe how ribosomes in the cell act to speed up production.*

Now that you have reviewed the steps in protein synthesis, in the following *Web Investigation* section you can explore how blocking these critical steps can cripple an organism—and in the process, save our lives.

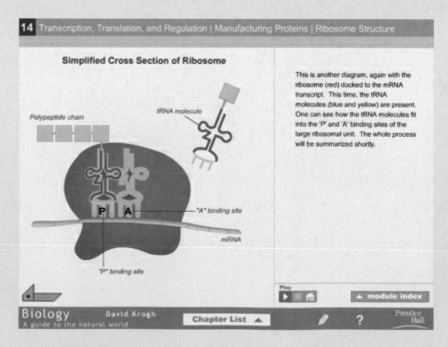

Investigation 1

Estimated time for completion = 5 minutes

If you still think that the steps in making a protein are confusing, maybe you would like to try a step-by-step, interactive demonstration. Select the Keyword **DNA WORKSHOP** on your CD or Website to visit a site that allows you to take part in coordinating the manufacture of proteins. Click on "DNA workshop activity" once you reach the site, and try your hand at using DNA in replication and in protein synthesis.

Investigation 2

Estimated time for completion = 5 minutes

You just saw two uses for DNA, replication and transcription. DNA to RNA to protein is called the central dogma because all cells use the same steps in protein synthesis. HIV, the virus that causes AIDS, is a retrovirus, meaning it carries its genetic material as RNA, not DNA. Click the Keyword **HIV** on your CD or Website to view a step-by-step overview of HIV infection. How can HIV replicate its genetic information or transcribe RNA without DNA?

Investigation 3

Estimated time for completion = 15 minutes

By now you understand how all proteins are made, but what would happen if you stopped protein synthesis—death? Yes, but a death you can rejoice in—the death of your enemies. Enemies in this case include opportunistic bacteria that can invade your tissues and cause disease or death. Many antibiotics work by stopping transcription or translation of proteins. Read about bacteria and antibiotics by selecting the Keyword **ANTIBIOTICS ATTACK** on your CD or Website, and you'll discover how a wide range of antibiotics act to defeat bacteria.

Exercise 1

Estimated time for completion = 5 minutes

Make a table of the similarities and differences between replication and protein synthesis that you observed in *Web Investigation 1*. Include the following: When are the two processes performed? What process occurs in both? Where? What enzymes are involved? What is the result in both cases?

Exercise 2

Estimated time for completion = 5 minutes

HIV uses a special viral enzyme called *reverse transcriptase* as a critical part of its life cycle. The popular anti-HIV drug AZT works by preventing reverse-transciptase functioning. Describe what would happen to a cell that was infected with HIV when AZT is used. AZT is a thymine analog (meaning a look-alike); the only difference is that it has a nitrogen atom on the important carbon in the ribose sugar that is needed to make covalent bonds between nucleotides. Why would this affect reverse transcription?

Exercise 3

Estimated time for completion = 15 minutes

You think your fever and sore throat is a streptococcal infection (strep throat). Your friend offers you penicillin left over from his last bacterial infection. Would it even work on your sore throat if you really did have an infection of streptococcus? How does penicillin work to kill bacterial cells, but not harm your cells? If you found out your infection was more serious (pneumonia) and the culprit was *Mycoplasma pneumoniae,* would the penicillin be effective? You decide to try a wide-spectrum antibiotic that kills many species of bacteria by preventing translation or transcription. What would your pharmacist and doctor recommend to block transcription, or translation by binding and inactivating ribosomes? Why doesn't this hurt your own translation and transcription?

15

The Future Isn't What It Used to Be
Biotechnology

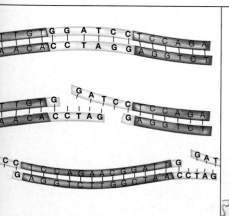

Restriction enzymes.
(Section 15.2, page 295)

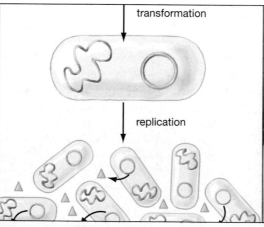

transformation

replication

Bacterial protein factories.
(Section 15.2, page 296)

Hello, Dolly!
(Section 15.3, page 298)

By manipulating the information contained in DNA, scientists are raising hopes—and some fears—on a grand scale.

During summertime in America, many a suburban and country resident takes pride in growing a lush, green lawn. The bane of weekend life for many of these same people, however, is that these lawns need to be mowed so often. For millions of Americans, one of the givens in life—right up there with paying taxes and dying—is the necessity of regularly cutting the grass.

Now, however, this given may be up for grabs. A giant American lawn products company currently is field testing a strain of grass that has been genetically engineered to make it grow slowly—so slowly that it might need to be cut only twice from spring through fall. And this "low-mow" trait may be just one of several changes in store for grass. By manipulating grass genes, it should be possible for grass to come in different *colors* (reds, blues, various greens) or have a different leaf shape. Then there is the possibility of making it resistant to droughts, so that it might need to be watered less.

When you hear the word "biotechnology," designer grass may not immediately come to mind. But in this one commercial product, it's possible to see most of the elements that make biotechnology seem both so promising and yet so threatening. Note that designer grass:

- Shakes up our notions of what is possible in the world.

- Has a big commercial potential. The market for low-mow grass has been estimated at a whopping $10 billion per year.

- Could have a personal impact. If brought to market, low-mow grass would affect not only businesses (think of golf courses) but personal lives as well. It would mean a change in how our lawns look and how we spend our time.

- Has critics who would like to see it go away entirely. Biotech adversary Jeremy Rifkin and the American Society of Landscape Architects have petitioned the U.S. Department of Agriculture to suspend low-mow field tests until the department determines that pollen from the grass won't mix or "hybridize" with strains of natural grasses. The fear is that low-mow traits could spread, thereby imperiling ecosystems that depend on natural grass.

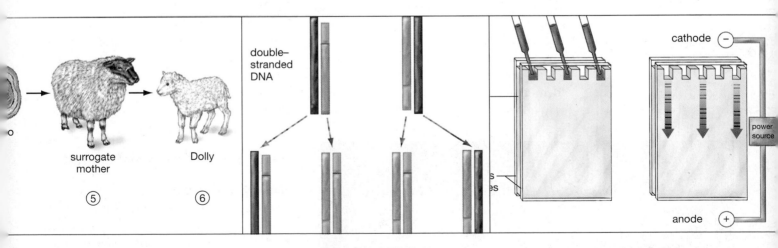

How it was done.
(Section 15.3, page 299)

A DNA copying machine.
(Section 15.4, page 301)

Visualizing DNA.
(Section 15.5, page 303)

On June 26, 2000, biotechnology had a glorious day in the sun when two American scientists—Craig Venter and Francis Collins—stood side by side with then-President Bill Clinton in announcing that the human genome had been sequenced. As of that day, human beings could read their own book of life; they could read the set of instructions contained as a series of A's, T's, G's, and C's along the human DNA molecule. It was a landmark day, but as the preceding example makes clear, biotechnology is bigger than even the human genome. Biotech has to do with manipulating genes within and among all kinds of organisms—from plants to bacteria to human beings (**see Figure 15.1**). But how can genes be manipulated within a single organism or transferred from one organism to another? And what real-world consequences stand to result from this work? This chapter is intended to serve as an introduction to these questions.

15.1 What Is Biotechnology?

In a strict sense, **biotechnology** can be defined as the use of living organisms to create products or to facilitate processes. A brief look at one real-world biotech product can show us what this notion can mean in practice.

Human beings grow to their full height under the influence of a hormone: human growth hormone (HGH), which normally is secreted by the human pituitary gland. HGH's role in promoting growth is, of course, most important during childhood and adolescence. A faulty pituitary gland can greatly reduce the amount of HGH young people have in their system, leaving them abnormally short. For years, the only way to get HGH was to laboriously extract it from the pituitaries of dead human beings, a practice that not only yielded too little HGH to go around but also turned out to be unsafe.

Enter biotechnology, which in the mid-1980s produced synthetic HGH in the following way. Using collections of human cells, the gene for HGH was isolated from the human genome, snipped out of it, and inserted into the *E. coli* bacterium. Each bacterium that took on the HGH gene thereupon began transcribing and then translating this gene, which is to say turning out a small quantity of HGH. These bacteria were then grown in vats by the billions. The result? Collectible quantities of HGH, clinically indistinguishable from that produced in human pituitary glands, manufactured by a biotech firm, and shipped to pharmacies worldwide.

In this account, you can see the essence of many biotech processes. First, genetic information must be understood. In this case, scientists had to find and *clone*, or make an exact copy of, the human HGH gene, and they had to know something about *E. coli* genetics as well. Second, gene sequences had to be manipulated, in this case with the splicing of the human gene into the *E. coli* genome. Third, a process that was invented in a small-scale laboratory had to be ramped up to an industrial scale.

But how do you cut DNA to get a human gene out of the human genome? How do you then get this gene coding for protein inside a group of bacterial mini-factories? These questions can be answered by taking a walking tour of some basic biotech processes. As a starting point, imagine that as with HGH, the goal is to produce quantities of a hypothetical human protein through the use of a living organism.

15.2 Some Tools of Biotechnology

Restriction Enzymes

In the early 1970s it became possible for scientists to cut genomes at particular places, with the discovery of **restriction enzymes**. These are enzymes, occurring naturally in bacteria, that are used in biotechnology to cut DNA into desired fragments. (In bacteria, they serve to cut up foreign DNA, such as that of invading viruses.) In isolating restriction enzymes, scientists found that many of them

Figure 15.1
Bigger Sooner, through Biotechnology
All the Atlantic salmon in the picture are about 14 months old, but the salmon on the left grew to 3 kilograms in this timeframe, while those on the right won't reach this "market weight" for another year. The difference? Scientists at the biotech firm Aqua Bounty Farms spliced a DNA sequence from another fish (an ocean pout) into the wild Atlantic salmon's genome. Addition of this "promoter" sequence changed the production site for growth hormone in the fish—from the pituitary gland to the liver and associated regions. The salmon seem to make more efficient use of growth hormones produced at the new sites and the result is faster growth at early ages. At maturity, however, the genetically engineered salmon will weigh no more than the wild variety. This technology stands to greatly improve production of salmon raised in enclosures for supermarket sale. Ecologists have voiced concerns, however, about the effect these "transgenic" salmon would have on natural salmon populations should the altered fish escape to breed with the wild variety.

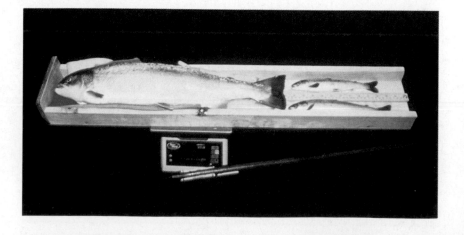

had a wonderful property: They didn't just cut DNA randomly; they cut it at very specific places. Here is how it works with an actual restriction enzyme called *Bam*HI. The two strands of DNA's double helix are complementary, as we've seen, and they also turn out to run in opposite directions. Thus, the sequence GGATCC would look like **Figure 15.2a** if we were viewing both strands of the helix.

Now, the *Bam*HI restriction enzyme will move along the double helix, leaving the DNA alone *until* it comes to this series of six bases, known as its **recognition sequence**, and here it will make identical cuts on both strands of the DNA molecule, always between adjacent G nucleotides (**see Figure 15.2b**). When another *Bam*HI molecule encounters another GGATCC sequence, it too makes cuts, which effectively will be like making a second cut in a piece of rope, giving us rope *fragments* (**see Figure 15.2c**).

*Bam*HI's recognition sequence may be GGATCC, but another restriction enzyme will have a *different* recognition sequence, and will make its cuts between a different pair of bases. Indeed, nearly 1,000 different restriction enzymes have been identified so far, which cut in hundreds of different places. The fact that they make cuts at so many specific locations has given scientists a terrific ability to cut up DNA in myriad ways.

Note that with *Bam*HI, each of the resulting DNA fragments has one strand that protrudes. Restriction enzymes that make this kind of cut are particularly valuable, for they produce "sticky ends" of DNA, so named because they have the potential to *stick to* other complementary DNA sequences. In the fragment on the left in Figure 15.2, for example, the protruding sequence CTAG could now easily form a base pair with *any* piece of DNA whose sequence is the complementary GATC. So great is the need for restriction enzymes that they can now be ordered from biochemical suppliers, much as a person might order a set of socket wrenches from a hardware store.

Another Tool of Biotech: Plasmids

From the overview of manufacturing human growth hormone, recall that the human gene for HGH was inserted into *E. coli* bacteria, which then started turning out quantities of this protein. The question is: How did this human gene get into a bacterium? Several

methods of transfer are available, but for now let us focus on a specific kind of DNA delivery vehicle. As it turns out, bacteria have small DNA-bearing units that lie *outside* their single chromosome. These are the **plasmids**, extrachromosomal rings of bacterial DNA that can be as little as 1,000 base pairs in length (**see Figure 15.3**). Plasmids can replicate independently of the bacterial chromosome; but just as important for biotech's purposes, they can *move into* bacterial cells.

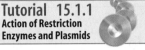

Tutorial 15.1.1
Action of Restriction Enzymes and Plasmids

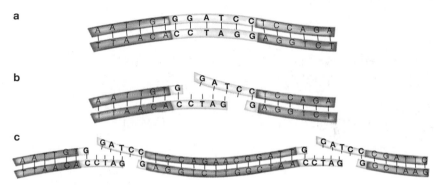

Figure 15.2
The Work of Restriction Enzymes

a A portion of a complementary strand of DNA has the highlighted recognition sequence GGATCC.

b A restriction enzyme moves along the DNA strand until it reaches the recognition sequence and makes a cut between adjacent G nucleotides.

c A second restriction enzyme makes another cut in the strand at the same recognition sequence, resulting in a DNA fragment.

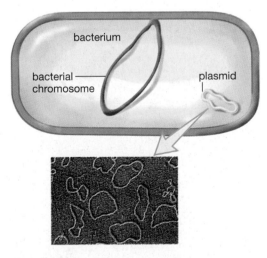

Figure 15.3
Transfer Agent

Plasmids are small rings of bacterial DNA that are not a part of the bacterial chromosome. They can exist outside bacterial cells and then be taken up by these cells. This artificially colored micrograph shows a type of plasmid, from *E. coli* bacteria, that is commonly used in genetic engineering.

How do plasmids do this? Bacteria are capable of taking up DNA from their surroundings, after which this DNA will function—that is, code for proteins—inside the bacterial cells. Appropriately enough, this process is known as **transformation**: a cell's incorporation of genetic material from outside its boundary. Some bacterial cells are naturally adept at transformation, while others, such as *E. coli*, can be induced to perform it by means of chemical treatment. Critically, plasmid DNA can be taken in via transformation and continue to function, as plasmid DNA, inside the bacteria.

Using Biotech's Tools: Getting Human Genes into Plasmids

At this point, you know about a couple of tools in the biotech tool kit: restriction enzymes and the transformation process involving plasmids. Let's now see how they work together. As you can see in **Figure 15.4**, the process starts with a gene of interest in the human genome. The first step is to use restriction enzymes on this human DNA. Knowing, say, the starting and ending sequence of the gene of interest, a restriction enzyme is selected that allows part of the genome to be cut into a manageable fragment

Tutorial 15.2.1
Getting Human Genes into Plasmids

Figure 15.4
How to Use Bacteria to Produce a Needed Human Protein

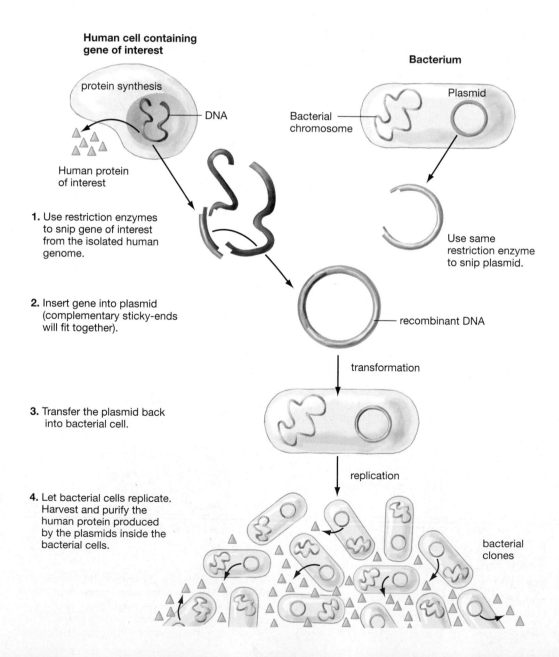

Human cell containing gene of interest

protein synthesis

DNA

Bacterium

Plasmid

Bacterial chromosome

Human protein of interest

1. Use restriction enzymes to snip gene of interest from the isolated human genome.

Use same restriction enzyme to snip plasmid.

2. Insert gene into plasmid (complementary sticky-ends will fit together).

recombinant DNA

transformation

3. Transfer the plasmid back into bacterial cell.

replication

4. Let bacterial cells replicate. Harvest and purify the human protein produced by the plasmids inside the bacterial cells.

bacterial clones

including this sequence of interest, preferably in a sticky-ended form.

Here's where the beauty of restriction enzymes really comes into play. If the same restriction enzyme is now used on the DNA of isolated *plasmids*, the result is *complementary* sticky ends of plasmid and human DNA. In other words, these segments of human and plasmid DNA, through sticky-ended base pairing, fit together like puzzle pieces.

When the DNA fragments are mixed with the "cut" plasmids, that's just what happens: Human and plasmid DNA form base pairs, and the human DNA is incorporated into the plasmid circle. With this, a segment of DNA that was once part of the human genome has now been *re-combined* with a different stretch of DNA (the plasmid sequence). Thus arises the term **recombinant DNA**, defined as two or more segments of DNA that have been combined by humans into a sequence that does not exist in nature.

Getting the Plasmids Back inside Cells, Turning out Protein

To this point, what has been done is to produce a collection of independent plasmids having a human gene as part of their makeup. Remember, though, the goal is to turn out quantities of whatever protein the human gene is coding for. To do this requires a vast quantity of plasmids working away, which means working away *back inside* bacterial cells, and it's here that transformation comes into play. If the plasmids are put into a medium containing compatible bacterial cells, a few of these cells will take up plasmids through transformation, after which the plasmids start *replicating* along with the bacterial cells themselves. As one bacterial cell becomes two, two become four, and so on, the plasmids are replicating away as well, and as the cell count reaches into the billions, collectible quantities of the protein begin to be turned out via instructions from the human gene inserted into the plasmid DNA.

A Plasmid Is One Kind of Cloning Vector

In this example, plasmids were the vehicle that served first to take on some foreign DNA and then to ferry it into working bacterial cells. Note, however, that plasmids are only one vehicle among several that can be used. Collectively, such vehicles are known as **cloning vectors**, meaning self-replicating agents that, in cloning, serve to transfer and replicate genetic material. Next to plasmids, the most common cloning vector is a type of virus that infects bacteria, known as a **bacteriophage**.

15.3 Cloning and the Wider World of Biotechnology

The word *cloning* in the preceding paragraph may seem at once familiar and yet foreign. Thanks to movies and recent real-life events, this word has taken on some sinister implications, as if biological cloning were inherently a Frankenstein-like procedure. But it need be no more threatening than the process just reviewed. To **clone** simply means "to make an exact genetic copy of." A clone is one of these exact genetic copies. In the biotech example you've been looking at, what's being cloned is a gene, copies of which were made in order to obtain quantities of the protein it codes for.

Cloning can involve, however, not just genes but whole organisms. Human beings actually have been making clones for centuries, though by low-tech rather than high-tech methods. A "cutting" taken from a plant and put into soil can sometimes grow into a whole new plant. When this happens, there is no mixing of egg and sperm. The new plant comes entirely from one individual—the original plant—and thus meets the definition of a clone, in that it is an exact genetic copy of another entity. (Every cell in the new plant has exactly the DNA of the parent plant.) Nature makes clones too. Stands of aspen trees amount to a whole series of clones of an original tree; each tree starts out as a "runner" that extends from an existing tree.

All this said, in recent years biotechnology has greatly expanded the range of what can be cloned. The potential for such cloning appears almost limitless—something that has inspired both great optimism and great concern. Biotechnology has now joined with various reproductive technologies to produce clones not just of genes or cells or plants, but of mammals, with the starting cells for these mammals coming from *adult* mammals. The most famous example is Dolly the sheep, cloned by researcher

Ian Wilmut and his colleagues in Scotland in 1997 (**see Figure 15.5**). This is **reproductive cloning**: cloning intended to produce genetically identical animals. Powerful in its own right, it gains added potential when combined with the basic biotech processes you've been reviewing.

Reproductive Cloning: How Dolly Was Cloned

Dolly the sheep is a clone as that term is defined: She is, to a first approximation, an exact genetic replica—in this case, of another sheep. Here's how she was produced (**see Figure 15.6**). A cell was taken from the udder of a six-year-old adult sheep and then grown in culture in the laboratory, meaning that the original cell divided into many "daughter" cells. While this was going on, researchers took an *egg* from a second sheep and removed its nucleus, meaning they removed all its nuclear DNA. Then they placed the udder cell (which had DNA) next to the egg cell (which did not have DNA) and applied a small electric current to the egg. This had two effects: It caused the two cells to fuse into one, and it mimicked the stimulation normally provided when a *sperm* cell fuses with an egg. With this, the udder cell DNA began to be reprogrammed. As a result of this reprogramming, the fused cell started to develop as an embryo. (Though the egg cell had its DNA removed, it still contained all kinds of egg-cell proteins whose normal function is to trigger development of an embryo. It was these factors still in the egg that reprogrammed the donor-cell DNA.) After the embryo had developed to a certain point, the researchers implanted it in a third sheep, which served as a surrogate mother. The result, 21 weeks later, was the lamb Dolly, who has since given birth to two sets of her own lambs.

Figure 15.5
Revolutionary Sheep
Dolly, the first mammal ever cloned from an adult mammal, is shown here as a young sheep with her surrogate mother. Note that Dolly is white-faced, like the sheep she was cloned from, while her surrogate mother is black-faced.

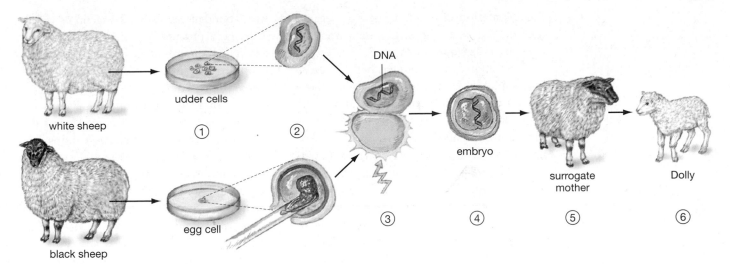

udder cells

① ②

DNA

③ embryo ④

surrogate mother

Dolly

⑤ ⑥

white sheep

egg cell

black sheep

Every cell in Dolly has DNA in its nucleus that is an exact copy of the DNA in the six-year-old donor sheep. Thus Dolly is a clone of that sheep. There was no second parent contributing chromosomes to Dolly, no mixing of genetic material at the moment of conception—just a copying of one individual from another.

Reproductive Cloning and Recombinant DNA

Ian Wilmut and his colleagues were not so much interested in cloning a sheep as they were in coupling the power of reproductive cloning to the techniques of *recombinant DNA* that you looked at earlier. Indeed, the workers at Wilmut's research institute then went on to produce Polly, who is not only a clone, but a clone who has had a human gene inserted into her genome. People who suffer from the disease hemophilia B are missing a blood-clotting protein called *factor IX*. Using the same kinds of DNA-cutting techniques described in this chapter, scientists snipped the gene that codes for factor IX out of some human DNA and inserted it into a sheep donor cell. This cell was then fused with an enucleated egg, just as with Dolly, and the result was Polly. Now, here's the importance of this: Polly secretes the human factor IX in her milk. She is expressing a needed human protein. You might say: Hasn't our protein "factory" now simply become a sheep instead of a bacterium? That's true, but because, compared to a bacterium, the sheep's genetic machinery is more like a human being's, some of

the proteins the sheep produces are in a more useful chemical form than those produced by bacteria.

Cloning and Xenotransplantation

This only scratches the surface of the potential for such "transgenic" clones (meaning clones that carry genes from other species). To give but one other example, thousands of people die each year because their liver or heart or kidneys failed and they could not find a compatible organ donor. In the United States, about a third of the people on organ-donor waiting lists die before they can receive a transplant. The essential problem is that the human immune system attacks any organ that it recognizes as "foreign," and many donated organs are recognized in this way.

Pig organs turn out to be about the same size as human organs, but they elicit an even stronger human immune-system response than do human-donated organs. Imagine, however, taking a pig donor cell and, by snipping out parts of its DNA, eliminating that cell's ability to produce the proteins that are perceived as foreign by the human immune system. Now fuse that cell with an enucleated egg, implant it in a surrogate mother, and the egg grows into a pig. Repeat this process enough times and you have lots of pigs. These are pigs, however, whose DNA was altered back when they were but single cells. Thus, cells making up the blood vessels of their organs do not have, protruding from their surfaces, proteins perceived as foreign by the human immune system. This means that these organs can be "harvested" for

Tutorial 15.2.4
Cloning Dolly

Figure 15.6
How Dolly Was Cloned

1. A cell was taken from the udder of a six-year-old white sheep and then allowed to divide many times in the laboratory. Meanwhile an egg was taken from a second black sheep.

2. One of the resulting udder cells was selected to be the "donor" cell for the cloning. Meanwhile, using a slender tube called a *micropipette*, researchers sucked the DNA out of the egg.

3. The donor cell and egg were put next to each other, and an electric current was applied to the egg cell.

4. This caused the two cells to fuse and prompted an activation that reprogrammed the donor-cell DNA. This caused the fused cell to start developing as an embryo.

5. After some incubation, the embryo was implanted in a third sheep, which served as the surrogate mother.

6. This mother gave birth to Dolly the sheep, which grew into an adult.

transplantation into human beings, because they do not elicit a strong human immune response (**see Figure 15.7**).

Work along these lines has been proceeding rapidly in the last few years, but it has a serious question to answer. Pigs carry a certain class of virus that seems to be able to move into human tissue. Viruses in this class have never been shown to be harmful to pigs or to humans, but what happens when they're introduced to humans through the unnatural route of **xenotransplantation**: the transplanting of organs from one species to another? One of the big questions for pig xenotransplantation (ZEE-no-trans-plan-ta-tion) is whether concerns about these viruses can be laid to rest. Some scientists think there is a more promising route to replacement organs: use of a special variety of human cells, called *stem cells*. You can read more about stem cells in Chapter 27 on page 628.

Human Cloning: Just Around the Corner?

Reproductive cloning may interest scientists because it has the potential to yield medicines or harvestable organs. It has captured the attention of the average person, however, because it raises the possibility of *human* cloning.

Such cloning is no longer the stuff of science fiction. It is probably achievable, most experts would say, through roughly the same process that was used to produce Dolly—take a cell from a human being, fuse it with an enucleated human egg cell, and implant the fused cell in the uterus of a woman who is willing to bring the resulting child to term. The fact that such a thing can be done doesn't mean it will be done, of course; but early in the 2000s, there

were abundant signs that it was on its way. In 2001, a U.S. physiology professor and an Italian fertility doctor announced that they would attempt to produce a human baby through cloning in the next two years in a "Mediterranean country." Meanwhile, a religious sect headquartered in Canada has made clear that it regards human reproductive cloning as a central tenet of its faith. With money, and with numerous female members willing to serve as egg donors and surrogate mothers, the sect seems well positioned to produce a human clone. And if this group doesn't do it, most observers say, somebody else will.

The prospect of a human clone is so dizzying that it's worthwhile to think about what such a person would represent in biological terms. He or she would be a genetic replica of the person who provided the donor cell with the DNA in it. Note that this donor does not have to be an adult, or even be alive. The Canadian sect just mentioned is working on behalf of an American couple whose 10-month-old child died in a hospital operation. The parents froze some cells from the boy while he was still alive, and it is these cells that may be used as the donor-DNA cells. Should the procedure succeed, the parents would have a baby who has exactly the same genetic makeup as their dead son.

One helpful way to think of this new child's biological status is in terms of a more familiar concept, that of an identical twin. Any set of identical twins shares an identical genetic makeup. Indeed, at one point, early in their mother's pregnancy, identical twins are a single organism—the single cell that was produced when their father's sperm fused with their mother's egg. In all pregnancies, this single cell divides—one cell becoming two, two becoming four, and so on—to produce a fully formed human being. In the case of identical twins, however, very early in this process, the separate cells that are produced through cell division go on to become separate human beings.

A human clone would, then, be like an identical twin of the person who provided the donor DNA. An identical twin develops, in a sense, from a "donor" cell—one of the cells that resulted from the original fertilized egg. A human clone likewise develops from a donor cell, only this cell is one that may be donated by an adult, taken from a child, and so forth.

This gives us some basis for thinking about the nature of a human clone. Identical

Figure 15.7
First Cloned Litter
In March 2000, the research company that funded the cloning of Dolly the sheep announced the cloning of five pigs: Millie, Christa, Alexis, Carrel, and Dotcom. Researchers are hopeful that cloned pigs may one day provide organs that can be transplanted into human beings.

twins may have some striking similarities, but they are not identical persons. This is so because no person is shaped solely by genes. Each of us is shaped by our "environment" as well, meaning everything from the position we have in our mother's uterus to the teachers we have in school. Identical twins often share very similar environments (in the uterus, in the home, and so forth), but a human clone might have wildly dissimilar environment from the DNA donor. Thus, on average, we would expect a DNA donor and his or her clone to be less alike than identical twins.

Be this as it may, the prospect of a human clone still has the power to stun us. Imagine a case in which the donor DNA comes not from a dead infant, but from a healthy adult. It is possible, of course, that such an adult might want to create a "Mini-Me," meaning a genetic replica of himself or herself. One of the motives more frequently cited for human cloning, however, is a situation in which two individuals wish to become parents and yet both are infertile. They decide they would like to have a child who carries the genes of at least one of them. Thus, one of the parents would donate DNA and have a child who is, in genetic terms, much like his or her identical twin.

Human cloning obviously is an area fraught with ethical considerations. To name just one of these, the reproductive cloning of any mammal is currently a hit-or-miss process. Hundreds of attempts may be required to produce a viable embryo that can be successfully implanted in a uterus and brought to term. And the clones that are produced often have physical defects or are abnormally large. Given this state of knowledge, is it ethical to even consider cloning a human being?

15.4 Other Biotechnology Processes: PCR

We'll now return to a review of some of the processes used in biotechnology, with an eye toward seeing how each of them is used in a real-world setting. Thus far, you've seen a couple ways of turning out human proteins through the use of recombinant DNA. Sometimes, however, the goal is not to get proteins, but to "amplify" DNA—to get many copies of a segment of DNA when the starting sample of DNA is small. The technique for doing so is called the **polymerase chain reaction**, or PCR for short (**see Figure 15.8**).

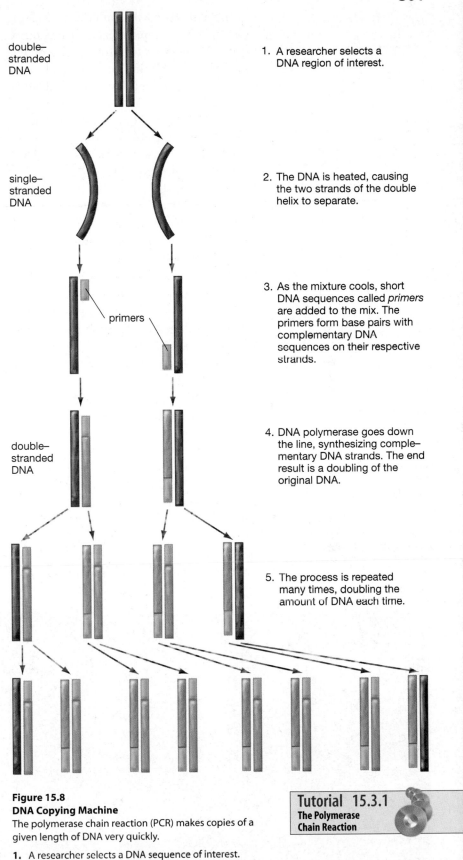

double–stranded DNA

single–stranded DNA

primers

double–stranded DNA

1. A researcher selects a DNA region of interest.

2. The DNA is heated, causing the two strands of the double helix to separate.

3. As the mixture cools, short DNA sequences called *primers* are added to the mix. The primers form base pairs with complementary DNA sequences on their respective strands.

4. DNA polymerase goes down the line, synthesizing complementary DNA strands. The end result is a doubling of the original DNA.

5. The process is repeated many times, doubling the amount of DNA each time.

Figure 15.8
DNA Copying Machine
The polymerase chain reaction (PCR) makes copies of a given length of DNA very quickly.

Tutorial 15.3.1
The Polymerase Chain Reaction

1. A researcher selects a DNA sequence of interest.

2. The DNA is heated, causing the two strands of the double helix to separate.

3. As the mixture cools, short DNA sequences called *primers* are added to the mix. The primers form base pairs with complementary DNA sequences on their respective strands.

4. DNA polymerase goes down the line, synthesizing complementary DNA strands. The result is a doubling of the original DNA segment of interest.

5. The process is repeated many times, doubling the amount of DNA each time.

The essence of this molecular copying machine is very simple: Heat a starting quantity of DNA until the two strands of its double helix separate, resulting in two single strands of DNA. Then add a collection of individual DNA nucleotides, and DNA polymerase (which, remember, goes down the line on single-stranded DNA affixing nucleotides to its available bases). Also add two DNA "primer" sequences—short sequences of single-stranded DNA that act as signals to DNA polymerase, saying "start base pairing here." As the mixture cools, primers will attach to each of the now-separate strands of DNA. Then the DNA polymerases go down the line, starting from the primers, linking nucleotides to the template DNA and thus producing strands that are complementary to the original strands. The result is *two* double-stranded lengths of DNA, both identical to the original double strand. In short, the DNA sample has been doubled in one copying "cycle." Then the entire process is repeated. Since each copying cycle takes only 1–3 minutes, by the time 90 minutes have passed, millions of copies of the original DNA strand can be created.

This technique is useful any time researchers want to quickly get a quantity of DNA, but it is particularly useful when the starting sample of DNA is very small. Crime scenes may contain tiny amounts of human blood or semen, for example, but larger amounts are needed for testing.

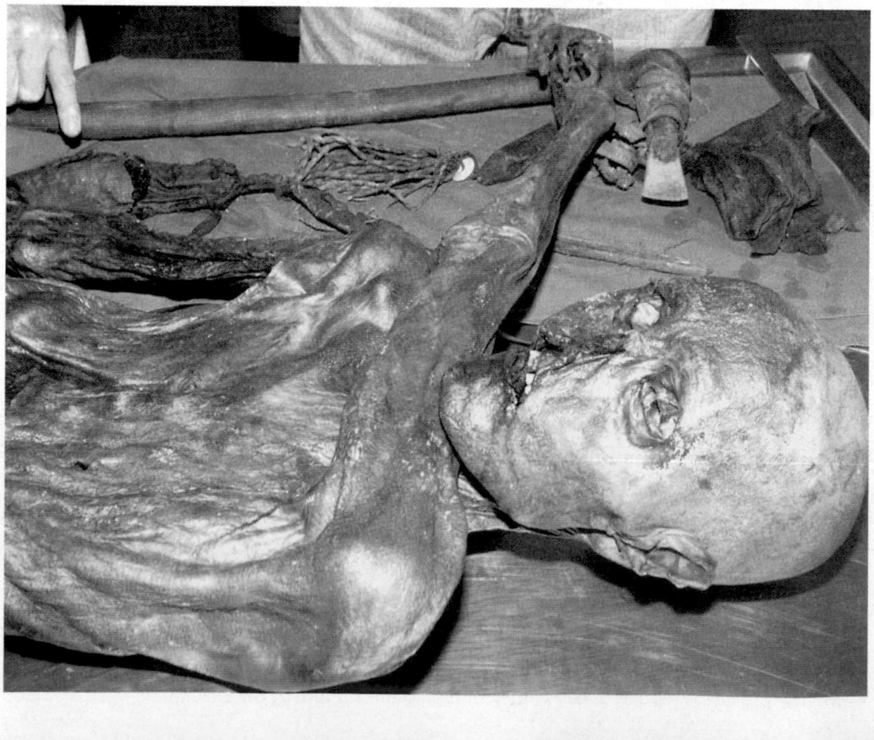

Figure 15.9
Making Ancient DNA Useful
The world's oldest mummy, the Italian Iceman, also known as Otzi, was found on a glacier in the Italian Alps in 1991. Though the Iceman lived some 5,200 years ago, scientists were able to extract small, intact samples of his DNA. Through the PCR technique, this DNA was copied or "amplified" sufficiently to allow scientists to learn something about his genetic makeup.

PCR's uses go way beyond criminal investigations, however. It can, for example, be used to check for infections by microorganisms. Lyme disease is caused by a bacterium, *Borrelia burgdorferi*, that is carried by deer ticks. But is a patient who complains of aching joints suffering from Lyme disease or from common arthritis? Doctors can get a sample of joint fluid from the patient and "PCR-up" the DNA within it, using primers for *B. burgdorferi* DNA sequences. If the DNA of the bacterium is detected, the patient has Lyme disease.

In an entirely different realm, back in 1991, the 5,200-year-old remains of a mountain traveler who came to be known as "Ice Man" were found in a glacier in Italy's Tyrolean Alps (**see Figure 15.9**). One of the intriguing questions about Ice Man was, whose ancestor is he? Who is he most closely related to among modern-day humans: northern Europeans, southern Europeans, or people from outside Europe altogether? Scientists were able to get small samples of his DNA, copy them through PCR, and determine that he is more closely related to central and northern Europeans than to people from any other region, including the Mediterranean. (The researchers compared one of Ice Man's DNA sequences to the same sequence from modern-day residents in different countries.)

Despite the fact that Iceman had been frozen in a glacier, it was not easy to get a usable DNA sample from him. When an organism dies, its DNA begins to degrade very quickly, and Iceman's DNA was badly degraded. That any sequence at all could be obtained from a 5,200-year-old mummy is, however, testament to PCR's power.

PCR was invented in 1983 by Kary Mullis, then a researcher at a California biotech firm. So important has the process become to science that in 1993, Mullis was awarded the Nobel Prize in Chemistry for his achievement.

15.5 Visualizing DNA Sequences

In working with DNA, scientists need a way to compare one segment of DNA to another. Does a given segment of DNA divide into three segments or five when a particular restriction enzyme is used on it? What is the exact sequence of bases *within* any of these segments? This question refers to DNA "sequencing," whose importance becomes apparent when you recall that, among the thousands of bases that make up the human hemoglobin gene, it is the substitution of *one base* (a T for an A) that brings about the disease sickle-cell anemia. Scientists can both sequence DNA and judge the sizes of larger DNA fragments through a process of visualization that is a kind of scientific "bar coding."

How can something as small as a segment of DNA be visualized? The key is that DNA carries a negative electric charge. As such, it is attracted to a positive charge. This fact helps scientists stage a kind of sprinting match of DNA fragments in order to separate them by size.

Running DNA through a Gel

There is a material used in biotechnology, called a gel, that in some ways is like the food gelatins we're used to. A quantity of this gel has running through it a series of microscopic spaces (or "pores"). Given this, the gel acts like a sifter: Substance such as DNA can pass through it, but only with some coaxing and at a controlled rate.

The starting length of DNA might be a whole genome's worth, but more likely will be some subset of a genome. Whatever the case, the process begins with using a given restriction enzyme to chop up a starting sample—from a collection of cells—into a quantity of fragments. Chopping up these segments with the same restriction enzyme results in a huge number of identical copies of fragment 1, a huge number of copies of fragment 2, and so on. As you know, restriction enzymes don't make evenly spaced cuts

throughout a length of DNA; they cut at specific recognition sequences. Thus fragment 1 will be a different size than fragment 2, which in turn will be a different size than fragment 3, and so on.

The question is, how do researchers learn the size of each fragment? The answer comes from the DNA sprint through the pores in the gel. As you can see in **Figure 15.10**, a tray is used that has two critical features. First, it has electrodes at each end—one negative and one positive—and second, it has a series of indentations, or "wells," located on the same end of the tray as the negative electrode.

The process begins with pouring gel into the tray. Then the tray is placed in a solution that conducts electricity. Then the collection of DNA fragments is poured into the wells, which lie on the end of the tray *opposite* that of the positively charged electrode. In electricity, opposites attract, and the negatively charged DNA fragments are thus drawn to the positively charged end of the tray. But to get there, they must make it through the gel. Given this obstacle, it is the *smallest* of the fragments that will travel the farthest in a given amount of time, while the largest of the fragments will travel the least amount of distance.

Learning the Size of the Fragments

What this means is that the fragments are separated out by size. A fragment that is, say, one thousand nucleotides long will make it farther through the gel than a fragment that is two thousand nucleotides long. Researchers can *see* where each fragment has ended up in the tray because, after putting the DNA fragments in the wells, they pour in

Tutorial 15.4.1
Gel Electrophoresis of DNA

Figure 15.10
Visualizing Lengths of DNA
In gel electrophoresis, negatively charged DNA fragments travel through the porous gel toward the positive end of the tray when a charge is applied to the gel. Large fragments, however, do not travel as far in a given amount of time as small fragments. Thus the fragments separate out by size. The first "stripe" at the top of a lane represents a collection of fragments of a given size, the second stripe a second collection of fragments of a smaller size, and so on.

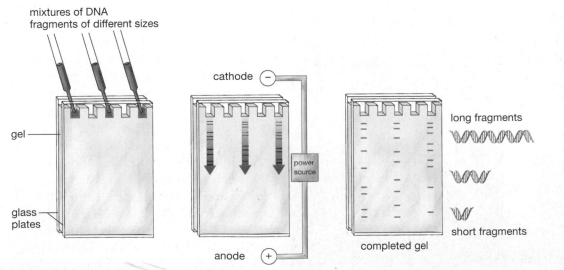

mixtures of DNA fragments of different sizes

gel

glass plates

cathode −

power source

anode +

long fragments

short fragments

completed gel

Figure 15.12
Great Day for Science
On June 26, 2000, President Clinton stood with Celera Genomics Corporation Scientist Craig Venter (left) and Human Genome Project leader Francis Collins to announce the completion of the first draft of the human genome.

a dye that stains the DNA. The result is a series of bands within the gel (similar to supermarket bar codes), with each band representing a huge collection of DNA fragments of about the same size. The size of each fragment is learned by putting, in one "lane" of the tray, a series of so-called reference fragments, meaning fragments of known size. If one of the reference fragments was 10,000 nucleotides long, and if one of the sample's fragments traveled far enough through the gel to line up perfectly with it, the researchers know that their sample fragment is roughly 10,000 nucleotides long as well.

This technique of sizing DNA fragments is called **gel electrophoresis**, a name that makes sense considering the gel that's used, the electricity, and the fact that *phoresis* is a Greek word meaning "to be carried."

Sequencing Requires a Different Operation

Characterizing a stretch of DNA by the number and size of its fragments is not, however, the same as sequencing. Knowing the size of a given fragment is not the same as knowing the sequence of A's, T's, C's, and G's *within* that fragment. The details of sequencing need not concern us here. Suffice it to say that there are two primary "manual" methods of sequencing small segments of DNA, both of which use gel electrophoresis in a way that produces fragments that differ in size by one nucleotide. The bands that result, corresponding to individual A's, T's, C's, and G's, can be "read" by a researcher, thus yielding the sequence of the DNA segment. The sequencing of larger stretches of DNA is now facilitated by automated sequencing machines, which analyze bases that have been "tagged" with fluorescent dyes (**Figure 15.11**).

Having learned how separation of fragments is done, you might well ask, to what end? Electrophoresis is used routinely in all kinds of scientific research. But to focus on one very specific use of it, imagine that two of the lanes of DNA fragments in Figure 15.10 came from two different suspects in a violent crime, while the third lane came from the person who indisputably did commit the crime—from blood spilled by the perpetrator at the crime scene. If a suspect's bands *line up* with those of the perpetrator, this can be evidence that the suspect *is* the perpetrator. You can learn about this application of gel electrophoresis by looking at "DNA in the Courtroom".

15.6 Decoding the Human Genome

When President Clinton stood at a podium with scientists Craig Venter and Francis Collins in June of 2000 to announce completion of the "first draft of the human genome," what was everyone celebrating (**see Figure 15.12**)? You've seen that DNA is composed of chemical building blocks called nucleotides, and that each nucleotide contains one of four chemical bases: A, T, G, or C (short for adenine, thymine, guanine, and cytosine). The entire complement of human DNA can be thought of as a single sequence that is about 3 billion bases long. By June of 2000, two rival teams of scientists—one led by Venter, the other

DNA *in the Courtroom*

A rape has been committed, but no fingerprints were left at the scene and there are no witnesses, apart from the victim. A suspect has been identified by the victim, but what physical evidence is there to corroborate her testimony? How can we tell whether the suspect is in fact the criminal?

In the mid-1980s, British geneticist Alec Jeffreys realized that the individual patterns each of us carries in our DNA might serve as just this kind of physical evidence. As a result of his work, courts in the United States have, since 1988, accepted so-called **forensic DNA typing**—better known as "DNA fingerprinting"—in criminal cases.

The essence of this work is to get two sets of physical samples: first, a sample left by the perpetrator at the crime scene in such forms as blood or semen; and second, a blood sample from the suspect. By comparing the DNA patterns of both samples, odds can be established as to whether the suspect *is* the perpetrator. (Conversely, it would be possible to use a victim's blood sample and test it against, for example, blood found on the suspect's clothing.)

This is evidence, however, that can be used both to convict and acquit. If the DNA patterns of suspect and perpetrator do not match up, a prosecution often ends its case right there, the innocence of the suspect having been established. (If the evidence reveals the presence of someone *else*'s semen, how could the suspect have been involved?)

So, what are these individual DNA patterns that each of us carries? Human genomes are filled with short, repeated stretches of DNA called *short tandem repeats*, or STRs. One of these might be the nucleotide sequence GGAGG repeated over and over. Such repeats don't code for any protein; as such, nature has put no premium on their remaining as they are. Over evolutionary time they have mutated—added or lost nucleotides—at very high rates. This gives them their usefulness in forensics: At a given location in the genome, one person might have 15 copies of an STR, while another person will have 30. Thus STR segments differ in *length*, and as you saw in the main text, different lengths of DNA can be visualized.

Forensic work begins by taking both a sample of the suspect's DNA and a sample of DNA from the crime scene. Thanks to the PCR copying process, only minute biological material from a perpetrator need to be found at crime scenes. Mere specks of semen or blood will provide enough DNA to work with. Next, DNA from both suspect and crime scene is chopped up with a restriction enzyme and then run through gel electrophoresis, thus separating DNA fragments in the samples by length. What ultimately results are two patterns of DNA fragments: The pattern

from the suspect and the pattern from the perpetrator. It is then a matter of comparing the two (**see Figure 1**). In crime cases, it is common to look not just at one length of DNA, but at several. The FBI looks at 13 DNA sequences, each about 100 to 600 bases long.

This is evidence that can be used to both convict and acquit.

Once this lab work is done, statistics plays a part in establishing guilt. Say the suspect's first STR pattern matched the pattern found in the crime-scene sample. Looking through the population at large, what fraction of people can be expected to *share* this pattern? The common way of establishing this number has been to analyze samples from

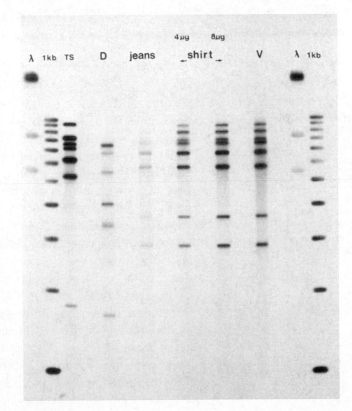

Figure 1
Telltale Stains
A defendant in this real-life murder case claimed that blood stains found on his clothing (see the rows under "shirt") came from his own blood rather than from the blood of a young woman who was stabbed to death. The DNA evidence said otherwise. Look at the pattern in the row under V (for victim) and see how closely it matches the pattern in the shirt rows. The probability that the blood on the defendant's clothing came from anyone but the victim was 1 in 33 billion. (Courtesy of Cellmark Diagnostics, Inc., Germantown, Maryland.)

continued

DNA *in the Courtroom* continued

blood banks. If a given STR pattern is found in 1 out of 25 samples, it's assumed that 1 person in 25 has this pattern. This is not the kind of statistic on which to convict a person of rape, but say you are looking at five DNA segments. Imagine that the matches between suspect and perpetrator continued with each of these segments, and that the odds of sharing in each of the remaining four were 1 in 50, 1 in 30, 1 in 50, and 1 in 25. Using the rule of multiplication, the odds of a person other than the suspect sharing in all *five* patterns can thus be calculated:

$$1/25 \times 1/50 \times 1/30 \times 1/50 \times 1/25 = 1/46,875,000$$

This simple calculation, then, seems to have established that only one person in 46.8 million could be expected to have the same five STR patterns as the suspect. Thus the claim, "sure the samples match, but it could have been someone else," seems to evaporate. (For more on the rule of multiplication, see "Proportions and Their Causes: The Rules of Multiplication and Addition" in Chapter 11, page 214.)

DNA fingerprinting has revolutionized law enforcement in more ways than one. It can be used, as just described, to link a known suspect to a known crime. Increasingly, however, genetic fingerprinting is being used to link *unknown* suspects to given crimes. It can do this because DNA fingerprints (or "profiles") can be stored and then retrieved later. Seven states now require all felons to provide DNA samples; another 25 states require such offenders as robbers, rapists, and killers to provide samples. DNA databanks result from this practice, and by using them, police can make what are known as *cold hits*. Imagine a situation in which a crime is committed, and no suspect was identified, but a DNA sample has been retrieved. A DNA profile will be created from this sample—a profile that can be matched against all those that are stored in a state's DNA databank. If DNA found at a crime scene matches a felon's stored profile, that in itself is sufficient evidence to make an arrest. The State of Virginia alone has thus far made more than 300 cold hits from its databank of about 135,000 felons.

DNA evidence has also worked on the other side of justice. By the summer of 2000, eight people in U.S. prisons had been freed from death row after having been exonerated through DNA testing.

by Collins—had sequenced enough of the human genome that they knew the order of about 90 percent of these bases. That was good enough for all concerned to say that a first draft of the human genome had been produced. (You may wonder whose genome they sequenced. Venter's team used DNA taken from five people: two Caucasian men, and three women—one African American, one Latino, and one of Chinese descent. Collins's team used a composite sequence from a dozen anonymous individuals.)

The sequencing of the human genome began back in 1990 as a single, unified effort. This was the publicly funded Human Genome Project, or HGP, which involved researchers from 16 laboratories in six countries. Under Collins's leadership, HGP seemed on track, by the late 1990s, to deliver a first draft of the human genome by 2005. Then, in 1998, Craig Venter, chief scientist of the private biotech company Celera Genomics, announced that his company would deliver the sequence of the human genome not in 2005, but in 2001. Thus was sparked a race between the two groups that ended with a draw of sorts when both Venter and Collins stood with President Clinton for the human genome announcement.

By February of 2001, both Celera and the HGP were ready to *publish* what they had learned about the human genome from their sequencing of it, and the groups did so through much-heralded scientific papers that appeared in separate scientific journals. Why the eight-month lag between sequencing and publishing? Recall that biotech can be big business. Celera intends to make money from its human genome sequence; it regards much of its human genome data as proprietary—as privately held information that has a commercial value. It expects to sell parts of its human genome database to pharmaceutical companies and other buyers. The publication of the human genome data was held up in part because of protracted negotiations about what kind of access the scientific community would have to Celera's genome data—a turn of events that illustrates the degree to which commercial concerns are intertwined with basic science in the world of biotech.

The Significance of Decoding the Human Genome

So, why should the average person care about the decoding of the human genome? An analogy may be helpful here. Imagine that you go to a library and ask for a particular book on the history of Russia, only to be told that the library has no idea whether it owns a copy of the book. The librarian is pretty sure that the library has books on Russian history, but really doesn't know where they are in the library's stacks. As you start walking around this library, you realize that there has been no orderly cataloguing of the library's holdings. You can find, in the library's electronic catalogue, some information about individual books, but you have no idea where one book is in relation to another.

This imaginary library bears comparison to the human genome and the knowledge we had of it until recently. For decades, scientists doing genetic research were confronted with a genetic *library* (the human genome) that had lots of books in it (genes), but little cataloguing of these books. In a real library, you might go to its electronic catalogue and type in: Huckleberry Finn, or even just Huckle*, after which you could find the book itself and look through it, word by word and letter by letter. If you had a counterpart *genetic* catalogue, all you would need to type in would be, say, a few dozen nucleotides from a DNA sequence of interest. With that information, the genetic catalogue would direct you to the "book" you're after, which is all the information that exists about the sequence: its location on a particular chromosome, its complete nucleotide sequence, and anything else scientists have learned about it.

This kind of genetic catalogue has now been created through the decoding of the human genome. Cancer researchers no longer have to hunt laboriously through the scientific literature to learn whether a gene they have identified is already known to other scientists; they can search human genome databases to find out. Pharmaceutical companies that have identified DNA strings that seem promising don't have to sequence long stretches of DNA that lie "upstream and downstream" from the

promising DNA. They can go to the Internet—where Human Genome Project data are posted—to find out what the surrounding sequences are. Basic scientists can look for *patterns* throughout the genome that might provide a key to how the different parts of the genome work together. In short, the decoding of the human genome has greatly accelerated the pace at which we are acquiring knowledge about human genetics. If you look at **Figure 15.13**, you can see one likely effect of this acceleration: Scientists are identifying disease-related genes at a faster rate with each passing year.

Surprises from the First Drafts

Completion of the first draft of the human genome allowed scientists from Celera and HGP to take a step back and consider the genome as a whole—an analysis that yielded several surprises. First, scientists learned that the human genome contains far fewer genes than most biologists thought. For years, the assumption was that the human genome contained about 100,000 genes. Celera and the HGP, however, came up with estimates of between 26,000 and 40,000 genes, with both teams agreeing that the final number is likely to be about 30,000. This figure actually holds a double surprise because it means that, contrary to our previous assumptions, human beings do not have five times or ten times as many genes as, say, fruit flies or roundworms.

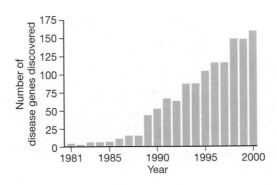

Figure 15.13
Finding Disease Genes at a Faster Rate
Through the year 2000, scientists had discovered 1,112 disease-related genes. Note, however, the increasing rate at which these genes are being discovered. This steady march upward can be partly attributed to the decoding of the human genome that began, in an organized way, in 1990. (Adapted from Leena Peltonen and Victor A. McKusick, *Science*, 16 February 2001, 1225.)

If you look at **Figure 15.14**, you can see that we humans, with our vision, speech, and 10 trillion cells, probably have only 11,000 more genes than the microscopic roundworm *C. elegans*, which is eyeless, speechless, and made up of 959 cells.

The 30,000-gene estimate for human beings has been questioned—some prominent researchers think 100,000 will still be more like the real number—but if 30,000 is accurate, the challenge to science is to understand how human beings do so much more with their 30,000 genes than *C. elegans* does with its 19,000. One part of the answer seems clear: Human beings have a relatively high proportion of multifunctional genes, meaning genes capable of coding for more than one protein. If you look at **Figure 15.15**, you can see the "alternative splicing" process by which organisms such as ourselves can get several proteins from a single gene. The upshot of this and other genetic factors is that, though we may have only 30,000 genes, we are likely to have hundreds of thousands of proteins functioning within us.

The unveiling of the human genome brought a second surprise that fits logically with the first: Very little human DNA codes for protein. From the 1970s forward, scientists have known that the noncoding portion of the genome dwarfs the protein-coding portion. For many years, however, scientists assumed that the coding portion made up 3–5 percent of the human genome. Not so, according to Celera and the HGP. At most, 1.5 percent of the genome codes for protein (HGP's estimate), and the figure may be as low as 1.1 percent (Celera's estimate). So, what function does the other 98.9 percent of the genome serve? For more on this, see "Making Sense of Junk DNA" on page 280 in Chapter 14.

Limitations of Human Genome Sequencing

Human beings may now possess nearly complete knowledge of the human genome sequence, but that does not mean we possess nearly complete knowledge of human genetics. What sequencing will provide is a kind of basic reference work for human genetics—a quick means of *getting to* more complete knowledge. An analogy may be helpful here. If you were interested in learning more about everyone you had gone to a summer camp with years ago, and you happened to run across a list of names and addresses of these people, that would be enormously *helpful* information. But it wouldn't begin to tell you what these people are doing now, whom they interact with, or any number of other things you might want to know about them. Just so with the genome: It is entirely possible to know that a gene exists in a certain place in the genome, what its sequence is, and what its protein product is, and yet not have a clue about what this protein *does*, or what other proteins it may interact with. Human genome sequencing has put genetic names and addresses together; coming to a complete understanding of all this catalogue's genes and their protein products is the work of the twenty-first century. It is now becoming clearer, however, what this work will entail.

Saccharomyces cerevisiae
(baker's yeast)

6,034

Drosophila melanogaster
(fruit fly)

13,061

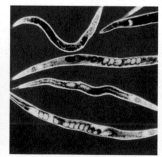

Caenorhabditis elegans
(roundworm)

19,099

Arabidopsis thaliana
(mustard plant)

25,000

Estimated number of genes:

Figure 15.14
Not as Much Difference as We Thought
At one time, scientists assumed that the human genome contained about 100,000 genes. Genome sequencing revealed, however, that the human genome probably contains about 30,000 genes. This is only about 11,000 more genes than the tiny roundworm *C. elegans*, and about 17,000 more than the *Drosophila* fruit fly. Scientists can make comparisons among genomes because, though the human genome has gotten most of the attention, the genomes of several other organisms have now been sequenced as well, including those shown here.

15.7 The Next Phase in Genetics: Genomics and Proteomics

Once an organism's genome is sequenced, the *next* steps in understanding its genetics involve two scientific disciplines you'll be hearing a good deal about in coming years. The first of these is **genomics**, meaning the study of sets of genes within or across genomes. The second is **proteomics**, meaning the study of sets of proteins with a focus on their functions and their interactions.

Genomics has to do with "mining" an organism's genome data to see what functions are served by the different DNA sequences. Which sequences are genes? Which sequences are related to, say, brain functioning? At what places in the genome do all individuals tend to be identical, and at what places do they differ?

The importance of proteomics is encapsulated in the metaphor used in Chapter 14: In living things, DNA is the cookbook, not the cook. It is *proteins* that actually carry out most of the myriad tasks that take place in each organism. Thus, to really understand how living things work, we must understand what their proteins do. Proteomics seeks to do this by looking at proteins both as they are expressed, and as they carry out their functions (which generally entails interacting with other proteins). Scientists now speak of an organism's *proteome*—all the proteins it expresses—in the same way they speak of its genome, meaning its entire complement of genetic information.

Biology and Computer Science

Genomics and proteomics turn out to be a natural fit with computer technology, for two reasons. First, it would be impossible to analyze something as complex as a genome *without* computers. Second, strange as it may seem at first glance, genetic information largely exists as digital information—the very kind of information computers are designed to handle.

You can appreciate the complexity of genetics by recalling that the human genome is made up of 3 billion DNA nucleotides and that, within this expanse of DNA, there are perhaps 30,000 genes. These genes may in turn code for hundreds of thousands of proteins, each one of which is likely to interact with half a dozen other proteins. As if that weren't enough, many researchers find themselves looking across *several* genomes—say, that of *Drosophila* and of human beings—to see if genes found in one organism have counterparts in another.

Fortunately for biologists, this unimaginable complexity is largely encoded in what can be thought of as digital information, if we don't stipulate that "digital" means "represented in numerals." To understand this, think of computers, with their strings of 0s and 1s encoding information. Each 0 or 1 is an individual unit of information (a bit) that is part of an informational sequence (a byte). Just so, information in DNA is encoded in discrete units, called bases, that are ordered into informational sequences called genes.

Information encoded in this way is tailor made for computer analysis. Because of this, and because of the complexity of the genomes being analyzed, genetic research has become intimately linked with computer science. As one sign of this linkage, consider that in August 2000, the IBM Corporation announced that biologists had replaced physicists as the main scientific users of supercomputers. A new subdiscipline of biology, called *bioinformatics*, is turning out researchers who are grounded in both biology and computer science.

What has this linkage meant in practice? To give but one example, for years the main method of hunting down genes linked to human disease was a technique called positional cloning. It entailed looking for families

Figure 15.15
Alternative mRNA Splicing
The human genome may contain only 30,000 genes, but that does not mean these genes code for only 30,000 proteins. Through the process of alternative mRNA splicing, it is possible for a single gene to code for several proteins. In the figure, a single mRNA transcript, containing five "exons" or protein-coding portions, has been transcribed from a single gene. The human genetic machinery can then splice these exons together in alternative ways to yield different proteins—A or B in the figure.

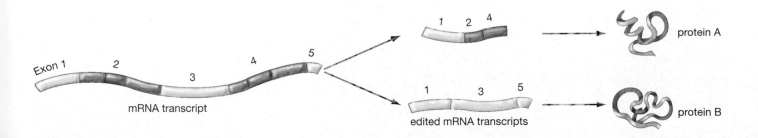

in which diseases seemed to be passed on over several generations, interviewing members of these families, cloning segments of their DNA, and then looking for mutations in these segments that might be causing the disease. This lengthy, laborious process is now being supplanted by something known as sequence-based discovery, in which scientists use computers to scour human genome databases, looking for sequences that match a "profile" of possible disease-causing genes.

15.8 Genetically Modified Foods

Every aspect of biotechnology has its critics, but no application of biotechnology has been as controversial as the production of food through genetic modification. Protesters have rallied worldwide against genetically modified (GM) foods; since 1998, the European Union has imposed a moratorium on the approval of any new GM foods; and a furor arose in the United States in September 2000 when a genetically modified strain of corn, intended for use only in animal feed, ended up in taco shells (**see Figure 15.16**). As one measure of the passions this issue engenders, consider that genetically modified "golden rice"—praised by some as having the potential to save millions of childrens' lives each year—is grown in a greenhouse in Switzerland that has been made grenade-proof because of fears it might be attacked by opponents of GM foods.

What's the furor about? As you might guess, a battle as protracted and intense as this one would not exist unless GM foods were perceived, by different camps, as both very threatening *and* very promising. Proponents of GM foods see in them the potential to feed a hungry world and to lessen environmental damage caused by human activity. Opponents see in GM foods the potential to harm human health and wreak havoc on Earth's ecosystems.

It's not hard to point to the promise of GM foods. Consider just a couple of examples.

- The aforementioned golden rice, which has genes from both a bacterium and the daffodil plant spliced into it, can produce its own beta carotene—something that regular rice does not do. The human body breaks beta carotene into vitamin A molecules. The World Health Organization estimates that 124 million children worldwide do not get enough vitamin A. The result is that half a million children go blind each year from vitamin A deficiency and one to two million children die from it. Golden rice appears to have the potential to reduce this scourge (**see Figure 15.17**).

- Bananas are a dietary staple in many underdeveloped countries. Scientists are hoping to insert genes into the banana genome that will allow the banana plant to resist fungal infections, thereby increasing crop yields. Even better, bananas are but one of the crops that have become candidates to produce edible vaccines. The concept—already demonstrated in raw potatoes—is to insert, into the banana genome, bacterial or viral genes that will allow the banana plant to produce the corresponding *proteins* of these microorganisms. Once ingested by humans, these proteins would serve the same function as any vaccine: They would set off a human immune system response that would render a person immune from future attacks by the microorganisms in question.

And what about the perceived threat of GM foods? Because there has been no evidence that any GM food has ever harmed human beings or damaged the environment, the fight over GM foods has to do with *potential* harm—to humans, to other animals, and to the environment in general.

Many GM food critics are convinced that these foods are, first and foremost, about corporate profits, not about feeding a hungry world. They note that the world already

Figure 15.16
Hoping to Stop Genetically Modified Food

Protesters in Montreal in January 2000 marched against genetically modified organisms (GMOs). The occasion was a United Nations convention on genetically modified food products and crops.

produces enough food to feed all its people; the problem is that this food is not distributed equitably. Beyond this, genetically modified seeds are patented products, technically "licensed" to farmers. GM food critics see in this development a move on the part of biotech firms to control the means of food production. If GM seeds bring higher yields at lower cost, then commercial farmers eventually will be forced to plant GM crops just to compete. In addition, some GM seeds contain genes that render the crops they produce incapable of producing their own fertile seeds. The result is that farmers can't use seeds that developed on last year's crop to produce *next* year's crop. Critics ask where such a technology leaves developing world farmers, who routinely save seeds from one year to the next.

Proponents of GM foods note that many of the GM plants in development are expressly aimed at aiding subsistence farmers. Indeed, the inventors of golden rice, researchers Ingo Potrykus and Peter Beyer, have put together a plan to *give* the rice, free of charge, to farmers who earn less than $10,000 a year, after which those farmers could save seeds from their crop and continue planting it year after year.

GM Foods Are with Us Now

Much of the debate over GM foods has to do with their future, but they actually are with us now in a big way. In 2000, according to the United States Department of Agriculture, 54 percent of all soybeans, 48 percent of all cotton, and 25 percent of all corn crops planted in the U.S. had been genetically modified in one way or another. Clearly, millions of Americans have been eating GM foods for years, and millions of acres of crops have been planted with GM seeds—all to no apparent harm. Let's take a closer look at one particular GM food technology in wide use today, however, to see why GM food crops can elicit both praise and alarm.

The bacterium called *Bacillus thuringiensis*, found naturally in the soil, produces proteins that are toxic to a number of insects. Collectively, these proteins constitute a natural insecticide known as Bt, which has been sprayed on crops for years, mostly by "organic" farmers, meaning those who do not use human-made pesticides or herbicides.

Enter biotechnology firms, which in the 1990s spliced *Bt* genes into crop plants, with the result that these plants—corn, cotton, and potatoes—now produce their *own* insecticide. The results have in some ways been an environmentalist's dream. In one survey conducted in the American Southeast, farmers who planted *Bt* cotton reduced the amount of chemical insecticides they applied to their fields by 72 percent. They did this, moreover, while increasing cotton yields by more than 11 percent (**see Figure 15.18**).

Despite such benefits, we can also see, in the use of *Bt* seeds, the qualities that make environmentalists uneasy about GM crops in

Figure 15.17
Crop of Plenty
Ingo Potrykus standing amidst stalks of the genetically modified golden rice he helped invent. Alone among rice varieties, golden rice produces its own beta carotene, which the human body converts into Vitamin A. Some GM food critics have challenged the nutritional benefits of golden rice, but Dr. Potrykus and others maintain the rice has the ability to lessen the death and disability that result in poor countries from Vitamin A deficiency.

Figure 15.18
Built-in Resistance to Pests
The cotton plants on the left have had genes spliced into them from the bacterium *Bacillus thuringiensis*. These genes code for proteins that function as a pesticide. Meanwhile, the cotton plants on the right are not "transgenic" in this way—they have not had genes from another organism spliced into them. Both strands of cotton were under equal attack from insects, but the Bt-enhanced plants fared much better.

general. Those few insects that survive in *Bt*-enhanced fields are likely to build up resistance to the natural toxin much faster than they would have before, with only periodic *Bt sprayings*. This raises the prospect of a valued natural insecticide losing its effectiveness against insects over the long run. So alarming is this possibility that the U.S. Environmental Protection Agency requires that at least 20 percent of any given farmer's crops must be non-*Bt* plants. The idea is that by creating non-*Bt* "refuges" near the *Bt* fields, *Bt*-resistant bugs will mate in these refuges and thereby continue to share genes with the many nonresistant bugs in them, thus dramatically slowing the buildup of resistance to *Bt*.

Other Concerns about GM Foods

Critics have raised lots of other concerns about GM foods, but two in particular deserve mention. One has to do with allergic reactions. The fear is that genes inserted into crop plants will cause the plants to produce proteins that will set off such reactions, at least in a small proportion of consumers. This was the fear that caused the furor over the corn (StarLink corn) that was intended for farm animals, but that ended up in taco shells. Researchers who subsequently looked into the issue found a "low probability" that any allergic reactions could have been triggered from StarLink, but that does not answer the question of whether GM foods *generally* pose a risk of setting off allergic reactions. Proponents of GM foods argue that genes transferred from one organism to another do not stand to make the second organism allergenic unless the first organism was allergenic. Allergenic risks, they say, can thus be assessed in any planned modification of food.

A second major concern about GM foods has to do with the environment. Here the fear is that traits engineered into GM crops will spread uncontrollably. Many plants are able to "hybridize" with others: Pollen from one species of plant can waft onto another species, and the resulting fusion of egg and sperm can produce fertile plants that have traits of both parent species. In the worst case, a crop plant genetically engineered to be, say, drought resistant might hybridize with a wild variety of that crop plant. The resulting new variety of plant—which could outcompete the plants around it, thanks to

its drought-resistant genes—would spread in nature with unknown consequences. Critics note that worldwide, 12 of the 13 major crop plants hybridize with wild varieties.

But do we have reason to believe that this theoretical scenario might come to pass? Results from a study published in 2001 were reassuring on this question, though no one claims that any final answers have been provided. Back in 1990, researchers in England planted varieties of corn, beets, potatoes, and oilseed rape that had been genetically modified either to produce insecticides or to resist herbicides. In the 12 locations in England where they were planted, not one of these GM plants spread beyond the 30-square-yard plot in which it was grown.

A Consensus Regarding Regulation?

Insofar as experts have weighed in on GM foods, there has been, and continues to be, a preponderance of opinion in favor of them. That said, a notable shift in attitude has taken place in the last couple of years. At one time, the view from the scientific community seemed to be that GM foods needed no more monitoring than did any other foods. When the National Academy of Sciences released a report on current GM foods in April of 2000, however, it recommended mandatory regulation of GM crops by the Environmental Protection Agency—something that is currently done only by informal agreement. In December 2000, the American Medical Association said there was no scientific justification for any special labeling of GM foods. It added, however, that there should be a systematic safety assessment of such foods and "continued research into the potential consequences to the environment of genetically modified crops." Even the researchers in England who found that GM crops were not spreading into their surrounding habitat were careful to say that they believe such crops need to be monitored on a case-by-case basis. In short, a consensus seems to be emerging for greater regulation of GM crops and food.

The Future of GM Foods

Early in the twenty-first century, there are differing views on the future of GM foods. One view holds that GM crops offer so many benefits, compared to natural varieties, that there is no stopping them in the long run.

The forces arrayed against them might just as well try to stop the use of electricity. The other view is that the tide has turned against these foods; that consumer resistance to them ultimately will shut down the entire GM food production chain—from research, to planting, to placement on a supermarket shelf. Strong opponents of GM crop production regard it as a dangerous and largely unplanned experiment that is using the living world as a test subject. Their hope is that the GM food genie can be put back in the bottle. Proponents of GM foods believe it would be a tragedy if progress on GM foods was stopped. They see GM foods as providing enormous benefits to humanity. The campaign to stop GM foods, they say, is not based on science; it is based on an ideology that sees all things "natural" as good and all things "corporate" as bad. Time will tell what lies in store for this aspect of biotechnology.

15.9 Ethical Questions in Biotechnology

When the Human Genome Project was started in the United States, policy makers decided that 3 percent of its budget would be used to fund research into the ethical ramifications of applied genetics. Subsequently, the proportion of HGP funds devoted to ethics was bumped up to 5 percent, making this expenditure "the largest investment in bioethics in the history of the world," in the words of HGP's Francis Collins. It is unusual, to say the least, for a science project to have an ethical study component built into it; but the action seemed appropriate given applied genetics' unprecedented potential for manipulating life-forms. Here is a brief look at a few of the applied genetics questions that have been raised.

Is It Inherently Unethical for Humans to Genetically Modify Themselves or Other Organisms?

Supporters of biotechnology note that the human race has been modifying itself for centuries through such means as better nutrition, vaccinations, dentures, and plastic surgery. Furthermore, humans have been modifying animals for this long through breeding (which is how something like a Dachshund has come about).

There is, however, an obvious difference between the insertion of dentures and the insertion of recombinant DNA into an organism. That difference raises a real ethical question: How far should we go in modifying organisms? In 1995, researchers at the University of Basel activated, in *Drosophila* fly embryos, a gene that brings about eye development. The twist was that they activated this gene in embryonic regions of the flies destined to become legs and antennae. The result was that the flies grew *eyes* on their legs and antennae. This does not mean that scientists are capable of modifying organisms in whatever way they wish. It does suggest, however, that we have not yet begun to learn the limits of what is possible. One observer has suggested a question that puts the ethical portion of this subject in sharp relief: It might be economical to produce an egg-laying chicken that had no eyes and no legs. If you could do such a thing, would you?

Then, of course, there are the possibilities raised by the cloning of Dolly, chief among them the prospect of cloning human beings. The question for the future is whether societies will decline to carry out a procedure that has become technologically feasible and for which there may be some demand.

Are Biotech Diagnoses Running Far Ahead of Biotech Treatments?

There is no question that this is the case. For example, there is a terrible affliction known as Huntington disease, which usually doesn't begin to cause problems for its victims until they are about 35. It is inevitably fatal, but its manifestations prior to death include spastic writhing and dementia. In recent years, it has become possible to test persons with a family history of the illness to tell whether they carry the single copy of the Huntington gene necessary to cause it. At present, however, we can do nothing to *treat* anyone who has this disease. Thus the dilemma: If you're at risk, do you want to know about something you are powerless to do anything about? A 20-year-old who carries the defective Huntington gene might well have 20 years of symptom-free life ahead; is it better to live those years in uncertainty, or in the certainty of a grim fate? If the answer is to live in uncertainty, does that mean not having children, or does it mean living in uncertainty about their fate as well?

Will Biotechnology Lead to "Genetic Discrimination"?

For many years, genetic testing was limited to the screening of unborn children for afflictions such as Down's syndrome, or to the screening of adults for certain rare genetic illnesses, such as sickle-cell anemia. The decoding of the human genome, however, has opened the door to the possibility of screening for diseases that are much more common—for cancer and heart disease, for example. In the future, physicians expect an era of "personalized medicine" in which each individual's entire genome will be screened for potential genetic problems.

The question is: Who gets to share this genetic information? You might think that health insurers would be interested in doing so, because they would not want to insure persons who are genetically disposed to serious illnesses. A study conducted in the United States has indicated, however, that insurance companies thus far seem uninterested in looking at the genetic profiles of potential clients.

The situation is not so clear with respect to employers. In February 2000, President Clinton signed an executive order prohibiting the federal government from making any decision to hire, promote, or dismiss a worker based on genetic information, and 23 states have likewise passed laws prohibiting employment discrimination based on genetics. Many experts believe, however, that private-sector employers still have great latitude not only to see the genetic information of current or potential employees, but to make employment decisions based on that information. Reports of actual discrimination based on genes are thus far anecdotal, but this may be because genetic tests currently can reveal very little about susceptibility to common diseases. What will happen, however, when these tests acquire a great predictive power? The fear is that employers, fearing high sick-leave and health care costs, will either not hire someone who appears to be at risk for genetic illness, or will fire a current employee for the same reason.

On to Evolution

In touring the strands of DNA's double helix, you've had many occasions to see how this microscopic molecule can profoundly affect our macroscopic world. DNA replicates, it mutates, it is passed on from one generation to the next. And some organisms will have more *success* in passing on their DNA—in reproducing, in other words—than will others. This latter fact has been critical in bringing about the huge variety of life-forms that Earth houses, from bacteria to bats to trees. Over billions of years, genetics has interacted with Earth's myriad environments to produce the grand story of life. That story is called evolution, and it is the subject of the next four chapters.

Chapter Review

Summary

15.1 What Is Biotechnology?

- Biotechnology is the use of living organisms to create products or to facilitate processes. Much of its power is derived from the manipulation of segments of DNA.

15.2 Some Tools of Biotechnology

- Restriction enzymes are proteins derived from bacteria that can cut DNA in specific places. Plasmids are small, extrachromosomal rings of bacterial DNA that can exist outside of bacterial cells and that can move into these cells through the process of transformation.
TUTORIAL 15.1.1: Action of Restriction Enzymes and Plasmids

- Human DNA can be inserted into plasmid rings. Scientists use the same restriction enzyme on both the human DNA of interest and the plasmids. Complementary "sticky ends" of the fragmented human and plasmid DNA will then bond together, thus splicing the human DNA into the plasmid. This produces recombinant DNA: Two or more segments of DNA that have been combined by humans into a sequence that does not exist in nature.
TUTORIAL 15.2.1: Getting Human Genes into Plasmids

- Once plasmids have had human DNA spliced into them, the plasmids can then be taken up into bacterial cells through transformation. As these cells replicate, producing many cells, the plasmid DNA inside them replicates as well. These plasmids are producing the protein coded for by the human DNA that has been spliced into them. The result is a quantity of the human protein of interest.

- Plasmids are one type of cloning vector, meaning a self-replicating agent that, in cloning, functions in the transfer of genetic material. Viruses known as *bacteriophages* are another common cloning vector.

15.3 Cloning and the Wider World of Biotechnology

- A clone is a genetically identical copy of a biological entity. Genes can be cloned, as can cells. Reproductive cloning is the process of making clones of whole, complex animals. Dolly the sheep is a reproductive clone.
TUTORIAL 15.2.4: Cloning Dolly

- Reproductive cloning can work in tandem with various recombinant DNA processes. Reproductive clones may produce proteins or even organs of interest. Reproductive cloning may be used in tandem with xenotransplantation, meaning the transplantation of organs from one species into another. There is a possibility that pig organs may someday be transplanted into human beings.

15.4 Other Biotechnology Processes: PCR

- The polymerase chain reaction (PCR) is a technique for quickly producing many copies of a segment of DNA. It is useful in situations in which a large amount of DNA is needed for some sort of analysis, yet the starting quantity of DNA is small.
TUTORIAL 15.3.1: The Polymerase Chain Reaction

15.5 Visualizing DNA Sequences

- Gel electrophoresis allows scientists to visualize the sizes of DNA segments and aids in the sequencing of lengths of DNA. Sequencing is the process by which scientists learn the exact order of bases in a length of DNA. Through use of electrical attraction, DNA fragments will move through a gel, with shorter fragments moving farther in a given period of time. Thus the fragments separate out by size.
TUTORIAL 15.4.1: Gel Electrophoresis of DNA

- In sequencing, gel electrophoresis is used to separate out individual DNA bases, whose order can then be "read" from the gel. Large-scale sequencing is facilitated today by sequencing machines.

15.6 Decoding the Human Genome

- The sequencing of the human genome is an effort whose goal, nearly complete now, is the identification and cataloguing of all 3 billion base pairs in the human genome. In June 2000 two competing teams of scientists announced that they had completed a "first draft" of the human genome. Sequencing of the entire human genome began in 1990 as a unified, international effort called the Human Genome Project (HGP). In 1998 a private biotech firm, Celera Genomics, announced it would finish decoding the human genome before HGP would, thus sparking a scientific race that ended with the June 2000 announcement by both teams.

- The decoding of the human genome has greatly accelerated the pace at which scientists are acquiring knowledge about human genetics. Knowledge of the genome's sequence will yield benefits in fields ranging from cancer research and pharmaceutical development to basic biology.

- Two surprises emerged from the first draft of the human genome. One was that there seem to be far fewer genes in the genome than had been thought—about 30,000, as opposed to the 100,000 that had been assumed. This in turn means that human beings have far fewer genes relative to other organisms than had been assumed. The second surprise was the low proportion of the genome that codes for proteins—perhaps as little as 1.1 percent, compared to the 3–5 percent that had been assumed.

- The sequencing of the human genome will not, by itself, provide knowledge of the function of genes or the proteins they code for. It will, however, provide a means of obtaining knowledge of these subjects.

15.7 The Next Phase in Genetics: Genomics and Proteomics

- Genomics is the study of genome data. Proteomics is the study of protein expression and function. Genomics has to do with "mining" an organism's genome data to see what functions are served by different DNA sequences. Proteomics seeks to understand what proteins an organisms expresses, what these proteins ultimately do, and how they interact with one another.

- Computer science has become an integral part of modern genetics for two reasons. First, the tremendous complexity of genetics can be analyzed only with computer technology. Second, genetic information is encoded largely in digital form: discrete units of information that combine to form larger units of information. Activities such as hunting for disease genes now are partly exercises in information science.

15.8 Genetically Modified Foods

- No aspect of biotechnology has been as controversial as the production of food through genetic modification. Critics charge that genetically modified (GM) foods are potentially dangerous to human beings, to other animals, and to the environment in general. Proponents of GM foods say these foods are of enormous potential to benefit humanity and the environment.

- Because there is no evidence that any GM food has ever harmed human beings or damaged the environment, the fight over GM foods has to do with the potential they have for causing harm.

- Genetically modified foods are ubiquitous in the United States today; in 2000, 54 percent of all U.S. soybeans, 48 percent of all cotton, and 25 percent of all corn crops had been genetically modified in one way or another.

- Two of the fears most commonly expressed about GM foods are that they stand to cause allergic reactions in some people, and that as crops they may spread uncontrollably, through hybridization with wild varieties of crop plants. Proponents of GM foods say that the risks of creating plants that cause allergic reactions are negligible. A study published in 2001 found that four GM crops planted 10 years earlier had remained within the confines of the plots of land on which they were grown.

- There seems to be a growing consensus in the scientific community that GM foods and crops need more regulation.

- The future of GM foods is unclear. One view is that society will continue to develop and use them because they are so useful. Another view is that consumer resistance to them may eventually end their development and use.

15.9 Ethical Questions in Biotechnology

- Biotechnology raises numerous ethical questions. One of them is whether it is inherently unethical for humans to genetically modify themselves or other organisms.

- Thanks to modern genetic research, the gap is growing between medical science's diagnostic capability and its therapeutic capability. This raises questions as to the ethics of learning about a condition that one may be powerless to treat.

- Genetic testing may be used in the near future to diagnose genetic susceptibility to common diseases, such as cancer or heart disease. There is concern that individuals diagnosed, through genetic testing, to be at risk of developing illness will be subject to genetic discrimination, either by insurance companies or by employers.

Key Terms

bacteriophage 297	polymerase chain reaction (PCR) 301
biotechnology 294	proteomics 309
clone 297	recognition sequence 295
cloning vector 297	recombinant DNA 297
forensic DNA typing 305	reproductive cloning 298
gel electrophoresis 304	restriction enzyme 294
genomics 309	transformation 296
plasmid 295	xenotransplantation 300

Understanding the Basics

Multiple-Choice Questions

1. How is biotechnology defined?
   a. the technical advances made by researchers in biology over the last decade
   b. the study of technical biology
   c. the use of living organisms in creating products and in facilitating processes
   d. the process of cloning genes
   e. the use of commercialization in biological research

2. Which sequence of events would be followed in the process of harvesting a protein of interest?
   a. Replicate vectors in host cells; isolate gene that codes for protein; insert gene into vector; harvest protein of interest; transform the vector into bacteria.
   b. Transform vector into bacteria; isolate gene that codes for protein; replicate vectors in host cells; insert gene into vector; harvest protein of interest.
   c. Harvest protein of interest; isolate gene that codes for protein; insert gene into vector; replicate vectors in host cells; transform the vector into bacteria.
   d. Isolate gene that codes for protein; insert gene into vector; replicate vectors in host cells; transform vector into bacteria; harvest protein of interest.
   e. Isolate gene that codes for protein; insert gene into vector; transform vector into bacteria; replicate vectors in host cells.

3. The process by which a cell takes up genetic material from outside its cell membrane is called
   a. genetic absorption
   b. transformation
   c. translocation
   d. transposition
   e. pinocytosis

4. If a criminal investigator wanted to amplify a DNA sample from a crime scene, the technique he or she would use would most likely be
   a. polymerase chain reaction
   b. transformation
   c. gel electrophoresis
   d. DNA sequencing
   e. xenotransplantation

5. Which of the following has biotechnology helped to accomplish?
   a. genetically engineered medicines
   b. constructing human family trees
   c. whole organism cloning
   d. genetically engineered crops
   e. all of the above

6. Which of the following is an example of recombinant DNA?
   a. human growth hormone gene introduced into a bacterial plasmid
   b. introduction of human gene for blood-clotting protein into sheep
   c. introduction of daffodil genes into rice genome

d. a and c

e. all of the above

7. Dolly the sheep was the result of
   a. applying an electric current to an egg and sperm in a laboratory
   b. taking an embryo from one sheep and implanting it in another
   c. taking a mammary cell from an adult sheep and fusing it with a sheep egg that had its DNA removed
   d. inserting new genes into an existing mammary cell
   e. getting sperm to fuse with egg outside a living sheep

8. The sequencing of the human genome has
   a. made human cloning possible
   b. allowed us to develop a genetic fingerprint of every human being
   c. provided us with the function of every protein coded for in the genome
   d. made xenotransplantation a possibility
   e. allowed us to locate and catalogue all the genetic information in the human genome

9. To date, genetically modified food crops have been modified to
   a. produce needed human nutrients
   b. resist human herbicides
   c. resist insect pests
   d. deliver vaccines
   e. all of the above

10. What is a clone?
    a. an exact genetic copy of a gene, cell, or an entire living thing
    b. a cell that can take up plasmids
    c. a segment of DNA of a specified length
    d. an offspring produced by test-tube fertilization
    e. a segment of DNA that can prime the polymerase chain reaction

Brief Review

1. Why is it useful to have organisms such as bacteria turning out human proteins?

2. In gel electrophoresis, how do the DNA fragments found at one end of a "lane" differ from the fragments found at the other?

3. Why is the polymerase chain reaction (PCR) so valuable?

4. The sequencing of the human genome has been regarded as one of the greatest advances in genetics to date. The effort does, however, have limitations. What are some of them?

5. Why has modern genetics become tightly linked with computer technology?

6. Why are restriction enzymes that cut with sticky ends useful? If the following piece of DNA were cut with *Bam*HI, what would the fragment sequences be?

   ATCGGATCCTCCG
   TAGCCTAGGAGGC

7. Why did Dolly the sheep end up looking like the sheep that "donated" the udder cell, rather than the sheep that donated the egg?

Applying Your Knowledge

1. Why are restriction enzymes important in recombinant DNA technology?

2. How is it possible for a single genetically engineered food crop to be both welcomed by environmentalists and feared by them?

3. One of the motives put forth for human cloning is that people want to replace children or other loved ones who have died. To what extent could a clone of a loved one be a replacement for that person? If the technique had then been available, should doctors in the nineteenth century have preserved the DNA of Abraham Lincoln for possible cloning?

4. You have recently been identified as having Huntington disease, which is a late-onset, lethal, and currently untreatable disorder. As troubling as this information is to you, how do you deal with it in relation to your family? Do you inform your siblings? What about your children? Do you also have them tested for the disease? Do you refrain from having any additional children?

5. One of the concerns mentioned in the chapter regarding the use of genetic technology centers around privacy issues such as "genetic discrimination." Once the ability to screen the human genome for diseases becomes more detailed, is there any level of testing that should be required by society as a means of reducing health care costs or reducing transmission of genetic diseases across generations? If a sophisticated genetic "profile" were available, would you get one of your own genome? Why or why not?

6. An extension of biotechnology techniques employed in this chapter leads to even more mind-boggling possibilities. What if whole organisms could be "engineered" with specific desirable traits in mind? Should society embrace the idea of "designer children"?

7. Should society demand that there be no risks to genetically modified foods before it allows them to be developed? Does any significant technology have no risks associated with it?

MediaLab

Holding Out Promise or Peril? Biotechnology

Biotechnology promises to create new pharmaceuticals (even replacement organs), improve livestock yields, and even reduce the need for pesticides and herbicides in agriculture. However, many people question whether scientists are proceeding at too great a rate, outdistancing careful, cautious evaluation. Using genetically manipulated organisms for food has met great resistance in Europe, where it is called Frankenfarming—a reference to Mary Shelley's Dr. Frankenstein—and the famous monster he created by combining human corpses. To learn more, enter the *CD-ROM Tutorial* for an animated review of the major techniques used to swap DNA between species. Don't be intimidated by the technology; the animations will unveil the simplicity of the techniques. In the *Web Investigation* and *Communicate Your Results* sections, you will evaluate and discuss different perspectives on the value of biotechnology—its promise, or perhaps its threat.

This *MediaLab* can be found in Chapter 15 on your CD-ROM (Tutorial 15.1.1) and Companion Website (http://www.prenhall.com/krogh3).

CD-ROM TUTORIAL

The process of manipulating DNA has little of the mystique displayed in science fiction. It is more akin to the mundane process of cooking. This *CD-ROM Tutorial* will lead you through some common techniques of biotechnology.

Activity

1. *First, you will learn about restriction enzyme digestion and how plasmids help introduce foreign DNA into bacterial hosts.*

2. *Next, you will learn about how the techniques of gel electrophoresis and polymerase chain reaction are used to create and organize DNA fragments.*

3. *Finally, you will investigate the real-life "science fiction" where these techniques are used in forensics and in cloning.*

Although the techniques are now commonplace, the result is not. Scientists are tinkering with the exact blueprint of an entire organism, of which only a small fraction is understood. Many people fear that scientists may make unforeseen changes that could irrevocably alter our lives.

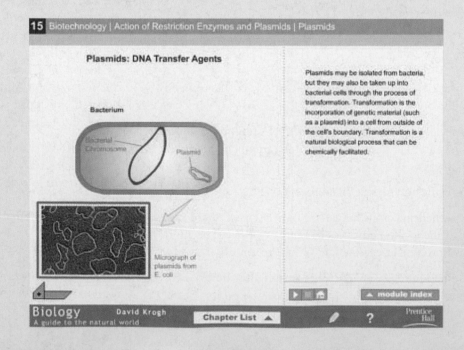

15 Biotechnology | Action of Restriction Enzymes and Plasmids | Plasmids

Plasmids: DNA Transfer Agents

Bacterium

Bacterial Chromosome

Plasmid

Plasmids may be isolated from bacteria, but they may also be taken up into bacterial cells through the process of transformation. Transformation is the incorporation of genetic material (such as a plasmid) into a cell from outside of the cell's boundary. Transformation is a natural biological process that can be chemically facilitated.

Micrograph of plasmids from E. coli

Biology
A guide to the natural world

David Krogh

Chapter List ▲

▲ module index

Prentice Hall

WEB INVESTIGATION

Investigation 1

Estimated time for completion = 5 minutes, followed by keeping a diet diary for 24 hours

Currently, 25 percent of corn, 54 percent of soybeans, and 48 percent of cotton grown in the United States are the product of genetic engineering, adding DNA from foreign species to create plants more resistant to insect pests or herbicides. Could you be using some of these foods? Select the Keyword **GE FOODS** from your CD or Website for this *MediaLab* to read a list of genetically engineered agricultural products that are currently marketed and sold. Then select the Keyword **GE DIET** to see if you are eating any of these products. Keep a diet diary for one day to see how many genetically engineered foods you may be eating.

Investigation 2

Estimated time for completion = 15 minutes

Today, biotech food is one of the most explosive environmental issues in Europe. Protesters there have destroyed crops they call Frankenplants, and they are insisting that food labels contain warnings if they contain genetically engineered products. Meanwhile, most Americans are already serving these foods for dinner. This may be the first time you have realized that fact, because the FDA has ruled that these genetically altered plants do not need to carry special labels warning consumers. Select the Keywords **GE PROTESTS** and **FDA GE POLICY** from your CD or Website for this *MediaLab* to read about both sides of the issue.

Investigation 3

Estimated time for completion = 5 minutes

Genetic engineering has already demonstrated its potential for great good, by alleviating diseases in countless hemophiliacs, diabetics, cancer sufferers, and immune-compromised children. These diseases are treated by pharmaceuticals manufactured in transgenic organisms—for example, pigs containing the human gene for human hemoglobin to be used in blood transfusions. Select the Keyword **CHOCOLATE COW** on your CD or Website for this *MediaLab* to test your knowledge of the techniques used to create transgenic organisms by creating cows that make chocolate milk. Then select the Keyword **TRANSGENIC HEMOGLOBIN** to visit a more serious site describing the creation of transgenic pigs for blood donation.

COMMUNICATE YOUR RESULTS

Now that you have been introduced to the level of genetically engineered products you come into contact with daily, use the following questions to analyze and communicate your feelings on the subject of biotechnology.

Exercise 1

Estimated time for completion = 5 minutes

In *Web Investigation 1* you created a record of your diet for 24 hours. What percentage of your diet could be from genetically engineered crops? How do you feel about it?

Exercise 2

Estimated time for completion = 20 minutes

The promise of lowering the use of pesticides and herbicides in growing genetically engineered food would appear to be a dream come true for organic farmers and environmental groups. However, environmental groups have organized a letter-writing campaign to force the FDA to inform consumers about genetically engineered food on ingredient labels. Do you think food labels should include information that identifies ingredients derived from genetically engineered foods? Form a group and compose a letter to the FDA supporting your position.

Exercise 3

Estimated time for completion = 30 minutes

What are some ethical concerns about using animals to replace human organs or to provide human medications? Write an essay defending or refuting this statement: It is not only ethically appropriate, but it would be unethical not to pursue creating transgenic organisms to provide blood substitutes as a way to save human lives. Consider these facts: Recombinant hemoglobin from livestock would cost less than human blood and would be free from HIV. Currently only 5 percent of Americans are blood donors; at this rate, it is estimated that by 2030 we will have a shortfall of 4 million units of blood. Over 60 million pigs are slaughtered for food each year, and our nation's donor blood needs could be entirely met by 1.2 million pigs.

16

An Introduction to Evolution

Charles Darwin, Evolutionary Thought, and the Evidence for Evolution

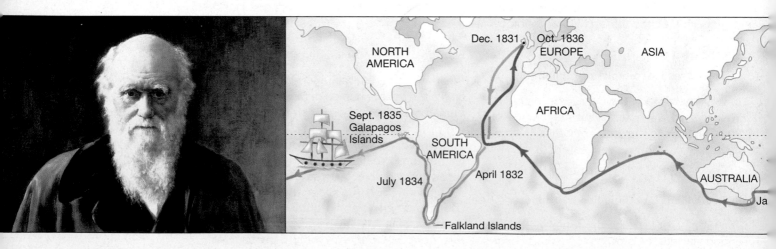

Charles Darwin.
(Section 16.2, page 323)

Fantastic voyage.
(Section 16.2, page 324)

Charles Darwin is the person singly most responsible for deepening our understanding of how life evolved on Earth. Evidence uncovered from his time to ours has confirmed his insights.

Standing in a coastal redwood grove in California, we see trees stretching hundreds of feet into the air and wonder: Why are they so tall? Watching a nature program on television, we observe the exquisite plumage of the male peacock, but are puzzled that such a *cumbersome* display could exist on a creature that has to be wary of predators. Leafing through a book, we discover that the color vision of honeybees is most sensitive to the colors that exist in the very flowering plants they pollinate. We wonder: Can this be an accident?

For as long as humans have existed, people have asked questions like these. Yet for most of human history, even the most informed individuals had no way to make *sense* of the variations that nature presented. Why does the redwood have its height or the peacock its plumage? Through the middle of the nineteenth century, the answer was likely to be either a shrug of the shoulders or a statement such as: "The creator gave unique forms to living things for reasons we can't understand."

Jumping ahead to our own time, modern-day biologists would give different answers to these questions. Their replies would be something like this:

- Redwoods (and trees in general) are so tall because, over millions of years, they have been in competition with one another for the resource of sunlight. Individual trees that had the genetic capacity to grow tall got the sunlight, flourished, and thus left relatively many offspring, which made for taller trees in successive generations.

- Extravagant plumage exists on male peacocks because it is attractive to *female* peacocks when they are choosing mating partners. Such finery is indeed cumbersome and carries a price (in lack of mobility), but this price is outweighed by the value of the feathers. Over time, the peacock ancestors with more attractive plumage mated more often and thus sired more offspring than peacocks who had less impressive plumage. This made for successive generations of peacocks with more elaborate plumage.

- It is not an accident that honeybee eyesight is most sensitive to the colors that exist in the flowering plants they pollinate. Both bees and flowering plants flourish because of their interactions with one another— bees for the food they derive, flowers for the pollination. It is likely that flower color, or bee vision, or both, were *modified* over time to maximize this interaction.

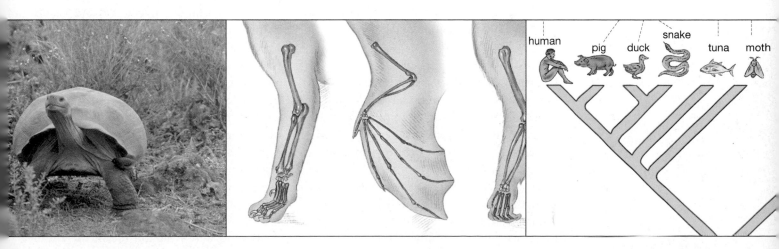

Island of discovery.
(Section 16.4, page 326)

How is a cat like a bat?
(Section 16.9, page 333)

Tracking evolution.
(Section 16.9, page 334)

Tutorial 16.1.1
Core Principles
of Evolution

Comparing contemporary answers such as these to that single nineteenth-century answer, you might well ask: What is it that came *in between*, so that we no longer must simply stand in puzzled wonderment before the diversity of nature?

16.1 Evolution and Its Core Principles

What intervened was a set of intellectual breakthroughs in the mid-nineteenth century that today go under the heading of the *theory of evolution*. Two principles lie at the core of this theory.

Common Descent with Modification

One of these principles has been labeled **common descent with modification**. It holds that particular groups, or species, of living things can undergo modification in successive generations, with such change sometimes resulting in the formation of new, separate species—one species separating into two, these two then separating further. If you had a videotape of this "branching" of species and then ran it *backward*, what you'd eventually see is a reduction of all these branches into one, meaning that all living things on Earth ultimately are descended from a single, ancient ancestor.

Evolution itself can have several definitions. In this chapter, we will think of it as synonymous with common descent with modification. In Chapter 17, you'll get a more technical definition of it.

Natural Selection

Descent with modification is joined to a second principle, that of **natural selection**: a process in which the fit of an organism with its environment selects those traits that will be passed on with greater frequency from one generation to the next. In the redwood example, trees that grew taller had a better fit with their environment than trees that didn't, in the sense that the taller trees got more sunlight. Accordingly, these trees left more offspring. Thus, the quality of tallness was *selected* for transmission to future generations. As you will see, there are other forces that shape evolution, but none is as important as natural selection.

The Importance of Evolution as a Concept

Evolution is not just important to biology; it is central to it. If biology were a house, evolution's principles would be the mortar binding its stones together. Why? Because all life on Earth has been shaped by evolution's key principles, natural selection and common descent with modification. This is easy to see once you think about any given organism. Take your pick: a fish, a tree, a human being? *All* of them evolved from ancestors, all of them ultimately evolved from a common ancestor, and natural selection was the primary force that shaped their evolution.

Given this, the theory of evolution provides a means for us to understand nature in all its buzzing, blooming complexity. It allows us to understand not just things that are familiar to us, such as the forms of the redwoods and peacocks, but all manner of *new* natural phenomena that come our way. When physicians observe that strains of infectious bacteria are becoming resistant to antibiotics, they don't have to ask: How could this happen? General evolutionary principles tell them that organisms evolve; that bacteria are capable of evolving very quickly (given their capacity to produce many generations a day); and that antibiotic resistance is simply an evolutionary adaptation.

Evolution Affects Human Perspectives Regarding Life

So far-reaching is the theory of evolution that its importance stretches beyond the domain of biology and into the realm of basic human assumptions about the world. Once persuasive evidence for the theory had been amassed, it meant that human beings could no longer view themselves as something *separate* from all other living things. Common descent with modification tells us that we are descended from ancestors, just like other organisms. Thus we sit not on a pedestal, viewing the rest of nature beneath us, but rather on one tiny branch of an immense evolutionary tree.

Beyond this, evolution meant the end of the idea of a *fixed* living world, in which, for example, birds have always been birds and whales have always been whales. Birds are the descendants of dinosaurs, while whales are the descendants of medium-sized mammals that walked on land. Life-forms only *appear* to be fixed to us, because human life is so short relative to the time frames in which species undergo change.

Finally, in evolution human beings are confronted with the fact that, through natural selection, life has been shaped by a force that has no mind, no goals, and no morals. Like a river that digs a canyon, natural selection shapes, but it does not design; it has no more "intentions" than does the wind. Trees that received more sunlight left more offspring, and thus trees grew taller over time. This is not a "decision" on the part of nature; it is an outcome of impersonal force. The interesting thing is that something that has no inherent direction has channeled living things down so many diverse paths.

If evolution has no consciousness, it follows that it can have no morals. This force that has so powerfully shaped the natural world is neither cruel nor kind, but simply indifferent.

16.2 Charles Darwin and the Theory of Evolution

How did such a far-reaching theory come into existence? Those who have read through the unit on genetics in this book know that the basic principles of that field were developed over the course of about a century by a large number of scientists. In contrast, a single person,—nineteenth-century British naturalist Charles Darwin—is credited with bringing together the essentials of the theory of evolution (**see Figure 16.1**).

Darwin's Contribution

Darwin's contribution was twofold. First, he developed existing ideas about descent with modification while providing a large body of evidence in support of them. Second, he championed the hypothesis, now regarded as correct, that natural selection is the primary force behind evolution.

Figure 16.1
Scientist of Great Insight
Charles Darwin, late in his life. (Painting by John Collier, 1883. London, National Portrait Gallery. © Archiv/Photo Researchers.)

Figure 16.2
What the *Beagle* Looked Like
A reconstruction of the HMS *Beagle*, sailing off the coast of Tierra del Fuego in South America.

To be sure, the study of evolution has been refined and expanded since Darwin. And it should be noted that before Darwin had ever published anything on his theory, a contemporary of his, fellow Englishman Alfred Russel Wallace, independently arrived at the insight that natural selection could be the driving force behind evolution. (This makes Wallace the co-discoverer of this principle.) Yet, there is near-universal agreement that it is Charles Darwin who deserves primary credit for providing us with the core ideas that exist in evolutionary biology even today.

Darwin's Journey of Discovery

Darwin was born on February 12, 1809 in the country town of Shrewsbury, England. He was the son of a prosperous physician, Robert Darwin, and his wife, Susannah Wedgwood Darwin, who died when Charles was eight. Young Charles seemed destined to follow in his father's footsteps as a doctor, being sent away to the University of Edinburgh at age 16 for medical training. But he found medical school boring, and his medical career came to a halt when, in the days before anesthesia, he found it unbearable to watch surgery being performed on children. His father then decided that he should study for the ministry. At the age of 20, Darwin set off for Cambridge University to spend three years that he later recalled as "the most joyous in my happy life." Darwin's happiness came in part from the fact that theology at Cambridge took a backseat to what had been his true passion since childhood: the study of nature. From his early years, he had collected rock, animal, and plant specimens and was an avid reader of nature books. His studies at Cambridge did yield a divinity degree, but they also gave Darwin a solid background in what we would today call life science and Earth science.

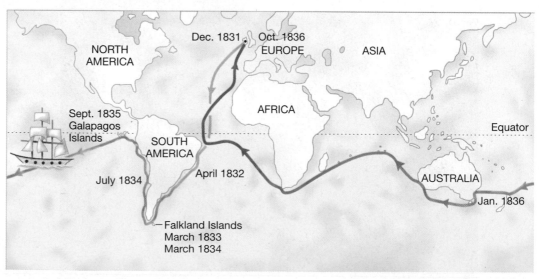

Figure 16.3
Journey into History
The main mission of HMS *Beagle*'s 5-year voyage of 1831–1836 was to chart some of the commercially promising waters of South America. Charles Darwin served on board the ship as a resident naturalist and companion to ship's captain Robert FitzRoy. Darwin observed nature and collected specimens throughout the ship's journey, but for purposes of the theory of evolution, the ship's most important stops came in 1835 on the Galapagos Islands, west of South America. The *Beagle* had a complex itinerary; for example, it stopped twice at the Falkland Islands off the southeast coast of South America, once in 1833 and again in 1834.

An Offer to Join the Voyage of the *Beagle*

Darwin's training, and the contacts he made at school, came together in one of the most fateful first-job offers ever extended to a recent college graduate. One of Darwin's Cambridge professors arranged to have him be the resident naturalist aboard the HMS *Beagle*, a ship that was to undertake a survey of coastal areas around the world (**see Figure 16.2**). If you look at **Figure 16.3**, you can trace the *Beagle*'s journey. In addition to numerous stops on the east coast of South America (the ship's primary survey site), the *Beagle* also stopped briefly at the remote Galapagos Islands, about 970 kilometers or 600 miles west of Ecuador—a visit you'll learn more about later.

Thus did Darwin spend time on a research vessel—five years in all—beginning in England two days after Christmas 1831 and ending back there in October 1836. Just 22 when he left, he was prone to seasickness; he had to share a 10-by-15-foot room with two other officers; he was not a traveler by nature; and the journey was dangerous (three of the *Beagle*'s officers died of illness during it). Yet Darwin was happy because of the work he was doing: looking, listening, collecting, and thinking about it all. Here he is writing about a part of coastal Brazil:

> A most paradoxical mixture of sound and silence pervades the shady parts of the wood. The noise from the insects is so loud that it may be heard even in a vessel anchored several hundred feet from the shore; yet within the recesses of the forest a universal silence appears to reign. To a person fond of natural history such a day as this brings with it a deeper pleasure than he can ever hope to experience again.

16.3 Evolutionary Thinking before Darwin

Darwin beheld these sights and sounds at a time when change was in the air with respect to ideas about life on Earth.

Charles Lyell and Geology

The single book Darwin took with him when he boarded the *Beagle* was the first volume of Charles Lyell's *Principles of Geology*, published in 1830 and bearing a message that even Lyell's fellow scientists found hard to accept: That geological forces *still operating* could account for the changes geologists could see in the Earth's surface. (As Darwin put it, "that long lines of inland cliffs had been formed, the great valleys excavated, by the agencies which we still see at work.") Under this view, Earth had not been put into final form at a moment of creation, but rather was steadily undergoing change. If such a thing were possible for the Earth itself, why not for the creatures that lived on it? (**See Figure 16.4.**)

a

b

Figure 16.4
Geologic Strata and What They Contain

a Several strata of sedimentary rock, with fossils embedded.

b A fossilized sea urchin found in Great Britain, dating from at least 65 million years ago.

Jean-Baptiste de Lamarck and Evolution

An idea along this line *had* been proposed, by the French naturalist Jean-Baptiste de Lamarck, in a book published in 1809. Lamarck believed that organisms changed form over generations through what has been termed the inheritance of acquired characteristics. Ducks or frogs did not originally have webbed feet, he said, but in the act of swimming they stretched out their toes in order to move more rapidly through the water. In time, membranes grew between their toes, effectively becoming webbing; critically, this characteristic was passed along to their offspring (**see Figure 16.5**). Over time, he believed, an animal would acquire enough changes that one species would diverge into two, with this branching extending all the way to human beings. In his work,

Figure 16.5
The Duck's Feet
Jean-Baptiste de Lamarck proposed that new species evolve through a branching evolution, which is correct. He also proposed, however, that one of the forces driving this evolution is the inheritance of acquired characteristics, which does not take place. Under Lamarck's view, common at the time, a duck got its webbed feet because of activities it carried out during its lifetime, and this change was then passed on to its offspring.

Lamarck lent support to a *means* of evolution that we know today is false. (Animals don't pass along traits in accordance with the activities they carry out.) Yet note what he got right: That organisms can evolve; that one kind of organism can be ancestral to a different kind of organism.

Georges Cuvier and Extinction

Another French scientist, Georges Cuvier, in examining the fossil-laden rocks of the Paris Basin early in the nineteenth century, provided conclusive evidence of the extinction of species on Earth. This was a radical notion at the time, because many Christians believed that the creator would never allow one of his creatures to perish. (One amateur fossil collector who believed this was Thomas Jefferson.) This much about extinction Cuvier got right. However, seeing in his rock layers seeming "breaks" in the sequence of animal forms—a layer of simple forms would be followed by a layer of more complex forms as he went from older to newer layers—he held that there had been a series of catastrophes, such as floods, that wiped out life in given areas, after which the creator had carried out *new* acts of creation, bringing more complex life-forms into being with each act.

If you were to survey a broad swath of scientific thought leading up to Darwin, you'd repeatedly find what you've seen with

Figure 16.6
The Galapagos Islands
Galapagos is Spanish for *tortoise*. A part of Ecuador, the islands are still an important site of evolutionary research. Here a giant tortoise moves through the Alcedo Volcano area on one of the Galapagos Islands, Isabela.

Lamarck and Cuvier: a given scientist getting things *partly* right and partly wrong (which is almost always the way in science). The result was a rich mix of scientific findings and fanciful speculation that existed in Darwin's world as he boarded the *Beagle*. He himself believed at the time that species were fixed entities; that they did not change over time.

16.4 Darwin's Insights Following the *Beagle's* Voyage

Darwin observed and collected wherever he went, but the most important stop on his journey took place nearly four years into it, when the *Beagle* stopped for a scant five weeks at the remote series of volcanic outcroppings called the Galapagos Islands (**see Figure 16.6**). There, in the dry landscape, amidst broken pieces of black lava, he saw strange iguanas and turtles and mockingbirds that varied from one island to another; he shot and preserved several of these mockingbird "varieties" for transport back home, along with a number of small birds that he took to be blackbirds, wrens, warblers, and finches.

Perceiving Common Descent with Modification

Docking in England in October of 1836, Darwin soon donated a good deal of his *Beagle* collection to the Zoological Society of London. If there was a single most important flash of insight for him regarding descent with modification, it came when one of the Society's bird experts gave him an initial report on the birds he had collected on the Galapagos. Three of the mockingbirds were not just the "varieties" that Darwin had thought; they were separate species, as judged by the standards of the day. Beyond that, the small birds he had believed to be blackbirds, finches, wrens, and warblers were all finches—separate species of finches, each of them found nowhere but the Galapagos (**see Figure 16.7**).

With this, Darwin began to see it: The Galapagos finches were related to an ancestral species that could be found on the mainland of South America, hundreds of miles to the east. Members of that ancestral species had come by air to the Galapagos and then, fanning out to separate islands, had diverged over time into separate species. Thus it was with Galapagos tortoises and iguanas and cactus plants as well; they had common mainland ancestors, but on these islands, had diverged into separate species. Darwin began to perceive the infinite branching that exists in evolution.

Perceiving Natural Selection

But what drives this branching? In England and set on a career as gentleman naturalist, Darwin married and settled in London for a time before moving to a small village south of London, where he would live out his days (**see Figure 16.8**). Two years after his homecoming, inspiration on evolution's guidance mechanism came to him not from looking at nature or thinking about the life sciences, but from reading a book on human population

Figure 16.7
One of Darwin's Finches
Geospiza fuliginosa feeds on sedge seeds in the Galapagos Islands.

Figure 16.8
Where *On the Origin of Species* Was Written
In this study in his house in rural England, Charles Darwin conducted scientific research and wrote his ground-breaking work, *On the Origin of Species by Means of Natural Selection*.

and food supply—*An Essay on the Principle of Population*, by T. R. Malthus. As Darwin later wrote in his autobiography:

> I happened to read for amusement Malthus on Population, and being well prepared to appreciate the struggle for existence which everywhere goes on . . . it at once struck me that under these circumstances favourable variations would tend to be preserved and unfavourable ones to be destroyed.

Thus did natural selection occur to him as the driving force behind evolution. You might think that, with two major insights in place—common descent with modification and natural selection—Darwin would have rushed to inform the world of them. In fact, more than 20 years were to elapse between his reading of Malthus and the 1859 publication of his great book, *On the Origin of Species by Means of Natural Selection*. In between, though incapacitated much of the time with a mysterious illness, Darwin published on geology, bred pigeons, and spent eight years studying the variations in barnacles—activities that each had relevance to his theory, and that yet constituted a kind of holding pattern for him as he contemplated informing the world of a theory that he knew would be controversial. Darwin finally got to work on what he thought would be his "big species book" in 1856.

16.5 Alfred Russel Wallace

Two years after Darwin began this labor, half a world away and unbeknownst to him, a fellow English naturalist lay in a malaria-induced delirium on an Indonesian island. Alfred Russel Wallace made a living collecting bird and butterfly specimens from the then-exotic lands of South America and Southeast Asia, selling his finds to museums and collectors. He had thought long and hard (and even published) about how species originate. Now, shivering with malarial fever in a hut in the tropical heat, with the question on his mind again, Wallace came upon the very insight that had come to Darwin 20 years earlier: Natural selection is the force that shapes evolution. Recovering from his illness, Wallace wrote out his ideas over the next few days and sent them to a scientific hero of his in England—Charles Darwin! He asked Darwin to read his manuscript and submit it to a journal if Darwin thought it worthy.

Darwin was stunned as he faced the prospect of another man being the first to bring to the world an insight he thought was his alone. Nevertheless, he would not allow himself to be underhanded with the younger scientist. He informed some of his scientist friends about his plight, and (without informing Wallace) they arranged for both Wallace's paper and some of Darwin's letters, sketching out his ideas, to be presented at a meeting of a scientific society in London on July 1, 1858. The readings before the scientific society turned out to have little immediate effect, but they did prod Darwin into working on what would become *On the Origin of Species*, published some 16 months later. This event had a great effect. Indeed, it set off a thunderclap whose reverberations can still be heard today.

16.6 Descent with Modification Is Accepted

Sparking both scorn and praise, Darwin's ideas were fiercely debated in the years after 1859. At least within scientific circles, however, it did not take long for common descent with modification to be accepted. Fifteen years or so after *On the Origin of Species* was published, almost all naturalists had become convinced of it, and it's not hard to see why. With evolution as a framework, so many things that had previously seemed curious, or even bizarre, now made sense. For example, years before *Origin* was published, scientists had known that, at a certain point in their *embryonic* development, species as diverse as fish, chickens, and humans all have structures known as pharyngeal slits (**see Figure 16.9**). In fish, these structures go on to be *gill* slits; in humans they develop into the eustachian tubes, among other things. The

Figure 16.9
Different Embryos, Same Structure
Pharyngeal slits, color-coded here, exist in the embryos of these five vertebrate animals: **(a)** sea lamprey, **(b)** pond turtle, **(c)** chicken, **(d)** domestic cat, and **(e)** human being. The common structure is evidence that all five evolved from a common ancestor. (Adapted from M. K. Richardson, 1997.)

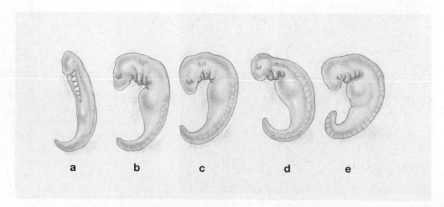

a   b   c   d   e

question was: Why would such different animals share this common structure in embryonic life? With evolution it became clear. All of them shared a common vertebrate *ancestor*, who had the slits. The vertebrate ancestral line had branched out into various species, yet elements of the common ancestor persisted in these different species in their embryonic stage.

In bringing together countless "loose ends" such as this, evolution became the mortar that unified the study of the living world. (It's interesting to think of biologists from this period as readers who start going over a detective novel a *second* time, saying, "Of course!" and "Why didn't I see that?") Given this, scientists had little trouble accepting the *fact* of evolution—that is, the occurrence of descent with modification. But the primary force that Darwin said was driving evolution, natural selection, was not embraced so quickly. Indeed, the twentieth century would be almost halfway over before it was generally accepted.

16.7 Darwin Doubted: The Controversy over Natural Selection

The essential stumbling block for the theory of evolution by natural selection was this. It asserted that traits can *vary* in ways that confer reproductive advantages on given individuals, and that these variations can be passed on from one generation to the next. With the trees considered earlier, the trait was height, and the variation that conferred a reproductive advantage was greater height. Thus in a given environment, added height represents a *difference* that allows some members of the species to leave more offspring than others, leading eventually to a taller species over all. But in the middle of the nineteenth century, no one could imagine how such differences could reliably be passed down over many generations. Even scientists of the time could not understand this, because they had no grasp of the field that we today call genetics.

Coming to an Understanding of Genetics

Those of you who have read the genetics section of this book may remember, from Chapter 11, that a trait such as tree height is likely to be governed by many genes working in tandem, with each of these genes coding for a protein that has some small, additive effect on height. Indeed, such multigene or "polygenic" inheritance is at work with most of the traits in living things. (See "Multiple Alleles and Polygenic Inheritance" in Chapter 11, page 220.) A tree's "height genes" will be shuffled in countless ways when tree egg and sperm are formed, and when they fuse in fertilization. The result is that gene combinations that yield particularly tall trees will come together in certain instances. In addition, outright mutations—alterations in genetic information—can take place, manifesting themselves as new physical traits, one of them being a taller tree. Most important for our purposes, whatever the genetic information is, it *persists*; it is passed on largely intact from one generation to another in the small informational packets we call genes.

Darwin could not have imagined a better mechanism for carrying out natural selection as he had envisioned it. The problem was that neither he nor anyone else in his time knew that things worked like this. Most nineteenth-century scientists believed that inheritance worked through a "blending" process akin to mixing different-colored paints. If you blend red and blue paint, for example, you get something that lies halfway between them, purple. Under such a system, differences in characters would not be preserved over time; they would be averaged into oblivion. Blending would work to make groups of organisms the *same*, not to allow differences among them. How could natural selection work if there were no persistent differences to select from?

Vindicating Natural Selection's Role in Evolution

The vindication of natural selection required, first, a demonstration (by Gregor Mendel) that inheritance doesn't work by blending. Rather, it works by the transmission of the stable genetic units we now call genes. Then, in the early part of the twentieth century, came the demonstration that, through polygenic inheritance, genes could account for very small physical differences that arise within a given population. Trees come in a *range* of heights, with the differences in this range governed by the combinations of individual genes described earlier. Genes are the mechanism for the process

Darwin envisioned: units of inheritance, persisting over generations, that can bring about small physical differences.

Darwin Triumphant: The Modern Synthesis

Advances in genetics such as this were later joined to other types of evolutionary research, such that in the period roughly from 1937 to 1950 there took place what is known as the **modern synthesis**: The convergence of several lines of biological research into a unified evolutionary theory. Taxonomists reported on how species are distributed throughout the world. Mathematical geneticists provided guidance on how evolution *could* work, given the shifting around of genes among organisms. Paleontologists studied evolution in the fossil record. Critically, these different types of scientists were involved in a process of "mutual education" with one another; the findings of geneticists, for example, were now known to taxonomists, and vice versa. The upshot was a greatly deepened understanding of how evolution works—an understanding that provided conclusive evidence for Darwin's assertions about the importance of natural selection. The architect of evolutionary theory had been vindicated nearly 100 years after publishing his great work. (For another example of evolution's explanatory power, see "Can Darwinian Theory Make Us Healthier?" on page 331.)

16.8 Opposition to the Theory of Evolution

So, why are Darwin's essential insights so widely believed to be correct? We'll now review some of the evidence for these insights, but first a word about why this is desirable. When you studied cells in this book, you did not review the "evidence for cells." Nor did you go over the "evidence for energy transfer" when studying cellular respiration, nor the "evidence for lipids" when studying biological molecules. What makes evolution different from these topics?

The False Notion of a Scientific Controversy

Evolution has an unusual status among major biological subdisciplines. Almost alone among them, its findings are regularly challenged as being unproven or simply wrong. For the average person, who can't be bothered with the details of such a controversy, these attacks make it appear that the theory of evolution is not a body of knowledge, solidly grounded in evidence, but rather a kind of scientific guess about the history of life on Earth.

What Is a Theory?

Several factors are critical in allowing the continued appearance of a "scientific debate" on the validity of the theory of evolution, when in fact none exists. The first of these has to do with a simple misunderstanding regarding terminology. You saw in Chapter 1 that, to the average person, the word *theory* implies an idea that certainly is unproven and that may be pure speculation. In science, however, a theory is a general set of principles, supported by evidence, that explains some aspect of the natural world. Accordingly, we have Isaac Newton's theory of gravitation, Albert Einstein's general theory of relativity, and many others. Lots of principles go into making up the theory of evolution, some of them on surer footing than others. Over time, however, some of these principles have achieved the status of established fact—we are as sure of them as we are sure that the Earth is round. Among these principles are the fact that evolution has indeed taken place and the fact that this has occurred over billions of years.

The Nature of Historical Evidence

A second factor that provides an opening for the opponents of evolution is the nature of the evidence for it. If you were called upon to provide the "evidence for cells" referred to earlier, a look through a microscope at some actual cells might be enough to stop this "debate" in its tracks. Conversely, one of evolution's most important manifestations—radical transformations of life-forms—can never be observed in this way, because these transformations take place over vast expanses of time. Asking to "see" the equivalent of a dinosaur evolving into a bird before you'll believe it is like asking to see the European colonization of North America before you'll acknowledge it.

Evolution is taking place right now, all around us, but the evidence for evolution is

Can Darwinian Theory Make Us Healthier?

Why do some pregnant women have "morning sickness"? Why do we like fats so much? Should we try to bring down the temperature of anyone who has a fever? As noted in this chapter, evolution is central to biology because it has such great explanatory power. In the 1990s, researchers began trying to harness that power to combat human disease and discomfort. The result is an emerging field called Darwinian medicine.

Two of the field's founders, Randolph M. Nesse (a medical doctor) and George C. Williams (an evolutionary biologist) note that their discipline doesn't seek to explain the direct causes of disease; rather, it aims to examine the physiological traits that make humans *vulnerable* to disease, thereby providing insight on how to treat it. The direct cause of morning sickness could perhaps be found at the level of molecules that are produced in some women early in pregnancy. But this does not tell us why so many women should be vulnerable to such a malady in the midst of something as common as pregnancy. More important, it does not tell us whether production of these molecules may be doing something *useful* as well as discomforting.

Darwinian medicine has a hypothesis about morning sickness. Put forth by the independent scientist Margie Profet, it is that morning sickness is an evolutionary *adaptation*; it is a means of minimizing the potential for exposing a growing fetus to harmful substances. Nauseated women are less likely to consume strong-tasting substances. Although few of these are likely to be harmful to a fetus, in something as important as reproduction, nature is likely to be operating under what Nesse and Williams call the "smoke-detector principle." In our homes, a smoke detector would be set off by an actual fire, of course, but it might also be set off by a burned piece of toast. The price of avoiding something as catastrophic as a fire is the existence of a super-sensitive alarm system. Just so, the female body may have a super-sensitive "alarm" on during early pregnancy to guard against the possibility of fetal damage. The price of avoiding catastrophe in pregnancy is an intermittent aversion to some foods, or to eating at all. Profet has made a first test of this thesis by examining the outcomes of a number of pregnancies. What she found was that women with more nausea were less likely to have a miscarriage.

With a result such as this, Darwinian medicine begins to assume some potential importance in the treatment of disease. Suppose that subsequent tests bear out Profet's hypothesis. We would then have to ask whether morning sickness should be "treated" at all, as it has been in the past. Much the same thing could be said of fevers—at least up to a certain level—because there is increasing evidence from humans and other animal species that higher body temperatures are a means of fighting off microbial invaders.

This is not to say that every baffling affliction that humans have is adaptive. Another hypothesis in Darwinian medicine is that near-sightedness may in part be a consequence of the modern practice of constantly looking at

Humans have gone from being hunters and gatherers to being computer programmers in only a few thousand years—the blink of an eye with respect to the evolution of physical traits.

objects that are close to us, such as books and computer screens—something that our evolutionary ancestors did not do. No one claims that near-sightedness serves any positive function; the idea is to combat it more sensibly by understanding its deeper cause.

Darwinian medicine holds that a consistent culprit in human physical affliction is the difference between the pace of cultural evolution on the one hand and physiological evolution on the other. Humans have gone from being hunters and gatherers to being computer programmers in only a few thousand years—the blink of an eye with respect to the evolution of physical traits. In near-sightedness, the evolution of our eyes hasn't had time to catch up with the change we've undertaken in how we spend our time. In another area, if we look at a typical "stone-age" diet, it turns out to be fairly close to the ideal diet recommended by doctors today: Low in saturated fats, high in fiber, high in vegetables and fruits. It was only with the rise of civilization that humans began consuming the large quantities of saturated fat that are a cause of heart disease. Yet fats taste good to us because our bodies have not evolved much from the time when fats were nutritional "prizes" that packed lots of energy into small quantities of food. In this area, as in many others, the assumption of Darwinian medicine is that to understand our present condition, we need to look at our past.

historical evidence to a degree that is not true of, say, the study of genetics or of photosynthesis. Any ancient historical record is fragmentary, and evolution's historical record is ancient indeed. The fossils that scientists analyze are what remain from tens of millions of years of weathering and decay. Such an incomplete record leaves room for a great deal of interpretation among scientists—far more than is the case in purely experimental science. This interpretation has to do, however, with the *details* of evolution, not with its core principles. It has to do with what group descended from what other; with the rate at which evolution has proceeded, or with the role that pure chance has played in evolution. It does not have to do with *whether* evolution has occurred.

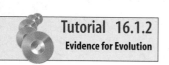

Tutorial 16.1.2
Evidence for Evolution

16.9   The Evidence for Evolution

So, what makes scientists so sure about the core principles of evolution? In any search for truth, we are more comfortable when lines of evidence are internally consistent and then go on to agree with *other* lines of evidence. The evidence for evolution satisfies both criteria. If we were to find even a single glaring inconsistency within or between lines of evidence, the whole body of evolutionary theory would be called into question. But no such inconsistencies have turned up yet, and scientists have been looking for them for about 140 years.

Radiometric Dating

One claim of evolutionary theory, as you have seen, is that evolution proceeds at a leisurely pace, with billions of years having elapsed between the appearance of life and the present. Yet how do we know that Earth isn't, say, 46,000 years old, as opposed to the 4.6 billion that scientists believe it to be? The conceptual bedrock upon which we have determined the age of the Earth and its organisms is **radiometric dating**, defined as a technique for determining the age of objects by measuring the decay of the radioactive elements they contain. As volcanic rocks are formed, they incorporate into themselves various elements that are in their surroundings. Some of these elements are radioactive, meaning they emit energetic rays or particles and "decay" in the process.

With such decay, one element can be transformed into another, the most famous example being uranium-238, which becomes lead-206 though a long series of transformations. The critical thing is that this transformation proceeds at a fixed *rate*; it is as steady as the most accurate clock imaginable. It takes 4.5 billion years for half a given amount of uranium-238 to decay into lead-206 (hence the term *half-life*). When such a transformation takes place in a cooled rock, the original or "parent" element is trapped within the rock, as is the "daughter" element, and the atoms of both elements can be counted. Therefore, comparing the *proportion* of a parent element to the daughter element in a rock sample provides a date for the rock with a fair amount of precision. This is so because of fixed rate, noted earlier: It takes a set amount of time for a given number of parent atoms to yield the number of daughter atoms found in a sample. There are now more than 40 different radiometric dating techniques used, each of them employing a different radioactive isotope. (See Chapter 2, page 23, for information on isotopes.) Some of these isotopes have very long half-lives (as with uranium-238), but some of have relatively short half-lives. Given the range of such "radiometric clocks," scientists have been able to date objects from nearly the formation of the Earth to the present, though there are some gaps in the picture.

Fossils

Today, one line of evidence for evolution is the similarity of fossil types by sedimentary layers. Looking at the same geologic layers of sediment worldwide, scientists find similar types of fossils, with a general movement toward more complex organisms as they go up through the newer strata. In Chapter 1, it was noted that scientific claims must be *falsifiable*; they must be open to being proved false upon the discovery of new evidence. The fossil record presents a falsifiable claim. For example, creatures called *trilobites* (**see Figure 16.10**) had a long run on Earth, existing in the ancient oceans from about 500 million years ago until some 245 million years ago, when they became extinct. By contrast, the evolutionary lineage that humans are part of, the primates, *began* about 60 million years ago. If scientists were to find, in a single fossil

Figure 16.10
Ancient Organism, Now Extinct
Shown is a fossilized trilobite, a kind of organism that became extinct long before the animals known as primates came to exist. This fossilized trilobite dates from about 370 million years ago and was found in present-day Morocco.

bed, fossils of trilobites existing side by side with those of early primates, our whole notion of evolutionary sequences would be called into question. No such incompatible pairing has happened, however, and the strong betting is it won't. Given this, the line of fossil evidence is internally consistent, but there's more. When we compare fossil placement with the dates we get from radiometric dating, we get excellent agreement *between* these two lines of evidence. We don't find trilobites embedded in sediments that turn out to be 60 million years old.

Comparative Morphology and Embryology

Morphology is the study of the physical forms that organisms take, while **embryology** is the study of how animals develop, from fertilization to birth. As it turns out, the evidence provided for evolution by comparative embryology was touched on earlier, in noting the pharyngeal slits that exist in the embryos of creatures as diverse as fish and humans.

In comparative morphology, some classic evidence for evolution is seen in the similar forelimb structures found in a very diverse group of mammals—in a whale, a cat, a bat, and a gorilla, as seen in **Figure 16.11**. Such features are said to be **homologous**, meaning the same in structure owing to inheritance from a common ancestor. Look at what exists in each case: one upper bone, joined to two intermediate bones, joined to five digits. Evolutionary biologists postulate that the four mammals evolved from a common ancestor, adapting this 1-2-5 structure over time in accordance with their varying needs.

Evidence from Gene Modification

In the genetics unit you saw that every living thing on Earth employs DNA and utilizes an almost identical "genetic code" (*this* triplet of DNA bases specifying *that* amino acid). At the very least, this means there is a unity running through all earthly life; it is also consistent with the idea that all life on Earth ultimately had a single starting point—the single common ancestor mentioned earlier.

In recent years, molecular biology has also provided another check on what scientists have long believed to be true about evolutionary relationships—about which species are more closely related and about how long it's been since they've shared a common ancestor. We know that similar genes function in different organisms. There is, for example, a gene called *hedgehog* that helps regulate embryonic development in the *Drosophila* fruit fly. As with any gene, *hedgehog* is composed of building blocks known as DNA nucleotides, represented as a series of "bases"— the A's, T's, C's, and G's along DNA's double helix. When we look at the base sequence of *hedgehog*, it turns out to be very similar to the base sequence of a developmental gene found in *mammals,* such as mice.

Though the *hedgehog* gene and its counterpart gene in mice are similar, we would not expect them to be identical. This is so because the base sequences of genes *change* over time through the process of mutation. If the *rate* of this change is constant, then base mutations are like a molecular clock that's ticking; with the passage of a given amount of time, you get a set number of mutations. Given this, the longer it has been since two

Figure 16.11
Four Animals, One Forelimb Structure
Whales, cats, bats, and gorillas are all descendants of a common ancestor. As a result, the bones in the forelimbs of these diverse organisms are very similar despite wide differences in function. Four sets of homologous bones are color-coded for comparison.

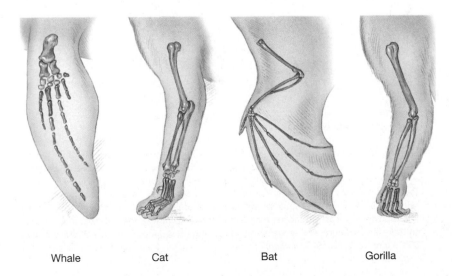

Whale   Cat   Bat   Gorilla

organisms shared a common ancestor, the greater the number of differences we should see in the base sequence of any gene they share.

To better understand this last point, it may be helpful to think in terms of something more familiar: natural language. The European settlers of Australia were predominately English. Once the Australians and English became fundamentally *separate*, however, their manner of speaking began to evolve independently. And the longer the Australians went on being separate from the English, the greater the number of differences there came to be between Australian speech and English speech. With respect to genes, far back in evolutionary history, mice and *Drosophila* flies had a common ancestor. Once the mouse and *Drosophila* evolutionary lines separated however, they began evolving independently, since the organisms in them no longer interbred. This meant that their *genes*

began to be modified independently (through random mutations, and so forth). And the longer it has been since such separate modifications began, the more differences we would expect to see in comparable genes.

The gene-modification hypothesis has been put to the test in connection with a gene that codes for an enzyme called cytochrome *c* oxidase, which exists in organisms as different as humans, moths, and yeasts. Going into this experiment, evolutionary theory predicted that there should be fewer DNA base-pair differences between the cytochrome *c* oxidase genes of, say, a human and a duck than between a human and a moth, because all the *other* evidence we had told us that humans and ducks share a more recent common ancestor than do humans and moths. If you look at **Figure 16.12**, you can see that evolutionary theory was fully borne out by the DNA sequencing. There are 17 sequence differences between a human and a duck, but 36 between a human and a moth. Indeed, all the differences between the species pictured fall into line with evolutionary theory. Thus we have another confirmation for evolution between lines of evidence—between DNA sequencing *and* the fossil record *and* radiometric dating *and* comparative morphology.

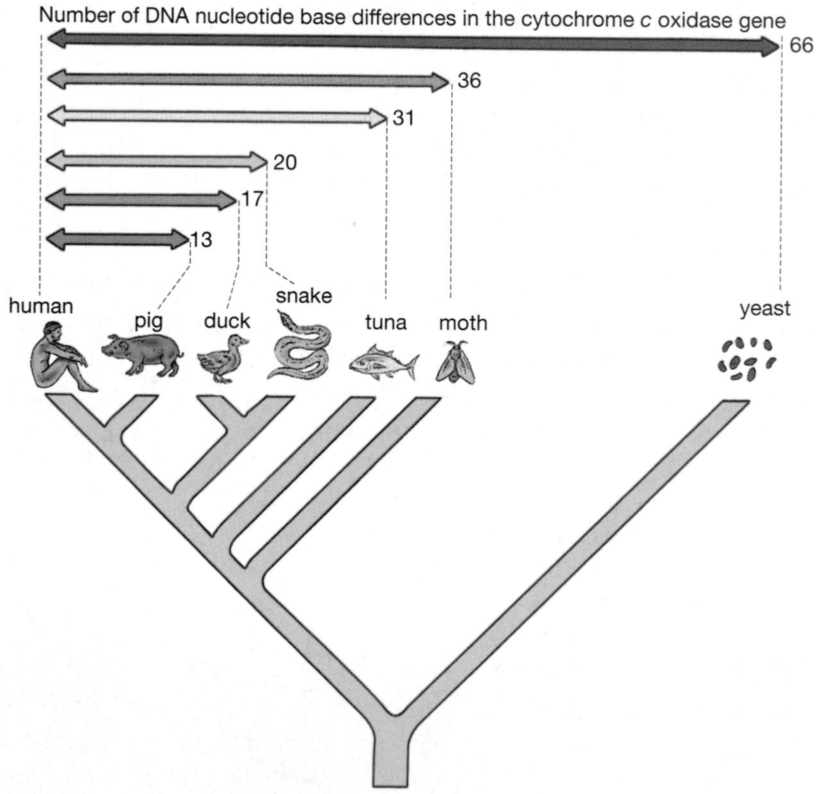

Experimental Evidence

Finally, much evidence for evolution has been provided in recent years by experiment and observation. At first glance, this may seem rather improbable because as you've seen, evolutionary change can often be perceived only over long stretches of time. Scientists have devised clever ways to catch evolution in the act, however.

Consider the experiments of John Endler, who believed that the male guppies he was studying were being pulled in two directions by natural selection. On the one hand, those who were larger and had brighter coloration were chosen more often by females for mating. On the other, these very characteristics made the males more vulnerable to predators. Endler saw that he could test natural selection by putting some guppies in a predator-free environment. In only a few generations, the males evolved brighter coloration and larger tails. When Endler then reintroduced predators into this population, things went the other way; over

Figure 16.12
Using Molecules to Track Evolution
Diverse organisms—such as yeast, moths, and pigs—all have genes that code for an enzyme called cytochrome *c* oxidase. These organisms inherited cytochrome *c* oxidase genes from a common ancestor many millions of years ago. Over time, however, the cytochrome *c* oxidase genes have undergone mutations that have altered the sequence of their DNA "building blocks," called bases. The longer it has been since any two species shared a common ancestor, the more differences there should be in their cytochrome *c* oxidase bases. There are 13 differences between the bases found in human cytochrome *c* oxidase genes and those found in pigs, but there are 66 differences between the human and yeast genes. Data from Whitfield, Philip. 1993. From *So Simple a Beginning: The Book of Evolution*. (New York: Macmillan: Maxwell Macmillan International)

several generations, the males evolved smaller tails and less brilliant colors. Both of these outcomes are precisely what evolutionary theory would predict. In both instances the guppies evolved in a direction that would maximize their reproductive success—more mates with the brighter colors in the predator-free environment, and a longer life (and hence more reproduction) with the drab coloration they evolved in the predator-laden environment.

On to How Evolution Works

Given the abundance of evidence for evolution, biologists long ago stopped asking whether it occurred. The really interesting questions for decades have been: Through what means has evolution proceeded? At what pace? In what direction, if any? These are the questions this text will address in the next three chapters.

Chapter Review

Summary

16.1 Evolution and Its Core Principles

- In the theory of evolution, descent with modification describes the process by which species of living things can undergo modification over time, with such change sometimes resulting in the formation of new, separate species. All species on Earth have descended from other species, and a single, common ancestor lies at the base of the evolutionary tree.
  TUTORIAL 16.1.1: Core Principles of Evolution

- Natural selection is the most important force in shaping evolution. Natural selection is a process in which the "fit" of an organism with its environment determines those traits that will be passed on with greater frequency from one generation to the next. Organisms whose traits better suit them to their environment will survive longer and leave more offspring than organisms with alternative traits. In this way, traits that help an organism survive and reproduce in its environment are selected for transmission to the succeeding generation.

- The theory of evolution has an importance beyond the domain of biology. Through it, human beings have become aware that (1) they are descended from other varieties of living things, and (2) the organisms that populate the living world are not fixed entities, but instead are constantly undergoing change.

16.2 Charles Darwin and the Theory of Evolution

- Charles Darwin deserves primary credit for the theory of evolution. He developed existing ideas about descent with modification while providing a large body of evidence in support of them. And he championed the hypothesis, now regarded as correct, that natural selection is the primary force behind evolution.

- Darwin's insights were inspired by the research he carried out during a five-year voyage he took around the world on the ship HMS *Beagle*, beginning in 1831.

16.3 Evolutionary Thinking before Darwin

- Some of Darwin's ideas can be traced to the work of Charles Lyell, Jean-Baptiste de Lamarck, and Georges Cuvier, who respectively noted the dynamic geological nature of the Earth, the possibility of descent with modification, and the extinction of species on Earth.

16.4 Darwin's Insights Following the *Beagle's* Voyage

- Darwin understood descent with modification for several years before he comprehended that natural selection was the most important force driving it. It was his reading of a work by Malthus that sparked his realization about natural selection.

16.5 Alfred Russel Wallace

- English naturalist Alfred Russel Wallace is the co-discoverer of the principle that evolution is shaped by natural selection.

16.6 Descent with Modification Is Accepted

- Descent with modification was accepted by most scientists not long after publication of Darwin's *On the Origin of Species By Means of Natural Selection* in 1859. Scientists accepted it because it explained so many facets of the living world.

16.7 Darwin Doubted: The Controversy over Natural Selection

- The hypothesis that natural selection is the most important force underlying evolution was not generally accepted until the middle of the twentieth century. Its acceptance hinged on a modern synthesis in the theory of evolution that brought together lines of evidence from genetics, the fossil record, and the distribution of organisms throughout the world.

16.8 Opposition to the Theory of Evolution

- Even today the theory of evolution is regularly challenged as being unproven or simply wrong. One factor leading to the appearance of a "scientific debate" is confusion over the meaning of the word *theory*. Though the average person may equate "theory" with speculation, in science, a theory is a general set of principles, supported by evidence, that explains some aspect of the natural world.

16.9 The Evidence for Evolution

- Five principal lines of evidence are consistent with the theory of evolution. First, radiometric dating indicates that the Earth is greater than 4 billion years old. Second, the placement of fossils is consistent with the theory of evolution and with radiometric dating. Third, the theory of evolution explains the common occurrence of homologous physical structures in different organisms. Fourth, the theory of evolution is consistent with variations found in the DNA sequences of various organisms. Fifth, experimental demonstrations of evolution have been carried out in the laboratory and in nature.

  TUTORIAL 16.1.2: Evidence for Evolution

Key Terms

common descent with modification 322	**morphology** 333
embryology 333	**natural selection** 322
homologous 333	**radiometric dating** 332
modern synthesis 330	

Understanding the Basics

Multiple-Choice Questions

1. In science, a theory is
   a. an untested hypothesis
   b. a general set of principles, supported by evidence, that explains some aspect of the natural world
   c. speculation
   d. an observation
   e. the first idea proposed to explain some aspect of the natural world

2. Which of the following are central ideas in the theory of evolution by natural selection? (Circle all that apply.)
   a. Organisms vary in characteristics that affect their survival and reproduction.
   b. Descent with modification occurs over generations.
   c. Evolution has a goal.

   d. Individuals with characteristics that allow them to survive and reproduce better than others are favored by natural selection.
   e. Characteristics that are acquired during the life of an individual contribute to evolution.

3. Which of the following observations provide evidence for evolution? (Circle all that apply.)
   a. Monkey and trilobite fossils are never found in the same fossil beds.
   b. Athletic training can produce an increase in muscle mass.
   c. DNA sequences of genes shared by various species vary in accordance with predictions about how related those species are.
   d. Almost all organisms use a common genetic code.
   e. Species whose adult forms look very different may have similar features in embryonic life.

4. During the formulation of his theory of evolution by natural selection, Darwin brought together ideas and results from several disciplines. Match the person with the phrase that describes his work.
   a. Lyell   Catastrophic extinction and new creations explain fossil record
   b. Wallace   Inheritance of acquired characters
   c. Lamarck   Natural selection is differential survival or reproduction
   d. Cuvier   Geological forces observable today caused changes in Earth

5. Some of the following structures are homologous with each other. Which ones are they, and which ones don't belong?
   a. whale flipper
   b. insect leg
   c. bat wing
   d. cat leg
   e. octopus tentacles

6. The discovery of _____ provided the evidence needed for natural selection to be widely accepted as the mechanism of evolution (select one):
   a. methods for radiometric dating
   b. extinctions in the fossil record
   c. genes as the cause of disease
   d. genes as the units of inheritance
   e. fossil bacteria

7. Important implications of the theory of evolution by natural selection include (select all that apply):
   a. All organisms, including humans, are descended from a common ancestor.
   b. The biological world is constantly evolving.
   c. Nature has an inherent morality.

d. The characteristics of organisms are molded by an impersonal force without a goal.

e. Humans are just a small part of the tree of life.

8. John Endler performed an experiment involving guppies in which he demonstrated evolution driven by natural selection that worked through a predator-prey relationship. Which of the following are true statements about the results of his experiments? (Select all that are correct.)

a. Males with longer tails and brighter coloration were more frequently eaten by predator fish.

b. Females preferred to mate with dull-colored males.

c. When predator fish were added to a population that had not previously had predators, males with dull coloration were favored by natural selection.

d. When predators were removed, the average brightness of male tails in the population increased.

e. When predators were removed, the average brightness of females increased.

9. When chemical insecticides were first used in the 1940s, very low doses of them were effective in killing most insects. Now, many pests are resistant to insecticides. Agriculturalists are working to control pests with reduced pesticide use. Given your knowledge of how evolutionary processes lead to the evolution of resistance, which of the following might help to reduce the speed at which resistance to a new chemical might evolve? (Circle all that apply.)

a. Apply chemicals only when absolutely necessary.

b. Apply chemicals as soon as a single pest insect is seen in a given field.

c. Use plant varieties that are resistant to insect attack when possible.

d. Try to shift planting to a time of year when the pest insect is not normally present.

e. Apply chemicals on a regular schedule, just to be sure.

Brief Review

1. What is homology, and why does it provide evidence for evolution?

2. How does the evidence presented in Figure 16.12 support the conclusion that humans are more closely related to pigs than to yeast?

3. Describe two examples in which concurrence of different lines of evidence provides evidence for evolution.

4. What is one observation that Darwin made in the Galapagos that influenced his thinking about evolution, and why was it important? Would he have been as likely to formulate his theory had he not had the opportunity to sail on the *Beagle*? Why or why not?

Applying Your Knowledge

1. Using evolutionary principles, explain why large ears might be expected to evolve in a terrestrial herbivore such as a rabbit or deer, but not in an aquatic mammal such as a seal.

2. Artificial selection is a process people use to decide which of a group of organisms will be the parents of the next generation, based on whether they possess particular desirable characteristics. Darwin used the differences that breeders had produced among breeds of pigeons as evidence for evolution. Describe how, using artificial selection, you would go about breeding a type of pigeon with a pronounced tuft of feathers on the head.

3. Explain the evolutionary steps by which bacteria may become resistant to antibiotics, using the core requirements for evolution by natural selection.

4. If all organisms had not descended from a common ancestor, and thus did not possess many common genes and mechanisms of development, would we be able to perform valid medical research using mice and in some cases, *Drosophila*? Why or why not?

17

The Means of Evolution
Microevolution

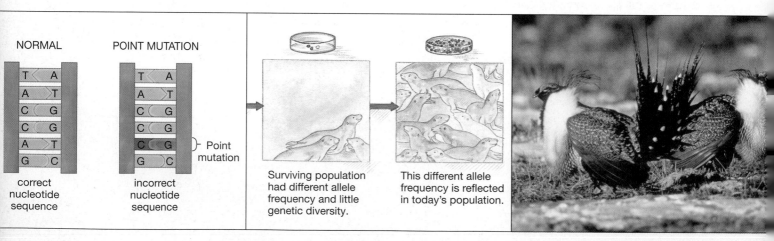

NORMAL   POINT MUTATION

T	A
A	T
C	G
C	G
A	T
G	C

correct nucleotide sequence

Point mutation

incorrect nucleotide sequence

Surviving population had different allele frequency and little genetic diversity.

This different allele frequency is reflected in today's population.

Route to new genetic information. (Section 17.3, page 342)

The bottleneck effect. (Section 17.3, page 344)

Ladies, take your pick! (Section 17.3, page 345)

Evolution at its most fundamental level is called microevolution. At root, all evolution is driven by the processes of microevolution.

When people think of the word *evolution*, what generally comes to mind is the grand sweep of evolution: The story of how microorganisms were the first life-forms; of how dinosaurs came and went; of how human beings arose in the last second of an evolutionary year. It is quite a story, to be sure, but to fully appreciate it we need to look first not at these *outcomes* of evolution, but rather at evolution's *processes*. How is it that generations of organisms become modified over time? That's the subject of this chapter. When you understand it, you'll be in a position to comprehend evolution on its grander scale.

17.1 What Is It that Evolves?

In approaching the topic of how organisms become modified over generations, the first question is: What is it that evolves? If you think about it, it's pretty clear that it is not individual organisms. A tree may inherit a mix of genes that makes it slightly taller than other trees, but if this tree is considered in isolation, it is simply one slightly taller tree. If it were to die without leaving any offspring, nothing

could be said to have evolved, because no persistent quality (added height) has been passed on to any group of organisms.

In Chapter 16 you were introduced to the idea of a **species**, which can briefly be defined as a group of organisms who can successfully interbreed with one another in nature, but who don't successfully interbreed with members of other such groups. You might think that it is species—such as horses or American elm trees—that evolve. Indeed, scientists often speak this way when they refer, for example, to an amphibious mammal having evolved into today's whales. But this is a kind of shorthand whose inaccuracy becomes apparent when species are considered as they live in the real world.

Populations Are the Essential Units that Evolve

Think of a hypothetical species of frogs living in, say, equatorial Africa in a single expanse of tropical forest. Suppose that a drought persists for years, drying up the forest such that this single lush range is now broken up into two ranges separated by an expanse of barren

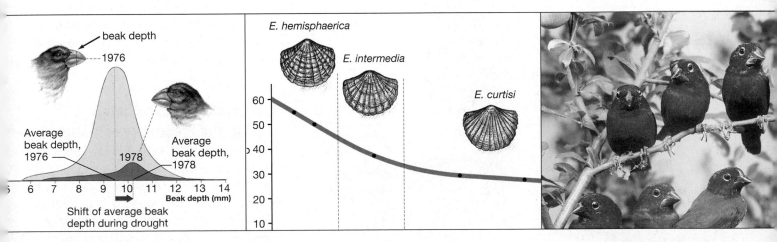

Surviving a drought.
(Section 17.4, page 348)

Seashells by the seashore.
(Section 17.5, page 350)

Big beak, little beak.
(Section 17.5, page 351)

terrain (**see Figure 17.1**). When the separation occurs, there is still but a single species of frog; but that species is now divided into two *populations* that are geographically isolated and hence no longer interbreeding. Each population now faces the natural selection pressures of its *own* environment—environments that may differ greatly even though they are nearly adjacent. One area may get more sun, while the other has a larger population of frog predators, for example. Thus, each population stands to be modified individually over time—to evolve. What is it that evolves? The essential unit that does so is a **population**, which can be defined as all the members of a species that live in a defined geographic region.

The question then becomes, *how do populations evolve?* The population of frogs that has more predators may, over many generations, evolve a coloration that makes them less visible to predators. But how does this slightly different coloration come about? Through genes. A given frog may be unable to change its spots, but *generations* of frogs can have theirs changed through the shuffling, addition, and deletion of genetic material.

Figure 17.1
Evolution within Populations

a A hypothetical species of frog lives and interbreeds in one expanse of tropical forest.

b After several years of drought, the forest has been divided by an expanse of barren terrain. The single frog population has thus been divided into two populations, separated by the barren terrain. The two frog populations now have different environmental pressures, such as the kinds of predators each faces, and they no longer interbreed; after many generations their coloration diverges as they adapt to the different pressures.

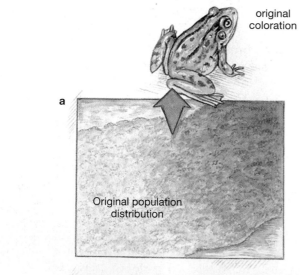

original coloration

a

Original population distribution

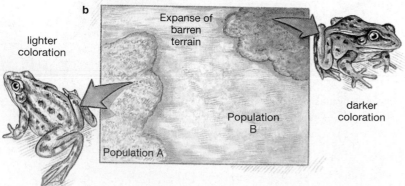

b

Expanse of barren terrain

lighter coloration

Population A

Population B

darker coloration

Genes Are the Raw Material of Evolution

Recall from Chapter 11 that the genetic makeup of any organism is its **genotype**—and that a genotype provides an underlying basis for an organism's **phenotype**, meaning any observable traits that an organism has, including its physical characteristics and behavior. Many genes are likely to be involved in producing a phenotype such as coloration. In sexually reproducing organisms, *each* of these coloration genes will likely come in two variant forms, called **alleles**, with offspring inheriting one allele from their father and one from their mother (**see Figure 17.2**). Though both alleles help code for coloration, one may result in slightly lighter or darker coloration than the other.

Genes don't just come in two allelic variants, however; they can come in many. In most species, no one organism can possess more than two alleles of a given gene, but a *population* of organisms can possess many allelic variants of the same gene (which is one reason human beings don't just come in one or two heights, but in a continuous *range* of heights.)

When the concept of a population is put together with that of genes and their alleles, the result is the concept of the **gene pool**: all the alleles that exist in a population. This gene pool is the raw material that evolution works with. If evolution were a card game, the gene pool would be its deck of cards. Individual cards (alleles) are endlessly shuffled and dealt into different "hands" (the genotypes that individuals inherit), with the strength of any given hand dependent upon what game is being played. (Survival in a drought? Survival against a new set of predators?)

17.2 Evolution as a Change in the Frequency of Alleles

Thinking of genes in terms of a gene pool gives us a new perspective on what evolution is at root: a change in the frequency of alleles in a population. This may sound a little abstract until you consider the frog coloration example. A frog inherits, from its mother and father, a coloration that allows it to evade predators slightly more successfully than other frogs of the same generation. It thus lives longer and leaves more offspring than the other frogs. It is successful in this way because of the advantageous set of alleles it inherited. Because this frog is more successful at breeding, its alleles are being passed to the *next* generation of frogs in relatively greater numbers than the alternative alleles carried by less successful frogs. Thus, this frog's alleles are increasing in frequency in the frog population. In looking at any example of evolution in a population over time, you would find this kind of change in allele frequency as its basis, assuming a stable environment.

With this perspective in mind, you're ready for a definition of **microevolution**: A change of allele frequencies in a population over a relatively short period of time. Why the *micro* in microevolution? Because evolution *within* a population is evolution at its smallest scale. This conception of microevolution allows you to understand the formal definition of evolution promised to you last chapter. **Evolution** can be defined as any genetically based phenotypic change in a population of organisms over successive generations. Taking a step back, the large-scale *patterns* produced by microevolution eventually become visible, as with the evolution of, say, mammals from reptiles. This is **macroevolution**, defined as evolution that results in the formation of new species or other groupings of living things.

17.3 Five Agents of Microevolution

So, what is it that causes microevolution? Put another way, what causes the frequency of alleles to change in a population? In the frog-coloration example, a familiar force was at work: natural selection. The frog's coloration allowed the frog to evade predators, thus helping the frog to produce more offspring than did other frogs. This particular coloration was thus *selected* for greater transmission to future generations, because this coloration had a better "fit" with its environment. With this selection, allele frequencies began to be altered in the frog population. Over generations, the alleles for protective coloration increased relative to other sets of alleles.

Natural selection is not the only force that can change allele frequencies, however. There are five "agents" of microevolution that can alter allele frequencies in populations. These are mutation, gene flow, genetic drift, nonrandom mating, and natural selection. You can see these forces summarized in **Table 17.1**. Let's now look at each of them in turn.

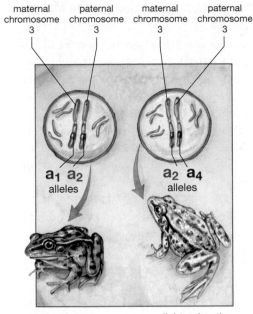

maternal chromosome 3   paternal chromosome 3   maternal chromosome 3   paternal chromosome 3

a₁ a₂ alleles   **a₂ a₄** alleles

dark coloration   light coloration

Tutorial 17.1.1
Genetic Basis of Evolution

Figure 17.2
Genetic Basis of Evolution
Many genes can produce a trait such as body coloration, and each gene often has many alleles or variants. Each individual in a population, however, can possess only two alleles for each gene, one allele inherited from its mother and one inherited from its father. The two frogs in the figure both have maternal and paternal copies of chromosome 3 that house genes for coloration. The chromosomes of the two frogs will differ, however, in the allelic variants they have of these genes. For example, the frog with dark coloration may possess alleles a₁ and a₂ of a chromosome-3 gene that codes for coloration, whereas the light-colored frog may possess alleles a₂ and a₄ of this same gene.

Table 17.1
Agents of Change: Five Forces That Can Bring about Change in Allele Frequencies in a Population

Agent	Description
Mutation	Alteration in an organism's DNA; generally has no effect or a harmful effect. But beneficial or "adaptive" mutations are indispensable to evolution.
Gene Flow	The movement of alleles from one population to another. Occurs when individuals move between populations or when one population of a species joins another, assuming the second population has different allele frequencies than the first.
Genetic Drift	Chance alteration of gene frequencies in a population. Most strongly affects small populations. Can occur when populations are reduced to small numbers (the bottleneck effect) or when a few individuals from a population migrate to a new, isolated location and start a new population (the founder effect).
Nonrandom Mating	Occurs when one member of a population is not equally likely to mate with any other member. Includes sexual selection, in which members of a population choose mates based on the traits the mates exhibit.
Natural Selection	Some individuals will be more successful than others in surviving and hence reproducing, owing to traits that give them a better "fit" with their environment. The alleles of those who reproduce more will increase in frequency in a population.

Mutations: Alterations in the Makeup of DNA

As you saw in the genetics unit, a mutation is any permanent alteration in an organism's DNA. Some of these alterations are heritable, meaning they can be passed on to future generations. Mutations can be as small as a change in a single base pair in the DNA chain (a point mutation) or as large as the addition or deletion of a whole chromosome or parts of it. Whatever the case, a mutation is a change in the informational set an organism possesses (**see Figure 17.3**). Looked at one way, it is a change in one or more alleles.

The rate of mutation is very low in most organisms; during cell division, it might be just a few DNA bases per billion. And of the mutations that do arise, very few are beneficial or "adaptive." Most do nothing, and many are harmful to organisms. Thus mutations usually are not working to *further* survival and reproduction, as the frog coloration alleles did. Given this, they generally are not likely to appear with greater frequency in successive generations. The upshot is that mutations are not likely to account for much of the change in allele frequency that is observed in any population.

But a few mutations occur that are adaptive. These genetic alterations are something like creative thinkers in a society: They are rare but very important. Such mutations are the only means by which *new* genetic information comes into being—by which new proteins are produced that can modify the form or capabilities of an organism. The evolution of eyes or wings had to involve mutations. No amount of shuffling of *existing* genes could get the living world from no eyes to eyes. Of course, no mutation can bring about a feature such as eyes in a single step; such changes are the result of many mutations, followed by rounds of genetic shuffling and natural selection, generally over millions of years.

Gene Flow: When One Population Joins Another

Allele frequencies in a population can also change with the mating that can occur after the arrival of members from a *different* population. This is the second microevolutionary agent, **gene flow**, meaning the movement of genes from one population to another. Such movement takes place through **migration**, which is the movement of individuals from one population into the territory of another. Some populations of a species may truly be isolated, such as those on remote islands, but migration and the gene flow that goes with it are the rule rather than the exception in nature. It may seem at first glance that migration would be limited to animal species, but this isn't so. Mature plants may not move, but plant seeds and pollen do; they are carried to often-distant locations by wind and animals (**see Figure 17.4**). Of course, for a migrating population to alter allele frequencies of another population, its gene pool must be different from that of the population it is joining.

Genetic Drift: The Instability of Small Populations

To an extent that may surprise you, evolution turns out to be a matter of chance. You can almost see the dice rolling in the third microevolutionary agent, genetic drift. Imagine a hypothetical population of 10,000 individuals. An allele in this gene pool is carried by one out of ten of them, meaning that 1,000 individuals carry it. Now imagine that some disease sweeps over the population, killing half of it. Say that

Tutorial 17.2.2
Changes in Allele Frequency: Mutations

Figure 17.3
Basis of New Genetic Information
A mutation is any permanent alteration in an organism's DNA. Examples of mutations include **(a)** point mutations, in which the nucleotide sequence is incorrect, and **(b)** deletions, in which part of a chromosome is missing.

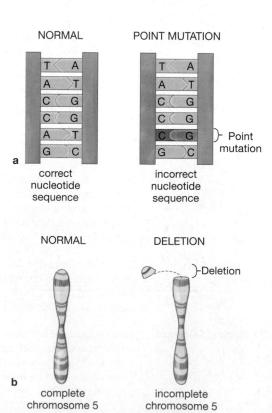

NORMAL

POINT MUTATION

T   A
A   T
C   G
C   G
A   T
G   C

T   A
A   T
C   G
C   G
C   G
G   C

} Point mutation

a

correct nucleotide sequence

incorrect nucleotide sequence

NORMAL

DELETION

} Deletion

b

complete chromosome 5

incomplete chromosome 5

this allele had nothing to do with the disease, so the illness might be expected to decimate the allele carriers in rough accordance with their proportion in the population. If this were the case, 5,000 individuals in the population would survive, and 1 in 10—or about 500 of them—could be expected to be carriers of this allele. Let us say, however, that just by chance, *550* of the allele carriers were killed, thus leaving the surviving population of 5,000 with only 450 allele carriers. In that scenario, the frequency of the allele in this population would drop from 10 percent to 9 percent (**see Figure 17.5a**).

Now for the critical step. Imagine the same allele, with the same 1-in-10 frequency, only now in a population of only *10*. There is now but a single carrier of the allele, and that individual may not be one of the 5 members of the population to survive the disease. In this case, the frequency of this allele drops from 10 percent to zero (**see Figure 17.5b**). It can be replaced only by a mutation (which is extremely unlikely) or by migration from another population. Assuming that neither happens, no matter how this population grows in the future, in genetic terms it will be a different population than the original, in that it lost this allele. The allele might be helpful or harmful, but its adaptive value doesn't matter. It has been eliminated, strictly through chance. This is an example of **genetic drift**: The chance alteration of allele frequencies in a population, with such alterations having the greatest impact on small populations. It is true that some genetic drift has taken place in the larger population, but look at how small the effect is:

The allele simply went from a 10 percent frequency to a 9 percent, with no loss of allele. Chance events can have much greater effects on small populations than on large ones.

Figure 17.4
Plant Migration
Brought about by volcanic eruptions, the Hawaiian Islands have always been surrounded by the Pacific Ocean. Therefore all the plant species existing on the islands today are descended from species that were introduced to the islands through one means or another—human activity, wind currents, water currents, or animal dispersal of seeds and pollen. **(a)** Hawaiian silverswords are derived from **(b)** a lineage of California plants commonly known as tarweeds.

Tutorial 17.2.3
Changes in Allele Frequency: Gene Flow

Figure 17.5
Genetic Drift

a In a hypothetical population of 10,000 individuals, 1 in 10 carries a given allele. The population loses half its members to a disease, including 550 individuals who carried the allele. The frequency of the allele in the population thus drops from 10 percent to 9 percent.

b A population of 10 with the same allele frequency likewise loses half its members to a disease. Because the one member of the population who carried the allele is not a survivor, the frequency of the allele in the population drops from 10 percent to zero.

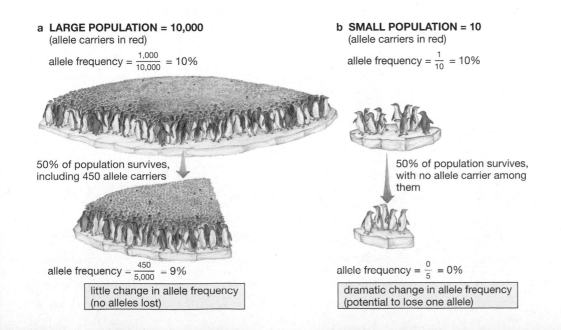

a LARGE POPULATION = 10,000
(allele carriers in red)

allele frequency $= \frac{1,000}{10,000} = 10\%$

50% of population survives, including 450 allele carriers

allele frequency $= \frac{450}{5,000} = 9\%$

little change in allele frequency (no alleles lost)

b SMALL POPULATION = 10
(allele carriers in red)

allele frequency $= \frac{1}{10} = 10\%$

50% of population survives, with no allele carrier among them

allele frequency $= \frac{0}{5} = 0\%$

dramatic change in allele frequency (potential to lose one allele)

Tutorial 17.2.4
Changes in Allele Frequency: Genetic Drift

Two scenarios, common in evolutionary history, produce the small populations that are most strongly affected by genetic drift. Populations can be greatly reduced through disease or natural catastrophe; or a small subset of a population can migrate elsewhere and start a new population. The first of these scenarios is called the *bottleneck effect*; the second is called the *founder effect*. Let's have a look at both of them in turn.

The Bottleneck Effect and Genetic Drift

The **bottleneck effect** can be defined as a change in allele frequencies in a population due to chance following a sharp reduction in the population's size. Real populations, or even species, can go through dramatic reductions in numbers. For example, northern elephant seals, which can be found off the Pacific coast of North America, were prized for the oil their blubber yielded and thus were hunted so heavily that by the 1890s, fewer than 50 animals remained. Thanks to species protection measures, the seals' numbers have rebounded somewhat in recent decades. But genetically, all the members of this species are very similar today, because they all descended from the few seals who made it through the nineteenth-century bottleneck. What occurs in these reductions is a "sampling" of the original population—the "sample" being those who survived the devastation (**see Figure 17.6**).

The reason allele frequencies change in such an event has to do with the nature of probability and sample size. Imagine a box filled with M&M's candies, with equal numbers of red, green, and yellow M&M's inside. You close your eyes and grab a handful of M&M's and pull out 12. With such a small sample, you might get, say, six reds, four greens, and two yellows, rather than the four-of-each-kind that would be expected from probability. If you pulled out *120* M&M's, conversely, your reds, greens, and yellows would be much more likely to approach the 40-of-each-kind that would be expected. In just such a way, a small sample of *alleles* is likely to yield a gene pool that's different from the distribution found in the larger population.

The Founder Effect and Genetic Drift

Genetic drift can also result from the **founder effect**, which is simply a way of stating that when a small subpopulation *migrates* to a new area to start a new population, it is likely to bring with it only a portion of the original population's gene pool. This is another kind of sampling of the gene pool, in other words; but in this case it's caused by the migration of a few individuals rather than the survival of only a few. This sample of the gene pool now becomes the founding gene pool of a new population. As such, it can have a great effect; whatever genes exist in it become the genetic set that is passed on to all future generations, as long as this population stays isolated.

The power of the founder effect can be seen most clearly when a founding population brings with it the alleles for rare genetic diseases. There is, for example, a very rare genetic affliction of the eyes called *cornea plana* that results in a misshapen cornea—the first structure of the eye through which light passes. The result can be impaired close-range vision and a general clouding of eyesight. Cornea plana is known to affect only 113 people worldwide. The strange thing is that 78 of these people are in Finland, most of them in an area in northern Finland. Current research

Figure 17.6
The Bottleneck Effect
Northern elephant seals were hunted so heavily by humans that in the 1890s, fewer than 50 animals remained. The population has grown from the few survivors (represented here by the three brown seals in the second frame), but the resulting genetic diversity of this population is very low. (Seal coloration for illustrative purposes only.)

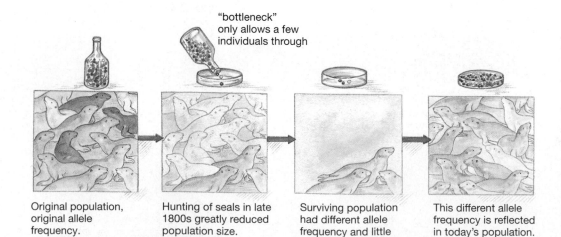

"bottleneck" only allows a few individuals through

Original population, original allele frequency.

Hunting of seals in late 1800s greatly reduced population size.

Surviving population had different allele frequency and little genetic diversity.

This different allele frequency is reflected in today's population.

indicates that about 400 years ago, a small population arrived in this isolated area, with at least one member of this population carrying the recessive allele that causes this affliction. Since then, this allele has continued to profoundly affect subsequent generations in the area. This will always be the case if the descendants of a founder population stay relatively isolated over time—that is, if the descendents breed mostly among themselves.

From this and the earlier examples, you may have perceived by now that there is an inherent value in genetic diversity; you can read about this in "Lessons from the Cocker Spaniel," on page 346.

Nonrandom Mating: When Mating Is Uneven Across a Population

You've looked at mutation, gene flow, and genetic drift as agents of change with respect to allele frequencies in a population. Now let's look at a fourth agent, nonrandom mating. Consider an imaginary population of eight animals—four males and four females—who pair off, mating for life and producing 16 offspring. Now imagine this same population of eight, but with *one* of its males mating with all four females while the other three males don't mate at all. It may be obvious that the allele frequencies of this population stand to differ depending on which of these two scenarios takes place since in the second scenario, a single male's alleles will be much more heavily represented in succeeding generations. This is an example of **nonrandom mating**, which is simply mating in which a given member of a population is not equally likely to mate with any other given member.

Some forms of nonrandom mating do not directly affect allele frequencies. Such is the case with **assortative mating**, which occurs when males and females that share a particular characteristic tend to mate with one another. Humans practice assortative mating in that short individuals tend to pick short mates, while tall individuals tend to pick tall mates. Such mating tends to bring similar alleles together ("short" with short, etc.) but it does not directly alter the *prevalence* of alleles in a population.

Nonrandom mating does affect allele frequency, however, when some members of a population mate *more* than others, as with the eight-animal example. In practice, this is mating based on phenotype, which you may recall is any observable trait in an organism, including differences in appearance and behavior. It is the *appearance* of particularly nice plumage on a male peacock that makes female peacocks choose it for mating, rather than another male while it is the *behavior* of strutting and chest-swelling in a male sage grouse that causes a succession of female grouse to mate with it rather than nearby competitors (**see Figure 17.7**). These are examples of **sexual selection,** defined as differential reproductive success, based on differential success in obtaining mating partners. Differential mating success among members of one sex in a species often is based on choices made by members of the opposite sex in that species. In general, it is females who are doing the choosing in these situations. Differential mating success can also, however, be based on differences in combative abilities that give individuals of one sex (generally males) greater access to members of the opposite sex. Sexual

Tutorial 17.2.6
Nonrandom Mating

Figure 17.7
Sexual Selection
Individuals in some species choose their mates based on appearance or behavior. Female sage grouses prefer to mate with males who put on superior "displays," which include sounds, a kind of strutting, and a puffing up of their chests. Here, two males display before females on a sage grouse breeding ground.

Lessons from the Cocker Spaniel: The Price of Inbreeding

Marry afar" is advice given by the pygmy men of Africa, by which they mean, "marry a woman who lives far from your home." The pygmies point to the hunting rights a man acquires in distant territory by marrying a woman from a remote region, but there is another benefit to this practice as well: It helps ensure genetic diversity among the pygmies. When pygmies from remote locations marry, it cuts down on the chances of "inbreeding," meaning mating in which close relatives produce offspring. Inbreeding can have harmful effects not only in humans, but in any species.

What are these effects? Well, consider what has happened in the United States with "purebred" dogs (**Figure 1**), which is to say dogs that over many generations have been bred solely with members of their own breed (cocker spaniels mating only with cocker spaniels, and so forth.). In the mid-1990s it was estimated that up to one-fourth of all U.S. purebred dogs had some sort of serious genetic defect—ranging from improper joint formation to heart defects, to deafness. We also have a good deal of experimental evidence from other animals on the effect of inbreeding. In one instance, a group of white-footed mice that had been inbred and then let out into the natural environment survived at only 56 percent the rate of a group of genetically diverse mice. As already noted, negative biological outcomes can appear in humans too; for that reason, many Western societies have made it illegal for first cousins to marry.

Why does inbreeding cause such trouble? As you've seen, every organism has a whole series of genes, almost all of which come in two variant forms, or alleles, one allele inherited from the father and one from the mother. In all organisms, a very small proportion of these alleles is potentially harmful or even lethal. But in most instances such alleles are recessive. *One* defective allele is not enough to cause trouble, because the second differs from it and provides sufficient information for the organism to function properly. Alleles that differ are said to be heterozygous, whereas alleles that are identical are homozygous.

In the mid-1990s it was estimated that up to one-fourth of all U.S. purebred dogs had some sort of serious genetic defect.

So how do these biological facts play out in inbreeding? Imagine two seals, brother and sister, that have mated together because their population has been so reduced. Imagine that their mother had a recessive allele that brings about a heart defect, and that this allele occurs in only one out of every hundred seals. There is a one-in-two chance that the mother will pass this allele on to any one of her offspring, and let's say it happened with the male and female here: Both brother and sister have inherited the defective allele. When *they* mate, the chance that either of

selection has something in common with natural selection, in that both processes result in some individuals passing along more of their genes to future generations than others. But while natural selection has to do with differential survival and reproductive capacity (and hence reproduction), sexual selection has to do with differential *mating* (and hence reproduction). In sum, sexual selection is a form of nonrandom mating that can affect allele frequencies in populations.

Natural Selection: Evolution's Adaptive Mechanism

You've already gone over a good deal about natural selection in this review of evolution, but it's time now to look at it as the last agent of change in microevolution. First, what is meant by natural selection? Here is a short

definition: **Natural selection** is a process in which the fit of an organism with its environment selects those traits that will be passed on with greater frequency from one generation to the next. What does this mean in practice? Here is what biologist Julian Huxley had to say about it almost 60 years ago.

Since there is a struggle for existence among individuals, and since these individuals are not all alike, some of the variations among them will be advantageous in the struggle for survival, others unfavorable. Consequently, a higher proportion of individuals with favorable variations will on the average survive, a higher proportion of those with unfavorable variations will die or fail to reproduce themselves.

them will pass along the defective allele to their offspring is likewise one out of two. More important, however, the chance that *both* of them will pass the allele along—producing an offspring with two defective alleles—is one out of four. If, on the other hand, the brother had mated with an unrelated member of the population, the chances of this union passing along two copies of this defective allele are one in *four hundred*.* Why the difference? Because the unrelated female had only a one-in-a-hundred chance of having the allele in the first place, whereas the sister had a one-in-two chance of getting it from her mother.

The essence of this lesson, as you may have realized, is that inbreeding has brought together identical alleles—the recessive homozygous alleles that are required for many genetic diseases. Random breeding, meanwhile, tends to keep homozygous alleles apart. It tends to preserve genetic diversity, to put it another way.

You might think from this that there are no uses for genetic *similarity*, but consider what human beings have been doing for centuries with dog breeding. Cocker spaniels did not come about by accident. Dogs that had cocker

*Remembering the rule of multiplication from Chapter 11, to calculate the odds of passing on two defective alleles in the random mating example, take the son's chances of passing along the allele (1 in 2, or 50 percent, or 0.5) × 0.01 (the chance of his female partner *having* the allele) × 0.5 (the chance of her passing the allele along). This works out to 0.0025, which equals 1 in 400. For the brother-and-sister mating, the figures are 0.5 (the chance of the brother passing it along) × 0.5 (the chance of the sister passing it along), which equals 0.25—or 1 in 4.

spaniel features were bred together for many generations ultimately giving us today's dog. It was the *novel features* of the spaniel that people wanted, and novel features are often the result of homozygous recessives coming together. While the breeding done to produce American cocker spaniels has resulted in dogs with cute, floppy ears, those ears are prone to infections, and the breed displays a "rage syndrome" that can result in unprovoked aggression, even against familiar humans.

We don't value the seal's heart condition the way we the spaniel's floppy ears, but looked at one way it too is a novel feature, made possible by the pairing of homozygous recessives.

Figure 1
Result of Generations of Controlled Breeding

By this process, the traits of those who are more successful in reproducing will become more widespread in a population—as the alleles that bring about these traits increase in frequency from one generation to the next.

A key concept here is that of **adaptation**, meaning a modification in the structure or behavior of organisms over generations in response to environmental change. Environmental change may come to a population (through streams drying up, for example), or a population may come to environmental change (through migration). Either way, natural selection is the force that pushes populations to adapt to new conditions. The frog population noted at the outset of the chapter adapted to a new predator, for example, by evolving a darker coloration.

Among the five agents of microevolution, natural selection is the only one that consistently works to adapt organisms to their environment. Genetic drift is random; it could as easily work against an adaptive trait as for it. Although mutations can have an adaptive effect, they more often have no effect or even a negative effect. Gene flow doesn't necessarily bring in genes that are better suited to a given environment. And nonrandom mating has to do only with success in securing sexual partners, not with matching individuals to environment.

Natural selection is, however, constantly working to modify organisms in accordance with the environment around them. As such, it is generally regarded as the most important agent in having shaped the natural world—in having given zebras their stripes and flowers their fragrance.

17.4 What Is Evolutionary Fitness?

Even the strongest supporters of natural selection as a shaping force would not maintain that it is some perfect mechanism working to produce perfect organisms. The concept of "fitness" is helpful here, if only to clear up some misconceptions. To a biologist, **fitness** means the success of an organism in passing on its genes to offspring *relative* to other members of its population at a particular time. An organism cannot be deemed "fit", even if it has 1,000 offspring. It can only have *more or less* fitness relative to other members of its population (who might have 900 or 1,100 offspring). This has to do, once again, with allele frequencies in a gene pool. No matter how many offspring an individual has, its allele frequencies will increase in a population only if it has *more* offspring than other members of its generation. Thus, fitness is a measure of impact on allele frequencies in a population.

This concept then gets us to the notion of "survival of the fittest," and the misunderstandings that arise from it. The phrase can be taken to imply the existence of superior beings; that is, organisms that are simply "better" than their counterparts, with images of being faster, more muscular, or smarter coming to mind. In fact, however, evolutionary fitness tells us nothing about organisms being *generally* superior, and it certainly tells us nothing about the value of any particular capacity, be it brawn or brain. All it tells us about are organisms who are better than others in their population at passing along their genes in a given environment at a given time. An accurate phrase, as others have pointed out, would not be "survival of the fittest," but rather "survival of those who fit—for now." Let's look at a real-life example of natural selection to see why this is true.

Galapagos Finches: The Studies of Peter and Rosemary Grant

When Charles Darwin stopped at the Galapagos Islands in 1835, some of the animal varieties he collected were various species of finches. Over the years, biologists kept coming back to "Darwin's finches" because of the very qualities Darwin found in them: They seemed to present a textbook case of evolution, with their 13 species having evolved from a single ancestral species on the South American mainland. Yet for more than 100 years after Darwin, it was a puzzle for scientists to figure out how Darwin's posited mechanism of evolution, natural selection, could have been at work with the finches. This changed beginning in the 1970s, when the husband-and-wife team of Peter and Rosemary Grant began a painstakingly detailed study of the birds.

Natural selection in the finches came into sharp focus in 1977, when a tiny Galapagos island, Daphne Major, suffered a severe drought. Rain that normally begins in January and lasts through July scarcely came at all that year. This was a disaster for the island's two species of finches; in January 1977 there were 1,300 of them, but by December the number had plunged to fewer than 300. Daphne's medium-sized ground finch, *Geospiza fortis*, lost 85 percent of its population in this calamity. The staple of this bird's diet is plant seeds. When times get tough, as in the drought, the size and shape of *G. fortis* beaks—their beak "morphology"—begins to define what one bird can eat as opposed to

Figure 17.8
A Product of Evolution on the Galapagos
A male of the finch species *Geospiza fortis*, which is native to the Galapagos Islands.

Tutorial 17.2.7
Natural Selection

Figure 17.9
Who Survives in a Drought?
A large percentage of the population of *Geospiza fortis* died on a Galapagos Island, Daphne Major, during a drought in 1977. Peter Grant observed in 1978 that individuals who survived the drought had a greater average beak depth than average individuals surveyed before the drought, in 1976. Individuals with larger beaks were better able to crack open the large, tough seeds that were available during the drought. The offspring of the survivors likewise had larger average beak size than did the population before the drought. Thus, evolution through natural selection was observed in just a few years on the island.

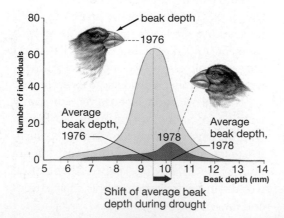

Shift of average beak depth during drought

another (**see Figure 17.8**). In *G. fortis*, larger body size and deeper beaks turned out to make all the difference between life and death in the drought of 1977. Measuring the beaks of *G. fortis* who survived the drought, the Grant team found they were larger than the beaks of the population before the drought by an average of some 6 percent. This was a difference of about half a millimeter, or roughly two-hundredths of an inch; by such a difference were the survivors able to get into large, tough seeds and make it through the catastrophe, eventually to reproduce (**see Figure 17.9**).

This is natural selection in action, but there is more to be learned from the Grant study, which is to say *evolution* made visible. The Grant team knew that beak depth had a high "heritability" in the finches, meaning that beak depth is largely under genetic control. As it turned out, the *offspring* of the drought survivors had beaks that were 4 percent to 5 percent deeper than the average of the population before the drought. In other words, the drought had preferentially preserved those alleles from the starting population that brought about deeper beaks, and the result was a population that evolved in this direction.

But the Grant study yielded one more lesson. In 1984–1985 there was pressure in the opposite direction: Few large seeds and an abundance of small seeds provided an advantage to *smaller* birds, and it was they who survived this event in disproportionate numbers.

Lessons from *G. fortis*

So, where is the "fittest" bird in all this? There isn't any. Evolution among the finches was not marching toward some generally superior bird. Different traits were simply favored under different environmental conditions. Secondly, there is no evolutionary movement toward combativeness or general intelligence here. Survival had to do with size—and not necessarily *larger* size at that. Looking around in nature, it's true that some showcase species, such as lions and mountain gorillas, gain success in reproduction by being aggressive. And it's true that our own species owes such success as it has had to intelligence. But in most instances, it is not brawn or brain that make the difference; it is something as seemingly benign as beak depth and its fit with the environment.

Finally, consider how imperfect natural selection is at the genetic level. Suppose the smaller *G. fortis* that disproportionately died

off in 1977 also disproportionately carried an allele that would have aided just slightly in, say, long-distance flying. In the long run this might have been an adaptive trait, but it wouldn't matter. The flight allele would have been *reduced* in frequency in the population (as the smaller birds died off) because flight distance didn't matter in 1977, beak depth did. Evolution operates on the phenotypes of whole organisms, not individual genes. As such, it does not work to spread *all* adaptive traits more broadly. Instead, the destiny of each trait is tied to the constellation of traits the organism possesses. Genes are "team players," in other words, that can only do as well as the team they came in on.

17.5 Three Modes of Natural Selection

In what directions can natural selection push evolution? As noted in Chapter 11, a character such as human height is under the control of many different genes and is thus **polygenic**. Such polygenic characters tend to be "continuously variable." There are not one or two or three human heights, but an innumerable number of them in a range. (See "Multiple Alleles and Polygenic Inheritance" on p. 220) When natural selection operates on characters that are polygenic and continuously variable, it can proceed in any of three ways. The essential question here is: Does natural selection favor what is average in a given character, or what is extreme? (**See Figure 17.10.**)

17.5 Three Modes
of Natural Selection

Tutorial 17.3.2
Three Modes of Natural Selection

Figure 17.10
Three Modes of Natural Selection
The bell curves represent the number of individuals in a population with a certain characteristic, in this case color intensity. Under stabilizing selection, the individuals that possess extreme values of a character—here the lightest and darkest colors—are selected against and die, or fail to reproduce. Over several generations, more and more members of the population move toward the average color of the population. Under directional selection, one of the extremes produces a better fit with the environment, meaning that individuals at the other extreme are selected against. Over several generations, the population thus moves in a direction—toward a darker coloration. Under disruptive selection, individuals with average coloration are selected against and die. Over many generations, half of the population becomes lighter, while the other half becomes darker.

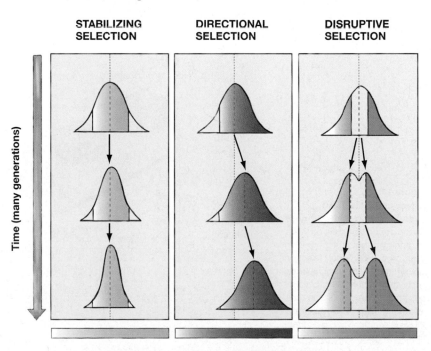

STABILIZING SELECTION   DIRECTIONAL SELECTION   DISRUPTIVE SELECTION

Time (many generations)

Range of a particular characteristic (such as color)

Stabilizing Selection

In **stabilizing selection**, intermediate forms of a given character are favored over extreme forms. A clear example of this is human birth weights. If you look at **Figure 17.11**, you can see, first, the weights that human babies tend to be. Notice that there are not only relatively few 3-pound babies, but relatively few 9-pound babies as well. A great proportion of birth weights fall at the average or "mean" of a little less than 7 pounds. Now look at the infant-mortality curve. Infant deaths are highest at both extremes of birth weight; low-birth-weight babies *and* high-birth-weight babies are more at risk than are average-birth-weight babies (though low birth weight poses the greater risk). Put another way, the children most likely to survive (and reproduce)

are those carrying alleles for intermediate birth weights. Thus, natural selection is working to make intermediate weights even more common. It is not working to move birth weights toward the extremes of higher or lower weights.

Directional Selection

When natural selection moves a character toward *one* of its extremes, **directional selection** is in operation—the mode in which we most commonly think of evolution operating. If you look at **Figure 17.12**, you can see an example of directional selection that took place over a very long period of time—about 10 million years—involving evolution toward smoothness of shells in certain species of brachiopods.

Disruptive Selection

When natural selection moves a character toward *both* of its extremes, the result is **disruptive selection**, which appears to occur much less frequently in nature than the other two modes of natural selection. This mode of selection is visible in the beaks of yet another kind of finch. A species of these birds found in West Africa (*Pyrenestes ostrinus*) has a beak that comes in only two sizes. Thomas Bates Smith, who has studied the birds, has observed that if human height followed this pattern, there would be some Americans who are 4 to 5 feet tall, and some who are 6 to 7 feet tall, but no one who is 5 to 6 feet tall (**see Figure 17.13**).

Figure 17.11
Stabililzing Selection: Human Birth Weights and Infant Mortality
Note that infant deaths are more prevalent at the upper and lower extremes of infant birth weights.

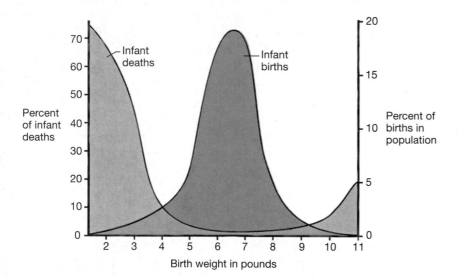

Figure 17.12
Directional Selection: Evolving toward Smoothness
Over a period of 10 million years, beginning about 415 million years ago, directional selection operated on these brachiopods, such that their shells became progressively smoother over time. (Modified from T. Dobzhansky et al., *Evolution*. New York: Freeman, 1977, p. 329.)

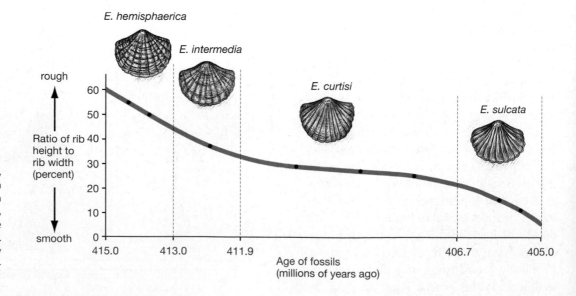

The environmental condition that leads to this mode of selection is, once again, diet. When food gets scarce, large-billed birds specialize in cracking a very hard seed, while small-billed birds begin feeding on several soft varieties of seed, with each type of bird being able to outcompete the other for its special variety. Given how these birds have evolved, it's probably safe to assume that a bird with an intermediate-sized bill would get less food than one with a bill of either extreme type. You might think that what really exists here are two separate species of bird, but large- and small-billed birds mate. Bill size seems to be under the control of a single genetic factor, so that bills are able to come out either large or small. This genotype was presumably shaped by natural selection over generations, such that any alleles for intermediate-sized bills were weeded out.

On to the Origin of Species

Stabilizing selection does what it says: It stabilizes given traits of a population, thereby keeping it a single entity. However, both disruptive and directional selection can serve as the basis for speciation—for bringing about the transformation of a single species into one or more *different* species. How is it, though, that such speciation works? And how do we classify the huge number of species that are the outcome of evolution's myriad branchings? These are the subjects of the next chapter.

Figure 17.13
Disruptive Selection: Evolution That Favors Extremes in Two Directions
Finches of the species *Pyrenestes ostrinus*, found in West Africa. Birds of this species have beaks that come in two distinct sizes— large and small.

Chapter Review

Summary

17.1 What Is It that Evolves?

- A population is all the members of a species living in a defined geographical area. Natural selection acts on individuals, but it is populations that evolve.
  TUTORIAL 17.1.1: Genetic Basis of Evolution

17.2 Evolution as a Change in the Frequency of Alleles

- Genes may be found in variant forms, called alleles. In most species, no individual will possess more than two alleles for a given gene, but a population may possess many such allelic variants. The sum total of alleles in a population is referred to as that population's gene pool.

- The basis of evolution is a change in the frequency of alleles within a population (a phenomenon that includes the appearance of new alleles through mutation). Evolution at this level is called microevolution.

17.3 Five Agents of Microevolution

- Five evolutionary forces can result in changes in allele frequencies within a population. These agents of microevolution are mutation, gene flow, genetic drift, nonrandom mating, and natural selection.
  TUTORIAL 17.2.2: Changes in Allele Frequency: Mutations

- Mutation happens fairly infrequently, and most mutations either have no effect or are harmful; yet adaptive mutations are vital to evolution in that they are the only means by which entirely new genetic information comes into being.

- Gene flow, the movement of genes from one population to another, takes place through migration, meaning the movement of individuals from one population into the territory of another.
  TUTORIAL 17.2.3: Changes in Allele Frequency: Gene Flow

- Genetic drift, the chance alteration of allele frequencies in a population, has its greatest effects on small populations.

Genetic drift works on small populations in two ways. The first of these is the bottleneck effect, defined as a change in allele frequencies due to chance during a sharp reduction in a population's size. The second is the founder effect, the fact that when a small subpopulation migrates to a new area to start a new population, it is likely to bring with it only a portion of the original population's gene pool.

TUTORIAL 17.2.4: Changes in Allele Frequency: Genetic Drift

- Nonrandom mating is mating in which a given member of a population is not equally likely to mate with any other member. Sexual selection is a form of nonrandom mating that can directly affect the frequency of alleles in a gene pool. It occurs when differences in reproductive success arise from differential success in mating.

TUTORIAL 17.2.6: Nonrandom Mating

- Some individuals will be more successful than others in surviving and hence reproducing, owing to traits that give them a better "fit" with their environment. This phenomenon is known as natural selection. Natural selection is the only agent of microevolution that consistently acts to adapt organisms to their environments. As such, it is generally regarded as the most powerful force underlying evolution.

17.4 What Is Evolutionary Fitness?

- The phrase "survival of the fittest" is misleading, because it implies that evolution works to produce generally superior beings who would be successful competitors in any environment. Evolutionary fitness, however, has to do only with the relative reproductive success of individuals in a given environment at a given time.

TUTORIAL 17.2.7: Natural Selection

17.5 Three Modes of Natural Selection

- Natural selection has three modes: stabilizing selection, directional selection, and disruptive selection. Stabilizing selection moves a given character in a population toward intermediate forms; directional selection moves a given character toward one of its extreme forms; and disruptive selection moves a given character toward two extreme forms.

TUTORIAL 17.3.2: Three Modes of Natural Selection

Key Terms

adaptation 347	**gene flow** 342
allele 340	**gene pool** 340
assortative mating 345	**genetic drift** 343
bottleneck effect 344	**genotype** 340
directional selection 350	**macroevolution** 341
disruptive selection 350	**microevolution** 341
evolution 341	**migration** 342
fitness 348	**natural selection** 346
founder effect 344	**nonrandom mating** 345

phenotype 340	**sexual selection** 345
polygenic 349	**species** 339
population 340	**stabilizing selection** 350

Understanding the Basics

Multiple-Choice Questions

1. In most populations of living things, each individual has__ copies of each gene, which may be the same or different allelic variants. A population may have _____ allelic variants for a given gene.
   a. 1, 2
   b. 2, 4
   c. 2, many
   d. 1, many
   e. 2, 2

2. Match the terms with their meanings:
   a. gene pool   a variant form of a gene
   b. allele   the set of all alleles in a population
   c. allele frequency   the genetic makeup of an organism
   d. genotype   exchange of genes between populations
   e. gene flow   the relative representation of a given form of a gene in a population

3. Which of the following statements is true of microevolution? (Select all that apply.)
   a. It is a change in allele frequency within populations.
   b. It is caused by inheritance of characters acquired during the life of an organism.
   c. It is seen when some allelic variants cause individuals to have increased survival or reproduction.
   d. It always leads to adaptation.
   e. It is caused only by natural selection.

4. Agents of change in allele frequency in populations include (select all that apply):
   a. genetic drift
   b. meiosis
   c. mutation
   d. segregation
   e. natural selection

5. Agents that consistently produce adaptive evolution (adaptation) include (select all that apply):
   a. genetic drift
   b. mutation
   c. meiosis
   d. natural selection
   e. gene flow

6. Mutations are (select all that apply):
   a. generally beneficial
   b. changes in the genetic material that may be inherited

c. always caused by changes in single base pairs of DNA
d. the ultimate source of new genetic variation
e. prone to occur with great frequency

7. In stabilizing selection, _____ individuals have the highest fitness:
a. large
b. intermediate
c. bright
d. extreme
e. small

8. Match the agent of evolution to its description:
a. mutation — chance alterations of allele frequencies in a small population
b. gene flow — a process in which the fit of an organism with its environment selects those traits that will be passed on with greater frequency from one generation to the next
c. natural selection — movement of alleles between populations by migration
d. genetic drift — genetic drift due to a few colonizing genotypes
e. founder effect — changes in the genetic material

9. When an insect population becomes resistant to insecticides, this is an example of a response to _____ selection.
a. disruptive
b. stabilizing
c. sexual
d. directional
e. random

10. Finches in the Galapagos experienced _____ selection on beak depth following a drought, while the African finch *P. ostranus* experiences _____ selection on beak size because it eats either very small, soft seeds or much larger, hard seeds.
a. stabilizing, directional
b. sexual, disruptive
c. directional, disruptive
d. disruptive, directional
e. stabilizing, disruptive

11. A large lizard population (1,000 individuals) on the coast contains mostly individuals that are plain brown, but a few have white spots. Coloration is genetically determined. One day during a storm, two spotted lizards hitch a ride on a piece of driftwood to a nearby island, where they join a population of 100 lizards, some of which are spotted and some of which are plain brown. The island's lizard population allele frequency has thus changed due to

a. natural selection
b. mutation
c. gene flow
d. nonrandom mating
e. inbreeding

Brief Review

1. Explain the statement, "Individuals are selected, populations evolve."

2. What is a gene pool, and why do changes in gene pools lie at the root of evolution?

3. What are the five causes of allele frequency changes (microevolution), and how does each work?

4. Why is evolutionary fitness always a measure of relative fitness in a population?

5. Why is natural selection the only agent of evolution that can produce adaptation?

6. Is the evolutionary fitness of an individual expected to be the same or different in different environments? How did the Grant's study of beak depth of finches after the drought and then again in 1984–1985 provide a real example of this?

Applying Your Knowledge

1. Cheetahs have long legs relative to other large cats. However, leg length in a cheetah population is more likely to be under stabilizing rather than directional selection. Why?

2. If a compound is known to cause mutations, people generally try to avoid it. Why is that so, even though we know that mutations are the raw material for adaptive evolution?

3. Explain why genetic drift may be important when captive populations of animals or plants are started with just a few individuals.

4. Two moth populations of the same species utilize different host plants. One rests on leaves and has evolved a green color that allows it to escape predation by blending in with the leaves. The other population rests on tree trunks, and is brown in color. The colors of these moths are genetically determined by different alleles of the same gene. What would be some consequences of migration of moths leading to gene flow between these two populations?

5. The text notes that natural selection is the force that pushes populations to adapt to environmental change. Is natural selection still going on in human populations? Why or why not?

MediaLab

Are Bacteria Winning the War? Natural Selection in Action

The last time your doctor prescribed an antibiotic for you, was it for a common cold or a real bacterial infection? If you had a bacterial infection, the first antibiotic might not have worked. Why not? You may have been infected with a strain of bacteria resistant to that antibiotic. The rise and spread of drug-resistant bacteria are consequences of natural selection—evolution in action. The *CD-ROM Tutorial* will help you review how environmental conditions can mold populations by natural selection. The following *Web Investigation* introduces the serious threat of evolving antibiotic-resistant bacteria. In the *Communicate Your Results* section, you can look for causes and solutions to this new problem.

This *MediaLab* can be found in Chapter 17 on your CD-ROM (Tutorial 17.3.2) and Companion Website (http://www.prenhall.com/krogh3).

CD-ROM TUTORIAL

In any population, organisms vary. In a group of bacteria, for example, there may be some rare antibiotic-resistant members. When the environment changes, let's say you take a prescribed antibiotic, and natural selection acts. All the sensitive bacteria die, leaving the antibiotic-resistant bacteria to thrive, thus changing the makeup of the population. This is just one of three types of natural selection that are animated in this activity.

Activity

1. *First, you will see how evolution, the modification of a trait in a population over time, is based in the change in frequencies of alleles for that trait.*

2. *Next, you will compare the three types of natural selection—directional, stabilizing, and disruptive selection—and see how they might act to change allele frequencies.*

3. *Finally, you will use a simulation of changing environmental conditions to see how the frequency of certain inherited variations change from one generation to another under the forces of natural selection.*

Use the Internet in the following *Web Investigation* to view modern examples of natural selection, including antibiotic resistance in bacteria, and see how natural selection affects us all.

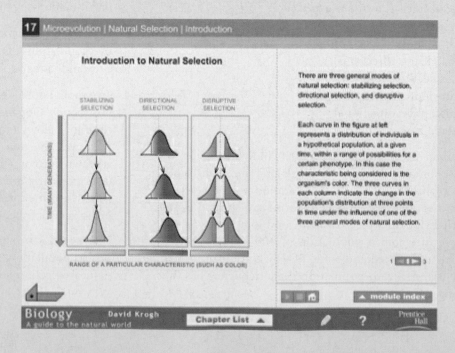

WEB INVESTIGATION

Investigation 1
Estimated time for completion = 10 minutes

Why should people care if antibiotic resistance occurs in bacteria? Read an online article by selecting the Keyword **ANTIBIOTIC RESISTANCE** on your CD or Website for this *MediaLab*. Note the biological causes of antibiotic (or antimicrobial) resistance, and the problems predicted. Two factors seem to be operating in this situation: antibiotic resistance by the bacteria in the context of natural selection, and the overuse (and misuse) of antibiotics by people.

Investigation 2
Estimated time for completion = 5 minutes

It's not just bacteria; many microbes have evolved to overcome the medical arsenals hurled against them. Americans were even beginning to see AIDS as a chronic but manageable disease after the introduction of the combination drug therapies that decimated the virus to nearly undetectable levels. Now,

however, many patients are finding these treatments no longer effective, and drug companies must develop new therapies. You can view the current state of several drug-resistant microbes by selecting the Keyword **CDC** or **HIV** on your CD or Website for this *MediaLab*.

Investigation 3
Estimated time for completion = 10 minutes

The curse of most infections isn't usually the direct damage inflicted by the virus. Rather, it is the body's defensive response and the accompanying symptoms like runny nose, sneezing, fever, pain, even diarrhea. From an evolutionary standpoint these symptoms are very beneficial. Select the Keyword **EVOLVED DEFENSES** on your CD or Website for this *MediaLab* to read about several examples from morning sickness to coughing.

Now that you have seen some real-life examples, do some of the following exercises to gain an understanding of the factors influencing natural selection and human diseases.

COMMUNICATE YOUR RESULTS

Exercise 1
Estimated time for completion = 5 minutes

What will happen when antibiotics are rendered useless against more and more antibiotic-resistant bacteria? What should we be doing to stop this trend? Using information from *Web Investigation 1*, compile a list of bacterial species that have developed antibiotic resistance, and indicate when. How has natural selection been a factor in the development and spread of antibiotic resistance? Write a list of common guidelines that you could use in explaining this problem to your friends and family. Include suggestions for stopping the spread of antibiotic resistance—and the consequences if we do not.

Exercise 2
Estimated time for completion = 5 minutes

In *Web Investigation 2*, you read about the emerging threat of drug-resistant microbes. Prepare a short presentation of one example to add to a list created by other students. Include the type of natural selection involved, the organism, and what science is doing about solving the problem.

Exercise 3
Estimated time for completion = 5 minutes

So you took some over-the-counter remedy for your common cold—but is it really best to completely obliterate those miserable symptoms? Using the article from *Web Investigation 3*, explain why or why not. Why might evolved defenses like runny noses, fevers, and diarrhea actually be beneficial for survival of our species? How did they evolve? Are the costs of these defenses actually worth it?

18

The Outcomes of Evolution
Macroevolution

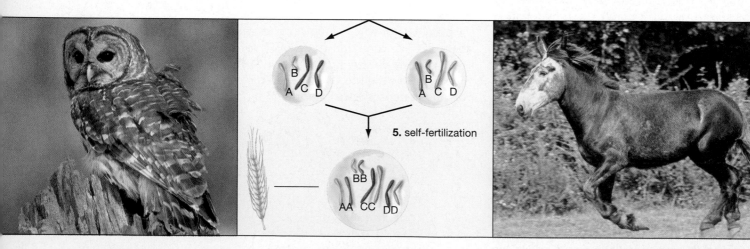

Endangered or not?
(Section 18.1, page 357)

5. self-fertilization

A new generation of wheat.
(Essay, page 362)

Not a horse, not a donkey.
What is it?
(Section 18.2, page 363)

New species come about when populations of existing species cease to breed with one another. Each new species that arises constitutes another branch on life's family tree. Scientists are trying hard to figure out what this tree looks like.

How many types of living things are there on Earth? How many varieties of life-forms are there that we can recognize as being fundamentally different from one another? It may surprise you to learn that we haven't the foggiest idea. The eminent naturalist Edward O. Wilson has pointed out that the number of species could be around 10 million, or it could be upwards of 100 million. Scientists simply have been unable to catalogue the vast diversity that exists on our planet.

Even if the 10 million figure is closer to the truth, isn't this number staggering? This seems particularly true given its starting point. A single type of organism arose some 3.6 billion years ago, branched into two types of organisms, and then the process continued—branches forming on branches—until at least 10 million different types of living things came to exist on Earth. Now here's the real kicker: These are just the *survivors*. The fossil record indicates that more than 99 percent of all species that have ever lived on Earth are now extinct. Branching indeed.

The questions to be answered in this chapter are: How does this branching work? How do we go from the microevolutionary mechanisms explored in Chapter 17 to the actual divergence of one species into two? And how do we classify the creatures that result from this branching?

18.1 What Is a Species?

There is a basic unit of living things you'll be reading about in this chapter, the species, that has a great importance to biology. If we allow that some fish and birds are endangered, that some bacteria are harmful to human beings while others are helpful, that some varieties of rice carry disease-resistance genes while others do not, then it follows that there must be some means of distinguishing one of these groupings from another (**see Figure 18.1** on **next page**). The whole notion of knowing something about the natural world begins to break down if we can't say *which* grouping it is that is endangered or harmful.

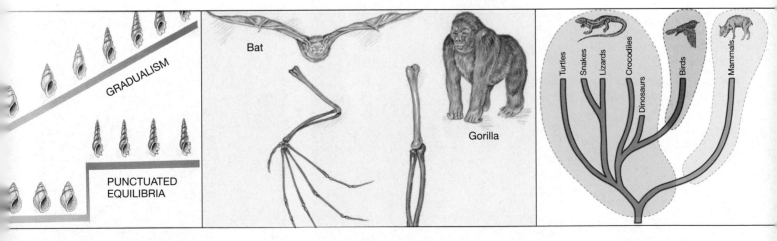

GRADUALISM

PUNCTUATED EQUILIBRIA

Jerky and smooth speciation. (Section 18.3, page 366)

Bat

Gorilla

How is a bat like a gorilla? (Section 18.5, page 370)

Turtles · Snakes · Lizards · Crocodiles · Dinosaurs · Birds · Mammals

Are birds reptiles? (Section 18.5, page 372)

Over the centuries, human beings have devised various ways of defining the groups now called species. But in science today, the most commonly accepted definition stems from what is known as the **biological species concept**, which uses breeding behavior to make classifications. We actually looked at a version of the biological species definition in Chapter 17. Here it is again, as formulated by the evolutionary biologist Ernst Mayr:

> Species are groups of actually or potentially interbreeding natural populations which are reproductively isolated from other such groups.

Note that the breeding behavior Mayr talks about can be real or potential. Two populations of finch may be separated from one another by geography, but if, upon being reunited in the wild, they began breeding again, they are a single species. Note also that Mayr stipulates that species are groups of *natural* populations. This is important because breeding may take place in captivity that would not in nature. No one doubts that lions and tigers are separate species; yet they will mate in zoos, producing little tiglons or ligers (depending on whether the father was a tiger or a lion). In natural surroundings, however, they apparently have never interbred, even when their ranges overlapped centuries ago.

You might think that, with the biological species concept in hand, scientists would be able to study any organism and, by discerning its breeding behavior, pronounce it to be a member of this or that species. Nature is so vast and varied, however, that this doesn't always work. We can't look at the breeding activity of bacteria, for example, because they don't *have* any breeding behavior; they multiply instead by simple cell division. (Microbiologists define bacterial groupings by sequencing their DNA or RNA and looking for the degree to which these sequences are the same.) Then there are separate species that sometimes interbreed in nature, producing so-called hybrid offspring as a result. Such mixing between species turns out to happen more in the plant world than the animal, but it does take place in both. If species are supposed to be "reproductively isolated" from one another, what are we to make of these crossings? (Mayr points out that it is *populations* in his definition that are reproductively isolated, not individuals, and that whole populations do tend to stay within their species confines.)

Despite these difficulties, the biological species concept provides a useful way of defining the basic unit of Earth's living things. Bacteria, some fungi, and some protists notwithstanding, most species carry out sexual reproduction, and relatively few

Figure 18.1
Separate Entities
The concept of a species can have very practical effects.

a This bird is a northern spotted owl (*Strix occidentalis*), which is an endangered species protected by federal law.

b This bird is a barred owl (*Strix varia*), which is not endangered.

a

b

species outside the plant world regularly produce hybrids. Moreover, this species concept is rooted in a critical behavior of organisms themselves—mating, which controls the flow of genes. And as you saw in Chapter 17, it is the change in the genetic makeup of a population (a change in its allele frequencies) that lies at the root of evolution.

18.2 How Do New Species Arise?

Having defined a species, let's now see how one variety of them can be transformed into another in the process called **speciation**: the development of new species through evolution. It is worth noting at the outset, however, that evolution does not always entail a multiplication of species.

Two Modes of Speciation: Cladogenesis and Anagenesis

As you've observed, a single species can diverge into two species, the "parent" species continuing while another branches off from it. This form of speciation, known as branching evolution or **cladogenesis**, is the central subject of this chapter (**see Figure 18.2**). It's also possible, however, for an entire species to undergo change without any branching taking place. If the change is extensive enough, a new species will evolve—one that presumably would be reproductively isolated from its ancestral species. This type of evolution is known as nonbranching evolution, or **anagenesis**.

Speciation Occurs When Populations Cease to Interbreed

With respect to branching evolution, the central question for scientists is: What brings it about? The answer has to do with the flow of genes reviewed in Chapter 17. As you saw then, evolution within a population means a change in that population's allele frequencies. Now imagine *two* populations of a single species of bird, with one being separated from the other—say, by one of them having flown to a nearby location. To the extent that they continue to breed with one another, with individuals moving between the locations, each population will *share* in whatever allele frequency changes are going on with the other population. Hence the two populations will evolve *together*, remaining a single

species. Now imagine that the migration stops between the two populations. Each population continues to undergo allele frequency changes, but it no longer shares these changes with the other population. Alterations in form and behavior may accompany allele changes, and these alterations may pile up over time. Coloration or bill lengths may change; feeding habits may be transformed. After enough time, should the two populations find themselves geographically reunited, they may no longer freely interbreed. At that point, they are separate species. Speciation has occurred.

The critical change here came when the two populations quit interbreeding. For scientists, the key question thus becomes: Why would this happen? What could drastically reduce interbreeding, and hence gene flow, between two populations of the same species?

The Role of Geographic Separation: Allopatric Speciation

In the preceding example, a very clear factor operated in reducing gene flow between the bird populations: They were separated geographically. And geographical separation turns out to be the most important starting point in speciation.

Populations can become separated in lots of ways. On a large scale, glaciers can move into new territory, cutting a previously undivided population into two. Rivers can change course, with the result that what was a single population on one side of the river may now be two populations on both sides of it. On a smaller scale, a pond may partially dry up, leaving a strip of exposed land between what are now two ponds with separate populations. Such environmental changes are not the only ways that populations can be separated, however. Part of a population might *migrate* to a

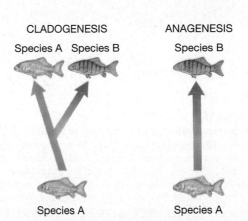

CLADOGENESIS

Species A   Species B

ANAGENESIS

Species B

Species A

Species A

Figure 18.2
Modes of Speciation
Speciation can occur either by the branching of one lineage into others (cladogenesis) or by gradual change along a single lineage (anagenesis).

remote area, as did the bird population, and in time he cut off from its larger population. For a real-life example of how migration can bring about speciation, **see Figure 18.3.**

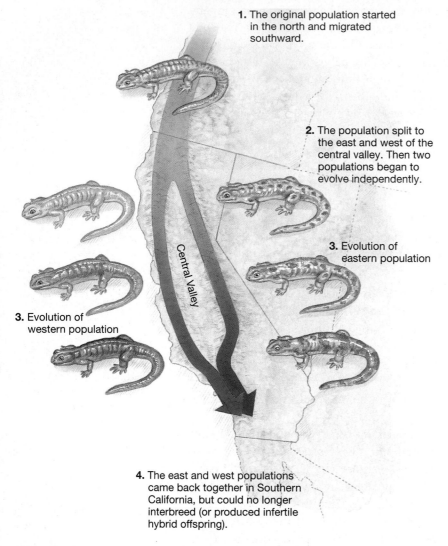

1. The original population started in the north and migrated southward.

2. The population split to the east and west of the central valley. Then two populations began to evolve independently.

Central Valley

3. Evolution of eastern population

3. Evolution of western population

4. The east and west populations came back together in Southern California, but could no longer interbreed (or produced infertile hybrid offspring).

Figure 18.3
Speciation in Action
Millions of years ago, the salamander *Ensatina eschscholtzii* began migrating southward from the Pacific Northwest. When it reached the Central Valley of California—an uninhabitable territory for it—populations branched west (to the coastal range) and east (to the foothills of the Sierra Nevada mountains). Over time, the populations took on different colorations as they moved southward. By the time the two populations were united in Southern California, they differed enough, genetically and physically, that they either did not interbreed or produced infertile hybrid offspring when they did. (Salamanders are falsely colored in drawing for illustrative purposes.)

When geographical barriers divide a population and the resulting populations then go on to become separate species, what has occurred is **allopatric speciation** (*allopatric* literally means "of other countries"; **see Figure 18.4**). Look along the banks on either side of the Rio Juruá in western Brazil, and you will find small monkeys, called tamarins, that differ genetically from one another in accordance with how wide the river is at any given point. Where it is widest, the members of this species do not interbreed, while at the narrow headwaters they do. Are the nonbreeding populations on their way to becoming separate species? Perhaps, but only if gene flow is drastically reduced between them, probably for a very long time.

Reproductive Isolating Mechanisms Are Central to Speciation

While geographical separation is the most important factor in getting speciation going, it cannot bring about speciation *by itself*. Following geographical separation, two populations of the same species must then undergo physical or behavioral changes that will *keep* them from interbreeding, should they ever be reunited. Allopatric speciation thus operates through a one-two process: first the geographic separation, then the development of differing characteristics in the two resulting populations—characteristics that will *isolate* them from each other in terms of reproduction.

Thus arises the concept of **reproductive isolating mechanisms**, which can be defined as any factor that, in nature, prevents interbreeding between individuals of the same species or

a

b

Figure 18.4
Geographical Separation—Leading to Speciation?
These two varieties of squirrel had a common ancestor that at one time lived in a single range of territory. Then the Grand Canyon was carved out of land in Northern Arizona, leaving populations of this species separated from one another, about 10,000 years ago. In the area of the Grand Canyon, **(a)** the Abert squirrel lives only on the Canyon's south rim, while **(b)** the Kaibab squirrel lives on the north rim. It's unclear whether these two varieties of squirrel would be reproductively isolated if reunited today—that is, whether they are now separate species—but the geographical separation they have experienced is the first step on the way to allopatric speciation.

of closely related species. Geographic separation is itself a reproductive isolating mechanism, because it is a factor that prevents interbreeding. (Populations that are physically separated cannot interbreed.) But, because the mountains or rivers that are the actual barriers to interbreeding are *extrinsic* to the organisms in question, geographic separation is called an **extrinsic isolating mechanism**. For allopatric speciation to take place, what also must occur is the second in the one-two series of events: The evolution of *internal* characteristics that keep organisms from interbreeding. Such factors are referred to as **intrinsic isolating mechanisms**: Evolved differences in anatomy, physiology, or behavior that prevent interbreeding between individuals of the same species or of closely related species.

What are these intrinsic mechanisms? A list of the most important ones is coming right up. An easy way to remember them is to think about what sequence of events would be required for fertile offspring to be produced in *any* sexually reproducing organism. First, organisms that live in the same area must encounter one another; if they don't, the "ecological isolation" mechanism is in place. If they do encounter one another, they must then mate in the same *time frame*; if they don't, then temporal isolation is in place. This sequence then continues on down through the list (see Table 18.1).

Six Intrinsic Reproductive Isolating Mechanisms

Ecological Isolation

Two closely related species of animals may overlap in their ranges and yet feed, mate, and grow in separate areas, which are called *habitats*. If they use different habitats, this means they may rarely meet up. If so, gene flow will be greatly restricted between them. Lions and tigers *can* interbreed, but they never have in nature, even when their ranges overlapped in the past. One reason for this is their largely separate habitats: Lions prefer the open grasslands, tigers the deep forests.

Temporal Isolation

Even if two populations share the same habitat, if they do not mate within the same time frame, gene flow will be limited between them. Two populations of the same species of flower may begin releasing pollen at slightly different times of the year. Should their reproductive periods cease to overlap altogether, gene flow would be cut off between them.

Behavioral Isolation

Even if populations are in contact and breed at the same time, they must choose to mate with one another for interbreeding to occur. Such choice, it turns out, is often based on specific courtship and mating displays, which can be thought of as passwords between members of the same species. Birds must hear the proper song, spiders must perform the proper dance, and fiddler crabs must wave their claws in the proper way for mating to occur.

Mechanical Isolation

Reproductive organs may come to differ in size or shape or some other feature, such that organisms of the same or closely related species can no longer mate. Different species of alpine butterfly look very similar, but their genital organs are different enough that one species cannot mate with another.

Tutorial 18.1.1
Reproductive Isolating Mechanisms

Table 18.1
Reproductive Isolating Mechanisms

EXTRINSIC ISOLATING MECHANISM	**GEOGRAPHIC ISOLATION** Individuals of two populations cannot interbreed if they live in different places (the first step in allopatric speciation).
	ECOLOGICAL ISOLATION Even if they live in the same place, they can't mate if they don't come in contact with one another.
	TEMPORAL ISOLATION Even if they come in contact, they can't mate if they breed at different times.
INTRINSIC ISOLATING MECHANISMS	**BEHAVIORAL ISOLATION** Even if they breed at the same time, they will not mate if they are not attracted to one another.
	MECHANICAL ISOLATION Even if they attract one another, they cannot mate if they are not physically compatible.
	GAMETE ISOLATION Even if they are physically compatible, an embryo will not form if the egg and sperm do not fuse properly.
	HYBRID INVIABILITY or INFERTILITY Even if fertilization occurs successfully, the offspring may not survive, or if it survives, may not reproduce (e.g., mule).

New Species through Genetic Accidents: Polyploidy

We have generally thought of speciation as something that takes place over many generations. Yet plants (and some animals) have a means of speciating in a single generation. It is called *polyploidy*, and it is very important in the plant world; more than 100,000 species of flowering plants in existence today are thought to have been brought about through it. Here is one version of how it works. From the genetics unit, you may recall that human beings have 23 pairs of chromosomes, *Drosophila* flies four pairs, and Mendel's peas seven. Whatever the *number* of chromosomes, the commonality among all these species is that their chromosomes come in *pairs*, which is the general rule for species that reproduce sexually.

As noted at the start of the chapter, plants are more adept than animals at producing "hybrids," with the gametes from two *separate* species coming together to create an offspring. Often these offspring are sterile, however, because when it comes time for them to produce *their* gametes (the counterparts to human sperm or eggs), the sets of chromosomes they inherited from their different parental species may not "pair up" correctly in meiosis, owing to differences in chromosome number or structure.

Now comes the accident. Suppose that, back as a single-celled zygote, the hybrid offspring carried out the usual practice of doubling its chromosome number in preparing for its first cell division. Now, however, the cell fails to actually divide; whereas it was supposed to put half its complement of chromosomes into one daughter cell and half into another, it doesn't do this. This single cell now contains *twice* the usual number of chromosomes. It then proceeds to undergo regular cell division, meaning that every cell in the plant that follows will have this doubled number of chromosomes. Critically, this plant will have doubled both its *sets* of chromosomes— the set it got from parental species A and the set it got from species B. If you double a set, by definition every member of that set now has a partner to pair with in gamete formation. Thus the roadblock to a hybrid producing offspring has been removed; the chromosomes can all pair up.

Many plants have employed a means of speciating in a single generation.

This pairing up yields gametes (eggs and sperm), which can then come together and fuse thanks to another capability of plants. Recall that many plants can *self-fertilize*. A single plant contains male gametes that can fertilize that same plant's female gametes. Thus the plant with the doubled set of chromosomes can fertilize itself, theoretically beginning an unending line of fertile offspring. This line of organisms is "reproductively isolated" from either of its parental species because it has a different number of chromosomes than either parental species. Were gametes from it to be "crossed" with gametes from either parent, sterile offspring would be the likely result. With reproductive isolation, we have a *separate species*; and with self-fertilization, we have a species that can perpetuate itself.

A multiplication of the normal two sets of chromosomes to some other set number is known as **polyploidy**; here we have speciation by polyploidy, which is one type of sympatric speciation. The importance of this in the plant

Gametic Isolation

Even if mating occurs, offspring may not result if there are incompatibilities between sperm and egg or between sperm and the female reproductive tract. In plants, the sperm borne by pollen may be unable to reach the egg lying within the plant's ovary. In animals, sperm may be killed by the chemical nature of a given reproductive tract, or may be unable to bind with receptors on the egg.

There is one form of gametic isolation, called *polyploidy*, that is very important in plants; indeed, at least 50 percent of all flowering plant species are thought to have come about through it. You can read about it in "New Species through Genetic Accidents."

Hybrid Inviability or Infertility

Even if offspring result, they may develop poorly—they may be stunted or malformed in some way—or they may be infertile, meaning unable to bear offspring of their own. A well-known example of such infertility is the mule, which is the infertile offspring of a female horse and a male donkey (**see Figure 18.5**).

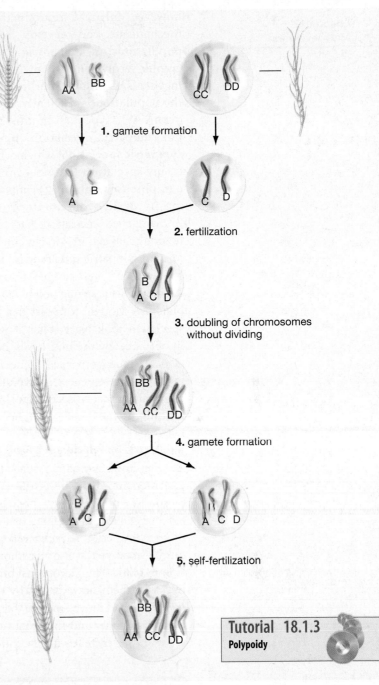

Many of our most important food crops are polyploid, including oats, wheat, cotton, potatoes, and coffee.

world is immense. Many of our most important food crops are polyploid, including oats, wheat, cotton, potatoes, and coffee. Indeed, the type of polyploid speciation we've looked at—which begins with a hybrid offspring—often produces bigger, healthier plants. As a result, breeders have developed ways to artificially induce polyploidy.

Figure 1
Polyploidy in Wheat
Two different species of wheat exist in nature, with slightly different "genomes" or complements of DNA.

1. Gametes (egg and sperm) are formed in the different species.

2. These gametes fuse, in fertilization, to form a zygote—a single cell that will develop into a new plant. Such a mixed-species or "hybrid" zygote generally develops into a sterile plant, because its chromosomes cannot pair up correctly when the plant produces its own gametes during meiosis. In this case, however, the zygote

3. doubles its chromosomes in preparation for cell division, but then fails to divide. With this doubling, each chromosome now has a compatible homologous chromosome to pair with during meiosis.

4. Gamete formation thus takes place in the plant, and these gametes from the same plant then

5. fuse, because this is a self-fertilizing plant. With this,

6. a new generation of wheat plant has been produced—one that is a different species from either parent generation, because each of the parent species has two pairs of chromosomes, while this new hybrid species has four pairs.

1. gamete formation

2. fertilization

3. doubling of chromosomes without dividing

4. gamete formation

5. self-fertilization

Tutorial 18.1.3
Polypoidy

Sympatric Speciation

Thus far, you have looked at the development of intrinsic reproductive isolating mechanisms strictly as a second step in speciation—one that follows a geographic separation of populations. As it happens,

Figure 18.5
Mules Are Infertile Hybrids
Even though mules cannot themselves reproduce, humans frequently cross horses and donkeys to produce mules in order to take advantage of their exceptional strength and endurance. Chromosomal incompatibilities between horses and donkeys leave the mules sterile.

however, intrinsic reproductive isolating mechanisms can develop between two populations in the *absence* of any geographic separation of them. If these isolating mechanisms reduce interbreeding between the populations sufficiently, speciation can occur. What occurs in this situation is not allopatric speciation, however; it is **sympatric speciation**, which can be defined as any speciation that does not involve geographic separation. (*Sympatric* literally means "of the same country.")

Sympatric speciation has been a contentious subject in biology for years. One form of sympatric speciation is the polyploidy reviewed in "New Species through Genetic Accidents." But setting polyploidy aside, most biologists thought for years that if sympatric speciation took place at all, it was of trivial importance. By the mid 1990s, however, new evidence had convinced some of the most prominent researchers in evolutionary biology that sympatric speciation does have significance in the broader picture of speciation.

Sympatric Speciation in a Fruit Fly

To see how sympatric speciation works, consider one of its best-studied examples, a species of fruit fly named *Rhagoletis pomonella*. Prior to the European colonization of North America, *R. pomonella* existed solely on the small, red fruit of hawthorn trees (**see Figure 18.6**). The Europeans brought *apple* trees with them, however, and by 1862 some *R. pomonella* had moved over to them. It seemed to at least one mid-nineteenth-century observer, however, that with the introduction of

apples there had arisen separate *varieties* of the flies—what might be called apple *R. pomonella* and hawthorn *R. pomonella*—with each variety courting, mating, and laying eggs almost exclusively on its own type of tree. A modern researcher named Guy Bush led the way, beginning in the early 1960s, in investigating whether this was the case, and the answer turned out to be yes. The two varieties of flies are separated from each other in all these ways. In one study undertaken on them, only 6 percent of the apple and hawthorn flies interbred with one another. The hawthorn and apple *R. pomonella* are not separate species yet, but they certainly give indication of being in transition to that status.

For our purposes, the first thing to note is that this separation has not come about because of *geographic* division. The apple and hawthorn trees that the two varieties of flies live on may scarcely be separated in space at all. Given this, how did this single species move toward becoming two *separate* species? Bush offers a likely scenario, based on a critical difference between the hosts the two species live on. Apples tend to ripen in August and September while hawthorn fruit ripens in September and October. In the summer, all fruit flies emerge as adults after wintering underground as larvae, after which they fly to their host tree to mate and lay eggs in the fruit.

Bush believes that about 150 years ago (when hawthorns were hosts to *all* the flies), some individuals in a population of *R. pomonella* experienced either a mutation or perhaps a chance combination of rare existing alleles. In either event, this change did two things: It caused these flies to emerge slightly *earlier* from their underground state than did most flies; and it drew these flies to the smell of *apples* as well as hawthorns. Because apples mature slightly earlier than the hawthorn fruit, these flies had a suitable host waiting for them. More important, these flies would have bred *among themselves* to a high degree. Recall that the other flies were emerging later, and the adult fly only lives for about a month. Thus, the variant alleles were passed on to a selected population in the next generation. Today, the apple *R. pomonella* flies indisputably do emerge earlier than the hawthorn flies. This is one of the things that ensures reproductive isolation between the two groups; their periods of mating don't fully overlap.

Figure 18.6
A Species Undergoing Sympatric Speciation?
The fruit-fly species *Rhagoletis pomonella*, pictured here on the skin of a green apple, may be undergoing speciation.

Such a lack of overlapping mating periods may sound familiar, because it is one of the intrinsic reproductive isolating mechanisms you looked at earlier. (It is temporal isolation.) You also can see ecological isolation in operation with these flies. Though the two types of flies live in the same area, they meet up with relatively little frequency, because they have different habitats (hawthorn vs. apple trees). In short, these populations have developed intrinsic reproductive isolating mechanisms without ever having been separated from one another geographically. They are headed toward speciation, in other words, but in this case it is sympatric speciation.

18.3 When Is Speciation Likely to Occur?

The horseshoe crab (**see Figure 18.7**) is not really a crab at all, but instead is distantly related to land-dwelling arthropods such as spiders and scorpions. (It lacks the antennae that real crabs possess, but it *has* the pincers around the mouth that all spiders and scorpions do.) Horseshoe crabs have been around in something like their modern form for more than 300 million years. They exist today in a scant four or five species, each one of which is pretty much like the others. The crabs live in the shallow oceans off North America and Asia, pushing through sand or mud to feed on everything from algae to small-bodied invertebrate animals.

Now consider the Galapagos (or Darwin's) finches you looked at in Chapter 17. There are *thirteen* of these species on the small Galapagos archipelago, all of them derived from a single species of South American

finch that arrived on one of the islands perhaps 100,000 years ago.

Think of the difference between horseshoe crabs and Darwin's finches. The former have remained almost unchanged for more than 300 million years throughout the world, while the latter diverged into 13 different species in the last 100,000 years in a small cluster of islands off South America.

Specialists and Generalists

Why the great differences in the rate of speciation here? Two general principles are at work. First, as Niles Eldredge has pointed out, horseshoe crabs are *generalists*: Their diet is extremely diverse; they will eat plants and small animals but will also scavenge for debris on the ocean bottom. By contrast, you saw in Chapter 17 how *specialized* some of the Galapagos finches are in their feeding behavior—particularly when food has become scarce because of drought conditions, as happens on the islands. In such times, species might exist on a single variety of plant seed. Species that are tied in this way to a particular food or environmental condition must adapt in connection with changes in them or face extinction. Think of how quickly the bills of the *Geospiza fortis* population on Daphne Major evolved toward greater depth when having a deeper bill meant the difference between life and death (see Chapter 17, page 348). By definition, this kind of adaptation means change—and change is what speciation is all about. By contrast, the horseshoe crab shifts from one food to another, depending on what is available, not adapting greatly in response to changes in any one food source.

a   b

Figure 18.7
Different Groups of Organisms Undergo Different Rates of Speciation

a Because horseshoe crabs look essentially the same today as they did 300 million years ago, some scientists have called them "living fossils." Compare the modern organism, left, with the fossil on the right, which is at least 145 million years old.

b By contrast, the Galapagos finches have diversified tremendously, forming 13 species in as little as 100,000 years.

New Environments: Adaptive Radiation

The second lesson offered by horseshoe crabs and finches concerns the kinds of environments that induce speciation. The Galapagos islands were formed by volcanic eruptions that brought the islands above the ocean's surface only about 5 million years ago. At that time, these volcanic outcroppings gave new meaning to the word *barren*. For a brief time at least the islands were utterly sterile, with no life on them. Very quickly, however, life did come to the islands, with bacteria, fungi, plant seeds, and tiny animals landing on the islands, all being borne by air or ocean currents. The South American finches obviously did not arrive until much later; but by then, with lots of large plant species well established, what these birds encountered was an environment rich with possibility, for there were *no birds of their kind* on the islands. Imagine that you are a graphic artist, working in a big city with lots of other graphic artists, most of whom *specialize* in this or that (magazines, Internet websites, etc.). Now you and a few other graphic artists move to a new city in which there are few graphic artists, but a good number of *possibilities* for graphic arts work. You would thus be able to specialize fairly easily—filling a *niche* or working role in this new environment—because many of these niches would not yet have been taken. Just so did Darwin's finches rush in to fill previously unoccupied niches on the Galapagos Islands—this plant seed, this insect, this mating environment.

Such a situation is ripe with possibilities for change (meaning speciation) because, while niches are in flux, there is a good deal of shaping of species to environment. But

more of this occurred on the Galapagos, because this niche-filling was taking place on 25 separate *islands*. The birds can fly from one island to another, but the islands nevertheless represent a geographic barrier to bird interbreeding, and you know what follows from this: allopatric speciation.

The Galapagos finches exemplify something known as **adaptive radiation**: the rapid emergence of many species from a single species that has been introduced to a new environment. The finches radiated out to fill new niches on the islands, with populations adapting to the environments over time. In summary, two conditions that are conducive to speciation are specialization (of food source or environment) and migration to a new environment, particularly when there are no closely similar species in that environment.

Is Speciation Smooth or Jerky?

Charles Darwin knew that in looking at large-scale evolutionary changes, such as the change from sea-dwelling to land-dwelling life-forms, what is striking is the lack of "transitional forms" in the fossil record. Not many creatures have been found that seem to lie in between the sea-dwelling and land-dwelling forms. Darwin thought that this was simply a problem of not having collected enough fossils in the right places; that in time, the transitional forms would show themselves.

Scientists have found some transitional-form fossils since Darwin's time, but not many. Indeed, the fossil record seems to speak against the mode of evolution as Darwin had envisioned it. He imagined evolution as a series of infinitesimally small changes accumulating in a slow, steady way in populations of a species until in time one species could be seen to have diverged into two.

What the fossil record exhibits again and again, however, is not evidence of slow, steady change in a species. Rather, there are indications of enormous lengths of time in which so-called **stasis** occurs—in which species stay exactly the same or undergo some minute modifications. After this long stasis, there appears to be an *abrupt* change to a new species.

In 1972, two young scientists proposed that it was time to take the fossil record at face value. Why isn't it plausible, they said, that species experience stasis for long periods

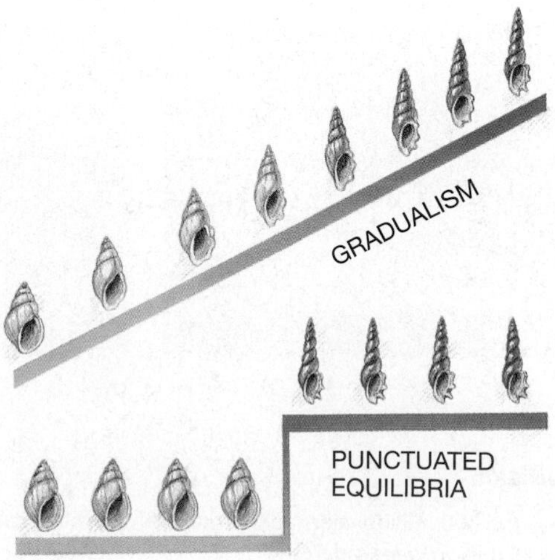

Figure 18.8
Is Speciation Smooth or Jerky?
According to the gradualism model, speciation occurs through gradual change over long periods of time. Contrast this with the punctuated equilibria model, which proposes that long periods of stasis may periodically be "punctuated" by rapid bursts of speciation.

GRADUALISM

PUNCTUATED EQUILIBRIA

of time, after which speciation then takes place in relatively brief bursts? This is the **theory of punctuated equilibria** proposed by Stephen Jay Gould and Niles Eldredge. (Periods of stasis are the equilibrium that species generally live in, with these then being "punctuated" by the speciation events.)

So, how rapid is speciation under this view? It can take place in thousands of years rather than the millions during which a species is in stasis. If that is the case, it would explain why there are so few transitional forms in the fossil record: They aren't around for long enough to leave many fossils (**see Figure 18.8**).

There has been great debate about the theory of punctuated equilibria ever since it was proposed. Some scientists have supported it, others have opposed it. Its detractors (sometimes called "gradualists") have noted that only the "hard parts" of species (bone and so forth) are preserved in the fossil record, while much evolutionary change must concern "soft parts." Other critics allege that the theory proposes nothing new, but merely puts terminology to ideas that already existed (in population genetics) about how rapidly species can evolve.

Some recent experimental evidence is consistent, however, with the idea of punctuated equilibria. Relying on the ability of bacteria to reproduce rapidly, scientists looked at 10,000 generations of them over a period of 4 years—freezing samples at various points in time. They found that these cells tended to stay the same size for hundreds of generations and then evolve to a larger average size very quickly. Other researchers have asked how much *genetic* change is required to bring about speciation and found that, in comparing two species of flower at least, the answer is very little. They found that changes in as few as eight genetic locations may account for the differences seen between two species of monkey flower that differ in color and accordingly attract different pollinators. If so few changes are required for speciation, then it could proceed more rapidly than if many changes were necessary.

One possibility is that speciation is both jerky (punctuated) and smooth (gradual), depending on the species and the characteristics being examined. Changes in "morphology" or physical structure are based on genetic changes, as you have seen. It could be that genetic changes are gradual, while the resulting morphological changes come about in spurts.

18.4 The Categorization of Earth's Living Things

This chapter began by noting the importance of the species concept—of being able to say that this organism is fundamentally separate from that. To have a species concept means, of course, that there has to be some means of *naming* separate species. You have probably noticed that most of the species considered in this chapter have been referred to by their scientific names; you didn't just look at a fruit fly, but rather *Rhagoletis pomonella*. To the average person, such names may be regarded as evidence that scientists are awfully exacting—or just plain fussy. Why the two names? Why the Latin? Can't we just say fruit fly?

To take these questions one by one, scientists can't just say "fruit fly" for the same reason a person can't say "the guy in the white shirt" while trying to identify a player at a tennis tournament. There are lots of guys in white shirts at tennis tournaments and there are lots of different kinds of flies in the world. It is important to be able to say *which* fly we are talking about. The importance of this can be seen with *Rhagoletis pomonella* itself, which is a major pest in the apple industry. Tell some apple growers that you have spotted flies on their apples and you may get a shrug of the shoulders; tell them you have spotted *R. pomonella* and you may get a very different reaction.

The Latin that is used stems from the fact that this naming convention was standardized by the Swedish scientist Carl von Linné in the eighteenth century, at a time when Latin was still used in the Western world in scientific naming. (In fact, Linné is better known by the Latinized form of his name—Carolus Linnaeus.)

Linnaeus recognized the confusion that can result from having several common names for the same creature, and thus devoted himself to giving specific names to some 4,200 species of animals and 7,700 species of plants—all that were known to exist in his time. Many of the names he conferred are still in use today.

The fact that there are *two* parts to scientific names—a **binomial nomenclature**—points to the central question of *groupings* of organisms on Earth. Consider the domestic cat, which has the scientific name *Felis domestica*. The first part of its name is its genus (plural *genera*), which designates a group of closely related, but still separate species. It turns out that, worldwide, there are five other species of small cat within the *Felis* genus, such as the small cat *Felis nigripes* ("black-footed cat") found in Southern Africa.

Taxonomic Classification and the Degree of Relatedness

The practical importance of the genus classification is that, if we know that two organisms are part of the same genus, then to know something about one of them is to know a good deal about the other. But what is the basis for placing organisms in such a category? Modern science classifies organisms largely in accordance with how closely they are *related*. In this context, "related" has the same meaning as it does when the subject is extended families. If people have the same mother, they are more closely related than if they shared only a common *grandmother*, which in turn makes them more closely related than if they had only a common great-great grandmother, and so on.

In the same way, species can be thought of as being related. Domestic dogs (*Canis familiaris*) are very closely related to gray wolves (*Canis lupus*), with all dogs being descended from these wolves. Indeed, by some reckonings dogs and gray wolves are the same species (*Canis lupus*), because they will interbreed in nature. Such differences as exist between them are the product of a mere 10,000 years or so of the human domestication of dogs. Scientists believe, then, that 15,000 years ago there was but *one* species (the gray wolf) that gave rise to both today's wolf and today's dog. A close relation indeed.

Domestic dogs are also related, however, to domestic *cats*; if we look far back enough in time, we can find a single group of animals that gave rise to both the dog and cat lines. This does not mean going back 10,000 years, however; it means going back perhaps 60 million years. Thus, there obviously is a big difference in how closely related dogs and wolves are, as opposed to dogs and cats. Establishing such *degrees* of relatedness is what the whole system of scientific classification is about. There is a field of biology, called **systematics**, that is concerned with the diversity and relatedness of organisms; part of what systematists do is try to establish the truth about who is more closely related to whom. They thus study the evolutionary history of groups of organisms.

Setting aside for the moment the difficulty in *determining* what such evolutionary histories are, there obviously is a tremendous cataloguing job here, given the number of living and extinct species mentioned earlier. Given this diversity, a grouping or **taxonomic system** is employed in order to classify every species of living thing on Earth. There are eight basic categories in use in the modern taxonomic system: **species, genus, family, order, class, phylum, kingdom**, and **domain**. The organisms in each category make up a **taxon**.

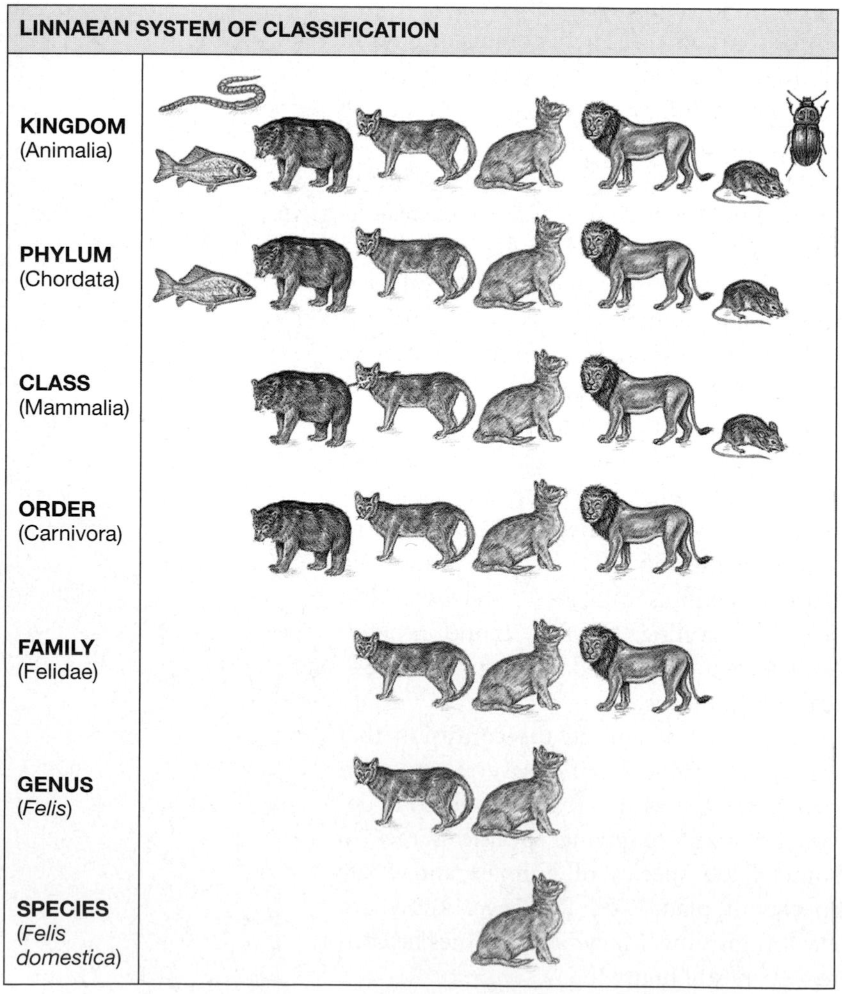

**Figure 18.9
Classifying Living Things**
The classification of the house cat, *Felis domestica*, based on the Linnaean system.

LINNAEAN SYSTEM OF CLASSIFICATION

| KINGDOM (Animalia) |
| PHYLUM (Chordata) |
| CLASS (Mammalia) |
| ORDER (Carnivora) |
| FAMILY (Felidae) |
| GENUS (Felis) |
| SPECIES (Felis domestica) |

A Taxonomic Example: The Common House Cat

If you look at **Figure 18.9**, you can see how this taxonomic system works in connection with the domestic cat. As noted, this cat is only one species in a genus (*Felis*) that has five other living species in it. The genus then is a small part of a family (Felidae) that has 17 other genera in it (panthers, snow leopards, and others). The family is then part of an order (Carnivora) that includes not only big and small "cats," but other carnivores, such as bears and dogs. On up the taxa we go, with each taxon being more inclusive than the one beneath it until we get to the highest category in this figure, the kingdom Animalia, which includes all animals. (Later you'll look at the "supercategory" above kingdom, called domain; but for simplicity's sake, kingdom is the highest-level taxon considered here.)

18.5 Constructing Evolutionary Histories: Classical Taxonomy and Cladistics

So how do systematists go about putting organisms in these various groups? What evidence do they use in constructing their evolutionary histories? The answer is radiometric dating, the fossil record, DNA sequence comparisons—all the things reviewed in Chapter 16 that are used to chart the history of life on Earth.

If you look at **Figure 18.10**, you can see the outcome of some of this work, an evolutionary "tree" for one group of organisms, in this case one of the major groups of mammalian carnivores. You can see that the tree is "rooted," about 60 million years ago, with an ancestral carnivore whose lineage then split two ways: to dogs on the one hand and everything else on the other. One interesting facet of this evolutionary history is how closely related bears are to the aquatic carnivores. All such histories are hypotheses about evolutionary relationships, with each such hypothesis known as a **phylogeny**.

Classical Taxonomy Looks for Similarities

The question is: How did systematists use the kinds of evidence noted in this chapter to construct this phylogeny? How do they interpret evidence to make conclusions about who is descended from whom and, in rough terms, at what time? There is a system for interpreting evolutionary evidence, handed down from the time of Darwin, that is sometimes referred to as *classical taxonomy*. In thinking about the organisms in this family tree, classical systematists would look first and foremost at the physical form or "morphology" of the animals in question, as preserved in the fossil record, and compare it to the morphology of modern animals. Skull shape, "dentition" (teeth patterns), limb structure, and much more would be considered. They would also look at *where* the ancient forms existed, compared to modern forms. More recently, they have used molecular techniques, such as examining DNA or protein sequences in different living species to determine the relatedness among them. In essence, classical systematists would look at how many similarities one group has with another and, on that basis, try to judge relatedness among them. The word *judge* is used advisedly here, because at the end of the day in the classical system, subjective judgments usually need to be made about who is more closely related to whom. It is a matter of weighing one piece of evidence against another, but who's to say which piece of evidence is more important?

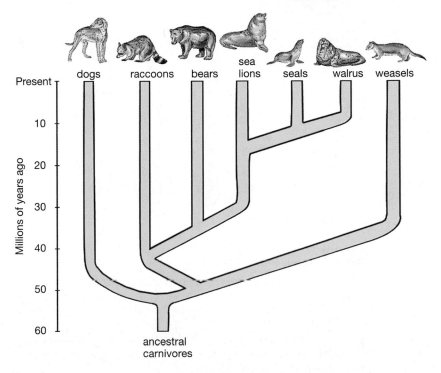

**Figure 18.10
A Family Tree for a Selected Group of Mammalian Carnivores**

Obscuring the Trail: Convergent Evolution
One of the things that makes classical interpretation of such evidence problematic is that similar features may arise *independently* in several evolutionary lines. A bedrock of

a HOMOLOGY: Common structures in different organisms that result from common ancestry

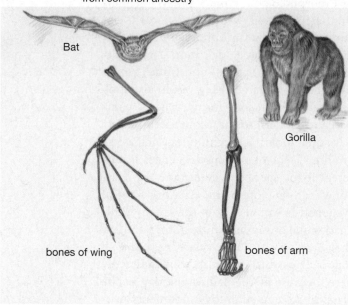

Bat

Gorilla

bones of wing

bones of arm

b ANALOGY: Characters of similar function and superficial structure that have *not* arisen from common ancestry

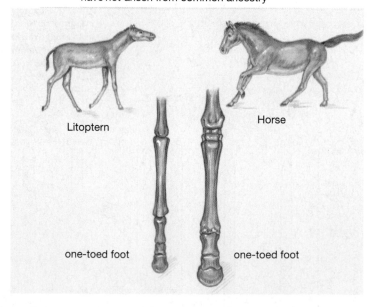

Litoptern

Horse

one-toed foot

one-toed foot

Figure 18.11
Related, as Opposed to Similar

a Bats and gorillas have the same bone composition in their forelimbs—one upper bone, joined to two intermediate bones, joined to five digits—because they share a common ancestor. Of course these forelimbs are used today for very different functions.

b Horses and the extinct litopterns have one-toed feet that serve the same function, but the one-toed condition was derived independently in the two creatures—in separate lines of descent.

systematic classification is the existence of **homologies** in organisms, which can be defined as common structures in different organisms, resulting from a shared ancestry. In Chapter 16 (Figure 16.12), a strong homology was noted in forelimb structure that exists in organisms as different as a gorilla, a bat, and a whale. If you look at **Figure 18.11**, you can see the gorilla and the bat again.

Now, however, consider an extinct group, the litopterns, that once lived in what is now Argentina. Like the modern-day horse, they roamed on open grasslands. Over the course of evolutionary time they developed legs that were extraordinarily similar to the horse's, right down to having an undivided hoof. In fact, litopterns are only very distantly related to the horse; but the leg similarities were enough to fool at least one nineteenth-century expert who thought he had found in litopterns the ancestor to the modern horse.

What is exhibited in the legs of the litoptern and horse is an **analogy**: A feature in different organisms that is the same in function and superficial appearance. Note the difference between it and a *homology*, which is the same structure in different organisms with a common ancestor. When nature has shaped two separate evolutionary lines in analogous ways, what has occurred is **convergent evolution**. (The litoptern and horse lines converged in the design of their legs.) But analogous features have nothing to do with common descent; they merely show that the same kinds of environmental pressures lead to the same kinds of designs. The problem for systematics is that analogy can be confused with homology; analogies can make us believe that organisms share common ancestry when in fact they do not. How common is analogy? To take one perhaps extreme example, eyes are thought to have evolved 38 separate times in animals.

Another System for Interpreting the Evidence: Cladistics

The confusion of analogy, and the subjectivity inherent in classical taxonomy, helped bring about the formation of another system for establishing relatedness. Developed in the 1950s by the German biologist Willi Henning, it is called **cladistics** (from the Greek *klados*, meaning "branch"). In practice, cladistics has become the core of most of the phylogenetic work going on today.

In **Figure 18.12** you can see a **cladogram**, which is an evolutionary tree constructed within the cladistic system. Note that there is a very simple branched line and that no time-scale is attached to it, as was the case with the evolutionary tree of carnivores. Cladistics concerns itself first with lines of descent—with the *order* of branching events. Once this order has been established, efforts can be made to fix events in time, but that is a secondary concern. First and foremost, this cladogram is a proposed answer to the question: Among the animals lizard, deer, lion, and seal, which two groups of animals have the most recent common ancestor? Then, which other two have the *next* most recent common ancestor, and so on. By extension, the question is: Who is more closely related to whom? Note that these questions are being asked about only four of the five animals listed. You'll get to the role of the hagfish in a moment.

Cladistics Employs Shared Ancestral and Derived Characters

The starting point for cladistic analysis is the difference between so-called ancestral characters and derived characters. If you look at any group of species (any taxon), its **ancestral characters** are those that existed in an ancestor common to them all. There are about 50,000 species of animals on Earth that possess a dorsal vertebral column (better known as a backbone). The ancestor to all fishes, reptiles, and mammals had such a feature, which makes the vertebral column an ancestral character for all these groups. On the other hand, only *some* vertebrates then went on to become animals that have four limbs (and are thus tetrapods). The tetrapod feature is thus a **derived character** of the vertebrate state; it is a character *unique* to taxa descended from a common ancestor.

In Figure 18.12, you can see a continuation of this process of ever-more selective grouping in that only some mammals are carnivorous feeders, and only some members of this taxon are *aquatic* mammalian carnivores. The presumption in cladistics is that shared derived characters are evidence of common ancestry and that the *more* derived characters any two organisms share, the more recently they will have shared a common ancestor,

compared to other organisms under study. At its most fundamental level, cladistics becomes a matter of counting shared derived characters. Seals and lions share *three* derived characters (tetrapod structure, mammary glands, and carnivorous feeding), while the lizard has only one derived character that it shares with all the other animals (tetrapod structure). Cladistics infers from this that the lion and the seal have a more recent common ancestor than do any other two groups in the cladogram—that their branching from this common line of descent came later than any other branching in the line. And the hagfish? It is a starting-point organism for the mammalian carnivores, known as an *outgroup*. If we are to say that the animals under consideration have derived characters, they must be derived *compared* to some other organism—in this case, the hagfish. The examples here employ visible, physical traits, but in practice modern cladistics relies heavily on comparisons among molecular sequences, such as the number of shared, derived base pairs in lengths of DNA or RNA.

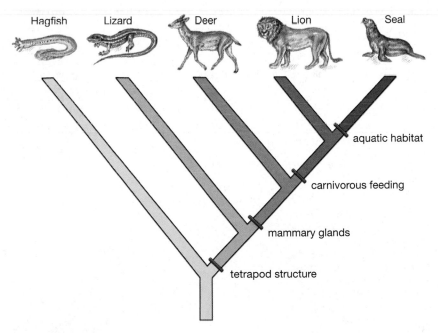

Figure 18.12
A Simple Cladogram
Among the lizard, deer, lion and seal, which two are most closely related? The tentative answer displayed in this cladogram is the lion and the seal. The basis for this assertion is the number of unique or "derived" characters these two creatures share compared to any other two organisms under consideration. (They share three: tetrapod structure, mammary glands, and carnivorous feeding.) Such a system pays no attention to superficial similarities among organisms. Seals swim in the water, like fish, but they share far more derived characters with lions than they do with fish—enough to persuade scientists that seals have returned to the water in relatively recent times, following a period in which their ancestors lived on land.

Note that this counting has done away with one of the problems of classical taxonomy: It has removed an element of subjectivity. There is a firm rule for inferring relatedness—the number of shared derived characters—which means that judgment need not play a part in this analysis.

Should Anything but Relatedness Matter in Classification?

Strict cladistic analysis can produce results that run headlong into commonsense notions of what is "different" and "the same." Which is more closely related: lizards and crocodiles on the one hand, or birds and crocodiles on the other? It may surprise you to learn that, in terms of evolution, it is birds and crocodiles. Birds split off from a dinosaur-crocodile line within the last 200 million years or so, while the split between the lizard and crocodile lines came far earlier. *Classical* taxonomy puts crocodiles, lizards, dinosaurs, and turtles into a single class (Reptilia), while putting birds in a different class (Aves; **see Figure 18.13**). Cladistic analysis rejects this classification, on grounds that crocodiles and birds have a more recent common ancestor than do crocodiles and lizards.

This gets us to an important point. Classical analysis holds that, in classifying creatures, something counts *besides* evolutionary relationships—in this case, the special qualities of birds (flight, warm-bloodedness)—while cladistics says the only criterion for placement is phylogeny. At root, then, cladistics is not about classification; it is about establishing lines of descent and only classifies species in accordance with the lines that emerge. Classical analysis is concerned with both evolutionary relationships and with classification.

On to the History of Life

The millions of species that exist on Earth today took billions of years to evolve. During that time, life went from being exclusively microscopic and single-celled to being the staggeringly diverse entity it is today. How did this happen over time? And how did we go from no life to life? And what is the order of appearance for life's living things? These are the subjects of Chapter 19.

Figure 18.13
Are Birds Reptiles?

a Classical taxonomy employs three classes, Aves, Mammalia, and Reptilia, for birds, mammals, and reptiles, respectively. Thus crocodiles and birds are in separate classes—a recognition on the part of classical taxonomists of the unique features of birds, such as feathers.

b In cladistic analysis, by contrast, birds are grouped very closely with crocodiles and dinosaurs, in recognition of the recent shared ancestry of the three groups. The only consideration in this system of classification is who is more closely related to whom.

a Classical view of relationships among tetrapods

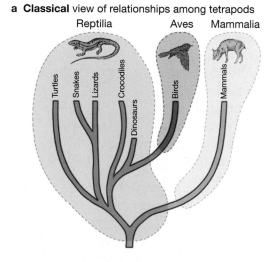

b Cladistic view of relationships among tetrapods

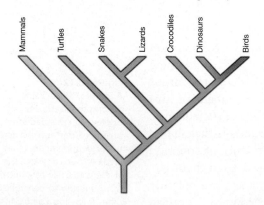

Chapter Review ▮▮▮▮▮▮▮▮▮▮▮▮▮▮

Summary

18.1 What Is a Species?

- The most accepted definition of a species is derived from the biological species concept: "Species are groups of actually or potentially interbreeding natural populations that are reproductively isolated from other such groups." This definition does not apply to asexually reproducing organisms, such as bacteria. Additionally, hybrids—the offspring of two different species—can occur with some frequency, especially among plants.

18.2 How Do New Species Arise?

- Speciation can occur through branching evolution (cladogenesis) or nonbranching evolution (anagenesis). Branching evolution occurs when a single species splits and gives rise to one or more daughter species, with each species reproductively isolated from the others. Nonbranching evolution occurs when a single lineage evolves into a new species, presumed to be reproductively isolated from its ancestral form.

- Speciation comes about when populations of the same species cease to interbreed.

- Branching evolution may occur through allopatric speciation, which always involves a geographic separation of populations of a species, or through sympatric speciation, which does not involve geographic separation.

- A central factor in speciation is the development of reproductive isolating mechanisms, defined as any factor that, in nature, prevents interbreeding between individuals of the same species or of closely related species.

- One form of reproductive isolating mechanism is the geographic separation that takes place in allopatric speciation. For allopatric speciation to occur, however, populations must also develop intrinsic reproductive isolating mechanisms—defined as differences in anatomy, physiology, or behavior that prevent interbreeding between individuals of the same species or of closely related species. Likewise, sympatric speciation always entails the development of intrinsic reproductive isolating mechanisms. At least six such mechanisms exist: ecological, temporal, behavioral, mechanical, gametic, and hybrid inviability or infertility.
  TUTORIALS: 18.1.1 Reproductive Isolating Mechanisms; 18.1.3 Polyploidy

18.3 When Is Speciation Likely to Occur?

- Speciation can occur rapidly with adaptive radiation into new niches, as with the migration of species to the Galapagos or Hawaiian islands.

- There is considerable debate among evolutionary biologists regarding the pace of evolution. The gradualist model holds that evolution proceeds at a steady rate, while the punctuated equilibria model holds that organisms undergo long periods of almost no change (stasis) followed by relatively brief episodes of rapid speciation.

18.4 The Categorization of Earth's Living Things

- The most important criterion for classifying species is their degree of relatedness. A binomial nomenclature is used in referring to any given species, with the first of its names identifying its genus and the second identifying the species within this genus (e.g., *Felis domestica* for the common house cat). Genus and species fit into a larger framework of *taxonomy*, meaning the classification of species. The most commonly used groupings in this taxonomy are species, genus, family, order, class, phylum, and kingdom. The highest-level category, domain, is sometimes also used.

18.5 Constructing Evolutionary Histories: Classical Taxonomy and Cladistics

- Homologous structures are common structures in different organisms that result from a shared ancestry. Homologous structures provide strong evidence regarding relatedness among species. One problem with their use is that they can be confused with analogous structures: similar features that developed independently in separate lines of organisms (as with the legs of modern horses and extinct litopterns).

- Classical taxonomy establishes evolutionary relationships among living and extinct organisms in accordance with such factors as physical form, distribution, and molecular similarities, but employs subjective judgments in deciding on what weight to give one piece of evidence as opposed to another. The phylogenetic system known as cladistics does away with this element of subjectivity by following a firm rule for inferring relatedness: It counts the number of shared derived characters two organisms have, meaning the features they uniquely share that are derived from another group of organisms.

- Cladistics is centrally concerned with establishing lines of descent and classifies species only as a byproduct of this effort, in accordance with the lines of descent that have emerged. Classical taxonomy holds that, in classifying organisms, another factor besides lines of descent ought to be taken into account: the special qualities that different groups of organisms possess, such as feathers on birds.

Key Terms

Understanding the Basics

Multiple-Choice Questions

1. In the biological species concept, the factor that defines a species is
   a. geographical separation
   b. reproductive isolation
   c. hybridization
   d. cladogenesis
   e. its behavior

2. In allopatric speciation (select all that apply):
   a. Allele frequency changes are shared between populations.
   b. Populations are geographically isolated.
   c. Gene flow between populations is stopped by physical barriers.
   d. Geographically isolated populations accumulate genetic changes over time that eventually block successful interbreeding.
   e. Populations are directly adjacent.

3. Factors that can be problems in determining the true relationships among organisms may include (select all that apply):
   a. homologous traits
   b. analogous traits
   c. convergent evolution
   d. poor fossil record
   e. shared derived traits

4. Possible intrinsic reproductive isolating mechanisms include (select all that apply):
   a. mating at different times of year or times of day
   b. geographical isolation
   c. incompatibility between eggs and sperm that prevent fertilization
   d. singing a different mating song than another species
   e. producing fertile hybrids

5. Match the reproductive isolating mechanism with how it works to reduce successful interbreeding:
   a. temporal isolation — populations live in different environments
   b. ecological isolation — populations are separated by physical barriers
   c. geographical isolation — mating occurs at different times of day or year
   d. gametic isolation — egg and sperm do not fuse
   e. hybrid sterility — progeny of a cross are unable to reproduce

6. In ocean-dwelling creatures like sea urchins or corals, sperm and eggs are released into the water by males and females and fertilization occurs externally. If several closely related species of coral live in the same location, what reproductive isolating mechanisms are likely to be effective at preventing interbreeding? (Select all that apply.)
   a. behavioral isolation
   b. gametic isolation
   c. temporal isolation
   d. ecological isolation
   e. hybrid infertility

7. Order the following categories into the appropriate biological hierarchy, with the most specific category first.
   a. family
   b. species
   c. phylum
   d. genus
   e. class

8. Adaptive radiations (select all that apply):
   a. are most likely to occur on mainlands
   b. are examples of rapid speciation that occurs when a number of ecological niches have not been filled
   c. concern the evolution of heat loss in vertebrates experiencing stressful climates
   d. are exemplified by the Galapagos finches
   e. often occur on islands

Brief Review

1. What is convergent evolution, and why is it a problem for phylogenetic analyses?

2. Contrast the gradual and punctuated equilibria models of long-term evolution. What type of evidence has been used to support the punctuated equilibria theory?

3. What is the difference between anagenesis and cladogenesis?

4. Describe one major difference between allopatric and sympatric speciation. What is one thing that they have in common?

5. State the biological species concept. Why is it not applicable to all species?

6. Why is the evolution of intrinsic reproductive isolation mechanisms required for two groups to be called separate species? Why isn't simple geographical isolation sufficient?

Applying Your Knowledge

1. Do you think that speciation and/or extinction are likely to be more frequent in species that are highly specialized in the resources they use than in more generalized species? Why or why not?

2. Populations of some kinds of organisms may be isolated by a barrier as small as a roadway, while other organisms require a much larger physical barrier to block gene flow. What features of organisms might determine how readily they become geographically isolated? What impact might these features have on how often or how rapidly speciation is likely to occur in these groups?

3. In an adaptive radiation, one species may colonize a previously empty environment, such as an island, and diversify evolutionarily into a set of closely related species that occupy a very wide range of habitats. For example, in the Darwin's finches on the Galapagos Islands, one species fills the role of woodpecker by using a stick to tap on trees and dig out insects. Do you think that such a species would have been likely to evolve in an area that already had woodpeckers? Why or why not?

4. Living things are classified into a set of hierarchical categories. Why is this more useful to biologists than simply giving everything a single part name (or a number) and eliminating some of the categories?

19

A Slow Unfolding
The History of Life on Earth

The history of life written in layers of rock.
(Section 19.1, page 379)

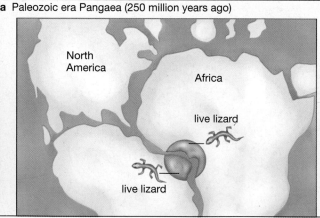

a Paleozoic era Pangaea (250 million years ago)

North America

Africa

live lizard

live lizard

So far apart but once so near.
(Essay, page 381)

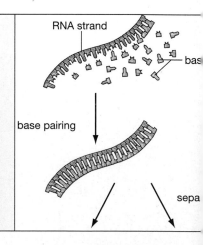

RNA strand

bas

base pairing

sepa

A critical step in the development of life.
(Section 19.2, page 385)

Life started out small at least 3.6 billion years ago and stayed that way for a long time. Eventually it exploded into a multitude of forms that came to include human beings.

Within 5 minutes of its birth, a newborn wildebeest on the plains of central Africa can be off and running with its herd (**see Figure 19.1**). By contrast, human infants probably will not be able to take a step by themselves until they are about a year old. Female lions reach sexual maturity at about 3 years, killer whales at about 7, but human males and females do not come to sexual maturity until they are past 10. Even in comparison to our fellow mammals, humans mature at a notably slow rate—a trait we share to some degree with our most closely related fellow primates, the chimpanzees.

At first glance, it might seem puzzling that primate evolution took this turn. From what we have learned of natural selection, wouldn't it make sense that survival would *decrease* for offspring that developed more slowly? What is the fitness payoff in having young that are essentially helpless for several years and then immature for many more? With respect to human beings, here is one possible answer. As a species, we have survived not so much because of our physical capabilities, but because of our wits. Our success is the product of our remarkable

capacity to *learn*. But learning takes time, and it is arguably best undertaken by minds that remain for a long time in a state of what might be called flexible immaturity. (Think of how easily young children acquire a second language in comparison with adults.) Among our human ancestors, it may be that those who survived tended to be those who could learn the most, and that those who could learn the most were those who took the longest to mature.

Figure 19.1
No Time to Lose
A newborn wildebeest struggles to its feet immediately after its birth on the plains of Africa. Wildebeest predators, such as hyenas, often concentrate their efforts on the newborn.

One of the earliest-evolving groups of plants.
(Section 19.6, page 389)

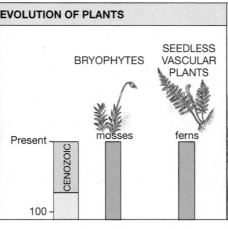

A tree of life for plants.
(Section 19.6, page 390)

EVOLUTION OF PLANTS

BRYOPHYTES

SEEDLESS VASCULAR PLANTS

Present

CENOZOIC

mosses

ferns

100

Neanderthal Life.
(Section 19.8, page 399)

Figure 19.2
Our Evolutionary Heritage
The gait exhibited by almost all four-limbed animals, including human beings, stems from the reversing S-pattern that our evolutionary ancestors used in getting around when they were making the transition from aquatic to terrestrial life. Here a salamander employs this same motion in getting from one place to another.

Under this view, we carry our evolutionary heritage with us in the form of delayed maturity. We did not leave this heritage behind in the African savanna, where human beings first evolved; we can see it every time we look at a 2-year-old. In the same way, we can see evolutionary vestiges in us from a much earlier time as well. Note the way we walk, our *left arm* swinging forward as our *right leg* does the same. Almost all four-limbed animals walk this way. One view is that we inherited this from the elongated, four-finned fish we evolved from. They came

Figure 19.3
A Timescale for Earth and Its Living Things

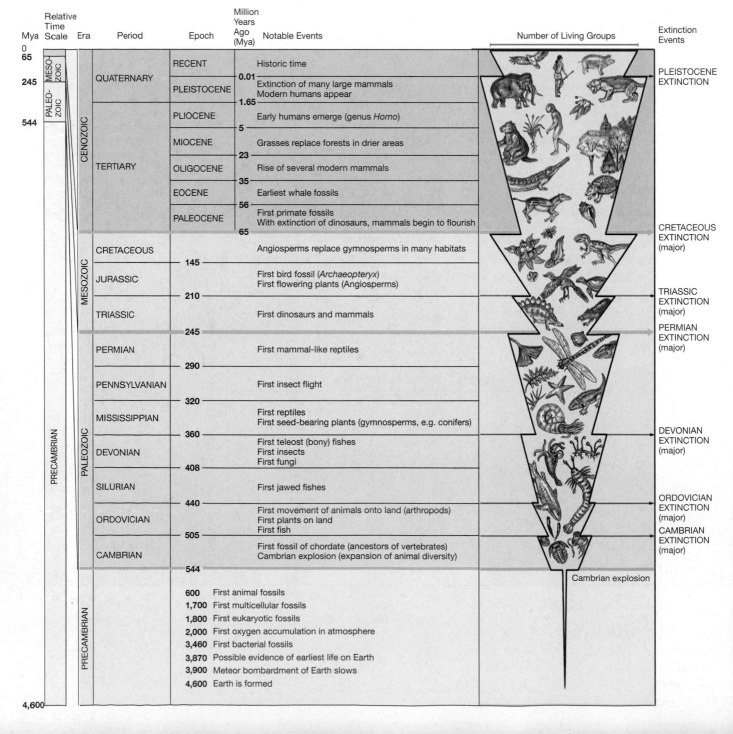

Mya	Relative Time Scale	Era	Period	Epoch	Million Years Ago (Mya)	Notable Events	Number of Living Groups	Extinction Events
0	MESOZOIC		QUATERNARY	RECENT		Historic time		PLEISTOCENE EXTINCTION
65		CENOZOIC			0.01			
245	PALEOZOIC			PLEISTOCENE		Extinction of many large mammals / Modern humans appear		
544					1.65			
				PLIOCENE		Early humans emerge (genus *Homo*)		
			TERTIARY		5			
				MIOCENE		Grasses replace forests in drier areas		
					23			
				OLIGOCENE		Rise of several modern mammals		
					35			
				EOCENE		Earliest whale fossils		
					56			
				PALEOCENE		First primate fossils / With extinction of dinosaurs, mammals begin to flourish		
					65			CRETACEOUS EXTINCTION (major)
		MESOZOIC	CRETACEOUS			Angiosperms replace gymnosperms in many habitats		
					145			
			JURASSIC			First bird fossil (*Archaeopteryx*) / First flowering plants (Angiosperms)		TRIASSIC EXTINCTION (major)
					210			
			TRIASSIC			First dinosaurs and mammals		
					245			PERMIAN EXTINCTION (major)
		PALEOZOIC	PERMIAN			First mammal-like reptiles		
					290			
			PENNSYLVANIAN			First insect flight		
					320			
	PRECAMBRIAN		MISSISSIPPIAN			First reptiles / First seed-bearing plants (gymnosperms, e.g. conifers)		
					360			DEVONIAN EXTINCTION (major)
			DEVONIAN			First teleost (bony) fishes / First insects / First fungi		
					408			
			SILURIAN			First jawed fishes		
					440			ORDOVICIAN EXTINCTION (major)
			ORDOVICIAN			First movement of animals onto land (arthropods) / First plants on land / First fish		
					505			CAMBRIAN EXTINCTION (major)
			CAMBRIAN			First fossil of chordate (ancestors of vertebrates) / Cambrian explosion (expansion of animal diversity)		
					544		Cambrian explosion	
		PRECAMBRIAN			600	First animal fossils		
					1,700	First multicellular fossils		
					1,800	First eukaryotic fossils		
					2,000	First oxygen accumulation in atmosphere		
					3,460	First bacterial fossils		
					3,870	Possible evidence of earliest life on Earth		
					3,900	Meteor bombardment of Earth slows		
4,600					4,600	Earth is formed		

onto the muddy land by slithering in an S-pattern, the left-front fin going back as the left-rear fin went forward, then the reverse (**see Figure 19.2**). Evolution may also be with us, some scientists speculate, in the lower-back pain that afflicts so many people in middle age and beyond. Our ancestors began walking upright probably no more than 5 million years ago; as this notion has it, we're still getting used to the change.

Note the sweep of evolution in all this. Maturation is with us from one period, walking style from another. If we look at the genetic code—*this* DNA sequence specifying *that* amino acid—we have within us a heritage that stretches back to the beginning of life on Earth, because this code is shared by nearly every living thing. What you will do in this chapter is trace some of these linkages and divergences. You'll look at how evolution has proceeded in its largest scale; how life began, and how it diversified into the major forms of life we see today.

19.1 The Geologic Timescale: Life Marks Earth's Ages

We can begin this inquiry by getting a sense of life's time line, which you can see in **Figure 19.3**. This is the geological timescale, which divides earthly history into broad **eras**, shorter **periods**, and shorter-still **epochs**. The scale begins with the formation of the Earth about 4.6 billion years ago and runs to "historic time," meaning time in the last 10,000 years.

Though this is a geologic timescale, you saw back in Chapter 16 that, in the days before radiometric dating, scientists measured the *relative* ages of geologic layers by looking at the fossils they contained. Looking at a certain layer and seeing fossils of a type of bony fish, a scientist knew he or she was looking at a Devonian layer, for example. Fossils, then, *defined* geologic layers or "strata" (**see Figure 19.4**).

Within this framework, what marks the boundary between one geologic era or period and the next is some sort of transition in living things. What may be surprising is what often lays the groundwork for these transitions: death on a grand scale in "major extinction events." Six events are noted in Figure 19.3, with all of them defining the end of an era or period. (The relatively recent Pleistocene Extinction, which brought about the demise of such large mammals such as the woolly mammoth, was not a major extinction event. One of the reasons it is notable, however, is that human beings played a part in it, through hunting.) The most famous extinction event, the **Cretaceous Extinction**, occurred at the boundary between the Cretaceous and Tertiary periods. This was the extinction, aided by the impact of a giant asteroid, that brought about the end of the dinosaurs, along with many other life-forms. The greatest extinction event of all, however, occurred at the end of the Paleozoic era. This was the **Permian Extinction**, in which as many as *96 percent* of all species on Earth were wiped out (see "Physical Forces That Have Shaped Evolution" on page 380).

Out of such extinctions, however, come new forms of life. With old forms gone, relatively rapid evolution can take place, as species fill depopulated niches (see Chapter 18, page 368). Such adaptive radiation can be seen with the evolution of our own lineage, the mammals. Before the extinction of the dinosaurs, we mammals existed in a few species, generally rat-sized insect hunters.

Figure 19.4
Revealing Layers
The history of life can be traced though fossilized life-forms found in layers of sediment, seen here in America's Grand Canyon.

Physical Forces that Have Shaped Evolution

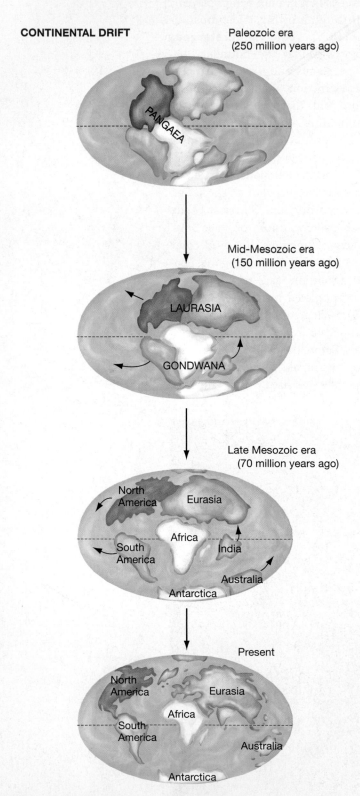

CONTINENTAL DRIFT

Paleozoic era
(250 million years ago)

PANGAEA

Mid-Mesozoic era
(150 million years ago)

LAURASIA

GONDWANA

Late Mesozoic era
(70 million years ago)

North America

Eurasia

Africa

South America

India

Australia

Antarctica

Present

North America

Eurasia

Africa

South America

Australia

Antarctica

Tutorial 19.1.1
Continental Drift and the
Geological Timescale

Figure 1
Continental Drift

During Earth's history, the tectonic plates that make up the surface layer of Earth have moved with respect to one another, causing the continents to spread apart and slam together.

Climate Change

The living world is very much affected by the physical forces operating on the Earth. You can see this in connection with the mass extinctions we noted in the main text. What has caused these extinctions? Evidence links at least four of those listed in Figure 19.3 to episodes of global climate change, specifically global cooling. (The four are the Ordovician, Devonian, Permian, and Cretaceous extinctions.) All kinds of things happen with global cooling. Glaciers develop, wiping out everything in their path and tying up huge amounts of water in ice. This reduces sea levels, thereby killing many organisms that live on the continental shelves. The oceans themselves get colder, particularly at greater depths, which is its own kind of killing event. As may be apparent, the factor of loss of habitat is intimately tied to the issue of global climate change. Global cooling is not the only factor involved in mass extinctions, but it is thought to be the most important factor.

Extraterrestrial Objects

In the public mind, the Cretaceous Extinction that finished off the dinosaurs had a single, nearly instantaneous cause: the explosive impact of an enormous asteroid, 10 kilometers (6.5 miles) in diameter, that struck Mexico's Yucatan Peninsula 65 million years ago. There is no doubt that this catastrophic event happened; what is in doubt is that it was the sole or even predominant cause of the Cretaceous Extinction. For one thing, fossil evidence indicates that many Cretaceous lifeforms, including the dinosaurs, were in severe decline *before* the asteroid struck. Secondly, as just noted, we have evidence that global cooling occurred in Cretaceous times. If there is a consensus view on this issue, it is that the asteroid probably was responsible for the nearly instantaneous demise of some species, but that in other cases it was merely the final blow in extinctions that already were under way.

At the very least, the Cretaceous Extinction shows that evolution can be affected by sudden, catastrophic events that have nothing to do with Earth's normal physical processes. A fascinating point is raised by this: Fitness, in the evolutionary sense, can change in an eyeblink. The qualities that made the dinosaurs so dominant in the Triassic Period were of no use once the asteroid struck. The game of survival had changed. Now *other* qualities came to the fore in terms of survival. Among these may have been small size, which mammals had, perhaps to their great benefit.

Continental Drift

As noted in Chapter 18, the primary engine driving speciation is the geographic separation of populations. Geographic separation was described as taking place in scales that were both small (a stream changing course) and large (a glacier's movement). It turns out that geographic separation can take place on a much larger scale yet, in that whole *continents* have come together and separated again during Earth's history.

If you look at **Figure 1**, you can see the land masses of the world as they have existed at four separate times, 250 million years ago, 150 Mya, 70 Mya, and the present. In the first of these, nearly all the Earth's continental mass was part of a single supercontinent, Pangaea, which stretched from the South to North Pole. By the Early Cretaceous, Pangaea was separating along the lines you see in the figure. This in turn gave way to our current division which, like its predecessors, is a configuration in flux.

These are only four "snapshots" in a process of **continental drift** that has produced lots of other alignments. How does such movement come about? In essence, the Earth's crust and part of its upper mantle are divided into a series of plates that move laterally over the globe. Hold two blocks together and slide them past each other; then move them away from each other, then *into* each other, and you get the idea of the kinds of movements that plates can have relative to one another.

The movement of the plates can be thought of in terms of a conveyor belt. Halfway between North America and Europe there is a long underwater mountain chain, the Atlantic Ridge. This ridge has in its middle a deep valley that forms the border between the North American and Eurasian Plates. Hot material in the Earth's interior is constantly being moved up into the valley from either plate, after which, as new crust, it moves eastward on one side of the ridge but westward on the other. What we have, then, is a spreading of the ocean floor in opposite directions. *Continents* sit atop this spreading material, and they are being moved in opposite directions—Europe to the east, North America to the west—as if they were objects on conveyor belts (**see Figure 2**).

The Impact of Continental Drift on Evolution

Continental drift has had profound effects on evolution. As noted, global cooling has been the most significant force behind mass extinctions. But what causes global cooling? One factor is the development of glaciers, which as it turns out can grow only when continents are located at or near the Earth's poles. This *was* the case 300 million years ago, then it was *not* the case 200 million years ago, and now it is again. By moving landmasses on and off the poles, continental drift has helped change the climate. As noted, glaciers lock up water, causing sea levels to drop, killing life on the continental shelves.

Continental drift has had a second profound effect on evolution, in that it has brought about the separation and mixing of living *populations*. For example, the breakup of the supercontinent Pangaea eventually meant the creation of island continents such as Australia and Antarctica. Until the coming of humans, Australia had scarcely any placental mammals—mammals whose embryos are nourished by a placenta (see Chapter 28). Since prehistoric times, however, Australia has had an abundance of marsupial (or "pouched") mammals, such as kangaroos. We can read evolutionary history in this distribution. Marsupials developed in Pangaea before its breakup, in what is now North America and Western Europe, and spread to Australia by way of what is now Antarctica. Meanwhile, placental mammals were spreading too; but few were able to reach Australia before it broke away as an island continent. Placental mammals seem to outcompete marsupials wherever the two varieties meet. This could not happen in Australia, however, because continental drift had *isolated* its marsupial populations, leaving them free to evolve into the diversity of species we see today, such as kangaroos, koalas, and wombats.

a Paleozoic era Pangaea (250 million years ago)

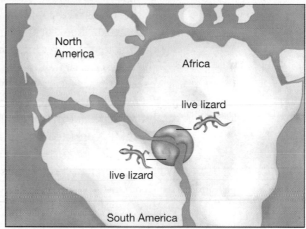

b Today

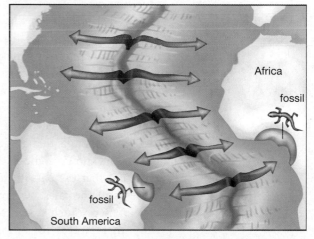

Figure 2
Similar fossils are found on the east coast of South America and the west coast of Africa. How can this be?

a 250 million years ago, the two continents were contiguous and formed part of the giant continent Pangaea.

b Animals that were alive then are now found as fossils on opposite sides of the ocean because the ocean floor has been spreading westward and eastward along the central Atlantic Ridge. The ocean floor is spreading because hot material from the Earth's interior is constantly being moved up into the central Atlantic Ridge, even today.

With the death of our fearsome enemies in the Cretaceous Extinction, we were able to evolve into the many forms that exist today.

Transition Does Not Always Mean Death: The Cambrian Explosion

Looking again at the eras in Figure 19.3, there is one exception to the transition-means-extinction rule. The transition between the Precambrian and Paleozoic eras marks not a major extinction, but a seeming explosion in the diversity of animal forms. The fossil record tells a tale of evolution proceeding at a dizzying pace for a short period that began about 544 million years ago (Mya), though this idea has been challenged, as you'll see. This "Cambrian Explosion" stands in marked contrast, however, to what preceded it. Look at the enormous amount of time that the Precambrian Era takes up in the timescale. In it, life evolved at such a slow pace that we scarcely have words for it. ("Glacial" greatly overstates things.) For more than *half* the time life has existed on Earth, it consisted of nothing but single-celled microscopic forms. Conversely, all the birds, reptiles, fish, plants, and mammals that exist today came about in the last 14 percent of evolutionary time.

What Is "Notable" in Evolution Hinges on Values

One final word about the timescale and the path you will follow in this chapter. Picking out the "notable events" in evolution is inevitably an exercise in making value judgments. The notable events covered in this chapter will lead you along a line that begins with microscopic sea creatures and ends with human beings. The value judgment that guides this path is that students have a great interest in knowing about their own evolution, and that the path to humans arguably presents the broadest sweep of evolutionary development. Such a course of study may leave the impression, however, that all of evolution amounts to a march toward the development of human beings, who then get to occupy the highest branch of an evolutionary tree.

In reality, we occupy one ordinary branch among a multitude of branches. Under notable events in the Cenozoic era,

the table could just as easily have listed milestones in the evolution of fish or birds, or the emergence of social insects such as bees. These creatures have continued evolving in the modern era, along with mammals, but in lines separate from us. Focusing on a line that leads to humans is like focusing on a railroad route that leads from, say, Baltimore to Denver, while ignoring the multitude of lines that intersect with it or that are separate from it. We regard the Baltimore-Denver route as special because we are *interested* in it, but that does not mean that it is fundamentally *different* from any other line. Some special qualities have evolved in human beings (notably our large brains), but the same could be said for almost any species. In phylogenetic terms, we simply lie at the tip of one evolutionary branch, while bees lie at another, and cactus plants lie at still another.

The Kingdoms of the Living World Fit into Three Domains

In tracing the history of life, you'll be looking to some extent at all the major life-forms that have evolved on Earth. Thus it will be helpful to have at the outset a sense of what these major life-forms are. This book employs a system of three *domains* of life, which then have within them "kingdoms" as noted in Chapter 18. The three domains are Archaea, Bacteria, and Eukarya.

Archaea is a domain populated entirely by microscopic, single-celled organisms, many of which live in "extreme" environments, such as boiling-hot vents on the ocean floor. **Bacteria** is likewise a domain of single-celled microscopic creatures, but these are the bacteria that are more familiar to us. The third domain, **Eukarya**, then includes all the organisms that are most familiar to us—plants, animals, and fungi—along with another group that is less familiar, the protists. (The protist kingdom has within it lots of single-celled water dwellers, but also includes some large organisms, such as giant sea kelp.)

19.2 Tracing the History of Life on Earth: How Did Life Begin?

Taking a historical approach to tracing life on Earth, the first question that arises is one of the toughest: How did life begin? Darwin

383

19.2 Tracing the History
of Life on Earth: How
Did Life Begin?

himself thought about this and imagined that life began in what he called a "warm little pond" in the early Earth. As it turns out, however, the early Earth had no warm little ponds; what it had was an environment more akin to hell on Earth.

You have seen that Earth was formed about 4.6 billion years ago. This took place in a process of "accretion," in which ever-larger particles clumped together: cosmic dust to gravel, gravel to larger balls, larger balls to objects the size of tiny planets. One consequence of such development was that the accretion now called Earth was periodically being slammed into by *other* large accretions—meteorites and comets, though some of them may have been as large as Mars. If we could look, then, at the Earth in its first 600 million years or so, we would see an enormously hot planet whose early seas were periodically vaporized to a depth of 100 meters or so because of the violent impact of the meteorites.

In What Kind of Environment Did Life Begin?

The meteorite bombardment of Earth seems to have slowed by about 3.9 billion years ago. With this change, we are left with a large question about what Earth was like. Scientists believe that the Sun shone some 20–30 percent less brightly then than it does now. But the Earth could have remained hot, or it could have been been very cold then, depending on how much heat-trapping carbon dioxide existed in Earth's atmosphere at that point. If it was cold, the Earth's seas may have been frozen to a depth of a few hundred meters.

These two views lead to different ideas about where life arose. Was it in very hot seas or in frigid seawater beneath a thick layer of ice? Most researchers in the field today seem to favor the hot-sea theory, with a critical piece of evidence being that the very earliest organisms seem to have lived in high-temperature environments.

What Was the Source of the Raw Materials for Life?

Whatever environment gave birth to life, it had to have in it the chemical raw materials that could be used in forming life's critical molecules. You saw in Chapter 3 that life's informational molecules (DNA and RNA) are composed of certain kinds of building blocks (nucleotides), while its proteins are composed of other kinds of building blocks (amino acids). The raw materials for such building blocks could have been the gases methane and ammonia. These substances certainly would have been spewed out (or "outgased") by Earth's early volcanoes; the question is whether they would have been quickly broken down in the ultraviolet light, coming from the Sun, that bathed the early Earth's atmosphere.

Some scientists believe that building blocks such as amino acids could have arrived ready-made, delivered by the meteorites and comets that smashed into the young Earth. It's clear, from meteorites that have landed on Earth in recent decades, that such objects *can* carry organic molecules (**see Figure 19.5**). Research has cast doubt, however, on whether such accidental "seeding" from outer space could have brought a sufficient *quantity* of organic materials to get life going.

Life May Have Begun in Very Hot Water

A third possibility for the supply of organic materials in the early Earth is that they came from the methane and hydrogen sulfide that gush out even today from deep-sea vents on the floors of the world's oceans. Interestingly enough, there are creatures living near these vents today. Some of these are microscopic archaea and bacteria that thrive at temperatures of 105° Celsius (247° Fahrenheit),

Figure 19.5
Space Travelers
Fragments of the Murchison meteorite, which landed on the township of Murchison in Australia in 1969. The meteorite contained water and organic molecules such as amino acids. Some scientists hypothesize that organic materials like these could have seeded life on Earth.

Figure 19.6
Hot Habitat
Material pours forth from a hot-water vent on the floor of the Atlantic Ocean. The fluid being emitted is mineral-rich enough to support bacteria and archaea that in turn form the basis for a local food web. A deep-sea crab can be seen at lower center.

Figure 19.7
Only Seemingly Inhospitable
The hot-water Grand Prismatic Pool in Yellowstone National Park, shown here in an aerial view, gets its blue color from several species of heat-tolerant cyanobacteria. The colors at the edge of the pool come from mineral deposits. The "road" at the top of the pool is a walkway for visitors.

which is above the temperature at which water boils (**see Figure 19.6**). To judge by their genetic makeup, these microbes come from truly ancient lineages of living things. Various lines of evidence, including the fact that these organisms live around a potential source of organic materials, leads to an idea that has a good deal of support among origin-of-life researchers. It is that life began in the "prebiotic soup" of hot-water systems—in the deep-sea vents or the kind of hot-spring pools that exist in Yellowstone Park (**see Figure 19.7**). Though life is fragile in some respects, it may be that it had its beginnings in hot, sulphurous water—quite a different scene than Darwin's warm little pond.

Wherever it began, life cannot have been created in a single step from basic compounds such as methane and hydrogen sulfide. There is a huge gap between these materials and organisms as complex as the archaea. If organic molecules did not come from outer space, early Earth had to be a kind of natural chemical lab in which ever more complex organic molecules were produced from simpler ones. A major portion of origin-of-life research involves creating, in the laboratory, simulations of early Earth environments and then watching to see if life's building blocks will come together in them. Scientists now have scenarios that can account for the creation of most, but not all, of these compounds.

The RNA World

Any living thing must have at least one critical feature: the ability to reproduce itself. Today, a key player in reproduction is DNA. In Chapter 14, you saw that our "genome," or inventory of DNA, can be thought of as a kind of cookbook containing recipes (DNA sequences) that lead to products (proteins) that undertake a wide variety of tasks. Yet, as noted there, though DNA is the cookbook, it is not the cook. When it comes time for DNA to be copied, the double helix cannot unwind itself or pair its bases with messenger RNA or edit the resulting mRNA "transcript." Instead, all these tasks are carried out by the *proteins* known as enzymes. Hence there is a chicken-and-egg dilemma: Enzymes can't be produced without the information contained in DNA, but DNA can't copy itself without the activity of enzymes. So, how did life get going if neither thing can function without the other?

In the late 1960s, researchers hit on a solution that has come to have a good deal of support. In the earliest life-forms, a *single* molecule performed both the DNA and enzyme roles. This molecule was RNA (ribonucleic acid), which was portrayed back in Chapter 14 as a kind of genetic middleman, ferrying DNA's information to the sites where proteins are put together. A critical piece of evidence about RNA's role in early life came in 1983, when researchers discovered the existence of enzymes that are composed of

385

19.2 Tracing the History of Life on Earth: How Did Life Begin?

RNA instead of protein. Called **ribozymes**, these are multitalented performers. They can encode information *and* act as enzymes (by, for example, facilitating bonds that bind RNA units together).

This leads researchers to the presumed existence of the "RNA world"—an early living world that consisted solely of self-replicating RNA molecules. The central question facing origin-of-life research today is this: How do we get to the RNA world from the chemical raw materials? RNA itself is so complex that many researchers believe it had to have taken over the role of a less complex precursor.

The Step at Which Life Begins

The critical step in life's development comes in imagining RNA or its precursor beginning to carry out replication of itself. *This* is the step at which life can be said to have begun. Given a molecule that can make copies of itself, a *line* of such molecules can come into existence; given the *mistakes* that are bound to take place in such copying, molecules can be produced that differ slightly from one another in being, say, more resistant to the elements. Molecules that have such an advantage will leave more daughter molecules, which might just as well be called offspring. What happens, in short, is natural selection among self-replicating molecules. This, of course, leads to evolution among these molecules and with evolution life begins to diversify into the innumerable forms we see today.

Lots of scenarios have been proposed for how this replication got going. A central assumption in most of them is that RNA or its precursor took part in the base pairing that takes place in contemporary nucleic acids (**see Figure 19.8**). If a *strand* of RNA exists, a *complementary* strand can easily be produced through base pairing. This is a reaction that takes place now, catalyzed by modern-day enzymes. Imagine in the early Earth, however, a ribozyme precursor that could carry out a similar sort of complementary pairing. With its enzymatic capability, this ribozyme-like molecule could link complementary bases together. It could thus form a strand of bases complementary to itself, then bring out separation of the resulting two strands, and then continue the process. The challenge for scientists today is to figure out what substances could have carried this replication out, and under what conditions.

Adding Life's Elaborations to Replication

There is a lot more to living things than replication, of course. Even the most primitive living things today are encased in the protective linings called plasma membranes; they have a sophisticated means of extracting energy from their environments; they can get rid of their wastes, and so forth. It may be that an elaboration such as an outer membrane developed along with the replication. But it is also possible that life first took the form of "naked RNA" or its precursor. Whatever the sequence of events, once life's elaborations are joined to replication, we move from simple molecules to the cellular ancestors of today's organisms.

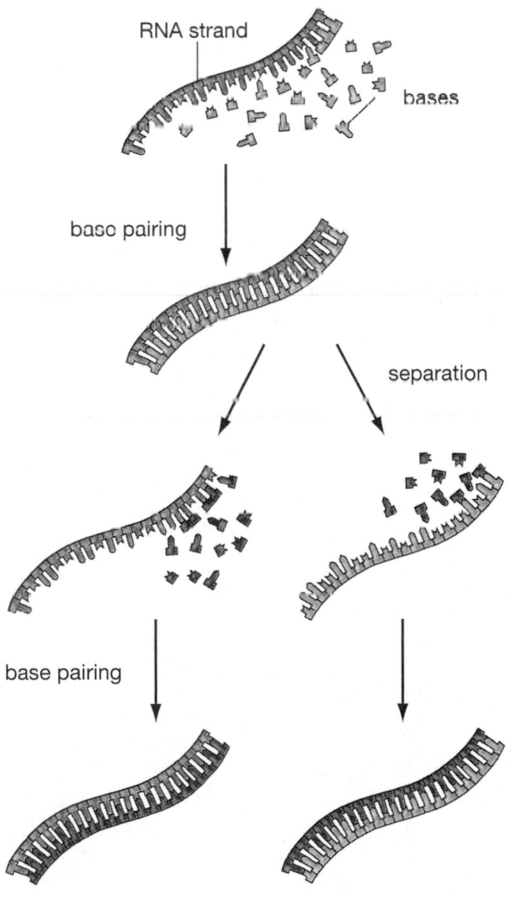

Figure 19.8
How Did Self-Replication Get Started?
A ribozyme-like precursor of RNA may have been the first molecule to replicate itself. Free nucleotide bases pair with their complementary bases to form a complementary strand of bases, linked together by the ribozyme's enzymatic capability. If the two strands then separate, the process can repeat itself and many RNA strands can be formed. This process occurs today with the help of enzymes, but the process may have occurred millions of years ago with only the RNA itself to catalyze the reactions.

Tutorial 19.2.1
RNA-Like Self Replication

19.3 The Tree of Life

Once life was established, how did it evolve? If you look at **Figure 19.9**, you can see the biggest picture of all. At the bottom you see a "universal ancestor," which is the organism— a successor to one of our replicating molecules—that gave rise to all current life. The evolutionary line that leads from the universal ancestor then branches out to yield, on the one hand, the domain Bacteria and, on the other, a line that leads to both the domain Archaea and the domain Eukarya. Archaea were long thought of as simply a different kind of bacteria, but recent research has shown these organisms to be distinct from bacteria in fundamental ways. Nevertheless, there are great similarities. Bacteria and archaea are essentially single-celled, and no archaeic or bacterial cell has a nucleus. Conversely, *all* the organisms in the domain Eukarya have nucleated cells, and are thus "eukaryotes."

As noted earlier, domain Eukarya has four "kingdoms" within it that you'll be learning about. Looking at the tree, you can see that the earliest kingdom in Eukarya is Protista—mostly one-celled, water-dwelling creatures—and it is from Protista that the other three eukaryotic kingdoms evolve. An unknown species within Protista gave rise to all the fungi that exists on Earth today; another grouping of protists called the *choanoflagellates* probably gave rise to all of the animal kingdom; and the protists we call green algae gave rise to all of today's plants.

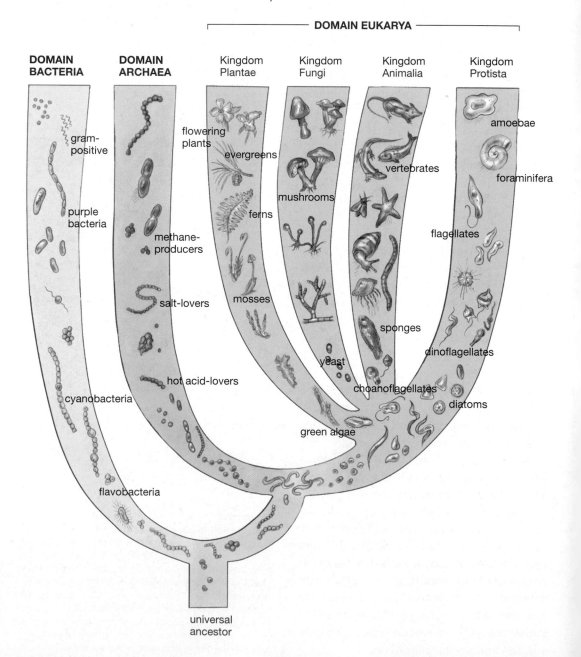

Figure 19.9
The Tree of Life
The universal ancestor at the base of the tree gave rise to two domains of organisms whose cells lack nuclei—Bacteria and Archaea. The third domain of life is the Eukarya, whose earliest kingdom was the protists. From these protists all plants, animals, and fungi evolved.

19.4 A Long First Period: The Precambrian

With this picture in mind, let's start tracing life from its cellular beginnings. As noted, living things started small and stayed that way for a long, long time.

The Slow Pace of Change in the Precambrian

Recall that the Earth was formed about 4.6 billion or 4,600 million years ago (Mya). The earliest undisputed evidence of life we have so far—bacterial microfossils from Australia—dates back 3,460 million years. There is some tentative physical evidence of life that dates back more than 3,800 million years, but let's stay with the sureness of the bacterial fossils and say that life appeared no *later* than 3,460 Mya. After this, more than 1,600 million years pass before we have evidence of *any other* form of life. The new form that does arise then is single-celled algae; after that another 100 million years pass before we get the first *multi*celled organisms, which are algae again. Evolution took its time in the Precambrian. The years rolled by in the millions, the land was barren, and the oceans were populated mostly by creatures too small to see with the naked eye.

Notable Precambrian Events

None of this is to say, however, that "nothing happened" in the Precambrian era. Those of you who have read Chapter 8 know how critical photosynthesis is for life on Earth. This capability first came about in the Precambrian era—and fairly early in the era at that.

The Initiation of Photosynthesis

Photosynthesis began in bacteria at about the time the first bacterial fossils left their imprint, which is to say about 3,500 Mya. This event was a turning point. Had this capacity not been developed, evolution would have been severely limited, for the simple reason that there would have been so little to eat on Earth. The earliest organisms subsisted mostly on organic material in their surroundings, but the supply of this material was limited. In photosynthesis, the Sun's energy is used to *produce* organic material—in our own era, the leaves and grasses and grains on which all animal life depends.

Large organisms require more energy and, through photosynthesis, a massive quantity of energy-rich food was made available.

Oxygen in the Atmosphere: A Dramatic Change on Earth

One particular kind of photosynthesis, again beginning in the Precambrian, had a second dramatic impact on evolution. The earliest undisputed fossils are called *cyanobacteria*. It turns out that these were the first creatures to produce *oxygen* as a by-product of photosynthesis (**see Figure 19.10**). For us the word oxygen seems nearly synonymous with the word life, but until about 2,000 Mya there was almost no oxygen in the atmosphere. When it finally did arrive, through the work of the cyanobacteria, its effect was anything but life-giving. For most organisms it was a deadly gas, producing what has been termed an "oxygen holocaust" and establishing a firm rule for life on Earth: Adapt to oxygen, stay away from it, or die.

Adapting to Oxygen: Bacteria and Eukaryotes Strike Up a Relationship

In the creatures that adapted to oxygen, we find one of the most interesting stories in evolution. Those of you who went through Chapter 7, on cellular energy harvesting, will recall that most of the energy that eukaryotes get from food is extracted in special structures within their cells, called *mitochondria*. Though mitochondria are tiny organelles within eukaryotic cells, they have characteristics of free-living *cells*, specifically free-living

Figure 19.10
Home to the Oldest Fossils
Layered sediments called stromatolites contain the fossilized remains of cyanobacteria that flourished more than 3 billion years ago in what is now Western Australia.

Figure 19.11
Adapting to Oxygen

The "powerhouses" of our cells—the organelles called mitochondria—have several characteristics of free-standing bacteria.

a There is general agreement among scientists that our mitochondria originally were bacteria that invaded eukaryotic cells many millions of years ago. These bacteria benefited from the eukaryote's cellular machinery, and the cell benefited from the bacteria's ability to metabolize oxygen.

b Over time, the bacteria came to be integrated into the host cells, replicating along with them. Two symbiotic organisms had now become a single organism.

c Today, the mitochondria in your cells turn the energy from food into a form that allows you to run, think, and turn the pages of this book. These artificially colored mitochondria functioned in a liver cell of a mouse.

bacterial cells. As it turns out, that is probably what they once were. Ancient bacteria that could metabolize oxygen took up residence in early eukaryotic cells and eventually struck up a mutually beneficial relationship with them (**see Figure 19.11**). The bacteria benefited from the eukaryotic cellular machinery, and the eukaryotes got to survive in an oxygen-rich world. In a second bacterial invasion, algae and their plant descendants got not only mitochondria, but the chloroplasts in which photosynthesis is carried out. (See "The Stranger within: Lynn Margulis and Endosymbiosis" in Chapter 4, page 83.)

One More Effect of Oxygen: The Blockage of Ultraviolet Light

The oxygen revolution had one more momentous consequence. Molecules of oxygen came together to form the gas called *ozone*, which rose through the atmosphere to form the ozone layer, giving Earth, for the first time ever, protection against the ultraviolet radiation that comes from the Sun. Marine creatures had been shielded from this radiation by the ocean itself; but prior to the formation of the ozone layer, there was no chance for life to develop on *land*.

19.5 The Cambrian Explosion: A Real Milestone or the Appearance of One?

As the Precambrian was coming to a close, life consisted of bacteria, archaea, and many kinds of protists (the earliest eukaryotes), all of which lived in the ocean. By about 600 Mya, however, we get the first evidence of *animals* in the seas. A formal definition of animals is given in Chapter 21. Suffice to say for now that animals have the characteristics you would expect. They all are multicelled, and they get their nutrition not from performing photosynthesis (like algae) but from consuming organic material or other organisms.

There currently is a debate about whether the earliest animal fossils represent a line of creatures that died out, or are instead related to today's animals. Whatever the case, these early animals can be seen as a kind of early stirring; a first blip on the screen bringing news of something big to come. To judge by the fossil record, about 70 million years after the emergence of these animals, we get it: A tidal wave of new animal forms, the like of which has never been seen before or since in evolution. This is the Cambrian Explosion, which began about 544 Mya.

Recalling the taxonomic system you went over in Chapter 18, you may remember that just below the "domain" and "kingdom" categories there is the **phylum**, which can be defined informally as a group of organisms that share the same body plan. There are at least 36 phyla in the animal kingdom today, some of these being Porifera (sponges), Arthropoda (insects, crabs) and Chordata (ourselves, salamanders). With one exception, the fossil record indicates, every single one of these came into being in the Cambrian Explosion.

a

eukaryotic cell

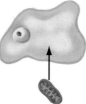

bacterial cell

b

eukaryotic cell

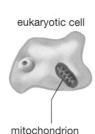

mitochondrion

c

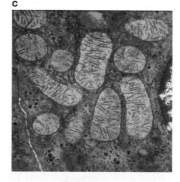

Figure 19.12
Product of the Explosion in Animal Forms

A model of an *Opabina regalis*, one of the now-extinct animals that arose in the flurry of evolutionary activity known as the Cambrian Explosion. *Opabina* had five eyes and a flexible proboscis beneath them.

This seeming riot of sea-floor evolution was actually more extensive than even this implies, in that a good number of phyla that *don't* exist today—phyla that have become extinct—also seem to have first appeared in the Cambrian. Some of these creatures are so bizarre, it's surprising that Hollywood hasn't mined the Cambrian archives for new ideas on monsters (**see Figure 19.12**). As if all this weren't enough, the Cambrian Explosion appears to have taken place in a very short period of time relative to evolution's normal pace—perhaps as little as 5 million years.

But did it? The Cambrian Explosion is so extreme that one of the primary challenges to evolutionary biology has been to explain why such a thing came about—or why it came about only once. Note the series of events: only a few animal forms before the Cambrian, then the explosion, then almost no new animal forms since. In the 1990s, several lines of evidence suggested something else: that animal forms actually began to diverge well back in the *Precambrian*. Under this view, all the Cambrian Explosion amounted to was an explosion of forms big and hard enough to leave *fossils*. The forms themselves appeared earlier and more gradually, this idea goes, but were too small and fragile to leave imprints behind.

If the Cambrian Explosion did in fact occur, what could have sparked it? Lots of ideas have been proposed; one of the best-received is that the explosion was triggered by the rise in atmospheric oxygen. To get bigger creatures, you need more oxygen, and levels of atmospheric oxygen may have reached a critical threshold about this time. Once larger forms *could* appear, adaptive radiation took place on a grand scale. So many niches were available that there was an explosion of forms. Once all the basic niches were taken, however, this frenzy of new forms not only came to a stop, it was pruned back (in the sense that some Cambrian phyla became extinct).

19.6 The Movement onto the Land: Plants First

The teeming seas of the early Cambrian period stand in sharp contrast to what existed on land at the time, which was no life at all. Earth was simply barren—no greenery, no birds, no insects. When life did come to the land, the first intrepid travelers were plants,

whose earliest fossilized remains on land date from some 460 Mya. Theirs was a gradual transition that began with marine algae in the ocean, then continued with freshwater algae, then freshwater algae that came to exist in *shallow* water, living partially above the waterline. When the transition to full-time living on land came, it was to damp environments.

Adaptations of Plants to the Land

Such a change required a lot of adaptation. Aquatic algae did not have to deal with water loss or the crushing effects of gravity, but land plants did. One of the plants' responses to the water problem was to evolve a waxy outer covering, called a *cuticle*, that could retain moisture. Meanwhile, an initial response to gravity was to stay low. Some of the most primitive land plants are mosses that often hug the ground like so much green carpet (**see Figure 19.13**).

Then there was the problem of reproduction. When green algae reproduce sexually, their gametes (eggs and sperm) float off from them as individual cells, after which they are brought together by ocean currents or the cells' own movement. Once this happens, the resulting embryos develop completely on their own. An embryo that matured this way on land, however, would dry out and perish. Plants adapted to the land by developing a protection mechanism for both embryos and gametes. We can see this clearly in some modern-day descendants of

Figure 19.13
Lying Low
Moss covering rocks along a stream in Tennessee's Great Smoky Mountains National Park. Mosses are members of the earliest-evolving group of plants, the bryophytes.

the early plants. Sperm develop in pollen grains, which have an outer coat, while eggs develop inside the parent plant and are fertilized there, producing an embryo that likewise begins development in the plant. The maturation of embryos *within* a parent is a key characteristic that separates plants from algae.

Another Plant Innovation: A Vascular System

All plants protect their embryos in this way, but plants then go on to diverge from each other in some very basic ways. If you look at **Figure 19.14**, you can see how the major divisions of plants evolved. The most primitive land plants, the **bryophytes** (represented by today's mosses), had no vascular structure—meaning a system of tubes that transports water and nutrients. Plants without such a system have very limited structural possibilities. They can grow out, but they lack ability to grow *up* very far, against gravity. By contrast,

the ancestors of today's ferns developed a vascular system; with this, over the ensuing 100 million years, a variety of **seedless vascular plants** evolved, including huge seedless trees such as club mosses, some of them 40 meters tall, which is about 130 feet.

Plants with Seeds: The Gymnosperms and Angiosperms

Even as the seedless vascular plants were reaching their apex, a revolution in plants was well under way as some of them developed *seeds*.

The Gymnosperms

With the first seed plants, called **gymnosperms**, offspring no longer had to develop within a delicate plant on the forest floor. Instead they developed within a seed, which can be thought of as a reproductive package that includes an embryo (brought about when sperm fertilized egg), food for this embryo, and a tough outer coat. In modern gymnosperms this is best exemplified by the common pinecone, which contains many seeds. Gymnosperms also represent the final liberation of plants from their ocean past. Mosses and ferns had sperm that could make the journey to eggs only through *water*. Not much water was needed; a thin layer would do, as with the last of a morning's dew on a fern leaf. With seed plants, however, the water requirement is done away with entirely. Sperm are encased inside pollen grains, which can be carried by the *wind*. This has obvious ramifications for the dispersal of such plants. Several types of gymnosperms developed beginning about 350 Mya. Easily the most widespread of these today are the pine and fir trees, which cover huge stretches of the Northern Hemisphere.

The Last Plant Revolution So Far: The Angiosperms

At the time the dinosaurs reigned supreme among land animals, the first flowering on Earth occurs, with the development of the flowering plants known as the **angiosperms**. Evolving about 165 Mya, the angiosperms eventually succeed the gymnosperms as the most dominant plants on Earth. Today there are about 700 gymnosperm species, but some 260,000 angiosperm species, with more being identified all the time. Angiosperms are not just more numerous than gymnosperms, they are vastly

Tutorial 19.1.2
Evolution of Plants

Figure 19.14
How Plants Evolved
The tree of life for bryophytes, seedless vascular plants, gymnosperms, and angiosperms.

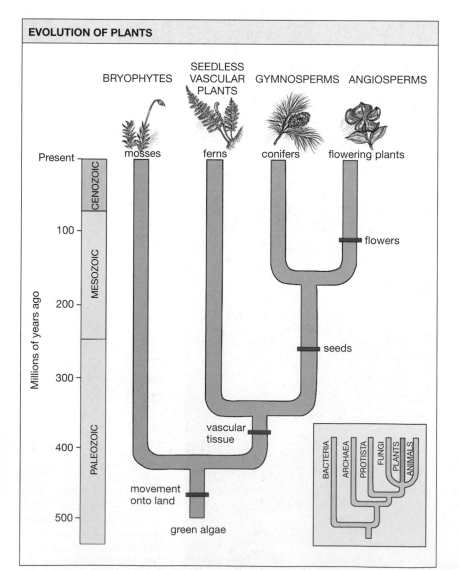

EVOLUTION OF PLANTS

BRYOPHYTES — mosses
SEEDLESS VASCULAR PLANTS — ferns
GYMNOSPERMS — conifers
ANGIOSPERMS — flowering plants

Millions of years ago

Present
CENOZOIC
MESOZOIC
PALEOZOIC

100
200
300
400
500

flowers
seeds
vascular tissue
movement onto land
green algae

BACTERIA ARCHAEA PROTISTA FUNGI PLANTS ANIMALS

more diverse as well. They include not only magnolias and roses, but oak trees and cactus, wheat and rice, lima beans and sunflowers.

19.7 Animals Follow Plants onto the Land

The movement of plants onto the land made it possible for animals to follow. With plants came food and shelter from the Sun's rays. The first animals moved to land about 20 million years after the first plants (back before there were even gymnosperms). Recalling the division of the animal kingdom into various phyla, it was the phylum of arthropods that first moved to land; about 440 Mya, a creature similar to a modern centipede laid down the oldest set of terrestrial animal tracks that have been found so far. It makes sense that arthropods were the first animals to come onto land, because the hallmark of all of them is a tough external skeleton (an "exoskeleton"), which can prevent water loss and guard against the Sun's rays.

Insects onto Land

Other kinds of arthropods, the insects, came on land some 50 million years after the first arthropod immigrants. The insect pioneers were wingless, perhaps being something like modern silverfish. Insect flight would not develop for tens of millions of years more, but when it did insects had the skies to themselves for 100 million years. The more primitive flying insects are represented today by dragonflies, whose wings cannot be folded back on their body (**see Figure 19.15**). The compactness that came with foldable wings gave later insects the advantage of being able to get into small spaces. (Think of a common housefly.) In both winged and wingless forms, the insects took to the land with a vengeance. About 1.4 million species

of living things have been identified on Earth to date, of which *half* are insects.

Vertebrates Move onto Land

At about the time the first insects were coming out of the sea, the first vertebrates slithered onto land.

Primitive Fish First

Vertebrates came in the form of primitive "lobe-finned" fishes. The lobed *fins* these creatures possessed are the precursors to the four *limbs* that are found in all tetrapod vertebrates, including ourselves (**see Figure 19.16**). It is a line of such tetrapods—the amphibians, the reptiles, and the mammals—that you'll follow for the rest of your walk through evolution. Here is the order of emergence among these forms. The lobe-finned fish give rise to amphibians, and early amphibians give rise to reptiles. Early reptiles then diverge into several lines, one of which leads to the dinosaurs and another of which leads to mammals. Dinosaurs eventually die out, of course, but before they do, a lineage that will live on—the birds—diverges from them

Figure 19.15
The First to Take Flight
Dragonflies, such as this dew-covered individual, are descendants of the insects that first developed the ability to fly. Flying insects that evolved later could fold their wings back, as with today's housefly.

Figure 19.16
Vertebrates onto Land

a Our ray-finned fish ancestors lacked bones in their fins.

b The lobe-finned fish that evolved from ray-finned fish had small bones in their fins that enabled them to pull out of the water and support their weight on tidal mudflats and sandbars.

c Finally, the early amphibians lost ray fins altogether, and had only bones in their limbs.

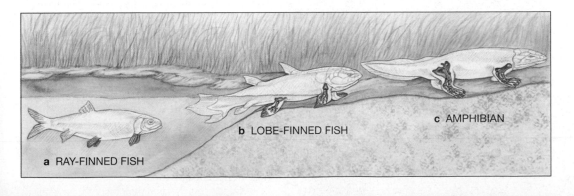

a RAY-FINNED FISH

b LOBE-FINNED FISH

c AMPHIBIAN

Tutorial 19.1.4
Evolution of
Vertebrates

Figure 19.17
How Vertebrates Evolved
A tree of life for lobe-finned fish, amphibians, reptiles, mammals, and birds.

(see Figure 19.17). Keep in mind the train-line metaphor employed earlier; in following these creatures, you are tracing only a couple of lines from among a huge number that exist.

Amphibians and Reptiles

In **Figure 19.18**, you can see an artist's rendering of *Ichthyostega*, the earliest known amphibian, which has a tail much like that of its lobe-finned ancestors. *Amphibian* literally means "double life," which is appropriate because in these creatures we can see the pull of both the watery world from which they came and the land onto which they moved. Modern-day frogs can serve as an example here. Many species of frog spend their adult lives as air-breathing land dwellers, but must return to the water to reproduce. Female frogs deposit eggs directly into the water, and the males then deposit sperm on top of them. The young that survive from this clutch go on to live as swimming tadpoles, complete with gills and a tail. After a time, however, both these features disappear, even as lungs start to develop, along with ears for hearing and legs for hopping. The tadpole has become a frog—a double life indeed.

A Critical Reptilian Innovation: The Amniotic Egg

In the transition from amphibian to *reptile*, there is a severing of the amphibian ties to the water through a new kind of protection for the unborn. Amphibian eggs require a watery environment; lacking any sort of outer shell, they dry out if taken out of the water, thus killing the embryo inside. The early reptiles solved this problem with the remarkable adaptation of the **amniotic egg**: an egg that has not only a hard outer casing, but an inner "padding" in the form of egg "whites" and a series of membranes around the growing embryo. These membranes help supply nutrients and get rid of waste for the embryonic reptile. With this hardy egg, the tie to the water had been broken; reptiles could move *inland*.

Amphibians were the sole terrestrial vertebrates for about 30 million years, beginning about 350 Mya. We think of these creatures now in terms of small frogs and salamanders, but the amphibians who roamed the Earth in the Pennsylvanian period were often the size of modern-day pigs or crocodiles. The carnivores among them probably fed on small early reptiles, but reptiles ultimately became the hunters rather than the hunted. Thriving everywhere from water to dry highlands, they grew in numbers and diversity.

One line of reptiles evolved, about 220 Mya, into the most fearsome creatures ever to walk the Earth—the dinosaurs (**see Figure 19.19**). For 155 million years, no other land creatures dared challenge them for food or habitat. Not all dinosaurs were huge, and many did not eat meat. Yet even a leaf-eater such as *Apatosaurus* must have commanded a good deal of respect, because it measured more than 26 meters, or 85 feet, from head to tail and weighed more than 30 tons. By comparison, a modern African elephant may weigh about 7 tons.

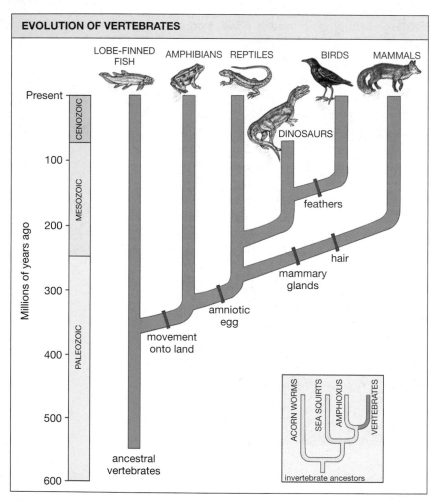

EVOLUTION OF VERTEBRATES

Figure 19.18
Transitional Animal
An artist's conception of *Ichthyostega*, the earliest known amphibian, seen at upper left climbing on the log and at center. Sometimes thought of as a "four-legged fish," *Ichthyostega* spent most of its time in water and probably used its back legs mostly for paddling. All land creatures with backbones, including human beings, ultimately trace their ancestry to tetrapod fishes such as *Ichthyostega*.

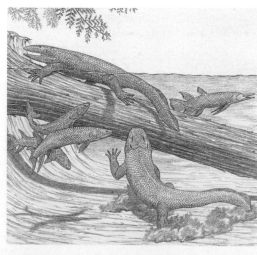

a

b

Figure 19.19
How Did *T. rex* Walk?
Previously, scientists thought *Tyrannosaurus rex* walked in the upright posture seen in photo (a). More recent research has persuaded scientists, however, that the "tyrant reptile" walked with the plane of its body parallel to the ground, as seen in (b). When standing still, *T. rex* may sometimes have raised itself into the upright position to look around.

From Reptiles to Mammals

Scurrying about in the underbrush when the dinosaurs reigned supreme were several species of insect-eating animals, none bigger than a rat. These animals fed their young on milk derived from special female mammary glands, however. In addition they probably possessed fur coats over their skins—an important feature for animals restricted to feeding at night, when the dinosaurs were less active. These were the mammals, the first of which appeared at about the same time as the first dinosaurs. In an example of being "born at the wrong time," these small mammals then lived in the shadow of the reptilian behemoths for more than 100 million years. Toward the end of the Cretaceous period, they began to radiate out into more environments and consequently evolved into more forms. At that point fate intervened: The asteroid that helped bring the reign of the dinosaurs to a crashing close 65 Mya also brought the mammals out of hiding. They radiated into niches far and wide, becoming the dominant form of large land animal, a status they retain to this day. Some of their members even returned to the sea, as with whales and seals. An obvious question here is: Why did the mammals survive the Cretaceous Extinction while the dinosaurs died out? Alas, we have no clear answers. Pure size may have had something to do with it; small creatures seem to survive extinction events better than large ones.

The Evolution of Mammals

So, how did mammals evolve from reptiles in the first place? As usual, the transition was a gradual one. Mammal-like reptiles had feet placed more nearly under their bodies (like mammals) as opposed to splayed out at angles (like reptiles), and they may have developed some body hair. This hair would become enormously useful in keeping some mammals warm. The transition to mammals also entailed something you've seen before: more protection for the unborn (recall that reptiles developed the amniotic egg). With the most advanced mammals, the young began developing entirely *inside* the mother before birth, sustained by the placental system in the womb.

The Primate Mammals

About five million years after the end of the dinosaurs, a species of placental mammals gave rise to the order of mammals called *primates*. If you look at **Figure 19.20**, you can see, in the

Figure 19.20
Primate Characteristics
These modern-day lemurs in Madagascar exhibit several characteristics common to most primates: Opposable digits that enable grasping, front-facing eyes that allow binocular vision, and a tree-dwelling existence.

a

b

c

Figure 19.21
Several Types of Primates

a A loris from India (*Loris tardigradus*) is representative of an early primate lineage, the prosimians, that includes lemurs.

b A wooly spider monkey (*Brachyteles arachnoides*) from the Atlantic rain forest in Brazil. Note the tail, which is capable of grasping objects. Only New World monkeys have such tails.

c A lowland silverback gorilla (*Gorilla gorilla*), native to Africa.

lemur of Madagascar, a modern-day descendant of these earliest primates. Three things characteristic of most primates are apparent in this picture: large, front-facing eyes that allow for the binocular vision (which enhances depth perception); limbs that have an opposable first digit, like our thumb (which makes grasping possible); and a tree-dwelling existence. There are about 230 species of primates living today (**see Figure 19.21**), which is a small

portion of the 4,600 species of mammals. It's a tiny number of species indeed compared to, say, the 60,000 species of molluscs, or the 750,000 species of insects. Yet because of one primate species, *Homo sapiens*, the primates have had an effect on living things all out of proportion to their numbers. **Figure 19.22** sets forth a possible primate family tree.

19.8 The Evolution of Human Beings

If you look to the upper right in Figure 19.22, you can see a fork that leads to chimpanzees on the one hand and human beings on the other. The animal at the branching point of this fork is a presumed "common primate ancestor" that gave rise to chimpanzees on the one hand, and the human family tree on the other, with this split occurring about 5 million years ago. **Figure 19.23** could be thought of as a kind of "close-up" of the line leading to human beings. It shows the human evolutionary line branching out in a fairly complex way, with one tip of this tree getting us to modern human beings, designated *Homo sapiens*.

Any such tree is merely an attempt to put a hypothesis about evolution into visual form, and this tree is no exception. Lots of other human family trees are under consideration; this one was chosen because it corresponds as closely as possible to a consensus view of how humans evolved. The researchers who study human evolution are anthropologists of various sorts—paleoanthropologists who study fossil evidence, for example, and biological anthropologists, who examine ancient and modern DNA sequences. These scientists are looking at a single body of evidence, but this evidence is fragmentary enough that it can be interpreted

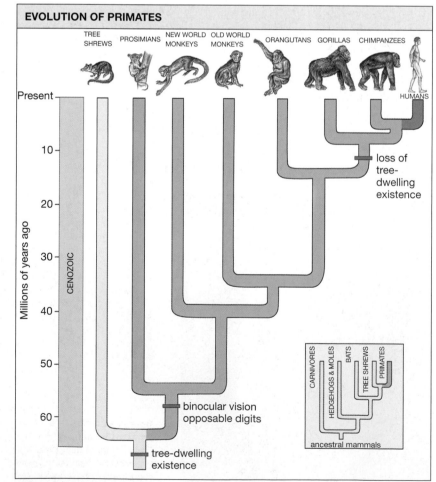

Figure 19.22
How Primates May Have Evolved

Tutorial 19.1.5
Evolution of Primates

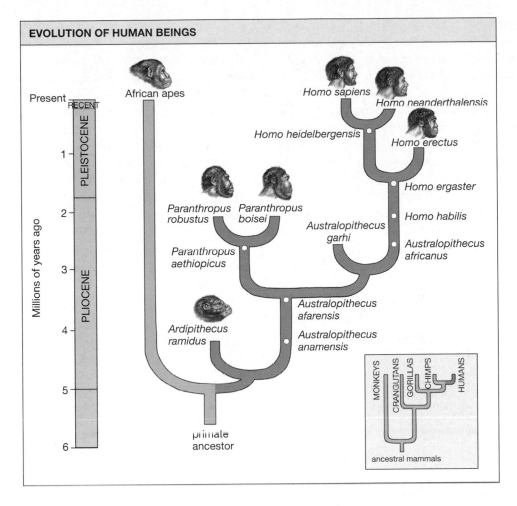

EVOLUTION OF HUMAN BEINGS

Present

RECENT

PLEISTOCENE

PLIOCENE

Millions of years ago

1

2

3

4

5

6

African apes

Homo sapiens

Homo neanderthalensis

Homo heidelbergensis

Homo erectus

Homo ergaster

Paranthropus robustus

Paranthropus boisei

Homo habilis

Australopithecus garhi

Paranthropus aethiopicus

Australopithecus africanus

Ardipithecus ramidus

Australopithecus afarensis

Australopithecus anamensis

MONKEYS
ORANGUTANS
GORILLAS
CHIMPS
HUMANS

ancestral mammals

primate ancestor

**Figure 19.23
A Possible Family Tree
for Human Beings**

In different ways (**see Figure 19.24**). As you'll see, these interpretations lead to very different views of who should be placed where on the human family tree.

All of the species pictured in Figure 19.23 are referred to as *hominids*, which is to say, members of the family of man-like creatures (*Hominidae*). Note that our own genus, *Homo*, is but one of four represented. The genus *Ardipithecus*, at the bottom of the tree, is represented by only one species, *Ardipithecus ramidus*, whose 4.4-million-year-old fossil fragments of jaw and teeth were found in Ethiopia in 1994. At the moment there is some controversy over whether *A. ramidus* deserves the "offshoot" status given in the figure or instead should lie directly between the common primate ancestor and the

**Figure 19.24
Types of Human Fossil
Evidence**
The fossils that paleoanthropologists unearth and interpret often are mere fragments of hominid skeletons, as with the fossils from the Gran Dolina site in Spain, at left. These are some of the remains of six different individuals—some of them children, some teenagers, and some adults, but all of them members of a hominid species named *Homo antecessor* that lived about 800,000 years ago. In contrast, a few unearthed fossils are remarkably complete, as can be seen in a picture of the skeleton of Turkana Boy, at right, so named because he appears to have been about 12 years old and was found near Kenya's Lake Turkana. About 90 percent of Turkana Boy's 1.6-million-year-old skeleton was recovered from the site. Hands and feet were the only major skeletal elements missing from this member of the *Homo ergaster* species.

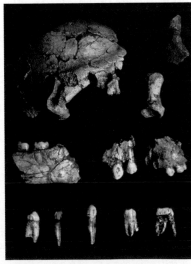

a *Homo antecessor* fossils

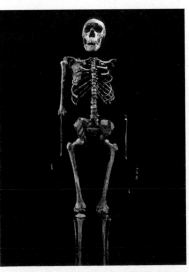

b Turkana Boy skeletal remains

next genus up, *Australopithecus*. Whatever the case, *A. ramidus* was very similar to the common primate ancestor, but had some hominid features as well. (It had teeth with thin dental enamel like that of great apes, but features at the base of its skull like those seen in later hominids.) *Australopithecus* is the genus that probably gave rise to the human genus *Homo* and to the *Paranthropus* genus, now extinct along with all the members of the *Australopithecus* genus.

With the possible exception of *A. ramidus*, all the creatures in this tree walked upright on two legs—they were all "bipedal"—as opposed to our closest non-hominid relatives, the chimpanzees and gorillas, who are "knuckle walkers." You can get some sense of how the various hominids differed by looking at the artist's renderings of several of them in **Figure 19.25**.

Interpreting Fossil Evidence

How do anthropologists know what characteristics a given hominid had? How, for example, do they know whether it stood upright or was a knuckle walker? Most clues are in skeletons, both ancient and contemporary. Modern chimpanzees and other knuckle walkers are, of course, putting weight on their arms when they walk. Not surprisingly, their arms have weight-bearing anatomical features. For example, there are ridges on their wrist joints that lock the wrist into position, meaning the wrist does not bend "backward" with the flexibility of our own wrists. In addition, chimpanzees have an adaptation in the upper arm bone that allows their elbow joints to lock in place.

Looking at the bones of a 4.2 million-year-old *Australopithecus anamensis* found in Northern Kenya, paleoanthropologists see only remnants of the wrist ridges, and no sign of the feature that would allow the elbow to lock into place. Looking at the lower half of the *A. anamensis* body, these scientists can see that its big lower-leg bone (its tibia) contained an amount of spongy, shock-absorbing tissue found only in two-legged walkers. These skeletal structures argue for bipedalism, then, but the most compelling evidence for upright walking comes from a series of footprints laid down in volcanic ash 3.6 million years ago in what is now Tanzania (**see Figure 19.26**). These footprints were made by a later-appearing hominid, *Australopithecus afarensis*. Two parallel tracks are in the stone, probably made by at least two individuals, one larger than the other (perhaps a female and a male). In over 30 meters of space with almost 70 footprints, there is no sign of the knuckle indentations, the bowlegged gait, or the ape-like footprint with a splayed-out big toe, just like the thumb on a hand. Instead, each footprint looks just like one of our own with all toes parallel and straight.

The name *Australopithecus afarensis* may not ring a bell, but one particular *A. afarensis* is perhaps the most famous hominid ever found. "Lucy," a three-foot-seven-inch *A. afarensis* who lived 3.18 million years ago, was discovered in Ethiopia by Donald Johanson and his colleagues in 1974.

The Descent from the Trees

As noted earlier, nearly all primates are tree dwelling (or "arboreal"). At some point, our hominid ancestors clearly abandoned the

**Figure 19.25
Extinct Members of the Hominid Family**

a An artist's conception of an *Australopithecus afarensis* couple.

b An artist's conception of *Homo erectus*.

a

b

arboreal lifestyle of the typical primate, coming down to dwell on the dry savannas of East Africa. At one time it was thought that the early hominids were *forced* to leave the trees by a climate change that dried up our forest homes. The problem with this hypothesis, modern research has shown, is that there was no environmental change at the time the hominids were appearing; the vegetation of East Africa remained fairly stable during that period. What seems likely is that bipedalism predated our descent from the trees, and that our ancestors lived for a time in both worlds.

Still, why was there a shift to bipedalism? A change this radical has costs as well as benefits. Upright posture requires a curvature in the spine, as pelvis, legs, and feet are brought parallel with the trunk. Such a shift leads to extraordinary pressure on the knees and spinal discs, resulting in the all-too-familiar joint and back pain. Against this, the benefits of upright posture are several. For a creature on the ground—as opposed to one still in the trees—standing is a more efficient way to pick fruit from low trees or to hold children or tools; it decreases exposure to sunlight while increasing exposure to cooling breezes; and the increased height that comes with it provides a better view of predators or rivals.

As the posture of our hominid ancestors was changing, so was their brain capacity. The cranial volume of *Australopithecus afarensis* ('Lucy') was about 450 cubic centimeters—about that of a chimpanzee—while that of modern *H. sapiens* is about 1,400 cubic centimeters. We can get a good idea of what evolution had produced with *A. afarensis* by noting that scientists have referred to Lucy as a "bipedal chimpanzee."

The Australopithecines are generally agreed to be ancestral to our own genus, *Homo*, but which Australopithecines stand in this line of descent? If you look at Figure 19.23, you can see two candidate species, *A. africanus* and the newly discovered 2.5-million-year-old *A. garhi* from Ethiopia. *A. garhi* has the right mix of characteristics, with long legs like *Homo*, but with ape-like long arms. More tellingly, it may also have been the first hominid to use tools to butcher animals. Stone tools—actually little more than sharp-edged flakes—were found with the *A. garhi* fossils, along with fossil remains of animals that had cut

marks consistent with butchering and extraction of bone marrow. Before this discovery, it was thought that the first toolmaker was *Homo habilis*.

Into the Genus *Homo*: *Habilis* and *Ergaster*

Homo habilis—"handy man" in honor of its toolmaking abilities—gets us to our own genus, *Homo*, as opposed to the earlier *Australopithecus* genus. But there is significant disagreement about whether *H. habilis* was *enough* like us that it deserves to be placed in our genus. Change to physical forms that are more like ours comes at several steps in the hominid line, but the most dramatic change by far comes with the rise of *Homo ergaster*, whose best-preserved remains, in the form of an adolescent called "Turkana Boy," date from about 1.6 Mya. Experts estimate that, had he grown to maturity, Turkana Boy would have reached a height of 2 meters (6 feet). His brain was 30 percent larger than that of *Homo habilis* and more than half the size of the average modern *Homo sapiens* brain. He had a much more modern face, long limbs typical of those humans who dwell today in arid climates in Africa, and also advanced tool technology.

Human Beings Emerged in Africa but Eventually Traveled

All of the hominids we've looked at so far lived and died solely in Africa. To put this another way, the whole human tree is rooted in Africa. The most important fossils in early human evolution have been unearthed in South Africa and in the Great Rift Valley that runs through much of eastern Africa. The

Figure 19.26
Evidence of Upright Walking
The footprints visible in the photo were laid down about 3.6 million years ago by at least two *Australopithecus afarensis* adults and possibly a child (who walked in the footsteps of one of the adults). The nature of the footprints strongly suggests that, by the time they were laid down, human ancestors had adopted upright or "bipedal" walking. The fossil bed was found in modern-day Tanzania by Mary Leakey in 1978.

portion of Africa that includes Kenya, Tanzania, and Ethiopia is assumed to be the cradle of the human species (**see Figure 19.27**). At some point, however, humans migrated out of Africa. Hominid fossils dated to 1.7 million years ago—and first found on the island of Java, in Indonesia—have now been located from China to southern Eurasia. These are fossils, then, of the first hominid transcontinental travelers, all of them belonging to the species labeled *Homo erectus*.

Out-of-Africa or Multi-Regionalism?

When we look at later-stage human evolution, we run into a controversy that has raged for years in the anthropological community. Here are the differing sides of this argument. One school of thought holds that various late-stage hominids, distinguishable by several anatomical features, left Africa at different times and then interbred in Europe and Asia. Over time, these differing hominids evolved into modern *Homo sapiens*—us, in other words. Under this view, there should be no distinction between *Homo heidelbergensis*, *Homo neanderthalensis*, or *Homo erectus*. All three actually were just different varieties of a single, primitive human species that together evolved into the modern human species. Because this view holds that modern humans evolved in numerous different regions of the globe, it has been dubbed the multi-regional hypothesis.

The contrasting view is represented by what you can see in Figure 19.23. *Homo heidelbergensis* was a distinct species that gave rise to two other distinct species. The first of these was *Homo neanderthalensis*; the other was *Homo sapiens*. Under this view, *Homo sapiens* fully evolved *in Africa*, no more than 200,000 years ago, and then spread throughout the globe perhaps 100,000 years ago. Hence, this view has been named the "out-of-Africa" hypothesis. Note that all views hold that human beings evolved in Africa if we simply go back far enough in time—to *H. ergaster* and earlier. The question is whether humans evolved to their fully modern form before *leaving* Africa.

The out-of-Africa hypothesis is sometimes called the "replacement" hypothesis and, by looking at *H. erectus* and *H. neanderthalensis* on the Figure 19.23 tree, you can understand why. Under this hypothesis, both species became extinct; they were "replaced" by modern humans, as the phrase has it, and this replacement was a fairly recent event in evolutionary terms. The Neanderthals survived until perhaps 25,000 years ago in Europe, and *H. erectus* hung on until perhaps 50,000 years ago in the Far East. At one time, the suspicion was that we *Homo sapiens*, with our supposedly superior tools and brainpower, moved out of Africa and proceeded to wipe out at least the Neanderthals. The consensus view among scholars today, however, is that there was no species warfare between modern humans and the Neanderthals. It's clear that, in at least some regions, humans and Neanderthals coexisted for tens of thousands of years. All this stands in contrast, however, to the multi-regional view, which holds that the Neanderthals never became extinct; they interbred with other hominids and all these varieties evolved into a single, modern human form.

The replacement hypothesis is the consensus view among anthropologists today. Strong pieces of evidence in support of it include the sequencing of both modern DNA (from living human beings around the world) and ancient DNA (recovered from the remains of three Neanderthals). This work showed, first, that certain DNA sequences of modern humans differ greatly from comparable sequences in the Neanderthals. The differences are so great, in fact, that they call into question the very thing asserted by the multi-regionalists: that modern humans and Neanderthals mated with one another. More recent work with contemporary DNA pointed to an African origin, some 170,000 years ago, for all human beings.

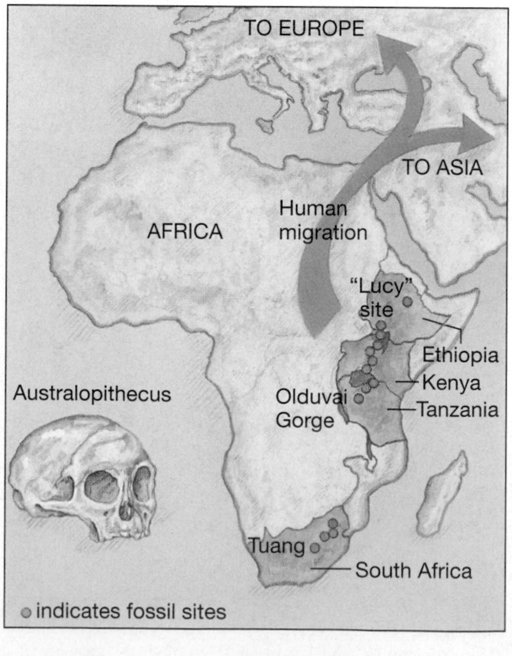

Figure 19.27
Africa Is the Cradle of the Humans
The oldest human fossils have been found in East and South Africa. Most scientists believe that humans originated in these regions, then migrated north and east into Europe and Asia, and eventually migrated into the Americas.

TO EUROPE

TO ASIA

AFRICA

Human migration

"Lucy" site

Ethiopia
Kenya
Tanzania

Australopithecus

Olduvai Gorge

Tuang

South Africa

o indicates fossil sites

In 2001, however, multi-regionalist proponents brought forth their own DNA evidence. Sequencing the DNA from the remains of an anatomically modern 62,000-year-old human, researchers in Australia found that *his* DNA contained a sequence not found in modern humans. The implication was that both this "Lake Mungo" man *and* the three ancient Neanderthals could have had DNA sequences that failed to be passed on to the modern human genome. This would allow Neanderthals to be genetically different from us (as the earlier DNA evidence indicates) and still be part of our biological lineage. At present, then, the debate between multi-regionalism and the replacement model shows no sign of abating.

Why Did the Neanderthals Die Out?

If we assume the replacement model is correct, one of the interesting questions it raises is: Why were the Neanderthals replaced? They had brains as big as ours, they were fine toolmakers, and they buried their dead, which is generally taken to be a sign of reverence for the deceased. Because of these attributes, most anthropologists have no trouble referring to Neanderthals as "people" (**see Figure 19.28**). So, if they were bright and they were not killed out right by modern humans, why did their lineage die out? One possibility is that modern humans brought about the Neanderthal's demise by outcompeting them. Humans and Neanderthals would have lived in the same kinds of habitats and eaten the same kinds of foods. If humans were consistently able to capture more of these resources, the Neanderthals may have died out slowly from a lack of them.

One More Possibility: *Homo antecessor*

Adding to this complexity, the remains of a tall, lanky species with a protruding brow and bulky lower jaw were unearthed in the mid-1990s in northern Spain and dated to about 800,000 years ago. This species, dubbed *Homo antecessor*, is bigger-brained than *Homo erectus* and lacks a distinctive central skull crest. Thus it is modern looking, but ancient, as its fossils stretch back to the time of *Homo heidelbergensis*. The question this discovery has raised is whether *H. antecessor*, rather than *H. heidelbergensis*, is the direct ancestor of human beings. Because the best-preserved *H. antecessor* skull is fragmentary and belonged to a juvenile—probably

a 10-year-old girl—most anthropologists have thus far been unwilling to endorse this idea, but the site from which *H. antecessor* was unearthed—the Gran Dolina site in Atapuerca, Spain—continues to yield fossils (**see Figure 19.29**). More clues to our past may be unearthed along with them.

Figure 19.28
Neanderthal Life
This scene, constructed by the American Museum of Natural History, shows three Neanderthals about 50,000 years ago at the Les Moustier Neanderthal fossil site in France. Neanderthals are known to have camped in the open as well as in caves and under rock overhangs such as this one. They used cooking hearths and almost certainly wore hide clothing of their own making. The young woman in the scene is scraping a hide with a stone tool while holding the hide with her teeth. The front teeth of Neanderthal fossils are usually heavily worn, suggesting that teeth were often used for work such as this.

Figure 19.29
Major Excavation Site
The Gran Dolina excavation site in northern Spain has yielded some important human fossil discoveries. The site has sediments some 18 meters (59 feet) in depth, with their geological strata ranging from 200,000 years old near the top to 900,000 years old at the bottom. The remains of *Homo antecessor*, believed by some to be the direct ancestor of human beings, were found at Gran Dolina in sediments dated to 800,000 years ago. (Some of these remains can be seen in Figure 19.24a.)

Modern *Homo Sapiens*

Whoever our direct ancestors were, we humans emerged in modern form fairly recently—no more than 200,000 years ago. The earliest fully modern *H. sapiens* fossils, uncovered in South Africa, date from 117,000 years ago. From our African origins we then spread out across the world. Whatever we may have in common with other species, it is certain that we have characteristics that make us unique: our unrivaled brain size, our superior toolmaking ability, and our capacity for language among them. These qualities evolved over much time, as you've seen. To get some perspective on *how* much time, consider that the history of Egyptian civilization stretches back for about 5,000 years. In contrast, the Neanderthals—latecomers in the human line—came to Europe 200,000 years ago and existed there until 25,000 years ago. Evolution within the entire human family has been going on for five million years.

Large as this latter figure is, it is the blink of an eye compared to the 3.6 billion years that have elapsed since the first microscopic living things took shape in Earth's ancient oceans. These ocean-dwelling species diversified, one variety of living thing evolving from another, with the result being the millions of life-forms that exist today. Whatever else evolution has wrought, however, only once has it produced, in human beings, a species capable of comprehending this story. Only once, to put it another way, has it produced a species capable of comprehending where it came from and how it got here.

On to the Diversity of Life

In reviewing the history of life in this chapter, you necessarily looked, in some small way, at almost all of Earth's major life-forms—bacteria, animals, plants, and so forth. In the two chapters coming up, you'll take a more detailed look at each of these life-forms. This is a tour that will make plain the astounding diversity that exists in the living world.

Chapter Review

Summary

19.1 The Geologic Timescale: Life Marks Earth's Ages

- Geological time is divided into eras, shorter periods, and shorter-yet epochs, which are defined by the kinds of life-forms that have existed in them.
  TUTORIAL 19.1.1: Continental Drift and the Geological Timescale

- What is notable in evolution hinges on values. Evolution can be conceptualized as a branching tree, with humans on one branch. It is not shaped as a pyramid with humans at the apex. The evolutionary process that led to human beings also led to millions of other organisms.

- Every living thing on Earth belongs to one of three domains of life: Archaea, Bacteria, or Eukarya. Both bacteria and the archaea are single-celled, microscopic life-forms; the domain Eukarya is composed of four kingdoms: plants, animals, fungi, and protists.

19.2 Tracing the History of Life on Earth: How Did Life Begin?

- Life on Earth began between 3.5 and 4 billion years ago with the origin of self-replicating molecules. There

currently is a debate regarding the conditions and sequences of events that led to the formation of these molecules. One hypothesis that has a good deal of support among researchers is that life began in a "prebiotic soup" of hot-water systems—in the deep-sea vents or the kind of hot-spring pools that exist in Yellowstone Park today.

- A critical question for orgin-of-life researchers had been to account for the existence in living things of both enzymes and information-bearing molecules (DNA and RNA), since neither kind of molecule can be synthesized without the other. The discovery of RNA molecules that have enzymatic capabilities helped solve this dilemma. The existence of such ribozymes provided evidence in support of a presumed "RNA world" in which the only living things were simple RNA molecules that could bring about their own replication.

- One of the primary challenges for origin-of-life researchers today is to discover a plausible sequence of events by which enzymatic RNA molecules, or their precursors, could have been synthesized from building-block organic molecules.
  TUTORIAL 19.2.1: RNA-Like Self Replication

19.3 The Tree of Life

- Life today can be traced to a universal ancestor that gave rise to the domains Archaea, Bacteria, and Eukarya. The earliest kingdom in Eukarya was the protist kingdom, whose organisms gave rise to the animal, plant, and fungi kingdoms.

19.4 A Long First Period: The Precambrian

- The oldest undisputed fossils we have are of bacteria dating from 3.4 billion years ago. During roughly half the time life has existed on Earth, it consisted of nothing but bacteria and archaea.

- Oxygen came to exist in the Earth's atmosphere through the activity of photosynthesizing organisms, originally photosynthesizing bacteria. Photosynthesis meant an abundance of food on the Earth. Further, a high concentration of atmospheric oxygen set a strict condition for all life: Adapt to oxygen, stay separate from it, or perish.

19.5 The Cambrian Explosion: A Real Milestone or the Appearance of One?

- The fossil record indicates a tremendous, rapid expansion in the number of animal forms in a "Cambrian Explosion" that began about 544 Mya, but this evidence has been challenged. It may be that the Cambrian Explosion was merely an explosion in the number of animal forms big and hard enough to leave fossils, and that the surge in animal diversity took place over a relatively long period of time beginning well back in the Precambrian era.

19.6 The Movement onto the Land: Plants First

- Plants, which evolved from green algae, made a gradual transition to land about 460 Mya, followed by animals. Major transitions in plant life came with the development of seed plants—first the gymnosperms (represented by today's conifers) and then the angiosperms (flowers, food crops, many tree varieties).
  TUTORIAL 19.1.2: Evolution of Plants

19.7 Animals Follow Plants onto the Land

- Four-limbed vertebrates (tetrapods) moved onto land about 400 Mya in the form of lobe-finned fishes, which gave rise to amphibians (represented by today's frogs and salamanders). Reptiles later branched off from amphibians, and mammals branched off from reptiles.
  TUTORIAL 19.1.4: Evolution of Vertebrates

- About 60 Mya, there arose an order of mammals called the primates, characterized by large, front-facing eyes, limbs with an opposable first digit (thumbs in today's human beings), and a tree-dwelling existence.
  TUTORIAL 19.1.5: Evolution of Primates

19.8 The Evolution of Human Beings

- Human evolution is studied by scientists called anthropologists, who specialize in various ways. Paleoanthropologists study fossil evidence, for example, while biological anthropologists examine ancient and modern DNA sequences.

- There is considerable debate about what the human family tree looks like. The body of evidence concerning human evolution is fragmentary enough that it can be interpreted in different ways by different anthropologists.

- All the members of the human family tree are referred to as hominids, meaning the family of man-like creatures. A common primate ancestor is thought to have given rise, about 5 million years ago, to both hominids and to chimpanzees. The genus of modern humans, *Homo*, is only one genus among four that have been part of the hominid family tree. The other three genera are *Ardipithecus*, *Australopithecus*, and *Paranthropus*. *Ardipithecus* lies near the base of the human family tree and was much like the common primate ancestor. *Australopithecus* is thought to have given rise to both *Paranthropus* and *Homo*. *Homo* is the only genus among these four that has not become extinct. The only remaining species in the genus *Homo* is our own species, *Homo sapiens*.

- Notable events in human evolution include the gradual abandonment of tree-dwelling existence and the switch to bipedalism, or two-legged walking.

- All early human evolution took place in Africa, more specifically in southern and eastern Africa. Though the hominid line began evolving about 5 million years ago, the first evidence we have of hominids living outside Africa dates to only 1.7 million years ago. These traveling hominids were members of the species *Homo erectus*.

- There are two competing hypotheses about the evolution of modern humans. One, called the multi-regional hypothesis, holds that several similar, but not identical, types of hominids left Africa and then interbred in Europe and Asia. Over time, these hominids evolved together into modern *Homo sapiens*. The second view, called the out-of-Africa hypothesis, holds that human beings fully evolved in Africa and the spread out over the globe, eventually replacing separate hominid species, such as *Homo neanderthalensis*, or the Neanderthals. Most experts believe the out-of-Africa hypothesis is correct.

Key Terms

amniotic egg 392	**era** 379
angiosperm 390	**Eukarya** 382
Archaea 382	**gymnosperm** 390
Bacteria 382	**period** 379
bryophytes 390	**Permian Extinction** 379
continental drift 381	**phylum** 388
Cretaceous Extinction 379	**ribozyme** 385
epoch 379	**seedless vascular plant** 390

Understanding the Basics

Multiple-Choice Questions

1. Put the following in the proper temporal sequence, earliest first.
   a. first insects
   b. first dinosaurs
   c. first bacteria
   d. first animals
   e. first seed-bearing plants
   f. first whales
   g. free oxygen accumulation in the atmosphere
   h. first land plants
   i. first birds

2. Periodic major extinctions (select all that apply)
   a. can eliminate the vast majority of species present at that time on Earth
   b. are usually caused by meteor impact
   c. open up new evolutionary opportunities
   d. can lead to major evolutionary transitions in the dominant forms of life on Earth
   e. have occurred only twice in Earth's history

3. The first life on Earth may have been (select all that apply)
   a. eukaryotic cells
   b. ribozyme-like molecules
   c. self-replicating DNA
   d. formed on land
   e. formed at extreme temperatures

4. Which of the following characteristics of animals and/or plants are adaptations to life on land? (Select all that apply.)
   a. multicellularity
   b. pollen and seeds
   c. amniotic egg
   d. muscles
   e. internal fertilization

5. Which of the following animals are tetrapods?
   a. birds
   b. frogs
   c. fish
   d. pigs
   e. insects

6. True or False: Dinosaurs were often hunted by humans.

7. Photosynthesis was a huge innovation in the history of life because (select all that apply)
   a. It allowed RNA to replicate itself.
   b. It put huge quantities of oxygen into the atmosphere.
   c. It resulted in the production of more food than had been previously available.
   d. It resulted in blockage of UV light, increasing habitability of the land for terrestrial organisms.
   e. It helped the Earth to thaw out.

8. Mitochondria and chloroplasts (select all that apply)
   a. are thought to have arisen as symbiotes
   b. are very similar to bacteria
   c. are very similar to algae
   d. are organelles within cells
   e. were early protists

9. Which of the following organisms have an amniotic egg? (Select all that apply.)
   a. grasshopper
   b. lizard
   c. salamander
   d. fish
   e. bird
   f. elephant
   g. turtle

10. The first lineage of terrestrial vertebrates was _____. The first lineage of vertebrates that was fully terrestrial was _____.
   a. lobe-finned fish, reptiles
   b. salamanders, frogs
   c. amphibians, reptiles
   d. amphibians, birds
   e. frogs, mammals

Brief Review

1. What are two differences between domain Bacteria and domain Eukarya?

2. What is the amniotic egg, and why was it a breakthrough for terrestrial animals? In which vertebrate class did the amniotic egg evolve?

3. What are two characteristics of primates that are not found in other mammals? Why are they important for the evolution of hominids?

4. Which two groups are more closely related, humans and chimps or humans and orangutans?

5. Were Neanderthals the direct ancestors of humans?

6. When did life begin, and how much time passed between the origin of the first bacterial cells and the colonization of land? What was the world like during that time?

7. Describe three characteristics that evolved in mammals that had not been present in earlier organisms.

Applying Your Knowledge

1. Mitochondria and chloroplasts within eukaryotic cells evolved from bacteria that were ingested and took up residence in early eukaryotic cells. How did this partnership benefit each party?

2. What was the Cambrian Explosion, and why was it notable in the history of life? What could have caused it, and why hasn't a similar explosion happened again?

3. How would Earth be different if photosynthesis had never developed in any organism?

4. Describe two problems experienced by plants that colonized land. What types of "solutions" evolved in plants for each of these problems? What problems did animals experience in colonizing land, and what types of solutions evolved in animal lineages?

20

Pond Dwellers, Log Eaters, and Self-Feeders
The Diversity of Life

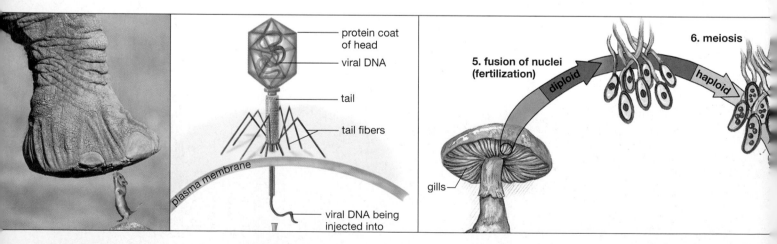

protein coat of head

viral DNA

tail

tail fibers

plasma membrane

viral DNA being injected into

5. fusion of nuclei (fertilization)

diploid

haploid

6. meiosis

gills

Creatures large and small.
(page 406)

Invader of living things.
(Section 20.1, page 408)

Life cycle of a fungus, part 2.
(Section 20.6, page 422)

Plant, fungus, microbe? Across its categories, life is stunningly diverse.

The smallest living things discovered to date are a type of bacteria that measure about two-hundredths of a micrometer in diameter—that is, two-hundredths of one-millionth of a meter in diameter. When we turn to the *biggest* living things, elephants or whales might come to mind, but it turns out that animals aren't even in the running for this title. At a maximum length of 27 meters (or about 89 feet), the blue whale is the largest animal that has ever existed, but it's small compared to California's coastal redwood trees, which might reach a height of 100 meters or about 330 feet. This is large indeed, but consider that a giant sea kelp once was found that was 274 meters in length, meaning it was about 900 feet long. It may be, though, that in terms of size, both kelp and redwoods have to take a back seat to a life-form normally thought of as quite small. In 1992, researchers reported finding a fungus growing in Washington State that runs underground through an area of 1,500 acres and that arguably is a single organism.

Now, how about the *deepest*-living things? The current record-holders are bacteria that were found in the early 1990s in a combination of oil drilling and scientific exploration—living almost 3 kilometers beneath the Earth's surface, which is about 1.9 miles down. Food and water are so scarce at such depths that these bacterial cells may live in a kind of suspended animation, dividing into "daughter" cells perhaps once a year, or even once a *century*. Turning to the highest living things, 12 species of bacteria have been found to be living in the Himalayas 8,300 meters or 27,000 feet above sea level, which is just a couple thousand feet lower than the peak of Mount Everest. A species of chickweed survives in the Himalayas at over 20,000 feet.

All kinds of life-forms—clams, tube worms, archaea—live kilometers beneath the surface of the oceans, near the "hot-water vents" that spew out water and minerals from the Earth's interior onto the ocean floor. It is the microbes called archaea of these deep-sea vents that set the pace for life in a hot environment. An average, midday temperature in the Sahara Desert would be so *cold* it would end the reproduction of the archaean *Pyrolobus fumarii*,

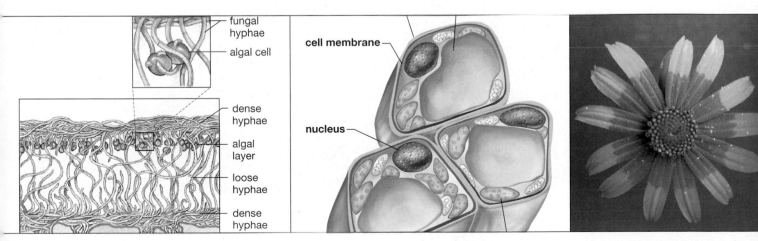

A mutually beneficial
arrangement.
(Section 20.6, page 423)

cell membrane

nucleus

An important type of cell.
(Section 20.7, page 425)

A bull's eye for bees
(Section 20.7, page 431)

fungal hyphae

algal cell

dense hyphae

algal layer

loose hyphae

dense hyphae

which reproduces within the hot-water vent walls at a temperature of up to 113° Celsius, which is hotter than the temperature at which water boils.

Finally, what about the oldest living things? Humans occasionally live past the age of 120, but this is an eyeblink compared to the lives of some plants. There is a Sequoia tree in California that was growing 3,500 years ago, when the pharaohs ruled Egypt. But even it is a youngster compared to the oldest living thing that is clearly a single organism, a bristlecone pine tree (*Pinus longaeva*) living in the dry slopes of Eastern California that has been dated at 4,900 years.

Looking at life this way—in its biggest or deepest or oldest extremes—is one way to get at the incredible diversity of it. But it does not begin to do justice to life's variety, because life is diverse in so many diverse *ways*. Honeybees do a "dance" to let their hive-mates know the location of a food supply. Corn plants, when attacked by army worms, can release an airborne substance that attracts parasitic wasps that prey on the worms. Pacific salmon live years in the ocean only to make one arduous, upstream journey to spawn in the fresh waters they were born in, after which they die. Look closely at almost any creature and you are likely to find something only slightly less dramatic (**see Figure 20.1**).

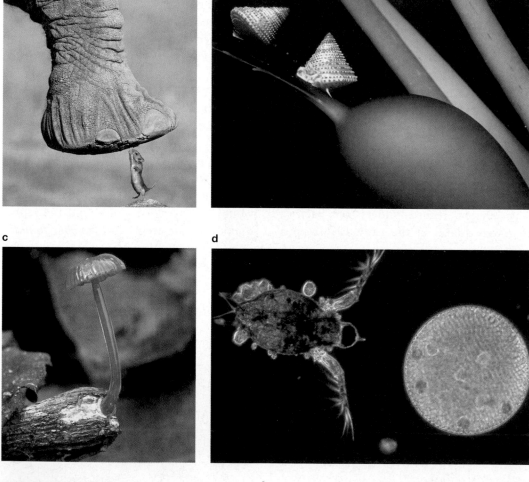

Figure 20.1
An Amazing Diversity in the Living World

a A mouse beneath the foot of an elephant.

b Aquatic snails, moving across kelp.

c A rain-forest fungus, growing from a fallen log.

d A tiny, aquatic animal, called a water-flea, swimming near a form of green algae called *Volvox*.

The Unsolvable Taxonomy Problem

This book uses a classification system that divides life first into three domains—Bacteria, Archaea, and Eukarya—and then into four kingdoms within the eukaryotic domain. But there are other ways of classifying all the creatures of the living world.

If you look at **Figure 1**, you can see both the three-domain structure (a) and another that employs six kingdoms (b). Note that this second system has no domains at all; instead, its highest taxa are the kingdoms Eubacteria, Archaebacteria, Protista, Plantae, Fungi, and Animalia, with Eubacteria corresponding to the domain Bacteria and Archaebacteria corresponding to the domain Archaea. It may be apparent that the Archaea and Bacteria have a lesser status in this second system. Rather than comprising two of life's three domains, they each comprise only a kingdom in a system that has six. Beyond this, the Archaea are thought of as bacteria in this latter system, whereas in the three-domain system they are a separate kind of organism.

Which system is "right?" Arguments can be made on both sides. No votes are taken on this issue, and there are highly respected biologists in both camps. Some specialists feel that the three-domain system best defines life's most fundamental distinctions, while other experts believe that the six kingdoms do a better job.

Another, more long-standing problem in taxonomy has to do with the catch-all kingdom Protista, whose members, mostly single-celled, are defined by what they are *not*—not plants, not animals, not fungi—which is a curious kind of definition. We would prefer to define a grouping by what its members *are*, which is to say what characteristics all its members share, but the Protista are so diverse that no one can come up with such a definition. By some definitions, green algae should be in the kingdom Plantae, but in this text all algae are included in Protista. Here there clearly is no "right" answer. The fundamental problem is that Earth's creatures are so diverse that it's impossible for all of them to fit neatly into categories devised by human beings. We can have sensible taxonomic systems, but we can't have perfect ones.

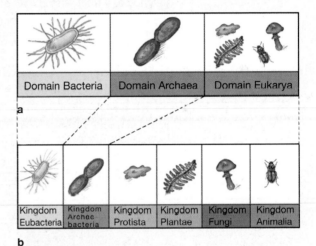

Figure 1
Different Ways of Categorizing Life

a The three-domain system.

b The six-kingdom system. Both systems take all living things into account, but group them differently.

The purpose of this chapter is to provide some sense of how this incredible diversity is divided up—to introduce you to life's large-scale categories. Those readers who went through Chapter 19 know all living things can be placed into one of three "domains"—Archaea and Bacteria, whose members are microscopic; and Eukarya, whose members range from the microscopic to the gigantic. Domain Eukarya is in turn divided into four "kingdoms," which are plants, animals, fungi, and another called protists. (See "The Unsolvable Taxonomy Problem," above, for a discussion of problems inherent in any such categorization.) In this chapter, you will take a brief look at the Archaea and Bacteria domains and then at three of the kingdoms within Domain Eukarya: protists, fungi, and plants. Animals are covered in Chapter 21. In beginning this tour of the living world, however, we'll start with a category of replicating entities that most biologists would say lies just outside of life, though they have a great effect on living things.

20.1 Viruses: Making a Living by Hijacking Cells

If a living thing is too small to be seen and can cause disease, most people would put it in the vague category known as "germs." But there is a fundamental distinction to be made between *two* kinds of infectious organisms. On the one hand there are bacteria, which may be very small but nevertheless are *cells*, complete with a protein-producing apparatus, a mechanism for extracting energy from the environment, a means of getting rid of waste—in short, complete with everything it takes to be a self-contained living thing. On the other hand there are viruses, which by themselves possess none of these features. Indeed, viruses can be likened to a thief who arrives at a factory he intends to rob possessing only two things: the tools to get inside and some software that will make the factory turn out items that he can use.

The factory is a living cell, and the software that the virus brings is its DNA or RNA, which it may inject into the "host" cell. Once inside a cell, most viruses first destroy the host's ability to use its own genetic machinery; then the viruses get to work expressing *their* genes with this machinery, meaning they turn out *viral* proteins that will facilitate the production of multiple copies of the virus. Some viruses accomplish this feat by splicing their genetic material into the host's DNA. Ultimately, hundreds or thousands of clones of the original invading virus are produced, which often go on to burst forth from the infected cell, killing it in the process. Once outside, random chemical motion brings these viral particles into contact with yet more cells, and the whole process starts over again. This is the so-called **lytic cycle**—meaning a cycle, used by some viruses to replicate themselves, in which the host cell is destroyed by viral rupture of its plasma membrane.

If you look at **Figure 20.2**, you can see an illustration and micrograph of one particular virus, known as T4, which preys on bacteria. (Viruses come in many shapes; it just so happens that this one looks like an invading spacecraft.) T4 attaches to its host via the tail fibers on its six spikes; its DNA is contained in its "head," which is sheathed in a protein coat. Its tail can contract, acting like a hypodermic needle to inject its DNA into the host cell. Once inside, this viral DNA uses the cell's own machinery (its enzymes) to turn out T4 virus clones in a multistep process. The viral DNA is replicated, even as new viral protein coats are being put together. Then these components are assembled to create the finished product: 100 or so clones of the virus (**see Figure 20.3**). Finally, a lytic enzyme is produced that causes the host cell's outer or plasma membrane to rupture—to **lyse**,—thus releasing these clones to the environment outside the cell. From the time T4 enters the cell to the time its clones burst forth, perhaps 20 minutes will have elapsed.

The life cycles of some viruses can take a course that doesn't involve lysis; in the so-called **lysogenic cycle**, the viral DNA integrates with the host DNA and then stays there, replicating each time the host cell divides. The virus ends this arrangement, however, if its host cell is threatened by, say, starvation. Then the virus "bails out," removing itself from the host genome and entering into the lytic cycle, its clones eventually bursting out of the cell.

Given how formidable viruses are, it is interesting to contemplate how little there is to them. They are so small that thousands of them can typically fit within a bacterial cell. They are so genetically simple that T4 has a DNA complement that is only about

Figure 20.2
One Variety of Virus

a The T4 virus looks like a spacecraft that has landed on the surface of a bacterium.

b A viral invader lands. An artificially colored T4 virus injects its DNA into an *E. coli* bacterium (in blue).

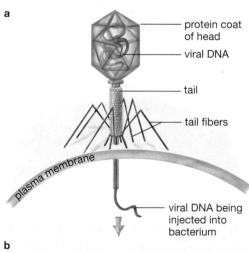

a

- protein coat of head
- viral DNA
- tail
- tail fibers
- plasma membrane
- viral DNA being injected into bacterium

b

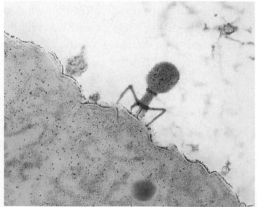

40,000 base pairs in length, compared to the *E. coli* bacterium's genome of 4 million base pairs, and the human genome of 3 billion base pairs. Some viruses amount to nothing *but* their genome and the protein coat that surrounds it. There are even virus-like entities, called **viroids**, that lack a protein coat; they are simply small strands of infectious RNA. There are more complex viruses, to be sure, but in no case is a virus complex enough to replicate by itself. Because they can carry on scarcely any of life's basic functions unaided, viruses aren't even classified as living things by most scientists.

The Trouble Viruses Cause

But what trouble from something that is not even alive! In humans alone, viruses cause smallpox, chickenpox, measles, mumps, rabies, polio, herpes, rubella, and some forms of cancer and hepatitis, to say nothing of common colds and flus. Then of course there is AIDS. And viruses can invade every life-form, from protists to fungi to plants to bacteria, not to mention all varieties of animals. In early 2001, hundreds of thousands of cattle and sheep had to be destroyed in Europe because they had been exposed to a virus that causes an affliction known as *foot and mouth disease.* From this, and the review of the techniques of viruses, it may seem as though all of them would be unstoppable. Yet we know that people get over colds and flus, that herpes doesn't kill us, and that polio has largely been conquered in the developed world. What puts the brakes on these cellular invaders? In human beings it is our immune system and the vaccines we have developed to prime this system's responses.

For the Defense: The Immune System and Vaccines

The human immune system provides two main lines of defense against viruses. One is the production of antibodies, which are proteins capable of recognizing virus particles and binding to them, thus blocking their life cycle. The other is our so-called T-cell complex, which is capable of recognizing *cells* that have been infected with viruses and then destroying these cells. Critically, both these arms of the immune system have a kind of memory. The immune system's *initial* response to an infection by a given virus may be relatively slow, but once the virus is cleared

from the body, the immune system retains clones of the antibodies and T cells it developed to fight the infection. This means that the *next* time this virus invades, the body can mount a rapid response, clearing the virus before harmful effects set in. This is why almost nobody gets chickenpox twice. The problem with this system is that some viruses do permanent damage on their first invasion (as with the poliovirus destroying the nerve cells that allow us to walk), while other viruses can remain hidden from the immune system, only to emerge later (as with herpes).

The theory behind vaccinations is that it is possible to elicit an initial immune response to a given virus without giving the disease

LIFE CYCLE OF A VIRUS

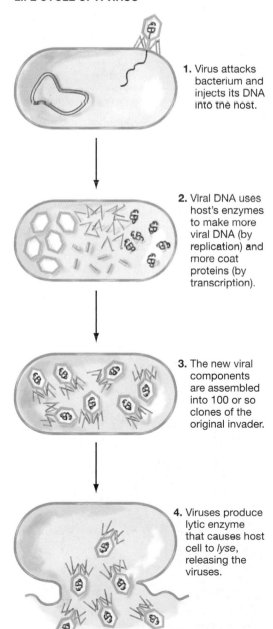

1. Virus attacks bacterium and injects its DNA into the host.

2. Viral DNA uses host's enzymes to make more viral DNA (by replication) and more coat proteins (by transcription).

3. The new viral components are assembled into 100 or so clones of the original invader.

4. Viruses produce lytic enzyme that causes host cell to *lyse*, releasing the viruses.

Tutorial 20.1.1
Life Cycle of a Virus

Figure 20.3
How Viruses Multiply
A lytic virus life cycle.

Not Alive, but Deadly: Prions and "Mad Cow" Disease

As the year 2000 turned into 2001, Western Europe was undergoing a crisis of fear about food and America was beginning to get nervous about it. Beef sales in the 15-country European Union were down by 27 percent in 2000. In January 2001, public health officials in Portugal ordered that 50,000 head of cattle be slaughtered and that their remains be sent not to supermarkets but to burial sites and incineration plants. Meanwhile, in the United States, health officials were doing everything they could to keep this European calamity from spreading across the Atlantic. In January 2001, the American Red Cross asked the U.S. Food and Drug Administration to prohibit blood donations in America by anyone who had lived in Western Europe for six months or more. Even as this was taking place, U.S. feedlots were on high alert regarding what they fed to their cattle.

Why this incredible flurry of activity? The fear of a new, seemingly transmissible human illness, commonly known as Mad Cow disease.

When we think of a disease that is transmissible, we normally assume that some sort of tiny living thing—a microorganism—is involved in transmitting it. Thus we have, for example, salmonella, meningitis, and tuberculosis: three diseases, each caused by a different bacterium that can move from one living "host" to another. And this is the case with most transmissible diseases; at root, they are caused either by microscopic living things, or by viruses, which always have the DNA or RNA that living things possess.

But Mad Cow disease is different. Almost alone among transmissible human diseases, it appears to have a *component part* of a living thing as its cause—a protein. Recall from Chapter 3 that much of the scaffolding, or structure, of living things is made from proteins; that the chemical tools called enzymes are made from proteins; and that some hormones are made from them. Clearly, proteins have many functions in living things. But proteins themselves are not living.

So, how could a protein cause a transmissible disease? A variety of protein called a prion protein exists in living tissue in a normal, harmless form. Such prion proteins also can come to exist, however, as "rogue" proteins that are *misshapen*. In their normal form in humans, prion proteins reside in the outer or "plasma" membranes of various cells, including brain cells. For reasons not yet understood, a prion protein will become deformed and then go on to deform *more* prion proteins when it comes into contact with them. The result is a chain reaction in which ever more proteins become deformed. Collections of these deformed proteins destroy brain cells; the ultimate result is a brain with sponge-like holes in it—and death for the victim. Rogue prion proteins, referred to as just-plain prions (PREE-ons), are now thought to be the infectious agent in Mad Cow disease. Other causes, such as slow-acting viruses, are still under consideration, but the evidence is increasingly pointing to prions.

A number of rare human illnesses are believed to be caused by prions, among them an affliction, identified in the 1920s, called Creutzfeldt-Jakob disease. However, even in its most common form, Creutzfeldt-Jakob disease is very rare, striking only about 1 in a million adults, for reasons that are still unknown. The wave of prion-linked human illness in Western Europe is something new, and it is linked to farm animals.

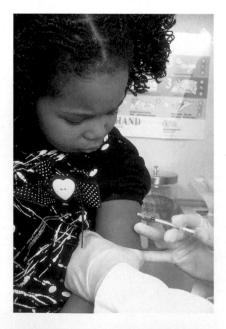

Figure 20.4
Protection for the Future
A four-year-old receives a vaccination.

itself to the person being inoculated (**see Figure 20.4**). How? By, for example, injecting *killed* virus particles—whose shape alone elicits the immune response—or *attenuated* viruses, meaning viruses that haven't been killed but are weakened so that they no longer can cause disease. The Salk polio vaccine is an example of killed virus, and the Sabin polio vaccine is an example of attenuated virus.

How Did Viruses Originate?

All the organisms you looked at in Chapter 19 fit into a grand evolutionary framework: Archaea and bacteria were the first to appear on Earth, followed by the protists, who gave rise to animals, plants, and fungi. So where do the viruses fit in? No one knows. There

In 1984, a single cow in England, began behaving erratically and then dropped dead of a disease that affected its nervous system. This turned out to be the start of an epidemic of prion-caused disease in British cattle—the actual Mad Cow disease, officially known as bovine spongiform encephalopathy (BSE). Scientists now believe that Mad Cow disease spread among more than 170,000 cattle in Britain because *prions* were transmitted through a kind of recycling: Cattle may have originally been infected by eating the remains of sheep who had developed a prion-caused disease called scrapie. More cattle were infected when they were fed the remains of ground-up *cattle*.

Then came the jump to human beings. People who ate prion-contaminated beef products likewise ingested prions, and these prions seem to be active in human beings. How is this possible? Mad Cow disease has been transformed into a human variant, dubbed New Variant Creutzfeldt-Jakob disease or nvCJD. Thus far, about 80 deaths have been reported across Europe from nvCJD (which most people continue to call Mad Cow disease).

Distressingly little is known about nvCJD, and what is known is not reassuring. The prions that appear to cause it cannot be eliminated from beef products by any known means: neither heat, nor radiation, nor common chemical treatments get rid of them. The extent to which the disease has spread in Europe is unknown; it is not clear how many Europeans have eaten prion-infected meat, nor is it clear whether doing so automatically brings on nvCJD. Not one instance of nvCJD has been reported in the United States; indeed, not one instance of Mad Cow disease has been reported among American cattle. U.S. health officials are taking no chances, however. No one knows whether deformed prions can be spread through blood transfusions, but the American Red Cross wants to guard against even this possibility by prohibiting blood donations from Western Europeans. Meanwhile, American feedlots are trying to make absolutely certain that cows are not being fed the ground-up remains of other animals. Only time will tell whether such measures will be enough to keep nvCJD from coming to America.

Figure 1
Mad Cow Quarantine
Health workers in Italy walk inside a barrier erected around some farmland. An animal on the farm was suspected of having Mad Cow disease.

are several hypotheses regarding the evolution of viruses, but all of them are weakened by a critical hole in the evidence, which is that there are no viral fossils. Viruses have no "hard parts" to make impressions, which severely hampers our ability to know when any virus first appeared. Another technique of evolutionary biologists, creating family trees by looking for similarities in DNA sequences among different organisms, turns out not to work well with viruses because their DNA (or RNA) often mutates so rapidly that no telltale patterns emerge.

These problems notwithstanding, various lines of evidence support a viral evolution hypothesis that probably is the most widely accepted among scientists today. It is that viruses represent "renegade" DNA or RNA. We know that genomes often contain stretches of DNA that can move from place to place within the genome. At a minimum these "mobile elements" always encode an enzyme that can snip them out of one location in a genome and patch them into another. Imagine, then, such an element breaking away from its genome entirely, and taking with it (or acquiring) the genes necessary to fabricate a protein coat. In this way, viruses may have come into being. (For a look at another agent that is infectious, but not living, see "Not Alive, but Deadly: Prions and "Mad Cow" Disease" above.)

Now let's move on to the first domain of life.

20.2 Domain Bacteria: Masters of Every Environment

Domain Bacteria is made up entirely of the microbes known as *bacteria*—single-celled prokaryotic organisms that, like viruses, are "germs" as people commonly understand that term. It is true that bacteria are responsible for lots of diseases: tuberculosis, syphilis, gonorrhea, cholera, tetanus, and leprosy, not to mention all manner of food poisonings and blood-borne infections. The bacterially caused bubonic plague, still with us today, wiped out a *third* of Europe's population in just four years in the fourteenth century. Yet in the larger picture, only a very small proportion of bacteria cause human disease. Most either don't affect us at all or are actually beneficial—not only to us, but to all of the living world. Indeed, it is hard to imagine how life on Earth could continue without bacteria.

Intimate Strangers: Humans and Bacteria

To look first at the human relationship with bacteria, from the time we travel down our mother's birth canal, we are bathed in these organisms, both outside and inside our bodies. Outside, we are covered in bacteria from head to toe, with heavy bacterial concentrations in the armpits and scalp. Inside, bacteria exist in the entire digestive tract, from mouth to anus. In the mouth alone, there are a presumed 600 *species* of bacteria, with the number of individual bacteria in a person's mouth probably exceeding the number of people who have ever lived. Further down, about half the contents of our colon are bacteria as are perhaps a quarter of our feces by weight. It may be discomforting to think about having been colonized in this way, but consider that some of the intestinal bacteria are producing sugars and vitamins that our bodies can use. Moreover, by making a living in various digestive environments, these bacteria are effectively occupying almost all the available bacterial niches, thus leaving few places for harmful bacteria to get started.

Bacterial Roles in Nature

In the larger living world, bacteria fill lots of roles. The element nitrogen is a component of all living things, being found in all the primary amino acids that make up proteins. But animals get their nitrogen only by eating plants, or by eating *animals* that have eaten plants. And how do plants get nitrogen? To a great extent from bacteria, which alone among organisms can take the *gas* nitrogen from the atmosphere, metabolize it, and produce nitrogen-containing compounds that can be taken up by plants.

Beyond this, it is not by magic that dead trees, the bones of dead animals, or for that matter orange peels, are broken down and then recycled back into the Earth. The Earth's refuse is broken down by *organisms*, and some of the most important of these "decomposers" are

Figure 20.5
Bacterial Purification
Waste water coming to this treatment plant goes first to the airless storage tanks in the background. There, a type of bacteria that thrive in an oxygen-less environment break the waste into gases and a humus-like material. The product then is transported to the ponds in the foreground, where another type of bacteria break down the solid material in it, yielding carbon dioxide and solids. The bacteria and solids are then filtered out of the water before it is discharged into a river or the sea.

bacteria. A spectacular meeting of waste and bacteria can be found in the sewage treatment plants of most cities, where a kind of production line of different kinds of bacteria is employed in treating solid and liquid waste (**see Figure 20.5**).

These are but a few examples of how bacteria are part of the fabric of life. We occasionally can't live with these tiny organisms, but we certainly can't live without them, which is a good thing because we have no choice. Bacteria exist in great numbers almost anywhere there is food and water, and they can withstand all kinds of hostile assaults on their environments. As long as life exists on Earth, bacteria will be among the living.

Common Features of Bacteria

If you look at **Figure 20.6**, you can see the three most common shapes that bacteria take—spherical, rod-shaped, and spiral-shaped. You can also see a size comparison of a bacterium called *E. coli*, the T4 virus you just looked at, and a protist you'll be looking at soon. Spherical bacteria are known as **cocci** (singular, *coccus*), which you may have heard of in connection with the *streptococcus* bacterium that causes strep throat. Rod-shaped bacteria are called **bacilli** (singular, *bacillus*). Spiral-shaped bacteria include the **spirochetes**, an example of which is the *Treponema pallidum* bacterium, which causes syphilis (see Figure 20.6).

Nobody knows how many species of bacteria come in these shapes and others. Using a DNA base-pairing technique, two Norwegian researchers looked at a single gram (about 1/28 ounce) of forest soil from their country and estimated that there were between 4,000 and 5,000 bacterial species living in it, a number that doubled when a gram of shallow marine sediment was tested. Given such abundance, it is useless to even speculate about how many bacterial species there may be in total.

So what characteristics do bacteria share? Here's a list.

- *No cell nucleus.* All bacteria are prokaryotes, meaning organisms whose cells lack a nucleus. As noted in Chapter 4, the nucleus is an internal compartment of a cell, bound by a membrane, that contains the cell's primary complement of DNA. Organisms whose cells have a nucleus are called *eukaryotes*, a category that includes all the organisms on Earth *except* bacteria and archaea.

- *No membrane-bound organelles and no cytoskeleton.* A nucleus is an example of an **organelle**: a highly organized structure within a cell that carries out specific cellular functions. You may remember from Chapter 4 that, in addition to a nucleus, eukaryotic cells have such organelles as mitochondria, which act as energy transfer centers, and lysosomes, which act as disposal and recycling centers. You may also recall that all these organelles are membrane-bound—their exterior surface is formed by a pliable

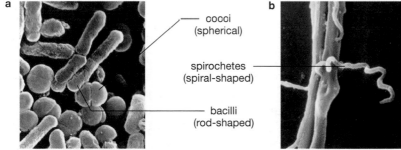

a

cocci
(spherical)

spirochetes
(spiral-shaped)

bacilli
(rod-shaped)

b

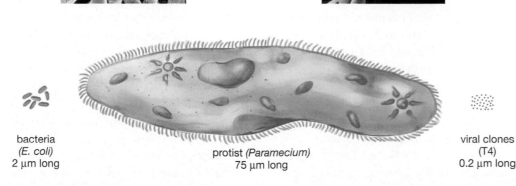

c

bacteria
(E. coli)
2 μm long

protist *(Paramecium)*
75 μm long

viral clones
(T4)
0.2 μm long

Figure 20.6
Types of Bacteria

a Both spherical cocci bacteria and rod-shaped bacilli bacteria can be seen in this micrograph, which shows *Streptococcus* bacteria, artificially colored orange, and *E. coli* bacteria, colored pink. These bacteria can cause infections, but they are also inhabitants of healthy human digestive and respiratory systems.

b Several of the spirochete bacterium *Treponema pallidum*, which causes syphilis, shown attaching to cell membranes in the testes, 22 hours after infection.

c Size comparison of *E. coli* bacteria on the left, with T4 virus clones on the right and a *Paramecium* protist in the middle.

membrane. By contrast, bacteria have only a single kind of organelle, the ribosome, and it is not membrane-bound. Bacterial cells also lack the sets of protein strands, collectively known as the cytoskeleton, that form the internal scaffolding of eukaryotic cells.

- *Single chromosome.* The DNA of bacteria exists in a single, circular chromosome, which makes bacteria **haploid** organisms: organisms that have but a single set of chromsomes By contrast, the chromosomes of eukaryotes generally exist in *pairs* for at least part of the eukaryote's life cycle, making eukaryotes **diploid** organisms during this time.
- *Asexual reproduction.* Bacteria reproduce by a simple cell-splitting or **binary fission**, with each pair of "daughter" cells being an exact replica of the parental cell.

Bacterial Organization and Biofilms

Bacteria usually are thought of as single-celled life-forms, and in one sense this is true. Each bacterial cell is a self-contained living thing, capable of carrying out most of the basic activities that, say, an animal does (acquiring food, getting rid of waste, and so forth). In recent years, however, biologists have been discovering that it is relatively rare for bacteria to actually live as isolated organisms. Rather, most bacteria come together in groupings, some of which can organize themselves in sophisticated ways. One form that such groupings take is that of a *biofilm*, meaning an organized collection of cells that adheres to a solid surface. Initially, the cells in a bacterial biofilm form a thin layer, adhering to their surface material by means of secreting a gummy carbohydrate.

Bacterial biofilms exist in a great many environments, forming on surfaces as diverse as plant roots and water pipes. It is their role in human disease, however, that has made them the subject of intense research activity in recent years. By one estimate, 65 percent of human bacterial infections involve biofilms. The bacterial "plaques" that form on human teeth generally are biofilms, and many ear and lung infections seem to involve biofilms. The "Legionnaire's Disease" that killed 29 members of the American Legion at a Philadelphia hotel in 1976 actually was the result of a biofilm that coated the insides of the hotel's air-conditioning system. (Pieces of the biofilm broke off from the system, thus becoming airborne.)

For our purposes, bacterial biofilms are important in that they provide a great example of organization in bacterial groupings. Biofilms get going when individual, free-swimming bacteria come together on a solid surface. If this process simply continued—more and more cells crowding into the same area—cells at the interior of the biofilm would start dying out. Nutrients couldn't get to them, but the toxic waste products of the other cells would be all around them. Biofilms, however, are capable of using the sticky carbohydrate they secrete to organize themselves into microscopic pillars that have water channels running between them. These channels bring nutrients to the biofilm cells and carry wastes away from them. This kind of organization is only possible if some sophisticated cell-signaling is going on among the cells that make up the biofilm. With all this in mind, it's easy to see that, though bacteria may be single-celled life-forms, they are capable of working together in cooperative ways.

Fighting Disease-Causing Bacteria with Antibiotics

In general, the medicines used to kill disease-causing bacteria are **antibiotics** meaning a substance produced by one organism that is toxic to another. There are antiviral and antifungal medicines that can properly be called antibiotics, but in the main when we speak of antibiotics, we are talking about medicines that kill bacteria but that do not work against viruses. Two things are worth observing about antibiotics. First, scientists do not design them from scratch; instead, they are derived from compounds produced by living things. This makes sense because organisms such as fungi have been involved in a war of survival with bacteria for hundreds of millions of years, and they have accordingly evolved antibacterial defenses. The best-known antibiotic in the world, penicillin, is the product of a mold, which is a type of fungus. Second, any antibiotic that humans use must exploit the *differences* between human and bacterial cells; otherwise the antibiotic would kill human cells along with the bacterial cells. With respect to penicillin, the critical difference is that bacterial cells have a cell wall, while human cells do not. Penicillin blocks a bacterial enzyme that helps construct this wall; the result is that, after penicillin has been administered, bacterial cells burst as they are being constructed, during cell division. One of the reasons biofilms are such a fierce medical enemy is that the sticky carbohydrate they secrete cannot be penetrated by some antibiotics. In addition, the cells in many biofilms can survive without

Modes of Nutrition: How Organisms Get What They Need to Survive

All living things need energy and nutrients. The question is, how do they get them? What is their "nutritional mode," as biologists would put it?

The most fundamental distinction in nutritional mode separates groups known as autotrophs from those called heterotrophs. **Autotroph** means "self-feeding." Autotrophs can manufacture their own food, defined as some form of organic (carbon-containing) molecule that can be broken down to yield energy. All organisms that are not autotrophs are **heterotrophs**, meaning "other feeders;" they cannot manufacture their own food, but must get it from elsewhere, as animals do.

The idea of heterotrophs is not strange to us—it's what we are, after all—but the notion of autotrophs may be a little more exotic. Those who have gone through Chapter 8 on photosynthesis will recall, however, that almost all *plants* are autotrophs in that they manufacture their own food, initially a sugar. To do this, they need a carbon source, because carbon is the "backbone" of the organic molecules we know as food, and they need an *energy* source that can drive the complex process of photosynthesis. The carbon source for plants turns out to be carbon dioxide (CO_2), obtained from the atmosphere, while the energy source is the rays of the Sun (the "photo" of photosynthesis). Thus do we see exemplified a *two-part* requirement for nutrition that holds for all organisms: on the one hand, a source of carbon and on the other, an energy source.

Within this framework there are four nutritional modes, which you can review in **Table 1**. The two most important of these modes are **photoautotrophy**, which is the nutritional mode of plants, and **chemoheterotrophy**, in which organic materials (better known as food) act as both carbon supplier *and* energy source. This is another way of saying that the cereal you ate this morning supplied you, as a chemoheterotroph, with both carbon and high-energy electrons.

There is also **chemoautotrophy**, in which organisms—some bacteria and archaea—get their carbon from carbon dioxide, like plants do, but power the production of food from this carbon by oxidizing (pulling electrons from) such inorganic materials as hydrogen sulfide and ammonia. A small number of bacteria and archaea practice **photoheterotrophy**, in which the Sun supplies the energy but the carbon comes from surrounding organic material.

Nutritional Mode	Carbon Source	Energy Source	Practiced by
Autotrophy			
Photoautotrophy	Carbon dioxide (CO_2)	The Sun's rays	Almost all plants, some bacteria, and many protists
Chemoautotrophy	Carbon dioxide	Inorganic compounds such as hydrogen sulfide and ammonia	Some bacteria and archaea
Heterotrophy			
Photoheterotrophy	Organic material	The Sun's rays	A few bacteria and archaea
Chemoheterotrophy	Organic material	Organic material	Almost all animals, all fungi, most bacteria, many protists, and a few plants

dividing; this makes them immune to an attack from an antibiotic such as penicillin, which does its work *while* cells are dividing.

Modes of Nutrition: Bacteria Do It All

Biofilms nothwithstanding, bacteria are primitive organisms in many ways; they multiply by simple division, they have only a single type of organelle (the ribosome), and they exist as unspecialized cells. Yet if the *diversity* of a grouping of organisms is thought of as an advanced characteristic, then bacteria are very advanced indeed. It is in the means by which they get their nutrition that the breadth of bacteria is most apparent; different kinds of bacteria utilize all the modes of nutrition known to living things, whereas humans utilize only one. You can read about these methods in "Modes of Nutrition," above.

Now let's proceed to the second of life's three domains, Archaea.

20.3 Domain Archaea: From Marginal Player to Center Stage

As little as 10 years ago, any biology text-book you picked up would likely have referred to the archaea as archae*bacteria*, and most would have mentioned the archaea as a kind of marginal life-form, existing in only a few extreme habitats, such as hot springs or salty lakes. Today, the archaea are recognized by many experts as constituting their own domain in life—standing alongside Domain Bacteria and Domain Eukarya—and they have belatedly been recognized as existing in abundance in lots of environments. The importance of the archaea stretches beyond this, however, in that, of all the organisms living today, archaea may have been the first to exist. Whether this means that archaea are ancestral to all other existing life-forms remains to be seen—many researchers have their doubts—but it is certain that archaea lie near the trunk of life's family tree.

Why did it take so long to recognize archaea as ancient, widespread, and in a class by themselves? With respect to their uniqueness, one problem was that archaea are superficially similar to bacteria: They are microscopic, basically single-celled, and their cells lack a nucleus (making them prokaryotes, along with the bacteria). In addition, it is hard to learn much about archaea because it is so difficult to "culture" them, meaning to grow populations of them in a laboratory. The lab environment is so different from the extreme environments inhabited by many archaea that cultures of them are quickly taken over by common bacteria.

Beyond these things, however, there was a long-term resistance among scientists to seeing archaea as anything but an odd type of bacteria. Indeed, a single individual, Carl Woese of the University of Illinois, had to wage a lonely battle lasting years before his claims about the uniqueness of the archaea were generally accepted (**see Figure 20.7**). In 1996 there came a triumphant confirmation of his ideas about the archaea with the sequencing of the entire genome of an archaean species, *Methanococcus jannaschii*. This work revealed that an amazing 56 percent of *M. jannaschii*'s genes were completely unknown to science—they were unlike anything seen in either bacteria or in eukaryotes. As for the remaining 44 percent of the genome, some *M. jannaschii* genes worked like those of eukaryotes, while others worked like those of bacteria.

All this speaks to the uniqueness and evolutionary place of the archaea, but what about the widespread distribution of them? In the 1980s, scientists developed a technique that allowed them to do a kind of census taking of archaea out in nature. This method revealed that archaeans live in extreme environments, such as the Yellowstone hot springs, in more numbers than had been thought. But the real surprise came when more expansive environments were tested; archaea turn out to account for almost a third of the microscopic organisms living in the surface waters off Antarctica, for example, and huge populations of them are believed to inhabit deep-ocean waters.

Prospecting for "Extremophiles"

The diverse species of archaea that do inhabit extreme environments have in recent years become the target of a kind of new-age prospecting, with "miners" being scientists from chemical and biotechnology firms. The archaea they are looking for are **extremophiles**—

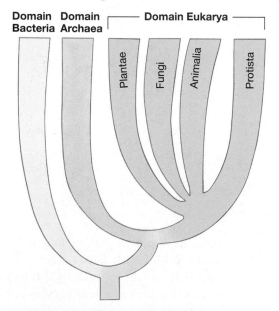

Figure 20.7
Carl Woese's Tree of Life
Woese has argued that the archaea are sufficiently distinct from both bacteria and eukaryotes to warrant the status of a third branch of life, the domain Archaea.

archaea or bacteria that flourish in conditions that would kill most organisms, such as high heat, high pressure, high salt, or extreme pH (**see Figure 20.8**). One of the most extreme of the extremophiles was reported on in 2000, when researchers revealed the existence of an archaean they found thriving in the acidic wastes of an abandoned copper mine. Dubbed *Ferroplasma acidarmanus*, this hardy microbe can live in liquid that has a pH of zero—a habitat more acidic than battery acid.

To live in tough environments, extremophiles ("extreme-lovers") must produce *enzymes* that function in these environments. And, through modern biotechnology processes, enzymes can be isolated from an organism and then turned out in quantity in factories. These can then be put to use in *manmade* extreme environments, such as the inside of a washing machine or the confines of a laboratory container. Lest you think these are fanciful examples, note that one U.S. company has already introduced a clothing detergent containing a cleaning additive that is simply an enzyme from a heat-loving extremophile.

20.4 Domain Eukarya: Protists, Plants, Fungi, and Animals

You have thus far looked at two of life's domains, Bacteria and Archaea, which leaves one more to review—Eukarya, meaning the eukaryotes of the world: protists, plants, fungi, and animals. As you've seen, all eukaryotes have cells with nuclei, and almost all have several other organelles. Beyond this, however, there is very little that is common to every eukaryote. Unlike bacteria or archaea, most eukaryotes are capable of sexual reproduction, but asexual reproduction is carried on to some extent in every eukaryan kingdom.

Domain Eukarya is special because of its incredible diversity of form and capabilities. Bacteria and archaea may be diverse at the molecular level, but as whole organisms they are all basically single-celled microscopic individuals. The eukaryotic domain has some single-celled microbes in it too, but it also includes birds and orchids and whales and mushrooms.

Figure 20.8
Looking for Life in Extreme Environments
These researchers from the Laboratory of the Molecular Biology of Extremophiles, located in France, are on a research outing in New Zealand prospecting for microscopic extremophiles. Their research group studies archaea and bacteria that live at temperatures close to 100° Celsius or 212° Fahrenheit—the temperature at which water boils.

As already noted, the eukaryotic domain is divided into four individual kingdoms: Protista, Plantae, Fungi, and Animalia. If you look at **Table 20.1**, you can see some of the major features of each kingdom. For the rest of the chapter, you'll be looking at three of these kingdoms individually—protists, fungi, and plants.

20.5 Kingdom Protista: An Undefinable Collection

As you progress through the eukaryotic kingdoms, you will see that we have fairly clear definitions of what a plant or an animal or a fungus is. Not so with the protists, which are a kind of grab-bag category of organisms that must be defined in terms of what they are *not*: They are eukaryotic organisms that do not have all the defining characteristics of plants or animals or fungi. About 100,000 species of protists are known to exist, most of them single-celled, and all of them living in environments that are at least moist, if not fully aquatic—oceans, lakes, damp forest floors. Many protists are parasites, as with the *Giardia* intestinal parasite much feared among campers.

Some protists perform photosynthesis (like plants), while others ingest food in order to live (like animals). Still others are capable of moving back and forth between the two nutritional modes, depending on the availability of sunlight and nutrients. Meanwhile, other protists reproduce more like fungi. In short, there are plant-like, animal-like, and fungus-like protists, which squares with what you learned in Chapter 19 about evolution of the eukaryotes: Plants evolved from one kind of protist (green algae), animals from another (probably a grouping called *choanoflagellates*), and fungi from still another (though scientists are not sure which). You can see two kinds of protists in **Figure 20.9**.

In an example of how protists can bend our ideas about categories of living things, **Figure 20.10** shows *Dictyostelium discoideum*, a protist that is part of a grouping called the cellular slime molds. *D. discoideum* has a fascinating multiple personality. At one point in its life cycle, this creature exists in the form of individual microscopic cells that live in soil, mostly by ingesting bacteria. When their food runs low, however, these cells begin aggregating, producing a migrating "slug" that has front and back ends and that moves toward light.

Table 20.1
Features of Kingdoms in the Eukaryotic Domain

Kingdom	Reproduction Method	Means of Nutrition	Multicellular?
Protista	Usually asexual, but many protists have both sexual and asexual reproductive phases	Both photoautotrophy (photosynthesis) and chemoheterotrophy (by ingestion of organic material) are common	Usually not, though both colonial multicellularity and true multicellularity exist
Fungi	Spores that are produced sexually or asexually	Always chemoheterotrophy by means of external release of enzymes that break down organic material, followed by absorption of this material	Usually, though unicellular forms also exist
Plantae	Usually sexual, though vegetative reproduction (for example, by cuttings) allows asexual reproduction	Almost always photoautotrophy	Always
Animalia	Generally sexual, but can be asexual	Always chemoheterotrophy, almost always by ingestion	Always

Eventually this aggregation forms a stalk-like reproductive structure that releases spores that will develop into new individual cells.

Means of Mobility in Protists

Many single-celled protists get to sunlight or nutrients by means of a flagellum (plural *flagella*), which is a long, whip-like extension of the cell that propels the cell by its movement. Other protists have cillia—hair-like cellular extensions that exist in profusion on the cell surface. Beating in rhythm, cilia move the cell toward food or away from noxious substances. Still other protists, the amoebas, move by extending a **pseudopodium** or "false foot," which is simply a portion of the cell cytoplasm that is cast out and then anchored, after which the rest of the cytoplasm streams into the extension. *Dictyostelium discoideum* is one of these amoeboid protists. More fearsome is the amoeba *Entamoeba histolytica*, which can contaminate drinking water, causing amoebic dysentery.

True Multicellularity: A Division of Labor

Some Protists Are Very Large: The Giant Kelp

Lest you think of all protists as microscopic, however, consider the giant kelp of California's coastal waters. These protists, which are one type of brown algae, are capable of *growing* 60 meters a year—almost 200 feet. You also saw at the start of the chapter that giant kelp have been found that can lay claim to being the world's longest organisms.

For an organism to grow to the size of kelp, obviously it must be composed of many cells, rather than just one. Bacterial cells are capable of coming together in several kinds of aggregations; yet it is only the eukaryotes, such as protist algae, that have true *multicellularity*. This means a form of organization among living things in which a single living thing is composed of cells that have different capabilities, resulting in a *division of labor* among cells. Your pancreas cells perform functions quite different from those of your brain cells. By contrast, none of the bacteria in the biofilms mentioned earlier is capable of performing functions different from any *other* bacteria in the structure. Such bacterial aggregation is merely colonial: A group of unspecialized cells is functioning as a single unit.

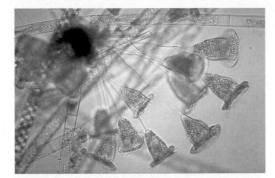

a

b

Figure 20.9
Protists Small and Large

a A microscopic protist called *Vorticella*.

b Towering examples of giant kelp off the California coast, with fish swimming nearby.

Figure 20.10
Multiple Identities

a The protist *Dictyostelium discoideum*, a cellular slime mold, exists for part of its life as a collection of individual cells that consume bacteria. Reacting to a reduced food supply, these cells start aggregating as shown here.

b The aggregation that forms takes on the appearance of a slug and can move across the forest floor.

c This organized collection of cells then transforms itself into a tower-shaped reproductive structure whose cap releases spores that will be dispersed to start life elsewhere as new individual cells. *Dictyostelium* is of great interest to biologists because it provides insight into how cells work together to form a multicellular organism.

a

b

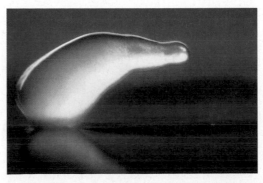

c

The general view among evolutionary biologists is that colonial protists were the bridge between unicellular and true multicellular life.

Single-Celled Algae: The Importance of Phytoplankton

Algae exist in single-celled as well as multicelled forms. The importance of the single-celled variety to life in general is great, because they form a large proportion of Earth's **phytoplankton**, meaning microscopic photosynthesizing organisms that drift in the upper layers of the oceans or freshwater bodies. Phytoplankton perform much of Earth's photosynthesis and occupy a key position in water-based food chains. Growing in great abundance, they are in turn eaten by **zooplankton**, which are the small animals and other heterotrophs that drift or move weakly in the upper layers of bodies of water. Phytoplankton and zooplankton (together called just plain **plankton**) are eaten by larger animals, which are eaten by still larger animals. Blue and humpback whales feed on shrimp-like plankton called *krill*, taking in great mouthfuls of ocean water and filtering out the krill that rush in.

Having looked at one kingdom in the eukaryotic domain, let's now move on to another.

20.6 Kingdom Fungi: Life as a Web of Slender Threads

If mushrooms are considered as an example, fungi may seem at first simply different kinds of plants. After all, they are **sessile**, or fixed in location, just as plants are, and they most often live in the soil. Yet there is a very basic distinction between plants and fungi: Almost all plants are photoautotrophic—they make their own food through photosynthesis—while all fungi are heterotrophic, meaning they must consume existing organic material in order to live.

When you think about what fungi are consuming, the term *nuisance* (or worse) may come to mind. Fungi cause mold, mildew, dry rot, ringworm, vaginal yeast infections, athlete's foot, and diaper rash for starters. The agricultural blights of corn smut and wheat rust are fungal infections. To state the obvious, many fungi are parasites, feeding on living hosts but not killing them, yet other fungi kill their hosts outright. Fungal infections can pose mortal threats to people, such as AIDS patients, whose immune systems have been weakened.

The Important Roles of Fungi in Nature

Their invader role notwithstanding, fungi are indispensable to humans and other living things. The mold that gives us penicillin is a fungus, as is the yeast we use to make bread rise and beer ferment. More generally, fungi join bacteria as the major "decomposers" of the living world, breaking down

Figure 20.11
Types of Fungi
Fungi are classified in accordance with the ways they carry out sexual reproduction. Some fungi, however, do not have a sexual phase in their reproductive cycle, or at least do not have one that has been observed. These fungi are classified as "imperfect" fungi. Pictured here are three of the four major types of fungi, including the imperfect fungi. The features and life-cycle of the fourth major grouping, the club fungi (which includes common mushrooms), are reviewed in Figures 20.12 and 20.14.

a Imperfect fungi. Shown greatly magnified (× 280) are reproductive branches from an imperfect fungus, *Aspergillus glaucus*, seen here feeding on some molding bread. The ball-like structures at the end of the branches eventually will rupture and release spores that will form new colonies of the fungus. In the *Aspergillus* genus, we can see both the benefits and threat of fungi. One species of *Aspergillus* is used to produce soy sauce and another helps ferment the alcoholic drink sake. Conversely, another species of *Aspergillus* produces a toxin that, when allowed to grow on foods, is a powerful cancer-causing agent.

b Sac fungi. Pictured is an edible morel mushroom, growing in Kansas.

c Bread molds or conjugation fungi. The picture shows a magnified view of the most common bread mold, *Rhizopus stolonifer*, which produces the familiar grayish patches on exposed bread.

a

b

c

organic material such as garbage or fallen logs and turning it into inorganic compounds that are recycled into the soil. Without this activity Earth would long ago have turned into a massive garbage heap of dead organic material. Beyond this, fungi are involved in a critical cooperation with plants. Some 80 percent of seed plants have fungi associated with their roots, something you'll learn more about shortly. In sum, we could say of fungi what we did of the bacteria: We could do without some of them, but we could not live without others. Now let's look at some of the qualities common to most members of this kingdom, which contains about 100,000 named species (**see Figure 20.11**).

- *Fungi largely consist of slender, barely visible filaments called* **hyphae**. Collectively these tube-like hyphae make up a branching web, called a **mycelium**. If you look at **Figure 20.12**, you can see the nature of this structure. You may have thought of a mushroom as the main part of a fungus, but in reality it is merely a reproductive structure that sprouts up from the larger web below.

- *The cells that make up hyphae tend to be very porous relative to one another, a quality that allows for rapid fungal growth.* Individual hyphae cells usually are separated from one another by structural dividers called *septa*, but these septa have microscopic openings that allow for a fairly free flow of cytoplasm between one fungal cell and the next. When combined with the mycelium structure, this quality brings about one of the most notable characteristics of fungi: They *grow toward* their food supply, and they can do so very rapidly. Because of the porous quality of the hyphae cells, a fungus can move a great many cellular resources, such as proteins, right to the point of growth. Did it ever seem to you that a mushroom sprouted in your yard overnight? It probably did.

- *Fungi get their nutrition by dissolving their food externally and then absorbing it.* A fungus spore lands on a log, begins to grow, and then starts releasing enzymes that break down the wood. This digestion takes place *outside* the fungus, after which the fungus absorbs the resulting nutrients. Note how this differs from animals which ingest their food by some means—engulfing it or biting off pieces of it—but then digest it *internally*. Some fungi attack prey almost as an animal would, their

hyphae taking the form of lasso-like loops that can be constricted around worms that pass through them (**see Figure 20.13**). Even here, however, the nutritional mode is to invade the trapped animal with special hyphae, discharge the dissolving enzymes, and then absorb the digested nutrients.

- *Fungi are almost always multicellular.* The exception to this is the yeasts, which are unicellular and thus do not form mycelia.

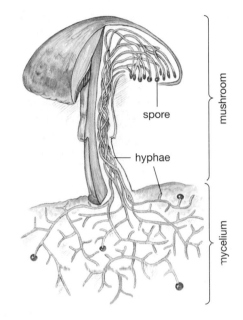

Figure 20.12
Structure of a Fungus
Fungi are composed of tiny slender tubes called *hyphae*. The hyphae form an elaborate network below ground, called a *mycelium*. The same hyphae also form a reproductive structure above ground, called a mushroom.

Figure 20.13
Trapped by a Fungus
A tiny worm, called a nematode, is trapped within the hyphae of a lasso fungus, and will soon be consumed by it.

Figure 20.14
Life Cycle of a Fungus

1. Spore formation Two haploid mushroom spores from different organisms are blown through the air and land on ground suitable for growth.

2. Germination The spores begin to germinate, which is to say, to grow haploid mycelia underground in their respective locations through simple cell division. These two fungi are of the of the same species, but they happen to have cells that are of complementary "mating types," meaning they are capable of fusing. (The two types are signified by the blue and red colors of their cell nuclei.)

3. Fusion Upon coming in contact with one another, hyphae of their respective mycelia then fuse, but note that it is only the cytoplasm of their cells that fuses, while the nuclei remain separate. Thus begins the dikaryotic phase of the life cycle—single cells with two nuclei. All the daughter cells of this original fusion event have this structure. Also note that the two organisms have now become one.

4. Mushroom formation The mushroom is a reproductive structure, initially made up entirely of dikaryotic cells.

5. Fusion of nuclei (fertilization) The accordion-like gills on the underside contain cells whose nuclei will fuse, producing cells that are not dikaryotic, but diploid.

6. Meiosis The diploid phase is short-lived, however; meiosis occurs quickly, yielding haploid cells that sprout out from the gills; these are spores that will drop off the bottom of the mushroom in tremendous numbers to be carried away by the wind and land on the ground, there to germinate and start the cycle again.

A Phase of Life Unique to Fungi: Dikaryotic Cells

Nearly all the cells in the human body, except for sperm and egg cells, are "diploid," meaning that they have *paired* sets of chromosomes within them. Conversely, many protists live all or most of their lives in a haploid condition—they have but a *single* set of chromosomes per cell. This diploid/haploid dichotomy runs throughout nature, and it exists in fungi as well. Fungi are "rule breakers" though, because in addition to having haploid and diploid phases in their life cycles, most of them go through a phase that lies *in between* these two states: a **dikaryotic** ("two-nuclei") phase in which each cell of a growing mycelium has *two* haploid nuclei within. Think of it this way. In human reproduction, when sperm and egg come together, the fusion of their cytoplasms is followed immediately by the fusion of their *nuclei*, which brings together the two sets of haploid chromosomes. In the mushroom, conversely, there is a pause between cytoplasm fusion and chromosome fusion—the dikaryotic phase, in which single cells have a common cytoplasm but *two* nuclei. This phase may be long-lived; it can go on for years, in fact, and ends only when the most visible manifestation of the mycelium, the mushroom, sprouts above ground.

If you look at **Figure 20.14**, you can see how this comes about. This is the life cycle of a mushroom, which is representative of one of four major types or "phyla" of fungi. If you looked at all four types, you'd see different reproductive structures and life cycles, but the mushroom cycle will give you a general idea of how fungi reproduce.

The four types of fungi are bread molds, the imperfect fungi, sac fungi, and club fungi. The mushrooms in Figure 20.14 are an example of club fungi, as are the rusts and smuts that cause crop damage. The best known of the bread molds is black bread mold, *Rhizopus stolonifer*, which grows on various kinds of food, including damp bread. The imperfect

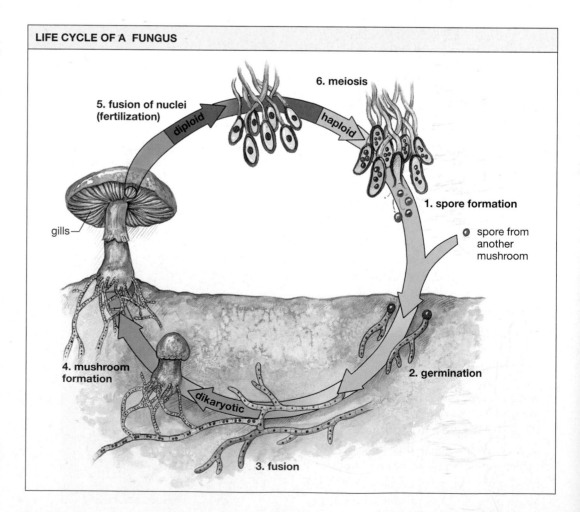

LIFE CYCLE OF A FUNGUS

5. fusion of nuclei (fertilization)

diploid

6. meiosis

haploid

1. spore formation

spore from another mushroom

2. germination

gills

4. mushroom formation

dikaryotic

3. fusion

fungi include some of the best-known fungi pests: *Trichophyton*, which causes athlete's foot, and *Candida albicans*, which causes yeast infections. Sac fungi include the highly prized morel and truffle mushrooms (see Figure 20.11). Fungi are put into one or the other of these categories on the basis of how they reproduce.

Fungal Associations: Lichens and Mycorrhizae

Lichens

Everyone has seen the thin, sometimes colorful coverings called *lichens* that seemingly can grow almost anywhere—on rocks as well as on trees; in Antarctica as well as in a lush forest (**see Figure 20.15**). As it turns out, lichens are not a single organism. They are always two organisms living as one: a fungus and then either an alga or a photosynthesizing bacterium that is nestled within the fungus. If you look at **Figure 20.16**, you can see that this association is structured as a kind of sandwich: An upper layer of densely packed fungal hyphae on top; then a zone of less dense hyphae that includes a layer of algal or bacterial cells; then another layer of densely packed hyphae on the bottom that sprout extensions down into the material the lichen is growing on (such as a rock). The example

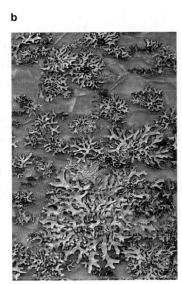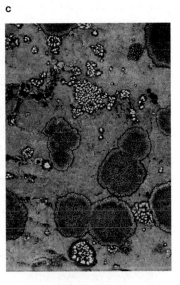

a   b   c

Figure 20.15
Flourishing in Different Environments
Lichens growing on **(a)** a boulder in the State of Washington's Olympic National Park, **(b)** a birch tree in Alaska's Kenai Peninsula, and **(c)** a rock in Alaska's Arctic National Wildlife Refuge.

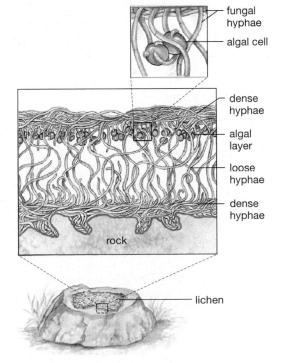

fungal hyphae
algal cell
dense hyphae
algal layer
loose hyphae
dense hyphae
rock
lichen

Figure 20.16
The Structure of a Lichen
A lichen is not a single organism. It is composed of a fungus and an alga (or sometimes a photosynthesizing bacterium) living in a mutually beneficial arrangement. The fungal hyphae form a dense layer on the top and the bottom and a loose layer in the middle, within which the alga is nestled. The fungus may provide a moist, protective environment for the alga, and the alga, which performs photosynthesis, provides food for the fungus.

here is an association between a fungus and an alga, which is the case in about 90 percent of lichens. In such a relationship, the fungal hyphae either wrap tightly around the algal cells or actually extend into them.

A lichen is mostly a fungus, then, but it is a fungus that could not grow without its algal partner. Why? Because the alga carries out photosynthesis and then proceeds to supply the fungus with some of the nutrients photosynthesis provides. It is generally assumed that what fungi do for algae in this relationship is keep them from drying out. One school of thought holds, however, that the fungi are doing nothing at all for the algae, in which case the fungi should be considered an algal parasite, rather than a partner.

Mycorrhizae

As noted earlier, about 80 percent of seed plants form an association with fungi. Here, both partners clearly are benefiting in a linkage that involves plant roots and fungal hyphae. What the fungi get from this is food, in the form of carbohydrates that come from the photosynthesizing plant; what the plants get is minerals and water, absorbed by the fungal hyphae. So important is this relationship to

plants that some species of trees, such as pine and oak, cannot grow without fungal partners. Other plants will have stunted growth without a fungal partner.

Such a relationship is technically an "infection" of plant roots by fungal hyphae, in that the hyphae generally grow into the plant roots. In a few instances the hyphae wrap around the root but do not penetrate it. In either case, the root-hyphae associations are known as **mycorrhizae** (**see** Figure **20.17**). The plant benefits from this relationship as the branched network of hyphae spreads out, thereby greatly increasing the volume of soil that the plant draws its nutrition from.

20.7 Kingdom Plantae: The Foundation for Much of Life

In your review of the living world so far, you've seen that human beings have reason to feel ambivalent about bacteria, fungi, and even protists in that there is much to appreciate in these organisms but a good deal to fear as well. When it comes to plants, however, it's hard to find much to fear—but easy to find much to appreciate. For starters, we humans are utterly dependent upon plants for food. Try naming something you eat that isn't a plant or that didn't come from creatures that do eat plants, and you will end up with a very short list. Then there is the atmospheric oxygen that plants produce as a by-product of photosynthesis. As you'll be seeing, these two critical contributions actually are just the start of how plants affect the world we live in.

What Are the Characteristics of Plants?

We all have a good intuitive sense of what plants are, though there are lots of plants that don't adhere to our preconceptions. Plants make their own food, but there are plants that do this while also being carnivorous, such as the Venus flytraps. Then there are a few plants that don't perform photosynthesis at all—for example, a plant commonly known as dodder—that live by parasitizing other plants. Allowing for such exceptions, plants generally share the following characteristics.

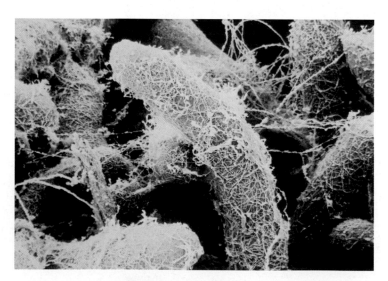

Figure 20.17
Underground Partners
Mycorrhizae are associations between plant roots and fungal hyphae that benefit both the plant and the fungus. Shown is an association between Aspen tree roots and thread-like mushroom hyphae that are wrapped around them. The hyphae help bring water and minerals to the tree, while the tree provides the fungus with food.

- *Plants are fixed in place, photoautotrophic, multicelled, and mostly land-dwelling.* We know that plants are fixed in one spot (sessile), and that they make their own food through photosynthesis (they are photoautotrophic). That they are multicelled follows from the fact that they develop from embryos, which are multicelled by definition. Recall from Chapter 19 that plants are descended from green algae, which live in water. As a group, then, plants made a transition to the land from water, and land is where most of them are found today—though some have made the transition *back* to water, as with water lilies.

- *Plant cells have cell walls, a specialized set of organelles, and contain a high proportion of water.* Both plant and animal cells have a plasma membrane around their periphery, but plant cells have something else as well: A **cell wall**, which surrounds the plasma membrane and is composed in large part of complex compounds called **cellulose** and **lignin** that help give cell walls their strength (**see Figure 20.18**). All cell walls help plants regulate their intake of water—they help them keep what they have, but not take in too much. This regulation is even more important to plants than animals; a typical animal cell may be 70 to 85 percent water, but for plant cells the water proportion is likely to be 90 to 98 percent. Plant cells have a set of organelles, called *plastids*, that animal cells lack; these plastids include the chloroplasts that are the actual sites of photosynthesis. (See Chapter 4 for details of plant-cell structure and Chapter 8 for photosynthesis.)

- *Successive generations of plants go through what is known as an alternation of generations.* This can best be explained by a comparison with human reproduction. Looked at one way, there is a haploid "phase of life" for human beings. The gametes (eggs and sperm) that human beings produce are haploid cells, meaning they have but a single set of chromosomes. When these haploid cells fuse in the moment of conception, what's produced is a *diploid* fertilized egg, meaning an egg with a *paired* set of

chromosomes. This egg gives rise to more diploid cells through cell division, eventually resulting in a whole new human being.

A typical *plant* in its diploid phase will, like human beings, produce a specialized set of haploid reproductive cells. Instead of these being the gametes that humans produce, however, what's produced are **spores**: single reproductive cells that, without fusing with anything, have the ability to grow into a new generation of plant strictly through cell division. There is thus a separate phase of life that in some types of plants is self-sustaining. This generation of plant eventually does produce gametes, which then fuse and create a diploid zygote, which in turn develops into a diploid plant

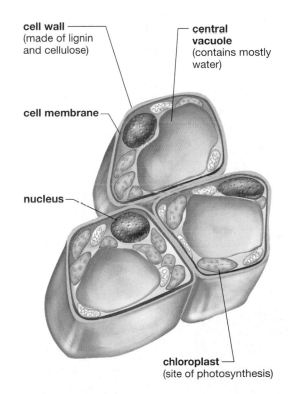

Figure 20.18
Characteristics of Plants
All cells have an outer membrane, but plant cells have a wall external to this membrane. The compounds cellulose and lignin, which help make up the cell wall, impart strength to it. Plant cells have a higher proportion of water in them than do animal cells, with much of this water located in an organelle called a central vacuole. The sites of photosynthesis in plants are the organelles called chloroplasts.

Figure 20.19
The Alternation of Generations in Plants, Compared to the Human Life Cycle

a Humans Almost all cells in human beings are diploid or 2n, meaning they have paired sets of chromosomes in them. The exception to this is human gametes (eggs and sperm), produced through meiosis, which are haploid or 1n, meaning they have a single set of chromosomes. In the moment of conception, a haploid sperm fuses with a haploid egg to produce a diploid zygote that grows into a complete human being through mitosis.

b Plants In plants, conversely, diploid (2n) plants—the multicellular sporophyte fern in the figure—go through meiosis and produce individual haploid (1n) spores that, without fusing with any other cells, develop into a separate generation of the plant. This is the multicellular gametophyte shown. This gametophyte-generation plant then produces its own gametes, which again are eggs and sperm. Sperm from one plant fertilizes an egg from another, and the result is a diploid zygote that develops into the mature sporophyte generation. The alternation between the sporophyte and gametophyte forms is called the alternation of generations.

like the one we started with (**see Figure 20.19**). Because this is a single species that alternates back and forth between two different forms, the plant life cycle manifests what

ALTERNATION OF GENERATIONS

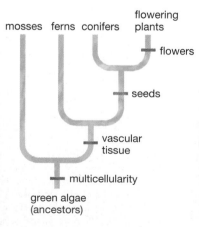

□ haploid (1n)
□ diploid (2n)

is known as the **alternation of generations**. In this life cycle, the generation that produces the spores is known as the **sporophyte generation**, while the generation that produces the gametes is the **gametophyte generation**. The fern used as an example in Figure 20.19 gives you one example of how physically different these generations can be, but the disparities actually can be much greater than this. In a massive organism such as a redwood tree, the tree that is familiar to us is the sporophyte generation. Meanwhile the redwood's gametophyte generation amounts to a microscopic, sperm-bearing pollen grain on the male side, and to a small collection of cells (that includes the egg) on the female side. When the sperm from the pollen grain fuses with the egg, this sets into motion the development of an embryo that can grow into a new redwood tree.

There Are Four Main Categories of Plants: Bryophytes, Seedless Vascular Plants, Gymnosperms, and Angiosperms

Botanists traditionally have partitioned the plant kingdom into a dozen "divisions," but for our purposes, it is convenient to separate the members of the plant kingdom into a mere four types (**see Figure 20.20**). These are the *bryophytes*, which include mosses; the *seedless vascular plants*, which include ferns; the *gymnosperms*, which include coniferous ("cone bearing") trees; and the *angiosperms*,

a Moss

b Ferns

c Conifers

d Flowering plants

Figure 20.20
Four Main Varieties of Plants
All four types are multicellular, but only the most primitive of them, **(a)** the bryophytes (moss in the figure), lack a fluid-transporting vascular system. **(b)** Ferns, representing the seedless vascular plants, have a vascular system but do not have seeds. **(c)** Gymnosperms (conifers in the picture) do utilize seeds. **(d)** Flowering plants (blossoming pear trees in the picture) produce seeds and are responsible for several other plant innovations, among them fruit.

a vast division of flowering plants—by far the most dominant on Earth today—that includes not only flowers such as orchids, but also oak trees, rice, and cactus. We'll look at all four types briefly here. Chapters 22 and 23 provide detailed descriptions of the workings of flowering plants.

Bryophytes: Amphibians of the Plant World

Look in low-lying, wet terrain and you are likely to see a carpet-like covering of moss; in countries such as Ireland, whole fields of peat moss grow and are harvested to be used as fuel for heating and cooking. Mosses are the most familiar example of a primitive type of plant that falls under the informal classification of **bryophyte**, which can be defined as a type of plant lacking a true vascular system. (**See Figure 20.21** for examples of several kinds of bryophytes.) What's a vascular system? A network of tubes within an organism that serves to transport fluids. You can get a good idea of what the bryophytes are like by looking at one type of them, the mosses.

Mosses are representatives of some of the earliest plants that made the transition from water to land, in the evolution of plants from green algae. As such, mosses can be conceptualized as plants that made only a partial break with aquatic living. In making this transition, they had to deal with something their fully aquatic relatives didn't, which is the effect of gravity. Lacking a vascular system, they could not transport water and other substances very *far* against the force of gravity; thus they must lie low, hugging the surface to which they are attached while spreading out horizontally to maximize their exposure to sunshine. Nevertheless, the mosses are hearty competitors in some tough environments, such as the arctic tundra and the cracks in sidewalks. In keeping with their aquatic origins, they have sperm that can get to eggs only by swimming through water. Not much water is necessary; a thin film left over from the morning dew will suffice. But some water must be present. Not surprisingly, bryophytes are most commonly found in moist environments. Bryophytes have no roots at all, but instead use single-celled extensions called *rhizoids* to anchor themselves to their underlying material, which may be soil or rock or wood. You might expect that, like roots, rhizoids would serve to absorb water, but they do this only to a very limited extent. Bryophytes take in water almost entirely through their above-ground exterior surface.

a

b

c

Figure 20.21
Three Kinds of Bryophytes
(a) Mosses, **(b)** liverworts, and **(c)** hornworts. These plants have no vascular tissue and thus tend to be small. The sperm of bryophytes must travel through water to get to eggs. For this reason bryophytes are found most often in moist environments.

Seedless Vascular Plants: Ferns and Their Relatives

What plants are the vascular plants? The answer is simple: all plants except the bryophytes. To put this another way, only the bryophytes lack the network of fluid-conducting tubes that make up a *vascular system*, meaning a kind of plumbing system for plants. One part of this system distributes water—upward, from root through leaf—and one part distributes the food produced in photosynthesis, along with hormones and other compounds.

Seedless vascular plants means just what it sounds like: Plants that have a vascular system, but that do not produce seeds as part of reproduction. Easily the most familiar representatives of the seedless vascular plants are ferns, with their often beautifully shaped leaves, called fronds. (**See Figure 20.22** for examples of several kinds of seedless vascular plants.) These plants have moved a step further in the direction of separation from an aquatic environment. The vascular system allows the plants to grow *up* as well as *out*—

something you'd expect would develop in organisms that are competing for sunlight—and it allows for roots that extend into the ground, where they serve their absorptive function. Despite this evolutionary innovation, the sperm of the seedless vascular plants, like that of the bryophytes, need to move through water to fertilize eggs.

The First Seed Plants: The Gymnosperms

There are two kinds of seed plants, the gymnosperms and the angiosperms, but putting things this way makes it sound as though these plants are on a kind of equal footing with the bryophytes and seedless vascular plants. In reality, the gymnosperms, which are nonflowering seed-bearing plants, took over from the seedless vascular plants about the time the dinosaurs came to dominance, only to find themselves replaced as the dominant plants when the angiosperms began flourishing, about 80 million years ago. Today there are only about 700 gymnosperm species; but the gymnosperm's presence is considerable, especially in the northern latitudes, where

Figure 20.22
Three Kinds of Seedless Vascular Plants
(a) Fall-colored ferns in New Hampshire, **(b)** horsetails, and **(c)** club mosses. Because these plants have vascular tissue, they are able to grow taller than most bryophytes. Like bryophytes, however, they do not produce seeds and are tied to moist environments.

they exist as vast bands of coniferous trees, including pine, Douglas fir, and white spruce. Conifers such as these provide most of the world's lumber. (**See Figure 20.23** for examples of several kinds of gymnosperms.)

Reproduction through Pollen and Seeds

When the gymnosperms are grouped together with the flowering plants as *seed* plants, it's apparent that they have been very good competitors, compared to the bryophytes and seedless vascular plants. Why should this be so? Part of the answer lies in how the male gametes (the sperm) of the gymnosperms get to the female gametes (the egg): Residing in tiny structures called pollen grains, sperm are carried through the air, rather than being limited to *swimming* to female gametes, as is the case with the seedless plants. Given this, these seed plants can propagate over great distances, which gives them an obvious competitive advantage over the seedless plants.

Another advantage is the seeds themselves, which can be thought of as tiny packages of food and protection. If you look at **Figure 20.24**, you can see one example of a gymnosperm seed, this one from a pine tree. **Figure 20.25** then shows you how this seed fits into the life cycle of the pine, which is a representative gymnosperm. Following the sequence through, once the sperm has reached the egg and fertilized it, an embryo begins to develop from it, in the same way that a human egg starts developing into an embryo once human sperm and egg have come together. In the pine tree example, however, this embryo is developing inside a seed. The seed has a tough coat that has packed inside it a food supply for the growing embryo—a kind of sack lunch of stored carbohydrates, proteins and fats. In sum, a **seed** is a plant structure that includes a plant embryo, its food supply, and a tough, protective casing.

Figure 20.25
The Life Cycle of a Gymnosperm
Pine trees have two kinds of cones. The small male cones produce pollen (the same yellow dust that causes many people with allergies to sneeze in the spring), and the larger female cones produce eggs. When the wind carries pollen onto the female cone, the sperm within the pollen fertilizes one of the eggs within the cone. An embryo then begins to develop inside a seed, which falls to the ground. Once conditions are suitable, the seed germinates and a whole new pine tree begins to grow.

Figure 20.23
Gymnosperms

a This spruce tree is a member of the grouping of gymnosperms known as conifers, which account for about three-quarters of all gymnosperm species. The conifers also include redwood, pine, juniper, and cypress trees.

b There are other types of gymnosperms as well. Pictured are the leaves and seeds of the maidenhair tree, *Ginkgo biloba*, which are used today in herbal preparations.

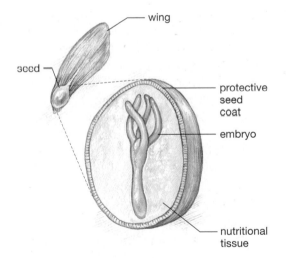

Figure 20.24
A Gymnosperm Seed
Seeds are tiny packages of food and protection. They come in many shapes and sizes, but they all contain an embryo, some food, and a protective seed coat.

wing

seed

protective seed coat

embryo

nutritional tissue

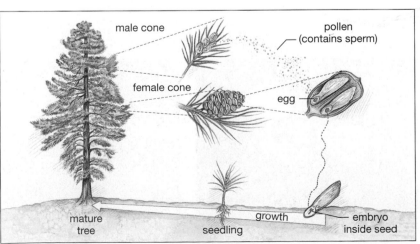

male cone

pollen (contains sperm)

female cone

egg

mature tree

seedling

growth

embryo inside seed

The two major groups of seed plants—the gymnosperms and the angiosperms—turn out to differ in the kind of seeds they have and the placement of these seeds within the larger plant reproductive structure. Angiosperm seeds come wrapped in a layer of tissue—called fruit—that gymnosperm seeds do not have. The details of this anatomical feature are presented in Chapter 23. For now, just be mindful that a **gymnosperm** can be defined as a seed plant whose seeds are not surrounded by fruit. The very name gymnosperm comes from the Greek words *gymnos*, meaning "naked," and *sperma*, meaning "seed."

Angiosperms: Nature's Grand Win-Win Invention

There are about 260,000 known species of the flowering plants known as angiosperms. The term *flowering plants* may bring to mind roses or tulips and, indeed, these flowers are angiosperms. But the angiosperm grouping includes all manner of other plants as well—almost all trees except for the conifers, all our important food crops, cactus, shrubs: the list is endless (**see Figure 20.26**). As noted, **angiosperms** are defined by an aspect of their anatomy. (Their seeds are surrounded by the tissue called fruit.) The details of angiosperm anatomy and physiology are covered in Chapters 22 and 23. In this chapter, we will look only look at why angiosperms are so important to human beings.

Pollination Can Be an Important Plant-Animal Interaction

For an angiosperm such as a new honeysuckle plant to come into existence, a two-part process must take place. First, sperm—developing inside a pollen grain—must get to an egg and fertilize it, which produces the embryo, encased in a seed coat. Second, this seed must land on a patch of soil somewhere and begin to germinate or sprout.

Now, how is it, in this angiosperm, that sperm gets to egg so that it can be fertilized in the first place? With gymnosperms such as pine trees, pollen grains are carried by *wind* to the female reproductive structure, residing in the familiar pinecone (see Figure 20.25). By contrast, the angiosperms developed a vast array of attractants that induce *animals* to carry pollen from one plant to another—to pollinate them. Not all angiosperms are pollinated in this way; some are pollinated by wind, and a few aquatic species rely on water currents. But most are pollinated with the help of animals, be they insects, birds, mammals, or even snails (**see Figure 20.27**).

The most important attractant that flowering plants use to encourage their animal couriers is nectar, which is essentially sugarwater. But angiosperms don't stop with nectar as an attractant. So important is pollination to them that they have developed a host of what might be called pollination marketing strategies. How about a *fragrance* that, say, bees are sensitive to? How about *colors* that they're attracted to? Insect-pollinated flowers generally have both sweet fragrances and coloration patterns, called nectar guides, that are nature's equivalent of homing signals and landing lights. They serve to get an animal to the flower in the first place and then into the right *position* for feeding and pollination (**see Figure 20.28**).

What's the value of all this for the flowering plants? The windborne pollination that gymnosperms practice can be thought of as a kind of scatter-shot approach: If the wind *happens* to blow a pollen grain onto the female reproductive structure, then

**Figure 20.26
Angiosperm Variety**

a Calla lilies on the California coast.

b Cholla cactus in Arizona.

c Corn in a field.

a

b

c

pollination occurs. Think how much more directed things are with animal pollination; what exists, essentially, is door-to-door service. This is one of the reasons that angiosperms are by far the dominant plants on Earth.

Seed Endosperm: More Animal Food from Angiosperms

The second big step in angiosperm reproduction is for the product of fertilization, the embryo (inside its seed), to begin germinating from the ground. All seeds contain food reserves for the growing embryo, but angiosperm seeds develop a special kind of nutritive tissue, called **endosperm**, that often surrounds the embryo (**see Figure 20.29**). It is this endosperm tissue that supplies much of the food for human beings throughout the world. Rice and wheat grains consist in large part of the endosperm meant to sustain the plant embryo. The white "meat" and liquid "milk" of the coconut are also forms of endosperm.

Fruit: An Inducement for Seed Dispersal

Some flowering plants then go on to add something else: tissue, whose function is to be attractive to animals as a means of getting the plant seeds *dispersed*. This is fruit not in the technical sense, but fruit as that term is commonly understood—the flesh of an apricot or

Figure 20.27
Animals Help Pollen Get from Here to There
Flowering plants were the first to take advantage of the mobility of animals to transport pollen from one plant to another. Relationships between flowers and pollinators often are species-specific, which helps to ensure that the pollen will be delivered to the right address.

a A pollen-covered honeybee on a dandelion.

b A purple-throated carib martinique pollinating a flower.

c A lesser long-nosed bat pollinating a saguaro cactus.

a This is what we see (normal sunlight).

b This is what the bee sees (ultraviolet light).

c This is what the bee "thinks."

Figure 20.28
Food Lies This Way
Animals may be attracted to a particular flower by its fragrance, color, or pattern of visual elements. Insects are guided into the nutrients in the center of the flower by color patterns called nectar guides, visible to an insect—because it perceives ultraviolet light—but not to us. **(a)** What a flower looks like to a human being. **(b)** What a flower looks like to an insect that can perceive ultraviolet light. The circular pattern is the nectar guide, and the light-colored flecks on the petals are pollen grains.

Figure 20.29
Food for the Seedling, Food for Humans
Endosperm is a food reserve in seed. It feeds the embryo before it can make its own food. It is this same endosperm that nourishes humans throughout the world in the form of corn, wheat, rice, and even coconut.

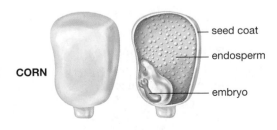

CORN

seed coat

endosperm

embryo

cherry, for example. As such, it represents one more piece of bounty that plants provide to animals. Yet plants get something out of this relationship as well. Imagine a bear who consumes a wild berry, which consists not only of the fruit flesh, but of the seeds it surrounds. The fruit is digested by the bear as food, but the *seeds* are tough; they are passed through its system intact to be deposited, with bear feces as fertilizer, at a location that may be very remote from the place where the bear *ate* the berry. *Voilá!* Seed dispersal at what may be a promising new location for a berry plant (**see Figure 20.30**).

Given pollination and the varieties of food that angiosperms produce, it's easy to see that, with the rise of the flowering plants, animals and plants entered into a much more interdependent relationship than had existed before, one that is still evolving today. Taking a step further back, if we ask which plants are the most important in shaping our world, the answer is the angiosperms.

On to a Look at Animals

In this chapter, you've looked at the microscopic life-forms known as bacteria and archaea, and at three of the kingdoms within the eukaryotic domain,—protists, fungi, and plants. This leaves one more eukaryotic kingdom to go. It is a kingdom that is arguably the most familiar of all, because we human beings are part of it. Yet it is a kingdom so diverse that it holds a good many surprises as well. It is the kingdom of animals.

Figure 20.30
Seed Carriers
Angiosperms have taken advantage of the mobility of animals not only to transfer pollen, but also to disperse seeds.

a Some seeds are wrapped in tasty fruit that is consumed by animals, such as bears.

b Other seeds come wrapped in burrs and spines that stick to the fur of animals and are carried away.

a

b

Chapter Review

Summary

- All living things on Earth can be classified as falling into one of three domains of life, Bacteria or Archaea, whose members are microscopic; or Eukarya, whose members range from the microscopic to the gigantic. Domain Eukarya is in turn divided into four kingdoms: plants, animals, fungi, and protists.

20.1 Viruses: Making a Living by Hijacking Cells

- There is controversy over whether viruses are living things. To carry out almost any of life's basic functions, viruses must use the structures of the cells they invade. Because viruses cannot carry out even replication by themselves, most scientists would not classify them as living things.

- Some viruses employ a lytic life cycle, while others employ both a lytic and lysogenic life cycle. In the lytic life cycle, clones of the invading virus burst forth from an infected cell and then go on to infect more cells. In the lysogenic life cycle, the virus may for a time integrate itself in the DNA of the host cell and then replicate with it, leaving the cell through lysis only when threatened.
TUTORIAL 20.1.1: Life Cycle of a Virus

- Viruses can invade all life-forms.

20.2 Domain Bacteria: Masters of Every Environment

- Though they can cause human disease, bacteria are indispensable to all life on Earth. They are largely responsible for moving the nutrient nitrogen into the living world and, along with fungi, are the major decomposers among living things. Bacteria are a single-celled life-form in one sense, but generally live in groupings, some of which display complex organization. Bacteria have only a single organelle, the ribosome, and reproduce through simple cell division.

20.3 Domain Archaea: From Marginal Player to Center Stage

- Archaea were once thought to be a form of bacteria but have recently been shown to differ greatly from bacteria in their genetic makeup and metabolic functioning. Like bacteria, they are single-celled and microscopic. Some archaea thrive in extreme environments, such as hot springs. The chemical products they produce are now being exploited by human beings.

20.4 Domain Eukarya: Protists, Plants, Fungi, and Animals

- Eukaryotes exhibit an incredible diversity of form and function; most reproduce sexually, though members of all eukaryotic kingdoms can reproduce asexually as well.

20.5 Kingdom Protista: An Undefinable Collection

- Kingdom Protista, part of the eukaryotic domain, is made up of an assortment of organisms so diverse that no one definition can be applied to all of them. Many protists are single-celled aquatic creatures, but large ocean-dwelling algae are protists, as are many parasites and soil-dwelling microbes.

- There are protists that perform photosynthesis (as do plants), protists that consume food in the manner of animals, and protists that reproduce in the manner of fungi. This makes sense in evolutionary terms because plants, animals, and fungi are all thought to have evolved from various forms of protists.

- The aquatic protists known as phytoplankton are important in photosynthesis and as the base of aquatic food chains.

20.6 Kingdom Fungi: Life as a Web of Slender Threads

- Fungi consist largely of slender tubes, called hyphae, that grow toward their food supply, releasing enzymes that digest the food externally, after which the fungus absorbs the nutrients. The network of hypae created by a fungus is called a mycelium, and it is often located underground.

- Along with bacteria, fungi are nature's primary decomposing organisms. With the exception of yeasts, all fungi are multicellular.

- Lichens are composite organisms, made up of both fungi and algae (or fungi and photosynthesizing bacteria). The relationship between fungi and algae in lichens is thought to be mutually beneficial to both kinds of organisms: Fungi derive food from the photosynthesizing algae, and the algae are shielded, by the fungi, from forces that would dry them out.

- Some fungi live in a mutually beneficial relationship with plants, the plants supplying the fungi with food produced in photosynthesis, the fungi using their hyphae to absorb water and nutrients for the plants. Associations of plant roots and fungal hyphae are called mycorrhizae.

TUTORIAL 20.1.2: Life Cycle of a Fungus

20.7 Kingdom Plantae: The Foundation for Much of Life

- Plants are the foundation for much of life on Earth, because they are responsible for much of the living world's food production and oxygen generation. All plants are multicelled; most are fixed in one spot and carry out photosynthesis. Plants reproduce through an alternation of generations.

- The four principal categories of plants are bryophytes, seedless vascular plants, gymnosperms, and angiosperms. Bryophytes include mosses; seedless vascular plants include ferns; gymnosperms include coniferous trees, such as pine; and angiosperms include a wide array of plants, such as orchids, oak trees, rice, and cactus.

- Bryophytes are representative of the earliest plants that made the transition from water to land. They lack a true fluid-transport or vascular system and thus tend to be low-lying. Bryophyte sperm can get to eggs only by swimming through water. Thus, bryophytes are most commonly found in damp environments.

- Seedless vascular plants have a vascular system but do not produce seeds in reproduction. Their sperm must move through water to fertilize eggs.

- Gymnosperms are seed-bearing plants whose seeds are not encased in tissue called fruit. There are only about 700 gymnosperm species; but their presence is considerable, particularly in northern latitudes where gymnosperm trees, such as pine and spruce, often dominate landscapes. The sperm of gymnosperms is encased in pollen grains, which are carried to female reproductive structures by the wind. Gymnosperms (and angiosperms) produce seeds in carrying out reproduction. Seeds are structures that include a plant embryo, its food supply, and a tough, protective casing.

- Angiosperms, or flowering plants, are seed plants whose seeds are encased in tissue called fruit. Angiosperms are easily the most dominant group of plants on Earth, with some 260,000 species having been identified to date.

- Sperm-containing angiosperm pollen grains often are transferred from one plant to another by animals such as bees and birds. Such animal-assisted pollination is unique to angiosperms. To induce animals to carry out this pollination, angiosperms produce nectar and have developed attractive colorations and fragrances.

- Angiosperm seeds contain tissue called endosperm, which functions as food for the growing embryo. Endosperm supplies much of the food that human beings eat. Rice and wheat grains consist largely of endosperm.

- Angiosperm seeds are unique in the plant world in being wrapped in a layer of tissue called fruit. Fruit that is edible functions in angiosperm seed dispersal, because animals will eat and digest the fruit but then excrete the tough seeds inside, often in a different location.

Key Terms

alternation of generations 426
angiosperm 430
antibiotic 414
autotroph 415
bacilli 413
binary fission 414
bryophyte 426
cell wall 425
cellulose 425
chemoautotrophy 415
chemoheterotrophy 415
cocci 415
dikaryotic 413
diploid 423
endosperm 414
extremophile 417
gametophyte generation 426
gymnosperm 429
haploid 414
heterotroph 415
hyphae 421`

lignin 425
lyse 408
lysogenic cycle 408
lytic cycle 408
mycelium 421
mycorrhizae 424
organelle 413
photoautotrophy 415
photoheterotrophy 415
phytoplankton 420
plankton 421
pseudopodium 419
seed 429
seedless vascular plant 428
sessile 421
spirochete 413
spore 426
sporophyte generation 426
viroid 409
zooplankton 420

Understanding the Basics

Multiple-Choice Questions

1. Characteristics shared by all bacteria include (select all that apply)
   a. rod shape
   b. no nucleus
   c. single chromosome
   d. sexual reproduction
   e. one organelle

2. What do all eukaryotes have in common? (Select all that apply.)
   a. multicellularity
   b. cell organelles
   c. nucleus
   d. sexual reproduction
   e. circular chromosomes

3. Important roles of fungi in life on Earth include (select all that apply)
   a. decomposition
   b. photosynthesis
   c. assist in water uptake in plants
   d. used in baking and brewing
   e. join with algae to make lichens

4. Match the organisms on the left with the descriptions on the right:
   a. hyphae — structures that are a combination of plant roots and fungal hyphae
   b. mycorrhizae — filamentous strands of a fungus
   c. cell walls — ancestors of animals, plants, and fungi
   d. archaea — aid plants in regulating water intake
   e. protists — often occupy extreme environments

5. Defining characteristics of plants include (select all that apply)
   a. alternation of generations
   b. hyphae
   c. cell walls
   d. form biofilms
   e. autotrophic nutrition

6. Which of the following is an angiosperm?
   a. moss
   b. a giant kelp
   c. a pine tree
   d. a rose
   e. a fern

7. If an organism is single-celled and that cell has no nucleus, the organism must be a
   a. virus or a bacterium
   b. protist or an archaean
   c. fungus or an archaean
   d. protist or a bacterium
   e. bacterium or an archaean

8. True or false: Most bacteria are harmful.

9. Only one of these organisms grows toward new sources of food. Which one is it?
   a. algae
   b. bacterium
   c. fungus
   d. flowering plant
   e. virus

Brief Review

1. What is an extremophile?

2. How do vaccines work to prevent viral diseases?

3. Why do most biologists classify viruses as nonliving?

4. Discuss two reasons why phytoplankton are so important to life on Earth.

5. Describe two common associations that fungi make with other organisms. What does each party in each of these associations do for the other, and what do they gain themselves?

6. How does nutrition in plants and fungi differ?

7. Given their structure, why are raw plants crunchy?

Applying Your Knowledge

1. What would life on Earth be like if there were no decomposers? Would evolution of life have followed the same course? What might have been different?

2. Discuss the statement, "If success is having the ability to eat anything or live anywhere, then bacteria are the most successful organisms on Earth."

3. If new viruses can form as "renegade" DNA that transposes out of eukaryotic genomes, might some viruses be more closely related to various eukaryotes than they are to other viruses?

4. What is the main difference between colonial multicellularity and true multicellularity? Why was this important for the evolution of eukaryotes?

21

Movers and Shakers:
The Animal Kingdom

The largest invertebrate. (Section 21.7, page 448)

Shedding an old skeleton. (Section 21.9, page 453)

A human tail. (Section 21.11, page 459)

The diversity of animals is one of the wonders of the natural world.

We human beings are, in a sense, intimate strangers with the bacteria and fungi that were reviewed in Chapter 20. We are intimate with these organisms in that we carry out our lives right along beside them; indeed, they are always on us or in us in great numbers. These creatures are strangers to us, however, in that, except for the occasional outbreak of athlete's foot or stomach flu, they never cross our minds. So small are bacteria and most fungi that, while they may be everywhere, for us they are nowhere. Not so with the animals that are the subjects of this chapter. Our cats and dogs and neighborhood birds are part of our conscious lives. We pay attention to a new spider web on our porch or to the lizard on our hiking path. We wonder about the circling hawk overhead.

Animals come from the long evolutionary line of *other* feeders: the heterotrophs, who cannot manufacture their own food (as plants do) but must get their nutrition from outside themselves. Bacteria performed this feat first, and fungi and protists did it as well. But animals, in their quest for food, added something that is unique to them: A nervous system; a system for transmitting complex messages over long pathways in the body. Once this happened, the living world was off to the races in terms of adding novel abilities. Think of sight, hearing, smell, flight, walking, singing, and reading. Then there are such things as the echo-location of bats and the waggle-dance of honey bees. For the average person, to be alive means to sense, to investigate, to respond, to move. In all these areas, animals reign supreme in the living world.

This chapter is about the broad diversity that exists in animals (**see Figure 21.1**).

Figure 21.1
Animal Diversity
An octopus, a feather star and a chimpanzee are very different creatures, yet all are members of the animal kingdom.

a Octopuses often are found crawling over ocean rocks, but they swim as well, by expelling a jet of water from a hose-like siphon, often propelling themselves "backwards" as with this one.

b Though they look like plants, ocean-dwelling feather stars such as this one are animals that are part of the same phylum as sea stars. They extend their arms to catch bits of food drifting in the ocean currents.

c Chimpanzees are the closest living relatives of human beings.

a

b

c

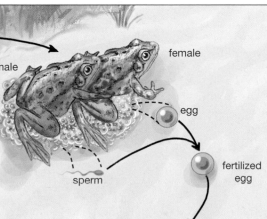

notochord   dorsal nerve cord

pharyngeal slits

male   female

egg

sperm   fertilized egg

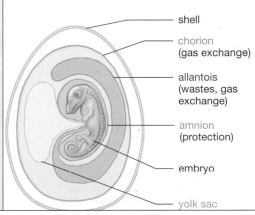

shell

chorion (gas exchange)

allantois (wastes, gas exchange)

amnion (protection)

embryo

yolk sac

One of our chordate relatives.
(Section 21.11, page 459)

A double life.
(Section 21.11, page 461)

An evolutionary innovation.
(Section 21.11, page 462)

Given that animals are part of our daily lives, some of what follows will be familiar. But there are likely to be many surprises as well.

21.1 What Is an Animal?

Our commonsense notions of animals serve fairly well in defining the animal kingdom—fairly well, but not well enough. We need to know what characteristics *all* animals have that other organisms *don't* have. Though it's fairly technical, there is a single feature that, by itself, is sufficient to set animals apart from all other living things.

Tutorial 21.1.1
Animal Classification

- Animals pass through something called a blastula stage in their embryonic development. A *blastula* is a hollow, fluid-filled ball of cells that forms soon after an egg is fertilized by sperm (see Chapter 27 for details). All animals go through a blastula stage, but no other living things do.

Three other characteristics are found in all animals, but they're found in other kinds of organisms as well. All animals:

- Are multicelled; there are no single-celled animals.

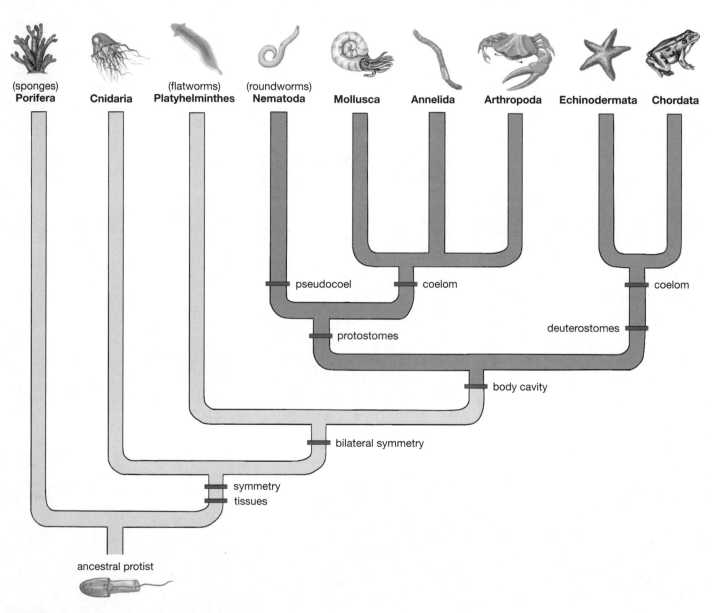

Figure 21.2
A Possible Family Tree for Animals
The animal kingdom is divided into groups called phyla. The members of each phylum have in common physical features that are evidence of shared ancestry. Nine of the estimated 36–41 animal phyla are shown here. All these phyla are regarded as having evolved from an ancestral species of protist, shown at lower left in the family tree. The red horizontal lines on the tree mark the points in animal evolution at which certain structural innovations appeared, such as tissues and bilateral symmetry. Human beings are members of phylum Chordata, at upper right.

- Are heterotrophs; they must get their nutrition from outside themselves.
- Are composed of cells that do not have cell walls. (The outer lining of animal cells is the plasma membrane. By contrast, all plants and most other creatures have a relatively thick lining—the cell wall—outside their plasma membrane.)

Apart from these universals, animals *usually* go on to share other characteristics. Animals tend to move, though there are animals that, for at least part of their lives, are as *sessile*, or fixed in one spot, as any plant. Except for the animal group that includes sponges, animal bodies are organized into **tissues**, each tissue being an assemblage of similar cells that serves a common function.

Then there are characteristics of animals that don't fit our preconceived notions. While it's true that animals generally are large relative to bacteria or protists, some animals are microscopic. We often think of animals as creatures with vertebral columns or "backbones," but there are far more **invertebrates**, or animals without vertebral columns, than animals with them. We generally think of animals as land creatures and, with a huge assist from insects, there certainly are a lot of land animals. However, in terms of animal *diversity*, as measured by basic body plans, there are more kinds of ocean-going animals than land animals, as you'll see. This makes sense, however, because animals existed in the sea for at least 100 million years before any of them came onto land.

21.2 Animal Types: The Family Tree

Basic body plans turn out to provide a sensible way of dividing up the animal kingdom. Depending on who is counting, there are between 36 and 41 animal phyla, with each phylum being a group of organisms that share a set of physical characteristics that stem from shared ancestry. The classification scheme for the living world introduced in Chapter 18—that of kingdom, phylum, class, order, family, genus, and species—provides the basis for a more formal definition of phylum. A **phylum** is a category of living things, directly subordinate to the category of kingdom, whose members share traits as a result of shared ancestry. (For more on classification, see Chapter 18, page 367.)

So, in what ways are the various animal phyla related to one another? To put this another way, what does the animal family tree look like? At the moment, this question is a matter of great debate among **zoologists**, meaning biologists who study animals. By measuring similarities in DNA sequences among living animals, these scientists have come up with entirely new conceptions of who is more closely related to whom. Pending the outcome of this debate, the family tree set forth here is fairly traditional. It posits animal life as moving from simpler to more complex through a series of *additions* to the characteristics found in more primitive animals. The twist is that only some varieties of animals evolved to get these additions, while others retained the primitive, "ancestral" condition.

You can see this on display in **Figure 21.2**, which is one possible phylogeny or family tree for animals. Not all the 36–41 animal phyla are shown in this tree; some of the phyla at the top represent groups of phyla. Let's walk through the tree now, focusing on features that were added to the animal kingdom through evolution.

Additions 1 and 2: Tissues and Symmetry

Looking down at the trunk of the tree in Figure 21.2, you can see that all animals have a common ancestor—probably a protist similar to the modern-day protists called choanoflagellates. Now notice that the tree splits above this common ancestor and yields, over to the left, a phylum called Porifera, which are sponges. Porifera are truly the outliers of the animal world in that, almost alone among animals, they lack tissues and a quality known as symmetry (defined shortly). If you look again at the split above the common ancestor, this time going to the right, you can see that the animals on this branch did get the additions of tissues and symmetry. Note also that this right branch leads to all the other animal phyla. Thus, all the animals that stem from this branch will have tissues and symmetry (which is what makes Porifera the outlier).

**Figure 21.3
Symmetry Is an Equivalence in Body Sections**
If an imaginary plane drawn through an animal can divide that animal into sections that are mirror images of one another, then that animal has symmetry.

a Radial symmetry The imaginary planes drawn through the jellyfish show that it has radial symmetry: a symmetry in which body sections are distributed evenly around a central point. Radial symmetry is characteristic of the phylum Cnidaria, which jellyfish are part of.

b Asymmetry The sponge has no symmetry—there is no plane that could be drawn through the sponge body that would yield sections that are mirror images of one another. Phylum Porifera, to which sponges belong, is the only major animal phylum that has no symmetry.

c Bilateral symmetry The dog has a kind of symmetry common to most animals: bilateral symmetry, in which the sides of an animal are mirror images of one another. More formally, bilateral symmetry exists when body sections on opposite sides of a sagittal plane are symmetrical to one another. Notice also the terms for different parts of the dog—dorsal and ventral, meaning bottom and top; and anterior and posterior, meaning front and back. These terms will be used fairly frequently throughout the chapter.

But what is this other addition, symmetry? If you look at **Figure 21.3a**, you can see that the jellyfish that's pictured can be thought of as being divided by imaginary planes, thus yielding body sections. These sections are mirror images of one another; they have **symmetry**, meaning an equivalence of size, shape, and relative position of parts across a dividing line or around a central point. The jellyfish, part of phylum Cnidaria, actually has a particular kind of symmetry. It has **radial symmetry**, meaning a symmetry in which body parts are distributed evenly around a central point. In simpler terms, the symmetry of jellyfish is the symmetry of pie sections.

To appreciate symmetry, consider what a lack of it is like. Look at the sponge in Figure 21.3b. Where is its symmetry? There isn't any, because there is no section of the sponge that is a mirror image of any other.

Symmetry can, however, come in several forms. Look at the dog in Figure 21.3c., divided by what is known as an imaginary "sagittal" plane. The dog obviously has symmetry, but it is a different kind of symmetry from that of the jellyfish. It has "bilateral" symmetry. We'll take a moment to consider this form of symmetry, as it constitutes the third addition to the animal kingdom.

Addition 3: Bilateral Symmetry

Animals usually are *different* front-to-back and top-to-bottom, but *symmetrical* side-to-side. Put another way, your head is different than your feet, and your chest is different than your back, but your left *side* is very similar to your right. This is what **bilateral symmetry** means: a bodily symmetry in which opposite sides of a sagittal plane are mirror images of one another.

In looking at the animal family tree, you can see that bilateral symmetry was an evolutionary innovation that affected all modern-day animals in the tree except the sponges and the phylum Cnidaria, which includes the jellyfish. Thus, bilateral symmetry is the general rule in the animal world.

The bilateral symmetry/radial symmetry divergence is in one sense the divergence between having a head and not having one. Jellyfish and sponges don't have one, but most other animals have at least a "cephalization," or concentration of nerve cells at one end of their bodies. Animals more complex than the cnidarian jellyfish evolved to sense their world primarily at one end of themselves—their heads—which is the end with which they move into their worlds. This is understandable because it is more advantageous to know something about the future (where you're going) than the past (where you've been).

Addition 4: A Body Cavity

Your stomach expands when you've just had a big meal, but then contracts when you're busy for a few hours. Your heart expands and contracts perhaps 70 times a minute. You bend over to tie your shoes and, unbeknownst to you, many of your internal organs slide out of the way. What allows you to do all this? The answer is an internal space you have; a large, fluid-filled body cavity. We humans are not alone in having such a cavity, or coelom (pronounced "SEE-lome"). If you look at Figure 21.2, you can see that there are only three animal groups pictured that *don't* have this cavity: the sponges, the cnidarians (jellyfish, etc.), and the members of phylum Platyhelminthes, which is made up of flatworms.

a Radial symmetry:
Symmetry around a central point

b Asymmetry:
No planes of symmetry

c Bilateral symmetry:
Symmetry across the sagittal plane

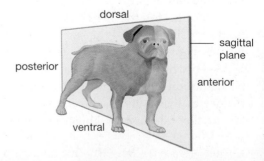

dorsal

posterior

sagittal plane

anterior

ventral

What's the value of such a cavity? Well, first, an expandable stomach has the same value as a gas tank: It allows you to go for a while without refueling. Then there is the fact that if a heart couldn't expand and contract, it couldn't work at all. Thirdly, a body cavity provides organs with protection from bodily blows and provides a large part of the body with flexibility.

In most instances the coelom surrounds another physical structure, the *digestive tract*—meaning the tube, functioning in digestion, that runs from the mouth to the anus. We can therefore think of the coelom as one tube that encircles another; the coelom is generally tube-shaped, and it surrounds the tube that is the digestive tract. (There are, however, lots of variations on this general principle.) **Figure 21.4** displays what it means to have no coelom, as in flatworms; a *pseudocoel*, as in roundworms; and a *true coelom*, as in earthworms.

The concept of a coelom is intimately linked to that of tissue layers. All animal embryos have what are known as germ layers, meaning layers of cells that *become* various types of tissue in adult animals. Most animals have three types of germ layers: endoderm, mesoderm, and ectoderm. Endoderm is initially an inner layer of cells in the embryo, mesoderm a middle layer, and ectoderm an outer layer. The general importance of these tissue layers will be reviewed later. For now, just note the linkage, displayed in Figure 21.4, between mesoderm and a coelom. This linkage provides a way of formally defining a **coelom**. It is a central body cavity in an animal that is lined with cells of mesodermal origin.

A Split in the Animal Kingdom: Protostomes and Deuterostomes

Our story of additions to animal complexity ends with the coelom. It's not that, after the coelom, animals ceased evolving more complex features. But look, in Figure 21.2, at the next divergence on the animal family tree. On the left is the grouping called protostomes. Among other animals, this group includes squid in the phylum Mollusca. Over on the right, meanwhile, the deuterostome grouping includes not only vertebrates such as ourselves (in phylum Chordata), but sea urchins as well. Now, which is more complex: a sea urchin or a squid? The answer is a squid; it is a much more sophisticated animal than a sea

urchin. As a *group*, then, deuterostomes are no more complex than protostomes. The protostome/deuterostome split simply represents two different evolutionary paths, rather than a transition from a simpler to a more complex form.

So, what is the difference between the protostomes and the deuterostomes? It actually is grounded in how the two types of animals develop in their early embryonic stages. Remember how you saw earlier that the chief defining characteristic of animals is that they all go through something called the blastula

a No coelom (acoelomate)

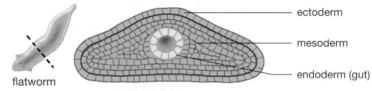

flatworm

ectoderm
mesoderm
endoderm (gut)

b Pseudocoel

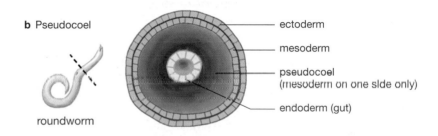

roundworm

ectoderm
mesoderm
pseudocoel
(mesoderm on one side only)
endoderm (gut)

c Coelom

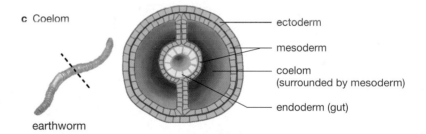

earthworm

ectoderm
mesoderm
coelom
(surrounded by mesoderm)
endoderm (gut)

Figure 21.4
An Important Space
Most animal bodies have a central cavity or coelom—an internal space that surrounds their digestive tract or other internal structures. A coelom gives an animal flexibility, protects its organs from external blows, and provides space for the expansion of such organs as the stomach. Only three of the phyla covered in the chapter lack a coelom.

a Phylum Plathyhelminthes, composed of flatworms, is one of the phyla that has no coelom. Note that from its gut to its exterior, the flatworm is composed of uninterrupted tissue—endoderm (inner tissue), mesoderm (middle tissue) and ectoderm (outer tissue).

b Roundworms, by contrast, have a central cavity, known as a "pseudocoel," that lies between their gut and mesodermal tissue layers. The pseudocoel develops as a continuation of the space that exists when an embryonic animal is in the blastula or hollow ball-of-cells stage.

c The earthworm is one of the many animals that has a "true" coelom (or just-plain coelom). Note that both the coelom *and* the gut of the earthworm are lined with mesodermal tissue layers (whereas in the roundworm, only the pseudocoel is lined by mesoderm). The true coelom develops from an animal's mesodermal tissue.

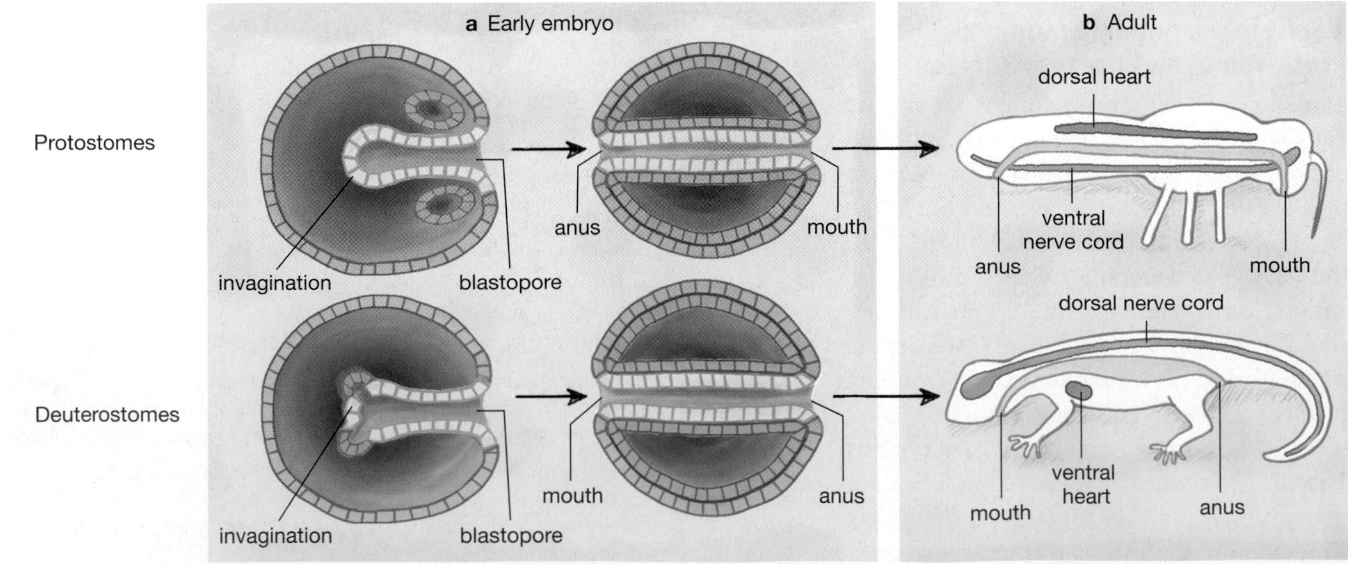

Figure 21.5
The Protostome-
Deuterostome Divide
(a) At one point in their embryonic lives, all animals pass through a so-called blastula stage of development, the blastula being a hollow ball of cells. In most animals, this blastula then invaginates to form a structure that develops into the animal's gut. The opening to this invagination is called a blastopore. In protostomes, the blastopore becomes the mouth. In deuterostomes, by contrast, the blastopore becomes the anus. **(b)** Developmental differences such as these have consequences in the adult. Protostomes, such as the insect shown, have their heart on their top or dorsal side and nerve cord on their bottom or ventral side. In deuterostomes, such as the reptile shown, this placement is reversed. The portions of the embryos colored yellow, red, and blue signify the endodermal, mesodermal, and ectodermal tissues noted in the discussion of the animal coelom.

stage in development? Well, in **Figure 21.5a**, you can see what happens to the hollow ball of cells called a blastula: It develops an invagination whose opening is called a blastopore. In protostomes, this blastopore becomes the mouth, but in deuterostomes it becomes the anus. (Protostome means "mouth first," while deuterostome means "mouth second.")

If you look at Figure 21.5b, you can see the effect that developmental differences such as this have on adult protostomes and deuterostomes. Note that the protosome has a dorsal heart and a ventral nerve cord, while in the deutorosome this arrangement is reversed.

With these broad-scale features in mind, we will now take a brief tour of the animal kingdom by walking through each of the nine phyla shown in Figure 21.2.

21.3 Phylum Porifera: The Sponges

Back in the 1780s, the United States was just that: a group of states that had just united into a single entity. The sponges that make up the phylum Porifera bear some comparison to this. Each sponge is a single, unified entity, but not by much. Sponges have no organs; no stomach, no heart. They don't really even have tissues. Each of their cells acquires its own oxygen and eliminates its own wastes. In experiments, scientists have strained some sponge species through a filter, "disaggregating" them into the individual cells they are made of. The scientists then watched as these cells came back together to

make up a single sponge once again. This is possible only in an organism in which each cell functions with a great deal of independence. If you were disaggregated into your individual cells, there is no chance they would come back together to re-form you. On the other hand, the cells in sponges clearly work together in a coordinated way. Sponges may not show much organization relative to fish, but they show a great deal of organization relative to, say, bacteria.

So, what manner of organism is a sponge? The simplest varieties have a layer of outer cells that is pockmarked with thousands of microscopic pores. Water flows in through these pores in the outer-layer cells and then is expelled out through a single large opening at the top of the sponge (**see Figure 21. 6**). Because sponges are fixed in one spot (generally on the sea floor), everything they need in the way of food and oxygen must come *to* them, while everything they need to get rid of must be washed away *from* them. The water that flows through them takes care of both needs. Food and oxygen wash in through the microscopic pores, and wastes wash out through the single large opening (called an *osculum*). Most sponges have elaborations on this theme, but the essential concept is the same: Move the water in, capture the food that comes with it, and move the water out, expelling wastes in the process.

Most sponges have skeletons composed of tiny, barbed structures called *spicules* that keep the sponge pores open and generally give the sponge structure. Some sponges have no spicules, but instead have skeletons made

a Early embryo

Protostomes

invagination   blastopore   anus   mouth

Deuterostomes

invagination   blastopore   mouth   anus

b Adult

dorsal heart

ventral nerve cord

anus   mouth

dorsal nerve cord

mouth   ventral heart   anus

of a special variety of the protein collagen, which can be a soft, pliable material. (Think of the collagen *injections* that people sometimes get to improve their looks.) Sponges with collagen skeletons are the sponges from which we get our everyday notion of a sponge, meaning a soft, absorbent object used in cleaning. Put another way, the sponges that people once used were the skeletal remains of ocean-dwelling animals (which were harvested by divers). Today our sponges generally are made of synthetic materials.

Depending on whose count you accept, between 5,000 and 10,000 species of sponge have been recognized so far, with probably lots more remaining to be identified. Almost all these species dwell in ocean water, though there are a few freshwater varieties (**see Figure 21.7**).

Sponges can reproduce merely by budding. A group of cells breaks off from an existing sponge and then develops into a new sponge. Sexual reproduction exists too, however, and here again the sponge puts water currents to use. Sperm are released from one sponge and make their way via water currents to another sponge, where they are taken up and transported to egg cells. An offspring that comes from sexual reproduction has a moving or "motile" stage. It swims for a time as a larva before landing on a solid surface—perhaps a rock or the body of a living animal—whereupon its outer layer of cells will secrete a substance that will allow it to attach to the material beneath it. Then it begins its life of filtering water.

21.4 Phylum Cnidaria: Jellyfish and Others

If the signature activity of sponges is filtering water to get food, the signature activity of cnidarians is stinging prey to get it. In the main, cnidarians (knee-DAR-ee-uns) are harpooning their prey with extensions that are not only barbed, but that may release poisons. These extensions are *tiny*, as each one springs from inside a single cell, but they are numerous enough that a single jellyfish may be able to immobilize and eat animals the size of small fish. Only a few species have a sting potent enough to cause real harm to human beings, but it makes sense to give all jellyfish a wide berth.

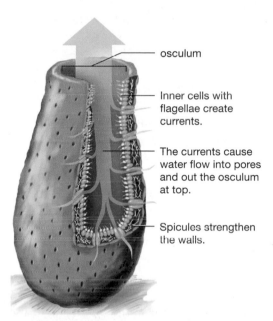

osculum

Inner cells with flagellae create currents.

The currents cause water flow into pores and out the osculum at top.

Spicules strengthen the walls.

Figure 21.6
A Body Plan for a Simple Lifestyle
Sponges filter water through themselves in order to live. Cells on their interior surface have hair-like extensions, called flagella, whose rapid back-and-forth movement draws water in through pores in the sponge's side walls. After filtering this water for nutrients, the sponge then expels it through a large opening at the top of its body called an osculum. Sponges lack true tissue layers, but possess many specialized cell types. The cells are held together in a stable arrangement by tiny, hard structures called spicules and by the flexible protein called collagen.

Figure 21.7
Sponge Diversity
Sponges differ greatly in size and shape, as you can see from these examples.

a A leaf sponge.

b A yellow, warm-water tube sponge.

a

b

Jellyfish are undoubtedly the best-known cnidarians, but the phylum also includes corals, whose calcified remains make up coral reefs; the delicate sea anemones, whose petal-like tentacles close up when touched; and hydrozoans, which can look more like sea plants than animals (**see Figure 21.8**). One type of hydrozoan is the hydra that you may see in your biology lab. There are about 9,000 identified species of cnidarians, nearly all of them ocean dwellers, with a few living in freshwater.

The basic body plan of a cnidarian is that of a sack, though it often is a sack that is turned upside down. If you look at **Figure 21.9**, you can see how this plays out in connection with a hypothetical jellyfish. In the familiar adult or "medusa" stage of a jellyfish, the sack is upside down; note that the mouth is on the *underside* of the body in this stage. Conversely, look at this jellyfish's immature "polyp" stage, when it is fixed to the ocean floor; now the mouth is on top. Medusa and polyp stages are present in many cnidarians, though corals, sea anemones, and hydra have only the polyp stage. The medusa stage predominates in jellyfish—misnamed, because they are not fish—while the polyp stage predominates in most hydrozoans. In all medusa and in many polyps, however, the basic body plan is the same: a single opening to the outside, serving as both mouth and anus, that is surrounded by tentacles that both sting prey and bring the prey to the mouth. Note also the material that lies between the jellyfish's gastrovascular cavity and its exterior. This is the **mesoglea**, a secreted, gelatinous material that makes up most of the medusa-stage jellyfish. Its consistency accounts for the

name jellyfish. The gastrovascular cavity functions in both digestion (*gastro-*) and in circulation (*vascular*).

Figure 21.10 shows you the life-cycle of a cnidarian, this one a hydrozoan known as *Obelia*. Medusa-stage male *Obelia* release sperm into the water while females release eggs; these unite, eventually producing a larval stage offspring that settles to the ocean-floor bottom, now becoming a polyp. This starting polyp can give rise to an entire colony of *Obelia* polyps through the process of *budding*: Groups of cells break off from one polyp and form another polyp. Note that in this polyp stage, some of the branch-like outgrowths are "feeding polyps" that have the familiar sack-like appearance of cnidaria (tentacles, mouth), while other outgrowths are entirely given over to reproduction, containing "medusa buds." These buds are immature medusa-stage *Obelia*, which eventually will be released to become free-swimming individual medusae. The *Obelia* life cycle is one among many that exist in cnidarians.

Cnidarians don't have any true organs but they are a long way from sponges in that they have a nervous system, and cells that function like muscle cells. This gives medusa-stage cnidarians the ability to swim. (Polyp-stage cnidarians sometimes can slide slowly along the ocean floor.) The typical jellyfish medusa contracts cells that ring its mesoglea, thus expelling a jet of water from its underside. This action both moves the animal in the opposite direction from the jet and pushes in the mesoglea. Like a compressed cylinder of foam rubber, the mesoglea then snaps back into shape, ready for the next contraction.

a

b

c

Special mention should be made of the reef-building corals, which have no medusa stage but live almost entirely as polyp-stage cnidarians. The fantastic ocean cities known as coral reefs take their names from the coral animals, which actually are close relatives of sea anemones. Coral reefs have a rock-like appearance and, indeed, they are composed mostly of calcium carbonate, better known as limestone. But this limestone comes from living organisms: The corals secrete it, a practice they share with a type of red algae with whom they share their habitat. The limestone forms an external skeleton for the corals, and when each coral polyp dies, its limestone skeleton remains. Thus coral reefs are composed largely of the stacked-up remains of countless generations of coral polyps, along with the remains of their neighboring red algae. But the thin outer veneer of each reef is composed of *living* algae and the latest generation of pinhead-sized polyps. If you looked up close at any one of these coral animals, you would see the basic features of a cnidarian polyp: stinging tentacles pointed upward, surrounding a mouth.

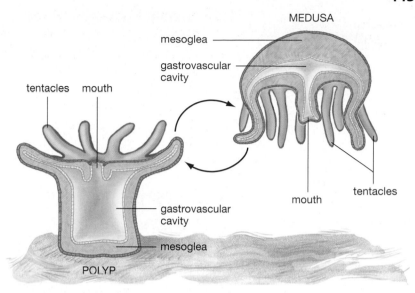

Figure 21.9
Two Stages of Life for Many Cnidarians
Many cnidarians go through both a polyp stage, in which they are attached to a solid surface beneath them, and a medusa stage, in which they swim freely through the water. Note that in the polyp stage, the cnidarian's mouth and tentacles face upward, while in the medusa stage, they face downward. The gastrovascular cavity of the cnidarian functions in both digestion and circulation. The mesoglea noted in the figure is a secreted, gelatinous material that makes up the bulk of the medusa stage in jellyfish cnidarians. All cnidarians use stinging tentacles to capture prey.

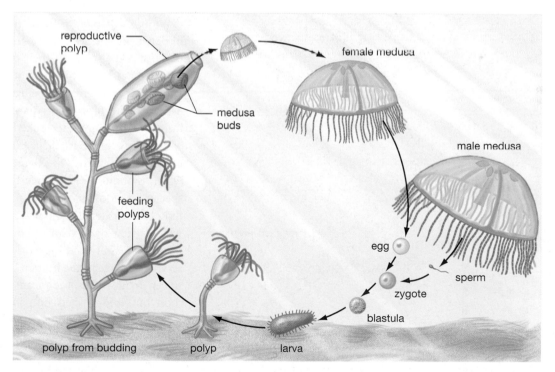

Figure 21.10
Cnidarian Reproduction
This hydrozoan cnidarian, known as *Obelia*, reproduces through both sexual and asexual means. Sexual reproduction can be observed at the right of the figure, where male and female medusae release sperm and eggs, respectively, that fuse to form the new generation of *Obelia*. The polyp that results from this fusion, however, can reproduce by asexual means. Cells bud off from the first polyp (center) and grow into another polyp, shown at left, which maybe the first of many that will develop this way. Polyps within these larger structures specialize: Some are involved in feeding, while others function in reproduction. The reproductive polyps have medusa developing inside them (the medusa buds), which are released to the surrounding water, starting the life cycle over again. The different stages of the life cycle are drawn at different scales.

21.5 Phylum Platyhelminthes: Flatworms

With the flatworms, you've arrived at animals that fit our everyday notions of what animals look like, in that they have something like a head and have the same side-to-side (bilateral) symmetry as, say, a mouse. Further, flatworms go beyond the tissues of the cnidarians and have **organs**, meaning highly organized struc-

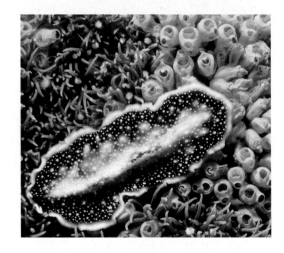

**Figure 21.11
Flatworms**
The flatworm in this picture, *Pseudoceros ferrugineus*, was photographed in the oceans off the coast of the Philippines. Most flatworms are free-living like this one, but there are also many parasitic flatworms that live for at least part of their lives inside a host.

a

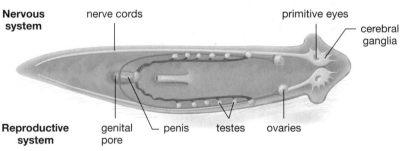

b

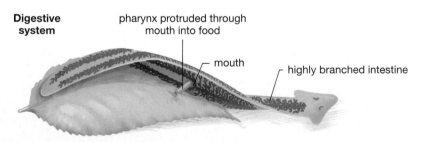

**Figure 21.12
Body Plan of a Flatworm**
Three sets of anatomical features are shown for *Dugesia*, a free-living flatworm.

a *Dugesia*'s nervous system includes primitive eyes and two collections of nerve cells, the cerebral ganglia, that connect to nerve cords that run the length of the animal. In reproduction, *Dugesia* is hermaphroditic, meaning it possesses male sex organs (testes and penis) as well as female sex organs (ovaries and other structures). When two *Dugesia* copulate, each projects its penis and inserts it in the genital pore of the other.

b *Dugesia* feeds by turning its muscular pharynx inside out and projecting it into the food source. Food taken up through the pharynx is distributed through the body by *Dugesia*'s highly branched intestine. Flatworms lack a true circulatory system and thus depend on their digestive system for the distribution of nutrients. Their digestive system lacks an anus, meaning *Dugesia* has only one digestive-tract opening, its mouth.

tures formed of several *kinds* of tissues. (Your heart is an organ that includes nerve and muscle tissues.)

Despite this complexity, flatworms lack a good many features that exist in almost all the phyla further up the complexity ladder. They have no coelom, meaning a central internal cavity. Thus, except for the tube of their digestive tract, they have very little internal space in them. This may sound like a good thing, but it actually limits their flexibility. Further, they have no system of blood circulation, meaning that every cell in a flatworm must get its oxygen directly from the environment around it, through diffusion from water or air outside the body. This is why flatworms are flat—they have to maximize the number of cells that are near an exterior surface, as opposed to being buried deep inside (**see Figure 21.11**). Those flatworms that are on land tend to be found under stones or rotting wood; many live at the bottom of the ocean, and a few live in freshwater. Some 25,000 species of flatworm have been identified.

Humans are unlikely to notice flatworms since, of the minority that live around us, most are very small. When they do come into our consciousness, it is often in the unpleasant role of parasites. Two of the three traditional classes of flatworms live strictly as **parasites**, meaning organisms that feed off their prey but do not kill them, at least immediately. Some of these worms can infect human beings. There are, for example, the tapeworms that can enter the human body in undercooked meat or fish. Tapeworms can be small, like most flatworms, but some reach up to 20 meters in length, which is about 66 feet. Then there are several species of another variety of flatworm, the flukes, that cause the disease schistosomiasis, which affects more than 200 million people in tropical areas around the world, often causing serious damage to the human bladder, liver, or spleen. With flukes, the means of entry is not food but unprotected human skin, as with people who are standing in fishing areas or flooded rice paddies.

The flukes are in one class of flatworms (Trematoda), while the tapeworms are in another (Cestoidea). There is then a third class of flatworms, the Turbellaria, whose members generally are not parasitic, but free-living. If you look at **Figure 21.12**, you can see something of the anatomy of one variety of tubellarian, the flatworm *Dugesia*, which is likely to be found crawling under rocks at the bottom of freshwater streams. About 1 centimeter or less than half an inch long, *Dugesia* has a "head," in that one end of it has not only a pair of primitive eyes (called *ocelli*) but also a concentration of nerve cells (the cerebral ganglia) that are connected to nerve trunks that run the length of the body.

Given this concentration of features in *Dugesia*'s head, the location of its mouth may come as a surprise. It's located midway down its body, on its ventral (bottom) side. To feed, it shoots a muscular structure called a pharynx outside its mouth, coats the prey with enzymes to break down its tissues, penetrates the prey with the pharynx, and then sucks the resulting semi-liquid material back in. (Your pharynx is the area at the back of your throat where your windpipe and digestive tract meet. Imagine being able to shoot it outside your mouth.) Food goes from the pharynx into an intestine, where it is digested and then simply diffuses into nearby tissues. Remember that flatworms have no blood circulation system. Thus, *Dugesia*'s intestine has to be highly branched because it is serving as both a food digestion and a food delivery system. Flatworms have no anus; their mouth is the single opening to their digestive tract, and waste material is expelled out of it.

Like most flatworms, each *Dugesia* has both ovaries and testes (the organs noted earlier). It is thus **hermaphroditic,** meaning a state in which one animal possesses both male and female sex organs. When two *Dugesia* copulate, each projects its penis and inserts it in the genital pore of the other. Most turbellarians can also reproduce asexually by a straightforward means: They break themselves in half. The posterior ("tail") part of the worm grasps the material beneath it while the anterior portion moves forward. The separated halves then grow whatever tissues they need to make themselves complete worms. Members of phylum Platyhelminthes, such as *Dugesia*, have organs thanks to a critical embryonic addition found in flatworms, and in all animals more complex than flatworms. Recall from the earlier discussion of the central body cavity (the coelom) that all animal embryos have what are known as germ layers, meaning layers of cells that *become* various types of tissue in adult animals. These layers are the endoderm, mesoderm, and ectoderm. If you looked at the embryos of sponges or jellyfish just past the blastula stage, you would see that they have only the endoderm and ectoderm. But Platyhelminthes and all the other more complex

phyla also have a third layer, the mesoderm, literally in the middle of the other two. (Endoderm is initially an inner layer of cells and ectoderm an outer layer; mesoderm comes in the middle.) Animals with the three germ layers are said to be **triploblastic**, while the sponges and cnidarians, with two layers, are **diploblastic**. The addition of mesoderm was an important event in animal evolution because the interactions between germ layers in embryos allow for the development of more complex structures. You may recall that the "true coelom" that is found in more complex animals develops from mesoderm. (For details on germ layers, see Chapter 27 on animal development.)

The Protostomes

21.6 Phylum Nematoda: Roundworms

Nematodes exist in such enormous numbers that it is impossible to imagine life without them. They fit into numerous ecological roles, such as the group of "decomposers" that also includes bacteria and fungi. Nevertheless, many farmers in the United States would *like* to imagine life without roundworms, because these animals are a major cause of damage to such crops as soybeans and corn. Farmers and research scientists have been united for decades in an entrenched battle against these pests. Adding to negative human views about roundworms is the fact that some them are human parasites. The disease trichinosis, usually brought on by eating undercooked pork, is caused by a roundworm; likewise there is hookworm, which affects hundreds of millions of people in warmer climates throughout the world (**see Figure 21.13**).

Figure 21.13
Roundworms
Most nematodes, or roundworms, are ecologically useful to human beings. However, a number of species are harmful to humans, either directly as parasites, or indirectly as crop pests. Pictured is an artificially colored micrograph of the nematode *Toxocara canis*, a parasite that infects dogs. Humans can become infected with *T. canis* eggs if they have contact with an infected dog or contaminated soil. The worm is magnified x 450.

Figure 21.14
Mollusc Diversity
Pictured are species in three of the most important classes of molluscs:

a Gastropods, represented here by a colorful terrestrial snail in the *Liguus* genus. These snails live in trees in southern Florida and through much of the Caribbean, though their numbers are dwindling in Florida.

b Cephalopods. The giant squid *Architeuthis dux*, shown in the picture, can reach a length of about 20 meters or 60 feet, but the example here, prepared for an exhibit at New York's American Museum of Natural History, measured a mere 25 feet in length and weighed 250 pounds. It was netted off the coast of New Zealand in 1997.

c Bivalves. Pictured are the bay scallops (*Argopecten irradians*) that are part of many seafood dishes. Note the small blue dots around the edge of the scallop. These are a few of the animal's many primitive eyes.

An acre of prime farmland might contain several hundred billion roundworms, but you don't have to go to farmland to find these animals. They exist in desert sands, polar ice, ocean bottoms, lakes—name the place where life is going on, and the nematode worms are likely to be among the living. Indeed, scientists have only the vaguest idea of how many *species* there are; about 15,000 have been catalogued, but 100,000 seems to be the minimum number that actually exist.

Most roundworm species are very small—the crop pests, which feed on roots, generally are microscopic—but there are parasitic roundworms that are several meters long. The roundworms that have been studied most intensely are those that cause disease and crop damage, and these have a characteristic appearance: transparent, smooth, C-shaped, and cylindrical with tapered ends (hence the term roundworm). Across the breadth of roundworm species, however, there is a great diversity in form.

In terms of animal features, Nematoda is notable because it is the first phylum we have gone over that has the important structure noted earlier, a central body cavity—technically a "pseudocoel" in the roundworms' case. Roundworms generally are of one sex or the other, meaning they are not hermaphroditic, as the flatworms are. Reproduction is always sexual; unlike flatworms, roundworms can't reproduce by pulling themselves apart. In contrast to the animals we've looked at so far, roundworms have a digestive tract with two openings—an anus as well as a mouth.

21.7 Phylum Mollusca: Snails, Oysters, and Squid and More

With the molluscs, we reach a phylum that has lots of members who are familiar to us. Everyone has seen a snail, many of us have eaten clams or mussels, and squid and octopus are probably known to us from television or books. As you think about these different kinds of molluscs, though, it's easy to see what a varied phylum they are part of. What do a snail and an octopus have in common?

Actually, molluscs are even more varied than these examples would suggest. Some molluscs are slugs and others worm-like; some have highly developed brains and others no brains at all; some are filter feeders and others fierce ocean predators. Some molluscs are very small, but one is a behemoth that is one of the enduring mysteries of the animal world. The giant squid, *Architeuthis dux*, reaches a length of about 20 meters or 60 feet and weighs up to a ton (**see Figure 21.14b**).

Depending on the expert consulted, there are between 50,000 and 100,000 described species of molluscs. They live in freshwater, saltwater, and on land. There actually are eight classes of molluscs, but here we will look only at the three most important classes, the **gastropods** (snails, slugs), the **bivalves** (oysters, clams, mussels), and the **cephalopods** (octopus, squid, nautilus).

What features unite this huge, diverse group as a single phylum? Outside the realm of technical anatomical details, there really is only one feature common to all molluscs. If

a

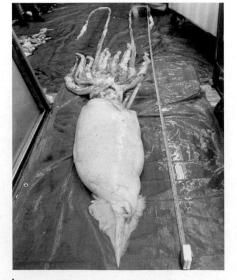

b

c

you look at **Figure 21.15a**, you can see an aquatic snail that displays this feature along with others. Note a layer near the dorsal (top) surface of the snail called a *mantle*. This is a fold of skin-like tissue that surrounds the upper body of the mollusc and that usually secretes material that forms a shell (though there are lots of molluscs without shells).

Past this universal feature, the snail displays a number of structures that are common to *most* molluscs. The mantle generally drapes over part of the mollusc body like a tablecloth, meaning there is a mantle cavity—a space underneath the mantle. This cavity generally is the site of gills, by which all aquatic molluscs obtain oxygen, and by which the filter-feeding bivalves (clams, mussels, oysters) obtain not only oxygen but food. In land-dwelling molluscs, such as garden snails, much of the cavity is a primitive lung that evolved from the gill.

Then there is the single or "unitary" foot as shown on the snail's ventral (bottom) side. Through wave-like muscular contractions, this foot can propel the mollusc along the surface underneath it, be it a leaf or the ocean bottom. Some sessile bivalves, such as oysters, have lost the foot altogether, however, and, as you'll see, in squid and octopus it has evolved into something else entirely. Note, in the blowup of the snail's mouth, there is a tooth-covered membrane called the *radula*. Present in most molluscs except the bivalves, this organ can be extended outside the mouth, where it serves as a kind of rasping file. Molluscs use it for scraping a surface or for cutting small prey; then they retract it back into the mouth, thus moving food into their digestive tracts.

Note that the snail has some features we haven't yet seen in our tour of animals, though we're very familiar with them in ourselves. These are a heart and a stomach. (It also has a kidney, though it is not shown.) These organs obviously aren't unique to snails (or even molluscs), and that's just the point. Along with gills and lungs, these organs are signs of the molluscs' advanced position in terms of structural specialization.

Why does something like a heart exist? Because creatures as large as molluscs need a *delivery system* for food and oxygen. Simple diffusion of food from a gut won't work anymore, because some tissues are buried so deep they would never be served. The delivery system's vehicle is blood, but that blood must be propelled by some force so that it can make a set of rounds through the animal, picking up oxygen from the gills and food from the digestive tract. The propelling force for this work is the beating of the heart.

The circulation of the cephalopods (squid, octopus) is like ours in that it is a **closed circulation system**—"closed" because the blood *stays within* the blood vessels, while the oxygen and nutrients diffuse out of the vessels to the surrounding tissues. Most molluscs, however, have an **open circulation system** in which blood flows out of the vessels altogether, into spaces or "sinuses" where the blood *bathes* the surrounding tissues. Veins collect the blood in these sinuses for a return trip to the gills and the heart. Gastropods such as snails can be either single-sex or hermaphroditic, and this is the case for the bivalves as well. All cephalopods, however, are single-sex.

Figure 21.15b gives you some idea of one set of variations evolution has wrought with the basic body plan of the mollusc. Remember how the unitary foot is a usual mollusc feature? Look at what has become of it in the squid; it has been divided into eight arms

21.7 Phylum Mollusca: Snails, Oysters, and Squid and More

Figure 21.15
The Body Plans of Molluscs

a Aquatic snails such as this one have features that are common to many molluscs. All molluscs possess a skin-like tissue called a mantle, which secretes material that can form a shell, whether it is an external shell (as in the snail) or an internal shell (as in a squid). Snails also possess a tooth-covered rasping structure for feeding, called a radula, and they are complex enough to have evolved some specialized organs that are familiar to us—the heart and the stomach. The unitary "foot" of the snail, which works through wave-like contractions, is another common mollusc feature.

b In the squid, the mollusc's unitary foot has evolved into a multisegmented appendage of eight arms and two tentacles. Note how other mollusc features have been modified in the squid. The mantle cavity houses not only gills, but a siphon the squid jets water out of, thus propelling itself. The radula now exists close to a parrot-like jaw.

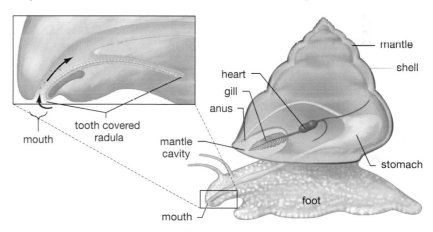

a Aquatic snail

mouth · tooth covered radula · mouth · mantle cavity · heart · gill · anus · mantle · shell · foot · stomach

b Squid

tentacles · arms · jaw · eye · stomach · internal shell · "foot" · radula · siphon · mantle cavity · mantle · gill · heart

(with suckers) and two longer tentacles. The radula is still there, but now it lies next to a set of jaws that operate something like a parrot's beak. Meanwhile, the mantle cavity now has another function; the siphon you can see at one end of it acts like a jetting water hose that not only helps propel squid movement—they expel water out of it—but also directs this movement by being aimed first one direction and then another.

Cephalopods deserve mention for another reason, which is that they are not only the biggest invertebrates in the world, they are the smartest. One researcher who worked for decades with octopus said that they had the intelligence of dogs, which is very smart indeed in the animal world. Many researchers would go further, however, saying that octopus and squid seem to plan and calculate in very sophisticated ways. These animals also are fast swimmers, have keen eyesight, can change color in an instant (to hide from predators), and as a last resort in defending themselves, have the ability to shoot out black "ink" from a special sac. Given their remarkable capabilities, the curious thing about these molluscs is that they live only a year or two; most animals this sophisticated live much longer lives.

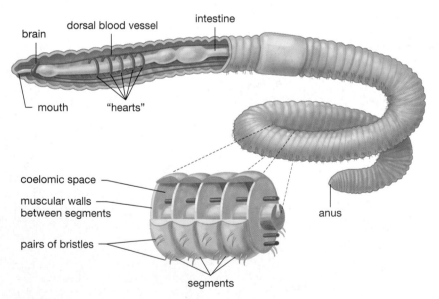

Figure 21.16
The Body Plan of an Earthworm
The segmentation of body parts—a widespread feature in the animal kingdom—is clearly seen in earthworms, whose segments are repeated hundreds of times. In general, such segmentation provides flexibility in combination with strength. In the earthworm, it also allows the actions being carried out in one segment to differ from those being carried out in another. Note, in the blow-up of the earthworm's body, the muscular walls between segments; these give the earthworm independent control of the segments and the bristles that can be extended from them. The earthworm also expands our notion of what a "heart" amounts to. It has a dorsal blood vessel that contracts, as do our human hearts, giving the main impetus to the circulation of red blood that moves through the worm's system. But it also has five pulsating vessels on each side of its intestine that serve as accessory hearts.

21.8 Phylum Annelida: Segmented Worms

More worms? You may have wondered why three of the nine phyla covered in this chapter are worms. Evolution resulted in a lot of worm or worm-like phyla—perhaps a dozen altogether—but the reason so many are being covered here is not because they are so numerous, but because they occupy interesting places in the spectrum of animal diversity.

Annelid worms occupy such a position, in that they provide the clearest example of a feature that exists in a lot of animals, including ourselves: **body segmentation**, which can be simply defined as a repetition of body parts in an animal. Just as a set of identical Lego blocks can be snapped together to make a larger structure, so can a repeated set of segments be connected to make a larger body. Think of the vertebrae that make up your backbone; 24 of them run in a line from your neck to the base of your back, each of them having a similar basic structure and a similar function (support and protection).

Segmentation is so useful in animals that it evolved independently in them at least twice: once in the protostome lineage that you've just been going over and then once again in the deuterostome lineage—more specifically in our own phylum, Chordata. Segmentation hasn't been mentioned until now because the annelid worms are the least complex animals to have it. You'll be seeing it again soon, however, in the arthropods (insects, spiders, crustaceans).

Though segmentation is widespread, nowhere in the animal kingdom is it more visible than with the annelid worms. Indeed, the name Annelida comes from the Latin word *anellus*, or "little ring." That pretty much describes what we see in the best-known annelid, the common earthworm, whose body seems to be composed of a series of little rings that have been joined together (**see Figure 21.16**).

Segmentation is valuable in earthworms for the same reason that individual rooms are valuable in a house: The activities being carried out in one room can *differ* from those being carried out in another. Each of the earthworm's 100–150 segments is separated from adjoining segments by a partition, and each segment can function independently of the others to some degree. The muscles in individual segments (or groups of segments) can contract independently. This, in turn, makes for independent control of needle-like bristles that the worm can extend from each segment.

These features account for the mesmerizing way in which earthworms move. One set of muscles toward the rear of the earthworm contracts, making the worm thick and compact in the area, even as the bristles in these segments are being pushed into the ground like so many stakes. Now, with its hindquarters fixed in place, the worm uses a different set of muscles to make the mid-portion segments *lengthen*, thinning out as they do, meaning this part of the worm has moved forward. Then the worm plants bristles in the front, anchors itself, and literally brings up its rear (see **Figure 21.17**). Note that this movement is possible only because of segmentation. If the *whole worm* lengthened, without part of it being compacted and fixed in one spot, it would get nowhere.

Segmentation is important in annelids, then, and is thought to have evolved in them precisely because of its power to help them move. (Earthworms use this power extensively, because they burrow in the dirt for hours at a time, swallowing soil for the nutrients it contains and eating leaves or other decaying vegetation.) In more general terms, segmentation is valuable to living things because it can confer flexibility while maintaining strength. We can see this in our own "backbones." Imagine how inflexible you would be if, instead of 24 moveable vertebrae, you had a single bone running from your skull to your posterior.

Unlike the other worms you've looked at, earthworms generally are helpful to human beings—and not just as fish bait. They churn up soil, aerating it and depositing their cylindrical "worm casting" waste, which enriches the soil. Earthworms have no lungs; they take

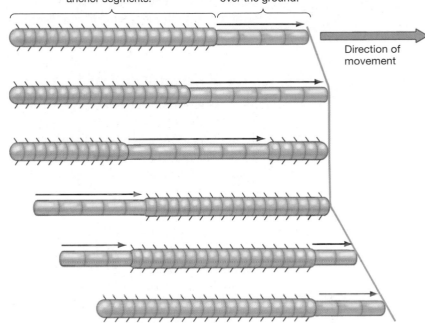

In these segments, one set of muscles contract, shortening segments and extending bristles to anchor segments.

In these segments, another set of muscles contract, elongating segments and withdrawing bristles. These segments advance forward over the ground.

Direction of movement

Figure 21.17
Earthworm Locomotion

in oxygen right through their skin. Their skin needs to be moist for this to happen, however. The great threat to an earthworm is drying out, which is why these "nightcrawlers" generally stay underground during the daytime. Like the flatworms you looked at earlier, earthworms are hermaphrodites. Each member of a copulating pair inseminates the other.

Though we've focused on earthworms, they actually represent only one small part of phylum Annelida, which contains about 17,000 named species. Most annelids actually are ocean-dwelling, and some of these colorful creatures do not fit in with our idea of drab worms (**see Figure 21.18**). There also are freshwater annelids, who are part of the same class as the earthworms but who tend

Figure 21.18
Annelid Diversity
Not all annelid worms are drab and not all live underground.

a Polychaetes are ocean-dwelling worms. Pictured is a Hawaiian Christmas tree worm in the genus *Spirobranchus*. These polychaetes stay in one spot, burrowed into tubes they themselves have constructed, generally within a coral. The bushy "Christmas trees" that extend from the worm's head are structures called radioles that it uses to filter-feed from the water.

b Leeches are a smaller class of mostly parasitic annelid worms. Pictured is a leech that is feeding from a human arm via suckers at both ends of its body. This particular leech, *Hirudo medicinalis*, serves a therapeutic function. These leeches are used today in connection with human reattachment and plastic surgery. Enzymes they secrete serve as anticoagulants—they keep blood from clotting—and help reduce pain at the site of surgery. A leech might be attached at the site of surgery for half an hour or so and then be removed, after which its enzymes will continue working for hours longer.

a

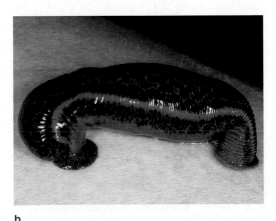

b

to be much smaller. Then there is the smallest class of the annelids, which is composed entirely of leeches. Most leeches live up to their reputation of being blood-sucking parasites. Though most live in freshwater, some live in the ocean while others dwell in damp-land habitats.

21.9 Phylum Arthropoda: So Many, but Why?

Phylum Arthropoda is large and varied indeed. There are lots of ways to group the animals in this phylum, but for our purposes, it will be convenient to think about them batched into three "subphyla." (**See Figure 21.19**.) These are:

- Subphylum Chelicerata (kay-LIS-e-RA-ta), which includes spiders, ticks, mites, and horseshoe crabs, among other organisms
- Subphylum Crustacea, which includes shrimp, lobsters, crabs, and barnacles, among other organisms
- Subphylum Uniramia, which includes insects, millipedes, and centipedes, among other organisms

Listed there among the arthropod types is one group that is so important it deserves special mention right away. This group is the insects.

Who can comprehend how many insects there are in the world? There are more species of them—about a million have been counted so far—than there are species in all the other groupings of animals combined. If we could get an accurate count of their numbers, we would probably find that there are *tens* of millions of insect species. To look at their numbers another way, for every person on Earth, there may be 200 million insects. The vertebrates in the animal world have been the biggest, fiercest creatures on land for the last 350 million years. But size and strength count for only so much in the face of the numbers the insects can put up. Here is the lead sentence from a September 2000 article in the Australian *Financial Review*:

> Extensive spraying could start as early as next week to try to control an impending locust plague which threatens to cause up to $500 million damage to Australia's forecast big grain harvest.

Had the *Financial Review* existed thousands of years ago, it could have made this same kind of report. Consider this passage on locusts from the Bible's Book of Joel, which dates from about 2,500 years ago:

> For a nation has invaded my land, powerful and innumerable . . . It has laid waste my vines, and splintered my fig trees . . .

The battles that insects and human beings have been engaged in for so long have involved not only plagues of locusts, but the bubonic plague (carried by fleas), malaria (transmitted by mosquitoes), and the everyday predations of lice, termites, and ants. On the other hand, we have insects to *thank* for a good deal of the plant growth we see in the world, in that insects pollinate plants. One of the real challenges to biology is to figure out why there are so many insects in the world, something you'll learn more about shortly.

Figure 21.19
Big Numbers, Much Diversity
The arthropods represent the largest animal phylum, in sheer numbers as well as in diversity. Shown are representatives of the three arthropod subphyla:

a Chelicerata, represented here by a web-building *Argiope* spider, this one found on the Snake River at the Idaho-Oregon border.

b Crustacea, represented here by a flame lobster from Hawaii. Note the well-defined arthropod characteristics: an exoskeleton and the legs that exist as paired, jointed appendages.

c Uniramia, represented here by a short-horned grasshopper native to western North America.

a

b

c

As noted, insects constitute just one part of the diverse arthropod phylum. What characteristics do all the arthropod animals share? First, all arthropods have an **exoskeleton**, meaning an external material covering the body, providing support and protection (**see Figure 21.20**). Second, all arthropods have what are known as **paired, jointed appendages,** which are just what they sound like: appendages, such as legs, that come in pairs and that have joints. (The word arthropod actually means "jointed leg.") The *combination* of the exoskeleton and the jointed appendages has a good deal to do with the terrific success of the arthropods.

Consider first the exoskeleton. It is made of a tough carbohydrate called chitin that comes embedded in protein. The crustacean arthropods (crabs, lobsters, and so forth) have, in addition, a calcified component to their skeleton that makes it very hard, as anyone who has ever "cracked" a lobster knows. Such a skeleton is not only protective but also serves to anchor arthropod muscles very securely. Your own muscles work by pulling against your *endo*skeleton—your internal skeleton. Arthropod muscles pull against a skeleton too, but it's a skeleton that surrounds their bodies.

Strong though it may be, the exoskeleton only works because it is also flexible. Like a knight's armor, the exoskeleton has plates that overlap and thin, flexible sections between these plates that can bend as necessary. Even with this flexibility, however, the exoskeleton presents a problem: The arthropod body can only grow so much before it expands right into the exoskeleton.

So how does the arthropod solve this problem? Its solution is **molting**, meaning a periodic shedding of an old skeleton. A crustacean such as a crab will retreat to a relatively safe place and begin to slip out of its old exoskeleton, as if it were Houdini trying to get out of a straightjacket (**see Figure 21.21**). Through chemical processes, the old skeleton has begun to split by this time. Since the new, developing exoskeleton lies just underneath the old one, you might think this process would leave the crab just as hemmed in as before. But at this stage the new skeleton can be *stretched*, and the crab proceeds to do just that by ingesting a large amount of water, thus inflating both its body and the skeleton to a new size. The new skeleton then hardens in its inflated size, the crab loses the extra water, and shrinks to its previous size. The result? It now has room to grow. Insects pull off this same feat, but stretch themselves with air instead of water.

The jointed appendages that arthropods possess come in many forms. There are legs for walking, claws for predation and defense, and wings for flying. These appendages are flexible and often come with sensory attachments, such as sensory "hairs," that keep the arthropod in touch with its environment.

Together, jointed appendages and an exoskeleton make for an animal that is often nimble and well defended—a winning combination. In addition, arthropods often have keen senses, such as sight and chemical perception (as with ants following a "pheromone" or chemical scent trail).

One other thing seems to have done much to bring about the success of at least the insects among the arthropods, though this factor is

Figure 21.20
Arthropod Features
Many arthropod bodies are made up of three large-scale sections: head, thorax, and abdomen. Like annelids, arthropods are segmented. This segmentation is clearly visible in the abdomen region of the grasshopper, but is less pronounced in the head and thorax because there the segments have fused. The entire arthropod body is covered in a rigid exoskeleton that is jointed to allow movement.

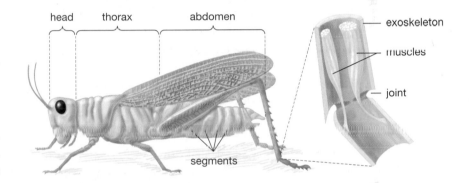

Figure 21.21
Old and New
Arthropods have an outer or exoskeleton of a fixed size. They are able to grow by periodically shedding, or molting, this stiff skeleton. The crab in the photograph has just finished molting; its old exoskeleton lies nearby.

only indirectly related to physical features. It is that many of the insects—beetles in particular—were rather like investors who happened to pick winning companies to buy stock in. At a certain point in *plant* evolution, some insects made a switch from feeding on the more primitive gymnosperm plants to feeding on a then-emerging group of plants, the angiosperms or flowering plants. (See Chapter 20, page 426, for the differences between these varieties.) Because the flowering plants are better competitors, they eventually became the dominant plants on Earth. Research done in the late 1990s showed that those lines of beetles that made the switch to angiosperms proceeded to evolve new species at a furious pace compared to beetles that stayed with the gymnosperms. The flowering plants were like companies whose superior competitive abilities allowed them to develop new "markets" for themselves—that is, new ecological niches, or ways of making a living in nature. The beetles that fed on these plants developed right along with them, diversifying in the new niches and going on to spur even more angiosperm speciation.

Insect Characteristics

If you look again at Figure 21.20, you can see the body parts of a typical insect. The three larger body regions featured in this grasshopper—head, thorax, and abdomen—are the rule in insects. In the abdomen, you can see a regular segmentation that looks more like what you saw with the annelid worms. Arthropods are segmented animals,

but evolution has made their segmentation less regular than that of the annelids.

The main cavities in the insect are the sinuses, or spaces, that its blood flows into. This blood bathes tissues in an "open" circulation system that almost all arthropods have. Sexes are separate in the insects and reproduction is generally sexual. We'll close our review of the arthropods with brief looks at a few groups of them other than insects.

Other Uniramians: Millipedes and Centipedes

Millipedes and centipedes are, with the insects, part of arthropod subphylum Uniramia (**see Figure 21.22**). Millipedes don't really have a thousand legs, despite their name, but hundreds of legs are possible, as these creatures have up to 100 body segments, each of which has two pairs of legs. This structure makes for movement that is slow but relatively powerful as millipedes make their way across the damp, dark environments they favor, which often means the forest floor.

Centipedes look much like millipedes, but are a different class of arthropod. They have only one pair of legs per segment and a pair of more prominent antennae extending from their head. Moreover, centipedes are carnivores, whereas millipedes feed largely on decaying plants. Centipedes come equipped for their predatory work as their front two legs have become modified into venom-injecting fangs. This may sound ominous, but only a few tropical species pose a threat to human beings.

Figure 21.22
Millipedes and Centipedes
Insects constitute only one of the three classes of arthropods in the arthropod subphylum Uniramia. Millipedes and centipedes make up the other two classes.

a Millipedes have two pairs of legs per segment and live mostly on decaying plant matter.

b Centipedes have only one pair of legs per segment and are carnivores. Shown is the Texas giant centipede, *Scolopendra heros*.

a

b

Subphylum Crustacea: Shrimp, Lobsters, Crabs, Barnacles and More

What do a crab and a barnacle have in common? Remember that in putting living things together in a category (or taxon), biologists are first and foremost concerned with shared *ancestry*, not with whether one creature has the same general appearance as another. At a crime scene, convincing evidence is likely to be found in small details, such as hair samples and fingerprints. The same thing holds true for evidence about the relatedness of groups of animals: the telling details are likely to involve small features.

This turns out to be the case with crustaceans. What all of them have in common is five pairs of appendages extending from their heads—two pairs of antennae and three pairs of feeding appendages. If you look at **Figure 21.23**, however, you can see the fantastic diversity that evolution wrought in crustaceans.

One of the crustaceans pictured in Figure 21.23, the water flea in the genus *Daphnia*, certainly is not familiar in the way that crabs or barnacles are, but it is representative of a variety of small, aquatic crustaceans that exist in huge numbers, mostly in the oceans, but also in freshwater bodies. As a whole, the crustaceans are mostly ocean-dwelling, but many live in freshwater, while a few live in damp-land

environments. Crustaceans are known to humans mostly as a food source, such as shrimp and lobsters, or as a marine pest, such as the barnacles that attach themselves to ships. It is not just humans who use crustaceans as a food source, however. Closely related to shrimp are the crustaceans known as krill. Generally no more than an inch long, these animals feed on the plant life that floats in the icy ocean waters off Antarctica. Traveling in huge, dense swarms, they then become an important part of the diet of whales, seals, and penguins.

Subphylum Chelicerata: Spiders, Ticks, Mites, Horseshoe Crabs, and More

With the Chelicerata, we have another example of small physical features being more important than general body structure in letting us know who is related to whom. You might think that horseshoe crabs should be categorized as crustaceans, along with the other crabs, but it turns out that horseshoe crabs aren't really crabs at all. If you looked at the underside of a horseshoe crab, near its mouth, you would see two small appendages called *chelicerae*, which have pincers and which are used by the crab to bring food to its mouth. If you looked near the mouth of a common spider, you would again see two chelicerae, only now each chelicera has

a

b

c

Figure 21.23
Crustacean Diversity
Diversity among the arthropod subphylum Crustacea is stunning.

a Barnacles are representative of one class of crustaceans. When seen on a pier or ship, barnacles may seem lifeless to us, but notice the feathery extensions coming from these barnacles in the south Pacific. These are cirri, which barnacles extend from their shell and then use to catch drifting plankton.

b Water fleas in the genus *Daphnia*, such as the one shown here, live in huge numbers in ponds and lakes. Though barely visible with the naked eye, they are complex animals. Under microscopes, their tiny hearts can be seen beating a furious rate. As they are not insects, they are not really fleas. They get their name from the jumping, jerky motions they make in the water. Note the clearly visible digestive tract in this water flea

c Shrimp are in the same crustacean class not only of lobsters and crabs, but of the pill bugs so often seen when we turn over rocks. Shown is the colorful fire shrimp, *Lysmata debelius*.

become modified to become a venom-injecting *puncturing* appendage, rather than a grasping appendage. Ticks, mites, and all the other chelicerates likewise have paired chelicerae, from which this subphylum gets its name. (**see Figure 21.24**).

Horseshoe crabs may be familiar to anyone who has spent time on the beaches of the Atlantic seaboard in May or June. Fearsome though they may look, they are harmless creatures that feed on worms, clams, or whatever else they can scavenge from the ocean bottom.

It is another grouping of chelicerates, however, that really inspires human fear: the arachnids, a class that includes mites, ticks, and scorpions, along with the more familiar spiders. Why "arachnophobia" should be so widespread in human beings is something of a mystery since, of the 34,000 identified species of spiders, only a handful cause any real harm to people. Part of the problem may be that spiders are blamed for wounds actually inflicted by other creatures. In one study of 600 suspected "spider bite" occurrences in southern California, 80 percent turned out to be the work of other arthropods—mostly ticks and some of the true bugs of the insect world (meaning bedbugs and other members of their order). Evolution gave spiders one of the world's great prey-snaring abilities, which is the spinning of their remarkable silk webs.

If we want to look for true causes of human misery, a better place to start would be with another arachnid, the microscopic dust mite, which is now known to be a major cause of asthma and other human allergic reactions. In the last decade the mite's fellow arachnid, the *Ixodes* tick, has become feared because of its role as a carrier of the bacterium that spreads Lyme disease (named after the town of Lyme, Connecticut, where the disease was first identified). If left untreated, this tick-borne disease can have long-term effects ranging from arthritis and numbness to facial paralysis.

The Deuterostomes

21.10 Phylum Echinodermata: Sea Stars, Sea Urchins, and More

With the echinoderms, we cross into the animal grouping noted earlier, the deuterostomes, that also includes our own phylum, Chordata. You might expect that, as our relatives, the echinoderms would have the kind of sophisticated features we've just been looking at—brains, well-developed sense organs, strictly sexual reproduction—but none of this is true in the best-known echinoderm, the sea star (**see Figure 21.25a**). It has no brain, its only sensory organs are the

a

b

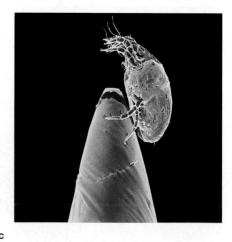

c

Figure 21.24
Chelicerate Diversity
Horseshoe crabs, spiders, and mites—all members of the arthropod subphylum Chelicerata—possess paired chelicerae, which are pincer-like appendages.

a Horseshoe crabs are a familiar sight on East Coast beaches in the springtime, as they gather there to mate and lay eggs. In this picture, a group of males is competing for access to a female.

b The much-feared but generally harmless tarantula, this one a dry-weather Chilean rose tarantula, *Grammostola spatulata*, that is often kept as a pet. Like all spiders, the tarantula is a member of the Chelicerata class Arachnida, from which we get the term arachnophobia.

c Dust mites actually are eight-legged arachnids as well. The dust mite shown is standing on the head of a sewing needle. Such mites exist in almost all homes in the millions. Body fragments and excrement from them actually make up a large portion of the "dust" that is visible when sunlight shines into a room. This material sets off allergic reactions in many people.

primitive eyespots found on the tips of its arms, and a whole new sea star can be regenerated from the arm and part of a central disk of an existing sea star. Even the bilateral symmetry that you've seen in every animal since the jellyfish is absent in the adult sea star: It has *radial* symmetry, like the jellyfish.

This is not to say, however, that the sea star and its echinoderm relatives are completely primitive. Most have a remarkable system for moving that involves forcing water into a series of suction-tipped tube feet that then extend, like so many inflated water balloons from the underside of the sea star's arms (**see Figure 21.26**). Moreover, the sea star evolved into radial symmetry *from* bilateral symmetry. How do we know? Sea star larvae have bilateral symmetry, a quality that is lost when the larvae develop into adults. Looked at together, these echinoderm features add up

to a phylum that is something of a mystery to zoologists—a mixture of characteristics from all over the animal map.

Sea stars aren't the only familiar echinoderms. Sea urchins, with their spine-covered spherical bodies, are echinoderms as well, as are sand dollars. Then there are the less familiar, but well-named sea cucumbers (**see Figure 21.25b**). There are about 7,000 identified species of echinoderms, all of them ocean-dwelling and nearly all of them inhabiting the ocean floor, though "floor" in this case may just mean the bottom of a tidepool. Some are sessile, but most move across the ocean floor at a slow pace. Their shared habitat and measured pace make sense when we realize that the entire phylum evolved from a group of sessile filter-feeders.

Despite their slow movement, echinoderms can be formidable predators. Sea stars

Figure 21.25
Echinoderm Diversity
The sea star, sea cucumber, and sea urchins shown here represent three of the five classes of the echinoderm phylum.

a Sea stars, undoubtedly the best-known, echinoderms, are fierce predators. Here a sea star is managing to open the shell of a bivalve known as a rock cockle.

b True to its name, this sea cucumber does look something like the vegetable, but sea cucumbers come in many shapes and colors. Like most echinoderms, sea cucumbers move slowly along the ocean floor.

c A group of sea urchins, foraging on some kelp.

a

b

c

Figure 21.26
The Body Plan of a Sea Star
The sea star is radially symmetrical in its adult form, and it has no brain. Its only sensory organs are the primitive eyespots found on the tips of its arms. It has, however, evolved a remarkable system for moving along the sea floor and grasping prey. It first takes in water from the surrounding sea and then channels this water into a series of water bulbs. When inflated with the water, these bulbs lengthen, functioning as suction-tipped "tube feet."

often feed on molluscs, such as oysters and mussels. In some cases they use the suction tubes on their feet to pry open a mollusc's shell just a little. Then they "evert" their stomach—turn it inside out—and slide it in through the narrow gap they have created between shell halves. Juices secreted by their stomach then start to digest the prey's tissues. Sea urchins, meanwhile, are some of the animal kingdom's most voracious algae eaters.

21.11 Phylum Chordata: Mostly Animals with Backbones

Say the word "animal," and most people think of a *vertebrate* animal—be it tiger, elephant, deer, lizard, or some other large and probably toothy creature. One thing that ought to be clear by now, however, is what a small portion of the animal world vertebrates make up. Nevertheless, there are a lot of vertebrate species. The 50,000 or so that have been identified are nothing compared to the 1 million species of insects, but there are more species of vertebrates than there are of sponges, or cnidarians (jellyfish, etc.), or flatworms, and the vertebrate numbers may match those of molluscs.

In other ways, too, the vertebrates have been a big success. All the largest and swiftest animals are vertebrates, whether in water or in air or on land. The "big, fierce" vertebrate land dominance noted earlier has gone on unchecked for 350 million years. And the sheer variety of sensory capabilities that vertebrates have—from echo-location to reading—is unrivaled anywhere in the living world. On the other hand, if one measure of an animal

grouping is its effect on *other* groupings, then the vertebrates need to be judged more harshly, as one vertebrate species—our own—has to be counted as one of the most destructive groups of living things ever to have arisen.

Large and important as they are, the vertebrates don't even constitute a phylum; instead they make up only a subphylum, Vertebrata. The larger phylum into which vertebrates fit, called Chordata, actually has two additional subphyla in it. One of these (Cephalochordata) is made up of only a single kind of animal, a small, eel-like creature called a lancelet, represented by only 25 or so species (**see Figure 21.27a**). The other subphylum (Urochordata) has about 3,000 ocean-going species in it, but is represented by only three classes of animals. The largest of these classes is made up of the tunicates, whose practice of expelling water earns them their alternative name of sea squirt.

What Is a Vertebrate?

What *distinguishes* vertebrates from these other chordates? Only the vertebrates have just what you'd expect: a **vertebral column**. Better known as a backbone, this flexible column of bones extends from the anterior to posterior end of an animal. Conversely, what *unites* all the chordates are four features, which you can see pictured in a lancelet in **Figure 21.28**. At some point in their lives, all chordates have

- A **notochord**, meaning a stiff, rod-shaped support structure, composed of cells and fluid and surrounded by a lining of fibrous tissue, running from the chordate's head to its tail

Figure 21.27
Chordate Diversity

a A single type of animal, the lancelet, is the sole representative of one of the three subphyla of chordates, Cephalochordata. Lancelets spend much of their time burrowed into the sand or mud of the shallow ocean floor, filter feeding on small organisms or passing food particles. Only a few centimeters long, the diminutive lancelets are chordates, but not vertebrates.

b The African cheetah shown here is a vertebrate, and thus a representative of another of the chordate subphyla, Vertebrata.

c Urochordates, representing the third chordate subphylum Urochordata, come in about 3,000 species, most of them similar to the vase-like tunicates shown here. These animals are filter feeders who spend their adult lives attached to rocks. Their common name, sea squirts, comes from their practice of suddenly ejecting a spout of water from an opening near the top of their bodies. This group was photographed in the ocean off the coast of Indonesia.

a

b

c

- A **dorsal nerve cord**, meaning a rod-shaped structure consisting of nerve cells, running from the chordate's head to its tail
- A series of **pharyngeal slits,** meaning openings to the pharyngeal cavity
- A **post-anal tail**, meaning a tail located posterior to the anus

At this point, you may be wondering where your *own* post-anal tail or pharyngeal slits went, so a little explanation of these features is in order. Although the word "notochord" sounds like "nerve cord," the notochord is a flexible but tough *support* structure, not a nervous system feature. A lancelet retains its notochord throughout life, but most vertebrates lose theirs in embryonic life. (Its outer portion develops into the backbone.)

The dorsal nerve cord may not seem like a feature unique to chordates—you've seen nerve cords in lots of other phyla—but it is the qualifier *dorsal* that makes the difference. The nerve cords you saw in the annelids and arthropods were ventral, meaning they were on the underside of these animals. The vertebrates among the chordates get not only the dorsal nerve cord, but a protective structure that develops around it, the vertebral column. The anterior (top) part of the vertebrate nerve cord undergoes tremendous development, becoming the brain.

The pharyngeal slits you see in the lancelet in Figure 21.28 take on more meaning when you remember that, in both humans and lancelets, the pharynx is a passageway just behind the mouth that can be thought of as the first chamber of the digestive tract. Pharyngeal *slits*, then, are perforations of the pharynx, and these serve different functions in different chordates. In the filter-feeding lancelets, they serve to trap food. Meanwhile, in embryonic fish, the slits *develop into* gills, which fish use to obtain oxygen. And in people? If you look at a month-old human embryo, you can see a series of pharyngeal clefts that for a time have openings in them. These pharyngeal slits, however, develop not into gills but into various other structures, including the space inside the middle ear.

Finally, what about that tail? If you look at **Figure 21.29**, you can see an 8-week-old human fetus with a very clearly developed tail. By the time the fetus is 10 weeks old, however, the tail has almost disappeared. All chordates have a tail at some point. In our oceangoing ancestors, it was used to help move them through the water, as is the case with modern fish. In human adults, though, all that remains of our embryonic tails is a small bone, called the coccyx, located at the base of our vertebral column.

Diversity among the Vertebrates

For the remainder of the chapter, we'll concentrate on the vertebrates among the chordates. The evolutionary relationships among these vertebrates were covered in Chapter 19. For a summary of who gave rise to whom, see Figure 19.17 on page 392.

Figure 21.28
Four Universal Chordate Features
All chordates possess four structures at some point in their lives: pharyngeal slits, a support structure called a notochord, a dorsal nerve cord, and a post-anal tail. In the lancelet, all these features are present in the adult. In humans, only the dorsal nerve cord is clearly evident in adults.

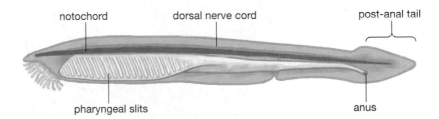

notochord   dorsal nerve cord   post-anal tail

pharyngeal slits   anus

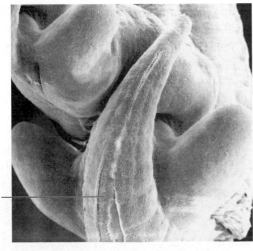

tail

a

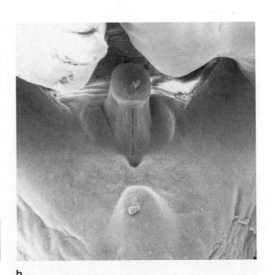

b

Figure 21.29
The Human Tail
(a) Eight weeks after conception, the human embryo has a very clearly developed tail. **(b)** Two weeks later, the tail has almost disappeared. In adult humans the only vestige of our embryonic tail is a small bone at the base of the vertebral column, the coccyx.

A signal shift in vertebrate evolution came with the development of a very familiar feature: jaws. There is general agreement that ancient, jawless vertebrates gave rise to the jawed variety. The development of jaws is considered perhaps the single most important event in vertebrate history, for it gave the vertebrates a great power to capture and consume prey. So useful was this feature that today all true vertebrates have jaws—with one exception.

A Jawless Vertebrate

The lamprey is a long, thin animal that looks something like an eel, but should not be confused with one. Eels are fish, complete with jaws; lampreys, by contrast, have a sucking disk at their anterior end with which they attach themselves to their prey, which are usually fish (**see Figure 21.30**). The lamprey's oral disk has numerous teeth around its edge. It combines these teeth with a rasping tongue to pierce the flesh of its prey. It then hangs on to the victim through suction, continuing to rasp away, ingesting the fluids and flesh that it obtains in this way.

Lampreys are best known in the United States as a destructive force in the Great Lakes. Once confined to the eastern Great Lakes, parasitic lampreys were able to move west by mid-century thanks to human alteration of a canal around Niagara Falls. By the 1950s, they had decimated a once-thriving trout fishing industry in the western Great Lakes. Human efforts to kill off lampreys while reintroducing trout have had only limited success. The main culprit seems to be human pollution, which is keeping stocked trout from reproducing.

Fish: Cartilaginous and Bony

All the early jawed vertebrates were fish, which today are a big success story in the vertebrate world. Of the 50,000 vertebrate species, more than half are fish. And of the 25,000 or so fish species, more than 24,000 are so-called bony fishes, whose name tells their story: with a couple of exceptions, their skeletons are made of bone. Another 750 fish species are cartilaginous fishes, meaning their skeletons are made of the more pliable connective tissue, called cartilage, that also gives shape to human ears.

By far the best-known cartilaginous fish are sharks, but not far behind are the graceful rays that seem to fly through the ocean (**see Figure 21.31**). The sharks' reputation for being an ancient life-form is well deserved. Their lineage dates back more than 350 million years, meaning they had been around for 130 million years by the time the first dinosaurs appeared. Unlike most of their bony fish relatives, sharks have no **swim bladder**, meaning an inflatable organ in a fish that the fish can fill with gas for optimum buoyancy. Fish that do have swim bladders can inflate them as needed to maintain a neutral buoyancy—that is, to float in the water at a given depth without expending any energy. By contrast, a shark must keep swimming or it will sink.

Sharks, rays, and their close kin aside, all the rest of the fish world is made up of bony fish, from goldfish to tuna to herring (**see Figure 21.32**). What accounts for their great numbers today? Their basic body plan seems to have allowed them to adapt to every kind of underwater environment. Some have

Figure 21.30
No Jaws, Powerful Predator
Lampreys, such as the one shown here, are jawless fish. Note the lamprey's sucking oral disk, which it uses to latch onto and then parasitize prey. This sea lamprey, *Petromyzon marinus*, is the species that, thanks to human intervention, was able to invade the western Great Lakes in the mid-20th century. It remains there today, although in numbers reduced from what they have been previously.

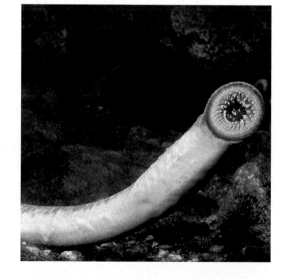

Figure 21.31
Cartilaginous Fish
Sharks and rays, such as the giant manta ray shown here, have cartilaginous, rather than bony skeletons. The ancient cartilaginous fishes represent only about three percent of all fish species. This ray and its accompanying diver were photographed off the coast of Mexico.

evolved a slender shape (eels), others a fantastic camouflage (anglerfish), and still others a biological "antifreeze" for frigid waters (the Antarctic icefish).

Almost all bony fish are so-called "ray-finned" fishes, so named because their dorsal fins are supported by straight-line structures that look like rays (**see Figure 21.32**). Four surviving species of bony fishes belong to another category, however, the lobed-finned fishes. These fish are important because it is their ancestors who brought vertebrates onto the land by struggling out of the swamps some 375 million years ago, using their muscular, lobed fins to propel themselves. Their lobed *fins* evolved into the four *limbs* of the **tetrapods**, meaning the four-limbed vertebrates (see Figure 19.16 on page 391 for this transition). All land vertebrates are tetrapods: amphibians, reptiles, birds, and mammals. Even snakes and whales are tetrapods; these animals simply lost the limbs their ancestors had. If you look at **Figure 21.33**, you can see a picture of one of the rare, living lobe-finned fish, the coelacanth (SEE-la-kanth).

Amphibians: At Home in Two Worlds

Amphibians were the first truly terrestrial vertebrates, having evolved from lobe-finned fishes, but this statement needs qualification. Are amphibians really land animals? Most of them actually live in two worlds: land and water. (Their very name means "double life.") If you look at **Figure 21.34**, you can see the life cycle of the best-known amphibian, the frog.

Figure 21.32
Bony, Ray-finned Fish
About 97 percent of fish species have bony (rather than cartilaginous) skeletons, and of the bony fish, almost all are "ray-finned," meaning that their dorsal fins are supported by stiff rays such as the ones visible on this yellowtail parrotfish, *Sparisoma rubripinne*.

Figure 21.33
Lobe-finned Fish
One of the few living lobe-finned (rather than ray-finned) fish is the coelacanth which was first discovered in modern times in the Indian Ocean near Madagascar. A fish similar to this one is thought to be the ancestor of all land vertebrates. Note its four fleshy lobe-fins. Fins such as these evolved, in the lobe-finned ancestors, into the four limbs of land vertebrates.

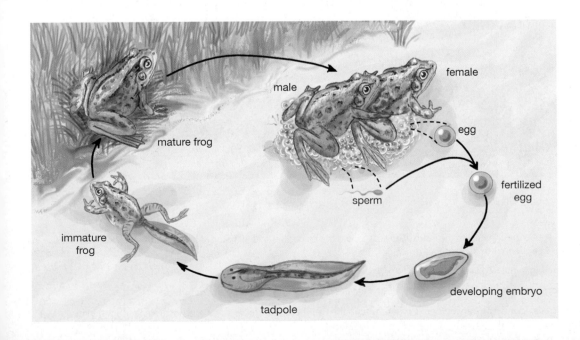

Figure 21.34
Amphibian Life Cycle
Amphibians (such as frogs, salamanders, and newts) require an aquatic, or at least moist, environment for at least part of their life cycle. Note the female frog in the figure is laying her eggs into the water. The male then releases sperm that falls on the eggs, fertilizing them. Most frogs dwell solely in the water through much of their development, first as embryos, then as tadpoles.

Even adult frogs must live in environments that are at least moist, because water can evaporate right through their skin. But frog offspring are fully aquatic for a time. They hatch from eggs that must be laid in water lest they dry out, and most young frogs spend the first part of their lives as swimming tadpoles. There are exceptions to the rule that amphibians must remain near standing water. Some frogs have become fully terrestrial, but even they must inhabit moist environments.

Apart from frogs, there are two other varieties of amphibians: salamanders and newts, collectively known as the tailed amphibians; and some wormlike, tropical amphibians known as caecilians. Salamanders may look like lizards, but their watery amphibian heritage betrays them. With some exceptions, they must live where it is moist, a fact that often keeps them hidden from human view.

Amphibians once were the dominant land animals; indeed, they were the only land vertebrates for some 75 million years. But about 250 million years ago they ceded their dominant status to the very group they gave rise to, the reptiles.

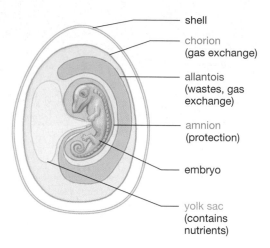

shell

chorion
(gas exchange)

allantois
(wastes, gas
exchange)

amnion
(protection)

embryo

yolk sac
(contains
nutrients)

Figure 21.35
An Egg for Many Environments
The amniotic egg was an evolutionary development that allowed vertebrates to move inland from the water. The egg's features meant that, in dry environments, the embryo would not "dessicate" or perish through a loss of fluids. The egg's tough shell is the first line of defense, as it limits evaporation of fluids. The amnion is one of several membranes that provide some cushioning protection for the embryo. The allantois serves as a repository site for the embryo's waste products. The chorion provides cushioning protection and works with the allantois in gas exchange, meaning the movement of oxygen to the embryo and the movement of carbon dioxide away from it. The yolk sac provides the nutrients the embryo will need as it develops.

Reptiles and Birds

It may seem strange to lump reptiles and birds together, but remember that scientific classification has mostly to do with shared ancestry. There is general agreement that birds are the direct descendants of the reptiles we know as dinosaurs, though this issue is not completely settled. If true, then the separate categories of "bird" and "reptile" make sense mostly in terms of describing *features* of the two kinds of animals.

a

b

c

Figure 21.36
Reptile Diversity
The three main groups of modern-day reptiles are the turtles, the lizards and snakes, and the crocodiles and alligators. Although some of these animals live in water, their scaly skin and amniotic eggs free them from dependency on it.

a A green turtle (*Chelonia mydas*) swimming in the Red Sea.

b A newborn green tree python (*Chondropython viridis*) in New Guinea. It will change into its adult color in six to eight months.

c An American crocodile (*Crocodylus acutus*) found from far-southern Florida, through the Caribbean, Central America, and northern South America.

One seemingly innocuous feature unites all reptiles and birds, but it is a feature that had far-reaching consequences. If you look at **Figure 21.35**, you can see an illustration of something known as the **amniotic egg**, named after a membrane (called an amnion) that aids the embryo inside of the egg. Recall that most amphibians must lay their eggs in water, lest the eggs dry out and the embryos inside them perish. The amniotic egg of reptiles and birds is different; it can be laid in environments ranging from moist forest floor to desert. Its tough shell keeps the egg from drying out, and the various membranes within it protect the embryo. The amniotic egg therefore gave reptiles a tremendous advantage over the amphibians in the kinds of locations the reptiles could call home; unlike the amphibians, reptiles could settle away from the water.

Characteristics of Reptiles

Several other things distinguish reptiles from amphibians. Looking back at the frog life cycle in Figure 21.34, you can see that the female frog lays her eggs in the water, after which the male spreads his sperm on top of them. This is external fertilization. By contrast, all reptiles employ internal fertilization—eggs are fertilized inside the female's body. A requirement for this, of course, is that males possess a copulatory organ that allows sperm to be deposited inside the female. Another difference between amphibians and reptiles is that reptiles have a tough, scaly skin that conserves water, as opposed to the thin amphibian skin that allows water to escape. Reptiles also have a stronger skeleton than amphibians, more efficient lungs, and a better developed nervous system.

There are three main varieties of modern-day reptiles: lizards and snakes; crocodiles and alligators; and turtles (**see Figure 21.36**). Then there is an ancient, extinct variety of reptiles that is as well known as the living examples. The dinosaurs first appeared about 220 million years ago and were certainly the most fearsome creatures ever to have walked on land. The last of them died out 65 million years ago.

Characteristics of Birds

Surprising as it may seem, one view is that "dinosaurs," properly defined, never did die out, but live on in their descendants, the birds. Indeed, many evolutionary specialists routinely think of birds as "avian dinosaurs." Two lines of evidence have convinced most scientists that birds descended from a line of bipedal (two-footed), meat-eating dinosaurs. One line concerns bone similarity. Relative leg sizes, types of hip bones, long, S-shaped necks: Only dinosaurs and birds share this list of characteristics. Second, there is the famous crow-sized fossil *Archaeopteryx*—a clear "transitional form" between dinosaurs and birds that had the dinosaur's teeth and claws, but the bird's feathers (**see Figure 21.37**). In the 1990s, scientists began filling in the transitional gaps on the other side, unearthing fossils of dinosaurs that were flightless but nevertheless had feathers.

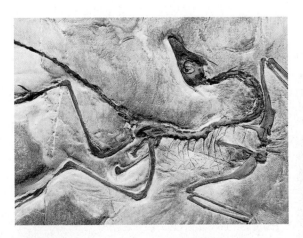 

Figure 21.37
Dinosaur and Bird
Scientists believe that this fossil, called *Archaeopteryx*, represents a transitional form between dinosaurs and birds. It has the teeth and claws of a dinosaur, but the unmistakable feathers of a bird. The drawing at right is an artist's interpretation of the fossil at left.

Birds are a vertebrate success. They exist all over the globe in great numbers, and their 9,100 recognized species exceeds the number of reptile species (7,000) or amphibian species (4,300). Though there are obvious differences between an eagle and a hummingbird, all flying birds have a very similar appearance. By contrast, think how different snakes are from crocodiles among the reptiles, or frogs are from salamanders among the amphibians. The similarity among birds is a consequence of the strict requirements of vertebrate flight: light, hollow bones, wings, and powerful flight muscles that attach to a breastbone; **see Figure 21.38**.

Mammals: Small Numbers, Big Impact

The story of mammals could be titled "Big Effects from a Small Group." There are only about 4,400 species of mammals—not even half the number of bird species, and scarcely more than the number of amphibian species. The total number of *individual* mammals is infinitesimal compared to the number of insects or roundworms. Yet it's the mammals that are the biggest creatures not only on land (the elephants) but in the sea (the whales). It's the mammals that are the fiercest creatures from the savanna (lions) to the arctic ice (polar bears). And, of course, it's the mammals, in the form of humans, who control the fate of so much of the rest of the living world.

But what is a mammal? Here are two mammal universals and two near-universals. All mammals have mammary glands and maintain a near-constant internal temperature. Most mammals have hair and have eggs that develop inside their mother's body. Here's some detail on each of these characteristics.

Mammals are named for the **mammary glands** that all of them possess: a set of glands that, in females, provide milk for the young. A feature that works closely with the mammary glands is something you can test out yourself, by slowly filling your mouth with water from a drinking fountain. Note that you can breathe through your nose while sucking the water in. Reptiles and amphibians cannot do this; their nasal passages feed into their mouths, but yours extend behind your mouth and feed into your windpipe (or trachea). Thus young mammals can suck from their mother's breast and keep breathing; likewise, cows can chew grass and never miss a breath.

Next, almost all mammals have hair, while no other animals have it. This feature is related to another universal mammal characteristic, which is that mammals are **endothermic**, meaning their internal body temperature is relatively stable and their body heat is generated internally—by their own metabolism. Conversely, amphibians and reptiles are **ectothermic**, meaning their internal temperature is controlled largely by the temperature of their environment.

Endothermy, sometimes misleadingly called warm-bloodedness, is rare in the animal world. In fact, other than mammals, the only creatures to possess it are birds. Endothermy and ectothermy are very important in determining which creature will live where in the animal world. After a cold night in the desert, the muscles of a lizard cannot function at full capacity until that lizard has warmed itself in the sun. By contrast, a desert rat is as functional at sunrise as it is at sunset. But there is a price to be paid for its readiness, and that price is energy expenditure.

Figure 21.38
Bird Similarity
Despite the obvious differences between the heron on the left and the bald eagle on the right, there are great similarities in the birds as well. Indeed, all flying birds display a similarity in body form. This is so because animals as large as birds must meet a strict set of requirements in order to fly. All birds have thin, hollow bones, tails to stabilize their flight, and powerful flight muscles that originate on their breastbone or sternum.

It takes a great deal of energy to maintain the constant, relatively high body temperatures of mammals. Pound per pound, the "basal" or resting metabolic rate of a rat may be four times that of a lizard. And it takes food to fuel this energy expenditure. The upshot is that, pound per pound, mammals and birds simply have to eat more than reptiles or amphibians.

The very quality of being able to maintain a stable body temperature has, however, allowed mammals to live where amphibians and reptiles cannot: in cold climates. Consider the fact that there are arctic hares, and arctic foxes, and polar bears, but there are no arctic lizards or frogs. Past a certain point, north or south on the globe, there is no place a lizard can go to warm its body enough to live. Mammals have their limits as well, but the range of climates they can adapt to is greater. Their hair is a great aid in maintaining temperature, as is the thick layer of fat that lies just beneath the skin of such creatures as whales and seals.

With two important exceptions, mammals are **viviparous**, meaning a condition in which fertilized eggs develop inside a mother's body. Contrast this with the **oviparous** condition, seen in all birds and many reptiles, in which fertilized eggs are laid outside the mother's body and then develop there. All embryos need nutrients and protection. In egg-laying (oviparous) animals, both things are provided by the egg itself (though parents also may provide protection). Conversely, a human egg (and then embryo) is protected by the mother's body and draws its nutrients from the mother in ways we'll look at shortly.

Reproduction turns out to be the defining feature of the three evolutionary lines of mammals: the monotremes, the marsupials, and the placental mammals (**see Figure 21.39**). **Monotremes** are egg-laying mammals, represented by the duck-billed platypus and spiny anteaters found in Australia. The monotremes do suckle their young on milk from mammary glands, but they have no nipples; thus the young must lap up the milk that diffuses onto the hair of the mother. Marsupials are represented by several Australian animals, including the kangaroo, and by several animals in the Western Hemisphere, including the North American opossum. **Marsupials** are mammals in which the young develop within the mother to a

a   b   c

Figure 21.39
Mammal Diversity
All mammals have mammary glands that provide milk for their young. However mammals differ in their reproductive strategies.

a Monotremes, such as this duck-billed platypus, lay eggs.

b Marsupials, such as this kangaroo, give birth to physically immature young, which then develop further within the mother's pouch. Pictured are a mother and her fairly mature "joey" in eastern Australia.

c The young of placental mammals, such as this grizzly bear (*Ursus arctos*) develop to a relatively advanced state inside the mother, with nutrients supplied not by an egg, but by a network of blood vessels and membranes called the placenta.

limited extent, inside an egg that has a membranous shell. Early in development, the egg's membrane disappears, and in a few days time, a developmentally immature but active marsupial is delivered from the mother. In the case of a kangaroo, the tiny youngster has just enough capacity to climb up its mother's body, into her "pouch," and begin suckling from a nipple there. We can think of the kangaroo as developing partly inside its mother's reproductive tract, and partly inside her pouch.

All other mammals are known as **placental mammals**. They are mammals nurtured before birth by the **placenta**; a network of maternal and embryonic blood vessels and membranes that allows nutrients and oxygen to diffuse *to* the embryo from the mother while allowing embryonic wastes to diffuse *from* the embryo to the mother. (For more on the placenta, see Chapter 28, page 650.) Thus, in placental mammals, the embryonic young derive their nutrition not from food stored in an egg, but directly from the circulation of the mother. Further, the embryos of placental mammals can develop for a very long time inside the mother. The human gestation period—the time from conception to birth—is about nine months, and in elephants it is almost two years. By contrast, the gestation time for a kangaroo is about a month, after which its "pouch life" might last six months.

The diversity of the placental mammals is impressive. There are flying mammals (bats), swimming ocean behemoths (whales, dolphins, and so forth), all manner of large herbivores (horses, deer, cows) and smaller herbivores (rodents, which make up more than half of all mammal species). Then, of course, there are the primates such as ourselves, defined by our forward-facing eyes, grasping hands, and arboreal or tree-dwelling habitat. Several species of primates, including our own, have come down from the trees to dwell on the ground (see Chapter 19, page 393). **Table 21.1** summarizes all of the animal phyla reviewed in this chapter.

On to Plants

In life's family tree, the animals you've just looked at are far removed from plants. Yet animals and plants are alike in one sense, in that they sit atop two great evolutionary lines. Plants occupy the higher branches of the line of "self-feeders": the *autotrophs* who make their own food by harnessing the power of the Sun. The first creatures who performed this feat were bacteria; then came the algae (which are protists). And then, evolving from the algae, came the plants, moving onto land and constructing their green edifice wherever they took root, thereby providing food and shade for all animals that came after them. In a sense, this sequence of events describes the relationship that animals and plants have: plants first, then animals. In a world without plants there would be no animals, but in a world without animals there would still be plants. The two chapters coming up describe the nature of the indispensable, green members of the living world, the plants.

**Table 21.1
Animal Phyla**

Phylum	Members Include	Live In	Characteristics
Porifera	Sponges	Ocean water (marine), with a few living in fresh water	Sessile (fixed in one spot) as adults, no symmetry. System of pores through which water flows serves to capture nutrients. Bodies do not have organs or tissues. Sexual and asexual reproduction.
Cnidaria	Jellyfish, hydrozoans, sea anemones, coral animals	Almost all marine; a few in fresh water	Radial symmetry, medusa (swimming) and polyp (largely sessile) stages of life. Use stinging tentacles to capture prey. Have tissues. Can reproduce sexually or by budding.
Platyhelminthes	Flatworms (flukes, tapeworms, tubellarians)	Marine, fresh water, or moist terrestrial environments	Bilateral symmetry, hermaphroditic, possess organs, no central cavity, most primitive animals with triploblastic tissue structure. Many are parasites. Can reproduce by dividing themselves.
Nematoda	Roundworms	Almost all environments	Small size in most, but some are several meters long. Many are crop pests, others animal parasites. Possess pseudocoelom. Reproduction always sexual. Have mouth and anus.
Mollusca	Gastropods (snails, slugs), bivalves (oysters, clams, mussels), cephalopods (octopus, squid, nautilus)	Marine, fresh water, moist terrestrial	Mantle that can secrete material that becomes shell; mantle cavity housing gills or lungs; unitary foot (lost in some sessile species, evolved into arms and tentacles in cephalopods).
Annelida	Segmented worms	Most are marine, some fresh water and moist terrestrial	Distinct body segmentation in varieties such as earthworms. Leeches are annelids.
Arthropoda	Vast group that includes insects, spiders, mites, crabs, shrimp, and centipedes	Many aquatic and terrestrial environments	External skeleton made of chitin; jointed, bilateral appendages; molting in many species; segmented bodies; generally an open circulation system.
Echinodermata	Sea stars, sea urchins, sea cucumbers	Always marine, most on ocean floor	Generally have radial symmetry as adults; tube feet in some species, slow ocean floor movement in most.
Chordata	Vertebrates, lancelets, sea squirts	Great variety of aquatic and terrestrial environments	At some point, all possess notochord, dorsal nerve cord, pharyngeal slits, post-anal tail. Vertebrates have vertebral column.

Chapter Review

Summary

21.1 What Is an Animal?

- A single physical feature is sufficient to define animals. All animals pass through a blastula stage in embryonic development, but no other living things do. A blastula is a hollow, fluid-filled ball of cells that forms once an egg is fertilized by sperm.

- Three other features are characteristic of all animals, but also are shared, to varying degrees, by members of other kingdoms in the living world. All animals are multicelled, are heterotrophs (they cannot make their own food), and are composed of cells that do not have cell walls.

21.2 Animal Types: The Family Tree

- The animal kingdom is divided into large-scale categories called phyla. There are between 36 and 41 animal phyla, each phylum being a group of organisms that share a set of physical characteristics that result from shared ancestry.
  TUTORIAL 21.1.1: Animal Classification

- Animals can be thought of as having evolved from simpler to more complex forms through a series of additions to the characteristics found in more primitive animals. These additions are tissues, radial symmetry, bilateral symmetry, and a central cavity or coelom.

- A fundamental split in animal evolution came with the divergence of the protostome and deuterostome animal lines. Protostome animals include roundworms, molluscs, annelid worms, and arthropods; deuterostome animals include echinoderms and chordates.

21.3 Phylum Porifera: The Sponges

- Sponges are simple animals, lacking organs, tissues, or symmetry. Nearly all sponges live in marine environments, though there are a few freshwater varieties.

- Sponges live by drawing water into themselves through a series of tiny pores on their exterior, and then filtering this water for food.

- Sponges can reproduce sexually, with eggs and sperm carried from one sponge to another by water currents; or they can produce asexually, through budding.

21.4 Phylum Cnidaria: Jellyfish and Others

- The defining characteristic of members of phylum Cnidaria is their use of stinging tentacles to capture prey.

- Cnidarians include jellyfish, sea anemones, hydras, and the reef-building coral animals whose skeletons make up the bulk of coral reefs.

- Many cnidarians have both an adult, medusa stage of life, which swims in the water; and an immature, polyp stage of life, which generally remains fixed to rocks, animals, or other solid surfaces. Some cnidarians have only the polyp stage. Cnidarians have no organs, but they do have tissues.

- Cnidarians have radial symmetry, and reproduction can be sexual or asexual.

21.5 Phylum Platyhelminthes: Flatworms

- The flatworms of phylum Platyhelminthes are mostly small creatures, dwelling either in aquatic or moist terrestrial environments. Some parasitic flatworm species can, however, reach enormous lengths.

- Flatworms have bilateral symmetry, but have no coelom or system of blood circulation. They have no anus, meaning the flatworm's mouth is the single opening to its digestive system.

- Flatworms are the least complex animals to be triploblastic: They have three embryonic germ layers—endoderm, mesoderm, and ectoderm—whereas sponges and cnidarians lack a mesoderm and are therefore diploblastic.

- Most flatworms are hermaphroditic—a single flatworm is likely to possess both male and female sex organs. Many flatworms can reproduce asexually by breaking themselves in half.

- Parasitic tapeworms, which can infect human beings, are flatworms, as are the flukes that cause the disease schistosomiasis.

21.6 Phylum Nematoda: Roundworms

- The roundworms of phylum Nematoda exist in enormous numbers in all kinds of habitats on Earth. Most are microscopic, though some are large.

- A number of roundworms are agricultural pests, and some are human parasites. The disease trichinosis is caused by roundworms, and the parasites known as hookworms are roundworms.

- Roundworms are the least complex animals to possess a central cavity—technically a pseudocoel instead of a true coelom. Their reproduction is always sexual and they are not hermaphroditic.

21.7 Phylum Mollusca: Snails, Oysters, and Squid and More

- Phylum Mollusca is a vast and extremely varied group of animals whose members range from sessile, brainless mussels to agile, intelligent squid. Mollusc habitats range from marine and freshwater to moist terrestrial.

- Three important classes of molluscs are the gastropods, which include snails and slugs; the bivalves, which include oysters, clams, and mussels; and the cephalopods, which include octopus, squid, and nautilus.

- All molluscs possess a fold of skin-like tissue, called a mantle, that can secrete material that forms a shell, though many molluscs do not have shells. Molluscs tend to have a mantle cavity that houses either gills (if the molluscs are aquatic) or lungs (if they are terrestrial); many have a unitary foot, whose wave-like contractions allow movement. Many molluscs have a tooth-lined membrane called a radula that can be extended outside the body for scraping and retrieving food.

- Cephalopods have a closed circulation system, in which blood stays within blood vessels, while oxygen and nutrients diffuse out of the vessels to the surrounding tissues. Most molluscs have an open circulation system, in which blood flows out of blood vessels into openings called sinuses, where the blood bathes tissues.

21.8 Phylum Annelida: Segmented Worms

- The worms of phylum Annelida provide a clear example of a physical feature that is widespread in the animal kingdom, body segmentation, meaning a repetition of body parts in an animal. Segmentation is generally valuable because it provides flexibility in tandem with strength.

- Earthworms are representative of the annelid or segmented worms; earthworm bodies display a clear segmentation, allowing the worm to exercise a degree of independent control over each segment or group of segments.

- Most annelids are marine, and some dwell in freshwater. All annelids exist in environments that are at least moist. Leeches are parasitic annelid worms.

21.9 Phylum Arthropoda: So Many, but Why?

- Phylum Arthropoda is enormous and extremely varied. There are more animal species in one class of arthropods, the insects, than there are in all the other groupings of animals combined.

- Arthropods can be divided into three subphyla: Chelicerata, which includes spiders, ticks, mites, and horseshoe crabs; Crustacea, which includes shrimp, lobsters, crabs, and barnacles; and Uniramia, which includes millipedes and centipedes along with insects.

- All arthropods have an external skeleton or exoskeleton, and paired, jointed appendages. Arthropods go through molting, meaning the periodic shedding of an old exoskeleton.

- All members of subphylum Crustacea have five pairs of appendages extending from their heads—two pairs of antennae and three pairs of feeding appendages. Crustaceans range from large lobsters to microscopic water fleas.

- All members of subphylum Chelicerata have in common a pair of appendages, called chelicerae, that serve various feeding functions—grasping in horseshoe crabs, for example, and puncturing in spiders.

21.10 Phylum Echinodermata: Sea Stars, Sea Urchins, and More

- Echinoderms include sea stars, sea urchins, sand dollars, and sea cucumbers, among others. All members of phylum Echinodermata are marine, and most inhabit the ocean floor.

- Echinoderms have a radial symmetry as adults. The sea stars among them evolved to this condition from bilateral symmetry; sea star larvae have a bilateral symmetry that is lost as they develop into adults. Sea stars have no brain and only one sensory organ, the primitive eyespots on the tips of their arms. A new sea star can be generated asexually from the arm and part of the central disk of an existing sea star.

- Echinoderms tend to move slowly across the sea floor; but many are formidable predators, often feeding on such molluscs as oysters or mussels.

21.11 Phylum Chordata: Mostly Animals with Backbones

- Phylum Chordata is made up of three subphyla: Vertebrata, which includes all the vertebrates, including human beings; Cephalochordata, made of up creatures called lancelets; and Urochordata, made up of three classes of animals, including the tunicates or sea squirts. Only the vertebrates have a vertebral column, meaning a flexible column of bones running from the anterior to posterior ends of an animal.

- All chordates possess, at some point in their lives, a rod-shaped support structure called a notochord; a nerve cord on their dorsal side; a post-anal tail; and a series of pharyngeal slits that develop into various structures, depending on the type of chordate.

- With the exception of lampreys, all true vertebrates have jaws. Today's jawed vertebrates developed from ancient jawless vertebrates.

- Fish account for more than half the 50,000 species of vertebrates, with the vast majority of fish being bony fish, as opposed to the more ancient line of cartilaginous fishes, such as sharks, whose skeletons are made of the connective tissue cartilage.

- Most bony fish are ray-finned fishes, named after the ray-like structures in their dorsal fins. One existing species of fish, the coelecanth, belongs to another category of fish, the lobe-finned fishes. These fish are important because an ancient variety of them is thought to have given rise to all tetrapods, meaning four-limbed vertebrates. All land vertebrates are tetrapods: amphibians, reptiles, birds, and mammals.

- Amphibians were the first truly terrestrial vertebrates. Today amphibians include frogs; the tailed amphibians, salamanders and newts; and some wormlike amphibians known as caecilians. All amphibians must live in environments that are at least moist, most employ external fertilization of eggs, and most live in aquatic environments for part of their lives.

- Reptiles evolved from amphibians. An important feature in their evolution was the development of the amniotic egg, which can keep embryos moist even in dry environments. The amniotic egg allowed the reptiles to move inland from water sources. Reptiles employ internal fertilization and have a tough, scaly skin that conserves water. There are three main varieties of modern-day reptiles: turtles, lizards and snakes, and crocodiles and alligators. Dinosaurs were reptiles.

- Most scientists believe that birds are the direct descendants of dinosaurs. Birds exist all over the globe in great numbers. The similar appearance of all flying birds stems from the strict requirements of flight for creatures as large as vertebrates: light bones, wings, and powerful flight muscles that attach to a breastbone.

- Though there are relatively few mammal species, mammals have had a great impact on the natural world. They account for the largest, fiercest creatures on land and sea, and one species of mammal, human beings, controls the fate of much of the living world.

- All mammals have mammary glands, meaning glands that deliver milk to the young. Most mammals have hair, and they have eggs that develop inside the mother's body (making them viviparous). All mammals are endothermic, meaning their internal body temperature is relatively stable and body heat is generated internally, by their own metabolism. Conversely, amphibians and reptiles are ectothermic, meaning their internal temperature is controlled largely by the temperature of their environment. Birds and mammals are the only true endothermic animals.

- There are three evolutionary lines of mammals: the monotremes, which are egg-laying (or oviparous) mammals, represented by the platypus; marsupials, in which the young develop within the mother only to a limited extent, represented by kangaroos; and placentals, in which young develop within their mothers, nurtured by a placenta—a network of maternal and embryonic blood vessels.

Key Terms

amniotic egg 463
bilateral symmetry 439
bivalves 448
body segmentation 450
cephalopods 448
closed circulation system 449
coelom 441
diploblastic 447
dorsal nerve cord 459

ectothermic 464
endothermic 464
exoskeleton 453
gastropods 448
hermaphroditic 447
invertebrate 439
mammary glands 464
marsupial 465
mesoglea 444

molting 453
monotreme 465
notochord 458
organ 446
open circulation
oviparous 465
paired, jointed
   appendages 453
parasite 446
pharyngeal slits 459
phylum 439
placenta 466

placental mammal 466
post-anal tail 159
radial symmetry 439
swim bladder 466
symmetry 439
tetrapods 461
tissue 439

triploblastic 447
vertebral column 458
viviparous 465
zoologist 439

Understanding the Basics

Multiple-Choice Questions

1. All animals (select all that apply)
   a. go through a blastula stage in development
   b. can see
   c. are multicelled
   d. can move
   e. all of the above

2. Bilateral symmetry is a condition in which
   a. No one animal species can overpower another.
   b. The top and bottom halves of an animal are symmetrical.
   c. Two animal species evolve along parallel lines.
   d. The sides of an animal are symmetrical.
   e. One side of an animal is as strong as the other.

3. A coelom is
   a. a central body cavity, not found in sponges, cnidarians, and flatworms
   b. a central body cavity, found only in deuterostomes
   c. a digestive organ, found in all animals except sponges
   d. an evolutionary offshoot of a species
   e. an offspring that arises from asexual reproduction

4. An animal has bilateral symmetry, but no central cavity. What phylum is it in?
   a. Mollusca
   b. Cnidaria
   c. Chordata
   d. Porifera
   e. Platyhelminthes

5. Members of the _____ phylum are, as a group, the most primitive animals; members of the _____ phylum are, as a group, the most advanced.
   a. Echinodermata, Mollusca
   b. Echinodermata, Chordata
   c. Cnidaria, Mollusca
   d. Porifera, Chordata
   e. Porifera, Mollusca

6. A marine animal with radial symmetry goes through both medusa and polyp stages in its life cycle. Which of the following features is it certain not to have? (Select all that apply.)
   a. a brain
   b. a heart
   c. a coelom
   d. legs
   e. a mouth

7. The world's largest invertebrates, the _____, always _____.
   a. crustaceans, have exoskeletons
   b. echinoderms, have radial symmetry
   c. cephalopods, are single-sex
   d. chordates, have a post-anal tail
   e. cephalopods, are filter-feeders

8. In earthworms, we can see a clear example of the widespread animal feature known as
   a. filter feeding
   b. body segmentation
   c. viviparous reproduction
   d. a post-anal tail
   e. a radula

9. Which of the following is not a physical feature of all vertebrates at some point in their lives? (Select all that apply.)
   a. pharyngeal slits
   b. a vertebral column
   c. a ventral nerve cord
   d. a post-anal tail
   e. a mantle cavity

10. If an animal has an exoskeleton, paired jointed appendages, and goes through the process of molting, it is:
    a. a bivalve
    b. a gastropod
    c. a tetrapod
    d. an arthropod
    e. a chordate

Brief Review

1. A central coelom amounts to internal space in an animal. Why is such a feature valuable?

2. Do all animals have a single ancestor, or do the different phyla of animals have different ancestors? What evidence supports your answer?

3. What are the benefits and the costs of endothermy? What animals are endothermic?

4. Why is it fair to characterize sponges as "primitive" animals?

5. What is a coral reef chiefly composed of?

6. What feature common to molluscs was modified through evolution to become the arms and tentacles of the squid?

7. Which animal is a spider more closely related to, a dust mite or a grasshopper?

8. A mother provides milk for her offspring, and yet those offspring are hatched from eggs. What kind of animal is the mother?

Applying Your Knowledge

1. The text notes the tremendous number of insects in the world—both the number of insect species and the number of individual insects. One of the most spectacular examples of insect numbers can be seen in the so-called periodical cicadas, which emerge in the Eastern and Central United States in groups called "broods" after completing an underground larval stage that lasts either 13 or 17 years. One 17-year brood that emerged in 1956 near Chicago was found in densities of 1,500,000 per acre in lowland forests—about 533 tons of cicadas per square mile over their entire range. Biologists often talk of the reproductive "strategies" of various species. These are the traits of a species that help ensure its continued reproduction, generation after generation. What is the reproductive strategy of the periodical cicadas? Does their long underground immature stage aid in this strategy?

2. Over evolutionary time, reptiles replaced amphibians as the biggest, most dominant land animals on Earth; with the extinction of the dinosaurs, mammals replaced reptiles in this role. Why should either event have occurred? Why did reptiles supplant amphibians in this regard, and why should mammals, rather than reptiles, hold this position today?

3. One of the distinguishing features of animals is that their cells do not have the thick outer lining known as a cell wall. Plant cells have a cell wall, as do fungi cells. Why would such a feature be advantageous for plants, but disadvantageous for animals?

4. Plants can have elaborate defense systems, and they can respond in sophisticated ways to their environment—for example, in preserving resources they go into a dormant state in winter. Given this, can plants have intelligence? Or is it only animals that have this trait, to one degree or another?

22

An Introduction to Flowering Plants

Heart medicine.
(Section 22.1, page 474)

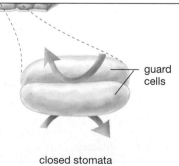

guard
cells

open stomata   closed stomata

Guard duty.
(Section 22.2, page 478)

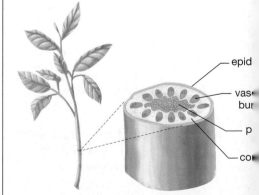

epid

vas
bun

p

co

A stem for support, storage, and transport.
(Section 22.2, page 479)

Plants are indispensable to most other living things, and are not the passive, defenseless entities they may appear to be. The flowering plants known as angiosperms are the most important plants of all.

About 500 years ago, Europeans first sailed to the Western Hemisphere and soon after began colonizing it. Now here's a question to ponder about this event: Why did things work out this way instead of, say, the Aztecs sailing to Europe and colonizing *it*? One obvious answer is that Europe was much more advanced technologically than was the New World: There were differences in ships, navigational equipment, weapons. This answer, however, just pushes the question back a step, prompting us to ask why this technological gap should have existed.

There are lots of possible answers to this question, but let's note one hypothesis that has been put forth by the UCLA physiologist Jared Diamond. It is that Europe developed faster in part because it got lucky in what might be called the plant sweepstakes.

22.1 The Importance of Plants

Human beings first developed agriculture—the systematic planting and care of food plants—about 10,000 years ago in what is known as the Fertile Crescent, meaning a crescent-shaped swath of territory stretching from modern-day Turkey through modern-day Iran. Wherever agriculture has developed, it has had a

tremendous effect on human culture, turning societies of hunters and gatherers into societies in which there is enough food that people can *specialize* in terms of vocation. This is critical because a culture that has metalworkers and accountants and soldiers advances technologically at a rapid rate compared to one that doesn't.

Now, as it turns out, farming not only developed first in the Fertile Crescent, but it spread relatively rapidly from it, as you can see in **Figure 22.1**. Within 3,000 years,

Figure 22.1
The Origins of Agriculture in the Old World
Farming originated almost 10,000 years ago in the "Fertile Crescent" east of the Mediterranean Sea. It then spread rapidly on a mostly east-west axis, helping to spur the rapid cultural evolution for which Europe is known. The lines indicate how far food production had spread by the dates shown. (Redrawn from Jared Diamond, "Location, Location, Location: The First Farmers," *Science*, November 14, 1997, p. 1243.)

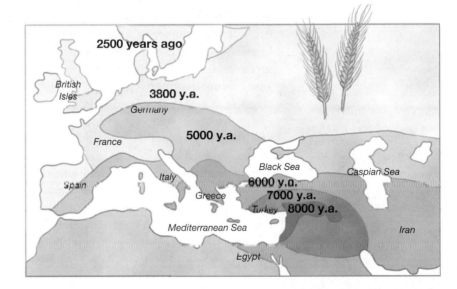

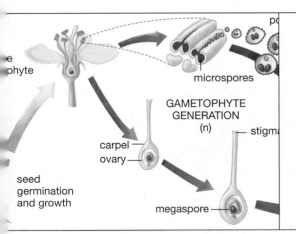

Part of a life cycle.
(Section 22.3, page 481)

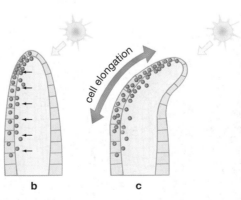

The power of light.
(Section 22.4, page 488)

The force of gravity.
(Section8 22.4, page 488)

agriculture that was based mostly on innovations made in the Fertile Crescent was being practiced throughout much of Europe. In the New World, by contrast, agriculture developed later and spread much more slowly. Why this difference?

First, genetic luck was involved. One of the "founder" crops of the Fertile Crescent was a form of wheat that, like almost all food crops, existed originally as a wild plant. As it happens, there are very few genetic differences between the wild variety of this wheat, which yielded little food, and the domesticated variety, which yielded a good deal more. By contrast, there was a large genetic difference between the wild variety of the main New World food crop, teosinte, and its domesticated descendant, maize (or corn). Take the members of two societies, one located in the Fertile Crescent and another in the New World, and let each group do the same thing with their respective wild plants, which is simply pick the grains from the most food-laden individual plants each year. As a result of this human activity, the *seeds* from these selected plants would then *germinate*, or sprout from the ground, in such places as the garbage dumps around human settlements. Both situations result in the rise of nutrition-laden strains of crops by means of this unintended cultivation. In the Fertile Crescent, however, these strains appear *quickly*—within a few hundred years—because of the short genetic distance between wild and domestic varieties. In the New World, the transition takes much longer because many more generations of hearty maize plants had to be "selected for" (picked) and "planted" (the excess discarded) to get the domesticated maize.

This explains why the Fertile Crescent was first with agriculture, but why the rapid *spread* of agriculture from it? Pure geographic luck, says Diamond. Europe has a largely east-west orientation, while the New World has a predominately north-south orientation. Plants can spread east to west without going through much climate change, meaning the same domestic strains can spread rapidly. The situation is different for the spread of plants either north or south, because they must adapt to changes in latitude (and temperature) as they go in either direction, which means a slower spreading.

In short, as this hypothesis has it, Europe won the technological race in part because it had a lucky head start in the agricultural race. This is a hypothesis, not established fact; but if it's true, think of its meaning. We need hardly note the tremendous effect that European colonization had on human affairs from the 1500s through the present. It may be that the fates of both colonizers and those who were colonized were determined in significant part by the ticket they held in the plant sweepstakes.

Figure 22.2
Uses Galore
Plants are used by human beings in innumerable ways.

a Concord grapes.

b Wood, being harvested from a forest in British Columbia.

c A foxglove plant, which yields the heart medicine digitalis.

a

b

c

Other Roles of Plants

Whatever their role may have been in history, plants are vital to almost all life at a much more basic level. They are the most important photosynthesizers on Earth, and it is photosynthesis that makes Earth a planet of living *surplus*—of biomass that is here for the harvesting by other life-forms. A single mature maple tree can produce over 100 kilograms (or 220 pounds) of leaves a year. As noted before, if you try to name something you eat that isn't a plant or that doesn't itself depend on plants for food, you'll come up with a very short list. There also is the oxygen that plants produce as a by-product of photosynthesis and, for humans, the lumber and paper that trees provide (**see Figure 22.2**).

Beyond this, plants are a kind of anchoring environmental force. Their roots prevent soil erosion, while their leaves absorb such pollutants as sulfur dioxide and ozone. One of the main greenhouse gases thought to be warming Earth is carbon dioxide, and plants absorb carbon dioxide in order to perform photosynthesis. Thus, there is not only a *production* side to atmospheric carbon dioxide—in significant part the pollution human beings produce—there is also a *consumption* side to it, in the amount plants absorb. One proposed way to fight greenhouse warming, therefore, is to plant trees. The maple tree mentioned earlier will absorb about 450 kilograms (about 1,000 pounds) of carbon dioxide from the atmosphere over the course of a single summer.

A Focus on Flowering Plants

Though the plant kingdom is vast and varied, it turns out that there are just four principal types of plants in it. These are the bryophytes, represented by mosses; the seedless vascular plants, represented by ferns; the gymnosperms, represented by coniferous (evergreen) trees; and the flowering plants, also known as angiosperms. Chapter 20, which introduced plants, contained information on the bryophytes, the seedless vascular plants, and the gymnosperms. In this chapter, we will focus strictly on flowering plants. Why pay so much attention to just one type of plant? Because of the overwhelming dominance of flowering plants within the plant kingdom. There are about 16,000 species of bryophytes, 13,000 species of seedless vascular plants, and 700 species of gymnosperms, but there are an estimated 260,000 species of flowering plant. The term "flowering plant" may bring to mind roses or orchids and these plants are indeed angiosperms. But food crops such as rice and wheat are flowering plants, as are all cacti, almost all the leafy trees, innumerable bushes, pineapple plants and cotton plants and ice plant—the list is enormous (**see Figure 22.3**). In this chapter, you'll get an overview of how flowering plants are structured, of how they function internally, and of how they respond to signals from the outside world. Chapter 23 will go into greater detail on angiosperm structure and function. A formal definition of angiosperms will be provided once you've learned a little about them.

a

b

c

**Figure 22.3
Angiosperm Variety**

a A California poppy surrounded by flowers called tidy tips.

b Organ pipe cacti.

c Corn in a field.

22.2 The Structure of Flowering Plants

Let's begin our tour of the angiosperms by looking at their component parts.

The Basic Division: Roots and Shoots

We'll first look at the larger-scale structures of the angiosperms, starting with a simple,

Figure 22.4
A Life-Form in Two Parts
Flowering plants live in two worlds, with their shoots in the air and their roots in the soil. Although flowering plants differ enormously in size and shape, they generally possess the external features shown.

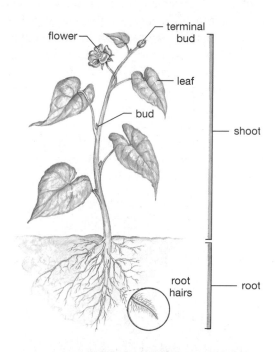

two-part division that rhymes: roots and shoots. Plants live in two worlds, air and soil, with their root system below ground and their shoot system above (**see Figure 22.4**).

The function of the root system is straightforward: Grow to water and minerals and absorb them from the soil and then begin transporting them up through the rest of the plant. (For more on the minerals plants absorb, see "What Is Plant Food?" on page 477.) Most roots also serve as anchoring devices for plants, and some act as storage sites for food reserves.

The function of the shoot system is more complex. Photosynthesis takes place in this system, primarily in leaves, which must thus be positioned to absorb sunlight. Plants are in *competition* with one another for sunlight, which is a primary reason so many plants are tall. The food that is produced in photosynthesis must be distributed throughout the plant. It is not just food production that is centered in the shoot system, however; it is plant reproduction as well. Within flowers and their derivatives, we find all the components of reproduction—seeds, pollen, and so forth. Now let's look in more detail at the components of the root and shoot systems.

Roots: Absorbing the Vital Water

If you look at **Figure 22.5**, you can see pictures of the two basic types of plant root systems, one a **taproot system** consisting of a large

a   b

Figure 22.5
Two Root Strategies

a Dandelions such as this one employ a taproot system—a large central root and a number of smaller lateral roots.

b The French marigold employs a fibrous root system—a collection of roots that are all about the same size.

Figure 22.6
Root Hairs
Root hairs enormously increase the surface area of a root. The greater the surface area, the more fluid absorption can occur. Shown here are root hairs on the taproot of a sweetcorn plant.

What Is Plant Food?

Plants are able to make their own food through photosynthesis if they have just four things: water, sunlight, carbon dioxide, and a few nutrients. Water, sunlight, and probably even carbon dioxide are no mystery to you by now, but what exactly are these nutrients you've been reading about?

A **nutrient** is simply a chemical element that is used by living things to sustain life. Recall from Chapter 2 that "elements" are the most basic building blocks of the chemical world. Silver is an element, as is uranium, but these are not nutrients because living things do not need them to live. In the broadest sense, carbon, oxygen, and hydrogen are nutrients. And it turns out that almost 96 percent of the weight of the average plant is accounted for by these three elements, which come to plants primarily from water and air.

Plants need at least another 13 elements to live, however, and not all of these are supplied by water and air. When we think of what is commonly called plant food, we are generally referring to those nutrients that plants can *use up* in their soil and that thus must be replenished to assure continued growth and reproduction. Of these, the most important three are nitrogen, phosphorus, and potassium, whose chemical symbols are N, P, and K. When you look at a package of an average plant fertilizer, you will see three numbers in sequence on it (for example, 10-20-10). What these refer to are the percentage of the fertilizer's weight accounted for by these nutrients. The growth of plants usually is enhanced by a fertilizer that has equal ratios of the three elements, but if increased *blooming* is your goal, an increased phosphorus ratio generally is recommended, as with the 10-20-10 example.

It is not just houseplants or lawns that benefit from fertilizer, but farm crops as well. Indeed, the planting of a crop on a parcel of land generally requires a large investment in fertilizer because a crop such as wheat removes a great deal of N, P, and K from the soil in a single season. Historically, fertilizers used on farms were **organic**, meaning they came from decayed living things, such as the fish that Native Americans famously taught the Pilgrims to use when planting corn. Nowadays, however, commercial fertilizers generally are **inorganic**, meaning they are mixtures of pure elements within binding materials, produced by chemical processes.

Nature is, of course, perfectly capable of producing a green bounty in such places as forests and marshes without the aid of any human-made fertilizer. Decaying plant and animal matter put nutrients back into the soil, and bacteria are able to fix nitrogen, meaning to absorb it from the air and transform it into a form that plants can take up. Looked at one way, however, houseplants and farm crops are *isolated* plants whose participation in this web of life is very limited, and thus they require a helping hand from humans in order to remain robust.

Figure 1
Needed Nutrients
Plants make their own food from sunlight, water, carbon dioxide, and a few nutrients, some of which come from the soil. The most important of these nutrients are nitrogen (N), phosphorus (P), and potassium (K). Because garden plants and houseplants are isolated from natural ecosystems, they must often get these nutrients in the form of fertilizer.

central root and a number of smaller lateral roots, and the second a **fibrous root system** consisting of a number of roots that are all about the same size. **Figure 22.6** then shows you not only the taproot of a young sweetcorn plant, but also another feature common to root systems. These are **root hairs**—threadlike extensions of roots that greatly increase their absorptive surface area. Each root hair actually is an elongation of a single outer cell of the root. Between roots and root hairs, the root structure of a given plant can be extensive. One famous analysis, conducted in the 1930s on a rye plant, concluded that its taproot and lateral roots alone, if laid end to end, would have totaled some 622 kilometers, which is about 386 miles. Meanwhile, the root hairs collectively were 10,620 kilometers long, meaning the plant had almost 6,600 miles of root hairs! This from a plant whose *shoot* stood 8 centimeters (about 3 inches) off the ground.

Why do plants lavish such resources on their roots? For one thing, water doesn't come to roots nearly as much as roots must come to water; roots must continually *grow to* new water supplies. Why such a great need for water? To perform photosynthesis, plants must have a constant supply of carbon dioxide, which enters the plants through microscopic pores, called *stomata*, that exist mostly on the underside of plant leaves. Open stomata, however, don't just let carbon dioxide in; they let the plant's water *out*, as water vapor. To accommodate this fact while still keeping vital tissues moist, plants continually pass water through themselves, from roots, up through the stem, and into the leaves (at which point it exits as water vapor). The evaporation of water from a plant's shoot is known as **transpiration**; through it, more than 90 percent of the water that enters a plant evaporates into the atmosphere. Because the roots of a single tall maple tree can absorb about 220 liters (or nearly 60 gallons) of water per *hour* on a hot summer day, the scale of transpiration obviously is immense. Thus do we see the importance of root development.

The food storage aspect of roots has a meaning for human beings. Sweet potatoes and carrots, which are roots, can represent stored food for people as well as for plants.

Shoots: Leaves, Stems, and Flowers

Now let's turn to the shoot system, looking first at the leaves within it.

Leaves: Sites of Food Production

The primary business of leaves is to absorb the sunlight that drives photosynthesis, which is why most leaves are thin and flat. This leaf shape maximizes the surface area that can be devoted to absorbing sunlight while minimizing the number of cells in leaves that are irrelevant to photosynthesis. Beyond this, through their tiny stomata, leaves serve as the plant's primary entry and exit points for gases. As noted, the most important gas that's entering is carbon dioxide, which is one of the starting ingredients for photosynthesis. What's exiting is the by-product of photosynthesis, oxygen, and the water vapor.

The broad, flat leaves that are so common in nature have in essence a two-part structure—a **blade** (which we usually think of as the leaf itself) and a **petiole**, more commonly referred to as the leaf stem. If you look at the idealized cross section of this leaf in **Figure 22.7b**, you can see that the blade

Tutorial 22.1.1
Sites for
Photosynthesis

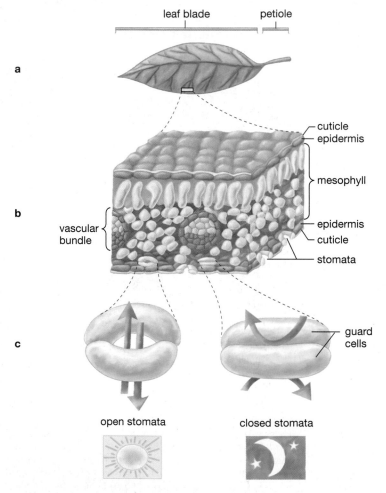

Figure 22.7
Site of Photosynthesis

a Leaves tend to be broad and flat to increase the surface area exposed to sunlight.

b Photosynthesis occurs primarily within the mesophyll cells in the interior of the leaf. The vascular bundles carry water to the leaves and carry the product of photosynthesis, sugar, to other parts of the plant. It is the pores called stomata, mostly on the underside of leaves, that allow for the passage of gases in and out of the leaf.

c Guard cells of the stomata control the opening and closing of the stomata. When sunlight shines on the leaf during the day, the guard cells engorge with water that makes them bow apart, thus opening the stomata. When sunlight is reduced, water flows out of the guard cells, causing the stomata to close.

can be likened to a kind of cellular sandwich, with layers of **cuticle**, or waxy outer covering, on the outside, and a layer of epidermal cells just inside them. In the leaf's interior there are **vascular bundles**, which bear some relation to animal veins in that they are part of a transport system. Then there are several layers of mesophyll cells. It is these cells that are the sites of most photosynthesis.

Now note, on the underside of the blade, the openings called **stomata** (singular, *stoma*). These are the pores, noted earlier, that let water vapor out and carbon dioxide in—but for most plants only during the day. When the Sun goes down photosynthesis can no longer be performed, and it is not cost-effective for a plant to lose water without gaining carbon dioxide. The stomata thus close up until photosynthesis begins again the following day. If you look at **Figure 22.7c**, you can see how this opening and closing of stomata is achieved: Two "guard cells," juxtaposed against one another like sides of a coin purse, are arranged around the stomata. When sunlight strikes the leaf, these cells engorge with water, which makes them bow apart. Then, in the absence of light, water flows out and the door closes again. A given square centimeter of a plant leaf may contain from 1,000 to 100,000 stomata.

Figure 22.8 gives you some idea of the varied forms of leaves. In addition to the so-called simple leaf we've been looking at, there are compound leaves in which individual blades are divided into a series of "leaflets." Pine needles and cactus spines are likewise actually leaves. The cactus, in particular, has reduced the amount of leaf surface it has as a means of conserving water. So how does it perform enough photosynthesis to get along? It carries out most of its photosynthesis in its stem.

Stems: Structure and Storage

We all generally understand what the stem of a plant is, though it is less generally appreciated that the trunk of a tree is simply one kind of plant stem. The main functions of stems are to give structure to the plant as a whole and to act as storage sites for food reserves. In addition, water, minerals, hormones, and food are constantly shuttling through (and to) the stem, with water on its way up from the roots and food on its way down from the leaves to the rest of the plant.

If you look at **Figure 22.9**, you can see a cross section of one type of plant stem, showing the vascular bundles or veins you first saw in the leaves and the outer, or *epidermal, tissue* just as the leaves had. You

Figure 22.8
Leaves Come in Many Sizes, Shapes, and Colors
Simple leaves have just one blade. In compound leaves, the blade is divided into little leaflets. Some leaves are so modified that the average person can hardly recognize them as leaves—for example, the spines of a cactus.

compound leaves

simple leaves

leaves modified as spines

highly reduced leaves (needles)

leaves modified as tendrils

Figure 22.9
The Stem and Its Parts
Stems provide support to the rest of the plant, act as storage sites, and conduct fluids. Vascular plants have a "plumbing" system that transports food, water, minerals, and hormones. The vascular bundles in the figure are groups of tubes, running in parallel, that serve this transport function.

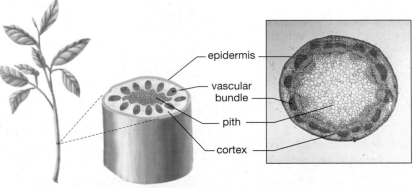

epidermis

vascular bundle

pith

cortex

can also see so-called *ground tissue* of two types: an outer cortex and an inner pith, both of which can play a part in food storage and wound repair and provide structural strength to the plant.

Flowers: Many Parts in Service of Reproduction

Flowers are the reproductive structures of plants. A single flower generally has both male and female reproductive structures on it, which might make you think that a given plant would fertilize *itself*. This is indeed the case with some plants, such as Gregor Mendel's pea plants, reviewed in Chapter 11. However, because the evolutionary benefit of sexual reproduction is to get the genetic diversity that comes with *mixing* genetic material from different organisms, natural selection has worked against self-fertilization by endowing many flowering plants with mechanisms to reduce the incidence of it. For example, the male pollen of a given plant might be genetically incompatible with that plant's female reproductive structures. Some of the illustrations you'll be seeing show a plant fertilizing itself, but this is done only for visual simplicity.

If you look at **Figure 22.10**, you can see the components of a typical flower. Taking things from the bottom, there is a modified stem, called a pedicel, which widens into a base called a receptacle, from which the flowers emerge. Flowers themselves can be thought of as consisting of four parts: sepals, petals, stamens, and a carpel. The **sepals** are the leaflike structures that protect the flower before it opens. (Drying out is a problem, as

are hungry animals.) The function of the colorful **petals** is to announce "food here" to pollinating animals.

The heart of the flowers' reproductive structures are the stamens and the carpel. If you look Figure 22.10, you can see that the **stamens** consist of a long, slender **filament** topped by an **anther**. These anthers contain cells that ultimately will yield sperm-bearing pollen grains. At maturity, these pollen grains will consist of three cells surrounded by an outer coat.

It is these pollen grains that will be released and then carried—perhaps by a pollinating bee or bird—to the carpel of another plant (or perhaps of the same plant). As Figure 22.10 shows, a **carpel** is a composite structure, composed of three main parts: the **stigma**, which is the tip end of the carpel, on which pollen grains are deposited; the **style**, a slender tube that raises the stigma to such a prominent height that it can easily catch the pollen; and the **ovary**, the area in which fertilization of the female egg and then early development of the plant embryo take place.

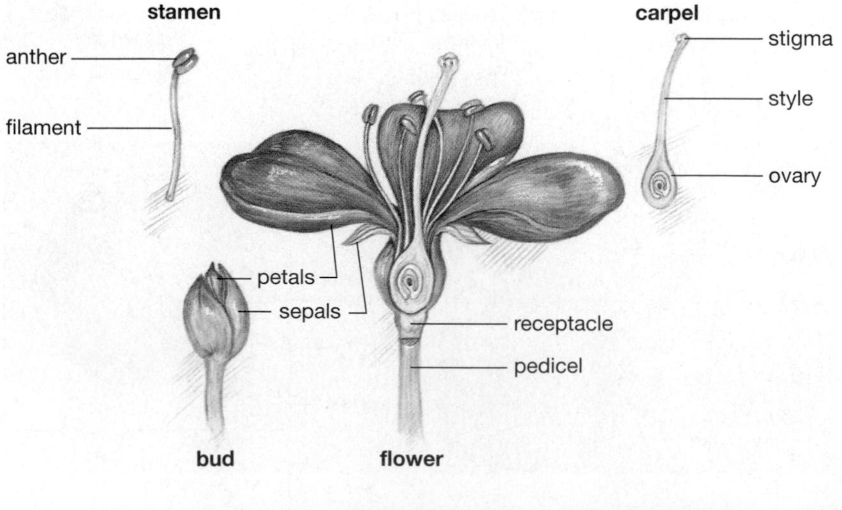

22.3 How Flowering Plants Function

Having looked at the structure, or anatomy, of the angiosperms, let's now go over the activities of some of these components. We'll think in terms of systems here: the reproductive system, the transport system, the hormonal system, the communication system, and the defense system. In addition, you'll look briefly at the nature of plant growth. Because you've just reviewed flower anatomy, let's start with reproduction.

Reproduction in Angiosperms

The function of a pollen grain is the fertilization of an egg. There are cells in the anthers of the original flower that undergo the type of cell division reviewed in Chapter 10, meiosis, thereby producing **microspores** that are haploid—that contain only a single set of chromosomes. As you can see in **Figure 22.11**, each pollen grain develops from one of the microspores inside the anther. In the type of plant pictured in the figure, the grain will consist, by the time it leaves the anther, of a tough outer coat and three cells: one **tube cell** and two **sperm cells**. Though the pollen grain never gets much more complex than this, it is

Figure 22.10
Parts of the Flower
Flowers are composed of four main parts: the sepals, petals, stamens, and a carpel. The sepals protect the young bud until it is ready to bloom. The petals attract pollinators. The stamens are the reproductive structures that produce pollen grains (which contain sperm cells). The carpel is the female reproductive structure; it includes an ovary that contains one or more eggs

not just a component of the original plant; it is its alternate generation on the male side. (For more on the alternation of generations in plants, see page 426 in Chapter 20.)

You can see in Figure 22.11 what the activities of the pollen grain's cells are in the context of the angiosperm life cycle. When the pollen grain leaves the anther, it is bound for the stigma of another plant (or its own plant). But for a grain to merely *land* on a stigma doesn't mean that anything has been fertilized. The tube cell in the grain must then begin to *germinate* on the stigma, forming a **pollen tube** that grows down through the style. Once this has taken place, one of the sperm cells in the grain travels through the tube, gets to the female egg, and only *then* is there fertilization. (The other sperm moves down along with the first, but then spurs the growth of food reserves for the growing embryo, about which you'll learn more in Chapter 23.)

And how does the egg arise that is fertilized by the sperm? Together with its supporting cells, the egg is the plant's alternate generation on the female side. Inside the plant's ovary, there is a cell that also goes through meiosis, thus producing a haploid **megaspore** (*mega*, because it's bigger than the male microspore). It in turn gives rise to several kinds of cells, one of which is the egg that the sperm from the pollen grain will fertilize.

Watch closely now, because once this happens—once sperm (from pollen grain) has fertilized egg (from megaspore)—it is the beginning of the kind of flowering plant we started with. We have alternated back to this generation of plant. Many a step remains before arriving at something that *looks* like the original plant: The fertilized egg (now called a zygote) must first have a tough covering develop around it. The combination of embryo, its food supply, and the covering is called a **seed**. This seed must be released and then land on a suitable patch of earth, there to germinate and grow to a full flowering plant. But the step at which this new generation appears is the point at which sperm fertilizes egg.

To put some terms to these generations, note that both the megaspore and pollen grain are haploid spores. Because *spores* are what the original, flowering plant produced, this flowering plant is the **sporophyte generation**. Meanwhile, the pollen and megaspore generation produced *gametes*—sperm or egg—and are thus the **gametophyte generation**. When the sperm and egg came together, they fused their haploid sets of chromosomes and thus produced a sporophyte embryo with a doubled, or diploid, set of chromosomes. A detailed account of angiosperm reproduction can be found in Chapter 23, beginning on page 512.

Recall that the angiosperm egg is fertilized and develops inside the structure called the ovary (see Figure 22.10). Also recall that as the fertilized egg develops into an embryo it becomes enclosed within a seed. As this is taking place, the ovary that surrounds the seed is also developing into something: a tissue called fruit. We are all familiar with fruit, of course, and sometimes the tissue surrounding the seed is fruit as we commonly understand that term—the flesh of an apricot, for example. But ovaries can mature into fruit in other forms. The pod that surrounds peas is fruit in this scientific sense, as is the outer covering of a kernel of corn. A **fruit** is simply the mature ovary of a flowering plant. All angiosperms have fruit in this sense, and it provides us with our definition of the flowering plants. An **angiosperm** is a flowering seed plant whose seeds are enclosed within the tissue called fruit.

Tutorial 22.1.2
Angiosperm Life Cycle

Figure 22.11
The Angiosperm Life Cycle
The mature sporophyte flower produces many male microspores within the anther, and a female megaspore within the ovary. The male microspores then develop into pollen grains that contain the male gamete, sperm. Meanwhile the female megaspore produces the female gamete, the egg. Pollen grains, with their sperm inside, move to the stigma of a plant. There the tube cell of the pollen grain sprouts a pollen tube that grows down toward the egg. The sperm cells then move down through the pollen tube, and one of them fertilizes the egg. This results in a zygote that develops into an embryo protected inside a seed. This seed leaves the parent plant and sprouts or "germinates" in the earth, growing eventually into the type of sporophyte plant that began the cycle.

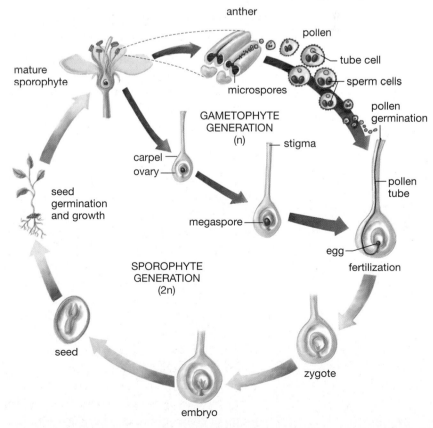

Figure 22.12
Reproduction without Sex
Individual Aspen trees often are clones of one another, reproducing without the fusion of eggs and sperm from different individuals. This stand of Aspen trees is in Colorado.

Asexual Reproduction in Angiosperms: Vegetative Reproduction

We are used to the human mode of reproduction, which—pending the arrival of cloned human beings—is always *sexual* reproduction: At some point, sperm must fertilize egg to produce a new embryo. But plants, including angiosperms, have another option—**vegetative reproduction**, which is *asexual* reproduction. This is familiar to us in the form of "cuttings" that can be taken from one houseplant to start another. A growing cutting represents a new plant that is an exact genetic replica of the original plant; no alternate generation or fertilization is required. One sporophyte plant has been grown from another.

Human beings are motivated to put such vegetative reproduction to use when nature produces a particularly *valuable* plant, generally through mutation. Navel oranges, for example, seem to have arisen through mutation on a conventional orange tree that was grown in Brazil in the nineteenth century. If pollen from that tree had helped fertilize an egg from another tree, we may well have *lost* the advantage this unique tree provided—fat, seedless navel oranges—through the shuffling of genetic material that comes with sexual reproduction. The trick was to use cuttings (or "grafts") from the original tree to produce one *identical* generation of orange tree after another. Vegetative reproduction is also common in nature. The roots of aspen trees, for example, produce a form of shoot known as a sucker which, if physically separated from the parent plant, will grow into a new aspen tree. Oftentimes, then, a stand of aspen trees amounts to a group of clones of one another (**Figure 22.12**).

Plant Plumbing: The Transport System

In looking at leaves and stems earlier, you saw that they have something called *vascular bundles* running through them. This term refers to collections of tubes through which fluid materials move from one part of the plant to another. This transport system bears obvious similarities with the circulation systems of animals, but there are differences as well. Many animals employ blood as a transport medium, and animals such as ourselves have a transport *pumping* device, called a heart. Plants do not have blood, and they have no pumping system. (In fact, plants have almost no "moving parts" at all.) Yet think of the transport job plants have to do. Water that is transpired from the top of a redwood tree may have made a journey of more than 100 meters straight up—more length than a football field! What kind of power is at work here?

There are two essential components to the plant transport system. First there is **xylem**, which can be defined as the tissue through which *water* and dissolved minerals flow in vascular plants. (**Tissue** means a group of cells that perform a common function.) Second there is **phloem**, the tissue through which the *food* produced in photosynthesis—mostly sucrose—is conducted, along with some hormones and other compounds. As noted before, water is making a directional journey from root through leaf, but the plant must be able to transport food and hormones everywhere within itself.

If you look at **Figure 22.13**, you can see an idealized view of a plant's transport system as it would appear within a stem. You'll also see that the vascular bundles noted earlier are composed of bundles of linked xylem and phloem tubes running in parallel.

Xylem is composed of two types of fluid-conducting cells, *vessel elements* and *tracheids*, which have different shapes as you can see. Upon reaching maturity, these cells do the rest of the plant the favor of dying. The cells' content is cleared out, leaving strands of empty cells stacked one on top of another—leaving tubes, in other words. The walls of these cells remain, however, and indeed are reinforced with the materials cellulose and lignin, which provide the *load-bearing* capacity that allows something as massive as a redwood tree to be so tall. (For more on the role of xylem in everyday houseplants, see "Keeping Cut Flowers Fresh," on the next page.)

Keeping Cut Flowers Fresh

As already noted, xylem is the plant tissue through which water moves *up*, from roots through leaves. The flowers we put in vases in our homes have lost their roots, of course, but they haven't lost their xylem, which continues to function long after the flower has been picked. Given this, flowers can last a long time indoors; but we can maximize their stay if we follow a few simple rules.

First, realize that the liquid in the xylem is under negative pressure—its natural tendency is to move up *into* the stem, not to flow out of it. As such, if the stems are cut when they are out of water, *air* gets sucked up into the cut ends, creating air bubbles (called *embolisms*) that can then get trapped in the xylem and keep water from rising up through it. When this happens, flowers can wilt, even when their stems are submerged in clean water. Recutting the stem under water (or under a steady stream from the faucet) can remove this blockage. Better yet, cut the stems under water the first time.

Beyond this, acidic sugar-water, such as can be found in citrus-flavored soft drinks, will prolong the life of some flowers by keeping bacterial growth down; changing water frequently is a good idea, particularly when it starts to look gummy or discolored. Keep your arrangement out of direct light or heat, and remove dead and dying flowers in the arrangement, because the hormone ethylene is given off by dying flowers and in many cases will hasten the demise of the healthy ones in the bunch. (For another role of ethylene, see "Ripening Fruit Is a Gas," page 485.)

Figure 1
Preserving Beauty
Following a few simple rules can prolong the life of cut flowers.

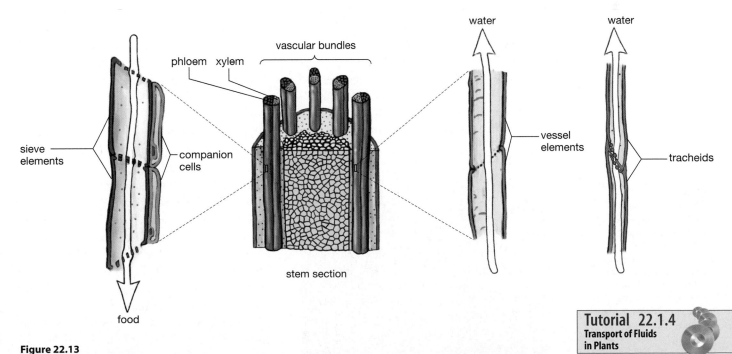

Figure 22.13
Fluid-Transport Structure of Plants
Fluids move through plants in sets of tubes called vascular bundles. Two types of tissue, running in parallel, can be found in each vascular bundle: xylem, the tissue through which water and dissolved minerals flow; and phloem, the tissue through which the food produced in photosynthesis flows. Xylem is composed of two kinds of fluid-conducting cells, vessel elements and tracheids. Phloem is composed of sieve elements and their companion cells.

Tutorial 22.1.4
Transport of Fluids in Plants

Table 22.1
Plant Hormones

Hormone	Major Functions	Where Found or Produced in Plant
Auxins	Suppression of lateral buds; elongation of stems; growth and abscission (falling off) of leaves; differentiation of xylem and phloem tissue	Root and shoot tips; young leaves
Cytokinins	Stimulate cell division; active in the development of plant tissues from undifferentiated cells	Roots
Gibberellins	Stem elongation, growth of fruit, promotion of seed germination	Seeds, apical meristem tissue, young leaves
Ethylene	Ripening of fruit, retardation of lateral bud growth, promotion of leaf abscission	Nearly all plant tissue
Abscisic Acid	Induces closing of leaf pores (stomata) in drought; promotes dormancy in seeds; counteracts growth hormones	Young fruit; leaves, roots

Phloem is also composed of two types of cells, but the arrangement here is very different. One type of phloem cell—a *sieve element*, which does the actual nutrient conducting—doesn't undergo cell death upon maturity, but it does lose its cell nucleus, meaning nearly all its DNA. Those of you who went through the genetics unit may wonder how any cell could function without its DNA information center, but there is an answer. Each sieve element cell has associated with it one or more *companion cells* that retain their DNA and that seem to take care of all the housekeeping needs of their sister sieve elements. So, why do sieve elements lose their nuclei? Apparently to make room for the rapid flow of food through them. In Chapter 23 you'll look at the means by which phloem and xylem conduct their respective materials.

Communication: Hormones Affect Many Aspects of Plant Functioning

We're so used to associating hormones with *animal* functioning that it may come as a surprise to learn that plants also have hormones. Taken as a whole, plant hormones do many of the same things that animal hormones do. Their most important roles are to regulate growth and development and to integrate the functioning of the various plant parts. In less abstract terms, hormones help buds grow and leaves fall and fruit ripen. This goes hand in hand with helping plants respond to their environments—to heat and cold, munching goats, and voracious insects.

Though most people have an intuitive sense of what hormones are, a definition might be helpful. **Hormones** are chemical messengers; they are substances that when released in one part of an organism, go on to prompt physiological activity in another part of that organism. In animals, hormones are generally synthesized in well-defined organs, called *glands*, whose main function is to produce hormones (for example, the thyroid or adrenal glands). In plants, however, hormone production is a more diffuse process, taking place not in glands, but in collections of cells that carry out a range of functions.

Five major plant hormones have been discovered to date. Three of these actually are *classes* of hormones, and two are individual hormones (**see Table 22.1** for a list of all these hormones). You can read about the hormone

Figure 22.14
Uniform "A" Shapes
The effects of apical dominance are clearly visible in this stand of blue spruce trees at a Michigan Christmas tree farm. The apical meristem tissue at the tip of the trees produces a hormone called IAA that inhibits the growth of lateral branches. Because the concentration of IAA is highest at the top near the apical meristem, and lowest at the bottom, the branches are very short near the top and longest at the bottom.

Ripening Fruit Is a Gas

That banana you had? The one you put in that plastic bag, thinking you'd make it part of your lunch? It already had a few black spots on it by the time it went in the bag, but then you forgot about it for lunch and, by the time you looked at it again, it was more black than yellow and something you were no longer interested in eating. The question is: Why did it go downhill so *fast*?

The answer is that those little black spots on the banana result from concentrations of a gas called *ethylene*—a plant hormone that can ripen lots of fruits. The bag you put the banana in became a kind of gas chamber, with the ethylene concentrations inside being very high. Moreover, ethylene stimulates its own production; the more ethylene there is, the more that gets produced. Given this contagious quality to ethylene, you can see why it is true that one bad apple can spoil the bunch.

Ethylene doesn't work on all fruits; grapes, strawberries, and cherries, for example, are unaffected by it. But apples, avocados, and tomatoes join bananas in being very sensitive to it. If you wish to ripen one of these latter fruits fast, putting it in a plastic bag will give you an edible fruit faster than putting it on the window sill.

But what does it mean for a fruit to be ripe? Let's match up our intuitive sense of this with the underlying physical changes. Compared to unripened fruits, ripe fruits tend to be sweeter (starches are converted to sugars); softer (a secretion of enzymes softens cell walls); and more fragrant (ripening releases compounds that our nose detects as sweet-smelling). Ethylene helps bring about all these changes, basically by upping the tempo at which a plant carries out its metabolic processes. The reason a green banana turns to yellow is that ethylene hastens the breakdown of the green chlorophyll in the banana, which allows us to see yellow pigments that had been there all along.

For years, commercial fruit growers have employed ethylene, or test-tube compounds that help release it, in order to *time* the ripening of their fruit. This is why we can pick green tomatoes at the source and then ripen them when they get closer to market. It's also possible to *suppress* ethylene production by storing fruit in a room that has high concentrations of carbon dioxide and low concentrations of oxygen, a combination that suppresses ethylene synthesis. In this way, a fruit such as apples can be kept on hold for months before being brought to ripeness by simply being exposed to normal air. Ever wonder why you can have fresh apples in the spring when they generally are picked in the fall? Ethylene control is part of the answer.

Figure 1
Spurring Their Own Ripening
These bananas ripen under the influence of a gas they produce, ethylene. Concentrations of ethylene produce the black spots on bananas.

ethylene in "Ripening Fruit Is a Gas." Let's go over just one other plant hormone, a member of the family of hormones known as *auxins*, to give you some idea of what plant hormones do.

An Auxin Gives the "A" Shape to Trees

Why do Christmas trees have their characteristic "A" shape? (**See Figure 22.14.**) They develop in this way thanks in large part to the effects of an auxin known as IAA. Unlike animals, plants do not grow globally, over their whole surface, but instead confine most of their vertical growth to special regions, called *apical meristems*, that exist at the *tips* of the roots and shoots. **Figure 22.15** shows you how this works out in connection with the

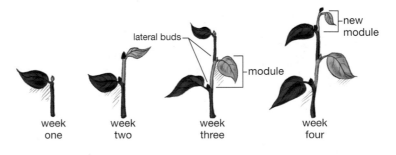

Figure 22.15
Modular Growth
Vertical growth in plants occurs in "modules." Each module includes a new portion of stem, at least one leaf, and a lateral bud that can give rise to a branch or a flower. Here alternating dark-green and light-green segments indicate modules, and we go from one module in week one to four modules in week four.

shoots; the shoot apical meristem gives rise to a series of growth *modules*: more stem, one or more leaves, and a lateral bud that forms in tandem with each leaf. These lateral buds have meristem tissue in them and can grow into new branches, but they normally don't *when they lie close to the apical meristem*. Why? The apical meristem is producing IAA, which works in tandem with other hormones to suppress the growth of these buds. But, the farther away from the apical meristem, the *smaller* the concentration of IAA that is available to the lateral buds. The result is a greater budding of branches at the base of a tree, and a tapering of this growth going from the base up through the tree's apex. This phenomenon is called **apical dominance**: a suppression of the growth of lateral branches through the activity of apical meristems. It is most pronounced in certain conifers, such as the Douglas firs often used for Christmas trees. Meanwhile, trees that don't have such a strict IAA gradient don't get the strict "A" shape. This maintenance of apical dominance is just one of the many things that IAA does. To give you some idea of the complexity of its function, IAA may serve to *suppress* the emergence of lateral buds, as noted, but it works to *promote* the cellular elongation that is important in plant growth.

Figure 22.16
The Basic Characteristics of Plant Growth
If a young boy drives a long nail into a tree and returns 10 years later, two things will be apparent: The nail is at the same height it used to be despite the increased height of the tree, and it is almost completely buried in the tree. What does this say about tree growth? Trees do not elongate throughout their whole length, but only at their tips in "primary growth." Further, they widen laterally in "secondary growth."

Plant Growth: Indeterminate and at the Tips

Having looked a little at how plant growth is affected by hormones, we can note some basic characteristics of plant growth. To visualize how it works, imagine a boy of about 8 years old who decides to drive a long nail into the trunk of a secluded tree. He doesn't drive the nail in completely, but lets it stick out a bit. After 10 years pass, the boy happens to walk by the same tree; upon seeing it, a couple of things occur to him (**see Figure 22.16**). First, he has grown but the nail is in the same place; he would have to bend down now to pull the nail out, whereas when he drove it in he was standing up straight. Second, he would have a hard time actually pulling the nail out now because not much of it is visible; the tree has grown around it. Third, if the nail's height is any indication, the trunk of the tree may not have moved up, but the *top* certainly has; it's much taller than he remembered it.

The lessons here are that, in their vertical growth, plants do not grow the way people do—throughout their entire length—but instead grow only at their tips (of both shoots *and* roots, though the boy couldn't see the roots). In addition, as the half-buried nail attests, this tree has carried out the *lateral* growth that botanists refer to as *secondary growth*, which you'll look at next chapter.

Beyond these things, it turns out that a plant's growth is indeterminate: In most plants it can go on indefinitely at the tips of roots and shoots. By contrast, animal growth is generally determinate: It comes to an end at a certain point in development. Imagine if, say, the tips of your fingers just kept growing long after your trunk and arms had reached their full extension.

Defense and Cooperation

Plants may be immobile and toothless, but they are not defenseless. Indeed, they have developed a formidable defense arsenal (**see Figure 22.17**). One variety of plant defenses, the structural defenses, are familiar to us because they're so visible—think, for example, of the spines on a cactus or the thorns on a rose. A second variety of defenses, the chemical defenses, are not so visible but are potent nevertheless. The chomping a beetle does on a potato plant induces the plant to produce substances (called *protease inhibitors*) that are believed to give the insect an ache in its gut that it won't forget. Other plants, when attacked by a fungus, release antifungal compounds that spread not only at the site of infection, but throughout the plant.

Beyond these self-defense mechanisms, plants can warn *other* plants of dangers in their midst. Airborne signaling compounds can be produced from the leaf surface of a plant under attack, waft to a neighboring plant, and prompt it to start producing compounds that are noxious or harmful to the would-be plant-eater. Some plants can even call in animal allies. Corn plants that are attacked by army worms release a substance that attracts parasitic wasps, which then devour the worms.

Plants also enter into cooperative relationships with other organisms. In Chapter 20 you saw that most plant roots are linked to the thin tubes, called *hyphae*, that extend from fungi. What the fungi get from this is nutrition from the photosynthesizing plant; what the plants get is a greatly expanded water and mineral absorption network. The combined root/hyphae associations are known as **mycorrhizae**. Other species, notably the legumes, such as soybeans, have developed partnerships with nitrogen-fixing bacteria, which is to say bacteria that are able to take nitrogen out of the atmosphere and put it into a form that living things can use. The bacterium infects the plant through a root hair, directs the construction of its own lodgings (called a *root nodule*), and then begins converting atmospheric nitrogen, while obtaining nutrition from the plant. This process must take place in the absence of oxygen, which the plant sees to by eliminating the available oxygen in the nodule. Lots of crops require applications of human-manufactured nitrogen, but legumes such as beans and alfalfa can do fine on their own (**see Figure 22.18**).

22.4 Responding to External Signals

Plants may lack a nervous system, but this doesn't mean they can't respond to their environment. They must do so, actually, because their life depends on it. They sense the march of the seasons by measuring available light and the surrounding temperature; they use internal clocks to synchronize their development and physical functioning; they respond to light, gravity, touch, heat, cold, and pH, among other things. Let's conclude the chapter by reviewing the plant's interactions with light, gravity, touch, and the seasons as examples of these environmental responses.

Responding to Light: Phototropism

Many plants will bend toward a source of light—meaning even a candle—with the value of this perhaps being obvious: Plants produce their own food, and that production depends on light. Thus a plant needs to respond when its sunlight becomes *blocked* by another plant or some other physical object

Figure 22.17
Silent and Fixed in One Spot, but Not Defenseless
The spiny bark of this Costa Rican tree, a member of a genus *Zanthoxylum*, protects it from plant predators.

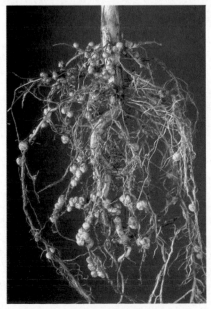

Figure 22.18
Working Arrangement
Plants need nitrogen to grow. Some get it through a cooperative relationship with bacteria that are nitrogen-fixing, meaning they can convert atmospheric nitrogen into a form living things can use. Shown are the root nodules (the small spheres) on the roots of a soybean plant. Bacteria make a home in the nodules, where they fix nitrogen that is then used by the plant.

Tutorial 22.2.2
Response of Plants to Light

in its surroundings. As it turns out, it is IAA, produced in the shoot-tips of growing plants, that controls this **phototropism**, meaning a curvature of shoots in response to light. When light strikes one side of the shoot, it causes IAA to migrate to the other side, where the IAA acts to promote the *elongation* of cells on this far side. The effect is to make the shoot curve toward the light (**see Figure 22.19**).

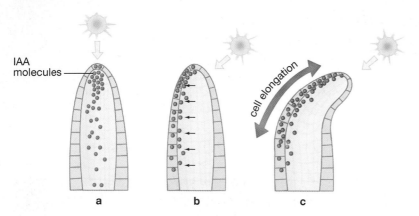

Figure 22.19
Plants Respond to Light

a When sunlight is overhead, the IAA molecules produced by the apical meristem are distributed evenly in the shoot.

b Once the sunlight shines on the shoot at an angle, the IAA molecules move to the far side and induce the elongation of cells on that side.

c Cell elongation results in the bending of the shoot toward the light.

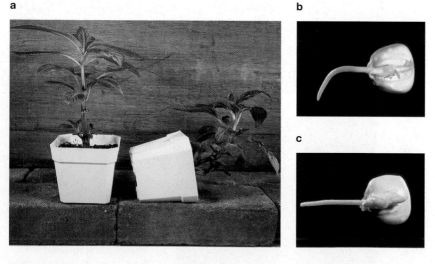

Figure 22.20
Plants Respond to Gravity

a Potted impatiens demonstrate the effects of gravitropism. The plant on the right was laid on its side 16 hours before the photo was taken, by which time the shoot had curved, beginning to grow upward.

b Similarly, the root of a germinating sweetcorn seed begins to curve downward only hours after being placed on its side. How do these plants sense which way is up and which way is down?

c One hint is that if the very tip, or "root cap" of the root is removed, the root no longer bends downward. Researchers believe that organelle "sedimentation"—the movement of certain organelles in response to gravity—is the sensing mechanism in gravitropism.

Responding to Gravity: Gravitropism

When a seed first starts to sprout, or "germinate," its roots and shoots must go in very specific directions: The root must go *down*, toward the water and minerals the plant needs, and the shoot must go *up* toward the sunlight. Imagine a seed that has simply fallen on the ground; it would not do for either root or shoot to grow *horizontally*; each must grow vertically. If you look at **Figure 22.20**, you can see dramatic evidence of a plant's ability to sense which way is up and which way down. The impatiens in Figure 22.20a was placed on its side and, within 16 hours, the shoot began curving upward, as you can see. Figure 22.20b shows the *root* of a germinating sweetcorn seed, likewise oriented horizontally (by researchers), but quickly beginning to bend downward. What's being exemplified here is **gravitropism**, meaning a bending of a plant's root or shoot in response to gravity. Figure 22.20c again shows the root of a sweetcorn plant, only this one has had its very tip, or *root cap*, snipped off. Here there is no bending in accordance with gravity, but rather a continuation of straight, horizontal growth. Conclusion? Cells or substances in the root cap are essential for root gravitropism.

But how do roots "know" which way is down, and how do they orchestrate a course correction when they need one? Logically, there must be at least two elements at work here: first, a gravitational *sensing* mechanism; and then a means of *responding* to gravitational cues, such that corrective bending points the root toward the center of Earth.

The consensus among botanists today is that the sensing mechanism in gravitropism is the "sedimentation" that various plant cell organelles perform in response to gravity. Recall that organelles are small but highly organized structures inside cells. (Mitochondria and ribosomes are organelles, for example.) Like the bubble in a carpenter's level, some plant-cell organelles can be seen changing position in response to gravity. The most important of these organelles is a group of starch-storing structures called *amyloplasts* (though some plants seem to employ alternate organelles). When these organelles move in response to gravity, it sets in motion the responding mechanism in gravitropism. The organelles "land" on other structures inside the cell (such as the cytoskeleton), and the resulting impact triggers a redistribution of IAA within the cell. This may sound familiar, because IAA redistribution also takes place in the phototropism you just looked at. And, just as in phototropism, this IAA redistribution results in differential growth within the plant—one *side* of the plant stem or root will grow more than the other side. With this growth, a root that is oriented horizontally will start bending toward the ground, while a stem that is oriented horizontally will start bending toward the sky.

Responding to Contact: Thigmotropism

We've all seen plants that manage to climb upward by encircling the stem of another plant. Whole stems can undertake such encircling, but it is often a thin, modified leaflet called a **tendril** that does so (**see Figure 22.21**). But how is a tendril able to wrap around another object? Once again through differential growth—more rapid growth on one side of a tendril than on the other. Contact with the object is perceived by outer, or epidermal, cells in the tendril. This sets into motion the differential growth, which probably is controlled by the hormones IAA and ethylene. This process is called **thigmotropism**, meaning growth of a plant in response to touch. Plants with this capability can piggyback on other plants in order to get more access to sunlight.

Responding to the Passage of the Seasons

The profusion of brightly colored leaves in the fall is one of the great seasonal markers in temperate climates such as exist in the American Midwest and most of the East Coast. As green leaves turn red, gold, and purple, we know that autumn has arrived (**see Figure 22.22**).

Dormancy in Winter

Trees that exhibit a coordinated, seasonal loss of leaves are called **deciduous** trees. But why should these broad-leafed trees lose their

Figure 22.21
Moving up by Curling around
This dodder plant is able to curl around its host plant because of thigmotropism—a plant's ability to grow in response to touch.

leaves, while the evergreen pines and firs keep the modified leaves known as needles?

First and foremost, there is relatively little water available in winter (most of it being frozen in the ground), and flat-leafed trees transpire more than do evergreens. In addition, because of the way wind flows over flat leaves, they lose more *heat* than do pine needles. Thus, the deciduous trees have evolved a strategy that is a matter of straight economics. Such winter photosynthesis as they might perform would not be worth the water loss that would result. The strategy thus becomes to lose the leaves, grow new ones next spring, and in

Figure 22.22
Seasonal Marker
Brightly colored fall leaves in Michigan's Ottawa National Forest. Plants have a number of mechanisms that allow them to respond to the passage of the seasons.

the meantime perform no photosynthesis but exist instead completely on food reserves. This state, in which growth is suspended and there is a low level of metabolic activity, is called **dormancy**. Evergreens exhibit dormancy too, in low temperatures, but deciduous trees are locked into it until the coming of spring.

Limitation of Loss in the Deciduous Strategy

The deciduous strategy, of first losing leaves and then growing them back, may at first seem very wasteful, akin to a company destroying much of its machinery each year, only to rebuild it months later. But consider how economical trees are about this. They *reclaim* the nutrients they have stored in their leaves. Proteins are broken down by enzymes and shipped, along with the nutrients, to storage cells in the stem and roots. By the time a leaf is ready for **abscission**, or detachment from the tree, it is little more than an empty shell of cell walls.

Before this happens, however, the change in coloration takes place. It may surprise you to learn there is less *adding* of new colors here than an *unmasking* of colors that existed in the leaves all along. With the approach of fall, green chlorophyll, which is the main pigment in photosynthesis, begins to break down in the leaf; with it gone, the yellow and orange colors provided by the carotenoid pigments become visible. New colors are added as well; blue and red come with the synthesis of pigments called anthocyanins. (These pigments are largely responsible for the stunning red color of many maples in fall.) The combination of carotenoids and anthocyanins gives us the final result, which is multihued leaves.

Photoperiodism

So, how does a deciduous tree sense when it's time to begin preparing for cold weather? In general it is an interaction between two factors. First there is cold itself, which brings about changes in metabolism. In addition, there is a phenomenon known as **photoperiodism**, which is the ability of a plant to respond to changes it is experiencing in the daily duration of darkness, relative to light. This is another example of the plant's ability to respond to its environment, but it should not be confused with the "phototropism" or *bending* toward the light that was discussed earlier. In photoperiodism,

plants are making seasonal adaptations based on the length of the nights they are experiencing. Most plants in the Earth's temperate regions and many in the subtropics exhibit photoperiodism, which can affect such processes as flowering as well as the onset of dormancy.

To get a feel for how photoperiodism works, consider ragweed, a plant that is the bane of many hay-fever sufferers. After sprouting in the spring and reaching a certain level of maturity, ragweed flowers only when it is in darkness for 10 hours *or more* per day. When would this threshold be crossed? As the nights begin to get longer. Ragweed thus joins a group of plants, called long-night plants, whose flowering comes only with an *increased* amount of darkness. As you can imagine, most of these are like ragweed in that they flower only in late summer or early fall. Their counterparts are the so-called short-night plants, which flower only when the nights get short enough—that is to say, when nighttime hours are *decreasing* as a proportion of the day. Predictably, these plants tend to flower in early to midsummer.

The mechanism by which photoperiodism works is still something of a mystery. Long-night plants, as their name implies, need a certain uninterrupted period of darkness to be able to flower. Shine a light of a certain wavelength on many of these plants at night, even for a couple of minutes, during the time of year in which we would expect them to flower, and the flowering will be suppressed. Botanists have long known of a pigment family called *phytochrome* that provides the plant with information about the relative amounts of daylight or darkness, yet its role is clearly not controlling in the process. For us, the important point to note is that plants have a way of sensing the seasons and reacting to them.

On to a More Detailed Picture of Plants

This chapter has served to introduce you to the basic features of plants—their varieties, structures, and the most important systems that operate in them. There is more to say, however, about many of the topics that we haven't gone over (such as how trees grow *out* as well as up). Chapter 23 goes into more detail on plant structure and functioning, including the function of reproduction.

Chapter Review

Summary

22.1 The Importance of Plants

- Plants are vital to almost all life on Earth. The photosynthesis they carry out indirectly feeds many other life-forms. The oxygen they produce as a by-product of photosynthesis is vital to many organisms. The lumber and paper that trees provide is important to human beings. Plants act as an anchoring environmental force by preventing soil erosion and absorbing carbon dioxide and pollutants.

- There are four principal varieties of plants: bryophytes, represented by mosses; seedless vascular plants, represented by ferns; gymnosperms, represented by coniferous (evergreen) trees; and the flowering plants or angiosperms. Of the four varieties, angiosperms are by far the most dominant on Earth.

22.2 The Structure of Flowering Plants

- Plants live in two worlds, above the ground and below it. As such, their anatomy can be conceptualized as consisting of the above-ground shoots and the below-ground roots.

- Roots absorb water and nutrients, anchor the plant, and often act as nutrient storage sites.

- Shoots include the plant's leaves, stems, and flowers.

- Leaves serve as the primary sites of photosynthesis in most plants. Leaves have a profusion of tiny pores, called stomata, that open and close in response to the presence or absence of light. In this way, the stomata control the flow of carbon dioxide into the plant and the flow of oxygen and water vapor out of the plant.

  TUTORIAL 22.1.1: Sites for Photosynthesis

- Stems give structure to plants and act as storage sites for food reserves.

- Flowers are the reproductive structures of plants, with most flowers containing both male and female reproductive parts. The male reproductive structure, called a stamen, consists of a slender filament topped by an anther. The anther's chambers contain the cells that will develop into sperm-containing pollen grains. The female reproductive structure, the carpel, is composed of a stigma, on which pollen grains are deposited; a tube called a style, which raises the stigma to such a height that it can catch pollen; and a structure called an ovary, where fertilization of the female egg and early development of the resulting embryo take place.

22.3 How Flowering Plants Function

- Pollen grains develop from cells called microspores inside the plant's anthers. At maturity, each pollen grain consists of two sperm cells, one tube cell, and an outer coat. When pollen grains from one plant land on the stigma of a second plant, they germinate, developing a pollen tube that grows down through the second plant's style. Sperm cells travel through the pollen tube, with one of the sperm cells reaching the egg in the ovary of the second plant and fertilizing it. Prior to this, the egg has developed from a cell called a megaspore within the ovary. Once it is fertilized by the sperm cell, the egg develops into an embryo that eventually will be surrounded by a tough outer covering. The combination of embryo, its food supply, and the outer covering is called a seed, which is capable of being implanted in the ground and growing into a new generation of plant. Angiosperms can be defined as plants whose seeds are surrounded by a layer of tissue called fruit. Fruit is the mature ovary of a flowering plant.

  TUTORIAL 22.1.2: Angiosperm Life Cycle

- Plants can reproduce asexually, through such means as grafting. This is known as vegetative reproduction.

- Fluid transport in plants is handled through two kinds of tissue: xylem, through which water and dissolved minerals flow; and phloem, through which the food the plant produces flows, along with hormones and other compounds. Xylem is composed of two types of fluid-conducting cells—vessel elements and tracheids—while phloem is composed of cells called sieve elements and their related companion cells.

  TUTORIAL 22.1.4: Transport of Fluids in Plants

- Plant hormones regulate plant growth and development and integrate the functioning of various plant structures. Many fruits ripen under the influence of the plant hormone ethylene, while the hormone IAA is important in controlling plant growth.

- Plants do not grow vertically throughout their length, but instead grow almost entirely at the tips of both their roots and shoots. Some plants, such as trees, thicken through lateral or "secondary" growth. The growth of most plants is indeterminate, meaning it can go on indefinitely.

- Plants have formidable defenses, both structural (such as cactus spines) and chemical (such as antifungal compounds).

- Plants enter into cooperative relationships with other organisms. Most plant roots are linked to fungal extensions called hyphae. This relationship brings added water and nutrients to the plant (from the hyphae) and food to the fungi (from the photosynthesis the plant performs). The combined root/hyphae associations are known as mycorrhizae. Some plants form cooperative relationships with nitrogen-fixing bacteria, the bacteria taking atmospheric nitrogen and transforming it into a form the plants can use, the plants providing the bacteria with nutrients.

22.4 Responding to External Signals

- Plants will bend toward a source of light through the process of phototropism, meaning a curvature of shoots in response to light.

  TUTORIAL 22.2.2: Response of Plants to Light

- Plants are able to sense their orientation with respect to the Earth and direct the growth of their roots and shoots accordingly—roots into the Earth, shoots toward the sky. This is called gravitropism.

- Some plants can climb upward on other objects by making contact with them and then encircling them in growth. This is thigmotropism, meaning growth of a plant in response to touch.

- Differential growth on one side of the root or stem makes possible phototropism, gravitropism, and thigmotropism.

- In temperate climates, deciduous trees exhibit a coordinated, seasonal loss of leaves and enter into a state of dormancy, existing on stored nutrient reserves in colder months.

- Plants can sense the passage of seasons and time their reproductive activities accordingly. One mechanism that assists in this process is photoperiodism, which is the ability of a plant to respond to changes it is experiencing in the daily duration of darkness, as opposed to light. Some plants that exhibit photoperiodism are long-night plants, meaning those whose flowering comes only with an increased amount of darkness—in late summer or early fall. Others are short-night plants, meaning those that flower only with a decreased amount of darkness—in early to midsummer.

Key Terms

abscission 490	**organic** 477
angiosperm 481	**ovary** 480
anther 480	**petal** 480
apical dominance 486	**petiole** 478
blade 478	**photoperiodism** 490
carpel 480	**phototropism** 488
cuticle 479	**pollen tube** 481
deciduous 489	**root hair** 477
dormancy 490	**seed** 481
fibrous root system 477	**sepal** 480
filament 480	**sperm cell** 480
fruit 481	**sporophyte generation** 481
gametophyte 481	**stamen** 480
gravitropism 488	**stigma** 480
hormone 484	**stomata** 479
inorganic 477	**style** 480
megaspore 481	**taproot system** 476
microspore 480	**tendril** 489
mycorrhizae 487	**thigmotropism** 489
nutrient 477	**tissue** 482

transpiration 478	**vegetative reproduction** 482
tube cell 480	**xylem** 482
vascular bundle 479	

Understanding the Basics

Multiple-Choice Questions

1. Which of the following is not a benefit that plants provide to other living things?
   a. act to decompose dead organic material
   b. build up food supply
   c. lock up carbon dioxide from the atmosphere
   d. produce oxygen
   e. prevent soil erosion

2. Plants take in _____ and _____through their roots, but take in _____ primarily through tiny pores on their leaves.
   a. minerals, carbon dioxide, water
   b. oxygen, carbon dioxide, water
   c. water, carbon dioxide, minerals
   d. oxygen, minerals, carbon dioxide
   e. minerals, water, carbon dioxide

3. Which of the following are not functions of roots?
   a. nutrient storage
   b. loss of water through transpiration
   c. loss of oxygen in photosynthesis
   d. grow to water supplies
   e. absorption of water

4. Flowers are the _____ structures of angiosperms.
   a. photosynthesizing
   b. carbon-dioxide absorbing
   c. reproductive
   d. oxygen-releasing
   e. food-producing

5. Pollen grains are
   a. hormonal messengers that travel from one plant to another
   b. tiny food packets that move from one plant to another
   c. sperm-bearing sporophyte plants
   d. toxin-bearing packets that stunt the growth of other trees
   e. sperm-bearing gametophyte plants

6. Plant growth
   a. occurs only in the midsection of any leaf, stem, or root
   b. serves almost solely to lengthen a plant, rather than widen it
   c. occurs, in vertical growth, only at the tips of roots or shoots
   d. must come to an end after a fixed number of months or years
   e. can go on as long as the plant is alive

7. The sucrose produced by photosynthesis is transported from the leaf to other parts of the plant through
   a. sieve tubes
   b. vessel elements
   c. tracheids
   d. rhizoids
   e. guard cells

8. Which of the following statements is true?
   a. Spores are diploid.
   b. Gametes are single-celled, haploid structures.
   c. Spores participate in fertilization directly, fusing to give rise to a diploid zygote.
   d. Gametes are diploid.
   e. Spores are the plant's genes.

9. Phototropism and gravitropism lead to bending of shoots or roots toward or away from the stimulus (light or gravity, respectively). Such bending is brought about by
   a. increased growth over the general surface of the root or shoot
   b. differential growth on opposite sides of the root or shoot
   c. decreased growth at the root or shoot apex, and increased growth at the base
   d. muscular contractions, leading to a curvature in the root or shoot
   e. all of the above

10. Photoperiodism allows some plants to respond to _____ by timing such things as _____ in accordance with the amount of _____ they are exposed to.
   a. predators, dormancy, moisture
   b. competition, growth, moisture
   c. the seasons, growth, toxins
   d. other plants, flowering, ethylene
   e. the seasons, flowering, darkness

Brief Review

1. The stomata that exist mostly on leaves allow important gas exchanges to take place in plants but these exchanges also present plants with a problem they must solve. What gases are exchanged through the stomata, and in which direction? What problem must plants solve because of these exchanges, and how do they solve it?

2. What functions do roots perform? What are the main types of roots produced by flowering plants?

3. List the main parts of a flower, and describe the functions performed by each part.

4. Contrast the way people grow, and explain how plants grow and develop over their lifetimes.

5. Give examples of mutually beneficial relationships between plants and bacteria, and plants and fungi.

6. Distinguish between the gametophyte and sporophyte plant generations.

Applying Your Knowledge

1. Consider the colonization of land by plants. What do you suppose were the biggest problems these plant pioneers faced in the terrestrial environment, compared to their formerly aquatic life? Would amphibians, the first vertebrates to invade the land, also have faced some or all of these problems?

2. The text makes clear that plants respond to their environments in sophisticated ways, even to the point of signaling one another during attacks by predators. Given this, can plants be said to be conscious beings, or are animals the only conscious beings?

3. How should we think of the gametophyte and sporophyte generations of a plant? As different organisms, or as different forms of the same organism?

4. Gardeners often "pinch off" the terminal shoot apex to stimulate bushiness in young plants. Explain the physiological basis for this common horticultural practice.

5. Why do broad-leaved plants lose their leaves in autumn in the colder parts of the world, and how do they "know" when it is time for them to shed leaves?

23

Form and Function in Flowering Plants

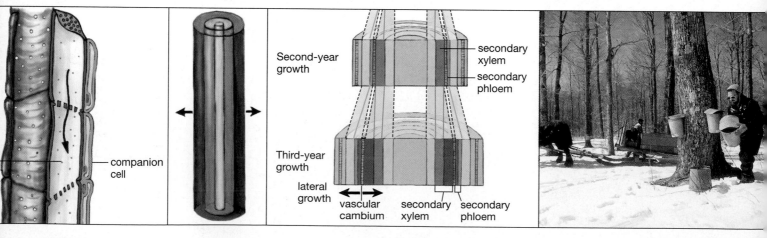

Plant tubes.
(Section 23.3, page 501)

Growing out.
(Section 23.5, page 505)

How woody plants grow wider.
(Section 23.5, page 506)

Pass the flapjacks.
(Essay, page 510)

Second-year growth

secondary xylem

secondary phloem

Third-year growth

lateral growth

vascular cambium

secondary xylem

secondary phloem

companion cell

Plants are fixed in one spot and have almost no moving parts, yet they manage to reproduce, transport fluids inside themselves, and keep growing for as long as they live.

Insects that eat plants are not much of a surprise, but plants that eat insects? About 500 species of flowering plants are carnivorous, with at least one variety, the tropical pitcher plant *Nepenthes*, able to capture animals up to the size of a small bird. But the most famous plant predator of them all undoubtedly is the Venus flytrap (*Dionaea muscipula*), which is native to the wetlands of North and South Carolina. So how does this plant, which has no muscles and is fixed in one spot, end up consuming insects?

Each trap is a single leaf folded in two, with the interlocking "fingers" of the two halves able to snap together in less than half a second, imprisoning any small wanderer (**see Figure 23.1**). But what keeps the leaf halves from wasting their energy by snapping together in pursuit of wind-borne blades of grass or other objects? The chemical reaction that puts the trap in motion is set off by a series of trigger hairs on the inside of each of the leaf's two sections. The trick is that two trigger hairs must be tripped in order for the halves of the leaf to snap shut. Insects wandering around inside the leaf are likely to trip two hairs, but blades of grass are not. And how does the plant slam its "jaws" shut without

muscles? When the trigger hairs are tripped, they set off a chemical reaction that engorges cells at the base of the leaf with water. This rapid movement causes the leaf halves to come together. Enzymes are then released that begin to digest the struggling insect.

It may surprise you to learn that Venus flytraps make their own food through photosynthesis, just like any other plant. So why do they need these side orders of insects? For nutritional supplements. Like many other carnivorous plants, Venus flytraps grow naturally in mineral-poor soil, and the insects they catch are a good source of nitrogen and phosphate.

Figure 23.1
Fatal Entry
A fly enters the leaf of a Venus flytrap. Once the fly pulls on two triggerhairs within the leaf, its opposing halves will snap shut, trapping the fly. Then the plant will begin to digest its prey.

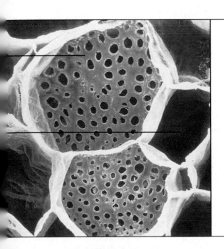

Food carriers.
(Section 23.6, page 511)

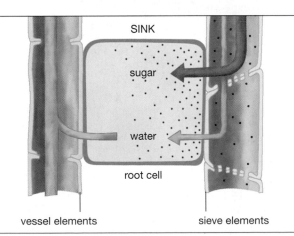

Food and water take a trip.
(Section 23.6, page 512)

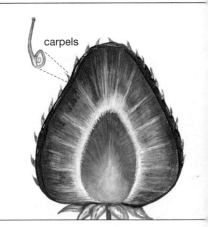

Where are the
strawberry seeds?
(Section 23.8, page 518)

Figure 23.2
Floating Platforms
A woman plays a violin while standing atop a Victoria water lily at the Missouri Botanical Garden in St. Louis at the turn of the century. The lilies are native to South America. Some South American Indians call the lilies *Yrupe*, which can be translated as "big water tray."

The world of plants is filled with species as unique in their own way as the Venus flytrap is. Queen Victoria water lilies (*Victoria regia*) have circular leaves that can be nearly 2 meters (about 6 feet) in diameter and that can serve as a floating platform for a person if pressed into service (**see Figure 23.2**). The seeds of orchids are so tiny that they are disseminated in nature by ants, while the seeds of the double coconut come wrapped in a fruit that might be 0.6 meter (about 2 feet) across.

Yet there is a unity to plant life, particularly in the case of the flowering plants, or angiosperms, that you'll be looking at in this chapter. Their common features and characteristics range from their cell types all the way to aspects of their reproduction.

The goal in this chapter is to cover some of these commonalties in four areas: cell and tissue types, growth, plant plumbing (meaning fluid transport), and reproduction.

You will be looking at flowering plants from the cell types to the larger systems. First, however, let's revisit the representation of the whole plant that you first saw in Chapter 22, so that you can "find your place" when reading about a plant's component parts. If you look at **Figure 23.3**, you can see a diagram of a typical plant, with its two-system division—roots and shoots—and the various parts of the plant that lie within these two systems. The function of each of these parts will become clear to you as you go through the details on them.

23.1 Two Ways of Categorizing Flowering Plants

The constituent parts of plants are often organized one way in a given kind of plant, but another way in a different type. So what are these types? Here are a couple of ways of *categorizing* plants, the significance of which will become apparent as you go along.

The Life Spans of Angiosperms: Annuals, Biennials, and Perennials

One important question that might be asked about any plant is: How long does it take for it to go through its entire life cycle—from being a seed germinating in the ground, through growth, flowering, seed dispersal, and then death? Plants that go through this cycle in one year or less are known as

Figure 23.3
Anatomy of a Flowering Plant (An Angiosperm)

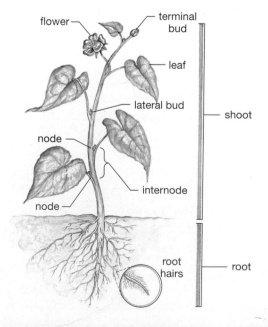

a

b

c

Figure 23.4
Categorizing Plants by Life Span

a Annuals such as tomato plants live for only one year.

b Biennials such as carrot plants have a two-year life span.

c Perennials such as this white oak tree live for many years, some for many hundreds of years. Plants with long life spans tend to be woody and larger than plants with short life spans.

annuals. Some of these are food crops, such as tomatoes and the commercial grains. Plants that go through the cycle in about two years are **biennials**, which include carrots and cabbage. (Have you ever seen the flowers of a carrot plant? Probably not; we *pick* the taproots called carrots after one year, meaning their shoots don't get a chance to flower. Most of us have seen the flowers of wild carrots, however, though these carrots go under another name: Queen Anne's lace.) Plants that live for many years are known as **perennials**, a category that includes trees, woody shrubs such as roses, and many grasses. **Figure 23.4** shows some representative annuals, biennials, and perennials.

A Basic Difference among Flowering Plants: Monocotyledons and Dicotyledons

A second distinction among the flowering plants has to do with their anatomy, meaning the arrangement of the structures that make them up. In this respect, there are two broad classes of flowering plants: monocotyledons and dicotyledons, which are almost always referred to as monocots and dicots. A cotyledon is an embryonic leaf, present in the seed, as you can see in **Figure 23.5**. **Monocotyledons** are plants that have one embryonic leaf; **dicotyledons** are plants that have two.

This distinction in the embryos then makes for major differences in the mature plants. Their roots are different, their leaves

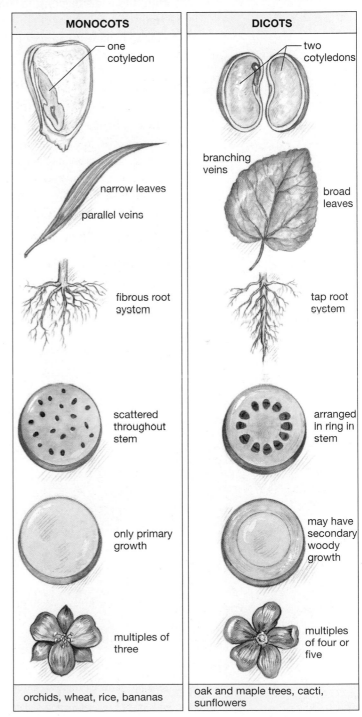

	MONOCOTS	DICOTS
Embryonic leaves	one cotyledon	two cotyledons
Mature leaves	narrow leaves — parallel veins	branching veins — broad leaves
Roots	fibrous root system	tap root system
Vascular bundles	scattered throughout stem	arranged in ring in stem
Type of growth	only primary growth	may have secondary woody growth
Flower parts	multiples of three	multiples of four or five
Examples	orchids, wheat, rice, bananas	oak and maple trees, cacti, sunflowers

Figure 23.5
A Cotyledon Is an Embryonic Leaf
Lots of differences result from whether a plant is a monocot (has one embryonic leaf) or a dicot (has two embryonic leaves).

are different, their transport tubes, or vascular bundles, are arranged differently (Figure 23.5). More than 75 percent of all flowering plants are the broad-leafed dicots. But given the 260,000 known species of flowering plants, this still leaves more than 50,000 species of narrow-leafed monocots, which include most of the important food crops. **Figure 23.6** shows you an example of both a monocot and a dicot.

With these distinctions in mind, you're now ready to start looking at the basic elements that go into making up flowering plants.

23.2 There Are Three Fundamental Types of Plant Cells

You saw in Chapter 4 that all living things are made up of cells, and plants are no exception to this rule. As it happens, there are three fundamental types of plant cells that, alone or in combination with each other, go on to make up most of the plant's **tissues**, meaning *groups* of cells that carry out a common function.

Parenchyma Cells

First, there are **parenchyma cells**, which have thin cell walls and are easily the most abundant type of cell in the plant, being found almost everywhere in it. Accordingly, these cells have a lot of different functions, forming the flesh of many fruits, the outer surface of most young plants, parts of leaves that are active in photosynthesis, and much of the "ground" tissue you'll be hearing about shortly. Further, parenchyma are alive at maturity, a quality that might hardly seem worth mentioning except that some plant cells are dead in their mature, functioning state.

Sclerenchyma Cells

This dead-at-maturity condition characterizes a second type of plant cell, **sclerenchyma cells**, which have thick cell walls that help the plant return to its original shape when it has been deformed by some force, such as the push of wind or animals. Sclerenchyma cells are found in parts of the plant that are mature—that will not be growing further, and that need the support that sclerenchyma can provide. Like all plant cells, sclerenchyma have, running through their cell walls, a compound called **cellulose** that can be likened to the steel bars in reinforced concrete. Most sclerenchyma cell walls also are infused with a tough compound called *lignin*. The combination of lignin and cellulose results in cells that can bear the crushing weight of massive trees (**see Figure 23.7**).

Collenchyma Cells

The third type of plant cell, **collenchyma cells**, can be thought of as support cells, like sclerenchyma cells, but with the function of stretching or elongating in parts of the plant that are growing, such as young leaves and stems. Collenchyma cells combine the properties of parenchyma and sclerenchyma cells, performing some of the functions of each of these cell types. They provide mechanical strength in areas undergoing active growth, and can also participate to some extent in photosynthesis and wound repair.

Parenchyma as Starting-State Cells

Both collenchyma and sclerenchyma are derived from parenchyma cells; that is, some parenchyma cells differentiate *into* collenchyma or sclerenchyma cells. Parenchyma are thus a kind of starting-state cell, but their capabilities are greater than this. They can be transformed "backward," in a sense, into the type of embryonic cells that give rise to the whole plant. You may have wondered in Chapter 22 how we can get a new plant by taking a cutting from an existing plant and placing it in the ground. The answer is that

a

b

Figure 23.6
Categorizing Plants by Physical Features

a Corn is one of several monocot plants that are food crops.

b Geraniums are an example of a dicot. See Figure 23.5 for a list of the anatomical differences between monocots and dicots.

when a plant is cut, parenchyma cells lying close to the cut are transformed into *growth* cells of a type you'll look at shortly, and these can sprout a new root. This obviously gives plants a flexibility not possessed by animals such as ourselves. If one of our fingers is cut off, we do not grow a new finger, to say nothing of a whole new person.

23.3 The Plant Body and Its Tissue Types

You saw earlier that tissues are groups of cells that carry out a common function. So what kinds of tissues are there in angiosperms? The short answer is dermal, ground, vascular, and meristematic, with each of these four types of tissue generally being composed of one or more of the cell types you just looked at. Dermal tissue can be thought of as the plant's outer covering, vascular tissue as its transport or "plumbing" tissue," meristematic tissue as its growth tissue, and ground tissue as almost everything else in the plant.

First: A Distinction Between Primary and Secondary Growth Tissue

Before proceeding further with this topic, we need to take note of a basic distinction in plant tissue that has to do with the way plants grow. Some plants are capable only of what is called **primary growth**, meaning growth at the tips of their roots and shoots that principally increases their *length*. In contrast, other plants exhibit not only primary growth, but something known as **secondary growth**, which can be thought of as the *lateral* growth, or thickening, that occurs in **woody plants**. Secondary growth exists in trees, for example, but not in orchids or strawberry plants. Nonwoody plants such as an orchid are known as **herbaceous plants**, meaning those that never develop wood (or bark) and that thus contain only primary tissue. It is this *primary* tissue that you will look at first, with a review of secondary tissue to come shortly.

If you look at **Figure 23.8**, you can see the location of all four kinds of primary tissue in a dicot. Looking at the stem in cross section, you can see that ground tissue is well named, because visually it forms a kind of background against which you can see the tubes of the vascular tissue. Dermal tissue then is at the periphery of the plant, forming a "skin" layer around the ground tissue. The

placement of the meristematic tissue is discussed shortly. First, let's look in a little more detail at dermal, ground, and vascular tissue.

Dermal Tissue Is the Plant's Interface with the Outside World

Plants can't move, they can't bite, and, as you saw in Chapter 22, they must carefully control their water supply. All this makes their **dermal tissue**, or *epidermis*, very important to them. This outer coat is generally only one layer of cells thick, but it serves to protect the plant and control its interaction with the

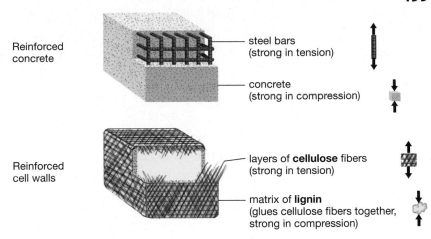

Figure 23.7
Plant-Cell Walls Are Composed of Composite Materials, as Is Reinforced Concrete
The cellulose fibers of plant-cell walls are very strong in tension, as are the steel bars in reinforced concrete. Lignin is similar to the matrix of concrete in that it binds the fibers together and resists compression. Both plant-cell walls and reinforced concrete are important in providing support for structures that are tall and heavy.

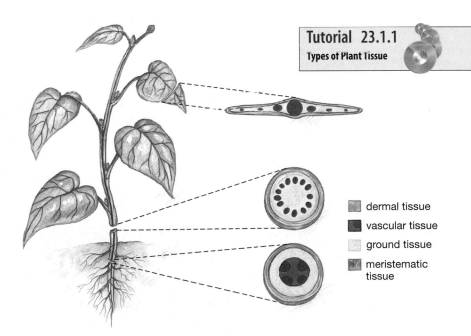

Tutorial 23.1.1
Types of Plant Tissue

- ▦ dermal tissue
- ■ vascular tissue
- ▨ ground tissue
- ▨ meristematic tissue

Figure 23.8
Four Types of Tissue in Primary Plant Growth
Dermal tissue (bright green) is found on the outside of the plant and is the interface between the plant and its environment. Ground tissue (pale green) is found throughout the plant and gives the plant its shape, stores food, and is active in photosynthesis. Vascular tissue (bright red) transports water, nutrients, and sugar throughout the plant. Meristematic tissue (blue) occurs at the tips of shoots and roots and at leaf bases, and gives rise to the entire plant, through the production of cells that develop into the other three tissue types.

outside world (**see Figure 23.9**). A plant's first line of defense against outside predators is a waxy coating called the **cuticle** that covers the epidermis of the plant's aboveground parts (its shoot). The cuticle is a kind of waterproofing, serving to keep water in, and an invader-proofing, helping to keep infecting bacteria and fungi out.

Plants need to exchange gases with their environments, however. They take in carbon dioxide and emit oxygen and water vapor through the microscopic pores called stomata, found mostly on the underside of their leaves. (You went over the functioning of stomata in Chapter 22, starting on page 479.) Stomata also represent a kind of weak point for plants, however, because they provide a passageway for microbial invaders and a conduit for excessive loss of moisture.

Dermal tissue also forms several kinds of extensions, called trichomes, that protrude from the plant's surface. You learned about one variety of them in Chapter 22: root hairs, meaning the thread-like outgrowths of individual epidermal root cells. As we noted there (page 477), these wispy outgrowths greatly increase the plant's ability to absorb water and minerals. Other types of trichomes serve not to absorb substances, but to secrete them; trichomes secrete the toxic chemicals that some plants use to ward off plant-eating animals.

Ground Tissue Forms the Bulk of the Primary Plant

Most of the primary plant is ground tissue, which can play a role in photosynthesis, storage, and structure (**see Figure 23.10**). Large parts of ground tissue may be simple parenchyma, but other parts are combinations of cell types.

Vascular Tissue Forms the Plant's Transport System

As you saw Chapter 22, the plant's vascular system has two main functions. First, it must transport water and minerals; second, it must transport the food made in photosynthesis. In line with this, there are two main components to the plant vascular system. First there is **xylem**, the tissue through which water and dissolved minerals flow. Second there is **phloem**, the tissue that conducts the food produced in photosynthesis, along with some hormones and other compounds (**see Figure 23.11**). The

Figure 23.9
Where Plant Meets Environment
Dermal tissue serves as the interface between a plant and the environment around it. The waxy cuticle and the single layer of cells in the epidermis work together to protect the plant and to control interactions with the outside world.

a Specialized epidermal cells called guard cells regulate the opening of the plant pores called stomata, thus controlling gas exchange between the plant and its environment. Here, guard cells have opened two stomata. In most plants, the majority of stomata are found on the leaves.

b Some epidermal cells have hairlike projections, called trichomes, that serve various functions. Two kinds of trichomes are visible in this rose plant. The larger of the two varieties, with the bulbous tips, help secrete chemicals that guard against plant-eating predators.

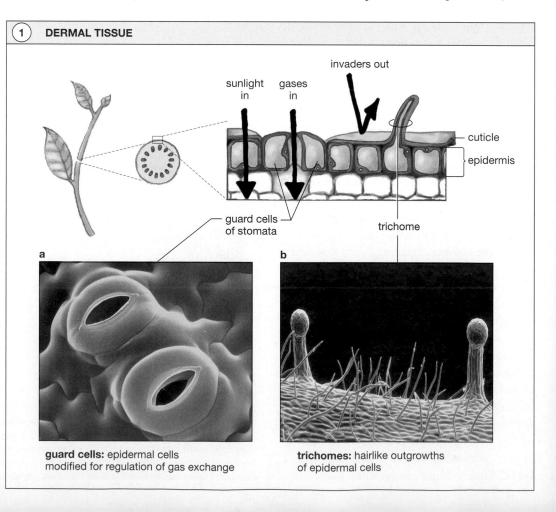

1 DERMAL TISSUE

sunlight in · gases in · invaders out

cuticle
epidermis
guard cells of stomata
trichome

a guard cells: epidermal cells modified for regulation of gas exchange

b trichomes: hairlike outgrowths of epidermal cells

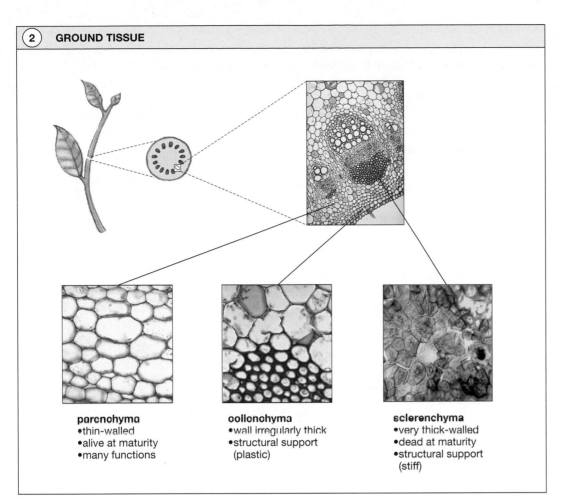

2 GROUND TISSUE

parenchyma
- thin-walled
- alive at maturity
- many functions

collenchyma
- wall irregularly thick
- structural support
(plastic)

sclerenchyma
- very thick-walled
- dead at maturity
- structural support
(stiff)

Figure 23.10
Ground Tissue
Most of a plant is made of ground tissue, which is composed of parenchyma, collenchyma, and sclerenchyma cells. Examples of these cell types shown here come from several different plants. In a cross section of a stem, thin-walled parenchyma can be found in most of the interior. Collenchyma—at the bottom of the collenchyma picture, surrounded by parenchyma are support cells that can elongate in growing parts of the plant. Very thick-walled sclerenchyma, such as these from a pear, can provide strength to tissues.

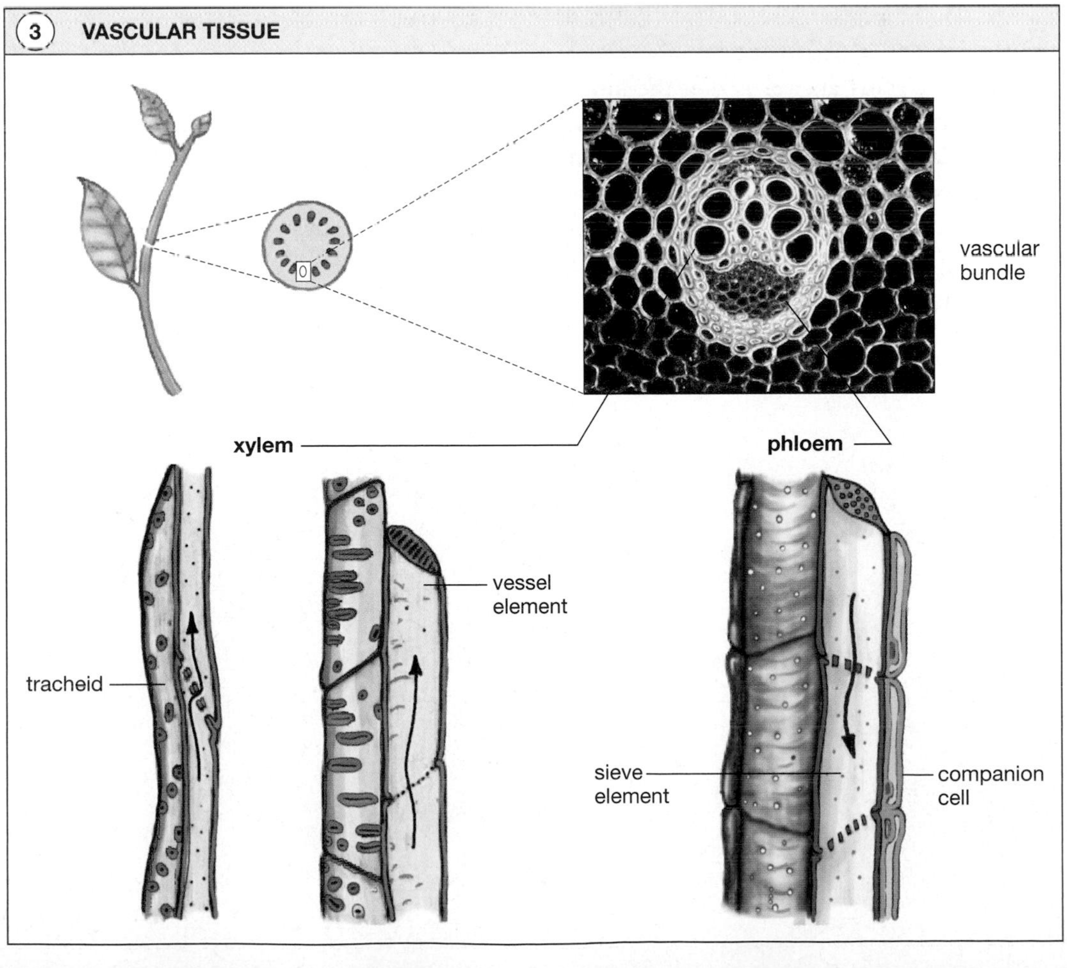

3 VASCULAR TISSUE

vascular
bundle

xylem

phloem

tracheid

vessel
element

sieve
element

companion
cell

Figure 23.11
Materials Movers
Vascular tissue transports fluids throughout the plant. Xylem is composed of cells called tracheids and vessel elements that transport water and dissolved minerals. Phloem is composed of sieve elements (and their companion cells), which transport the food produced during photosynthesis. These cell types are stacked upon one another to form long tubes.

movement of water through xylem is directional—from root up through stem and then out through leaves as water vapor. Meanwhile, the food made in photosynthesis must travel through phloem to every part of the plant, though, as a practical matter, in temperate climates the net flow is downward in summer (from food-producing leaves) but upward in early spring (from root and stem storage sites.)

Xylem and phloem are arranged in **vascular bundles**, which is to say collections of xylem and phloem tubes that run together in parallel in the stem—xylem tubes toward the inside of the stem, phloem tubes toward the outside. The bundles are arranged in different ways within a stem, depending on whether the plant is a monocot or a dicot; the dicot bundles are configured in a circle, the monocot in an irregular pattern. This is why dicot ground tissue often is conceptualized as existing in two regions, which you can see in **Figure 23.12**: a **cortex** is outside a ring of the vascular bundles, and a **pith** is inside it, with the pith cells being specialized for storage.

Meristematic Tissue and Primary Plant Growth

So you've seen three types of primary plant tissues. The question now is: How do these arise? How is it that a plant gets its vertical growth? The short answer is that the fourth type of plant tissue, meristematic tissue, gives rise to all the other tissue types. In considering the role of meristematic tissue, recall from Chapter 22 that, in their vertical growth, plants do not grow the way people do—throughout their entire length—but

instead grow only at the tips of their roots and shoots. In addition, a plant's growth is *indeterminate*: In most plants it can go on indefinitely at the tips of roots and shoots, as opposed to animal growth, which generally comes to an end when the animal matures.

23.4 How a Plant Grows: Apical Meristems Give Rise to the Entire Plant

Putting indeterminate growth together with the concept of growth at the tips, it follows that there must be plant cells at root and shoot tips that remain perpetually young— or more accurately, perpetually embryonic. That is, these cells are able to keep dividing, thereby giving rise to cells that then differentiate into all the primary cell and tissue types you've been reading about. Indeed, everything in plants (both their primary and secondary tissues) develops ultimately from these cells, which collectively form the **apical meristems** of plants.

If you look at **Figure 23.13**, you can see the apical meristem locations in a typical plant. Note that in this plant with a taproot, each lateral root tip has its own apical meristematic tissue that is capable of giving rise to yet more roots. In the shoot, plants confine their growth to the shoot apices that lie at the tip of each stem. Why? The better to compete for precious sunlight. It is the **shoot apical meristem** that gives rise to all the cells that allow this vertical growth. Note, however, that there is a second location for meristematic tissue in the shoot: the area nestled between leaf and stem. It's

Figure 23.12
Two Types of Flowering Plants

These cross sections show the organization of primary tissues in a monocot and a dicot.

a Monocots (in this case an onion) have vascular bundles arranged in an irregular pattern within ground tissue.

b Dicots (in this case a buttercup) have vascular bundles arranged in a ring that separates the ground tissue into pith on the inside of the bundles and cortex on the outside.

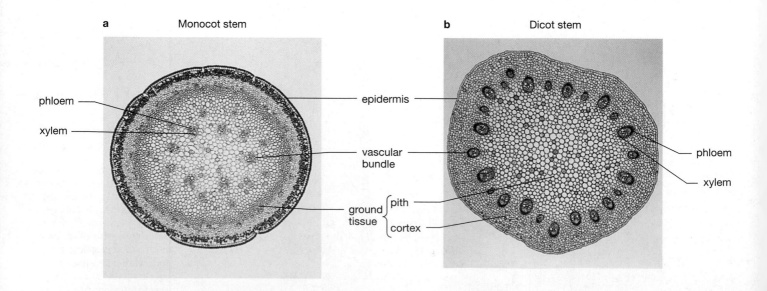

a Monocot stem   **b** Dicot stem

phloem · xylem · epidermis · vascular bundle · ground tissue { pith · cortex } · phloem · xylem

④ MERISTEMATIC TISSUE

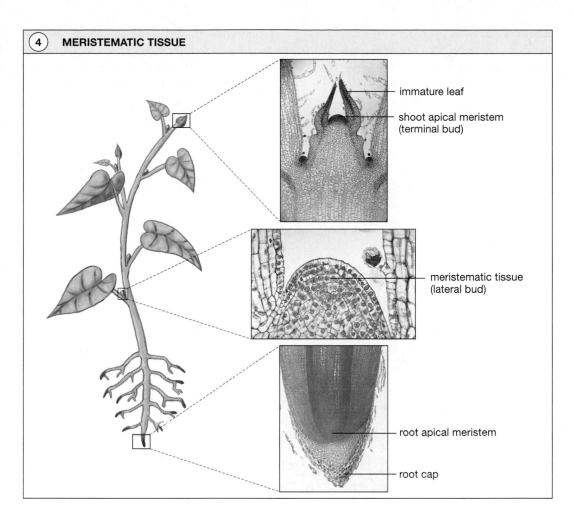

- immature leaf
- shoot apical meristem (terminal bud)
- meristematic tissue (lateral bud)
- root apical meristem
- root cap

Figure 23.13
Meristematic Tissue
Growth originates from meristematic tissue. The shoot and root elongate at the apical meristems, while new branches originate at the lateral buds. Meristematic tissue gives rise to more dermal, ground, and vascular tissue, as well as more of itself.

here that we find lateral buds (sometimes called *axillary buds*). Any **bud** is an undeveloped shoot, composed mostly of meristematic tissue. A **lateral bud** is meristematic tissue that may give rise to a branch or a flower (which obviously ends up growing laterally from the stem). Lateral buds also serve, however, as a kind of insurance policy for the plant in that they can switch roles. Should the plant's apical meristem become damaged, one of the lateral buds steps in to assume the role of the shoot apex, thus allowing the plant to maintain its vertical growth. As long as the original apical meristem is intact, however, it generally will produce hormones that *suppress* the growth of the lateral buds near it, leaving them dormant. The shoot apex is itself sometimes referred to as a **terminal bud**, particularly when *it* is dormant, as with trees in winter.

We can think of plant growth in terms of growth modules. Each module consists of an internode (the stem between leaves), a node (the area where a leaf attaches to stem), and one or more leaves (**see Figure 23.14**).

A Closer Look at Root and Shoot Apical Meristems

Now let's look a little closer at these important collections of cells from which all else comes, the shoot and root apical meristems. The right-hand pictures in Figure 23.13 show you what shoot apical meristem, lateral bud, and root apical meristem look like up close.

Tutorial 23.1.7
How Plants Grow

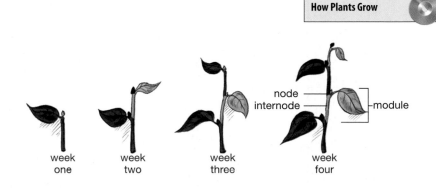

week one | week two | week three | week four

node
internode
module

Figure 23.14
Vertical Growth in Plants Occurs in Modules
Each growth module (indicated here by alternating dark green and light green segments) consists of an internode, a node, and one or more leaves. This plant moves from one module in week one to four modules in week four.

Shoot Meristems

As you can see in the topmost picture, the shoot apical meristem lies atop a dome-shaped collection of cells. Overlapping it are two budding leaves that serve a protective

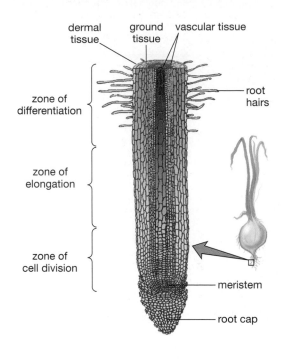

Figure 23.15
Tissue Development from Apical Meristems
Apical meristems give rise to the primary tissues. In development, cells first are mostly engaged in division, just above the apical meristem, then elongation, and finally differentiation. The result is three fully formed kinds of tissue.

function. Next, in the middle picture, you can see a second shoot location for meristematic tissue—the lateral bud, nestled between leaf and stem. These buds are collections of meristematic tissue that are left behind, in a sense, by the apical meristem tissue as it continues its growth. (If you look again at the topmost picture, you can see two small collections of darker colored cells on each side of the apical meristem dome. These are collections of meristem tissue that have recently been left behind; they will become lateral buds.)

Root Meristems

Because roots are pushing their way into the ground, root apical meristems are located not at the very tip of the root, but just up from the tip. The root apical meristems are adjacent to a collection of cells called a **root cap** that both protects the meristematic cells and secretes a lubricant that helps ease the way for the root in its progress. Toward the middle of the apical meristem, there is a collection of cells called the quiescent center. These slowly dividing cells amount to another kind of insurance policy for the plant. Should the meristematic cells become damaged, quiescent center cells will start dividing more rapidly, producing new meristematic cells. Cells in the quiescent center are, in short, a reserve that goes into action in times of trouble. They are well suited for this because, in their dormancy, they are more resistant than apical meristem cells are to injury from such influences as toxic chemicals.

How the Primary Tissue Types Develop

You've looked so far at three of the plant's primary tissues—the ground, dermal, and vascular tissues. And you know that the fourth type of primary tissue, meristematic tissue, gives rise to all these. So how do plants get from meristem to the others? Not in a single step, it turns out—not directly from apical meristem cells to, say, fully functioning vascular tissue. Instead, there is a transition involving three main changes: *More* cells are produced through cell division; cells *elongate* through cell growth; and cells *differentiate* into the three kinds of tissue. These three changes are overlapping—cells are elongating even as they are differentiating—but events do take place in roughly this order. You can see this by looking at **Figure 23.15**. In a gradual transition with no sharp boundaries, the apical

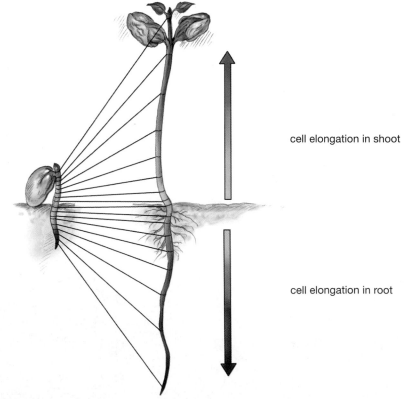

Figure 23.16
Growth by Cell Elongation
Much of the elongation of the shoot and root in a young plant results from cell elongation rather than from cell division. Thus it is cells just above the root tip, in the zone of elongation, that provide most of the root's growth, as shown here. The tip of the root (where cell division occurs) has elongated very little, and the cells in the upper root have completed most of their elongation. This same pattern holds true for the shoot.

meristem gives way to a **zone of cell division**, then there is a **zone of elongation**, and finally a **zone of differentiation**, which is where the ground, dermal, and vascular tissues fully take shape. Note that root hairs don't appear until the zone of differentiation. Though the three kinds of tissues *finish* taking shape in the zone of differentiation, they actually begin this process just above the apical meristem (in the zone of cell division). It's accurate to think of the tissues in the zone of cell division as being precursors to the three tissue types.

If you look at **Figure 23.16**, you can see the practical effect of a zone of elongation: It is where most of the primary plant growth occurs. There is some growth with the addition of more cells down near the apical meristem, but most of the vertical expansion of the plant takes place because of cell elongation. Such elongation has the effect of extending the roots into the soil and the stem into the air.

Intercalary Meristems Keep Grasses Growing

There is one important variation on primary growth through apical meristems. What about plants such as grasses that are constantly losing their tops through such means as grazing or mowing? Grasses get their growth from tissue known as **intercalary meristems**, which are found at the base of *each node*. There thus exists a series of vertical growth tissues that are intercalated, or interspersed, between regions of nondividing cells as we go up the plant. One practical effect of this structure is that when we mow the top half of a blade of grass, it grows back. A second is that, because these plants are growing at *many* points along their length (rather than just their apex), they can manifest very rapid growth. We can easily see nodes in the giant grass known as bamboo, which exhibits intercalary growth (**see Figure 23.17**).

23.5 Secondary Growth Comes from a Thickening of Two Types of Tissues

So far, you've been looking at a plant's vertical or primary growth. In the herbaceous plants that's all the growth there is. But you've seen that some plants—the woody plants—grow out as well as up and down; they have primary *and* secondary growth, in other words (**see Figure 23.18**). So how does this secondary growth come about? Once again, through cells that are meristematic. In essence, secondary growth yields three new kinds of tissue: a different type of xylem and phloem— secondary xylem and secondary phloem— and several layers of outer tissue that collectively go under the familiar name of *bark*.

At the start, it's worth noting that secondary growth in plants is related to two of the plant characteristics that we reviewed at the start of the chapter. First, secondary growth almost always takes place in perennials, rather than annuals (though plenty of perennials exhibit only primary growth, as with grasses.) Second, with very few exceptions, the monocots described earlier have only primary growth. Meanwhile, dicots are more often woody, though many are herbaceous.

Secondary Growth through the Vascular Cambium: Secondary Xylem and Phloem

You saw earlier that the plant's vascular tissues are arranged into bundles of tube-like vessels, with the food-carrying phloem tubes toward the outside of the stem and the water-carrying xylem tubes toward the inside. In a cross section of a dicot plant stem, these

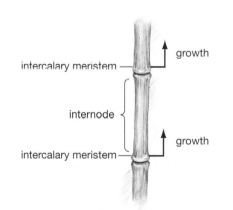

Figure 23.17 Intercalary Meristems Keep Grasses Growing One reason bamboo grows so rapidly is that growth occurs at several points along its length at once. The intercalary meristems responsible for this growth are located at the base of each internode.

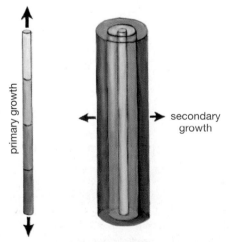

Figure 23.18 Growing Up versus Growing Out Herbaceous plants grow at their tips, mostly vertically, through primary growth. Woody plants, however, grow not only vertically but also laterally, through what is called secondary growth. Dicots can be either woody or herbaceous, but almost all monocots are herbaceous.

tubes form concentric rings. If you look at the plant another way—through its entire *length*—you can see that these rings take the shape of cylinders that are nested, as if these layers of tissue were a series of open-ended cans, one inside the next (see **Figure 23.19**).

Now notice that woody plants develop a thin layer of tissue *between* the primary xylem and primary phloem cylinders. This is the **vascular cambium**, one of the additional types of meristematic tissue that brings about secondary growth. The vascular cambium has one group of cells in it, called *ray initials*, that produce exactly what their name indicates: *rays* of parenchyma cells that carry water and dissolved compounds through the *width* of the plant. (If you've ever looked at the stump of a tree, you've probably noticed fine lines that seem to radiate out, from near the center to the periphery of the trunk. These

are the tree's transport rays, made up of parenchyma cells that form streaks of living cells in an expanse of dead xylem. You can see an example of them in Figure 23.22.)

More important for our purposes, the vascular cambium is something like a cell factory that is continually pushing out two different kinds of cells to either side of itself. The result is two new cylinders of tissue: **secondary xylem** to the inside of the vascular cambium, and **secondary phloem** to the outside. Secondary xylem and phloem initially do just what primary xylem and phloem do: transport, through the *length* of the plant, water through xylem and food through phloem. **Figure 23.20** shows you how the vascular cambium pulls off its feat of two-way xylem/phloem production. Note in the figure that the cells of the vascular cambium divide, producing one cell that differentiates into either xylem or phloem and another cell that remains vascular cambium. This process continues over and over again, causing the stem of the plant to thicken.

Secondary Xylem Is Responsible for Most of a Plant's Widening

It is the secondary xylem that is responsible for most of the widening of trees and other woody plants. Indeed, another term for secondary xylem is **wood**, a material that comprises about 90 percent of an average tree. After an initial period of carrying water, the secondary xylem cells will cease doing this, but are still valuable to the tree because of their strength.

Almost everyone is familiar with the **annual rings** of trees in temperate climates such as ours. But what is each visible ring? **Figure 23.21** tells the tale. Each ring represents the abrupt change between the secondary xylem of late summer and the secondary xylem of the following spring. In late summer, when plants are getting little water, they produce xylem cells with smaller interiors and proportionately thicker cell walls. Because cell walls are *darker* than the cell interiors, the summer xylem cells are relatively dark. Come next spring, with its plentiful water, the cells that are produced are much larger and thus have lighter color. Thus, the "line" of any given ring is made up of darker, late-summer cells. We can look just to the outside of any line and say "spring began here." To learn more about what we can tell by looking at a cross section of a felled tree, see "A Tree's History Can Be Seen in Its Wood."

Figure 23.19
Vascular Cambium and Secondary Growth
Vascular cambium is meristematic tissue that produces secondary xylem tissue to its interior and secondary phloem tissue to its exterior. Over the years, the thickening of a tree comes about primarily because of the growth of secondary xylem.

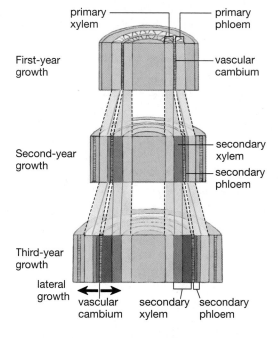

Figure 23.20
How Vascular Cambium Thickens the Plant
The cells of the vascular cambium divide, producing one cell that differentiates into either xylem or phloem and another cell that remains vascular cambium. This process repeats, causing the stem of the plant to thicken.

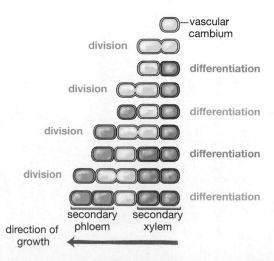

A Tree's History Can Be Seen in Its Wood

Everyone knows about looking at the cross section of a felled tree trunk and counting its "rings" to gauge the age of the tree. But what other stories does this kind of tree wood have to tell?

First, are the rings wider in some years than others? To the extent that the tree grew more in a given year, it will have a wider ring. The wider the ring, the more likely it is that optimal conditions for growth existed during that year: plentiful rainfall, enough sun, a lack of pests. A series of unusually narrow rings probably means just the opposite (**see Figure 1**).

Next, are the rings perfectly concentric, or are they *eccentric* at some point in the tree's growth? Is a given set of circles wide on, say, the right as you look at the trunk, but narrower on the left, as in Figure 1? If so, it may be because some force—for example, pressure from another tree—was bending the tree trunk toward the right at that point in its history. Because trees prosper by being as tall as they can, they will resist being pushed over. They do so by growing extra wood on one *side* of themselves. The wider areas in a given set of circles represent this extra wood, known as **reaction wood**.

Look at the center of any large log, and the wood you see there is almost always darker than the surrounding wood. Why? The inner wood—which like all wood is secondary xylem—has long since closed down with respect to its water-transport function. The cells that make up the water columns were always dead in their mature, water-transporting state; but eventually each column was severed altogether, meaning it quit conducting water. The problem is that an *empty* xylem column is a *dangerous* column as far as the tree is concerned, because a fungus could grow right up through it. Therefore, trees have developed a plugging mechanism: Surrounding living cells push material into the columns, sealing them off. Over time, these cells produce oils, lignin, and microbe-fighting substances and then die themselves. The result is a center

of a tree that is completely dead—and dark in coloration—but strong in its load-bearing strength and resistant to bacteria and fungi. This is the tree's **heartwood**. Its counterpart is the tree's **sapwood**—the lighter, outer wood whose xylem tubes are still conducting xylem sap.

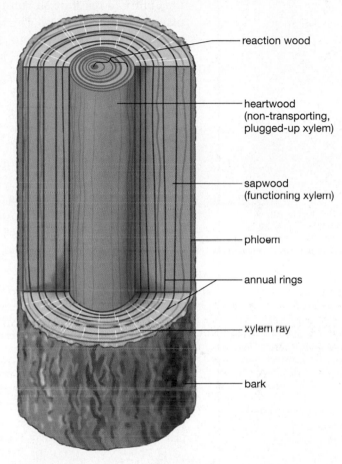

reaction wood

heartwood (non-transporting, plugged-up xylem)

sapwood (functioning xylem)

phloem

annual rings

xylem ray

bark

Figure 1
The Anatomy of Wood

Figure 23.21
Annual Rings Tell Time and More

a The thick secondary xylem (the wood) of a tree contains many concentric rings, with the dark line of each ring representing the end of one year's growth. The xylem rays transport water and nutrients laterally through the trunk.

b This close-up of the annual rings from an actual tree shows how growth can vary with environmental conditions. The earlier growth rings on the left are thicker than the rings on the right. The tree grew less in each of the later years because it was being affected by acid rain.

c This micrograph of a pine tree shows the structural basis for annual rings. On the left—toward the inside of the trunk—are smaller cells that grew in late summer and early fall, when water was less plentiful. The very narrow cells at left-center create the dark line of an annual ring. Immediately to the right of these densely packed cells are the larger cells that grew the following spring, when water was again plentiful. These larger cells are visible as the lighter wood in the trunk.

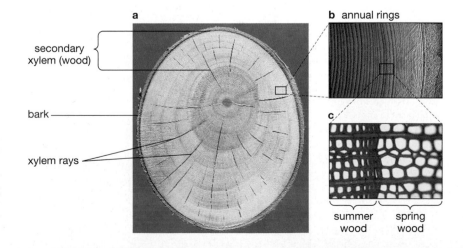

a

secondary xylem (wood)

bark

xylem rays

b annual rings

c

summer wood

spring wood

Secondary Growth through the Cork Cambium: The Plant's Periphery

With the secondary xylem cells pushing out from the interior of the tree or shrub, the plant's original outer covering (its epidermis) can't last. It splits apart, eventually to be shed altogether, usually within a year. Before it goes, however, the plant develops a new kind of outer covering, one that has the ability to renew itself constantly as the plant continues to grow. This quality may sound familiar by now, and indeed **cork cambium** represents the final type of meristematic tissue you'll be encountering—one that gives rise to the plant's outer tissues (**see Figure 23.22**).

The cork cambium itself is a product of other tissue in the stem—in a mature tree, the secondary phloem tissue you just looked at. Cork cambium is once again secondary meristemic tissue, which is like a cell factory that is continually pushing out different kinds of cells to either side of itself. The main product of this activity goes to the *outside* of the cambium in the form of **cork** cells. As these cells mature, they go on to infuse their cell walls with a waxy substance that acts as nature's own waterproofing and invaderproofing. But cork cells are born to die, in the sense that their genetic blueprint brings about a so-called programmed cell death. It is in this dead, mature state that they serve their protective function. In a single growing season, the cork cambium can produce layer upon layer of cork cells, the result being a well-protected tree. To its interior, the cork cambium can produce, in a few species, a layer of parenchyma cells called phelloderm.

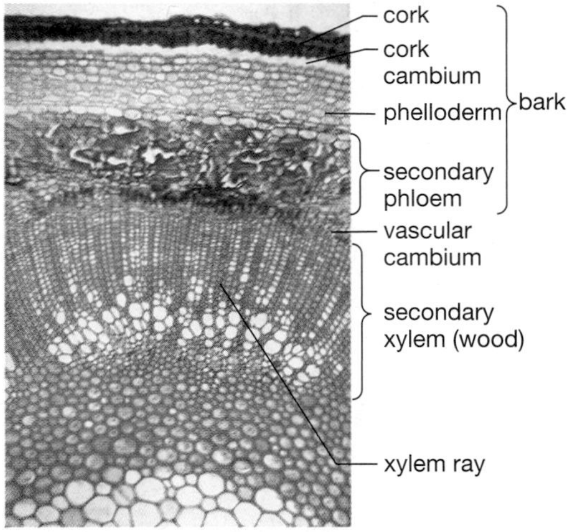

What Is Bark?

So, woody plants have cork cambium and its two products (cork and phelloderm). Then *inside* this three-part structure, there is the secondary phloem. Put together, all four parts constitute a region of a woody plant that goes by a familiar name: **bark**. Put another way, bark is everything outside the vascular cambium. Figure 23.22 shows all the layers of tissues in secondary growth.

If a tree is "ringed" by having a strip removed around its entire circumference—as hungry animals might do in winter—it might live without its cork, and the cork cambium interior to it, and the phelloderm inside that. But if this cutting continues into the younger, food-conducting secondary phloem, that would be the end of the tree, because it could no longer move its energy stores to where they're needed. Indeed, this kind of cutting describes the "girdling" that is a common way to kill unwanted trees (**see Figure 23.23**).

To conclude this section on plant growth, **Figure 23.24** shows you how the entire plant grows, with the starting point for both secondary and primary growth being apical meristem tissue.

23.6 How the Plant's Vascular System Functions

You have had ample opportunity already to look at the plant's vascular or "plumbing" system as it relates to plant growth, but now it's time to take a closer look at how the vascular system itself works, starting with the water-conducting xylem and then continuing to the food-conducting phloem.

Figure 23.22
Secondary Growth Tissues
The secondary xylem, the wood with which we are all familiar, is responsible for water transport and the structural support of the plant. Xylem rays are lateral conduits for water. Bark is composed of four layers of tissue: secondary phloem, phelloderm, cork cambium, and cork.

cork
cork cambium
phelloderm ⎫ bark
secondary phloem
vascular cambium
secondary xylem (wood)
xylem ray

Figure 23.23
Death from Lack of Nutrients
This tree has been girdled by someone who intends to kill it. The bark was removed all the way down to the secondary phloem, the tissue layer that transports food from one part of the plant to another.

How the Xylem Conducts Water

Remember the statement at the top of the chapter that a good many plant cells are dead in their mature, functioning state? You saw this once already with cork cells; here is another variety of these cells, which form most of the xylem. Two types of cells make up the water-conducting portions of both the primary and secondary xylem. These are **tracheids** and **vessel elements**. Upon reaching maturity, these cells die, and their contents are then cleared out. What's left is a strand of empty, thick-walled cells stacked one on top of another—tubes, in other words (**Figure 23.25**). The walls of these xylem cells remain, however; indeed, *these* are some of the sclerenchyma cells noted earlier that have the strong lignin reinforcement in their cell walls. The upshot is not only a transport system but a *load-bearing* system, which is why something as massive as a tree can grow so tall. As noted earlier, this load-bearing function remains in a given group of xylem cells long after they have ceased to act as a conduit for the water and dissolved nutrients that are sometimes called **xylem sap**. (To get an idea of one of the uses human beings have for this sap, see "The Syrup for Your Pancakes Comes from Xylem" on page 510.)

A tube metaphor is more accurate for the vessel elements than for the slender, tapered tracheids, because the vessel elements lose not only their internal contents, but may also lose part or all of the cell-wall barriers between adjoining cells. The result is a great capacity to transport water relative to the tracheids, whose cell walls remain intact. Any two tracheid cells will have matched pairs of perforations that line up with one another, allowing water to move from one tracheid to another.

The Value of Having Tracheids and Vessel Elements

You may remember from Chapter 22 that, through evolutionary time, flowering plants took over as the dominant variety of plant from the gymnosperms (which include coniferous trees). One reason this happened was the development of vessel elements; almost all angiosperms have vessel elements *and* tracheids, while almost all woody gymnosperms possess only tracheids. As a result, angiosperms can simply transport more water than can gymnosperms. You may wonder, though: If

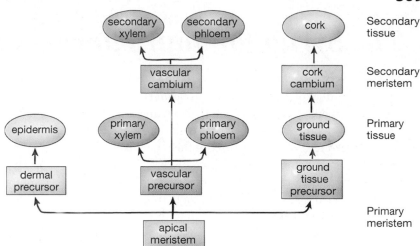

vessel elements are so much more efficient than tracheids at transporting water, why has evolution preserved tracheids in angiosperms? As it turns out, compared with vessel elements, tracheids are less susceptible to trapping air bubbles, or "embolisms," that can arise in flowing water. Given this, tracheids may be another kind of insurance policy for plants as they face various environmental changes.

Suction Power Moves Water through Xylem

In some trees, water coming from the roots goes through a vertical journey of up to 100 meters, or about 300 feet, before it gets to the topmost leaves of the tree. What kind of force can move water such a long way? Plants spend no energy at all in this process. Instead, it is **transpiration** itself—the loss of water from a plant, mostly through the leaves—that moves water through xylem. As water vapor evaporates into the air, it creates a region of low pressure, and this pulls a continuous

Figure 23.24
Summary of Primary and Secondary Growth in the Stem of a Vascular Plant
The "precursor" tissues that arise from the apical meristem are those that take shape in the zones of cell division, elongation, and differentiation noted earlier in the chapter.

Figure 23.25
Water Carriers
The fluid-conducting cells of the xylem are tracheids and vessel elements. Both kinds of cells stack up to form tubes.

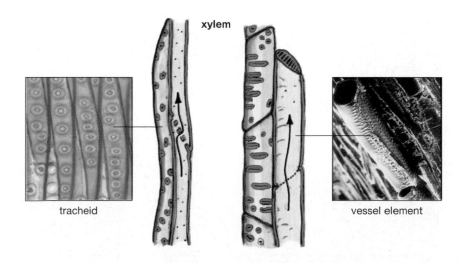

The Syrup for Your Pancakes Comes from Xylem

In early March, when winter is winding down in New England and Eastern Canada, farmers who own sugar maple trees are kept busy in a game of drilling and watching. What they are drilling is a series of two-to-three-inch holes, about three of them per maple tree. What they are watching is the weather, because they need a series of freezing nights followed by warmer days to get what they want. Such nights prompt the maple trees to convert nutrients they have stored into sugar; the warmer days then bring about a flow of this sugar, upward in sap through the tree's xylem. This material can then be "tapped" through spigots inserted into the drilled holes. The clear white sap that drips out—as much as 150 liters (about 40 gallons) of it per tree per year—is the starting material for pure maple syrup. Vermont is the chief maple-syrup-producing state in the United States, with an output in 1996 of just over 2 million liters (about 550,000 gallons). It is Canada, however, that is the world's largest maple syrup producer.

The "maple syrup" that most Americans buy in stores actually has very little maple syrup in it (perhaps 2–3 percent, the bulk being made up of corn syrup and syrup from regular sugar). This is understandable, because pure maple syrup is expensive, costing about $25–$35 per gallon. One reason for this expense is that harvesting the syrup is a labor-intensive business; each tap must be put in by hand, just as it was hundreds of years ago (**see Figure 1**). Beyond this, it takes about 50 gallons of sap to make a single gallon of maple syrup. The sap must be boiled down, until its sugar concentration reaches 66.5 percent.

Too many taps can kill a tree, but experienced syrup harvesters know how to tap the same tree year after year—potentially for hundreds of years—without harming it. The 150 liters of sap a tree can yield up sounds like a lot, but it may account for only 10 percent of a tree's total supply of sugar.

In the text, you've read much about the plant's food supply flowing through *phloem*, not xylem. But in sugar maples and some other deciduous trees, the food produced in summer's photosynthesis is stored as starch in xylem cells, to be broken down into sugar and released to the growing upper portions of the tree in early spring.

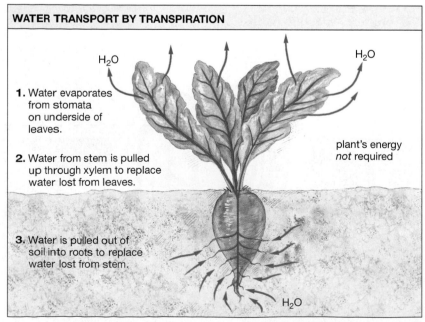

Figure 1
Source of Maple Syrup
Maple trees in Vermont being tapped for their xylem sap, which will become the key ingredient in maple syrup.

Tutorial 23.2.1
Water Transport through Xylem by Transpiration

WATER TRANSPORT BY TRANSPIRATION

H₂O

H₂O

1. Water evaporates from stomata on underside of leaves.

2. Water from stem is pulled up through xylem to replace water lost from leaves.

plant's energy *not* required

3. Water is pulled out of soil into roots to replace water lost from stem.

H₂O

column of water upward (**see Figure 23.26**). Ultimately, the energy source for this movement is the Sun, whose warm rays power the evaporation of water at the leaf surface.

Food the Plant Makes Is Conducted through Phloem

Now let's turn from water-and-nutrient transport to food transport. What is the nature of the food that flows through phloem? It's mostly sucrose—better known as table sugar—that is dissolved in water. Hormones, amino acids, and other compounds move through phloem as well, but sucrose and

Figure 23.26
Suction Moves Water through Xylem
When water evaporates from leaves, a low pressure is created that pulls more water up to fill the void. Water is so cohesive that it moves up in a continuous column. The Sun's energy, rather than the plant's energy, fuels this process.

water are the main ingredients of this material, which is known as **phloem sap**.

The phloem cells that this sap moves through are called **sieve elements**. Unlike xylem cells, sieve elements are alive at maturity, only in a curiously altered state. Each one has lost its nucleus, which you may remember is the DNA-containing information center of a cell. Sieve elements continue to function as living cells, however, because each one has associated with it one or more **companion cells** that retain their DNA and that seem to take care of all the housekeeping needs of their sister sieve elements (**see Figure 23.27**). The association between the two kinds of phloem cells is so intimate that when sieve elements die out, as they often do at the end of a growing season, their companion cells die as well. The reason sieve elements lose their nuclei seems to be to make room for the rapid flow of nutrients through them. The reason they are called *sieve* elements is that the ends of adjoining cells are studded with small holes, much like a sieve. Stacked on top of each other, these cells form the plant's food pipes, the **sieve tubes**.

Sugar-Pumping and Water Pressure Move Nutrients through the Phloem

The motive power for the movement of food through phloem is different from that of water through xylem; with nutrient transportation, the plant does not get a free ride on transport (via transpiration). Instead, to load the sucrose it makes into the phloem, the plant must expend its own energy (**see Figure 23.28**). After this, a principal player in moving fluid through the phloem is a phenomenon you've encountered before—osmosis.

Osmosis and the Movement of Phloem Sap

You can review the process of osmosis in Chapter 5 (page 102), but here's a brief summary. Any time there is a membrane, such as the lining of a cell, that is immersed in water and is semipermeable—some molecules can pass through it but others can't—this creates the conditions for osmosis. There is a net movement of water across the membrane to the *side* in which there is a greater concentration of solutes, or substances that can be dissolved in liquid (**see Figure 23.29**, on page 512). In plants, the solute is sucrose, which arrives at the cell walls of the sieve elements and is then *pumped* into them through special membrane channels. (The pumping is what

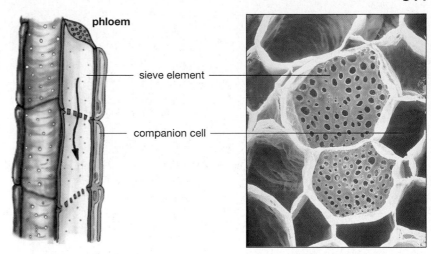

phloem

sieve element

companion cell

requires energy from the plant.) The result is a higher concentration of sucrose inside the cells than before, resulting in a net movement of *water* into the cells because of osmosis. The effect is the same as water coming into a garden hose from a spigot: The water moves through its path of least resistance—through the hose and out the nozzle. Just so is this solution of water and sucrose moved through the sieve tubes under the power of this pressure flow (**see Figure 23.30**, on page 512). The name given to this hypothesis about phloem sap movement is the **pressure-flow model**, first developed in the 1920s by German plant researcher Ernst Münch.

The loading site of this system is called a **source**, and the unloading site is called a **sink**, but these sites could also be called the "producer" and the "consumer," or perhaps "producer" and "bank." Sources tend to be sugar-producing leaves, while sink sites tend

Figure 23.27
Food Carriers
The micrograph at right shows a cross section through some sieve elements (looking at them from above), surrounded by a number of companion cells. A plant's phloem sap moves from one sieve element to another through the perforations visible in the cells.

Tutorial 23.2.2
Sugar Transport through Phloem by Pressure Flow

Figure 23.28
How Food Moves through the Plant
Sucrose is produced in the leaves (source) by photosynthesis, after which the plant expends its own energy to load the sucrose into the phloem. Osmosis is then a key factor in transporting the sucrose through the plant for use and storage (sink).

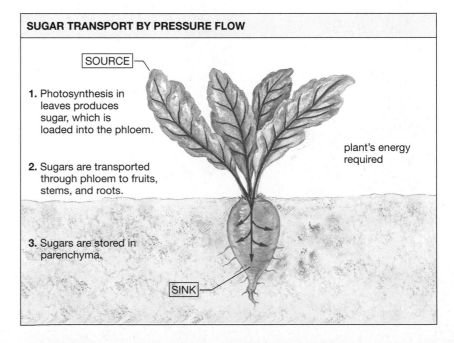

SUGAR TRANSPORT BY PRESSURE FLOW

SOURCE

1. Photosynthesis in leaves produces sugar, which is loaded into the phloem.

2. Sugars are transported through phloem to fruits, stems, and roots.

3. Sugars are stored in parenchyma.

plant's energy required

SINK

Figure 23.29
Osmosis in Action

a An aqueous solution divided by a semipermeable membrane has a solute—in this case salt—poured into its right chamber.

b As a result, though water continues to flow in both directions through the membrane, there is a net movement of water toward the side with the greater concentration of solutes in it.

Tutorial 23.2.3
Phloem-xylem Linkage and Food Transport

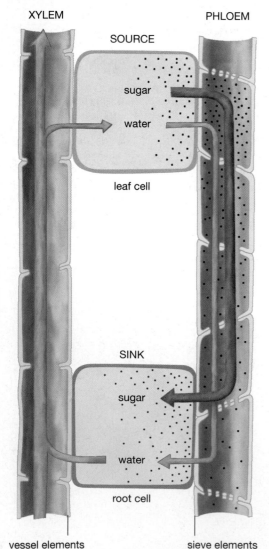

XYLEM   PHLOEM

SOURCE

sugar

water

leaf cell

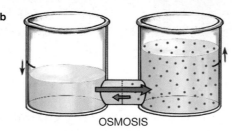

1. Sugar is actively transported into phloem (requires ATP).

2. Water follows by osmosis.

3. Pressure gradient moves fluid down phloem.

SINK

sugar

water

root cell

4. Sugar moves by active or passive transport into root cell (may require ATP).

5. Water follows by osmosis.

vessel elements   sieve elements

to be storage tissue such as roots, along with growing fruits and stems. At the sink, sucrose moves passively or is actively pumped from sieve element cells to sink cells (the roots, for example). This triggers osmosis once again; only this time the net water flow is *out* of the sieve tube, because there has been an increase in the solute concentration outside the sieve tube. This flow creates a pull that, like the push at the source, helps move the phloem sap along.

Water Flows from Xylem to Phloem and Back Again

You saw earlier that phloem and xylem tubes are stacked together in bundles. These transport systems are not sealed off from each other, however; but instead are closely interconnected. Remember the phloem sources, where water flowed into the sieve tube? Well, where did that water come from? From the adjacent xylem. And in the phloem sinks, when water flows out of the sieve tubes, most of it flows into adjacent xylem, where it joins water coming from the roots and moves on up with it (see Figure 23.30). With this, let's take our leave of transport in plants in order to cover another critical topic, reproduction and development.

23.7 Sexual Reproduction in Flowering Plants

As almost everyone knows, the human reproductive cells called eggs and sperm fuse—in a moment often referred to as conception—thereby creating a single cell that starts dividing, eventually producing a new human being. Imagine, however, that instead of eggs and sperm, human beings

Figure 23.30
Food Transport through Phloem and Phloem-Xylem Linkage

In food transport, on the right, (1) sugar produced within leaves is actively pumped by the plant into the sieve elements of the phloem. (2) Water then "follows" the sugar in, under the power of osmosis. This creates (3) a pressure gradient that moves the sugar-water compound through the phloem. At the sink, (4) sugar moves either passively or through pumping, this time out of phloem cells, with (5) water following, again through osmosis. Because water flows into the phloem at the leaves and out of the phloem at the root, leaves are regarded as the "source" and roots as the "sink" of the phloem. The phloem and xylem systems are linked. Water returns to the xylem from the phloem (bottom of the figure) and is transported back to the leaves. Meanwhile, it is water transported to the *phloem* from the *xylem* (top left of the figure) that provides the water that flows through the phloem.

produced *different* kinds of reproductive cells; ones that—on their own, without coming together—grew into a separate form of human life that looked very different from the humans we're used to. Then imagine that *this* generation produced the eggs and sperm (the gametes) that give rise to human beings. Beyond this, imagine that the human eggs and sperm were brought together not by sexual intercourse between humans themselves, but by the wind, or by other animals. In a nutshell, this describes some of the major differences between plant and animal reproduction. Because plant reproduction has an added step—an intervening generation that produces gametes—it is a bit more complicated than animal reproduction, but fascinating to look at precisely because it differs from what we're used to.

Flowers Are the Reproductive Parts of Plants

To begin, it may be helpful to look again briefly at the reproductive parts of plants, their flowers, which you can see in **Figure 23.31**. You may recall from Chapter 22 that the heart of the flowers' reproductive structures are the stamens and the carpel. A *stamen* consists of a long, slender filament that is topped by an *anther*, whose chambers contain cells that give rise to the male pollen. The carpel, meanwhile, is a composite structure, composed of three parts: the *stigma*, which is the tip end of the carpel, on which pollen grains are deposited; the *style*, a slender tube that raises the stigma to such a prominent height that it can easily catch the pollen; and the *ovary*, the area in which fertilization of the

female egg and then development of the next generation of plant will take place.

Flowering Plants Reproduce through an Alternation of Generations

Critical to an understanding of how a new generation of maple trees or roses arises from an existing generation is the alternate generation described earlier. Maple trees and rose bushes don't produce gametes (eggs and sperm) that go on to fuse and then give rise to a new generation that looks like the old one. Instead, they produce **spores**—single cells that, without fusing with anything, grow into their own generation of plant. It is *this* generation of plant that produces the gametes that will once again result in a maple tree or rose bush.

Our trouble in grasping this is that the generation growing from spores is a *tiny* generation compared to the parent generation. Further, it is a generation that spends almost its entire life in (or on) the parent generation of plant. In short, from a human perspective this separate generation looks like some small component of the original generation.

The back-and-forth between types of generations in plants gives us a term that was reviewed in Chapter 20—the **alternation of generations** (see page 425). Because the generation that is familiar to us—the maple tree or rose bush—produces spores, it is known as the sporophyte generation. These spores then give rise to the gametes (eggs and sperm) that, upon fusing, will develop into maple trees and rose bushes once again. Hence, this is the gametophyte generation. For a view of how sporophyte and gametophyte generations differ not just in angiosperms but in all

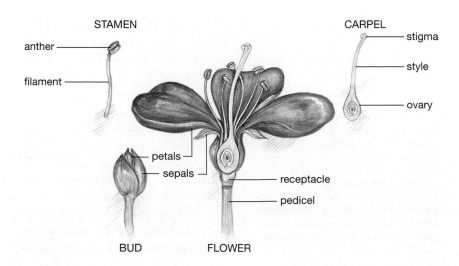

STAMEN

anther

filament

petals

sepals

BUD

FLOWER

CARPEL

stigma

style

ovary

receptacle

pedicel

**Figure 23.31
Anatomy of a Flower,
the Reproductive Structure
of an Angiosperm**

Figure 23.32
Generational Dominance Varies with the Type of Plant
In more primitive plants, the gametophyte generation is larger and more dominant. In mosses, for example, the gametophyte generation houses the sporophyte generation and supplies it with nutrition. By contrast, in the most recently evolved plants, the angiosperms, the gametophyte generation has been reduced to a microscopic entity that is almost wholly dependent on the sporophyte generation (in this example a tree). Though not plants, green algae are included for comparative purposes.

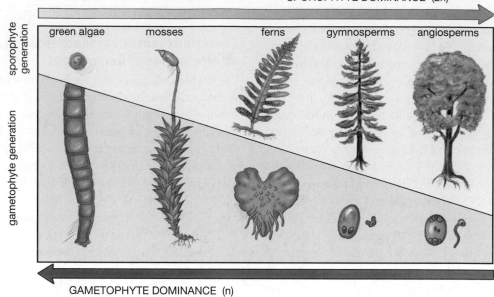

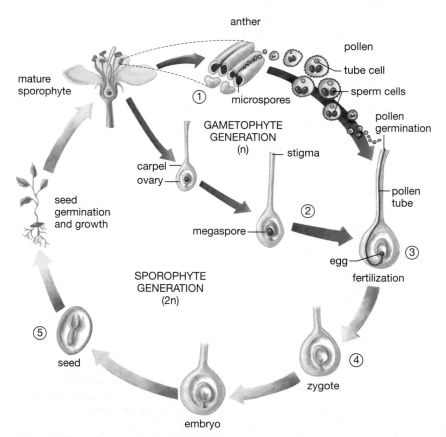

Figure 23.33
Overview of Angiosperm Reproduction
1. The mature sporophyte flower produces many male microspores within the anther and female megaspore within the ovary. The male microspores then develop into pollen grains that contain the male gamete, sperm. **2.** Meanwhile, the female megaspore produces the female gamete, the egg. **3.** Pollen grains, with their sperm inside, move to the stigma of a plant. Moving through a pollen tube, the sperm reaches the egg and fertilizes it. **4.** This results in a zygote that develops into an embryo protected inside a seed. **5.** This seed leaves the parent plant and sprouts or "germinates" in the earth, growing eventually into the type of sporophyte plant that began the cycle. In this diagram, the plant is self-fertilizing: a sperm from the plant is fertilizing an egg in this same plant. This does occur in nature, but it is the exception rather than the rule. More often, pollen from one plant will fertilize one or more eggs in a different plant.

types of plants, see **Figure 23.32**. In the more primitive plants, such as mosses, it is the gametophyte generation that is larger and more dominant, but things reverse with movement up the evolutionary line. By the time we reach angiosperms, the sporophyte generation of, for example, a tree might be 50 meters (about 165 feet) tall, while its gametophyte generation is microscopic.

Look now at **Figure 23.33** to get an overview of angiosperm reproduction. As you proceed through the story, you'll be looking at various pieces of this figure in greater detail.

The Sporophyte and Gametophyte Generations Are Genetically Different
Apart from being different in size and appearance, the sporophyte and gametophyte generations are genetically very different. Once again, there is a generational back-and-forth. Let's look at this first on the male side, starting with the sporophyte of a typical flowering plant (**see Figure 23.34**). Students who have been through the genetics section will recall that the normal state for the cells of most complex organisms is diploid, meaning that cells have chromosomes that come in matched *pairs*—a state sometimes referred to as 2n. All the cells in the sporophyte generation are diploid, or 2n. But special cells called **microspore mother cells,** which lie within the anthers of the sporophyte plant, will undergo the special kind of cell division called meiosis (reviewed in Chapter 10). Meiosis in the microspore

mother cells will result in *microspores* that have only a single set of chromosomes each (making them haploid or n). It is these microspores that will develop into pollen grains. Keeping an eye on the forest as well as the trees, it's worth noting that the line between one generation and the next is crossed after the microspore mother cell has undergone meiosis. Once the microspores have been produced, the result is the gametophyte generation.

Chromosome reduction takes place in the production of the female gametophyte as well. Here meiosis, as well as development of the female gametophyte, take place entirely inside the ovary lying at the base of the plant's carpel. In this case a single diploid cell—called a **megaspore mother cell**—undergoes meiosis, yielding female *megaspores*. (Why is the female spore "mega" compared to the male's "micro"? Because the female spore is bigger.)

As you can see in Figure 23.34, both the male and female gametophyte generations are multicelled structures (though just barely), with these cells arising by mitosis, or common cell division, from the original single haploid spores. Because mitosis is a process that makes exact copies of cells, this means that *every cell* in the gametophyte generation is haploid. In sum, the gametophyte generation is haploid (n),while the sporophyte generation is diploid (2n).

Tutorial 23.3.1
Development of Male &
Female Gametophytes

Figure 23.34
Development of the Male and Female Gametophyte Generation

Male gametophyte The male gametophyte generation takes shape inside the chambers of the pollen sacs, where the diploid microspore mother cells will undergo meiosis, each one of them thereby giving rise to four haploid microspores. It is these microspores that represent the new, haploid generation of the plant on the male side. The single cell in each microspore goes through cell division, thereby producing two cells, a tube cell and a generative cell. Before or during this time, a protective coat develops around the microspore. The combination of the cells and protective coat is the pollen grain. At some point, the generative cell in the grain divides into two sperm cells. (In the plant pictured, this happens before the pollen grain leaves the anther.) With this cell division a mature male gametophyte has developed, consisting of an outer coat plus three cells—two sperm cells and a tube cell. The pollen grain is then released from the anthers to make its way to a stigma.

Female gametophyte For the female gametophyte generation, development begins inside a structure of the parent sporophyte plant called an ovule, located deep within the carpel of the plant. From a central mass of cells in the ovule, a single diploid megaspore mother cell undergoes meiosis, producing four haploid megaspores. This marks the beginning of the female gametophyte generation. Three of the megaspores then die, leaving a single megaspore. The megaspore undergoes mitosis, eventually producing six cells with a single nucleus each and one large cell with two nuclei. These seven cells form the embryo sac, which is the mature female gametophyte. Note that one of the seven cells in the embryo sac is the egg that will undergo fertilization by one of the sperm in the pollen grain.

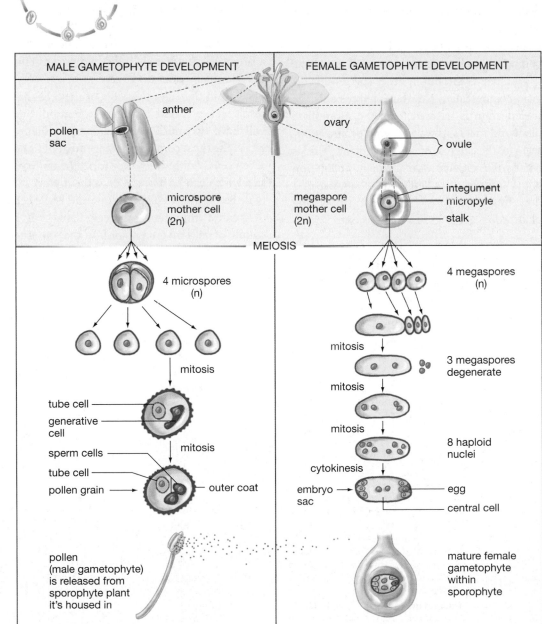

MALE GAMETOPHYTE DEVELOPMENT	FEMALE GAMETOPHYTE DEVELOPMENT

anther

pollen sac

ovary

ovule

microspore mother cell (2n)

megaspore mother cell (2n)

integument
micropyle
stalk

— MEIOSIS —

4 microspores (n)

4 megaspores (n)

mitosis

mitosis

3 megaspores degenerate

mitosis

mitosis

8 haploid nuclei

tube cell

generative cell

sperm cells

tube cell

pollen grain

outer coat

cytokinesis

embryo sac

egg

central cell

pollen (male gametophyte) is released from sporophyte plant it's housed in

mature female gametophyte within sporophyte

Development of the Male and Female Gametophyte Generation

So, for our purposes, the starting point for plant reproduction is the mature sporophyte plant that gives rise to both male and female gametophyte generations. Figure 23.34 also shows you *how* the male and female gametophyte generation develop in angiosperms.

The Male Gametophyte

On the male side, things are pretty simple. Meiosis in each microspore mother cell yields four haploid microspores (the start of the gametophyte generation). Each spore develops into a pollen grain that initially consists of three things: an outer coat, a tube cell, and a generative cell. This pollen grain can be thought of as an *immature* male gametophyte plant. At some point, the generative cell divides once, resulting in two *sperm* cells. (In the type of plant you're looking at, this division comes before the pollen grain is released from the sporophyte plant, but in other species this division comes later.) Now having three cells, the pollen grain is released from the sporophyte plant and makes its way, by wind or bird or insect, to the stigma of another plant. This is what pollination is. More formally, **pollination** is the transfer of pollen from the anther of a flowering plant to the stigma of a flowering plant. You'll get to its story in a second.

The Female Gametophyte

On the female side, note in Figure 23.34 that there develops from the ovary wall a structure called an *ovule*—a part of the sporophyte plant—that will enclose the female gametophyte generation. Note that the ovule consists of the protruding stalk, a central core of cells, and a couple layers of tissue called **integuments**. The integuments wrap around the central cells, almost coming together except for a tiny opening they leave, called a **micropyle**. From the ovule's central mass of diploid cells, a single cell—the megaspore mother cell—undergoes meiosis, thereby producing four megaspores. This action represents the crossover into the gametophyte generation. Then three of the megaspores will die, leaving the one that remains as an immature female gametophyte plant.

The remaining megaspore now undergoes three rounds of mitosis, but of a curious sort: There are several rounds of doubling of the cell's DNA (in the nucleus), but no subsequent cell division, as is usually the case. The result is a single cell that has eight haploid nuclei. This cell eventually does undergo cell division, but once again there is a twist. Six of the cells that result will each have a single nucleus, but one will have two nuclei. You can see, in Figure 23.34, the most common arrangement of these seven cells. Collectively, they form the **embryo sac**, which can be thought of as the mature female gametophyte plant. At one end is the cell of greatest importance, the egg, flanked by a couple of cells on either side. On the opposite end there are three other cells; and in the middle, a very large central cell, which is the cell that has two nuclei.

Tutorial 23.3.2
Double Fertilization in Angiosperms

Figure 23.35
Double Fertilization in Angiosperms
When the male pollen (shown greatly enlarged) lands on the stigma, the tube cell within it sprouts a pollen tube, through which the two sperm pass in moving to the ovule. One sperm cell fertilizes the egg, forming a zygote—the new sporophyte plant. The other sperm cell fuses with the two nuclei of the central cell, forming the endosperm that will later nourish the embryonic plant.

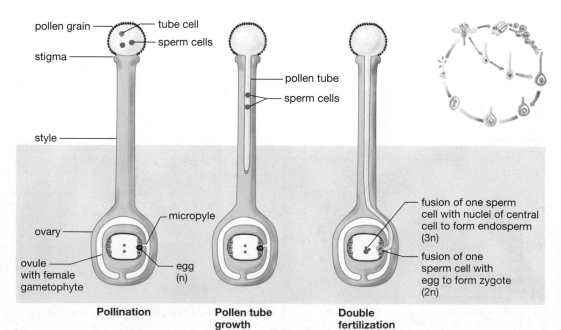

pollen grain — tube cell
stigma — sperm cells

style

ovary

ovule with female gametophyte
micropyle
egg (n)

Pollination

pollen tube
sperm cells

Pollen tube growth

fusion of one sperm cell with nuclei of central cell to form endosperm (3n)

fusion of one sperm cell with egg to form zygote (2n)

Double fertilization

Fertilization of Two Sorts: A New Zygote and Food for It

The stage is now set for male to come into contact with female. The pollen grain, with its two sperm cells and one tube cell, now lands on the stigma of a plant with an egg and embryo sac matured down in the ovule (**see Figure 23.35**). The tube cell then sprouts something: a pollen tube, which tunnels its way down through the style, eventually gaining access to the female reproductive cells within the ovule. The two sperm cells travel down this tube to achieve fertilization. (Why *two* sperm cells? Stay tuned.) In sum, in its most mature state, the male gametophyte generation consists of three cells, one of which grows a tube and two of which are sperm cells—quite a different thing from the sporophyte generation with its billions of cells.

The pollen tube grows through the ovule's micropyle and ejects its two sperm cells into the embryo sac. One of these sperm fuses with the egg, yielding the zygote that is the beginning cell in the new generation of plant—another sporophyte generation like the maple tree or rose bush we started with. This is no different in principle than what takes place with *human* life, where egg and sperm fuse, producing a single-celled zygote that develops into a whole human being. And, as with human egg and sperm, remember that the plant egg and sperm were both haploid, containing a single set of chromosomes each. Their fusion produces a diploid zygote—one with a *paired* set of chromosomes—that will divide over and over, resulting in a diploid sporophyte plant.

And what of the other sperm? Here is one of the beauties of flowering plants. This sperm enters the central cell with its two nuclei, producing a cell with *three* nuclei (making it triploid or 3n.) This too is a fertilization, only what it sets in motion is the development of *food* for the embryo—endosperm, which is tissue rich in nutrients that can be used by the embryo that will now develop from the zygote. Thus two fertilizations have taken place. This is known as **double fertilization** in angiosperms—a fusion of gametes and of nutritive cells at the same time. Double fertilization is an innovation that belongs almost solely to the flowering plants. As it happens, it has been a boon to all kinds of *animals*, including people. It is endosperm that supplies much of the food we human beings consume. The rice and wheat grains we're familiar with consist in large part of endosperm meant to sustain the plant embryo (**see Figure 23.36**).

23.8 Embryo, Seed, and Fruit: The Developing Plant

The Development of a Seed

Egg fertilization produces a single-celled zygote. The question is, how does the plant develop from this zygote into an embryo-containing *seed* that falls to the ground, eventually giving rise to a growing sporophyte plant? Many steps are involved here; what follows are only the highlights. First, recall that the female gametophyte, the embryo sac, was nearly surrounded by ovule tissue called *integuments*. These will now develop into the **seed coat** that will protect the embryonic plant until it germinates in the ground. What once was an ovule with a gametophyte inside is now a seed: a reproductive structure of a plant that includes an outer seed coat, an embryo, and nutritional tissue for the embryo (the endosperm). This is a reproductive structure that bears similarities to one employed by some animals, the egg (**see Figure 23.37**).

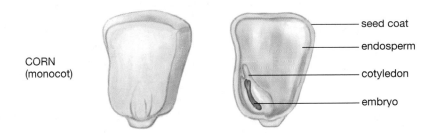

Figure 23.36
Endosperm Is Food for the Embryo
In monocots such as corn, endosperm persists in the mature seed, providing nutrition for the embryo inside. The cotyledon in the figure is the corn plant's embryonic leaf.

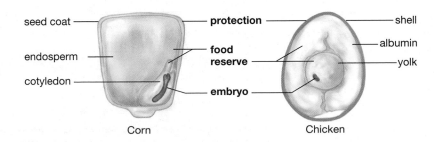

Figure 23.37
A Kernel of Corn and a Chicken Egg Are Similar in Some Ways
Both are home to an embryo, both contain plenty of food for development, and both offer protection for the embryo in dry, land environments.

Figure 23.38
Fruits Come in Many Forms
A fruit, the mature ovary of a flowering plant, surrounds a seed or seeds. Thus **(a)** the flesh of an apricot fits this definition of fruit, but so does **(b)** the pod of a pea, which has several seeds inside. The structure we commonly think of as **(c)** the strawberry fruit actually is the strawberry plant receptacle, with each receptacle having many fruits on its surface. What are commonly thought of as strawberry "seeds" are tiny strawberry plant carpels, each complete with its own fruit and seed. Fruits are important not only because they protect the seeds but also because they attract animals that disperse the seeds.

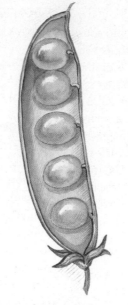
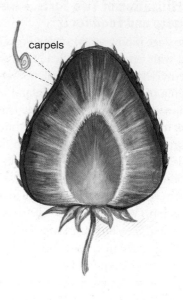

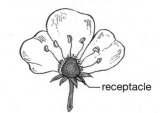

a Apricot
(one carpel, one seed)

b Pea
(one carpel, many seeds)

c Strawberry
(many carpels, many seeds, one receptacle)

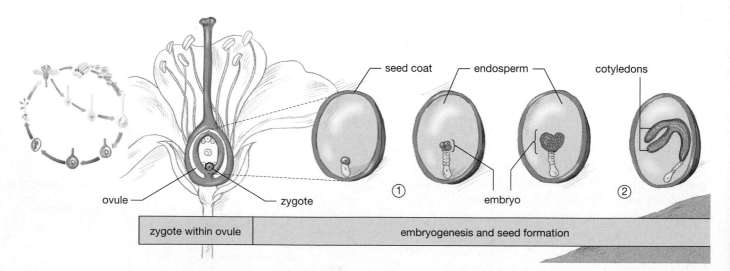

Figure 23.39
Development of a Plant Embryo, Germination of a Plant Seed
Egg and sperm have come together in this dicot to produce a single-celled zygote, which divides repeatedly in developing as the plant embryo. The zygote is surrounded by the plant ovule, which develops into a seed.

1. The zygote divides into two cells, one of which is the embryo (green in the figure). The other cell (yellow in the figure) develops into a structure that will push the embryo into the endosperm.

2. The embryo starts to develop its cotyledons or embryonic leaves. Most of the endosperm nutrients eventually are taken up by these cotyledons. (Because this plant is a dicot, it has two cotyledons.)

The Development of Fruit

Something more will now develop *outside* the seed coat. Recall that the ovule was surrounded by the ovary of the flower. The ovary wall now develops into fruit. Strictly speaking, **fruit** is the mature ovary of any flowering plant. Given this, *all* flowering plants develop fruit, but it's clear that most of the 260,000 species of flowering plants are not fruits as we commonly understand that term. It follows that the botanical definition of fruit differs from the common definition; under the botanical definition, fruits can take on lots of forms. The pod of a pea plant is a fruit, for example, as is the outer covering of a kernel of corn (though here fruit has fused with the corn seed coat, leaving scarcely anything separate that can be recognized as fruit). On the other hand, fruits can take exactly the form we're used to, as with the flesh of an apricot or cherry. To add another twist, where is the fruit in a strawberry? The answer, shown in **Figure 23.38**, may surprise you. Fruit develops in all angiosperms, and it develops *only* in angiosperms. Thus, angiosperms can be defined by this anatomical feature. An **angiosperm** is a plant whose seeds develop inside the tissue called fruit.

Fruits Serve in Protection and in Seed Dispersal

What is common to many fruits is that they serve a protective function for the underlying seed, and many act as *attractants* for seed dispersal. As noted in Chapter 20 (page 431), the wild berry that a bear eats consists not only of the fruit flesh, but of the seeds it surrounds. The fruit is digested by the bear as food, but the seeds pass through the bear's digestive tract. Thus has this plant engineered seed dispersal at what may be a promising new location for a berry plant. Plants with such attractants *time* their fruit ripening, so that the fruit is not attractive before the seed is ready for dispersal.

Development of the Embryo and Germination of the Seed

While fruit and seed coat are maturing, the embryo inside them is also developing. **Figure 23.39** shows you the outline of the development of a plant embryo, this one a dicot plant. Recall that the process starts with a single cell, the zygote, encased within the structure that will become the seed, the ovule. The zygote divides, yielding two cells, one of which becomes the embryo. The

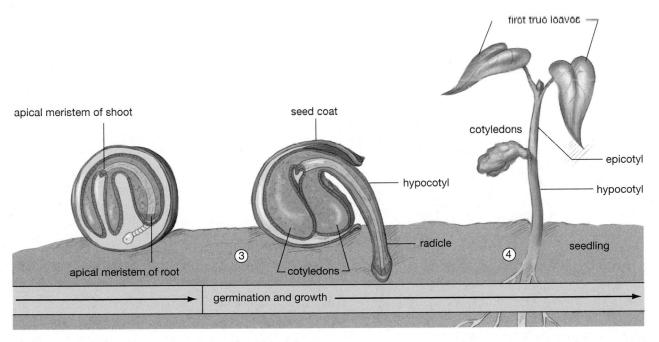

3. Having made its way to the ground, the seed is ready to "germinate" or start sprouting. The seed coat splits, allowing the emergence of the hypocotyl (tissue below the cotyledons), which includes the plant's radicle, or early root structure.

4. Root and shoot have both sprouted. The plant's cotyledons, having completed their function, will wither away. The plant sprouts its first true leaves and begins the process of photosynthesis. The reproductive cycle is ready to begin again.

Tutorial 23.3.3
Plant Embryo and Seed Germination

other cell gives rise to a paddle-shaped structure that pushes the growing embryo into the endosperm.

Eventually, the embryonic tissue develops two **cotyledons**, or embryonic leaves. When the embryo is still tightly encased in the seed coat, the cotyledons can take on an important role you've seen before, which is to provide nutrients to the growing embryo. You may say: But isn't this the function of the endosperm? In most monocots (for example, the grains) the endosperm retains this function, but in most dicots the nutrients stored in endosperm tissue are *taken up* by the cotyledons as they develop. The foods we know as beans are comprised of embryos with large cotyledons (**see Figure 23.40**).

The seed that encases the embryo eventually separates from the sporophyte parent plant and makes its way into a suitable patch of earth, there to germinate, or to sprout both roots and shoots. First to emerge from the seed is the root structure or radicle, which is attached to **hypocotyl**, which can be thought of as all the plant tissue below the cotyledons. Then there emerges the **epicotyl**, meaning the tissue above the cotyledons. This epicotyl gives rise to the first true leaves of the plant. The seed has sprouted into a new sporophyte plant.

Figure 23.40
Cotyledons Serve a Nutritive Function
In a dicot such as this bean plant, the endosperm food reserves are taken up by the embryonic leaves or cotyledons of the embryo, which then serve as the mature seed's food reserves.

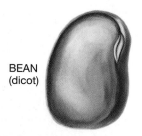
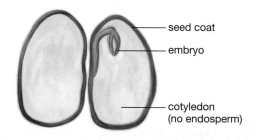

BEAN
(dicot)

seed coat

embryo

cotyledon
(no endosperm)

Seed Dormancy Can Be Used to a Plant's Advantage

One of the survival strategies of plants is the use of dormancy, or a prolonged low metabolic level in seeds. Seeds can *postpone* germinating until they have favorable conditions around them, such as the proper temperature or amount of light. What generally sets seed germination in motion is an uptake of water from the surrounding environment. It is, however, the triggering mechanisms for this water uptake that tell the tale in dormancy. Some seeds require a scraping or abrasion before they will take up water; others simply need to dry out, once they've shed the fruit from around them; still others require the action of enzymes from an animal's gut, or even charring by fire.

Once the plant has germinated and matured a little, you could take a cross section of its stem and find concentric circles of the tissue types mentioned near the start of this chapter: dermal, vascular, and ground. Meanwhile, meristematic tissue could be found at the tips of the root and shoot, allowing the plant to grow further. With this, we have come full cycle: Another generation of plants is maturing.

On to Animals

In Chapters 22 and 23, you have looked at a group of living things, the plants, whose members are silent, fixed in one spot, and have almost no moving parts. Now it's time to shift gears. There are animals that are silent and animals that are fixed in place, but you'll also find with animals nature's only sound-makers and its supreme travelers. The next five chapters are concerned with the functioning of animals in general, but they focus on one animal that in some ways is very familiar to us—the human animal.

Chapter Review

Summary

23.1 Two Ways of Categorizing Flowering Plants

- Plants can be categorized by how long it takes them to go through a cycle from germination to death. Those that go through this cycle in a year or less are annuals; those that go through in about two years are biennials; those that live for many years are perennials.

- A cotyledon is an embryonic leaf, present in the seed. Angiosperms are classified according to how many cotyledons

they have—one in the case of monocotyledons, two in the case of dicotyledons. This distinction in embryos results in major differences in the anatomy and functioning of monocot and dicot plants.

23.2 There are Three Fundamental Types of Plant Cells

- There are three fundamental types of cells in plants that, alone or in combination, make up most of the plant's tissues, meaning groups of cells that carry out a common function. These three cell types are parenchyma, sclerenchyma, and collenchyma.

23.3 The Plant Body and Its Tissue Types

- Some plants are capable only of primary growth, meaning growth at the tips of their roots and shoots that primarily increases their length. Plants that exhibit only primary growth are herbaceous plants, composed solely of primary tissue. Other plants exhibit both vertical growth and lateral, or secondary growth. These are the woody plants, composed of primary and secondary tissue.

- There are four tissue types in the primary plant body: dermal, vascular, meristematic, and ground. Dermal tissue can be thought of as the plant's outer covering, vascular tissue as its "plumbing," meristematic tissue as its growth tissue, and ground tissue as almost everything else in the plant.
  TUTORIAL 23.1.1: Types of Plant Tissue

23.4 How a Plant Grows: Apical Meristems Give Rise to the Entire Plant

- The entire plant develops from meristematic cells in regions called apical meristems. Meristematic cells remain perpetually embryonic, able to give rise to cells that differentiate into all the plant's tissue types.

- Shoot apical meristems give rise to the entire shoot of the plant. In addition to providing for vertical growth, shoot apical meristems produce meristematic tissue called lateral buds at the base of leaves that can give rise to a branch or flower. Root apical meristems are located just behind a collection of cells at the very tip of the root, called the root cap.

- The plant's tissue types develop in stages from meristematic cells. This development takes place in a series of regions adjacent to the apical meristem. In a gradual transition, the apical meristem gives way to a zone of cell division, followed by a zone of elongation (in which developing cells lengthen) followed by a zone of differentiation (in which cells fully differentiate into different tissue types).
  TUTORIAL 23.1.7: How Plants Grow

23.5 Secondary Growth Comes from a Thickening of Two Types of Tissues

- Secondary growth in plants takes place through the division of cells in two varieties of meristematic tissue: the vascular cambium, which gives rise to secondary phloem and xylem; and cork cambium, which gives rise to the outer tissues of woody plants. Secondary xylem, also known as wood, is responsible for most of a tree's widening.

- Looking at a tree from the secondary phloem outward to the tree's periphery, four tissues constitute the tree's bark: secondary phloem, phelloderm, cork cambium, and cork. The cork cells are dead in their mature state and provide layers of protection for the tree.

23.6 How the Plant's Vascular System Functions

- Two types of cells make up the water-conducting portions of xylem tissue. These are tracheids and vessel elements, both of which are dead in their mature, working state. Vessel elements, which exist almost solely in angiosperms, conduct more water than tracheids. Their existence in angiosperms is one of the reasons for the angiosperms' dominance in the plant world.

- Water movement through xylem is driven by transpiration, meaning the loss of water from a plant, mostly through the leaves. As water evaporates into the air, it pulls a continuous column of water upward through the plant.
  TUTORIAL 23.2.1: Water Transport through Xylem by Transpiration

- Sucrose is the main product that flows through phloem. The fluid-conducting cells in phloem are called sieve elements, which lack cell nuclei in maturity. Each sieve element has associated with it one or more companion cells, which retain their nuclei and seem to take care of the housekeeping needs of their related sieve elements.
  TUTORIAL 23.2.2: Sugar Transport through Phloem by Pressure Flow

- Plants must expend their own energy to load the sucrose they produce into the phloem. Dissolved sucrose then moves through the plant through the power of osmosis and the fluid pressure that results from it.
  TUTORIAL 23.2.3: Phloem-xylem Linkage and Food Transport

23.7 Sexual Reproduction in Flowering Plants

- All plants, including angiosperms, reproduce through an alternation of generations. A sporophyte generation (the familiar tree or flower) produces haploid spores that develop into their own generation of plant, the gametophyte generation. The male gametophyte is the pollen grain, consisting in maturity of an outer coat, two sperm cells, and one tube cell. The female gametophyte consists at maturity of an embryo sac composed of seven cells, one of which is the egg. The female gametophyte is housed inside a structure of the parent sporophyte plant called an ovule.
  TUTORIAL 23.3.1: Development of Male & Female Gametophytes

- Fertilization of the egg by sperm requires that a pollen grain land on the stigma of a plant. The tube cell of the pollen grain then germinates, sprouting a pollen tube that grows down through the sporophyte plant's stigma and style, eventually reaching the female reproductive cells inside the ovule. The sperm cells inside the pollen grain travel down through the pollen tube, and one of the sperm cells fertilizes the egg in the ovule, producing a zygote. With this, the new sporophyte generation of plant has come into being.

- The second sperm cell in the pollen grain enters the central cell in the embryo sac, setting in motion the development of food for the embryo, endosperm tissue. This second fertilization completes the process of double fertilization—a fusion of gametes and of cells producing nutritive tissue.
  TUTORIAL 23.3.2: Double Fertilization in Angiosperms

23.8 Embryo, Seed, and Fruit: The Developing Plant

- With fertilization, the ovule integuments that surrounded the embryo sac begin to develop into the seed coat that will surround the growing sporophyte embryo. The ovary that surrounded the ovule then starts to develop into a layer of tissue that will surround the seed—fruit, which is defined as the mature ovary of a flowering plant. Under this definition the pod of a pea plant is fruit, as is the flesh of an apricot.

- The seed with its fruit covering eventually separates from the sporophyte parent plant and then germinates in the earth, sprouting both roots and shoots.
  TUTORIAL 23.3.3: Plant Embryo and Seed Germination

Key Terms

Understanding the Basics

Multiple-Choice Questions

1. Dermal tissue _____, while vascular tissue _____.
   a. protects the plant, conducts fluids
   b. provides strength, facilitates growth
   c. facilitates reproduction, protects the plant
   d. facilitates growth, provides strength
   e. protects the plant, facilitates reproduction

2. Biennials, like cauliflower, radishes, and beets,
   a. go through their entire life cycle in one season
   b. go through their entire life cycle in about two years
   c. grow and bloom each season, for many years
   d. have the shortest life span, among annuals, biennials, and perennials
   e. go through their entire life cycle twice each year

3. Sclerenchyma cells
   a. lack cellulose
   b. function in metabolic processes such as photosynthesis
   c. are generally dead at maturity
   d. are commonly found in rapidly growing parts of the plant
   e. are found exclusively in roots

4. Secondary growth
   a. leads to increased height of the plant
   b. is more widespread in herbaceous plants than in woody shrubs and trees
   c. leads to an increase in the girth of the plant
   d. is brought about through cell divisions in the cuticle
   e. is growth that occurs in a plant's second year

5. Which of the following statements about the ascent of water in tall trees is true?
   a. Plants must expend metabolic energy to power the ascent of water into treetops.
   b. The evaporation of water from the leaf surface is required to lift water from the roots to the canopy of a tall tree.
   c. The ascent of water in the xylem is fastest when the stomata are closed.
   d. The formation of embolisms, or air bubbles, in xylem elements increases the speed of ascent of water.
   e. All of the above are true.

6. Phloem sieve tubes develop a positive pressure because
   a. Osmotic uptake of water occurs when sugars are loaded into the sieve elements.
   b. Sieve tubes lack a nucleus.
   c. Sieve tubes are dead at maturity.
   d. Sucrose is a bulky molecule that exerts a pressure against the sieve tube walls.
   e. All of the above are true.

7. Endosperm is formed when
   a. A sperm nucleus from the pollen fuses with one haploid nucleus from the embryo sac.
   b. Two nuclei from the embryo sac fuse with each other.
   c. A sperm nucleus from the pollen fuses with the egg nucleus in the embryo sac.
   d. An egg nucleus from the pollen fuses with a diploid nucleus in the embryo sac.
   e. A sperm nucleus from the pollen fuses with two haploid nuclei from the embryo sac.

8. Which of the following statements about spores is true?
   a. Spores are diploid structures that give rise to gametes.
   b. Spores are multicellular, haploid structures that give rise to gametes.
   c. Spores are part of the sporophyte generation.
   d. Spores give rise to multicellular haploid gametes.
   e. Spores give rise to single-celled diploid gametes.

9. Which of these is a gametophyte plant? (Select all that apply.)
   a. petal
   b. stigma
   c. embryo sac
   d. pollen grain
   e. sepal

10. The difference between a seed and a fruit is that
    a. Fruit is composed of triploid tissue; seeds are diploid.
    b. Seeds are always hard; fruits are always soft.
    c. Seeds develop from ovules while fruit tissue is derived from ovary walls.
    d. Annuals produce seeds while perennials produce fruits.
    e. Fruits are annuals while seeds are perennial.

Brief Review

1. Compare monocots and dicots with respect to their anatomy, morphology, and seed structure.

2. What are the three main cell types in a plant, and what distinct functions are performed by each?

3. What is a meristem? Describe the organization of the meristem, and the zonation in cell activity, in a root tip.

4. Compare the structure and function of xylem- and phloem-conducting elements.

5. How are growth rings in a tree trunk created? Why is it that you can tell the age of a tree by counting the number of growth rings?

6. Describe the process of double fertilization in plants.

Applying Your Knowledge

1. If you were to regularly shear a lilac bush down to a stub, you would probably kill it. Yet, mowing after mowing, the grass in your lawn continues to grow. Is there a fundamental difference in the location of growing points in the lilac and the grass? Explain.

2. Many grocery stores try to improve on Mother Nature by dipping fruits and vegetables in natural and synthetic waxes to prolong their shelf life. What is the role of the waxy cuticle in plants? Would you expect desert plants to have a thicker or thinner cuticle than plants that occur in moist areas?

3. If you were to pull out the "strings" on the ridges of a celery stick, you'd be peeling off strands of almost pure collenchyma cells. Why is this cell type especially prevalent in large but tender (non-woody) parts of the plant, such as the leaf stalks of rhubarb and celery?

4. Beavers often kill young trees by girdling them, that is, removing a strip of bark from the circumference of a tree trunk. Why does this bark removal kill a tree? Which part of the tree would die first, the root or the shoot?

5. Vessels are wider than tracheids, and therefore more efficient as water conduits. Even though flowering plants have evolved vessels, they also produce a good number of tracheids, producing even greater numbers of them late in the season in temperate areas. Can you think of a reason why this might happen?

MediaLab

Why Do We Need Plants Anyway? The Importance of Plant Diversity

Evidence indicates that a meteorite struck the Earth near the Gulf of Mexico about 65 million years ago. After that event, there is no fossil record of any animal species over 65 pounds surviving. Most animals were probably not killed by rocks or tidal waves, but died when Earth's plant life could no longer support their needs. A significant meteor impact would send millions of tons of dust and debris into the atmosphere, blocking the Sun's rays from reaching Earth's surface, significantly reducing photosynthesis, and ultimately leaving little food to support animal life. This event illustrates the essential role of plants in creating an environment that supports animal life. We simply couldn't survive without them. To learn more, the *CD-ROM Tutorial* will introduce you to the basics of plant structure and growth. In the *Web Investigation*, you will see how plants have adapted these basic structures to specific habitats and conditions. Finally, in the *Communicate Your Results* section, you will synthesize and apply your knowledge to the plant life around you.

This *MediaLab* can be found in Chapter 23 on your CD-ROM (Tutorial 23.4.1) and Companion Website (http://www.prenhall.com/krogh3).

CD-ROM TUTORIAL

Plants, because of their ability to trap solar energy in a chemical form, are a food source for all other living organisms. They are pivotal members of every ecosystem and have adapted their basic body parts to suit these various environmental conditions.

Activity

1. First, review the four principal anatomical structures of flowering plants—roots, stems, leaves, and flowers.

2. Then, cover the process of plant growth and reproduction.

3. Finally, follow the development of the next generation of plants from seed through germination.

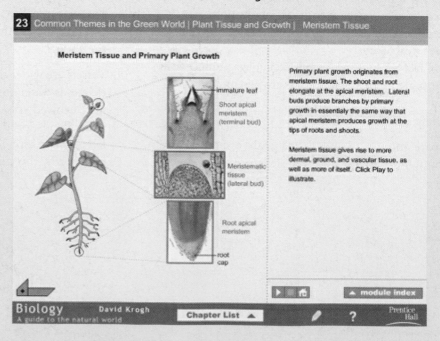

Investigation 1

Estimated time for completion = 15 minutes

There are two major groups of flowering plants, the monocots and the dicots, whose lineages diverged early in their evolutionary history. To contrast how these two groups of plants are constructed, select the Keywords **DICOT** and **MONOCOT** on your CD or Website for this *MediaLab*. Then compare the architecture and anatomy of the tomato plant, a dicot, and the rice plant, a monocot. How many differences can you observe in these two plants? Examine the roots, stems, leaves, and flowers.

Investigation 2

Estimated time for completion = 5 minutes

Flowering plants appeared on Earth about 150 million years ago. Since then, they have moved into an extraordinary number of habitats. Over generations, plant species have altered their form and function to become better adapted to living in varying habitats. Select the Keyword **SILVERSWORD** on your CD or Website for this *MediaLab* to view a striking example of adaptation by a plant species found in Hawaii. How do the different forms of these plants suit their environment?

Investigation 3

Estimated time for completion = 5 minutes

The structures of flowering plants are the culmination of millions of years of plant evolution. Features such as vascular tissue and enclosed seeds have appeared in various plant groups at various times in the evolutionary history of plants. Select the Keyword **FOSSIL PLANTS** on your CD or Website for this *MediaLab* to determine when these plant features first appeared in the fossil record.

Now that you've gained some understanding of plant structures and functions by doing the preceding activities, consider these questions.

Exercise 1

Estimated time for completion = 15 minutes

Use your observations from *Web Investigation 1* to construct a chart illustrating the differences between the architecture and anatomy of dicot and monocot plants. You could start with something like Figure 23.5, but you should be able to greatly expand on the differences. To do this exercise most efficiently, you may want to systematically work your way from flowers to roots, or any other order you would prefer.

Exercise 2

Estimated time for completion = 20 minutes

Using the adaptations you observed of the silversword species to environments in the Hawaiian Islands, answer this question: How would you construct a plant to be well adapted to the following environment?

- *Poor-quality soil, frequently disturbed by rock slides*
- *Growing season about 8 months long*
- *Annual rainfall of about 18 inches, most of it during the first third of the growing season*
- *About 8 hours of sunlight at the start of the growing season; about 16 hours by the last half*

After deciding how to construct this plant so that it will be adapted to its environment, draw your hypothetical plant. Then, list how its adaptations help the plant cope with its environmental hardships.

Exercise 3

Estimated time for completion = 10 minutes

Using the information you obtained from *Web Investigation 3*, construct a graph illustrating when the various plant features first appeared in the fossil record. Your chart should extend from the time when the first land plants appeared to the present day.

24

Introduction to Animal Anatomy and Physiology

The Integumentary, Skeletal, and Muscular Systems

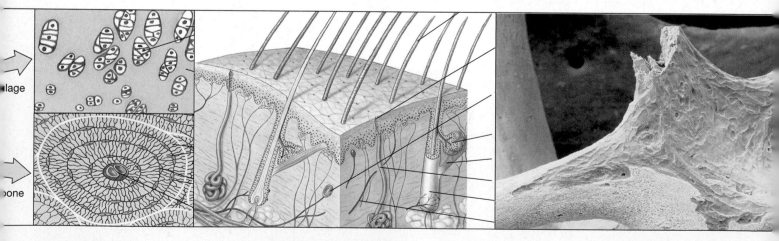

lage

one

This helps to support us.
(Section 24.4, page 531)

Our outer wrap.
(Section 24.6, page 537)

A common bone disease.
(Essay, page 540)

Just as several individuals may make up an office, and several offices a department, so our bodies are organized into a number of small working units that go on to make up larger ones. Three of the larger units involve skin, bone, and muscle.

Almost all human beings are constantly in contact with other human beings, of course. As a result, most of us take for granted such basic human capabilities as speaking, walking, reasoning, or listening. Anyone who has ever tried to devise a *machine* that can carry out these human activities would tell you, however, that human capabilities should be looked at with more wonder than indifference. At tremendous cost of time and money, IBM devised a computer capable of being a world-class chess player, but how about devising a computer that can play chess *and* recognize faces *and* speak in fluid sentences? How about devising a computer, in other words, that had a few of the capabilities of an average 10-year-old? There are differing schools of thought about what limits there might be to machine or "artificial" intelligence. One view, however, is that computers will never be able to think as humans do, in significant part because of what computers lack: a body. Think of what the body provides in the way of learning. First, wonderful sensory inputs—such as seeing and hearing—that *allow* us to learn in diverse ways. Second, a set of needs—for food, warmth, and so forth—that *drive* us to learn from the moment of our birth.

As remarkable as any one human capability may be, the truly wondrous thing is the body as a whole. Our bodies have a system for acquiring energy (a mouth, a digestive tract) a system for eliminating waste (a bladder, etc.), a system for movement (muscles), a system for moving materials around internally (blood vessels), and lots more. It is these systems that you will be reading about in the next five chapters. The idea is to give you some sense of the workings of the human machine.

24.1 The Sciences of Anatomy and Physiology

Zoology is the study of animals, and one way of thinking about animals is to consider the makeup and functioning of their bodies. Anatomy (literally "to cut up") is the study of the body's internal and external structure and the physical relationships between the body's constituent parts. Physiology is the study of how these various structures work. Anatomy tells us, for example, where the heart is, how it is structured, and how it is situated in relation to other organs and tissues. Physiology describes to us the *mechanisms* of the heart's functioning.

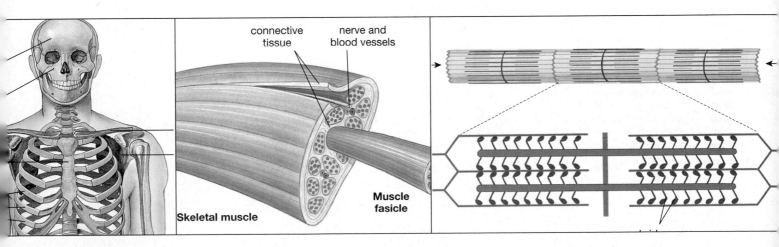

Some of our 206 bones.
(Section 24.7, page 541)

connective tissue

nerve and blood vessels

Skeletal muscle

Muscle fasicle

Muscles help us move.
(Section 24.8, page 542)

How does a muscle contract?
(Section 24.8, page 543)

Animals vary tremendously in size and complexity. There are sponges, corals, various types of worms, and then vertebrates such as sharks, birds, and humans. Because human beings are the most-studied and best-understood animal in the animal kingdom, you'll primarily be looking at humans in this discussion of animal anatomy and physiology.

24.2 What Are the General Characteristics of Humans?

There are obvious differences between a human being and, say, a bird; but there are also some similarities. Indeed, human beings have a good number of characteristics in common with many other animals. Three of these shared characteristics are an internal body cavity, an internal skeleton, and internal temperature regulation.

Internal Body Cavity

Vertebrates and most invertebrates have an internal body cavity, or **coelom**, which is not open to the external environment. Humans and other mammals have two primary body cavities. First, there is a dorsal (back) cavity that is further divided into cranial and spinal cavities, into which our brain and spinal cord fit (**see Figure 24.1**). Second,

there is a ventral (front) cavity that includes a thoracic or "chest" portion that contains the lungs and heart, and a lower abdominopelvic cavity that contains such organs as the stomach, liver, pancreas, and intestines. Between the thoracic and abdominopelvic cavities lies a muscular sheet called the *diaphragm*. Body cavities allow for flexibility; with them, internal organs can expand and contract without disrupting the activities of neighboring organs.

Internal Skeleton

Humans have an internal skeleton that supports the body and allows for continuous body growth up to maturity. We humans are described as **vertebrates**, because our internal skeleton contains a backbone, or vertebral column, made up of individual bones called *vertebrae*. The vertebrae protect the spinal cord by forming the spinal cavity, while the skull protects the brain by forming the cranial cavity.

Internal Temperature Regulation

Most fish, amphibians, and reptiles are **ectothermic**, meaning that their internal body temperature is controlled largely by the temperature of their environment (*ecto* means "outside," and *therm* means "heat"). These animals are unable to regulate their

**Figure 24.1
Body Cavities**

a A side or "lateral" view of the two main cavities in the human body—the dorsal (back) and ventral (front) cavities—and the smaller cavities within them. The dorsal body cavity is bounded by the bones of the skull and vertebral column and includes both the cranial and spinal cavities. The ventral cavity includes a thoracic (chest) cavity and an abdominopelvic cavity, which is separated from the thoracic cavity by a muscular diaphragm.

b A front or "ventral" view of the body's cavities.

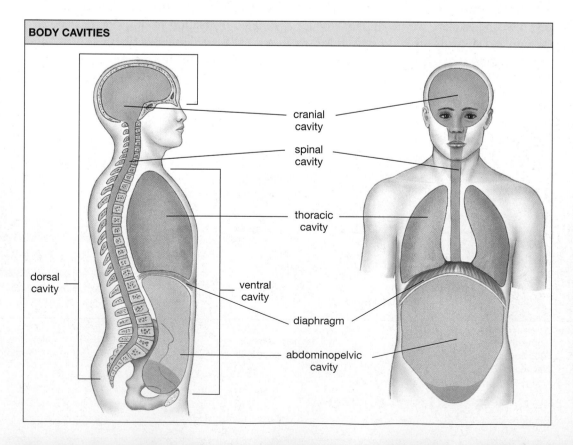

BODY CAVITIES

cranial cavity

spinal cavity

thoracic cavity

dorsal cavity

ventral cavity

diaphragm

abdominopelvic cavity

body temperature and so exist at the same temperature as the air or water surrounding them. Thus a lizard may warm itself for the day's activities by lying in the sun. The term *cold-blooded*, which is often applied to ectotherms, is misleading because ectotherms may have warm body temperatures—if they are in warm environments. Likewise, they can maintain a *stable* internal temperature, but only if their environment does.

In contrast, humans, other mammals, birds, and some fish and insects warm their bodies primarily through the activities of their own metabolism. These animals are referred to as **endothermic**, meaning an animal whose internal body temperature is relatively stable and whose body heat is generated internally, by its own metabolism. There are both costs and benefits to endothermy. One benefit is that endotherms can be fully functional in lots of environments. Imagine that, like lizards, we had to warm ourselves in the sun before we could be fully active. The price we pay for this functionality, however, is energy usage. Pound per pound, in a resting state a human being burns up many more calories than would, say, a crocodile. And because endotherms burn up proportionately more calories, they must obtain proportionately more food.

24.3 Animal Architecture and Organization

One distinguishing characteristic of animals is that they are multicellular. Put another way, there is no such thing as a single-celled animal. With multicellularity comes size, and with size comes complexity. As animals became larger through evolutionary time, they also became more structurally complex. In a sense they had to because, as you learned in Chapter 4 ("The Size of Cells," page 72), as anything grows in size, its volume increases more than its surface area. With large size, then, not all the cells of an organism can have access to the outside world. Thus, new *systems* had to evolve that would allow cells far from the surface to get needed nutrients and gases and to expel their own wastes safely. In addition, most large organisms need muscles in order to move and some type of skeleton for general support and for muscle attachment.

This division of labor resulted in the formation of several types of specialized **tissues**, meaning groups of cells that have a common function. In more complex animals, tissues are in turn arranged into various types of **organs**, meaning complexes of several kinds of tissues that perform a special bodily function. (The heart is an organ that has muscle tissue within it, but it also has another type of tissue called epithelial tissue.) Organs then go on to form **organ systems**, which are groups of interrelated organs and tissues that serve a particular set of functions in the body. Contractions of the heart push blood into a network of blood vessels. The heart, blood, and blood vessels form the cardiovascular *system*, which is one of 11 systems the human body has.

Tissues, organs, and organ systems are three levels of organization within animals. You may remember from Chapter 1 that the levels of organization in living things do not *start* with tissues. Tissues are made of cells, cells have within them tiny working structures called *organelles*, and organelles are in turn made of the building blocks called *molecules*. At the bottom of this chain is the fundamental unit of matter, the atom. In this chapter, the focus is on the levels of organization that run from tissues through organ systems. **Figure 24.2** shows the various levels of structural organization in the human body, using the cardiovascular system as an example.

24.4 The Animal Body Has Four Basic Tissue Types

So, what kinds of tissues are there in animals? You'll be reading about a lot of different tissue types in this chapter, but it turns out there are only four *fundamental* tissue types: epithelial, connective, muscle, and nervous. All the other tissue types are subsets of these fundamental types.

Figure 24.2
Levels of Organization in the Human Body
Atoms form molecules, which in turn combine to form organelles, which combine to form cells, such as heart muscle cells. A group of cells that performs a common function is a tissue; here cells have formed heart muscle tissue (whose common function is heart muscle contraction). Two or more tissues combine to form an organ such as the heart. The heart is one component of the cardiovascular system; other parts of the system are blood and blood vessels. All the organ systems combine to create an organism, in this case a human being.

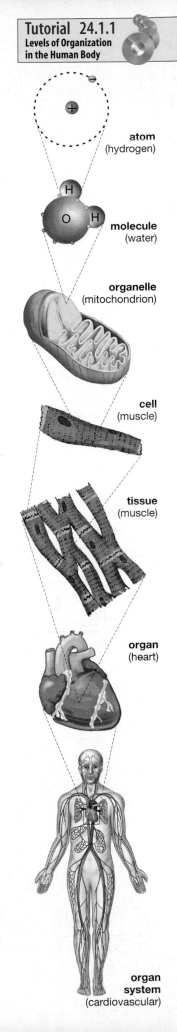

atom
(hydrogen)

molecule
(water)

organelle
(mitochondrion)

cell
(muscle)

tissue
(muscle)

organ
(heart)

organ
system
(cardiovascular)

Epithelial Tissue

Epithelial tissue covers all the exposed body surfaces of animals. The surface of your skin is epithelia, for example. But epithelial tissue also lines your internal passageways—such as your digestive tract—and forms **glands**, which are organs or groups of cells that secrete one or more substances.

Epithelial tissue may consist of a single layer (a simple epithelium) or be made up of two or more layers (a stratified epithelium). In their shape, epithelial cells may be squamous, meaning flat; cuboidal, meaning square-shaped; or columnar, meaning rectangular. **Figure 24.3** shows some of the epithelial tissues in the human body. Because the exterior surface of an epithelial cell may be exposed, its *interior* surface needs to be firmly anchored to deeper body tissues. Epithelial cells are attached to what is called a *basement membrane*, which is made up of a dense network of protein fibers and filaments. This network also keeps large molecules in the underlying connective tissue from reaching the epithelium.

Epithelial tissues have several functions in humans and other animals. All materials that enter or leave the body must contact epithelial cells. Why? Because these cells form the internal and external *surfaces* of animal organs. In skin, epithelial cells can serve as an effective barrier to the outside world, in large part because they fit together tightly, with portions of their cell membranes interlocking. This is one of the body's key means of keeping invaders and harmful substances out. In addition, in land vertebrates such as ourselves, epithelial cells that lie on our body's surface produce the waterproofing protein keratin.

Epithelial glands in animals as diverse as flatworms, earthworms, and snails produce mucus secretions that aid body movement and, in humans, ease the movement of foods through the digestive tract.

Glands made up of epithelial tissue come in two main types: **Exocrine glands**, which secrete their materials through tubes or "ducts" onto an epithelial surface; and ductless **endocrine glands**, which release their materials directly from cells into surrounding tissues or into the bloodstream. The perspiration on your skin comes through a duct; thus your sweat glands are exocrine glands. Meanwhile endocrine glands secrete many **hormones**—substances that, when released in one part of an organism, go on to prompt physiological activity in another part of the organism.

Connective Tissue

The second type of animal tissue, connective tissue, is unlike epithelial tissue in that it is never exposed to the external environment. **Connective tissue** has two key properties: It supports and protects other tissue and its cells secrete an extracellular material that they are then embedded in. Unlike epithelial tissues, which consist almost entirely of cells, connective tissue cells are separated from each other by the extracellular ("outside the cell") material they produce. Indeed, a prime function of most connective tissue cells is to produce such materials, which are composed of protein fibers and a "ground substance" that varies in consistency from a fluid to a solid. See **Figure 24.4** for the different types of connective tissues in the human body.

Loose Connective and Adipose Tissue

Not surprisingly, connective tissues are classified according to their surrounding extracellular material. Loose connective tissue is composed of a fluid-like ground substance and fibers of a protein called collagen, which

Figure 24.3
Some Epithelial Tissues in the Human Body

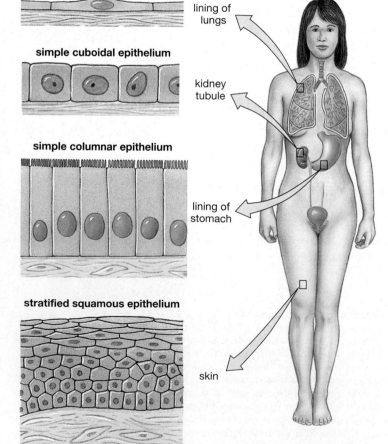

simple squamous epithelium

simple cuboidal epithelium

simple columnar epithelium

stratified squamous epithelium

lining of lungs

kidney tubule

lining of stomach

skin

together provide strength and flexibility. This tissue is the packing material of the body. It fills spaces between organs, provides cushioning, and supports the epithelia. **Adipose tissue**, or fat, is a type of loose connective tissue that contains large numbers of fat cells. It too provides cushioning, but it also insulates the body and stores energy.

Fibrous and Supporting Connective Tissue

Containing less ground substance and packed with collagen fibers, fibrous connective tissue is tough and strong. Such tissues form tendons that attach skeletal muscle to bones, ligaments that connect one bone to another, and capsules that surround organs and enclose joint cavities.

Supporting connective tissues, the most rigid connective tissues, make up bone and cartilage. The extracellular material of these tissues contains a dense ground substance and closely packed fibers. In humans, **cartilage** makes up the outer portions of the nose and ear, where it provides support and flexibility. It is also found between the bones of the spine and at the ends of limb bones,

where it protects the bones by reducing friction and acting as a shock absorber. Then there is **bone**, which is a supporting connective tissue whose ground substance contains mineral deposits primarily made of calcium (thus making it calcified). These minerals, along with flexible collagen fibers, make bone both strong and resistant to shattering.

Fluid Connective Tissue

It may be difficult to think of blood as a tissue at all, but it is one more type of connective tissue—a fluid connective tissue. Blood is composed of a distinctive population of cells (blood cells) suspended in a watery ground substance (called *plasma*) in which tissue proteins and other materials are dissolved. In humans, blood contains red blood cells, which carry oxygen; white blood cells, which are important to the immune system; and platelets, which are cell fragments that aid blood clotting.

Some fluid is constantly pushed out of the smallest blood vessels and into the surrounding body tissues, where it is known as **interstitial fluid**. At a certain point this fluid enters special

**Figure 24.4
Connective Tissues in the
Human Body**

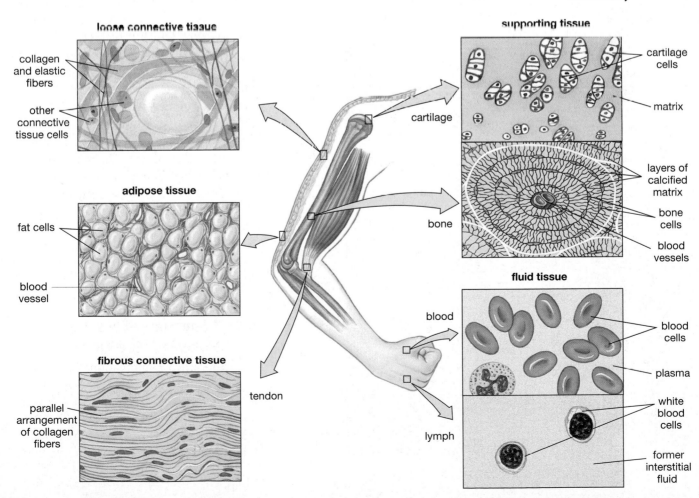

vessels, called **lymphatics**, that return it to the blood circulation. While in the lymphatics, the fluid constitutes a second kind of fluid connective tissue, **lymph**. Before being returned to the blood, lymph is checked for signs of injury or infection by cells of the immune system.

Muscle Tissue

Muscle tissue is tissue that is specialized in its ability to shorten, or contract. It does so through the interaction between two contractile protein elements or "fibrils," actin and myosin. Muscle in which the fibrils are regularly arranged in stripe-like patterns is called striated muscle, while muscle lacking such an arrangement is called nonstriated muscle. Striated muscle cells form **skeletal** and **cardiac muscle tissues**, and nonstriated muscle cells form **smooth muscle tissue**.

Most of the cells in the living world contain a single nucleus, and this is the case with smooth muscle tissue and the striated cardiac muscle tissue. Skeletal muscle tissue, conversely, is made up of long cells, each of which typically contains hundreds of nuclei. In humans and other vertebrates, skeletal muscle is always associated with bones and is responsible for movement of the body—it is skeletal muscle that makes up the biceps, for example. Meanwhile, smooth muscle is responsible for contractions in blood vessels, air passageways in the lungs, and hollow organs such as the uterus. As might be apparent from this list of tasks, smooth muscle is not under voluntary control. Cardiac muscle tissue is well named, because it is found only in the heart; its function is to pump blood throughout the body. In **Figure 24.5**, you can see illustrations of these three muscle tissue types in humans. You'll get more detail on how muscles work later in this chapter.

Nervous Tissue

Nervous tissue is tissue that is specialized for the rapid conduction of electrical impulses. It is made up of two basic types of cells: **neurons** and some associated helper cells called **neuroglia**, which you can see in **Figure 24.6**. Neurons are the functional units of nervous tissue, while neuroglia physically support the neurons, aid their nourishment and insulation, and defend nervous tissue from infection. A typical neuron has a cell body, which contains the nucleus; a single long extension called an **axon**, along which information generally is transmitted *from* the neuron; and many extensions called **dendrites**, along which information generally is transmitted *to* the neuron. In humans, axons may reach a meter in length.

24.5 A Summary of the Organ Systems of the Human Body

As noted earlier, combinations of the four tissue types just reviewed go on to form various organs in the body, and these organs then work in concert to form the body's organ systems. **Table 24.1** summarizes the 11 organ systems of the human body, while

Figure 24.5
Muscle Tissue in the Human Body
The three types of muscle tissue are cardiac, smooth, and skeletal.

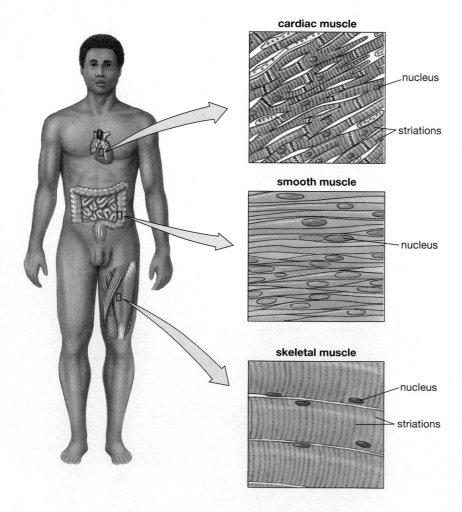

cardiac muscle

nucleus

striations

smooth muscle

nucleus

skeletal muscle

nucleus

striations

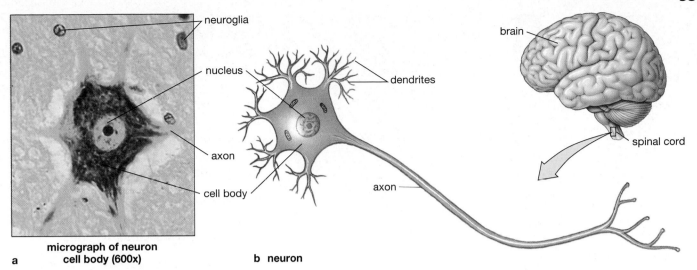

micrograph of neuron
cell body (600x)

a

b neuron

neuroglia

nucleus

dendrites

axon

cell body

brain

spinal cord

axon

Figure 24.6
Nervous Tissue in the Human Body

a A micrograph of a neuron and associated neuroglia.

b The functional unit of nervous tissue is the neuron, which contains a single axon that carries signals from the cell body; and dendrites, which carry signals to the cell body. Most neurons are surrounded by an array of supporting neuroglia cells. This neuron happens to be located in the spinal cord.

Table 24.1
Organ Systems in the Human Body

Organ System	Composed of	Main Functions
Integumentary	Skin, hair, nails, associated exocrine glands	Protection, water and temperature insulation, fat storage
Skeletal	Bone and cartilage	Support, mineral storage, red blood cell production
Muscular	Skeletal muscles	Movement, posture, balance, maintenance of body temperature
Nervous	Brain, spinal cord, sense organs and interconnecting nerves	Sensing, communicating, activating
Endocrine	Endocrine glands, hormone-producing cells	Maintenance of internal state, regulation of development, maintenance of sex characteristics
Lymphatic	Lymphoid organs (e.g., thymus gland) and lymphatic vessels	Protection against invading organisms
Cardiovascular	Heart, blood, blood vessels	Transport of nutrients, gases, and hormones; waste removal
Respiratory	Lungs and air passageways	Delivery of oxygen to the blood, removal of carbon dioxide from the blood
Digestive	Mouth, stomach, small and large intestine, digestive (exocrine) glands	Digestion of food, elimination of waste
Urinary	Kidneys, ureters, urethra	Elimination of liquid wastes; regulation of water balance
Reproductive	Testes, penis, associated glands in men; ovaries, uterine tube, uterus, vagina and associated glands in women	Reproduction

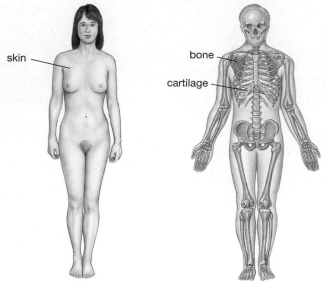

skin

a The Integumentary System

bone

cartilage

b The Skeletal System

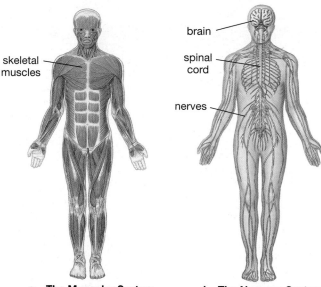

skeletal
muscles

c The Muscular System

brain

spinal
cord

nerves

d The Nervous System

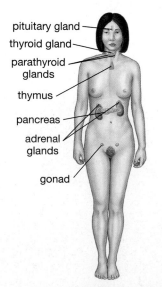

pituitary gland

thyroid gland

parathyroid
glands

thymus

pancreas

adrenal
glands

gonad

e The Endocrine System

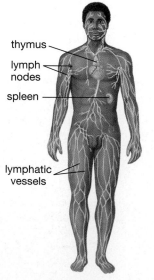

thymus

lymph
nodes

spleen

lymphatic
vessels

f The Lymphatic System

Figure 24.7
The Organ Systems of the Human Body

Figure 24.7 gives you a schematic view of the systems. What follows is a brief look at each of these systems, grouped according to the functions they perform. The balance of this chapter is then devoted to a more detailed look at the three systems in the first group—the integumentary, skeletal, and muscular systems. In Chapters 25, 26, and 28, you'll be taking more detailed looks at the systems in group 2 and group 3.

Organ Systems 1: Body Support and Movement—The Integumentary, Skeletal, and Muscular Systems

- The **integumentary system**, or integument (Figure 24.7a), is made up of the skin and associated structures, such as glands, hair, and nails. This system protects the body from the external environment and assists in the regulation of body temperature.

- The **skeletal system** (Figure 24.7b) forms an internal supporting framework for the body. This is a complex framework; humans have about 206 bones in their body, along with a large number of cartilages. This system also includes the connective tissues and ligaments that connect the bones at various joints. The individual bones store minerals, and some bones, such as those of the skull and ribs, also protect delicate tissues and organs.

- The **muscular system** (Figure 24.7c) includes all the skeletal muscles of the body that are under voluntary control—about 700 of them. It is the contractions of skeletal muscle that produce movement of the body. Muscles also maintain posture and balance, support soft tissues, and help maintain body temperature, in that our human bodies are largely warmed by the heat given off from muscle contraction. The smooth and cardiac muscle tissues noted earlier are part of different systems; cardiac muscle, for example, is part of the cardiovascular system.

Organ Systems 2: Coordination, Regulation, and Defense—The Nervous, Endocrine, and Lymphatic Systems

- The nervous *tissue* reviewed earlier is made up of neurons and their helper neuroglial cells. This tissue then goes on to make up the **nervous system** (24.7d), which includes the brain, spinal cord, sense organs such as the eye and ear, and all the nerves that inter-connect these organs and link the nervous system with other systems.

- The **endocrine system** (Figure 24.7e) bears comparison with the nervous system, in that both systems are in the business of sending signals throughout the body. The endocrine system, however, works through the substances we call hormones, many of which are produced in specialized glands such as the thyroid or adrenal glands. Hormones generally travel through the bloodstream and often function to maintain stability in the body—by conserving water or maintaining temperature, for example.

- The **lymphatic system** (Figure 24.7f) has a great overlap with the immune functions of the body. The system consists of a widespread network of lymphatic vessels that collect extracellular fluid and deliver it, as the fluid called lymph, to the blood vessels of the cardiovascular system. While the fluid is in the lymphatic system, it is checked for signs of infection. The lymphatic system also contains lymphoid organs, which produce or support specialized immune cells that monitor and destroy disease-causing invaders. These organs include the small lymph nodes and the larger thymus gland, tonsils, and spleen.

Organ Systems 3: Transport and Exchange with the Environment—The Cardiovascular, Respiratory, Digestive, and Urinary Systems

- The **cardiovascular system** (Figure 24.7g) consists of the heart, blood, and blood vessels. It also includes the bone marrow, where red blood cells are formed. The cardiovascular system transports nutrients, dissolved gases, and hormones to tissues throughout the body. It also carries waste products from the body's tissues to the site of their filtration, the kidneys.

- The **respiratory system** (Figure 24.7h) includes the lungs and the passageways that carry air to them. These passageways begin at the nasal cavities, and continue through the pharynx (throat), larynx (voicebox), and trachea (windpipe). Through this system, oxygen comes into the body and carbon dioxide is expelled from it.

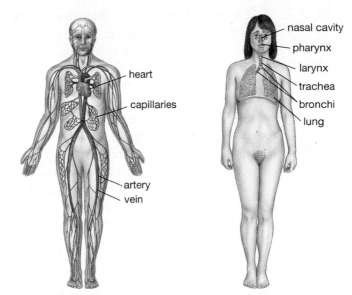

g The Cardiovascular System

h The Respiratory System

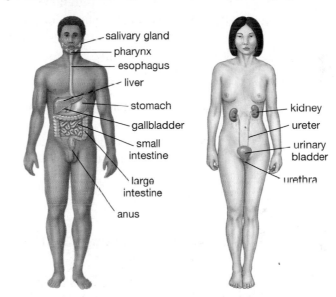

i The Digestive System

j The Urinary System

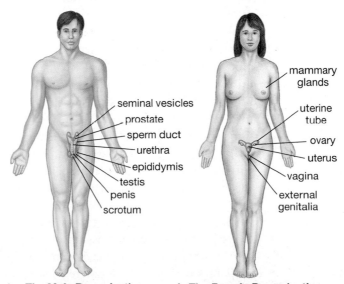

k The Male Reproductive System

l The Female Reproductive System

- The central feature of the **digestive system** (Figure 24.7i) is the digestive tract, a long tube that begins at the mouth and ends at the anus. Along the way, there is also the stomach, the small intestine, and the large intestine. Associated with the tract through its length are a number of accessory glands, including the salivary glands, the liver, and the pancreas. These glands, along with the stomach and the small intestine, secrete digestive enzymes that help break down food so that it can be absorbed and used by the body.

- The major function of the **urinary system** (Figure 24.7j) is the elimination of waste products from the blood through the formation of urine. The system is made up of the kidneys, where waste filtration occurs; tubes called the ureters, which carry urine to another part of the system, the urinary bladder; and the urethra, which conducts urine from the bladder to the exterior of the body.

- Humans have two **reproductive systems**—one for males and another for females (Figure 24.7k–l). The two are linked, of course, in that together they produce offspring. Both systems are discussed in detail in Chapter 28.

24.6 The Integumentary System: Skin and More

Now let's look in greater detail at the first three organ systems, grouped under "body support and movement," starting with the integumentary system. The primary component of this system is the organ we call skin. Just as it may have been difficult to think of blood as a tissue, it may also seem strange to think of the skin as an organ, because it is not the kind of small, easily definable structure we associate with the term. However, skin actually constitutes the largest and most visible organ of the human body. Together with its related structures (hair, nails, and a variety of exocrine glands) it makes up the integumentary system.

Skin does more than just cover the body. It also protects underlying tissues and organs from infection or puncture; it controls the evaporation of body fluids and regulates heat loss; and it stores fat and synthesizes vitamin

D. Meanwhile the exocrine glands that are accessories to the skin excrete materials such as salt, water, oils, and milk; and specialized nerve endings in the skin detect touch, pressure, pain, and temperature.

The Structure of Skin

The **skin** is an organ in two parts: A thin, outer epithelial covering, the **epidermis**, and a thicker underlying **dermis**, composed mostly of connective tissue. Beneath the dermis—and not part of the skin, proper—is the subcutaneous layer, or hypodermis, made up of loose connective tissue that attaches the skin to deeper structures, such as muscles or bones. You can see a cross section of human skin in **Figure 24.8**. One way to look at the skin is as a complex organ composed of all four basic tissue types described earlier: epithelial, connective, nerve, and muscle. You'll learn shortly how all four tissue types work together in skin.

The Outermost Layer of Skin, the Epidermis

The outermost of our skin layers, the epidermis, itself consists of several different cell layers that form a stratified squamous epithelium. (Remember a stratified epithelium is one with more than one layer, and squamous epithelial cells are *flat* cells.) Because the outermost skin layer, the epidermis, is continually being rubbed or scraped by objects, its cells are worn away as a result. This means that epidermal cells must be constantly replenished. As cells divide in the epidermis, they are pushed outward, toward the surface, to replace cells that have worn away. As these new epidermal cells are pushed farther outward by even newer cells, they increase their production of the water-resistant protein **keratin**. The uppermost, or surface, layer of the epidermis is made up of flattened, dead epithelial cells that are tightly joined and filled with large amounts of keratin—an effective barrier for water control and invader protection.

It takes about 14 days for a cell to move outward from the deepest to the outermost layer of the epidermis. The dead, surface cells usually remain in their exposed position for an additional two weeks before they are shed or washed away.

A Skin Layer beneath the Epidermis, the Dermis

The skin's deeper layer, the **dermis**, is filled with accessory structures derived from the epidermis, such as hair follicles and sweat glands. The uppermost region of the dermis consists of loose connective tissue that supports and nourishes the epidermis in several ways. For example, it contains the capillaries and nerves that supply the surface of the skin. Nerve fibers control blood flow, adjust gland secretion rates, and monitor the sensory receptors of touch, pain, pressure, and temperature that are scattered throughout the dermis.

Integumentary Structures Associated with Skin

Hair

Hair projects above the surface of the skin almost everywhere except over the sides and soles of the feet, the palms of the hands, the sides of the fingers and toes, the lips, and portions of the external genital organs.

Hairs originate in tiny complex organs called **hair follicles.** The epithelium at the base of a follicle surrounds the **hair papilla**, a peg-like structure of connective tissue containing capillaries and nerves. Ribbons of smooth muscle (arrector pili muscles) extend from the upper dermis to the hair follicle. When stimulated, the arrector pili can pull on the follicles and elevate the hairs. Contraction of the arrector pili may be caused by emotional states, such as fear or rage, or as a response to cold, producing the characteristic "goose bumps" associated with shivering. As you might guess, the arrector pili muscles are *involuntary* muscles, which is why we can neither "give ourselves" goose bumps nor stop them from forming.

Glands

The skin contains two types of duct-containing or exocrine glands, sebaceous glands and sweat glands (see Figure 24.8). **Sebaceous glands** produce a waxy, oily secretion called *sebum*, which is secreted into hair follicles and directly onto the surface of the skin, where it lubricates the hair shaft and inhibits bacterial growth in the surrounding area.

The **sweat glands** of the skin come in two varieties: *apocrine* sweat glands and *merocrine* (or eccrine) sweat glands. Apocrine sweat glands are found in the armpits,

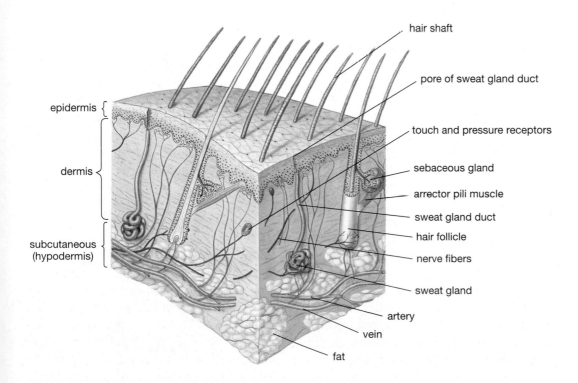

**Figure 24.8
Our Outer Wrap: Components of the Integumentary System**

epidermis

dermis

subcutaneous (hypodermis)

hair shaft

pore of sweat gland duct

touch and pressure receptors

sebaceous gland

arrector pili muscle

sweat gland duct

hair follicle

nerve fibers

sweat gland

artery

vein

fat

around the nipples, and in the groin. Like sebaceous glands, apocrine sweat glands also empty into the hair follicles. At puberty, these coiled tubular glands begin discharging a thick, cloudy secretion that becomes odorous when broken down by bacteria. In other mammals this odor is an important form of communication; for us, whatever function body odor may have is usually masked by deodorants. The milk-secreting mammary glands of the breast are modified apocrine sweat glands.

Merocrine (eccrine) sweat glands are far more numerous and widely distributed than apocrine glands. Merocrine sweat glands produce the clear secretion called *sweat*, also known as perspiration. The primary function of this secretion is to cool the surface of the skin and to reduce body temperature. The perspiration produced by merocrine glands is more than 99 percent water, but it contains a mixture of other compounds, including salts—which give sweat its salty taste—and waste products such as urea.

Nails

Nails form over the tips of the fingers and toes, where they protect the exposed tips. The body of the nail covers the nail bed, but nail growth occurs from the nail root, an epithelial fold not visible from the surface. The deepest portion of the nail root lies very close to the bone of the fingertip. A portion of the outer epidermal layer of the fold extends over the exposed nail nearest the root, forming the cuticle. Underlying blood vessels give the nail its pink color.

24.7 Body Support and the Skeleton

Now let's turn from skin to bone and cartilage. The human skeletal system (composed of both bone and cartilage) performs more functions than just the obvious one of support. For example, the bones serve as storage sites for valuable minerals such as calcium and phosphate. In addition, some bones store lipids within their internal cavities and produce red blood cells and other blood elements. Most of the human skeletal system is bone, but cartilage, which does not contain blood vessels, exists in certain places—at the end of our noses and ribs and in our outer ears, for example. Some vertebrates, such as sharks, have a skeleton that is made entirely of cartilage.

Structure of Bone

Bone is largely composed of osseous tissue (bone tissue), which is one of the two types of supporting connective tissue noted earlier. But each bone of the skeleton is considered to be an organ because it contains not only connective tissue but other tissue types as well. Like other connective tissues, osseous tissue is made up not only of specialized cells but also of an extracellular material these cells secrete. Bone tissue contains large deposits of hard calcium compounds that give the tissue its characteristic solid, stony feel. The remaining bone mass is mainly collagen fibers, with bone cells and other cells providing only around 2 percent of the mass of a bone.

Large-Scale Features of Bone

The typical features of a long bone such as the humerus of a limb are shown in **Figure 24.9**. A long bone has a central shaft, or diaphysis, and expanded ends, or epiphyses (singular, epiphysis). The diaphysis surrounds a central cavity filled with **marrow**, which is a loose connective tissue found in the internal cavities of bones. Marrow comes in two forms: red marrow, which is blood-forming tissue; and yellow marrow, which is made up of fat. The ends (epiphyses) of adjacent bones form joints with each other and are covered by slippery cartilages called *articular cartilages*. There are two types of bone tissue. Compact bone is relatively solid, whereas **spongy bone** is porous and less dense. Both types are present in long bones, with compact bone forming the diaphysis, and spongy bone filling the epiphyses (see Figure 24.9a).

Small-Scale Features of Bone

Dropping down to the microscopic level and looking at bone *cells*, three different types are involved in the growth and maintenance of bone. **Osteoblasts** are immature bone cells that are responsible for the production of new bone. They secrete the organic material that becomes the bone matrix. Once surrounded by this material, the osteoblasts reduce their production of it. This marks their transition to osteocytes, the second type of bone cell. **Osteocytes** are mature bone cells; they *maintain* the structure and density of normal bone by continually recycling the calcium compounds around themselves. Then there is the third type of bone cell, the **osteoclasts**, which could be thought of as the demolition team of bone tissue. They release enzymes into the bone matrix that break it down, thus liberating the stored minerals in it. This process, essential in bone growth, also helps regulate calcium and phosphate concentrations in body fluids. Osteocytes are the most numerous bone cells in a person whose bone growth is complete, but bone is a very dynamic tissue. At any given moment, osteoclasts are removing bone and osteoblasts are adding to it. Beginning in middle age, the body may start removing more bone than it adds, a condition that particularly affects women. You can read more about this in "Doing Something

about Osteoporosis While You Are Young," on page 540.

Within both compact and spongy bone, osteocytes exist in microscopic cavities called lacunae (which means "little lakes"). Within compact bone, the basic functional unit is the **osteon** or Haversian system (Figure 24.9c). The essence of bone growth is that, within each osteon, osteoblasts produce layer after layer of concentric cylinders of bone—much like layers of insulation wrapped around a pipe. These layers surround what could be thought of as the hole in the pipe, which is a central or Haversian canal. These canals parallel the long axis of the bone and have running through them a kind of support system for the bone—blood vessels and nerves. Unlike compact bone, spongy bone has no osteons. It consists of an open network of interconnecting, calcified rods or plates. These open areas of spongy bone provide space for red marrow.

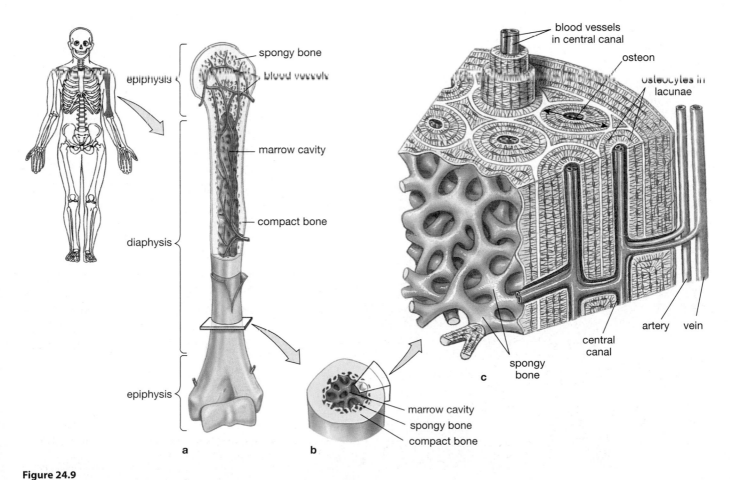

Figure 24.9
Large- and Small-Scale Features of a Typical Bone
The building units of compact bone, the osteons, run parallel to the long axis of the bone. Spongy bone does not contain osteons, but instead is composed of an open network of calcified rods or plates.

Doing Something about Osteoporosis While You Are Young

Osteoporosis is a thinning of bones that leaves its victims susceptible to bone fractures—breaks or cracks in bones, especially bones of the hips, spine, and wrist (**see Figure 1**). It is generally thought of as a condition that affects women, and indeed 80 percent of the Americans who suffer from it are female. It is further thought of as a condition related to age, and this too is true in a sense, in that it is menopause—with its related hormonal changes—that generally brings about the onset of osteoporosis.

Most of the attention paid to osteoporosis has to do with prevention of it *in middle age*, which is to say prevention therapies that are undertaken beginning with menopause. Activities that generally keep us healthy—exercise, proper diet, not smoking—can play a part in this therapy. But the most well-known means of treating osteoporosis is one that involves replacing the hormones whose levels drop off with menopause. In hormone replacement therapy (HRT), women generally take pills or use skin patches that contain estrogen and a synthetic form of the hormone progesterone.

There is no doubt that HRT significantly reduces the risk of developing osteoporosis, and it seems to have a wealth of other benefits as well, probably including a reduction in the risk of heart disease. Against this, HRT appears to *raise* the risk of developing breast cancer. Given these complexities, there is no blanket recommendation about HRT. Instead, decisions need to be made on the basis of individual risk factors. (Is there a history of heart disease or breast cancer in the family, for example?)

There is another side to the prevention of osteoporosis, however—one that can involve *young* women, rather than middle-aged women. The main text in this chapter notes that bone is a dynamic tissue; at any given moment, the bone cells called osteoclasts are removing bone while the cells called osteoblasts are adding to it. It is the *balance* in this activity that changes through life. In a 15-year-old, there is much more bone being added than removed; in a 30-year-old these two activities are roughly balanced out; in a postmenopausal woman there is more bone being removed than being added.

Human females attain almost all their growth in height by late adolescence, but they are still gaining in bone density up until about the age of 30. The critical thing is that the entire period until age 30 represents a *window of opportunity* to develop bone mass. And the more bone mass a woman has by 30, the less risk she will have of osteoporosis after menopause. Once the window of opportunity has closed, however, there is no possibility of raising *peak* bone density, meaning the highest bone density a woman will ever have. Whatever a woman achieves in this regard, she will achieve by age 30.

Most bone is deposited by late adolescence. But medical researcher Robert Recker and his colleagues at Creighton University found in the early 1990s that a group of women they tracked over several years added almost 13 percent to their bone mass during their 20s. Looking for lifestyle factors that might make a difference in adding bone, the researchers found three: physical exercise, intake of calcium and protein, and use of oral contraceptives. None of the women studied were heavy exercisers or had particularly high calcium intake. Thus, the research suggested that simple, modest increases in exercise and calcium during a woman's 20s might yield benefits decades later, when menopause begins and bones begin to thin.

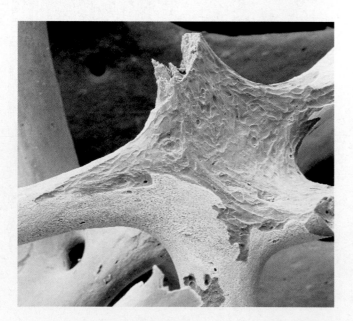

Figure 1
Thinning Bone
Pictured is the fractured thighbone (femur) of a patient with osteoporosis. Note that the lighter areas of the bone have a grainy, pock-marked quality. The dark spots in these areas are places that have lost bone mineral. (×85)

The Human Skeleton

Taking a step back from the microscopic world, you can see in **Figure 24.10** that there are a lot of bones in the skeletal system—206 of them in all. (Only the major bones are visible in the figure.) There are two main divisions to the skeletal system: The **axial skeleton**, whose 80 bones include the skull, the vertebral column, and the rib cage that attaches to it; and the **appendicular skeleton**, whose 126 bones include those of our paired appendages—

the arms and the legs, along with the pelvic and pectoral "girdles" to which they are attached.

Joints

Joints, or articulations, exist wherever two bones meet. Each joint represents a compromise between the need for strength and the need for mobility. When movement is not required, joints can be very strong. For example, joints such as the immovable sutures in the skull hold its separate elements together

Figure 24.10
Solid Framework
A view of the human skeleton, which has two divisions, axial and appendicular. The major bones of the axial division are listed on the left, and the major bones of the appendicular division on the right.

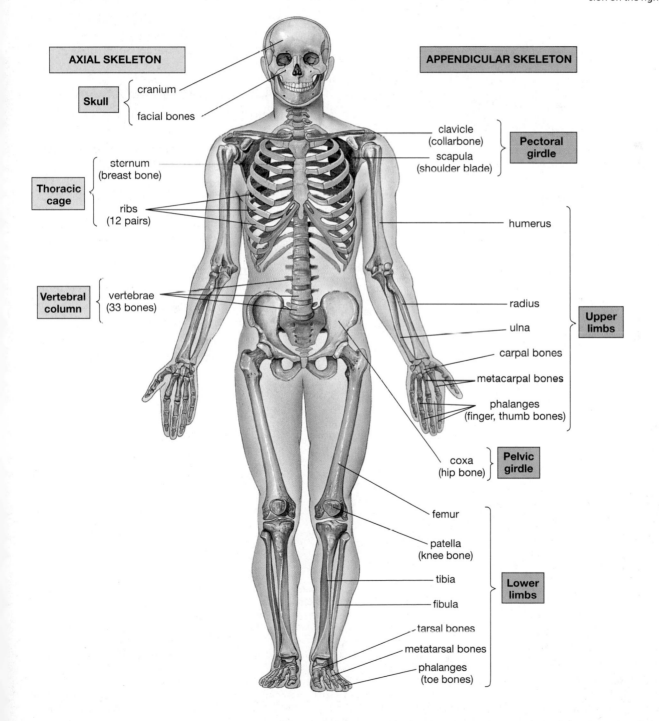

AXIAL SKELETON

APPENDICULAR SKELETON

Skull { cranium / facial bones

clavicle (collarbone)
scapula (shoulder blade) } **Pectoral girdle**

Thoracic cage { sternum (breast bone) / ribs (12 pairs)

humerus

Vertebral column { vertebrae (33 bones)

radius

ulna

carpal bones

metacarpal bones

phalanges (finger, thumb bones)

} **Upper limbs**

coxa (hip bone) } **Pelvic girdle**

femur

patella (knee bone)

tibia

fibula

tarsal bones

metatarsal bones

phalanges (toe bones)

} **Lower limbs**

Figure 24.11
A Highly Moveable Joint
A simplified view of the human knee joint. Note that the two bones never touch each other because they are covered with articular cartilage. The joint has shock-absorbing cartilage in the form of the menisci and a fat pad that protects the articular cartilage.

as if they were a single bone. Conversely, the ball-and-socket joint at the shoulder permits a wide range of arm movement that is limited more by the surrounding muscles than by joint structure. The shoulder joint itself is relatively weak, however, and as a result shoulder injuries are rather common.

Highly movable joints are typically found at the ends of long bones, such as those of the arms and legs. You can see a view of such a joint, the human knee joint, in **Figure 24.11**. Under normal conditions the bony surfaces of such a joint do not contact one another, for

they are covered with special articular cartilages. Such a *hinge joint* is surrounded by a fibrous articular capsule, and the inner surfaces of the joint cavity are lined with a fluid-producing membrane that provides additional lubrication. The knee joint has additional padding in the form of cartilage that lies between the opposing articular surfaces. Such shock-absorbing cartilage pads are called menisci. Also present are fat pads, which protect the articular cartilages and act as packing material. When the bones move, the fat pads fill in the spaces created as the joint cavity changes shape. Additionally, ligaments that join one bone to another may be found on the outside or inside of the articular capsule.

24.8 Muscles and Movement

Bones may provide a scaffolding for the body, but how is it that this scaffolding is capable of movement? The answer, of course, is muscles, specifically skeletal muscles, which are numerous and large enough that they account for about 40 percent of our body weight. (When we talk about the "meat" in, say, cattle, we are mostly talking about skeletal muscle.) Like bones, skeletal muscles are organs in that they contain a variety of tissues—connective and neural tissues as well as skeletal muscle tissue. At each end of a muscle, the fibers of the outer layer come together to form **tendons** that attach skeletal muscles to bones. You can see a typical skeletal muscle in **Figure 24.12**.

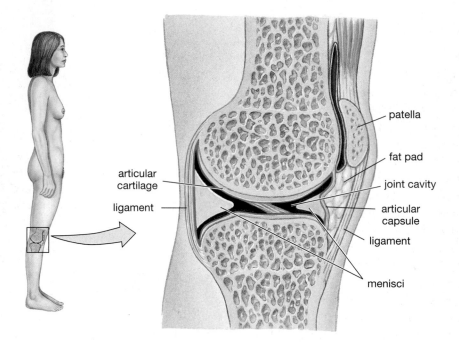

articular cartilage

ligament

patella

fat pad

joint cavity

articular capsule

ligament

menisci

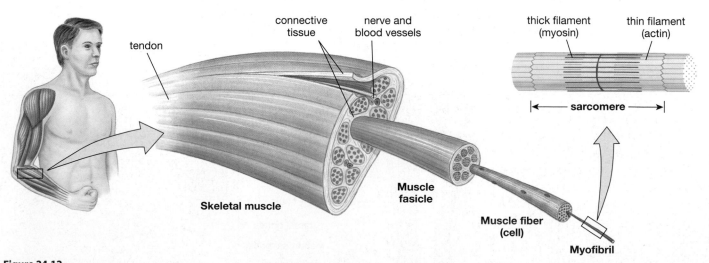

tendon

connective tissue

nerve and blood vessels

thick filament (myosin)

thin filament (actin)

sarcomere

Skeletal muscle

Muscle fasicle

Muscle fiber (cell)

Myofibril

Figure 24.12
Structure of Skeletal Muscle
Elongated muscle cells or "fibers" are grouped into oval-shaped bundles called fascicles. Each fiber is composed of many thin myofibrils, and each myofibril has within it a collection of alternating thick and thin protein strands called filaments. The thin filaments are made of the protein actin, the thick filaments of the protein myosin. Each myofibril is divided up into a series of repeating functional units called sarcomeres.

Makeup of Muscle

Each muscle has within it a number of oval-shaped bundles called fascicles. Inside each fascicle boundary there is a collection of muscle cells, but the term *cell* here may be a little misleading, because any one of these cells can be as long as the muscle itself—a gigantic length relative to the strictly microscopic dimensions of most cells. Because skeletal muscle cells are elongated, they are referred to as **fibers**. A look inside one of these fibers reveals that it is composed of more strands—hundreds of long, thin contractile structures called **myofibrils** that run the length of the cell (see Figure 24.12.) Each myofibril, in turn, has a large collection of strands inside it; these are assemblies of two kinds of protein strands that alternate with one another: thin filaments made of the protein *actin* and thick filaments made of the protein *myosin*. Along their length, these fibers are then divided into a set of repeating units, called **sarcomeres**, that are the functional units of the muscle.

How Does a Muscle Contract?

The structure of the sarcomere provides a key to understanding how a muscle fiber contracts. Thick filaments lie in the *center* of the sarcomere. Thin filaments are attached to either *end* of the sarcomere and extend toward the center, where they overlap with the thick filaments (**Figure 24.13**). In contraction, the thin filaments *slide toward the center* of the sarcomere, causing the unit to shorten—which is to say, causing the muscle to contract.

Tutorial 24.1.2
How Muscles Contract

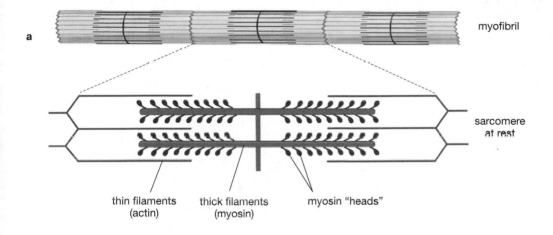

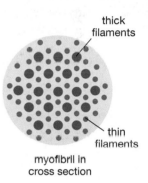

thick
filaments

thin
filaments

myofibril in
cross section

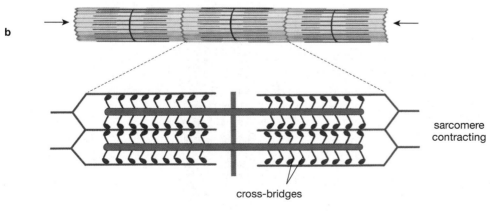

cross-bridges

Figure 24.13
How a Skeletal Muscle Contracts

a The myofibril contains several sarcomeres, one of which is shown at rest. The thin filaments within the sarcomere, made of actin, overlap with adjacent thick filaments, made of myosin. The thick filaments have club-shaped outgrowths, called heads. The myofibril is also shown in cross section, at a location where thick and thin filaments overlap.

b Muscle contraction is a matter of the thin actin filaments sliding together within a sarcomere. This comes about when the myosin heads attach to the actin filaments and pull them toward the center of the sarcomere. The myosin heads pivot at their base and are capable of alternately binding with, or detaching from, the thin filaments.

544

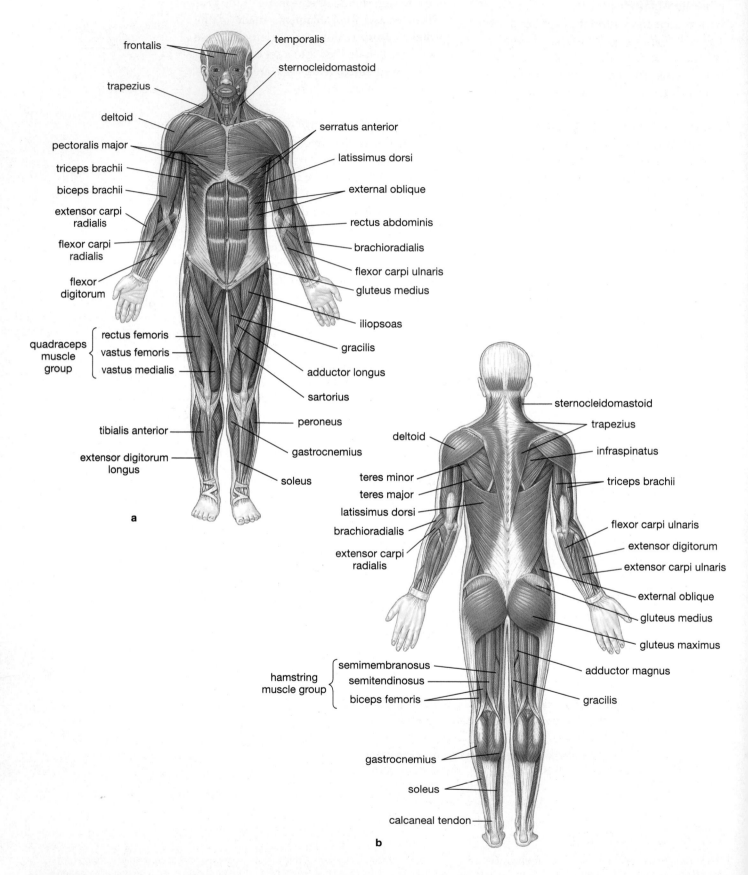

frontalis

temporalis

sternocleidomastoid

trapezius

deltoid

serratus anterior

pectoralis major

latissimus dorsi

triceps brachii

biceps brachii

external oblique

extensor carpi
radialis

rectus abdominis

brachioradialis

flexor carpi
radialis

flexor carpi ulnaris

flexor
digitorum

gluteus medius

iliopsoas

quadraceps
muscle
group

rectus femoris

vastus femoris

gracilis

vastus medialis

adductor longus

sartorius

tibialis anterior

peroneus

extensor digitorum
longus

gastrocnemius

soleus

a

sternocleidomastoid

trapezius

deltoid

infraspinatus

triceps brachii

flexor carpi ulnaris

extensor digitorum

extensor carpi ulnaris

teres minor

teres major

latissimus dorsi

external oblique

brachioradialis

gluteus medius

extensor carpi
radialis

gluteus maximus

adductor magnus

gracilis

hamstring
muscle group

semimembranosus

semitendinosus

biceps femoris

gastrocnemius

soleus

calcaneal tendon

b

Figure 24.14
Major Superficial Muscles of the Body

a Anterior or "front" view.

b Posterior or "back" view.

But what enables the thin filaments to slide? The thick myosin filaments have numerous club-shaped "heads" extending from them that are capable of alternately binding with or detaching from the adjacent thin actin filaments, creating so-called *cross-bridges*. The myosin heads pivot at their base, as if they were on a hinge. Here's the order of their activity: Attach to the actin filament, pull it toward the center of the sarcomere, detach from the actin, reattach to it and pull again. This whole process is powered by ATP.

Major Muscles of the Body

You can see an overview of the muscular system in **Figure 24.14**. The general appearance of the skeletal muscles provides clues to their primary function. Muscles involved with locomotion and posture work across joints, producing skeletal movement. Those that support soft tissue form slings or sheets between relatively stable bony elements.

On to the Nervous, Endocrine, and Immune Systems

Having looked at the basic characteristics of the body and at three of its organ systems, we're now ready to move on to three other systems, which function in communication, regulation, and defense against invaders. These are the nervous, endocrine, and immune systems of the body.

Chapter Review

Summary

24.1 The Sciences of Anatomy and Physiology

- Anatomy is the study of the body's internal and external structure and the physical relationships between the body's constituent parts. Physiology is the study of how the body's various structures work.

24.2 What Are the General Characteristics of Humans?

- Human beings share a number of characteristics with many other animals. Three of these are an internal body cavity, called a coelom; an internal skeleton; and internal temperature regulation.

- Like all mammals, human beings are endothermic, meaning that human bodies are warmed primarily by their own metabolism. Conversely, most fish, amphibians, and reptiles are ectothermic, meaning that their primary source of body heat is external to them.

24.3 Animal Architecture and Organization

- Animal bodies have several levels of organization. Tissues are groups of cells that have a common function. In some animals, tissues are arranged into organs: complexes of several kinds of tissues that perform a special bodily function. Organs in turn form organ systems, such as the cardiovascular system. The human body has 11 organ systems.
  TUTORIAL 24.1.1: Levels of Organization in the Human Body

24.4 The Animal Body Has Four Basic Tissue Types

- There are four fundamental tissue types in the human body: epithelial, connective, muscle, and nervous.

- Epithelial tissue covers all the exposed body surfaces of animals and also forms glands, which are organs that secrete one or more substances.

- Glands come in two main types: exocrine glands, which secrete materials through ducts; and endocrine glands, which release their materials directly from cells into surrounding tissues or the bloodstream.

- Connective tissue is active in the support and protection of other tissues. The cells of connective tissue are embedded in material that they have secreted. A prime function of most connective tissue cells is to produce this extracellular material, composed of protein fibers and a ground substance that varies in consistency from a fluid to a solid.

- Connective tissues are classified according to their surrounding extracellular material. There is loose connective tissue, which provides strength and flexibility as a packing material of the body. Fibrous connective tissues form tendons and ligaments, along with capsules, which surround organs and enclose joint cavities. Supporting connective tissues make up bone and cartilage. Fluid connective tissue includes blood—blood cells suspended in a watery ground substance called plasma—and lymph, which functions primarily in bodily defense.

- Muscle tissue is specialized in that its protein elements or fibrils have the ability to contract. There are three primary types of muscle tissue: skeletal muscle, cardiac muscle, and smooth muscle. Skeletal muscle is always attached to bone and is under voluntary control; it is the muscle that is responsible for movement of the body. Smooth muscle is responsible for contractions in blood vessels, air passages, and hollow organs such as the uterus; it is not under voluntary control. Cardiac muscle exists only in the heart and is responsible for its beating.

- Nervous tissue is specialized for the rapid conduction of electrical impulses. It is made up of two basic types of cells: neurons, which are the functional units of nervous tissue; and neuroglia, which are support cells for neurons.

24.5 A Summary of the Organ Systems of the Human Body

- The integumentary, skeletal, and muscular systems function in body support and movement. The integumentary system is made up of skin and such associated structures as glands, hair, and nails. The skeletal system forms an internal supporting framework for the body. The muscular system includes all the skeletal muscles of the body that are under voluntary control.

- The nervous, endocrine, and lymphatic systems function in coordination, regulation, and defense of the body. The nervous system includes the brain, spinal cord, sense organs such as the eye and ear, and all the nerves that interconnect these organs and that link the nervous system with other systems. The endocrine system functions in bodily regulation, working through hormones. The lymphatic system has a great overlap with the immune functions of the body; it consists of a network of lymphatic vessels that collect extracellular fluid and deliver it, as lymph, to the blood vessels. While in the lymphatic vessels, lymph is checked for signs of infection. The lymphatic system also produces cells that function in defense of the body. Lymphatic system organs include the lymph nodes, thymus gland, tonsils, and spleen.

- The cardiovascular, respiratory, digestive, and urinary organ systems function in transport and in exchange with the environment. The cardiovascular system consists of the heart, blood, blood vessels, and the bone marrow—where red blood cells are formed. The respiratory system includes the lungs and the passageways that carry air to them. Through this system, oxygen comes into the body and carbon dioxide is expelled from it. The central feature of the digestive system is the digestive tract, a tube that begins at the mouth and ends at the anus. Other digestive system organs include the stomach, the small and large intestines, and such accessory glands as the salivary glands, the liver, and the pancreas. The urinary system functions to eliminate waste products from the blood through the formation of urine. It is made up of the kidneys, the ureters that carry urine to the urinary bladder, and the urethra, which conducts urine from the bladder to the exterior of the body.

- Humans have two reproductive systems—one for females and one for males.

24.6 The Integumentary System: Skin and More

- The primary component of the integumentary system is the organ called skin. Integumentary structures associated with skin are hair, nails, and a variety of exocrine glands.

- In addition to covering the body, skin protects underlying tissues and organs, controls the evaporation of body fluids, regulates heat loss, stores fat, and synthesizes vitamin D. Exocrine glands associated with skin excrete such materials as water, oils, and milk. Specialized nerve endings in the skin detect touch, pressure, pain, and temperature.

- Skin is an organ in two parts: a thin outer epithelial covering, the epidermis; and a thicker underlying dermis, composed mostly of connective tissue. Epidermal skin cells must constantly be replenished. The surface layer of the epidermis is composed of tightly interlocked cells that produce a water-resistant protein called keratin. The dermis is filled with accessory structures, such as hair follicles and sweat glands, and contains capillaries and nerves that support the surface of the skin.

- Hairs originate in tiny organs called hair follicles. The skin contains two types of exocrine glands: sebaceous glands, that produce an oily secretion; and sweat glands. Sweat glands come in two varieties: apocrine glands—found in the armpits, around the nipples, and in the groin—which produce a thick, cloudy secretion that empties into hair follicles; and the more numerous merocrine glands, which produce perspiration.

- Nails form over the tips of the fingers and toes, where they protect the exposed tips.

24.7 Body Support and the Skeleton

- The human skeletal system is composed of bone and cartilage. It functions in support, as a storage site for minerals, and as a production site for red blood cells and other blood elements.

- Each bone of the skeleton is considered to be an organ because it contains not only connective tissue—mostly the supporting connective tissue called osseous tissue—but other tissues as well. Like other connective tissues, osseous tissue is made not only of specialized cells, but of extracellular fibers and a ground substance.

- The typical bone features a long central shaft, called a diaphysis, and expanded ends, called epiphyses. The diaphysis surrounds a central cavity filled with a loose connective tissue called marrow that comes in two forms: red marrow, which is blood-forming; and yellow marrow, which is made up of fat.

- There are two types of bone tissue: compact bone, which is relatively solid; and spongy bone, which is porous and less dense. Compact bone forms the diaphysis of long bones; spongy bone fills in the epiphyses.

- Three different types of cells are involved in bone growth and maintenance. Osteoblasts are responsible for the production of new bone; osteocytes are mature bone cells that maintain the structure and density of normal bone; and osteoclasts release enzymes into the bone matrix that breaks it down.

- In compact bone, the basic functional unit is the osteon. Osteoblasts produce layers of concentric cylinders of bone, which are wrapped around a central canal of the osteon. The central canal, which parallels the long axis of the bone, has blood vessels and nerves running through it.

- The human skeleton has 206 bones in all, arranged in two main divisions: an axial skeleton, whose 80 bones include the skull, the vertebral column, and the rib cage; and the appendicular skeleton, whose 126 bones include those of the arms and legs and the pelvic and pectoral girdles to which they are attached.

- Joints or articulations exist wherever two bones meet. Normally, the bony surfaces of highly movable joints do not meet directly, but are instead covered with special cartilages and pads that provide shock absorption and lubrication.

24.8 Muscles and Movement

- A given skeletal muscle cell can be as long as the muscle itself; because of their elongation, these cells are called fibers. Muscle fibers are composed of thin structures called myofibrils that are in turn composed of assemblies of two kinds of protein strands that alternate with one another: thin filaments made of the protein actin and thick filaments made of the protein myosin. These filaments are divided into a set of repeating units called sarcomeres.

- In contraction, thin muscle filaments slide toward the center of the sarcomeres, causing the unit to shorten. The thick myosin filaments have numerous club-shaped heads that can bind with and detach from the adjacent thin actin filaments. The myosin filaments thus bring about contraction by attaching to the actin filaments, pulling them toward the center of the sarcomeres, detaching, and then pulling again.

TUTORIAL 24.1.2: How Muscles Contract

Key Terms

adipose tissue 531	**endothermic** 529	
appendicular skeleton 541	**epidermis** 536	
axial skeleton 541	**epithelial tissue** 530	
axon 532	**exocrine gland** 530	
bone 531	**fiber** 543	
cardiac muscle tissue 532	**gland** 530	
cardiovascular system 535	**hair follicle** 537	
cartilage 531	**hair papilla** 537	
coelom 528	**hormone** 530	
connective tissue 530	**integumentary system** 534	
dendrite 532	**interstitial fluid** 531	
dermis 536	**keratin** 536	
digestive system 536	**lymph** 532	
ectothermic 528	**lymphatic** 532	
endocrine gland 530	**lymphatic system** 535	
endocrine system 535	**marrow** 538	

muscle tissue 532	**respiratory system** 535
muscular system 534	**sarcomere** 543
myofibril 543	**sebaceous gland** 537
nervous system 532	**skeletal muscle tissue** 532
nervous tissue 532	**skeletal system** 534
neuroglia 532	**skin** 536
neuron 532	**smooth muscle tissue** 532
organ 529	**spongy bone** 538
organ system 529	**sweat gland** 537
osteoblast 538	**tendon** 542
osteoclast 539	**tissue** 529
osteocyte 539	**urinary system** 536
osteon 539	**vertebrate** 528
reproductive system 536	

Understanding the Basics

Multiple-Choice Questions

1. Which of the following organs is not found in the abdominopelvic cavity?
   a. stomach
   b. liver
   c. lungs
   d. pancreas
   e. intestines

2. The four basic types of tissue in the human body are
   a. epithelial, connective, muscle, nervous
   b. epithelial, blood, muscle, nervous
   c. skin, bone, muscle, nervous
   d. skin, connective, heart, nervous
   e. epithelial, organ, connective, muscle

3. Endocrine glands differ from exocrine glands in that
   a. they have ducts
   b. they are ductless
   c. they produce only perspiration
   d. they consist of epithelia and muscle tissue
   e. they are found only in the liver

4. Large muscle fibers that are multinucleated, striated, and voluntary are found in
   a. cardiac muscle tissue
   b. skeletal muscle tissue
   c. smooth muscle tissue
   d. rough muscle tissue
   e. all four types of muscle tissue

5. The skeletal system includes 206 bones and several
   a. tissues
   b. muscles
   c. nerves
   d. organs
   e. cartilages

6. The lungs and the passageways that carry air make up the
   a. cardiovascular system
   b. respiratory system
   c. lymphatic system
   d. urinary system
   e. connective system

7. The following structures are all associated with the integumentary system except
   a. nails
   b. sweat glands
   c. hair
   d. dermis
   e. salivary glands

8. The two types of bone tissue are
   a. spongy bone and cartilage
   b. dense bone and compact bone
   c. compact bone and spongy bone
   d. compact bone and periosteum
   e. cartilage and ligaments

9. The cells that maintain bone structure are
   a. osteoblasts
   b. osteoclasts
   c. osteocytes
   d. osteons
   e. osteolytes

10. The appendicular skeleton consists of the bones of the
    a. pectoral and pelvic girdles
    b. skull, thorax, and vertebral column
    c. arms, legs, hands, and feet
    d. limbs, pectoral girdle, and pelvic girdle
    e. arms, legs, and neck

11. A skeletal muscle contains
    a. connective tissues
    b. blood vessels and nerves
    c. skeletal muscle tissue
    d. all of the above
    e. none of the above

Brief Review

1. Beginning with the molecular level, list in correct sequence the levels of organization of the human body from the simplest level to the most complex.

2. What fluid connective tissues are found in the human body?

3. What is keratin?

4. List three functions of the skeletal system.

5. List three primary functions of the muscular system.

6. Which organ system is composed of ductless glands whose secretions coordinate cellular activities of other organ systems?

7. Why does the epidermis continually shed its cells?

8. A sample of bone shows concentric layers surrounding a central canal. Is this sample from the shaft (diaphysis) or the end (epiphysis) of a long bone?

9. Why does skeletal muscle appear striated when viewed under a microscope?

Applying Your Knowledge

1. Why is the same stratified epithelial organization found in the mucous membranes of the pharynx, esophagus, anus, and vagina?

2. Lack of vitamin C in the diet interferes with the body's ability to produce collagen. What effect might this have on connective tissue?

3. The repair capability of cartilage is low compared with that of bone. What accounts for this difference?

25

Control and Defense
The Nervous, Endocrine, and Immune Systems

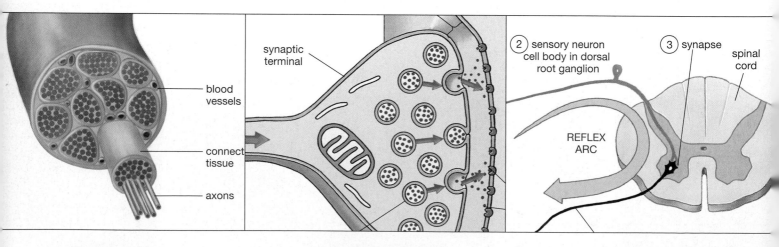

The nature of a nerve.
(Section 25.1, page 553)

A nerve impulse is transmitted.
(Section 25.2, page 555)

A knee-jerk reaction.
(Section 25.3, page 557)

Communication within the human body takes several forms. Nervous-system communication is fast and specific, while hormonal communication is slower and more general. Our bodies have an elaborate system set up to fight off invaders.

One of the central qualities that separates living beings from inanimate objects is that living beings have the ability to respond to their external environment and to maintain a relatively stable internal environment. To be alive means to monitor what's around us, to pursue what we need, and to adapt to change. We vertebrate animals are able to do these things largely because of the complex and marvelous workings of three bodily systems. One is the nervous system, with its ability to respond almost instantaneously to external signals or internal desires; another is the endocrine or hormonal system, which allows us to maintain a physical steady state over time; a third is the immune system, which allows us to deal with the microscopic invaders that mount their attacks on us every day. These three systems are the subject of this chapter.

25.1 Overview of the Nervous System

The basic function of the nervous system is to monitor an animal's internal and external environment, integrate the sensory information the animal receives in this process, and coordinate the voluntary and involuntary responses of many systems throughout the body. All of these functions are performed by **neurons**—the key functional cells of the nervous system—with major assistance from **neuroglia** cells, which surround, support, and protect neurons.

The major subdivisions of vertebrate nervous systems are shown in **Figure 25.1** (page 552). The two major divisions are, first, the **central nervous system (CNS)**, consisting of the brain and spinal cord, which are responsible for integrating and coordinating sensory data and issuing action or *motor* commands. The brain is also the seat of higher functions, such as intelligence, memory, and emotion. All communication between the CNS and the rest of the body occurs via the nervous system's second major subdivision, the **peripheral nervous system (PNS)**, which includes all of the neural tissue outside the CNS.

The peripheral nervous system then has an **afferent division**, which brings sensory information to the CNS. It also has an **efferent division** that carries motor commands to muscles and glands, which collectively are

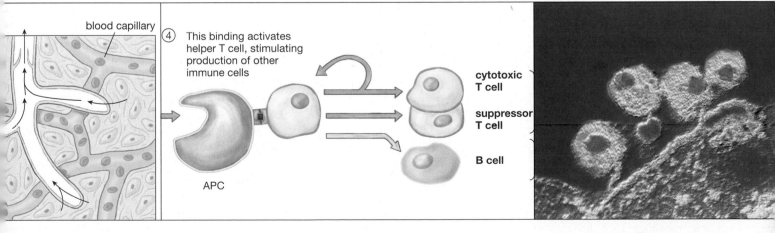

Fluids on the move. (Section 25.12, page 577)

Central to the immune system. (Section 25.14 page 579)

④ This binding activates helper T cell, stimulating production of other immune cells

blood capillary

cytotoxic T cell

suppressor T cell

B cell

APC

Budding killers. (Section 25.16, page 581)

Figure 25.1
Divisions of the Nervous System

Information about the body and its environment comes to the nervous system from its sensory receptors (for example, cells in the ears), which are part of the peripheral nervous system. This information then goes through the afferent division of the peripheral nervous system to the brain and spinal cord, which constitute the central nervous system. The brain and spinal cord process information and issue motor commands through the peripheral nervous system's efferent division. These commands go to "effectors" such as skeletal muscles and glands. The peripheral nervous system's efferent division has two systems within it: The somatic nervous system, which provides voluntary control over skeletal muscles; and the autonomic system, which provides involuntary regulation of smooth muscle, cardiac muscle, and glands. The autonomic system is further divided into sympathetic ("fight-or-flight") and parasympathetic ("rest-and-digest") divisions.

known as the body's **effectors**. Within the peripheral system's efferent division, there are two subsystems: the **somatic nervous system**, which provides voluntary control over skeletal muscles; and the **autonomic nervous system**, which provides involuntary regulation of smooth muscle, cardiac muscle, and glands. The autonomic nervous system is further subdivided into the **sympathetic division** (which generally has stimulatory effects) and **parasympathetic division** (which generally has relaxing effects).

Lest all this seem too confusing, let's think of it in terms of a real-world occurrence: a bell going off for a high-school class on a day when you had to give a report in front of the class. Sensory receptors in your ears (cells that are part of the peripheral nervous system) perceived the sound of the bell and transmitted a message through afferent auditory nerves to your brain (part of the central nervous system), which processed the information and transmitted motor commands through the peripheral nervous system's efferent division to two places. One was to the effectors we call *skeletal muscles* (in your legs and arms), commanding them to move so you could get to

class; the other was to the effectors called *adrenal glands*, prompting them to produce adrenaline (because this was a public performance). It was the somatic nervous system that issued the command to move, but it was the autonomic nervous system that issued the command for adrenaline release.

Cells of the Nervous System

All the functions of the nervous system involve neurons communicating—either with one another, or with cells of other types, such as muscle or hormonal cells. The helper *neuroglial cells* surround the neurons, providing a supporting framework for neural tissue, and acting as defenders of it. Although neuroglia are much smaller than neurons, they outnumber neurons by about 10 to 1.

On the basis of the functions they perform, we can distinguish three types of neurons: sensory neurons, motor neurons, and interneurons.

Sensory neurons do just what their name implies: They sense conditions both inside and outside the body and convey information about these conditions to neurons inside the CNS. Sensory information from the external environment includes touch, temperature, and pressure sensations, as well as the more complex senses of sight, smell, hearing, and taste. Internal sensory information concerns such things as the positions of skeletal muscles, the activities of the digestive, respiratory, and urinary systems, and sensations of deep pressure and pain.

Motor neurons carry instructions from the CNS to other tissues, organs, or organ systems. The targets of information for motor neurons are the effectors noted earlier, such as skeletal muscles; they are called effectors because they can effect *change*, as with skeletal muscle contracting after it receives a neural signal.

The third class of neurons, **interneurons** (or association neurons), are located entirely within the brain and spinal cord. As their name implies, they interconnect other neurons. They are responsible for the analysis of sensory inputs and the coordination of motor outputs.

Anatomy of a Neuron

Figure 25.2 shows a typical motor neuron. Its cell body contains all the basic elements of cells in general—a nucleus, organelles such as mitochondria, and so forth. Extensions projecting from the cell body include a variable number of **dendrites**, which receive signals coming *to*

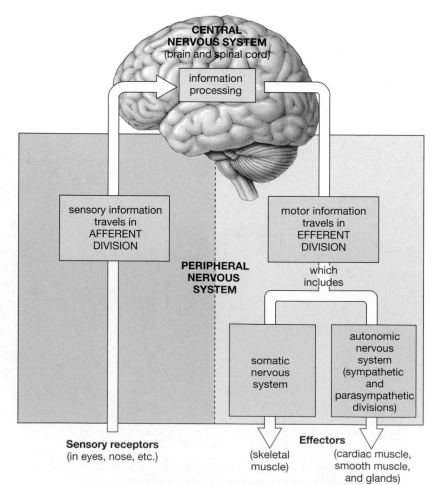

CENTRAL NERVOUS SYSTEM (brain and spinal cord)

information processing

PERIPHERAL NERVOUS SYSTEM

sensory information travels in **AFFERENT DIVISION**

motor information travels in **EFFERENT DIVISION**

which includes

somatic nervous system

autonomic nervous system (sympathetic and parasympathetic divisions)

Sensory receptors (in eyes, nose, etc.)

Effectors
(skeletal muscle)
(cardiac muscle, smooth muscle, and glands)

the cell body; and a single, large **axon** that carries signals *from* the cell body. Each neuron has an outer plasma membrane, just like any other cell; what sets the neuron's membrane apart is the kind of sensitivity it has to chemical, mechanical, or electrical stimulation. Such stimulation often leads to the generation of electrical signals—meaning nerve impulses—that are conducted along the axon. At the tips of an axon or its branches, there may be structures called **synaptic terminals**, which are part of a **synapse**—a site where a neuron communicates with another cell. Neurons can be extraordinarily *long* cells; the axon that allows the movement of your little toe starts in a cell body that lies partway up your spinal cord.

The Nature of Neuroglia

Neuroglia, which are found in both the CNS and PNS, have no information processing ability of their own. In the CNS, one type of neuroglia is protective in that it can ingest invaders or foreign molecules. A second group of neuroglia covers tiny blood vessels within the brain, forming what is known as the **blood-brain barrier**, which keeps some substances from passing from general circulation into the brain. Other neuroglia wrap their cell membranes around the axons of neurons in the brain and spinal cord. Such a covering is called **myelin**, and an axon covered in this way is said to be myelinated.

Axons that are myelinated carry nerve impulses faster than those that are not. Myelin is fat- or lipid-rich, and areas of the brain and spinal cord containing myelinated axons are glossy white. Thus do we get the term the **white matter** of the CNS, which contains mostly axons, whereas areas dominated by neuron cell *bodies* are **gray matter**. If you look at **Figure 25.3**, you can see what a myelinated axon looks like. In the PNS, the most important neuroglia are called *Schwann cells*, which wrap around the axons of neurons, making them myelinated.

Nerves

A **nerve** is a bundle of axons in the PNS that transmit information to or from the CNS. Any given bundle may stem from a collection of sensory neurons, a collection of motor neurons, or a mixture of the two types of neurons. Like muscles, nerves are covered by and contain three layers of connective tissue; they also contain blood vessels that supply nutrients and

oxygen. In a dissection, nerves are whitish string-like structures, easily visible to the naked eye.

25.2 How Does Nervous-System Communication Work?

Nerve impulses travel not only between a given neuron and other neurons but also between neurons and effectors, such as muscle cells or gland cells (as Figure 25.2 indicates). But nerve signals must also travel *within* a given neuron—along the length of its axon. This communication occurs through changes in an electrical property of the neuron's outer lining or plasma membrane.

Nerve Signal Transmission within a Neuron

Those who read Chapter 5 will recall that the plasma membrane of any cell regulates the passage of substances in and out of the cell.

Figure 25.2
The Neuron
Structure of a motor neuron, in this case innervating a skeletal muscle.

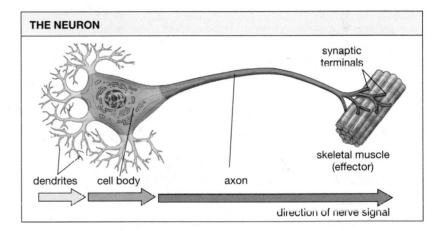

THE NEURON

synaptic terminals

skeletal muscle (effector)

dendrites   cell body   axon

direction of nerve signal

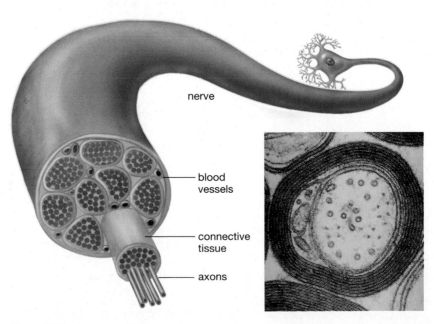

nerve

blood vessels

connective tissue

axons

Figure 25.3
Looking at an Axon
The micrograph at right shows a cross section of a single vertebrate axon that is wrapped in multiple layers of myelin covering. As seen at left, a nerve is a bundle of axons.

Because the plasma membrane is not permeable to charged particles (such as ions), the membrane separates charged particles that lie inside the membrane from those that lie outside.

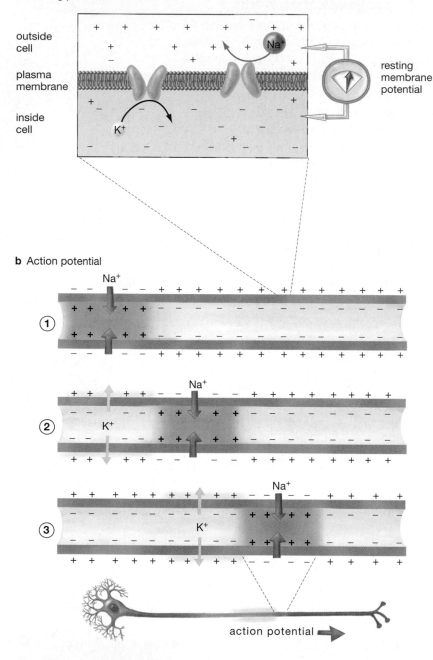

a Resting potential

outside cell

plasma membrane

inside cell

resting membrane potential

b Action potential

action potential

Figure 25.4
Nerve Signal Transmission within a Neuron

a Electrical energy is stored across the cell membranes of resting neurons because there is an excess of positively charged compounds on the outside of the membrane and an excess of negatively charged compounds on the inside. This stored energy is called a membrane potential. Protein channels (shown in green) that can allow the movement of electrically charged ions across the membrane remain closed in a resting cell, thus maintaining the membrane potential.

b **1**. Nerve signal transmission begins when, upon stimulation, some protein channels open up, allowing a movement of positively charged sodium ions (Na⁺) into the cell. For a brief time, the inside of the cell becomes positively charged. **2**. The Na⁺ gates close and the gates for positively charged potassium ions (K⁺) open up, allowing a movement of K⁺ out of the cell. With this, there is once again a net positive charge outside the membrane. The influx of Na⁺ at one point in the cell membrane then triggers the same sequence of events in an adjacent portion of the membrane—note the influx of Na⁺ next to the outflow of K⁺. **3**. The nerve signal continues to be propagated one way along the axon by means of this action potential.

There are, however, protein channels in the membrane that can open or close like gates, allowing for the controlled flow of charged particles across the membrane. A resting cell has a greater positive charge outside itself than inside; this is so because there is an excess of positively charged compounds (ions) on the outside and an excess of negatively charged compounds (ions and protein molecules) on the inside. Because oppositely charged particles attract, this difference in overall charge across the two sides of the membrane amounts to a form of stored energy that can be used in transmitting nerve signals; it is potential energy that, fittingly enough, is called a membrane potential (**see Figure 25.4**).

In response to stimulation a neuron receives from other neurons, its permeability to charged particles can change. Recall that the cell membrane has protein "gates" that can open or close. Upon stimulation, positively charged sodium ions (Na^+), which lie in abundance outside the cell, are let inside in greater numbers through gates that open up. After a very brief time, the influx of Na^+ is great enough that the inside of the cell is more positive than the outside. This very activity then causes the Na^+ gates to shut and a group of gates for another positively charged ion, potassium (K^+), to open up more fully. However, K^+ ions initially exist in greater numbers *inside* the plasma membrane than outside. With the opening of their gates, they rush out of the cell and the positive charge that existed on the inside of the cell begins to decline. Eventually, enough K^+ ions have rushed out that the cell returns to its resting state, with a net positive charge outside the membrane and a net negative charge inside. This entire process takes place in a few thousandths of a second (milliseconds).

Transmission along the Whole Axon: The Action Potential

You have just looked at the alteration of membrane potential at *one spot* along an axon. But how does a nerve signal get transmitted down the *length* of the axon? The key to such impulse transmission is that an influx of sodium ions at an initial location on the membrane triggers reactions that cause *neighboring* portions of the membrane to permit sodium ions to rush into the cell in their location (see Figure 25.4). What occurs, in other words, is a chain reaction that moves down the entire length of the axon membrane, again in milliseconds. This is an **action potential**: a transient reversal of electric potential across a cell membrane that results in a conducted nerve impulse. Action potentials have been compared to what happens with a lighted fuse: The heat of the spark causes the neighboring section of the fuse to catch fire, thus moving the spark along.

We expect a fuse spark to move in one direction, of course, and the same is true of an action potential. In the case of a fuse, this one-way movement comes about because the burnt part of the fuse cannot be burnt further. In the case of nerve signals, the portion of an axon that has potassium (K^+) ions flowing out of it cannot simultaneously have its sodium (Na^+) channels opened up. In effect, there is a short *refractory* period in which a given portion of an axon is incapable of transmitting a nerve signal. Thus the signal always moves away from the original point of stimulation.

All action potentials are of equal strength and duration. An all-or-none principle applies to neurons: A given stimulus triggers either a full-blown action potential or none at all.

Communication between Cells: The Synapse

Having traveled the length of an axon, an action potential then reaches the synaptic terminals (the branched tip of the axon). How does the signal then get to the *next* neuron (or muscle or gland cell)? This question was once very intriguing to biologists, because it was clear that a neuron does not touch the downstream cell. The axon of one neuron comes *close* to touching the dendrites or cell body of another neuron (within about 20 millionths of a meter), but there is inevitably a small intervening space, or gap, between them. There are thus three entities involved in cell-to-cell transmission: the "sending" cell, the "receiving" cell, and the gap between them. The location where all three come together is the area in which cell-to-cell communication takes place, and it is called a **synapse** (see Figure 25.5).

Structure of a Synapse

Every synapse has three basic parts: (1) the **presynaptic neuron**, which is the transmitting neuron; (2) the **synaptic cleft**, which is the space between presynaptic and postsynaptic neurons; and (3) the **postsynaptic neuron** (or other effector cell type), which receives the action potential. Upon arrival of

the action potential at a synaptic terminal, the presynaptic neuron releases a chemical, called a **neurotransmitter**, which is stored in small sacs or vesicles in the synaptic terminal. When released into the synaptic cleft, the neurotransmitter diffuses over to the receiving cell and binds to receptors in that cell's membrane. This binding stimulates the opening of sodium channels, thereby allowing positively charged sodium ions to cross the *post*synaptic cell membrane, setting in motion the generation of an action potential in that cell. (See Figure 25.5 for an example involving the neurotransmitter acetylcholine.) So, why doesn't the neurotransmitter *keep* stimulating the postsynaptic cell? One means of control is enzymes, which the postsynaptic cell releases into the synaptic cleft to breakdown or inactivate the neurotransmitter.

The Importance of Neurotransmitters

There are many different neurotransmitters. Some, like acetylcholine, almost always have stimulatory effects on postsynaptic cells. Others, like dopamine in the CNS, are usually inhibitory. The importance of neurotransmitters is immense. As just one example of their power, you may have heard of a widely prescribed class of drugs called SSRIs (selective serotonin reuptake inhibitors) that are aimed

Tutorial 25.1.2
Structure and Function of a Synapse

Figure 25.5
Structure and Function of a Synapse
As a nerve impulse reaches the synapse, molecules of a neurotransmitter (such as acetylcholine) are released into the synaptic cleft from synaptic terminals of the presynaptic neuron. This release occurs when small membrane-bound vesicles—each filled with thousands of neurotransmitter molecules—fuse with the outer membrane of the presynaptic neuron, and in so doing release their contents to the synaptic cleft. The neurotransmitter molecules then move across the cleft and stimulate receptors in the receiving cell's membrane, thus initiating a nerve impulse within it.

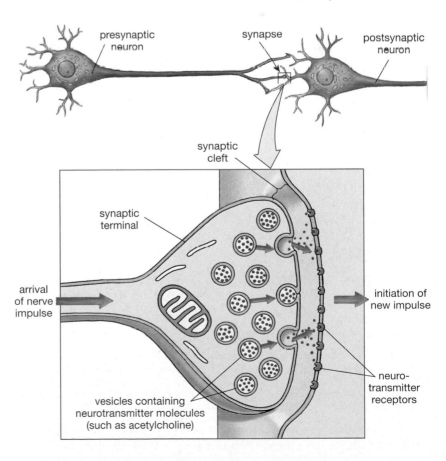

presynaptic neuron   synapse   postsynaptic neuron

synaptic cleft

synaptic terminal

arrival of nerve impulse

initiation of new impulse

neurotransmitter receptors

vesicles containing neurotransmitter molecules (such as acetylcholine)

at fighting depression in human beings. (The drugs Prozac, Zoloft, and Paxil are in this class.) Serotonin is a neurotransmitter found in the brain. Presynaptic cells have means of taking serotonin back into themselves once they have released it into the cleft. SSRIs work by reducing this "reuptake." The result is an increased amount of serotonin in the synaptic cleft—and perhaps a reduction in depression.

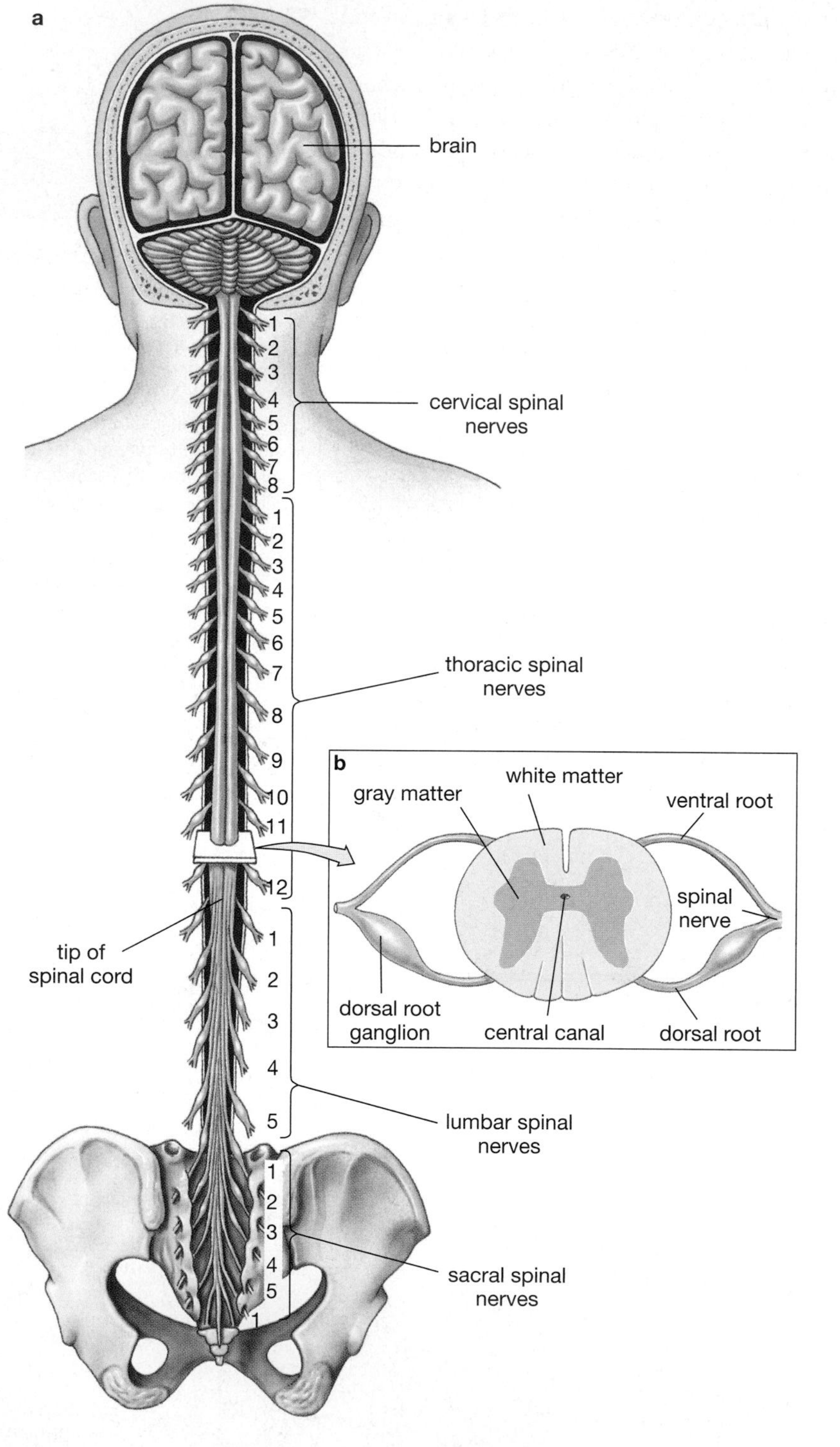

25.3 The Spinal Cord

Now that you've looked at how nerve impulses are transmitted at the cellular level, it's a good time to take a step back and look at the nervous system in its larger dimensions. Let's start by examining the basic structure and function of the spinal cord.

The spinal cord serves two major functions. First, it can act as a communication center on its own, receiving input from sensory neurons and directing motor activities in response, with no input from the brain. These are our so-called reflexes. Second, the spinal cord is the major pathway for sensory impulses going to and from the brain.

Extending from the base of the brain to an area just below the lowest rib, the spinal cord measures approximately 45 cm (18 inches) in length and about 14 mm or (0.5 inches) in width (see Figure 25.6a). The entire spinal cord consists of 31 segments, each connected to a pair of **spinal nerves**—one nerve for the left side of the body, one for the right. The spinal nerves can be grouped into several classes according to the region of the body they serve, as Figure 25.6 illustrates.

The most striking feature of a cross section of the spinal cord is a rough H- or butterfly-shaped area around the narrow central canal. This H is the gray matter of the spinal cord, composed mostly of the cell bodies of neurons and their supporting neuroglial cells. Surrounding the spinal cord's gray matter is its white matter. White matter, as you've seen, is largely myelin-coated neural *axons*, and this is the case with the spinal cord. The **central canal** of the spinal cord that you see in Figure 25.6b is a space filled with **cerebrospinal fluid**, which circulates not only in the spine but in the brain, supplying nutrients, hormones, and white blood cells and providing protection against jarring motion.

The Spinal Cord and the Processing of Information

Sensory input is, of course, coming *to* the spinal cord from various sensory receptors in the body, while motor commands must go *from* the spinal cord to various effectors. The motor neurons that are sending the motor

Figure 25.6
Structure of the Spinal Cord

a The numbers to the right of the spinal cord identify 31 pairs of spinal nerves, which can be grouped according to the body part they control. The eight cervical nerves control the head, neck, diaphragm and arms; the 12 thoracic nerves control the chest and abdominal muscles; the five lumbar nerves control the legs; and the five sacral nerves control the bladder, bowels, sexual organs, and feet.

b Cross section through the lower thoracic region of the spinal cord, showing the arrangement of gray and white matter, the central canal, the ventral roots, the dorsal roots, and the dorsal root ganglia.

commands lie in the gray matter of the spinal cord, as do the interneurons discussed earlier. The axons of these motor neurons leave the spinal cord through the ventral roots you can see in Figure 25.6b.

Now consider sensory neurons, which are responsible for transmitting the information that goes to the spinal cord. Their cell bodies lie *outside* the spinal cord in the **dorsal root ganglia** you can see in Figure 25.6b. (A **ganglion** is any collection of nerve cell bodies in the PNS.) The axons of these sensory neurons may extend all the way to distant sensory receptors in the body—say, to a touch receptor located in your arm. The receptor itself may be a dendrite that feeds into the axon, or it may be a specialized cell that communicates with the axon across a synapse. Either way, once stimulated, the sensory neuron axon transmits a message to the sensory neuron cell body in a dorsal root ganglion. This message then goes to the spinal cord, through the dorsal root you can see in Figure 25.6b. In sum, motor commands leave the spinal cord through a ventral root, while sensory information comes to the spinal cord through a dorsal root. Inside the spinal cord, sensory and motor neurons may link up directly (through synapses) or they may link up with interneurons. Note that both dorsal and ventral roots come together, like fibers being joined in a single cable, to form a given spinal nerve.

Quick, Unconscious Action: Reflexes

The sensory/motor division of labor is clearly reflected in simple body **reflexes**, which can be defined as automatic nervous-system responses, triggered by specific stimuli, that help us avoid danger or preserve a stable physical state. If you accidentally touch a hot stove, you automatically pull back your hand. No conscious thought of "Oh, I've touched a hot stove; I'd better pull my hand back," is necessary; the reaction is automatic. The neural wiring of a single reflex is called a reflex arc. A reflex arc begins at a sensory receptor, runs through the spinal cord, and ends at an effector, such as a muscle or gland. In the simplest reflex arc, a sensory neuron links directly to a motor neuron. Because only one synapse is involved in such a situation, simple reflexes control the most rapid motor responses of the nervous system. **Figure 25.7** shows one of the best-known examples, the knee jerk, or patellar reflex, in which a properly placed sharp rap on the knee produces a noticeable kick. Note that

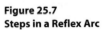

Tutorial 25.1.4
Steps in a Reflex Arc

Figure 25.7
Steps in a Reflex Arc

1. Stimulus (tapping) arrives and a receptor is activated.

2. The signal from the receptor reaches a sensory neuron cell body, in the dorsal root ganglion near the spinal cord; the signal moves to a sensory neuron/motor neuron synapse in the spinal cord.

3. Information processing of the event takes place, prompting a signal to be sent through the motor neuron.

4. The motor neuron signal stimulates the effector (the quadriceps muscle) to contract. Note that CNS processing for this reaction was handled entirely in the spinal cord; the brain was not involved.

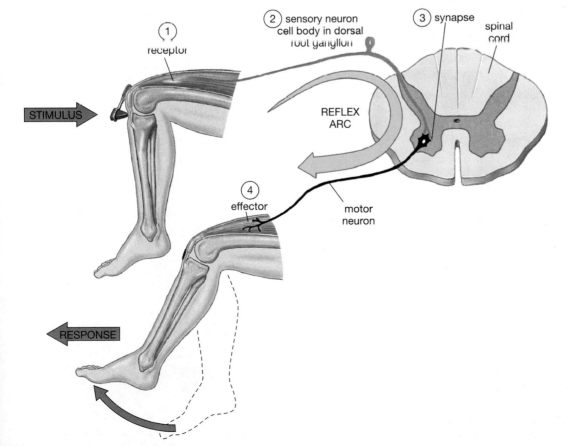

Figure 25.8
Involuntary Control of Bodily Functions
The autonomic nervous system has two divisions, the sympathetic and the parasympathetic, which exercise automatic control over the body's organs, generally in opposing ways. Note how the parasympathetic aids in digestion, while the sympathetic controls "fight or flight" responses, such as an increase in heart rate. Axons of the parasympathetic division emerge not only from the spinal cord, but from the brain as well.

this exemplifies what you saw earlier: A sensory receptor is stimulated, thus prompting a signal to move through the axon of a sensory neuron and then into the spinal cord. There the sensory neuron is linked via a synapse to a motor neuron, which issues a command that is carried out by an effector—in this case a skeletal muscle.

Many other reflexes may have at least one interneuron placed between the sensory receptor and the motor effector. Due to the larger number of synapses, these reflexes have a longer delay between a stimulus and response. However, they can produce far more complicated responses because the interneurons can control several different muscle groups simultaneously.

25.4 The Autonomic Nervous System

Effectors, such as muscles, respond to motor commands that come through the "outgoing" or efferent division of the PNS. But such commands can be either voluntary or involuntary. As noted earlier, there is a division of the PNS over which we have conscious control; it is called the **somatic nervous system**, and it is responsible for the voluntary control of skeletal muscle. The division of the PNS over which we do *not* have conscious control is called the **autonomic nervous system**, and it is very important; imagine having to consciously control your breathing. This system controls the involuntary regulation of smooth muscle, cardiac muscle, and glands.

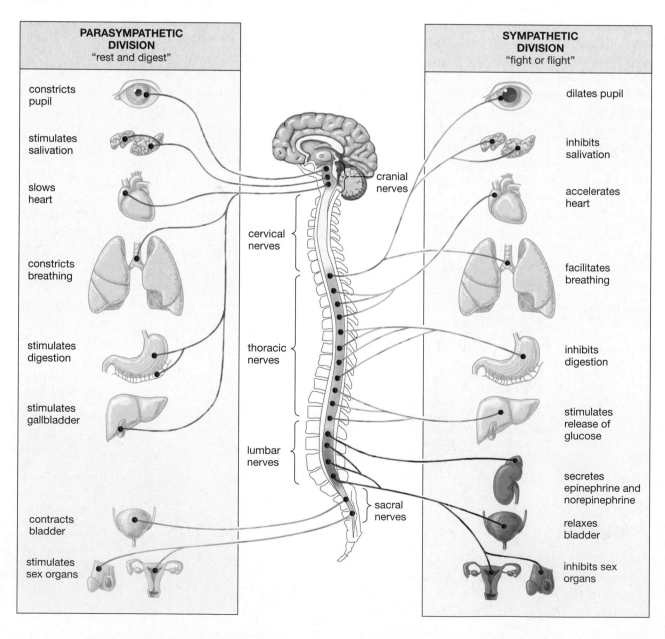

PARASYMPATHETIC DIVISION
"rest and digest"

constricts pupil
stimulates salivation
slows heart
constricts breathing
stimulates digestion
stimulates gallbladder
contracts bladder
stimulates sex organs

cranial nerves
cervical nerves
thoracic nerves
lumbar nerves
sacral nerves

SYMPATHETIC DIVISION
"fight or flight"

dilates pupil
inhibits salivation
accelerates heart
facilitates breathing
inhibits digestion
stimulates release of glucose
secretes epinephrine and norepinephrine
relaxes bladder
inhibits sex organs

The Sympathetic and Parasympathetic Divisions of the Autonomic Nervous System

The autonomic nervous system exercises its control through two "divisions," the sympathetic and parasympathetic. The **sympathetic division** is often called the *fight-or-flight* system, because it usually stimulates tissue metabolism, increases alertness, and generally prepares the body to deal with emergencies. The **parasympathetic division** is often regarded as the *rest-and-digest* system, because it conserves energy and promotes activities such as digestion. In **Figure 25.8**, you can see the effects these two divisions have on various bodily functions. Although some organs are innervated by one division or the other, most vital organs receive dual innervation, meaning instructions from both autonomic divisions. Where dual innervation exists, the two divisions often have opposing effects, keeping the body's stability mechanisms working in a dynamic balance.

25.5　The Human Brain

The adult human brain is far larger and more complex than the spinal cord. It contains almost 98 percent of the neural tissue in the body. An average adult brain weighs 1.4 kg (a little over 3 pounds), but brain size varies considerably between individuals. The brains of males are generally about 10 percent larger than those of females (because of the males' greater average body size), but there is no correlation between brain size and intelligence.

The brain consists of gray and white matter. **Gray matter** is made up of nerve cell bodies that exist as individual working groups called *nuclei*, and as a layer at the surface of the brain called the **cerebral cortex**. **White matter**, made up of bundles of axons, forms "tracts" that surround the nuclei and connect them with the cerebral cortex. Like the spinal cord and its spinal nerves, the brain can communicate directly with other body tissues and organs by means of cranial nerves that extend from the brain.

Six Major Regions of the Brain

There are six major regions in the adult brain: (1) cerebrum, (2) thalamus and hypothalamus, (3) midbrain, (4) pons, (5) cerebellum, and (6) medulla oblongata. **Figure 25.9** shows the locations of these brain regions and their general functions. Now let's take a look at each of them.

- The **cerebrum** is divided into large left and right **cerebral hemispheres**, the outer portion of which is the cerebral cortex. Conscious thought processes, sensations, intellectual functions, memory storage and retrieval, and complex motor patterns originate in the cerebral hemispheres.
- The **cerebellum** refines ongoing movements on the basis of sensory information and stored memories of previous movements. Its activities help us adjust postural muscles, maintain balance and equilibrium, and reproduce learned movements such as swinging a golf club and playing the piano.

In looking at the brain, if you take away the cerebrum and the cerebellum, what remains is the brain stem, whose structures are shown in Figure 25.9b. The brain stem contains important processing centers and relay stations for information passing to or from the cerebrum or cerebellum. The brain stem also controls vital functions such as breathing, and activities of the heart and digestive system. Here are its constituent parts:

Figure 25.9
The Brain

a Lateral view of the left surface of the brain, showing the cerebrum and cerebellum.

b The structures of the brain stem, and the endocrine system's pituitary gland. The medulla oblongata connects the brain with the spinal cord.

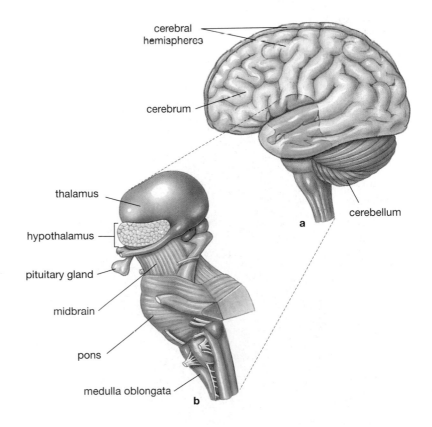

cerebral hemispheres

cerebrum

thalamus

hypothalamus

pituitary gland

midbrain

pons

medulla oblongata

cerebellum

a

b

- The **thalamus** contains relay and processing centers for incoming sensory information. Lying below it is the **hypothalamus**, which contains cell bodies involved with emotions, autonomic function, and hormone production. A narrow stalk connects the hypothalamus to the pituitary gland, which is part of the endocrine system. The hypothalamus is the primary link between the nervous and endocrine systems.
- The **midbrain** processes seeing and hearing information and generates involuntary motor responses related to muscle tone and posture. It also contains centers involved with maintaining consciousness.
- The term **pons** refers to a bridge, and the pons serves this function, connecting, with relay centers in it and axons that run through it, several other structures of the brain. The pons also contains cells involved with the involuntary control of breathing.
- One structure the pons is connected to is the **medulla oblongata**, the segment of the brain that is attached to the spinal cord. The medulla oblongata relays sensory information to the thalamus and brain-stem centers; it also contains major centers concerned with the regulation of autonomic functions, such as heart rate, blood pressure, respiration, and digestive activities.

The brain and spinal cord contain internal cavities filled with the cerebrospinal fluid noted in connection with the spinal cord. In the brain, this fluid provides cushioning for delicate neural structures and support—the brain essentially floats in the cerebrospinal fluid. Though the brain weighs about 1.4 kilograms (about 3 pounds) in air, it weighs only about 50 grams (about 2 ounces) when supported by the cerebrospinal fluid. As noted earlier, the cerebrospinal fluid also transports nutrients, chemical messengers, and waste products.

25.6 The Nervous System in Action: Our Sense of Vision

If you wanted to evoke the richness of life in as short a space as possible, you could do worse than to write down just five words: seeing, touching, hearing, smelling, and tasting. By themselves, these words seem capable of reminding us of how sensual our world is. What does it mean to be alive? For many people, a large part of the answer is: It means we can sense the world around us. Our sensory capabilities work through the nervous system you have just been reading about. What follows could be thought of as a detailed example of how the nervous system works in one area, that of perception.

The idea of "five senses" is so ingrained in us that it may come as a surprise to learn that we actually have more sensory capabilities than the famous five of vision, touch, smell, taste, and hearing. We have a sense of balance, for example, but it isn't one of these five senses. We have another little-recognized sensory capability that helps us do such things as dance, catch softballs, and simply sit up straight. To demonstrate this capability to yourself, extend one arm out, parallel to the ground, and close your eyes. Now touch your index finger to the tip of your nose. How were you able to do this? You have sensory neurons that monitor such things as the position of your joints and the tension of the tendons in your body. These neurons provide a sense, called *proprioception*, that in this instance helped guide your finger to your nose. (It sent out signals that said "on the right course" or "correction needed.") How do you know if you're sitting up straight? Proprioception is part of the answer.

**Figure 25.10
Anatomy of the Human Eye**

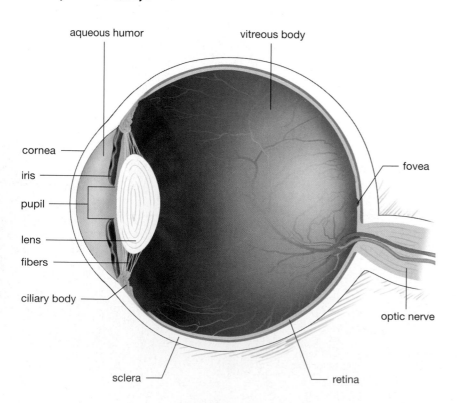

Taking these and other sensory capabilities into account, it becomes apparent that our "five senses" really amount to the senses we are consciously aware of. It is understandable that these senses get our attention, however, because they are the sources of such varied information about the world around us. Let's look now at the most-studied of these senses, vision, to see how this one complex sensory capability works in human beings.

Three Tasks for Vision

Our visual system has to accomplish three central tasks. First, it has to capture light from the outside world and focus this light at a very precise location—on a layer of tissue at the back of our eyes. Second, it has to take the light signals it receives and convert them into nervous-system signals—the signals you have come to know as action potentials. Third, it has to make sense of the visual information it has received. The most simple image has many elements to it: shape, orientation, color, and movement among them. The brain first has to register these elements *as* elements—as "elongated," "moving," and "green," for example. It then has to integrate these elements into a coherent message, which in this case might be "snake up ahead." Let's see how the visual system takes care of all its major tasks.

Task 1: Capturing and Focusing the Image

If you look at **Figure 25.10**, you can see the anatomy of part of our visual system, the eye and the optic nerve that extends from it. At the eye's front or anterior portion, you can see its **cornea**, a curved membrane through which light first passes upon entering the eye. The cornea actually is the transparent portion of the tissue called *sclera* that you can see lining the eye's perimeter. Over most of its area, the sclera is familiar to us as the "whites" of the eyes. Behind the cornea, you can see the **iris**: a colored structure of the eye, composed partly of smooth muscle and capable of contracting and relaxing. The iris has a central opening, called the **pupil**, which grows smaller or larger in response to contractions or relaxations of the iris muscles. Then, behind the pupil, you can see the **lens**: a transparent structure of the eye, lying behind the iris, that serves to focus incoming light. Note that the eye's lens is attached by slender fibers to something called ciliary bodies.

These bodies actually are rings of muscle whose contractions allow the lens to change its shape in ways you'll look at shortly. If you look at the back of the eye, inside the sclera, you can see a layer of tissue called the **retina**: an inner layer of tissue in the eye that contains cells that convert light signals into neural signals. The retina is the destination of the light that first enters through the cornea. Most of the volume of the eye is made up of a jelly-like material called the *vitreous body*. In between the lens and the cornea is a fluid called the *aqueous humor*.

So how do all these substances and structures work together in performing the visual system's first task of receiving and focusing light? Light reception by the eye is not surprising; the cornea and lens are transparent, after all, and the pupil is an opening. The focusing that these structures carry out is a little more complicated. If you look at **Figure 25.11a**, you can see what it means to have light focused properly on the retina: parallel rays of light need to converge on a small area of the retina. This is no different in principle

Figure 25.11
Focused and Unfocused
Sharp vision requires that incoming light rays converge just as they reach the retina.

a In normal vision, the eye's cornea and lens bring about this convergence; they refract the incoming light at an angle that matches the length of the eye.

b People who are farsighted have a mismatch between the length of their eye and the amount of refraction provided by their cornea and lens: Their eye is too short for the amount of refraction provided. Thus, light rays converge behind their retina. As a result, these people cannot see nearby objects clearly.

c People who are nearsighted have the opposite problem: Their eye is too long for the amount of refraction provided, with the result that light rays converge in front of their retina. These people cannot see distant objects clearly. Both nearsighted and farsighted vision can be corrected.

a Normal vision
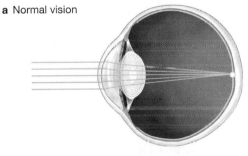
light rays converge on the retina

b Farsighted vision
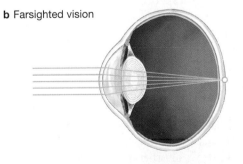
light rays converge behind the retina

c Nearsighted vision
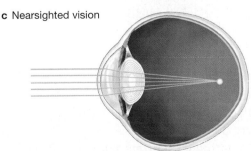
light rays converge in front of the retina

than what happens with a camera. We take a picture of an object (say a statue) that ends up as a small image on a frame of film. Just so, objects we view with our eyes end up as *very* small images on our retinas. (How small? At fractions of a millimeter, some of them are microscopic.) To go from large objects to small images of them, the rays of light reflected by the objects have to converge onto a small space on the back of the eye.

For this convergence to take place, parallel rays of light, entering from a distance, need to be bent or *refracted*, as you can see in Figure 25.11. We've all seen refraction at work when we look at a spoon sitting in a glass that's half-filled with water. The spoon seems to bend abruptly just as it enters the water. This is so because *light* can bend when it passes from one transparent "medium" into another. Light coming into the glass gets refracted when it passes from the medium of air into the medium of water. In the case of our eyes, light gets refracted first when it passes from air into the cornea, and again when it passes from the cornea into the lens. This light refraction creates an *inverted* image on the retina. A friend who we see waving to us ends up as a tiny, *upside-down* image of a friend on our retinas. Our brains then have to correct this inversion.

If you look at Figures 25.11b and c, you can see what happens when there is a mismatch between the length of the eye and the amount of refraction provided by the cornea and the lens. The light ends up being focused not on the retina, but on an area behind it or in front of it. If the eyeball is too short, the rays converge after they get to the retina, making the person farsighted or unable to see nearby objects clearly. If the eyeball is too long, rays of light converge before they get to the retina, making the person nearsighted or unable to see distant objects clearly. Different kinds of glasses and contact lenses can correct either problem.

There is another visual affliction, called *presbyopia*, that generally comes with middle age. Here the culprit is not the length of the eyeball, but instead the aging of the eye's lens. Note again in Figure 25.10a how the lens is attached to the muscular ciliary bodies through sets of slender fibers. It turns out that for us to focus properly on objects that are close to us, the lens needs to go from flatter to rounder—from more plate-shaped to more ball-shaped. This feat is accomplished

by a relaxation of the fibers in response to a contraction of the ciliary muscles. The problem is that, as we get older, the lens hardens and therefore becomes less able to "fatten up" when the fibers relax. The result is an inability to focus on nearby objects—and our first pair of reading glasses.

One final thing about the eye's outer structures. Because the iris is a muscular structure, it can serve the well-known function of controlling the amount of light that comes into the eye. Outside on a sunny day, one set of involuntary muscles in the iris contracts, thus reducing the size of the pupil and the amount of light that enters the eye; in a darkened movie theater another set of iris muscles contracts, allowing the pupil to dilate or open up, thus letting more light in.

Task 2: Converting a Light Signal into a Neural Signal

With light captured and properly aimed, we now have an image sharply focused on the retina at the back of the eye. How does the eye turn the light that falls on the retina into a nervous system signal that our brain processes? If you look at **Figure 25.12b,** you can see what retinal tissue looks like when greatly magnified. Light comes through the cornea, back through the eye, and then passes through the two layers of cells you see at the front of the retina, ganglion and connecting cells. It then arrives at the two kinds of cells you can see at the back of the retina, rods and cones. These are the cells in which information contained in light is converted into information expressed as neural signals. Rods and cones collectively are known as *photoreceptors* because they are sensitive to photons, which are individual packets of light.

Rods and cones divide up the work of converting light energy into neural signals. The numerous rods (about 120 million of them) predominate around the periphery of the retina, while the less numerous cones (about 7 million of them) are concentrated at the center of the retina. If you look at Figure 25.10, you can see a small structure called the fovea, at the back of the eye, that represents an especially dense concentration of cones. The fovea has been called the most important square millimeter of tissue in the body, because it contains the cones that provide our sharpest vision. We employ the foveal cones whenever we need to see very fine detail in an object. Thread a needle and you are forming an image of the needle's eye directly on the foveal cones.

Cones require more light to function than do rods. Cones work well whenever the light source is bright—as with sunlight or a neon sign—but they do not function in dim-light situations, such as might exist on a residential street after dark. It is rods, not cones, that provide our dim-light vision, because rods are more sensitive to light. Anyone who is out on a darkened night can attest, however, that our night vision is *black-and-white* vision. This is so because rods do not register differences in the color of light. Only cones do this, but cones are not active in dim light. Sit in your yard by day and you have color vision provided by cones; stay in the yard as day turns into night and, at a certain point, rods will take over as your sole working photoreceptors, giving you black-and-white vision. The rod-cone difference explains why you can perceive dim stars better with your peripheral vision than by looking at them directly. Rods predominate in the periphery of the retina, and only rods respond to the dim light of such stars.

With all this in mind, you can appreciate formal definitions of rods and cones. **Rods** are photoreceptors that function in low-light situations, but that do not provide color vision. **Cones** are photoreceptors that respond best to bright light, that allow us to see minute detail in objects, and that provide color vision.

Our eyes contain three different kinds of cones, commonly referred to as red, blue, or green cones because each kind responds most strongly to red, green, or blue light. Any object that we look at produces a specific pattern of activity across all three kinds of cones—a fuchsia object creates one pattern of activity, a teal object another pattern, and so forth. So varied are these patterns that humans with normal vision can discriminate among seven million different color shades.

One final thing about rods, cones, and light reception. Note that there are no rods or cones—indeed, there is no retina—at the back of the eye where the optic nerve begins. The outcome of this is simple: No rods or cones, no perception of images. The area where each optic nerve begins is your "blind spot," which

Figure 25.12
Turning Light into Neural Signals
It is photoreceptor cells called rods and cones that convert incoming light into neural signals.

a Incoming light moves to the retina at the back of the eye. The blow-up of the retina shows its three layers of cells: ganglion cells, connecting cells and rods and cones. Light reaches the rods and cones and stimulates a molecular change in them that results in a signal being sent back out the other direction—through the connecting cells and then the ganglion cells. With this series of events, a light signal has been converted to a neural signal that will go, via ganglion cell axons, to the optic nerve. The anatomy of both rods and cones can be seen at right. Light is absorbed by millions of pigment molecules embedded in the flattened membranes of rods and cones. Only the cones have pigments that respond to color.

b A color-enhanced micrograph of rods and cones. The light blue structure is a cone.

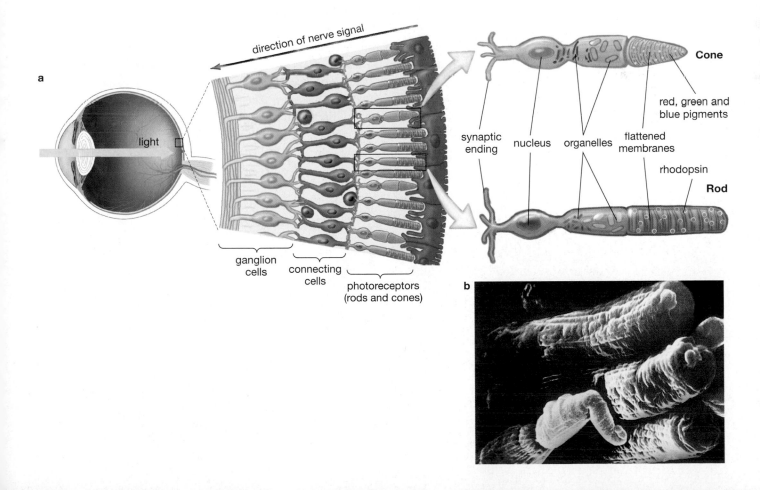

you can locate by closing one eye and slowly moving a narrow, vertical object, such as a pencil, horizontally across your visual field.

Rods and cones function through *pigments*: chemical compounds that absorb light. Rods have a single type of pigment, called *rhodopsin*, while each cone has one of three types of pigments: red, green, or blue. If you look at Figure 25.12, you can see an illustration of rods and cones accompanied by a micrograph of them. Note that all the usual structures we associate with cells—the nucleus, various organelles—lie at the anterior portion of rods and cones (toward the front of the eye). However, the posterior portions of these cells (toward the back of the eye) have extensions that house hundreds of thousands of stacked-up, flattened membranes. Embedded within these membranes are millions of pigment molecules that do the actual light absorption.

Now let's see how this works. Remember that light has come through the pupil, back through the eye, and past two outer layers of cells in the retina; it now lands, say, on rhodopsin pigments embedded within one of the stacked membranes in a rod. Stimulation of these rhodopsin molecules by the light causes a molecular change that sets into motion a cascade of chemical reactions. These reactions culminate in the rod sending a signal back *out* through the two layers of cells anterior to it. The rods pass signals to connecting cells and connecting cells, in turn, pass these signals along to the ganglion cells. These signals say, in effect, "light stimulation here, in this part of the retina." With

this, the visual system has converted light into a neural signal, coursing through the nervous system. Note, in Figure 25.12, that both rods and cones have synaptic terminals. The nerve signal coming from the photoreceptors works like any other nerve signal—by means of neurotransmitters shuttling between neurons.

Having come from the photoreceptors, the nerve impulses then are channeled to the brain. How? The ganglion cells that received the photoreceptor input have axons that go to the same place. These axons are like tributaries coming together to form a river—only this river is the optic nerve, composed of about a million axons stemming from ganglion cells. Because we have two eyes, we thus have a *pair* of million-fiber optic nerves carrying visual information from the eyes into the brain, where most of our visual system's processing will take place.

Task 3: Making Sense of Visual Signals

If you look at **Figure 25.13**, you can see the large-scale path for most of the visual information that comes into the brain. Signals from the eye are routed first to a kind of relay station in the brain's thalamus called the *lateral geniculate nucleus*. There, the nerve signals make synaptic connections and are taken, through a new set of axons, to an area at the back of the brain called the **primary visual cortex.** This is the main site of our visual processing.

The activity that goes on in the primary visual cortex and adjacent structures is studied as much by psychologists as by biologists. This visual processing is far too complex to describe here, but a central lesson from this complexity is vital to an understanding of vision. It is that the human visual system does not work like a camera; it does not passively record bits of light and register a collection of these bits as images. Rather, the brain *constructs* images as much as it records them. You've already seen a couple of examples of this construction. Remember how the eye's cornea and lens invert the images we perceive, such that the image that strikes the retina is upside down compared to the real-world image? If the brain were passively recording nerve impulses based on light, we would see the world upside down, but we don't because the brain turns things right side up. Second, recall that each eye has a blind spot. But do we

Figure 25.13
Visual Pathway
An object that we perceive stimulates rods and cones in the retina, producing a neural signal that goes through the optic nerve to a relay station, called the lateral geniculate nucleus, that lies in the brain's thalamus. The signal is then sent to an area at the back of the brain called the primary visual cortex—our primary site of visual processing.

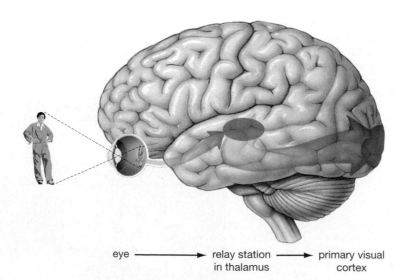

eye ⟶ relay station ⟶ primary visual
in thalamus   cortex

ever perceive "holes" in our vision in everyday life? We do not, because the brain "fills in" the perceptual gaps that exist.

Beyond these examples, scientists have created scores of so-called "subjective figures" that catch the brain in the act of constructing visual images. **Figure 25.14** is one of these subjective figures. In it, you see a V, formed out of bumps and a dot-like indentation. Now turn the book upside down and look at the figure. The bumps have become indentations and vice versa. Why? Note that when the book is right side up, most dots in the V are shaded in such a way that light seems to be coming from above them—there are "shadows" on the lower part of the dots. Light generally comes from above us, whether from a lamp or from the Sun. This is so common that evolution has produced a genetically based "rule" in our visual systems that says: "When shadows are in the lower part of an object, that object is raised above a surface."

Our visual system evolved to make sense of the world around us. Ancestors of ours who could perceive, say, a snake more *quickly* than other indivduals survived longer and thus left more offspring. What mattered in evolution was not whether we perceived objects "correctly." What mattered was whether we perceived objects in ways that helped us survive and reproduce. Thus, we can see the value of the many genetically based visual rules that exist: They provide a series of "best bets" for quickly telling us what we are looking at. (If an object's lower half is shaded, the best bet is that it's raised above a surface.) The idea that our visual system is often creating perceptions, rather than simply recording them, may seem strange, but that's the reality of visual perception.

25.7 The Endocrine System: Hormones and How They Work

Let's now shift the focus from the nervous system to the endocrine system. Though the two systems are separate, they work closely together, as you'll see; and they are alike in function in that both are concerned with coordination and control in the body.

Endocrine-system cells secrete chemicals called **hormones** that travel mostly through the circulatory system and affect the activities of *other* cells. Hormonal production and

secretion take place to a significant extent within specialized organs called **endocrine glands**: glands that release their materials directly into the bloodstream or into surrounding tissues, without using ducts. The major endocrine glands are shown in **Figure 25.15**. Some of these glands, such as the pituitary gland, secrete nothing but hormones;

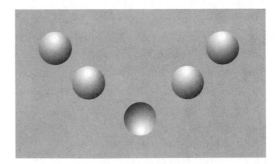

Figure 25.14
Seeing Is Believing
Look at the figure and then turn the book upside down and look at it again. With the shift, dots that once appeared to be raised above the surface of the drawing now appear to be indented, and vice versa. Our visual system has a rule that causes us to perceive objects that are shaded at the bottom as protruding from a surface, while objects that are shaded at the top appear to be indented. This is but one example of how our visual system creates visual images, rather than simply recording them.

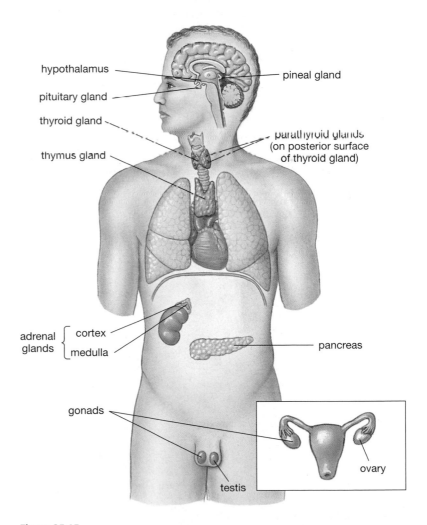

Figure 25.15
The Major Hormone-Secreting Glands of the Body
The hypothalamus is part of the nervous system, but releases hormones and plays a central role in endocrine regulation.

others, such as the pancreas, have other functions in addition to hormone secretion. Not all hormone-producing structures are glands. For example, the heart, kidneys, stomach, liver, small intestine, placenta, and adipose (fatty) tissue also secrete hormones.

Types of Hormones

Most hormones are composed of amino acids, the building blocks of proteins. These amino-acid-based hormones range in size from single amino acid molecules to long chains that form proteins. **Steroid hormones**, on the other hand, are lipid-based hormones, built from cholesterol molecules. They are released by the reproductive organs and the outer layers (or cortex) of the adrenal glands, which sit atop the kidneys. Another group of lipid-based hormones is called *prostaglandins*. Produced from fatty acids by most tissues of the body, these hormones coordinate cellular activities and affect body processes, such as blood clotting, that take place in extracellular fluids. Unlike most hormones, the prostaglandins do not travel throughout the body; thus they affect only nearby cells. For this reason, they are sometimes called *local hormones*.

Tutorial 25.2.1
Hormones and Target Cells

Actions of Hormones

You've noted that hormones are substances that, having been released by cells in one location, then travel short or long distances through the body (generally through the circulatory system) to affect the activities of cells in *other* locations. Thus, each hormone is said to have specific **target cells**. Most target cells have receptor molecules protruding from their outer membranes; it is these receptors to which specific amino-acid-based hormones attach, as you can see in **Figure 25.16**. (The steroid-based hormones work by passing through a cell's plasma membrane, after which they bind with a receptor protein inside the cell.) Because a given *type* of target cell may occur in multiple locations throughout the body, a single hormone can simultaneously affect the activities of multiple tissues and organs.

As noted in Chapter 3, proteins have a great diversity of functions in living things, and it is this diversity that enables living things to be so complex. Structural proteins determine the general shape and internal structure of a cell, and the special proteins called enzymes speed up cellular activities, in effect enabling the cell's chemical processes to go forward. Hormones alter the workings of a target cell by changing the identities, activities, or quantities of its important enzymes and structural proteins. As an example, the steroid hormone testosterone—found in differing amounts in males and females—stimulates the production of enzymes and proteins in skeletal muscle fibers, thus increasing muscle size and strength.

25.8 How Is Hormone Secretion Controlled?

Hormones are major players in a general process through which the body maintains a stable state. Like people in a large city, the cells, tissues, and organ systems of the body exist together in a shared environment. And just as city dwellers would suffer if the air in their environment became highly polluted or too hot or cold, so cells would be harmed if the *fluid* environment in which they exist became too hot or cold or acidic. The body flourishes in a stable environment, and it has developed an elaborate

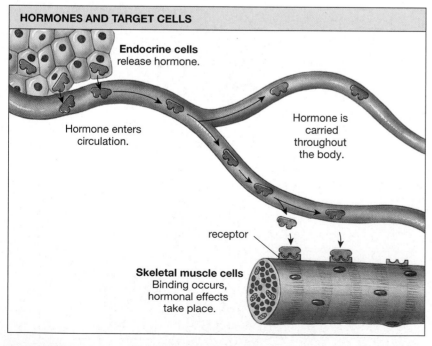

HORMONES AND TARGET CELLS

Endocrine cells
release hormone.

Hormone enters
circulation.

Hormone is
carried
throughout
the body.

receptor

Skeletal muscle cells
Binding occurs,
hormonal effects
take place.

Figure 25.16
Hormones and Their Target Cells
For a hormone to affect a target cell, that cell must have receptors that can bind the hormone. The binding of hormone to receptor then initiates a change in the target cell's activity. The figure shows an amino-acid-based hormone that affects skeletal muscle tissue. The hormone would not affect, for example, nervous tissue, because only muscle tissue has the appropriate receptors to respond to this hormone.

set of mechanisms to maintain stability. **Homeostasis**, derived from the words *homeo* ('unchanging') and *stasis* ('standing'), refers to an organism's tendency to maintain a relatively stable internal environment in the midst of a constantly changing *external* environment. And it turns out that a single process is almost entirely responsible for maintaining homeostasis in the body.

Negative Feedback

Most people are aware of how a home heating system works. Falling temperature causes a thermostat to turn on a furnace. To look at this another way, there was a **stimulus** (cold air) that brought about a response (furnace operation). The *product* of this response is hot air. When enough of this hot air circulated to raise the temperature, the thermostat sensed this and shut the furnace down. Now let's think of your body, which has—as an example—a certain amount of calcium circulating through it in the bloodstream. When levels of calcium in the blood fall too low (stimulus), some glands in your body, the parathyroid glands, sense this and secrete a hormone that causes your bones to release calcium that they have stored up (response; see **Figure 25.17**). The product of *this* stimulus-response chain is released calcium, and when enough of it is circulating, the parathyroid glands sense this and stop releasing their calcium-liberating hormone. Thus, with both the thermostat and the parathyroid glands, their responses to a stimulus bring about a decrease in their activity. In other words, their responses feed back on their activity in a negative way—they reduce it. This is **negative feedback**, which can be defined as a process in which the elements that bring about a response have their activity reduced by that response. The calcium example is only one of the countless negative feedback loops that exist in the body.

Positive Feedback

Far less frequently, the body employs **positive feedback**, in which the initial response *magnifies* the stimulus that gave rise to it. For example, the primary stimulus that begins labor and delivery in childbirth is the stretching of the womb, or uterus, by the growing fetus. This stretching stimulates a control center in the mother's brain,

and the result is the release of a hormone that brings about muscle contractions in the wall of the uterus. These contractions begin moving the fetus toward the birth canal. The product of the stimulus is thus movement, but this product causes *more* uterine stretching, which brings about additional stimulation of the control center. This kind of cycle, a positive feedback loop, can be broken only by some external force or process—in this instance, the delivery of the newborn infant. Is it obvious why, in physiology, negative feedback is so much

Tutorial 25.2.3
Negative Feedback and Calcium Levels

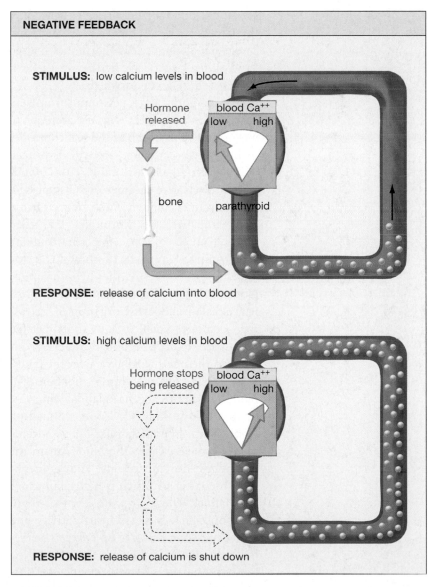

Figure 25.17
Calcium Levels: An Example of Negative Feedback
Calcium levels in the bloodstream are regulated in part by the parathyroid glands. When these glands sense that calcium levels are too low (stimulus in the upper drawing), they secrete a hormone that causes the bones to release calcium. In the lower drawing, the stimulus is high calcium levels, once again sensed by the parathyroid glands. These increased calcium levels cause the parathyroid glands to halt their production of the calcium-liberating hormone.

more common than positive feedback? Positive feedback brings about instability, which rarely benefits organisms; negative feedback supports homeostasis, which almost always benefits organisms.

A Hierarchy of Hormonal Control: The Hypothalamus

Though negative feedback is the ultimate control mechanism in the release of hormones, it's accurate to view a good part of the endocrine system as a hierarchy that has a structure in the brain, called the *hypothalamus*, sitting on top. In this system, the hypothalamus releases hormones that control a "master" gland, called the *pituitary gland*; the pituitary then releases hormones that control several other endocrine glands. These glands then release hormones that go to work on target cells, be they in the kidneys, the muscles, or elsewhere.

The hypothalamus gets input from several sources, which is why it's accurate to view negative feedback as the real controller of hormonal secretion. Sensory nerves feed into the hypothalamus, and it senses bloodstream levels of a number of substances in the body. In response to these inputs, the hypothalamus releases hormones that then control much of the endocrine system's functioning. **Figure 25.18** shows you the systems governed by the hypothalamus. Note all that the hypothalamus does. First, it acts as an endocrine organ itself, releasing hormones you'll look at shortly called *ADH* and *oxytocin*. Second, it exercises control, via the nervous system, over the endocrine cells of the inner portion (or medulla) of the adrenal glands, which sit on top of the kidneys. Upon stimulation from the hypothalamus, the adrenal medulla releases into the bloodstream the hormones norepinephrine and epinephrine, the latter of which is better known as adrenaline. These are the fight-or-flight hormones that increase heart activity and blood pressure, and free up energy reserves for immediate use. Beyond these activities, the hypothalamus secretes regulatory hormones. These are the special hormones that regulate the endocrine secretions in the pituitary gland—specifically the anterior pituitary gland. There are two classes of regulatory hormones that come from the hypothalamus: releasing hormones (RH), which *stimulate* the production of pituitary hormones; and inhibiting hormones (IH), which *prevent* the production and secretion of pituitary hormones.

The Pituitary Gland Link

The pituitary gland itself is a small, oval structure that is suspended from the brain's hypothalamus by a slender stalk (see Figure 25.9). It is divided into two distinct regions, the anterior and posterior regions, which differ greatly in function.

The pituitary secretes eight different hormones. Two of these are secreted by the **posterior pituitary**: an endocrine gland that receives its hormones directly from the hypothalamus and then stores these hormones and later releases them. The other six are released by the **anterior pituitary**: an endocrine gland that releases two hormones that work directly on target cells and four other hormones that regulate the production of hormones by other endocrine glands. In sum, the anterior pituitary synthesizes its *own* hormones, and releases them under control of the hypothalamus, while the posterior pituitary is simply a storage-and-release site for hormones synthesized by the hypothalamus. Figure 25.18 shows the hormones released by the pituitary gland and the target organs for them.

The Anterior Pituitary Gland

The endocrine cells of the anterior pituitary are surrounded by a network of the tiny blood vessels called *capillaries*. Move up to the hypothalamus, and it too has a capillary bed in its lower portion. Running between these two capillary beds are a small number of blood vessels. Here's how the hypothalamus-pituitary system works. Nerve cells in the hypothalamus synthesize hormones—in vanishingly small amounts—that travel down axons and are released into the hypothalamic capillaries; from there they travel down the linking blood vessels to the anterior pituitary capillaries, and then move into the pituitary gland itself. In the pituitary these hormones—sometimes called *releasing factors*—stimulate pituitary cells to produce hormones. Various *cell types* in the pituitary, each synthesizing a different hormone, are responsible for the number of hormones that this master gland produces. These hormones then circulate out into the

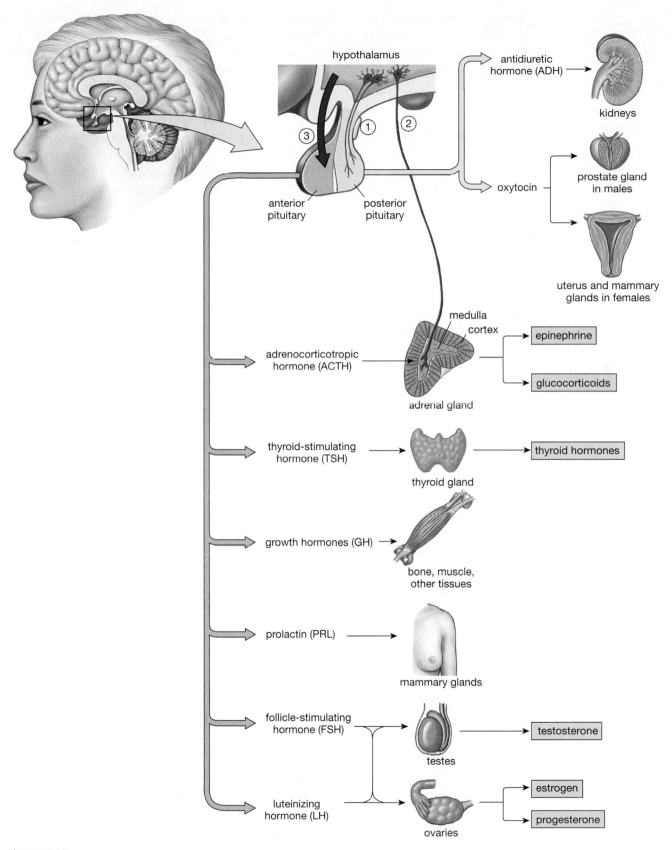

Figure 25.18
Pivotal Player in Hormone Secretion
The structure of the brain called the hypothalamus directs much of the body's endocrine activity.

1. It secretes antidiuretic hormone (ADH) and oxytocin, which are released by the posterior pituitary gland.

2. It exercises direct control over the release of epinephrine and norepinephrine from the medulla of the adrenal glands.

3. It releases regulatory hormones that either stimulate or prevent the production of several hormones by the anterior pituitary gland. The functions of the body's various hormones are summarized in Table 25.1.

bloodstream and head toward their target cells. Here are the six hormones that the anterior pituitary releases:

1. **Thyroid-stimulating hormone (TSH)**, which targets the thyroid gland and triggers the release of thyroid hormones.

2. **Adrenocorticotropic hormone (ACTH)**, which stimulates the release of steroid hormones by the outer cells of the adrenal gland (the adrenal cortex). In particular, ACTH stimulates the cells producing hormones called glucocorticoids.

3. **Follicle-stimulating hormone (FSH)** promotes egg development in women and stimulates the secretion of **estrogens**, steroid hormones produced by cells of the ovary. In men, FSH production supports sperm production and testosterone secretion in the testes.

4. **Luteinizing hormone (LH)** induces ovulation (egg release) in women and promotes the ovarian secretion of estrogens and the **progestins** (such as progesterone) that prepares the body for possible pregnancy. FSH and LH are called gonadotropins, because they regulate the activities of the male and female gonads or sex organs. In males, LH stimulates testosterone production.

5. **Prolactin**, which means "before milk," stimulates the development of the mammary glands and their production of milk in the female and stimulates cell growth in both sexes.

6. **Growth hormone**: Human growth hormone (hGH), or somatotropin, stimulates overall body growth through cell growth in a two-step linkage. Human growth hormone stimulates the production in the liver of another hormone, somatomedin, which directly spurs on cell growth. Human growth hormone has its greatest effects on muscular and skeletal development, especially in children, via the actions of sonatomedin, which is often referred to as IGF (for insulin-like growth factor).

The Posterior Pituitary Gland

The hormones the posterior pituitary stores and releases are produced by two different groups of nerve cells within the hypothalamus.

One group manufactures antidiuretic hormone (ADH), while the other produces oxytocin. These hormones are transported within nerve fibers that extend to the posterior pituitary, which, in turn, secretes the hormones into blood capillaries.

Antidiuretic hormone (ADH) is released when there is a rise in the concentration of solutes in the blood or a fall in blood volume or pressure. ADH acts to decrease the amount of water lost at the kidneys. It would most definitely be released if we had to make a long, hot walk with no water.

In women, **oxytocin** (a name that means "quick childbirth") stimulates smooth muscle cells in the uterus and special cells of the mammary glands. The stimulation of uterine muscles by oxytocin helps maintain and complete normal labor and childbirth. After the infant is born, oxytocin triggers the release of milk from the mother's breasts. In the male, oxytocin stimulates the smooth muscle contraction in the walls of the prostate gland. This action may be important prior to ejaculation. **Table 25.1** provides a summary of the body's most important hormones and their effects.

25.9 The Immune System: Defending the Body from Invaders

It's time to turn now to the third organ system reviewed in this chapter, the immune system. It may go without saying that the human body is constantly under attack by microbes. There are viruses, bacteria, fungi, and single-celled life-forms such as the *Giardia* that is so well known for contaminating the water in outdoor areas. The human body has multiple means of defending itself against such invaders. (The military metaphor is irresistible when talking about the immune system because the actions it's involved in are so much like war.)

Two Basic Types of Defense: Nonspecific and Specific

The body's defensive strategies fall into two basic types. First, there are general nonspecific defenses that do not discriminate between one invader and the next. These provide the body with a defensive capability known as *nonspecific resistance*. The body's second main strategy, its specific defenses,

Table 25.1
Hormones of the Endocrine System: Their Sources and Effects

Gland/Hormone	Effects
Hypothalamus	
Releasing hormones	Stimulate hormone production in anterior pituitary
Inhibiting hormones	Prevent hormone production in anterior pituitary
Anterior Pituitary	
Thyroid-stimulating hormone (TSH)	Triggers release of thyroid hormones
Adrenocorticotropic hormone (ACTH)	Stimulates adrenal cortex cells to secrete glucocorticoids
Follicle-stimulating hormone (FSH)	Female: promotes egg development; stimulates ovaries to produce estrogen Male: promotes sperm production
Luteinizing hormone (LH)	Female: produces ovulation (egg release); stimulates ovaries to produce estrogen and progesterone Male: stimulates testes to produce androgens (e.g., testosterone)
Prolactin (PRL)	Stimulates mammary gland development and production of milk
Growth hormone (GH)	Stimulates cell growth and division
Posterior Pituitary	
Antidiuretic hormone (ADH)	Reduces water loss at kidneys
Oxytocin	Stimulates contraction of smooth muscles of uterus and release of milk in females; stimulates prostate gland contraction in males
Thyroid	
Thyroxine	Stimulates general rate of body metabolism
Calcitonin	Reduces calcium ion levels in blood
Parathyroid	
Parathyroid hormone (PTH)	Increases calcium ion levels in the blood
Thymus	
Thymosins	Stimulate development of white blood cells (lymphocytes) in early life
Adrenal Cortex	
Glucocorticoids	Stimulate glucose synthesis and storage (as glycogen) by liver
Mineralocorticoids	Cause the kidneys to retain sodium ions and water and excrete potassium ions
Androgens	Produced in both sexes, but functions uncertain
Adrenal Medulla	
Epinephrine (E)	Also known as adrenaline; stimulates use of glucose and glycogen and release of lipids by adipose tissue; increases heart rate and blood pressure
Norepinephrine (NE)	Also known as noradrenaline; effects similar to epinephrine
Pancreas	
Insulin	Decreases glucose levels in blood
Glucagon	Increases glucose levels in blood
Testes	
Testosterone	Promotes production of sperm and development of male sex characteristics
Ovaries	
Estrogens	Support egg development, growth of uterine lining, and development of female sex characteristics
Progesterones	Prepare uterus for arrival of developing embryo and support of further embryonic development
Pineal Gland	
Melatonin	Delays sexual maturation; establishes day/night cycle

provides protection against *particular* invaders. A specific defense may protect against infection by one type of bacteria, but ignore other bacteria and viruses. Specific defenses are dependent upon the immune system's ability to develop a kind of memory regarding invaders it has faced. What these defenses produce is a state of long-lasting protection known as specific resistance, which is better known as **immunity**.

Before going into details, here are a couple of terms that you'll see repeatedly in this review of the immune system. One is the word **antigen**, which is any foreign substance that elicits an immune-system response. A bacterial cell is, of course, a foreign substance; but it is likely to be certain *proteins* on the surface of the bacterial cell that set off the human immune response. In general, antigens are the molecules that are part of a foreign body—generally proteins or polypeptide chains. Thus, a single invader could "present" more than one antigen to the immune system.

Also note that the immune system is reliant on several types of cells and proteins to get its work done. Among cells, several varieties of **white blood cell** are critical to immune function. You'll be hearing the terms *T cell, B cell, neutrophil*, and *macrophage*, among others. All these are different varieties of white blood cell. **Table 25.2** presents some of the cells and proteins involved in the body's defense.

25.10 Nonspecific Defenses of the Immune System

In dividing the immune system into its nonspecific and specific defense functions, it makes sense to start with the nonspecific variety, because it is the body's first line of defense. Here's a list of the body's nonspecific defenses:

- Physical barriers
- Phagocytic cells
- Immunological surveillance
- Interferons
- Complement
- Inflammation
- Fever

Nonspecific defenses generally deny entrance to, or limit the spread of, microorganisms or the harmful substances—called toxins—that they can secrete.

Table 25.2
Cells and Proteins Active in the Human Immune System

White Blood Cells	Other Cells Involved in Defense	Proteins Involved in Defense

• Phagocytes (cells that can ingest other cells)
—Microphages ("small eaters")
  Neutrophils
  Eosinophils
—Macrophages ("large eaters"), which can become antigen-presenting cells (APCs)
• Lymphocytes
—Natural killer cells
—B cells, which specialize into:
  Plasma cells
  Memory B cells
—T cells, which specialize into:
  Helper T (CD4) cells
  Cytotoxic (killer) T cells
  Suppressor T cells
	Mast cells (connective tissue cells that release histamine and heparin)	• Lysozymes (destructive enzymes found in bodily fluids)
		• Complement proteins
		• Antibodies (receptors on surface of B cells; later released by plasma cells derived from B cells)
		• Interferons (released by immune cells and infected cells; interfere with viruses; stimulate other parts of the immune system)
		• Interleukins
		—Interleukin-1 (released by antigen-presenting cells; helps activate helper T cells)
		—Interleukin-2 (released by helper T cells; spurs activation of B cells and cytotoxic T cells)

Physical Barriers

To cause trouble, a disease-causing organism—known as a **pathogen**—must enter the body tissues; and that means crossing an *epithelium*, or tissue that covers an exposed body surface. The epithelial covering of the skin, described in Chapter 24, has multiple layers, a waterproof keratin coating, and a network of tight seams that lock adjacent cells together. These specializations create a very effective barrier that protects underlying tissues.

The exterior surface of the body is also protected by hairs and glandular secretions that provide some protection against mechanical scraping (or "abrasion"), especially on the scalp. Secretions from sebaceous and sweat glands flush the surface of the skin, washing away microorganisms and chemical agents. Some secretions also contain bacteria-killing chemicals and destructive enzymes, called *lysozymes*, which can be found in tears and saliva.

The epithelial tissue lining the digestive, respiratory, urinary, and reproductive tracts is more delicate; but it is equally well defended. Mucus bathes most surfaces of the digestive tract, and the stomach contains a powerful acid that can destroy many potential pathogens. Mucus moves across the lining of the respiratory tract, urine flushes the urinary passageways, and glandular secretions do the same for the reproductive tract.

Phagocytes

As noted earlier, white blood cells play an important role in the immune system. Some of these white blood cells are **phagocytes**, meaning cells that can ingest other cells, parts of cells, or other particles. In body tissues, phagocytes remove not only parts of the body's own cells that have become cellular debris but also pathogens, by means of phagocytosis or "cell eating." There are two general classes of phagocytic cells. **Microphages**, or "small eaters," are the white blood cells called *neutrophils* and *eosinophils*. Though normally found in the circulating blood, these phagocytic cells can leave the bloodstream and enter tissues subjected to injury or infection. **Macrophages**, or "large eaters," are large cells derived from white blood cells called *monocytes*. Almost every tissue in the body shelters resident or visiting macrophages.

Immunological Surveillance

The immune system generally ignores normal cells in the body's tissues, but attacks and destroys abnormal cells. The constant monitoring of normal tissues is called immunological surveillance. This surveillance primarily involves a type of cell known as a **natural killer (NK) cell**, a special variety of lymphocyte, which is a variety of white blood cell. NK cells are sensitive to the presence of antigens on the body's own cells. These antigens in effect say "abnormal cell membrane here." When NK cells encounter these antigens on a cancer cell or a cell infected with viruses, they secrete proteins that kill the abnormal cell by destroying its cell membrane. So, why do we get cancer? Some cancer cells avoid detection by the system; they can multiply and spread without interference by NK cells.

Interferons

Interferons are small proteins released by activated lymphocytes and macrophages, and by tissue cells infected with viruses. These proteins cause normal, uninfected cells to produce antiviral compounds that interfere with viral replication. They also stimulate macrophages and NK cells into action.

Complement

Complement refers to a group of proteins in the blood that "complements," or supplements, the action of *antibodies*, a group of immune-system proteins that will be reviewed shortly. Complement activation enhances phagocytosis, destroys the membranes of invading cells, and promotes the defensive reaction known as *inflammation*.

Inflammation

Inflammation is a localized tissue response to injury. It produces the well-known sensations of swelling, redness, heat, and pain. Inflammation can be produced by any stimulus that kills cells or damages loose connective tissue. When stimulated by local tissue damage, special connective tissue cells called **mast cells** release—into the interstitial fluid—special substances called *histamine* and *heparin* that initiate the process of inflammation. **Figure 25.19** follows the different actions that occur during inflammation of the skin. These actions serve to slow the spread of pathogens at sites away from the injury, to prevent the entry of additional pathogens, and to set into motion a wide range of defenses that can overcome the pathogens and make permanent tissue repairs. The repair process is called *regeneration*. Note that one of the later actions in the inflammation process is "activation of specific defenses." You'll learn more about these defenses shortly; for now, just note that the immune system almost always works together as an integrated whole. Though inflammation is a nonspecific defense, a skin puncture that causes inflammation will activate specific defenses as well.

Fever

A **fever** is a continued body temperature greater than 37.2°C (99°F). The hypothalamus contains groups of nerve cells that regulate body temperature and act as the body's thermostat. Macrophages, pathogens, and bacterial toxins, among other things, can reset the thermostat and cause a rise in body temperature. Within limits, a fever may be beneficial, because high body temperatures speed up the activities of the immune system. However, high fevers (over 40°C, or 104°F) can disrupt or damage many different organ systems.

25.11 Specific Defenses of the Immune System

Interferons and natural killer cells do not discriminate between one invader and the next. However, immunity or *specific resistance* is the

Tutorial 25.3.1
The Inflammation Process

Figure 25.19
The Inflammation Process

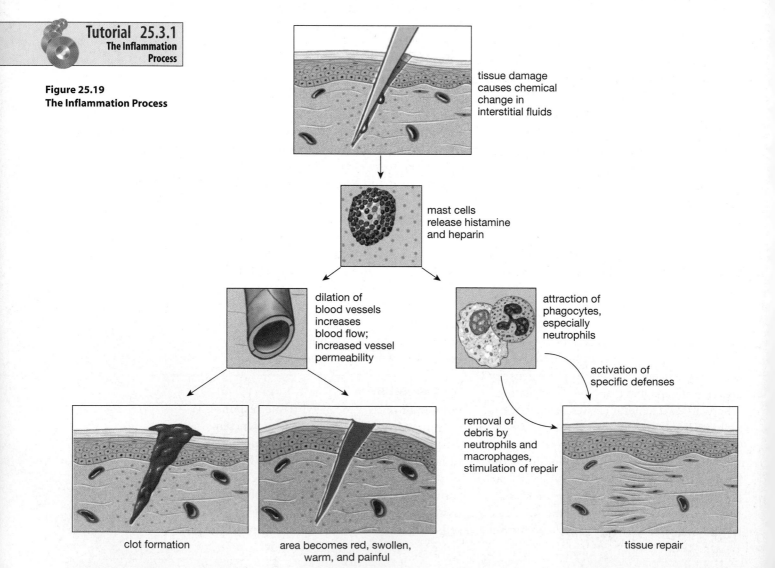

tissue damage causes chemical change in interstitial fluids

mast cells release histamine and heparin

dilation of blood vessels increases blood flow; increased vessel permeability

attraction of phagocytes, especially neutrophils

activation of specific defenses

removal of debris by neutrophils and macrophages, stimulation of repair

clot formation

area becomes red, swollen, warm, and painful

tissue repair

resistance to injuries and diseases caused by *specific* foreign chemical compounds and pathogens. In the inflammation response, some parts of the immune system took actions that they would have employed against any invader. But while a skin puncture may be a general type of injury, *specific* viruses, or bacteria, or fungi would take advantage of it. Thus, the body's response to such an injury involves not only nonspecific immunity, but specific immunity. Specific immunity comes in two varieties, innate and acquired. Innate immunity is inherited; it is present at birth and has no relation to previous exposure to the antigen involved. Such inborn immunity is what protects us from diseases that may infect farm animals or pets. For example, humans are not subject to the same diseases as goldfish, and vice versa. In contrast, an acquired immunity is an acquired resistance to infection. It may develop in either of two ways, passively or actively, as you'll now see. The relationships between innate immunity and the different forms of acquired immunity are diagrammed in **Figure 25.20**.

Passively acquired immunity is acquired by the administration of disease-fighting substances produced by another individual. This occurs naturally when substances produced by the mother provide protection against infections that could attack a fetus. Such substances may also be *administered* to fight or prevent infection. For example, antivenoms are used to treat bites or stings suffered by a person, while immunoglobulin (antibody) shots are given to increase resistance to possible infection. Importantly, these means of immunity provide only short- or medium-term resistance to harm—lasting as much as several years—because disease-fighting compounds produced *outside* our own bodies are eventually degraded in our system. Truly long-lasting immunity is something that our bodies develop with their own resources.

Actively acquired immunity is developed as a result of natural exposure to an antigen in the environment, or from deliberate exposure to an antigen. It is the means by which we get long-lasting protection against invaders. Such immunity normally begins to develop after birth, and it is continually updated and broadened as individuals encounter pathogens or other antigens that they have not been exposed to before. One form of actively acquired immunity is produced artificially by the immunization shots given to children and adults to protect them against specific diseases, such as smallpox. Let's now begin to look at how actively acquired immunity works.

Figure 25.20
Types of Immunity: Acquired and Innate

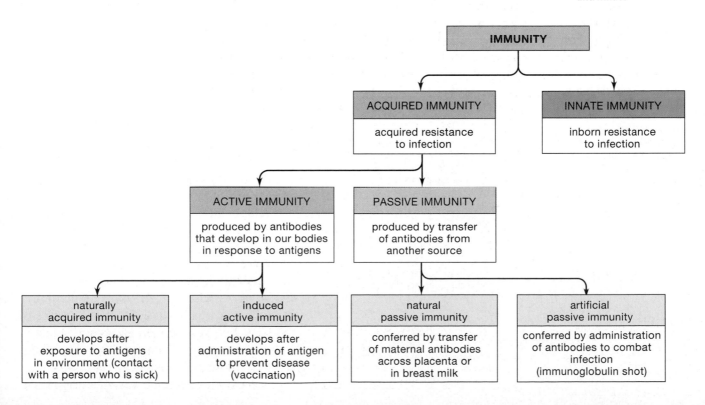

25.12 Antibody-Mediated and Cell-Mediated Immunity

A key to understanding the body's actively acquired immunity lies in what takes place with vaccination against a disease like polio. In vaccination, a person is not being treated with a long-lasting substance that is toxic to viruses or bacteria. Rather, what is being injected or swallowed is an antigen—at least part of the disease-causing organism itself. So why is it that, having been inoculated in this way, a person is unlikely to contract polio? The answer is that as far as the body's immune system is concerned, the person *did* contract polio when he or she was inoculated. The immune system was presented with polio antigens, and it then proceeded to mount an attack on this invader. What "getting polio" means, then, is not having the poliovirus come into our bodies; it means being *harmed* by the virus before the immune system can clear it from our system. This is what can happen when the full-blown virus invades our system by natural means before we are inoculated against it.

The critical thing is that once we are inoculated with a vaccine, the poliovirus is nearly powerless to harm us. Not powerless to come into our bodies, mind you; powerless to harm us because the body is now *ready*. As a product of the original "attack" (the vaccination), the immune system is put into a permanent heightened state of readiness with respect to the poliovirus. What vaccines do, then, is elicit readiness. But how does the immune system stay ready for specific invaders? For that matter, how does it work to defeat them when they come? That is the subject of this section on immune-system coverage.

There are two major components to the body's specific resistance system. The first is called **antibody-mediated immunity**, meaning an immune-system capability that works through the production of proteins called antibodies. The second is called **cell-mediated immunity**, meaning an immune-system capability that works through the production of cells that destroy infected cells in the body. Both systems begin in the same place, in that the white blood cell called the *lymphocyte* is the basis for cells that are critical to both systems. Like all white blood cells, lymphocytes originate in the bone marrow. Some lymphocytes, however, migrate to the body's thymus gland (located just above the heart), which is where these lymphocytes differentiate into the main cells of *cell-mediated* immunity, the **T-lymphocyte cells** or just **T cells** (with the *T* standing for "thymus" cell).

The Body's Lymphatic System

Meanwhile, lymphocytes that remain in the bone marrow can develop into the central cells of *antibody*-mediated immunity, the **B-lymphocyte cells** or **B cells**. Mature T and B cells are most often found in the body's lymphatic system organs—the spleen and the lymph nodes, for example—where they are prepared to latch onto invaders. The development of T and B cells is tracked in **Figure 25.21**. The body's **lymphatic system** is laid out for you in **Figure 25.22**. Think of it as a system of vessels that picks up fluids (interstitial fluids) that have leaked from blood vessels into areas outside the body's cells. Eventually, these fluids are channeled back into the circulatory system. While in the lymphatic system, however, these fluids pass through lymphoid organs, which are packed with immune-system cells. This is

Figure 25.21
Development of T Cells and B Cells
Key players in the body's specific defense system are T cells and B cells, both of which start out as a type of white blood cell, called a lymphocyte, in bone marrow. Some of these lymphocytes migrate to the thymus gland, where they differentiate into T cells (thymus cells). Others that remain in the bone marrow develop into B cells. Most T and B cells then migrate to lymphoid tissues, such as lymph nodes and the spleen, where they will serve their immune function.

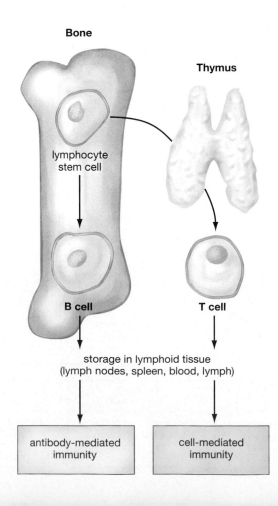

Bone

Thymus

lymphocyte stem cell

B cell

T cell

storage in lymphoid tissue
(lymph nodes, spleen, blood, lymph)

antibody-mediated immunity

cell-mediated immunity

why, when you get sick, your "lymph glands" are swollen—they are filled with an abnormally large number of cells that are busy fighting off the infection.

From this point forward, the antibody-mediated and cell-mediated immune operations are treated separately. The two systems work together to a degree, but for now let's look at them in isolation. After separate tours of the two systems, you'll take a step back and see how the immune system functions as a whole. Let's start now with antibody-mediated immunity.

25.13 Antibody-Mediated Immunity in Detail

As you might imagine, a key player in antibody-mediated immunity is the **antibody**, which can be defined as a circulating immune-system protein that binds to a particular antigen. From this definition, it may be apparent that antibodies and antigens are closely linked. (So closely linked that one is named for the other; the word *antigen* is short for "antibody generating.") To a first approximation, a given antibody will bind to a specific antigen and to no other antigen. In their binding capacity, you will see antibodies in two guises: first as *receptors* on the surface of B cells, extending in a generalized Y-shape from the cells' outer membranes; second as *free-standing* molecules produced by B cells, moving through the bloodstream. In their receptor identity, antibodies are also known as antigen receptors.

The Fantastic Diversity of Antibodies

It's common knowledge that there is no such thing as the common cold; rather, there are countless *varieties* of common cold, each caused by a different virus. Then there are all the other potential invaders, such as bacteria, fungi, and so forth. How can the immune system cope with such a variety of foes? One of its key strengths is that it produces millions of different kinds of B cells, each having a type of antigen receptor on it that is specific to only one antigen. Look at one B cell and it will have antigen receptors that latch onto an antigen only on *this* virus; look at a *different*

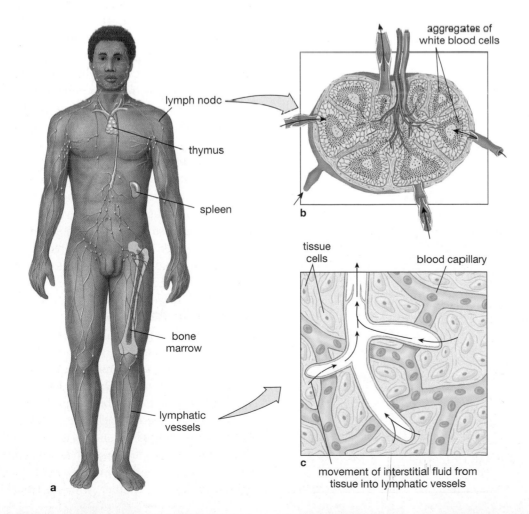

aggregates of
white blood cells

lymph node

thymus

spleen

bone
marrow

lymphatic
vessels

tissue
cells

blood capillary

b

c

movement of interstitial fluid from
tissue into lymphatic vessels

a

**Figure 25.22
The Human Lymphatic
System**

a Fluid moving through lymphatic vessels passes through lymphatic organs, such as lymph nodes or the spleen, where the fluid is inspected by an array of immune-system cells.

b Detail of fluid moving through a lymph node.

c Detail of interstitial fluid moving into lymphatic vessels.

578

Figure 25.23
B Cells and Antibody-Mediated Immunity

1. The antigen receptors (antibodies) of a particular B cell allow it to bind with a virus displaying a specific antigen.

2. This action spurs the B cell to produce a huge number of clones of itself by dividing rapidly. These B cells differentiate into two types: memory cells and plasma cells. Memory cells remain in the body long after the infection has passed, ready to quickly produce more B cells of the same type, should the same invader come again.

3. Plasma cells secrete free standing antibodies, which enter the bloodstream.

4. Antibodies produced by the plasma cells attack the invader.

B cell and it has a *different* set of receptors. This mind-boggling complexity is made possible by DNA arranging that takes place back in the bone marrow, when the precursors of B cells are being formed. The result is millions of varieties of B cells, each expressing its own specific antigen receptors (antibodies).

A good many of these "soldiers" will never see battle, for the simple reason that the body is never invaded by a pathogen displaying their type of antigen. Some B-cell types are called on to fight, however; and when they do, they have the marvelous capacity not only to make more soldiers like themselves, but to create a permanent set of sentries for a given invader, as you'll now see.

The Cloning of Selected Cells in Antibody-Mediated Immunity

Lying in readiness in the lymph nodes or circulating through the bloodstream, a B cell encounters, say, a virus, one that presents an antigen that can be bound by this B cell's antigen receptors (meaning antibodies). If

you look at **Figure 25.23**, you can see what happens next. This very binding causes the B cell to start dividing very rapidly. Each new B cell of this type gives rise to more identical B cells, creating a selected "clone," or a huge number of identical cells of this type. Such numbers are necessary because the *virus* will be multiplying rapidly as well.

Fighters and Sentries: The Differentiation of B Cells

After a time, these B cells then differentiate into two types of cells. One variety is called a **plasma cell**, which exists in the lymph nodes and is specialized to produce antibodies. These antibodies are not the cell-surface receptors you've been looking at, however. Instead, they are freestanding antibodies that leave the plasma cells by the millions—they are *exported* from the B cells—after which they move through the bloodstream and fight the invader directly. The other type of B cell is called the **memory B cell**, and it is the body's permanent sentry. Long after the initial war with this invader is over, this specific variety of memory B cell remains in the body. Should this invader come in a *second* time, these memory cells will be ready to produce more plasma cells to mount a very quick defense. This is why almost no one gets chicken pox twice; it's also why being inoculated with antigens from the poliovirus protects us permanently from the damage caused by this crippler.

If you look at **Figure 25.24**, you can see the typical time-course of exposure to antigen and production of antibodies. What is known as the "primary immune response" has reached a substantial level seven days out from first exposure. Look at the second exposure, however, to see what a difference this first exposure has made. Now there is a much greater response in a much shorter time.

Action of the Antibodies

But what is it that the circulating antibodies do to combat the invader? First, because they are binding to the invader's antigens, the antibodies can sometimes prevent the invader from attaching to anything *else*, which stops its spread. They can also cause an "agglutination" of antibodies and antigens in an antigen-antibody "complex" that renders the invaders inactive. Finally, the antibody binding can trigger the activities of another immune-system

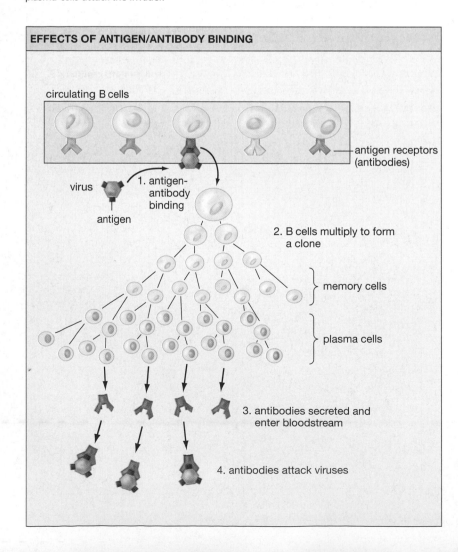

EFFECTS OF ANTIGEN/ANTIBODY BINDING

circulating B cells

antigen receptors (antibodies)

virus

antigen

1. antigen-antibody binding

2. B cells multiply to form a clone

memory cells

plasma cells

3. antibodies secreted and enter bloodstream

4. antibodies attack viruses

component, the complement proteins you looked at earlier. Complement can bind to the antibody at the same time the antibody is bound to the antigen. Critically, complement binds to an antibody *only* when the antibody is bound to an antigen. This dual binding sets into motion the production of a cascade of complement proteins, the final members of which can poke holes in the cell walls of bacteria.

25.14 Cell-Mediated Immunity in Detail

It may go without saying at this point that antibody-mediated immunity is quite spectacular. But it also turns out to be limited in that it works strictly on foreign organisms and molecules. The problem with this is that many viruses (as an example) are successful in invading the body's *own cells*, which then become part of the problem. To deal with this threat, the body has developed another line of acquired immunity, the cell-mediated immunity mentioned earlier.

You may recall that the central player in cell-mediated immunity is the *T*-lymphocyte or T cell. T cells come in several varieties, three of which are **cytotoxic T cells** (also known as *killer T cells*), **helper T cells**, and **suppressor T cells**. Let's now go through the central actions of the cell-mediated immunity system, taking as a starting point a body that has been infected with a virus.

Macrophages Bearing Invaders: Antigen-Presenting Cells

A given body cell that has been infected with a common virus puts out one of the interferon proteins, noted earlier in the discussion of nonspecific immunity. This protein acts as a chemical signal saying, in effect, "infected cell here," and the result is an attack by the natural killer and macrophage cells noted earlier. The macrophages actually serve two functions at this point. First, they ingest infected cells by engulfing them; second, they then display *fragments* of proteins from the invading virus on their own surface. You won't be surprised to learn that these fragments will now serve as antigens. When macrophages play this role, they are sometimes referred to as **antigen-presenting cells** or **APCs**.

If you look at **Figure 25.25**, you can see how the next interaction plays out. The

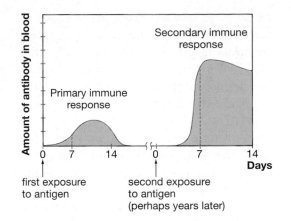

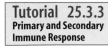
Tutorial 25.3.3
Primary and Secondary Immune Response

Figure 25.24
Prepared for an Invasion
The memory cells produced by the body during a first attack by an invader allow it to mount a faster, more vigorous defense should the same invader attack a second time.

Tutorial 25.3.4
T Cells and Cell-Mediated Immunity

Figure 25.25
T Cells and Cell-Mediated Immunity

1. A cell-eating macrophage ingests a virus.

2. The virus is displayed as an antigen within one of the macrophage's receptors, making the macrophage an antigen-presenting cell, or APC.

3. A helper T cell binds to the antigen-APC binding site.

4. This binding produces activated T cells that put several other processes in motion: the production of clones of helper and cytotoxic T cells specific to the invader, the production of suppressor T cells, and the production of B-cell clones.

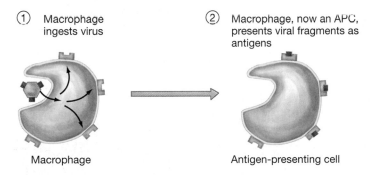

① Macrophage ingests virus

Macrophage

② Macrophage, now an APC, presents viral fragments as antigens

Antigen-presenting cell

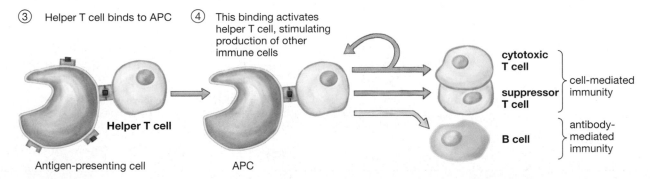

③ Helper T cell binds to APC

Helper T cell

Antigen-presenting cell

④ This binding activates helper T cell, stimulating production of other immune cells

APC

cytotoxic T cell
suppressor T cell
cell-mediated immunity

B cell
antibody-mediated immunity

helper T cell now locks onto the APC; but as usual with the immune system, there is great specificity in this. The APC is displaying the viral protein fragment within one of its own membrane proteins, and the helper T cell cannot bind with the APC unless this dual condition is met: A viral antigen must be bound up with what we can call a self-protein of the APC. This makes an antigen/self-binding site that is unique in two ways: Only this specific invader would prompt the site's creation, and only one variety of the body's tens of millions of varieties of helper T cells would bind with it. In short, we once again have a tremendously specific immunity, this time mediated by cells. After binding with the helper T cell, the APC also puts out a protein, interleukin-1, that helps activate helper T cells.

Activated Helper T Cells Are Central to the Immune Response

The "activated" helper T cells now secrete a protein, interleukin-2, that puts several other processes into motion. First, this secretion spurs production of clones of helper and cytotoxic T cells. These clones include not only active helper and cytotoxic T cells but also *memory* cells, for future invasions. Second, the interleukin-2 also stimulates production of suppressor T cells, which multiply more slowly than do cytotoxic T cells. Ultimately

the suppressor cells will inhibit the production of cytotoxic T cells, once the immediate infection battle has passed. Third, the interleukin-2 helps activate B-cell immunity (by stimulating clone development).

By now you may be saying: So where's the invader-fighting in all this? The answer is in the clone of cytotoxic T cells, which are called "killer" T cells for a good reason. Using their own special receptors, they latch onto infected cells in the body and puncture their outer membranes, causing them to "lyse" or burst. Note that cytotoxic cells are not binding with the APCs, as the helper cells did. These killers are binding with the body's own infected cells. From this, it may be apparent why this is called *cell-mediated* immunity. In the end, it is a housecleaning involving not the antibodies described before, but several sets of cells.

Taking a step back from cell- and antibody-mediated immunity, in **Figure 25.26** you can see how the two systems work together, along with nonspecific defenses.

25.15 Allergies and Autoimmune Disorders

Clearly, a critical factor in cell-mediated immunity is the body's ability to distinguish between "self" (its own uninfected cells) and "non-self" (the infected cells). Otherwise,

**Figure 25.26
An Overview of Cell-Mediated and Antibody-Mediated Defenses and Their Link with Nonspecific Defense**

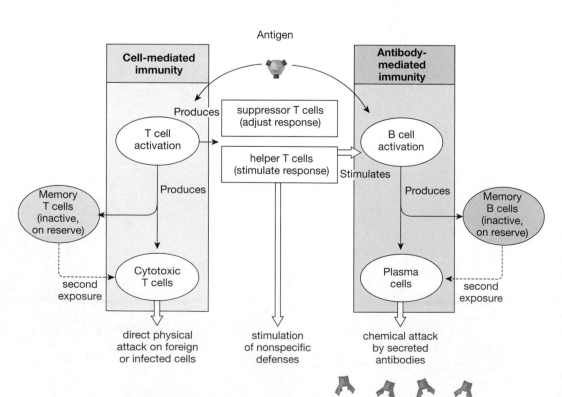

the immune system would attack the body's own healthy cells. Despite safeguards built into the system, the immune system does attack either harmless substances or the body itself with disturbing frequency.

T cells learn to distinguish self from non-self during the embryonic or newborn phases of life. Developing during this period in the thymus, some T cells may never "learn" to distinguish correctly between self and non-self. Conversely, it may be that an invading virus is able to incorporate some of the body's own proteins into itself, meaning that "self" would now be read as "non-self." For whatever reason, the body does sometimes attack itself—with devastating consequences. If you look at **Figure 25.27**, you can see the results of the disease known as *rheumatoid arthritis*, which may be a result of helper T cells attacking the wrong targets. There is also multiple sclerosis, in which the immune system starts attacking the axon-covering myelin you read about when looking at the nervous system.

Beyond this there are **allergies**: immune-system overreactions to antigens. Any immune-system attack is triggered by substances that the body perceives as foreign. The foreign substances that trigger allergic reactions are called **allergens**. These substances cause the body to start releasing the histamine and heparin noted earlier in connection with inflammation. Allergic reactions can run from life-threatening (asthma) to merely annoying (sneezing and sniffling). Allergens need not be living things, but almost always are at least derived from living things, because it is protein molecules that are serving as the antigens setting off the immune attack. Thus pollen, dust mites, foods, and fur are common allergens. Because allergens commonly elicit histamine release, it makes sense that a common treatment for allergies is the class of substances called *antihistamines*.

25.16 AIDS: Attacking the Defenders

The disease called acquired immunodeficiency syndrome, or **AIDS**, is devastating precisely because it attacks the very immune system that normally protects the body. Within the immune system, it is the helper T cells noted earlier that are decimated by the human immunodeficiency virus, or **HIV**, which is the cause of AIDS.

HIV works by attaching itself to a helper T cell, injecting its genetic material into it, and then hijacking the T cell's own protein-making machinery. The cell then begins turning out *HIV* proteins; eventually, complete virus particles are assembled within the T cell. These then leave the cell, destroying it in the process. They then latch onto other helper T cells, infecting them. If you look at **Figure 25.28**, you can see a picture of AIDS virus particles emerging from an infected T cell.

You saw earlier that helper T cells are critical not only in getting cell-mediated (T-cell) immunity going, but in stimulating the antibody-mediated system as well (B-cell immunity). As a result, when the

Figure 25.27
When the Body Attacks Itself
Rheumatoid arthritis has so disfigured this person's hands that simply signing a greeting card has become a difficult act. Rheumatoid arthritis is an autoimmune condition.

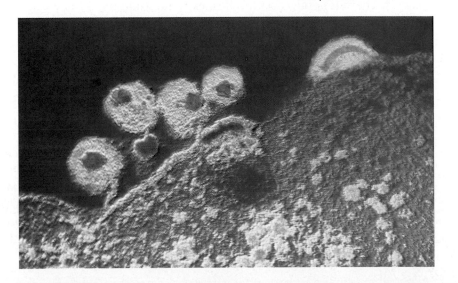

Figure 25.28
The Invaders Emerge
Artificially colored AIDS virus particles bursting forth from a stricken helper T cell. The cell will be killed by this budding of virus particles from it.

helper T cell "count" drops too low, the AIDS victim falls prey to a host of afflictions that the immune system normally would make quick work of. A fungal infection that most people wouldn't even realize they had can become life-threatening to an AIDS sufferer. Many cells have so-called cell identity markers on their surface—proteins that serve to identify what type of cell they are. Helper T cells have what are known as CD4 glycoproteins on their surface, and this is the nomenclature that you will hear employed in connection with AIDS. A critical question becomes: What is the patient's CD4 count? From your review of the immune system, you probably can see why this question would be so important.

On to Transport and Exchange

Having looked at the amazing capabilities of the nervous, endocrine, and immune systems, next you'll look at four other systems that are just as important in maintaining a healthy body. Chapter 26 is concerned with blood, breathing, digestion, and elimination.

Chapter Review

Summary

25.1 Overview of the Nervous System

- The key functional cells of the nervous system are neurons, which are supported by neuroglial cells.

- The two major divisions of the human nervous system are the central nervous system (CNS), consisting of the brain and spinal cord; and the peripheral nervous system (PNS), which includes all the neural tissue outside the CNS. The PNS has an afferent division, which brings sensory information to the CNS; and an efferent division, which carries motor (action) commands to muscles and glands—the body's effectors. Within the PNS efferent division are two subsystems: the somatic nervous system, which provides voluntary control over skeletal muscles; and the autonomic nervous system, which provides involuntary regulation of smooth muscle, cardiac muscle, and glands. The autonomic system is further divided into the sympathetic division, which generally has stimulatory effects; and the parasympathetic division, which generally has relaxing effects.

- There are three types of neurons. Sensory neurons sense conditions inside and outside the body and convey information to neurons inside the CNS. Motor neurons carry instructions from the CNS to other tissues, organs, or organ systems. Interneurons, located entirely within the CNS, interconnect other neurons.

- Neurons have extensions called dendrites that receive signals coming to the neuron cell body, and a single large extension, called an axon, that carries signals from the cell body. At the tips of an axon there may be a structure called a synaptic terminal that is part of a synapse—a site where a neuron communicates with another cell.

- Neuroglia have no information processing ability, but serve to protect neurons and to facilitate communication by them.

- A nerve is a bundle of axons in the PNS that transmits information to or from the CNS.

25.2 How Does Nervous System Communication Work?

- An electrical charge, called a membrane potential, exists across the plasma membrane of neurons. It represents a form of stored energy that is used in transmitting nerve signals. The basis of nerve signal transmission is that the plasma membrane's permeability to charged substances (ions) can change upon stimulation.

TUTORIAL 25.1.1: Signal Transmission within a Neuron

- Transmission along the entire axon of a neuron takes place through an action potential, in which a change in membrane (ion) permeability at one point along the axon changes the permeability at an adjacent point. The action potential moves in one direction along the axon, away from the initial point of stimulation.

- A nerve signal moves from one neuron to another (or to a muscle or gland cell) across a synapse, a site of nerve cell communication that includes a "sending" presynaptic neuron, a "receiving" postsynaptic neuron or effector cell, and a synaptic cleft, which is the space between the two cells. Communication takes place via a chemical, called a neurotransmitter, released into the synaptic cleft by the presynaptic neuron.

TUTORIAL 25.1.2: Structure and Function of a Synapse

25.3 The Spinal Cord

- The spinal cord can receive input from sensory neurons, and it can direct motor activities in response; it is also the major pathway for impulses going to and from the brain.

- Spinal cord motor neurons, which are sending motor commands to various effectors in the PNS, have cell bodies that lie within the gray matter of the spinal cord. The axons of the motor neurons leave the spinal cord through ventral roots. Sensory neurons, which transmit information to the spinal cord, have their cell bodies outside the spinal cord, in the dorsal root ganglia.

- Sensory neurons are stimulated via sensory receptors, which may be dendrites that feed into an axon of a sensory neuron, or may be specialized cells that communicate with the axon across a synapse. Upon stimulation, a message goes from sensory receptor, through the axon of the sensory neuron, through a dorsal root ganglion and the dorsal root, and finally to a site within the spinal cord where the sensory neuron links either with a motor neuron or with an interneuron.

- Reflexes are automatic nervous-system responses, triggered by specific stimuli, that help us avoid danger or preserve homeostasis.

- The neural wiring of a single reflex, called a reflex arc, begins with a sensory receptor, runs through the spinal cord, and ends at an effector—such as a muscle or gland.
  TUTORIAL 25.1.4: Steps in a Reflex Arc

25.4 The Autonomic Nervous System

- The division of the PNS over which we do not have conscious control, the autonomic nervous system, has sympathetic and parasympathetic divisions. The sympathetic division is often called the fight-or-flight system, because it generally activates bodily functions. The parasympathetic division is often called the rest-and-digest system, because it conserves energy and promotes digestive activities. Most bodily organs receive input from both systems.

25.5 The Human Brain

- The brain contains almost 98 percent of the human body's neural tissue. It consists of gray matter, made up of nerve-cell bodies and existing primarily as a layer at the surface of the brain called the cerebral cortex; and of white matter, made up of bundles of nerve-cell axons.

- There are six major regions in the adult brain: the cerebrum, thalamus and hypothalamus, midbrain, pons, cerebellum, and medulla oblongata. The cerebrum is divided into right and left cerebral hemispheres and is the seat of conscious thought processes, intellectual functions, and memory storage and retrieval.

- The brain stem is all the structures of the brain except the cerebrum and cerebellum. The four primary structures of the brain stem are active in controlling involuntary bodily

activities, in relaying information, and in processing sensory information.

25.6 The Nervous System in Action: Our Sense of Vision

- Human sensory capabilities work through the nervous system. The common notion that human beings possess "five senses" is incorrect. Human beings have more sensory capabilities than this.

- The human visual system must accomplish three central tasks. It has to gather and focus light reflected by objects in the outside world; it must convert light signals into nervous-system signals (action potentials), and it must make sense of the visual information it has received.

- Light enters the eye first through the cornea and then passes through the lens and various materials on its way to a layer of tissue called the retina at the back of the eye. Light is bent or refracted by the cornea and the lens in such a way that it ends up as a tiny, sharply focused image on the retina.

- If the eyeball is too long, rays of light entering the eye converge before they get to the retina, causing nearsightedness or an inability to see distant objects clearly. If the eyeball is too short, rays of light converge after they get to the retina, causing farsightedness, or an inability to see nearby objects clearly. Presbyopia is caused, generally in middle age, by a hardening of the eye's lens that leaves it unable to change shape when viewing nearby objects.

- Light signals are converted to nervous-system signals by cells in the retina called photoreceptors that come in two varieties: rods and cones. Rods function in dim light but are not sensitive to color; cones function best in bright light but are sensitive to color. Any given cone is most sensitive to one of three colors of light: red, blue, or green. Any object we look at produces a specific pattern of activity across all three kinds of cones. These specific patterns allow us to perceive any hue in the color spectrum.

- Rods and cones function through pigments: chemical compounds that absorb light. Stacks of flattened membranes at the posterior portion of each rod and cone contain millions of pigment molecules that absorb the light coming from the outside world. This absorption sets into motion a chemical cascade that results in a signal being sent from rods and cones to the connecting cells lying anterior to them in the retina. With this, light energy has been converted into a nervous-system signal.

- The signals from the photoreceptors are channeled first to connecting cells and then to ganglion cells, whose axons come together to form the two optic nerves in the body. Visual information is carried by the optic nerves first into a portion of the thalamus and then, by different axons, to an area at the back of the brain called the primary visual cortex, the brain's primary site of visual processing.

- The brain does not passively record the visual information that comes to it. Rather, the brain constructs images as much as it records them. The visual system operates through a series of genetically based "rules" that allow us to quickly make sense of what we perceive. Evolution shaped our vision in this way, to be maximally useful to us in survival and reproduction.

25.7 The Endocrine System: Hormones and How They Work

- Endocrine system cells secrete chemicals called hormones that travel mostly through the circulatory system and that affect the activities of other cells. Hormonal production and secretion takes place to a significant extent within endocrine glands.

- Most hormones are amino-acid-based hormones; but there are also steroid hormones, built from cholesterol molecules. Some hormones, called local hormones, do not travel throughout the body, but affect only nearby cells.

- Each hormone has specific target cells upon which it works. Amino acid hormones link to their target cells via receptors that protrude from the target cells' outer membranes. This binding produces changes in the target cell's production of proteins. Steroid hormones pass through a cell's plasma membrane, after which they bind with a receptor protein inside the cell.
  TUTORIAL 25.2.1: Hormones and Target Cells

25.8 How Is Hormone Secretion Controlled?

- Homeostasis refers to the body's tendency to maintain a relatively stable internal environment in the midst of a constantly changing external environment.

- Homeostasis is maintained primarily through negative feedback, which is a process in which the elements that bring about a response have their activity reduced by that response.
  TUTORIAL 25.2.3: Negative Feedback and Calcium Levels

- Infrequently, the body employs positive feedback, in which the initial response magnifies the stimulus that gave rise to it. Positive feedback promotes instability, while negative feedback promotes stability.

- The endocrine system's release of hormones is fundamentally controlled by negative feedback; but a hierarchy of control exists for much of the system, in which the brain structure called the hypothalamus (1) acts as an endocrine organ; producing its own hormones, which are released by the posterior pituitary gland; (2) exercises control, via the nervous system, over the hormonal activity of the medulla of the adrenal glands; and (3) releases regulatory hormones that control secretion of hormones by the anterior pituitary gland. The hypothalamus is the central link between the body's nervous and endocrine systems. The pituitary is regarded as the master gland of the endocrine system.

25.9 The Immune System: Defending the Body from Invaders

- The body's immune system protects it against invading organisms and toxic substances.

- The body has two primary defensive strategies: nonspecific defenses that do not discriminate between one invader and the next, and specific defenses that provide protection against particular invaders. Specific defenses can produce immunity, defined as a state of long-lasting protection against particular invaders.

- An antigen is any foreign substance—often a protein from an invading organism—that elicits an immune-system response.

- Several varieties of white blood cell are important to immune function; these include T cells, B cells, and macrophages.

25.10 Nonspecific Defenses of the Immune System

- The body's nonspecific defenses include physical barriers, phagocytic ("cell-eating") cells, immunological surveillance, interferon, complement, inflammation, and fever.
  TUTORIAL 25.3.1: The Inflammation Process

25.11 Specific Defenses of the Immune System

- Specific immunity can be either innate or acquired. Innate immunity is present at birth and has no relation to exposure to an antigen. Acquired immunity can develop either actively or passively. Passively acquired immunity is achieved through the administration of disease-fighting substances produced by another individual. Actively acquired immunity develops as a result of natural or deliberate exposure to an antigen. One form of actively acquired immunity is produced artificially by immunization shots.

25.12 Antibody-Mediated and Cell-Mediated Immunity

- There are two major components to actively acquired immunity: antibody-mediated immunity and cell-mediated immunity.

- White blood cells called lymphocytes are fundamental to both types of resistance, because lymphocytes develop into both T cells (which are central to cell-mediated immunity) and B cells (which are central to antibody-mediated immunity).

- Mature T and B cells are most often found in organs of the body's lymphatic system, which is a system of vessels that captures interstitial fluids, subjects these fluids to scrutiny by immune-system cells, and returns the fluids to the circulatory system.

25.13 Antibody-Mediated Immunity in Detail

- An antibody is a circulating immune-system protein that binds to a particular antigen. Antibodies exist as (1) receptors on the surface of B cells that bind to specific antigens, and (2) as freestanding invader-fighting proteins that are secreted by B cells and move through the bloodstream.

- Antibody-mediated immunity begins with the binding of an antigen to a B-cell receptor (antibody). This binding produces a cascade of effects, including the production of a B-cell clone—a huge number of identical copies of the B cell that are bound to the antigen.
  TUTORIAL 25.3.2: B Cells and Antibody-Mediated Immunity

The B cells produced differentiate into plasma cells, which produce infection-fighting antibodies, and memory B cells, which remain in the body long after the initial infection from the specific invader has passed, providing permanent immunity against it.

TUTORIAL 25.3.3: Primary and Secondary Immune Response

25.14 Cell-Mediated Immunity in Detail

- Cell-mediated immunity provides protection in instances in which the body's own cells are harmful after having become infected by an invader.

- The central player in cell-mediated immunity is the T cell, which comes in three varieties: cytotoxic (or killer) T cells, helper T cells, and suppressor T cells.

TUTORIAL 25.3.4: T Cells and Cell-Mediated Immunity

- An infected cell secretes an interferon protein that elicits an attack by natural killer and macrophage cells. Macrophages activated in this way ingest infected cells and display fragments of proteins from the invader, bound with macrophage receptors, on the macrophage's surface. Macrophages that play this role are known as antigen-presenting cells, or APCs.

- Helper T cells, binding with the combined antigen/APC binding site, activate the T cells, bringing about a cascade of activities that includes the production of helper, cytotoxic, and suppressor T cells, and the stimulation of B-cell clone development. The T-cell activity produces memory T cells for future invasions by the specific pathogen. Cytotoxic T cells bind with infected cells and puncture their outer membranes.

25.15 Allergies and Autoimmune Disorders

- The immune system attacks the body's own healthy tissues with disturbing frequency, producing such conditions as rheumatoid arthritis and multiple sclerosis.

- Allergies often are an immune-system disorder in which a foreign substance, termed an allergen, causes the body to release infection-fighting histamine and heparin.

25.16 AIDS: Attacking the Defenders

- AIDS is devastating because it attacks the very immune system that normally protects the body.

- The cause of AIDS, the human immunodeficiency virus or HIV, attaches itself to helper T cells, making copies of itself inside and destroying the helper T cells in the process. Because helper T cells are central to specific immunity, the result can be a great vulnerability to common pathogens.

Key Terms

Understanding the Basics

Multiple-Choice Questions

1. The peripheral nervous system (PNS) is made up of
   a. the brain and spinal cord
   b. the brain and cranial nerves
   c. all of the nervous tissue outside of the CNS
   d. all of the nervous tissue inside the CNS
   e. all of the above

2. Which of the following is *not* a response of the sympathetic division of the autonomic nervous system?
   a. dilation of the airway passages
   b. dilation of the pupils
   c. increased heart rate
   d. increased digestive activity
   e. All of the above are responses to the sympathetic division of the ANS.

3. Conscious thought processes and all intellectual functions originate in the
   a. cerebrum
   b. cerebellum
   c. thalamus
   d. medulla oblongata
   e. hypothalamus

4. The regulation of autonomic function, such as heart rate and blood pressure, occurs in the
   a. cerebrum
   b. cerebellum
   c. thalamus
   d. pons
   e. medulla oblongata

5. Action potentials work by means of
   a. hormonal surges
   b. the unwinding of DNA molecules
   c. the storing of neurotransmitters
   d. transient reversals of electrical potential across a cell membrane
   e. long-lasting reversals of electrial potential across a cell membrane

6. Light coming into our eyes is focused by the (more than one answer may be correct)
   a. cornea
   b. retina
   c. primary visual cortex
   d. lens
   e. optic nerve

7. We can perceive millions of colors because
   a. We have a different kind of cone for each color.
   b. Our rods and cones combine colors in ways that allow this variety.

   c. Our brain can recognize different colors irrespective of the visual input it gets.
   d. Three kinds of cones respond to each reflected color in a specific pattern to yield the full color spectrum.
   e. We have a different kind of rod for each color.

8. Cells of different tissues and organs respond to the same hormone if they have the same kind of _____ molecules.
   a. receptor
   b. target
   c. DNA
   d. protein
   e. all of the above

9. Steroid hormones are produced from
   a. amino acids
   b. fatty acids
   c. proteins
   d. cholesterol molecules
   e. plasma cells

10. What is the ultimate source of control in the endocrine system?
    a. the hypothalamus
    b. genes
    c. the pituitary gland
    d. positive feedback
    e. negative feedback

11. Inflammation
    a. aids in the temporary repair at an injury site
    b. slows the spread of pathogens
    c. produces swelling and redness
    d. produces heat and pain
    e. all of the above

12. B cells can be activated only by
    a. pathogens
    b. exposure to a specific antigen
    c. fever
    d. cells infected with viruses
    e. bacteria

Brief Review

1. Three functional groups of neurons are found in the nervous system. What role does each fill?

2. What advantage does dual innervation of a bodily organ provide?

3. List the important functions of cerebrospinal fluid.

4. What is the primary difference in the way the nervous and endocrine systems communicate with their target cells?

5. How are the body's specific defenses different from its nonspecific defenses?

6. The medulla oblongata is one of the smallest regions of the brain, yet damage there can cause death, whereas similar damage in the cerebrum might go unnoticed. Why?

7. Describe the path that light and its resulting nervous-system signals take in the human visual system.

8. What is the difference between negative feedback and positive feedback?

9. Why is the pituitary gland referred to as the "master gland" of the human body?

10. How would a lack of helper T cells affect the antibody-related immune response?

11. What types of cells would be affected by a decrease in the monocyte-forming cells in the bone marrow?

Applying Your Knowledge

1. A police officer has just stopped Bill on suspicion of driving while intoxicated. Bill is first asked to walk the yellow line on the road and then asked to place the tip of his index finger on the tip of his nose. How would these activities indicate Bill's level of sobriety? What part of the brain is being tested by these activities?

2. Julie is pregnant and is not receiving any prenatal care. She has a poor diet consisting mostly of fast food. She drinks no milk, preferring soft drinks instead. How will this situation affect the activity of Julie's parathyroid glands?

3. What kind of immunity protects the developing fetus, and how is it produced?

4. It's often said that seeing is believing. Is it reasonable to put this much faith in the human visual system?

MediaLab

How Does Your Body Fight the Flu? Understanding the Immune System

In 1972, a child named David Vetter was born, and then spent his 12 short years encased in a plastic bubble. The bubble protected him from the millions of microbes we all battle daily. David had a disease called SCID (severe combined immune deficiency, found in 1 in 100,000 people born each year), that affects the body's ability to produce a functional immune system. This *CD-ROM TUTORIAL* will help you explore how a normal immune system functions to spare us a similar fate. In the *Web Investigation*, you'll gather information concerning the actions of antibodies and their success in fighting two of these biological battles, against influenza and HIV. In the *Communicate Your Results* section, you will discuss and analyze how well our bodies may be fighting these battles.

This *MediaLab* can be found in Chapter 25 on your CD-ROM (Tutorial 25.3.4) and Companion Website (http://www.prenhall.com/krogh3).

CD-ROM TUTORIAL

There are two main defensive mechanisms of our immune system, the antibody-mediated and cell-mediated immune responses. This *CD-ROM TUTORIAL* will illustrate the actions of the two types of white blood cells, B- and T-lymphocytes, which coordinate these responses.

Activity

1. First, you'll learn how the immune system is activated during the inflammation response and how this stimulates B-lymphocytes to produce antibodies.

2. Then, you'll compare the primary and secondary immune response and why we develop immunity only after a primary infection.

3. Finally, you'll learn the role of the T-lymphocytes and cell-mediated immunity that allows our bodies to destroy cells that may be harboring a foreign invader.

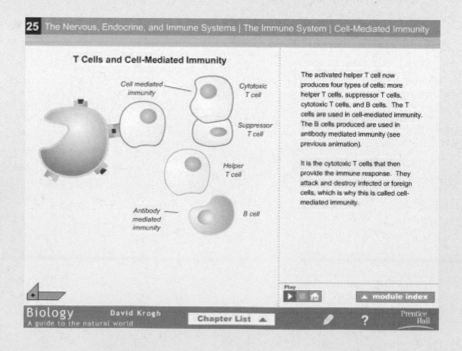

WEB INVESTIGATION

Investigation 1

Estimated time to complete = 15 minutes

In your everyday experiences, you know that getting the flu is at least an inconvenience, but you may not be aware that influenza can be deadly. Select the Keyword **PANDEMIC** on your CD or Website to read an article describing the 1918 influenza outbreak that killed over 20 million people. Then, click the Keyword **FLUNET** to visualize the development and extent of flu outbreaks in the United States and around the world. Click on "epidemic activity" on the left side of the screen; you can select the geographic area you want data from, the specific type of virus (leave this at "Epidemiological Activity" for now), the format for presenting the data, and the time period for the data. Go back in time and try to find any patterns that indicate a worldwide flu outbreak like the pandemic of 1918.

Investigation 2

Estimated time to complete – 10 minutes

Select the Keyword **ANTIBODIES** on your CD or Website for this *MediaLab* to read an article describing the general structure and activities of antibodies. Pay particular attention to how the structure of antibodies helps them recognize a huge number of different antigens, and what functions are associated with the five different classes of antibodies.

Investigation 3

Estimated time to complete = 20 minutes

By this time you've probably heard at least a little about the HIV virus that causes AIDS. Select the Keyword **CDC** on your CD or Website for this *MediaLab* to access data for HIV and AIDS. To see a report of cumulative AIDS cases in the United States, click on "Cumulative Cases." How many have there been in the United States? How are these cases distributed through different age groups, or races, in the United States? Go to "Exposure Categories" for data on how the virus has been spread. What is the most frequent mode of transmission in the United States? Is it the same for all age groups? Races? Now click on the Keyword **EUROPE** or **ASIA** to compare the U.S. data with data from other parts of the world.

COMMUNICATE YOUR RESULTS

Exercise 1

Estimated time to complete = 10 minutes

Going back through the years, can you find a pattern (from the websites in *Web Investigation 1*) that indicates a worldwide flu season? Are there flu seasons within regions of the world? Is the outbreak of flu as widespread from year to year? Recently, which years have had the worst flu outbreak? Fast-forward now, about 10 years. You are working for a government agency that has been charged with creating policy to stem the spread of a new influenza strain, which has been rapidly spreading to pandemic proportions during the past four months. What is your plan?

Exercise 2

Estimated time to complete = 10 minutes

Use the description of antibody function from *Web Investigation 2* to compose a 100-word description of the immune system's response when presented with an antigen such as a particular influenza strain the first time, or a second or third time. What types of antibodies are produced at each exposure? What antibodies regulate the robust response to the second exposure? Why does influenza repeatedly attack human populations, frustrating our immune systems?

Exercise 3

Estimated time to complete = 15 minutes

Construct bar graphs illustrating differences in the occurrence of HIV/AIDS in the United States with its occurrence in 10 other countries around the world. Use the bar graphs to answer the following questions: Are the differences you find as striking between two developed countries versus a developed country and an undeveloped country? How about between Western countries and countries in Asia? Where is the occurrence of HIV/AIDS the highest? How likely are you to meet someone with HIV/AIDS in your own age group in the United States versus one of the other countries?

26

Transport, Nutrition, and Exchange
Blood, Breath, Digestion, and Elimination

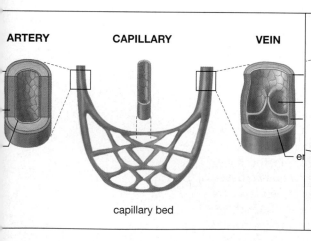

ARTERY   CAPILLARY   VEIN

capillary bed

Body pipes.
(Section 26.1, page 594)

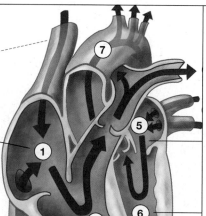

A hearty pumper.
(Section 26.2, page 595)

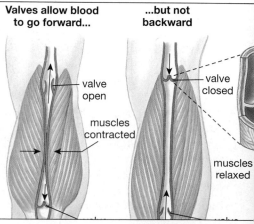

Valves allow blood to go forward...   ...but not backward

valve open

muscles contracted

valve closed

muscles relaxed

One-way street.
(Section 26.4, page 598)

Our bodies have a circulation system that transports materials to where they are needed, a respiratory system that acquires oxygen and gets rid of carbon dioxide, a digestive system that breaks down food, and a urinary system that gets rid of some materials while recycling others.

Many times in this text, you have had occasion to think of life in terms of its building blocks, with one set of those blocks being the cells that help make up organisms. One problem with this metaphor is that cells are living things and thus, unlike real building blocks—bricks or cinder blocks—they have ongoing maintenance requirements. Cells have to obtain energy and get rid of wastes; in many cases they have to obtain oxygen. They have to have a source of nutrients such as nitrogen and phosphorus. When life consisted of nothing but single-celled creatures, every cell had access to the world outside itself, which meant that food could enter, waste could be expelled, and gases could be "exchanged" in a fairly simple manner. In a human being, conversely, many of our 10 trillion or so cells spend their entire existence wrapped inside layers of tissue which themselves are wrapped inside other layers of tissue.

During the course of evolution, this access problem, as it might be termed, was dealt with by the development of physiological *systems*: a special bodily apparatus for delivering nutrients, another for delivering oxygen, and another for removing waste, for example. The capabilities of these three systems—digesting food, taking in oxygen, and getting rid of waste—are subjects of this chapter. These systems are all involved in carrying out various exchanges of materials between a living thing and its surroundings. One additional system then links all three of these systems together. It is the cardiovascular system, which—through blood—picks up gases, nutrients, and wastes from one location in the body and brings them to others. The essential subject of this chapter is how the body's transport and exchange operations function as a coherent life-support system. If you look at **Figure 26.1** on page 592, you can see a representation of how these systems are linked. Let's start this review with the cardiovascular system.

26.1   The Cardiovascular System and Body Transport

Those of you who went through Chapter 7, on energy, know that the reason oxygen is necessary to organisms such as ourselves is

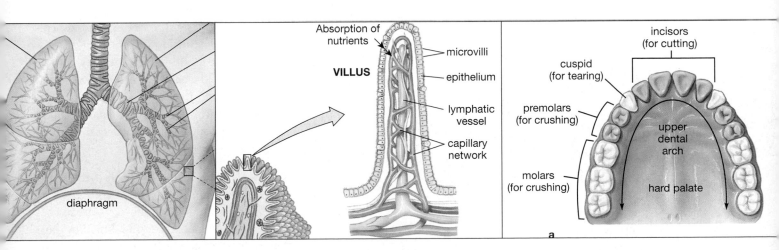

Breathe deeply.
(Section 26.5, page 599)

An absorbing story.
(Section 26.7, page 603)

Open wide.
(Section 26.7, page 604)

that oxygen is in the energy transfer business. Large living things need large amounts of energy, and oxygen is a critical element in helping them obtain it. But what is the distribution mechanism for oxygen? After entering the body through the lungs, how is it delivered to every cell in the body?

The answer is the cardiovascular system, which can be compared to the cooling system of a car. A car uses antifreeze and water as its circulating fluids; the cardiovascular system has blood. A car may use a water pump to circulate its fluids; the cardiovascular system uses a pump called the heart. A car has an assortment of conducting pipes; the cardiovascular system does too, in the form of blood vessels.

Apart from oxygen, the cardiovascular system also transports other materials needed and excreted by the body's cells. These materials include nutrients, vitamins, carbon dioxide, hormones, waste products, and molecules and cells specialized for body defense. Let's start this tour of the cardiovascular system by looking at the material that runs through it—blood.

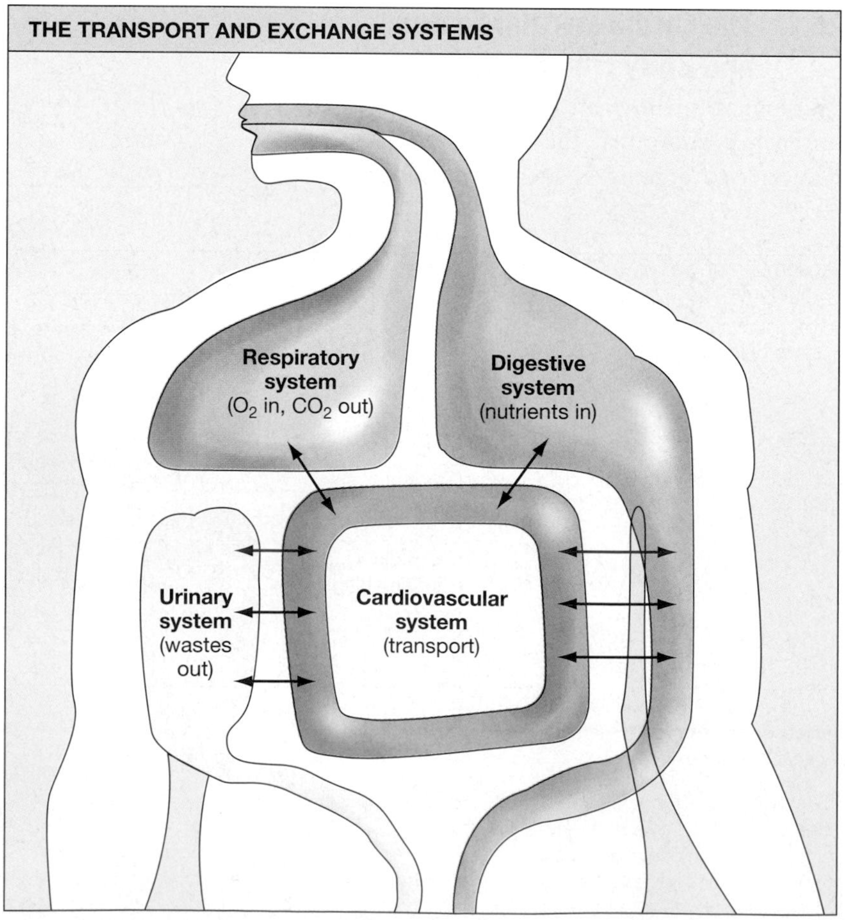

Figure 26.1
The Transport and Exchange Systems of the Body
The respiratory system moves oxygen into and carbon dioxide out of the body. The digestive system puts the food we eat into a form that can be transported throughout the body. The urinary system filters bodily fluids, removing waste while conserving water and other materials. The cardiovascular system is central to all the others, in that it transports materials to and from them.

THE TRANSPORT AND EXCHANGE SYSTEMS

Respiratory system (O_2 in, CO_2 out)

Digestive system (nutrients in)

Urinary system (wastes out)

Cardiovascular system (transport)

Blood Has Two Major Components: Formed Elements and Plasma

The volume of blood in the body varies from 4 to 5 liters (about 1 to 1.3 gallons) in an adult woman to 5 to 6 liters in an adult man. The temperature of the blood in the body is roughly 38°C (100.4°F), slightly higher than normal body temperature. Blood pH averages 7.4 and is therefore slightly alkaline or "basic," as opposed to acidic. Blood is also sticky, cohesive, and somewhat resistant to flowing freely. Blood is about five times "thicker," or more viscous, than water.

Collect a sample of blood from a human being, prevent it from clotting, and spin it in a centrifuge and it will separate into two layers. The lower layer, accounting for some 45 percent of the total volume, consists of blood *cells* and cell fragments, which are known as the **formed elements** of blood (**see Figure 26.2**). The yellowish, upper layer, which makes up 55 percent of the volume of blood, is called **plasma**, meaning the fluid portion of blood, which is mostly water. In the body, the formed elements are suspended in the plasma.

Formed Elements

Formed elements include **red blood cells (RBCs)**, which transport oxygen to, and carbon dioxide from, every part of the body; the far less numerous **white blood cells (WBCs)**, which are components of the immune system; and **platelets**—small fragments of cells that have pinched off from larger, complete cells. Platelets contain cellular material that includes enzymes and other substances, collectively called **clotting factors**, that are important to the blood-clotting process.

The formed elements (including platelets) are produced through a process called *hemopoiesis*, which means "to make blood." Almost all blood cells and platelets are formed within bone marrow.

Red Blood Cells   Red blood cells (RBCs) are also known as **erythrocytes**—from *erythros*, which means "red." Red blood cells are red because they contain the iron-containing protein **hemoglobin**, which is the element in the red blood cell that oxygen binds to. Blood as a whole is deep red because RBCs constitute 99.9 percent of all the formed elements in it. Indeed, the *number* of RBCs in the blood of an average person staggers the imagination. One cubic millimeter of blood—an amount that might scarcely be visible—contains roughly 4.8 to 5.4 million RBCs. So numerous are these cells that they make up roughly one-third of all the cells in the human body.

Numerous though they may be, RBCs are much different from most cells. A mature red blood cell in a mammal primarily consists of a cell membrane surrounding a mass of hemoglobin. Viewed another way, an RBC has neither the nucleus nor many of the organelles that are standard equipment in almost all eukaryotic cells. Because of this difference, the RBC has a life span of about 120 days, which is relatively short. About 1 percent of the circulating erythrocytes are replaced each day, meaning that approximately 180 million new erythrocytes enter the circulation *each minute*.

White Blood Cells White blood cells are also known as **WBCs** or **leukocytes** (*leukos* means "white"). They are quite different from RBCs because they have nuclei, but lack hemoglobin. Those of you who read about the immune system in Chapter 25 were introduced there to the pivotal role of white blood cells in the body's defense system.

A typical cubic millimeter of blood contains 6,000 to 9,000 white blood cells—a tiny number relative to the red blood cells present in such a volume. Most of the WBCs in the body are, however, found in body tissues; those in the blood represent only a small fraction of the total population. For WBCs, the bloodstream is used primarily for transportation to sites of invasion or injury. WBCs also move under their own power by extending parts of their cytoplasm as a kind of cellular foot. This allows them to move along the walls of blood vessels and, when outside of the bloodstream, through surrounding tissues. When problems are detected, these cells leave the circulation and enter the affected area, getting out of the bloodstream by squeezing between the cells that form the blood-vessel wall. Once the WBCs reach the problem area, they engulf pathogens, cell debris, or other materials by **phagocytosis** or "cell eating." You went over the types of WBCs and their functions in Chapter 25 (see pages 572–573).

Platelets Platelets can be thought of as enzyme-bearing packets. They are not cells, but fragments of cells that have broken away from a set of unusually large cells that exist in the bone marrow. The function of the enzymes in platelets is to aid in blood clotting at sites of injury. Apart from releasing these enzymes, the platelets facilitate clotting by clumping together at the site of an injury, forming a temporary plug that slows bleeding while the process of clotting continues.

Blood's Other Major Component: Plasma

All the formed elements you've just read about are suspended in blood's other major component, plasma, which is 92 percent water. Proteins and a mixture of other materials are dissolved in this fluid as well. Most blood proteins are too large to cross through the walls of the tiny blood vessels called *capillaries*, and they thus remain trapped within the circulatory system. This large concentration of plasma proteins is one of the chief differences between plasma and the **interstitial fluid** found outside the circulatory system.

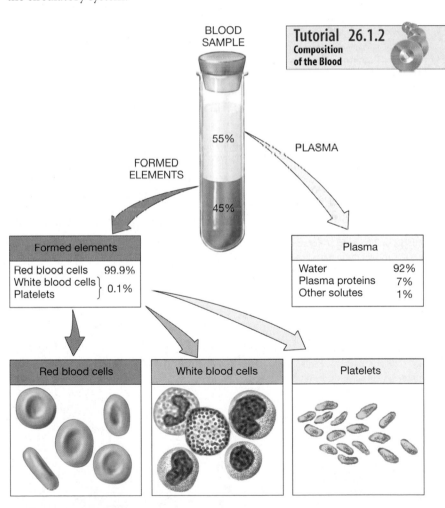

BLOOD SAMPLE

Tutorial 26.1.2
Composition of the Blood

55%

45%

PLASMA

FORMED ELEMENTS

Formed elements	
Red blood cells	99.9%
White blood cells }	0.1%
Platelets	

Plasma	
Water	92%
Plasma proteins	7%
Other solutes	1%

Red blood cells

White blood cells

Platelets

Figure 26.2
The Composition of Blood
Blood that is spun in a centrifuge separates into its two primary constituent parts, formed elements and plasma. Formed elements are cells and cell fragments that help make up the blood—red blood cells, white blood cells, and the cell fragments called platelets. Blood plasma, the liquid in which the formed elements are suspended, is composed mostly of water, but also contains proteins and other materials.

There are three primary classes of plasma proteins:

- **Albumins**, which constitute roughly 60 percent of the plasma proteins, are synthesized by the liver and function in the transport of both hormones and the components of fats known as fatty acids.
- **Globulins**, which make up 35 percent of the dissolved proteins, come in two forms that have important functions. The immunoglobulins or **antibodies** that you saw so much of in Chapter 25 attack foreign proteins and pathogens (disease-causing organisms). Meanwhile, the **transport proteins**, synthesized by the liver, carry small ions, hormones, or other compounds. Some of these transport proteins carry lipids that normally do not dissolve in water. These proteins are called **lipoproteins**; that name may sound familiar, because an important consideration in health is the relative abundance of the high-density lipoproteins (or **HDLs**) a person has circulating compared to low-density lipoproteins (or **LDLs**). For more on these proteins, see Chapter 3, page 60.
- **Fibrinogen**, the third type of plasma protein, functions in blood clotting. It, too, is made by the liver. Under certain conditions, fibrinogen molecules interact to form large, insoluble strands of fibrin, the basic framework molecule in a blood clot.

The remaining 1 percent of the plasma consists of such compounds as electrolytes (ions), organic nutrients, organic wastes, and hormones. The two major ions dissolved in plasma are sodium (Na^+) and chloride (Cl^-), whose combined form (NaCl) is better known to us as table salt. Some of the other ions important for normal cellular activities include potassium (K^+), calcium (Ca^{2+}), and iodide (I^-). Nutrients include glucose, fatty acids, and amino acids. They are used by cells for production of the energy-currency molecule, ATP, as well as for growth and maintenance. Wastes dissolved in the plasma are carried to sites of excretion. For example, the **urea** produced by the liver's breakdown of amino acids is carried to the kidneys.

Blood Vessels

Having looked at what moves through the cardiovascular system, let's now look at the plumbing that the blood moves through, our network of blood vessels.

Blood pumped by the heart is carried within tubes of varying diameter to all parts of the body. Blood vessels carrying blood *away* from the heart are the **arteries**, while the vessels carrying blood back *to* the heart are the **veins**. The farther from the heart these vessels are, the smaller they tend to be. The smallest of all blood vessels, the **capillaries**, connect the arteries with the veins. It is through the thin walls of the capillaries that gases, nutrients, and wastes are exchanged between the blood and the body's tissues.

The walls of arteries and veins contain three distinct layers (**see Figure 26.3**). The innermost layer is made up of a group of flat epithelial cells, an endothelium. The middle layer contains smooth muscle tissue. When these smooth muscles contract, the vessel decreases in diameter (an action known as **vasoconstriction**); when they relax, the diameter increases (an action termed **vasodilation**).

This mechanism plays a part in regulating our blood pressure; as diameter decreases, pressure increases. The outer layer of arteries and veins consists of a stabilizing sheath of connective tissue. These multiple layers give arteries and veins considerable strength, and the muscular and elastic components permit controlled alterations in diameter.

Arteries and Veins

The largest arteries are extremely resilient, elastic vessels with diameters of up to 2.5 centimeters or about 1 inch. The smallest arteries, the **arterioles**, have an average diameter of about 30 micrometers (μm), which is about a thousandth of an inch. The vessel walls of veins are thinner than arteries, because veins do not have to withstand much blood pressure. Most of this pressure has been spent by the time blood arrives at the veins from the capillaries.

Figure 26.3
The Structure of Blood Vessels
Both arteries and veins have three layers to them: an inner endothelium, a layer of smooth muscle, and a layer of connective tissue. By contracting or relaxing, the smooth muscle can change the diameter of the vessels. Many veins contain valves that prevent a backflow of the blood that is moving through them toward the heart. Arteries and veins are linked by intervening capillary beds. A capillary generally consists of a single layer of endothelial cells.

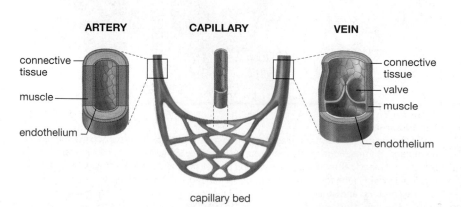

ARTERY   CAPILLARY   VEIN

connective tissue
muscle
endothelium

connective tissue
valve
muscle
endothelium

capillary bed

Capillaries

Capillaries are the only blood vessels whose walls are thin enough to permit exchanges between the blood and the surrounding interstitial fluids. A typical capillary consists of a single layer of endothelial cells and has an average diameter of only 8 μm. Capillaries exist in interconnected networks called **capillary beds**, as you'll see shortly.

Systems of Circulation

Vertebrates such as ourselves employ the blood vessels you've just reviewed in what is called a **closed circulation** pattern of blood transport—one in which blood is delivered from one set of vessels to another. In contrast, many invertebrates have an **open circulation** that lacks capillaries. In such systems, arteries carry blood into open spaces, called *sinuses*. Blood from such sinuses flows around cells and is then returned to a heart or hearts through openings in veins.

26.2 The Heart and the Circulation of Blood

What makes the blood flow through this system of closed circulation? The answer is the muscular pump we call the heart. A small organ, roughly the size of a clenched fist, the heart actually is shaped something like a valentine heart in its major outline. It lies near the back of the chest wall, directly behind the bone known as the *sternum*, and sits at an angle with its blunt tip pointed downward and to the left side of the body.

As noted earlier, a central task of the circulatory system is to get oxygen to all the various bodily tissues. For blood to carry oxygen to tissues, the blood first has to *take up* the oxygen that is brought into the body through the lungs. So imposing is this task of oxygenating blood that the body has an entire circulation loop devoted to this function. Once oxygenated in this loop, blood then returns to the heart, where it is pumped out to the rest of our tissues in the body's second major circulation loop.

To put these things more formally, there are two general networks of blood vessels in the cardiovascular system. One of these is the **pulmonary circulation**, in which the blood flows between the heart and the lungs. Then there is the **systemic circulation**, in which the blood is moved between the heart and the rest of the body. Now let's see how this two-network system works (**see Figure 26.4a**).

The human heart contains four muscular chambers, two associated with the pulmonary circulation and two associated with the systemic circulation. These are the right atrium and right ventricle (associated with

Tutorial 26.1.3
The Two Circulatory Networks

Figure 26.4
Two Circulatory Networks and How the Heart Supports Them

a Veins in the systemic circulation bring blood into the right side of the heart; it is pumped out via the pulmonary circulation to the lungs, where it is oxygenated, and then returns to the left side of the heart. It is then pumped out from the left side of the heart, through the systemic circulation, to the rest of the body.

b Deoxygenated blood enters (1) the right atrium of the heart through the superior and inferior venae cavae. The right atrium pumps the blood into (2) the right ventricle, which contracts, pumping the blood to the lungs, via (3) the pulmonary arteries. After being oxygenated in the lungs, the blood returns to heart through (4) the pulmonary veins and moves into (5) the left atrium, which pumps the blood into (6) the left ventricle. The powerful contractions of the left ventricle pump the blood out through (7) the aorta, after which the blood is distributed to the rest of the body.

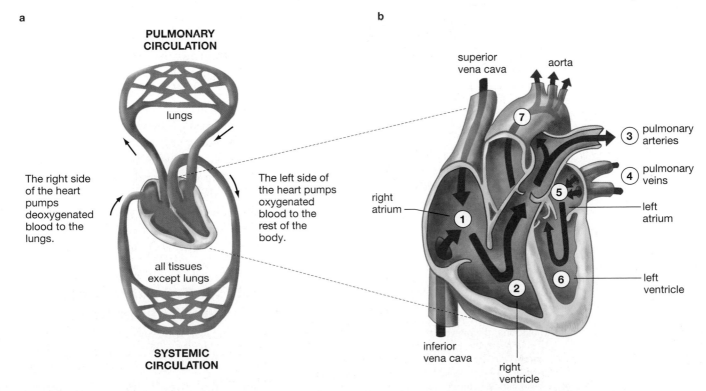

a

PULMONARY CIRCULATION

lungs

The right side of the heart pumps deoxygenated blood to the lungs.

The left side of the heart pumps oxygenated blood to the rest of the body.

all tissues except lungs

SYSTEMIC CIRCULATION

b

superior vena cava

aorta

7

3 pulmonary arteries

4 pulmonary veins

right atrium

1

5

left atrium

left ventricle

2

6

inferior vena cava

right ventricle

the pulmonary circulation) and the left atrium and left ventricle (associated with the systemic circulation). A look at the path of blood through these chambers will give you a clear idea of how blood moves through the body.

Following the Path of Circulation

Let's take as the starting point the blood entering the right atrium (on the left side of **Figure 26.4b**) from the veins called the *superior* and *inferior venae cavae*. These veins receive the blood returning to the heart from the upper and lower portions of the body, respectively. This is blood that is returning to the heart after having distributed oxygen and other blood-borne materials to the rest of the body. As such it is deoxygenated blood, actually dark red in color, but indicated by the blue color of the veins it flows through. After the right atrium receives this blood, it pumps it the short distance to the right ventricle, which contracts, pushing the blood into the pulmonary arteries. (These are arteries instead of veins, because blood moves *away* from the heart through them.) This blood is now bound for the lungs, where it will pick up oxygen. Having done so, it returns to the heart through the pulmonary veins—veins that in this case are carrying oxygen-*rich* blood, which is bright red. These veins both empty into the left atrium. The blood then is pumped into the left ventricle, which contracts and sends the blood coursing up into the enormous artery called the **aorta**. This vessel has branches stemming from it that will carry the blood to all the tissues of the body except the lungs.

Taking a step back from this, you can see that the right side of the heart is pumping blood to the lungs, while the left side of the heart is pumping blood to all the body's other tissues, as shown in Figure 26.4a. All four chambers of the heart pump blood, but the right and left *atria* are essentially blood collection sites that pump blood only into their adjacent ventricles. The right *ventricle* is pumping blood to the lungs, so there is some greater contraction power needed there. But the *left* ventricle is pumping to all the rest of the body, and its contractions are by far the most powerful. When we "take our pulse," what we are feeling is the surge of blood produced by the contraction of the left ventricle.

Control of Blood Flow through Valves

Because both ventricles are contracting with such force, you may wonder why blood doesn't go back into their respective atria with each beat. The answer is that there are valves between the atria and ventricles that allow the blood to flow only one way—from atrium to ventricle. Occasionally, because of disease or genetic predisposition, people do have this kind of backflow into the atria. The fluid turbulence that results can be heard as a kind of gurgling sound, known as a "heart murmur" (which is not necessarily a sign of a serious heart problem). Valves also exist to prevent backflow between the ventricles and the arteries they pump blood into. The heart's set of valves is responsible for the familiar "lub-dub" sound of the heartbeat. The "lub" is the sound of the valves closing between the atria and ventricles, while the "dub" is the sound of the valves closing between the ventricles and the arteries.

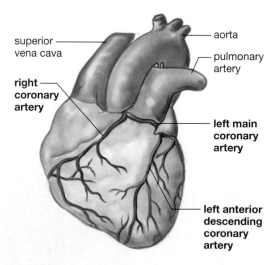

superior vena cava
aorta
pulmonary artery
right coronary artery
left main coronary artery
left anterior descending coronary artery

Figure 26.5
Critical Vessels
Damage to the coronary arteries brings about the heart attacks and heart conditions that are responsible for about half of all deaths in the United States. The coronary arteries begin with a right and left coronary artery that branch off from the aorta. These arteries then undergo branching themselves, thus supplying blood to the whole heart. Blockage of the left anterior descending coronary artery, shown in the drawing, figures in almost half of all heart attacks; this vessel supplies blood to heart muscle in the left ventricle.

26.3 The Heart's Own Blood Supply: What Is a Heart Attack?

The heart is in some ways a unique organ; but like any other organ, it too needs a supply of blood and the oxygen and nutrients that come with it. Indeed, the heart is a voracious user of these materials because it essentially is a set of muscles that never stop working. Blood does not come to the muscles of the heart through any of the chambers you've been reviewing. (Think about the two ventricles; one is sending blood to the lungs, the other to the rest of the body, but not directly to the heart itself.) The way blood does come to the heart is through two arteries that branch off from the aorta just after it emerges from the left ventricle. Because these arteries encircle or "crown" the heart before they start branching, they are known as the **coronary arteries** (see Figure 26.5).

It is difficult to overstate the importance of the coronary arteries, because about half of all deaths in the United States today are caused by a blockage of one or more of them. Such blockages generally are initiated by a buildup of small masses of fatty material called *plaque* that accumulate between the smooth muscle and inner lining of an artery. As a result of the blockage, the blood supply can be cut off altogether to given collections of heart-muscle cells. Small or large groups of these "myocardial" cells can die when this happens; the technical term for this event is a myocardial infarction, but it is better known as a **heart attack**. Death can come suddenly from such an event in several ways. The heart muscles may no longer be able to contract with sufficient force to pump blood through the body; or a wild, irregular rhythm, called a *ventricular fibrillation*, can be set off.

The terrible consequences that can result from having a blockage of one small vessel within the body ought to make us think twice about high-fat diets, smoking, and lack of exercise, all of which are contributing factors to heart disease. It's worth noting that while heart attacks are generally events of middle or old age, the buildup of fatty plaques seems to proceed from childhood forward. In other words, it is never too early to adopt a healthy lifestyle.

26.4 Getting the Goods to and from the Cells: The Capillary Beds

Having looked at the players involved, you're now ready to see how the cardiovascular system actually carries out its central functions of bringing materials to tissues and taking materials away from them. Propelled by the heart through the aorta, blood flows through the tube-like arteries. The large-diameter arteries branch repeatedly, gradually decreasing in size until they become the smaller-diameter **arterioles**. Arterioles then branch into the delivery vehicles of the cardiovascular system, the capillary beds. If you look at **Figure 26.6a**, you can see a representation of one of these beds. Note that at one end, the bed extends from the body's arterial system, but that at the other it is feeding into the body's system of *veins*—the venous system that is bringing blood *back* to the heart. Looking at the venous system as it begins with the capillaries, you can see that it initially consists of the tiny **venules**, which then merge into veins; smaller veins eventually merge into larger ones that finally come together into the largest veins of all, the two vena cavae that feed directly into the heart.

This explains "blood-out" (through the arteries) and "blood-back" (through the veins), but the real action in the circulatory system takes place in the capillaries that lie *between* the arteries and the veins. Capillaries are so small in diameter that red blood cells must squeeze through them in single file; and as you've seen,

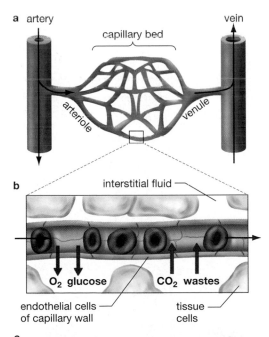

a artery | vein

capillary bed

arteriole | venule

b interstitial fluid

O₂ glucose | CO₂ wastes

endothelial cells of capillary wall | tissue cells

c

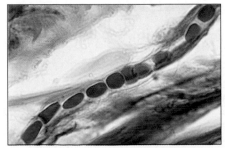

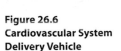

Tutorial 26.1.5
Capillary Beds and Exchange of Materials

Figure 26.6
Cardiovascular System Delivery Vehicle
Capillary beds are the sites of the exchange of materials between blood vessels and the body's tissues.

a An idealized view of how blood flows from arteries to arterioles, through capillary beds (losing oxygen along the way) and then into the venules and veins that return it to the heart.

b Blood pressure moves materials out of the capillaries near the arterial end of capillary beds, while osmotic pressure moves materials into the capillaries near the venous end of the beds.

c Micrograph of red blood cells moving in single file through a capillary.

each capillary is composed of but a single layer of cells. Think now about a line of oxygenated red blood cells, suspended in plasma and moving through a capillary along with hormones, glucose, ions, and other materials needed by the cells *outside* the capillary. The oxygen, glucose, and all the rest move out through the thin wall of the capillary itself and into the **interstitial fluid**—the liquid in which both the capillary and the cells around it are immersed. It is from this fluid that these needed materials will make their way to the surrounding cells. The distance the materials must travel to the cells is not far, however. So extensive is the body's capillary network that no cell is farther than two or three cells away from a capillary. Oxygen leaves the red blood cells and diffuses across the capillary lining and into the interstitial fluid. (The red blood cells themselves are too big to pass out.) Glucose and small ions such as sodium likewise are distributed in this manner. Meanwhile, carbon dioxide and nitrogenous wastes are flowing *into* the capillaries, with this movement, like

that of oxygen and other materials, driven by a concentration gradient. (For more on concentration gradients, see Chapter 5, page 102.)

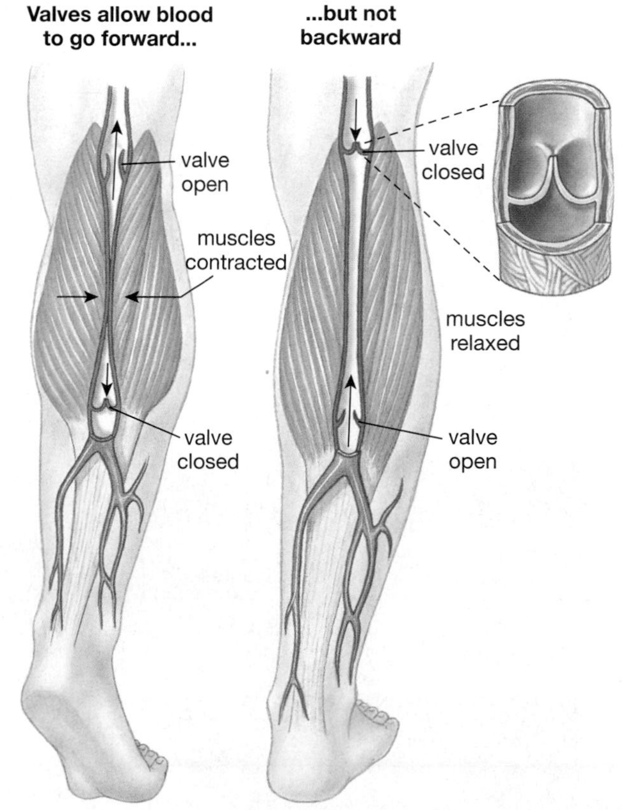

Forces That Work on Exchange through Capillaries

Water and some of the small molecules in it tend to flow *out of* a capillary bed at its arterial end, but *into* a bed at its venous end. This is the result of two opposing forces at work at any given point along a capillary. Blood pressure, which results from the heart's contractions, tends to move materials out of the capillary. At the arterial end of a bed, this pressure is relatively strong—strong enough to bring about a net movement of water out of the capillary. But the narrowness of the capillaries means that this pressure is largely spent by the time blood approaches the venous end of a capillary bed. At that point another force, osmotic pressure, tends to move materials *into* the capillary: Proteins too big to leave the capillary result in an osmotic gradient that pulls on fluids (for more on osmosis, see Chapter 5, page 102). At the venous end of a capillary, then, the force of osmosis overcomes the force of blood pressure. Most of the water that leaves the capillary beds is returned to them in this way, though some remains as interstitial fluid to be taken up by the vessels of the body's lymphatic system.

Muscles and Valves Work to Return Blood to the Heart

Because blood pressure has dropped to such low levels by the time the blood gets to the venous system, you may wonder how it can get back to the heart. The answer is that our muscles are essentially doing double duty. In contracting, they squeeze the veins in a way that moves the venous blood along. Blood can move only *toward* the heart in this system, because the veins have a series of valves in them that block movement of blood away from the heart (**see Figure 26.7**).

Valves allow blood to go forward...   **...but not backward**

valve open

muscles contracted

valve closed

valve closed

muscles relaxed

valve open

Figure 26.7
One-Way Flow to the Heart
Skeletal muscle contraction is the primary force that drives blood back to the heart through the venous circulation. A system of valves guards against the backflow of this blood.

26.5   The Respiratory System and the Exchange of Gases

One-celled organisms and other simple lifeforms (such as flatworms) do not need to breathe in the way that we do. They rely instead on the process of diffusion in order to bring oxygen in and move carbon dioxide out of themselves. Larger organisms have evolved various specialized organs, within which oxygen and carbon dioxide are exchanged with their environment. In aquatic habitats, gills serve this purpose for several different groups of animals (segmented worms, arthropods, mollusks, and fishes). Land-dwelling vertebrates, meanwhile, have evolved various types of lungs for the exchange of gases.

The Functions of Respiration

People can live perhaps a couple of months without food, and generally a few days without water. But if they go 5 or 6 minutes without breathing, death is usually the result. This is because, as you've seen, oxygen is in the energy transfer business; without oxygen, our cells simply don't have enough energy to function.

Cells may need to obtain oxygen to live, but they *generate* carbon dioxide as a consequence of living and need to dispose of it. Thus do we see the two central functions of respiration: capturing oxygen and disposing of carbon dioxide. Additional functions of respiration include defending the body against invasion by microorganisms, producing sounds for speaking, assisting in the regulation of blood volume and pressure, and assisting in the control of body fluid pH.

Structure of the Respiratory System

The lungs are one part of a **respiratory system** that also includes the nose, nasal cavity, and sinuses; the pharynx (throat), the larynx (voice box), and the trachea (windpipe); and the conducting passageways, called *bronchi* and *bronchioles*, that lead to the lungs (**see Figure 26.8**). The end points of the bronchioles are a profusion of tiny air sacs called **alveoli**. Alveoli

Figure 26.8
Components of the Respiratory System

a In the magnification of the alveoli, the capillary network that surrounds all alveoli has been cut away, on the right, to reveal the structure of the alveolar sacs.

b Artificially colored micrograph of a section of a human lung showing alveoli at the left and a larger bronchiole at right.

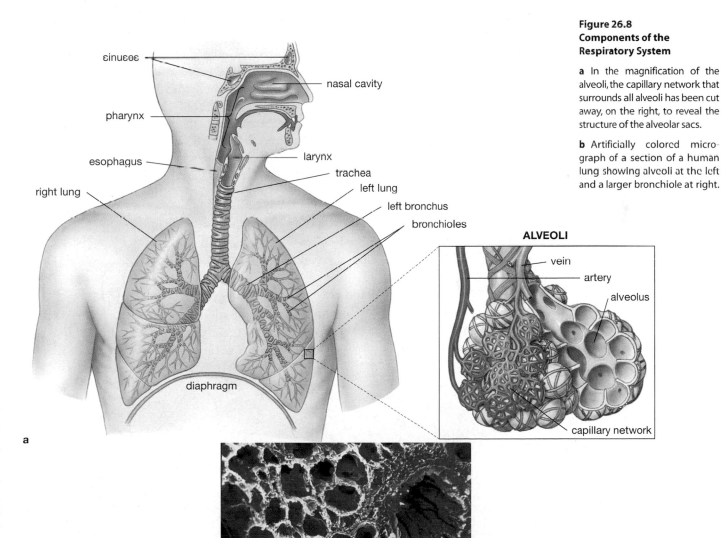

are like capillaries in a sense, because they are structures that lie at the ends of conducting passageways and are specialized for diffusion. Like the capillaries, the alveoli are thin enough to allow the diffusion of materials in and out of themselves.

Steps in Respiration: Breathing

So how does respiration work? First there is **pulmonary ventilation**, or breathing, meaning the physical movement of air into and out of the lungs. A single breath, or *respiratory cycle*, consists of an inhalation—the movement of air into the lungs; and an exhalation—the movement of air out of the lungs. This movement of air is controlled by differences in air pressure between the lungs and the external environment. At the start of a breath, pressures inside and outside the lungs are identical and there is no movement of air. When the chest or "thoracic" cavity enlarges, the pressure inside the lungs drops, and air enters the respiratory passageways (**see Figure 26.9**). Enlargement of the thoracic cavity involves the contractions of the diaphragm, aided by muscles of the rib cage (the *external intercostal muscles*). In exhalation, the downward movements of the rib cage and upward movement of the diaphragm reduces the size of the lungs. Pressure inside the lungs now exceeds atmospheric pressure, and air moves out of the lungs.

During quiet breathing, as when you're sitting and reading, exhalation is a passive process. That is, when the respiratory muscles used during inhalation relax, the stretched elastic fibers of the lungs recoil and air is forced out. When you are breathing heavily, contractions of another set of rib cage muscles (the *internal intercostal muscles*) and the abdominal muscles assist in exhalation.

Steps in Respiration: Exchange of Gases

Pulmonary ventilation is just one step in respiration. Next there is the exchange of respiratory gases, which occurs in two general processes. *External respiration* is the diffusion of gases between the alveoli and the blood. Think of it; the oxygen you inhale is needed in all the internal tissues of the body. How is it going to get there? It diffuses through the thin layer of alveoli and into an adjacent capillary; inside the capillary, oxygen is bound to one of the red blood cells that are so numerous within the blood. Thus locked up in the blood, the oxygen is ready to move with it into the heart, which pumps the oxygenated blood out to the rest of the body. At some point, the oxygen is dropped off, to be used by a bodily tissue.

Meanwhile, at the same capillary-alveoli interface, carbon dioxide that has come *from* the body's tissues moves into the alveoli from the capillaries. From the alveoli, it is expelled into the environment outside the body (**see Figure 26.10**). Carbon dioxide is moving to the external environment in this process, while oxygen is moving from the external environment; hence, these are forms of external respiration.

But there is also *internal respiration*, meaning the two-way diffusion of gases between the blood that flows through the capillaries and the interstitial fluid that lies outside the capillaries. Because cells are immersed in the interstitial fluid, they also can use the two-way movement of gases between the capillaries and interstitial fluid to obtain the gases they need and to dispose of their waste gases.

In both internal and external respiration, oxygen and carbon dioxide diffuse independently of each other—the exchange of one is

**Figure 26.9
How We Breathe**

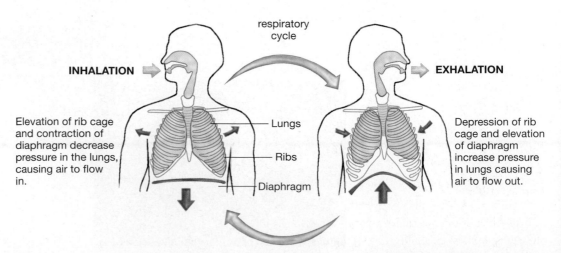

respiratory cycle

INHALATION

Elevation of rib cage and contraction of diaphragm decrease pressure in the lungs, causing air to flow in.

EXHALATION

Depression of rib cage and elevation of diaphragm increase pressure in lungs causing air to flow out.

Lungs

Ribs

Diaphragm

not linked to the exchange of the other. The gases are moving down their concentration gradients, which is to say moving from an area of their higher concentration to an area of their lower concentration. In the external respiration of carbon dioxide, for example, more carbon dioxide exists inside the capillaries than in the alveoli; thus the carbon dioxide is "following" its concentration gradient in diffusing out of the capillaries and into the alveoli.

Steps in Respiration: Oxygen and Carbon Dioxide Transport

So how does blood transport the two gases you've been looking at? In other words, how do these gases get loaded into and out of blood? Recall that blood is made up of both formed elements (cells and cell fragments) and plasma (a watery, protein-filled fluid). Almost all the oxygen that moves from the alveoli to the blood in the capillaries is bound to molecules of the protein hemoglobin (Hb) that exist in red blood cells.

In contrast to this simplicity, the carbon dioxide that is being transferred *to* the alveoli is transported by the blood in three different ways after being picked up from the interstitial fluid. As **Figure 26.11** shows, 93 percent of the carbon dioxide that comes into capillaries enters red blood cells, meaning that only 7 percent remains dissolved in the blood plasma. In the red blood cells, some of the carbon dioxide binds directly to hemoglobin, but most (70 percent) is rapidly converted within the red blood cells to bicarbonate ions (HCO_3^-). These newly formed ions then move out of the red blood cells and into the blood plasma. Hydrogen ions that are produced by this reaction are absorbed by hemoglobin molecules. Upon arrival at the capillaries in the lungs, this process is reversed. The bicarbonate and hydrogen ions combine and reform carbon dioxide, which diffuses into the alveoli; the hemoglobin-bound carbon dioxide is released, diffusing into the alveoli; and the carbon dioxide that had been dissolved in the plasma diffuses into the alveoli.

26.6 The Digestive System

Let's shift gears now, moving from respiration to digestion. The digestive system is in one sense very simple. Its central feature is a muscular tube that passes through the body from the mouth to the anus. The system consists of the tube, the **digestive tract**, and various accessory organs, including the salivary glands, gallbladder, liver, and pancreas. Through the food we eat, the digestive system supplies the energy and chemical building blocks for growth and maintenance.

The Digestive Process in Overview

The central function of the digestive system is to get the foods the body ingests into a form the body can use; the system must also collect and then rid the body of the waste that remains once the useful material from

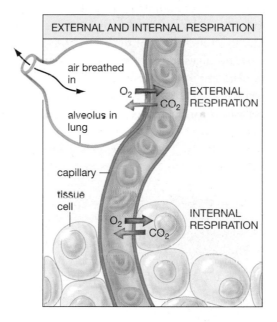

Figure 26.10
How Carbon Dioxide Is Transported in the Blood

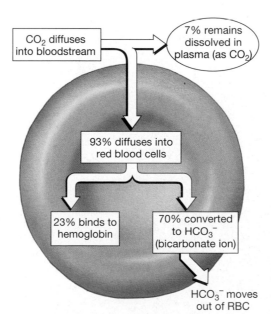

Figure 26.11
External and Internal Respiration
External respiration involves the movement of gases between capillaries and the alveoli—oxygen moving from the alveoli to the capillaries, and carbon dioxide moving to the alveoli from the capillaries. Internal respiration involves the exchange of gases between capillaries and the cells outside them.

the food has been removed. You've seen already that nutrients are delivered to tissues throughout the body via the circulatory system. So the real question for the digestive system is: What does it need to do to food to get it into a form that can *move into* the circulatory system? For carbohydrates and proteins, the pathway is straightforward. Partway through the organs called the *small intestines*, carbohydrates and proteins have been broken down enough that they move out through the inner lining of the intestine and into adjacent capillaries. This network of capillaries then feeds into a common vein that goes to the liver. Most nutrients, then, are carried straight from the small intestine to the liver. In a controlled release, the liver then sends these nutrients out (through another vein) to the heart, which pumps them out to all the tissues of the body.

The digestion of fats takes a somewhat more complicated route, as you'll see. But the principle remains the same; fats must be broken down to such a point that they can leave the digestive tract for the rest of the body. For all foods, "breakdown" means just what it sounds like: a transition from large to small. The transition goes from big, visible bites of food to smaller bits, and eventually from big *molecules* of food to the building blocks that make them up. It is only these smaller molecules—simple sugars, amino acids, and fatty acids—that are ready to leave the digestive tract for the rest of the body. Now let's take a look at the steps in digestion and the components of the digestive system, after which you'll follow the path of digestion. The functions of the digestive system involve a series of interconnected steps:

- *Ingestion.* The entry of food into the digestive tract through the mouth.
- *Mechanical processing.* The physical manipulation of solid foods, first by the tongue and the teeth and then by swirling and mixing motions of the digestive tract.
- *Digestion.* The chemical breakdown of food into small molecules that can be absorbed by the digestive lining.
- *Secretion.* The release of water, acids, enzymes, and buffers from the digestive tract lining and accessory organs, aiding digestion.
- *Absorption.* The movement of small organic molecules, electrolytes, vitamins, and water across the digestive inner lining—or epithelium—into capillaries or lymphatic vessels.
- *Excretion.* The elimination of waste products from the body. In the digestive tract, these waste products are compacted and discharged through defecation.

The lining of the digestive tract also plays a defensive role by protecting surrounding tissues against the corrosive effects of digestive acids and enzymes, and pathogens that are either swallowed with food or reside within the digestive tract.

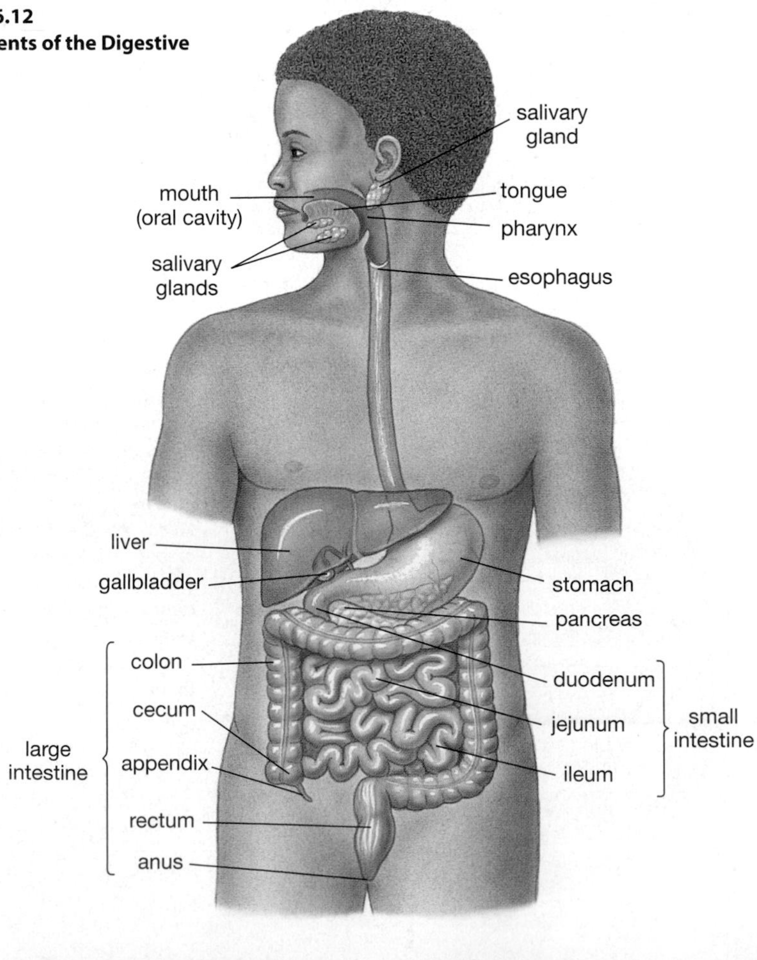

26.7 Components of the Digestive System

Figure 26.12 shows the locations and functions of the accessory glands and different subdivisions of the digestive tract. The digestive tract begins with the oral cavity and continues through the pharynx, esophagus, stomach, small intestine, and large intestine before ending at the rectum and anus. (The small and large intestines are further divided into various regions that are described later.) Other terms used to describe the digestive tract are **alimentary canal** and **gastrointest-**

**Figure 26.12
Components of the Digestive System**

inal tract. Generally, the gastrointestinal tract, or **GI tract**, refers to the lower portion of the digestive tract—the stomach and intestines.

The Digestive Tract in Cross Section

Thinking again of the digestive tract as a muscular tube, let's look at a cross section of this tube, within the small intestine, to see how it is structured.

As you can see in **Figure 26.13**, the outermost layer of the tract is connective tissue known as the serosa, inside of which there is a layer called the muscularis externa, which is composed of two sets of smooth muscle. In a process called **peristalsis**, these muscles take turns contracting, which produces waves of contraction that push materials along the length of the digestive tract. Inside the muscularis externa there is the submucosa, a layer of loose connective tissue that contains large blood vessels and a network of nerve fibers, sensory neurons, and motor nerve cells. This nerve tissue helps control and coordinate the contractions of the smooth muscle layers, and also helps regulate the secretions of digestive glands. Then there is the epithelial layer that makes up the internal lining of the tract. It is called the mucosa, because it is constantly bathed in mucous secretions. Along most of the length of the digestive tract, the mucosa has a large number of folds that increase the surface area available for absorption and permit expansion like an accordion—helpful after a large meal. Additionally, in the small intestine, the mucosa forms fingerlike projections called **villi** (singular *villus*) that greatly increase the area for absorption. It is the capillary and lymph vessels that run through the villi, just outside the epithelium, that nutrients move into after having been sufficiently digested.

The Oral Cavity

Let's now look at the process of digestion from start to finish, beginning with the part that's most familiar to us, the mouth.

The mouth opens into the oral cavity, the part of the digestive tract that ingests, or receives, food. Food is, of course, tasted and broken up into smaller pieces in the oral cavity through the mechanism of chewing. This mechanical processing is joined by a lubrication, through mucus and salivary secretions, that also takes place in

**Figure 26.13
Structure of the Digestive Tract**
The digestive tract is a tube formed of several layers of tissue, each performing a different function. Note that the absorption of nutrients takes place when food that has been sufficiently digested moves across the epithelium and is taken up by either a capillary or a lymphatic vessel inside a villus.

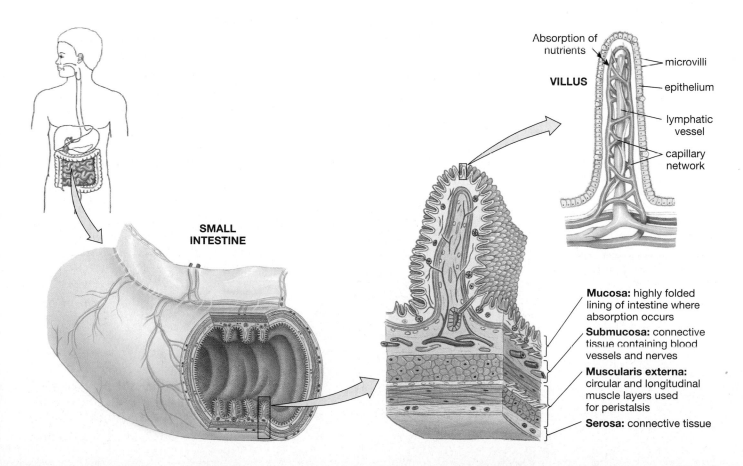

Absorption of nutrients

VILLUS

microvilli

epithelium

lymphatic vessel

capillary network

SMALL INTESTINE

Mucosa: highly folded lining of intestine where absorption occurs

Submucosa: connective tissue containing blood vessels and nerves

Muscularis externa: circular and longitudinal muscle layers used for peristalsis

Serosa: connective tissue

the oral cavity. The digestion of food—the breakdown by chemical means—starts here as well, as enzymes in the saliva start to break down carbohydrates.

Three pairs of salivary glands produce up to 1.5 liters, or almost half a gallon of saliva each day. Saliva is more than 99 percent water, but it also contains ions, pH neutralizers, waste products, antibodies, and the digestive enzymes already mentioned. Between meals, small quantities of saliva are continually released onto the surfaces of the oral cavity. These secretions clean the oral surfaces and, through the action of a bacteria-destroying enzyme, reduce the numbers of bacteria.

The Structure and Function of Teeth

The action of chewing breaks down tough connective animal tissues and plant fibers, and helps saturate the materials with salivary lubricants and enzymes. Adult teeth are shown in **Figure 26.14a**. There are four different types of teeth. Incisors, the blade-shaped teeth found at the front of the mouth, are useful for clipping or cutting. The pointed cuspids, also called *canines* or *eyeteeth*, are used for tearing or slashing. Bicuspids (or premolars) and molars have flattened crowns with prominent ridges for crushing and grinding food.

During infant development, two sets of teeth begin to form. The first to appear are the deciduous teeth, also known as *baby teeth*. When a child reaches the age of five or six, these teeth are gradually replaced by the adult permanent teeth. As we age, three additional teeth appear at the back of each side of the upper and lower jaws, extending the length of the tooth rows and bringing the permanent tooth count to 32. The third molars, or

wisdom teeth, may not erupt before age 21 because of abnormal positioning or inadequate space in the jaw. To avoid problems later in life, wisdom teeth are often surgically removed.

A cross section of a typical adult tooth is shown in **Figure 26.14b**. The bulk of each tooth consists of a mineralized material called dentin. It surrounds the connective tissue that makes up the central pulp cavity. The pulp cavity receives blood vessels and nerves through a narrow root canal at the tooth's base, or root, which sits within a bony socket. The neck of the tooth marks the boundary between the root and the crown. The dentin of the crown is covered by a layer of enamel—the hardest substance manufactured by living things.

The Pharynx and Esophagus

The **pharynx**, or throat, transmits solid food, liquids, and air. The pharyngeal muscles cooperate with muscles of the oral cavity and esophagus to initiate the process of swallowing. The muscular contractions during swallowing force the food along the esophagus and into the stomach.

The **esophagus** is a muscular tube that begins at the pharynx and ends at the stomach. It is approximately 2 centimeters or about 0.8 inch in diameter.

The Stomach

The stomach's main functions include (1) the temporary storage of ingested food, (2) the mixing of ingested foods, and (3) the breaking of chemical bonds (the digestion) of proteins through the action of acids and enzymes. What leaves the stomach is a mixture of food and gastric juices; this thick, soupy material is called **chyme**.

**Figure 26.14
Adult Teeth**

a Permanent teeth of an adult, showing the upper dental arch (roof) of the mouth.

b A typical adult tooth.

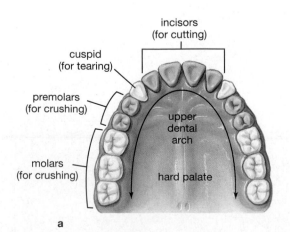

incisors
(for cutting)

cuspid
(for tearing)

premolars
(for crushing)

molars
(for crushing)

upper
dental
arch

hard palate

a

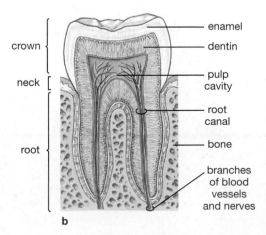

enamel

dentin

crown

pulp
cavity

neck

root
canal

root

bone

branches
of blood
vessels
and nerves

b

Figure 26.15 shows the muscular stomach. The **pyloric sphincter** muscle at its lower end regulates the flow of chyme between the stomach and small intestine. The dimensions of the stomach are extremely variable. When empty, it resembles a tube with a narrow cavity. When full, it can expand to contain 1 to 1.5 liters or up to almost 0.5 gallon of material. (The fact that it can hold material in this way is the reason we don't have to eat all the time.) This degree of expansion is possible because the stomach wall contains thick layers of smooth muscle, and the mucosa of the relaxed stomach contains numerous folds, called *rugae*. As the stomach expands, the smooth muscle stretches and the folds gradually disappear.

The Stomach's Digestive Juices

The stomach is lined by an epithelium filled with mucous cells. The mucus these cells produce helps protect the stomach lining from the acids, enzymes, and abrasive materials it contains. Shallow depressions called gastric pits open onto the internal surface of the stomach. Secretions from underlying gastric glands pass through the openings of the gastric pits. Each day, these glands produce about 1,500 milliliters or 45 ounces of gastric juice, which contains hydrochloric acid (HCl) and pepsin, an enzyme that breaks down proteins. The hydrochloric acid lowers the pH of the gastric juice, kills microorganisms, and breaks down cell walls and connective tissues in food.

The Small Intestine

The small intestine is about 6 meters (almost 20 feet) long and ranges in diameter from 4 centimeters (1.6 inches) at the stomach to about 2.5 centimeters (1 inch) at the junction with the large intestine. Although longer than the large intestine, it is called the small intestine because its overall diameter is smaller. It is made up of three regions, as you can see in **Figure 26.16**. The *duodenum* is the 25 centimeters (10 inches) of the small intestine closest to the stomach. It receives chyme from the stomach and secretions from the pancreas and liver. The *jejunum*, the next region, is about 1 meter (3 feet) in length. Most of the chemical digestion and nutrient absorption occurs in the jejunum. The third segment, the *ileum*, is the longest and averages 2 meters (6 feet) in length. The ileum ends at a sphincter muscle that controls the flow of chyme from the ileum into the large intestine. On average it takes about 5 hours for chyme to pass through the small intestine.

Most of the important digestive processes are completed in the small intestine, where the final products of digestion—simple sugars, fatty acids, and amino acids—move out of the digestive system, along with most of the system's water content. Approximately 80 percent of all absorption takes place in the small intestine, with the rest divided between the stomach and the large intestine.

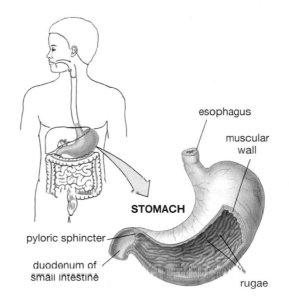

**Figure 26.15
External and Internal Views
of the Stomach**

esophagus

muscular wall

STOMACH

pyloric sphincter

duodenum of small intestine

rugae

**Figure 26.16
Structure of the Small
Intestine**

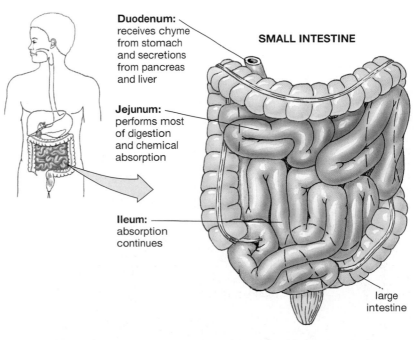

Duodenum: receives chyme from stomach and secretions from pancreas and liver

SMALL INTESTINE

Jejunum: performs most of digestion and chemical absorption

Ileum: absorption continues

large intestine

The Pancreas

The pancreas, shown in **Figure 26.17**, lies behind the stomach, extending laterally from the duodenum. It is an elongated, pinkish-gray organ about 15 centimeters (6 inches) long that is primarily an exocrine organ—one that secretes its materials through ducts. However, about 1 percent of its cells are endocrine cells that secrete substances directly into the bloodstream. One of these substances, insulin, moves blood sugar out of the bloodstream and into cells. The other substance, glucagon, releases glucose from storage in skeletal muscles and the liver. The exocrine secretions of the pancreas are digestive enzymes and buffers that are carried to the duodenum by the *pancreatic duct*, which passes through the duodenal wall along with the common bile duct from the liver and gallbladder. Pancreatic exocrine enzymes are broadly classified according to their intended targets. *Lipases* work on lipids, *carbohydrases* digest sugars and starches, and *proteases* break proteins apart.

The Liver and Gallbladder

Setting aside the skin, the liver is the largest organ in the body, being about the size of a football and weighing about 1.5 kg (3.3 pounds). In addition to its digestive role, this large, firm, reddish-brown organ provides a wide range of other essential services vital to life.

The liver lies directly under the diaphragm. As Figure 26.17 shows, it is divided into two large lobes, the left and right. Lodged within a recess under the right lobe of the liver is the **gallbladder**, a muscular sac that stores and concentrates a material called *bile*. Produced by the liver, **bile** is a digestive substance that aids in the digestion of fats. Bile can move into the small intestine through either of two routes. It always passes through passageways, called *hepatic ducts*, within the liver until it eventually leaves the liver through the common hepatic duct. The bile in this duct may either (1) flow into the small intestine's duodenum through the common bile duct, or (2) enter the cystic duct that leads to the gallbladder and only later flow into the small intestine (through the common bile duct). Why would it take one course or the other? A sphincter at the intestinal end of the common bile duct stays opens only at mealtimes. At other times, bile backs up through the cystic duct for storage within the expandable gallbladder before moving to the small intestine. Liver cells produce roughly 1 liter (about a quart) of bile each day.

Functions of the Liver

The liver plays a central role in regulating body metabolism. It can do this because all the blood carrying nutrients from the digestive tract is channeled into a single blood vessel, called the *hepatic portal vein* that flows into the liver. This enables liver cells to extract absorbed nutrients or toxins from this blood before they are sent to the rest of the body, and also to monitor and adjust the circulating levels of nutrients. Because the liver is the first stop for much digested material, it is also the first stop for toxins, such as alcohol (which is one of the few substances to move directly from the stomach to the liver without going through the small intestines). It's not surprising, then, that people who have abused their bodies with too much alcohol can over time find themselves with damaged livers.

The liver also functions as the largest blood reservoir in the body. In addition to nutrient-laden blood arriving through the hepatic portal vein, the liver also receives about 25 percent of the cardiac output. Phagocytic cells in the liver constantly remove aged or damaged red blood cells, debris, and pathogens from the body's circulation.

The Large Intestine

The large intestine begins at the end of the third section of the small intestine, the ileum, and ends at the anus. As you saw in Figure 26.12, the large intestine lies below the stomach and liver and almost completely frames the small intestine. If you look at **Figure 26.18**, you can see the structure of this organ in greater detail. The main function of the large

**Figure 26.17
Accessory Glands in
Digestion: The Liver,
Pancreas, and Gallbladder**

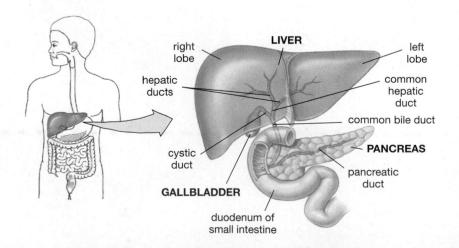

right lobe
LIVER
left lobe
hepatic ducts
common hepatic duct
common bile duct
cystic duct
PANCREAS
pancreatic duct
GALLBLADDER
duodenum of small intestine

intestine is to hold and compact the refuse produced by the rest of the digestive system (as solid waste, or feces). It also absorbs water and important vitamins produced by resident bacteria. The large intestine is called "large" only because it has a wider diameter than the small intestine, but it is only about 1.5 meters (5 feet) long—about a quarter the length of the small intestine. Often called the *large bowel*, it is divided into three major regions: (1) the cecum; (2) the colon; and (3) the rectum. Let's look at these in order.

Three Regions of the Large Intestine: Cecum, Colon, and Rectum

Material (chyme) arriving from the ileum of the small intestine first enters an expanded chamber, called the *cecum*. A muscular sphincter, the ileocecal valve, forms the connection between the ileum and the cecum. The cecum usually has the shape of a rounded sac, and the slender, hollow appendix attaches to it, below the ileocecal valve. The appendix walls contain lymphatic tissue, and it is infection of this tissue that produces the symptoms of appendicitis. The longest portion of the large intestine is the **colon**. Visible bands of longitudinal muscle produce the series of pouches characteristic of the colon. The lower portion of the colon empties into the rectum.

The reabsorption of water is an important function of the large intestine. Although roughly 1,500 milliliters (45 ounces) of watery material arrives in the colon each day, some 1,350 milliliters of water move back out of it, leaving only about 150 milliliters (4.5 ounces) that is ejected with the feces. In addition to water, the large intestine absorbs vitamins produced by bacteria that live within the colon.

Bacterial action breaks down protein fragments remaining in the feces into various compounds. Some of these are reabsorbed, processed by the liver into relatively nontoxic compounds, and eventually excreted at the kidneys. Others are responsible for the odor of feces or result in the generation of hydrogen sulfide (H_2S), a gas that produces a "rotten egg" odor.

Some carbohydrates are not digested and arrive in the colon intact, where they provide a nutrient source for bacteria. The metabolic activities of these bacteria are responsible for intestinal gas, or flatus. Beans often trigger gas because they contain a high concentration of indigestible polysaccharides for the bacteria to work on. Movement of chyme through the large intestine occurs very slowly, allowing hours for the reabsorption of water.

The **rectum** forms the end of the digestive tract. The rectal chamber is usually empty except when powerful peristaltic contractions force fecal materials out of the lower part of the colon. Stretching of the rectal wall then triggers the defecation reflex. **Defecation**, the excretion of solid wastes, occurs through the rectal opening, the anus.

26.8 Different Digestive Processes for Different Foods and Nutrients

A typical meal contains a mixture of carbohydrates, proteins, lipids, water, electrolytes (ions), and vitamins. Water, electrolytes, and vitamins can be absorbed directly into circulation, but large organic molecules, such as starches and proteins, must be broken down through digestion before absorption can occur. Once absorbed, these molecules can

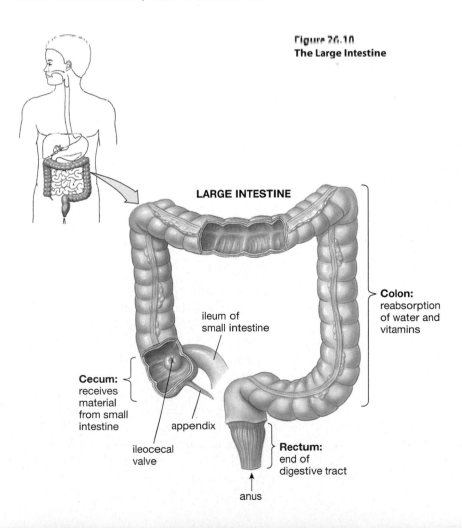

Figure 26.10
The Large Intestine

LARGE INTESTINE

Colon: reabsorption of water and vitamins

ileum of small intestine

Cecum: receives material from small intestine

appendix

ileocecal valve

Rectum: end of digestive tract

anus

be used by the body as building blocks for other molecules or to generate the body's "energy currency" molecule, ATP.

Organic molecules are usually complex chains of simpler molecules. A complex carbohydrate, such as a starch, is made of simple sugars. In a protein, meanwhile, the building blocks are amino acids, and in lipids they are fatty acids and glycerol molecules. Digestive enzymes split the bonds between the different building-block molecules in a process involving water, called **hydrolysis** ("to split with water.") (For more on carbohydrates, lipids, and proteins, see Chapter 3, beginning on page 47.)

Carbohydrates

Carbohydrate digestion begins in the mouth through the action of salivary amylase. Amylase breaks down complex carbohydrates into smaller fragments, producing a mixture primarily composed of two- and three-sugar molecules. Salivary amylase travels with the meal and continues to digest the starches and glycogen in it for an hour or two before stomach acids render the amylase inactive. In the small intestine, the remaining complex carbohydrates are broken down through the action of pancreatic amylase.

Further digestive activity occurs on the surfaces of the intestinal microvilli, where digestive enzymes reduce the two- and three-sugar molecules to simple single sugars. *Now* the carbohydrates exist in a form that is ready to pass into circulation. The simple sugars pass through the cells of the intestinal wall into the interstitial fluid of the villi. They then enter the intestinal capillaries and are transported directly to the liver by the hepatic portal vein.

Proteins

Proteins are relatively large and complex molecules, and different regions of the digestive tract cooperate to take proteins apart. Stomach cells secrete acid and the protein-splitting enzyme pepsin. The strongly acidic environment of the stomach provides the proper pH for pepsin to begin the process of reducing the large protein molecules into smaller fragments.

After the chyme enters the small intestine and the pH has risen (become less acidic), pancreatic enzymes come into play. Working together, different protease enzymes from the pancreas break the protein fragments into a mixture of short amino acid chains and individual amino acids. Enzymes on the surfaces of the microvilli complete the process by breaking the short amino acid chains into single amino acids. In this form, they are absorbed by the intestinal cells and then diffuse into intestinal capillaries and are carried to the liver.

Lipids

The most common dietary lipids are **triglycerides**. A triglyceride molecule consists of three fatty acids attached to a single molecule of glycerol. Triglycerides and other dietary fats are relatively unaffected by conditions in the stomach and enter the duodenum in the form of large, lipid drops. Bile breaks the large drops apart in a process called **emulsification**, creating tiny droplets that now present enough surface area for pancreatic lipase to begin breaking the triglycerides apart. Although the resulting lipid fragments are absorbed by the cells of the intestinal wall, they are too large to then enter intestinal capillaries. Instead, they are attached to proteins that aid their transport. The lipid-protein packages enter small lymphatic vessels that underlie the intestinal villi (Figure 26.13). From these vessels, the lipids proceed along the larger lymphatic vessels, and finally enter the venous circulation.

Water and Vitamins

Each day 2 to 2.5 liters (or 2.1 to 2.65 quarts) of water enter the digestive tract in the form of food and drink. The salivary, gastric, intestinal, and accessory gland secretions provide another 6 to 7 liters (6.3 to 7.4 quarts). The water total, then, is about 9.5 liters or 2.5 gallons a day; of this a remarkably small amount, about 150 milliliters or 4.5 ounces, goes out of the body with the fecal wastes. All the rest of the water is absorbed into the body from the digestive system.

Vitamins are organic compounds required in very small quantities for critical chemical reactions. There are two major groups of vitamins: water-soluble vitamins (those that dissolve in water; the C and the B vitamins) and fat-soluble vitamins (those that dissolve in fat; A, D, E, and K).

There is plenty of fat (or lipid) material in the body for the fat-soluble vitamins to dissolve into, such as the plasma membranes of cells, which are largely lipid. This means that our bodies can *store* substantial amounts of the fat-soluble vitamins. This can be very helpful, because we might be able to go for months at a time on stored reserves. It also means, however, that we can get too *much* of a fat-soluble vitamin such as A or D, with potentially harmful consequences. (Too much A can cause liver damage; too much D can cause calcium deposits.) There's little danger of this with the water-soluble vitamins, because they readily move between fluid stores and the bloodstream, with excess amounts excreted in the urine.

26.9 The Urinary System in Overview

Here our subject changes once again (for the last time in this chapter), as your tour of the body's systems moves from digestion to the elimination of liquid wastes. All organisms create waste materials simply by living; the question every living thing confronts is how to get rid of them. Single-celled creatures can simply dump their waste materials into their external environment through exocytosis. (See Chapter 5, page 107.) In all highly organized animals, excretory *organs* take over this task. If these organs are small and numerous, and scattered throughout the body, they are known as *nephridia*. Conversely, higher invertebrates and vertebrates have a single compact set of excretory organs that performs this function. These organs—the kidneys, the bladder, and others—are what you'll be looking at in this review of the body's urinary system.

Though gathering and discharging waste is an important function of the urinary system, it has other, equally important roles. For example, it regulates blood volume—how much blood we have within us—a function whose necessity becomes obvious when we think about consuming lots of liquids. Suppose the body simply incorporated all the water in these liquids into circulating blood? We would eventually swell and burst like a water balloon. Instead, the body *regulates* how much water it retains and how much it excretes, with the kidneys being a critical player in this process. The urinary system also controls ion concentration and maintains pH balance in the body. In short, it is important in preserving homeostasis. We

almost never think of the urinary system in this regard, essentially because it works so well. But a person who suffers kidney failure and must rely on a kidney "dialysis" machine to filter wastes knows the value of the urinary system in a personal way.

In carrying out its activities, the urinary system is very much involved in the *conservation* of bodily resources. Indeed, the kidneys are as much in the conservation business as the urination business. They are conserving water, amino acids, ions, sugars—all kinds of things that the body needs. Consider that the kidneys will process about 180 liters (or almost 48 gallons) of fluid every day. Yet our urination is only about 1.8 liters, or about *half* a gallon daily. More than 47 gallons cycled back into the body each day! Conservation indeed. Now let's see how the kidneys and the larger urination system manage this feat.

Structure of the Urinary System

The organs of the urinary system are shown in **Figure 26.19**. The pivotal organs in the system are the two **kidneys** that produce urine. Urine leaving the kidneys travels along two tubes, the left and right **ureters**, to the **urinary bladder** for temporary storage. When urination occurs, contraction of the muscular bladder forces the urine through the **urethra** and out of the body.

**Figure 26.19
The Urinary System
in Overview**

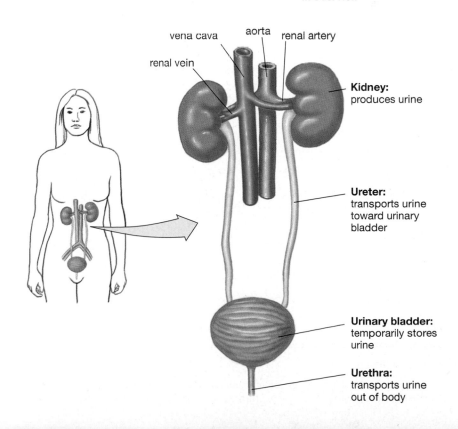

vena cava
aorta
renal artery
renal vein

Kidney:
produces urine

Ureter:
transports urine
toward urinary
bladder

Urinary bladder:
temporarily stores
urine

Urethra:
transports urine
out of body

26.10 How the Kidneys Function

The kidneys are located on either side of the vertebral column, at about the level of the eleventh and twelfth ribs. Not surprisingly, they are shaped like kidney beans. A typical kidney is about 10 centimeters (4 inches) long and about 3 centimeters (1 inch) thick.

In considering how the kidneys work, it's helpful to think first in terms of input and output. What comes into the kidney in the way of fluids? Blood from the circulatory system arrives by way of the renal arteries. As you can see in Figure 26.19, these branch off from the aorta, one renal artery going to the left kidney, one to the right. (The word *renal* comes from *renes*, which means "kidneys"; you'll be seeing a lot of "renal" this and that.) Now, what comes *out* of the kidney? As you've seen, urine, which exits each kidney through its ureter. However, blood is also flowing out of the kidneys, through the renal *veins*. Thus, each kidney has one input (the renal artery) but *two* outputs (the ureter and the renal vein). This is a clue to what's happening inside. Blood is flowing into the kidney through the renal artery and out through the renal vein; but inside the kidney, some of the materials that come in with the arterial blood (the waste materials) are being channeled out of the blood vessels and into a *separate* system of vessels—the system whose output tubes are the ureters. Now let's see how this happens inside the kidney.

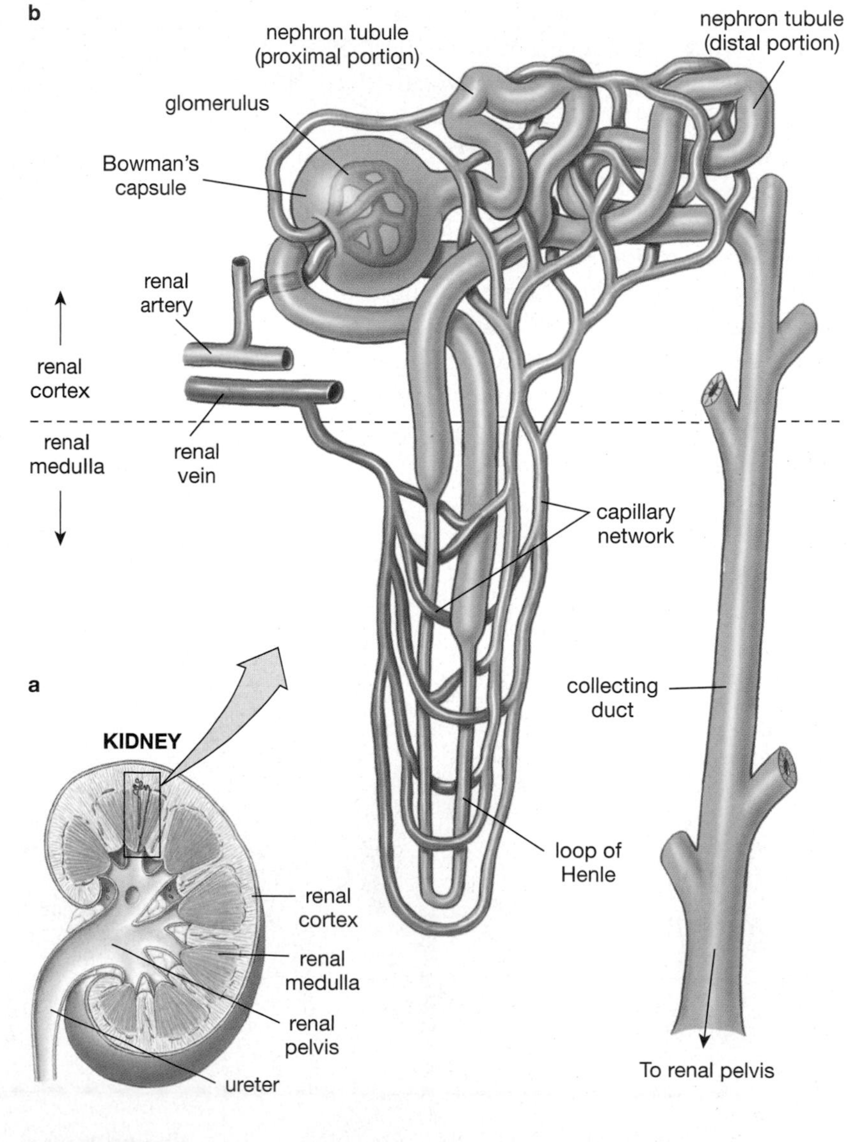

Seen in cross section in **Figure 26.20a**, the kidney can be divided into an outer *renal cortex*, an inner *renal medulla*, and a chamber called the *renal pelvis* whose numerous branches collect the urine produced by the kidneys and channel it into the ureter.

The work of the kidney begins in its outer portion, or cortex. If you look at **Figure 26.20b**, you can see the kidney's basic working unit, called a *nephron*. Each kidney has within it a million or so nephrons, each serviced by a single renal arteriole that has branched off from the entering renal artery. In the figure you can see that an arteriole branching off the renal artery enters a hollow chamber, called the **Bowman's capsule**, and then promptly expands into a knotted network of capillaries, called the **glomerulus**, that is surrounded by the capsule. *Emerging* from this network, there is again a single arteriole; but it once again branches into a network of capillaries that surrounds a convoluted nephron tubule that has both proximal ("near") and distal ("distant") portions. The **nephron**, then, is the functional unit of the kidneys, composed of a nephron tubule, its associated blood vessels, and the interstitial fluid in which both are immersed. At the "output" end of the nephron, the tubule is one of several that are emptying into a common **collecting duct**. Eventually, many of these ducts will themselves merge to feed into the renal pelvis, which feeds into the ureter.

Figure 26.21 uncomplicates things a little by showing you what the nephron tubule structure looks like without the surrounding blood vessels. You can see blood input at the

Figure 26.20
Structure of the Kidney and Its Working Unit

a A kidney in cross section.

b A nephron in the kidney, composed of a tubule and its associated blood vessels, which are immersed in interstitial fluid.

Bowman's capsule, then the proximal portion of the nephron tubule, known properly as the **proximal tubule**. Then there is the U-shaped loop of Henle, which ends with the **distal tubule** that feeds into the collecting duct. Note that the Bowman's capsule is in the renal cortex, but that the loop of Henle dips deep into the renal medulla. Now, what happens in this structure?

First Kidney Function: Filtration in Bowman's Capsule

Blood flows into the glomerulus capillaries under high pressure. As a result, some of the plasma component of the blood moves out of the capillaries and into the Bowman's capsule. Water, amino acids, ions, and glucose contained in the blood—along with a waste product called *urea*—are among the substances

Tutorial 26.3.3
Physiology of the Nephron

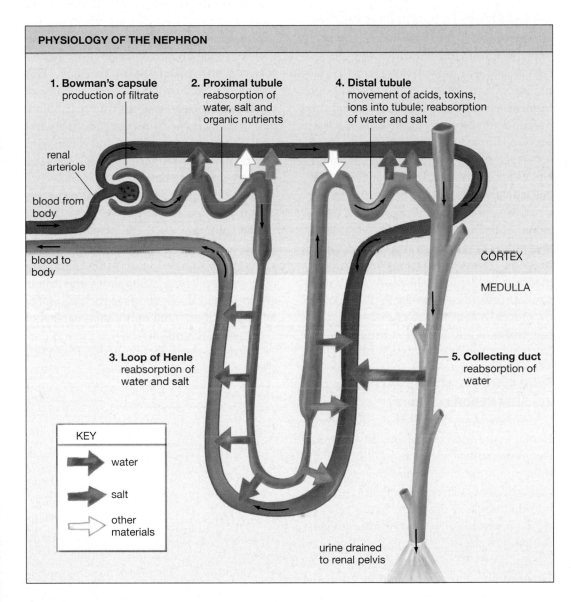

PHYSIOLOGY OF THE NEPHRON

1. Bowman's capsule
production of filtrate

2. Proximal tubule
reabsorption of water, salt and organic nutrients

4. Distal tubule
movement of acids, toxins, ions into tubule; reabsorption of water and salt

renal arteriole

blood from body

blood to body

CORTEX

MEDULLA

3. Loop of Henle
reabsorption of water and salt

5. Collecting duct
reabsorption of water

KEY

water

salt

other materials

urine drained to renal pelvis

Figure 26.21
Physiology of the Nephron

1. Blood flows into the glomerulus within the Bowman's capsule under pressure, driving some plasma components of the blood out of the glomerulus and into the capsule.

2. Conservation begins, as water, glucose, and other needed materials move from the proximal tubule back into the circulatory system.

3. Water and then salt move out of the tubule in the loop of Henle, producing an osmotic effect that moves more water out of the tubule, and into the interstitial fluid, at the collecting duct.

4. Acids, toxins, and ions move from capillaries into the distal tubule, even as water and salt move from the tubule back into the bloodstream.

5. Water moves from the collecting duct into the interstitial fluid and from there back into the circulatory system, thus preserving the water. Urine that remains drains to the renal pelvis.

that will make this transition. Together, they constitute a fluid, called **filtrate**, that is funneled from the capsule into the proximal tubule. In short, a transfer has been made from the blood vessels to the kidney's nephron tubule. What is not part of this **filtration** process are materials that were too big to pass out of the glomerulus capillaries: red blood cells, proteins, and lipid molecules.

Second Kidney Function: Reabsorption from the Nephron Tubule

Remember how you saw earlier that the kidneys are in the conservation business to a significant extent? Well, this conservation begins immediately after the filtrate moves into the proximal tubule. Cells in the tubule actively pump sodium (Na^+) from the filtrate into the interstitial fluid surrounding the tubule. Chloride (Cl^-) and water follow the sodium, and all these substances move into the network of capillaries surrounding the tubule, along with amino acids and sugars, which also are pumped out of the tubule. What is happening, then, is **reabsorption** of valuable materials back into the bloodstream. Indeed, about two-thirds of the water—and nearly all the glucose and amino acids that initially became part of the filtrate—are now moving back to the circulatory system.

Third Kidney Function: Secretion into the Nephron Tubule

If you now jump ahead (past the loop of Henle) to the *distal* tubule, you can see a similar movement of water and salt out of the tubule and into the bloodstream. Notice, however, that substances are also moving into the tubule at this point: Hydrogen ions (H^+) and potassium ions (K^+) are among the compounds that will move from the capillaries into the distal tubule, in a process known as **secretion**. (Substances coming from the capillaries are being taken up by tubule cells, which are secreting the substances into the tubule.) Some poisons, and medicines such as penicillin will move into the tubules at this point, to be excreted in the urine.

Fourth Kidney Function: The Loop of Henle, the Collecting Duct, and the Conservation of Water

As noted earlier, at the proximal tubule about two-thirds of the water in the filtrate moved back to the circulatory system. What

about the other third of the water? This is where the loop of Henle and the collecting duct come into play. The makeup of the loop allows for the selective movement of water and salt (NaCl) out along its length. In this process, the loop functions to maintain a concentration of salt in the interstitial fluid—a concentration that increases with depth in the medulla. This salt gradient exists in the fluid surrounding both the loop of Henle and the collecting duct. The result is a greater solute concentration outside the loop and collecting duct than inside them, particularly at the lower part of the loop. This means that water moves by osmosis out of the collecting duct, into the interstitial fluid, and from there back into the circulatory system. The arrows you see around the loop in Figure 26.21 show the movement of water and salt along its length.

When the filtrate reaches the collecting duct, it is called urine; with its loss of water in the collecting duct, the urine that flows through the duct may have as much as four times the concentration of dissolved solvents as does blood plasma. In short, this whole system allows the body to rid itself of a large amount of waste with only a minimal loss of water and nutrients.

To recap, the kidneys function by undertaking four main processes:

- *Filtration.* Portions of the blood plasma move from the glomerulus capillaries into the nephron tubule by way of the Bowman's capsule.
- *Reabsorption.* Water and nutrients move from the tubule back to the blood vessels, while waste is retained within the tubule.
- *Secretion.* Hydrogen ions and various compounds (for example, toxins) move from the capillaries into the tubules.
- *Concentration.* The final solute concentration of the urine is increased through removal of water from the filtrate in the collecting duct, thus conserving water.

Hormonal Control of Water Retention

The kidneys don't just go along producing the same amount of urine in simple accordance with the amount of filtrate that flows into the Bowman's capsule. Rather, the kidneys are able to *control* how much water there is in urine. This means they control how much water we retain or urinate, with this regulation in turn being under hormonal control. This capacity can be critical for people who are, say, lost in the desert. In that situation, the body needs to *retain* as much water as possible. How does it do it? The heart has receptors in it that sense blood volume, while the brain's hypothalamus has receptors that sense the solute concentration in the blood. The concentration of solutes in our desert travelers' blood is likely to rise, while their blood volume will fall because they have been steadily losing water (in breathing, perspiring, and so forth). When these things happen, the hypothalamus orders the release of **antidiuretic hormone (ADH)** from the pituitary gland. ADH works on the distal nephron tubule and the collecting duct, having the same effect on both: It increases their permeability to water. This means that more water moves out of the tubule and collecting duct and thus back *into* circulation. ADH has thus acted to reduce the amount of water lost by the kidney. Looked at another way, it means that the urine becomes more concentrated because it has less water in it. Should our travelers be saved, however, and then not be able to resist a fourth glass of lemonade, the pituitary's production of ADH will be shut down. The tubule and collecting duct will *retain* more water, meaning that more of it will be urinated. Along these lines, if you've ever wondered why 12 ounces of beer makes a person have to urinate more than does 12 ounces of lemonade, it's because the alcohol in the beer suppresses the production of ADH.

26.11 Urine Transport, Storage, and Excretion

Having seen how the kidneys produce urine, let's now look at the steps by which urine is then excreted from the body. Recall that urine is transported to the urinary bladder by the pair of tubes called the **ureters**—one stemming from each kidney. The ureters are muscular tubes that extend for a distance of about 30 centimeters, or 12 inches. The wall of each ureter contains an inner expandable epithelium, a middle layer of smooth involuntary muscle, and an outer connective tissue covering. Starting at the kidney, about every 30 seconds a peristaltic wave of muscle contraction squeezes urine out of the renal pelvis, along the ureter, and into the urinary bladder. Occasionally, solids composed of calcium deposits, magnesium salts, or crystals of uric acid develop within the collecting tubules, collecting ducts, or ureters. Such kidney stones can be very painful; in addition to obstructing the flow of urine, they may also reduce or eliminate filtration in the affected kidney.

The Urinary Bladder

The **urinary bladder**, shown in **Figure 26.22**, is a hollow muscular organ that stores urine. Its dimensions vary with the volume of stored urine, but a full urinary bladder can contain up to 800 milliliters of urine, which is about 27 ounces. The area surrounding the urethral entrance, called the *neck* of the urinary bladder, contains the internal sphincter muscle that provides involuntary control over the discharge of urine from the bladder.

The Urethra

The urethra extends from the urinary bladder to the exterior of the body. Its length and functions differ in males and females. In the male (shown in Figure 26.22), the urethra extends to the tip of the penis and is about 18 to 20 centimeters (7 to 8 inches) in length. The initial portion of the male urethra is surrounded by the prostate gland. In addition to transporting urine, the male urethra carries sperm cells and the reproductive secretions from the prostate and other glands. In contrast, the female urethra is very short, extending 2.5 to 3.0 centimeters (about 1 inch), and transports only urine. In both sexes, the urethra contains a circular band of skeletal muscle that forms an external sphincter. Unlike the internal sphincter, the external sphincter is under *voluntary* control.

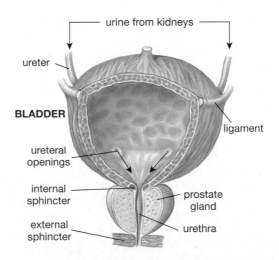

urine from kidneys

ureter

BLADDER

ligament

ureteral
openings

internal
sphincter

prostate
gland

external
sphincter

urethra

**Figure 26.22
The Urinary Bladder in a Male**

Urine Excretion

The urge to urinate usually appears when the bladder contains about 200 milliliters (about 7 ounces) of urine. We become aware of this need through nerve impulses sent to the brain from stretch receptors in the wall of the urinary bladder. The stimulation of these receptors also results in involuntary contractions of the urinary bladder (which is why we "have to go" to the bathroom). Such contractions increase the fluid pressure inside the bladder, but urine ejection cannot occur unless both the internal and external sphincters are relaxed. We control the time and place of urination by voluntarily relaxing the external sphincter. When this sphincter relaxes, so does the internal sphincter. If the external sphincter does not relax, the internal sphincter remains closed, and the bladder gradually relaxes. A further increase in bladder volume begins the cycle again, usually within an hour. Once the volume of the urinary bladder gets large enough, sufficient pressure is generated to force open the internal sphincter. This leads to an uncontrollable relaxation in the external sphincter, and urination occurs despite voluntary opposition. Normally, less than 10 milliliters (about a third of an ounce) of urine remain in the bladder after urination.

As noted earlier, an average of about 1.8 liters (half a gallon) of urine is excreted daily. Normal urine is a clear, bacteria-free solution with a yellow color.

On to Development and Reproduction

Animals breathe and move and fight off invaders in ways that have been reviewed over the last three chapters. Each generation of animals gives rise to another, of course, through the workings of a specialized system, the reproductive system. Between the moment of conception and the moment of birth, an animal must also go through an amazing transformation—in the case of humans, from a single cell to a baby in full cry. How do single sets of male and female cells interact to form a new living thing? How does this new life then make the transition from one-celled existence to trillion-celled existence? These are the subjects of the next two chapters.

Chapter Review

Summary

26.1 The Cardiovascular System and Body Transport

- The cardiovascular system transports materials to and from the body's cells. These materials include oxygen, carbon dioxide, nutrients, vitamins, hormones, waste products, and molecules and cells specialized for the body's defense. These materials are transported within blood.

- The two major components of blood are formed elements and blood plasma. Formed elements include red blood cells, white blood cells, and cell fragments called platelets. Blood plasma is the liquid medium in which the formed elements are suspended.
  TUTORIAL 26.1.2: Composition of the Blood

- Red blood cells constitute 99.9 percent of all the formed elements in blood and are responsible for transporting oxygen to the body's tissues and carbon dioxide from them. White blood cells come in many varieties and are active in defending the body from invaders. Platelets—cell fragments broken off from larger cells—assist in blood clotting at injury sites.

- Plasma is 92 percent water, but it also contains a number of dissolved materials, including proteins. There are three primary classes of plasma proteins: albumins, which transport hormones and fatty acids; globulins, which aid the immune system and serve as transport proteins; and fibrinogen, which functions in blood clotting. Other plasma compounds include electrolytes, organic nutrients, organic wastes, and hormones.

- Blood vessels carrying blood away from the heart are arteries; blood vessels returning blood to the heart are veins. The smallest of the blood vessels, the capillaries, connect the arteries with the veins.

- Vertebrates such as humans utilize closed circulation—circulation in which blood is delivered from one set of vessels to another. This contrasts with the system of open circulation, used by many invertebrates, in which arteries carry blood into open spaces, called sinuses. The blood then flows around cells, and is returned to a heart or hearts through openings in veins.

26.2 The Heart and the Circulation of Blood

- Blood is transported out to the various tissues of the body through the muscular contractions of the heart.

- There are two general networks of blood vessels in the cardiovascular system. One is the pulmonary circulation, in which blood flows between the heart and the lungs. The other is the systemic circulation, in which blood circulates between the heart and the rest of the body.
TUTORIAL 26.1.3: The Two Circulatory Networks

- The human heart contains four muscular chambers—two associated with pulmonary circulation, and two associated with systemic circulation. The right atrium and right ventricle are associated with pulmonary circulation, while the left atrium and left ventricle are associated with systemic circulation.

- Blood returning to the heart from the systemic circulation flows into the right atrium from veins called the superior and inferior vena cava. It then flows into the right ventricle, from which it is pumped into the pulmonary arteries, which carry it to the lungs. It returns to the left atrium through the pulmonary veins and then flows into the left ventricle, from which it is pumped to the rest of the body.

- A series of valves that open and close ensures that blood flows only one way through the heart.

26.3 The Heart's Own Blood Supply: What Is a Heart Attack?

- About half of all deaths in the United States today are caused by the blockage of one of the coronary arteries, which supply heart tissue with blood. Such blockages can cut off the blood supply to collections of heart muscle cells, which can die if deprived of blood for long enough. The result is a heart attack.

- Heart cells that have died are replaced by scar tissue, which cannot take part in the heart's pumping function. High-fat diets, smoking, and lack of exercise all are contributing factors to heart disease.

26.4 Getting the Goods to and from the Cells: The Capillary Beds

- Arteries near the heart branch into smaller-diameter arterioles, which feed into the delivery vehicles of the cardiovascular system, the capillary beds. The capillary beds then feed back into the body's system of veins that returns blood to the heart. The venous system begins with venules that merge into larger veins.

- Materials carried by arteries that are needed by the body's tissues move directly out of the capillaries (through their thin walls) and into the interstitial fluid—the liquid in which both the capillary and the cells that surround it are immersed. Meanwhile, carbon dioxide and wastes from cells flow into capillaries from the interstitial fluid.

- The movement of materials out of the capillaries occurs primarily near the arterial end of the capillary beds and is driven by blood pressure; movement of materials into the capillaries occurs primarily near the venous end of the capillary beds and is driven by osmosis.
TUTORIAL 26.1.5: Capillary Beds and Exchange of Materials

- Blood pressure is at low levels by the time blood has moved through the capillaries. Blood returns to the heart by the contraction of muscles, which squeeze the veins in a way that moves the venous blood toward the heart. A system of valves in the veins blocks venous movement away from the heart.

26.5 The Respiratory System and the Exchange of Gases

- The respiratory system functions primarily to capture oxygen and to dispose of carbon dioxide. It also aids in immune functions, in producing sounds for speaking, in regulating blood volume and pressure, and in controlling body fluid pH.

- The respiratory system includes the lungs; the nose, nasal cavity, and sinuses; the pharynx (throat); the larynx (voice box); the trachea (windpipe); and the conducting passageways, called bronchi and bronchioles, that lead to the lungs. The end points of the bronchioles are the air sacs called alveoli, which are the sites of oxygen and carbon dioxide exchange in the lungs.

- Steps in respiration include, first, breathing or pulmonary ventilation, meaning the movement of air into and out of the lungs. The second step is the exchange of respiratory gases that occurs in two general processes: external respiration, or the diffusion of gases between the alveoli and the blood; and internal respiration, meaning the two-way diffusion of gases between capillary blood and interstitial fluid.
TUTORIAL 26.2.2: Respiration

- In both internal and external respiration, gas exchange occurs through diffusion down concentration gradients; the gases are moving from areas of their higher concentration to areas of their lower concentration.

- Almost all the oxygen that moves from the alveoli to the blood in the pulmonary capillaries is bound to molecules of the protein hemoglobin, which exists within red blood cells. Carbon dioxide is transported within the blood in several different ways; some dissolves in blood plasma, some binds to hemoglobin in red blood cells, and some is converted to bicarbonate ions that move from red blood cells into the blood plasma.

26.6 The Digestive System

- The central function of the digestive system is to get the foods the body ingests into a form the body can use; the system must also collect and then rid the body of the waste that remains once the useful material from the food has been absorbed.

- The process of digestion is a process of the breakdown of foods—from large pieces of food to smaller pieces, and then from relatively large molecules of food to smaller molecules. It is these smaller molecules that leave the digestive system for transport through the rest of the body.

- The central feature of the digestive system is a muscular tube, the digestive tract, that passes through the body from the mouth to the anus. The functions of the digestive system involve a series of interconnected steps: ingestion, mechanical processing, digestion, secretion, absorption, and excretion.

26.7 Components of the Digestive System

- The mouth opens up into the oral cavity, which receives or ingests food that is then broken up into smaller pieces through the mechanical action of chewing.

- There are four different types of teeth: incisors (used for clipping and cutting), cuspids (tearing, slashing), bicuspids, and molars (crushing and grinding).

- The pharynx (throat) transmits solid food, liquids, and air. The esophagus is a muscular tube that begins at the pharynx and ends at the stomach.

- The stomach is a muscular digestive organ whose functions include the temporary storage of food, the mixing of ingested foods, and the breaking of chemical bonds of foods through chemical action. Gastric glands lining the stomach produce quantities of gastric juice that aid in digestion and kill microorganisms. The stomach empties chyme—a mixture of food and gastric juices—into the small intestine.

- The small intestine is the part of the digestive tract that receives chyme from the stomach, and completes the digestion of most of the food in it. Absorption is the movement of digested food out of the digestive tract; about 80 percent of all absorption takes place in the small intestine, with the rest divided between the stomach and large intestine. The small intestine is aided in its digestive function by the digestive secretions it receives from the pancreas and the liver.

- The pancreas is an organ that lies behind the stomach and that secretes both blood-borne hormones (insulin and glucagon) and substances that move through the pancreatic duct (digestive enzymes and pH buffers).

- The liver is a large organ that lies directly under the diaphragm and plays a central role in regulating the body's metabolism. All the blood leaving the absorptive areas of the digestive tract flows through the liver before going into general circulation. Thus the liver is the first stop for much digested material and can monitor and adjust circulating levels of nutrients. The liver also functions as the largest reservoir of blood in the body and serves to detoxify harmful substances before they reach the rest of the body.

- Lodged under the liver, the gallbladder is a muscular sac that stores and concentrates a substance produced by the liver, bile, which aids in the digestion of fats. Bile can move into the small intestine either from the gallbladder or directly from the liver.

- The large intestine begins at the small intestine and ends at the anus. The main function of the large intestine is to hold and compact the refuse produced by the rest of the digestive system (as solid waste, or feces). It also absorbs water and important vitamins produced by resident bacteria. The final portion of the colon, the rectum, forms the end of the digestive tract. The rectal chamber is usually empty except when peristaltic contractions force fecal materials out of the lower part of the colon. Stretching of the rectal wall then triggers the defecation reflex.

26.8 Different Digestive Processes for Different Foods and Nutrients

- A typical meal contains a mixture of carbohydrates, proteins, lipids, water, electrolytes (ions), and vitamins. Water, electrolytes, and vitamins can be absorbed directly into circulation, but large organic molecules, such as starches and proteins, must be broken down through digestion before absorption can occur. Different types of nutrients are digested in different ways.

- Both carbohydrates and proteins are broken down into their component molecules (simple sugars and amino acids, respectively) and move through intestinal cells, diffusing into intestinal capillaries. They then move through the capillaries to the hepatic portal vein, which carries them to the liver. Conversely, because of their large size, lipids do not enter intestinal capillaries. Instead, they enter lymphatic vessels after becoming attached to proteins. From these vessels, lipids enter the larger lymphatic vessels, after which they enter venous circulation.

- Vitamins are organic compounds required in small amounts for chemical reactions in the body. There are two major groups of vitamins: water-soluble vitamins (those that dissolve in water; the C and the B vitamins) and fat-soluble vitamins (those that dissolve in fat; A, D, E, and K). Humans can store fat-soluble vitamins for significant periods of time, but we cannot store water-soluble vitamins. This means that we can consume an excess of fat-soluble vitamins, with potentially harmful consequences.

26.9 The Urinary System in Overview

- Vertebrates and higher invertebrates have a single, compact set of excretory organs—the kidneys. The kidneys filter bodily fluids, removing waste in the process, but conserving such bodily materials as amino acids, ions, sugars, and water. By regulating fluid retention, the kidneys serve to control the volume of blood in the body.

- The urinary system consists of two kidneys that produce urine; the left and right ureters that the urine travels through upon leaving the kidneys; the muscular urinary bladder, which temporarily stores urine; and the urethra, through which urine passes out of the body.

26.10 How the Kidneys Function

- The kidneys undertake four principal processes: (1) filtration, in which portions of the blood plasma move from the bloodstream to the kidney's nephron tubules; (2) reabsorption, in which water and nutrients move back into the blood vessels, while waste is retained within the tubules; (3) secretion, in which hydrogen ions, toxins, and other compounds move from the capillaries into the tubules; and (4) concentration, in which the final solute concentration of the urine is increased through the removal of water from the collecting duct, thus conserving water.

  TUTORIAL 26.3.3: Physiology of the Nephron

- Retention of water by the kidneys is regulated by the hypothalamus' control of the secretion of antidiuretic hormone (ADH) by the pituitary gland. An increased secretion of ADH means that more water will move out of the kidney's tubules and collecting ducts and back into circulation.

26.11 Urine Transport, Storage, and Excretion

- Frequent peristaltic waves of muscle contraction squeeze urine out of the renal pelvis, through the ureter and into the urinary bladder. The urethra extends from the urinary bladder to the exterior of the body.

- The area surrounding the urethral entrance contains the internal sphincter muscle that provides involuntary control over the discharge of urine. The urethra also contains a band of skeletal muscle under voluntary control, the external sphincter. We become aware of the need to urinate through nerve impulses sent to the brain from stretch receptors in the wall of the urinary bladder. Urine ejection cannot occur unless both the internal and external sphincters are relaxed; relaxation of the external sphincter prompts the internal sphincter to relax.

Key Terms

Understanding the Basics

Multiple-Choice Questions

1. The formed elements of blood include
   a. albumins, globulins, fibrinogen
   b. red blood cells, white blood cells, platelets
   c. plasma, fibrin, serum
   d. plasma, globulins, and lacteals
   e. red blood cells, white blood cells, fat cells

2. The amount of blood in the average person is in the range of _____ liters.
   a. 1 to 2
   b. 2 to 4
   c. 4 to 6
   d. 8 to 10
   e. 12 to 14

3. The two-way exchange of substances between blood and body cells occurs only through
   a. capillaries
   b. arteries
   c. veins
   d. arterioles
   e. valves

4. The oxygen-carrying protein within an erythrocyte is
   a. fibrin
   b. hemoglobin
   c. fibrinogen
   d. plasmid
   e. globulin

5 A heart attack can best be described as
   a. a sudden harmful contraction of the heart muscles
   b. an invasion of heart tissue by pathogens such as bacteria
   c. a death of cells in an area of the heart caused by a reduction of blood supply
   d. a failure of the nervous system to cause heart contractions
   e. a rapid, harmful expansion of blood volume supplied to the heart

6. The arteries of the pulmonary circulation differ from those of the systemic circulation in that they carry _____.
   a. oxygen, carbon dioxide, and nutrients
   b. oxygen and nutrients
   c. deoxygenated blood
   d. oxygenated blood
   e. platelets

7. The movement of air into and out of the lungs is
   a. external respiration
   b. internal respiration
   c. cellular respiration
   d. pulmonary circulation
   e. pulmonary ventilation

8. The part of the digestive tract that plays the primary role in digestion and absorption of nutrients is the
   a. large intestine
   b. cecum and colon
   c. stomach
   d. small intestine
   e. esophagus

9. An adult who has a complete set of permanent teeth has _____ teeth.
   a. 20
   b. 26
   c. 32
   d. 36
   e. 40

10. The urinary system includes all the following except
    a. kidneys
    b. gallbladder
    c. ureters
    d. urinary bladder
    e. urethra

Brief Review

1. What are the major types of functions performed by blood?

2. Why do capillaries permit the diffusion of materials, whereas arteries and veins do not?

3. What is external respiration?

4. Name and describe the four layers of the digestive tract, beginning with the innermost layer.

5. What is the primary function of the urinary system?

6. If you drank three glasses of fruit juice just because it tasted so good, how would you expect your body to react with respect to its production of antidiuretic hormone (ADH)?

7. In what form is most of the carbon dioxide transported to the lungs?

8. How is the small intestine adapted for the absorption of nutrients?

9. Why don't blood cells and large protein molecules in the plasma appear in the filtrate under normal circumstances?

Applying Your Knowledge

1. Which of the formed elements would you expect to see increase after donating a pint of blood?

2. Why is the wall of the left ventricle more muscular than the right ventricle?

3. Why do long-haul trailer truck drivers frequently experience kidney problems?

27

An Amazingly Detailed Script
Animal Development

GASTRULATION

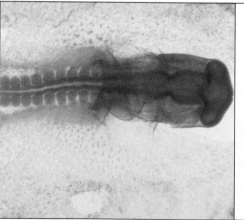

a

animal pole

vegetal pole

blastocoel

blastula

Crucial directions. (Section 27.1, page 623)

What are somites? (Section 27.1, page 625)

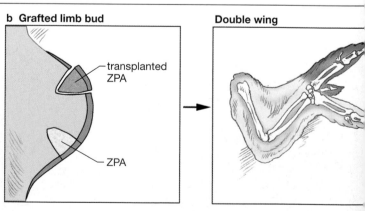

b Grafted limb bud

Double wing

transplanted ZPA

ZPA

Sprouting two sets of digits. (Section 27.2, page 625)

How can a living thing start out as a single cell and end up as a fully formed animal? Through the process of development.

Of all the marvels in the natural world, the most spectacular may be a complex process that goes under a simple name: development. The set of steps by which, for example, a human embryo goes from being a speck of microscopic cells to a newborn cradled in its mother's arms is so remarkable that even the old adage about "seeing is believing" is turned on its head. People who view pictures of such a work in progress can scarcely believe what they're seeing. It's not uncommon, months after the birth of a child, for parents to find themselves looking at the baby and then at each other in amazement, still wondering how such a thing is possible.

This process is all the more notable because it takes place through self-assembly. There is no outside hand here; no master artisan watching over each step and intervening where necessary. There is instead a starting set of instructions supplied by maternal and paternal DNA, and a starting set of proteins and other molecules that come in the maternal egg. From these elements, an embryo fashions itself, aided only by the nurturing environment of egg or womb.

How can such an event take place? As you'll see, in species as diverse as frogs, insects,

and human beings a common, core set of steps is followed in development. Moreover, nature has provided a fairly small number of "tools" that are employed in development, and these, too, tend to be utilized across the different species.

Because nature can fashion an entire organism from the instructions contained in a single egg, scientists long wondered: Would it be possible to regenerate *parts* of the human organism by harnessing the power of development? Could we grow new knee joints for arthritis sufferers, for example, or new cardiac tissue for heart attack victims? In just the past few years, researchers have taken some giant steps toward this goal. As you'll see, the key to this progress has been a deepened understanding of some developmental tools known as stem cells.

27.1 General Processes in Development

Let's now begin by looking at the developmental progression that is common to much of the animal world, noting first *what* happens in development, and then going over *how* these things happen.

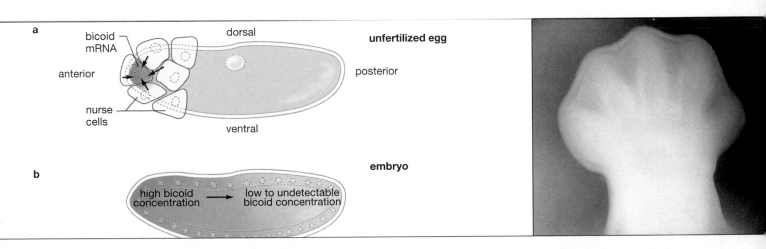

Where does the head go?
(Section 27.2, page 627)

Fish, fowl, or human?
(Section 27.3, page 628)

Two Cells Become One: Fertilization

Those who read about meiosis in Chapter 10 saw there that, in the formation of egg and sperm cells, a *halving* of the cell's usual genetic complement takes place. In humans, each resulting egg or sperm contains 23 chromosomes, rather than the 23 *pairs* of chromosomes that exist in most cells in the body. Going forth with this reduced, or *haploid* set of chromosomes, egg and sperm then fuse in the moment generally referred to as **fertilization**, more poetically known as conception. Because egg and sperm each bring a haploid set of chromosomes to this union, the fertilized egg contains a doubled, or *diploid*, chromosome set. The fertilized egg, also known as a **zygote**, is a cell, although a very special one: An entire human being will develop from it. Development at its simplest level is a process by which this single, original cell divides to become two cells, after which the two become four, the four eight, and so on. Once the zygote starts dividing, we call it an **embryo**.

Three Phases of Embryonic Development

Embryonic development in animals can be divided into three phases.

- *Cleavage*. Cellular division results in a ball of cells with an interior cavity.
- *Gastrulation*. These cells rearrange themselves into three layers that will give rise to specific organs and tissues.
- *Organogenesis*. Organs begin to form.

First Phase: Cleavage

In **cleavage**, the zygote is repeatedly divided into smaller and smaller individual cells through cell division. The reason the cells get progressively smaller is that none of the usual cell growth occurs between these cell divisions.

The product of this process of cleavage is a **morula** (from the Latin for "mulberry"), a tightly packed ball of cells. This configuration soon changes, however, as the cells arrange themselves in one or more layers around a liquid-filled cavity. Several hundred to several thousand cells make up this structure, which is called a **blastula**, with the cavity inside known as a **blastocoel** (see Figure 27.1).

The original egg contains a quantity of yolk, which in development means both the yellow "yolk" and the surrounding "whites." This yolk helps give the zygotes of many species a *polarity*, meaning that one end of the zygote is different from the other. Many zygotes have a **vegetal pole**, which is to say an end that contains a relatively greater proportion of yolk, and an **animal pole**, which has relatively less yolk and which lies closer to the cell's nucleus. As you'll see, this factor helps define the way the embryo develops.

Second Phase: Gastrulation

Anyone who has ever watched a play in rehearsal probably has witnessed a phenomenon that bears some comparison with the second phase in embryonic development—gastrulation. It's the moment in which the director says to the assembled cast, "Take your places." What happens then is a carefully directed movement of actors; some move only a little, whereas others go from one side of the stage to the other. Whatever the case, after all this movement is done, the actors—having assumed their places—are ready to take on their assigned roles.

Gastrulation is likewise a process of directed movement, in this case of groups of cells that move to particular places in the developing embryo. Once this movement is completed, the cells will have formed themselves into three different layers: endoderm, mesoderm,

Figure 27.1
Cleavage of a Frog Egg

a Repeated cell division of the zygote results in a tightly packed ball of cells called the *morula*. This configuration gives way to a blastula, in which the cells are arranged around a liquid-filled cavity, the blastocoel. The yolk at the vegetal end of the zygote impedes cleavage, so that fewer but larger cells result at the vegetal pole than at the animal pole of the morula and blastula. The figure at far right shows a cross section of the blastula, with its blastocoel visible.

b A micrograph of cleavage in a frog egg, when the egg is at the eight-cell stage.

a

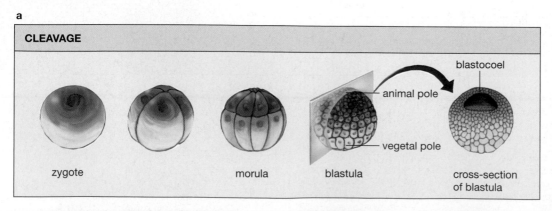

CLEAVAGE

blastocoel

animal pole

vegetal pole

zygote   morula   blastula   cross-section of blastula

b

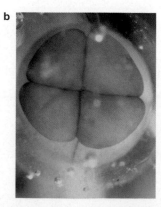

and ectoderm. These are "germ" layers of cells, meaning layers that *lead to* other kinds of structures (**see Table 27.1**). As it turns out, endoderm, mesoderm, and ectoderm have a rough kind of "inside/middle/outside" quality to them, and this is reflected in the organs and tissues that develop from them. In gastrulation, endoderm takes its place as the innermost layer, mesoderm as the middle layer, and ectoderm as the outer layer. In line with this, endoderm gives rise to interior tissues (the lungs, for example), mesoderm to tissues exterior to these (skeletal systems), and ectoderm to tissues that are more exterior yet (nails, hair).

One Example of Gastrulation: The Sea Urchin

Let's now look at the process of gastrulation in one organism, the sea urchin (**see Figure 27.2**). The sea urchin blastula amounts to a few thousand cells, arranged in a single layer around the liquid-filled blastocoel. Gastrulation begins at the blastula's vegetal pole. Some cells at the pole detach from the ring of blastula cells and move into the blastocoel. These are called the *primary mesenchyme cells*, and with their loss, the vegetal pole begins to pinch inward. This process, called *invagination*, eventually will carry the vegetal pole as much as halfway toward the blastula's animal pole. The structure that starts forming around this indentation is called the *archenteron*, or primitive gut. Cells growing at

the tip of the archenteron begin to send out slender extensions of themselves that adhere to selected cells that lie "above" them on the rim of the blastocoel. The contraction of these extensions then effectively *pulls* the archenteron toward the animal pole, eventually moving the archenteron all the way to the pole itself.

The cells with the extensions are called *secondary mesenchyme cells*; once they have gotten the archenteron to the top of the blastocoel, they move back down into the blastocoel cavity. There they will form the mesoderm, or middle layer, of the sea urchin—eventually its skeleton and muscles. Meanwhile, the mouth of the sea urchin will form in the place where

Table 27.1
Three Layers of Cells That Are Assembled in Gastrulation and Some of the Structures They Give Rise to in Mammals

Cell Layer	Gives Rise to Organs or Types of Tissue
Endoderm	Lungs, liver, pancreas, lining of digestive tract, lining of bladder, and several glands
Mesoderm	Much of the body mass of higher animals, including most of the muscle and skeletal systems; blood, blood vessels, and bone marrow; the uterus and the sperm- and egg-producing gonads; the kidneys, and the cortex of the adrenal gland
Ectoderm	Entire nervous system; nails, hair, lenses of eyes, tooth enamel, several glands, the outer or epidermal layer of skin

Tutorial 27.1.2
Three Phases of Embryonic Development

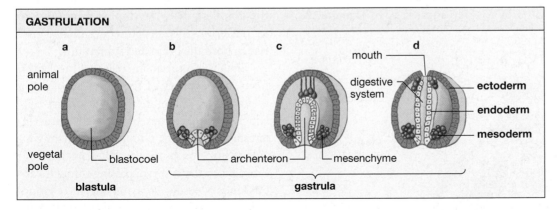

Figure 27.2
Gastrulation in the Sea Urchin

a The sea urchin blastula amounts to a few thousand cells, arranged in a single layer around the liquid-filled blastocoel.

b Some cells at the blastula's vegetal pole move into the blastocoel, prompting the vegetal pole to pinch inward, moving toward the animal pole. The structure that starts forming around this indentation is the archenteron, or primitive gut.

c Mesenchyme cells, growing at the tip of the archenteron, send out slender extensions of themselves that adhere to cells on the rim of the blastocoel. These extensions contract, moving the archenteron to the animal pole.

d The mouth of the sea urchin forms where the archenteron contacted the blastocoel lining. A digestive tube then runs from this mouth to the rim of the vegetal pole, which develops into an anus. These cells constitute the inner layer or endoderm of the sea urchin. Mesenchyme cells move back down into the blastocoel, forming the middle layer or mesoderm of the sea urchin. The outer cells of the blastula form the sea urchin's outer layer or ectoderm.

Tutorial 27.1.2
Three Phases of Embryonic Development

Figure 27.3
Body Planes in a Frog and Formation of the Frog Central Nervous System

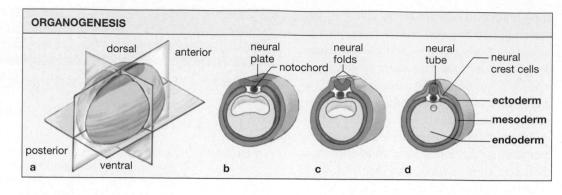

ORGANOGENESIS

the archenteron moved up to meet the blastocoel lining. This mouth will fuse with the archenteron, creating a tube that runs the length of the sea urchin, from the mouth right down to the original vegetal pole. There, the rim of the invagination you saw earlier will develop into an anus. This entire feeding and digestive apparatus is the inner cell layer or endoderm of the sea urchin. Finally, the outer cells of the blastula form the outer cell layer or ectoderm of the organism, eventually becoming its skin and related organs.

Third Phase: Organogenesis

In vertebrates, an interplay of two of these three germ layers marks the first steps on the path to the development of organs—in this case, organs of the nervous system.

In talking about development in any organism, an important concept is *body planes*—imagined divisions that pass through the center of an animal, dividing it into two parts of roughly equal size. In **Figure 27.3a**,

you can see that a frog embryo has been divided with three such planes. The first yields dorsal and ventral portions of the embryo; the second yields anterior and posterior ends. In the case of a frog, dorsal eventually will mean its back, ventral its belly; anterior will mean its head, and posterior its hindquarters.

Looking at Figures 27.3b and 27.3c, you can see what takes place toward the frog embryo's dorsal surface as gastrulation is nearing completion. First, a temporary rod-shaped support organ, the **notochord**, develops. Apart from support, one of its functions is to *induce*, or bring about, development in the ectodermal tissue that lies above it, toward the dorsal surface. (You'll hear more about such induction shortly.) This development begins with an elongation of the ectodermal cells to form a flattened surface called the *neural plate*, which curls up and eventually folds all the way over on itself to form an enclosed, hollow **neural tube** that will lie below the dorsal surface. Developing from anterior to posterior, this structure eventually gives rise to both the brain and spinal cord.

Equally important are cells that break away from the periphery of the neural tube as it is folding over. These are the **neural crest** cells, which exist only in vertebrates and have two interesting properties: Groups of them migrate along specific routes to different locations throughout the embryo, passing through various kinds of tissues along the way; these groups then go on to develop into very different organs or tissues, depending on where they come to rest. For example, the medulla or inner portion of the adrenal glands (which in humans sit atop the kidneys) is derived from neural crest cells. You can see the routes of these migrations and some of their outcomes in **Figure 27.4**. One consequence of a failure of migrating neural crest cells to come together is a gap that occasionally occurs between the nose and mouth in humans, which is known as a cleft lip.

Figure 27.4
Neural Crest Migration in a Frog Embryo
This cross section through a frog embryo shows some of the pathways of neural crest cells as they migrate to various locations, prompting the development of different tissues and organs.

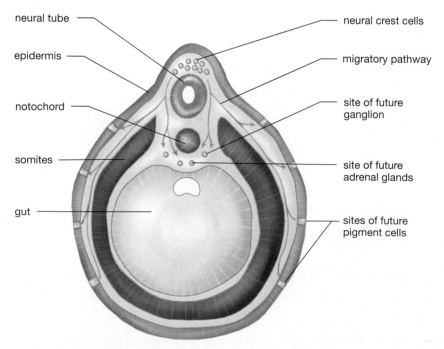

As this ectodermal movement is taking place, mesodermal tissue on both sides of the notochord is developing into blocks called **somites** (**see Figure 27.5**) that will give rise to muscles as well as to the vertebrae that enclose the spinal cord.

A Theme in Development: From the General to the Specific

As complicated as all this may seem, it amounts to a brief look at important, but early periods of development in animals, with an emphasis on vertebrates. Looking at the larger picture, it may be obvious that in the early stages of development, there is a movement from the general to the specific. The animal goes from a small group of superficially similar cells, to three *layers* of cells, to tissues that proceed from these layers (notochord, somites) to actual organs formed from these tissues. (Many of these organs are formed from several different cell layers.)

27.2   What Factors Underlie Development?

Now that you've taken a brief look at what happens in early development, the next question is: *How* do these things happen? How do vegetal pole cells "know" they are vegetal pole cells? How do neural crest cells take different routes through an embryo and end up serving as different kinds of cells?

The Process of Induction

You saw earlier that the notochord induces development of the ectodermal tissue that lies above it. Such **induction** is an important player in development in general. It can be defined as the capacity of some embryonic cells to direct the development of other embryonic cells. If you look at **Figure 27.6**, you can see how induction works in the well-studied

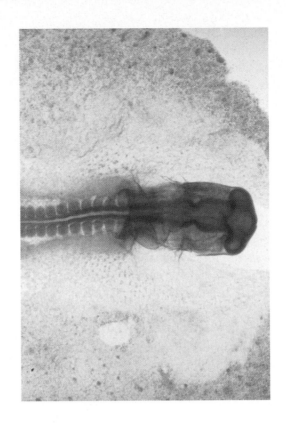

Figure 27.5
Somites
These regular, repeating blocks of mesodermal tissue, visible in this case on either side of the notochord of a chick embryo, will give rise to muscles and the vertebrae that enclose the spinal cord.

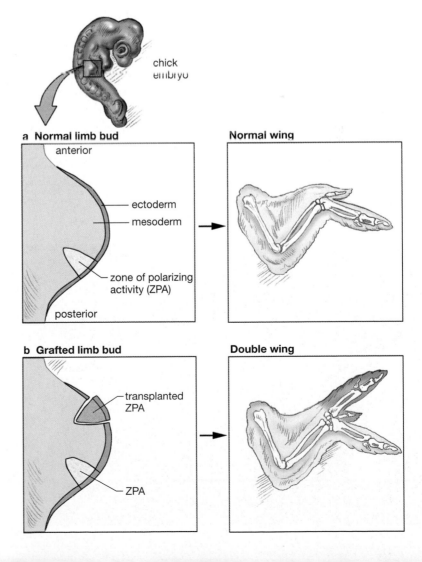

chick embryo

a Normal limb bud
anterior
ectoderm
mesoderm
zone of polarizing activity (ZPA)
posterior

Normal wing

b Grafted limb bud
transplanted ZPA
ZPA

Double wing

Figure 27.6
Cells Can Induce Development in Other Cells
Transplantation experiments with chick embryos revealed that a group of cells on the posterior portion of the chick's limb bud give out positional information for the development of digits in the chick wing.

a When cells from a zone of polarizing activity (ZPA) were left in their normal posterior location, the chick wing developed as usual.

b When limb bud ZPA cells were transplanted to the anterior portion of a second limb bud, the result was mirror images of two sets of developing wing digits, much like two sets of human fingers extending from a joined hand.

example of chick wings. In essence, there is a relatively small group of cells that gives out positional information for the development of a portion of the wing. Researchers learned this by working with the "limb bud" that is transformed into the chick wing. They took cells from a posterior "zone of polarizing activity" (ZPA) in one limb bud and transplanted them to the *anterior* portion of a second limb bud. The result was mirror images of *two* sets of developing digits in wings, spreading out like two sets of human fingers developing from joined hands. In short, the cells in the ZPA induced the development of other cells, sending out a message that said, in effect, "begin digit development here."

The Interaction of Genes and Proteins

The induction process just described raises the question of what lies at the root of development. There are lots of cells in an embryo. How can some of them (such as the ZPA cells) control the development of others? It turns out that, at its most basic level, development is controlled by the interaction of genes and proteins—the genetic regulation reviewed in Chapter 14. Biologists' understanding of how this interaction works in development has grown dramatically in the period from the early 1980s to the present. And the key "model" organism in this research has been the tiny fruit fly, *Drosophila melanogaster*. So important has this work been that three developmental investigators who worked with *Drosophila*—Edward Lewis

and Eric Wieschaus of the United States, and Christiane Nüsslein-Volhard of Germany— were awarded the Nöbel Prize in Physiology or Medicine in 1995 for their discoveries (**see Figure 27.7**). Let's look at a little of what the *Drosophila* research has uncovered.

Two Lessons in One Gene

There is a *Drosophila* gene, called *bicoid*, that codes for a protein that is also called bicoid. (It is a convention that the gene and protein have the same name; note, however, that the gene name is italicized while the protein is not.) This single gene offers a clear example of two principles that operate in development:

- A critical early step in development is the establishment of positional information in the embryo.
- Substances called *morphogens* are important in establishing this positional information.

Look at the unfertilized *Drosophila* egg pictured in **Figure 27.8** to see how these principles play out with respect to the bicoid protein. Recall first, the body planes described earlier. In Figure 27.8a, the top of the egg is its dorsal portion and the bottom its ventral portion, while the left side corresponds to its anterior portion and the right side to its posterior. Eventually, the anterior portion will be the head of the fly, the posterior its hindquarters, the dorsal portion its back, and the ventral portion its belly.

While the egg is still in the mother, a group of "nurse cells" surrounds its anterior portion and pumps molecular products into it. These products include bicoid messenger RNAs (mRNAs)—genetic instructions copied from the *bicoid* gene. Eventually, these mRNA instructions will lead to the bicoid protein. Looking at the figure, you can see that the *bicoid* mRNAs are found at the anterior end of the egg.

After the egg is fertilized, it begins dividing into many cells, of course; but these cells in the early *Drosophila* embryo can be thought of as part of an undivided egg. Once the *bicoid* mRNA molecules are translated into proteins, these proteins diffuse through roughly 70 percent of the length of the egg, starting from anterior and stretching toward posterior. However, as you can see in Figure 27.8b, this bicoid protein exists in a *concentration gradient*, with the greatest concentration being at the

**Figure 27.7
Advancing Human
Knowledge of Development**
Edward B. Lewis of the United States (left), Christiane Nüsslein-Volhard of Germany (center), and Eric F. Wieschaus of the United States (right), in Stockholm after receiving Nobel Prizes for research they did in developmental biology.

anterior end, a lesser concentration toward the center, and no detectable amounts found at the posterior end.

Positional Information

With this general background, you can now see how *bicoid* exemplifies one of the two principles of development noted earlier. What does the bicoid protein do? It imparts positional information to the embryo and is critical in setting in motion the development of structures in the anterior and center portions of the embryo. Recall that the far-anterior portion of the egg eventually becomes the head of the *Drosophila* fly. In experiments, however, scientists have taken bicoid-laden cytoplasm from the anterior portion of one egg, placed it at the *center* of an egg with no bicoid in it, and watched as the head and related structures developed at the center. Meanwhile, an egg with no bicoid protein gives rise to a fly that never develops a head, though its posterior structures develop normally.

Morphogens

The bicoid protein is a **morphogen**: a diffusable substance whose local concentration affects the course of local development in an organism. Several other morphogens are operating in *Drosophila* in early development. Indeed, several positional "systems" are at work in the fly. What's important to note is that, with these systems, a three-dimensional *grid* is set up in which the destiny of various cells is determined by their position within this grid. How? Their exposure to different morphogens varies according to their position.

So, how do morphogens work? You may remember learning in Chapter 14 about proteins that feed back on DNA, controlling its transcription, which is to say turning genes on or off. These DNA-regulating proteins are called *transcription factors*. All of the *Drosophila* morphogens you've been reading about here are transcription factors. These morphogens, in other words, are proteins that, once brought into being by genes, then go on to help turn *other* genes on or off.

All this provides a basis for understanding the chick wing development noted earlier. Recall that ZPA cells in the limb buds were able to induce the development of digits in the chick wing. These cells can play their controlling role because they produce a

morphogen (called sonic hedgehog) that acts on surrounding cells, regulating their genes and thus inducing their development.

A Hierarchy of Genetic Regulation

Once you begin to see the cascade of events that are controlled through genetic regulation, it's easier to understand how development moves from the general to the specific. Genes operate in a hierarchy in development. The first level of genes specifies broad positional patterns across the embryo (as with *bicoid*). Going down the hierarchy, this positioning gets finer-grained—now development is being specified *within* sections of the embryo. Finally, control is exercised over the development of specific body *structures*, such as wings or antennae. Remember how development goes from cells, to layers of cells, to specific structures such as a mouth and a gut? The genetic cascade explains this pattern.

27.3 Developmental Tools: Sculpting the Body

To this point, you've been considering development at the level of genes. Ultimately, however, the shaping of an organism takes place through the actions of complete cells. You may wonder how a hand, for example, can be shaped from a seemingly formless mass of cells. The answer lies in cell *capabilities* that allow for this sort of development. Three examples of these capabilities are

- Cell movement
- Cell adhesion
- Cell death

Figure 27.8
***Drosophila* Egg and Bicoid Gradient**
A fly begins to develop. **(a)** Nurse cells surround the anterior portion of an unfertilized *Drosophila* egg. The cells pump messenger RNA molecules for the bicoid protein into the egg. **(b)** The messenger RNAs are translated into bicoid proteins that diffuse through the now-fertilized egg. These proteins exist, however, in a concentration gradient—in high concentration at the anterior end of the embryo, but in low to undetectable concentration at the posterior end. The bicoid protein is a morphogen that imparts information regarding development of anterior and center-position structures in the fly. An egg without bicoid in it gives rise to a fly that never develops a head.

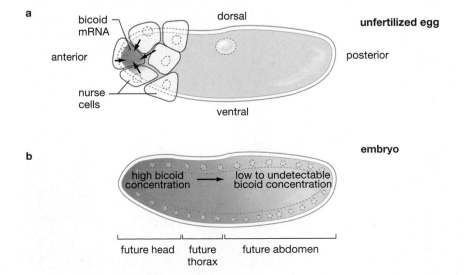

a

bicoid mRNA

dorsal

unfertilized egg

anterior

posterior

nurse cells

ventral

b

embryo

high bicoid concentration → low to undetectable bicoid concentration

future head | future thorax | future abdomen

You saw examples earlier of cell movement and cell adhesion when reviewing the process of gastrulation in the sea urchin. Recall that cells growing at the tip of the primitive gut were first able to move into the blastocoel, after which they sent out slender extensions of themselves that pulled the gut toward the "top" of the blastocoel. These extensions were able to do this, however, only after *adhering* to selected cells that lay "above" them in the blastocoel. Several families of proteins, collectively called *cell adhesion molecules*, allow cells to carry out this trick of selectively gluing themselves onto other cells.

Cell death may not sound like a promising means of developing an organism, but consider that a sculptor may work not by adding material, but by chipping away at an initial mass of it. Thus it is in development, where cell death works to create spaces that define shapes. An example of this is the human hand which, but for the death of large numbers of cells between future fingers, would look more like the webbed foot of a duck. Cells that die in this fashion are not attacked by other cells, nor do they perish for lack of nutrients. Instead, they die through their own metabolic processes in a cellular suicide known as *programmed cell death* (**see Figure 27.9**).

27.4 The Promise of Stem Cells

Having reviewed a couple of examples of wings or heads being induced to grow where they normally wouldn't, you may have the impression that groups of cells are infinitely adaptable in what they can become; that given the right developmental factor (such as a morphogen), a cell could be induced to become anything at any point. For most cells, however, this is not the case. As you'll see, most cells are fated to become a given kind of cell, and they cannot become any other kind of cell. A relatively small number of cells, however, do have the capacity to *give rise to* different kinds of cells. These cells are called stem cells, and only in recent years have scientists begun to learn how adaptable some of them can be. This is not just a matter of academic interest, however. Research into stem cells has exploded in recent years because of the medical potential they hold. So promising are stem cells that scientists now routinely express the hope that, through them, we might someday be able to completely repair the damage done by, for example, spinal-cord injuries or heart attacks. To understand the importance of stem cells, let's look first at how they differ from most cells in the body.

Cell Fates: Determined and Committed

In thinking about the average cell, it might be helpful to consider an analogy with the human acquisition of language. A newborn baby can learn to speak Chinese, English, or Dutch with equal ease. If the child is raised in an English-speaking household, English will be the lifelong language, barring some intervention. A switch could be made to Dutch or Chinese, and until about the time of puberty, this can be done with relative ease. Once the child passes puberty, however, the window of adaptability slams shut; a new language can be acquired only though hard, slogging work. Most cells in the developing body are like this. They are adaptable in early embryonic life, but then the window of adaptability begins to close as development proceeds.

This pattern is understandable because, as you've seen, development proceeds from the general to the specific. Early in development, a given type of cell needs to be adaptable enough to go down several different developmental pathways. Mesodermal cells, for example, need to be able to give rise not only to muscle tissue, but to bone tissue as well. Once the body is fully formed, however, all its specialized tissues and organs are in place, meaning there is less need for cells that have this kind of flexibility. The result is that most cells in the adult human body have undergone

Figure 27.9
Soon to Be Different
The webbed hand of a human embryo, 40 days after conception. The programmed death of cells in the hand will bring about the emergence of five separate fingers.

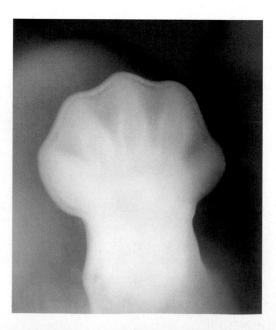

what is known as **commitment**: A developmental process that results in cells whose role is completely determined. Almost all your liver cells, for example, have undergone commitment in that they can be nothing but liver cells and can give rise to nothing but liver cells.

From a human perspective, the problem with this sequence of events is that the tissues of fully developed human beings can become damaged or lost altogether. They break down or become diseased; they are attacked by the body's own immune system; they are injured in car accidents or fires. And many damaged tissues will not regenerate themselves. A person whose spinal cord is severed does not generate a new spinal cord. Thus, for medicine, the question for decades has been: Could we use developmental processes to generate new tissue? One obvious way to go about this would be to harness the power of the cells that, in nature, generate new kinds of tissue—early embryonic cells. It turns out, however, that there is also another possibility As development proceeds, a small proportion of *adult* cells retain the ability to produce specialized cells. Both the adult and the embryonic cells that have this generative capacity are known as *stem cells*. Scientists have known about both adult and embryonic stem cells for years. What's new is our ability to find and manipulate these cells.

The Breakthrough in Embryonic Stem Cells

Remember that the very early embryo is a fertilized egg that has begun to undergo cell division. In mammals, up through the eight-cell embryo stage, any one of these cells can give rise to a whole new mammal. Indeed, it is these early cells that produce identical twins in human beings. A cell that can give rise to every kind of cell in an organism—and hence to a whole new organism—is said to be **totipotent**.

Now go forward in development a little to the embryonic structure you looked at earlier called the blastula—the hollowed-out ball of cells with a fluid-filled center. In mammals, the blastula is known as a **blastocyst**, but it is very similar to the blastula seen in Figure 27.1 (see page 622). It turns out that some of the cells in the human blastocyst are **pluripotent**: They are capable of giving rise to almost all of the specialized cells or tissues in the body. They do not have the flexibility of totipotent cells—they can't develop into a whole new embryo—but they are very flexible indeed. The 200-cell human blastocyst fits into a particular stage of human reproduction. One section of its cells becomes the placenta that will help nurture a growing embryo, while another section of cells, called the *inner cell mass*, constitutes the embryo itself. It is these latter blastocyst cells that are pluripotent. (For a look at the human blastocyst, see Figure 28.9 in the human reproduction chapter, on page 649.)

For years scientists could not imagine doing much with human blastocyst cells, because it was thought that these cells could exist only for fleeting periods of time *in blastocysts*. Despite many efforts, no one could first retrieve the pluripotent cells from human blastocysts and then get them to grow and divide *in the laboratory*. In other words, scientists could not "culture" these cells, or get them to maintain themselves in laboratory petri dishes. This changed in 1998 when a research team led by James Thomson of the University of Wisconsin announced it had been able to manage this feat. (Almost simultaneously, a team led by John Gearhart of Johns Hopkins University announced that it had cultured pluripotent cells from a later-stage embryonic source.) Pluripotent blastocyst cells now figure regularly in news stories, but they are referred to by a different name: **embryonic stem cells**. These can be defined as cells from the blastocyst stage of an embryo that are capable of giving rise to almost all of the cells or tissues in the human body.

A Further Advance: Adult Stem Cells

The breakthrough with embryonic stem cells led to a flurry of activity regarding stem cells from the *other* source already noted: adult tissues. In one important piece of work, a research team led by Margaret Goodell at Baylor College of Medicine took adult mice, isolated stem cells from their muscle tissues, and watched as these muscle cells developed into various types of *blood* cells. Prior to such advances, scientists had known that adult tissue contained stem cells. It was thought, however, that these stem cells had a very limited capability: Liver stem cells could give rise only to liver cells, bone stem cells to bone cells, and so forth. The new research made clear that one kind of adult stem cell could give rise to *other* kinds of cells. With this, science was off to the races in locating sources of **adult stem cells** and then seeing what these cells would differentiate into. Most of this research has been done with mice cells, but in one particularly welcomed result, scientists at UCLA and the University of Pittsburgh isolated what appear to be stem cells from human fat tissue. The researchers then got these cells to differentiate into muscle, bone, and cartilage tissue. Imagine using liposuction to harvest fat tissue from a person who suffers from, say, arthritic knees, and then coaxing the stem cells in this fat to differentiate into bone tissue. Imagine then implanting this bone tissue into this same person's knees, thus helping restore them.

The Bigger Picture of Stem Cells

All of this leads to the more general concept of a **stem cell**, which can be defined as any cell with the capacity for prolonged self-renewal that can produce at least one type of differentiated daughter cell. In this definition, we actually can see *three* valuable qualities of stem cells. First, in dividing they can produce specialized cells—muscle cells, nerve cells, and so forth, or their precursors. Second, while doing this, they keep on producing more stem cells. A stem cell division will, like any other cell division, result in two daughter cells. One of these may be a specialized cell, but the other will be another stem cell. Third, stem cells have the capacity to *keep on* producing both specialized cells and more stem cells. They don't die out after a few dozen generations, as is usually the case with cells. Indeed, some embryonic stem cell lines appear to be "immortal," meaning that they never die out. This is important because the idea is to have stem cells become tissue factories, and to be useful, factories need to keep producing for long periods of time.

The first step in stem cell research is to isolate and culture stem cells in such a way that they maintain themselves solely *as* stem cells. (Keeping embryonic stem cells from specializing as cells can be difficult, because that is what they are programmed to do.) The second step is to use various procedures and chemical treatments to coax some stem cells from this original line to differentiate as needed. The third step is to get the resulting specialized cells working away back inside a human being.

The Ethical Debate over Embryonic Stem Cells

Scientists cannot generate stem cells from scratch. Investigators need a source of both adult and embryonic stem cells in order to conduct research on them. Obtaining adult stem cells can be difficult, depending on the tissue being looked at, but this is a technical issue. Human *embryonic* stem cells, meanwhile, present science with an ethical issue.

To state the obvious, human embryonic stem cells come from human embryos—from the collections of cells that have the potential to become fully formed human beings. Where do researchers get the embryos that are the source of embryonic stem cells? In most cases from fertility clinics, which each year discard thousands of excess embryos that have been produced for in vitro fertilizations. (A couple undergoing fertility treatments generally will produce many more embryos than they need to bring about a pregnancy.) With the couples' consent, excess embryos that would have been discarded are used instead in scientific research. When scientists extract stem cells from these early-stage embryos, however, the embryos are destroyed and this is a matter of great concern to many people. For them, the destruction of an embryo is a destruction of human life. If research using such embryos is to go forward at all, these critics say, the government should not fund it.

Given such objections, you might think that scientists could simply stop using embryonic stem cells and instead work solely with adult stem cells. There is a consensus among experts, however, that while adult stem cells may be flexible, they are not *as* flexible as embryonic stem cells. They will not give rise to as many kinds of cells as will embryonic stem cells. Moreover, adult stem cells may have shorter life spans than embryonic cells, and they generally are hard to isolate in the first place. If stem cell research is to proceed as rapidly as possible, experts say, embryonic stem cells must continue to be used along with adult cells.

The practical consequence of this debate has to do with research funding. Private biomedical companies are pressing ahead with research in which they obtain early-stage human embryos and then retrieve the stem cells from them. However, for years the federal government—the largest single source of biological research funds—has not funded any research in which embryos have been destroyed. This stipulation holds even if the embryos in question would be destroyed anyway, as in the case of the embryos that come from fertility clinics. The question before political leaders today is: Should the federal government fund research that merely *involves* embryonic stem cells, even if no embryos were destroyed during this research? Could a researcher obtain embryonic stem cells from a private-sector scientist, and then receive government funding for research on the embryonic cells received? In August 2001, President Bush decided that the answer is a qualified "yes." Federal funding will be provided to researchers who work with cells from any of the 64 stem-cell lines cultured before the date of the President's announcement. The decision meant a change in U.S. policy: The federal government will now fund some research that is linked to the destruction of embryos. It is unclear, however, what effect the strict limitation on the number of eligible cell lines will have on stem-cell research in the U.S.

The Potential of Stem Cells

No human disease has yet been cured with stem cells; so far, these cells have demonstrated therapeutic benefits mostly in mice. Nevertheless, all parties seem agreed that stem cells hold out enormous potential. Consider just a partial list of the conditions that stem cells stand to alleviate or cure altogether: Parkinson's disease, heart disease, spinal-cord injuries, severe burn injuries, arthritis, Alzheimer's disease, multiple sclerosis, osteoporosis, and diabetes. Now consider just the last of these afflictions. People who suffer from "Type I" diabetes have lost the capacity to produce the hormone insulin, which moves blood sugar into cells. They have lost this capacity because the cells in their pancreas that produce insulin have been destroyed. As a result, they have to inject themselves with insulin each day to keep their blood sugar from building up to toxic levels. Even with this, they suffer from a host of diabetes-related afflictions, particularly as they get older. In 2001, researchers announced that embryonic stem cells they had retrieved from mice had begun secreting insulin when transplanted back into diabetic mice. Imagine a diabetic child being treated not with lifelong insulin injections, but with a similar cell transplant. Imagine scenarios like this involving *all* the conditions listed earlier, and you can see why scientists are so excited about stem cells.

On to Human Reproduction

The steps of development that you have just looked at form a kind of middle phase of the process you'll be looking at next chapter, which is reproduction—specifically human reproduction. Coming *before* the development you've reviewed are the formation of the egg and sperm that come together to create a new human being. Coming *after* it are late-stage human embryonic growth and then birth itself. With this chapter, you know something about development; in Chapter 28, you'll see how development fits into the larger scheme of human reproduction.

Chapter Review

Summary

27.1 General Processes in Development

- Development in animals begins with the creation of a zygote or fertilized egg. Once a zygote starts to divide, it is called an embryo.

- There are three phases of embryonic development. The first is cleavage, which produces a ball of cells, called a blastula, with an interior cavity. The second is gastrulation, in which these cells rearrange themselves into three layers that give rise to specific organs and tissues. The third is organogenesis, in which organs begin to form.
TUTORIAL 27.1.2: Three Phases of Embryonic Development

- The first product of cleavage is a tightly packed ball of cells called a morula. The cells in the morula rearrange themselves into a structure composed of one or more layers of cells surrounding a liquid-filled cavity. This structure is the blastula; its interior cavity is called the blastocoel.

- Gastrulation produces an embryo composed of three layers of cells, the endoderm, the mesoderm, and the ectoderm. The endoderm gives rise to interior organs and tissues, such as the liver and bladder in mammals. The mesoderm gives rise to much of the body mass of higher animals, including most of the muscle and skeletal systems. The ectoderm gives rise to the entire nervous system of mammals, as well as various external tissues, such as hair and tooth enamel.

- In early organogenesis in vertebrates such as frogs, a rod-shaped support organ, called a notochord, develops near the dorsal surface and induces the development of ectodermal tissue that lies above it. A neural plate develops, folding over on itself to form a hollow neural tube that gives rise to both the brain and the spinal cord. Groups of neural crest cells break away from the periphery of the neural tube during its development and then migrate to different locations throughout the embryo. These groups go on to develop into different organs or tissues.

- The early stages of development represent a transition from the general to the specific. A small group of superficially similar cells gives rise to three different layers of cells, which give rise to tissues and then to specific organs.

27.2 What Factors Underlie Development?

- Induction is the capacity of some embryonic cells to direct the development of other embryonic cells.

- Development is fundamentally controlled by the interaction of genes and proteins. A key model organism in scientific understanding of this interaction is the fruit fly, *Drosophila melanogaster*.

- A morphogen is a diffusable substance whose local concentration affects the course of local development in an organism. For example, the bicoid protein is a *Drosophila* morphogen that imparts positional information to the embryo, directly affecting the development of its anterior and center portions. Morphogens generally are transcription factors, meaning proteins, brought into production by genes, that then regulate the activity of other genes.

- Development works through a hierarchy of events controlled by genetic regulation. A first level of genetic instruction specifies broad positional patterns in an embryo (for example, the head and thorax portions of *Drosophila*). Subsequent levels of genetic instruction specify development within these body sections. Finally, control is exercised over the development of specific structures, such as organs.

27.3 Developmental Tools: Sculpting the Body

- Three cell capabilities help shape the animal body in development. The first is cell movement. The second is cell adhesion, in which several families of proteins produced by cells have the capability to selectively adhere to other cells. The third is selective cell death, in which tissues take shape through the destruction of cells, often through programmed cell death.

27.4 The Promise of Stem Cells

- Most cells in the adult body have fates that are committed. A cell has undergone commitment when its role has become completely determined.

- Cells in the early embryo have not yet undergone commitment and can give rise to various kinds of cells. A relatively small number of cells in the adult body have this same capability. Both the embryonic and adult cells that have this capability are called stem cells. Stem cells can be defined as cells that have the ability to generate both more stem cells and at least one variety of specialized cell (such as heart or nerve cells or their precursors). Stem cells are the subject of intense research interest today because of their potential to alleviate or cure conditions such as spinal-cord injuries and heart attacks.

- One source of human stem cells is early embryonic structure known as the blastocyst. Cells from the blastocyst's inner cell mass are pluripotent, meaning they can give rise to nearly all the different cells or tissues in the human body. A research breakthrough came in 1998 when a team of researchers announced they had been able to "culture" or grow pluripotent human blastocyst cells in the laboratory. Cultured, pluripotent blastocyst cells are commonly referred to as embryonic stem cells.

- The advance in embryonic stem cells sparked advances in the isolation and culturing of adult stem cells.

- An ethical controversy exists over the use of embryonic stem cells in scientific research, since embryos must be destroyed for the embryonic stem cells to be retrieved. In August 2001, President Bush decided to allow federal funding of research on a tightly restricted number of stem-cell lines.

- Stem cells have an enormous medical potential. A partial list of the medical conditions they stand to alleviate or eliminate includes Parkinson's disease, heart disease, spinal-cord injuries, severe burn injuries, arthritis, Alzheimer's disease, multiple sclerosis, osteoporosis, and diabetes.

Key Terms

adult stem cell	630	**morphogen**	627
animal pole	622	**morula**	622
blastocoel	622	**neural crest**	624
blastocyst	629	**neural tube**	624
blastula	622	**notochord**	624
cleavage	622	**pluripotent cell**	629
commitment	629	**somite**	625
embryo	622	**stem cell**	630
embryonic stem cell	629	**totipotent cell**	629
fertilization	622	**vegetal pole**	622
gastrulation	622	**zygote**	622
induction	625		

Understanding the Basics

Multiple-Choice Questions

1. In what order do the following developmental steps occur?
   a. organogenesis
   b. fertilization

c. blastula formation
d. morula formation
e. gastrulation

2. All of the following tissues are mesodermal derivatives **except**
   a. blood
   b. skin
   c. kidneys
   d. muscle
   e. bone marrow

3. Which of the following statements best describes the gastrula?
   a. hollow ball of cells
   b. newly fertilized zygote
   c. embryo with fully formed organs
   d. embryo with three distinct layers of cells
   e. embryo with a vegetal and animal pole

4. The process of gastrulation involves
   a. massive cell death
   b. little or no growth of the embryo
   c. cell reorganization and movement
   d. cell enlargement
   e. organogenesis

5. Organogenesis is accomplished, in part, by the interaction between various regions of the three germ layers. This interaction is referred to as
   a. reduction
   b. transformation
   c. translation
   d. induction
   e. fusion

6. The cells that move away from the developing notochord and eventually form many different organs and tissues are
   a. neural crest cells
   b. archenteron cells
   c. primary mesenchyme cells
   d. secondary mesenchyme cells
   e. endoderm cells

7. Substances, like the bicoid protein, that are essential in controlling local development in an embryo are called
   a. nurse cells
   b. mRNAs
   c. morphogens
   d. receptor proteins
   e. enzymes

8. Many proteins, like bicoid, that help to establish positional information function as
   a. neural crest cells
   b. transcription factors
   c. cell adhesion molecules
   d. blastulas
   e. mRNAs

9. A stem cell is valuable because it has not yet undergone _____ and can _____.
   a. gastrulation; move through the embryo
   b. commitment; give rise to specialized cells
   c. programmed cell death; still function
   d. division; produce morphogens
   e. organogenesis; stop dividing

10. A breakthrough occurred with human embryonic stem cells in 1998 when scientists
    a. learned of their existence
    b. got them to fuse with each other
    c. cultured them in a laboratory setting
    d. followed them through organogenesis
    e. were able to locate them

Brief Review

1. Briefly describe the order of events that occur after fertilization to produce an embryo with three layers of cells.

2. Outline the steps involved in the production of the chick wing.

3. How do cell adhesion molecules influence early events during development?

4. What is a morphogen concentration gradient, and how does it affect early development of the fruit fly, *Drosophila*?

5. Describe the role cell death plays during the growth and development of an organism.

6. Why have scientists become so enthusiastic about stem cells?

Applying Your Knowledge

1. Explain how the bicoid protein acts as a morphogen. What would happen to the developing *Drosophila* embryo if bicoid mRNA was introduced into the posterior region of the egg prior to fertilization?

2. The text listed numerous examples of medical conditions that stand to be cured or alleviated through the use of stem cells. But is the potential of stem cells broader than even this list would indicate? Can you name a single medical condition that would *not* benefit from the addition of new, functional cells as replacements for malfunctioning, diseased, or missing cells?

3. Cell death is a common feature in the development of many structures in the growing embryo. This process of cell death, if unregulated, could cause severe problems. What types of problems might be observed if the cell death process is altered?

28

How the Baby Came to Be
Human Reproduction

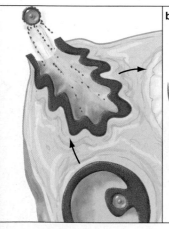

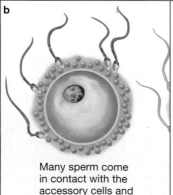

Many sperm come in contact with the accessory cells and are capacitated.

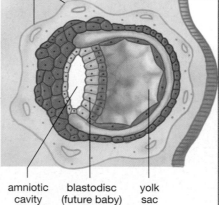

amniotic cavity   blastodisc (future baby)   yolk sac

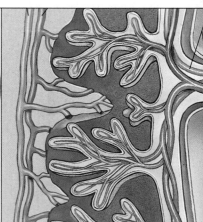

An egg is formed. (Section 28.2, page 638)

Fertilization begins. (Section 28.4, page 647)

Implantation. (Section 28.5, page 649)

What's a placenta? (Section 28.5, page 650)

To produce a child, human males and females must first produce sperm and eggs, after which sperm and egg must come together in the act of fusion known as conception. The fertilized egg that results will, over the next 38 weeks, develop into a fully formed baby.

Memorable events mark every human life, but for parents, one event that is sure to be indelibly etched into memory is the birth of a child. The mother's labor, the final "pushing" that brings the baby forth, the first real look at the infant and the infant's first look at the world; these are moments that, once experienced, are never forgotten.

The birth of a child is a beginning, of course, and yet it is also a culmination. Human beings must grow to a certain level of biological maturity before being capable of reproduction. After this, finding a partner may be difficult, and getting pregnant more difficult yet. Eventually, however, two very special "seeds"—*this* egg and *this* sperm, out of all the millions a man and a woman produce in a lifetime, fuse to create this child.

In this chapter you'll trace the biology of human reproduction, which might be defined as all the biological steps that must fall into place before a child can be born. The process of development you studied in Chapter 27 represents part of what you'll be looking at, but other elements are involved as well.

28.1 Overview of Human Reproduction and Development

Taking this story from the beginning means picking up a narrative that was begun in Chapter 10, when you went over the formation of the reproductive cells called **gametes**—eggs in females, sperm in males. In this chapter the focus widens, taking into account the physical *setting* in which this formation takes place and the steps that bring egg and sperm together. Finally, you will look at human development from conception to birth. For information on some of the methods humans use to *prevent* conception, see "Methods of Contraception" on page 645.

The human male and female reproductive systems are functionally alike up to a certain point, in that they first develop their respective gametes and then deliver them to a location where they can fuse in the act known as fertilization or conception. However, the female reproductive system is called on to do all that the male system does *and more* in that, once fertilization takes place, the female system must then provide the environment for the embryo to develop, after which it must deliver the baby into the world. This process takes place over a period of 38 weeks, on average.

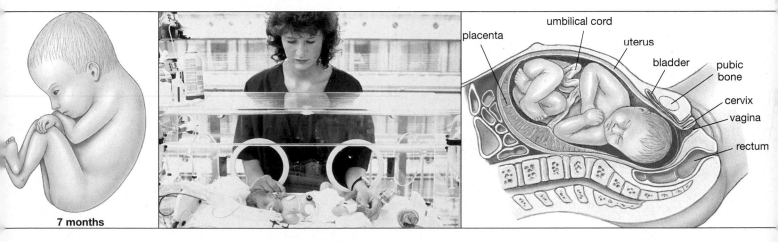

7 months

Lungs develop, eyelids open.
(Section 28.5, page 651)

Getting special treatment.
(Section 28.5, page 652)

placenta
umbilical cord
uterus
bladder
pubic bone
cervix
vagina
rectum

Ready for delivery.
(Section 28.6, page 653)

Reproduction in Outline

The first part of reproduction, the formation and delivery of egg and sperm, is a straightforward story in outline. All the eggs a woman possesses exist from birth in a precursor form, called oocytes, that lie within two walnut-shaped structures called ovaries. These exist just right and left of center within the female pelvic area (**see Figure 28.1**). Oocytes sit in the ovaries within nurturing complexes of cells and fluids called **follicles**. On average, once every 28 days one follicle-oocyte complex is brought to a state of maturity such that the oocyte and some accessory cells that surround it rupture from their ovary and begin a slow journey to the uterus. This trip takes place inside the structure called the **uterine** (or **Fallopian**) **tube**. The release of the oocyte is called **ovulation**. If, during this trip, the oocyte encounters a male sperm, the two may fuse, in which case conception has occurred.

Sperm in men are produced, in unfinished form, in sets of tubules in the two male testes (**see Figure 28.2**). They then are transported to a structure called the epididymis (one for each testis) for further development. Then, with sexual excitation, sperm are transported in a loop: up and over the urinary bladder through two ducts, each called a **vas deferens**. The contents of these ducts then empty into a single duct, called the urethra, from which they are ejaculated into the female vagina. Along the way, materials secreted by several glands will join the sperm in a process that can be likened to tributaries feeding into a common stream. The resulting mixture of sperm and glandular materials is called **semen**.

Whereas the female releases an average of one oocyte per month, the male releases an average of 250 million sperm per ejaculation. Upon entering the vagina, sperm are able to propel themselves up into the uterus and from there into the uterine tubes. Of the hundreds of millions of sperm that begin the journey, a few dozen may encounter the female oocyte as it continues its journey from ovary to uterus (**Figure 28.3**). The first sperm to breach the coating that surrounds the

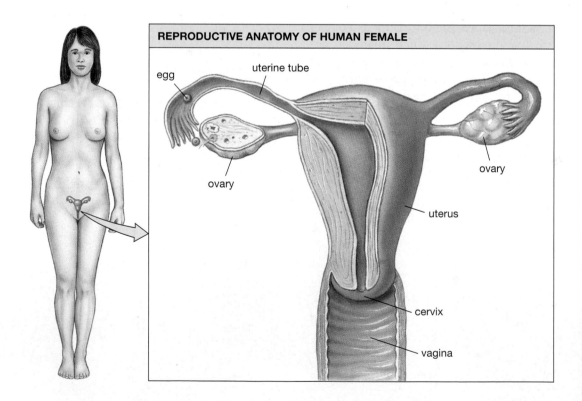

REPRODUCTIVE ANATOMY OF HUMAN FEMALE

egg

uterine tube

ovary

ovary

uterus

cervix

vagina

Figure 28.1
Reproductive Anatomy of the Human Female
Eggs develop in the ovaries within nurturing complexes called follicles. Once every 28 days, on average, an egg is expelled from one ovary or the other and journeys through the uterine tube to the uterus. If the egg encounters sperm on the way, pregnancy will result when egg and sperm fuse.

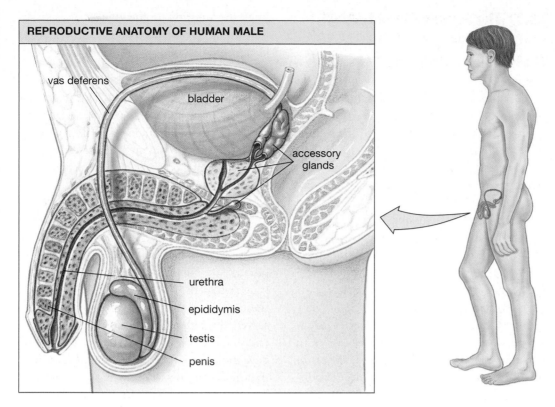

Figure 28.2
Reproductive Anatomy of the Human Male
Sperm are produced in the testes and transported to the epididymis, where they mature and are stored. With orgasm, the sperm are transported through the vas deferens and then the urethra, from which they are ejaculated. Along the way, they are joined by materials secreted by several accessory glands.

oocyte fuses with it, while the rest are shut out through a kind of gate-slamming mechanism that operates almost instantaneously. When this one sperm has joined with an oocyte, conception has occurred. The now-fertilized egg still floats free in the uterine tube, where it begins to develop in the ways described in Chapter 27: One cell becomes two, two become four, and so on. Four days after fertilization, the developing embryo enters the uterus; a day or so more and it implants itself in the uterine wall.

Both the male and female reproductive systems are greatly influenced by the actions of hormones—chemical signaling molecules that often move through the bloodstream. Indeed, the reproductive organs are primary sites of production for some of the most important of these hormones, testosterone in men and estrogen and progesterone in women.

a

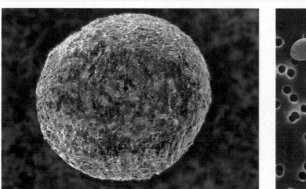

b

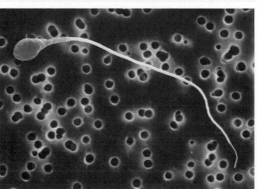

Figure 28.3
Egg and Sperm
Artificially colored images of **(a)** a human egg and **(b)** a human sperm.

28.2 The Female Reproductive System

The Female Reproductive Cycle

Although hormones operate in both sexes, in females they bring about a reproductive *cycle* that repeats itself about once every 28 days (though it can range from 21 to 42 days). In males, conversely, there is no cycle, but instead a steady production of sperm and a destruction or removal of the sperm that are not ejaculated. Why a cycle for females and not for males? The answer lies partly in the two roles the female reproductive system must fulfill. It must form an egg *and* create an environment in the uterus in which a fertilized egg can develop. An important part of this uterine environment is essentially built up and then dismantled once every 28 days—except in those rare instances in which an oocyte is fertilized. The end result of the dismantling of the uterine environment is **menstruation**, which is a release of blood and the specialized uterine tissue it infuses, the **endometrium**. This is tissue that lines the interior of the uterus; it is tissue that *would have* housed an embryo, had one been implanted there.

Why Does Menstruation Exist?

From this, it's reasonable to ask: Why this monthly cycle of buildup and destruction, rather than a one-time formation of a lasting embryonic environment? After all, menstruation has some obvious costs attached to it; apart from pain and discomfort, there are what might be called species-survival costs: a loss of fertile days and a loss of blood and other tissue. (On average women lose about 40 milliliters—or about 1.4 fluid ounces—of blood in a menstrual cycle, and about that much other tissue as well.) So, in the face of these costs, why has menstruation been preserved throughout human evolution? Scientists have no firm answers here, but we do have a couple of intriguing hypotheses. One is that menstruation serves to clear the uterus of harmful bacteria that can be transported in on sperm. (These bacteria exist on male and female genitalia and are able to attach

Tutorial 28.1.1
Steps in Ovulation and Pregnancy

Figure 28.4
Steps Leading to an Egg
An egg develops from precursor cells called oocytes that exist near the surface of the ovary. Oocytes join with nurturing complexes of cells to form ovarian follicles. On average, only one of these follicles matures—once every 28 days—through the stages of primary, secondary, and tertiary follicle. The oocyte contained in that follicle is expelled at the end of the process for a journey through the uterine tube. With its expulsion, the tertiary follicle is transformed into the corpus luteum, which for a time produces hormones that can aid in pregnancy.

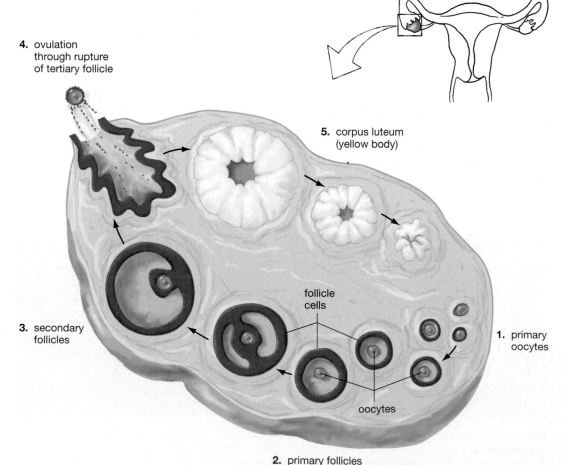

4. ovulation through rupture of tertiary follicle

5. corpus luteum (yellow body)

follicle cells

1. primary oocytes

3. secondary follicles

oocytes

2. primary follicles

themselves to ejaculated sperm.) The other hypothesis is that, costly as menstruation may be, in terms of expended energy it is not *as costly* as the alternative of maintaining a permanent embryonic environment in the uterus.

How Does an Egg Develop?

As noted earlier, eggs develop from precursor cells called **oocytes**. If you look at **Figure 28.4**, you can see where oocytes develop: in the outer portion of each ovary. Each oocyte is surrounded by a sphere of accessory cells, called *follicle cells*, that eventually provide the oocyte with nutrients. The complex of oocyte and follicle cells is called an **ovarian follicle**; it is the basic unit of oocyte development.

Follicle Development and Ovulation

Every oocyte that matures into an egg must go through the developmental steps you see in Figure 28.4. This is a monthly process of selection, running from a large starting set of follicles, through fewer *primary follicles* that develop from them, to a greatly reduced number of *secondary follicles* that continue

development, and finally to a single *tertiary follicle* that nurtures an oocyte through its time in the ovary. Responding midway through the menstrual cycle to hormonal influence, the tertiary follicle is squeezed to the extent that it ruptures, thus expelling the oocyte and its surrounding accessory cells not only from the follicle but from the ovary as well. With this event, ovulation has occurred. It takes place in this way in *one* ovary each month, with the ovary selected through an unknown process.

Movement through the Uterine Tube

Freed from its ovarian housing, but still surrounded by accessory cells, the oocyte—technically not yet an egg—is now swept into the uterine tube through the actions of the finger-like *fimbriae* lying at the tube's funnel-shaped end (**see Figure 28.5**). Propelled by liquid currents and the uterine tube's whip-like cilia, the oocyte makes a journey of about 10 centimeters (4 inches) through the uterine tube over the next four days. While within the tube, the oocyte may encounter sperm and become fertilized prior to entering the uterus.

Tutorial 28.1.1
Steps in Ovulation and Pregnancy

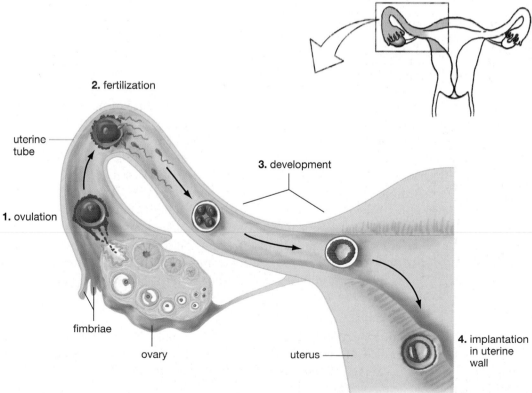

Figure 28.5
Journey to Pregnancy
Surrounded by accessory cells, an oocyte is expelled from the ovary into the uterine tube, where it encounters sperm and is fertilized by one of them. Now a fertilized egg (or zygote), it begins the process of development and completes its journey with implantation in the uterine wall.

Hormones and the Female Reproductive Cycle

One of the more interesting aspects of human reproduction is that in females it is carried out through a *cycle*, which is to say a series of events that repeat themselves regularly—on average, once every 28 days. But how can the female reproductive system "keep time" like this? What is the clock that governs the cycle?

A real clock is actually a good analogy for what happens. If you look at an old-fashioned mechanical watch, what you'll see is a central spring that exerts a force through a series of gears. Given these parts, it takes 12 hours for the physical mechanism of the "small hand" on the watch to make its way through a complete cycle—to end up where it began. In women, the substances called *hormones* set in motion a group of physical processes that take about 28 days to complete. The last of these processes leaves the body right where it began, at the start of another cycle.

We actually should speak of female reproductive *cycles*, in the plural, because several are going on in reproduction, all of them interlinked. There is the ovarian cycle, which has to do with the development and release of oocytes; the menstrual or "uterine" cycle, which concerns the regular buildup and then breakdown of tissue in the inner lining of the uterus; and several hormonal cycles, which help govern both menstruation and egg creation.

If you look at **Figure 1**, you can see a time line for all these cycles and what occurs during them. To give you a feel for the processes at work, let's follow just the menstrual cycle, whose "first day," over on the far left, is defined as the first day of menstruation. What the graph tracks is the state of the tissue (the endometrium) lining the inside of the uterus. As you can see, beginning on day 1, this tissue goes through a breakdown that lasts three or four days, after which it begins a steady buildup that continues all the way to the end of the cycle—at which point it starts to break down again. Now look at what is going on in the ovarian cycle. This part of the graph is following the single "successful" oocyte (out of many that begin development) through its progression into the oocyte that bursts from the ovary on day 14.

The real lesson, however, is to think about the menstrual and ovarian cycles as they relate to *each other*. The endometrium is built up in the uterus as a means of creating an environment in which an egg could implant itself, should one be fertilized. If such an egg begins its journey on day 14 and then takes another 6 days or so to implant itself in the uterus, you can see what stage the *endometrium* will be in at this point: near its maximum thickness (and at a highly productive stage of synthesizing a blood-vessel system). Such an environment is just what an implanted embryo would need.

And what of the figures over on the right side of the ovarian cycle? They represent the corpus luteum, the "remains" of the follicle that the successful oocyte broke out of. The corpus luteum is much more than a spent vessel at this point, however, because hormones secreted by it will promote the growth of the endometrium. Without a pregnancy occurring, the corpus luteum carries out such secretion only for about a week. After that it begins to degenerate, and *this* means the degeneration of the endometrium, which washes out, along with some of the blood that supplied it, in the process called *menstruation*.

What hormones are secreted by the corpus luteum? Look now at the ovarian hormone cycle, and you'll see a line for "progestins"—a family of hormones whose most important member is progesterone. As you can see, the level of progestins takes a sharp turn upward just as the corpus luteum is formed—no surprise, because it is the corpus luteum that is secreting them.

There is a circularity to all this hormonal control, with the hormones involved influencing each other in different ways as the cycle goes on. Let's arbitrarily take as a starting point, however, first the development of a follicle and then the release of an egg from it.

One important hormone in the cycle is produced neither in the ovaries nor the uterus, but in the brain. A brain structure called the *hypothalamus* releases vanishingly small amounts of a hormone called **gonadatropin-releasing hormone**, whose function is spelled out in its name: it controls the release of *other* hormones, namely the gonadatropins.

continued

Meanwhile the ruptured tertiary follicle develops into a body, called the **corpus luteum** ("yellow body"). This structure secretes hormones that help prepare the reproductive tract for pregnancy. If pregnancy does not occur, the corpus luteum begins degenerating after about 12 days. With the corpus luteum, you can see again the part that hormones play in the monthly female reproductive cycle. For details on this hormonal regulation, see "Hormones and the Female Reproductive Cycle," above.

Changes through the Female Life Span

In between the creation of oocytes and the journey of one oocyte down the uterine tube lies a multistep process that could be characterized by the phrase: "Many are called, but few

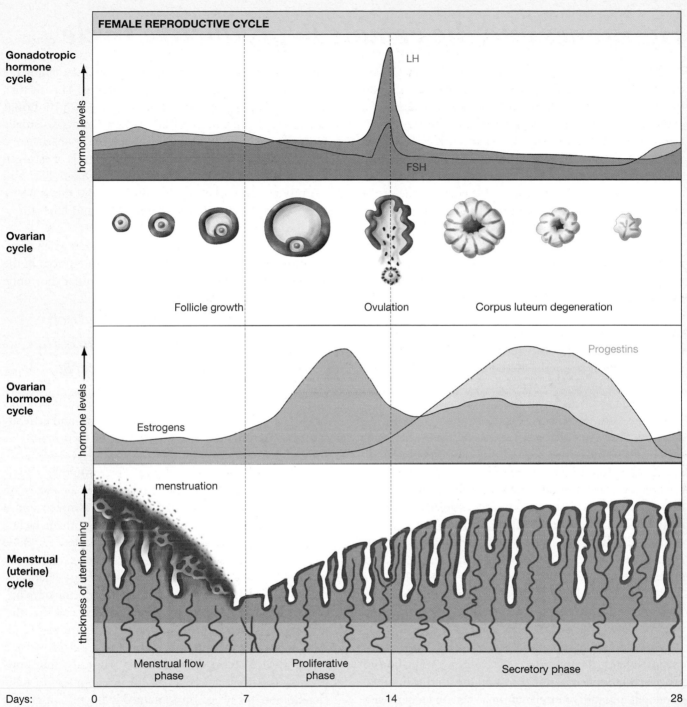

Figure 1
The Female Reproductive Cycles
Female reproduction is controlled by the interrelated secretions of a number of hormones.

Tutorial 28.1.3
The Female Reproductive Cycle

are chosen." Recall from Chapter 10 that all the oocytes a woman will ever possess are produced in her prior to birth; these oocytes in turn become associated with the nurturing follicle complexes. A 6-month-old female fetus has in its ovaries about 7 million follicles that are capable of becoming eggs. At birth, this number has been reduced to about 1 million, primarily because of the process of *atresia*, or the natural degeneration of follicles. By puberty, the number has declined to about 250,000. An unknown number of these follicles are "recruited" monthly for further growth through hormonal action, but only *one* of these will, each month, go on to develop into the oocyte that bursts forth from the ovary. The only exception to this rule comes in the 1 percent or so of ovarian cycles that result in

Hormones and the Female Reproductive Cycle *continued*

Two gonadatropins are spurred into production in this way, **follicle-stimulating hormone (FSH)** and **luteinizing hormone (LH)**, both of them produced in a gland located beneath the hypothalamus, the *anterior pituitary gland*. You can see both FSH and LH levels charted in the gonadatropic hormone cycle in Figure 1.

After being secreted by the anterior pituitary, both FSH and LH have their primary early-cycle effects at the site of the ovarian follicles. FSH promotes the development *of* the follicles, while LH stimulates the synthesis of a hormone *by* the follicles, **estrogen** (which actually is a family of hormones). The more the follicles develop, the more estrogen they secrete. The result, as you can see in Figure 1, is that estrogen levels slowly rise through the first week or so of the cycle. One thing estrogen is doing is promoting the growth of the endometrium in the uterus; another, however, is controlling the release of both FSH and LH through *feedback* mechanisms, which is to say mechanisms in which the level of one thing controls the level of another.

Estrogen feedback operates in two ways at two different phases of the cycle. Early on, the relatively low (but increasing) levels of estrogen act to *decrease* FSH—not enough to stop development of the follicles, but enough to slightly reduce FSH levels over the first 10 days or so of the cycle, as you can see in Figure 1. In the second week of the cycle, however, estrogen output is greatly increased thanks to the development of the tertiary follicle, whose oocyte will burst forth from the ovary. At this level, estrogen acts to *increase* not only FSH production but LH production as well. The result, as seen in Figure 1, is the "spike" of LH production just before day 14 of the cycle. This hormonal surge is what causes the oocyte to break out from both follicle and ovary.

Now the former tertiary follicle has a new identity, the corpus luteum. In this role it has begun production of progesterone, which as noted acts on the uterine lining to develop it for embryo implantation. Then, in tandem with estrogen, progesterone has another effect. It acts on the hypothalamus and pituitary to *lower* FSH and LH levels. But it turns out that the corpus luteum needs LH in order to remain healthy. In the absence of pregnancy, falling LH levels spell the end for the corpus luteum. This in turn causes a decline in progesterone levels, which brings about the degeneration of the endometrium, essentially by depriving it of a blood supply. Now, in the absence of a corpus luteum or developed follicles, the levels of estrogen and progesterone are *low*. Such levels *stimulate* FSH and LH production in the anterior pituitary, and this is where we started: The cycle begins again, as LH and FSH stimulate the growth of another set of follicles.

So what keeps this cycle from continuing when pregnancy does occur? The implantation of an embryo in the endometrium causes release of yet another hormone

How can the female reproductive system "keep time"? What is the clock that governs the cycle?

(called human chorionic gonadotropin), which acts like LH in that it keeps the corpus luteum robust and generating progesterone and estrogen, which in turn maintains the endometrium.

With this knowledge under your belt, you're in a position to see what it is that birth control pills do. They are compounds that contain both synthetic estrogen and a synthetic progestin. Women generally take them beginning five days after the start of menstruation for three weeks (and during the fourth week take a chemically functionless "spacer" or nothing). Think about what is required at the beginning of a cycle to prompt the development of follicles: *low* levels of estrogen. (Recall that this low level stimulates the hypothalamus and pituitary to secrete follicle-stimulating hormone.) The pill provides a *higher* level of estrogen than exists naturally, thus suppressing the release of FSH and the development of follicles. Second, the progestin contained in the pill suppresses LH release by the pituitary, which prevents the *ovulation* of any follicle that might manage to develop, because the LH surge is required for ovulation.

multiple ovulations, in which case fraternal twins, triplets, and other multiple births are possible. In sum, the oocyte that begins the journey down the uterine tube each month has been selected from more than 7 million initially created, and it is one of about 500 that will make this trip in a woman's lifetime (about 13 a year for 35 or 40 years).

The Consequences of Follicle Loss

Given that a woman begins puberty with about 250,000 follicles, you might think that her supply would never run out. This is not the case, because the process of atresia continues throughout the lifetime, not only for "recruited" follicles, but for other follicles as well. By the time a woman reaches her early fifties, perhaps a thousand follicles remain; this scarcity is the primary factor that brings about **menopause**, or the cessation of the monthly ovarian cycle. Because follicles not only nurture oocytes, but are primary sites of estrogen synthesis as well, the loss of follicular activity helps cause the physical problems associated with menopause: the loss of muscle tone, the thinning of the vaginal walls, and a loss of bone deposition, which has been associated with **osteoporosis**,

or a decrease in bone mass. One aid in combating these problems is hormone replacement therapy (HRT), which is the use of synthetic hormones by women who are past menopause. Now let's look at the male reproductive system.

28.3 The Male Reproductive System

Figure 28.6 provides another view of the male reproductive system. Here you see the large-scale structures of the system. There is the

Tutorial 28.2.3
Sperm Development and Fertilization

Figure 28.6
Development and Delivery of Sperm

a After the sperm develop in the testes and mature in the epididymis, orgasm prompts them to be transported through the vas deferens and then the urethra. Secretions from accessory glands (the seminal vesicles, the prostate gland, and the bulbourethral glands) join with the sperm to form semen before ejaculation.

b How sperm mature. Sperm development begins within seminiferous tubules, located in the testes. Sperm start out as spermatogonia at the periphery of the tubules and end up as immature sperm in the interior cavity of the tubules. From there, they are transported to the epididymis for further development and storage.

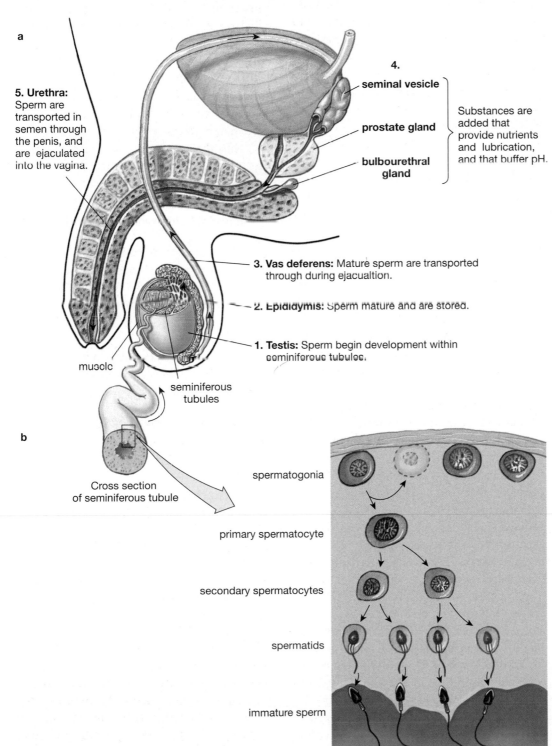

a

5. Urethra: Sperm are transported in semen through the penis, and are ejaculated into the vagina.

4.

seminal vesicle

prostate gland

bulbourethral gland

Substances are added that provide nutrients and lubrication, and that buffer pH.

3. Vas deferens: Mature sperm are transported through during ejacualtion.

2. Epididymis: Sperm mature and are stored.

1. Testis: Sperm begin development within seminiferous tubules.

muscle

seminiferous tubules

b

Cross section of seminiferous tubule

spermatogonia

primary spermatocyte

secondary spermatocytes

spermatids

immature sperm

testis, in which sperm begin development; the **epididymis**, the tubule in which sperm mature and are stored; and the **vas deferens**, the tube through which sperm move in the process of ejaculation. (All three of these structures exist in pairs, though the second member of each pair is not visible in the figure.) Then there is the single **urethra**, which each vas deferens empties into and through which sperm continue their journey. (Eventually the urethra is channeled through the penis.) There are also several accessory glands whose contributions are added to sperm as it proceeds: the **seminal vesicles**, the **prostate gland**, and the **bulbourethral glands**. (The seminal vesicles and bulbourethral glands also exist in pairs.)

Structure of the Testes

The testes are surrounded by a muscle whose contraction helps regulate the temperature of the sperm inside. Sperm development requires temperatures somewhat cooler than those found in the rest of the body; this is why the testes are *external* to the body's torso. Sperm cannot develop in temperatures that are too cold, however. Thus, when a male steps into, say, a cold pond, the muscle surrounding the testes will contract, drawing the testes up closer to the torso—and warming the sperm.

Figure 28.6a shows that the testes are divided into a small series of lobes, and that within each lobe there are a number of highly convoluted tubes, the **seminiferous tubules**. These are the sites of initial sperm development. If you now look at a single seminiferous tubule in cross section, in Figure 28.6b, you'll see how this development proceeds: from the outside of each tubule—where sperm precursors are in their *least* developed state—toward the interior cavity of each tubule, where sperm are in a *more* developed state. Eventually, the immature sperm are released into the interior cavity, and are transported up it, moving then to the epididymis for further processing and storage. Thus, there is an assembly-line quality to sperm production, starting with initial development at the periphery of the tubules, continuing in the interior of the tubules, and ending with storage in the epididymis. Older sperm that are not ejaculated are destroyed (by other cells) or eliminated in the urine.

Male and Female Gamete Production Compared

The production of sperm bears an interesting comparison with the development of eggs. The cells that give rise to sperm are called **spermatocytes**. In going back to the precursors of *these* cells, however, we see a significant difference between males and females. The cells that give rise to primary spermatocytes are called **spermatogonia**; they exist at the periphery of the seminiferous tubules and, through normal mitotic division, they actually lead to *two* kinds of cells. One is the primary spermatocyte, which goes on to develop into a sperm; the other, however, is another spermatagonium. Thus, spermatogonia are "sperm factories," if you will, that give rise not only to sperm, but to more sperm *factories* as well.

This, then, is the reason males *keep* producing sperm throughout their lives while, as we've noted, females have a fixed quantity of oocytes. Although male sex hormone production lessens with age, it is possible for male reproductive capability to remain intact until death. In contrast, as you've seen, female reproductive capability ends with menopause, generally when a woman is in her fifties, primarily because of a lack of oocytes.

If you look at Figure 28.6b, you can see how the male sperm progression works in the testes: from spermatagonium, through spermatocyte to spermatid, and finally to immature sperm.

Further Development of Sperm

As you can see in Figure 28.6a, the seminiferous tubules eventually feed into the epididymis, a single convoluted tubule (one for each testis) that is about 7 meters (about 23 feet) long. This structure serves as the site of storage and final sperm development and as a kind of materials recycling site for damaged sperm. By the time sperm arrive at the bottom portion of the epididymis, they are fully matured and "motile," or capable of movement.

Time Sequence, Selection, and Sperm Competition

How long does this whole process of sperm development take? About 2.5 months from start to finish. Perhaps 250 million sperm are produced in this assembly line each day, which you may recall is about the number

Methods of Contraception

Selected methods of contraception are presented here. The rate of effectiveness for each method refers to effectiveness with correct use during every instance of sexual intercourse. In practice, effectiveness for each method tends to be lower, because use of the method may be incorrect or inconsistent. See "Hormones and the Female Reproductive Cycle" on page 640 for information on how birth control pills work. The type of birth control pill most commonly used has an effectiveness of 99.9 percent.

Table 28.1
Methods of Contraception

Condom

Made of latex, plastic, or animal tissue, condoms are put on the erect penis prior to intercourse and catch sperm after it is ejaculated. Condoms are the only method of contraception that helps protect both partners from some sexually transmitted diseases.

Effectiveness: 97 percent.

Diaphragm

A soft, circular rubber barrier inserted into the woman's vagina before sex. Covering the cervix, it prevents sperm from reaching the egg. Along with the related cervical caps, diaphragms offer females some protection against some sexually transmitted diseases, such as gonorrhea and chlamydia.

Effectiveness: about 95 percent, when used with a spermicide (a chemical compound that immobilizes sperm).

Intrauterine device (IUD)

A small device made in some instances of plastic and copper that is inserted into a woman's uterus by a physician. These IUDs can be left in place for several years. Other IUDs are made of plastic and contain a supply of a progestin hormone that is continuously released. These IUDs must be replaced once a year. Both types of IUDs help prevent fertilization of egg by sperm, in a process that may involve altering the movement of each type of gamete. Both types of IUDs also work to prevent implantation in the uterus of any embryo that might be produced.

Effectiveness: copper IUD, 99.4 percent; hormone-releasing IUD, 98.5 percent.

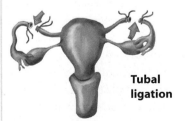

Tubal ligation

A surgical procedure that generally results in the sterilization of a woman, meaning permanent infertility. In it, the uterine tubes are first cut and then tied, thus blocking eggs from moving to a position in which they could be fertilized by sperm. Does not affect feminine characteristics, hormonal production, or sexual performance. Very rarely, tubes will reconnect themselves, which accounts for the (low) failure rate of the procedure. It is possible to reverse the operation in some instances.

Effectiveness: 99.6 percent.

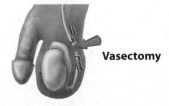

Vasectomy

A surgical procedure that can result in the sterilization of a man, though vasectomies are sometimes reversible. In the procedure, the vas deferens are cut and tied, thus blocking the movement of sperm during ejaculation. Does not affect male hormonal production, physical capacity for erection, or ability to ejaculate. (Sperm make up a small portion of the semen that is ejaculated.) In rare instances, vas deferens will grow back together again, which accounts for the (low) failure rate of the procedure.

Effectiveness: 99.9 percent.

that are ejaculated with each orgasm. (This latter number should be taken as a very rough average, however. It varies somewhat from person to person and greatly in accordance with how much time has elapsed since the last ejaculation.)

Why Are So Many Sperm Produced?

As noted, the watchword for oocyte selection could be "many are called but few are chosen." How much more extreme is the selection that goes on with sperm. The number that are expelled with *each ejaculation* vastly exceeds the number of oocytes a woman ever has stored in her ovaries. On average, however, only a single sperm per ejaculation will be able to fertilize an egg—assuming an egg is present in the uterine tube at the right time.

Why is there such an enormous difference between the number of sperm produced and the number that will fertilize eggs? As with the reason for menstruation, we can't say for sure, but we do have an interesting hypothesis put forward by two British scientists. It is that most sperm don't function to fertilize eggs; they function to block or destroy the sperm of *other males* (thus helping ensure the reproductive success of the male they came from). Part of the evidence supporting this idea is that perhaps half of all mature sperm are so ill-shapen that they seem incapable of making the long journey to the uterine tube, and are perhaps better suited to serve as "blockers" of other sperm.

From Vas Deferens to Ejaculation

Once they have matured, human sperm are stored in both the epididymis and the vas deferens. With sufficient sexual stimulation, ejaculation takes place through the contractions of muscles that surround the penis at its base. This contraction pushes the sperm through the urethra and out the urethral opening at the tip of the penis—at a speed of about 10 miles an hour.

Supporting Glands

As noted earlier, several other kinds of materials join with sperm before ejaculation to form the substance known as **semen**. Indeed, given the number of sperm ejaculated, it may be surprising to learn that sperm don't account for much of the volume of semen; some 95 percent of the ejaculated material comes from the accessory glands already mentioned: two seminal vesicles, the prostate gland, and two bulbourethral glands. It's helpful to think of sperm as getting "outfitted" for their travels by these glands just as they are exiting from the body; the glands supply both nutrients (which sperm will use to power their movement) and alkaline and lubricating substances (which neutralize the acidic environment of the vagina and facilitate transportation).

One of these supporting glands may sound familiar. The **prostate gland** surrounds the urethra near the bladder and contributes a substantial amount of material to semen. Later in life, a significant percentage of men develop either enlarged prostate glands or, worse, the growth of cancerous cells in the prostate, which is to say prostate cancer. Given that the prostate encircles the urethra—and that the urethra transmits not only semen but urine as well—you can see why a warning sign of these conditions is reduced urine flow.

28.4 The Union of Sperm and Egg

To this point, you have reviewed the process by which male sperm and female egg are first created and then *positioned* so that they can come together in the process called fertilization. Now let's look at this moment of fusion in somewhat greater detail.

When the oocyte is expelled from the ovaries, there still is no solitary oocyte in transit, but rather an oocyte surrounded by a group of accessory cells and connective substances.

How Sperm Get to the Egg

Now consider the sperm (each one of which is known as a *spermatozoon*). As you can see in **Figure 28.7a**, each spermatozoon has a "head" that contains not only the nucleus and its vital complement of chromosomes, but an outer compartment, called an **acrosome**, which contains enzymes. These enzymes are released *externally*, so that they break down materials lying in their path, meaning the layers of accessory cells that surround the oocyte. To release these enzymes, the sperm need to be *capacitated*, which means they need to have a protective coating that covers the acrosome stripped away. This is accomplished by the action of substances secreted by the outer layers of cells surrounding the oocyte.

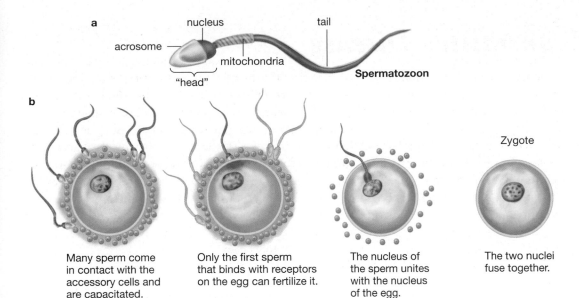

a

acrosome — nucleus — tail

mitochondria

"head"

Spermatozoon

b

Many sperm come in contact with the accessory cells and are capacitated.

Only the first sperm that binds with receptors on the egg can fertilize it.

The nucleus of the sperm unites with the nucleus of the egg.

Zygote

The two nuclei fuse together.

Figure 28.7
Anatomy of a Spermatozoon and Steps in Fertilization

a The "head" of each sperm includes not only its complement of chromosomes but also an acrosome, which contains enzymes that, when released externally, break down the layers of cells and other materials surrounding the egg. This release comes when the acrosome is "capacitated" by its contact with these surrounding cells. Behind the sperm head are the mitochondria that supply the energy for the motion of the tail that propels the sperm.

b The process by which sperm fertilize an egg.

Tutorial 28.2.3
Sperm Development and Fertilization

Many sperm are capacitated, but only one of them completes the next step, which is to latch onto the oocyte (**Figure 28.8**). This happens through an extension of the sperm cell that binds with receptors on the surface of the oocyte (Figure 28.7b). If you've ever wondered what separates one species of animal from a closely related, but different species, this is an important control point. Only sperm extensions and egg receptors that are chemically matched can bind with one another—and if there is no binding, there will be no offspring. Once this binding has occurred, the plasma membranes of sperm and oocyte fuse; the sperm is inside the oocyte, and sperm and egg nuclei combine. Fertilization is complete.

How Latecomers Are Kept Out

But what of the other dozens of sperm surrounding the egg? The fusion of the plasma membranes of egg and sperm slams a kind of gate shut faster than the blink of an eye. Membrane channels in the egg open up, allowing sodium ions to rush in; this changes the electrical potential of the egg membrane, effectively shutting out any other sperm. This process, called the *fast block to polyspermy*, is a kind of first line of defense against *multiple* sperm entering the egg. It's critical that a block like this exists, because if several sperm did enter the egg, there would be no development of a zygote. (How could there be, since an egg would have, say, one and a *half* complements of genetic material within it?) The sexual activity that is so important to human reproduction also carries with it the risk of sexually transmitted disease, which you can read about in the essay on page 648.

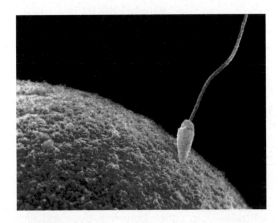

Figure 28.8
Near the Moment of Conception
A human sperm and egg at the moment the sperm binds the egg. Very soon after this, the sperm will enter the egg and its nucleus will fuse with the egg nucleus, producing a human zygote.

Sexually Transmitted Disease

Given the necessity of sex for the continuation of human life, it may seem surprising that so much risk can be associated with it, but such is the case. The root of the problem is that microorganisms that literally couldn't stand the light of day—that is, that couldn't exist for long in an exterior environment—are quite well adapted to life inside human tissue. More important, these organisms can be transmitted from one human being to another during the act of sex.

The risk of sexually transmitted diseases (STDs) varies in accordance with lots of things: the disease itself, the type of sexual activity a person engages in, and the precautions that are taken to avoid trouble. But one rule that holds true for all STDs and all behaviors is that as the number of partners goes up, so does risk. Here are the most common STDs, their prevalence in the United States, and the methods of treatment for them.

Disease	Caused by	Possible Consequences	Estimated Cases in United States	Treatment
Chlamydia	Chlamydia trachomatis organism—has features of both virus and bacteria	Spread to uterine tubes in women, causing pelvic inflammatory disease (PID), which can lead to sterility	3 million cases per year	Curable with timely use of antibiotics
Gonorrhea	Gonococcus bacterium	Arthritis, pelvic inflammatory disease	650,000 per year	Curable with timely use of antibiotics, though antibiotic-resistant strains now exist
Syphilis	Spirochete bacterium	Heart, nervous system, bone damage if allowed to progress	35,000 cases of primary and secondary infection per year	Curable with timely use of antibiotics, though antibiotic-resistant strains now exist
Genital herpes	Virus	Recurrent skin lesions, eye damage, pregnancy complications	1 million per year	Occurrence and severity of outbreaks can be lessened with treatment, but no cure exists
Genital warts	Human papilloma virus	Cervical infection in women, association with increased risk of cervical and penile cancer	5.5 million per year	Several treatments exist for removing warts, but virus may persist
AIDS	Human immunodeficiency virus (HIV)	Death, dementia, injury from a variety of opportunistic infections	46,400 new AIDS cases in 1999	No cure; treatments to reduce symptoms exist

28.5 Human Development Prior to Birth

Having seen how sperm and egg form and get together, you have reached a point in human reproduction that should be familiar: development, meaning the process by which a fertilized egg is transformed, step by step, into a functioning organism. As you'll see, processes you looked at in Chapter 27 in connection with frogs and flies have counterparts in human development. What may be surprising is how *early* in human development many of the critical events of development occur. Here is a list of some of the key developmental processes discussed in Chapter 27.

- Formation of the blastula, or hollowed-out body of cells (called a *blastocyst* in humans and other mammals)
- Gastrulation, in which the three primary layers of cells are formed
- Formation of the neural tube (which gives rise to the brain and nervous system)
- Organogenesis, in which the organs of the body start to take shape

Now, keeping in mind that a human pregnancy lasts an average of 38 weeks, here is the time line for these major developmental events. Blastocyst formation begins 5 days after conception; gastrulation begins 16 days after conception; neural tube formation begins in the third week after conception; and organogenesis begins in the fourth week. By the end of 12 weeks, immature versions of all the major organ systems have formed. To place some of these events in the context of what the mother experiences, organogenesis is occurring about the time the first menstrual period is missed, and formation of the immature organ systems is being completed at about the same 12-week point at which morning sickness generally begins.

The embryo's development over time can be contrasted with its *growth* over time. By the end of the third week—after blastocyst formation and gastrulation—a human embryo is perhaps 2 millimeters (or less than a tenth of an inch) in diameter. Skip to the twelfth week after conception, when major portions of organogenesis are concluding, and the fetus may still be only about 13 centimeters, or a little over 5 inches in length.

The message here is that development is concentrated early in pregnancy, while the simple growth of what *has* developed comes later. So pronounced are the changes during development that biologists use different terms for the growing organism over time. The original cell—the fertilized egg—is a **zygote**. In humans, about 30 hours after the zygote is formed, it undergoes its first cell division; at this point, the developing organism is referred to as an **embryo**, a name that will be applied to it through the eighth week of development. From the ninth week of development to birth, the organism is referred to as a **fetus**. The typical 9-month pregnancy is divided into periods of roughly three months each, called **trimesters**. The first trimester ends at about the time organogenesis does.

Early Development

You saw earlier that the zygote/embryo moves from ovary to uterus during the first four days after fertilization. It arrives in the uterus on day four as the mulberry-like *morula*, or tightly packed ball of cells noted in Chapter 27 (see Figure 27.1, page 622). Over the next two days the morula becomes the blastocyst, which implants itself in the endometrial lining. Actually, this "implantation" is more like an invasion. The blastocyst carves out a cavity for itself by releasing enzymes that effectively digest endometrial cells (**see Figure 28.9**). Implantation normally occurs in the dorsal wall of the uterus. Occasionally, however, a

blastocyst attaches itself inside the uterine tube, or sometimes to the **cervix**, which is the lower part of the uterus that opens onto the vagina. This is an **ectopic pregnancy**, which the embryo usually does not survive.

Tutorial 28.3.1
Implantation of Embryo in Uterus

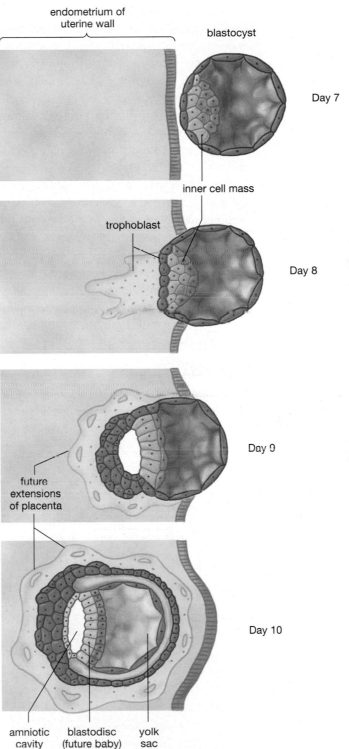

Figure 28.9
Implantation of the Embryo in the Uterus
Seven days after conception, the embryo, now in the blastocyst stage of development, arrives at the uterine wall. It is composed of both an inner cell mass, which develops into the baby, and a group of cells called a trophoblast, which carve out a cavity for the embryo in the uterine wall and then begin establishing physical links with the maternal tissue. The outgrowth of this linkage is the placenta.

Two Parts to the Blastocyst

Figure 28.9 shows you the arrangement of cells in the blastocyst. On one side of the blastocyst cavity is a group of cells called the **inner cell mass**; these are the cells that develop into the baby. The periphery of the blastocyst, however, is formed by cells called the **trophoblast**. These are the cells that etch their way into the uterine wall; with implantation, they begin extending farther into the endometrium, establishing indirect links with the maternal blood supply. In time, these cells grow into the fetal portion of a well-known structure in pregnancy, the **placenta**. This structure is a complex network of maternal and embryonic blood vessels and membranes. It allows nutrients and oxygen to diffuse *to* the embryo (from the mother), while allowing carbon dioxide and waste to diffuse *from* the embryo (to the maternal blood system). See **Figure 28.10** for a look at how the placenta functions within a uterine environment as it exists at about two months. The tissue that links the fetus with the placenta is the **umbilical cord**, which houses fetal blood vessels that move blood between the fetus and the placenta.

The Origins of Twins

Having learned a little about fetal development, you're now in a position to understand something about the formation of twins. Identical twins can develop at any time in the very early stages in pregnancy—from the time the zygote undergoes its first division up until about the 15th day of pregnancy, after the blastocyst has implanted in the uterine wall. In all cases, identical twins result from two *parts* of the embryo developing into separate human beings. This may be a matter of the two cells produced by the zygote's first division each developing as separate embryos; or it may entail the inner cell mass of the blastocyst separating into *two* masses, with each then giving rise to a different person. In contrast, **fraternal twins** (or triplets, and so on) are twins who are produced first through multiple ovulations in the mother, followed by multiple fertilizations from separate sperm of the father, and then multiple implantations of the resulting embryos in the uterus. From this, you can see the difference between fraternal and identical twins: Fraternal twins are like any other pair of siblings, except that they happen to develop at the *same time*. Meanwhile, **identical twins** are twins who develop from a single zygote; strictly speaking, they are a single organism at one point in their development and as such have exactly the same genetic makeup.

Development through the Trimesters

If you look at the series of pictures and diagrams that make up **Figure 28.11**, you can see how an embryo and then fetus develops at selected points in the 38 weeks of pregnancy. Here's an overview of what happens through time.

Figure 28.10
Nutrient and Gas Exchange in the Placenta

a Oxygen- and nutrient-rich blood comes from the mother through the maternal arteries and moves, through smaller arterial vessels, into the maternal blood pools in the placenta. Oxygen-poor blood from the embryo, transported through the umbilical cord's fetal arteries, does not flow into the maternal blood pools, but instead stays within the fetal blood vessels, which project into the maternal blood pools. Oxygen and nutrients in the maternal blood pools move into the fetal blood vessels, however, through such processes as diffusion. The resulting oxygen- and nutrient-rich blood is carried back to the embryo through the umbilical cord's fetal vein. Meanwhile, carbon dioxide from the embryo moves the other way—into the maternal blood pools—to be carried away through maternal veins.

b Larger structure of the placenta and uterine environment.

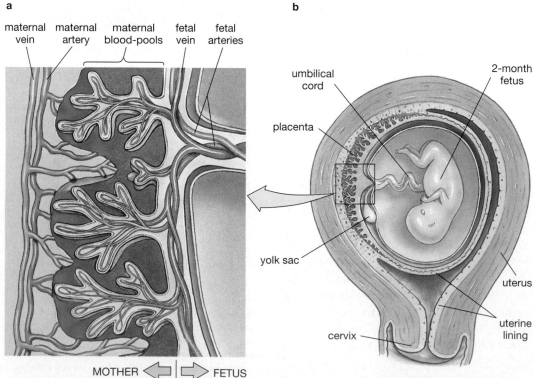

a

maternal vein · maternal artery · maternal blood-pools · fetal vein · fetal arteries

MOTHER ⟸ | ⟹ FETUS

b

umbilical cord
placenta
yolk sac
2-month fetus
uterus
uterine lining
cervix

a EMBRYO DEVELOPMENT:

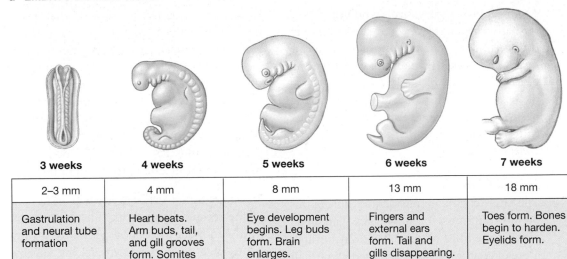

3 weeks	4 weeks	5 weeks	6 weeks	7 weeks	8 weeks
2–3 mm	4 mm	8 mm	13 mm	18 mm	30 mm
Gastrulation and neural tube formation	Heart beats. Arm buds, tail, and gill grooves form. Somites form.	Eye development begins. Leg buds form. Brain enlarges.	Fingers and external ears form. Tail and gills disappearing.	Toes form. Bones begin to harden. Eyelids form.	Arms bend at elbows. Genitals begin to develop.

b FETUS DEVELOPMENT:

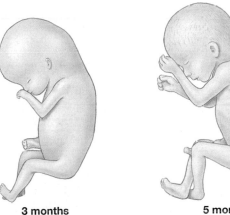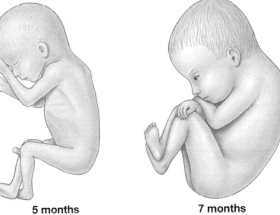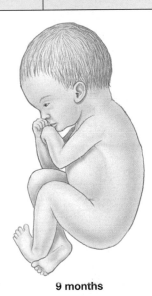

3 months	5 months	7 months	9 months
87 mm	190 mm	270 mm	350 mm
Well-defined neck appears. Genital formation complete. Sucking reflex appears.	All major organs have been formed. Head and body hair appear. Movements felt by mother.	Lungs and lung circulation develop. Eyelids open. Fat deposited under skin. May be viable if born.	Fetus usually viable if born. Body hair lost, head hair well developed. Most senses are well developed.

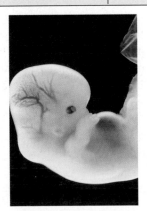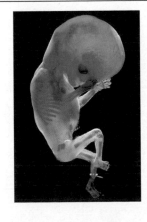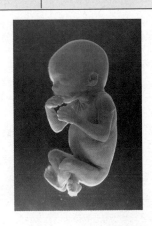

Figure 28.11
Development through the Trimesters
(a) Development of the human embryo (3 weeks to 8 weeks). **(b)** Development of the human fetus (3 months to 9 months). Photos of a human embryo at 6 weeks (left), 11 weeks (middle), and 5 months (right).

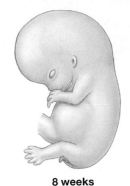

Tutorial 28.3.2
Human Fetus Development

First Trimester

As noted, blastocyst, gastrula, and neural tube formation all take place within the first month of conception; but much more happens during this period as well. By the 25th day, a primitive heart has formed in the human embryo and has begun pumping blood; the "somites" or repeating segments you read about in Chapter 27 have begun to take shape for muscular and skeletal formation.

Second Trimester

The fetus lengthens considerably during this period, but most of its weight gain takes place in the third trimester. Midway through a pregnancy, a fetus has achieved about half of its birth *length* (which is about 50 centimeters or 20 inches on average), but only about 15 percent of its birth *weight* (which averages 3.5 kilograms or 7.5 pounds).

The heartbeat of a fetus can be heard through a stethoscope during the second trimester, and the mother can feel the baby begin kicking by the end of the fourth month. Kicking is only one of several movements the fetus is capable of at this point; others include facial movements and the sucking reflex. By the end of the trimester, about half of all fetuses can be startled by loud sounds. A fine, white downy hair, called the *lanugo*, covers the body of the fetus by the end of the fifth month, probably as an aid to getting an oily, protective coating to adhere to the skin.

Third Trimester

Along with the terrific growth that takes place during the third trimester, fetuses are running through a series of activities that can be thought of as practice for life outside the uterus (or womb). The fetus is surrounded by a protective and nutritive **amniotic fluid**, which fills its lungs. It therefore doesn't breathe, but it periodically exercises its breathing muscles during this period by moving amniotic fluid in and out of them.

Premature Babies and Lung Function The end of the second trimester and the start of the third also mark the critical period for a baby being able to survive *outside* the uterus. Clinically, any baby born before 37 weeks of gestation is *premature*; but many babies born between 35 and 37 weeks after conception require little special care. The *more* premature a baby is, however, the greater its risk of having difficulties with physical and cognitive development, at least for the first few years of life. In very premature babies, survival itself is called into question (**see Figure 28.12**). In general, babies cannot survive without having spent at least 22 weeks inside the uterus.

A critical factor in whether a premature baby can live outside the uterus is lung function. Working against the survival of very early "preemies" is the fact that, when they exhale, the air sacs in their lungs tend to close (or "collapse") and then stick together. Why doesn't this happen with the lungs of a full-term 38-week fetus? Because, beginning in the 28th week after conception, fetuses begin synthesizing a lung *surfactant*—a soap-like fatty substance that acts as a lubricant to keep lung tissue from sticking together. One of the great (and strangely unheralded) medical advances of the 1980s was the development of a synthetic surfactant by John Clements and his colleagues at the University of California, San Francisco. Infant deaths due to so-called respiratory distress syndrome (RDS) were cut roughly in half by this advance, thus saving the lives of thousands of premature babies.

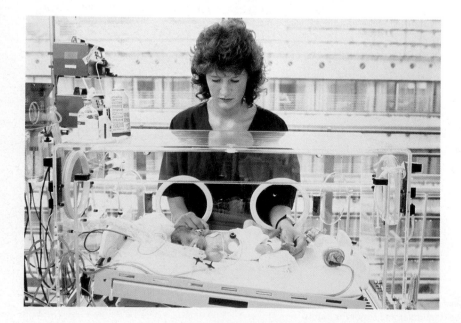

Figure 28.12
Tough Going Early on
A pediatric nurse attends to a premature baby in an incubator.

28.6 The Birth of the Baby

As the fetus grows, the uterus becomes an increasingly cramped place, and the movements the fetus undertakes become much more restricted. In normal circumstances, late in the pregnancy the fetus's head will lodge near the base of the mother's spine (**see Figure 28.13**). In this "upside-down" position, the fetus is ready to make its entrance into the world. What triggers this journey? The immediate cause is **labor**: the regular contractions of the uterine muscles that sweep over the fetus from its legs to its head. The pressure this generates opens or "dilates" the cervix (ultimately to approximately 10 centimeters or about 4 inches). This creates an opening large enough for the baby to pass through. These things must occur in stages, however. The first phase of the uterine contractions is given over to cervical dilation; then comes the expulsion of the baby and finally expulsion of the placenta (also called the *afterbirth*, for obvious reasons).

These events take place in response to a complex interplay of hormones. Interestingly, research on sheep and monkeys indicates that, in them at least, this hormonal cascade begins not with the mother but with the fetus. In sheep, it is fetal stress hormones (the "fight-or-flight" hormones) that trigger the release of the maternal hormones that in turn bring on labor.

On to Ecology

In this chapter and the four that preceded it, you followed the workings of the cells and tissues that allow animals to function in their marvelously diverse ways. But animals—and for that matter all the other varieties of living things—are not self-contained units. Rather, they are constantly *interacting* with each other. In addition, all the creatures of the living world are influenced by the physical forces that surround them—climate and soil characteristics and ocean temperature, to name a few. In the next two chapters, you'll be looking at life at the level of this integrated web. What you will be studying is ecology.

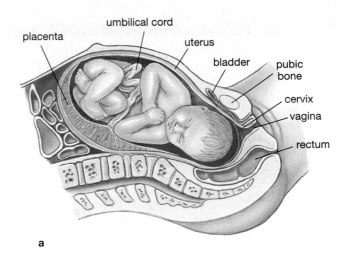
a

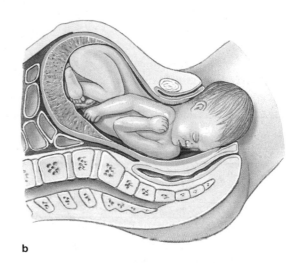
b

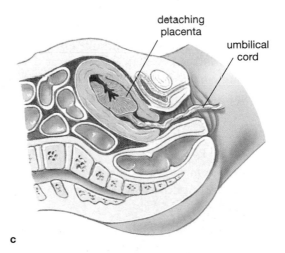
c

Figure 28.13
Birth of the Baby
With uterine contractions, the cervix dilates, allowing room for the baby to pass through the birth canal and to the outside world. The placenta and its associated fluids and tissues are expelled shortly afterward, as the "afterbirth."

Chapter Review ▰▰▰▰▰▰▰▰▰▰▰

Summary

28.1 Overview of Human Reproduction and Development

- Both the human male and female reproductive systems must produce their respective gametes (sperm and eggs) and then deliver them to a place where they can fuse, in the moment of fertilization.

- The precursors to eggs, called oocytes, exist in the ovaries, which lie just left and right of center in the female pelvic area. Oocytes mature within nurturing complexes of cells and fluids called follicles. On average, once every 28 days, one follicle-oocyte complex is brought to a state of maturity. The egg and some of its accessory cells then rupture from the follicle and ovary and begin a journey through the uterine tube, to the uterus. The release of the oocyte is called ovulation. Conception can occur if the oocyte encounters sperm on the way to the uterus.

- Sperm are produced in the male testes and then pass through two types of ducts, the vas deferens and the urethra, after which the sperm are ejaculated from the penis into the female vagina. Materials secreted by three separate sets of glands will join with the sperm before ejaculation. The resulting mixture of sperm and glandular materials is called semen. The first of the sperm to reach the oocyte moving through the uterine tube may fertilize it. Development of the fertilized egg, called a zygote, then ensues.

28.2 The Female Reproductive System

- A portion of the female reproductive system is built up and then dismantled in a reproductive cycle that takes approximately 28 days to complete. The result of the dismantling is menstruation, which is a release of blood and the specialized uterine tissue it infuses, the endometrium. The endometrium, which lines the uterus, is tissue that an embryo implants itself in when pregnancy occurs.

- The basic unit of oocyte development is the ovarian follicle, a complex that includes the oocyte that develops into an egg and some surrounding accessory cells.

- On average, a single follicle each month will develop into a tertiary follicle that ruptures, expelling the egg and its accessory cells from the ovary; this complex of cells then begins a journey through the uterine tube to the uterus. The ruptured tertiary follicle develops into the corpus luteum, which secretes hormones that prepare the reproductive tract for pregnancy.
  TUTORIAL 28.1.1: Steps in Ovulation and Pregnancy

- All the follicles a woman will ever possess are produced in her prior to birth. This number steadily decreases throughout her lifetime, primarily through the process of follicle degeneration called atresia. The scarcity of follicles a woman has in middle age is the primary factor that brings about menopause, the cessation of the monthly ovarian cycle.
  TUTORIAL 28.1.3: The Female Reproductive Cycle

28.3 The Male Reproductive System

- Sperm begin development in the testes and continue development and are stored in the adjacent epididymis. Sperm move from the epididymis through the vas deferens and then the urethra in ejaculation. The accessory glands that contribute material to the sperm before ejaculation are the two seminal vesicles, the prostate gland, and the two bulbourethral glands.
  TUTORIAL 28.2.3: Sperm Development and Fertilization

- Within the testes, sperm development takes place in the seminiferous tubules. Development begins at the periphery of a given tubule, with sperm becoming more developed toward the interior of the tubule. Eventually, developing sperm are transported from the interior of a tubule to the epididymis for further development and storage.

- Sperm development in the testes begins with spermatogonia, which develop into spermatocytes, which give rise to spermatids and then to immature sperm. Spermatogonia give rise not only to spermatocytes, but to more spermatogonia. This self-generation of sperm precursors is the reason men keep producing sperm throughout their lifetimes. About 250 million sperm are produced each day—about the number of sperm ejaculated with each orgasm.

- The materials secreted by the seminal vesicles, the prostate gland, and the bulbourethral glands make up 95 percent of semen. These materials provide nutrients, pH neutralizers, and lubricating fluids for the sperm.

28.4 The Union of Sperm and Egg

- Sperm release enzymes externally that break down layers of accessory cells that surround the oocyte. To release these enzymes, the sperm must be capacitated, meaning they need to have a protective coating that surrounds their "head" stripped away. This happens through action of substances secreted by cells surrounding the oocyte.

- Extensions of the sperm cells bind with receptors on the surface of an oocyte, after which the nuclei of the two cells will fuse. This is the moment of conception. Fusion brings about an almost instantaneous change in the electrical potential of the egg membrane, which effectively shuts out other sperm. This process, called the fast block to polyspermy, is a defense against multiple sperm entering the egg.
  TUTORIAL 28.2.3: Sperm Development and Fertilization

28.5 Human Development Prior to Birth

- Human development proceeds through the stages of animal development noted in Chapter 27: formation of the blastocyst, gastrulation, formation of the neural tube, and organogenesis or the formation of organs. Many of the most important stages in development are completed by the end of 12 weeks of pregnancy. The developing organism is termed a zygote at conception, an embryo from the zygote's first division through the eighth week of pregnancy, and a fetus from the ninth week of pregnancy to the birth of the baby.

- The blastocyst implants itself in the endometrial lining of the uterus during the first week of pregnancy.
  TUTORIAL 28.3.1: Implantation of Embryo in Uterus

- The blastocyst initially consists of an inner cell mass, which develops into the baby, and a group of cells called a trophoblast. The trophoblast is active in uterine implantation and the production of cells that extend into the endometrium, establishing links with the maternal blood supply. These trophoblast cells grow into the fetal portion of the placenta, a complex network of maternal and embryonic blood vessels and membranes. The placenta allows for the movement of nutrients, wastes, and gases between the developing embryo and the mother.
  TUTORIAL 28.3.2: Human Fetus Development

28.6 The Birth of the Baby

- The immediate cause of birth is labor, meaning the regular contractions of uterine muscles that sweep over the fetus from its legs to its head.

- The pressure that results from these contractions opens or dilates the cervix, thus creating an opening large enough for the baby to fit through.

Key Terms

acrosome 646	gonadatropin-releasing hormone 640
amniotic fluid 652	identical twin 650
bulbourethral gland 644	inner cell mass 650
cervix 649	labor 653
corpus luteum 640	luteinizing hormone (LH) 642
ectopic pregnancy 649	menopause 643
embryo 649	menstruation 638
endometrium 638	oocyte 639
epididymis 644	osteoporosis 643
estrogen 642	ovarian follicle 639
Fallopian tube 636	ovulation 636
fetus 649	placenta 650
follicle 636	prostate gland 646
follicle-stimulating hormone (FSH) 642	semen 636
fraternal twin 650	seminal vesicle 644
gamete 635	seminiferous tubule 644

spermatocyte 644	umbilical cord 650
spermatogonia 644	urethra 644
testis 644	uterine tube 636
trimester 649	vas deferens 636
trophoblast 650	zygote 649

Understanding the Basics

Multiple-Choice Questions

1. The precursors of the eggs produced in the ovary are referred to as
   a. gametes
   b. sperm
   c. oocytes
   d. small eggs
   e. oviducts

2. The structure that transports sperm up and over the urinary bladder to be delivered during sexual excitation is the
   a. epididymis
   b. oviduct
   c. penis
   d. testes
   e. vas deferens

3. The tertiary follicle ruptures midway through the menstrual cycle, expelling
   a. one oocyte along with some accessory cells
   b. two oocytes along with some accessory cells
   c. one mature egg, ready to be fertilized
   d. one mature egg with some accessory cells
   e. the fourth follicle

4. Once the oocyte and its accessory cells have been expelled from the ovary, the fingerlike fimbriae act to
   a. move the incoming sperm toward the oocyte
   b. protect the oocyte from premature fertilization
   c. move the oocyte into the uterine tube
   d. force the oocyte into the uterus
   e. implant the newly fertilized egg in the uterus

5. The natural degeneration of follicles during a woman's lifetime is referred to as
   a. menopause
   b. ovulation
   c. puberty
   d. atresia
   e. apoptosis

6. Sperm development begins in the
   a. bone marrow
   b. epididymis
   c. seminiferous tubules
   d. vas deferens
   e. seminal fluid

7. These sperm factories give rise not only to sperm but also to more sperm factories, called
   a. spermatocytes
   b. spermatogonia
   c. oocytes
   d. vas deferens
   e. prostate glands

8. The portion of the sperm that contains the enzymes used to degrade materials found on the outside of the egg is the
   a. acrosome
   b. nucleus
   c. tail
   d. mitochondria
   e. chromosomes

9. The portion of the blastocyst called the trophoblast eventually forms the
   a. head of the baby
   b. legs of the baby
   c. entire baby
   d. nervous system of the baby
   e. placenta

10. Birth of identical twins results from unusual events during
   a. egg production
   b. sperm production
   c. early cell divisions after fertilization
   d. first trimester
   e. second trimester

Brief Review

1. Describe the hormonal changes that regulate the female reproductive cycle.

2. Discuss how the testes act to protect the developing sperm.

3. Menstruation represents a costly biological process. Why do scientists believe it occurs so frequently?

4. Compare and contrast the tertiary follicle and the corpus luteum.

5. What is the fast block to polyspermy? Why is it necessary?

6. Outline the events in human development that occur prior to implantation.

7. Define the term *premature* with respect to human development. What complications may arise during a premature birth?

8. How do birth control pills work?

Applying Your Knowledge

1. When a couple is having trouble conceiving a child, the first thing fertility specialists generally will look at is the fertility of the prospective father. What do you think a specialist would be looking for?

2. The fast block to polyspermy prevents more than one sperm from fertilizing the egg by quickly changing the ionic environment of the egg after the first sperm has successfully fertilized the egg. There is also a slow block to polyspermy. This block involves the production of a hard outer coating around the newly formed zygote. Why is this necessary?

3. Birth control pills manipulate hormone levels in the female to prevent ovulation. Could alterations in hormone levels be used to manipulate sperm production in males? Why? Why not?

4. Compared to molecules such as O_2 and CO_2, viruses are rather large and more complex. In order for a virus to enter a cell or cross a blood vessel, it must be transported across. This transportation process is specific to certain cells. The virus that causes AIDs, HIV, can be transported only into a particular type of white blood cell called the helper T cell. Based on this information, do you think that an HIV-positive mother will pass this virus to her child through the placenta?

MediaLab

Are Test Tube Babies the Solution? Understanding Reproductive Problems

Where do babies come from? This question is usually followed by parents squirming through the old-standby talk about birds and bees. Imagine, though, the difficulty some modern parents will have in describing their use of assisted reproductive technology, mechanical methods that remove the sex from sexual reproduction—like *in vitro* fertilization, egg and sperm donors, even sperm injection. These techniques seem to be the innocent product of a late twentieth-century desire to use scientific advances to make childbirth possible for couples who could never conceive naturally. But is there a more sinister reason for the increase in reproductive technology? Are we becoming less fertile as a species? To learn more, the *CD-ROM Tutorial* will help you review normal female and male reproduction, so that in the following *Web Investigation* and *Communicate Your Results* sections, you can evaluate how these processes might be sensitive to our modern lifestyle.

This *MediaLab* can be found in Chapter 28 on your CD-ROM (Tutorial 28.4.4) and Companion Website (http://www.prenhall.com/krogh3).

CD-ROM TUTORIAL

Human sexual reproduction involves multiple, highly coordinated steps. In addition to the events of meiosis to create haploid gametes, fertilization and embryonic development occur internally. Humans thus need not only accessory structures to direct sperm and egg fusion but also structures to allow the embryo to access oxygen and food and get rid of wastes. This *CD-ROM Tutorial* leads you through the reproductive systems of males and females, and examines the early internal development of a human embryo.

Activity

1. First, you will review the female menstrual cycle: the coordinated release of an oocyte and preparation for the female body to receive and nourish a developing embryo.

2. Next, you will learn about the formation and maturation of sperm, and the accessory glands needed to activate sperm for their journey to fertilize the egg.

3. Finally, you will learn about the voyage of the fertilized egg into the uterus and the creation of the membranes used to access nutrients from the mother: the placenta, amnion, and umbilical cord.

Failure in any one of these steps can result in an inability to conceive or to carry the fetus to term; a problem for increasing numbers of couples dealing with infertility. Could factors from our environment be responsible for this problem? The following *Web Investigation* and *Communicate Your Results* sections are meant to help you evaluate that question.

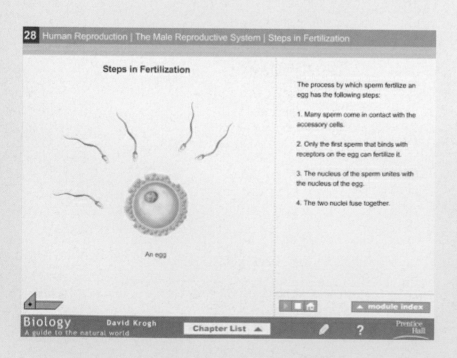

28 Human Reproduction | The Male Reproductive System | Steps in Fertilization

Steps in Fertilization

The process by which sperm fertilize an egg has the following steps:

1. Many sperm come in contact with the accessory cells.

2. Only the first sperm that binds with receptors on the egg can fertilize it.

3. The nucleus of the sperm unites with the nucleus of the egg.

4. The two nuclei fuse together.

An egg

Biology
David Krogh
A guide to the natural world
Chapter List ▲
Prentice Hall

WEB INVESTIGATION

Investigation 1

Estimated time for completion = 5 minutes

The environmental movement in the past 30 years has awakened public concern over the effects of pollution on animal health and reproduction. Now, many are questioning whether the same human-made chemicals—including pesticides, plastics, and detergents—that interfere with normal hormone function in laboratory animals are causing a decline in human sperm counts. Select the Keywords **SPERM CONTROVERSY 1** or **2** on your CD or Website for this *MediaLab* to read two contradictory articles on declining sperm counts.

Investigation 2

Estimated time for completion = 5 minutes

What have fertility specialists concluded about the major causes of infertility in most couples? Select the Keyword **INFERTILITY** on your CD or Website for this *MediaLab* to view a chart of the major causes of infertility. Then select the Keyword **ART** to view a government site posting background information on couples seeking help from Assisted Reproductive Technology (ART).

Investigation 3

Estimated time for completion = 10 minutes

Many pregnant women fear birth defects caused by environmental substances (teratogens that interfere with normal development). Although only about 5 percent of birth defects are attributed to exposure to environmental substances, these environmental agents should be clearly recognized and avoided. However, information on the identity of teratogens is often inaccurate, alarming, and not supported by scientific studies. Select the Keyword **PREGNANCY** on your CD or Website for this *MediaLab,* and list the substances that are suggested to be teratogens and are thus to be avoided during pregnancy. Then, select the Keyword **TERATOGENS** to view a list of known teratogens.

COMMUNICATE YOUR RESULTS

Now that you have been introduced to several sources of suspicion that the environment is adversely affecting human reproduction, use the following questions to evaluate the validity of these claims.

Exercise 1

Estimated time for completion = 15 minutes

Although members of the scientific community disagree over whether sperm counts are declining worldwide, they generally agree about an increase in testicular cancer and congenital birth defects in male reproductive systems. In this exercise, form a group and propose a hypothesis to explain these observations. Your group should search the Internet to find supporting studies, and use the Message Board at the Website for your textbook to post the Internet addresses.

Exercise 2

Estimated time for completion = 15 minutes

Test-tube babies, sperm injections, frozen eggs and sperm are all commonly covered in news stories; but we often hear more about technology than about patients. Using information from the websites in *Web Investigation 2,* write a 250-word portrait of the average woman (or man) who is using ART. Include answers to some of these questions: What is the most common procedure? How many ART patients are not pregnant after one try or cycle? How many couples experience multiple births after one cycle? What is the average age of females receiving this treatment?

Exercise 3

Estimated time for completion = 5 minutes

Hopefully it can be said that for every inaccurate Internet site there is a well-informed, scientifically credible site. Compare the list of chemicals to avoid during pregnancy that you made for *Web Investigation 1* with the information from the site listing human teratogens. In 50 words, express your conclusions about the first article.

29

An Interactive Living World
Populations and Communities in Ecology

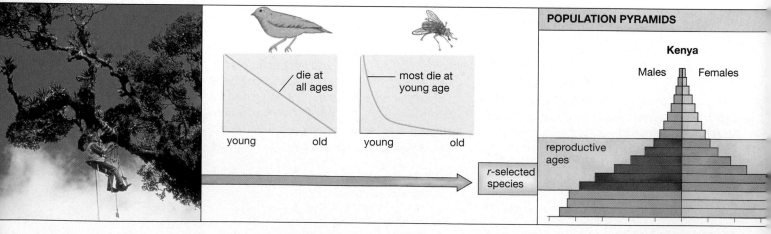

Ecologist on the job.
(Section 29.1, page 662)

Why do most flies die young?
(Section 29.3, page 670)

A population pyramid.
(Section 29.4, page 671)

Ecology is the study of how living things interact—with each other and with their physical environment. One way of studying these interactions is to focus on populations, meaning all the members of a single species living in one area. Another way of studying these interactions is to look at communities, meaning all the populations of all the species living in one area.

For those who already have been through some earlier units of this book, a word that is a favorite in the humanities by now may have come to mind in connection with biology. It is *microcosm*, which comes from the Greek *mikros* (tiny) and *kosmos* (universe): a tiny universe, which is also to say a world within a world.

What are biology's microcosms? Well, for one, in all living things, a *system* exists whereby DNA contains information that is then used for the production of proteins. Once put together, these proteins may make up some structure (your hair, for example) or get busy on some specific task (such as breaking down sugar). Critically, one of these tasks is turning on or off the very DNA that called the proteins into being in the first place. Thus it is a *self-regulating* system; it exists within a whole living being, but it has some of the characteristics of a tiny world of its own—a microcosm.

Now take a step up in scale and consider the parts of a human body. Drink some lemonade and your body will start producing insulin, which will help store away the blood sugar that is now coursing through your veins. *Stop* eating for awhile and your body will produce a hormone called glucagon that helps bring these stored food supplies *out* of reserve. This self-regulating system is thus a kind of world of its own—one that works in concert with, and yet that is in some ways separate from, the circulatory system, the nervous system, and on and on.

By moving up another level in scale, you arrive at a whole organism—a person in the example just given, but it could be a bat or a tree. You might be tempted to think that this is the end of the line for microcosms; that here you have reached the *macrocosm*, containing all microcosms. Think again, however, because the life of, say, a tree is profoundly influenced by *other* trees around it as well as by other kinds of organisms (insects, fungi). Under this view, a living thing is a kind of microcosm that exists within the larger world of living *things*.

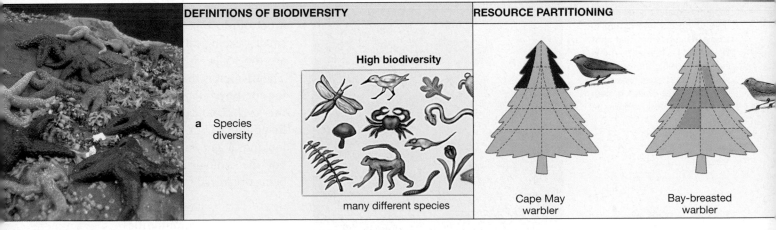

| DEFINITIONS OF BIODIVERSITY | RESOURCE PARTITIONING |

High biodiversity

a Species diversity

many different species

Cape May warbler

Bay-breasted warbler

What is a keystone species? (Section 29.5, page 674)

Biodiversity. (Section 29.5, page 675)

That's my branch. (Section 29.6, page 677)

Even after taking this larger world into account, there is still one more jump to make in scale. The whole of the living world exists within the context of the *nonliving* world and universe: the Earth's climate and gases, its sea currents and volcanoes, as well as the energy from the Sun. Life doesn't exist apart from these things. Instead, it is very much conditioned by them—though as you'll see, some of these things are also conditioned by life.

29.1 The Study of Ecology

In your walk through biology so far, you have made it through the molecular world (DNA, for example) and through the world of organ systems (circulation, for example) and to some extent through whole organisms. Now it is time to take the last steps up; in this chapter and the one that follows, you will consider biology's largest realm of all, **ecology**, defined as the study of the interactions that living things have with each other and with their environment. The final chapter in the book, on animal behavior, is also placed within the ecology unit because it too is concerned with interactions in the living world.

Ecology Is Not Environmentalism

In common speech the word *ecology* has become synonymous with the conservation of natural resources or with the "environmental movement" in all its political dimensions. Under this view, you might assume that the job of an ecologist is to help preserve the natural world. In fact, however, the job of an ecologist is to *describe* the natural world in its largest scale; to say what is, rather than what should be. Are ecologists environmentalists? Most are, but this shouldn't be surprising. We would no more expect ecologists to be indifferent to the environment than we would expect art historians to be indifferent to paintings. Moreover, there is a branch of biology, conservation biology, that is directly concerned with preserving natural resources. But the business of ecology per se is to tell us about the large-scale interactions that go on in the natural world. This work has an important use in that it provides the *information base* that society can use to make decisions about the environment. Does species diversity make for healthier natural communities? How many gray whales are there in the world? How important are prairie dogs to the prairie? To answer these questions, ecologists end up working in some places you might expect—in tents near the Arctic Circle or at treetop level in tropical rain forests (**Figure 29.1**). They also work in some locations you might not expect, however, such as in offices, hunched over computers, seeing how mathematical models of the natural world work out.

Path of Study

In this chapter and the next one, you'll follow a bottom-to-top approach by moving from smaller ecological units to larger ones. What are the scales of life that concern ecology? The smallest is that of the physical functioning, or **physiology**, of given organisms. At this level, ecologists are examining individual living things, but at all other scales they are looking at *groups* of organisms or species. For the ecologist, life is conceptualized as being organized into:

> Populations
> Communities
> Ecosystems
> The Biosphere

From Individuals to a Population

The smallest level of group organization is a **population**, defined in Chapter 17 as all the

Figure 29.1
An Ecologist on the Job
Ecologists spend a good deal of time going out into nature to study the conditions that exist in various ecosystems—however inaccessible those ecosystems may be. Here a researcher is studying tree branch growth in the rain-forest canopy in Costa Rica.

members of a single species that live together in a specified geographic region. The North American bullfrog (*Rana catesbeiana*) is a species that can be found from southern Canada to central Florida, but all the *R. catesbeiana* in a given pond constitute a single *population* of this species. Of course it's possible to define our population to be all the bullfrogs in *two* ponds, or in two states, or two countries—whatever geographic region is most useful for the question an ecologist is asking.

From Populations to a Community

Going up the scale, if you take the populations of *all* species living in a single region, you have a **community**. Often the term community is used more restrictively, to mean populations in a given area that potentially interact with one another.

From a Community to an Ecosystem

If you add to the community all the *non*living elements that interact with it—rainfall, chemical nutrients, soil—you have an ecosystem. More formally, an **ecosystem** is a community of organisms and the physical environment with which they interact. Ecosystems can be of various sizes; you'll be looking at a range that goes from fairly small (a single field, for example) up through the enormous ecosystems called *biomes* that may take up half a continent.

From Ecosystems to the Biosphere

The largest scale of life is the **biosphere**, which can be thought of as the interactive collection of all the Earth's ecosystems. Given what you've reviewed about ecosystems, this means all life on Earth and all the nonliving elements that interact with life. Sometimes, however, the term biosphere is used purely in a territorial sense—to mean that portion of the Earth that supports life. If you look at **Figure 29.2**, you can see a graphic representation of ecology's scales of life. We'll now begin to look at ecology through these levels of biological organization, starting small, with the population.

29.2 Populations: Size and Dynamics

So what is it ecologists want to know about a population? Well, they need to know how to count it, how and why it is distributed

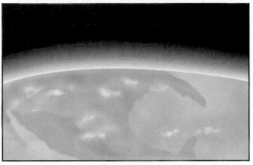
Biosphere

Ecosystem

Community

Population

Organism

Figure 29.2
The Scales of Life That Concern Ecology
In this chapter, you will consider how organisms of one species are associated in populations, and how populations of different species are associated in communities. In Chapter 30, you will look at ecosystems, which include not only living community members but the non-living factors that interact with them (such as the rainfall in the ecosystem panel above). Chapter 30 also reviews the large-scale ecosystems called biomes and the biosphere, which is the interactive collection of all the Earth's ecosystems.

over its geographical area, and how and why its size *changes* over time—what its population dynamics are, to use the term employed by ecologists.

Estimating the Size of a Population

The reason ecologists want to count the members of a given population is straightforward: Without such a count, there is no way to answer a question such as how much territory a group of cheetahs must have in order to flourish, or how fast a population of finches recovers from a drought. (How would you know about the latter unless you could compare the population after the drought with the population before it?)

With large, immobile species such as trees, taking a census can sometimes be easy; just mark off an area and count. Things become more difficult when the area under consideration is so large that not all species in it can be counted, but rather must be estimated based on a population counted in a smaller representative area. Estimating becomes more difficult yet when the individuals are numerous and mobile, as with birds. Ecologists employ various means to estimate such populations; they count animal droppings within a defined area or survey bird populations as they migrate, for example.

Growth and Decline of Populations over Time

How is it that populations *change* size? As you begin to think about this question, it's worth going over a more general concept, which is the way a population of *anything* might increase in number. If you look at, say, the number of cars coming off the end of a production line, there might be 1,000 on Monday and then 1,000 on Tuesday, and so forth. The important thing to note is that the number of cars produced on Tuesday is not related to the number produced on Monday. The increase in the number of cars is thus an **arithmetical increase**: Over each interval of time (a day in this case), an unvarying number of new units (cars) is added to the population.

Now contrast this with what happens to living things. In most cases, each new unit (each living thing) is capable of playing a part in giving rise to *more units*, which certainly is not the case with cars. Population increases for living things are thus *proportional to* the number of organisms that already exist. Thus the increase in a population of organisms comes about through a different sort of increase, an **exponential increase**, which occurs when, over an interval of time, the number of new units added to a population is proportional to the number of units that exist. **Table 29.1** gives an example of the difference between the two kinds of growth over a period of weeks, using cars for arithmetic growth and the tiny water flea *Daphnia* for exponential growth. Let's start out at the end of day 1 with the 1,000 cars produced that day and with 1,000 water fleas existing in an optimal laboratory environment.

As you can see, *Daphnia* is a relatively slow starter, but its population quickly overwhelms that of the car population—not surprising, because the *Daphnia* population doubles every 3 days. If you look at **Figure 29.3**, you can see what these two kinds of growth look like when plotted on a graph.

Population Growth in the Real World

As noted, the *Daphnia* were growing in an "optimal" laboratory environment—plenty of food, habitat kept clean, no predators. Thanks to human intervention, this was a kind of paradise for water fleas, in other words. But could any population ever grow like this in the real world? In all cases the answer is "not forever," and in most cases the answer is "not for long," but some population growths can be dramatic while they last.

Table 29.1
Arithmetic Growth versus Exponential Growth

	Day 1	Day 4	Day 7	Day 10	Day 13	Day 16	Day 19	Day 22	Day 25
Arithmetic Growth: Cars	1,000	4,000	7,000	10,000	13,000	16,000	19,000	22,000	25,000
Exponential Growth: *Daphnia*	1,000	2,000	4,000	8,000	16,000	32,000	64,000	128,000	256,000

In 1859, Thomas Austin released 24 wild European rabbits (*Oryctolagus cuniculus*) onto his estate in southern Australia, near Melbourne, in order to provide more game for sport hunting. His was not the first attempt to introduce rabbits to Australia, but it was by far the most successful—or perhaps the most fateful. This alien species spread north and west in Australia like wildfire; within 16 years it had expanded its range through the entire latitude of the continent, a distance of almost 1,800 kilometers or 1,100 miles. And this invasion was not a matter of a rabbit here and a rabbit there; the European rabbits became a scourge that only recently has shown signs of being brought under control, through use of a virus (**see Figure 29.4**). Here is Eric Rolls' description in *They All Ran Wild* of the rabbit problem east of Peterborough in South Australia in 1887:

. . . for over a hundred miles most trees and plants had been killed. Outalpa head-station was a thousand square miles of rabbits. . . . Rabbits crawled like possums in branches several feet off the ground. They had burrowed under the beautiful granite homestead; they had taken possession of the cellar; they were in the turkeys' coop; they ran in and out of the men's huts apparently without fear; they took no notice of the dogs nor the dogs of them.

There are lots of instances of such seemingly uncontrolled growth. Charles Darwin devoted some thought to the issue as a theoretical matter. He wrote that "There is no exception to the rule that every organic being naturally increases at so high a rate, that, if not destroyed, the Earth would soon be covered

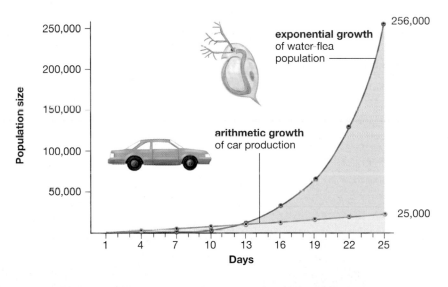

Tutorial 29.1.3
Exponential Growth

Figure 29.3
Arithmetic and Exponential Growth
When the same number of objects is produced in a given interval of time—in this case cars from a factory each day—arithmetical growth is at work. By contrast, the populations of most organisms, such as the water flea *Daphnia*, can exhibit exponential growth, at least for a time, meaning the population grows in proportion to its own size.

Figure 29.4
Rabbit Control in Australia
Rabbits run free on one side of a "dingo fence" in Australia. Note the contrast between the barren land where the rabbits feed and the grassland, to the left, that the rabbits have been kept out of.

by the progeny of a single pair." To put a little finer point on this, a single pair of carrion flies (which are a little larger than a housefly) could in theory give rise to enough flies in a single year to cover the state of New Jersey to a depth of 1 meter (3 feet).

The Shape of Growth Curves

If you look at **Figure 29.5a**, you can see what growth of this type, **exponential growth**, looks like when plotted on a graph. This is called the **J-shaped growth form** because it can be so extreme that when plotted it looks like the letter J. It may look familiar, because it is the growth form you just saw with *Daphnia*. As with *Daphnia*, there is relatively slow growth at first, then faster growth, then faster yet. Growth occurs, in other words, in which the *rate* of increase keeps accelerating. In the real world, the curve eventually drops off, often sharply, for reasons you'll soon go over.

Now look at **Figure 29.5b** to see a second model of growth for natural populations, the **S-shaped growth form**, which tracks **logistic growth**. This *starts* like exponential growth, but eventually the rate of growth slows and

finally ceases altogether, stabilizing at a certain level, denoted here as *K* (which you'll learn the significance of shortly).

What has intervened to account for the difference between the dizzying increase of the J-shaped form and the moderate increase of the S-shaped form? All the forces of the environment that act to limit population growth, which in ecology are known as **environmental resistance**. Organisms will run out of food or have their sunlight blocked; greater numbers of predators will discover the population; the wastes produced by the organisms will begin to be toxic to the population.

There are populations whose dynamics over time look something like either the J- or S-shaped curve, but in the real world, population size is likely to vary in more complex patterns. In **Figure 29.5c**, you can see one example of such a pattern.

Calculating Exponential Growth in a Population

You've seen that ecologists have ways to count natural populations; but is it possible for them to *predict* the size of populations, given certain pieces of starting information? The answer is yes, as you'll now see. Let's look first at a simplified means of calculating exponential growth, meaning the J-shaped curve. For now, assume the population is isolated—that is, that no individuals are moving in or out of it.

What first needs to be done is to pick a time period to evaluate population change. This period may be 40 minutes (for bacterial populations) or 50 years (for elephants), but some unit of time needs to be chosen for the analysis. The next step is to learn the difference between the birth rate and the death rate in the population over the time period you've picked. This, of course, raises the question: What's a birth rate? It is the number of individuals born in a given time period, expressed as a proportion of the whole population. Thus, if a population of 1,000 had 100 births in a year, its birth rate would be 0.10 or 10 percent per year.

Figuring the death rate is similar; if a population of 1,000 had 80 deaths per year, its death rate would be 0.08 or 8 percent per year.

The *difference* between these birth and death rates, of course, is 2 percent. This number is the population's growth rate in an optimal setting, denoted as *r*—standing for

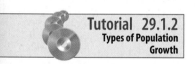

Tutorial 29.1.2
Types of Population Growth

Figure 29.5
Three Models of Growth for Natural Populations

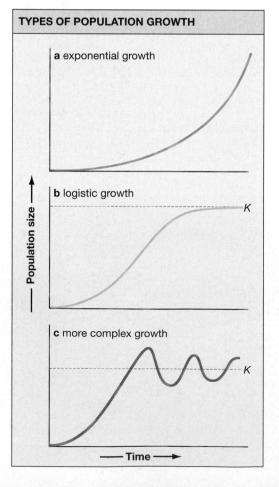

TYPES OF POPULATION GROWTH

a exponential growth

b logistic growth
K

c more complex growth
K

Population size →

← Time →

the *intrinsic rate of increase*—and it is very important. With it in hand, you can predict what the population's size will be in the future. Growing at 2 percent per year, the population of 1,000 will increase to 1,020 in the first year, and to about 1,480 in 20 years.

Growth rate (r) is critical because a small change in it can mean a big change in population over time. Take a population of another species, once again with 1,000 individuals in total, but this time with an r of 6 percent rather than 2 percent. In the first year this population would grow by 60 individuals, giving it 1,060 members as opposed to the first population's 1,020. Go 20 years out, however, and this population has about 3,200 individuals in it—it's now more than twice the size of the first population even though both were the same at the start. Meanwhile, if a population's r is less than zero—that is, if the number of deaths exceeds the number of births—the population is shrinking. If r is zero on the nose, meaning that births exactly match deaths in a given period, the population is at **zero population growth** (see Figure 29.6).

As noted, the r figure goes by another name, which is a population's **intrinsic rate of increase**: The rate at which a population would grow if there were no external limits on its growth. You could think of it as a population's potential for growth. Importantly, populations of each species have characteristic rates of increase. It won't surprise you to learn that this rate is higher for *Daphnia* populations than it is for whale populations; you can see some examples of an important factor in rate of increase, the time it takes to produce a new generation, in **Figure 29.7**.

Logistical Growth of Populations: Reality Makes an Appearance

You've now observed exponential or J-shaped growth, but you've seen that such growth never occurs for long in nature. Environmental resistance begins to assert itself. In some instances there will be a flattening out of the population increase, and in time perhaps a complex pattern with the curve moving above and below the line denoted earlier as K. But what is K? What is this point around which the population hovers? It is the maximum population density of a given species that a defined geographical area can sustain over time. The term for this measure is an area's **carrying capacity**.

Tutorial 29.1.4
Intrinsic Rate of Increase in a Population

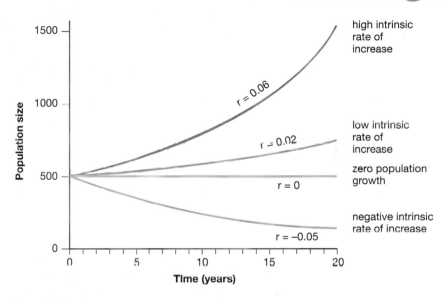

Figure 29.6
A Population's Potential for Growth: The Intrinsic Rate of Increase (r)
If the birth rate exceeds the death rate, then the population size will increase over time ($r = 0.02$, $r = 0.06$ in the example). If the birth rate equals the death rate, then the population size will stay constant ($r = 0$). Finally, if the birth rate is lower than the death rate, the population size will decrease over time ($r = 0.05$). Note what dramatic effects a small change in r can have over time, in this case on a population whose starting size is 500 individuals.

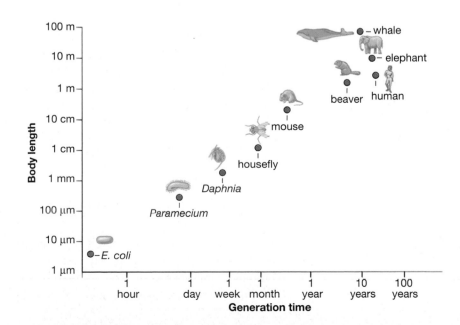

Figure 29.7
How Long between Generations?
An important factor in the population dynamics of any species is its "generation time," meaning the time that elapses between the birth of one generation and the birth of the next. Shown on the graph are the minimum generation times for a few selected species, along with a measurement of the length of each organism at the time it reproduces. The larger a species is, the longer its generation time tends to be. (Redrawn from J. T. Bonner, *Size and Cycle: An Essay on the Structure of Biology* [Princeton University Press, 1965].)

A Real-Life Example of Carrying Capacity

There is an island in the San Francisco Bay called Angel Island, just a short boat ride from the Golden Gate side of the Bay, that had a small population of deer introduced to it by humans early in the twentieth century. The problem was that the island lacked any natural *predators* for the deer, with the result that the deer population proliferated way beyond what the island's vegetation could support. Had this taken place in a purely natural environment, the deer population would have shrunk in accordance with the amount of available vegetation. But there were human hikers on Angel Island and, seeing emaciated deer, they did what you might expect: They fed the deer with what they brought along in their backpacks. Surviving through this artificial means, however, *more* deer came to live on the island and stripped it of its vegetation, thus reducing its carrying capacity.

The main lesson here is that there are limits to how many members of a given species an expanse of territory can support. Note also, however, that carrying capacity is not fixed; in the case of Angel Island, it was lowered as the deer, artificially fed by human beings, stripped the island of vegetation. In other instances carrying capacity might be raised for given periods (perhaps through abundant rainfall). The environments that many species live in tend to be stable over time, however, meaning that their carrying capacities will tend to be stable as well.

The *K* line can generally be established for a given population by estimating its numbers over an extended period of time. For many species, their population may exceed or "overshoot" *K* at times, but this excess will be offset by periods during which the population is under *K*. Once *K* is known, it is valuable in predicting the change in a population's size, because it often serves as a *factor* that can be used in mathematical equations to calculate the limits to population growth. The higher a population is relative to *K*, the more severely its growth will be limited.

29.3 *r*-Selected and *K*-Selected Species

It's well known that for human beings, about 9 months will elapse between conception and birth; but it's less well known that for an elephant this same process takes almost 2 years. For any given human or elephant female, then, births are relatively few and far between. And once the offspring come, the amount of attention lavished on them is great indeed. It's not unusual for an elephant calf to be nurtured by its mother and a larger kinship group for at least the first 10 years of its life. In the first year of this period, a mother might not ever let her calf wander farther than 20 meters away from her. Meanwhile, humans may lavish 18 years of constant care and concern on a child before the domestic bonds are finally broken.

By contrast, the common housefly can produce a new generation once a month, and the parental generation provides no care whatsoever to the offspring.

Houseflies and elephants lie at opposite ends of a continuum of *reproductive strategies*, by which is meant characteristics that have the effect of increasing the number of fertile offspring an organism bears. A plant or animal obviously does not consciously *decide* on a strategy for producing offspring; rather, such a strategy is handed to it by virtue of its species' evolution.

So how does the strategy of elephants differ from the strategy of houseflies? This may be apparent from what you've observed already. For elephants the strategy is to bring forth few offspring, but lavish attention on them. For houseflies it is to bring forth a multitude of offspring, but give no attention to any of them. These strategies are in turn related to other characteristics of these species, as you'll now see (**see Figure 29.8**).

K-Selected, or Equilibrium, Species

Elephants will not seek out a totally new environment; they experience their environment as a relatively *stable* entity, and compete among themselves and with other species for resources within it. Given this stability, species like elephants are known as **equilibrium species**. In line with this, the elephant *population* stays relatively stable compared to a fly population; its numbers will fluctuate in a relatively narrow range above and below the environment's carrying capacity. This is another way of saying that the population regularly bumps up against carrying capacity (*K*), which is the density of population that a unit of living space can support. Two things follow from this. First, elephants are said to be a *K-selected species*: one whose population sizes tend to be limited by carrying capacity. The second point, which follows from this definition, is that the pressures on the elephant population are **density dependent**: As the density of the population goes up, the factors that limit the population—food supply, living space—assert themselves ever more strongly.

r-Selected, or Opportunist, Species

In contrast to elephants, houseflies are known as an **opportunist species**—a species whose population size tends to fluctuate greatly in reaction to variations in its environment. Should favorable weather suddenly arrive, or a food supply suddenly appear, the fly population in the area will skyrocket. When the food is gone, or if the temperature suddenly changes, this same population will plunge in number. In short, environment tends to be highly variable for fly populations, and the flies have a high population

growth rate that allows them to take advantage of environmental opportunities. For these animals, the strategy is to produce a multitude of offspring very fast. Flies are therefore said to be an **r-selected** species, meaning one whose population sizes tend to be limited by reproductive rate.

This definition implies that fly population numbers have nothing to do with fly population *density*, at least for a time. The pressures on the population are thus said to be **density independent**. The forces at work in these populations tend to be *physical* forces—frost, temperature, rain—as opposed to the *biological* forces that operate in density dependent control (competition for food, disease). Given this, the *r*-selected strategy of many offspring/little attention is understandable. When you exist at the whim of environmental forces, "life is a lottery and it makes sense simply to buy many tickets," as R. M. May and D. I. Rubinstein have phrased it.

CHARACTERISTICS OF *K*-SELECTED AND *r*-SELECTED SPECIES

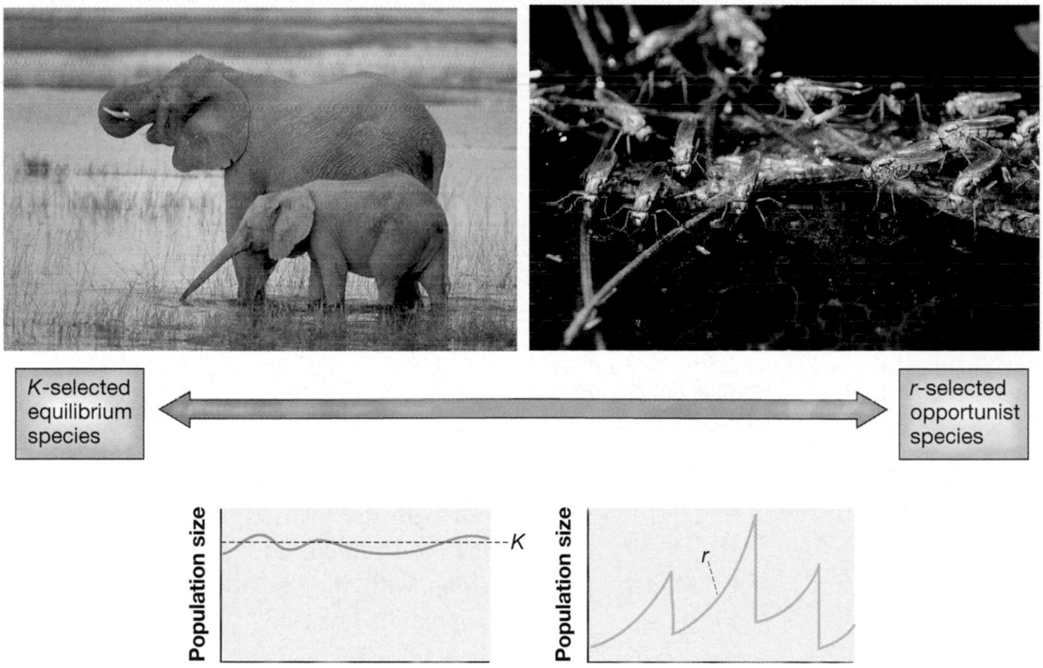

K-selected equilibrium species ⟷ *r*-selected opportunist species

population size:
- limited by carrying capacity (*K*)
- density dependent
- relatively stable

organisms:
- larger, long lived
- produce fewer offspring
- provide greater care for offspring

population size:
- limited by reproductive rate (*r*)
- density independent
- relatively unstable

organisms:
- smaller, short-lived
- produce many offspring
- provide no care for offspring

Figure 29.8
***K*-Selected and *r*-Selected Species**
The elephants on the left are a *K*-selected species. Here, a mother stays close to her calf, providing the kind of careful attention typical of *K*-selected species. The pond flies on the right are an *r*-selected species, bearing many young but giving no attention to them after birth.

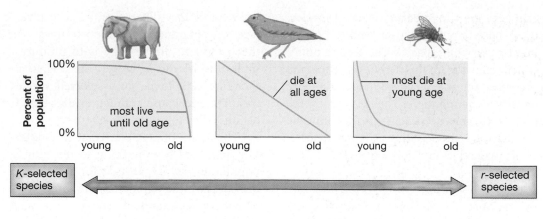

Figure 29.9
When Is Death Likely to Come?
An organism's chance of living a long life is related to the reproductive strategy of its species, as reflected in these survivorship curves. Elephants and humans (late-loss species) produce few young, but most survive until old age. At the other extreme, flies (early-loss species) produce many young, but most die young. In between are constant-loss species; they produce a moderate number of young, which die at a fairly constant rate during a typical life span.

The *r*- and *K*-selected groupings are categories invented by human beings and, as usual, nature refuses to fit neatly into such boxes. It is true that small, short-lived species tend to lie at the *r*-end of the continuum, and that large, long-lived species lie at the *K*-end, but sea turtles are long-lived and they give no care at all to their newborns. Meanwhile flies can be limited in a density dependent way for at least part of their population cycle—by predators and by competition among themselves. All sorts of variations exist on this theme. You've looked at animals as examples of *r*- and *K*-selection, but the concept applies to plants as well. A plant such as ragweed is at the *r*-selected extreme in the plant world, while oak trees and cacti are at the *K*-selected extreme.

Survivorship Curves: At What Point Does Death Come in the Life Span?

K- and *r*-strategies are related to the concept of "survivorship curves," which, as you can see in **Figure 29.9**, are thought of in terms of three ideal types: late loss, constant loss, and early loss (sometimes referred to as types I, II, and III, respectively). Humans and most of our fellow *K*-selected mammals fall into type I, because we tend to survive into old age (our lives are lost late). Insects and many amphibians are type III, because their death rates are very high early in life, but level out thereafter. Other types of living things, such as birds, fall into type II, because they die off at a nearly constant rate through their life span.

29.4 Thinking about Human Populations

We'll look at the final elements in population dynamics in connection with the human population. The concepts involved could be applied to any population of living things, but here the focus is on human beings, so that you can consider population principles along with the real-world issue of human population growth.

Survivorship Curves Are Constructed from Life Tables

The survivorship curves you just looked at are constructed by developing what are known as **life tables** for the species in question. To create these, scientists divide the species' life span into suitable units of time—days for fruit flies, years for human beings—and see how likely it is that an average

Table 29.2
A Life Table for the United States

Taking a hypothetical group of 100,000 persons born in the United States in 1996, the table shows the proportion likely to still be living at the ages indicated and the average remaining lifetime for persons at each age. Note that life expectancy does not drop 10 years for each 10 years lived. The average person surviving to age 70 can expect to live until age 84, whereas the average person at age 10 could expect to live only to about age 77.

At Age	Number Still Living	Average Remaining Lifetime in Years
10	99,022	66.9
20	98,519	57.2
30	97,487	47.7
40	95,881	38.4
50	92,946	29.5
60	86,630	21.2
70	73,056	14.1
80	49,276	8.4
85 +	33,629	6.1

Source: National Center for Health Statistics

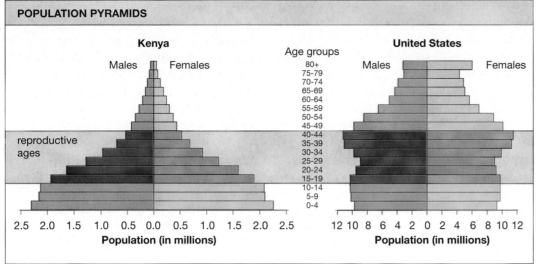

POPULATION PYRAMIDS

Kenya — Males / Females

United States — Males / Females

Age groups: 80+, 75-79, 70-74, 65-69, 60-64, 55-59, 50-54, 45-49, 40-44, 35-39, 30-34, 25-29, 20-24, 15-19, 10-14, 5-9, 0-4

reproductive ages

Population (in millions)

Figure 29.10
Age Structure in Populations: Kenya and the United States, 2000
Each bar on the graphs represents a five-year age grouping of the population. Note the greater proportion of the Kenyan population that is of "pre-reproductive" age—an indication of greater population increases in the future than is the case with the United States. (Data from the United States Census Bureau.)

29.4 Thinking about Human Populations

species member will be alive after a given number of days or years. The technique for doing this was not invented by biologists who wanted to know about the survival of flies, but by a nineteenth-century British actuary—a person trained to calculate probabilities—who knew the information would be useful for life insurance companies. How likely is it that a person will be alive at 10, or 20, or 80 years of age? If you look at Table 29.2, you can see the answer for a hypothetical group of 100,000 persons born in the United States in 1996.

Population Pyramids: What Proportion of a Population Is Young?

You've seen that population growth in the natural world can be exponential (for a time) because a population of living things grows in proportion to its own size. However, all members of a population do not count equally in calculating this growth. If scientists want to peer into the future of a given population, what they want to know is: What proportion of the population is *past* the age of reproduction as opposed to the proportion that is, or will be, of reproductive age? If you look at **Figure 29.10**, you can see the answer to this question, expressed as a "population pyramid" for two countries in the year 2000. Each bar on the graph represents a 5-year age grouping, or "cohort," of the population (those who are 0–4 years old, 5–9, and so on), with the length of each bar representing the size of the cohort. The reproductive age-range for humans generally is considered to be between 15 and 45. If you look at the graph for Kenya, you can see that it has many

more individuals of reproductive and pre-reproductive age than of post-reproductive age. What does the future hold for Kenya? In all probability, a huge growth in the population. In contrast, the age-structure diagram for the United States shows that a much greater proportion of its population is at or beyond reproductive age.

Immigration and Population Change: The United States

Thinking globally, there is but one human population, of course. But to continue thinking along lines of population by country, two factors that haven't been mentioned come into play—the movement of individuals into a population (**Immigration**) and the movement of individuals out of a population (**emigration**). Immigration is expected to have a great impact on the population of the United States, as you can see in **Figure 29.11**.

Figure 29.11
Immigration Can Have a Big Impact on Population Size
Current projections are that without immigration, the United States population would level out and begin to decrease by the middle of this century. With its influx of immigrants, the U.S. population is expected to continue growing into the foreseeable future. (Data from *The New Americans* [National Research Council, 1997], Table 3.3.)

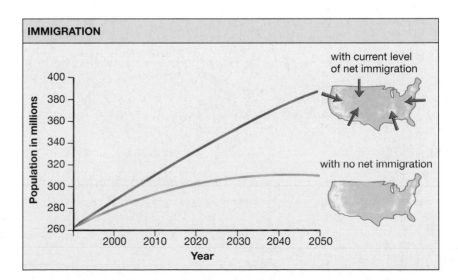

IMMIGRATION

with current level of net immigration

with no net immigration

Population in millions / Year

According to projections released in 1997 by the National Research Council, if the United States allowed no net legal or illegal immigration, its 1995 population of 263 million would grow to about 307 million by the year 2050. At current net levels of immigration, however (about 800,000 people per year), the U.S. population is projected to grow to 387 million people by 2050—80 million more people than if there were no immigration, and an increase of 47 percent over the 1995 population. Note also the longer-term trends suggested in the figure. With no immigration, the U.S. population actually would be declining after 2040, while with immigration the trend will be upward for many more years.

The Increase in World Population

Large as the U.S. growth figure is, it pales in comparison with the growth projected for many other parts of the world. Between 1990 and 1995, the annual growth rate for the more developed countries such as the United States was 0.4 percent, but for the less developed countries it was 1.88 percent. In parts of Africa, meanwhile, it approached 3.0 percent (though this high rate is now being drastically reduced in some African countries because of AIDS). Remember how, as noted earlier, small differences in the rate of growth stand to have tremendous impact *over time*? In 1950, developed countries accounted for 33 percent of the world's population; by 2025 they are expected to account for 15 percent of the total. By 2025, Bangladesh is likely to have a total population of 196 million people—this in a country about the size of Iowa, whose population in 1994 was about 2.8 million people (**see Figure 29.12**).

Human Population Growth over Time

If you look at **Figure 29.13**, you can see how human population has grown over time. For centuries our numbers grew relatively little, but then began an upward climb about 1700. This increase was nothing, however, compared to the rise that occurred beginning about 1950. Improved sanitation, better medical care, and increases in the food supply came together to produce the rate of growth you see. The Earth's human population didn't pass the 1-billion mark until 1804; it then took 123 years to double to 2 billion (in 1927), then 48 years to double to 4 billion (in 1974), and has now exceeded 6 billion. The curve you see that resulted from this growth may look familiar, because it is a more extreme version of the J-shaped curve, though greatly elongated on its left side. Were we looking at any species other than our own and saw this kind of growth curve, we would predict that this kind of increase could not go on for long; that environmental resistance would assert itself and that the population might even crash in one catastrophic sense or another.

The Earth's population is projected to grow from its present 6.1 billion to between 7.3 and 10.7 billion by 2050. So, can the Earth support a population of 8 or 10 billion people? Do current famines and disease outbreaks represent environmental resistance to such population sizes? Do we see on the horizon any other signs of environmental resistance? These questions are a matter of spirited debate among ecologists, economists, and demographers. As may be apparent, this is a discussion about *carrying capacity*—in this case, the Earth's carrying capacity for human beings. Some scholars think we have come nowhere near this capacity, while others think we have effectively exceeded it already.

Current Population Impacts on Quality of Life and the Environment

It may be that the notion of planetary carrying capacity is beside the point; that what we ought to be focusing on is the impact human population increase is having right now on the quality of human life and on the quality of Earth's environment. The relationship of population growth to the quality of human life is complex; suffice it to say here

Figure 29.12
Small Area, Big Population
By 2025, Bangladesh, which is about the size of Iowa, may have 196 million people, compared to 331 million in all of the United States.

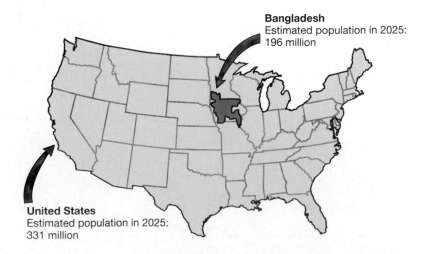

Bangladesh
Estimated population in 2025:
196 million

United States
Estimated population in 2025:
331 million

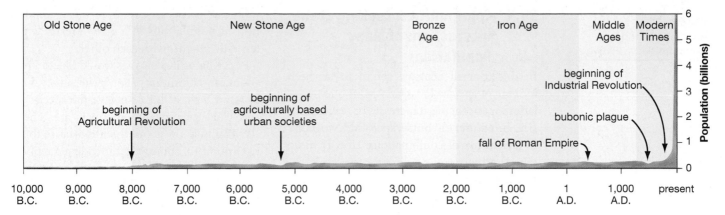

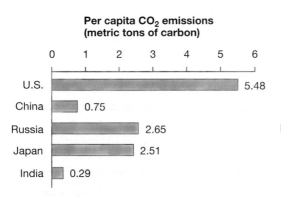

that there is great disagreement about whether growth per se works against human well-being.

With respect to the environment, some scientists would argue that there is no greater single environmental threat than the continued growth of the human population. The basis for this argument is that population affects so many environmental issues: the use of natural resources, the amount of waste that is pumped into the environment daily, the reduction of species habitat, the decimation of species through hunting and fishing. Look at almost any environmental problem and you are likely to find human population growth playing a part in it.

Other experts would argue that it is not population per se, but rather the use of resources *per person* that is of most pressing concern. Consider that the United States used less water in 1995 than it did in 1980, even though the U.S. population grew by 16 percent during the period. Another perspective on this is provided by the use of resources in more-developed, as opposed to less-developed, countries. If you look at **Figure 29.14**, you can get a sense of the differences that exist. Carbon dioxide (CO_2) is a gas that is found naturally in the atmosphere,

but that is also put into the atmosphere when human beings burn "fossil fuels," such as coal or gas, to power car engines or run power plants. CO_2 produced by these human activities is now regarded by most scientists as a cause of global warming—the rise in Earth's surface temperature that has occurred over the past century. Figure 29.14.a shows the amount of CO_2 produced per person by human activity in five countries. Note how little CO_2 is produced per person in less-developed countries, such as India or China, compared to that produced in the more-developed countries, such as the United States. Now, in Figure 29.14.b note the effect this per capita difference has on total CO_2 emissions. The United States put 64 percent more carbon into the atmosphere than did China in 1997, even though China had nearly five times as many people as the United States (1.3 billion to the U.S.'s 265 million).

With this, you can begin to see why environmentalists are worried about the strain that economic development will put on the Earth's environment *irrespective* of population growth. What will happen when people in China start driving cars and using air conditioners at the rate that Americans do?

Figure 29.13
Human Population Increase through the Centuries
How many people can the Earth sustain?

Figure 29.14
Use of Resources per Person
The use of natural resources per person (or "per capita") varies greatly from one country to the next. The average resident of a developed country, such as the United States, uses far more resources than the average resident of a less-developed country, such as India, and this has environmental consequences. (a) One measure of resource use is per-capita carbon dioxide (CO_2) emissions: the amount of CO_2 a country releases into the atmosphere through human-caused activities, divided by the number of people that country has. The graph displays per-capita CO_2 emissions for five countries in 1997, as measured in metric tons of carbon. (b) The differences in per-capita emissions have significant consequences for total emissions within a country. In 1997, the United States released 64 percent more CO_2 into the atmosphere than did China, even though the U.S. had only one-fifth as many people as China.

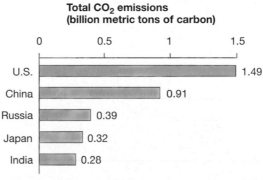

29.5 Communities: Looking at the Interactions of Many Populations

In this tour of ecology, you've so far been looking at species in isolation—at one population or another as it increases or decreases in size. In nature, of course, species live together in a rich mix of combinations. How rich? To take the high end of diversity, which is the tropical rain forest, Smithsonian biologist Terry Erwin found an estimated 1,700 species of beetle in a single tree in Panama. Not 1,700 beetles, mind you; 1,700 *species* of beetle.

In beginning to think about the interactions of these diverse populations, a helpful concept is that of the community. An **ecological community** generally is defined as all the populations that inhabit a given area; all the plants, animals, fungi, bacteria, protists—every living thing, in other words. The term also is used, however, to mean a collection of populations in a given area that potentially *interact* with each other.

So what is it ecologists would like to know about any community? For starters, they'd like to know what at least some of the species are and what their relative numbers are. They'd also like to know about the importance to the community of some of its individual members.

Large Numbers of a Few Species: Ecological Dominants

There's a tremendous variability in the mix of species found in different communities, but many communities tend to be dominated by only a few species. Forests tend to be populated by certain kinds of trees, stretches of prairie by certain kinds of grasses. The few species that are abundant in a given community are called **ecological dominants** (see Figure 29.15). These generally are plants, but they can be other life-forms, as is the case with the tiny animals known as corals that are the ecological dominants in coral reefs.

Importance beyond Numbers: Keystone Species

Ecologists have long recognized that there are also species that may not be numerous in a given area, but whose *absence* would bring about significant change in the community. These organisms are called **keystone species** (named for the pressure-bearing "keystone" in an arch). The concept of the keystone species was introduced in the 1960s by marine ecologist Robert Paine, who went with his students to a shallow-water zone of the Pacific Ocean in Washington State and for six years regularly removed all sea stars in the genus *Pisaster* from a small area. Such a sea star may sound harmless enough, but *Pisaster* was, in fact, the **top predator** in the area: It preyed on other species, but no species preyed upon it (see Figure 29.16). The impact of the *Pisaster* removal was big: Before the change there had been 15 species in the area; after it there were 8. One species of mussel, freed from its former predator's control, took over much of the attachment space in the area, crowding out other animals, such as barnacles.

Over time, the keystone species concept has undergone some modification. Where once scientists thought of keystones as always sitting at the top of food chains, they now recognize that organisms in other positions

Figure 29.15
Ecological Dominants

a A forest dominated by spruce trees in Bavaria, Germany.

b Prairie in Kansas dominated by tall grass.

a
b

Figure 29.16
Keystone Species
When the predatory sea star *Pisaster ochraceus* was removed from a small area of rocky shore along the Pacific coast of the United States, the species composition of the community changed drastically, more so than if one of the other species had been removed. Pictured are several *Pisaster* sea stars on the Pacific Northwest coast.

can take on a keystone role—beavers building dams, for example, or even lichens that are critical in getting communities going in the desert. Beyond this, it turns out that there are communities without keystones; remove any one species from such a community and its role will be taken over by another species.

Variety in Communities: What Is Biodiversity?

Apart from ecological dominants and keystones, a third element that ecologists pay attention to in any community is the *range* of species it has in it. This touches on the more general concept of **biodiversity**, which can be defined as variety among living things. In everyday speech, what biodiversity means is a diversity *of species* in a given area. This is an important measure of diversity, but it is only one among several. You can see why from an imaginary experiment that has been noted by ecologist Paul Ehrlich.

Suppose that you could get a few members of every species on Earth, but that you restricted each species to a single population housed somewhere (in a single zoo or an aquarium or botanical garden). Species *diversity* would not drop at all in such a scenario—you'd still have some of each kind of creature, after all—but the Earth quickly would become barren, because what's needed is a rich *distribution* of species in populations across the planet. This geographical distribution of populations is the second measure of biodiversity. The third measure exists *within* populations or species. It is genetic diversity, meaning a diversity of "alleles" or variants of genes in a population. Without such diversity, populations are vulnerable to disease, their members may die young or suffer from a variety of inherited mental and physical afflictions. (See "Lessons from the Cocker Spaniel" in Chapter 17.) In summary, biodiversity means species diversity, geographic diversity, and genetic diversity (**see Figure 29.17**).

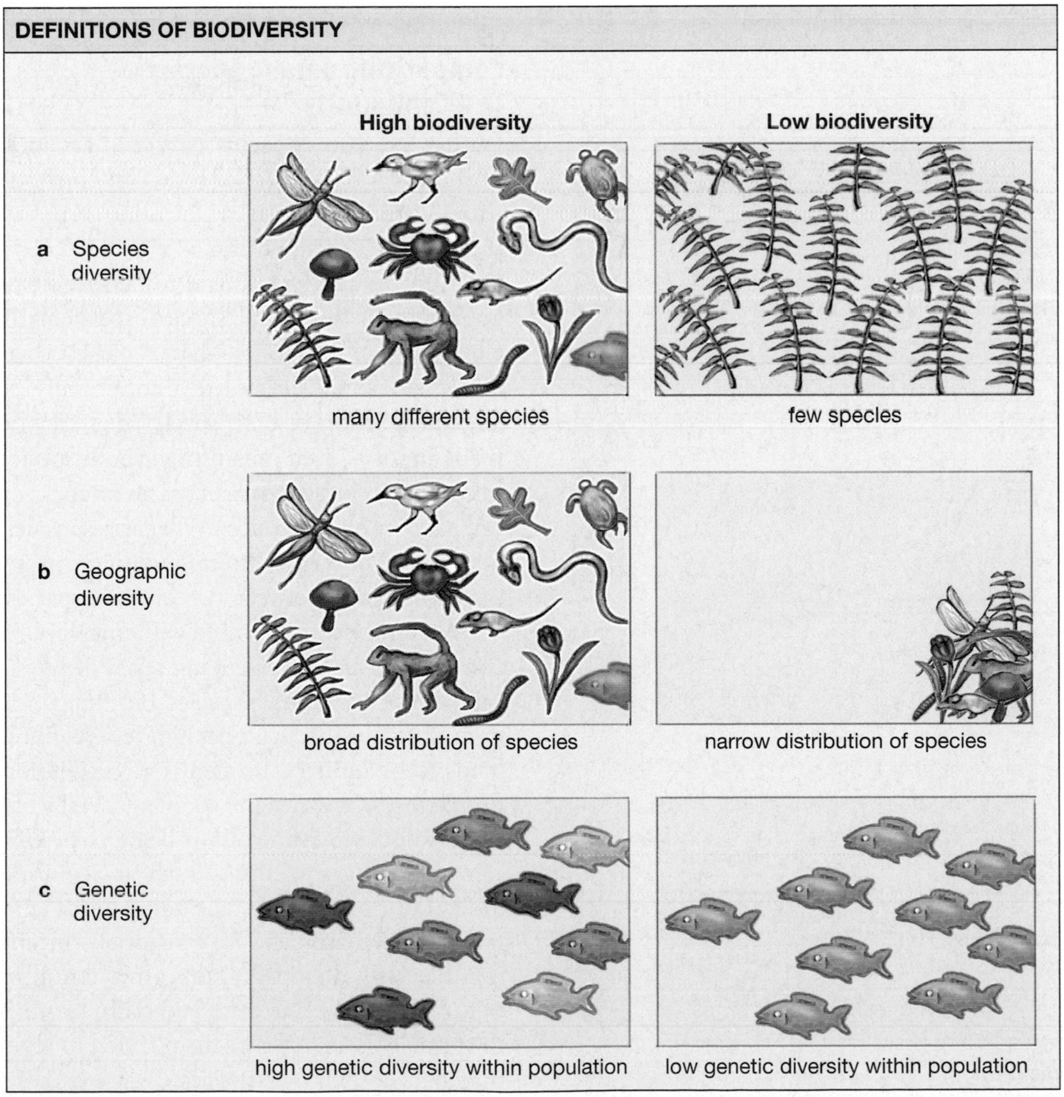

DEFINITIONS OF BIODIVERSITY

High biodiversity | Low biodiversity

a Species diversity — many different species | few species

b Geographic diversity — broad distribution of species | narrow distribution of species

c Genetic diversity — high genetic diversity within population | low genetic diversity within population

Figure 29.17
Three Types of Biodiversity

Figure 29.18
Competition for Resources among Species
Laboratory experiments by G. F. Gause showed that competition for resources between two species can have two possible outcomes.

a When two species compete for the same limited, vital resource, one will always drive the other to local extinction—as the paramecium *P. aurelia* did to the paramecium *P. caudatum*. This is the competitive exclusion principle at work.

b Conversely, when Gause put *P. aurelia* together with another paramecium, *P. bursaria*, the two species divided up the habitat and both survived. This is a demonstration of resource partitioning.

29.6 Types of Interaction among Community Members

Having considered some general issues regarding communities, let's now turn to the subject of how members of a community might interact with each other. Here's a short list of ways:

> Competition
> Predation and parasitism
> Mutualism and commensalism

Before continuing with the exploration of these modes of interaction, let's go over a couple of concepts that will apply to them all.

Two Important Community Concepts: Habitat and Niche

A **habitat** is the physical surroundings in which a species can normally be found. Though two populations of a species may be widely separated, they can generally be found dwelling in similar natural surroundings. A habitat is sometimes described as a species' "address," but a more accurate metaphor might be a species' preferred type of neighborhood.

The word **niche** has been defined in several ways, but it is useful to think of it in terms of a simple metaphor: A niche is an organism's occupation. How and where does the organism make a living? What does it do to obtain resources? How does it deal with competition for these resources? The horseshoe crab has found a niche walking the bottom of shallow coastal waters, feeding on food items that range from algae to small invertebrates. Note that this is about more than what the horseshoe crab eats. It includes specific surroundings (shallow ocean waters), specific behaviors (ocean-floor crawling), and perhaps seasonal or daily feeding times among other things. If you were to specify all the things that define a horseshoe crab's niche, the odds are that no other organism would exactly fit into it.

Competition among Species in a Community

With these two concepts covered, let's look at the ways organisms interact in communities, starting with a familiar type of interaction—competition.

Even though niches tend to be specific to given organisms, some species—particularly closely related species—have niches that *overlap* to some degree in a community. A large proportion of both species' diet may be made up of a given organism, or both species may occupy similar kinds of spaces on rocks or branches or pond surfaces. What arises in such instances is **interspecific competition**, meaning competition between two or more species.

What may come to mind with interspecific competition is a never-ending series of physical battles between two species, but things seldom work like this. For one thing, competition tends to be indirect. It often is a competition for *resources*, in which the winner generally triumphs not by fighting but by being more efficient at doing something, such as acquiring food. Lots of animals, including birds and many large mammals, are **territorial**—meaning they will attempt to keep other creatures out of a territory they have laid claim to—but this usually is a matter of them trying to repel members of their own species.

a Competitive exclusion

Time (days)

b Resource partitioning

Time (days)

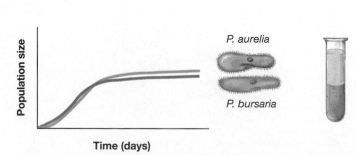

No Two Species Can Share the Same Vital Resource for Long

When the niches of two species greatly overlap, it's unlikely that you'd see *long-standing* competition among them, for the simple reason that one species or the other is likely to *win* such a competition in fairly short order. Because the competition is for nothing less than vital resources, the result is that the losing species will be driven to local extinction.

It was a laboratory experiment, performed by the Russian biologist G. F. Gause in the 1930s, that pointed the way on this latter principle. Gause grew two species of *Paramecium* protists, *P. caudatum* and *P. aurelia*, in culture and found that, in time, *P. aurelia* was always the sole survivor. True to what was just noted, *P. aurelia* wasn't eating or wounding *P. caudatum*; instead, it grew faster and thus used more of the surrounding resources. Nevertheless, the outcome was always the death of all the *P. caudatum*. This led Gause to formulate what came to be called the **competitive exclusion principle**: When two populations compete for the same limited, vital resource, one will always outcompete the other and thus bring about the latter's local extinction (**see Figure 29.18a**).

Ecologists wouldn't expect to witness the competitive exclusion principle operating much in nature, however. This is because, as noted, the *Paramecium* scenario is likely to be played out quickly.

Nevertheless, competitive exclusion has been observed in nature, often when humans have a hand in things. For example, humans introduced a Southeast Asian vine, the kudzu (*Pueraria lobata*), to the American South on a large scale in the 1930s. Growing at up to 30 centimeters or 1 foot per day, kudzu has now taken over millions of acres in the South, locally eliminating many plants in its path (**see Figure 29.19**).

Resource Partitioning Is Common in Natural Environments

Competitive exclusion notwithstanding, there are instances in nature in which two related species will use the same kinds of resources from the same habitat over a long period of time. So, why isn't one of them eliminated? The answer is contained in another experiment conducted by Gause. He took the successful species from the earlier experiment, *P. aurelia*, and placed it in a test tube with a different paramecium, *P. bursaria*. This time, neither species was eliminated (see Figure 29.18b). Instead, the two species divided up the habitat, *P. aurelia* feeding in the upper part of the test tube and *P. bursaria* flourishing in the lower part. (*P. bursaria* had an advantage in the lower, oxygen-depleted water because it has symbiotic algae that grow with it, producing oxygen.) In a nutshell, this result describes a situation that often exists in nature: **coexistence** (meaning a sharing of habitat) through a practice called **resource partitioning**, which can be defined as a dividing up of scarce resources among species that have similar requirements. If you look at **Figure 29.20**,

Figure 29.19
Kudzu Vines Run Wild
The kudzu plant was introduced in the American South in the 1930s and has since spread at a rapid rate, locally eliminating many plants in its path. This is competitive exclusion in action. Here kudzu has overgrown an abandoned house in Mississippi.

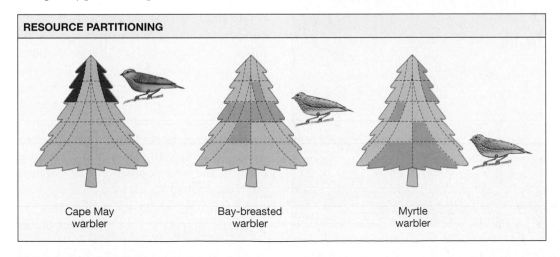

RESOURCE PARTITIONING

Cape May warbler

Bay-breasted warbler

Myrtle warbler

Figure 29.20
Resource Partitioning
Ecologist Robert MacArthur spent long stretches of time over several years in the 1950s observing the feeding patterns of several species of warblers. All of them ate caterpillars, but from substantially different, though overlapping, parts of the tree.

you can see how this works among some species that are a little more familiar—several varieties of warbler. The main message here is that species often flourish by specializing, deriving resources by feeding from different locations (or during different times) within the same habitat.

Other Modes of Interaction: Predation and Parasitism

It is one thing for two species to compete for resources; it is another for one species to *be* a resource for another—to be eaten or used by another species. Thus do ecologists distinguish the first mode of interaction in communities, competition, from the second, predation, and a variety of it called parasitism.

Predation can be defined as one freestanding organism feeding on parts or all of a second organism. Note that the prey here can be plants, protists, animals—whatever is preyed upon. Predation is generally thought of in terms of animals killing other animals, but note that its definition includes such things as animals grazing on plants. **Parasitism** is a variety of predation in which the predator feeds on prey, but does not kill the prey immediately—and may not ever kill it. Parasites that come to mind tend to be animals, but there are an estimated 3,000 species of parasitic plants, one of which can be seen in **Figure 29.21**.

Figure 29.21
Plants Parasitizing Plants
The orange tendrils in the picture belong to a species of dodder, *Cuscuta pentagona*, a parasitical plant that survives by tapping into the food reserves of other plants, thus feeding off them. The dodder is orange, rather than green, because it lacks the chlorophyll that is used in photosynthesis.

The Value of Predation and Parasitism

Predation and parasitism are the features of nature that many nature lovers can't stand. Why this should be so is a fascinating question, though beyond the scope of this book. If one values more complex organisms (such as ourselves), predation and parasitism have had a clear value, in that they have spurred on evolution by culling the less capable members from populations and by stimulating the "arms race" that has resulted in evolutionary adaptations such as vision and flight. (One organism's predatory adaptation spurs the development of another organism's defensive adaptation, and vice versa.) Whatever we may think of it, predation is simply a fact of nature, and it is impossible to imagine life without it. But it's also true that to admire the beauty of nature is in significant part to admire the handiwork of predation, since it has done so much to shape the living world. You can read about a very familiar predator, the house cat, in "Purring Predators: Housecats and Their Prey" on page 679.

Predator-Prey Dynamics

It may be obvious that although predators attack prey, predators are also *dependent* on prey as a food source. An ongoing question in ecological research is how tight this linkage is. An attractive theory is that when the prey population flourishes, the resulting bounty of available food makes the *predator* numbers grow; this, however, means that more predators will hunt, which will bring the prey population's numbers down. When this happens, there may be a steep decline in the predator population, because a large part of its food supply is now gone. The end-result would be a series of linked population ups and downs, or cycles. If you look at **Figure 29.22**, you can see a series that has occurred with one linked pair of species, the moose of Michigan's Isle Royale National Park and the wolf that preys on it. At first glance, these figures seem to provide evidence for the scenario just described, because the moose and wolf numbers move generally in opposite directions. In reality, however, things are more complex than this. The huge crash in the wolf population that began in 1981 was caused in part by a canine virus that invaded the island and perhaps by a lack of genetic diversity among the wolves. Suffice

Purring Predators: Housecats and Their Prey

Housecats are famous for bringing home animals they've killed or captured, of course; intrigued by the carnage brought into their own homes in the 1980s, Peter Churcher and John Lawton decided to make a scientific study of the predatory behavior of the domestic cat. To do so, they enlisted the help of 172 households in the small English village of Bedfordshire, where Churcher lived. The two researchers asked the locals to "bag the remains of any animal the cat caught," and turn the evidence over to them once a week. This process went on for a year with a high degree of cooperation from the cat owners.

Some of the results were not surprising. Young cats hunt more than older cats; small mammals, such as mice, are the favored prey; and cat hunting is not based purely on hunger—all the cats were fed by their owners and yet hunting was widespread.

What was remarkable, however, was the *scale* of the killing. Concentrating on the village's house sparrow population, Churcher and Lawton found that cats were responsible for between one-third and one-half of all sparrow deaths in the village, a figure they believed no other single predator could match. When they looked at all animals killed, and projected the village figures onto the whole of Great Britain, the researchers calculated that cats were responsible for about 70 million deaths a year.

What was remarkable was the scale of the killing.

Given that birds account for somewhere between 30 and 50 percent of these kills, this means that cats kill at least 20 million birds a year in Britain. The "at least" here may be an important qualifier. An American biologist, the researchers note, found that cats bring home only about half of the food they catch.

Figure 1
Predator on the Loose
A domestic cat carrying its prey.

it to say that population cycles generally are the result of many factors, only one of which is a given predator-prey relationship.

Parasites: Making a Living from the Living

Let's look now at the special variety of predation called *parasitism*. Some familiar parasites, such as leeches and ticks, have a straightforward strategy: Get on, hold on, and consume material from the prey, known as the **host**. But the relationship between parasites and hosts can be much more sophisticated than this. Take, for example, the barely visible roundworm *Strongyloides stercoralis*, which lives throughout the tropics and has human beings in the southeastern United States as one of its hosts. The worm comes into the human body from the soil in a threadlike larval form, then makes its way into the veins, which it travels through to get

Tutorial 29.2.3
A Simple Mathematical Predator-prey Model

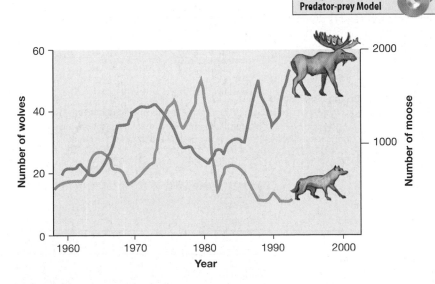

Figure 29.22
Predator and Prey Populations
The population dynamics of a predator species (the wolf) and its prey (the moose) in Michigan's Isle Royale National Park over several decades. (Redrawn and adapted from an illustration in *Science*, August 27, 1993, p. 1115.)

to the lungs. From there it moves up the trachea and then down into the digestive system, ending up in the small intestine. There it lays eggs that develop into larvae, that this time move *out*, with feces, back into the Earth to start the whole life cycle over again. *Strongyloides* is not just using humans as an incubator, however; it can worm its way, you might say, directly from the small intestines into the bloodstream, bringing about a continuous cycle of infection whose symptoms range from severe diarrhea to lung problems. (Lest all of this seem too disturbing, take heart; there is a medicine that can clear this pest from human bodies.) So, note the complexity here: *Strongyloides* not only feeds on a host, but uses the host in its reproduction cycle, something that actually is fairly common among parasitic worms and protozoans.

Were you to start examining the *range* of host-parasite dynamics, you would find lots of interactions as elaborate—and chilling—as this. To give but one example, Nancy

Figure 29.23
Parasitism Can Be Very Specialized
When laying its eggs in a caterpillar, the parasitoid wasp *Cotesia congregata* also injects a special virus that coats the eggs, preventing the caterpillar's immune system from destroying them. The wasp larvae feed on the caterpillar while inside it, eventually bursting forth from the caterpillar, which then dies.

Figure 29.24
Fooling Predators about Prey
The predatory frogfish uses a modified spine resembling a tasty worm to lure its prey. Here a tasseled frogfish clearly shows its lure while swimming near Edithburg, South Australia.

Beckage and her colleagues have been studying a species of wasp, *Cotesia congregata*, that lays its eggs inside its host, in this case a caterpillar. There's nothing special about this; there are legions of these "parasitoid" species whose offspring *develop* in the host, feeding on it and ultimately killing it (**see Figure 29.23**). The difference here is that the eggs the wasp lays are coated with a virus that cripples the caterpillar's immune system, thus allowing the eggs to develop free of a caterpillar immune system attack. This virus is produced in the female wasp's ovary, with the genetic instructions for its production coming from the wasp's own DNA.

The Effect of Predator-Prey Interactions on Evolution

With the interactions between *Cotesia* and the caterpillar, you can begin to see some *evolutionary* effects of predator-prey interactions. Through evolution, *Cotesia* developed a new weapon that conferred greater reproductive success on it. Such an adaptation is part of what was referred to earlier as the arms race in evolution, meaning the continual adaptation of species in response to pressure from other species. Warm-water "frogfish" have evolved a spine on their dorsal fins that is tipped with a piece of flesh that looks for all the world like a small worm floating free in the ocean. When a would-be predator tries to snatch up this "worm," however, it finds *itself* the prey (**see Figure 29.24**). On the prey side, look at **Figure 29.25** to see the amazing kinds of camouflage that animals use to keep themselves hidden from the eyes of predators. Needless to say, predator-prey evolution goes on within all the kingdoms of life. Plants have not only developed spines, thorns, and other protective structures but also a vast array of chemical compounds—more than 15,000 have been characterized so far—that protect them from predators.

Mimicry Is a Theme in Predator-Prey Evolution

The ability to fool an opponent by faking is as valuable in nature as it is in sports, as you've seen with the frogfish. Another way such deception takes place in nature is through **mimicry**, a phenomenon through which one species has evolved to assume the appearance of another. One form of mimicry involves at least three participants: a model, a mimic, and a dupe. If you look at **Figure 29.26**, you can see a model on the left, the yellowjacket wasp, which obviously can provide a painful lesson to any animal that tries to eat it. On the right you can see the mimic, the clearwing moth, which is harmless but *looks* a great deal like a wasp. For the dupe, you could select any predator species that sees a clearwing moth and passes it by, believing it to be a dangerous yellowjacket. This is so-called **Batesian mimicry**: the evolution of one species to resemble a species that has a superior protective capability.

In a second type of mimicry, **Müllerian mimicry**, several species that *have* protection against predators come to resemble each other (**see Figure 29.27**). This creates a visual warning that becomes known to an array of predators. The benefit is that the experience a would-be predator has with an individual from one species will keep it away from *all* the individuals in any look-alike species, meaning that fewer individuals in any of these species will be disturbed or killed.

Beneficial Interactions: Mutualism and Commensalism

The competitive, predatory, and parasitic modes of interaction you've reviewed have an each-organism-against-all quality to them, but other kinds of relationships exist in nature as well. Consider **mutualism**, meaning an interaction between individuals of two species that is *beneficial* to both individuals. There's a great example of this in Chapter 20 (page 424), with the linkage between plant roots and the slender, below-ground extensions of fungi called *hyphae*. What the fungi get from growing into the plant roots is food that comes from the photosynthesizing plant; what the plants get is minerals and water, absorbed by the network

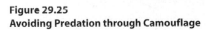

Figure 29.25
Avoiding Predation through Camouflage

a A spanworm, looking like a twig, on a maple tree.

b A casque-head chameleon against a tree in Kenya.

Figure 29.26
Harmless, but Looking Dangerous
In an example of Batesian mimicry, the clearwing moth (on the right) has no sting but has evolved to look like an insect that does, the yellowjacket wasp (left). The yellowjacket is the model and the moth is the mimic.

Figure 29.27
Müllerian Mimicry
(a) A yellowjacket and **(b)** a cicada-killer wasp are two of the many varieties of stinging insects that have similar patterns of warning coloration. This form of mimicry sends a message to predators that they should keep away from all organisms that have this color pattern—or pay a price.

a

b

a

b

Figure 29.28
Mutually Beneficial

a Several oxpecker birds sit atop a black rhinoceros, ridding the rhino of ticks and other pests while the rhino provides a safe habitat for the birds. This is a demonstration of mutualism—an interaction between two species that is beneficial to both.

b The orange-spotted shrimp goby, on the right, and the snapping shrimp at lower left also exhibit mutualism. Here the shrimp goby stands guard near the snapping shrimp, which digs out the burrow on the sea floor that the two creatures share.

of hyphae. Some plants, such as orchids, cannot grow without this association. Perhaps the most famous example of mutualism is between the rhinoceros and oxpecker birds. Sitting on the back of the rhino, the oxpecker removes parasites and pests (ticks and flies), thus getting food and a safe place to perch, while the rhino gets relief from its tiny adversaries (**Figure 29.28**).

There is also **commensalism**, meaning an interaction between two species in which one benefits while the other is neither harmed nor helped. Birds can make their nests in trees, and benefit from this, but generally don't affect the trees in any way. **Table 29.3** summarizes all the kinds of interactions you've gone over in this chapter.

Coevolution: Species Driving Each Other's Evolution

Earlier you read about the effect that predator-prey interactions have had on the evolutionary arms race among species. Yet it's clear that organisms have affected each other's evolution through mutualism as well. Indeed, species that are tightly linked in any way over a long period of time are likely to shape one another's evolution. Perceiving this, scientists Paul Ehrlich and Peter Raven developed the concept of **coevolution**, meaning the interdependent evolution of two or more species. Some of the clearest examples of coevolution involve flowering plants and the animals that pollinate them. Recall from Chapter 20 that many flowers have both fragrances that attract insects and ultraviolet color patterns (invisible to us) that guide the insects to the proper spot for pollination. These features are nature's own homing signals and landing lights, communicating the messages "Food lies this way" and "Land here" (**see Figure 29.29**). On the other side of the coin, consider that the color vision of honey bees is most sensitive to the colors that exist in the very flowering plants they pollinate. What seems likely is that honeybee eyesight evolved in response to plant coloration, and that plant coloration evolved to be maximally attractive to the pollinating insects. These groups of organisms coevolved, in other words.

Having considered how community members interact with one another, let's now go on to consider how communities change over time.

29.7 Succession in Communities

When Washington State's Mount St. Helens volcano erupted in May of 1980, it first collapsed inward, thus sending most of the

Table 29.3 Community Interactions		
In This Type of Interaction . . .	**One organism . . .**	**While the other . . .**
Competition	Is harmed	Is harmed
Predation and parasitism	Gains	Is harmed
Mutualism	Gains	Gains
Commensalism	Gains	Is unaffected

mountain's north face sliding downhill in the largest avalanche in recorded history. Then came the actual eruption, which sent a huge volume of rock and ash hurtling not straight up but out at an angle, toward the north. Some stands of forest in the path of this blast were instantly incinerated down to bare rock; another 35,000 hectares (86,000 acres) of trees were snapped in two like so many twigs. Then came the mudslides caused by the vast expanse of snow and ice melted in the explosion; then came the fall of hundreds of millions of tons of ash, some of it landing as far away as Wyoming. After viewing the devastated area around the blast, President Jimmy Carter said that it "makes the surface of the moon look like a golf course."

But today? No one would claim that the Mount St. Helens area has returned to anything like its former state, but look at the pictures in **Figure 29.30** to get an idea of the transition that has occurred at one location around Mount St. Helens.

The exact form of this rejuvenation may surprise us, but don't we intuitively expect something like this? Just from our everyday experience of watching, say, an abandoned urban lot becoming progressively weed-filled, don't we expect this would generally be the case? As it turns out, our intuition is right, because almost any parcel of land or water that has been either abandoned by humans or devastated by physical forces will be "reclaimed" by nature, at least to some degree.

The question is, how does this reclamation proceed? Areas that are rebounding don't just get one type of vegetation and then retain it. Instead, the vegetation changes through time—one type of growth *succeeds* another—with a general movement toward more and larger greenery as time goes by. Ecologists call this phenomenon **succession**, meaning a series of replacements of community members at a given location until a stable final state is reached. Within this framework, there are two kinds of succession. The first is **primary succession**, meaning succession in which the starting state is one of little or no life and a soil that lacks nutrients. Then there is **secondary succession**, meaning succession in which the final state of a habitat has been disturbed by some force, but life remains and the soil has nutrients. The classic example of this is land that has first been cleared for farming but later abandoned by the farmer.

The relatively stable community that develops at the end of any process of succession is called a **climax community**. There will be some shifts over time in a climax community, just as there are in any other. A prolonged drought will cause some change in the mix of the community's animal and plant life, for example. But, with some exceptions, what does not happen is a shift to a *fundamentally* different mix of life-forms, barring a major change in climate. Grassland stays grassland, and forest stays forest.

29.7 Succession in Communities

Figure 29.29
Coevolution of Plants and Their Pollinators
This flower has an ultraviolet color pattern that is not normally visible to humans, but that attracts bees. It is likely that these color patterns evolved in the flowers because they aided in attracting the bees, while the bees developed vision that was sensitive to the colors exhibited by these same plants.

a what we see

b what a bee sees

Figure 29.30
Rebound from Disaster
The Lang Ridge, 7 miles from Mount St. Helens crater, as it looked in August 1980, 3 months after the blast, and in 1999. This is an example of ecological succession in action.

Succession can be fairly predictable within small geographical areas. Two abandoned farm plots that lie close together will have very similar kinds of succession—so much so that local farmers or naturalists can sometimes tell to within a few years how long it's been since a field was abandoned. As you may be able to tell, succession generally is thought of in terms of the various *plant* communities that succeed one another, though modern ecology is attempting to bring animals into the picture as well.

An Example of Primary Succession: Alaska's Glacier Bay

One of the best examples of primary succession comes from Alaska's Glacier Bay, where glaciers have been retreating for some 200 years. Such withdrawal initially leaves behind a rocky terrain that is completely devoid of organic matter—no fallen leaves or branches, no decomposing animals. Indeed, what's left is "till" (not soil) that has a high pH and no available nitrogen—not a promising environment for life to establish itself. But establish itself it does. Glacier Bay's succession particulars—

what species succeeded what—are not important to us; what's important is the general process that occurred. Here's what happened in Glacier Bay after the glacier retreated (**see Figure 29.31**).

The first "pioneer" species to come were mosses and a non-woody plant that had the ability to grow essentially on rocks. Then came more plants, including an Arctic herbal rose, *Dryas*, that grows in a mat. *Dryas* and other pioneer plants have a capability that is critical not only to their own growth in these environments, but to the growth of plants that come after them. These pioneers harbor bacteria in their roots that can *fix nitrogen*—that can take nitrogen from the air and convert it into a form that plants can use. This is how the plant community could get started on the till in Glacier Bay, which *lacked* nitrogen.

Willows that hug the ground followed the pioneers, making a combined *Dryas* and willow mat that began to build up the decayed organic matter that helps produce soil. This community was succeeded by a plant called the alder bush that excluded almost

Figure 29.31
One Example of Primary Succession: Glacier Bay, Alaska

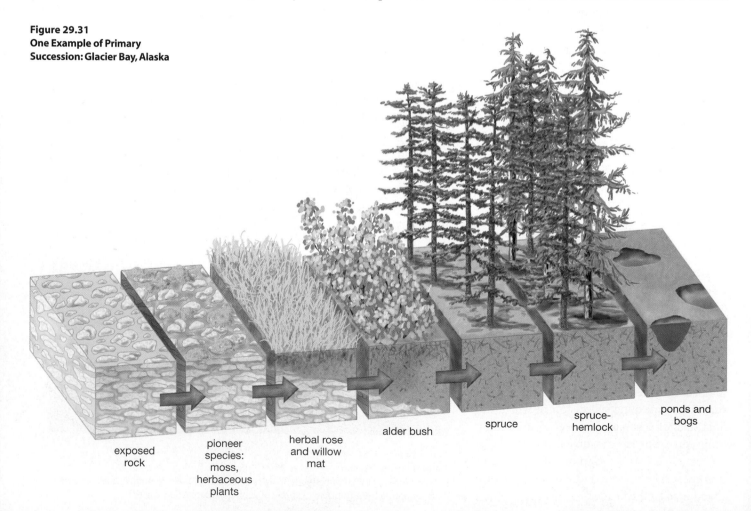

exposed rock

pioneer species: moss, herbaceous plants

herbal rose and willow mat

alder bush

spruce

spruce-hemlock

ponds and bogs

everything else—it was a highly successful dominant, in other words. Fifty years after the glacier's retreat, alders were growing in stands that were 10 meters, or about 33 feet tall. With their growth, the terrain got a great deal of nitrogen and, through acidic alder leaves that fell to the ground, a rapid lowering of pH.

The alder sowed the seeds of its own elimination, however. With the lowered soil pH, an extensive population of Sitka spruce trees could develop; when they did, they "shaded out" the smaller alders by blocking their sunlight. Some 120 years after the glacier's retreat, a spruce forest existed—one that, within 30 more years, started becoming a combined spruce-hemlock forest.

You might think that this would be the end of things; that the spruce-hemlock trees and the foliage around them would constitute the climax community. Instead, it was this forest that was replaced. Water-absorbing mosses eventually started to grow and, with them, the forest floor became watery, the tree roots couldn't absorb the oxygen they needed, and the trees died. What eventually was left was land that was filled with ponds and watery bogs. Whether *this* is the climax community will become known only with the passage of more time.

Common Elements in Primary Succession

Successions often are presented as ending with a mixture of tall forests and undergrowth, but there obviously are lots of different climax communities, of which bogs and ponds may be one. In the Glacier Bay succession you can, however, see developments that apply to many primary successions:

- An arrival of wind-borne "pioneer" plants in the form of seeds (or spores in other cases).
- A steady increase in biomass, compared to the starting state. This increase eventually stops in all successions; in Glacier Bay it actually was reversed with the invasion of the late-stage mosses.
- A general movement toward longer-lived species.
- A general trend toward an increase in species diversity, though not without some reversals (the alder bushes and the late-stage moss community).
- The facilitation of the growth of some later species (the Sitka spruce here) through the

actions of earlier species (the shedding of leaves by the alder bush). More formally, **facilitation** is a process within ecological succession in which the qualities or actions of some early-arriving species enable the success of later-arriving species.
- The competitive driving out of some species (the alder bush) by the actions of later species (the blocking of sunlight by the spruce trees).

Lessons in Succession from Mount St. Helens

Ecologists have been studying succession for about 100 years now, but it remains a field filled with basic questions. Spirited debate has gone on for years about what factors hold true in all instances of succession, and what models of development best describe succession. (Is facilitation more important than competition?)

It turns out, however, that in just the past 20 years, Washington's Mount St. Helens has provided a treasure trove of insight into how primary succession works. If there is a silver lining behind the Mount St. Helens explosion, it is that it has given ecologists an unparalleled opportunity to understand succession. This is so because the Mount St. Helens succession began on a crystal-clear starting date (May 18, 1980) and proceeded from such well-defined starting states—obliteration of some habitats, and major or minor damage to others.

So, what have we learned from Mount St. Helens? Perhaps the most important lesson is this: What is left from the past is crucial to the future. When trees were knocked over in the blast, their roots were yanked above-ground. In this moment of destruction, however, these roots literally carried the seeds of rejuvenation with them. Soil, small plants, and seeds were pulled up with the roots, and they stayed on them—above the deadening layer of ash below. Dead wood, left strewn about after the blast, provided habitat and food for wood-boring insects. These insects in turn attracted birds, who were able to build nests in "snags," meaning trees that were dead but standing. Gophers survived the blast in large numbers because they live below ground, where they feed on roots. They actually flourished in the years following the explosion, and their activity not only turned over the soil—mixing fertile soil below with ash on top—but also provided a kind of subway system for other animals who were migrating among patches of viable habitat (see Figure 29.32). So important were the gophers, snags, and surviving seeds to renewal at Mount St. Helens that a collective term for them has now entered the ecological lexicon. They are examples of **biological legacies**: living things, or products of living things, that survive a major ecological disturbance.

Succession at Mount St. Helens has yielded some other surprises as well. "Classical" primary succession, as outlined in the Glacier Bay example, assumes that life will come to a devastated area from the outside world, by means of seeds and spores being carried in on the wind. Though this happened at Mount St. Helens, in the main recovery was internal: thousands of small patches of habitat rejuvenated without being seeded from the outside. When organisms did come in from the outside, many were the "pioneer" species predicted by classical succession theory. However, ecologists also found solitary tree seedlings—supposedly late comers in succession—growing right out of some of the most devastated terrain in the blast area, with no pioneer species setting the stage for them. The second main lesson from Mount St. Helens, then, is that succession is more varied and habitat-specific than our theories had predicted. Put another way, we still have a lot to learn about succession.

On to Ecosystems and Biomes

In this chapter, you have looked at populations and communities. In populations, you saw the building blocks of communities. In communities, you looked through a lens of the interactions of living things. But it's obvious that *nonliving* forces and factors—weather, soil chemistry—interact with communities. When living things and nonliving factors are considered together, the resulting whole is called an ecosystem—the subject of the first part of Chapter 30. Very large-scale ecosystems are called biomes; a review of them forms the concluding portion of Chapter 30 and will bring to a close this book's coverage of ecology.

Figure 29.32
Important Survivor at Mount St. Helens

The pocket gopher (*Thomomys talpoides*) played a role in the restoration of the Mount St. Helens environment. The gopher's underground habitat allowed it to survive the blast, after which it continued to feed on roots, turning over soil in the process. In addition, its tunnels provided a passageway for other animals who, in the years following the explosion, were making their way to viable habitats. This gopher is in Wyoming.

Chapter Review

Summary

29.1 The Study of Ecology

- Ecology is the study of the interactions living things have with each other and with their environment.

- Ecology is not the same thing as environmentalism. The function of ecology is to describe interactions that affect the living world, not to work on behalf of environmentalism.

- There are five scales of life that concern ecology: physiology, populations, communities, ecosystems, and the biosphere. A population is all the members of a single species that live together in a specified geographical area; a community is all the species living in a single area; an ecosystem is a community and all the nonliving elements that interact with it; the biosphere is the interactive collection of all the Earth's ecosystems.

29.2 Populations: Size and Dynamics

- Ecologists employ several means to estimate the size of populations of living things, among them counting animal droppings or surveying bird populations as they migrate.

- An arithmetical increase occurs when, over a given interval of time, an unvarying number of new units is added to a population. An exponential increase occurs when the number of new units added to a population is proportional to the number of units that exists. Populations of living things are capable of increasing exponentially, because living things are capable of giving rise to more living things.
TUTORIAL 29.1.3: Exponential Growth

- The rapid growth that sometimes characterizes living populations is referred to as exponential growth, or as the J-shaped growth form. Populations that initially grow, but whose growth later levels out, have experienced logistic growth, sometimes referred to as the S-shaped growth form.
TUTORIAL 29.1.2: Types of Population Growth

- The size of living populations is kept in check by environmental resistance, defined as all the forces of the environment that act to limit population growth.

- Exponential growth in living populations can be calculated by subtracting a population's death rate from its birth rate, which yields the population's growth rate. Denoted as *r*, this rate is also known as the population's intrinsic rate of increase; it can be thought of as the population's potential for growth.
TUTORIAL 29.1.4: Intrinsic Rate of Increase in a Population

- Carrying capacity, denoted as *K*, is the maximum population density of a given species that can be sustained within a defined geographical area over an extended period of time.

29.3 *r*-Selected and *K*-Selected Species

- Different species have different reproductive strategies, meaning characteristics that have the effect of increasing the number of fertile offspring a species member will bear.

- Some species are said to be *K*-selected, or equilibrium, species. These species tend to be large, to experience their environment as relatively stable, and to lavish a good deal of attention on relatively few offspring. The pressures on *K*-selected species tend to be density dependent, meaning that as a population's density goes up, factors that limit the population's growth assert themselves ever more strongly.

- Other species are said to be *r*-selected or opportunist species. These species tend to be small, to experience their environment as relatively unstable, and to give little or no attention to the numerous offspring they produce. The pressures on *r*-selected species tend to be density independent, meaning pressures that are unrelated to the population's density.

- Survivorship curves describe how soon species members tend to die within the species' life span. There are three idealized types of survivorship curves: late loss, constant loss, and early loss.

29.4 Thinking about Human Populations

- Survivorship curves are created from life tables, which set forth the probabilities of a member of a species being alive after given intervals of time.

- An important step in calculating the future growth of human populations is to learn what proportion of the population is at or under reproductive age. A population pyramid displays this proportion. Populations whose pyramids are heavily weighted toward younger age groups are likely to experience relatively large growth.

- Immigration can have a large impact on a population. Because of immigration, the population of the United States is projected to grow much more rapidly in the coming decades than it would without immigration.

- The populations of less-developed nations are in general growing at a much faster rate than the populations of developed nations.

- The Earth's human population has increased greatly in the last 50 years and now stands at about 6 billion people.

There is a debate over whether the Earth has reached its carrying capacity for the human population.

- Some scientists believe that there is no greater single threat to the environment than the continued growth of the human population. Others argue that a more important concern is the use of natural resources per person.

29.5 Communities: Looking at the Interactions of Many Populations

- An ecological community is all the populations that inhabit a given area, though the term can be used to mean a collection of populations in a given area that potentially interact with one another.

- Most communities tend to be dominated by only a few species; the few species that are abundant in a given area are called ecological dominants.

- Keystone species are species that may not be numerous in a given area, but whose absence would bring about significant change in the community.

- Some keystone species are top predators in a community, meaning species that prey upon other species but that are not preyed upon themselves. Keystone species need not be top predators, however, nor does every community have a keystone species.

- Biodiversity is variety among living things. It can take three primary forms: a diversity of species in a given area; a distribution of species across the Earth; and genetic diversity within a species.

29.6 Types of Interaction among Community Members

- There are three primary types of interaction among community members: competition; predation and parasitism; and mutualism and commensalism.

- Habitat is the physical surroundings in which a species can normally be found.

- Niche can be defined metaphorically as an organism's occupation, meaning what the organism does to obtain the resources it needs to live.

- The competitive exclusion principle states that when two populations compete for the same limited, vital resource, one always outcompetes the other and thus brings about the latter's local extinction.

- There are numerous instances in nature in which two related species use the same kinds of resources from the same habitat over an extended period of time. Neither species undergoes local extinction in this interaction, because neither is superior in competing for all resources; this leads to coexistence through resource partitioning.

- Predation is defined as one free-standing organism feeding on parts or all of a second organism. Parasitism is a variety of predation in which the predator feeds on prey, but does not kill it immediately, if ever.
TUTORIAL 29.2.3: A Simple Mathematical Predator-prey Model

- Predation has done much to shape the natural world.

- The population dynamics of a predator and its prey can be linked, but predator-prey interaction generally is only one of several factors controlling the population level of either group.

- The prey of a parasite is known as the host. Hosts may be used by parasites not only for food, but to facilitate parasite reproduction.

- Mimicry is a phenomenon by which one species has evolved to assume the appearance of another. Species that have evolved in this way generally suffer less predation as a result. Batesian mimicry is the evolution of one species to resemble a species that has a superior protective capability. In Müllerian mimicry, several species that have protection against predators come to resemble each other.

- Mutualism is an interaction between individuals of two species that is beneficial to both individuals. Commensalism is an interaction in which an individual from one species benefits while an individual from another species is neither harmed nor helped.

- Coevolution is the interdependent evolution of two or more species. Flowers have evolved colors and fragrances that attract bees, for example, while bees have evolved vision that is most sensitive to the colors in the flowers they pollinate.

29.7 Succession in Communities

- Parcels of land or water that have been abandoned by humans or devastated by physical forces will almost always be reclaimed by nature to some degree. The process by which this takes place is called succession: a series of replacements of community members at a given location until a stable final state is reached.

- Primary succession proceeds from an original state of little or no life and soil that lacks nutrients. Secondary succession occurs when a final state of habitat is first disturbed by some outside force, but life remains and the soil has nutrients. The final community in any process of succession is known as the climax community.

- A common set of developments occurs in most instances of primary succession, including the facilitation of the growth of some later species through the actions of earlier species, and the competitive driving out of some species by the actions of later species.

- The rejuvenation of the Mount St. Helens area that has occurred since 1980 has provided ecologists with a wealth of information regarding both primary and secondary succession. One of the chief lessons learned concerns the importance of biological legacies, meaning living things, or products of living things, that survive a major ecological disturbance. Biological legacies have been critical to the rapid renewal of habitat at Mount St. Helens.

Key Terms

arithmetical increase 664
Batesian mimicry 681
biodiversity 675

biological legacies 686
biosphere 663
carrying capacity (*K*) 667
climax community 683
coevolution 682
coexistence 677
commensalism 682
community 663
competitive exclusion principle 677
density dependent 669
density independent 669
ecological community 674
ecological dominants 674
ecology 662
ecosystem 663
emigration 671
environmental resistance 666
equilibrium species 668
exponential growth 666
exponential increase 664
facilitation 685
habitat 676
host 679

immigration 671
interspecific competition 676
intrinsic rate of increase (*r*) 667
J-shaped growth form 666
keystone species 674
K-selected species 669
life table 670
logistic growth 666
mimicry 681
Müllerian mimicry 681
mutualism 681
niche 676

opportunist species 669
parasitism 678
physiology 662
population 662
predation 678
primary succession 683
resource partitioning 677
r-selected species 669

secondary succession 683
S-shaped growth form 666
succession 683
territorial 676
top predator 674
zero population growth 667

Understanding the Basics

Multiple-Choice Questions

1. A(n) _____ consists of members of a single species that live in a specific geographic region, while a _____ consists of the populations of all species that live in a single region.
   a. ecological dominant, community
   b. ecological dominant, biosphere
   c. community, ecosystem
   d. population, community
   e. population, ecosystem

2. A community and all the nonliving elements that interact with it (sunlight, nutrients, soil, etc.) is called
   a. a population group
   b. a biosphere
   c. an ecosystem
   d. a biome
   e. a social system

3. When population growth increases in proportion to the number of members of a population, _____ growth is said to have occurred.
   a. density-dependent
   b. density-independent
   c. *K*-selected
   d. arithmetical
   e. exponential

4. The use of pesticides in many agricultural communities has led to a great reduction in many "targeted" insects. However, many other "nontarget" insects have declined as well. In particular, ecologists and fruit growers have noticed that certain species of bees that pollinate valuable plants have declined, leading to a decrease in many bee-pollinated species. In communities where great change has occurred due to the decline in this ecosystem function, bees may be
   a. ecological dominants
   b. bottleneck species
   c. keystone species
   d. top predators
   e. parasitoids

5. An interaction between two species in which one species benefits while the other is neither harmed nor helped is called
   a. parasitism
   b. mutualism
   c. mimicry
   d. commensalism
   e. cooperation

6. In the process of nitrogen fixation, plants contribute food to bacteria that are capable of fixing atmospheric nitrogen gas, and the bacteria can supply the plants with a usable form of nitrogen. This is an example of
   a. parasitism
   b. mutualism
   c. fertilization
   d. mimicry
   e. cooperation

7. The process by which one species evolves to resemble another species is called
   a. mimicry
   b. mutualism
   c. behaviorism
   d. parasitism
   e. natural deception

8. The largest scale of life, which can be thought of as the interaction of all of Earth's ecosystems, is called
   a. the biome
   b. the ecosphere
   c. the biosphere
   d. the environment zone
   e. the ionosphere

Brief Review

1. Name the four levels of group organization of living things, in order of increasing size.

2. Compare and contrast *r*-selected and *K*-selected species, and give an example of each. Are humans *r*-selected or *K*-selected?

3. Explain what is meant by environmental resistance and its relationship to population growth.

4. The red-spotted newt *Notophthalmus viridescens* secretes toxins from its skin that make predators avoid it. The red salamander *Pseudotriton ruber* has no such secretions but resembles the red-spotted newt and gains protection from this resemblance. What is this an example of? Which is the model, the mimic, and the dupe?

5. A volcano erupted in Hawaii, and lava covered land that had been farmed for centuries. Eventually, new lichens, and then plants, grew on the lava. Is this an example of primary or secondary succession?

Applying Your Knowledge

1. Few species can continue exponential growth for very long, because environmental resistance will assert itself. However, the human population has been growing exponentially in the last few decades. Do you think that human populations are truly without an upper bound, or do you think that carrying capacity (*K*) is just always rising? What evidence do you see that humans might eventually reach an absolute upper limit (other than that of pure physical space)? How does the emergence of disease, or epidemics, factor into this discussion?

2. What is a keystone species? How do you think we can tell if a particular endangered species is a keystone species—such as the spotted owl in the Pacific Northwest, or the California condor? Should a species' role as a keystone species in a community make us increase efforts to protect it? Explain, and give several arguments for (or against) laws that protect individual species.

3. Some scientists argue that no single factor poses a greater threat to the environment than the continued growth of the human population, while others believe the use of resources per person is a more critical factor. What arguments can you think of for or against either proposition?

4. In steep, mountainous terrain, logging of forests is thought to cause an increase in the chance of landslides. When landslides occur, most of the topsoil and parent rock fragments slide downhill. Is the regrowth on this newly exposed terrain an example of primary or secondary succession? Would you expect the vegetation to grow faster or slower than the nearby land that was harvested, but that did not experience a landslide?

MediaLab

Can Earth Support All of Us? Population Growth Patterns

Nineteenth-century economist Thomas Malthus noted that populations increase geometrically until they encounter a factor that limits continued population growth. We know some populations grow rapidly, others slowly. Since 1800 the number of people on Earth has increased dramatically, with more yet to come. How can we accurately predict the effect of this continued growth? This *CD-ROM Tutorial* reviews concepts about population growth. The following *Web Investigation* lets you examine a variety of population growth situations and the problems they cause. In *Communicate Your Results*, you can apply these concepts.

This *MediaLab* can be found in Chapter 29 on your CD-ROM (Tutorial 29.1.2) and Companion Website (http://www.prenhall.com/krogh3).

CD-ROM TUTORIAL

Species can differ greatly in the type of population growth they can experience. In this *CD-ROM Tutorial*, you'll investigate the factors controlling growth and how to use these factors to predict future population increases.

Activity

1. *First, you will use an interactive program where you can enter factors like births and deaths and determine a population's growth rate.*

2. *Then, you will look at the three different models of growth seen in natural populations.*

3. *Finally, you will see what factors are important for achieving a negative, low, or high intrinsic growth rate or even zero growth.*

The following *Web Investigation* presents examples of population growth in both human and animal populations, so that you can see how the characteristics of a species' growth influence how slowly or rapidly its population increases.

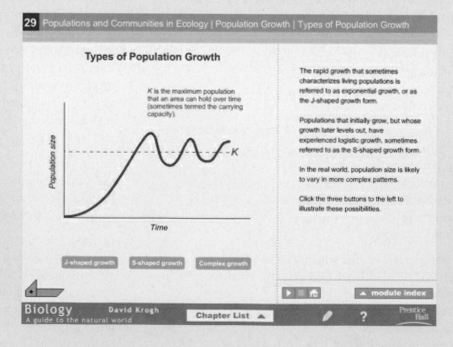

WEB INVESTIGATION

Investigation 1

Estimated time for completion = 15 minutes

Select the Keyword **GROWTH RATE** on your CD or Website for this *MediaLab* and read about the population growth rate. Notice how the human population has changed over time. Would you say human population is growing arithmetically (straight line) or geometrically (curved line)? What factors are increasing or slowing the rate of increase?

Investigation 2

Estimated time for completion = 10 minutes

Tour an online exhibit about population growth by selecting the Keyword **BIRTH RATE** on your CD or Website for this *MediaLab*. What factors affect the number of children a woman can have in your area of the planet? Sign in and choose another continent, and see how geography might affect the number of children a woman might have.

Investigation 3

Estimated time for completion = 20 minutes

"So what—we have more people on the planet. That doesn't concern me!" Perhaps it should. Select the Keyword **OVERPOPULATION** on your CD or Website for this *MediaLab* and read the project description. This project describes some issues that ought to concern all of us, no matter what country we are living in. Consider recent news reports. Are there any signs that population pressures, as detailed at this site, might be factors in recent aggression?

Now that you've seen how populations grow and how increasing populations might be factors in social and ecological problems, do the following exercises to determine the severity of increasing populations.

COMMUNICATE YOUR RESULTS

Exercise 1

Estimated time for completion = 5 minutes

You have seen how populations grow. Plot the following data on a sheet of graph paper (or enter it into a program).

Year	Estimated population (in millions)
8000 BC	5
1750 AD	75
1950	2,500
1975	3,700
1990	5,600

From the *CD Activity*, you will recall that humans are a *K*-selected species. Does the graph you produced look like a *K*-selected species? What developments or events might account for the shape of the human population graph?

Exercise 2

Estimated time for completion = 5 minutes

Since the 1970s, the Chinese government has promoted a policy of one child per family. Apply what you learned in the first two sections of this *MediaLab* to reconstruct the reasons why the Chinese, who traditionally valued children and family, undertook such a harsh policy. Would that policy work where you live? What measures would be needed to make such a policy work?

Exercise 3

Estimated time for completion = 5 minutes

What should be done about overpopulation? Birth rates are declining in most countries, so perhaps there is no reason to worry. Get together with some of your classmates, divide into teams, and take "sides" about human population growth, presenting evidence in a limited debate.

30

An Interactive Living World
Ecosystems and the Biosphere

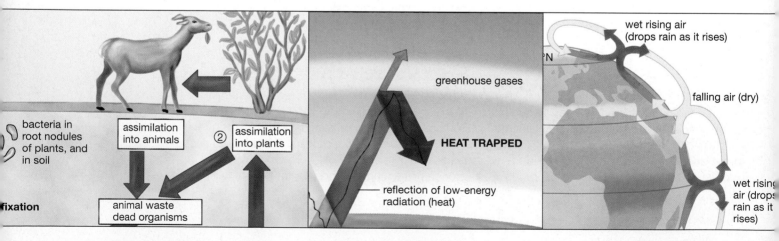

The nitrogen cycle.
(Section 30.2, page 696)

Greenhouse effect.
(Section 30.4, page 709)

Making weather.
(Section 30.4, page 711)

The living world is affected by the larger physical world around it—and vice versa. The interconnected web of living things is being damaged in many ways by human activity.

The word "system" is much in use today. We hear about a given country's political system or its economic system. The term was first used in this context by academics who needed a way to refer to complex entities composed of many interacting parts. This way of looking at things is applicable in ecology as well as in economics or politics. In Chapter 29, when you went over ecological communities, you actually were studying something in isolation: the interactions of all the *living* things in a given area. It goes without saying that these living things also interact with many nonliving entities. To get a complete picture of how the living world functions, then, it's necessary to understand how living things interact not only with each other, but with these nonliving elements.

30.1 The Ecosystem Is the Fundamental Unit of Ecology

What do nonliving elements add to life? All communities need an original source of energy, generally meaning the Sun; they all need a supply of water and a supply of nutrients, such as nitrogen and phosphorus; and all organisms need to exchange gases with their environment (think of your own breathing). The Sun's energy is making a one-way trip through the community, ultimately being transformed into heat; but water and nutrients are being *cycled*—from the Earth into organisms and then back again into the Earth. Looking at this, we can begin to perceive the reality of ecological *systems*: of functional units that tightly link both living or **biotic** factors and nonliving or **abiotic** factors to form a whole.

Given this, the fundamental unit of ecology is the **ecosystem**, which can be defined as a community of organisms and the physical environment with which they interact. This chapter begins with a look at small-scale ecosystems and concludes with a look at the very large-scale ecosystems called biomes. In today's world, it's impossible to note the ways in which ecosystems function without also noting the ways in which human activity is damaging them.

The taiga.
(Section 30.5, page 715)

King penguins off Antarctica.
(Section 30.6, page 720)

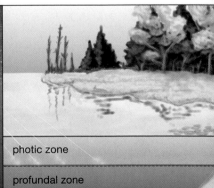

photic zone

profundal zone

Zones in a lake.
(Section 30.6, page 723)

30.2 Abiotic Factors Are a Major Component of Any Ecosystem

In Chapter 29 you learned a good deal about two of the biotic factors of the ecosystem, namely individual populations of living things and the communities that are made up of these populations. This chapter focuses on some of the abiotic factors in an ecosystem. In overview, these abiotic factors fall into two categories. First are the *resources* that exist in the ecosystem, such as water and nutrients; second are the *conditions* in which an ecosystem exists, such as average temperature. Resources will be considered first in this chapter and conditions later.

The Cycling of Ecosystem Resources

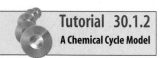

Tutorial 30.1.2
A Chemical Cycle Model

Those of you who went through Chapter 2, on chemistry, know what a chemical **element** is: a substance that is "pure" in that it cannot be broken down into any other component substances by chemical means. The 92 stable elements include such familiar substances as gold and helium, but only 30 or so of these elements are vital to life and are thus called **nutrients**. Some nutrients, such as iron or iodine, are needed only in small or "trace" quantities by living things, but others are needed in large quantities, among them carbon, oxygen, nitrogen, and phosphorus. Living things also have a great need for water, which is not an element but a molecule composed of two elements, hydrogen and oxygen. Water and nutrients move back and forth between the biotic and abiotic realms, with the term for this movement being a mouthful: **biogeochemical cycling.** Let's look now at the cycling of two elements, carbon and nitrogen, and then at the cycling of water.

Carbon as One Example of Ecosystem Cycling

A certain amount of cosmic debris makes its way to Earth's surface in the form of meteorites, but this material is pretty sparse. What else comes to Earth from space? The Sun's rays, to be sure, as well as light from distant stars, and some other forms of radiation. But when we're talking about chemical elements, the Earth really is a spaceship: It carries a fixed amount of resources with it. Nothing comes to Earth *from* the outside, and relatively little leaves Earth *for* the outside. For an ecologist, this self-containment simplifies things: What we possess in terms of elements is all we'll ever possess. Thus the question becomes not so much what we have, but where we have it.

In recent years there has been considerable concern about the buildup of the heat-trapping gas carbon dioxide (CO_2) in the Earth's atmosphere, as this increase is now regarded by most scientists as a cause of global warming. Well, if Earth has this fixed quantity of elements—in this case the carbon and oxygen that make up CO_2—how could there be a buildup of carbon dioxide in the atmosphere? The answer is that carbon has been *transferred* from one place on Earth to another. It has moved from the "fossil fuels" of coal and oil, where it was stored, into the atmosphere.

Storage and transfer; these two concepts are fundamental to biogeochemical cycling. When we look a little deeper into carbon cycling, we find other players involved in carbon's storage and transfer. Plants need CO_2 to perform photosynthesis, and they take in great quantities of it for that purpose, producing their own food as a result. While the plants are alive, they use part of the carbon they take in to grow—to build up leaves and stems and roots. And as long as these structures exist, they will store carbon within them, just as coal and oil do. (Thus, one of the proposed solutions to global warming is to grow a huge number of trees, each of which could be thought of as a kind of piggy bank for carbon.) Eventually the plants die, of course, after which their decomposition by bacteria and fungi releases carbon into the soil and atmosphere, again as CO_2. With this, you can begin to see the whole of the carbon cycle in the natural world (**see Figure 30.1**).

Some plants will, of course, be eaten by *animals*, who need carbon-based molecules for tissues and energy, just as plants do. The difference is that animals get these molecules *from plants*—in the form of seeds and leaves and roots—rather than from the air. These animals will likewise lock up carbon for a time, only to return it to the soil with their death and decomposition. Note, however, the critical difference between animals and plants: Carbon comes into the living world only through plants (and their fellow photosynthesizers algae and some bacteria). An animal might take in some carbon dioxide in breathing, but nothing life-sustaining happens as a result. Conversely, a plant that takes in carbon dioxide can *incorporate* it into a life-sustaining molecule, in this case a sugar that is the initial product of photosynthesis. With this sugar as a food source, plants grow, and this bounty ultimately feeds everything else. Look at this simplified cycle:

atmospheric CO$_2$ → plants → animals →
decomposers → atmospheric CO$_2$

What does it track, the path of carbon or the path of food? The answer is both. This is one of the reasons that biologists refer to life on Earth as "carbon-based."

How Much Atmospheric CO$_2$? In following the carbon trail, it's probably apparent that, just as money is a medium of exchange in our economy, CO$_2$ is the living world's medium of exchange for carbon; photosynthesizers take *in* carbon as CO$_2$, and decomposers yield *up* carbon as CO$_2$. (All organisms also release CO$_2$ as a by-product of carrying out life's basic processes, as with your own exhalations in breathing.) Given how much **biomass** or living material there is in the

Tutorial 30.1.1
The Carbon Cycle

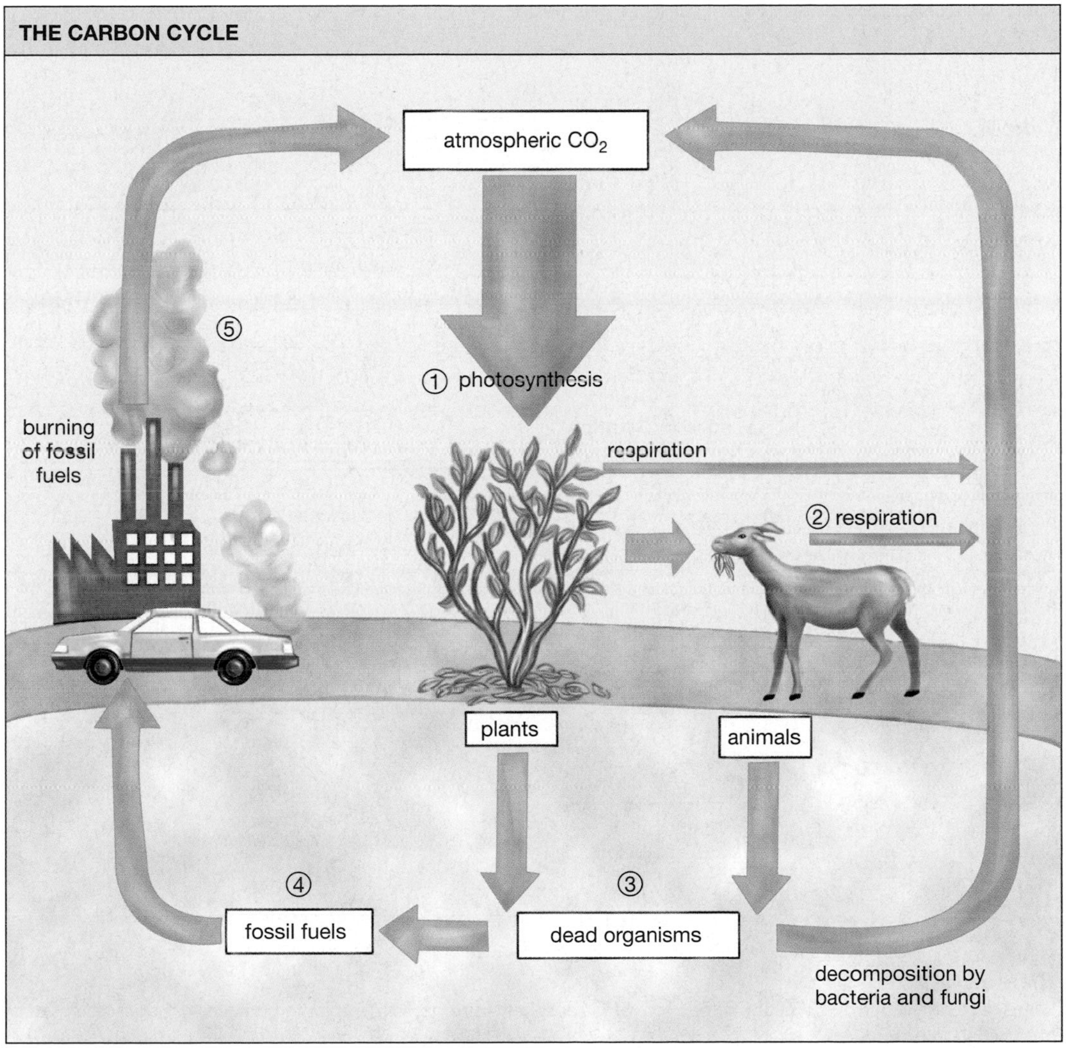

THE CARBON CYCLE

Figure 30.1
The Carbon Cycle

1. Plants and other photosynthesizing organisms take in atmospheric carbon dioxide (CO$_2$) and convert or "fix" it into molecules that become part of the plant.

2. The physical functioning or respiration of organisms converts the carbon in their tissues back into CO$_2$.

3. Plants and animals die and are decomposed by fungi and bacteria. Some CO$_2$ results, which moves back into the atmosphere.

4. Some of the carbon in the remains of dead organisms becomes locked up in carbon-based compounds such as coal or oil.

5. The burning of these fossil fuels puts this carbon into the atmosphere in the form of CO$_2$.

world, you might think there would have to be a great deal of CO_2 in the atmosphere, but as a proportion of atmospheric gases, CO_2 is a bit player, making up about 0.035 percent of the atmosphere. But what an important player it is! It's critical in feeding us all, and it's likely that relatively small changes in its atmospheric concentration are now leading to significant changes in global temperature.

The Nitrogen Cycle

Like carbon, nitrogen is an element; and like carbon, nitrogen cycles between the biotic and abiotic domains. Nitrogen makes up only a small proportion of the tissues of living things, but it is a critical proportion because it is required for DNA, RNA, and all proteins. All living things need this element, then. The question is, how do they get it?

Tutorial 30.2.1
The Nitrogen Cycle

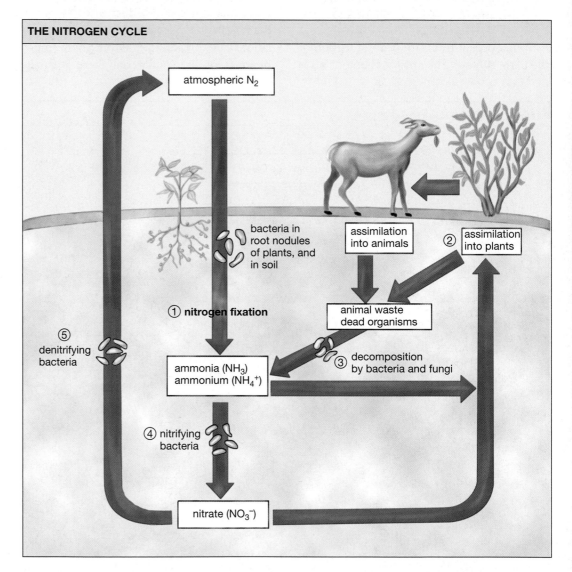

Figure 30.2
The Nitrogen Cycle

Some bacteria "fix" nitrogen, meaning they convert atmospheric nitrogen (N_2) into an organic form that can be used by other living things.

1. Nitrogen-fixing bacteria convert N_2 into ammonia (NH_3), which converts in water into the ammonium ion (NH_4^+). The latter is a compound that plants can assimilate into tissues. In the diagram, bacteria living symbiotically in plant root nodules have produced NH_4^+, which their plant partners have taken up and used. Meanwhile, free-standing bacteria living in the soil have likewise produced NH_4^+.

2. Other plants take up NH_4^+ that has been produced by soil-dwelling bacteria and assimilate it. Animals eat plants and assimilate the nitrogen from the plants.

3. Animal waste and the tissues of dead animals and plants are decomposed by fungi and by other bacteria, which turn the organic nitrogen back into NH_4^+.

4. Other "nitrifying" bacteria convert NH_4^+ into nitrate (NO_3^-), which likewise can be assimilated by plants.

5. Some nitrate, however, is converted by "denitrifying" bacteria back into atmospheric nitrogen, completing the cycle.

The source for all the nitrogen that enters the living world is atmospheric nitrogen, which exists in great abundance. You've seen that carbon dioxide makes up less than 1 percent of the atmosphere, and it turns out that oxygen makes up about 21 percent. But nitrogen makes up 78 percent of the atmosphere. So rich is the atmosphere in this element that a typical garden plot has tons hovering above it. The problem? All this nitrogen is in the form of pairs of nitrogen atoms (N_2) that have a great tendency to stay together, rather than combining with anything else. Thus, in a case of so near and yet so far, plants have no direct access to atmospheric nitrogen. So how do they take it up? In the natural world, the answer essentially is: through the actions of bacteria. It is bacteria that are carrying out the process of **nitrogen fixation**, meaning the conversion of atmospheric nitrogen into a form that can be taken up and used by living things.

The Process of Bacterial Fixation of Nitrogen

You can see nitrogen fixation and the rest of the nitrogen cycle diagrammed in **Figure 30.2**. The essence of the cycle is that several types of nitrogen-fixing bacteria take in atmospheric nitrogen (N_2) and convert it into ammonia (NH_3). This ammonia then is converted (either in water or by other bacteria) into two types of nitrogen-containing compounds that plants can *assimilate*—can take up and use. And as you've seen, what comes into the plant world will come into the animal world, when animals eat plants. The two usable nitrogen compounds produced in the cycle are the ammonium ion (NH_4^+) and nitrate (NO_3^-). Some of the nitrate that results from this process is used by yet another kind of bacteria, denitrifying bacteria, which can convert nitrate back into atmospheric nitrogen, completing the cycle.

Nitrogen as a Limiting Factor in Food Production

If you were to look at the nitrogen story 100 years ago, there would be little more to it than what you've just gone over. In other words, through most of human history, nitrogen was fixed solely by bacteria (with a small assist from lightning, which can produce a usable form of nitrogen that falls to the Earth in rain). Human beings had a problem with this single route to nitrogen, however, because nitrogen is critical for all plant growth,

**Figure 30.3
Nutrients Beyond What
Nature Provides**
A helicopter applies fertilizer to a sugar beet crop in California.

including *agricultural* plant growth. Indeed, for centuries a lack of nitrogen was a primary limiting factor in getting more food from a given amount of land. Long before anyone knew what nitrogen was, farmers were trying to get more of it to crops in two ways. One was by applying organic fertilizer, such as rotting organic material, to their fields. (Remember how the Pilgrims were taught by the Indians to bury dead fish around their corn plants?) The second was by planting crops that carried their own nitrogen-fixing bacteria within them. This is the case with certain legumes, such as soybeans, which carry symbiotic bacteria in their root nodules.

Early in this century, however, a momentous change came about when the German chemist Fritz Haber developed an *industrial* process for turning atmospheric nitrogen into ammonia. In essence, human beings became nitrogen fixers and, by the 1960s, had taken up this activity on a grand scale. In 1990 about 80 million metric tons of biologically active nitrogen were manufactured for use as fertilizer; in addition, human beings planted crops that resulted in the production of another 40 million tons (**see Figure 30.3**). This human-driven production was roughly equal to the amount of active nitrogen that all of nature produced on its own.

It is hard to overstate the importance of the industrial fixation of nitrogen. Put simply, the world's human population could not be fed without such fixation. Only with synthetic nitrogen can the Earth's arable land produce enough food to feed the Earth's burgeoning human population.

The massive human intervention in the nitrogen cycle has a downside as well, however. First, remember that human-manufactured nitrogen ends up being used in a very *concentrated* way; it is poured onto the relatively small patches of earth we call farms. A good deal of this nitrogen does not end up in the crops it was intended for, but instead departs as runoff. Later in this chapter you'll consider a phenomenon called eutrophication, in which lakes can contain too many nutrients (such as nitrogen), sometimes resulting in enormous fish kills. Nitrogen application can also make the soil too acidic, and it contributes to smog and to global warming.

The Cycling of Water

Now let's turn from nutrient to water cycling. Those who went through the material on water in Chapter 3 know how important water is to life. The human body is about 66 percent water by weight, and any living thing that does not have a supply of water is

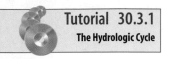

Tutorial 30.3.1
The Hydrologic Cycle

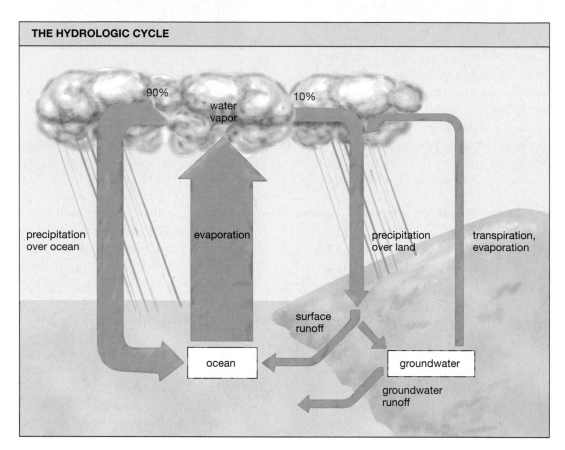

Figure 30.4
The Hydrologic Cycle
More than 95 percent of Earth's water is stored in the oceans. When water evaporates from the ocean, 90 percent returns to the ocean directly by way of precipitation. The other 10 percent falls on land. There, the water either runs back into the ocean, is stored in such structures as glaciers, or is moved by transpiration and evaporation back into the atmosphere.

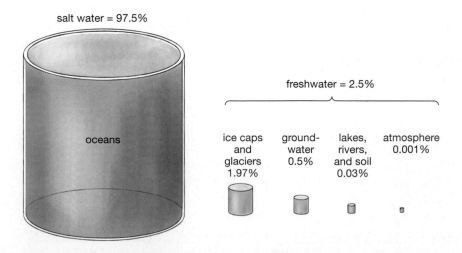

Figure 30.5
Earth's Water
Although there is a tremendous amount of water on our planet, only a small fraction of it is available to us as freshwater.

doomed in the long run. As with nitrogen or carbon, all water is either being cycled or is locked up. Carbon may be stored in coal or trees, but water can be stored in ice—in glaciers or in polar ice, for example.

The oceans hold more than 95 percent of Earth's water, but thankfully for us, when the oceans' salt water evaporates, it falls back to Earth as freshwater. We can also be grateful that not all ocean water falls back to Earth on the oceans; instead, about 10 percent falls on land (**see Figure 30.4**). This water represents about 40 percent of the precipitation that land areas get; the other 60 percent comes from a form of cycling reviewed in Chapter 22: *transpiration*, meaning water that is taken up by the roots of land plants, then moved up through their stems and out through their leaves as water vapor. After evaporating into the atmosphere, water returns to Earth as rain, fog, snow—all the forms of precipitation possible. The driving force behind the hydrologic cycle is the energy of the Sun; its heat powers both evaporation and transpiration.

With more than 95 percent of Earth's water in the oceans, perhaps as little as 2.5 percent of the Earth's water is freshwater at any given time. Of this freshwater, more than 75 percent is locked up in glaciers and other forms of ice (**see Figure 30.5**). Thus any snapshot we would take of Earth would show that as little as 0.5 percent of its water is available as liquid, freshwater; but even here we have to qualify things. About 25 percent of this freshwater is **groundwater**, meaning water that moves down through the soil until it reaches porous rock that is saturated with water. In line with this, about 20 percent of the water used in the United States comes from groundwater. The porous rock that groundwater is stored in is called an **aquifer**, which you can see in **Figure 30.6**. Major aquifers can be enormous, as you can see from the map of the High Plains Aquifer in **Figure 30.7**.

Human Use of Water

Given the sheer number of people on Earth and the importance of water to each of them, it will not surprise you to learn that human beings are laying claim to a great deal of Earth's freshwater; civilization now uses about half the world's accessible supply. But from what you've seen so far, think of what this means. There is no less *water* than there ever was. Like nitrogen or carbon, all the water we have is either being

recycled or is stored. What's changed is that the human population is using an ever-greater proportion of the water that's available. Human beings fight among themselves for water, of course, because it is a scarce commodity in

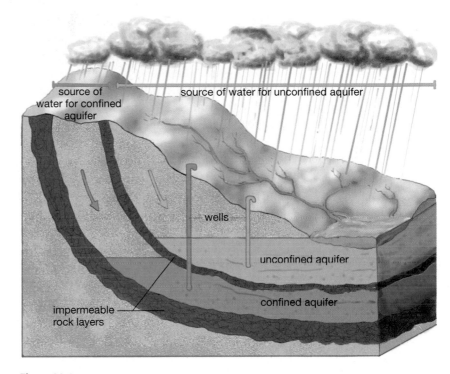

Figure 30.6
Aquifers Store Groundwater
When water seeps into the ground, it moves freely through layers of sand and porous rock but moves either extremely slowly or not at all through layers of impermeable rock. Water thus becomes trapped in different underground layers, from which humans can draw water. Unconfined aquifers receive water directly from a large surface area and therefore tend to be more vulnerable to pollution by chemicals such as fertilizers and pesticides. Confined aquifers are located between layers of impermeable rock. They are replenished more slowly but tend to contain purer water.

Figure 30.7
Enormous Store of Underground water
The High Plains Aquifer underlies an expanse of land stretching across eight states. Billions of gallons of water are withdrawn from it each day, mostly for agriculture. (Modified from Gutentag, E. D., Heimes, F. J., Krothe, N. C., Luckey, R. R., and Weeks, J. B., 1984. Geohydrology of the High Plains Aquifer in Part of Colorado, Kansas, Nebraska, New Mexico, Oklahoma, South Dakota, Texas, and Wyoming. U.S. Geological Survey Professional Paper 1400-B.)

locations throughout the world (**see Figure 30.8**). Close to home, parts of California's Owens Valley that once were farming communities now are desert-like because of the diversion of Owens River water to Los Angeles.

One part of the human water problem, water expert Peter Glieck has noted, is that humans make such poor use of the water they capture. Mexico City's leaky water system *loses* enough water to meet the needs of a city the size of Rome. If the toilet in your house is more than 10 years old, chances are it uses six gallons of water with each flush. If it was built in the 1990s, following imposition of new federal standards for low-flow toilets, it probably uses less than a third of that—a mere 1.6 gallons.

Human beings are, of course, capable of diverting water from species that have no say in this division of resources. More than 20 percent of all freshwater fish species are estimated to be threatened or endangered precisely because their water has been dammed or diverted for human use. California's natural populations of coho salmon have been reduced by some 94 percent since the 1940s, in significant part through diversions of the freshwater streams that the salmon swim up in order to spawn.

If you look at **Figure 30.9**, you can get a quick idea of humanity's impact on the three resources just reviewed—carbon, nitrogen, and water.

Human Beings Are Not Separate from the Earth They Live on   One final thing about water and nutrient cycling. The average person, who has no knowledge of these cycling processes, may regard human beings as a kind of independent line that stretches back in time. Under this view, water and food may pass through us, but they are as separate from us as gasoline is from a car engine.

From your reading of this section, however, it may now be apparent to you that this is not true. Some of the carbon in the bread you ate this morning could have been contained in a tree that decomposed in China 100 years ago. After many cycles, carbon from this tree was taken up by a wheat plant in America whose seeds were ground up, yielding flour that ended in the bread on your shelf. This carbon may be burned by your body to supply the energy it needs, it's true, but it may also become part of your bones or your nerves. Then, when you die, the material you are made of will go *back* to the Earth to be recycled in the ways you've been reading about, perhaps to be part of a tree or a person in a future generation. We are intertwined with the Earth in some basic, physical ways.

30.3   How Energy Flows through Ecosystems

Having looked at the abiotic elements of nutrients and water, let's now turn to another important component of any ecosystem, energy. Those of you who read through Chapter 6, on energy, know that one of energy's inflexible laws—the first law of thermodynamics—is that energy is never gained or lost, but only transformed. The Sun's energy is not used up by green plants; rather, some of it is *converted* by plants into chemical form—initially into the chemical bonds of the sugar that plants make in photosynthesis.

The second law of thermodynamics is that energy spontaneously flows in only one direction: from more ordered to less ordered. The carbohydrate molecules in bread are very ordered things—so many atoms of carbon, oxygen, and hydrogen in a precise spatial relationship to one another. Contrast this with another form of energy, heat, which is the *random* motion of molecules—clearly a disordered form of energy compared to a carbohydrate. Indeed, heat is the last stop on the energy line, because it is the least-ordered form of energy. And, because all energy spontaneously moves toward less order, heat is the ultimate *fate* of all energy. Energy relentlessly ratchets down toward the form of heat because, like a casino that must get a "cut" of

**Figure 30.8
Freshwater Is a Limited Resource**

Women carry drinking water through a polluted slum in Port-au-Prince, Haiti.

each bet laid down, heat gets a cut of each energy transaction. The chemical bonds in gasoline can be broken through combustion and thus be used to power a car engine, but not all the energy released from the combustion drives the engine. Some dissipates as heat, and this happens *every* time energy is used, whether we're talking about a piston firing or a cell dividing.

All of this provides a framework to conceptualize an important part of ecosystems, which is how energy flows through them. Think of the Sun's energy, with the energetic rays that leave it as a starting point; now think of the ultimate destiny of all this energy, which is heat, randomly dispersed in the universe. Looked at one way, life on Earth *intervenes* in this flow. Life is an enormous energy collection and storage enterprise, gathering some of the Sun's energy and locking it up for a time in the form of chemical bonds. These bonds can then be broken—think of digesting a muffin—which *releases* this stored energy so that an organism can grow and reproduce. Life does not stop the march of the Sun's energy into heat, but it does intervene in it by transforming it into chemical bonds that have order and stability.

Once we see life in these terms, we can begin to think of the *flow* of energy through it. That's what you'll be looking at in the sections that follow.

Producers, Consumers, and Trophic Levels

One way to look at food and energy is in terms of production and consumption. Plants and other photosynthesizers (algae, some bacteria) are an ecosystem's **producers**, while the organisms that eat plants are one kind of **consumer**: an organism that eats other organisms, rather than producing its own food. But as you know, there are consumers *of* consumers—grass is eaten by zebras, but zebras are eaten by

lions. Thus do we get to the concept of feeding *levels* or, as ecologists put it, trophic levels. More formally, a **trophic level** is a position in an ecosystem's food chain or web, with each level defined by a transfer of energy between one kind of organism and another. Producers are one level, and then there are several levels of consumers. Here's a list of trophic levels through four stages.

First trophic level:	**Producers—** (photosynthesizers)
Second trophic level:	**Primary consumers—** plant predators (herbivores)
Third trophic level:	**Secondary consumers—** herbivore predators (carnivores)
Fourth trophic level:	**Tertiary consumers—** organisms that feed on secondary consumers (carnivores)

It is the Sun's energy that is locked up in leaves and stems, and it is this energy that is being passed along through these trophic levels. Note that the levels are not just categories that could be put in any order. We must have primary consumers to have secondary consumers; sharks cannot exist on algae (**see Figure 30.10**). You may remember that an animal that eats only plants or algae is a

Figure 30.9
Human Impact on Resources

a Nearly a quarter of the current atmospheric carbon dioxide concentration is produced by human activity, primarily through the burning of fossil fuels.

b Over half of terrestrial nitrogen fixation comes about because of human activity, including the manufacture of fertilizer for agriculture.

c Over half of the Earth's accessible surface water is now diverted for use by humans, mostly for agriculture. Percentages are approximate.

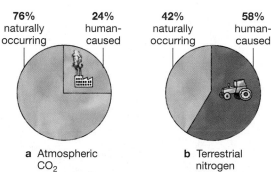

76% naturally occurring	24% human-caused	42% naturally occurring	58% human-caused	46% available	54% used

a Atmospheric CO$_2$ concentration   **b** Terrestrial nitrogen fixation   **c** Accessible surface water

Figure 30.10
Trophic Levels
Plants, algae, and some bacteria are called producers because they use the Sun's energy to produce their own food through photosynthesis. Organisms at all other trophic levels ultimately derive their energy from these photosynthesizers.

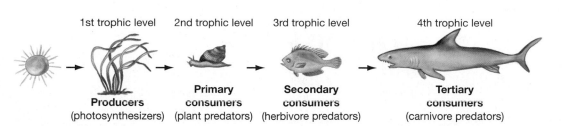

1st trophic level   2nd trophic level   3rd trophic level   4th trophic level

Producers (photosynthesizers)   **Primary consumers** (plant predators)   **Secondary consumers** (herbivore predators)   **Tertiary consumers** (carnivore predators)

herbivore; then there are **carnivores**, which eat only meat, and **omnivores**, which eat both plants and meat. Of course, many organisms cannot be assigned to just a single trophic level. Most human beings, for example, are primary, secondary, and tertiary consumers.

You have thus far thought in terms of food chains, which is to say trophic *lines* in which a single organism follows another. In reality, of course, nature is much more complex than this; what really exists are feeding patterns called *food webs*, one example of which you can see in **Figure 30.11**.

A Special Class of Consumers: Detritivores

To this scheme of trophic levels, ecologists add a special class of consumers, the **detritivores**. These are consumers who feed on *detritus*, which in normal usage simply means a collection of debris. In ecology, however, detritus is the remains of dead organisms or cast-off material from living organisms (waste, a fallen branch from a living tree). A worm or a dung beetle feeding on such material would fit this description of a detritivore.

Of particular interest, however, is a special kind of detritivore: A **decomposer**, which is an organism that, in feeding on dead or cast-off organic material, breaks it down into its inorganic components, which then can be recycled through the ecosystem. *Inorganic* here can be thought of as building blocks; more technically, it means an element or compound that does not contain carbon bound covalently to hydrogen. The most important decomposers are fungi and bacteria. A number of detritivores usually are responsible for breaking up organic material, first into pieces and then into its inorganic components (**see Figure 30.12**).

Stepping back from this, it's worth noting that if a factory made such complete use of its materials as nature does, that factory would be considered a wonder of the world. Think of it this way: What is thrown away in the feeding and decomposing chain you've been looking at? Nothing, which is all to the good, because if there were any natural garbage of this sort, spaceship Earth would sooner or later be filled up with this useless material. This very thought ought to give us some pause about

Figure 30.11
A River Food Web
The diagram indicates who eats whom along a portion of the Eel River in Northern California. The arrows have been color-coded by trophic level. Note that there are up to five trophic levels in the web. (Adapted with permission from an original drawing by Mary E. Power, University of California, Berkeley.)

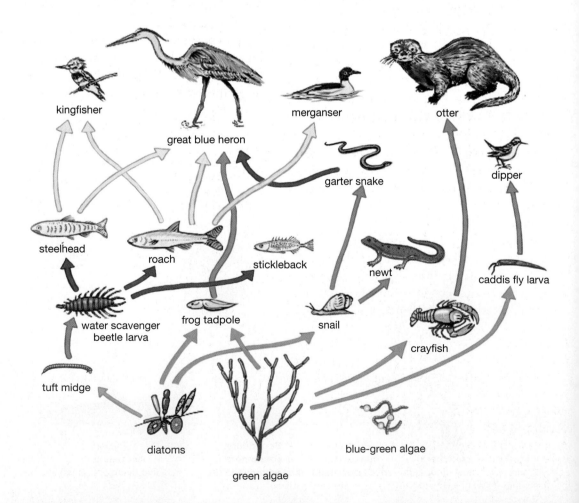

the compounds human beings currently are putting into the environment that are not **biodegradable**, meaning capable of being decomposed by living organisms. Of particular concern here are plastics, many of which will remain intact for hundreds of years following their production.

Accounting for Energy Flow through the Trophic Levels

All material substances eventually are recycled in ecosystems, but this is not true of energy, which is just passing through, on its way to heat. Two interesting questions arise from this: How *much* of the Sun's energy does the living world collect in the first place, and then how much of the collected energy makes it through to each successive trophic level?

This may seem like a routine set of questions, but the first ecologists who thought in these terms brought about a revolution in ecology. Working in the 1940s at Minnesota's Cedar Bog Lake, ecologist Raymond Lindeman was the first to conceptualize ecosystems as units in which energy is first captured by given organisms and then transferred on to others. This conceptualization is today known as the **energy-flow model** of ecosystems. Part of the impetus for Lindeman's research was to answer a very down-to-earth question, succinctly expressed by ecologist Paul Colinvaux: Why are big, fierce animals rare? (If you doubt that they are, think of this. It's utterly routine to see plants, insects, or small birds, but it's often the experience of a lifetime to see a bear or a shark.) The energy-flow model explained the small numbers of fierce animals, as you'll see, but it then went on to shed light on all kinds of ecological relationships. Why? Because energy is a kind of currency in the natural world; all organisms use it, and it is transferred among them. Moreover, it can be *measured*, usually in units called **kilocalories** (the amount of heat it takes to raise 1 kilogram of water one degree Celsius). If we know that a field mouse must use 68 percent of the kilocalories it consumes just to stay alive—to move, to keep a constant body temperature—but that a weasel must use 93 percent of its kilocalories for this purpose, that gives us a meaningful basis for comparison between these two animals. Beyond this, weasels *eat* field mice, and with the energy-flow model we have some basis for *tracking* this important commodity as it moves through an ecosystem.

Available Energy Is Greatly Reduced from One Trophic Level to the Next

So, what has energy-flow research told us? One critical insight is that the organisms in each succeeding trophic level bear a comparison with children who inherit far more than they will ever pass along. Recall that the Sun is the source of almost all energy on Earth, and that photosynthesizers are the entry point for this energy into the living world. So how much of the Sun's energy do plants

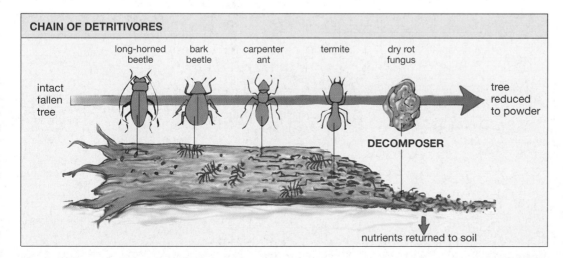

CHAIN OF DETRITIVORES

Figure 30.12
How Organic Material Is Broken Down
Different kinds of detritivores consume the organic material in a branch that has fallen from a tree. Most of these animals are consuming pieces of the log, but the detritivore called a decomposer (a fungus in the figure) is breaking organic materials into their inorganic components, thus returning nutrients to the soil.

capture? As you can see in **Figure 30.13**, very little; only about 2 percent of the Sun's energy that falls around plants is taken in for photosynthesis. This raises an important point, which is noting the difference between the energy that *comes to* an organism and what the organism *does with* that energy.

The amount of material that a plant produces as a result of photosynthesis is known as its **gross primary production**. Those of you who have been through Chapter 8 know that the initial product of photosynthesis is a sugar that serves as a starting "building block" for all the material in the plant. Every time a molecule of this sugar is produced in photosynthesis, the plant has added to its gross primary production. But plants don't get to retain all the material they produce in this way. Why? Because plants incur energy *costs* in just staying alive. It takes energy to perform photosynthesis, to move sap from one end of the plant to the other, and so forth. Such costs are subsumed under the heading of "cellular respiration," but you could also think of them as the plant's overhead; they are the price of doing business. Apart from this, we know that, in every energy transaction, some of the energy is lost as heat. After subtracting these various costs, the plant is left with its **net primary production**: the amount of material a plant accumulates as a result of photosynthesis. This is the material that serves to build the plant up—its stems and leaves and roots.

One question that arises from this framework is: What fraction of the energy a plant receives from the Sun ends up as net primary production? How efficient are plants, in other words, at turning solar energy into stems and leaves and roots? Of the solar energy that plants receive, somewhere between 30 and 85 percent is transferred into net primary production. In other words, at the low end of this range, only 30 percent of the solar energy a plant assimilates is transferred into material that is even potentially available to consumers.

Up a Level, to Primary Consumers

Now, what happens when you jump up one trophic level, to those primary consumers? Well, in the first place, they can't *get to* a lot of the tissue that's been produced—it's in roots that are buried or in leaves that are too high to reach. A good deal of the material is inedible in any event, except by detritivores. Critically, most of the energy these scavengers take in is making an *exit* from the trophic chain; it won't be available to primary consumers, which means it won't be available to secondary or tertiary consumers either. Adding all this up, you can begin to see that of the energy locked up by plants, very little is assimilated by organisms at the next step up—the primary consumers.

This pattern of energy reduction is then repeated in the successive trophic levels. But the reductions tend to get more severe, because animals have higher energy costs than plants. They excrete a lot of the material they consume, and the warm-blooded or "endothermic" animals we're most familiar with spend a great deal of energy keeping a constant body temperature—producing heat, which obviously is energy lost to the food chain. So, while a plant may transfer 50 percent of the energy it receives into new material, for endothermic

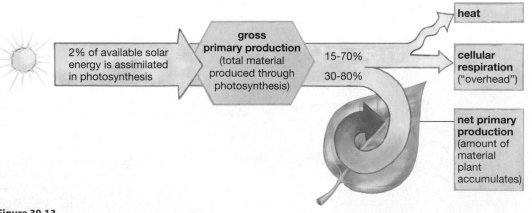

Figure 30.13
Absorption and Use of Solar Energy by Plants
Only 2 percent of the solar energy reaching the Earth's surface is taken in by plants for photosynthesis. Of this energy, only a fraction ends up being contained in plant material, such as tissues. It is this material (the plant's net primary production) that is available to the next trophic level in a food chain. The remaining energy is either dissipated as heat or used by photosynthesizers to power their own physical processes.

animals the figure is about 2 percent, meaning they use up 98 percent of the energy they assimilate just to stay alive, rather than to produce tissue with energy locked up inside it.

All of this leads to what you see in **Figure 30.14**: A reduction in available energy, and, in most cases, a reduction in the number of organisms at each trophic level. The gist of this is that each succeeding trophic level has a great deal less energy available to it than did the level before it. The amount that is passed on varies by community, but there is a rule of thumb in ecology that for each jump up in trophic level, the amount of available energy drops by 90 percent. Take all the energy converted into tissue by, say, a second trophic level herbivore, and of this, only 10 percent ends up being converted into tissue in a *third* trophic level carnivore. Hence the comparison to children's inheritance. There is a great variability in conversion rates, however, and 10 percent probably is more like a maximum passed on.

The Effects of Energy Loss through Trophic Levels

So, what are the consequences of this energy reduction at trophic levels? First, *this* is the reason that big, fierce animals are rare. If you had a field that could feed 100 field mice, how many weasels could the field support, if those weasels were existing solely on the mice? Remembering that, at most, only 10 percent of a trophic level's energy is likely to be passed along to the next level, you might guess that 10 weasels could make a living in the field, but even this number would be too high. For 10 predators to make a living from 100 mice, these predators would have to be no *bigger* than the mice, which is certainly not true of weasels. Extending this further, not even a single hawk could survive on the available weasels in the field. Why are big, fierce animals rare? Because of the loss of available energy at each step in the trophic chain. Because weasels are not about to start eating the same things as mice, weasels are locked into a trophic level with a more limited energy supply. When there's only so much energy, there can be only so many weasels.

Beyond this, the energy-flow model allows us to see the *limits* to top predators. Why couldn't we have a land carnivore bigger and fiercer than a grizzly bear? Well, imagine that you found some awesome creature that fit this bill somewhere on Earth, and that you

plopped it down in Alaska. What would happen? If it could feed only on grizzly bears, it would starve to death. There simply is not enough energy locked up in grizzly bears for there to be a trophic level above them. This is why, in most food webs, there are so few trophic levels above primary producers; the energy simply runs out. This question of energy loss with each trophic level has relevance for a human issue, which is eating meat that comes from grain-fed cattle, as you can see in "A Cut for the Middleman," on page 706.

Primary Productivity Varies across the Earth by Region

Primary productivity concerns the production of *biomass* or living material. It will come as no surprise that some areas of the Earth produce more biomass than others. If you look at **Figure 30.15**, you can get an idea

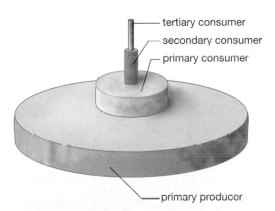

Figure 30.14
Energy Pyramid
Only a small fraction (10 percent or less) of energy at each trophic level is available to the next higher trophic level.

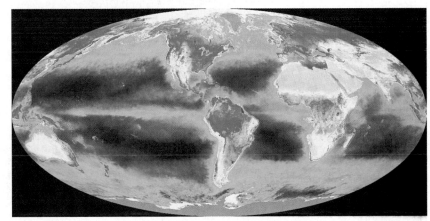

Figure 30.15
Primary Productivity by Geographic Region
Primary productivity around the globe. This composite satellite image shows how the concentration of plant life varies on both land and sea. On land, the areas of lowest productivity are tan-colored, as with the enormous swath of land that begins with the eastern Sahara Desert in North Africa. More productive land areas are yellow, more productive yet are light green, and the most productive of all are dark green. On the oceans, the gradient runs from dark blue (the least productive), though lighter blue, green, yellow, and orange-red (the most productive). The ocean measurements are of concentrations of phytoplankton—the tiny, photosynthesizing organisms that drift on the water and form the base of the marine food web. Note the high productivity of the oceans in both the far northern and southern latitudes. Antarctica itself is gray because the imaging technique did not work well for it.

A Cut for the Middleman: Livestock and Food

It may be difficult to think of cows in a field as occupants of one of the trophic levels you've been looking at, but that's just what they are, with some provocative consequences for human beings and food. People can't survive on grass the way cows and sheep can, and it follows that these animals are turning food that is useless to human beings into products that are use*ful* to human beings—meat, wool, and milk. In this sense, then, these animals are a boon to the efficiency of worldwide agriculture in that they allow us to get more human food and other products from the same amount of land.

There is another side to this story, however, having to do with the *loss* of food that occurs as we move up through the trophic levels. Recall that as a rough rule of thumb, 90 percent of the energy that is locked up in one trophic level is lost to the organisms in the next trophic level. If cattle and other domesticated animals were doing nothing but *foraging*—if they were being fed solely on the grass available in pastures—this would not be a matter of concern, because the grass is not useful to humans as a food source anyway. But in developed countries such as the United States, much of the diet of domesticated animals comes from *grain* that is grown on farms. A large portion of this animal feed could go directly to human beings, but it instead passes through the trophic level of animals, with a predictable result: It takes 9 pounds of feed to produce 1 pound of beef ready for human consumption. What does this mean in practice? A harvest of 2,000 pounds of grain will feed five people for a year. Take roughly this same harvest, feed it to livestock, and then feed the *livestock* products to people, and you will feed only *one* person for a year.

> *It takes 9 pounds of feed to produce 1 pound of beef ready for human consumption.*

This calculation takes on added importance when considering the scale of the grain-fed operation. In the feedlot, while being fattened for slaughter, an average steer consumes up to 2,700 pounds of grain. It may be apparent that such consumption adds up quickly; some 70 percent of the grain produced in the United States is fed to livestock. There are about 70 million cattle in the United States and Canada alone.

But does this mean that human hunger would be reduced if domestic animals were eliminated as a "middle-man"? Not necessarily. In a simple calculation of calories-produced versus calories-needed, the world *already* produces enough food to feed all its people. The current problem, then, is one of food distribution, not food production. As an often-used phrase has it, "Hunger is real, scarcity is not." The burden that grain-fed animals pose may have to do with the future. As countries get richer, their human populations will eat more meat, which would mean less total food available for humans worldwide.

Figure 1
Fattening at the Feedlot
About 40,000 head of cattle are fed for meat production at this commercial feedlot in Imperial Valley, California.

of the differences around the globe in connection with plant biomass. The reasons for these differences will be explored in this chapter's section on the large ecosystems called biomes. For now, suffice it to say that much of what you see in Figure 30.15 is intuitive. Productivity is very low in such desert areas as North Africa and very high in such tropical rain-forest areas as equatorial Africa. Yet why should one part of the Earth be a rain forest and another part a desert? The answer has to do with Earth's large-scale physical environment.

30.4 Earth's Physical Environment

Earth's Atmosphere

Earth itself exists within an environment, called the **atmosphere**, which is the layer of gases surrounding the Earth. One of the main things to realize about Earth's atmosphere is how it differs from outer space: There's something *to* Earth's atmosphere (a mix of gases) while outer space really is *space*; there's almost nothing in it.

The atmosphere is divided into several layers, but here you need be concerned with only two of them, which you can see in **Figure 30.16**. The lowest layer of the atmosphere, called the **troposphere**, starts at sea level and extends upward about 12 kilometers (or 7.4 miles), and it contains the bulk of the gases in the atmosphere. Nitrogen and oxygen make up 99 percent of the troposphere, but carbon dioxide exists there in small amounts, as do some other gases, such as argon and methane. After a transitional zone, the next layer above the troposphere is the **stratosphere**. Of greatest concern in it is a gas that reaches its greatest density at about 20 to 35 kilometers (13 to 21 miles) above sea level. The oxygen we breathe comes in the form of two oxygen molecules bonded together (O_2). When ultraviolet sunlight strikes O_2 in the stratosphere, however, it can put it in a form of three oxygen molecules bonded together (O_3). This is the gas known as **ozone**. Created initially as a *product* of life—since atmospheric oxygen came from photosynthesis—this ozone layer ultimately came to protect life by blocking some 99 percent of the ultraviolet (UV) radiation the Sun showers on the Earth. It was this blockage that allowed life to come onto land some 460 million years ago, and it is this blockage that protects life now; the UV radiation that pours from the Sun can wreak havoc on living tissue, bringing about cancer and immune-system problems in people, and damaging vegetation as well.

The Worrisome Issue of Ozone Depletion

Given the importance of the ozone layer, it is sobering to contemplate how fragile it is. Some human-made chemical compounds have the effect of destroying stratospheric ozone, and such destruction went on unchecked for years until atmospheric chemists Sherwood Rowland and Mario Molina revealed, in 1974, that compounds called chlorofluorocarbons posed a direct threat to the ozone layer. Chlorofluorocarbons or **CFCs**—found at one time in spray cans, refrigerators, and plastic foams—are undoubtedly the most famous of the ozone-depleting

compounds, but they are by no means the only chemicals that have this effect. Other harmful compounds include methyl bromide, which is used as a pesticide, notably on tomato and strawberry crops; and bromine, which is found in a group of fire-extinguishing chemicals.

The damage that these compounds have done to stratospheric ozone has brought about the spectacle of annual reports on how big the "ozone holes" are over the North and South Poles, along with a concern about a general thinning of the ozone layer. In 2000, the spring ozone hole over the Antarctic grew to a record 11 million square miles—an area three times the size of the United States. So large was this hole that, for the first time, it extended over a center of human population, the city of Punta Arenas in southern Chile. The residents of this city thus joined life-forms in the Antarctic in being seasonally exposed to potentially harmful levels of ultraviolet radiation.

Despite the ominous nature of this news, the long-term outlook on ozone depletion actually is good. In an agreement signed in Montreal in 1987, many of the world's nations pledged to phase out production of ozone-depleting chemicals, and this agreement seems

Figure 30.16
Earth's Atmosphere
The two lowest layers of the atmosphere are shown, with Mount Everest (the tallest mountain) and the Concorde (the highest flying passenger airplane) included for vertical scale. The troposphere contains most of the atmosphere's gases—mostly nitrogen and oxygen, with some carbon dioxide, methane, and other gases. The stratosphere contains the ozone layer, which blocks 99 percent of the Sun's harmful ultraviolet radiation.

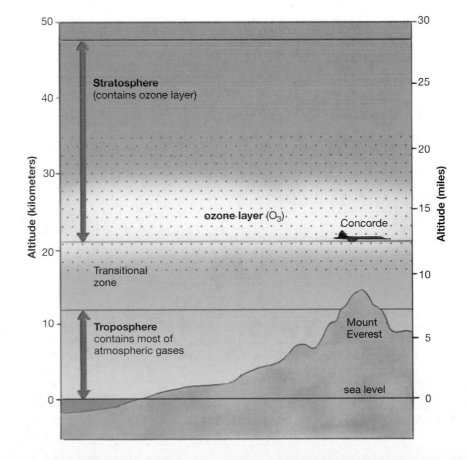

to be working. Production of CFCs was banned in the United States even before the Montreal agreement was signed and has dropped dramatically throughout the world in recent years. Meanwhile methyl bromide—most of which is injected into fields to kill pests—will be banned in the United States by 2005.

Critically, the reduction in the use of these compounds has had exactly the effect intended. Overall levels of ozone-depleting compounds are now dropping in the stratosphere. Atmospheric scientists reported in 2001 that if current progress continues, the ozone hole could be repaired within 50 years. This outcome is far from assured, and the ozone layer may get thinner before it gets thicker, but in the large scheme of things, the trend looks good.

The Worrisome Issue of Global Warming

A second environmental issue involving the atmosphere is that of global warming. The essential questions regarding this issue are straightforward: Is the Earth getting warmer? If so, to what degree is human activity responsible for this warming? And, if the Earth is warming, what consequences will this have? Put another way, what will a warmer Earth be like?

Is the Earth Getting Warmer?

At one time, all of these questions were contentious within the scientific community. In the last few years, however, near-unanimity seems to have been reached regarding one of them—*whether* the Earth is warming. Almost all parties are now agreed that Earth's surface temperature is indeed rising. It probably has increased by about 0.6° Celsius (or about a degree Fahrenheit) in the past century, with much of this increase coming in just the past 20 years. It's likely that, globally, the 1990s was the warmest decade since the middle of the nineteenth century. For the Northern Hemisphere, at least, the 1990s appear to have been the warmest decade in the last thousand years.

Is Human Activity Responsible?

This leads to the second question, which is whether human activity has caused this increase. Here, while there is still disagreement, a consensus seems to have emerged that the answer is yes: Human activity is at least partly to blame for global warming. In 2001, the United Nations Intergovernmental Panel on Climate Change (IPCC)—which received input from hundreds of scientists around the globe—issued a report stating "there is new and stronger evidence that most of the warming observed over the past 50 years is

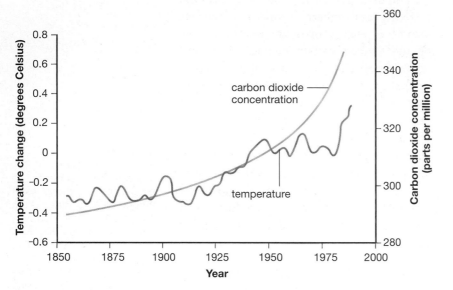

Figure 30.17
Global Warming
Atmospheric carbon dioxide concentration is increasing, and global temperature appears to be as well. It is difficult to be certain that the CO_2 increase is driving the increase in temperature because, over long periods of time, there are great natural fluctuations in global temperature. Most experts have now concluded, however, that rising levels of greenhouse gases such as CO_2 are making the world warmer.

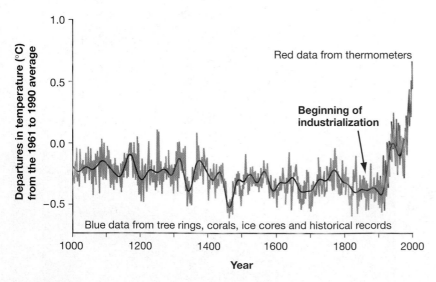

Figure 30.18
Earth's Temperature during the Past 1,000 Years
Using data from such sources as tree rings and ice cores, scientists have estimated average temperatures in the Northern Hemisphere over the last 1,000 years. They concluded that the increase in Earth's temperature in the twentieth century seems to have been greater than the increase that occurred in any other century during the last 1,000 years. Further, the 1990s appear to have been the warmest decade—and 1998 the warmest single year—over the past 1,000 years. (Adapted from *The Third Assessment Report of Working Group I of the Intergovernmental Panel on Climate Change* (IPCC) 2001, *Summary for Policymakers*, Figure 1b.)

attributable to human activities." What activities are these? The burning of fossil fuels, such as coal and oil, and the deforestation of the Earth. Fossil-fuel burning has put more carbon dioxide and other "greenhouse" gasses into the atmosphere, while deforestation has meant that less carbon is locked up in trees and other plants. The result is an increased quantity of greenhouse gases in the atmosphere. Warming has resulted from this increase because greenhouse gases trap heat, as you'll see.

It's important to note that there are respected scientists who disagree with this view of global warming. From their perspective, such warming as has taken place may represent nothing more than the random temperature variations that occur on Earth over very long time scales. Under this view, society should not rush to reduce greenhouse gas emissions because it doesn't yet know whether these emissions have anything to do with global warming.

While conceding that this view is supported by some valid arguments, most experts have become persuaded that human activity is indeed playing a part in global warming. What makes them think this is so? If you look at **Figure 30.17**, you can see that a strong correlation exists between increasing temperatures and rising atmospheric carbon dioxide (CO_2) levels over the last 150 years. If you then look at **Figure 30.18**, you can see temperature trends in the Northern Hemisphere for a much longer period, the last thousand years. Note the strong uptick in temperature that took place beginning after 1900, as civilization was becoming more industrialized, and the even more prominent upturn in the past few decades. This longer-term temperature trend is likewise consistent with longer-term CO_2 trends. The IPCC report notes that atmospheric concentration of CO_2 has increased by 31 percent since 1750 and that the Earth's current CO_2 concentration "has not been exceeded during the past 420,000 years and likely not during the past 20 million years."

A firm rule in science is that correlation is not causation; the fact that two trends are correlated does not mean that one caused the other—that rising CO_2 levels caused global warming, in this case. But by constructing theoretical models of what might be at play, IPCC scientists concluded that the most plausible explanation for the rise in Earth's temperature is that human activity has combined with natural forces, such as solar variation and volcanic activity, to produce the warming we have experienced.

What Is the Greenhouse Effect?

Why should the concentration of gases such as CO_2 have anything to do with a warmer planet? Sunlight that is not filtered out by ozone comes to Earth in the form of very energetic, short waves that can easily pass through the atmosphere (**see Figure 30.19**). Once this energy reaches the land and ocean, most of it is quickly transformed into heat that does all the things you've just read about: warms the planet, drives the water cycle, and so forth. This heat ultimately is radiated back toward space. But heat is not short-wave radiation; it is *long*-wave radiation, and it can be *trapped* by certain compounds, among them carbon dioxide and methane.

What Are the Likely Consequences of Global Warming?

All of this leads to the third question that frames the issue of global warming: What will its consequences be? The 1-degree rise recorded so far in Earth's temperature may seem tiny, but consider the effects that *already* appear to have been caused by this small increase. Arctic sea-ice has shrunk by about 6 percent since 1978; ice cover on lakes and rivers in some northern latitudes now lasts about two weeks less than it did 150 years ago; in the European Alps, some plant species have been migrating to higher altitudes at the rate of 4 meters per decade; in Europe and North America migratory birds

Tutorial 30.4.1
Greenhouse Effect

Figure 30.19
The Greenhouse Effect
The high-energy rays of the Sun can easily penetrate the layer of gases in the troposphere. However, the lower-energy radiation (heat) that reflects from Earth's surface cannot penetrate the layer of gases as easily. Carbon dioxide and methane thus take on the role of glass panes in a greenhouse: They let solar energy in, but do not let heat out.

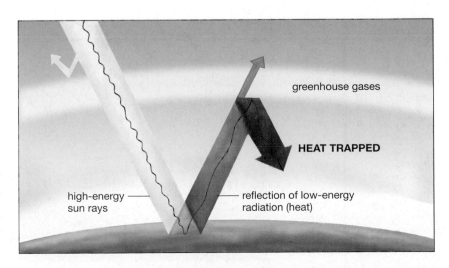

greenhouse gases

HEAT TRAPPED

high-energy
sun rays

reflection of low-energy
radiation (heat)

are now arriving earlier in the spring and leaving later in the fall; and general snow coverage in the Northern Hemisphere appears to have shrunk by 10 percent since 1972. Glaciers around the world, from the Himalayas to the Andes, are shrinking because of higher temperatures (**see Figure 30.20**). Meanwhile, the ice cap on top of Africa's Mount Kilimanjaro is melting at such a rapid rate that it is expected to disappear in less than 15 years.

Given such changes, if the temperature increase of the past hundred years were merely repeated in the *next* hundred years, we might worry about what is in store for the Earth. But IPCC research has predicted an increase not of 1 degree Fahrenheit for the coming century, but of 2.7 to 10 degrees. What would such a change mean? A picture of total calamity might come to mind, but the effects actually stand to be mixed. A second IPCC report notes that crop yields in some mid-latitude regions probably will increase because of the greater warmth, as would timber yields in some areas. More water might become available in areas, such as parts of Southeast Asia, where water is scarce now. On the other hand, low-lying parts of the globe may see entirely too *much* water, since melting polar ice is likely to cause a rise in sea levels that stands to displace tens of millions of coastal residents. Insect-borne diseases, such as malaria, can be expected to move both north and south from equatorial regions into formerly temperate areas. Australia and New Zealand are expected to experience a drying trend. Crop yields in current warm-weather regions can be expected to fall. In general, it is the warmer-weather regions of the Earth—and the mostly underdeveloped countries in them—that stand to lose the most from a warmer Earth.

What Can Be Done about Global Warming?

One obvious way to combat global warming is to cut back on emissions of greenhouse gases. This is much easier said than done, however. It costs money to build cleaner-burning power plants, and people want to drive cars. If the answer to global warming is to impose a "carbon tax" on such things as gasoline consumption, the economic impacts would be severe right where consumers are most sensitive to them—at the cash register. As one measure of this problem, consider that the United States was a party to a 1997 treaty, the Kyoto Protocol, under which 38 developed nations committed themselves to reducing their greenhouse gas emissions. But such dire economic predictions followed the signing of the treaty that the U.S. Senate never ratified it and, in 2001, President Bush decided to withdraw from it altogether.

Other ideas have been proposed for dealing with global warming. The Kyoto Protocol actually allowed participant nations to meet their CO_2 reduction targets not just by reducing gas emissions, but by planting trees. Remember from the carbon cycle that plants lock up carbon—they are carbon "sinks," while smokestacks are carbon "sources." By some estimates, new trees could take care of a large proportion of the Earth's added greenhouse gas emissions. But uncertainties abound in these estimates. Suffice it to say that it is far from certain that humanity will be able to plant its way out of global warming.

Figure 30.20
Warming Planet, Disappearing Glaciers
Global warming is melting glaciers around the world. Peru's Quelccaya ice cap, in the southern Andes, has shrunk by at least 20 percent since 1963. One of the main glaciers flowing from the cap, Qori Kalis, is shown here as it existed in 1978 and then in 2000. A 10-acre lake now exists where the glacier once extended. Qori Kalis' rate of retreat has reached 155 meters or 509 feet per year. This is three times greater than its rate of shrinkage from 1995 to 1998.

a The Qori Kalis Glacier 1978

b The Qori Kalis Glacier 2000

a

b

Earth's Climate: Why Are Some Areas Wet and Some Dry, Some Hot and Some Cold?

Did you ever look at a globe of the Earth and wonder why it is tilted? Earth exists in space, after all, so what could it be tilted against? When we say that a rod stuck in the ground is tilted, we mean that it is not *perpendicular* to the ground. Here, of course, we're thinking of the ground as a flat surface—a plane. Earth, too, can be viewed as existing on a plane, in this case the plane of its orbit around the Sun. (We generally think of this orbit as looking like a large hula hoop with the Sun as a yellow ball in the middle.) Now think of the spherical Earth on this plane as having a rod sticking through it in the form of its north-south axis—the imaginary line that runs from the North Pole straight through to the South Pole. The critical thing is that this rod does not stick straight up and down with respect to the plane of the Earth's orbit; it is *tilted*, at an angle of 23.5°. This tilt dictates a good deal about Earth's climate, and Earth's climate dictates a great deal about life on Earth. Once again, in other words, the Sun determines the basic conditions for life.

You can see the effects of Earth's tilt in **Figure 30.21**: In June, our Northern Hemisphere tilts toward the Sun, while in January it tilts away from it. Thus the Sun's rays strike us more directly in June, and the days are warmer (and longer) than days in January. The angle at which the Sun's rays strike a given portion of the Earth is very important. Relative to, say, Brazil, sunlight strikes the Arctic at an indirect or "oblique" angle. Thus, the far north gets less of the solar energy that powers photosynthesis.

The Circulation of the Atmosphere and Its Relation to Rain

The variation in temperature caused by these sunlight variations is the most important factor in the *circulation* of Earth's atmosphere. Near the equator Earth simply gets more warmth and, critically, warm air *rises*. Warm air also can retain more moisture than cold air. You've seen an example of this whenever you've looked at the beads of "sweat" that form on the outside of a cold glass in summer. Moisture-laden warm air comes into contact with the cooler air immediately around the glass; thus cooled, the air cannot hold as much moisture and releases it onto the glass.

This same thing happens with the warm air that rises on both sides of the equator. Because of the heat of the tropics, air is rising in quantity from the tropical oceans; because this air is warm, it is carrying with it a great deal of moisture. The air cools as it rises, however, and then drops much of its moisture on the tropics, which is why they're so wet.

Following this a step further, this volume of air is now cooler, drier, and moving toward the poles in both directions from the equator (**see Figure 30.22**). At about 30° north and

Tutorial 30.4.2
Seasons

Figure 30.21
Earth's Tilt and the Seasons
The Northern Hemisphere gets more sunlight in June and less in January, while the reverse is true for the Southern Hemisphere. This is the reason for seasonal climate variations over large portions of the globe.

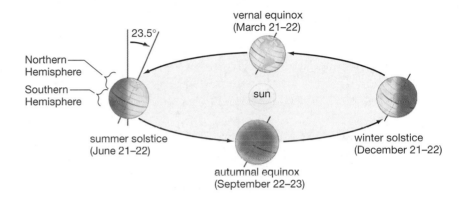

vernal equinox
(March 21–22)

23.5°

Northern Hemisphere

Southern Hemisphere

sun

summer solstice
(June 21–22)

winter solstice
(December 21–22)

autumnal equinox
(September 22–23)

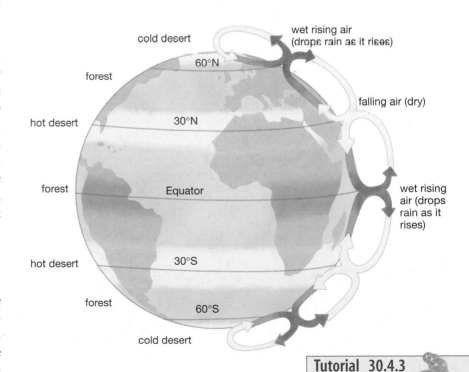

cold desert

wet rising air
(drops rain as it rises)

60°N

forest

falling air (dry)

hot desert

30°N

forest

Equator

wet rising air (drops rain as it rises)

hot desert

30°S

forest

60°S

cold desert

Tutorial 30.4.3
Atmospheric Circulation

Figure 30.22
Earth's Atmospheric Circulation Cells
Because the Earth receives its most direct sunlight at the equator, this region is warm—and warm air holds moisture and rises. This air moves north and south from the equator, cooling as it rises, and then loses its moisture as rain, with the result that the land and ocean immediately north and south of the equator get lots of rain. As the now-dry air moves farther north and south, it cools and sinks again at about 30° in both hemispheres. Thus, the land at these latitudes tends to be dry. Two other bands of circulation cells exist at higher latitudes, operating under the same principles.

south of the equator, it descends, warming as it drops and actually absorbing moisture *from* the land. The land will be dry at these latitudes because this is where the dry, hot air descends.

The air that has descended now flows in two directions from the 30° point. In the Northern Hemisphere this means north toward the North Pole and south toward the equator. Traveling at a fairly low altitude, this air eventually picks up moisture, rises, and deposits its moisture, this time at about 60° north and near the equator, with the same events happening at the same locations in the Southern Hemisphere.

What you get from this rising and falling is what you see in Figure 30.22. A set of inter-related "circulation cells" of moving air, each existing all the way around the globe at its latitude, and each acting like a conveyor belt that is dropping rain on the Earth where it rises but drying the Earth where it descends.

The Impact of Earth's Circulation Cells
The full meaning of this becomes obvious only when you look at a map such as the one in **Figure 30.23** (or better yet, a globe). Draw your finger across the equatorial latitude on the map and look at how much green there

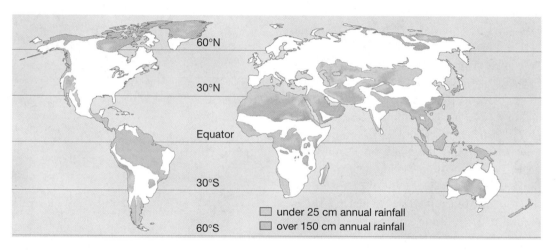

Figure 30.23
Earth's Atmospheric Circulation Cells and Precipitation
The wettest regions of the Earth lie along the equator, while the driest regions lie along the latitudes 30° N and S. Compare these regions to the circulation cells shown in Figure 30.22. The variations of rainfall patterns within cells are explained by factors such as the presence of mountain ranges.

Figure 30.24
Rain Shadows
Mountain ranges force air to rise. This cools the air, causing it to lose its moisture on the windward side of the range. This creates a rain shadow, meaning a lack of precipitation, on the leeward side of the range.

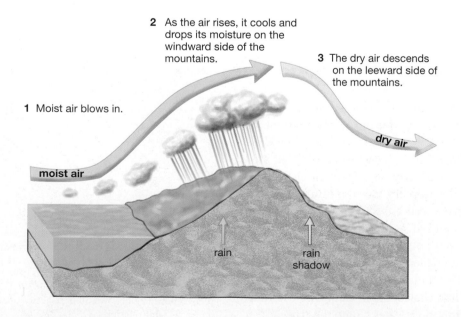

2 As the air rises, it cools and drops its moisture on the windward side of the mountains.

3 The dry air descends on the leeward side of the mountains.

1 Moist air blows in.

dry air

moist air

rain

rain shadow

is—Southeast Asia, equatorial Africa, the Amazon. This is no surprise, because we expect the equator to be wet and green. Now, however, do the same thing at the dry 30° north latitude. Look at how much desert there is at this latitude all around the globe—it's where we find the Sahara, the Arabian, and the Sonoran deserts, among others. And there's more desert territory at 30° south, with the Australian desert and Africa's Namib. From this you can see that the climatic fate of Earth's various regions is largely written in our globe's large-scale wind patterns.

Mountain Chains Affect Precipitation Patterns

Latitude doesn't tell the whole story about precipitation on Earth, however. Mountain ranges force air upward and it too cools as it rises, dropping its moisture on the windward side of the range. Then the air descends on the opposite (leeward) side of the range, only this time it is dry air and is thus *picking up* moisture from the ground, rather than depositing it (**see Figure 30.24**). A dramatic example of this "rain shadow" effect can be seen on the Pacific Coast of the United States in southern Oregon, where the western side of the Cascade Mountains is lush with greenery all year round, while the basin below the eastern slopes is a desert.

Beyond this, we know that it's possible to be in a warm latitude, such as that of Ecuador in South America, and still be very cold—if you go to the top of the Ecuadorian Andes. As altitude increases, temperature drops, in other words. The changes in climate that occur as you move from flatland to mountain peak are similar to the changes that occur as you move from the equator to one of the poles.

The Importance of Climate to Life

The factors you have just reviewed are among the most important in shaping the large-scale **climates** or average weather conditions we find on Earth. It's obvious that climate has a great deal of influence on life, since we never see polar bears in Panama or monkeys in Montana, but this influence is even greater than you might think. Looking over the globe, you see different forms of dominant vegetation in different areas—grassland in America's Great Plains, rain forest in northern

Brazil. So strongly does climate affect this pattern that Earth's vegetation regions essentially are defined by Earth's climate regions. Far in Canada's north, there is a fairly distinct boundary at which Canada's vast coniferous forest gives way to the mat-like vegetation called *tundra*. North of this line there is tundra; south of it are the forests. What is this line? In summer it is the far southern reach of a mass of cold air called the *arctic frontal zone* (**see Figure 30.25**).

Vegetative formations generally overlap one another to a greater extent than is seen with Canada's tundra and coniferous forest, so that the closer we get to a vegetation boundary the less distinct it looks. Nevertheless, there are discrete regions of vegetation on Earth whose boundaries are largely determined by climate—particularly by temperature and rainfall. Vegetative exceptions *within* these regions tend to be mountains, whose climates can vary greatly from base to peak.

30.5 Earth's Biomes

When these large, land vegetation formations are looked at as *ecosystems*, complete with animals, fungi, nutrient cycling, and all the rest, they go by the name **biomes**. Grasslands, whether found in Russia or in the American Midwest, are ecologically very similar to one another and thus constitute

Figure 30.25
Line of Transition
In Alaska's Denali National Park, a region of forested taiga vegetation gives way to a region of grassy tundra vegetation in a fairly abrupt way.

one type of biome. The tropical rain forests found in Africa, South America, and Asia constitute another. The world can be divided into any number of different biome types—this is a human classification system overlaid on nature—but six are recognized as the minimum: tundra, taiga, temperate deciduous forest, temperate grassland, desert, and tropical rain forest. See **Figure 30.26** for an overview of locations of these biome types throughout the globe. Polar ice and mountains often are recognized as separate biomes as well, as is another grassland variation, the savanna, found most famously in equatorial Africa. There is also chaparral, a shrub-dominated vegetation formation found in "Mediterranean" climates such as those in Southern California's coastal region.

The fact that two biomes are of the same type does not mean that they are identical. Two areas that are broadly the same in climate can end up with some significant differences in life-forms. Let's look briefly now at each of Earth's major biome types and then at its aquatic ecosystems.

Cold and Lying Low: Tundra

Along with polar ice, **tundra** is the biome of the far north, stretching in a vast, mostly frozen ring around the northern rim of the world. So inaccessible is tundra that the average person may never have heard of it, yet it

occupies about a fourth of Earth's land surface. The word *tundra* comes from a Finnish word that means "treeless plain," and the description is apt. Its flat terrain stretches out for mile after mile with little change in the vegetational pattern of low shrubs, mosses, lichens, grasses, and the grass-like sedges (**see Figure 30.27**).

Tundra is a paradoxical biome in that it averages about 25 centimeters (or 10 inches) of rain per year, which almost puts it in the desert category. Yet it is marshy in summer, with numerous shallow lakes dotting its surface. How can both things be true? Once the summer rain falls, it essentially has no place to go. No more than a meter below the surface, the ground is permanently frozen down to a depth of as much as 500 meters or a quarter-mile. This is the **permafrost**, or permanent ice, which acts as a boundary beyond which neither water nor roots can go. Because water cannot drain far into the soil, it sits on top of the tundra in summer, which makes for the bogs and small lakes. The low-lying tundra vegetation makes the most of a short growing season—lasting as little as 50 days—by producing a year's worth of food through photosynthesis in only a few weeks.

The tundra has a good deal of animal life, particularly in the summer when migrating animals move north to feed on the vegetation. Large mammals such as caribou

Figure 30.26
Distribution of Biomes across the Earth

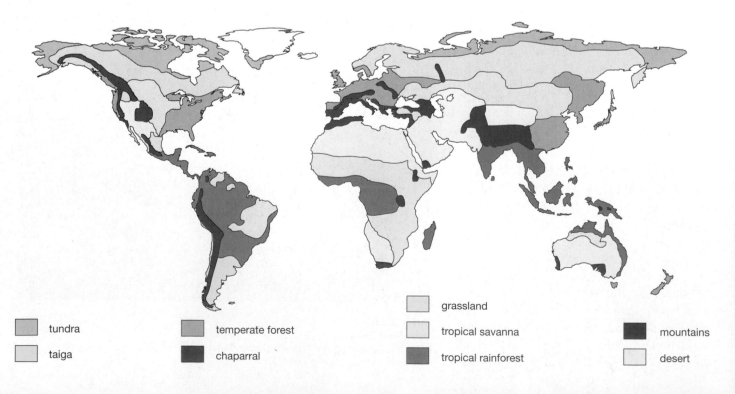

tundra

taiga

temperate forest

chaparral

grassland

tropical savanna

tropical rainforest

mountains

desert

and musk oxen live there. Arctic hares and foxes are year-round residents, as are arctic lemmings—herbivores that are key to much of the tundra's animal food web. (Contrary to popular belief, these lemmings do not commit suicide by jumping off cliffs. They merely *migrate* in great numbers at certain times.)

Northern Forests: Taiga

The **taiga** or boreal (northern) forest includes the enormous expanse of coniferous trees mentioned earlier that juts up against the tundra. Conditions are still fairly dry and cold, and the growing season is short in this northern biome, but in all three respects the taiga is less severe than the tundra. Note in Figure 30.26 that, as with tundra, taiga is found almost solely in the Northern Hemisphere. This is so because there is little land in the Southern Hemisphere at the extreme latitudes of these biomes. This in turn is related to a more general point, which is that most of Earth's landmass is in the Northern Hemisphere.

The taiga is a model of species uniformity as only a few types of trees—spruce, fir, and pine—are the ecological dominants, but they are present in great numbers. The taiga's vegetation supports lots of animals, including large mammals such as moose, bear, and caribou (**see Figure 30.28**). When thinking of the fur trade, you might have in mind some freezing, coniferous territory in the far north. The taiga turns out to be it. Sables, minks, beavers, arctic hares, wolverines: This harsh biome has an abundance of these furred creatures.

Hot in Summer, Cold in Winter: Temperate Deciduous Forest

Temperate deciduous forest is a biome familiar throughout much of the United States as it exists roughly from the Great Lakes nearly to the Gulf of Mexico and from the western Great Lakes to the Atlantic Ocean. The pine forests that cover large parts of the southeastern United States generally are considered to be part of the temperate deciduous biome because the pines are a successional stage *leading to* deciduous forest. (Remember, from Chapter 29, the concept of ecological succession, under which species in a community are replaced by other species over time until a final, stable

state is reached.) Meanwhile, the warm Gulf Coast is not temperate deciduous forest per se, but a variant on it called *temperate evergreen forest*. Much of Europe and eastern China fit within the temperate deciduous classification, though large parts of Europe have few forests left, and China has very little.

Figure 30.27
Tundra
Grizzly bears forage for fall berries in Alaska in the shrub-and-grass biome known as the tundra.

Figure 30.28
Taiga
Alaskan caribou in winter in the northern-forest biome known as taiga.

Temperate deciduous forests are the forests of maple and beech and oak and hickory (**Figure 30.29**). The "deciduous" in the name refers to trees that exhibit a pattern of loss of leaves in the fall and regrowth of them in the spring. Trees do not dominate these forests to the extent they do the taiga, however. Making use of the good soil, a robust "understory" of woody and herbaceous plants will spring up on the floor of the deciduous forest. Some of this activity occurs early in the growing season, before the trees have "leafed out" and blocked the sunlight.

Temperate forests grow where there is plenty of water: between 75 and 200 centimeters (or 30 to 78 inches) per year, with the lower amount being three times what the tundra gets.

It's worth noting that while the tundra has no reptiles or amphibians and the taiga has only a few, the temperate deciduous forest has an environment that is wet and warm enough to support a variety of both animal forms. Many species of the large mammals that once populated the deciduous forests in the United States have either been greatly reduced (black bears, wild pigs) or nearly eliminated altogether (wolves), though in the case of the black bear, fairly robust conservation and restoration efforts are under way.

Dry but Sometimes Very Fertile: Grassland

There is an irregular line in western Indiana that forms the dividing line between deciduous forest and prairie. What is "prairie" in the United States is known as "steppes" in Russia, "pampas" in Argentina and Uruguay, and "veldt" in South Africa. These are all names for the biome that ecologists call *temperate grassland* (**Figure 30.30**).

Looked at one way, grasslands are what often literally lie between forests and deserts. Whereas the precipitation range for deciduous forests is from 75 to 200 centimeters per year (30 to 78 inches), for grassland the range is from 25 to 100 centimeters per year (10 to 39 inches). The North American grassland has its own internal division based on the amount of rainfall. It ranges from tall-grass prairie in Illinois to a drier mixed-grass prairie that begins at about the middle of Kansas and Nebraska—this is where the Great Plains begin—to a short-grass prairie that lies in the rain shadow of the Rocky Mountains. (This same tall-grass/short-grass gradient exists in the steppes of Asia, only there the gradient runs from North to South.) Other American grassland areas can be found in California's Central Valley and eastern Washington and Oregon.

A type of *tropical* grassland, the **tropical savanna**, exists in Australia, Africa, and South America. Tropical savanna is characterized by seasonal drought, small changes in the generally warm temperatures throughout the year, and stands of naturally occurring trees that punctuate the grassland. This is the homeland of many of the great mammals we associate with Africa, such as lions and zebras.

When Europeans first beheld the American grasslands, they could scarcely believe what they saw: A territory that had essentially no trees, but instead a sea of grasses that could grow up to 3 meters or 10 feet tall in the wetter tall-grass prairie. Prairie soil turned out to be some of the most fertile on Earth, however, and that fact eventually spelled the end of natural prairie in the United States and most of the rest of the world. Almost all the tall-grass and mixed-grass prairie in the

Figure 30.29
Temperate Deciduous Forest
Plenty of water and seasonally warm temperatures lead to abundant plant and animal life in the biome known as temperate deciduous forest. This forest is in the Great Smoky Mountains National Park in Tennessee.

United States eventually was plowed under for cultivation, while the short-grass prairie is now used for crops and cattle grazing. Prior to the coming of Europeans, 22 million acres of Illinois was prairie; today that figure stands at slightly more than 2,000 acres. From the former prairie, however, the United States feeds itself and some of world outside its borders. Illinois and Kansas and Nebraska are still grasslands, but the grasses that grow there now are not the native big bluestem or needlegrass; they are corn and wheat.

The American grasslands once were home to huge herds of one of the largest of Earth's grazing animals, the American bison (*Bison bison*). Better known as the buffalo, this animal was shot nearly to extinction—hunted is not the right word—but it is now rebounding somewhat in protected areas.

Chaparral: Rainy Winters, Dry Summers

Chaparral is not grassland, but instead is a biome dominated by evergreen shrub vegetation that is dotted with pine and scrub oak trees. Chaparral is found in areas that have a Mediterranean climate: around the Mediterranean Ocean and in western Australia, parts of Chile and South Africa, and coastal Southern California. What most of these areas have in common is that they are on the west coast of a landmass. This placement results in mild, wet winters and very dry summers.

The Challenge of Water: Deserts

Deserts are defined in various ways, one of them being: an area in which rainfall is less than 25 centimeters or 10 inches per year. Note that temperature has nothing to do with this definition. Thus, there can be temperate deserts (the Mojave in Southern California), cold deserts (the Gobi in China) and hot-weather deserts, like the one in **Figure 30.31**. There are areas in some deserts, such as Death Valley in California, that achieve the dubious distinction of having very little life in them, but this is rare. Most deserts harbor a variety of life-forms, and what we think of as desert can shade imperceptibly into dry grassland.

Precipitation is not only low in the desert, it is sporadic; that is, most of the moisture a desert gets will come in just a few days during the year. Given this, desert life has relentlessly been shaped by one overriding requirement: Collect water and then conserve it. The flowering plant grouping called bromeliads have a species, *Tillandsia straminea*, that is blown around on the deserts of coastal Peru much like a dead tumbleweed; only this is a living plant, nearly rootless, that gets all its moisture from *fog* that it is able to absorb.

Desert animals are no less remarkable in this respect. The kangaroo rat (genus *Dipodomys*) can be found in many parts of the American Southwest. It comes out of its

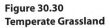

Figure 30.30
Temperate Grassland
A white-tailed deer fawn stands among the grass and flowers in a tall-grass prairie in Missouri. Tall-grass is one kind of growth found in the biome known as temperate grassland.

Figure 30.31
Desert
When water gets sparse, life does too. Shown is a desert biome in Monument Valley, Arizona.

underground burrow only at night when temperatures fall, it has no sweat glands, and it generally does not drink standing water (there being almost none available). Instead, the Kangaroo rat gets by on the water produced by metabolizing the seeds it eats.

The Issue of Desertification

Earlier you read about the loss of natural grasslands and temperate deciduous forests. With deserts, however, the concern is just the opposite. Some of them may be growing *larger* in the process known as **desertification**, meaning the transformation of an area into a desert, sometimes through human activity. Desertification is thought to be taking place, for example, on land to the north of the American Chihuahuan Desert in western Texas and eastern New Mexico. Where once this territory was a grassland community, for decades it has been turning into a mesquite-filled desert like the land to the south of it. Cattle grazing, introduced to the region in the 1880s, is thought to be a primary cause of this transformation.

Conversely, what is undoubtedly the world's best-known example of "desertification" seems now to be nothing but the random fluctuations of nature in a dry area. For decades it was generally accepted that Africa's Sahara Desert was moving south, taking over a very long, bordering strip of arid land known as the Sahel, with cattle grazing and other human activities thought to be at fault once again. In 1998, however, detailed satellite imaging of the area revealed that the Sahara isn't moving south at all. The Sahara-Sahel border does shift around—by as much as 300 kilometers

(about 186 miles) over some periods since 1980. But this is a forward-then-*back* movement of the desert controlled by rainfall, not a steadily forward one caused by human activity. The clear message here is that we need to be wary of what we "know" to be true.

Lush Life, Now Threatened: Tropical Rain Forest

Many people have an idea that the Earth's **tropical rain forests** are important, but they're not quite sure why. There are two main reasons. First, along with coral reefs and algal beds, they are the most productive biome type in the world, meaning they produce more biomass per square meter of territory than any other. Second, no other ecosystem can touch them in terms of species diversity. As noted earlier, for mile after mile in the taiga, trees might be limited to one or two species. In the tropical forests of the Amazon, more than *three hundred* species of trees have been identified in a single hectare of land, which is about 2.5 acres. In Chapter 29, you read about a researcher who found an estimated 1,700 species of beetle in one tree in the Peruvian rain forest. Indeed, half of all the identified plant and animal species on Earth reside in tropical rain forests. This despite the fact that, compared to the other biomes you've looked at, they don't occupy a lot of territory: about 7 percent of the Earth's surface.

The problem, as most people know, is that the rain forests are disappearing, by being burned down or cut down. How great is the loss? One estimate is that more than 35 hectares—about 86 acres—are lost each minute. South America has what is by far the world's largest tropical rain forest, the Amazon Basin Forest that runs from Colombia through Bolivia and at one point nearly from the Atlantic to the Pacific Ocean. Significant rain forests also can be found in Central America, equatorial Africa, and regions of Southeast Asia near the equator.

The rain forests are lands of rain and stable, warm temperatures (**see Figure 30.32**). The norm for rainfall is between 200 to 450 centimeters or about 78 to 175 inches a year, but some areas get up to 1,000 centimeters annually, which is 390 inches.

Poor Soil, Bountiful Covering

You've seen that much of the richness of the temperate grasslands can be found in their soil. In the tropical rain forest, almost all the

Figure 30.32
Tropical Rain Forest
Abundant rain and warm weather mean abundant growth and a great diversity in life-forms in the biome known as the tropical rain forest. Shown is a lowland rain forest on the Segama River in Borneo.

richness is on display aboveground (much of it 20 to 40 meters—65 to 130 feet—aboveground in the extensive forest "canopy"). The rain forest has such a large and efficient group of detritivores that organic material dropping to the forest floor is decomposed by them and absorbed almost immediately by plant roots rather than by the surrounding soil. This means the soil never gets built up; it is nutrient-poor and often acidic.

How can this be squared with this biome's fantastic productivity? Note what is going on in the example: The nutrients are simply bypassing the soil, going from plant to detritivore and back to plant. Some rain-forest trees have roots that grow *up*, onto tree trunks, rather than down into the soil, giving them the ability to absorb nutrients from the water that washes down the trunks. Farmers on the Illinois prairie could plow under the natural vegetation and grow crops there because perhaps 90 percent of the prairie's nutrients are held in the soil. What happens, though, when a rain forest is cleared for agriculture? The nutrient-poor land will support a year or two of crops and then that's it: No more rain forest, no more crops. (The new crops don't build up the soil, because nutrients are leached out by the heavy rains.) With the loss of rain forest comes the loss of habitat for its animals, resulting in a greatly elevated rate of extinction in this biome.

Why Worry about the Rain Forests?

Why should we care about the loss of rain forests? That is, so what if we lose a lot of plant material and a lot of species? Regarding the first question, the enormous biomass of the rain forests locks up *carbon*. Cutting down the forests, therefore, results in an increase of the greenhouse gas, carbon dioxide. Losses from the Brazilian Amazon alone are thought to put hundreds of millions of tons of CO_2 into the atmosphere each year. Overall, the concern is that, beyond greenhouse warming, loss of rain forest will affect global climate, perhaps in ways that we can't even anticipate yet.

On the question of species diversity, there is a value in preserving plant, fungal, and animal species purely as a matter of self-interest. This is so because of the medicines these organisms provide. It is very rare for scientists to create a medicine from scratch. What they do is survey the world for *natural* substances that show signs of being effective in fighting disease, after which these substances are tested.

The cancer-fighting drug Taxol, derived from yew tree bark, was found in just this way.

Beyond these practical considerations, to some degree, large or small, aren't we irreversibly poorer every time we lose a species? There is some obvious loss every time a finely crafted work of art is destroyed. Is there any less loss when something living, shaped over millions of years, disappears forever?

30.6 Life in the Water: Aquatic Ecosystems

Marine Ecosystems

Because we humans are land creatures, the ocean is mostly unknown to us; thus it might be good to get our bearings by thinking about how the ocean can be divided up in terms of the life in it. The ocean's **coastal zone** extends from the point on the shore where the ocean's waves reach at high tide to a point out at sea where an ocean-floor formation called the continental shelf drops off (**see Figure 30.33**). Beyond this point there is the **open sea**. *Within* the coastal zone, there is an area bordered on one side by the ocean's low-tide mark and on the other by its high-tide mark. This is the **intertidal zone**.

If you look next at the ocean strictly in its vertical dimension, all of the water from the ocean's surface to its floor is called the **pelagic zone**, while the ocean floor itself is called the **benthic zone**. *Within* the pelagic zone, from sea level down to a depth of at least 100 meters (about 330 feet), the Sun's rays can penetrate strongly enough to drive photosynthesis. This zone of photosynthesis is known as the **photic zone**.

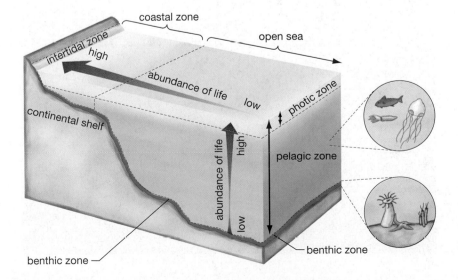

**Figure 30.33
Ocean Zones**

a

b

c

More Productive Near the Coasts, Less Productive on the Open Sea

Each of these zones marks off an area that is meaningful in terms of life. Start first at the intertidal zone. This is a world *in between*, we might say, because the creatures in it live part of the time submerged in the ocean and part of the time exposed to the air. It turns out that the intertidal zone is extremely productive for a reason that you might think would make it *unproductive*: the ocean waves, which bring in nutrients, carry away wastes, and expose more of the surface area of photosynthesizers to sunlight.

Ocean life continues to be abundant out past the intertidal zone and in general remains so all the way to the end of the coastal zone. In short, ocean life is most productive in its shallower depths (**see Figure 30.34**). Once past the continental shelf and into the open ocean, productivity can drop off to levels that are less than those of a desert. Why should this be so? Remember that sunlight strong enough to power photosynthesis penetrates ocean waters only to a depth of 100 meters or so. Meanwhile, the average depth of the ocean is about 3,000 meters or 1.8 miles. Thus there is a vast depth of open ocean that simply isn't very productive. In addition, coastal areas benefit from higher levels of nutrients, primarily as a result of nutrient runoff from the land.

One measure of the coastal zone's importance is that it is the location of all the world's great fisheries. At one time, it was thought that these areas were places of such endless bounty that they could never be depleted. The 1990s proved this wrong, however. Modern industrial fishing techniques have left one commercial fish species after another seriously depleted—haddock and bluefin tuna, Atlantic cod and swordfish. Depletion is not the same as endangerment, but it should be of real concern if only from a self-interested point of view: Almost 20 percent of the animal protein that humans consume worldwide comes from fish. For a perspective on environmental issues, see "Good News about the Environment," on the next page.

Figure 30.34
Diversity in Ocean Life

a A man and two children investigate tidepools, a part of the ocean's intertidal zone. The deeper water beyond them can be seen both above and below the surface.

b King penguins swim underwater near a breeding colony in the fertile waters off the coast of Antarctica.

c Fish called sweepers and fairy basslets feed in the currents above a coral plate covered with animals called crinoids. The structure is off the coast of Papua New Guinea.

Good News about the Environment

A sense of gloom about the environment may arise from between the lines of these pages as the list of current or potential environmental problems piles up: global warming, nitrogen buildup, overuse of water resources, tropical rain forest destruction. These are real problems, but their existence should not blind us to the fact that there has been a great deal of good news about the environment in the last 30 years, much of it showing how effective government can be when prodded into action by its citizens.

If there was a time for unrestrained gloom about the environment in the United States, it probably would have been in the late 1960s, when the country had plenty of endangered species but no Endangered Species Act; when levels of smog in Los Angeles were almost twice what they are now (**see Figure 1** on page 722); when there was plenty of household trash to be recycled, but almost no recycling going on; when only a third of the bodies of water in the country were safe for fishing and swimming, whereas today two-thirds are.

As a symbol of change, consider Cleveland, Ohio, whose Cuyahoga River was so polluted in 1969 that it *caught fire*, while southwestern Lake Erie, into which the Cuyahoga flows, was essentially a dead zone for fish during the late summer. The lake came to this low state in part because of the nearly 80 metric tons of phosphorus—then regularly added to washing detergents—that were dumped into it each day. That figure has now been reduced by perhaps 85 percent, and Lake Erie is biologically very active and a major freshwater fishery. Meanwhile the Cuyahoga River has been cleaned up enough that an area near the river's mouth has been transformed

Environmentalism has entered not only our lawbooks but, more important, our consciousness.

into an entertainment district complete with restaurant tables near the water's edge—not bad for a small river running through a big city. This does not mean that the Cuyahoga now looks as it did when Europeans first saw it; toxic sediment from the river, mostly the legacy of the past, still flows into Lake Erie. And the lake itself is now beset by another environmental calamity in the form of an invasion of a small, voracious mollusc known as the zebra mussel (*Dreissena polymorpha*). But if we look at the *trend*, there is no doubt that things have gotten better in Cleveland, not worse.

This same thing could be said of most of the United States. Simply put, there has been a profound change with respect to the environment over the past 30 years. From pollution controls to species protection to personal recycling, environmentalism has entered not only our lawbooks

continued

More Productive Near the Poles, Less Productive Away from Them

Apart from the shallow-to-deep transition, the ocean also has one other gradient with respect to the amount of life it harbors. In the pelagic zone, there is relatively more ocean life toward the poles, with this concentration decreasing with movement toward the equator—the exact opposite of what we find with land life. The oceans that surround Antarctica are filled with life. **Phytoplankton**, meaning floating, microscopic photosynthesizers such as algae and cyanobacteria, are the base producers for it all. These are consumed by the tiny floating animals known as **zooplankton**. Shrimp-like crustaceans called krill feed mostly on the phytoplankton and serve as the principal source of food for the whales and seals we associate with this area.

Cities of Productivity: Coral Reefs

Despite the richness of the colder oceans, the tropical oceans do boast marine communities that are as productive as any in the ocean—indeed, as any on Earth. These are the **coral reefs**, which lie in shallow waters at the edge of continents or islands at tropical latitudes. Coral reefs actually are the piled-up remains of many generations of coral animals, with each individual animal, known as a polyp, being a tiny relative of the more familiar sea anemones. Coral polyps secrete calcium carbonate, better known as limestone, a practice they share with a type of red algae with whom they share their habitat. The limestone forms an external skeleton for the corals and, when each coral dies, its limestone skeleton remains. Thus, coral reefs are composed largely of the stacked-up remains of countless generations of coral polyps,

Good News about the Environment *continued*

but, more important, our consciousness. And support for the environment has grown not just in America but in most developed countries. The cleanup of Lake Erie required the cooperation of the United States and Canada; the heartening trend in preservation of the Earth's ozone layer is the product of an international agreement.

None of this is to say that mechanisms are in place such that we need not worry about the environment. Far from it; largely because of the growth in human population and the greater per-capita use of natural resources, the environment is threatened on thousands of fronts and will remain so as far out as we can see into the future. But the record of the last 30 years is not just one of environmental loss; it is one of great environmental accomplishment as well. Perhaps the period's most important lesson is that there is nothing inevitable about environmental degradation. Concerned, hard-working people have made a great deal of difference in this issue.

2000 ← Pounds of hydrocarbons produced by an average new passenger car during 100,000 miles of driving.

50

1965   1998

Figure 1
Moving in the Right Direction
Over the last 30 years, the amount of car emissions in California has decreased 40-fold, as measured in pounds of hydrocarbon emissions produced by an average new passenger car in California during 100,000 miles of driving. One result is that cities like Los Angeles have experienced dramatic decreases in smog levels, despite steep increases in population.

along with the remains of their neighboring red algae. But the thin outer veneer of each reef is composed of *living* algae and the latest generation of pinhead-sized corals.

You can get an idea of how many generations of coral are piled up by considering that most reefs are between 5,000 and 8,000 years old—and some are millions of years old. What the reef as a whole creates is *habitat*: nooks and crannies and tunnels and surfaces that are home to an amazing array of living things. So rich is this habitat that it covers only 2 percent of the ocean's floor, but is home to perhaps 25 percent of the ocean's species. One coral reef, Australia's Great Barrier Reef, also is one of the largest structures created by any living thing, including humans. It measures some 2,000 kilometers or 1,200 miles in length.

Sadly, coral reefs are now imperiled cities of biological productivity. By one estimate, 27 percent of the world's coral reefs have now been lost as functioning ecosystems, though with coral reefs, loss of activity is not always permanent. The greatest single source of this destruction was the El Niño-caused ocean warming of 1997–1998, which brought about a phenomenon known as *coral bleaching*, meaning a discoloration of the reefs due to the death of symbiotic algae that live within the coral animals. Temporary weather events aside, the longer-term threat to coral reefs is human activity. Oil is dumped into the ocean, sewage and runoff from cities harms the reefs, divers may trample on them. Perhaps most disturbing, thousands of pounds of the poison sodium cyanide are dumped each year on coral reefs in the Philippines and other parts of Asia as a means of capturing colorful aquarium fish. (The cyanide stuns the fish, but kills the coral animals.) Global warming has now been tied to the threat to coral reefs as well; both ocean warming and increased ocean CO_2 concentrations are harmful to the reefs.

Freshwater Systems

Whereas the oceans cover almost three-quarters of Earth's surface, freshwater ecosystems—inland lakes, rivers, and other running water—together cover only about 2.1 percent of Earth's surface. Like the ocean, lakes can be divided into biological zones. The most productive zone in a lake is the shallow-water area along its edge, the **littoral zone**, whose outer boundary is defined as the point at which the water is so deep that rooted plants can no longer grow. Then there is the **photic zone**, which starts at the surface of the lake and extends down to the point at which sunlight no longer penetrates strongly enough to drive photosynthesis. Finally, there is the area beneath the photic zone, the **profundal zone**, in which photosynthesis can't be performed (**Figure 30.35**).

Nutrients in Lakes

Lakes can be naturally **eutrophic**, meaning nutrient-rich, or **oligotrophic**, meaning nutrient-poor. A lake that has few nutrients will have relatively few photosynthesizers, and this in turn means few animals. Oligotrophic lakes generally are clear (and deep), while eutrophic lakes often have an abundant algal cover.

Nutrient enrichment can happen naturally over time, or it can take place because of human intervention. The latter variety is known as **artificial eutrophication**, a process that sometimes is undertaken intentionally just to increase the yield of fish from a lake or pond. There is such a thing as having too *many* human-added nutrients in a lake, however, and this can be big trouble for this ecosystem. An overabundance of such nutrients as phosphorus or nitrogen brings on an overabundance of algae—a so-called **algal bloom**—and this eventually means an overabundance of

dead algae falling to the bottom of the lake (**see Figure 30.36**). When this happens, decomposing bacteria flourish, and they are using *oxygen*—so much of it that the fish in a lake can suffocate. How do these harmful levels of phosphorus or nitrogen get into aquatic ecosystems? Fertilizers from lawns or agriculture, construction that disturbs bedrock, sewage, and detergents are some of the sources (though in many states, phosphate-containing detergents have been banned for decades). This is one form of human environmental impact that in some instances is reversible in a straightforward way: Reduce the nutrients that are flowing into the ecosystem.

Estuaries and Wetlands: Two Very Productive Bodies of Water

There is one important type of aquatic water system that always straddles the line between a freshwater and saltwater habitat. This is the **estuary,** an area where a stream or river flows

Figure 30.35
Zones in a Lake
The littoral zone starts at water's edge and extends out to the point at which rooted plants can no longer grow. The photic zone starts at the lake's surface and extends down to the point at which sunlight no longer drives photosynthesis. The profundal zone is the area beneath the photic zone in which photosynthesis cannot be performed.

photic zone

profundal zone

littoral zone

Figure 30.36
Too Many Nutrients
A pond with an overabundance of nutrients has experienced an overgrowth of algae—an algal bloom—that may be detrimental to other life-forms in the pond.

into the ocean. Estuaries rank with tropical forests and coral reefs in being the most productive ecosystems on Earth. The cause of this productivity is the constant movement of water—the same force at work in the ocean's intertidal zone. Ocean tides and river flow are constantly stirring up estuary silt that is rich in nutrients. Plants and algae thus get an abundance of these substances, grow in abundance, and pass their bounty up the food chain. In **Figure 30.37** you can see an image of an important estuary in the United States, the Chesapeake Bay.

Another important type of aquatic ecosystem is **wetlands**, which are just what they sound like: lands that are wet for at least part of the year. "Wet" here covers a variety of conditions, from soil that is merely temporarily waterlogged—a prairie "pothole" in Minnesota, for example—to a cover of permanent water that may be several feet in depth (the bayous of Louisiana). Wetlands go by such specific names as swamps, bogs,

marshes, and tidal marshes; the overwhelming majority of them are freshwater **inland wetlands**, rather than the ocean-abutting **coastal wetlands**, which can be freshwater or salt water (**see Figure 30.38**). Wetlands are sites of great biological productivity, and they serve as vital habitat for migratory birds. It's been estimated that the amount of wetlands in the United States has been reduced by 55 percent since the arrival of Europeans—120 million of the 215 million acres that were originally wetlands have been drained or paved over. Currently, the state of Florida and the federal government are working on one of the most far-reaching and expensive habitat restoration efforts in history, as they try to bring the largest expanse of wetlands in the nation, the Florida Everglades, to a healthier condition over the next 20 years.

Life's Largest Scale: The Biosphere

You've been reviewing life on a very large scale, which is to say the aquatic ecosystems and biomes of the Earth. But there is one larger scale of life yet, which is the collection of *all* the world's ecosystems, the **biosphere**, which is also thought of as that portion of the Earth that supports life. Much could be said about the biosphere, but here's just one observation about life on this global scale. The Earth is about 40,000 kilometers or a little more than 24,000 miles in circumference, with life existing along any line you could draw all around the globe. If we look at life in its *vertical* dimension, however, it has been found to exist only about 3 kilometers or 1.8 miles below Earth's surface and 8.3 kilometers or about 5.2 miles above sea level (in the upper Himalayas). This means that all of life on Earth exists in a band that is at most about 10 kilometers or 6.2 miles thick, with the vast bulk existing in an even narrower span. Actually, this is all the life that we know of in the *universe*, so that if we could imagine ourselves traveling by cosmic elevator, we could potentially come up from the molten center of Earth, pass through the bacteria lodged between underground rocks, and then move through a tissue-thin layer of a green living world, after which we might never see anything like it again, no matter how far out into the universe we traveled. Taking this perspective, the really remarkable thing is not how much life there is, but how little.

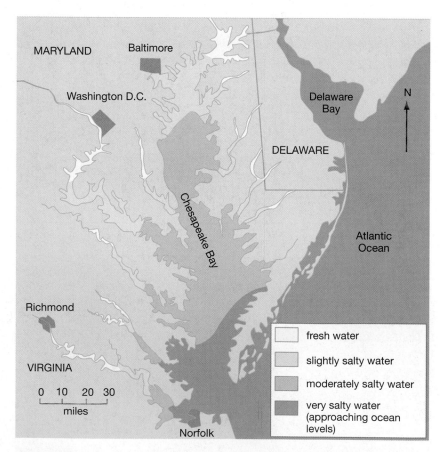

Figure 30.37
Where Salt Water and Freshwater Meet
The Chesapeake Bay is one of the largest and most productive estuaries in the world. In it, the freshwater of 19 principal rivers meets the salt water of the ocean, forming zones of different degrees of saltiness. Each zone supports specific communities of organisms. (Redrawn from A. J. Lippson and R. L. Lippson, *Life in the Chesapeake Bay*. Baltimore: Johns Hopkins University Press, 1984.)

On to Animal Behavior

Animals have figured prominently in the ecology you've been reading about in the last two chapters. The chapter coming up, however, will put a magnifying glass to one aspect of animal life, animal behavior. In the living world, only animals possess the communications network known as a nervous system. The internal instant-messaging that comes with this system has given animals an unrivaled ability to move, sense, seek, and react. In addition, more than any other living things, animals seemingly are able to *choose* from among an array of different possible actions. So, why do animals behave as they do? That is the subject of Chapter 31.

a

b

Figure 30.38
Inland Wetland and Coastal Wetland

a A bird's nest in a prairie wetland in the United States.

b A view from an observation tower across the sawgrass prairie of Florida's Everglades National Park.

Chapter Review

Summary

30.1 The Ecosystem Is the Fundamental Unit of Ecology

- An ecosystem is a community of organisms and the physical environment with which these organisms interact.

- Very large-scale ecosystems are known as biomes, which are defined by the types of vegetation within them.

30.2 Abiotic Factors Are a Major Component of Any Ecosystem

- The approximately 30 chemical elements that are vital to life are known as nutrients. Along with water, nutrients move back and forth between abiotic (nonliving) and biotic (living) domains on Earth in a process called biogeochemical cycling.
TUTORIAL 30.1.2: A Chemical Cycle Model

- Elements can be stored in living things, transferred between them, or transferred between them and the abiotic domain.

- Carbon comes into the living world through the plants, algae, and bacteria that take in atmospheric carbon dioxide to perform photosynthesis. Animals obtain their carbon from these photosynthesizing organisms. Carbon moves back into the atmosphere in the form of carbon dioxide, which plants and animals give off as a product of their respiration, and as a product of their decomposition after death.
TUTORIAL 30.1.1: The Carbon Cycle

- Carbon dioxide makes up a small but critical proportion of the Earth's atmosphere. It is vital to life and may greatly affect global temperature.

- Prior to this century, nitrogen entered the biotic domain mostly through the action of certain bacteria that have the ability to convert atmospheric nitrogen into forms that can be used by living things. Other bacteria have the ability to convert this organic nitrogen back into atmospheric nitrogen, completing the cycle. Human beings now use industrial processes to convert nitrogen into a biologically usable form, thus greatly expanding agricultural productivity.

  TUTORIAL 30.2.1: The Nitrogen Cycle

- As with carbon or nitrogen, all of Earth's water is either being cycled or is being stored—in such forms as glaciers or polar ice. As little as 0.5 percent of Earth's water is available as fresh, liquid water, with about 25 percent of this being groundwater.

  TUTORIAL 30.3.1: The Hydrologic Cycle

30.3 How Energy Flows through Ecosystems

- Plants and other photosynthesizers are an ecosystem's producers, while the organisms that eat plants are its consumers. Every ecosystem has a number of feeding or trophic levels, with producers forming the first trophic level and consumers forming several additional levels.

- A detritivore is a class of consumer that feeds on the remains of dead organisms or cast-off material from living organisms. A decomposer is a special kind of detritivore that breaks down dead or cast-off organic material into its inorganic components, which can then be recycled through an ecosystem.

- The energy-flow model of ecosystems provides ecologists with a powerful analytical tool; it measures energy as it is used by and transferred among different members of an ecosystem.

- Very little of the energy that a given trophic level receives is passed along to the next trophic level. A rule of thumb in ecology is that for each jump up in trophic level, the amount of available energy drops by 90 percent. This explains why there are many plants on Earth relative to the number of large animals.

30.4 Earth's Physical Environment

- The atmosphere is the layer of gases surrounding the Earth. The lowest layer of the atmosphere, the troposphere, contains the bulk of the atmosphere's gases. Nitrogen and oxygen make up 99 percent of the troposphere, with carbon dioxide and small amounts of such gases as argon and methane making up the rest.

- The gas called ozone screens out 99 percent of the Sun's potentially harmful ultraviolet radiation. Human-made compounds such as chlorofluorocarbons (CFCs), used in various consumer and industrial products, can destroy ozone. International action taken on this issue in the 1980s seems to be reducing the amount of ozone-destroying compounds being put into the atmosphere.

- There is near-unanimity among scientists that global warming is taking place and a consensus has emerged that human activity is at least partly responsible for this warming. The greenhouse effect is one mechanism that underlies global warming. In it, higher atmospheric concentrations of gases such as carbon dioxide—released through fossil-fuel burning—trap additional quantities of the heat that comes to the Earth from the Sun. Deforestation is also thought to be a cause of global warming, as trees serve as storage sites for carbon.

  TUTORIAL 30.4.1: Greenhouse Effect

- The Earth appears to have warmed by 1° Fahrenheit in the past century. The effects of this warming can be seen in such phenomena as the shrinkage of Arctic sea-ice and the migration of mountain plant species to higher altitudes. Scientists are now predicting an increase of between 2.7 and 10° Fahrenheit in the coming century, with the consequences of this warming ranging from beneficial in some areas to catastrophic in others.

- Earth is tilted at an angle of 23.5° relative to the plane of its orbit around the Sun, a fact that dictates much about climate on Earth, which in turn dictates much about life on Earth.

  TUTORIAL 30.4.2: Seasons

- Sunlight strikes the equatorial region of the Earth more directly than the polar regions. The differential warming that results produces a set of enormous interrelated circulation cells of moving air, each existing all the way around the globe at its latitude. Each of these cells drops rain on the Earth where it rises, but dries the Earth where it descends. This is why some regions of the Earth get so much more rainfall than others.

  TUTORIAL 30.4.3: Atmospheric Circulation

- More rain will be deposited on the windward side of a mountain range than the opposite leeward side. This rain-shadow effect can cause opposite sides of a mountain range to have dramatically different vegetation patterns.

- A climate is an average weather condition in a given area. Large vegetative formations essentially are defined by climate regions.

30.5 Earth's Biomes

- A biome is an ecosystem defined by a large-scale vegetation formation.

- Six types of biomes are recognized at a minimum: tundra, taiga, temperate deciduous forest, temperate grassland, desert, and tropical rain forest. Polar ice and mountains often are recognized as separate biomes, as are the tropical grasslands called tropical savannas, and the dry, shrub-dominated formations called chaparral.

- Tundra is the biome of the far north, frozen much of the year, but with a seasonal vegetation formation of low shrubs, mosses, lichens, grasses, and the grass-like sedges.

- Taiga is another biome of the north; it includes the enormous expanse of coniferous trees that lies south of the tundra at northern latitudes. The taiga exhibits a great deal of species uniformity because only a few types of trees—spruce, fir, and pine—are the ecological dominants. The region supports large populations of fur-bearing animals.

- Temperate deciduous forests grow in regions of greater warmth and rainfall than is the case with tundra or taiga. These forests, existing over much of the eastern United States, are composed of an abundance of trees such as maple and oak, complemented by a robust understory of woody and herbaceous plants.

- Temperate grassland goes by several names around the world, including prairie and steppes. This biome is characterized by less rainfall than that of temperate forest and by grasses as the dominant vegetation formation. Such regions can be very fertile agricultural land.

- Chaparral is a biome dominated by evergreen shrub vegetation. It is found in a few, relatively small regions of the world that have a Mediterranean climate. These regions have mild, rainy winters and very dry summers.

- Deserts are characterized by rainfall that is low and sporadic. Deserts may be hot, cold, or temperate, but all desert life is shaped by the need to collect and conserve water. Desertification refers to the process by which an area may be transformed into a desert, sometimes through human activity.

- The tropical rain forest biome is characterized by warm, stable temperature, abundant moisture, great biological productivity, and great species diversity. Rain-forest productivity is concentrated above the forest floor, often in the canopy high aboveground. Found in Earth's equatorial region, rain forests are being greatly reduced through cutting and burning.

30.6 Life in the Water: Aquatic Ecosystems

- Ocean or marine ecosystems are most biologically productive near the coasts, with the deep open oceans having a productivity that can be less than that of deserts. Coastal areas benefit from wave actions that bring in nutrients, carry away wastes, and expose more of the surface area of photosynthesizers to sunlight.

- Modern fishing techniques have seriously depleted many once-abundant commercial fish species.

- There is relatively more pelagic ocean life toward the poles, with this concentration decreasing as one moves toward the equator. The food webs in the oceans surrounding Antarctica are based on photosynthesizing phytoplankton, meaning floating microscopic organisms such as algae and cyanobacteria.

- Coral reefs are warm-water marine structures composed of the piled-up remains of generations of coral animals. Coral reefs provide a habitat that results in a rich species diversity.

- Freshwater ecosystems, which include inland lakes, rivers, and other running water, cover only about 2.1 percent of Earth's surface.

- Freshwater lakes are most productive near their shores and near their surface. Lakes can be naturally eutrophic, meaning nutrient-rich, or oligotrophic, meaning nutrient-poor. The human activity known as artificial eutrophication can sometimes introduce too many nutrients into a lake, which can cause some species to flourish (algae, bacteria) while harming other species (some fish).

- Estuaries are areas where streams or rivers flow into the ocean. They are characterized by high biological productivity because of the constant movement of water within them.

- Wetlands, also known as swamps or marshes, are lands that are wet for at least part of the year. Wetland soil may merely be waterlogged for part of the year or under a permanent, relatively deep cover of water. Wetlands are very productive and are important habitats for migratory birds. The amount of wetlands in the United States is estimated to have been reduced by 55 percent since the arrival of the Europeans.

Key Terms

abiotic 693	**gross primary production** 704
algal bloom 723	
aquifer 699	**groundwater** 699
artificial eutrophication 723	**herbivore** 702
atmosphere 706	**inland wetlands** 724
benthic zone 719	**intertidal zone** 719
biodegradable 703	**kilocalorie** 703
biogeochemical cycling 694	**littoral zone** 723
biomass 695	**net primary production** 704
biome 713	**nitrogen fixation** 697
biosphere 724	**nutrients** 694
biotic 693	**oligotrophic** 723
carnivore 702	**omnivore** 702
CFCs 707	**open sea** 719
climate 713	**ozone** 707
coastal wetlands 724	**pelagic zone** 719
coastal zone 719	**permafrost** 714
consumer 701	**photic zone** 719
coral reef 721	**phytoplankton** 721
decomposer 702	**primary consumer** 701
desertification 718	**producer** 701
detritivore 702	**profundal zone** 723
ecosystem 693	**secondary consumer** 701
element 694	**stratosphere** 707
energy-flow model 703	**taiga** 715
estuary 723	**tertiary consumer** 701
eutrophic 723	**trophic level** 701

Understanding the Basics

Multiple-Choice Questions

1. In nature, it is _____ that are primarily responsible for converting atmospheric nitrogen into a form that living things can use.
   a. plants
   b. algae
   c. bacteria
   d. animals
   e. fungi

2. In recent years there has been substantial concern about global warming, which is caused to a great extent by the build-up of the heat-trapping gas
   a. oxygen
   b. nitrogen dioxide
   c. sulfur dioxide
   d. carbon dioxide
   e. neon

3. You walk in on the middle of a nature program on television and see aerial shots of huge expanses of one or two types of coniferous trees. The type of biome you are looking at probably is
   a. chaparral
   b. taiga
   c. tundra
   d. savanna
   e. temperate forest

4. There is a simple food chain in an eastern Oregon desert grassland: Mice are consumed by several species of snakes, which themselves are consumed by hawks. The hawk is a _____ and occupies the _____ trophic level in this ecosystem.
   a. tertiary consumer, fourth
   b. secondary consumer, fourth
   c. tertiary consumer, third
   d. secondary consumer, third
   e. primary consumer, third

5. A dung beetle (can you guess what it eats?) is an example of a
   a. detritivore
   b. omnivore
   c. carnivore
   d. autotroph
   e. homeotherm

6. The so-called greenhouse effect is caused by
   a. incoming long-wave radiation that is trapped as short-wave radiation by the atmosphere
   b. short-wavelength light that is re-radiated as long-wave radiation and is trapped by certain gases
   c. heat that is stored as short-wave radiation by rough surfaces of the Earth and then re-radiated as long-wavelength light
   d. cold air masses that mix with the warm gases and produce warm atmospheric gases
   e. reactions in the stratosphere between these greenhouses gases and ozone, producing long-wave heat

7. Donna is measuring rainfall over a large elevation gradient in the Cascade Mountains of Oregon. As she walks up the west side, she notices that rainfall increases and that when she reaches the top and walks down the east side, it is much drier. This is because
   a. Dry air from the equator flows east at this latitude, dropping moisture on the west side.
   b. Wet air from the tropical equator flows east at this latitude, dropping moisture on the west, and then becomes dry by the time it reaches the eastern side.
   c. Air is flowing from the west and drops moisture as it rises and cools, then absorbs moisture from the land as it descends and warms.
   d. Warm air from 30° latitudes picks up moisture as it travels to 60° north, then drops this moisture as it cools.
   e. The rain-shadow effect means that cool, moist air drops rainfall as it warms as elevation rises and the air travels closer to the Sun.

8. Microscopic floating photosynthesizers such as algae that are the base producers in the oceans are known as
   a. pelagic producers
   b. secondary producers
   c. littoral consumers
   d. zooplankton
   e. phytoplankton

9. A eutrophic lake will have
   a. high nutrient levels, large phytoplankton populations, and low oxygen levels at depth
   b. high levels of nutrients, low phytoplankton levels, high oxygen levels in surface waters
   c. low nutrient levels, large phytoplankton populations, and low oxygen levels at depth
   d. low nutrient levels, low phytoplankton populations, and high oxygen levels at depth
   e. low nutrient levels, large zooplankton populations, and low phytoplankton levels

Brief Review

1. In a large reserve in Idaho, Forest Service wildlife scientists measured wolf populations as well as elk populations, and found that the wolf population was much lower than that of the elk population. If wolves feed on elk, why aren't the population sizes similar?

2. Explain why most deserts across the globe occur at about 30° north and south of the equator.

3. What relationship does climate have to the large-scale vegetation patterns that are seen on Earth?

4. A great many dead fish are found floating on the surface of a pond. What might be the cause of such a fish kill?

5. Name several ways in which the tundra and taiga biomes differ.

Applying Your Knowledge

1. You have learned that ozone is essential to protect life against ultraviolet radiation, and that CFCs deplete ozone, with the result that CFCs have been banned in order to protect ozone. But in Los Angeles and other large cities, there are strict air-emission laws to stop excess production of ozone. What is going on here?

2. One solution that has been proposed to the problem of global warming is to "grow trees." What is the basis of this argument? Knowing that there are different possible fates for different trees, explain whether growing trees is a reasonable solution to global warming.

3. Why have some ecologists argued that organic fertilizer ("green manures"—plant debris, compost, or animal wastes) are better for the soil than Haber-process fertilizer?

4. Slash-and-burn agriculture is common in tropical regions: Farmers cut rain-forest trees, burn them, plant crops for 1–2 rotations, and then move on. Why is this so common: In other words, why are the trees burned and not sold, and why must the farmers then move on?

MediaLab

El Niño and the Greenhouse Effect: How Climate Affects Our Weather, Food, and Water Supplies

It is almost impossible for us to imagine what the world will be like 50 or 100 years from now. Yet that is exactly the challenge facing government agencies and scientists who are planning for world food and water supplies in the next century. Will Earth's biomes, and our food and water supplies, change with forces such as global warming and climatic fluctuation over the next hundred years? To learn more, enter the *CD-ROM Tutorial* for an animated review of Earth's weather and climate. In the *Web Investigation*, you'll examine the forces that shape climate from year to year and from decade to decade. Then, in *Communicate Your Results*, you'll use that information to develop projections of future conditions.

This *MediaLab* can be found in Chapter 30 on your CD-ROM (Tutorial 30.4.2) and Companion Website (http://www.prenhall.com/krogh3).

CD-ROM TUTORIAL

This *CD-ROM Tutorial* reviews concepts about the physical conditions that influence Earth's weather and climate, and explores their impact on the abundance of life.

Activity

1. First, you'll investigate how carbon and other chemicals cycle through ecosystems, and then how human alteration of these cycles can change climate.

2. Next, you will investigate how the relationship between the Sun and the Earth cause seasonal change.

3. Finally, you will examine the implications of geographic and geologic landforms and learn how they influence precipitation patterns and thus climate and weather around the globe.

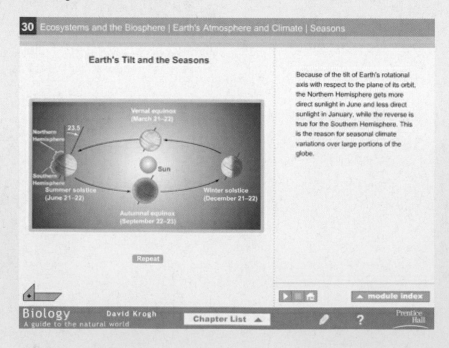

WEB INVESTIGATION

Investigation 1

Estimated time for completion = 15 minutes

What is El Niño, and how does it affect our climate? To begin your investigation, select the Keyword **EL NIÑO** on your CD or Website for this *MediaLab*. Read about the history and influence of the El Niño Southern Oscillation (ENSO). Describe the warm and cold extremes of this phenomenon. How frequently does the warm phase of ENSO occur, and how does it affect different parts of the world? Then select the Keyword **CROP PROSPECTS** to view the United Nations Food and Agriculture Organization maps of current and past years' unfavorable crop prospects and food supply shortfalls in different parts of the world. How might global food supply problems be helped or worsened by ENSO phenomena?

Investigation 2

Estimated time for completion = 15 minutes

Why have El Niño events become more frequent in recent decades? The answer may lie in natural long-term climatic forces, which create a "conveyor belt" of ocean water circulation that last for many years. Select the Keyword **CLIMATE** on your CD or Website for this *MediaLab* to read an article from Oregon State University on this process. Do the phenomena described in the article represent a permanent global shift or a cyclic phenomenon? What is thought to cause the "conveyor belt" of ocean circulation? Is it linked to, or separate from, global warming effects?

Investigation 3

Estimated time for completion = 10 minutes

What will Earth's natural environment look like in 50 or 100 years from now? Select the Keyword **PROBABILITY** on your CD or Website for this *MediaLab* to read an article on the probability of sea-level rise. Have the climate predictions of the 1970s and 1980s been borne out by recent sampling? How large a difference exists between the low and high ends of the reported range of climate estimates? To what do you attribute these differences? Some of the world's largest cities lie at or near sea level. Under the most pessimistic forecasts, how might low-lying areas such as these be affected? Select the Keyword **MAPS** to access a wide range of map resources on the Web.

COMMUNICATE YOUR RESULTS

Exercise 1

Estimated time for completion = 15 minutes

Using the information you have gleaned about the global effects of the ENSO phenomenon, suggest some ways that an understanding of ENSO can be used to protect global food and water supplies from year to year or over longer time periods.

Exercise 2

Estimated time for completion = 15 minutes

Which of the phenomena described in the *Web Investigation* are short term? Which are reversible? Write a 250-word summary of the relative effects of ENSO, oceanic circulation, and global warming on Earth's climate. In your view, what would constitute a worst-case combination of these phenomena? What would be a best-case combination? Can you suggest a different perspective, perhaps from a different part of the world or culture?

Exercise 3

Estimated time for completion = 10 minutes

How certain are climate forecasts? Use the discussion of predictive models and uncertainty from *Web Investigation 3* to discuss the challenges of understanding and forecasting climate accurately. Does the level of certainty change if the prediction is short term (say, 1 to 5 years) rather than long term (say, 20 to 100 years)? Why or why not? Are models useful or misleading in planning for global water and food supply? Debate this question in your class.

31

Animal Behavior

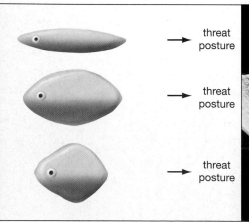

Rivals are red.
(Section 31.3, page 739)

Ready to go.
(Section 31.3, page 741)

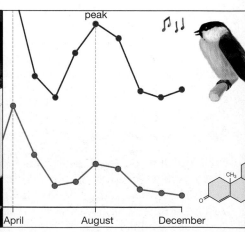

Sex and Songs.
(Section 31.3, page 741)

Genetics, environment, and learning all work together to produce the behavior we see in animals.

My six-year-old daughter Tessa says she would like to be a cat, but I've been trying to talk her out of it. Sure, I say, cats get to lie around all day, and they seem to live pretty interesting lives in some ways. But from my point of view, cats have a real problem: They appear to live in fear to a degree that I think human beings would find intolerable. I point out to Tessa that our cat, Ursa, is afraid of any loud sound, of being turned over on her back, and of traveling in a car on the way to the veterinarian's. Topping everything on Ursa's fear scale, however, is the sound of the garbage truck that stops in front of our house every Monday morning. The noise the truck makes is not particularly loud, but when Ursa hears it, she runs under the bed or pleads to be let out of the house. *Don't let it trap me here!* she seems to feel. The reason I would not like to be a cat, I tell Tessa, is that Ursa is indeed trapped. She is trapped in her fears. She is incapable of learning *not* to be afraid of dozens of harmless things.

But *why* is Ursa afraid of such things? And why does she react to, say, the garbage truck with such specific responses (running under the bed or heading for the door)? Does the garbage truck sound like a predator? How specific does this sound have to be? If the disposal company got a different truck, would it set Ursa off as well? Alas, I don't have any firm answers about Ursa's behavior. But there is a field of scientific study that delves into questions exactly like these.

31.1 The Field of Behavioral Biology

The field is **behavioral biology**, defined as the study of the behavior of animals, and there is no end of questions in it. Why do birds sing? How do they learn their songs? How do sea turtles, when migrating to lay eggs, manage to navigate thousands of miles through a featureless ocean to land on the very beach on which they were hatched? How does each honeybee "know" what to do in a hive that may contain 50,000 workers? (And does behavioral biology apply to human beings? See "Are Men 'Naturally' Promiscuous and Women Reserved?" on page 744.)

Tutorial 31.1.2
The Beewolf

Tutorial 31.1.4
Honeybees

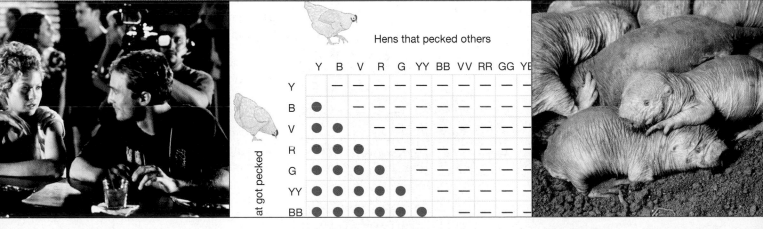

Genetic predispositions
in action?
(Essay, page 744)

To peck or be pecked?
(Section 31.6, page 749)

Truly social.
(Section 31.6, page 750)

Behavioral Biology Asks What, Why and How

Like biology in general, behavioral biology asks three fundamental types of questions. The first of these are "what" questions. What does a given animal actually do? What behaviors does it exhibit? The answers to these questions are interesting in and of themselves, but they are indispensable for answering the field's other two types of questions, the "why" and "how" questions: Why does an animal behave as it does, and how does it manage to exhibit these behaviors?

Answering "what" questions requires observation. Scientists must look at the animals, listen to them, or otherwise measure their behavior in some way. This can be fairly easy to do with animals that can be studied in a laboratory, but difficult and time-consuming for animals that can be studied only in the wild. There is the difficulty of *getting to* the habitats of some of these latter animals—think of whales or moles—and then there is the problem of observing them without disturbing them (**see Figure 31.1**). Scientists who study eagles often must be hidden away in treetop observation huts, while those who study chimpanzees might have to spend years around a given group before the animals will accept them as a part of their surroundings.

Behavioral biology has been answering "what" questions for hundreds of years, but only since the 1930s has it come into its own as a branch of science by systematically pursuing "why" and "how" questions. Two researchers, Konrad Lorenz of Austria and Niko Tinbergen of the Netherlands, led the way in this effort by observing animals in natural conditions and then *manipulating* these conditions as a means of getting at the truth. Tinbergen, for example, observed a species of wasp called a beewolf, so named because it paralyzes honeybees with its sting and then drags the bees down into its nest, where they will serve as food for the wasp offspring. The beewolf's nest lies at the end of a tunnel that it digs in the sand, and the wasps Tinbergen watched made a round trip of about a mile to get the bees and bring them back home.

Watching all this, Tinbergen hit on a question that fascinated him: When returning with prey, how does the beewolf pick out *its* nest from among the multitude that exist? The nests themselves are underground, after all, and the beewolves cover up the entrances to them upon departing to look for bees. He resolved to find out, but this meant he had to have some means of telling one wasp from another. He therefore caught wasps, one by one, and marked them with colored dots. Doing so gave him a feeling that many behavioral biologists have experienced. As he wrote:

> It was remarkable how this simple trick of marking my wasps changed my whole attitude to them. From members of the species *Philanthus triangulum* they were transformed into personal acquaintances, whose lives from that very moment became affairs of the most personal interest and concern to me.

Tinbergen had observed that beewolves circled their nests in ever-widening and ever-higher loops before leaving on a hunting expedition. With this in mind, he hit on a possible explanation for how they found their nests upon returning. As they circled, perhaps they were putting into memory the position of their nests relative to surrounding landmarks. If so, he reasoned, *disturbing* these landmarks after a wasp left should interfere with its ability to locate its nest upon return. Tinbergen did just this, moving the sticks and rocks around a given nest, while leaving the nest entrance unaltered. When the beewolves returned to these changed surroundings, they were disoriented; four

Figure 31.1
Observing without Disturbing
A biologist studies mountain goats, but from a distance so as not to interfere with their behavior.

feet or so above the ground they would stop, zigzag back and forth, hang in the air, and then circle in a wide loop, starting the whole process over again. This test and others convinced Tinbergen that the wasps were locating their nests through the use of visual landmarks.

Setting aside Tinbergen's specific finding, note how this one piece of research answered all three types of questions fundamental to behavioral biology. *What* does the beewolf do? (Circles its nest before leaving.) *Why* does the beewolf circle its nest before leaving? (So it can find its nest upon returning.) *How* does the beewolf know where its nest is? (It associates landmarks with its nest.) From Tinbergen's time to the present, observation has been joined to ever more sophisticated experimental methods. Scientists now videotape their animal subjects, sequence their DNA, and track them with Earth-orbiting satellites (**see Figure 31.2**).

Proximate and Ultimate Causes

When we look at the "why" questions in behavioral biology, a helpful concept is that of proximate and ultimate causes of behavior. "Proximate" can be thought of as any cause of behavior that involves the *physiology* or physical functioning of an animal. It is an "immediate" or "triggering" cause of a behavior. Meanwhile, while the ultimate cause of a behavior generally is linked to survival and reproduction—to evolution in the final analysis, as you'll see.

To take one example of this proximate/ultimate distinction, consider the behavior of the small "Belding's" ground squirrels (*Spermophilus beldingi*) found in the Sierra Nevada mountains of California (**see Figure 31.3**). About two months after they are born, the males among these squirrels leave their burrow and never return to it for the rest of their lives. The *sisters* of these travelers, meanwhile, remain close to their "natal" burrow for their whole lives. So, why do the males leave while the females pretty much stay put? To find out, researchers Kay Holekamp and Paul Sherman tested a number of hypotheses, but the evidence they sifted through pointed to only two factors: hormonal influence early in life and growing to a certain minimal body size. To test the first idea, they injected a group of *female* squirrels with the "male" hormone

testosterone soon after birth, after which these squirrels and their mother were returned to the wild. The result? The injected females showed the same kind of leaving-home behavior as males. With respect to the body-size idea, the researchers found that males tended to leave about the time their weight reached a minimum of 125 grams, which presumably allowed them to survive on their own. (Remember the observation made earlier about how difficult fieldwork can be in behavioral biology? Consider that Holekamp, Sherman, and their colleagues spent nearly *six thousand hours* looking through binoculars at squirrels they had marked.)

On one level, then, what seems to cause male Belding's squirrel dispersal is early hormonal influence and the achievement of a certain minimum weight. But these are proximate causes in that they are "triggering" and

Figure 31.2
High-tech Tracking
An Atlantic loggerhead turtle, outfitted with a satellite radio transmitter, is prepared for release into the ocean. Researchers use satellite transmitters to track the migratory movements of animals.

Figure 31.3
Much-studied Squirrel
An adult Belding's ground squirrel surrounded by three young on the eastern slopes of California's Sierra Nevada mountains. Young males of the species leave home at an early age while young females remain near the burrows in which they were born.

involve the physiology of the squirrels (hormones, weight). They don't tell us much about the *ultimate* causes of the dispersal. In looking at the latter issue, the researchers tested a number of hypotheses—that the males could avoid food shortages by leaving, or that they would have better mating prospects away from home—but none of these ideas was supported by the evidence.

Another possibility, however, was that the males were dispersing to avoid the possibility of "inbreeding," meaning mating with close relatives. Here the evidence was supportive. The movement of males away from the home burrows resulted in a complete absence of mating among kin, the researchers observed, and, after breeding, the males who mated most moved farthest away. Why do juvenile male Belding's squirrels disperse? The *ultimate* cause may be to avoid inbreeding, thereby gaining a reproductive benefit.

Natural Selection and Ultimate Cause

This account may make sense to readers who have been through the evolution section of this book (Unit 4), but other readers may need one more string tied together. Male Belding's squirrels obviously do not sit around and think: "Well, I'm about two months old now; I'd better leave the burrow or face the possibility of breeding with one of my relatives." (For an account of the harmful consequences of inbreeding, see "Lessons from the Cocker Spaniel" on page 346.) In the *absence* of such conscious thought, how can something like "the consequences of inbreeding" motivate squirrels to leave home? Here's how.

Through evolutionary time, squirrel ancestors who mated with close kin left offspring who suffered the consequences: physical maladies and perhaps an earlier death, meaning these offspring would leave fewer offspring of their own. Meanwhile, the squirrels who did not engage in inbreeding had relatively *more* offspring who went on to have pups of their own.

Now, what behaviors would help keep male squirrels from mating with relatives? How about moving into new environments in a less fearful manner? How about displaying an interest in foraging in these new environments? Squirrels that exhibited these behaviors would be led farther from their natal homes at younger ages and would thus be

less likely to mate with close relatives. Genes can help influence the development of such behaviors, and, as noted, squirrels with these genes would leave relatively more offspring than squirrels without them. Thus, genes for leaving home would be *selected* for inclusion in the squirrel genome and, as a result, everlarger portions of the squirrel population would come to possess these genes over time. Through such a process, an entire species can become predisposed to carry out certain behaviors. These *behaviors*, then, are selected for inclusion in the animal's repertoire because they aid in reproduction. What is at work here, in short, is **natural selection**, defined as a process in which the "fit" of an organism with its environment determines those traits that will be passed on with greater frequency from one generation to the next. Those male squirrels that leave home have a better fit with their environment, in that they have more offspring. Thus, through natural selection, a given trait (leaving home) will be passed on with greater frequency to offspring than an alternative trait (not leaving home).

As may be obvious, natural selection is not just important in squirrels. It is critical to behavioral biology because it *generally* lies at the root of ultimate causes. So strong is its influence that, when behavioral biologists are looking for the ultimate cause of this or that behavior, the first question they will ask is: How does this behavior stand to improve the chances that this organism will pass on more of its genes to future generations?

31.2 The Web of Behavioral Influences

Natural selection may lie at the heart of ultimate causes, but it is linked to many factors that produce behavior. There are, for example, environmental influences. The tide comes in, winter begins, an animal is cared for by its mother; all these things influence behaviors in animals. Likewise, there are internal influences, such as the hormones at work in the Belding's squirrels. Then there is learning, which all animals are capable of to some degree. When describing the motivational influences on animal behavior, it is convenient to think of influences as falling into categories such as these, and there will be some of this compartmentalizing in what follows. First, however, let's take a look at why, in real life, motivational factors are very difficult to separate from one another.

A young spider that builds a circular or "orb" web will, on its very first try, build a web that is perfect, and this is true even of spiders that, from birth, have been isolated from other spiders. We might surmise from this that web-building is purely a matter of "instinct," to use a term that behavioral biologists rarely employ anymore. After all, a spider can build its web without learning anything about this task from other spiders, and it can do so from the start without making mistakes. If we think a little more closely about this, however, we can see that, from the moment of its birth, each spider is engaged in a learning process that will yield benefits when it comes time to build a web. Simply by virtue of being alive, a spider moves about, it gets oriented to its environment, it manipulates objects with its legs, and so forth. It learns by doing these things, and what it learns can be applied to web-building.

Following on this thought, consider what happens when a spider involved in web-building is transferred to a web that is less far along. It generally will go back to the *earlier* stages of building necessary to construct a proper web. In this instance, the spider is acting partly on the basis of sensory input—it senses that the web lacks some early-stage construction and acts accordingly. Thus we can see that web-building is not a rote activity, hardwired in the spider's genes, that must be carried forward in the same way from beginning to end each time.

Broadening this idea further still, the behavioral biologist P. J. B. Slater has noted that baby rats will not urinate for the first time unless their genital area is stimulated, usually by their mother licking them soon after they are born. You might think that there could be few activities more instinctual than urination, but a baby rat who is isolated before its genital area is stimulated will hold its urine until its bladder bursts. Now think about the influences on the baby rat's behavior. Its physiology makes it capable of urinating, and it does not "learn" to urinate by practicing or watching others. Rather, there is an environmental influence (genital licking) that is seemingly required to set this behavior in motion.

These examples serve to make the point that, if we examine any animal behavior closely enough, we are likely to see a complex interplay at work involving genetic predisposition, learning, and environmental influence. To be sure, not all of these factors are equally at work in every behavior. Genetic predisposition will be strongly at work in some behaviors, while learning will play a relatively larger part in other behaviors. In line with this, we can think of behaviors as existing in a spectrum that runs from very rigidly defined (through genetic influence) to very "plastic," or open to alteration.

We will now begin looking at what might be called the component parts of animal behavior, starting with influences on behavior that can be thought of as largely internal. Once we have reviewed these influences, we will look at animal learning, which inevitably has a strong external component to it (animals learn from the behavior of other animals, they learn by interacting with their environment, and so forth). We will then conclude this part of the chapter by looking at an example of how, in the real world, these various motivational components work together as inseparable parts of a whole.

31.3 Internal Influences on Behavior

Reflexes

Any time you step on a sharp object with a bare foot, you instantly lift your foot up, of course, but something else happens as well. Your *other* leg will straighten into "locked" position, so that you automatically support yourself with it. This behavior is one of many **reflexes**, meaning rapid, automatic responses to stimuli. Reflexes are the epitome of an internal influence on behavior, as animals do not need to be taught anything about them and they are completely **stereotyped**, meaning they are performed precisely the same way each time by individuals of the same species.

Action Patterns

Reflexes are very simple actions. Looking beyond them to more complex behaviors, it is possible to see *patterns* of action in animal behavior that seemingly are as stereotyped as reflexes. The time-honored example of such a pattern is the egg-rolling behavior of the greylag goose, which was studied by both of the founders of behavioral biology, Niko Tinbergen and Konrad Lorenz. For a female greylag trying to hatch chicks, an egg *outside* the nest is a matter of great importance, because eggs that are not properly incubated won't hatch. As such, females have a very specific means of getting wayward eggs back inside the nest's perimeter: They reach just beyond the egg with their bills and roll the egg back toward the nest until it safely retrieved. (**see Figure 31.4**). Now, here's the interesting part. Any small, round object placed just outside the greylag's nest will elicit this behavior. Tinbergen and Lorenz constructed "eggs" much larger than any the goose would ever have laid and used beer bottles to boot. No matter; placed just outside a greylag's nest, these objects would prompt the same behavior. Indeed, no object at all needed to be present for the goose to carry out this action. Tinbergen and Lorenz snatched eggs away from geese that were in the middle of their rolling, and they still continued the behavior right through to the end (meaning until they got their beaks inside the nest).

At one time, an important concept in behavioral biology was that of a *fixed action pattern*, meaning a stereotyped behavior that, once triggered by a stimulus, is always carried through to its conclusion. The greylag's egg-retrieval behavior is a paradigm of the fixed action pattern in that geese can do it on their first try, it is stereotyped (all greylags do it the same way), and it is always carried through to its conclusion. A more informative way to state this last phrase might be: It is always carried through to its conclusion *regardless of sensory input*. Remember it doesn't matter whether the goose's senses tell her an egg is under her bill or not.

In recent years, the "fixed" in "fixed action pattern" has fallen from favor among behavioral biologists, because there seem to be very few behaviors that are absolutely "fixed" in this way. Nevertheless, the concept is useful because it sets up a standard against which behaviors that are not so rigid can be measured.

Releasers in Action Patterns

Given how specific action patterns can be, it is perhaps not surprising that they are often set off by very specific stimuli. A key concept here is that of a **releaser**, meaning the critical element in an action or object that triggers an action pattern. Consider yet another species studied by Niko Tinbergen, the stickleback fish. When breeding season arrives, the heads and undersides of the male three-spined stickleback take on a red coloration. Male sticklebacks establish mating territories, and they do not welcome the presence of *other* male sticklebacks within them. Thus, a male, perceiving another male on the border of his territory, will go into an aggressive pose called the head-down threat posture to warn the potential rival away.

This clearly is a stereotyped behavior, but more to the point, it is triggered by a very specific stimulus: the color red. By building clay models of male sticklebacks, Tinbergen was able to establish that a life-like image of a male stickleback is not required to induce the head-down threat posture; indeed, some of Tinbergen's models lacked gills, fins, even a tail. No matter; as long as the undersides of these crude models were painted red, the head-down posture was triggered (**see Figure 31.5**). Tinbergen even observed sticklebacks adopting the posture when red mail trucks drove past his laboratory window. The releaser for the stickleback's action pattern happened to be visual, but odors, touch sensations, tastes, and sounds can also stimulate behavior.

Figure 31.4
Action Pattern
When a female greylag goose discovers an egg outside the nest, she reaches just beyond the egg with her bill and rolls the egg back toward the nest. Once initiated, this action pattern is always carried through to its conclusion.

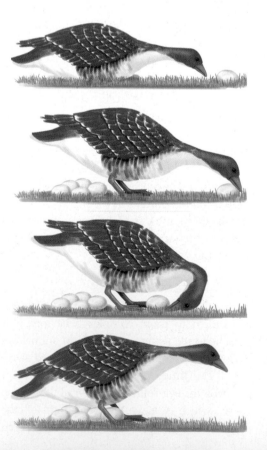

Orientation Behavior: Taxis

Animals constantly need to put their bodies in the correct orientation relative to their environment, after which they often will move in connection with this orientation. Orientation is sometimes partly learned—as it was with the beewolf orienting itself to its nest —and this can be an important part of the long-distance navigation of some species, as you'll see.

Orientation can also be influenced by genetic predisposition. There are numerous examples of animals being genetically predisposed to move toward or away from a stimulus, a phenomenon known as **taxis**. A positive taxis is a movement toward the stimulus; a negative taxis is a movement away from the stimulus. Certain moths will fly away from any sound in a given high-frequency range—the range of tones emitted by the bats who prey on these moths. This is an example of a negative taxis. Conversely, many moths also have a well-known tendency to fly toward artificial light sources, which is a positive taxis. Other kinds of stimuli that elicit taxis are odors, the direction of water movement, and the Earth's gravitational field.

Biological Rhythms: The Internal Clock

Many squirrels hibernate each year, many birds migrate south for the winter, and crickets begin their mating calls each day near sunset. But what drives these cycles of behavior? Animals have internal clocks that help govern behavior on time scales that run from a few minutes to a year.

To establish that internal clocks exist, it is necessary to show that periodic behaviors really are internal—that they are not being triggered by external "cues," such as the fading of daylight or the coming of colder temperatures. To tease out the truth, behavioral biologists have often moved animals to the laboratory, where they can be isolated from all environmental cues. The male *Teleogryllus* cricket, when confined to a lab where temperature is held constant and light kept on around the clock, will still chirp almost 11 hours per day, and will begin each new bout of calling about 25 or 26 hours after the end of the previous bout. The first lesson here is that the chirping is prompted by an internal clock; chirping will continue in a fairly fixed way in the absence of any environmental cues.

The second lesson has to do with how internal clocks and environmental cues work together to produce a behavior. Note that the cricket's internal clock does not keep a strict 24-hour time, but instead takes 25 or 26 hours to complete a full cycle. However, if *Teleogryllus* crickets are subjected in the laboratory to 12 hours of light followed by 12 hours of darkness, they will eventually adopt a very fixed 24-hour pattern of chirping. This indicates that an environmental cue—the change from light to dark—is serving to **entrain** or initiate each new cycle of the cricket's internal clock. In the real world, of course, the cricket's internal clock is being entrained by the setting of the Sun, which means the clock is being synchronized each day to Earth's 24-hour cycle of light and darkness.

Internal clocks have a wide distribution in the living world. Indeed, when we look not just at behavior but at metabolism as well, it may be that all living things have such clocks. Internal cycles, such as those of the cricket's, that last about a day are called circadian rhythms, from the Latin *circa*, meaning "about" and *dies*, meaning "a day." More formally, **circadian rhythms** are biological cycles that function independently of environmental cues and that are roughly synchronized to Earth's 24-hour rotation. Sunlight (or its absence) is the most important environmental cue in entraining circadian rhythms, but various types of biological rhythms are entrained by such factors as

What stimulates the threat posture?

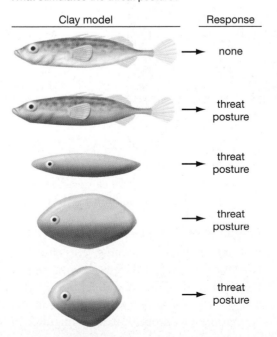

Figure 31.5
Stimulating an Action Pattern

Niko Tinbergen constructed various clay models of sticklebacks to determine what stimuli would trigger the threat posture. He discovered that crude models, bearing little resemblance to a fish, would elicit much stronger reactions than exact stickleback replicas as long as the ventral surfaces of the models were painted red.

Biological Rhythms and Sports

We human beings have our own internal clock that functions much as internal clocks do in the rest of the animal world. The circadian rhythms this clock controls can have some surprising effects on our functioning.

Scientists have known for decades that a human internal clock exists, but they have made great strides in recent years in figuring out its detailed mechanisms. The primary governor of the human master clock is a small group of cells, called the suprachiasmatic nucleus or SCN, located in an area of the brain called the hypothalamus. SCN cells have genes that can turn on or off, thus regulating a cascade of bodily functions having to do with sleepiness, and with general physical functioning. In the complete absence of external cues, the human internal clock takes an average of 24.2 hours to complete one full cycle. As with the circadian clocks of other animals, however, the human clock is entrained by an external cue—the presence or absence of sunlight.

Here's how one part of this system works. The presence of sunlight is signaled to the SCN through pathways coming from the eyes. When sunlight is not present, the SCN sends a signal to the pineal gland, located in the brain, to raise its production of the hormone melatonin. This hormone acts to lower body temperature and bring on sleepiness. When sunlight reaches our eyes the next day, it entrains a signal to the pineal gland that essentially says, "Day is here again; time to be alert; shut down melatonin production."

This system, the product of millions of years of evolution, runs into problems when humans undertake the very modern practice of travel by jet airplane. Getting off a plane, a traveler is in a different location, where daylight may begin, say, three hours earlier than in the old locale. The critical thing is that it takes several days for the timing of sunlight in a new locale to reset our internal clock. Thus, a person whose alarm clock is saying, "get up" may have an internal clock that is saying, "still time for sleep." This is what it means to be jet-lagged.

We normally think of jet lag as some minor annoyance that makes us feel "off" for a few days in a new location. But it can have practical effects as well. Research has shown that athletic performance tends to peak in late afternoon, as measured by such things as reflex speed and quick muscular bursts. In line with this, athletes tend to report that they perform best from about 3 to 6 p.m. and worst before 9 in the morning or after 9 at night. So what happens when an athlete *travels*? Researchers looked at the results from the National Football League's "Monday Night Football" and found that the outcome of the games was significantly skewed in favor of West Coast teams, who won 71 percent of their home games, while the East Coast teams won only 44 percent of theirs. These results held up even when Las Vegas point spreads were taken into account. Why should this be? Monday night games begin at 6 p.m. Pacific Coast time, which means 9 p.m. East Coast time. The East Coast athletes, then, are starting to perform at one of the worst times of the day in terms of their circadian rhythms.

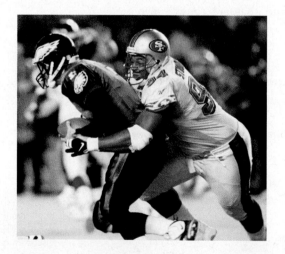

Biological Rhythms and Football
Monday Night Football games played on the West Coast start at 6 p.m. pacific time, but for players from East Coast teams, this is effectively 9 p.m. Human athletic performance starts to diminish about 9 p.m., and this drop in performance is reflected in the scores of Monday Night games. Their outcome has been skewed significantly in favor of West Coast teams.

temperature, tides, and even the phase of the moon. We human beings have our own circadian rhythms, which you can read about in "Biological Rhythms and Sports," above.

Longer Internal Cycles: Annual Clocks

Not all biological rhythms are circadian rhythms; some internal clocks keep time over the course of a year. Consider the garden warbler (*Sylvia borin*), which breeds in the springtime in Europe, but then migrates in the fall to central and southern Africa, where it stays until the following spring. During the summer and winter, the garden warbler is active in the daytime. Like many migratory birds, however, it *migrates* only at night.

So what do warblers do in the absence of environmental cues? When they are kept in cages with a constant, artificial day-night cycle over a whole year, they start becoming active at night just at the times of the spring and fall when they would normally be setting off on their migrations. Moreover, closely related species of warbler will stay in this state of "migratory restlessness" in rough accordance with how long they would normally migrate; longer-distance migrants remain restless for longer than shorter-distance migrants. Remarkably, at about the same time that free-flying warblers will be changing direction in an actual migration, caged warblers will change their orientation and end up facing the same direction as their flying counterparts. At least 20 species of birds have been shown to have annual internal clocks, as have some fish, reptiles, amphibians, and mammals. If you look at **Figure 31.6**, you can see one of the means by which scientists can determine the preferred orientation of caged birds during different times of the year. (To learn how animals manage to navigate over long distances, see "How Do Sea Turtles Find Their Way?" on page 755.)

The Effects of Hormones

Hormones are substances that, when released in one part of an organism, go on to prompt physiological activity in another part of the organism. (See page 565 for more on hormones.) What effect can hormones have on behavior? Male cichlid fish perform a "fanning" behavior that keeps sufficient oxygen flowing around eggs their mates have laid. Fanning is normally prompted by environmental cues, such as the amount of dissolved gases in the water around the eggs. Males that have been injected with the hormone prolactin, however, will engage in fanning behavior whether or not they have any eggs to fan. In another example, male songbirds generally increase their singing in the spring and reduce it in the fall. The proximate cause is a change in the birds' testosterone levels. If you look at **Figure 31.7**, you can see how close the relationship is between testosterone and singing in one bird species, the willow tit.

31.4 Learning and Behavior

We now turn to another influence on animal behavior, **learning**, which can be defined as the acquisition of knowledge through experience. Learning is a behavioral influence that has a relatively strong external component to it, in that animals learn by receiving sensory input from the outside world. (In contrast, an influence such as an internal clock can function in the absence of sensory input.) It's important to

note, however, that learning is never purely a matter of experience; it always has a genetic component to it. As you'll see, animals can be predisposed by genetics to learn—or not learn—certain things.

Establishing Relationships: Imprinting

There is a type of learning that actually was brought to public attention because of research done in behavioral biology. It is called

Figure 31.6
Ready to Fly
Some birds display a "migratory restlessness" during their species' normal migration season, even when kept in captivity away from their natural environment and its behavioral cues. Such birds may also orient themselves in the same direction as they would during migration. This cage has an inkpad on the floor and cone-shaped sides lined with paper. Scientists can judge the bird's preferred orientation by looking at the ink marks the bird lays down on the pad with its feet.

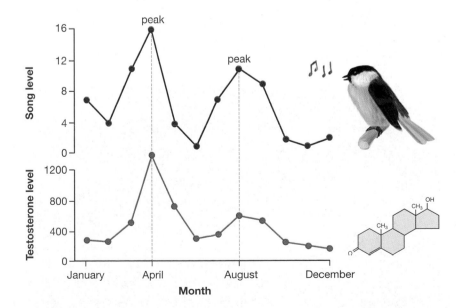

Figure 31.7
Hormonal Inspiration
The close correlation, visible in the graph, between testosterone levels and singing in the male willow tit suggests that testosterone release is the proximate cause of singing in these birds.

imprinting: a process of learning that results in one animal preferentially associating with another. As famously manifested in Konrad Lorenz and his geese, imprinting is very simple in concept. Early in their lives, goslings must, for their own safety, follow their parents on journeys outside of the nest. Lorenz discovered, however, that goslings could be made to follow *him*, rather than their parents, if he raised the youngsters from the time they were hatched and if he walked away from them mimicking the proper goose call (**see Figure 31.8**). Thus, the goslings were genetically predisposed not to follow "parents" per se, but to follow any large organism that (a) was near them early in life and that (b) exhibited some specific behaviors. Other experiments have shown that the object of attachment need not even be an organism; ducklings can be made to imprint on a moving decoy, and other birds have imprinted on toy balloons.

The Sensitive Period

If you look at **Figure 31.9**, you can see a representation of one of the key elements in the process of imprinting, the **sensitive period**, which can be defined as the time-interval during which an animal can learn to respond to a given stimulus. Ducks in this study learned to follow a model duck after a single 30-minute exposure to it early in life. As you can see, the "window" of their sensitive period lasted only about a day—the first day of their lives—with their sensitivity to imprinting peaking at 15 hours after hatching.

Imprinting is seen in mammals as well as birds, and it need not be offspring who are imprinted. In goats, a mother will imprint on her kid during a sensitive period that peaks in the hour following the kid's birth. During this critical first hour, a *different* kid can be placed with the mother and she will raise it as if it were her own.

Other Forms of Learning

Learning can take many other forms, only a few of which will be briefly noted here. A very simple form of learning employed by all animals—and many other living things—is **habituation**, meaning a reduction in a response, based on repeated exposure to a stimulus that has no positive or negative consequences. Humans experience habituation all the time. When a fire alarm goes off, we may jump and turn our heads in panic. A minute later, however, we are walking calmly out of the building accompanied by what is now merely an irritating noise. Habituation is learning that allows an organism *not* to react to a stimulus; it is valuable in that it saves an animal time and effort.

More complex forms of learning come into play when experience teaches animals either to associate a new stimulus with an existing stimulus (classical conditioning) or to associate one of their own actions with a particular positive or negative outcome (operant conditioning).

The time-honored example of classical conditioning involves psychologist Ivan Pavlov's dog, who was taught to associate

Figure 31.8
Imprinting in Geese
By raising these goslings from the time they were hatched, and by walking away from them mimicking the proper goose calls, Konrad Lorenz was able to make the goslings imprint on him rather than on their mother. This resulted in the goslings following him, just as they would have their mother.

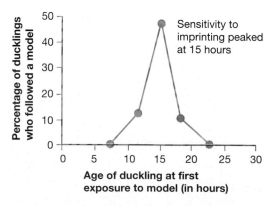

Figure 31.9
The Sensitive Period
Animals tend to be capable of imprinting only during relatively brief time frames, known as sensitive periods. The graph shows the proportion of ducklings who learned to follow a model duck after a single 30-minute exposure to it. The ducklings were most sensitive to imprinting when they were 15 hours old.

meat powder it was given with the sound of a ringing bell. Meat powder alone would cause a salivation reflex in the dog, but Pavlov wanted to see what would happen if he rang a bell just before delivering the meat powder. He found that, after five or six pairings of bell and meat powder, the sound of the bell alone would elicit salivation in the dog. Thus **classical conditioning** is learning to respond in a customary way to a new stimulus that has been paired with an existing stimulus. In the case of Pavlov's dog, a customary response (salivation) was elicited when a new stimulus (the sound of a bell) was paired with an existing stimulus (food delivery). In nature, such learning may allow animals to respond more quickly to events in their environment. Should, for example, a certain posture in one animal be a stimulus that signals "aggression likely" to a second animal, this stimulus can set off responses in the second animal—such as hormonal releases—that will allow it to prepare for the upcoming encounter.

Closely related to classical conditioning, is **operant conditioning,** which is learning that occurs when the animal's own behavior brings about a response that has either negative or positive consequences. A hungry rat is enclosed in a cage that has at least one lever that can be pushed. Accidentally doing this during the course of its wanderings, the rat finds that a pellet of food is delivered into a small hopper (**see Figure 31.10**). It will not be long before the rat learns that *every* push of the lever will produce the food. Each time the food is delivered, the behavior of lever-pushing is reinforced. This is positive reinforcement because the rat is getting something it likes from the behavior, but negative reinforcement is possible as well; rats might receive, for example, a mild electric shock upon pressing a lever. Note the difference between this form of learning, operant conditioning (or trial-and-error learning), and classical conditioning. In classical conditioning, two *stimuli* are paired—food and a bell, for example—while in operant conditioning it is a *behavior* of the organism that elicits an effect (lever-pressing elicited food delivery in the example just given).

Certain animals are capable of more sophisticated forms of learning yet; they can learn through **imitation,** meaning they can copy a behavior they are observing in another animal. Finally, there is **insight learning**, meaning a form of learning in which an animal makes associations between objects or events that it has previously regarded as unrelated. Put more simply, insight learning is the ability to reason. The most famous example of it comes from experiments Wolfgang Kohler did with chimpanzees in the early twentieth century. Presented with bananas suspended above their heads, chimps could learn either to put sections of a pole together, which enabled them to whack the bananas and bring them down, or to stack boxes one on top of another and reach the bananas with their hands (**see Figure 31.11**). In a separate experiment, the bananas were moved higher and the chimps were given access both to boxes and to pole sections. The result? They put the pole sections together *and* stacked the boxes, after which they climbed the boxes and whacked the bananas with the pole.

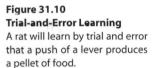

Figure 31.10
Trial-and-Error Learning
A rat will learn by trial and error that a push of a lever produces a pellet of food.

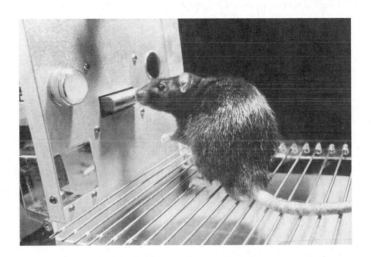

Figure 31.11
Insight Learning
In his famous experiment of the early twentieth century, Wolfgang Kohler observed chimpanzees manipulating boxes and pole sections in order to reach bananas suspended above their heads. To do so, he realized, they had to be carrying out insight learning, which requires the ability to reason.

Are Men 'Naturally' Promiscuous and Women Reserved?

We're the gas; they're the brakes.

—Male character from the movie *Ed tv*, talking about the differing sex roles of men and women.

How different are human beings from other animals? Are we really the free agents we imagine ourselves to be? Or, like other animals, are we driven by genetic influences that are beyond our conscious understanding? One school of thought holds that the limits on human free will are essentially the limits imposed by culture—the sum of what we learn from those around us. Another school of thought holds that our evolutionary heritage shapes our behavior in ways we are only beginning to understand. It holds that we are genetically predisposed to act in certain ways. There is, in fact, a scientific discipline dedicated to investigating this idea. It is called evolutionary psychology.

Evolutionary psychologists maintain that the effects of evolution can be seen in many human behaviors. One clearly defined hypothesis in the field has to do with sexual activity. It is that, when we compare human males and females, males have a genetic predisposition to be sexually promiscuous while females are genetically predisposed to be sexually reserved. Under this view, human beings are much like the many animal species in which the norm is male sexual eagerness and female sexual choosiness (a phenomenon known as "female choice").

To support its claim, evolutionary psychology has a lot of hurdles to clear. First it has to demonstrate that human males *are* promiscuous relative to females. Second, this behavioral difference must be universal—it must be carried out in nearly every culture, instead of just here and there. (If it weren't we'd write it off to the effects of culture, rather than genetics.) Finally, note that this is *evolutionary* psychology; it posits an evolutionary basis for behaviors that people are disposed to. That means these behaviors must, at one time at least, have served an *adaptive function*: They must have helped foster reproductive success in human beings.

So, does the "sexual difference" hypothesis pass these tests? Take the last test first. Why would human females benefit, in an evolutionary sense, from being sexually reserved while males would benefit from being sexually eager? Evolutionary psychologists would say: Because the two sexes must pursue different "strategies" to be as reproductively successful as possible. This has partly to do with the different "investments" each sex has in a given offspring. Females carry fertilized eggs to term, which takes some time; and many will then nurse and care for the young they bear, which takes a lot of time. A human female usually bears only one child at a time, takes nine months to bring it to term, and may nurse it for years. During all this time, she is unlikely to ovulate and thus is unlikely to get pregnant again. This means that sexual activity during this period will yield her nothing in terms of passing on her genes; no matter how much she copulated, she would only bring one child into being. She thus has a great investment in this one child and is much better served by being *selective* about who its father is. You might say that she benefits by concentrating on quality, rather than quantity.

Now consider the human male. Compared to the female, his minimum investment in each offspring is very small. For him, there is no carrying a child to term or breast feeding. Theoretically, he can sire a huge number of children by securing more mating partners. This is not to say, however, that for our male ancestors the optimal strategy was simply to copulate with as many partners as possible. Remember that reproductive success means not simply having children, but having children who themselves go on to have children. Given the considerable needs of human infants, it seems likely that males had to join with females in providing care for offspring, if those offspring were going to survive long enough to have children themselves. The lasting pair-bond, institutionalized today as marriage, seems to be a universal feature in human cultures. In the view of some evolutionary psychologists, the optimal strategy for our human male ancestors would have been to help raise children within a marriage bond while simultaneously pursuing "low investment" mating possibilities with other women. Thus males would be eager to copulate with other women, even if these women were already bonded to other men. (Through such matings, it would be possible for a man to pass along his genes while having another man expend the time and effort necessary to raise the child.)

For females, however, there was no such thing as a low-investment mating because of the huge minimum investment a female has to make in connection with each child. Our female ancestors may have pursued outside affairs, but unlike the case with males, not just any man would do for these affairs. Women had to be conscious of the quality of potential mates. They had to be looking out for the same kinds of things that would have benefited them most, in a genetic sense, within the marriage bond: men who showed signs of being good providers (of such things as food and protection) or who showed signs of having "good genes," in the sense of being physically robust.

If we accept this theoretical underpinning, the question is whether it is reflected in *modern* male and female mating behavior. As a starting question, are contemporary human males promiscuous compared to females? A number of tests of this hypothesis have been carried out on U.S. college students, and all of them point in the same direction. In one, involving college students in Hawaii, students were approached by a good-looking member of the opposite sex and asked one of three questions: "Would you like to go out on a date with me tonight?" "Would you go back to my apartment with me tonight?" or "Would you have sex with me tonight?" Of the women who were approached, 50 percent agreed to the date, 6 percent agreed to go to the apartment, and 0 percent agreed to have sex. Of the men who were approached, 50 percent agreed to the date, 69 percent agreed to go back to the woman's apartment, and 75 percent agreed to have sex. (The men who declined sex pleaded excuses such as prior commitments or fiancées.) But is a disparity like this universal, or is it restricted to U. S. college students? Well, consider that prostitution exists throughout countless cultures around the world, and the fact that, with very few exceptions, it is men who are seeking casual sex from women in this arrangement and not the other way around.

Along another line, researchers have asked male and female college students about standards of acceptability they have for casual sex partners, as compared to potential marriage partners. These studies indicate that the standards men have for marriage partners do not apply when it comes to casual sex partners. Men greatly relax their standards in connection with such qualities as charm, education, independence, and generosity, while women relax their standards only slightly. This is consistent with the idea, noted earlier, that our female ancestors always needed to be concerned about quality in a sex partner, while our male ancestors only sometimes did.

Critics of evolutionary psychology are unimpressed by such arguments. Looking at the Hawaiian college study, they respond that it is no surprise that women wouldn't want to have sex with a stranger. Men are *stronger* than women and any woman who accepted the offer could be putting her life

Different Strategies?
Evolutionary psychology holds that men and women are genetically predisposed to differ in their approach to sexuality. Pictured is a scene from the movie *Ed tv*.

at risk. And what about such things as prostitution? As the science writer Natalie Angier has written, "Women are said to have lower sex drives than men, yet they are universally punished if they display evidence to the contrary . . . a woman can't sleep around without risking terrible retribution, to her reputation, to her prospects, to her life." In short, differences in sexual conduct can be explained by culture.

Evolutionary psychology has been hailed in some quarters as the dawning of a new day in that it holds promise of elucidating aspects of the human condition that everyone has observed for millennia but that no one could ever explain. Why is there a bias toward older-man/younger-woman coupling, and not the other way around? Why do children face a greater risk being killed by step-parents than by biological parents? Evolutionary psychology says that there are answers to these questions and that the answers have to do with human evolution. Note that no one is saying that human beings have no free will in certain areas. To say that human beings have a tendency to behave in a certain way is not to say that they must behave in these ways.

Critics of evolutionary psychology hold that it is a dangerous discipline in that, on a personal level, it can provide an excuse for bad behavior, while, on a mass scale, it can breed a sense of fatalism about societal conditions. (Husbands who cheat on their wives are simply "doing what their genes told them to.") To this, evolutionary psychologists would reply that society can only begin to deal with its problems when it understands all the forces that are at work in generating them—including the forces of evolution.

Figure 31.12
Bird Song Made Visible
The songs of birds can be represented in a visual form called a sonogram. Pictured are sonograms for two birds, both of them members of a subspecies of white-crowned sparrow that makes its home year-round on the central and northern California coast (*Zonotrichia leucophrys nuttalli*). Sonograms display the pitch of bird songs—the highness or lowness of their sounds—as measured in kilohertz (kHz), meaning thousands of vibrations per second. They also show exactly when, in time, each element of the song took place. Note that the entire white-crown song lasts about two seconds. These particular sonograms also provide information about something else: Regional dialects that exist in white-crown songs. One of these birds was recorded in the Ft. Mason area of San Francisco, on the city's northern edge. The other bird was recorded about two miles away, on Alcatraz Island. Note that the first two large-scale elements of the birds' songs are very similar, but that the songs differ significantly in their middle and later elements. By making many such recordings, researchers discovered that all the white-crowns in a given location share a distinct dialect and that these dialects are passed on from one generation to the next. (Adapted from Andrea Joan Jesse, *Cultural Homologues and Insular Dialects in Nuttall's White-Crowned Sparrow*, 1997.)

31.5 Behavior in Action: How Birds Acquire Their Songs

Having seen some internal influences on behavior, and having looked some modes of learning, let's see how internal influences and learning work together to create one particular type of animal behavior, singing in birds.

Looking at song acquisition across all bird species, there is a great deal of variation in the degree to which internal influences or learning seem more important. A bird called the eastern phoebe, when reared by scientists completely away from other birds, will go on to develop the normal song for its species, even when deafened at an early age by the scientists. Moreover, when young eastern phoebes are exposed to specially edited tapes of adult phoebes—edited to yield an unusual song—the student phoebes are unaffected by what's on the tapes, meaning they will still sing the standard phoebe song. In short, singing in eastern phoebes seems to be rigidly defined, mostly by genetic predisposition. At the other end of the spectrum, there are birds that, when deafened while young, go on to produce mere screeches rather than their species' songs.

In all the species of the birds that we think of as "songbirds," learning is very important to song development. Indeed, song is clearly transmitted from one generation to the next in songbirds and a given species of songbird may have local or regional dialects of its song just as, say, Italians have regional dialects of their language.

Bird song is a fleeting thing, of course, with most songs lasting no more than a few seconds. If you look at **Figure 31.12**, you can see how scientists are able to discern the subtle differences between one song and another. What you see is a sonogram: a visual representation of a bird's song. By comparing sonograms, it's easy to see how two songs are alike or different in their minute details.

Most songbirds hatch in the spring and then merely listen to the songs of adult male birds until sometime in the later summer or fall, when the adults stop singing, not to resume until perhaps February of the following year. (It is usually male birds that are doing the singing in latitudes such as ours, though female singing is common in the tropics.) In general, young songbirds do no singing of their own until nearly a year after their birth. With the coming of their second spring, their testosterone levels rise and this in turn prompts them to begin singing, with their song development following a predictable pattern over a period of weeks. At first, their songs may be a quiet, jumbled series of chirps and whistles. Over time, they begin to use the "syllables" of their species' songs, though the order in which these syllables appear will vary. Finally, their songs "crystallize" into the clear, orderly songs of their species, with the number of songs that adults sing varying from one to 1,000.

There is a songbird, called the white-crowned sparrow (*Zonotrichia leucophrys*), whose song development follows this general script while providing some variations that

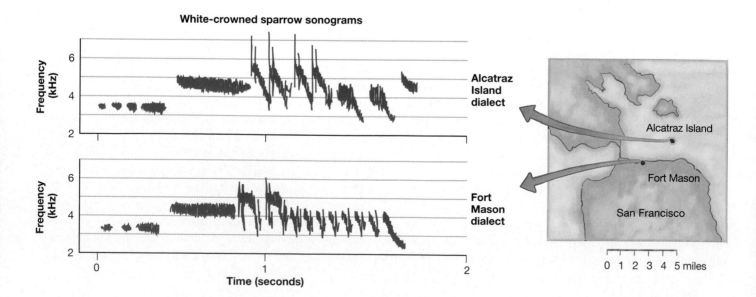

are instructive about the interplay of internal influences and learning in bird song (**see Figure 31.13**). You need not pay attention to the *details* of learning in the white-crowned sparrow; the details of song learning will vary greatly from one species to the next. Just note the intricate way in which internal influences and learning interact in one behavior in one species.

White-crowned sparrows raised *in captivity* will follow the pattern of song acquisition just described: they listen to songs in their first spring and summer but don't themselves begin singing until they are perhaps six months old. In nature, however, things are different. A type of white-crown found year-round in the San Francisco area begins singing within six weeks or so of birth and may progress to fully crystallized song as early as three months after birth, meaning about September. Why would there be a difference between singing in nature and singing in the lab? The pressures of nature. As year-round residents, the San Francisco white-crowns don't fly into an area in the spring and then establish territories. Rather, they establish territories as early as their first autumn. One function of bird song is to announce, "I have a territory here." Young white-crowns will, like many species, extend this practice by "countersinging," meaning a male, upon hearing the song of a nearby male of its species, will repeat the exact song he has heard, thus setting off a back-and-forth duel, like two children in an argument, each of them saying, "I'm still here."

Internal influences and learning are also on display in white-crowns in the way they acquire their songs. You saw earlier that there is often a so-called sensitive period for animal learning—a kind of window in which an animal is able to acquire certain skills or information. In laboratory-raised white-crowns, the sensitive period starts at about 10 days after birth and extends until about 50 days after birth. A white-crown deafened prior to the opening of the sensitive period eventually will sing individual notes, but it will never learn to sing its species' song. Meanwhile, white-crowns raised through part of their sensitive period in nature, and then taken to the laboratory, will begin singing the following winter in the dialect of the area in which they were hatched. Two points are worth observing about this. First, note that these birds are *learning* the white-crown song months before they ever start *practicing* it themselves. Indeed, the learning window will be closed completely (in their first summer) before these lab-reared birds ever sing a note (the following winter). Second, learning is important enough in song acquisition that white-crowns learn not just their species' song, but local or regional variants of it, which they are able to recall months after last hearing them.

But what about internal influences? Interestingly, all white-crowns who are reared in isolation from birth eventually sing nearly identical versions of a kind of standard white-crown song. In other words, there seems to be a built-in version of the white-crown song that becomes modified with local dialects only when birds are raised in the wild. Beyond this, isolated white-crowns who are exposed to tapes of *other* species' songs will ignore the other birds' songs entirely and go on to sing the basic white-crown song. White-crowns are thus genetically disposed to learn their own song while ignoring the songs of others.

31.6 Social Behavior

Animals vary greatly in their living arrangements. Some live lives of almost complete isolation, while others are in constant contact with other members of their species. At one extreme, consider the female mason wasp (genus *Manobia*), which in her few weeks of life has exactly one moment of contact with another adult member of her species—the moment in which she mates with a male wasp. Other than this, she

Figure 31.13
What Factors Produce Birdsong?
Male white-crowned sparrows are genetically predisposed to sing a certain type of song—the white-crowned sparrow song. However, males will sing varying dialects of their species' song in accordance with the area in which they were raised. Thus, learning is important in white-crowned song acquisition as well.

Figure 31.14
Social Animal
Zebras are social animals: they live in groups. The members of a zebra family groom each other and warn one another of predators—behaviors that benefit each individual. Living in groups also has costs, such as the spread of disease. These zebras are part of a herd in Kenya.

Figure 31.15
Defense through Cooperation
Group defense can be one of the benefits of group living. Male musk oxen form a circle like this one when predators threaten. These oxen are in Canada.

spends her whole life working alone in the service of her offspring. She lays eggs in hollowed-out plant stems, paralyzes caterpillars, and provisions the egg sites with the caterpillars (which will serve as food for her young). You might think this would leave her with at least the possibility of having some contact with her offspring once they have matured, but she will die before they emerge from their plant-stem homes.

On the opposite end of the spectrum are zebras, which are never far from members of their species. Plains zebras travel in "family" and "bachelor" groups, with the families being composed of a dominant stallion, up to six mares, and offspring (**see Figure 31.14**). Young males leave these families to join bachelor groups of up to 10 individuals. These two basic zebra units may then become part of herds of zebras that number into the tens of thousands. Zebras do more than just live in close proximity, however. Family members groom one another (for up to 30 minutes at a time), play with one another when young, keep lookout for predators to protect sleeping family members, and show what appears to be great loyalty to one another. (Zebras move at the pace of the *slowest* family member, never leaving it behind, and they may attempt rescue missions for family members who have become separated from a group that is under attack.)

Why Live Alone—or Together?

From the human perspective, the life of the mason wasp seems lonely and difficult compared to that of a zebra, but of course we have no indication that wasps or zebras would see things this way. It will come as no surprise that the unsentimental logic of natural selection is once again at work in channeling animals toward solitary or social living. A species will exhibit social behavior to the degree that such behavior aids in the survival and reproduction of individuals in that species. Social behavior always has *costs* as well as benefits; animals will be social to the extent that the costs are outweighed by the benefits. There is a cost to many animals in merely living together, in that parasites and diseases can move easily among them because of the physical contact they have. This very factor is decimating North American populations of honeybees, *Apis mellifera*, who pass among themselves a tiny mite that makes its home in their respiratory tubes, suffocating them. Beyond this, think of a hungry zebra moving to a better grazing territory. It incurs a cost by having to move at the speed of the slowest member of its family.

Against the costs, many benefits can come from group living. There is group defense, which you can see a vivid example of in the musk oxen pictured in **Figure 31.15**. One of the oxen's predators, meanwhile, is the wolf, which benefits from *hunting* in groups. The V-pattern that geese maintain while migrating reduces the amount of energy each animal needs to expend in flight. Pigs and penguins conserve heat by keeping close together in winter.

Dominance Hierarchies

One form of organization that often arises among social animals is the **dominance hierarchy**: a persistent power ranking in an animal population that gives those of higher rank the ability to control some aspect of the behavior of those of lower rank. This control often results in higher-ranking individuals gaining better access to resources, such as food and mates. If you look at **Figure 31.16**, you can see an exact accounting of a dominance hierarchy in chickens. This is, literally, a pecking order, documented by a researcher who observed a group of 12 hens. It is indeed well-ordered in that the hen at the top, "Y," pecked any and all beneath her while the hen at the bottom, "BR," pecked no one but *got* pecked by everyone.

Dominance hierarchies usually are not this rigid. There may be a single dominant individual and then "all others" who have no ranking among them; animal A may be dominant over B, but not over C, and so forth. Wolves (*Canis lupus*) employ a caste system with each wolf pack composed of a dominant male and female that get privileged access to food and that are the only members of the group who breed. Then, in the middle, come adults that do not breed. Finally, at the bottom, are other adults that can literally live on the periphery of the group.

Arrangements such as this may seem unfair from the human point of view, but fairness doesn't count in the struggle for survival and reproduction. At root, dominant hierarchies seem to exist because certain individuals in a group—the dominant ones—benefit from such hierarchies and can enforce them.

Territoriality

Territoriality can be defined as the effort an animal makes to keep other animals out of a given area. In general, territoriality refers to efforts to keep members of one's *own* species from entering an area. Why would a robin be more concerned about another robin than about a rabbit? Because fellow robins will be competing for the same resources—food, nesting space, and mating partners. Members of the same species are not necessarily territorial, however; any two may inhabit so-called "home ranges," meaning relatively large, and sometimes overlapping, areas that neither lays claim to.

The term *territoriality* may bring big, fierce creatures to mind, but most songbirds are territorial, as are animals as small as aphids. Territoriality can be seasonal; birds that are territorial during mating season may congregate in large flocks at other times of the year, completely unconcerned about "turf." Mating partners or larger groups of animals may defend territories, but in general it is individual males that carry out the practice. The songs that male songbirds use to attract females often serve the dual purpose of warning other males to stay away.

Territoriality and dominance hierarchies are mutually exclusive to a certain degree. A male bird who has a territory is master of his own domain with respect to his species. There is no dominance hierarchy within his territory, in other words. A given species, however, can move back and forth between territoriality and dominance hierarchies. Some lizard species defend territories during breeding season, but only until their population reaches a certain density level; then they abandon territorial defense and shift to a dominance hierarchy.

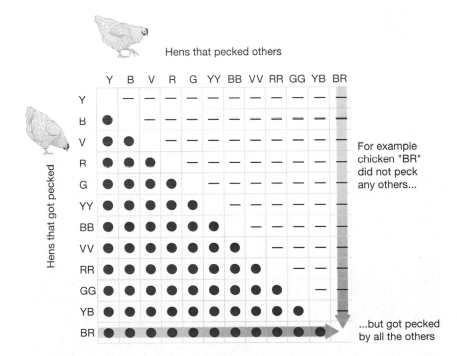

Hens that pecked others

For example chicken "BR" did not peck any others...

...but got pecked by all the others

Figure 31.16
Pecking Order
By recording which hens pecked others in a group of 12 hens, A. M. Guhl documented a pecking order. It is a "linear" pecking order, because dominance in it runs in a straight line: Not one hen ever pecked a hen above her in the hierarchy, but all hens pecked hens beneath them. Thus, hen "Y" pecked hen "B" and "V" and so on, down the line, but got pecked by no one. Meanwhile hen BR, at the bottom, got pecked by every hen in the group while pecking no one. Pecking orders are one form of dominance hierarchy; it is rare in nature, however, for a dominance hierarchy to be this rigid. Adapted from Aubrey Manning and Marian Stamp Dawkins, *Introduction to Animal Behaviour* (5th ed.), London: Cambridge University Press, 1998.

Eusociality: Life in Animal Societies

Some animals live in such highly organized, cooperative groups that they are referred to as "eusocial" or "truly social" species. Ants and termites, along with some bees and wasps, provide the most familiar examples of eusocial species. However, one species of mammal, an unpleasant-looking East African rodent called the naked mole rat, has

Figure 31.17
Extremely Social Society
Naked mole rats are unusual mammals in that they have a eusocial system similar to that of bees and some other insects. In each colony, a single dominant female mates with a few males. The remaining members of the colony forage for food and maintain the burrow system. Reproduction in subordinate females may be suppressed by pheromones in the dominant female's urine.

Figure 31.18
Honeybee Society
Out of thousands of females in a honeybee colony, only one bee, the queen, is likely to bear offspring. Each colony is composed of numerous female worker bees, a small number of male drones—whose sole function is to mate with a queen—and the queen bee herself. Shown is a group of workers surrounding a queen in a North American hive.

been shown to be eusocial (**see Figure 31.17**). Lots of species exhibit complex cooperative relationships, but eusociality means something more than this. To be **eusocial**, a species must be organized into a caste system in which there is a division of labor—different members of a population will consistently perform different tasks. Further, the young of a eusocial species must be raised through cooperative care, meaning care provided by many members of the group. Let's see how eusociality plays out in a colony of honeybees (*Apis mellifera*).

Honeybees as a Female Society

Honeybee colonies, like ant colonies, are essentially societies of females. In a hive of perhaps 20,000 bees, only a few hundred will be male "drones," and they are around only in the spring or summer—long enough to rise to treetop level in a comet-like swarm, chasing after one of the queen bees that have assembled from various hives at a mating site. Of the many drones assembled, only 10–15 will actually mate with a queen during one of her "nuptial flights." Each drone that is successful dies in the process, however, and a similar fate awaits drones who aren't successful; once mating is done, they will be kicked out of their hives or killed.

The sperm the queen receives in her week of nuptial flights serves her for a lifetime of prodigious egg laying; she will produce up to 2,000 fertile eggs a day for years. Nearly all of the offspring that hatch from these eggs are female; they are the hive's "workers," and they are well named, for it is they who will maintain the hive, forage for food, store the food away, care for newly laid eggs, and more. It is they who will do everything for the colony, in other words, except lay eggs and mate with the queen (**see Figure 31.18**).

How Does a Worker Know What To Do?

Over their brief adult lives of perhaps six weeks, every worker bee takes on, in a predictable order, nearly all the worker tasks a hive has to offer. For the first three days of her life, a worker is primarily a cleaner of the cells the bee larvae are stored in. As the days pass, she becomes primarily a larvae feeder, then a hive construction worker, then an entrance guard and food storer, and finally a forager, going out to secure nectar, pollen, and water for the colony. Within this structure, however, a worker's life is one of surprising flexibility. After becoming a construction worker, for example, she still engages in some cell cleaning; and throughout her life, she spends a good deal of time resting and "patrolling" the hive.

Importantly, there is no chain of command in a colony—no group of workers communicating the message "more food needed now" or "cell cleaning needed over here." How, then, does all this work get organized among tens of thousands of bees? Bees are prompted to act either because of environmental conditions (the temperature of the hive, for example) or because of signals or cues they receive from other bees. The "signals" are explicit acts of communication, as with the famous "waggle-dance" bees perform to inform their fellow workers of the location of food sites. Quite often, however, bees are reacting to "cues" they get from other bees that simply *imply* a given condition. Take, as an example, a cue that researcher Thomas Seeley confirmed that has to do with "unloading time" at the hive.

In a well-fed hive, forager bees gather food only from flower patches that have lots of nectar. When a hive is near starvation, however, the foragers aren't so choosy; then low-yield flower patches will do. So, how does a forager know whether to be choosy or not? How is she informed of the nutritional status of the colony, in other words? Her informational source is the length of time it takes her to unload her food. Providing the cues are the "food-storer" bees, which receive the food the foragers bring back and then process it into honey and pack it away in the hive. It takes a returning forager a relatively long time to make contact with a food-storer bee in a well-fed hive, but a relatively short time in a starving hive. Why? Because in a well-fed hive, the food storers have plenty to keep them busy—there is plenty of food to store away. If, however, a forager can make contact with a food storer within 15 seconds of entering the hive,

the forager knows the colony is low on food and will start paying visits to low-yield sites. This is but one example of how life in the colony is self-organizing; each bee's behavior is shaped by the behavior of other bees.

31.7 Altruism in the Animal Kingdom

In a world in which survival is so tough, why would one animal do anything to help another? Certain animals do this. Zebras come to the aid of other members of their group, and a honeybee will sting any intruder into the nest, dying in this act of group defense. It turns out that honeybees and other eusocial species hold clues to explaining **altruism**, meaning a costly or risky behavior carried out by one animal for the benefit of a second. To help explain altruism, we'll look here at the fire ants (genus *Solenopsis*) that exist throughout the southeastern United States, to the consternation of so many residents there (**see Figure 31.19**).

Fire ants are much like honeybees in that they generally live in colonies in which there is but a single queen that lays all the colony's eggs, most of which hatch as females. As with honeybees, these female fire ants are the workers of the colony, while the relatively few males have no function in the colony except to mate with the queen. The reproductive difference between fire ants and honeybees, however, is that a *single* male is father to all the ants in a colony headed by a single queen (rather than the several males that will mate with a queen bee in her nuptial flights). The female workers in the ant colony are completely sterile—they don't even have ovaries—meaning a worker will play no direct role in reproduction during her entire lifetime. Rather, like a honeybee worker, an ant worker *tends to* the queen's offspring.

Figure 31.19
Fire Ant Society
Like honeybees, fire ants live in societies that are organized around the care of the one breeding female, the queen, and the eggs she lays.

Well, given all you've seen in this chapter about how natural selection favors those who produce the most offspring, how can we account for the behavior of these worker ants? They have abandoned reproduction altogether and instead are caring for the offspring of others. To see what sparks this extreme form of altruism, let's look a little more closely at fire ant reproduction.

Fire ants are unusual in that the males among them are born from eggs that are not fertilized by male sperm. The queen can choose to fertilize a given egg—from the store of sperm she collected during mating—or not to fertilize it. Critically, the males that result from this process are haploid: They have but a *single* set of chromosomes, because they have no father. Meanwhile, the queen and all the workers are diploid; just like humans, they have *two* sets of chromosomes, one set from each parent.

What does this mean for ant offspring? If we consider two worker sisters in a colony, each of them will get exactly the same genes from her father as the other. To simplify, think of their father as having only a single chromosome, chromosome A, rather than a set of chromosomes. Inheritance here is simple: Each of the father's sperm contains chromosome A, meaning both sisters will inherit chromosome A (**see Figure 31.20**). Contrast this with the case of two *human* sisters. Now the father starts with two sets of chromosomes, because he is diploid. To simplify, think of these as chromosomes A and B. When the human father's sperm are formed, any given sperm may get chromosome A, or it may get chromosome B. Thus, while two ant sisters will inherit a given paternal chromosome *every* time, two human sisters are likely to inherit a given paternal chromosome only *half* the time.

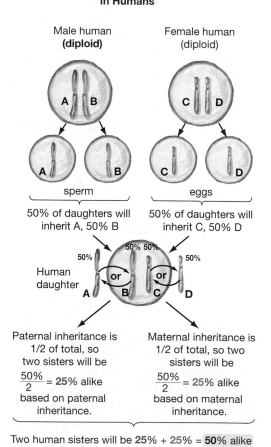

In Fire Ants

Male fire ant **(haploid)**   Queen fire ant (diploid)

Parent generation

Gametes

A   A
sperm

C   D
eggs

100% of daughters will inherit A

50% of daughters will inherit C, 50% D

Fire ant daughter

Paternal inheritance is 1/2 of total, so two sisters will be
$$\frac{100\%}{2} = \mathbf{50\%}\text{ alike}$$
based on paternal inheritance.

Maternal inheritance is 1/2 of total, so two sisters will be
$$\frac{50\%}{2} = \mathbf{25\%}\text{ alike}$$
based on maternal inheritance.

Two Fire ant sisters will be **50% + 25% = 75%** alike

In Humans

Male human **(diploid)**   Female human (diploid)

Parent generation

Gametes

A   B
sperm

C   D
eggs

50% of daughters will inherit A, 50% B

50% of daughters will inherit C, 50% D

Human daughter

Paternal inheritance is 1/2 of total, so two sisters will be
$$\frac{50\%}{2} = \mathbf{25\%}\text{ alike}$$
based on paternal inheritance.

Maternal inheritance is 1/2 of total, so two sisters will be
$$\frac{50\%}{2} = \mathbf{25\%}\text{ alike}$$
based on maternal inheritance.

Two human sisters will be 25% + 25% = **50%** alike

Figure 31.20
Altruism and Degrees of Relatedness
Why do female worker ants forgo reproduction and instead care for their developing sisters? The answer lies in the unusually close genetic relationship between sister ants. Because their father is genetically haploid, sister fire ants end up sharing an average of 75 percent of their genes. The drawing on the left illustrates why this is so. Meanwhile, the drawing on the right illustrates why *human* sisters end up sharing only 50 percent of their genes on average. Female worker ants forgo reproduction because, by tending to their larval-stage sisters, they can pass on more of "their genes"—the genes they have in common with their sisters.

The upshot of this, as you can see in Figure 31.20, is that fire ant sisters are likely to have 75 percent of their genes in common. So what? Well, think of this figure in relation to the genetic similarity between a regular diploid mother and her daughter—say, a human mother and her daughters, or an ant mother and hers. A daughter gets 50 percent of her genes from her mother (and 50 percent from her father). Mothers and daughters, then, have 50 percent of their genes in common. Meanwhile, *fire ant sisters* have 75 percent of their genes in common. This means that a worker ant has more in common genetically with one of her sisters than she would with any daughter she might give birth to. Why should evolution act to make worker ants forgo having children of their own? Because more of "their genes"—the genes they have in common with their sisters—can be passed on by tending to their larval-stage sisters than by having their own offspring.

Most biologists think that this general phenomenon accounts for a good deal of the altruism seen in the animal world. It shows up in its most vivid form—animals forgoing reproduction entirely—in species like ants that have the odd genetic twist of having one haploid parent. But as you'll see, it exists in less dramatic form in other species. The essential story here is that individuals can increase their genetic representation in future generations by aiding the reproduction of their *relatives* as well as by fostering their own reproduction.

This concept of increasing genetic representation may ring a bell with readers who have been through the evolution section of this book. An important idea introduced there is that of evolutionary *fitness*, meaning the relative genetic contribution of an individual to a succeeding generation in a population. If the average organism in a population has 10 offspring, but one individual in this population has 15 offspring, that individual's genes will become slightly more prevalent in the next generation. This individual thus has a greater evolutionary fitness than the average individual in the population. Research that began in the 1960s fostered the recognition, however, that evolutionary fitness can include more than *individual* fitness. The behavior of animals can be influenced by **inclusive fitness**:

an individual's relative genetic contribution to a succeeding generation, made both through itself *and* through relatives that have reproduced because of assistance the individual has provided.

The evolutionary mechanism that underlies inclusive fitness is **kin selection**, meaning the selection of genes that increase the inclusive fitness of an individual. A fire ant will attack an invader to the colony but may well die in doing so. The genes that influence the altruistic defense behavior have been *selected* for inclusion in the ant's genome because they increase the inclusive fitness of the ants—they increase the total genetic contribution the ant will make, indirectly, to future generations through the survival and reproduction of her sisters. A grizzly bear is not nearly so likely to make this kind of selfless sacrifice for her sisters. Why? Because sibling bears will not have as much in common genetically as sibling ants; given this, an altruistic bear would *reduce* its inclusive fitness by dying for its sisters.

These concepts, first articulated by William Hamilton in the mid-1960s, are important because they helped solve a riddle that had puzzled biologists since the time of Darwin: Given natural selection, why would one animal do anything that helped further the reproduction of another? The answer is that, in most altruistic acts, animals are only *incidentally* helping other animals. At root, most altruistic acts are a way for animals to pass on more of their own genes, only this time through relatives.

Inclusive Fitness at Work

Apart from the altruism evidenced by such species as ants, how do we know that inclusive fitness actually is at play in the world? Consider the numerous animal species that are cannibals—that eat members of their own species. Paper wasps (*Polistes fuscatus*) eat offspring of other paper wasps, but they prefer to eat offspring of distant relatives, rather than offspring of their sisters. Then there is an animal called the spadefoot toad, which, as a tadpole, eats members of its own species. The spadefoot begins its predation, however, by merely nipping at a fellow tadpole. If the prey tadpole is a sibling, the predator tadpole releases it unharmed; but if the prey is not a relative, it will be eaten (see Figure 31.21). Such behavior requires, of course, that animals can recognize how related they are to animals they encounter—an ability that has now been demonstrated in a number of species.

Reciprocal Altruism

Does altruism in animals always involve relatives? No, but if relatives are not in the picture, altruism does require a reasonable prospect of a return favor. This you-scratch-my-back-and-I'll-scratch-yours phenomenon is known as **reciprocal altruism**: an exchange of altruistic acts by individuals over time.

Reciprocal altruism was predicted originally by theoreticians and has subsequently received support in studies of a number of species. Where would we expect to see it? In species that rely on each other for such things as feeding or defense and that live fairly long lives—so that favors can be returned and "cheaters," who take favors but do not return them, can be denied favors in the future. To take but one example, the vampire bat (*Desmodus rotundus*) feeds solely on blood. Vampire bats are not always successful in getting meals, however, and can starve to death if they go three successive nights without feeding. Researchers have found that vampire bats who have fed will regurgitate blood to feed starving bats, but will be particularly inclined to provide this favor to bats who have fed *them* in the past.

Reciprocal altruism does not seem to be common in the animal world, which is not surprising since so many conditions have to be in place for it to work. In addition to living long lives in close proximity, members of a group have to face the same kinds of challenges. In the case of bats, it turns out that *any* vampire bat can be unsuccessful in getting food for a few days and thus stands to need the help of its neighbor at some point. If this were not the case—if some vampire bats were simply bad at getting food—then reciprocal altruism would make no sense, because some bats would be chronic "takers" while others would be chronic "givers."

Learning about Animal Navigation

How are animals able to migrate long distances in order to end up end up at very precise locations? For that matter, how have scientists been able to learn about the ways in which animals navigate? For more on this, see "How Do Sea Turtles Find Their Way?" on page 755.

On to ... the Rest of Life

Every chapter of this book has ended with a little preview of what is to come in the next chapter. This time there is no such thing as a next chapter, because you have reached the end of the book. Hopefully, this animal behavior chapter has conveyed the same general message as the other chapters: Life has interesting rules, and the scientists called biologists are having some success in figuring out what those rules are. Readers who have been through large portions of this book know, however, that our ignorance of the living world at least matches our knowledge of it. The good news is that biologists are acquiring knowledge at a faster pace now than ever before. The average college student eventually will understand nature in ways that his or her grandparents could not have imagined. Stay tuned for what's coming up.

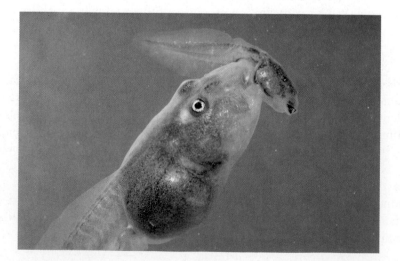

Figure 31.21
If It's Kin, Don't Eat It
Spadefoot toads are cannibalistic in their tadpole stage, but they avoid eating their siblings. This large predator tadpole is just tasting the smaller prey tadpole, which will be released if it's a sibling.

How Do Sea Turtles Find Their Way?

Animals must navigate over distances large and small. It is a matter of life and death for a ground squirrel to know precisely how to get to its burrow from wherever it may be. It is long-distance animal navigation, however, that has long been an object of human wonder. The gray whale (*Eschrichtius robustus*) feeds from April through November in the arctic waters near Alaska and then migrates almost 10,000 kilometers (6,200 miles) south to Mexico's Baja California for breeding. A shorebird called the red knot (*Calidris canutus*) stays in northern Canada, above the Arctic Circle, during spring and summer and then flies nearly the length of the Western Hemisphere, to the southern tip of South America, as weather gets colder.

These species are practicing **migration**, meaning a regular movement from one location to a distant location. The costs of migration are high, but the usual payoffs—seasonally rich feeding grounds and a more hospitable climate—make the arduous journeys worth it. In migrating, animals must practice **navigation,** meaning the use of various cues to enable movement toward a desired location. The cues that animals use in navigating are varied. Some steer a course by employing sunlight; others use the Earth's magnetic field, sound, geographical features, and the location of stars in the sky. In general, animals use multiple cues to make their way.

In recent years, scientists have discovered that another kind of animal, a species of sea turtle, is itself an ultra-long-distance migrator. Indeed, some Pacific populations of the loggerhead turtle *Caretta caretta* seem to migrate farther than either the red knot shorebird or the gray whale. DNA sequencing done in the mid-1990s indicated that loggerheads found off the Baja California coast actually had been born in the coastal waters of Japan. Proof of

this hypothesis came in 1994, when scientists marked a loggerhead and released it into Baja's Pacific waters. Four hundred and seventy-eight days later, this indefatigable swimmer was pulled out of the water by a fisherman off the coast of Kyushu, Japan.

Turtles such as this one are returning to the area—perhaps even the exact beach—in which they were hatched. What route do they take? If you look at **Figure 1**, you can see the answer for one loggerhead, which in 1996 was outfitted with a satellite signaling device and put into Baja California's Pacific waters after having lived in captivity for 10 years. Wallace J. Nichols and his colleagues, then at the University of Arizona, and Antonio Resendiz and his colleagues at the Mexican Secretary of Fisheries then tracked this animal for the next 368 days over the route you see in the figure. (The route's pattern near Japan may be the result of the turtle having been hauled aboard a Japanese fishing boat at the end of its journey. Sadly, Nichols and his colleagues do not know its fate.) This was a big animal, weighing more than 95 kilograms or about 210 pounds, and it moved at the pace of a "slow walk," as Nichols characterized it. On its journey of more than 11,500 kilometers (or about 7,100 miles) its average speed was less than 1 mile per hour.

But how do loggerheads manage this kind of long-distance navigation? The hypothesis most favored by scientists today is that, as with many species, loggerheads use an internal magnetic compass to steer—to point them in the correct direction as they make their journey.

continued

Figure 1
Long-Distance Swimmer
The line in the figure traces the route followed by a single loggerhead sea turtle as it swam from Baja California to Japan over the course of 368 days in 1996–1997. Tracked by satellite, the turtle may have been captured by Japanese fishermen as it neared the end of its journey.

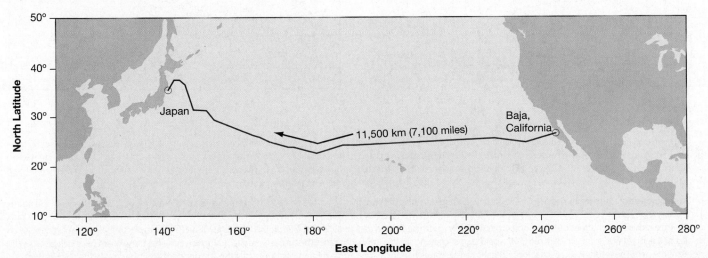

How Do Sea Turtles Find Their Way? *continued*

This "directional sense" is then combined with "positional sense," meaning that the turtles have a sense of where they want to go. Much of our information on this topic comes from loggerheads that have been studied on the other side of North America, in the Atlantic Ocean.

Earth's magnetic field exists at an angle to the Earth that varies in accordance with latitude. At the magnetic equator, the field is parallel to the Earth's surface, while at the magnetic poles, the field lines are perpendicular to the Earth's surface. Can turtles respond to variations in this angle? Yes. As part of a long-running research program on turtle navigation, University of North Carolina scientists Kenneth and Catherine Lohmann measured the preferred orientations of young loggerheads tethered to an arm in a laboratory tank (**see Figure 2**). Using a magnetic coil system, the Lohmanns were able to create an *artificial* magnetic field around the tank, varying the field's inclination angle as they desired. Critically, they knew that, in the wild, loggerheads from Florida must remain through the first 5 or 10 years of their lives within an enormous mass of moving ocean water called the North Atlantic Gyre. Turtles that venture too far north in the gyre are swept into the frigid waters of the North Atlantic

near Europe, where they die from cold. Turtles that venture too far south risk being swept out of their normal range in the other direction. In their experimental tank, the Lohmanns created a magnetic field whose angle matched that of the northern extreme of the North Atlantic Gyre. The result? The turtles turned south-southwest and started swimming. When confronted with an extreme southern magnetic field, the turtles oriented themselves the other way—north-northeast, back toward what they perceived as the safety of the gyre.

Any long-distance navigator must have a sense not only of the direction it is headed, but of the location it wants to reach. But how could, for example, the Baja California loggerheads be directed to a location as specific as a Japanese island? How do they acquire this positional sense? It may be that they *imprint* on the magnetic coordinates of the place they want to go to. Just as Konrad Lorenz's geese knew where they wanted to be (with Lorenz) through a process of visual imprinting, so loggerhead turtles in, for example, Baja California might know where they want to go (Japan) through a process of magnetic imprinting. In the loggerheads' case, their imprinting would be completed when they are young, before leaving Japanese waters for the coast of Baja California.

a Loggerhead orientation in an experimental setting

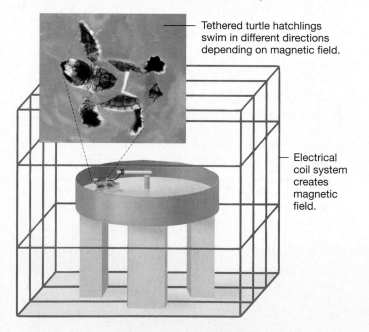

Tethered turtle hatchlings swim in different directions depending on magnetic field.

Electrical coil system creates magnetic field.

b Loggerhead navigation in nature

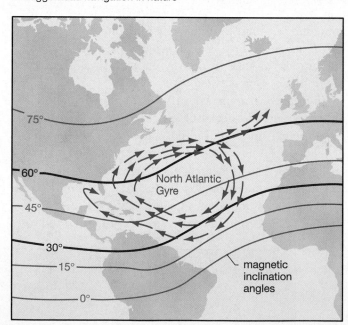

75°

60°

North Atlantic Gyre

45°

30°

15°

magnetic inclination angles

0°

Figure 2
How Do Turtles Navigate?

a Loggerhead orientation in an experimental setting. Researchers Kenneth and Catherine Lohmann constructed a tank to test the directions that loggerhead turtles preferred to swim in under different experimental conditions. Using an electrical coil system (the red lines around the tank), the Lohmanns created an artificial magnetic field that could be varied to mimic the magnetic fields that loggerheads would encounter in different parts of the Atlantic Ocean. Tethered to an arm in the experimental tank, turtles were free to orient themselves in response to magnetic field changes.

b Loggerhead navigation in nature. The loggerheads studied in the experiment live for the first 5 to 10 years of their lives in the Atlantic Gyre, shown in the figure. Turtles dare not venture either too far north or too far south in the Gyre, lest they be swept out to inhospitable habitats. The Lohmanns found that turtles in their tank who were exposed to a magnetic inclination angle of 60 degrees (the northern dark red line in the figure) would start swimming southwest, while turtles exposed to an angle of 30 degrees (the southern dark red line) would start swimming northeast. In both cases, the turtles were swimming toward what they would have perceived as the safety of the gyre. The results indicate that loggerheads can perceive changes in the Earth's magnetic field and that they respond to these changes in their navigation.

Chapter Review ▐

Summary

31.1 The Field of Behavioral Biology

- Behavioral biology tries to answer three questions: What do animals do, why do they do it, and how do they manage to do it? Observation of animals is critical to answering all three types of questions.

- In the 1930s, pioneering behavioral biologists Niko Tinbergen and Konrad Lorenz began developing hypotheses for why animals behave as they do.

- Most animal behaviors can be conceptualized as having both proximate and ultimate causes. Proximate causes involve the physiology of an animal and are triggering or immediate causes. Ultimate causes are linked to survival, reproduction, and evolution.

- Natural selection is critical to behavioral biology because it generally lies at the root of ultimate causes. When any given behavior is observed, behavioral biologists are likely to ask how it might be linked to survival and reproduction, and hence to natural selection.
  TUTORIALS: 31.1.2 The Beewolf; 31.1.4 Honeybees

31.2 The Web of Behavioral Influences

- Genes, learning, and an animal's environment almost always work together to produce a given behavior. Though it may be convenient to conceptualize each of these influences on behavior as being separate, in real life each of these factors almost always works in concert with the others.

31.3 Internal Influences on Behavior

- Reflexes are rapid, automatic responses to stimuli. They are the epitome of an internal influence in that animals do not have to learn anything to be able to perform them and they are stereotyped, meaning they are performed precisely the same way each time by individuals of the same species.

- Action patterns are linked sets of behaviors that are rigidly defined and largely influenced by genes, but that generally can be modified by sensory input. A releaser is the critical element in an action or an object that triggers an action pattern.

- Orientation skills can be learned but often have a strong genetic component to them. A genetic predisposition to move toward or away from a stimulus is called taxis. In a positive taxis an animal moves toward a stimulus; in a negative taxis, it moves away from a stimulus.

- Animals have internal clocks that can function independently of environmental signals and that help regulate such behaviors as migration and mating. Internal cycles that last about a day are called circadian rhythms. Some animals also have annual internal clocks.

- Hormones are substances that, when released in one part of an organism, go on to prompt physiological activity in another part of that organism. Hormones can be a powerful internal influence on animal behavior.

31.4 Learning and Behavior

- Learning is the acquisition of knowledge through experience. Though learning is based on experience that comes to an animal from the outside world, it always has a genetic component to it. Animals can be genetically predisposed to learn, or not learn, certain things.

- Imprinting is a process of learning that results in one animal preferentially associating with another. In it, a young animal often learns to associate with a parent figure. Imprinting generally takes place during a sensitive period, meaning the time-interval during which an animal can learn to respond to a given stimulus.

- One simple form of learning is habituation: a reduction in a response, based on repeated exposure to a stimulus that has no positive or negative consequences. Habituation is learning that allows an organism not to react to a stimulus.

- A more complex form of learning is classical conditioning: learning to respond in a customary way to a new stimulus that has been paired with an existing stimulus. Another complex form of learning is operant conditioning or trial-and-error learning, in which animals learn to associate their own actions with a particular outcome.

- Animals also can learn through imitation, meaning they copy a behavior they are observing in another animal; and they can practice insight learning, meaning learning in which an animal makes associations between objects or events that it has previously regarded as unrelated.

31.5 Behavior in Action: How Birds Acquire Their Songs

- Some bird songs are produced largely under genetic influence, but most songs are acquired in part through learning.

- Songbirds always utilize learning in producing their songs. A given species of songbird may have local or regional dialects of its song.

- Song development follows a predictable pattern in most songbirds: Males generally do the singing, but, in the first few months of their lives, they merely learn songs without performing them. In the spring following their birth, when their testosterone levels rise, males begin singing. Their songs, at first jumbled and incoherent, eventually crystallize into the recognizable songs of their species.

- White-crowned sparrows provide an example of the interplay between learning and internal influences in bird-song acquisition. White-crowns learn to sing regional dialects of their species' song by listening to previous generations of birds.

When reared in isolation, however, they will eventually sing a standard white-crown song, even if the only "model" song they are exposed to is sung by a bird from another species.

31.6 Social Behavior

- Some animals live in almost complete isolation from other members of their species, while other animals are never far from other members of their species.

- Animals who live with other members of their species are exhibiting social behavior. Such behavior always has costs as well as benefits: Disease may be transmitted easier among members of a species living together, but benefits accrue to individuals in these groups as well, such as group defense or group predation. A species will exhibit social behavior to the degree that such behavior aids in the survival and reproduction of individual members of that species.

- A common feature of social living is the dominance hierarchy: a persistent power ranking in an animal population that gives those of higher rank the ability to control some aspect of the behavior of those of lower rank. Higher ranking often provides an animal with greater access to such resources as food and mating partners. Dominance hierarchies seem to exist because certain individuals in a group—the dominant ones—benefit from such hierarchies and can enforce them.

- Territoriality refers to efforts on the part of one animal to keep other animals from entering a given area. Animals who exhibit territoriality generally are trying to keep members of their own species from entering a given area. Territoriality can be constant or seasonal, depending on the species; it is generally individual males who are territorial.

- Eusocial or "truly social" species are organized into a caste system, in which there is a division of labor in a population and in which the young of the species are raised through cooperative care, meaning care provided by many members of the group. Eusocial species include honeybees and many species of ants, and at least one species of mammal, the naked mole rat.

- Honeybees are a good example of a eusocial species. Only one permanent colony member, the queen, reproduces, aided by male drones from another colony who mate with her during a series of "nuptial flights." Nearly all colony members are sterile female workers whose functions in the colony change over the course of their lives.

- Behavior in a honeybee colony is self-organizing. Workers are prompted to carry out certain behaviors either through environmental conditions (such as a change in the temperature of the hive) or through signals or cues they receive from other bees. Signals are explicit acts of communication, such as the waggle-dance that shows the location of food; cues simply imply a given condition to which the bees need to react.

31.7 Altruism in the Animal Kingdom

- Some animals clearly display altruism, which is costly or risky behavior carried out by one animal for the benefit of a second. Altruism is conspicuous in eusocial species, in which most individuals forgo reproduction altogether and instead care for the offspring of others (queens and the males they mate with).

- Altruism in the animal kingdom is correlated with how closely related two individuals are: The more closely they are related, the more likely it is that they will act altruistically toward each other. Certain eusocial species provide the clearest example of this phenomenon. Because they have an unusually close genetic relationship, female fire ant workers can pass on more of their genes by caring for sisters than by having their own offspring.

- Inclusive fitness is an individual's relative genetic contribution to a succeeding generation, made both through itself and through relatives that have reproduced because of assistance the individual has provided. The evolutionary mechanism underlying inclusive fitness is kin selection, meaning the selection of genes that increase the inclusive fitness of an individual.

- Animals occasionally exhibit reciprocal altruism: an exchange of altruistic acts by individuals over time. Reciprocal altruism tends to be exhibited only in species whose members depend on each other for support and who stay together long enough for altruistic acts to be "paid back."

Key Terms

altruism 751	learning 741
behavioral biology 733	migration 755
circadian rhythms 739	natural selection 736
classical conditioning 743	navigation 755
dominance hierarchy 749	operant conditioning 743
entrain 739	reciprocal altruism 754
eusocial 750	reflexes 737
habituation 742	releaser 738
imitation 743	sensitive period 742
imprinting 742	stereotyped 737
inclusive fitness 753	taxis 739
insight learning 743	territoriality 749
kin selection 753	

Understanding the Basics

Multiple-Choice Questions

1. Behavioral biology asks (select all that apply):
   a. How can the behavior of living things be changed?
   b. How does the behavior of animals compare to that of bacteria?
   c. What do animals do and why do they do it?
   d. How do animals manage to exhibit the behaviors they do?
   e. What is the nature of the animal genome?

2. In animal behavior a proximate cause is _____; an ultimate cause is _____.
   a. one that happens right away; one that happens hours earlier
   b. one that explains only part of a behavior; one that explains all of a behavior
   c. one that may be true; one that has been proven to be true
   d. one that is a triggering and physiological cause; one that is linked to survival and reproduction
   e. one that is based on observation; one that is based on theory

3. What generally lies at the root of ultimate causes?
   a. operant conditioning
   b. female choice
   c. natural selection
   d. learning
   e. hormonal influences

4. A(n) _____ is stereotyped, meaning that _____.
   a. reflex; all members of a species perform it the same way
   b. trial and error learning; all members of a species learn the same things
   c. ant colony; all members are genetically identical
   d. dominance hierarchy; it can have many leaders
   e. eusocial species; all members act the same way

5. Which of the following is true? (select all that apply):
   a. A circadian rhythm is an internal biological cycle that lasts about a day.
   b. Circadian rhythms generally are entrained by environmental cues.
   c. Birds are the only animals that possess annual internal clocks.
   d. Animals are the only living things that possess internal clocks.
   e. Circadian rhythms most often are entrained by temperature.

6. Konrad Lorenz' geese followed him because:
   a. An orientation reflex prompted them to.
   b. They learned to through trial and error.
   c. Geese will follow the largest object they see.
   d. They had imprinted upon him after hatching.
   e. He provided the right cue for their circadian rhythms.

7. Pavlov's dog learned to associate food with the sound of a bell, while a bird will learn to associate trying to eat a wasp with getting stung. What type of learning is represented in each instance?
   a. imitation, classical conditioning
   b. insight learning, imitation
   c. operant conditioning, imitation
   d. insight learning, classical conditioning
   e. classical conditioning, operant conditioning

8. Male songbirds generally:
   a. memorize their species' songs months before first performing them
   b. will learn any song presented to them early in life
   c. must be in the presence of females before starting to sing
   d. have a repertoire of two songs
   e. cannot distinguish between one song they hear and another

9. Animals are social to the degree that:
   a. There is no cost to it.
   b. They do not have to submit to dominance hierarchies.
   c. The benefits of social living outweigh its costs.
   d. They can share equally in all resources.
   e. They are inclined to take care of one another.

10. Much of the altruism displayed among animals probably can be explained as:
    a. a genuine concern for one another
    b. a lack of understanding of its consequences
    c. imprinting on the behavior of others
    d. the reproductive benefits that can come through aiding relatives
    e. fear of cannibalism

Brief Review

1. The increase in singing that male birds exhibit in the spring seems to be prompted by an increase in their testosterone levels. But are testosterone levels a proximate or ultimate cause of the singing? If they are not an ultimate cause, what is?

2. What is a circadian rhythm?

3. Describe the difference between classical and operant conditioning.

4. Is singing in birds generally a learned behavior or a genetically controlled behavior?

5. Why don't all animals live in groups?

6. What is the ultimate cause of most animal altruism?

7. Some authorities say that all animal learning amounts to trial-and-error learning. Why is this a plausible argument?

Applying Your Knowledge

1. Is it possible to imagine any animal behavior that is completely learned—one that has no genetic component to it, in other words?

2. Eusociality might be described as a form of animal living in which the individual means nothing, while the group means everything. Explain why life in a honeybee colony provides an example of this.

3. The evolutionary biologist J. B. S. Haldane once said in a British pub that he would be willing to die "for two brothers, four uncles, or eight cousins." Explain what he meant in light of kin selection and inclusive fitness.

Appendix

Metric-English System Conversions

Length

1 inch (in.) = 2.54 centimeters (cm)
1 centimeter (cm) = 0.3937 inch (in.)
1 foot (ft) = 0.3048 meter (m)
1 meter (m) = 3.2808 feet (ft) = 1.0936 yard (yd)
1 mile (mi) = 1.6904 kilometer
1 kilometer (km) = 0.6214 mile (mi)

Area

1 square inch (in.^{2}) = 6.45 square centimeters (cm^2)
1 square centimeter (cm^2) = 0.155 square inch (in.^{2})
1 square foot (ft^2) = 0.0929 square meter (m^2)
1 square meter (m^2) = 10.7639 square feet (ft^2) = 1.1960 square yards (yd^2)
1 square mile (mi^2) = 2.5900 square kilometers (km^2)
1 acre (a) = 0.4047 hectare (ha)
1 hectare (ha) = 2.4710 acres (a) = 10,000 square meters (m^2)

Volume

1 cubic inch (in.^{3}) = 16.39 cubic centimeters (cm^3 or cc)
1 cubic centimeter (cm^3 or cc) = 0.06 cubic inch (in.^{3})
1 cubic foot (ft^3) = 0.028 cubic meter (m^3)
1 cubic meter (m^3) = 35.30 cubic feet (ft^3) = 1.3079 cubic yards (yd^3)
1 fluid ounce (oz) = 29.6 milliliters (mL) = 0.03 liter (L)
1 milliliter (mL) = 0.03 fluid ounce (oz) = 1/4 teaspoon (approximate)
1 pint (pt) = 473 milliliters (mL) = 0.47 liter (L)
1 quart (qt) = 946 milliliters (mL) = 0.9463 liter (L)
1 gallon (gal) = 3.79 liters (L)
1 liter (L) = 1.0567 quarts (qt) = 0.26 gallon (gal)

Mass

1 ounce (oz) = 28.3496 grams (g)
1 gram (g) = 0.03527 ounce (oz)
1 pound (lb) = 0.4536 kilogram (kg)
1 kilogram = 2.2046 pounds (lb)
1 ton (tn), U.S. = 0.91 metric ton (t or tonne)
1 metric ton = 1.10 tons (tn), U.S.

Metric Prefixes

Prefix	Abbreviation	Meaning
giga-	G	$10^9 = 1,000,000,000$
mega-	M	$10^6 = 1,000,000$
kilo-	k	$10^3 = 1,000$
hecto-	h	$10^2 = 100$
deka-	da	$10^1 = 10$
		$10^0 = 1$
deci-	d	$10^{-1} = 0.1$
centi-	c	$10^{-2} = 0.01$
milli-	m	$10^{-3} = 0.001$
micro-	μ	$10^{-6} = 0.000001$

Answers

CHAPTER 1

Multiple Choice
1. c, 2. b, 3. b, 4. d, 5. e.

Brief Review
1. Science is a body of knowledge and a formal process for acquiring that knowledge. It is thus a collection of unified insights about the physical world, and also a way of learning. Like belief systems such as religion, science is an attempt to understand and explain the physical reality of the world around us and our experience of it. Science differs from religious belief, however, in that it deals exclusively with physical phenomena and does not claim to describe a non-material or metaphysical world. Indeed, the existence of such worlds is not a question taken up by science. Further, science does not accept descriptions of physical reality that are based on faith, nor does it recognize any entity as an unquestionable authority. Scientific information is not immutable; it is open to challenge by all, and is regularly challenged by its own practitioners, scientists.
2. An experiment is said to be controlled when the experimenter controls the factors in it that will change and those that will be kept constant. Suppose a scientist wanted to test the effect of the insecticide DDT on the reproductive development of lab rats. In this case, the experiment might consist of feeding DDT-laced food to one group of rats, while another similar group—called the control group—would receive the same food without DDT. It is important that the two groups of rats be as closely matched as possible in regard to all other factors. For instance, both groups must consist of the same strain of lab mice; both must be the same gender or mix of genders; they must be given the same amount of food and water, and housed in identical environments. Suppose differences were observed in, for example, the size of the offspring born to one group of rats as opposed to the other. Were the starting conditions not controlled in the experiment, there would be no way to tell whether the DDT or some other factor (such as different housing) was responsible for these differences.
3. Pasteur investigated a widely held notion of the time that living things can arise spontaneously from basic chemicals. He hypothesized that instances where life appeared to arise in medium believed to lack it could be readily explained as the result of contamination from unseen airborne microscopic organisms. He sterilized meat broth in a flask with a long, S-shaped neck, by heating both the broth and the flask. Although the mouth of the long tube remained open, any incoming dust particles laden with microscopic organisms would be trapped in the bent neck and could not gain access to the meat broth. Under these conditions, the broth remained free of any visible signs of life. In other tests, the S-shaped neck was broken off or the flask tilted so that broth made contact with the neck of the flask; in these instances, the broth came "alive" with teeming bacteria and fungi. Pasteur inferred from these experiments that the appearance of life forms depended on whether the broth was contaminated by airborne microscopic organisms or not. When the airborne organisms had no access to the broth, however, it remained sterile. There was no evidence, in other words, of life coming about through spontaneous generation.
4. All living things, as we know them on this planet, acquire, store, and use energy through metabolism; they respond to stimuli in their environment; they maintain a relatively constant internal environment (homeostasis); they reproduce, using DNA to transmit hereditary information; they exhibit cellular organization; and they show evidence of having evolved, or changed over time.
5. Life can be analyzed at the minutest level of organization—atoms and molecules—or at successively larger and more complex levels. Atoms and molecules are components of organelles, which in turn are constituents of cells, the basic functional unit of life. In multicellular organisms, cells are organized into distinct tissue types. Collections of various tissues compose the organ systems of plants and animals. The total assemblage of organ systems constitutes an individual organism. All the individuals of the same species living in a defined area are referred to as a population of that organism. Populations of different types of organisms living in a defined area compose a community. All the communities of the Earth, together with their physical environment, make up the biosphere.

CHAPTER 2

Multiple Choice
1. b, d, 2. b, 3. c, 4. a, 5. d, 6. b, 7. e, 8. a.

Brief Review
1. The forms are called isotopes. Isotopes of an element differ from one another in accordance with the number of neutrons they have. A regular carbon atom has 6 neutrons in its nucleus, while a carbon-14 atom has eight neutrons.
2.

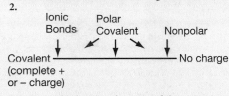

3. Free radicals exist when one of the atoms in a covalently bonded molecule has an unpaired electron. One free radical can produce many more free radicals in a chain reaction. Free radicals can do damage by irritating artery walls or by causing mutations in DNA.
4. An atom's electrons move through volumes of space outside the atom's nucleus. An entire atom is about 100,000 times larger than its nucleus. Thus, most of an atom is the space through which electrons move.
5. When the valence (outer) shell is filled, the atom is at a lower, more stable energy state.

CHAPTER 3

Multiple Choice
1. e, 2. a, 3. e, 4. c, 5. e, 6. b, 7. b, 8. d, 9. a.

Brief Review
1. Water absorbs tremendous amounts of heat from the Sun, and releases this heat slowly, with the onset of night's colder temperature. The perspiration human beings throw off carries with it a great deal of heat, which has been absorbed by the water.
2. LDL is associated with deposition of cholesterol in the arteries, so you should be concerned about high levels. HDL is associated with carrying cholesterol to the liver, away from the arteries where it does damage. A high HDL count relative to LDL count is in general a good sign.
3. Table 3.2 lists several protein functions, including enzymes, transport proteins, structural proteins, contractile proteins, and so forth. Proteins can do many different kinds of jobs because they take on so many different shapes. Each shape allows the protein to do a different job.
4. The water molecules cling together so tightly that when each molecule evaporates from a plant leaf, it draws up a chain of water molecules that extends all the way down to the root.
5. Strong hydrogen bonds help explain why water is so extremely cohesive, has such a strong surface tension, and has such a high specific heat.

CHAPTER 4

Multiple Choice
1. e, 2. d, 3. a, 4. b, 5. c, 6. d, 7. b, 8. e.

Brief Review
1. The cell would have difficulty obtaining materials from the external environment because its surface-to-volume ratio would be too low.
2. Eukaryotes have a nucleus, bound within a thin membrane, that contains almost all their DNA. The DNA of prokaryotes is localized (in a "nucleoid" region), but it is not bound within a membrane. Eukaryotes are usually much larger than prokaryotes. Eukaryotes are also quite often multicelled organisms, while prokaryotes are single-celled. Most eukaryotes need oxygen to exist, whereas many prokaryotes can survive without oxygen.
3. The DNA code is copied into an RNA molecule, which leaves the nucleus through the pores in the nuclear membrane. The RNA is then "read" and translated into a molecule similar to insulin in a ribosome on the rough ER. This molecule is dumped into the cisternae of the ER, where it folds and may be modified. This molecule is then encased in a membrane for shipment to the Golgi complex. The Golgi complex further modifies the molecule and packages it for export out of the cell in a vesicle.
4. Cilia and flagella are cell extensions formed by microtubules that are either a means of propulsion or, in the case of some cilia, a means of moving material around a cell. Cilia exist as a profusion of hair-like growths extending from cells that move back and forth very rapidly, while only a few flagella sprout from a given cell—indeed, there is often just a single, tail-like flagellum. In the human body, cilia are found in the nasal passages and in the lungs. The only animal cell that has a flagellum is a sperm cell.
5. Each cell would be isolated from each other cell by two thick cell walls. Thus plants could not actually be an organism if there were no plasmodesmata. Each cell would have to be pretty much independent of the rest.
6. A gap junction consists of clusters of protein structures that shoot through the plasma membrane of a cell from one side to the other. When these structures line up in adjacent cells, they form a channel for passage of small molecules and electrical signals.

CHAPTER 5

Multiple Choice
1. c, 2. a, 3. b, 4. e, 5. d, 6. b, 7. b, 8. c, 9. a, 10. c.

Brief Review
1. There are receptor proteins on the surface of the taste-bud cell that recognize sugar molecules. When this recognition occurs, the cell changes in a way that allows a message to travel to the brain and convey that what you ate was sweet. Because salt receptors are not on that cell, it responds to sugar but not salt.
2. Because steroids are lipids, they pass easily across the phospholipid bilayer by dissolving in it. Thus they can affect every single cell of the body.
3. Because the phospholipid "heads" are polar or hydrophilic, they are attracted to the polar water molecules on the outer edges (outside and inside) of the membrane. Because the "tails" are very hydrophobic, they tend to be pushed away from the water to the inside of the membrane.
4. Receptor proteins are not the same for every cell. Thus a cell can be "tailored" to respond to a specific hormone or group of hormones by the specific receptor proteins on its surface.

5. You would not expect that communication and recognition, and possibly transport, would be as important inside the cell as on the outside. Thus the glycocalyx would be unnecessary, and the proteins in the membranes would serve different functions.

6. When you shake the tray gently, the marbles move until they are fairly evenly spaced across the tray. In diffusion, molecules or other small particles such as ions hit against each other until they eventually bump each other to places that are roughly equidistant. This is purely a random process. Because temperature is a measure of the speed of molecules and is the result of the energy of the molecules, shaking the tray is roughly like changing the temperature of molecules—the harder you shake, the faster the rate of diffusion.

7. Your stomach cells carry out active transport, which requires energy. In this process, the cells may also need a pumping protein in the membrane.

CHAPTER 6

Multiple Choice
1. a, 2. c, 3. d, 4. d, 5. e, 6. e, 7. a, 8. c.

Brief Review
1. Plants abide by the second law of thermodynamics in that, in building themselves up, they are contributing to the total amount of entropy in the universe; in building a leaf from carbon dioxide and water, heat is released. A plant leaf represents only a local increase in order.

2. This is a coupled reaction in that the breakdown of ATP to ADP is an exergonic or downhill reaction that powers an endergonic or uphill reaction—the movement of sodium ions against their concentration gradient to the outside of the cell.

3. The energy stored and released by ATP comes from food. ATP is an energy "currency" molecule in that it facilitates an organism's expenditure of energy on various processes. Just as the value inherent in labor is put into a medium of exchange—money—that can be easily expended, so the energy in food needs to be put into the form of ATP to be easily used by living things.

4. Enzymes are involved with every aspect of cellular life. They are necessary for breaking things down, but they are also used to make larger molecules out of smaller ones.

5. Allosteric regulation of an enzyme occurs when the product of a reaction facilitated by an enzyme binds to the enzyme itself and changes its shape, thus making the enzyme unable to bind to its substrate.

CHAPTER 7

Multiple Choice
1. b, 2. d, 3. e, 4. a, 5. d, 6. b, 7. d, 8. a.

Brief Review
1. To get a trampoline to bounce you into the air, you put energy into it by jumping first. The trampoline stops bouncing when you quit jumping. The electrons from glucose are losing energy (just like you do when you jump). As they lose energy, some of it is transferred to ADP to make ATP (just as the energy from your jump is transferred to the trampoline, and from there to your body, to shoot you back into the air).

2. Muscles require a lot of energy. Thus they need mitochondria to carry out the Krebs cycle and electron transport, which extract the most energy from glucose.

3. We cannot make ATP without a source of energy. Foods supply the energy so that we can make ATP.

4. Water is produced in respiration at the very end of the process, when the electrons from the ETC are passed to oxygen, which then combines with hydrogen to make water.

5. Like NAD^+, FAD serves to carry electrons (and the energy stored in them) from the Krebs cycle to the ETC. Therefore, if your supply of FAD is inadequate, the Krebs cycle and electron transport cease, and you will be able to get only small amounts of energy out of food.

CHAPTER 8

Multiple Choice
1. e, 2. a, 3. c, 4. d, 5. e, 6. c, 7. b.

Brief Review
1. Yes, plants are as dependent on cellular respiration as we are. Plants make food—carbohydrates—in photosynthesis. Just as we have to break down the food we eat to get the energy out and into ATP, so plants have to break down the food they make to get the energy out and into ATP.

2. The sunlight that makes it to the Earth's surface is composed of a spectrum of energetic rays (measured by their "wavelengths") that range from very short ultraviolet rays, through visible light rays, to the longer and less energetic infrared rays. The photosynthesis we are concerned with is driven by part of the visible light spectrum—mainly by blue and red light of certain wavelengths.

3. In a photosystem, several hundred pigment (chlorophyll a and accessory pigment) molecules serve as antennae to capture light energy. The molecules pass their excited electrons, still in the excited state, to two chlorophyll a molecules, called the reaction center. These pass the excited electrons to a receptor molecule, called the primary electron acceptor. All of these parts are necessary: If there were no antennae molecules, the light would have to strike the reaction center directly each time, slowing photosynthesis significantly. The reaction center is essential because its molecules pass the electrons on. Without the electron acceptor, energy would not be made available to make sugars.

4. There would soon be no life, at least as we know it. Because most organisms on Earth depend on the energy harnessed in photosynthesis, soon all animals and humans would die off. Algae would die along with plants. Fungi and many bacteria would eventually die when all the other dead organisms had been decomposed.

5. In a bright, moderately hot environment, the C_4 plant would grow most rapidly. But the C_3 plant would grow fastest in cloudy or shady conditions, in which the C_4 plant would be spending too much energy moving compounds around in the plant leaf. The CAM plant, which always photosynthesizes slowly, would grow most rapidly in the desert. In other words, each plant grows most rapidly in the area to which it is best adapted.

6. (1) Rubisco adds CO_2 to RuBP to form two molecules of 3PGAL. (2) Energy is added to convert to G3P. (3) G3P is used to supply energy and/or matter to the plant for all the processes that plants carry out. (4) Most of the G3P is used to make more RuBP so that the cycle can continue. The first two steps add matter (CO_2) to the energy and hydrogens brought over from the light-dependent reactions. The third step allows the plant to use this energy. The fourth step allows the process to continue. Energy is like money to the cell; plants and animals are very unlikely to waste it. Occasionally, cells waste energy because there is no simple way to avoid it, but typically organisms guard their use of energy carefully.

CHAPTER 9

Multiple Choice
1. a, 2. d, 3. d, 4. c, 5. b, 6. a, 7. d, 8. b, 9. b, 10. e, 11. c.

Brief Review
1. DNA contains information for the production of proteins. These proteins are "workers" that carry out a vast array of tasks in organisms such as ourselves. Through the work of proteins, our bodies develop and carry out everyday functions.

2. The cell is in metaphase. There are eight chromatids present in this stage, and each daughter cell will have four chromosomes.

3. Mitosis serves to equally divide a cell's genetic material, while cytokinesis divides the cell's cytoplasm.

4. Binary fission

5. Interphase—This is the longest phase of the cell cycle. During this time, the cell may be in a period of "rest," or cell growth without division, called

G_1. If the cell is an actively dividing cell, chromosome duplication will occur during the S phase, followed by additional cell preparations for mitosis in the G_2 phase.
Prophase—Duplicated chromosomes condense and become visible, the centrosome duplicates and centrosomes migrate to the poles, the nuclear membrane breaks down, and the mitotic spindle is formed.
Metaphase—Chromosome pairs align along the cell mid-line or metaphase plate and attach to the microtubules extending from the cellular poles.
Anaphase—Separation of sister chromatids paired along the metaphase plate, and their migration to the separate poles of the cell.
Telophase—The final stage of mitosis, involving the unwinding of the condensed chromosomes at their respective poles, and the formation of new nuclear membranes. At this point, division of genetic material is complete. The cell is ready for cytokinesis.

6. Plant cells have cell walls; animal cells do not. In plants, membrane-lined vesicles migrate to the center of the dividing cell, near the metaphase plate. These vesicles form the flat cell plate. As the flat cell plate grows toward the original cell wall, the membrane of the flat cell plate fuses with the membrane of the original cell. This action completes cell division, resulting in two new daughter cells and a new cell wall.

7. The 23 pairs of human chromosomes are matched (or homologous) in the sense that the two members of each pair contain information about similar functions, such as hair color, metabolic processes, and so forth. The one exception to the matched-pair rule comes in human males, who have 22 pairs of matched chromosomes, but then one X chromosome and one Y chromosome, which are not matched. Human females have 23 matched pairs, including two X chromosomes.

8. Prokaryotes only have one circular chromosome, while eukaryotes have several linear chromosomes.

9. (a) four different nitrogen-containing bases (adenine, thymine, guanine, and cytosine); (b) in the nucleus; (c) "messenger"; (d) mRNA is used as a cellular "messenger," copying the DNA information in the nucleus and carrying it to the cytoplasm, where the message is used by ribosomes to assemble amino acids into proteins; (e) amino acids; (f) in the cytoplasm.

CHAPTER 10

Multiple Choice
1. b, 2. a, 3. c, 4. a, 5. b, 6. a, 7. e, 8. d, 9. a, 10. e.

Brief Review
1. Chromosome reduction takes place during meiosis I.

2. No. Asexual reproduction also occurs in single celled prokaryotes, as well as many eukaryotes outside the plant kingdom. Some animals are capable of a form of asexual reproduction called regeneration.

3. The cells in males that go through meiosis, giving rise to immature sperm, are called primary spermatocytes and are themselves the product of cells called spermatogonia. These spermatogonia can give rise not only to primary spermatocytes but to more spermatogonia as well.

4. Meiotic cytokinesis differs from mitotic cytokinesis in the selective shunting of cytoplasm. In mitosis, cytokinesis divides the cytoplasmic contents equally; in meiotic cytokinesis, cytoplasm is preferentially shunted to the cell destined to be the ovum, and away from the polar bodies.

5. Somatic cells are diploid cells that undergo mitosis. They include every cell in the body except the reproductive cells. Gametes are haploid reproductive cells, resulting from the process of meiosis: eggs and sperm in the female and male, respectively.

6. (a) metaphase I, meiosis I; (b) anaphase II, meiosis II; (c) anaphase, mitosis; (d) metaphase II, meiosis II; (e) metaphase, mitosis.

7. Meiosis ensures genetic diversity in two ways. In the process called recombination, homologous chromosomes swap sections with one another. Then, in the process called independent assortment, homologous chromosome pairs line up in a random way, relative to one another, along the metaphase plate.

8. Anaphase of meiosis II is most like anaphase of mitosis, because both involve the separation of sister chromatids—in anaphase of meiosis I, homologous chromosomes separate, but not sister chromatids.

9. Sperm contribute only DNA, packaged in chromosomes, to the succeeding generation. In addition, they must make a journey to an egg in order to fertilize it, thus making small size an advantage. By contrast, eggs need to have stored within them a rich mixture of nutrients and other materials to support the newly formed offspring in its early stages.

10. (a) gamete: a reproductive cell (sperm or eggs); (b) recombination (or crossing over): a process during meiosis in which homologous chromosomes exchange reciprocal parts with one another; (c) asexual reproduction: production of a new organism without a fusion of gametes; (d) haploid: the state of possessing only a single set of chromosomes (as opposed to diploid, in which two sets of chromosomes exist).

CHAPTER 11
Multiple Choice
1. a, 2. a, 3. e, 4. c, 5. d, 6. b, 7. b, 8. c, 9. d, 10. c.
Brief Review
1. (a) Phenotype is the representation of visible traits in an organism (that is, yellow vs. green peas), while genotype is the genetic composition responsible for the phenotype of a trait (that is, YY or Yy = yellow peas, yy = green peas). (b) Dominant is used to refer to a trait that is equally expressed in either the homozygous or heterozygous genotype (i.e., YY or Yy = yellow peas). Recessive is the term used to describe a trait that will be expressed only in the homozygous condition. Recessive traits are masked in the heterozygous genotype (i.e., Yy = yellow peas). (c) Codominance refers to the condition in which a heterozygote produces an equal amount of product from each allele it carries. This is exhibited in blood types, where an AB individual produces both A and B cell surface molecules. Incomplete dominance is displayed when the heterozygote displays a phenotype that is an intermediate between the dominant and recessive phenotypes. An example is the generation of pink F_1 flowers resulting from a parental cross of one red-flowering and one white-flowering plant.

2. True.

3. The fragile-X syndrome demonstrates the genetic principle of pleiotropy, in which one gene has more than one effect. This differs from what Mendel observed in that, in his pea plants, one gene had one effect. There was a gene for flower color, for example, that came in alleles that brought about either purple or white flowers. But that gene did not then go on to affect other characters in the plant. By contrast, the gene connected to fragile-X syndrome has multiple effects—on intelligence and various physical characteristics.

4. In Mendelian genetics, two different genotypes may represent a dominant phenotype: YY and Yy.

5. Fifty percent of the progeny will be yellow peas.

6. Mendel's "element of inheritance" is now referred to as the gene.

7. Mendel's first law (law of segregation): Members of a gene pair will segregate from each other during gamete formation such that each gamete receives one and only one of the original two genes from the gene pair. Mendel's second law (law of independent assortment): During gamete formation, gene pairs assort independently of one another.

8. (a) $Aa \times aa$; (b) $Zz \times Zz$; (c) $Bbcc \times bbCc$; (d) $EeFf \times EeFf$.

9. (a) Two characters—yellow and white, and supersweet and not-so-sweet—are clearly different alleles of two different genes, one for color (Y for yellow and y for white) and the other for sugary taste (Su for starchy and su for supersweet). The alleles segregate from each other during meiosis in accordance with Mendel's Law of Independent Assortment so that you can get all possible combinations of traits. Thus, you should be able to get the yellow supersweet variety you were striving for. The two parental plants were true-breeding, yellow starchy, $YYSuSu$, and white supersweet, $yysusu$. The F_1 hybrids, $YySusu$, were heterozygotes and thus showed the dominant alleles, yellow and starchy. When the two F_1s are crossed together, the F_2 generation should show Mendel's expected ratios of 9/16 yellow and starchy, 3/16 yellow and supersweet, 3/16 white and starchy, and 1/16 white and supersweet. Here is how this works out in a Punnett square:

	YSu	Ysu	ySu	ysu
YSu	$YYSuSu$	$YYSusu$	$YySuSu$	$YySusu$
Ysu	$YYSusu$	$YYsusu$	$YySusu$	$Yysusu$
ySu	$YySuSu$	$YySusu$	$yySuSu$	$yySusu$
ysu	$YySusu$	$Yysusu$	$yySusu$	$yysusu$

(b) Yellow and starchy, 563; white and starchy, 187; yellow and supersweet, 187 (three shaded boxes); white and supersweet, 63.
(c) As you can see from the Punnett square, 3/16 of the offspring should be yellow and supersweet, but of these, only 1/3 are homozygous for the yellow allele ($YYsusu$), so only 1/3 will be true-breeding. The other 2/3 ($Yysusu$) will give a mixture of yellow (3/4) and white (1/4) supersweet offspring.

10. The Chestnut gene C shows an incomplete dominance pattern of inheritance, so that the homozygous recessive (cc) is white, the homozygous dominant (CC) is Chestnut, but the heterozygotes (Cc) are Palomino, as illustrated in Table A-1.
Below are three Punnett squares that will help you understand the outcomes of the last three crosses in the question.

Palomino (Cc) x Palomino (Cc) = 1/4 White, 1/4 Chestnut, 1/2 Palomino:

	C	c
C	1/4 CC Chestnut	1/4 Cc Palomino
c	1/4 Cc Palomino	1/4 cc white

Palomino (Cc) x Chestnut (CC) = 1/2 Chestnut, 1/2 Palomino:

	C	C
C	1/4 CC Chestnut	1/4 CC Chestnut
c	1/4 Cc Palomino	1/4 Cc Palomino

White (cc) x Palomino (Cc) = 1/2 White, 1/2 Palomino:

	C	c
c	1/4 Cc Palomino	1/4 cc white
c	1/4 Cc Palomino	1/4 cc white

CHAPTER 12
Multiple Choice
1. c, 2. e, 3. a, 4. d, 5. e, 6. d, 7. c, 8. e, 9. d, 10. a.
Brief Review
1. (a) autosomal dominant; (b) chromosomal deletion; (c) aneuploidy; (d) aneuploidy; (e) X-linked; (f) autosomal recessive; (g) aneuploidy; (h) X-linked; (i) X-linked.

2. Male. Sex-linked diseases usually are associated with the X chromosome, which females have two copies of. A female can carry one faulty copy of an X-chromosome gene and still be protected by a functional copy on her second X chromosome. Males, however, have only a single copy of the X chromosome. Should a gene on it be faulty, they have no second, protective copy of it.

3. The genes for red and green color vision are very closely aligned on the X chromosome, while the gene for blue vision is on chromosome 7. It would be possible to have dysfunctional genes for red and green but a functional gene for blue.

4. Deletions—A portion of the chromosome is completely lost.
Inversions—A portion of the chromosome is broken off and rejoins the original chromosome, but in a reversed order.
Translocations—Occur when two chromosomes that are not homologous exchange pieces, leaving both with improper gene sequences.
Duplications—A portion of a chromosome is duplicated, most often by gaining a chromosome piece during abnormal crossing over.

5. (a) non-disjunction; (b) polyploidy.

6. Because people who have Klinefelter syndrome have a Y chromosome, which confers the male sex, but they also have two copies of the X chromosome, rather than the one copy that males normally possess. Females normally have two copies of the X chromosome. Thus it makes sense that those with Klinefelter syndrome would exhibit some female characteristics.

7. A heterozygous genotype for hemoglobin would be advantageous in areas of mosquito infestation, while neither homozygous genotype would be advantageous in such areas. Those who are homozygous for hemoglobin A would not have the increased resistance to malaria; those who are homozygous for hemoglobin S would inherit sickle-cell anemia. Those who are heterozygous would have malaria resistance and yet would not inherit sickle-cell anemia.

Genetics Problems
1. (a-I) 50 percent; (a-II) 0 percent; (b-I) $Xx \times XY$; (b-II) $xY \times XX$; (c-I) 50 percent; (c-II) 100 percent.

2. (a) $bbEe \times bbEe$; (b) black, 0 percent; brown, 75 percent; golden, 25 percent; (c) no black puppies, 9 brown, 3 golden.

3. (a) deafness. For pedigrees, it is often useful to make a hypothesis that the trait is passed on by any of the possible patterns of inheritance, and then ask if the data fit. If even one cross does not fit with the hypothesis, then that pattern of inheritance must be excluded. So, let's try all the possibilities: X-linked recessive—No, the data do not fit. If the disease-causing allele of this gene was X-linked recessive, and a mother was affected with the trait, then all of her sons would also express the trait, because sons receive the X chromosome from the mother. Individual II4 is a female with the condition, but only some of her sons show the trait,

Table A-1				
Parents				
Female	**Genotype**	**Male**	**Genotype**	**Progeny Colors**
White	(cc)	White	(cc)	all White (cc)
Chestnut	(CC)	Chestnut	(CC)	all Chestnut (CC)
White	(cc)	Chestnut	(CC)	all Palomino (Cc)
Palomino	(Cc)	Palomino	(Cc)	1/4 White, 1/4 Chestnut, 1/2 Palomino
Palomino	(Cc)	Chestnut	(CC)	1/2 Chestnut, 1/2 Palomino
White	(cc)	Palomino	(Cc)	1/2 White, 1/2 Palomino

so it cannot be an X-linked recessive. In addition, for a father to pass the trait on to his daughter, the mother would also have to be a carrier, and the chance that both unrelated individuals I2 and II2 are carriers for a trait that is usually found in only 1 in 10,000 people is highly improbable.

X-linked dominant—No, the data do not fit. Sons inherit their X chromosome from their mothers. If the disease-causing allele of this gene was X-linked dominant, then affected sons should have affected mothers, in all cases. Individual III3 is affected, but his mother is not, so the trait cannot be X-linked dominant.

Autosomal dominant—Yes, this is clearly the most likely hypothesis. In an autosomal dominant, if a parent is affected, then about half the children will be affected, regardless of sex. It also should make no difference if it is the mother or father who passes on the trait.

Autosomal recessive—It's possible that the trait is recessive, but this hypothesis is highly unlikely for several reasons. For a parent affected with an autosomal recessive trait to pass it on to their offspring, the other parent must either have the trait or be a carrier for the trait. In this case, all three of the following unrelated individuals must have been carriers for the trait to pass it on: I2, II2, and II3. Most autosomal recessive alleles are incredibly rare, so it is highly improbable that all three would be carriers.

(b) muscle atrophy. Again, let's examine all possibilities:

X-linked recessive—No, the data do not fit. If the disease-causing allele of this gene were X-linked recessive, and a mother was affected with the trait, then all of her sons would also express the trait, because sons receive their only X chromosome from their mother. Individual I1 is a female with the condition, but only half of her sons show the trait, so it cannot be an X-linked recessive. In addition, for parents to pass the trait on to their daughter, the father must have the trait. Daughter III3 has the trait, but her father does not, therefore ruling out X-linked recessive inheritance.

X-linked dominant—No, the data do not fit. Sons inherit their X chromosome from their mothers. If the disease-causing allele of this gene was X-linked dominant, then affected sons should have affected mothers. Individuals III4 and III8 are both affected, but neither mother is, so the trait cannot be X-linked dominant.

Autosomal dominant—No, the data do not fit, because individuals III3 and III4 both have the trait while neither of their parents does. Dominant alleles are expressed if they are present.

Autosomal recessive—This is clearly the most likely hypothesis. In an autosomal recessive, both male and female children can be affected even if their parents are not. This is clearly the case with individuals III3 and III4. It is interesting that this hypothesis requires that children inherit the trait from both parents, so clearly I2, II1, II2, and II6 must have all been carriers.

(c) ichthyosis (extremely dry skin). Again, let's examine all possibilities:

X-linked recessive—This is clearly the most likely hypothesis. The hallmark of an X-linked recessive allele is that mostly sons are affected, and the trait is passed from the mother to her sons. Individual II2 had a father with the trait; therefore she would have inherited the allele on one of the Xs. She doesn't have the trait, but one of her sons does. Sons receive their only X chromosome from their mother, so if she was a carrier, then her sons should be affected—and one is. None of her daughters are affected, because daughters receive one X from each parent, and II1 did not have the disease.

X-linked dominant—No, the data do not fit. Daughters inherit the father's X chromosome; that's why they are female. If the disease-causing allele of this gene is X-linked dominant, then

affected fathers should always have affected daughters. Individuals II2 and IV7 are both unaffected, but both fathers were, so the trait cannot be X-linked dominant.

Autosomal dominant—No, the data do not fit. Individual I1 has the trait, but his daughter does not, so clearly she did not inherit a dominant disease-causing allele—if she had, she would have the condition. But her son does have the disease, so if he inherited it from his mother, it must have been recessive.

Autosomal recessive—It's possible that the trait is autosomal recessive, but it's unlikely. With an autosomal recessive trait, you expect to see equal numbers of male or female offspring affected, but in this case you see only males affected.

4. **(a)**

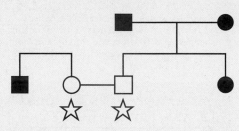

(b) Dominant. We know that Barry's mother and father are both affected by the condition. If it were a recessive condition, both parents would have to be homozygous for the trait, meaning all their offspring would have to inherit the condition as well. Yet they have produced one offspring (Barry) who is not affected. **(c)** genotypes: Barry's mom, *Aa*; Barry's sister, *AA*, or *Aa*, Barry *aa*; Jan: *aa*. **(d)** There is no chance that Barry and Jan will have a child with the syndrome.

5. **(a)** If the wife were a carrier for the X-linked recessive allele that causes Hunter's syndrome, half her sons would be likely to inherit the allele and therefore have the disease. But if she were a carrier for the autosomal recessive allele that causes Hurler's syndrome, she would have less cause to worry. With autosomal recessive traits, both parents need to be carriers so that their offspring have a 25 percent chance of inheriting the recessive allele from both parents. Because her husband had no history of the disease in his family, the chances would be quite good that he would not carry the allele, and they would have no chance of having a child with the condition. **(b)** The mucopolysaccharidosis condition the woman fears was not likely caused by Hunter's syndrome (an X-linked recessive condition), because the only affected relative was her aunt, individual II1. For the parents of this aunt to pass the trait on to her, the mother, I2, would have to be a carrier for it; but the father, I1, would have the condition, because males have only one X chromosome. Since the father I1 doesn't have the condition, it seems more likely that the mucopolysaccharidosis is correctly diagnosed as Hurler's syndrome, but a genetic counselor would need to do a genetic test to determine if III1 or III2 are carriers.

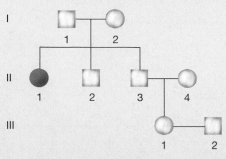

6. Mother, AO. (Her blood type can't be AA; if she were, the child would have to have at least one A allele.) Father, BO. (Same reasoning as above.)

Child, OO. (A person who is type O cannot have inherited either the A or B allele.)

7. **(a)** Debbie is *hh*, June is probably *Hh*. (It's likely that Debbie and June's mother is *hh*, because this is a rare disease, but one can't be sure. Thus, June could be *HH*.) **(b)** Debbie 0 percent, June 100 percent; **(c)** Debbie 0 percent, June 50 percent.

8. **(a)** 25 percent; **(b)** 50 percent; **(c)** 25 percent (males only); **(d)** 100 percent if either parent is homozygous dominant, 75 percent if both parents are heterozygous.

9. **(a)** Person I: *A* or *a*; Person II: *a*; Person III: *aB* and *ab*; Person IV: *AB*, *Ab*, *aB*, *ab*; **(b)** Person I with person II: *Aa × aa* will produce 50 percent *Aa* and 50 percent *aa*. Person III with person IV: *AaBb × aaBb* will produce about 13 percent *AaBB*, 25 percent *AaBb*, 13 percent *Aabb*, 13 percent *aaBB*, 25 percent *aaBb*, and 13 percent *aabb*.

10. **(a)** autosomal recessive, *aa*; **(b)** X-linked recessive, *xY*.

11. The parental generation cross is *TT × tt*. The F₁ cross is *Tt × Tt*. Remembering Mendel's 3:1 ratio, the outcome would be 150 tall plants (*TT*, *Tt*, *Tt*) and 50 short plants (*tt*). There would be no intermediate height plants, because there was no incomplete dominance in Mendel's crosses.

CHAPTER 13
Multiple Choice
1. b, 2. e, 3. d, 4. e, 5. c, 6. a, 7. b, 8. e, 9. d, 10. b.
Brief Review
1. "Something old, something new" indicates that the cell replicates its DNA in such a way that each of the two helixes formed contains one "old" handrail, and its associated bases and one "new" complementary set of bases and a handrail.
2. A point mutation may arise through an error in DNA replication. For example, a G base may pair with a T base across the DNA helix (rather than with a C base). Conversely, a C-G base pair may be inserted in a sequence where an A-T pair normally resides.
3. False—DNA polymerases have an editing function, moving backward to check for and correct mistakes.
4. False—A small proportion of mutations act to improve or change protein function, therefore contributing to the evolutionary process.
5. Even though there are only four bases, the order they occur in as they line up to form the DNA can vary extensively, thereby allowing for the production of a huge number of proteins.
6. Huntington disease and sickle-cell anemia are examples of human diseases caused by point mutations. In the case of sickle-cell anemia, the substitution of one T base for one A base is responsible for the condition.

CHAPTER 14
Multiple Choice
1. d, 2. d, 3. c, 4. c, 5. a, 6. e, 7. e, 8. d, 9. a, 10. b.
Brief Review
1. The replication of DNA and the production of proteins from DNA.
2. **(a)** mRNA (messenger RNA), which takes DNA information to ribosomes; **(b)** rRNA (ribosomal RNA), which helps make up the ribosomes that act at the site of protein synthesis; and **(c)** tRNA (transfer RNA), which brings amino acids to the ribosomes and binds with mRNA codons there.
3. Transcription is the process by which the information encoded in DNA sequences is put into the form of mobile mRNA sequences. Translation is the process by which the information encoded in mRNA sequences is put into the form of amino acid sequences; this results in the production of a protein.
4. Viruses require the use of host genetic machinery in order to function and replicate. If we used a different genetic code than viruses, we would be immune to virus invasion. As it is, we are susceptible because of the common code.

5. Introns are the sequences of DNA and mRNA that do not code for amino acid sequences. Existing only in eukaryotic cells, introns are edited out of the original mRNA sequence prior to the transport of mRNA out of the nucleus. Exons are the sequences of DNA and mRNA that code for amino acids. They are the sequences that constitute the mRNA transcript after its introns have been edited out.

6. It is inaccurate to conceptualize DNA as autonomously controlling a cell's production of proteins, for two reasons. First, while DNA contains information that is used to produce proteins, it carries out none of the actual steps of protein synthesis. All these actions are carried out by proteins. Second, some of the very proteins that are encoded in DNA sequences then feed back on DNA, binding with it to regulate its activity.

7. Three, one.

8. Because the genetic code—this DNA triplet coding for that amino acid—is nearly universal in all living things, we infer that it must have been operating in an ancient organism that is ancestral to all the organisms alive today. Because DNA encodes information, and because we share in the genetic code, we share in an informational linkage that stretches back billions of years.

CHAPTER 15
Multiple Choice
1. c, 2. e, 3. b, 4. a, 5. e, 6. e, 7. c, 8. e, 9. e, 10. a.
Brief Review
1. It is useful because some human beings cannot turn out sufficient quantities of their own supply of needed human proteins; this is so with some diabetics and the protein insulin, for example. Meanwhile, the protein may be available only in limited quantity—or with limited purity—from other sources. Thus it can be useful for human beings to manufacture supplies of given proteins through the processes of biotechnology.

2. The DNA fragments differ in their lengths. Those fragments that make it farthest through the gel are the shortest fragments, while those that go least far through the gel are the longest fragments.

3. PCR is valuable because it allows scientists to quickly make a large quantity of a given segment of DNA from a very small starting sample. This is useful in producing enough crime-scene DNA for testing, or for producing quantities of DNA from small amounts extracted from ancient artifacts.

4. Some limitations of the sequencing of the human genome are that (a) it provides the means of getting more complete knowledge; and (b) it may identify genes, but it does not identify their function or their effect on other genes.

5. Computers and modern genetic research are tightly linked because genomes and proteomes are so complex that they could not be analyzed without computers. In addition, genetic information is encoded as digital information, which computers are designed to work with.

6. Sticky ends are useful in restriction enzymes because they allow for the linking of specific sequences of DNA with one another.
   Fragment sequences:
   ATCG   GATCCTCCG
   TAGCCTAGGAG   GC

7. Dolly looked like the sheep that donated the udder cell, rather than the sheep that donated the egg, because the udder cell retained the DNA in its nucleus while the DNA in the egg was removed. Thus, all the instructions for Dolly's development ultimately came from the DNA in the udder cell.

CHAPTER 16
Multiple Choice
1. b; 2. a, b, d; 3. a, c, d, e; 4. Lyell—Geological forces observable today caused changes in Earth. Wallace—Natural selection is differential survival or reproduction. Lamarck—Inheritance of acquired characters.

Cuvier—Catastrophic extinction and new creations explain fossil record. 5. Homologous: a, c, d; don't belong together: b, e. 6. d; 7. a, b, d, e; 8. a, c, d; 9. a, c, d.
Brief Review
1. The term *homology* describes structures that are similar due to common ancestry. Homology provides evidence for evolution because it shows descent with modification.

2. Figure 16.12 shows that, in a gene shared by pigs, moths, and yeast, there are fewer DNA base differences between the forms of the gene found in humans and pigs than between the forms found in humans and moths. This is what you would expect if humans and pigs shared a common ancestor more recently than did humans and moths.

3. Congruent lines of evidence that are consistent with the theory of evolution: (a) The placement of fossils found in geological strata is consistent with the ages of those strata as determined by radiometric dating. (b) Evidence from molecular clocks is consistent with fossil and other evidence regarding the degree of relatedness among species.

4. Darwin brought back many birds that he thought were the blackbirds, wrens, and warblers he was familiar with in Britain, but he found they were in fact all species of finches. Darwin perceived this as an example of descent with modification. Without taking the *Beagle* voyage, Darwin would not have been exposed to such extravagant examples of branching evolution, nor would he have made various other biological observations that spurred his thinking about evolution by natural selection.

CHAPTER 17
Multiple Choice
1. c; 2. gene pool—the set of all alleles in a population; allele—a variant form of a gene; allele frequency—the relative representation of a given form of a gene in a population; genotype—the genetic makeup of an organism; gene flow—exchange of genes between populations; 3. a, c; 4. a, c, e; 5. d; 6. b, d; 7. b; 8. mutation—changes in the genetic material; gene flow—movement of alleles between populations by migration; natural selection—a process in which the fit of an organism with its environment selects those traits that will be passed on with greater frequency from one generation to the next; genetic drift—chance alterations of allele frequencies in a population, with such alterations having greatest impact on small populations; founder effect—genetic drift due to a few colonizing genotypes; 9. d; 10. c; 11. c.
Brief Review
1. An individual may be selected for survival—and hence greater reproduction—because of the qualities that individual has, such as coloration or height. But this individual's characteristics do not change because it possesses these qualities. An individual does not evolve, in other words. An individual can pass on more of its genes relative to other individuals in a population, however; this causes the population's allele frequencies to change, meaning the population has evolved.

2. A gene pool is the sum total of alleles that exist in a population. The inheritable characteristics of living things are based on the alleles they possess. Thus, for a population to change in a way that is passed on from one generation to the next—for the population to evolve—the mixture of alleles in that population must change. Thus the evolution of a population is based at root on changes in that population's gene pool.

3. The five causes of microevolution are (1) Mutation—Changes in the form of the genetic material that result in new alleles. (2) Gene flow—Brings alleles from other populations into the target population. (3) Nonrandom mating—Mating in which a given member of a population is not equally likely to mate with any other given member. Can change the genetic contribution individuals make to a succeeding generation. (4) Genetic drift—The chance alteration of allele frequencies in a population. May occur if a new

population is founded by only a few individuals (founder effect), or if the size of a population is decreased suddenly (the bottleneck effect). (5) Natural selection—Differential survival and hence reproduction of individuals based on their fit with a particular environment. If the phenotypic variation is inherited, the population will evolve to be like the individuals that survive longest and reproduce most.

4. An individual in a population has to have greater reproductive success relative to other members of the population in order to contribute more to the gene pool of the succeeding generation.

5. Mutation and genetic drift alter allele frequencies in random directions, and hence have nothing necessarily to do with adaptation. Gene flow doesn't necessarily bring in alleles that will be suited to the new environment, and nonrandom mating concerns mate choice, rather than an organism's fit with its environment. Through natural selection, however, phenotypes of individuals that survive and reproduce best in a given environment spread through the population.

6. Evolutionary fitness—the relative success in contributing genes to the next generation—depends crucially on the environment. It is the fit between environment and organism that determines which organisms will have the greatest success in survival and reproduction. Thus, the evolutionary fitness of an individual could be expected to vary greatly from one environment to another. In the Grants' study of the Galápagos finches, a deeper bill was favored after the drought, but this situation later reversed.

CHAPTER 18
Multiple Choice
1. b; 2. b, c, d; 3. b, c, d; 4. a, c, d; 5. temporal isolation—mating occurs at different times of day or year; ecological isolation—populations live in different environments; geographical isolation—populations are separated by physical barriers; gametic isolation—egg and sperm do not recognize one another; hybrid sterility—offspring are unable to reproduce; 6. b, c, e; 7. species, genus, family, class, phylum; 8. b, d, e.
Brief Review
1. Convergent evolution is the evolution of a very similar character in two unrelated organisms, caused when the organisms experience similar types of natural selection in their environments. For example, many unrelated species of desert plants have succulent leaves that store water. But these similar characters are not homologous (they are analogous), so they cannot be used to determine relationships in a phylogenetic study.

2. In the gradualistic model, evolutionary changes are steady and gradual over time. In the punctuated equilibria model, organisms go through long periods of stasis (remaining the same) and then enter into relatively brief periods of rapid evolutionary change. Some of the evidence for punctuated equilibria is that bacteria, which can be observed over tens of thousands of generations in a laboratory, do seem to evolve in spurts. Secondly, it appears that, in some species at least, changes in reproductively important traits can come about as a result of a small number of genetic changes.

3. Anagenesis is nonbranching evolution, in which a species evolves, giving rise to a single species that presumably would be reproductively isolated from the ancestral species. Cladogenesis is branching evolution, in which a single species may evolve into two or more species, with the ancestral species continuing while others branch off from it.

4. A first requirement of allopatric speciation is that two populations of the same species must become separated by a geographical barrier that prevents or restricts gene flow between them. This allows the forces operating in microevolution—natural selection, genetic drift, mutation, and so forth—to work separately on the two populations. Eventually, the populations may develop intrinsic

reproductive isolating mechanisms, such that, should the geographical barrier between them be removed, they would no longer interbreed. In sympatric speciation, intrinsic isolating mechanisms evolve in the absence of geographical separation. The quality that both sympatric and allopatric speciation have in common is the development of intrinsic reproductive isolating mechanisms.

5. In the biological species concept, two organisms are in the same species if they actually or potentially interbreed successfully in nature, while being reproductively isolated from other groups.

6. Without the evolution of intrinsic isolating mechanisms, two populations would simply interbreed, remaining one species, if the physical barrier to interbreeding were removed. Thus, they can't be called separate species unless they possess intrinsic reproductive isolating mechanisms.

CHAPTER 19
Multiple Choice
1. c, g, d, h, a, e, b, i, f; 2. a, c, d; 3. b, c, d; 4. b, c, e; 5. a, b, d; 6. False; 7. b, c, d; 8. a, b, d; 9. b, g; 10. a.
Brief Review
1. Eukaryotes can be single-celled or multi-celled, but in all cases, these cells have nuclei. Bacteria are always single-celled, with these cells not having nuclei.

2. The amniotic egg evolved in the Reptilia. It is an egg in which the embryo lives in a watery environment (albumin, the "white") while being enclosed in a protective leathery or hard shell. Special membranes facilitate gas exchange while protecting the embryo from drying out. This was a huge breakthrough for terrestrial animals, because it allowed them to lay eggs outside an aquatic environment without risking drying out.

3. Primates have forward-looking eyes and an opposable thumb. This facilitates depth perception and the ability to pick up and manipulate objects, paving the way for the use of tools.

4. Humans and chimps.

5. No.

6. Life may have begun as early as about **3.9** billion years ago. Land was not colonized until about 460 million years ago. Between those times, all life existed in the oceans, the land having no life on it.

7. Mammals have hair, which helps in heat retention. Most have internal development of young in the uterus, and they have mammary glands that enable them to feed their offspring with milk.

CHAPTER 20
Multiple Choice
1. b, c, e; 2. b, c; 3. a, c, d, e; 4. hyphae—filamentous strands of a fungus; mycorrhizae—structures that are a combination of plant root and fungal hyphae; cell walls—aid plants in regulating water intake; archaea—often occupy extreme environments; protists—ancestors of animals, plants, and fungi; 5. a, c, e; 6. d; 7. e; 8. False; 9. c.
Brief Review
1. Extremophiles are bacteria or archaea that live in environments that are too harsh for ordinary organisms. Such environments can exhibit extremes of heat, cold, pH, salt, or other factors.

2. Viruses stimulate the immune system to produce specific antibodies and T cells to recognize and eliminate a specific virus. Even after the infection is cleared, some of these highly specific proteins and cells remain. If the same virus invades again, it is wiped out before it has a chance to become established. Vaccines made from killed virus or from weakened virus can provoke the immune response and thus offer protection from the virus.

3. Viruses cannot replicate themselves without being in a host cell. The ability to self-replicate is generally considered to be a fundamental, defining feature of life.

4. Phytoplankton do much of the photosynthesis on Earth, and they are at the base of many aquatic food chains.

5. Some fungi interact with plant roots, assisting them in their water uptake. Fungi obtain food from the plant in this process. Other specialized fungi are able to form a close association with algal or bacterial cells in forming a lichen. In a lichen, the fungus obtains food from photosynthesis, and the algae or bacteria obtain protection from drying out.

6. Plants make their own food through photosynthesis; they are photoautotrophs. Fungi do not make their own food; they are chemoheterotrophs that digest food outside their bodies.

7. Raw plants have cell walls made out of cellulose, and each cell has a large, water-filled vacuole. Chewing breaks the cell walls, releases the water, and makes the plant crunch.

CHAPTER 21
Multiple Choice
1. a, c; 2. d; 3. a; 4. e; 5. d; 6. a, b, c, d; 7. c; 8. b; 9. c, e; 10. d.
Brief Review
1. A central cavity provides an animal with flexibility, allows for the expansion and contraction of organs such as the heart and stomach, and helps protect an animal's organs from being damaged by external blows.

2. All animals are thought to have evolved from a single, common ancestor, probably a protist called a choanoflagellate. The evidence for this is that all animals share several physical features. All go through a blastula stage in development—whereas no other life-forms do—and all are composed of cells that lack cell walls. It would be difficult to imagine a feature as specific as a blastula stage of development arising more than once in evolution.

3. The benefit of endothermy is that animals that possess it are fully functional in a wider range of temperatures than are ectotherms, as endotherms do not have to rely on their external environment to control their internal temperature. Endothermic animals also are able to live in colder climates than ectothermic animals. The cost of endothermy is higher energy expenditure; to obtain the energy they need, endothermic animals have to eat more, per unit of body weight, than do ectothermic animals. Only mammals and birds are truly endothermic.

4. Sponges have no brains or organs of any sort, no tissues, and each of their cells functions with a great deal of independence. Thus, the cells in a sponge have a lower level of specialization and integration with one another than is the case in more advanced animals.

5. Coral reefs are the skeletal remains of many generations of a variety of cnidarian—the reef-building coral—along with the remains of an associated type of algae.

6. The mollusc's unitary foot evolved over time into the arms and tentacles of the squid.

7. A dust mite; both the spider and mite are part of the arthropod subphylum Chelicerata, while the grasshopper is part of the arthropod subphylum Uniramia.

8. The mother must be a monotreme mammal, because only mammals provide milk for their young, and only monotreme mammals are hatched from eggs.

CHAPTER 22
Multiple Choice
1. a; 2. e; 3. b, c; 4. c; 5. e; 6. c, e; 7. a; 8. b; 9. b; 10. e.
Brief Review
1. The stomata allow the movement of carbon dioxide into the plant and oxygen out. These same openings also allow water to escape from the plant as water vapor. The plant solves the resulting problem of water loss by moving large quantities of water up through the roots and stem and out through the stomata.

2. Roots serve to anchor the plant (usually in soil), and they are the main absorptive organs of the plant. They absorb water and minerals from their surroundings and make them available to the above ground parts of the plant. In some species, roots also store substantial reserves of food. Taproots and fibrous roots are the two main types of roots produced by flowering plants.

3. Parts of a flower and main functions of each:

Flower Part	Function
pedicel	Flower stalk; lifts flower up to make it noticeable to pollinators.
sepals	Leaf-like structures that protect flower buds before they open; commonly, also augment food supplies because they are usually green and photosynthetic.
petals	Generally most conspicuous part of the flower; often brightly colored; sometimes fragrant; attract pollinators.
stamens	Male reproductive structures; consist of a stalk called a filament, which holds aloft the pollen-producing sacs called anthers.
pollen	The male gametophyte; the vehicle that conducts male sperm to the female reproductive structures.
carpel	Female reproductive structure; consists of the ovary, style, and stigma—the ovary produces and protects the egg-containing female gametophyte, and after fertilization turns into the fruit; the stigma is the receptive surface that captures pollen and promotes germination of the right kind of pollen; the slender style raises the stigma to make it accessible to pollinators.

4. While people and other animals grow over their whole body, plants confine their vertical growth to distinct zones at the tips of their roots and shoots. Because these zones can be active throughout the life span of a plant, plants continue to grow all through their lives. Animals, in contrast, typically cease growth upon reaching maturity.

5. Plants enter into cooperative arrangements with members of nearly all the kingdoms of life. The roots of some plants—those of legumes such as beans and peas, for example—harbor nitrogen-fixing bacteria; the plant supplies carbohydrate to the bacteria, which in return manufactures hard-to-obtain nitrogen-containing compounds that the plant can use. In nature, the roots of most plants are colonized by certain types of fungi (mycorrhizal fungi) whose hyphae increase the plant's ability to absorb water and minerals. In return, these fungi receive carbohydrates produced by the plant.

6. The gametophyte is the multicellular haploid phase of a plant; that is, each cell in the gametophyte has just one set of chromosomes. The gametophyte produces gametes (egg and sperm). The sporophyte is the multicellular diploid phase of the plant; the sporophyte produces the spore that gives rise to the gametophyte.

CHAPTER 23
Multiple Choice
1. a; 2. b; 3. c; 4. c; 5. b; 6. a; 7. e; 8. b; 9. c; d; 10. c.
Brief Review
1. Differences between monocots and dicots:

	Monocots	Dicots
number of cotyledons (food-storing parts of the embryo)	One	Two
arrangement of flower parts	Sets of three	Sets of four or five
leaf vein pattern	parallel veins	net-like veins

arrangement of vascular bundles	Scattered	In a discrete ring
type of root system	Fibrous root	Taproot

2. The three main cell types in a plant and the functions performed by each:

Cell Type	Function
parenchyma	Living cells; perform a wide variety of metabolic functions such as photosynthesis, food storage, wound repair.
collenchyma	Living cells with irregularly thickened walls; perform some metabolic functions such as photosynthesis, but also impart mechanical strength to tissues experiencing active growth.
sclerenchyma	Usually dead at maturity; main role is structural support for the plant body, and water conduction in some cases.

3. Meristems are tissues specialized for cell division alone. After embryonic development, meristems are the source of almost all new cells that contribute to the growth of the plant body. Meristematic cells do not acquire new, more specialized cell functions. Meristems are perpetually maintained. Apical meristems at the tips of shoots and roots contribute to the length or height of a plant. Lateral meristems are of two main kinds—the vascular cambium gives rise to secondary xylem and phloem, and cork cambium creates a protective layer of cells known as bark; both types contribute to an increase in the girth of the plant.

4. *Xylem-conducting elements*—tracheids and vessel elements—are dead at maturity. They have thick, lignified walls. Tracheids have many small perforations, vessels with one or a few relatively large perforations. Xylem conducting elements provide mechanical support, and conduct water and dissolved minerals to all parts of the plant. *Phloem-conducting elements*—sieve tubes—are living, but lack nuclei. They have relatively thin walls. Sieve tubes are composed of several sieve tube members (individual cells) connected through cytoplasmic channels. Each sieve tube member is closely associated with, and totally dependent on, living cells called companion cells. Sieve tubes conduct carbohydrates, mainly sucrose, from sites of production (sources) to locales where they will be utilized or stored (sinks) within the plant. Some other compounds, such as amino acids and plant hormones, are also transported in phloem sap.

5. Growth rings are created by abrupt changes in the size and structure of xylem elements laid down earlier in the growing season versus later in the growing season. The first-formed wood, when the growing season resumes, is called early or spring wood. Early wood is composed of xylem elements with a wide diameter. It is more porous and less dense, and appears lighter because there's proportionately more air space relative to thick cell walls. Late wood is formed later in the season, when soil water becomes depleted and trees may experience water shortage. The tracheids and vessels produced then have a narrower bore and proportionately thicker walls. Late wood therefore appears denser and darker. Each band of early and late wood constitutes an annual ring, and represents the wood deposited in a single growing season. In temperate regions, that means a single year's worth of secondary growth. Because the dark, late wood of the current season lies next to the following season's lighter, early wood, the annual rings produced in successive seasons are easy to see. Thus the total number of annual rings in a tree trunk is equivalent to the number of seasons, or years, that tree has been growing.

6. Double fertilization refers to the two fusion events that take place when pollen tubes deliver the two sperm cells to the embryo sac (haploid female gametophyte, in the ovule). One of the two sperm cells fuses with the egg to give rise to the diploid zygote and future embryo. The second sperm cell fuses with a set of two haploid nuclei that are suspended within a large central cell; this fusion event gives rise to a triploid structure, the endosperm.

CHAPTER 24
Multiple Choice
1. c; 2. a; 3. b; 4. b; 5. e; 6. b; 7. e; 8. c; 9. c; 10. d; 11. d.
Brief Review
1. The sequence of the levels of organization of the human body—from the simplest level to the most complex—is molecular, cellular, tissue, organ, organ system, organism (the individual).
2. Blood and lymph.
3. Keratin is a durable, water-resistant protein. It fills the outer epidermal cells, making a waterproof skin. (It also makes up the nails of the hand and foot, and forms the outer layer of hairs.)
4. Three functions of the skeletal system are support, storage of minerals and lipids, and blood-cell production.
5. Three primary functions of the muscular system are to produce skeletal movement, to maintain body posture and body position, and to maintain body temperature.
6. The endocrine system is made up of ductless glands whose secretions (hormones) are carried throughout the body, where they alter the activities of organ systems.
7. Cells are lost through abrasion (scraping), and the new cells that are formed through cell division push up on the layers above them.
8. Concentric layers of bone around a central canal indicate a Haversian system (osteon) that makes up compact bone. Because the ends (epiphyses) of long bones are mostly spongy bone, this sample most likely came from the shaft (diaphysis).
9. Skeletal muscle appears striated, or striped, under a microscope because it is composed of thin (actin) and thick (myosin) protein filaments arranged in such a way that they produce a banded appearance in the muscle.

CHAPTER 25
Multiple Choice
1. c; 2. d; 3. a; 4. e; 5. d; 6. a; d; 7. d; 8. a; 9. d; 10. e; 11. e; 12. b.
Brief Review
1. The three functional groups are sensory neurons, motor neurons, and interneurons. The sensory neurons carry information about the internal and external environments to the central nervous system (CNS). Motor neurons carry information from the CNS to other tissues, organs, or organ systems that can change their activities. Interneurons analyze the sensory information and coordinate motor responses.
2. Dual innervation of a bodily organ by both the sympathetic and parasympathetic divisions of the autonomic nervous system produces opposite effects on organ function. This helps maintain relatively rapid control of the organ in question, stimulating or relaxing it.
3. The cerebrospinal fluid provides cushioning and support for the brain. It also transports nutrients, chemical messenger molecules, and waste substances.
4. In communication by the nervous system, the source and destination are quite specific and the effects are short-lived. In endocrine communication, the effects are slow to appear, and they often persist for days. A single hormone can alter the metabolic activities of multiple tissues and organs simultaneously.
5. The nonspecific defenses are present at birth and do not discriminate between one threat and another. Specific defenses provide protection against threats on an individual basis.
6. Although the medulla oblongata is small, it contains many vital reflex centers including those that control breathing and regulate the heart and blood pressure. Damage to the medulla oblongata can result in cessation of breathing or changes in heart rate and blood pressure that are incompatible with life.
7. Light enters through the cornea and passes through the opening in the iris (the pupil), then through the lens and vitreous body on its way to the retina. It passes through two layers of retinal cells—the ganglion and connecting cells—before arriving at the photoreceptors (rods and cones) at which its energy will be converted into nervous-system signals. These signals pass back out from the photoreceptors to the connecting cells and then to the ganglion cells, whose axons come together to form the optic nerve. Neural signals pass through the optic nerve to the thalamus and then, through different axons, to an area at the back of the brain called the primary visual cortex.
8. In negative feedback, elements that bring about a response have their activity decreased by that response. In positive feedback, elements that bring about a response have their activity increased by that response.
9. The pituitary gland releases six hormones from its anterior portion and two from the posterior portion. Many of the pituitary's hormones stimulate, or "turn on," other endocrine glands. As a result, the pituitary is called the "master" endocrine gland.
10. Helper T cells promote B-cell division, the maturation of plasma cells, and the production of antibodies by the plasma cells. Without the helper T cells, the antibody-mediated immune response would take much longer to occur and would not be as efficient.
11. A decrease in the number of monocyte-forming cells in the bone marrow would result in a decreased number of macrophages in the body, because all the different macrophages are derived from monocytes.

CHAPTER 26
Multiple Choice
1. b, 2. c, 3. a, 4. b, 5. c, 6. c, 7. e, 8. d; 9. c; 10. b.
Brief Review
1. Transportation—of dissolved gases, nutrients, hormones, and metabolic wastes. Regulation—pH and electrolyte composition of interstitial fluids throughout the body, stabilization of body temperature, and restriction of fluid losses through damaged vessels. Protection—defense against toxins and pathogens.
2. Capillaries are only one cell thick. Small gaps between adjacent endothelial cells permit the diffusion of water and small solutes into the surrounding interstitial fluid but prevent the loss of blood cells and plasma proteins. The walls of arteries and veins are several cell layers thick and are not specialized for diffusion.
3. External respiration is the diffusion of gases between the blood in the lung capillaries and air in the alveoli of the lungs.
4. The four layers of the digestive tract, beginning with the innermost layer, are the mucosa, submucosa, muscularis externa, and serosa.
5. The urinary system performs vital excretory functions and eliminates the organic waste products generated by cells throughout the body, while conserving needed materials.
6. Drinking this much fluid would likely cause your body to reduce its production of antidiuretic hormone, thus allowing the excess fluid to be retained by the kidney tubules and flow into the bladder.
7. Most of the carbon dioxide released by body tissues enters red blood cells, where it's converted into bicarbonate ions that then diffuse into the blood

plasma. At the lungs, the bicarbonate ions are converted back into carbon dioxide within the red blood cells and released into the alveoli of the lungs.

8. The small intestine has several adaptations that increase surface area to increase its capacity to absorb food. Its internal wall consists of folds that are covered by fingerlike projections of tissue called villi. The cells covering the villi, in turn, have smaller fingerlike projections called microvilli. In addition, the small intestine has a rich blood and lymphatic supply to transport the absorbed nutrients.

9. Blood cells and large protein molecules are too large to pass out of the circulatory system and into the filtrate in the kidneys.

CHAPTER 27
Multiple Choice
1. b, d, c, e, a; 2. b; 3. d; 4. c; 5. d; 6. a; 7. c; 8. b; 9. b; 10. c.
Brief Review
1. After fertilization, the cleavage stage of development begins. This results in the formation of a hollow ball of cells. The embryo does not change in size during cleavage. Next, these cells undergo rearrangement and movement, in the process called gastrulation. Now the embryo has a body cavity and three layers of cells—the endoderm, ectoderm, and mesoderm.

2. After gastrulation in the chick embryo, a small group of cells is induced to form the wing of the chick. This primitive wing, the limb bud, is transformed into the wing. A portion of the developing wing, called the zone of polarizing activity (ZPA), sends out a morphogen that directs the proper order of digit production in the chick's wing.

3. Cell adhesion molecules underlie all the processes of development that involve cell movement and reorganization. Gastrulation and neural-crest cell migration are just two examples of processes that rely on this class of proteins.

4. A concentration gradient refers to the distribution of a molecule within an embryo in a pattern of high concentration to low. The morphogen bicoid is distributed in *Drosophila* flies in a concentration gradient with high concentrations in the anterior of the embryo and progressively lower concentrations toward the posterior regions. High concentrations of bicoid result in the production of head structures.

5. Cell death is a normal process in the development of most organisms. In humans, cell death "trims" down the tissue between each finger, so that the fingers are no longer webbed when the baby is born.

6. Scientists are enthusiastic about stem cells because these cells hold the potential to produce tissues that could cure or help alleviate many kinds of human illnesses, such as heart disease and diabetes.

CHAPTER 28
Multiple Choice
1. c; 2. e; 3. a; 4. c; 5. d; 6. c; 7. b; 8. a; 9. e; 10. c.
Brief Review
1. Gonadotropin-releasing hormone, which is released by the hypothalamus, stimulates the release of follicle-stimulating hormone (FSH) and lutenizing hormone (LH). Both of these hormones are released by the anterior pituitary gland. FSH promotes the development of the follicles, and LH stimulates the production of estrogen. As estrogen levels increase, FSH levels at first begin to decrease, but as estrogen levels begin to peak due to the development of the tertiary follicle, both FSH and LH levels spike. This results in ovulation. The corpus luteum then begins to synthesize progesterone, which acts on the hypothalamus to reduce FSH and LH to original levels, resulting in degeneration of the endometrium and menstruation.

2. Sperm development is dependent on temperature. If things get too hot, then sperm production is impaired. Similarly, cold temperatures also prevent sperm development. The structure

of the testes acts to keep sperm production at the correct temperature. The sperm are protected from the high body temperatures because of the location of the testes outside the body. In addition, a muscle that surrounds the testes contracts when temperatures drop, so that the testes are pulled in closer to the warm body.

3. Scientists hypothesize that menstruation serves to remove harmful pathogens, such as bacteria, that may have entered along with sperm during intercourse. Also, the cost of maintaining an embryonic environment may exceed the monthly buildup of uterine tissue.

4. The tertiary follicle and the corpus luteum are variations of the same structure. Once ovulation has occurred and the tertiary follicle releases the oocyte, the tertiary follicle then becomes the corpus luteum. Both structures release hormones.

5. The fast block to polyspermy occurs when the first sperm has fertilized the egg. Membrane channels in the egg open, allowing positively charged sodium ions to enter the egg. This changes the electrical potential of the egg, which prevents additional sperm from fertilizing the egg. This is necessary to prevent the introduction of additional sperm, with their additional sets of genetic information.

6. Prior to implantation, the fertilized egg divides to form a morula. This tightly packed ball of cells continues to divide, forming a hollow ball of cells called the blastocyst. The blastocyst will then implant in the uterine wall.

7. Any baby born before 37 weeks is premature. Premature births may have many complications due to the incomplete development of organs such as the lungs.

8. Birth control pills work using a combination of synthetic estrogen and progestin. Normally, low estrogen levels trigger the beginning of follicle development. The pill creates high estrogen levels that suppress the release of FSH from the hypothalamus. The progestin in the pill suppresses the release of LH in order to prevent ovulation of any follicle that might develop in the presence of high estrogen levels.

CHAPTER 29
Multiple Choice
1. d; 2. c; 3. e; 4. c; 5. d; 6. b; 7. a; 8. c.
Brief Review
1. Population, community, ecosystem, biosphere.

2. *K*-selected species experience their environment as relatively stable, spend most of their lives in populations that are close to carrying capacity (*K*), and provide a relatively large amount of care to a relatively small number of offspring. In contrast, *r*-selected species experience their environment as relatively unstable, have populations whose numbers are limited largely by reproductive rate, and provide little or no parental care to a relatively large number of offspring. Humans and elephants are *K*-selected species; flies and ragweed are *r*-selected species.

3. *Environmental resistance* is the term used to describe all the environmental forces that act to limit population growth. Organisms will run out of food or have their sunlight blocked; temperatures will plunge or rain will cease; greater numbers of predators will discover the population; the wastes produced by the organisms will begin to be toxic to the population.

4. This is an example of Batesian mimicry, in which one species (the red salamander) has evolved to acquire the appearance of a species with superior protective capabilities (the red newt). The red newt is the model, the red salamander is the mimic, and the dupe is any predator species that would avoid the red salamander, believing it to be a red newt.

5. Because this succession begins from a state of little or no life and no soil, it is an example of primary succession.

CHAPTER 30
Multiple Choice
1. c; 2. d; 3. b; 4. a; 5. a; 6. b; 7. c; 8. e; 9. a.
Brief Review
1. According to the energy-flow model of ecosystems (and the second law of thermodynamics), the energy contained in each trophic level is much less than the energy contained in the trophic level below it, because much of the energy in each trophic level is used for metabolic processes (running, staying warm) and much energy from these processes is lost from the system as heat. In addition, not all of the prey is eaten by the predators.

2. Warm air (a) rises and (b) can carry more moisture than cold air. Near the equator, the Earth gets warm due to the direct rays coming from the Sun. As this warm air rises, it cools and drops moisture. This cooler air moves toward the poles in both directions, and descends at about 30° north and south, warming as it goes, thus absorbing moisture from the land, which is dry as a result.

3. Climate essentially defines the large-scale vegetation patterns that are seen on Earth. Variations within large-scale climate regions tend to be mountains, whose climates can vary greatly from base to peak.

4. It is likely that eutrophication has occurred in the pond. If so, large algal blooms can result from the excess nutrients. As this dead organic matter falls to the lower depths, decomposers consume the organic matter but deplete the water of oxygen, thus killing fish.

5. Permafrost makes the tundra wet during summer months, even with low rainfall, and it is too cold for trees to grow. Taiga has a less severe climate and a longer growing season, and no permafrost; coniferous trees grow slowly there and support a larger animal community.

CHAPTER 31
Multiple Choice
1. c, d; 2. d; 3. c; 4. a; 5. a, b; 6. d; 7. e; 8. a; 9. c; 10. d.
Brief Review
1. A proximate cause. The ultimate cause is attracting females and warning other males away from a territory, which brings reproductive benefits.

2. A circadian rhythm is a biological cycle that functions independently of environment and is roughly synchronized to Earth's 24-hour rotation cycle. Such a cycle may be metabolic (when does an animal get sleepy?) or behavioral (when does a cricket start chirping?).

3. Classical conditioning exists when an animal learns by pairing one stimulus that has positive or negative consequences with another; operant conditioning exists when an animal's behavior brings about a response that has positive or negative consequences for the animal.

4. Bird song generally has both innate and learned components to it. Many songbirds will learn variations on songs in accordance with what they have been exposed to when young, but nevertheless have an innate tendency to learn the songs of their own species, rather than the songs of other species.

5. Because there are costs as well as benefits to group living; animals will live in groups to the extent that the costs are outweighed by the benefits.

6. The proximate cause is that animals have genes that prompt them to carry out the altruistic behavior. The ultimate cause is inclusive fitness—animals can often gain reproductive benefits by aiding their relatives, particularly close relatives.

7. Here is the argument: Habituation is trial-and-error learning in which the trials are the many stimuli that fail to result in a positive or negative consequence, and the errors are the initial fright responses. In classical conditioning, an animal learns, through a trial-and-error process, which stimuli (such as nectar) go with which other stimuli (such as a flower of a certain color). Insight learning is trial-and-error learning in which the trials go on inside the head of the learner.

Glossary

1n Haploid; having one chromosome of each type per cell instead of a pair of chromosomes of each type.

2n Diploid; having two chromosomes of each type per cell.

abiotic Pertaining to nonliving things.

abscission The detachment of plant parts, such as leaves, petals, or fruits, from the main plant.

accessory pigment A type of pigment found in chloroplasts that, along with chlorophyll *a*, absorbs some wavelengths of sunlight in the first step of photosynthesis.

acid Any substance that yields hydrogen ions when in solution. An acid has a number lower than 7 on the pH scale.

acrosome A structure located on the front end of the head of a vertebrate sperm cell; contains enzymes that help the sperm penetrate the accessory cells surrounding the oocyte.

action potential A transient reversal of electric potential across a cell membrane that results in a conducted nerve impulse.

activation energy The energy required to initiate a chemical reaction. Enzymes lower the activation energy of a reaction, thereby greatly speeding up the rate of the reaction.

active site The area of an enzyme, usually a groove or pocket, that binds the substrate and changes its shape to affect its reactivity.

actively acquired immunity Immunity resulting from natural exposure to an antigen in the environment or from deliberate exposure to an antigen through such means as an immunization shot.

active transport Transport of materials across the plasma membrane in which energy must be expended. Through active transport, solutes can be moved against their concentration gradient. The sodium-potassium pump is an example of active transport.

adaptation An evolutionary modification in the structure or behavior of organisms over generations that makes them better suited to their environment.

adaptive radiation The rapid emergence of many species from a single species that has been introduced to a new environment. The different species specialize to fill available niches in the new environment.

adipose tissue Also known as fat, this type of loose connective tissue insulates the body and stores energy in addition to providing cushioning.

adrenocorticotropic hormone (ACTH) A hormone, secreted by the anterior pituitary, that stimulates release of steroid hormones from the outer cells of the adrenal gland.

adult stem cell A cell from a tissue of an adult that is capable of giving rise to many other types of cells or tissues.

afferent division The division of the peripheral nervous system that carries sensory information toward the central nervous system, having gathered information about the body or environment.

AIDS Acquired immunodeficiency syndrome, a disease caused by the human immunodeficiency virus (HIV) that attacks the body's own immune system and leaves it unable to fight off even mild infections.

albumins With globulins and fibrinogen, one of the three primary classes of proteins suspended in the fluid or plasma portion of the blood. Albumins function in the transport of hormones and fatty acids.

alcoholic fermentation The process by which yeasts produce alcohol as a byproduct of glycolysis they perform in oxygenless environments.

algal bloom An overabundance of algae in a lake, resulting from an excess of nutrients. The many dead algae that fall to the bottom allow decomposing bacteria to flourish, using up so much oxygen that fish can suffocate.

alimentary canal Another term for the digestive tract, the muscular tube that runs through the body from mouth to anus and through which food passes.

alkaline Basic, as in solutions. Alkaline (basic) solutions have numbers above 7 on the pH scale.

allele One of the alternative forms of a single gene. In pea plants, a single gene codes for seed color, and it comes in two alleles—one codes for yellow seeds, the other for green seeds.

allergen A foreign substance that triggers an allergic reaction. These substances are usually derived from living things, including pollen, dust mites, foods, and fur.

allergies Immune system overreactions to antigens.

allopatric speciation Speciation that involves the geographic separation of populations. Most speciation involves geographic separation, followed by the development of intrinsic isolating mechanisms in the separated populations.

allosteric Denoting control of a chemical reaction through the actions of a second binding site on an enzyme. Allosteric reactions are important chiefly in negative feedback, when the product of a reaction binds to the enzyme that hastened the reaction, thereby changing the enzyme's shape and temporarily halting its ability to bind to more substrate.

alpha helix A common secondary structure of proteins that resembles a corkscrew or spiral staircase. Hair, nails, and horns are proteins made up almost entirely of alpha helices.

alternation of generations Plant life cycle in which a haploid generation alternates with a diploid generation. A haploid gametophyte plant produces gametes through mitosis; fertilization of the female gamete (egg) with the male gamete (sperm) produces a diploid zygote that grows into a diploid sporophyte plant. Through meiosis, this plant produces haploid spores that grow into haploid gametophytes.

altruism A costly or risky behavior carried out by one animal for the benefit of another animal.

alveoli (singular, *alveolus*) The many tiny air sacs that form the terminal branches in the lungs. Gases are exchanged with the capillaries through the walls of the alveoli.

amniocentesis A medical procedure in which amniotic fluid is withdrawn from a mother's abdomen with a needle and the fetal epidermal cells contained in the fluid are tested for genetic abnormalities.

amniotic egg An egg with a hard outer casing and an inner series of membranes and fluids that form a padding around the growing embryo. The evolution of the amniotic egg, in reptiles, freed them from the constraint of having to reproduce near water.

amniotic fluid A protective and nutritive fluid that surrounds the fetus of mammals, including filling the lungs.

anagenesis A form of evolution in which a species changes over time, eventually having changed enough that it would be reproductively isolated from its ancestral species. In this nonbranching evolution, the original species changed into the new species and itself disappeared.

analogy A structure found in different organisms that is similar in function and appearance, but is not the result of shared ancestry. Analogies must be distinguished from homologies to get a true picture of evolutionary relationships.

ancestral character A character that existed in the common ancestor of a group of organisms. Cladistics distinguishes ancestral from derived characters and uses these characters to determine evolutionary relationships.

aneuploidy A condition in which an individual organism has either more or fewer chromosomes than is normal for its species, so that the individual has the wrong number of chromosomes in its set. Down syndrome is the result of aneuploidy—generally three copies of chromosome 21.

angiosperm A flowering seed plant whose seeds are enclosed within the tissue called fruit. Angiosperms are the most dominant and diverse of the four principal types of plants. Examples include roses, cacti, corn, and deciduous trees.

animal pole The end of a zygote with relatively less yolk and lying closer to the cell's nucleus. The location of the egg's poles defines the orientation in which the embryo develops.

annual A type of plant that goes through its entire life cycle—from germination of the seed through growth, flowering, and death—in one year or less. Many crop plants, such as tomatoes and grains, are annuals.

annual ring The ring formed in wood—shows the abrupt change between the secondary xylem of late summer, when cells get less water and thus are smaller and have proportionately larger cell walls, and the secondary xylem of early spring, when the cells get more water, are larger, and have proportionately smaller cell walls.

anterior pituitary An endocrine gland that releases two hormones that work directly on target cells and four other hormones that regulate the production of hormones by other endocrine glands.

anther The part of a flower that produces pollen grains. The anther is on top of a filament, and together they are called the stamen.

antibiotic A substance produced by one organism that is toxic to another. When speaking of antibiotics, we usually mean medicines directed against disease-causing bacteria.

antibody A protein of the immune system, found on the surface of a B cell or circulating free in the plasma, that is formed in response to a particular antigen and reacts with it specifically, deactivating it.

antibody-mediated immunity An immune system capability that works through the production of proteins called antibodies.

anticodon The end of the transfer RNA molecule that can form a base pair with a particular codon on the mRNA transcript.

antidiuretic hormone (ADH) Hormone, released by the posterior pituitary, that acts to decrease the amount of water lost by the kidney, helping to conserve water.

antigen Any foreign substance that elicits a response by the immune system. Certain proteins on the surface of an invading bacterial cell, for instance, act as antigens that trigger an immune response.

antigen-presenting cell (APC) A macrophage that has ingested an infected cell and displays fragments of proteins from the invader on its surface. The APCs bind with helper T cells and activate them, stimulating secretion of interleukin-2, which has multiple functions in mounting further attacks against the invader.

aorta The major artery in the circulatory system. The aorta has branches that send blood to all of the body's other tissues except the lungs.

apical dominance Suppression of the growth of the lateral branches of a plant through the activity of apical meristems.

apical meristem The group of plant cells at the tips of the roots and shoots that are able to keep dividing, and from which all tissues ultimately develop.

appendicular skeleton The division of the skeletal system consisting of the bones of the paired appendages, including the pelvic and pectoral girdles to which they are attached.

aqueous solution A solution in which water is the solvent. Because living things contain so much water, nearly all solutions important in biological systems are aqueous solutions.

aquifer The porous underground rock in which groundwater is stored.

Archaea With Bacteria and Eukarya, one of three domains of the living world, composed solely of microscopic, single-celled organisms superficially similar to bacteria but genetically quite different. Many of the Archaea live in extreme environments such as boiling-hot vents on the ocean floor.

arithmetical increase An increase in numbers by an addition of a fixed number in each time period.

arteriole The smallest type of artery, which branches into the capillary beds, where gases and nutrients are delivered to tissues.

artery A blood vessel that carries blood away from the heart.

artificial eutrophication Nutrient enrichment of a lake through human intervention. This can be intentional, or it can occur accidentally when, for example, fertilizers from lawns or agriculture are washed into the lake.

asexual reproduction Reproduction that occurs without the union of two reproductive cells (sexual reproduction). Offspring of asexual reproduction are genetically identical to their parents.

assortative mating Mating in which males and females that share a particular characteristic, such as short height, tend to mate with one another. Such mating tends to bring similar alleles together but does not directly alter allele frequencies in the population.

atmosphere The layer of gases that surrounds the Earth. The atmosphere is nonliving, but its presence enables life to exist on Earth.

atom The fundamental unit of matter; once thought to be the smallest particle in nature, but now known to be made up of many smaller units, including electrons, protons, and neutrons.

atomic number The number of protons in the nucleus of an atom. Gold has 79 protons in its nucleus and thus has the atomic number 79. All elements are ordered on the periodic table according to atomic number.

atomic weight The combined weight (mass) of protons and neutrons in an atom's nucleus.

ATP synthase The enzyme that catalyzes the synthesis of ATP in the electron transport chain.

autonomic nervous system A subdivision of the efferent division of the peripheral nervous system that provides involuntary regulation of the smooth muscles, cardiac muscles, and glands.

autosomal dominant disorder A genetic disorder caused by a single faulty allele located on an autosomal (non-sex chromosome). Huntington disease is one example.

autosomal recessive disorder A recessive dysfunction caused by faulty alleles on an autosome (non-sex chromosome). Sickle-cell anemia is one example.

autotroph Any organism that manufactures its own food. Almost all plants and algae, and certain bacteria, are autotrophs.

axial skeleton The division of the skeletal system that forms the central column, including the skull, vertebral column, and rib cage.

axon A single, large extension of the cell body of a neuron that carries signals away from the cell body toward other cells.

bacilli (singular, bacillus) Rod-shaped bacteria.

Bacteria With Archaea and Eukarya, one of three domains of the living world, composed solely of single-celled, microscopic organisms that superficially resemble archaea but are genetically quite different.

bacteriophage A type of virus that infects bacteria; used in biotechnology to put foreign DNA into bacterial cells, therefore serving as a cloning vector.

ball-and-stick model A diagram showing the three-dimensional structure of a molecule, with the atoms drawn as balls and the bonds between them drawn as sticks. This type of representation clearly shows the spatial relationship of the atoms to each other.

bark Everything outside the vascular cambium in woody plants; bark consists of cork cambium, cork, phelloderm, and secondary phloem.

base Any substance that accepts hydrogen ions in solution. A base has a number higher than 7 on the pH scale.

Batesian mimicry A type of mimicry in which one species evolves to resemble a species that has superior protection against predators.

behavioral biology The study of the behavior of animals.

bell curve A distribution of values that is symmetrically largest around the average.

benthic zone The ocean floor, where bottom-dwellers live.

beta pleated sheet A common secondary structure of proteins that resembles the folds of an accordion. Silk is a protein made up almost entirely of beta pleated sheets lying stacked like pancakes on top of each other.

biennial A type of plant that goes through its life cycle in about two years, flowering in the second year.

bilateral symmetry A bodily symmetry in which opposite sides of a sagittal plane are mirror images of one another. Animals generally are bilaterally symmetrical.

bile A substance produced by the liver that aids in the digestion of fats. Bile can enter the small intestine through ducts directly from the liver, or from a storage organ, the gallbladder.

binary fission The form of reproduction carried out by prokaryotic cells in which the chromosome replicates and the cell pinches between the attachment points of the two resulting chromosomes to form two new cells. In this type of simple cell-splitting, each pair of daughter cells is an exact replica of the parental cell.

binding site An area of a protein or glycoprotein, located at the surface of the plasma membrane, with a particular conformation that recognizes only one molecule or a closely related group of molecules. Binding sites enable specificity of cell communication.

binomial nomenclature The system of naming species that uses two names (genus and species) for each species. This system helps identify groupings among living things.

biodegradable Capable of being broken down by living organisms.

biodiversity The diversity of living things. Includes species diversity, diversity of distributions, and genetic diversity.

biogeochemical cycling The movement of water and nutrients back and forth between biotic (living) and abiotic (nonliving) systems.

biological legacy A living thing, or product of a living thing, that survives a major ecological disturbance. Biological legacies proved to be crucial in the process of succession that occurred at Washington State's Mount St. Helens following its eruption in 1980.

biological species concept A definition of species that relies on the breeding behavior of populations in nature. Accordingly, groups of populations that actually or potentially interbreed in nature and are reproductively isolated from other such groups constitute a species.

biology The study of life.

biomass Living material, generally measured by weight.

biome An ecosystem dominated by a large vegetation formation, whose boundaries are largely determined by climate. The same biome type can occur on two different continents and have different species, but the two regions will bear striking similarities.

biosphere The interactive collection of all the world's ecosystems. Also thought of as that portion of the Earth that supports life.

biotechnology The use of living organisms to create products or facilitate processes.

biotic Pertaining to living things.

bivalves A class of molluscs that includes mussels, clams, and oysters, among other organisms.

blade In plants, the major, broad part of a leaf.

blastocoel The fluid-filled cavity in a blastula.

blastocyst Hollow, fluid-filled ball of cells that is formed in the early stages of the embryonic development of humans and other mammals. In non-mammalian animals, the blastocyst is known as the blastula.

blastula Hollow, fluid-filled ball of cells that is formed in the early stages of the embryonic development of animals other than mammals. In mammals, the blastula is known as the blastocyst.

blood-brain barrier The covering over tiny blood vessels in the brain, made of neuroglia cells, that prevents some substances from passing from the general circulation into the brain. This barrier helps protect the brain from potentially damaging substances.

B-lymphocyte (B cell) A type of lymphocyte that is central to antibody-mediated immunity. B cells bearing a specific antibody divide rapidly, then produce both antibodies to fight a specific invader and permanent lines of memory cells sensitive to the invader, should it appear in the body again.

body segmentation A repetition of body parts in an animal. An example can be found in the vertebrae that make up the human vertebral column or backbone.

bone A type of supportive connective tissue that contains mineral deposits, primarily calcium, which—along with flexible collagen fibers—makes the tissue both strong and resistant to shattering.

bottleneck effect A change in allele frequencies in a population due to chance, following a sharp reduction in the population's size. One of the factors that potentiates genetic drift.

Bowman's capsule A hollow structure in the kidneys of vertebrates that receives the filtrate from the blood and passes it into the proximal tubule of the nephron. This structure is the point where wastes are transferred from the blood into the kidney.

bryophyte A plant that lacks a vascular (fluid transport) structure. Bryophytes are the most primitive of the four principal varieties of plants. Mosses are the most familiar example.

bud An undeveloped plant shoot, composed mostly of meristematic tissue.

buffering system A solution that helps to keep pH at optimal levels. These solutions are generally weak acids or bases that can work to neutralize sudden infusions of acid or base by accepting or donating hydrogen ions.

Buffering systems help living beings function by keeping pH at optimal levels.

bulbourethral gland A gland that, in human males, contributes fluids to the semen as it proceeds through the urethra in the process of ejaculation.

bundle-sheath cells Special cells, wrapped around leaf veins, that receive carbon dioxide from a 4-carbon molecule in the C_4 pathway. In C_4 plants, the Calvin cycle takes place in these cells.

Calvin cycle The set of steps in photosynthesis in which an energy-rich sugar is produced by means of two essential processes—the fixing of atmospheric carbon dioxide into a sugar, and the energizing of this sugar with the addition of electrons supplied by the light-dependent reactions of photosynthesis. Named for the biochemist Melvin Calvin, who elucidated its steps. Also known as the C_3 or Calvin-Benson cycle.

CAM metabolism A type of photosynthesis used by succulents and some other plants from dry climates. In it, stomata stay open at night, allowing CO_2 to enter the plant, but then close during the day, thereby conserving water. Carbon thus "banked" during the night is then used the following day in the plant's completion of photosynthesis.

capillary The smallest type of blood vessel, connecting the arteries and veins in the body's tissues. Gases, nutrients, and wastes are exchanged between the blood and the tissues through the thin walls of capillaries.

capillary bed An interconnected network of capillaries that lies in the body's tissues. The network brings the capillaries close to each part of the tissues and provides lots of surface area for more efficient exchange.

carbohydrate A class of biological molecule that contains carbon, oxygen, and hydrogen, usually with twice as much hydrogen as oxygen. Carbohydrates are composed of monosaccharide (simple sugar) building blocks. Starch, cellulose, sucrose, and chitin are carbohydrates.

carcinogen Any environmental influence that can contribute to the onset of cancer.

cardiac muscle The type of striated muscle tissue that forms the muscles of the heart.

cardiovascular system The organ system that transports nutrients, dissolved gases, and hormones to tissues throughout the body, and carries wastes to sites of filtration. This system consists of the heart, the blood vessels and blood, and the bone marrow where blood cells are formed.

carnivore An animal that eats meat.

carpel The female reproductive structure of a flower, consisting of an ovary, a style, and a stigma.

carrier A person who does not suffer from a recessive genetic debilitation, but who carries genes for it that can be passed along to offspring.

carrying capacity The maximum population of a species that can be sustained in a given geographical area over time. In ecology, this is often denoted as K.

cartilage A type of supportive connective tissue that provides support and flexibility, as in the outer portions of the nose and ear, between bones of the spine, and between limb bones.

catalyst A substance that retains its original chemical composition while bringing about a change in a substrate. Enzymes are catalysts in chemical reactions; one enzyme can carry out hundreds or thousands of chemical transformations without itself being transformed.

cell The functional unit of life. A highly organized, generally microscopic structure whose component parts always include an outer membrane, genetic material in the form of DNA, and a host of accessory molecules. All living things either are single cells or are composed of collections of cells.

cell cycle The repeating pattern of growth, genetic duplication, and division that many types of cells undergo. The cell cycle has two main phases—interphase, in which cells grow, duplicate chromosomes, and perform routine functions; and mitotic phase, in which they divide.

cell junction A linkage between animal cells. A gap junction is a type of cell junction that facilitates cell communication.

cell-mediated immunity An immune system capability that works through the production of cells that destroy infected cells in the body.

cell plate   A membranous sac that, in cell division, is the precursor to the plant cell wall.

cell wall   A structure composed mostly of cellulose and sometimes of lignin that surrounds the plasma membrane in plants. The cell wall provides structure and helps regulate water intake.

cellular respiration   The process of extracting energy from a food molecule to form ATP or other energy transfer molecules. Three stages are included: glycolysis, the Krebs cycle, and the electron transport chain.

cellulose   A complex carbohydrate that is the largest single component of plant cell walls. Cellulose is dense and rigid and provides structure for much of the natural world. Mammals cannot digest cellulose, so it serves as insoluble dietary fiber that helps move food through the digestive tract.

central cell   Cell with two nuclei in the embryo sac (mature female gametophyte) in a flowering plant. Upon fertilization, this cell will develop into the endosperm, the food for the embryo.

central nervous system (CNS)   The portion of the nervous system consisting of the brain and spinal cord.

central vacuole   An organelle in a plant cell, containing mostly water but also other substances, that may occupy up to 90 percent of the volume of the cell. The vacuole stores nutrients, retains or degrades waste, and is involved in metabolism.

centrosome   A cellular structure that acts as an organizing center for the assembly of microtubules. A cell's centrosome duplicates prior to mitosis and plays an important part in the development of the cell's mitotic spindle.

cephalopods   A class of molluscs that includes squid, octopus, nautilus, and cuttlefish.

cerebellum   Region of the vertebrate brain that adjusts ongoing movements on the basis of sensory information and memory of past movements. The cerebellum helps us both maintain balance and remember how to carry out complex motor tasks.

cerebral cortex   The highly convoluted layer of gray matter at the surface of the brain of vertebrates. This layer consists largely of cell bodies of sensory and motor neurons.

cerebral hemispheres   The two halves (left and right) of the cerebrum of the vertebrate brain. This is the site of many functions, including conscious thought, sensations, and memory.

cerebrospinal fluid   The fluid that circulates in the spine and brain, supplying nutrients, hormones, and white blood cells, and cushioning the nervous tissue from jarring injury.

cerebrum   The top portion of the vertebrate brain, divided into two cerebral hemispheres. Here, conscious thought, memory, sensations, and complex motor patterns originate.

cervix   The lower part of the uterus, a narrow neck that opens into the vagina.

character   A genetically determined attribute of an organism, such as stem length, seed color, or seed shape in Mendel's pea plants. Each character can manifest in two or more forms, called traits. The character of seed color in Mendel's plants manifested in the traits of yellow and green seeds.

chemical bonding   General term for a bond created when electrons of two atoms interact and rearrange into a new form that allows the atoms to become attached to each other. Ionic, covalent, and hydrogen bonds are all chemical bonds.

chemoautotrophy   Nutritional mode in which an organism gets its carbon from carbon dioxide, but gets the energy to make food from this carbon by oxidizing inorganic materials. Some bacteria and some archaea are chemoautotrophs.

chemoheterotrophy   Nutritional mode in which an organism gets its carbon and energy from organic materials, referred to as food. Animals are chemoheterotrophs.

chitin   A complex carbohydrate that gives shape and strength to the external skeleton of arthropods, including insects, spiders, and crustaceans.

chlorofluorocarbons (CFCs)   A class of human-made chlorine compounds that destroys the atmospheric ozone that protects life on land from damaging ultraviolet radiation.

chlorophyll *a*   The primary pigment of chloroplasts, found embedded in its membranes. Together with the accessory pigments, chlorophyll *a* absorbs some wavelengths of sunlight in the first step of photosynthesis.

chloroplast   A type of plastid (an organelle) in a plant cell that contains chlorophyll and is the site of photosynthesis.

chorionic villus sampling (CVS)   A medical procedure to test for genetic abnormalities in a fetus. Cells derived from the embryo are suctioned from the villi (extensions of the chorionic membrane surrounding the placenta) and examined. This alternative to amniocentesis can be done somewhat earlier, and results are faster; but it is riskier for the fetus.

chromatid   One of the two identical strands of chromatin (DNA plus associated proteins) that make up a chromosome in its duplicated state.

chromatin   Substance consisting of DNA plus the protein material around which DNA is wrapped. Chromosomes are composed of chromatin.

chromosome   Structural unit containing part or all of an organism's genome, consisting of DNA and its associated proteins (chromatin). The human genome is made up of 23 pairs of chromosomes, or 46 chromosomes in all.

chyme   The soupy mixture of food and gastric juices that is passed from the stomach to the duodenum (upper portion of the small intestine) during digestion.

cilia (singular, *cilium*)   Hair-like extensions of a cell, composed of microtubules. Many cilia occur on the surface of a given cell, and they move rapidly back and forth to propel the cell or to move material around the cell.

circadian rhythms   Biological cycles that can function independently of environmental cues and that are roughly synchronized to Earth's 24-hour rotation.

cisternae (singular, *cisterna*)   Flattened sacs of the rough endoplasmic reticulum (ER) in which polypeptide chains, transferred in from ribosomes, fold into their protein conformations.

cisternal space   Space in the middle of the cisternae (flattened sacs) of the rough endoplasmic reticulum. Polypeptide chains, newly synthesized on the ribosomes, enter the cisternal space and then fold into their protein conformations.

citric acid cycle   Another name for the Krebs cycle, one of the three main sets of steps in cellular respiration; named for the first product of the cycle, citric acid.

cladistics   The branch of systematics that uses shared derived characters to determine the order of branching events in speciation and therefore which species are most closely related. Cladistics is concerned only with evolutionary relationships, not classification.

cladogenesis   A form of evolution that occurs by means of a single species diverging into two species, with the original species and the new species both persisting. Also referred to as branching evolution.

cladogram   An evolutionary tree constructed using the cladistic system.

class   A taxonomic grouping subordinate to phylum and superordinate to order. Humans are in the class Mammalia.

classical conditioning   A form of learning in which animals learn to respond in a customary way to a new stimulus that has been paired with an existing stimulus. In the case of Pavlov's dog, a customary response (salivation) was elicited when a new stimulus (the sound of a bell) was paired with an existing stimulus (food delivery).

cleavage   The cellular division of a zygote during animal development, resulting in a ball of smaller cells which then differentiate.

climate   The average weather conditions, including temperature, precipitation, and wind, in a particular region.

climax community   The relatively stable community that develops at the end of any process of ecological succession.

clone   An exact genetic copy. Also, used as a verb, to make one of these copies. A single gene or a whole, complex organism can be cloned.

cloning vector   A self-replicating agent that, in the cloning process, serves to transfer and replicate genetic material. Cloning vectors generally are bacterial plasmids or the viruses known as bacteriophages.

closed circulation system   A type of circulatory system, found in all vertebrates and in some invertebrates, in which blood stays within vessels.

clotting factor   A group of substances, found in blood platelets, that are important in the process of blood clotting.

coastal wetlands   Lands adjacent to the ocean that are wet for at least part of the year. These can be fresh or salt water wetlands.

coastal zone   The region lying between the point on shore where the ocean's waves reach at high tide to the point off shore where the continental shelf drops off.

cocci (singular, *coccus*)   Spherical bacteria.

codominant   The condition in which two alleles in a heterozygous organism are both expressed. Blood type in humans (A, B, AB, O) is an example of codominance.

codon   An mRNA triplet that codes for a single amino acid or a start or a stop command in the translation stage of protein synthesis.

coelo   A central body cavity, found in animals, that is lined with cells of mesodermal origin.

coenzyme   A type of accessory molecule that is part of the active site of an enzyme and allows it to function. Many vitamins are important coenzymes.

coevolution   The interdependent evolution of two or more species. Coevolution can benefit both species, as in flowering plants and their animal pollinators, or it can be an arms race between species, as in a plant and its predators.

coexistence   The condition in which two species can live in the same habitat, dividing up resources in a way that allows both to survive.

collecting duct   A tube in the nephron that receives urine from several nephron tubules. Some additional reabsorption, especially of water, takes place in the collecting duct, resulting in more concentrated urine.

collenchyma cell   The type of plant cell that provides support to allow stretching and elongating the growing parts of the plant.

colon   The longest region of the large intestine, between the cecum and the rectum.

commensalism   An interaction between two species in which one benefits while the other is neither harmed nor helped.

commitment   A developmental process that results in cells whose role is completely determined. Most cells in the adult body have undergone commitment, meaning they are fated to function as a particular kind of cell, and give rise only to more such cells.

common descent with modification   The process by which species of living things undergo modification in successive generations, with such modification sometimes resulting in the formation of new, separate species. The lineage of all species traces back to a single common ancestor.

community   All the populations of living things that inhabit a given area. The term also is used to mean a collection of populations in a given area that potentially interact with each other.

companion cell   In plants, cells that are closely associated with sieve elements in phloem. The companion cells provide housekeeping needs of the sieve elements, which have lost their nuclei to provide room for faster conduction of phloem sap.

competitive exclusion principle   When two populations compete for the same limited, vital resource, one will always outcompete the other and thus bring about the local extinction of the latter species.

complement   A group of proteins in the blood that supplements the action of antibodies by enhancing phagocytosis, destroying the membranes of invading cells, and promoting inflammation.

complex carbohydrate   A carbohydrate built of many monosaccharides. Also known as polysaccharides, these compounds include starch, glycogen, cellulose, and chitin.

concentration gradient   A gradient within a given medium defined by the difference between the highest and lowest concentration of a solute. The solute will have a natural tendency to move from the areas of higher concentration to lower, thus diffusing.

cones   In human vision, photoreceptors that respond best to bright light, that allow organisms to see minute detail in objects, and that provide color vision.

connective tissue   In humans, one of the four principal types of tissue. A tissue, active in the support and protection of other tissues, whose cells are surrounded by a material that they have secreted.

consumer Any organism that eats other organisms rather than producing its own food.

continental drift The lateral movement of continental plates over the globe, allowing continents to divide and rejoin in different patterns. This process can separate populations of organisms, providing the geographic barriers that can result in speciation.

control A comparative condition in an experiment. A control is typically a treatment in an experiment in which no parts are varied, so that other treatments can be compared to this standard and the effect of the variables can be determined.

convergent evolution Evolution that occurs when similar environmental influences shape two separate evolutionary lines in similar ways.

coral reef Ocean structure, found in shallow, warm waters, that consists primarily of the piled-up remains of many generations of the animals called coral polyps. Such reefs provide habitat for a rich diversity of marine organisms.

cork Cells, dead in their mature state, that form the outermost covering of woody plants. These cells are infused with a waxy substance that protects the plant from drying out and from invaders.

cork cambium Secondary meristematic tissue in woody plants that forms the outer living covering of woody plants. This cambium produces the cork cells that, in their dead, mature state, protect the outside of the trunk and branches.

cornea A curved membrane through which light first passes upon entering the eye.

coronary artery An artery that delivers oxygenated blood to the muscles of the heart. Blockage of coronary arteries causes heart attack.

corpus luteum The structure that develops in the mammalian ovary from the ruptured follicle after ovulation. The corpus luteum secretes hormones that help prepare the reproductive tract for pregnancy. If pregnancy does not occur, this body degenerates.

cortex In dicotyledonous plants, the area of ground tissue in the stem that is outside the ring of vascular bundles.

cotyledon An embryonic leaf. A major division in plants is between the monocots, which have one embryonic leaf, and the dicots, which have two embryonic leaves.

coupled reaction The situation in which an exergonic and an endergonic reaction are brought together, with the exergonic reaction releasing energy that is used to power the endergonic reaction.

covalent bond A type of chemical bond in which two atoms are linked through a sharing of electrons.

Cretaceous Extinction A mass extinction event that occurred at the boundary between the Cretaceous and Tertiary periods. This event, which included an asteroid impact, resulted in the extinction of the dinosaurs along with many other organisms.

cross-pollinate To pollinate one plant with pollen of another plant. Mendel used this technique in conducting his experiments to uncover rules of heredity.

crossing over The exchange of pieces of chromosomes that occurs when homologous chromosome pairs intertwine during prophase I of meiosis. This process, also called recombination, accounts for some of the diversity in individual gametes.

cuticle In plants, the waxy outer covering of leaves and shoots. The cuticle forms a protective layer over the dermal tissue, keeping water in and infecting bacteria and fungi out.

cytokinesis The stage of cell division in which a cell's cytoplasm divides.

cytoplasm The region of a cell inside the plasma membrane and outside the nucleus. Usually, this region is filled with the jelly-like cytosol containing the cell's extra-nuclear organelles.

cytoskeleton The set of protein strands that function as internal scaffolding for a cell. These protein strands give the cell shape, anchor its structures, act as monorails for particles moving within the cell, and allow whole cells to move. Microfilaments, intermediate filaments, and microtubules are all parts of the cytoskeleton.

cytosol The protein-rich, jelly-like fluid in which a cell's organelles that are outside the nucleus are immersed.

cytotoxic T cell A type of T cell, active in cell-mediated immunity, that attaches to infected cells in the body and punctures their outer membranes, causing them to rupture. Also known as killer T cells.

deciduous Refers to plants that show a coordinated, seasonal loss of leaves. This strategy allows plants to conserve water during a time they could perform little photosynthesis anyway.

decomposer A type of detritivore that, in feeding on dead or cast-off organic material, breaks it down into its inorganic components. Most decomposers are fungi or bacteria.

defecation The excretion of solid wastes through the rectal opening (anus).

deletion A chromosomal condition in which a piece of a chromosome has been lost. This occurs when a chromosomal fragment that breaks off does not rejoin any chromosome.

denatured Pertaining to a protein that has lost its shape, and therefore also its function.

dendrite An extension of the cell body of a neuron that generally receives information that it transmits to the neuron. Typically, a neuron has many dendrites.

density-dependent In ecology, effects on a population that increase or decrease in accordance with the size of that population. Density-dependent effects tend to involve biological factors. The level of surrounding toxins will change in a population, for example, in accordance with the density of that population.

density-independent In ecology, effects on a population that are not related to the size of that population. Density-independent effects tend to involve physical forces, such as temperature and rain.

deoxyribonucleic acid (DNA) A molecule in living things that contains the information for the replication of itself and for the synthesis of proteins and RNA.

derived character A character unique to groupings of organisms (taxa) descended from a common ancestor.

dermal tissue The epidermis, or outer layer of cells of a plant. In addition to covering the plant, dermal tissue forms trichomes such as root hairs.

dermis In certain animals, the thick layer of the skin—composed mostly of connective tissue—that underlies, nourishes, and supports the epidermis.

desertification The transformation of an area into a desert, sometimes through human activity.

detritivore An organism that feeds on the remains of dead organisms or the cast-off material from living organisms.

dicotyledon A type of plant that has two embryonic leaves within the seed. More than three-quarters of all flowering plants are dicotyledons.

diffusion The movement of molecules or ions from areas of their higher concentration to areas of their lower concentration. Over time, the random movement of molecules will result in the even distribution of the material.

digestive system The organ system that transports food into the body, secretes digestive enzymes that help break down food to allow it to be absorbed by the body, and excretes waste products. This system consists of the esophagus, stomach, and large and small intestines, plus the accessory glands that produce the enzymes along the way.

digestive tract The muscular tube that runs through the body from mouth to anus, through which food passes.

dihybrid cross An experimental cross in which the plants used differ in two of their characters.

dikaryotic Having two haploid nuclei in one cell. A dikaryotic phase occurs in the life cycle of fungi because there is a delay between cytoplasm fusion and fusion of the nuclei, so all cells of a growing mycelium are dikaryotic.

diploblastic In animals, the state of developing from two layers of germ cells.

diploid Possessing two sets of chromosomes. All human cells are diploid with the exception of human gametes (eggs and sperm), which are haploid. Such haploid cells possess only a single set of chromosomes.

directional selection The type of selection that moves a character toward one of its extremes. Compare to **stabilizing** and **disruptive selection.**

disruptive selection The type of selection that moves a character toward both of its extremes, operating against individuals that are average for that character. This type of selection seems to be less common in nature than either stabilizing or directional selection.

distal tubule In the nephron tubule of the kidney, the convoluted portion that lies between the loop of Henle and the collecting duct. Some reabsorption of salt and water occurs here.

DNA polymerase An enzyme that is active in DNA replication, separating strands of DNA, bringing bases to the parental strands, and correcting errors by removing and replacing incorrect base pairs.

domain The highest-level taxonomic grouping of organisms. There are only three domains: Archaea, Bacteria, and Eukarya.

dominance hierarchy A persistent power ranking in an animal population that gives those of higher rank the ability to control some aspect of the behavior of those of lower rank. This control often results in higher-ranking individuals gaining better access to such resources as food and mates.

dominant Expressed in the heterozygous condition. When an organism possesses two different alleles of the same gene, but only one allele of the two is expressed, that allele is said to be dominant.

dominant disorder Genetic conditions in which a single faulty allele can cause damage, even when a second, functional allele exists.

dormancy A state in which growth is suspended and there is a prolonged low level of metabolic activity. Dormancy allows organisms to conserve energy during times of unfavorable environmental conditions.

dorsal nerve cord A rod-shaped dorsal structure consisting of nerve cells, running from the chordate animal's head to its tail.

dorsal root ganglion The collection of cell bodies of the sensory neurons that send their axons into the spinal cord through the dorsal root, carrying sensory information into the central nervous system.

double bond Type of covalent bond formed when two pairs of electrons are shared between atoms.

double fertilization In plants, the fusion of one sperm with the egg and another sperm with the central cell; the first of these fertilizations results in the zygote, the other in endosperm to provide food for the embryo. Double fertilization occurs almost exclusively in flowering plants.

Down syndrome A disorder in humans in which affected individuals usually have three copies of chromosome 21. Individuals with this syndrome have short stature, shortened life span, and low IQ.

ecological community All the populations of living things that inhabit a given area. The term is also used to mean a collection of populations in a given area that potentially interact with each other.

ecological dominant A species that is abundant and obvious in a given community. In any community, a few species, usually plants, will dominate in numbers.

ecology The study of the interactions that living things have with each other and with their environment.

ecosystem A community of living things and the physical environment with which they interact.

ectopic pregnancy An abnormal pregnancy in which the blastocyst attaches inside the uterine tube or to the cervix, rather than in the normal position in the dorsal uterine wall. The embryo usually does not survive in this type of pregnancy.

ectothermic Having an internal temperature that is controlled largely by the temperature of the external environment. For example, lizards are ectothermic and often bask in the sun to warm up.

effector A muscle or gland that responds to instructions received from a motor neuron, thus effecting change in the body.

efferent division The division of the peripheral nervous system that carries motor commands from the central nervous system (CNS) toward the effectors (muscles and glands).

electrical charge A quantity of unbalanced electricity. For a molecule, the electrical charge is written as one or more times that of an electron. Electrical charge enables some types of chemical bonds to occur.

electron A basic constituent of an atom that has negative electrical charge. Electrons are distributed in an atom at a distance from the nucleus. Electrons interact to form chemical bonds between atoms.

electron transport chain (ETC)   The third stage of cellular respiration, occurring within the inner membrane of the mitochondria, in which most of the ATP are formed. In this process, electrons are transported along a chain of molecules, providing energy to pump hydrogen ions against their concentration gradient. When these hydrogen ions return down their concentration gradient, the energy released drives the synthesis of up to 32 ATP per molecule of glucose.

electronegativity   The measure of the strength of attraction an atom has for electrons. An atom with higher electronegativity will tend to pull electrons away from atoms with lower electronegativity.

element   A substance that cannot be reduced to any simpler set of components through chemical processes. An element is defined by the number of protons in its nucleus.

embryo   A developing organism. In humans, the developing organism from the time a zygote undergoes its first division through the end of the eighth week of development.

embryo sac   The mature female gametophyte plant, consisting of seven cells produced by the megaspore. These cells include the egg and the central cell, which contains two nuclei.

embryology   The study of how animals develop, from fertilization to birth. Comparing embryology in different organisms can provide insights into their evolutionary relationships.

embryonic stem cell   A cell from the blastocyst stage of an embryo that is capable of giving rise to almost all of the cells or tissues in the body.

emigration   The movement of individuals out of a population.

emulsification   In digestion, the breaking of a large droplet of lipid material into many small droplets, greatly increasing the surface area on which digestive enzymes can act.

endergonic   A type of chemical reaction in which the products contain more energy than the reactants, so that energy is stored in the reaction. Energy must be put into such a reaction to make it go.

endocrine gland   A gland that releases its materials directly into surrounding tissues or into the bloodstream, without using ducts. Many hormones are produced by endocrine glands.

endocrine system   The organ system that sends signals throughout the body using hormones, which generally travel through the blood. This system consists of the endocrine glands and other organs that produce hormones.

endocytosis   The process by which the plasma membrane folds inward and pinches off, bringing relatively large materials into the cell enclosed inside a vesicle.

endomembrane system   The network of organelle membranes and the membrane-bound transport vesicles that bud from them and fuse with them.

endometrium   The tissue lining the interior of the uterus in mammals, which thickens in response to progesterone secretion during ovulation and is shed during menstruation. If pregnancy occurs, this tissue houses the embryo.

endosperm   The nutrient tissue that surrounds an angiosperm embryo in the seed. The rice and wheat grains that we eat consist mostly of endosperm.

endothelium   The innermost layer of blood and lymph vessels, made up of a type of flat epithelial cells. Capillary walls consist of one layer of endothelial cells, and exchange of gases, nutrients, and wastes occurs through these cells.

endothermic   Having a relatively stable internal body temperature and body heat that is generated internally, by the organism's own metabolism. Mammals are endothermic.

energy   The capacity to bring about movement against an opposing force.

energy-flow model of ecosystems   A conceptualization of ecosystems as units in which energy is first captured by given organisms and then transferred to other organisms. Because this capture and transfer can be quantified, the energy-flow model has been of great value in elucidating the workings of ecosystems.

entrain   To initiate a new cycle of an organism's internal clock. Many biological rhythms are entrained by such environmental cues as sunlight and temperature.

environmental resistance   All the forces in the environment that act to limit the size of a population. Limited food or sunshine, low temperature or rainfall, or predators are some components of environmental resistance.

enzyme   A chemically active type of protein that speeds up, or in practical terms enables, chemical reactions in living things.

epicotyl   All tissue of an embryonic or seedling plant above the cotyledons. The epicotyl gives rise to the first true leaves.

epidermis   The outermost layer of skin in animals, or the outermost cell layer in plants.

epididymis   The mass of tubules near the testis in which sperm complete their development and are stored.

epithelial tissue   In humans, one of the four principal types of tissue. A tissue that covers exposed surfaces and also forms glands. In humans, skin is an epithelial tissue, as is the lining of the digestive tract.

epoch   A division of a geologic period, which is in turn a division of an era. The Pleistocene is an epoch.

equilibrium species   A species whose population size stays relatively stable and at or near its carrying capacity. Also known as a *K*-selected species.

era   A large division of geologic time. The Precambrian, Paleozoic, Mesozoic, and Cenozoic are eras. An era includes several periods.

erythrocyte   A blood cell, also known as a red blood cell, that transports oxygen to and carries carbon dioxide from every part of the body.

esophagus   The muscular tube that begins at the pharynx and ends at the stomach, forming a portion of the digestive tract.

estrogen   A class of hormones, produced primarily by cells of the ovary, that supports egg development, growth of uterine lining, and development of female sex characteristics.

estuary   An area where a river flows into the ocean, bringing freshwater and saltwater habitat together. Estuaries are among the most productive ecosystems on Earth.

Eukarya   With Bacteria and Archaea, one of three domains of the living world, composed of four subordinate kingdoms: plantae, animalia, fungi, and protistia. All eukaryotes are composed of cells that have a nucleus.

eukaryotic cell   A cell possessing a membrane-bound nucleus that contains most of the cell's DNA and generally containing numerous other membrane-bound organelles. All organisms except bacteria and archaea either are eukaryotic cells or are composed of them.

eusocial species   A species of animal that is organized into a caste system in which there is a division of labor, such that different members of a population will consistently perform different tasks, and in which the young are raised through cooperative care, meaning care provided by many members of the group.

eutrophic   Pertaining to a lake that has many nutrients, and therefore sustains many photosynthesizers. Eutrophic lakes often have abundant algal cover, cutting down visibility in the water.

evolution   Any genetically based phenotypic change in a population of organisms over successive generations. Evolution can also be thought of as the process by which species of living things can undergo modification over successive generations, with such modification sometimes resulting in the formation of new species. Evolution is of central importance in biology because every living thing has been shaped by it.

exergonic   Denoting a chemical reaction in which the reactants contain more energy than the products, so that energy is released in the reaction.

exocrine gland   A gland that secretes its materials through ducts (tubes). For example, sweat glands conduct perspiration through ducts to the skin.

exocytosis   The process in which a transport vesicle fuses with the plasma membrane of a cell and the contents of the vesicle are ejected outside the cell. Exocytosis can be used to expel waste products in single-celled organisms or to export protein products in multicelled organisms.

exon   A segment DNA, or the messenger RNA transcript complementary to it, that encodes information for the sequencing of amino acids in a protein. Exon is short for an "expressed sequence" of mRNA or DNA. Compare to **intron.**

exoskeleton   An external material covering the animal body, providing support and protection.

exponential growth   Growth in numbers that occurs from exponential increase, in which the number produced is proportional to the number already in existence. This type of growth in the size of a population can produce huge numbers of individuals over time.

exponential increase   An increase in numbers that is proportional to the number already in existence. This type of increase occurs in populations of living things, and it carries the potential for enormous growth of populations.

extremophile   Type of organism that flourishes in extreme conditions such as high heat, high pressure, high salt, or extreme pH. Many archaea and some bacteria are extremophiles; the enzymes that allow them to thrive in their environments can be used by biotechnologists to make products useful to humans.

extrinsic isolating mechanism   A barrier to interbreeding of populations that is not an inherent characteristic of the organisms in the populations. Geographic barriers such as rivers are extrinsic isolating mechanisms.

facilitated diffusion   The passage of materials through the cell's plasma membrane, aided by a concentration gradient and a transport protein.

facilitation   In primary succession, the actions or qualities of earlier species that in some way assist growth of later species.

fallopian tube   The tube that transports the ovulated egg from the ovary to the uterus in humans. Fertilization occurs inside this vessel, also known as the uterine tube.

family   A taxonomic grouping of related genera. This category is subordinate to order and superordinate to genus. Humans are in the family Hominidae.

fatty acid   A lipid molecule composed of a long chain of carbon and hydrogen atoms that terminate with an acid carboxyl (COOH) group.

fertilization   The fusion of two gametes to form a zygote. In humans (and most other organisms), the gametes known as sperm and egg fuse, resulting in a zygote.

fetus   An organism at a later stage of development. In humans, the developing organism from the start of the ninth week of development to the moment of birth.

fever   A sustained, elevated body temperature that speeds up the activities of the immune system and may therefore help to rid the body of an infection. In humans, fever is defined as a continued body temperature greater than $37.2°C$ (99°F).

fiber   In skeletal muscle, a single elongated muscle cell, containing hundreds of long, thin myofibrils that run the length of the cell.

fibrinogen   A type of protein in blood plasma that, under certain conditions, forms into long strands of fibrin, the framework molecule of a blood clot.

fibrous root system   A plant root system that consists of many roots, all about the same size.

filament   The part of a stamen (male reproductive part of a flower) that is shaped like and functions as a stalk, and has an anther at the top.

filtrate   The fluid that, in the kidney, is passed from the blood in the glomerulus into the Bowman's capsule of the nephron in the first stage of urine formation.

filtration   The process of passing fluid from the blood into the kidney at Bowman's capsule. Materials such as red blood cells, proteins, and lipids are too large to pass out of the capillaries of the glomerulus, and thus remain in the blood.

first filial generation   The offspring of the parental generation in an experimental cross.

first law of thermodynamics   Energy cannot be created or destroyed, but only transformed from one form to another. The process of transformation of energy is never totally efficient, and some energy is always converted into heat.

fitness   The success of an organism, relative to other members of its population, in passing on its genes to offspring. Fitness is a relative concept only; some organisms are better than others at passing on their genes in a given environment at a given point in time.

fixation   The process of a gas being incorporated into an organic molecule.

flagella (singular, *flagellum*)   Long, whip-like cell

extensions, composed of microtubules, that help certain cells move. Only one or a few flagella will occur on any given cell. The tail of a human sperm cell is a flagellum that enables it to swim to an unfertilized egg.

fluid-mosaic model The widely accepted view of the cell membrane as a mosaic of proteins moving laterally within the fluid that is the phospholipid bilayer.

follicle In the vertebrate ovary, the complex of the oocyte (developing egg) and the cells and fluids that surround and nourish it.

follicle-stimulating hormone (FSH) A hormone, secreted by the anterior pituitary, that promotes egg development and stimulates secretion of estrogens in women, and supports sperm production and testosterone secretion in men.

forensic DNA typing Popularly known as genetic fingerprinting, this technique compares the number of copies of short, repeated stretches of DNA between DNA samples. These DNA patterns are highly variable among individuals, so a DNA crime sample can be matched to a DNA sample from a suspect.

formed elements Cells and cell fragments that form the nonfluid portion of blood. Formed elements include red blood cells, white blood cells, and platelets. Contrast with **plasma,** the fluid portion of blood, consisting mostly of water.

founder effect The phenomenon by which an initial gene pool for a population is established by means of that population migrating to a new and isolated area. One of the conditions that potentiates genetic drift.

fraternal twins Twins who are produced first through multiple ovulations in the mother, followed by multiple fertilizations from separate sperm of the father, and then multiple implantations of the resulting embryos in the uterus.

free radical A molecule with an unpaired electron, usually existing for only a very brief time. Although free radicals are natural products of biological processes, these molecules can be destructive in living tissues.

free ribosome A type of ribosome that is located in the cytosol of a cell, not attached to the endoplasmic reticulum. Generally, proteins that will reside in the cell's membranes and proteins that will be exported out of the cell are synthesized on the free ribosomes.

fruit The mature ovary of any flowering plant. Many fruits protect the underlying seeds, and many attract animals that will eat the fruit and disperse the seeds.

gallbladder A muscular sac, located beneath the right lobe of the liver in humans, that stores and concentrates bile, a digestive substance produced by the liver.

gamete A haploid reproductive cell, either egg or sperm.

gametophyte generation The generation in a plant's life cycle that produces gametes (sperm or egg). This generation is microscopic in angiosperms. Contrast with **sporophyte generation.**

ganglion Any collection of nerve-cell bodies in the peripheral nervous system.

gap junction A type of cell junction that facilitates cell communication, made up of clusters of protein structures extending from one side to the other of the cell's plasma membrane. When these structures line up in adjacent cells, the channel formed allows small molecules and electrical signals to pass through.

gastrointestinal (GI) tract Generally refers to the lower portion of the digestive tract (the stomach and intestines).

gastropods A class of molluscs that includes snails and slugs, among other organisms.

gastrulation The stage of animal development in which an embryo's cells migrate in a programmed fashion to form three layers of cells, each of which gives rise to specific organs or tissues.

gel electrophoresis A technique used to separate DNA fragments by size, using electrical charge to run the fragments through a gel. Smaller fragments move faster, and the different lengths of fragments arrange in bands according to size.

gene A segment of DNA that brings about the transcription of a segment of RNA. A gene may contain regulatory chemical sequences, information for the synthesis of one or more proteins, or sequences that bring about the production of RNA used at such sites as the ribosomes.

gene flow The movement of genes from one population to another.

gene pool The entire collection of alleles in a population.

genetic code The inventory that specifies which nucleotide triplets code for which particular amino acids, or for start or stop commands in protein synthesis. With few exceptions, the genetic code is universal in living things.

genetic drift The chance alteration of allele frequencies in a population, with such alterations having greatest impact on small populations.

genetics The study of heredity.

genome The complete haploid set of an organism's chromosomes.

genomics The study of sets of genes within or across genomes.

genotype The genetic makeup of an organism, including all the genes that lie along its chromosomes.

genus A taxonomic grouping of related species. This category is subordinate to family and superordinate to species. Humans are in the genus *Homo.*

germ-line cell A cell that becomes an egg or a sperm cell.

gland An organ or group of cells that secretes one or more substances.

globulins With albumins and fibrinogen, one of the three primary classes of proteins suspended in the fluid or plasma portion of the blood. Globulins include antibodies, which attack foreign invaders; and transport proteins, such as the lipoproteins that transport lipids.

glomerulus The ball of capillaries that sits inside the Bowman's capsule of a vertebrate kidney. Filtrate passes from the blood in the glomerulus into Bowman's capsule in the first stage of urine formation.

glyceride A type of lipid, consisting of a head made of an alcohol (usually glycerol) attached to fatty-acid subunits.

glycerol A particular kind of alcohol, frequently found in the "head" of glyceride.

glycocalyx An outer coat for a cell, consisting of carbohydrate chains protruding from the proteins and phospholipids of the membrane.

glycogen A complex carbohydrate that serves as the major form of carbohydrate storage for animals.

glycolysis The first stage of cellular respiration, occurring in the cytosol. For some organisms, glycolysis is the sole means of extracting energy from food. In most organisms, it is a means of extracting some energy and a necessary precursor to the other two stages of cellular respiration, the Krebs cycle and the electron transport chain.

glycoprotein A molecule that combines protein and carbohydrate. Glycoproteins play important roles as cell receptors and some types of hormones, among other functions.

Golgi complex The organelle in the cytosol that receives proteins via transport vesicles from the rough endoplasmic reticulum, processes them by adding or trimming side groups, and sorts and sends them to the correct destinations. The Golgi complex thus acts as a processing and distribution center for protein products of the cell.

gonadotropin-releasing hormone A hormone released in tiny amounts from the brain's hypothalamus that stimulates the anterior pituitary to release two other hormones (follicle-stimulating hormone and luteinizing hormone) important in reproduction.

grana Stacks of flattened vesicles (thylakoids) inside a chloroplast. The light-dependent reactions of photosynthesis occur here.

gravitropism The bending of a plant's roots or shoots in response to gravity. This capability helps a plant orient roots and shoots properly—roots toward the center of the Earth, shoots away from it.

gray matter The tissue within the central nervous system that appears gray and consists largely of cell bodies plus neuroglia.

gross primary production The amount of material that a plant produces as a result of photosynthesis.

groundwater Water that is found underground in porous rock saturated with water.

growth hormone In humans, this hormone stimulates overall body growth through cell growth. Also known as somatotropin.

gymnosperm A seed plant whose seeds do not develop within a fruit or carpel. One of the four principal varieties of plants, gymnosperms reproduce through wind-aided pollination. Coniferous trees, such as pine and fir, are the most familiar examples.

habitat The surroundings in which individuals of a species are normally found.

habituation A simple form of learning, consisting of a reduction in a response, based on repeated exposure to a stimulus that has no positive or negative consequences.

hair follicle A tiny complex organ, formed from epidermis, from which hair grows.

hair papilla A peg-like structure of connective tissue, containing capillaries and nerves, at the base of a hair follicle.

half-life The time needed for one-half of a given amount of a radioactive element to decay into another isotope or element. This time differs among elements but is uniform for each element. Half-life can be used, along with measured proportions of elements, to radiometrically date a substance.

haploid Possessing a single set of chromosomes. Human gametes (eggs and sperm) are haploid cells, because they have only a single set of chromosomes. All other cells in the human body are diploid, meaning they possess two sets of chromosomes. Some organisms are strictly haploid. Bacteria, for example, have only a single chromosome, making them haploid.

HDL High-density lipoprotein. A lipoprotein, carried in the blood, that has a relatively high ratio of protein to lipid. HDLs carry cholesterol away from outlying tissues to the liver; a high proportion of HDL is a predictor of keeping a healthy heart.

heart attack An episode of heart failure in which, due to blockage of a coronary artery, the blood supply is cut off to a group of cardiac muscles. The interruption in blood supply can cause the muscles to stop beating or to beat in a wild, irregular rhythm.

heartwood An area of dead, darker colored, inner wood in a tree trunk, where xylem tubes no longer conduct sap and are filled with oils, lignin, and microbe-fighting substances for strength and resistance to bacteria and fungi.

helper T cell A type of immune system cell, active in both cell-mediated and antibody-mediated immunity, that helps spur production of various types of T cells and one type of B cell. Helper T cells are also known as CD4 cells.

hemoglobin The iron-containing protein in red blood cells that binds to oxygen. Hemoglobin gives our blood its red color.

herbaceous plant A plant that never develops wood (secondary xylem), and therefore has relatively thin and weak shoots. Herbaceous plants do not undergo secondary growth (thickening).

herbivore An animal that eats only plants.

hermaphroditic A state in which one animal possesses both male and female sex organs.

heterotroph A type of organism that cannot manufacture its own food, but must instead get food from elsewhere. Animals are heterotrophs.

heterozygous Possessing two different alleles of a gene for a given character.

HIV Human immunodeficiency virus, the causative agent of the disease AIDS. This virus attacks the body's immune system, eventually leaving it unable to fight off even minor infections.

homeostasis A process that maintains a relatively stable internal environment of an organism, even when the outside environment varies. An example of homeostasis is that humans maintain a body temperature of about 98.6°F, even when the outside temperature fluctuates.

homologous In anatomy, having the same structure owing to inheritance from a common ancestor. Forelimb structures in whales, bats, cats, and gorillas are homologous.

homologous chromosomes Chromosomes that are the same in size and function. Species that are diploid (have two sets of chromosomes) have matching pairs of homologous chromosomes; one member of each homologous pair is inherited from the male, the second member of each homologous pair is inherited from the female. Human beings have 46 chromosomes—22 pairs of homologous chromosomes and either two homologous X chromosomes (in the case of females) or one X and one Y chromosome (in the case of males).

homology A structure that is shared in different organisms owing to inheritance from a common ancestor. Homologies are used to help decipher evolutionary relationships.

homozygous Having two identical alleles of a gene for a given character.

homozygous dominant Having two identical alleles of a gene, with the allele being dominant, so that this allele is the one expressed in the heterozygous condition.

homozygous recessive Having two identical alleles of a gene, with the alleles being recessive, meaning not expressed in the heterozygous condition.

hormone A substance that, when released in one part of an organism, goes on to prompt physiological activity in another part of the organism. Both plants and animals have hormones.

host The prey in a parasitic relationship.

hydrocarbon A compound made of hydrogen and carbon. Hydrocarbons are nonpolar covalent molecules and therefore are not easily dissolved in water.

hydrochloric acid The acid, secreted in the vertebrate stomach, that lowers the pH of the gastric juice, kills microorganisms, and breaks down cell walls and connective tissues in food.

hydrogen bond A chemical bond that links an already covalently bonded hydrogen atom with a second, relatively electronegative atom. The hydrogen atom can be linked to another atom in the same molecule, in which case the hydrogen bond helps hold a molecule in a certain shape; or it can be linked to an atom in another molecule, thus weakly holding the two molecules together.

hydrolysis The splitting of bonds in a molecule by reaction with water.

hydrophilic Property of a compound indicating that it will interact with water. Table salt (NaCl), which dissolves readily in water, is hydrophilic.

hydrophobic Property of a compound indicating that it will not interact with water. Oil is hydrophobic and will not readily dissolve in water.

hydroxide ion The OH^- ion. Compounds that yield hydroxide ions are strongly basic, so they can be used to counteract acids and shift solutions toward neutral or basic on the pH scale.

hypertonic solution A solution that has a higher concentration of solutes than another.

hyphae (singular, *hypha*) The slender filaments that make up the bulk of most fungi. Cellular material moves through the hyphae, which grow toward sources of food. The network of hyphae produced by a fungus is called a mycelium.

hypocotyl All tissue of an embryonic or seedling plant below the cotyledons. The hypocotyl includes the plant's radicle, or early root structure, and is first to emerge from a seed when it starts to sprout.

hypothalamus Small section of the vertebrate brain, lying below the thalamus and above the pituitary. The hypothalamus is linked by a narrow stalk to the pituitary, making it the main link between the nervous and endocrine systems. Cell bodies involved in emotions, autonomic functions, and hormone production are also located here.

hypothesis A tentative, testable explanation of an observed phenomenon.

hypotonic solution A solution that has a lower concentration of solutes than another.

identical twins Twins who develop from a single zygote; strictly speaking, identical twins are a single organism at one point in their development and as such have exactly the same genetic makeup.

imitation A sophisticated form of learning in which one animal copies behavior it observes in another animal.

immigration The movement of individuals into a population.

immunity Specific resistance, or long-lasting protection that results from reaction of the specific defenses of the immune system to foreign chemical compounds or pathogens.

imprinting A process of learning that results in one animal preferentially associating with another.

inclusive fitness An individual's relative genetic contribution to a succeeding generation, made both through itself and through relatives who have reproduced because of assistance the individual has provided.

incompletely dominant The condition in which neither of two alleles in a heterozygous organism is completely dominant to the other, so that the phenotype expressed is different from what the phenotype would be if the individual were homozygous for either allele.

independent assortment The random alignment of homologous chromosomes at the metaphase plate during metaphase I of meiosis. This random alignment assures that different gametes will have different alleles, accounting for some of the diversity among sexually reproducing organisms.

induction The capacity of some embryonic cells to direct the development of other embryonic cells. For example, the notochord induces the tissue above it to form the neural tube. Induction helps organize development.

inflammation A localized tissue response to injury. Inflammation interferes with the spread of pathogens and initiates a wide range of defenses to overcome the pathogens and repair damaged tissues.

inland wetlands Lands away from the ocean that are wet at least part of the year. The overwhelming majority of wetlands are inland and freshwater.

inner cell mass In mammalian development, the group of cells in the embryo that will develop into the baby, rather than into the placenta.

inorganic Pertaining to the chemistry of compounds that do not contain carbon.

insight learning A sophisticated form of learning in which an animal makes associations between objects or events that it has previously regarded as unrelated. In short, the ability to reason.

integral protein A protein of the plasma membrane that is attached to the membrane's hydrophobic interior. Such proteins may extend from one or both sides of the membrane and may be involved in cell recognition, cell communication, and transport of materials through the membrane.

integumentary system The organ system that protects the body from the external environment and assists in regulation of body temperature. This system consists of the skin and associated structures, such as glands, hair, and nails.

integuments The outer layers of tissue of an ovule, surrounding the embryo sac in a flower. These tissues develop into the seed coat that protects the embryonic plant.

intercalary meristem Meristematic tissue found at the base of each growing node in grasses. These multiple growing regions permit fast growth, and also allow grasses to keep growing even when their tops are removed by grazing or mowing.

interferon A small protein that causes normal, uninfected cells to produce compounds that interfere with viral replication. Interferon is released by activated lymphocytes and macrophages, and by tissues infected with viruses.

intermediate electron carrier A molecule in the electron transport chain that serves to accept and in turn donate electrons down the energy hill.

intermediate filament The protein filaments of the cytoskeleton that are intermediate in size. The positions of the nucleus and organelles are stabilized by intermediate filaments.

interneuron A type of neuron, located only within the brain or spinal cord, that connects other neurons. These neurons are responsible for the analysis of sensory inputs and the coordination of motor commands.

interphase The stage of the cell cycle in which a cell simultaneously carries out its work, grows, and duplicates its chromosomes. Cells spend most of their time in interphase.

interspecific competition Competitive interaction between individuals of two different species.

interstitial fluid Fluid between cells in the bodily tissues of animals.

intertidal zone The region within the coastal zone of the ocean that extends from the ocean's low-tide mark to its high-tide mark.

intraspecific competition Competitive interaction between individuals of the same species.

intrinsic isolating mechanism A difference in anatomy, physiology, or behavior that prevents interbreeding between individuals of the same species or of closely related species. One or more intrinsic isolating mechanisms must exist for two populations of the same species to begin the process of evolving into separate species.

intrinsic rate of increase The rate at which a population would grow if there were no external limits on its growth. In ecology, this is often denoted as *r*.

intron A segment of DNA, or the messenger RNA transcript complementary to it, that does not encode information for the sequencing of amino acids in a protein. Introns are spliced out of mRNA transcripts before the transcripts move to the cell's cytoplasm to take part in protein translation. Intron is short for an "intervening sequence" of mRNA or DNA. Compare to **exon**.

inversion Flipping of the orientation of a chromosome fragment in meiosis before it rejoins the chromosome it came from, so the order of the fragment's chemical sequence is reversed from normal. Inversion can affect the phenotype.

invertebrate An animal without a vertebral column.

ion An atom that has positive or negative charge because it has fewer or more electrons than protons. Ions will be attracted to other ions with the opposite charge, thus forming ionic compounds.

ionic bonding A linkage in which two or more ions are bonded to each other by virtue of their opposite charge.

ionic compound A substance in which two or more atoms are held together through ionic bonding. Table salt (NaCl) is a familiar example.

iris A colored structure of the eye, composed largely of smooth muscle and capable of contracting and relaxing, thus controlling the amount of light that passes into the eye through the iris' central opening, the pupil.

isomer A form of a molecule that has the same chemical formula but a different spatial arrangement than another molecule. Because of their different shapes, different isomers can have different properties.

isotonic solution A solution that has the same concentration of solutes as another.

isotope A form of an element as defined by the number of neutrons contained in its nucleus. Different isotopes of an element have varying numbers of neutrons but the same number of protons.

J-shaped growth In ecology, exponential growth in a population. When plotted on a graph, this type of increase in numbers can be so extreme that it looks like the letter J.

karyotype A pictorial arrangement of a full set of an organism's chromosomes. A karyotype can be helpful in diagnosing chromosomal abnormalities.

keratin A tough, fibrous, water-resistant protein formed by some epidermal tissues. In skin, keratin keeps water in and invaders out. Also found in claws, feathers, hair, and hooves.

keystone species A species whose absence would bring about a significant change in a community of organisms.

kidney In vertebrates, the organ that regulates water balance in the blood and excretion of metabolic wastes as urine.

kilocalorie The amount of energy it takes to raise one kilogram of water one degree Celsius. Food consumption is measured in kilocalories, usually written as Calories.

kin selection In natural selection, the selection (preferential preservation) of genes that increase the inclusive fitness of an individual animal.

kinetic energy The energy of motion. A rolling rock has kinetic energy.

kingdom A taxonomic grouping superordinate to every other grouping except domain. There are four kingdoms in domain Eukarya: protista, fungi, animalia, and plantae.

knot In a tree trunk, a branch that died out when the tree was younger and then was surrounded by secondary xylem.

Krebs cycle The second stage of cellular respiration, occurring in the inner compartment of mitochondria. This cycle of reactions yields 6 NADH and 2 $FADH_2$ that will enter the electron transport chain, plus two ATP per molecule of glucose. The Krebs cycle is the major source of electrons for the electron transport chain.

K-selected species A species that tends to be relatively long-lived, that tends to have relatively few offspring for whom it provides a good deal of care, and whose population size tends to be relatively stable, remaining at or near its carrying capacity (*K*). Also known as an equilibrium species.

labor The regular contractions of the uterine muscles that sweep over the fetus, creating pressure that opens the cervix and expels the baby and the placenta.

lactate fermentation The process in animal cells by which pyruvic acid, the product of glycolysis, accepts electrons from NADH to form lactic acid. Lactic acid is a waste product for animals.

lactic acid A waste product formed in animal cells when pyruvic acid, the product of glycolysis, accepts electrons from NADH. Lactic acid buildup causes muscle burn when a person's exertions outstrip that person's capacity to transfer energy through aerobic respiration.

lateral bud An undeveloped plant shoot, located between the stem and a leaf, that may give rise to a branch, a flower, or even take over the role of shoot apex if the apical meristem becomes damaged. All buds are composed mostly of meristematic tissue.

law of conservation of mass In a chemical reaction, matter can neither be created nor destroyed.

Law of Independent Assortment Mendel's insight that during gamete formation, gene pairs (alleles) assort independently of one another. The physical basis for this law is the independent assortment of pairs of homologous chromosomes at the metaphase plate during meiosis.

Law of Segregation Mendel's insight that organisms have two genetic elements (alleles) that separate when gametes are formed. The physical basis for this law is the separation of homologous chromosomes during meiosis.

LDL Low-density lipoprotein. A lipoprotein, carried in the blood, that has a relatively low ratio of protein to lipid. LDLs carry cholesterol to the outlying tissues; a high proportion of LDL is a risk factor for heart attack.

learning The acquisition of knowledge through experience.

lens A transparent structure of the eye, lying behind the iris, that serves to focus incoming light.

leukocyte Another term for white blood cell. This class of blood cells is critical to immune function.

life cycle The repeating series of steps that occur in the reproduction of an organism.

life science All scientific disciplines dealing with living things; for example, biology, medicine, and forestry are life sciences.

life table A table showing how likely it is for an average species member to survive a given unit of time.

light-dependent reactions The first set of steps in photosynthesis, in which energy from sunlight produces the energetic electrons and ATP that will be used in the second set of steps in photosynthesis, the light-independent reactions. The light-dependent reactions produce atmospheric oxygen as a by-product.

light-independent reactions The second set of steps in photosynthesis, in which carbon dioxide is brought together with high-energy electrons and ATP, supplied by the light-dependent reactions of photosynthesis, to produce the high-energy sugar G3P.

lignin A polymer that is often a major component of the cell walls of plants. Lignin provides stiffening by binding and supporting cellulose fibers.

lipid A type of biological molecule made up of carbon, oxygen, and hydrogen; lipids have much more hydrogen relative to oxygen than do the carbohydrates. Pure lipids are not readily broken down in water. Oils, fats, cholesterol, and some hormones are lipids.

lipoprotein A molecule that combines lipid and protein. Lipoproteins transport fat molecules through the bloodstream to all parts of the body.

littoral zone The shallow-water zone in a lake that extends from the lake's edge to the point where water is too deep for rooted plants to grow.

logistic growth A type of growth, occurring in natural populations, in which exponential growth slows and then stops in response to environmental resistance. Also known as S-shaped growth.

luteinizing hormone (LH) A hormone secreted by the anterior pituitary; in women, LH induces ovulation

and stimulates the ovary to secrete estrogens and progestins to prepare the body for possible pregnancy. In men, LH stimulates the testes to produce androgens such as testosterone.

lymph The fluid carried by the lymphatic vessels. This fluid is checked by cells of the immune system for signs of injury or infection before being returned to the blood.

lymphatic Vessel of the lymphatic system, through which lymph flows.

lymphatic system The organ system that collects interstitial fluid, transports it as lymph through lymphatic vessels, checks the fluid for infection, and delivers the fluid to blood vessels.

lysogenic cycle A viral life cycle in which a virus invades a cell and integrates its DNA into the host's DNA; the virus then remains inside the cell, replicating along with it, until the cell is threatened, at which point the virus may lyse or burst the cell, departing from it.

lysosome Membrane-bound organelle in an animal cell that breaks down and then exports, stores, or recycles the resulting materials back into the cytosol.

lytic cycle A viral life cycle in which a virus invades a cell, makes copies of itself inside the cell, and then lyses (bursts) the cell, thereafter invading new cells.

macroevolution Evolution that results in the formation of new species or other groupings of living things. The underlying basis of macroevolution is microevolution, defined as a change of allele frequencies in a population over a short period of time.

macrophage A large type of phagocyte (a type of white blood cell) that can be found in most tissues of the body, capable of engulfing infected cells and cell fragments.

mammary glands A set of glands that, in female mammals, provide milk for the young.

marrow A loose connective tissue, found in the internal cavities of bones, that comes in two forms: red marrow, which is blood-forming tissue; and yellow marrow, which is made up of fat.

marsupial A type of mammal in which the young develop within the mother to a limited extent, inside an egg having a membranous shell. Early in development the egg's membrane disappears, after which the mother delivers a developmentally immature but active marsupial. Kangaroos are one example of a marsupial.

mass A measure of the resistance of an object to being moved, taking into account the object's density and volume. For most biological purposes, mass is equivalent to weight.

mast cell A special type of connective-tissue cell that releases histamines and heparin into the interstitial fluid, thus initiating the process of inflammation.

mechanical work The work done in moving an object through a space. Muscle contraction is mechanical work.

medulla oblongata The section of the vertebrate brain that is connected to the spinal cord. This structure contains major centers involved in autonomic functions, including heart rate and respiration, and relays sensory information to the brain stem and thalamus.

megaspore A haploid cell that results from meiosis inside the ovary of a flower. The megaspore then gives rise to several types of cells, including the egg.

megaspore mother cell A single diploid cell in the ovary of a flower that undergoes meiosis, producing four megaspores, one of which develops into the embryo sac containing the egg.

meiosis A process in which a single diploid cell divides to produce haploid reproductive cells. Meiosis produces the gametes for sexual reproduction.

memory B cell A type of B cell, produced in antibody-mediated immunity, that remains in the body, ready to produce more plasma cells and quickly mount a defense against another invasion by the same invader.

menopause The cessation of the monthly ovarian cycle that occurs when women reach about 50 years of age. The scarcity of follicles remaining in the ovary is the main factor that brings about menopause.

menstruation The release of blood and the specialized uterine lining, the endometrium; occurs about once every 28 days in human females, except when the oocyte is fertilized.

mesoglea A secreted, gelatinous material that makes

up most of the medusa-stage jellyfish. Its consistency accounts for the name jellyfish.

messenger RNA (mRNA) A type of RNA that encodes, and then carries to the ribosomes from the cell's nucleus, information used in the synthesis of proteins. This information is encoded initially in DNA and is encoded onto mRNA in the process of transcription.

metabolic pathway A sequential set of enzymatically controlled reactions in which the product of one reaction serves as the substrate for the next. Metabolic pathways carry out complex processes with many steps.

metabolism The sum of all chemical reactions carried out by a cell or larger organism.

metaphase plate A plane located midway between the poles of a dividing cell.

microevolution A change of allele frequencies in a population over a short period of time. Microevolution is the basis for all large-scale or macroevolution, meaning evolution that results in the formation of new species or other groupings of living things.

microfilament The smallest-sized component of the cytoskeleton, composed of the protein actin. Actin is a support structure of cells, helps a cell grow rapidly at one end to move or capture prey, and, (along with the protein myosin) enables muscles to contract.

micrograph A picture taken with a microscope.

micrometer (μm) A millionth of a meter.

microphage A small type of phagocyte (a type of white blood cell) that generally circulates in the blood but leaves the bloodstream to enter injured or infected tissues. Neutrophils and eosinophils are the two types of microphages.

micropyle A tiny opening in the integuments that cover the ovule in the female gametophyte of a flowering plant. The pollen tube grows through the micropyle in the process of fertilization of the egg lying within the ovule.

microspore A haploid cell that results from meiosis in the anthers of flowering plants. Microspores develop into pollen grains, the male gametophyte plant.

microspore mother cell A diploid cell type, found within the anthers of flowering plants, that undergoes meiosis to produce the haploid microspores that develop into pollen grains.

microtubule The largest-sized component of the cytoskeleton. Microtubules help determine the shape of the cell, act as monorails along which transport vesicles travel, form the mitotic spindle, and form cilia and flagella for propulsion.

midbrain The part of the vertebrate brain that processes information about sight and hearing, generates involuntary motor responses to maintain posture, and has centers that help maintain consciousness.

migration A regular movement of animals from one location to a distant location. Also, the movement of individuals from one population into the territory of another population. Migration is the basis of gene flow among populations.

mimicry A phenomenon in which one species evolves to resemble another species.

mitochondria (singular, *mitochondrion*) Membrane-bound organelles that convert energy to a form easily used by cells. Mitochondria are the power plants of cells.

mitosis The stage of cell division in which a cell's duplicated chromosomes separate.

mitotic phase The stage of the cell cycle in which the duplicated chromosomes separate and the whole cell splits in two.

mitotic spindle The microtubules active in cell division, including those that align and move the chromosomes.

modern synthesis The unified evolutionary theory that resulted from the convergence of several lines of biological research between 1937 and 1950. Advances in taxonomy, genetics, and paleontology all contributed to this deeper understanding of how evolution and natural selection work.

molecular biology The study of life at the level of its individual molecules, their constituent parts, and how they interact.

molecular formula A notation specifying the elements in a molecule, with the number of each shown as a subscript. This notation reveals the constituent elements of a molecule, but not the spatial arrangement of those elements.

molecule A structure with a defined number of atoms in a particular spatial arrangement. The atoms and their arrangement determine how the molecule interacts with other molecules.
molting A periodic shedding of an old skeleton.
monocotyledon A type of plant that has one embryonic leaf within the seed. Although comprising only one-quarter of all flowering plants, most important food plants are monocotyledons.
monohybrid cross An experimental cross in which the organisms used differ in only one of their characters.
monomer Small molecules that are building blocks for larger molecules, known as polymers.
monosaccharide The building block (monomer) of carbohydrates. Also known as simple sugars; glucose and fructose are familiar examples.
monotremes Egg-laying mammals, represented by the duck-billed platypus and spiny anteaters found in Australia.
monounsaturated fatty acid A fatty acid with one double bond between the carbon atoms of its hydrocarbon chain.
morphogen A diffusible substance whose concentration in a region of an embryo affects development in that region.
morphology The study of physical forms that organisms take. Comparing these forms can provide insights into the evolutionary relationships of organisms.
morula A tightly packed ball of cells that results from repeated cleavage of the zygote, before gastrulation begins during animal development.
motor neuron A neuron that carries instructions from the central nervous system to an effector (muscle or gland) in the body.
Müllerian mimicry A type of mimicry in which several species that have protection against predators evolve to look alike. All of the individuals in the species benefit, because once a predator learns what one distasteful or harmful individual looks like, it will avoid individuals in all the similar-looking species.
multiple alleles Three or more alleles or alternative forms of a gene occurring in a population. In most species, any given individual can possess only two of these alleles at most. The existence of multiple alleles within a population can provide a range of variation for different traits.
muscle tissue In humans, one of the four principal types of tissue. A tissue that has the ability to contract.
muscular system The organ system that produces movement and maintains posture of the body. This system consists of all the skeletal muscles of the body that are under voluntary control.
mutation A permanent alteration in a DNA base sequence. Mutations can come about as random, spontaneous events during the process of DNA replication. Mutation rates, however, can be affected by environmental influences such as radiation, chemicals, and viruses.
mutualism An interaction between two species that is beneficial to them both.
mycelium A web of fungal hyphae that makes up the major part of a fungus.
mycorrhizae (singular, *mycorrhiza*) Associations of plant roots and fungal hyphae. The fungal hyphae absorb minerals, growth hormones, and water that are then available to the plant, and the fungus gets carbohydrates from the photosynthesizing plant. The association is so important that some plants cannot live without their fungal partners.
myelin The white, fatty covering over axons of vertebrates that enables faster transmission of nerve impulses. Myelin is made of neuroglia cells called Schwann cells, which wrap their cell membranes around the axons.
myofibril The contractile structure of a skeletal muscle cell, containing alternating thick and thin filaments arranged in sarcomeres.
nanometer (nm) A billionth of a meter.
natural killer (NK) cell Type of lymphocyte (a type of white blood cell) that kills abnormal cells by secreting proteins that destroy their cell membranes.
natural selection A process in which the fit of an organism with its environment determines those traits that will be passed on with greater frequency from one generation to the next. Organisms whose traits better suit them to their environment will survive longer and leave more offspring than organisms with alternative traits. In this way, traits that better suit an organism to its environment are selected for transmission to the succeeding generation.
navigation The use of various cues to enable movement toward a desired location. The cues that animals use in navigating are varied. Some steer a course by employing sunlight; others use the Earth's magnetic field, sound, geographical features, and the location of stars in the sky. In general, animals use multiple cues to make their way.
negative feedback A system of control in which the product of a process reduces the activity that led to the product.
nephron The functional unit of the mammalian kidney, consisting of a nephron tubule, its associated blood vessels, and the interstitial fluid in which both are immersed.
nephron tubule A convoluted tubule in the kidney of birds and mammals. Filtrate passes through the tubule, where reabsorption and secretion occur during the formation of urine.
nerve A bundle of axons in the peripheral nervous system that transmits information to or from the central nervous system.
nerve cord A rod-shaped structure consisting of nerve cells, running from an animal's head to its tail.
nervous system The organ system that monitors an animal's internal and external environment, integrates the sensory information received, and coordinates the animal's responses. This system consists of all the body's neurons, plus the supporting neuroglia cells. In humans, it includes the brain, spinal cord, sense organs such as the eye and ear, and all the nerves that interconnect these organs and link the nervous system with other systems.
nervous tissue In humans, one of the four principal tissue types. A tissue specialized for the rapid conduction of electrical impulses.
net primary production The amount of material a plant (or other photosynthesizing organism) accumulates through photosynthesis. Net primary production is gross primary production minus energy lost to heat and the plant's energy expenditures on its own maintenance.
neural crest Embryonic cells, unique to vertebrates, that break away from the periphery of the neural tube as it is folding over during development. These cells migrate to different locations, where they develop into different organs and tissues.
neural tube An enclosed, hollow tube that forms from ectodermal cells in the developing animal embryo. Eventually, this structure gives rise to the brain and spinal cord.
neuroglia Associated helper cells of neurons, providing physical support, nourishment, insulation, and defense from infection.
neuron A nerve cell, typically possessing a cell body with a nucleus, a long extension called an axon that transmits information from the neuron, and many dendrites that transmit information to the neuron. Neurons are the functional units of nervous tissue, which is specialized for rapid conduction of electrical impulses.
neurotransmitter A substance that is released into a synaptic cleft by a presynaptic neuron and binds to receptors on the postsynaptic neuron or effector cell. The neurotransmitter allows an action potential to be transmitted from one cell to another.
neutron A basic constituent of an atom, possessing no electrical charge and found in the atom's nucleus. Isotopes are defined by the number of neutrons in an atom.
niche A characterization of an organism's way of making a living that includes its habitat, food, and behavior.
nicotinamide adenine dinucleotide (NAD^+) The most important intermediate electron carrier in cellular respiration, NAD^+ oxidizes food molecules, picking up an electron and a hydrogen atom from them to form NADH, after which it donates the electrons it received to other molecules in a subsequent step of respiration, thereby returning to the NAD^+ form.
nitrogen fixation The conversion of atmospheric nitrogen into a form that can be taken up by living things. Bacteria fix nitrogen, which is essential to life.
nondisjunction The failure of homologous chromosomes or sister chromatids to separate during meiosis, resulting in unequal numbers of chromosomes in the daughter cells. Nondisjunction results in aneuploidy.
nonpolar covalent bond A type of covalent bond in which electrons are shared equally between atoms.
nonrandom mating A type of mating in which a given member of a population is not equally likely to mate with any other given member.
notochord A rod-shaped support structure, composed of cells and fluid and surrounded by a lining of fibrous tissue, running from a chordate's head to its tail. At some point in their lives, all chordates have a notochord.
nuclear envelope The concentric, double membrane that forms the outer layer of a cell's nucleus. The nuclear envelope contains nuclear pores that enable substances to pass in and out of the nucleus.
nuclear pore A channel through the nuclear envelope surrounding the cell's nucleus. Materials pass in and out of the nucleus through the thousands of pores in the nuclear envelope.
nucleolus A structure within the cell's nucleus that functions in the synthesis of ribosomal RNA, one of the main constituent molecules in ribosomes.
nucleotide The building block of nucleic acids, including DNA and RNA, consisting of a phosphate group, a sugar, and a nitrogen-containing base.
nucleus of atom The central core of an atom, containing its protons and neutrons. Nearly all of the mass of the atom resides in its nucleus.
nucleus of cell The membrane-bound organelle in a eukaryotic cell that contains nearly all of its DNA.
nutrient A chemical element that is used by living things to sustain life.
oil Fat in liquid form. Polyunsaturated fatty acids naturally occur as oils at room temperature.
oligotrophic A state of having few nutrients, generally thought of in connection with lakes.
omnivore An animal that eats both plants and animals.
oocyte The precursor cell of eggs in the vertebrate ovary. Diploid primary oocytes develop from oogonia and in turn give rise to haploid secondary oocytes in the process of meiosis.
oogonium (plural, *oogonia*) The diploid cell that is the starting female cell in gamete (egg) production. Diploid oogonia develop into diploid primary oocytes, which give rise to haploid secondary oocytes. A limited number of secondary oocytes then continue through the process of maturation.
open circulation system A type of circulatory system, found in many invertebrates, in which arteries carry blood into open spaces called sinuses. The blood then bathes surrounding tissues and is channeled into veins, through which it flows back to the heart.
open sea The region of the ocean that begins where the continental shelf drops off and extends out to sea. The open sea includes the entire ocean except the coastal zone.
operant conditioning A complex form of learning that occurs when experience teaches animals to associate one of their own actions with a particular outcome. Such learning occurs when an animal's own behavior brings about a response that has either negative or positive consequences.
opportunist species A species that tends to be relatively short lived, that tends to produce relatively many offspring for whom it provides little or no care, and whose population size tends to fluctuate widely in reaction to an environment that it experiences as highly variable. Also known as an *r*-selected species.
order A taxonomic grouping subordinate to class and superordinate to family. Humans are in the order Primates.
organ A highly organized functioning unit within an organism, performing one or more functions, that is formed of several kinds of tissue. Kidneys, heart, lungs, and liver are all familiar examples of organs in humans.
organelle A highly organized structure within a cell that carries out specific cellular functions. Almost all organelles (meaning "tiny organs") are bound by membranes. The lone exception is the ribosome, which has no membrane and is the single organelle possessed by prokaryote cells (bacteria and archaea). Organelles in

eukaryotic cells include the cell nucleus, mitochondria, lysosomes, chloroplasts, and ribosomes.

organic   Pertaining to the chemistry of compounds that contain carbon.

organic chemistry   The chemistry of compounds that contain carbon. This branch of chemistry is directly relevant to biology because life on Earth is based on carbon compounds.

organism   A living being. An organism can be one-celled, like a bacterium, or can contain many cells, like a dog or a rose bush.

organismal biology   The study of whole organisms.

organ system   A group of interrelated organs and tissues that serve a particular set of functions in the body. For example, the digestive system consists of mouth, stomach, and intestines, and functions in digesting food and eliminating waste.

osmosis   The net movement of water across a semipermeable membrane from an area of lower solute concentration to an area of higher solute concentration.

osteoblast   An immature bone cell that secretes organic material that becomes bone matrix, thus producing new bone.

osteoclast   A type of bone cell that dissolves bone matrix, thus liberating the minerals stored in it.

osteocyte   Mature bone cell that maintains the structure and density of bone by continually recycling calcium compounds around itself.

osteon   The functional unit of compact bone, consisting of osteoblasts and the concentric layers of bone produced by them, forming a cylinder of bone with a hole in the center, and paralleling the long axis of the bone.

osteoporosis   A medical condition stemming from a decrease in bone mass that occurs when old bone is broken down faster than new bone is produced. Osteoporosis is especially common in postmenopausal women.

ovarian follicle   The complex of the oocyte (developing egg) and a set of accessory cells that surrounds it.

ovary   In flowering plants, the area, located at the base of the carpel, where fertilization of the egg and early development of the embryo occurs. In animals, the female reproductive organ in which eggs develop.

oviparous   A condition, seen in all birds and many reptiles, in which fertilized eggs are laid outside the mother's body and then develop there.

ovulation   The release of an oocyte from the ovary in animals.

oxidized   The state of having lost one or more electrons to another substance.

oxytocin   A hormone secreted by the posterior pituitary; stimulates smooth-muscle contractions in the uterus and special cells of the mammary glands in women, and in the walls of the prostate gland in men.

ozone   A gas in the Earth's atmosphere consisting of three oxygen atoms bonded together (O_3) that serves to protect living things from the Sun's ultraviolet radiation.

paired, jointed appendages   Appendages, such as legs, that come in pairs and have joints. Such appendages are characteristic of arthropods. Grasshoppers, for example, have paired, jointed appendages.

parasites   Organisms that feed off their prey but do not kill them, at least not immediately.

parasitism   A type of predation in which the predator gets nutrients from the prey, but does not kill the prey immediately and may never kill it. A parasite can be a plant or an animal.

parasympathetic division   The division of the autonomic nervous system that generally has relaxing effects on the body.

parenchyma cell   The most abundant type of cell in plants. These cells have thin cell walls, are usually alive at maturity, and serve numerous functions within the plant, including giving rise to the other plant-cell types.

parental generation   The generation that begins an experimental cross between organisms. Such a cross is used to study genetics and heredity of traits.

parthenogenesis   A type of reproduction in which offspring develop from an egg that has not been fertilized by sperm.

passively acquired immunity   Immunity acquired by the administration of disease-fighting substances produced by another individual. This can occur naturally, as when substances produced by the mother provide protection against infections that could attack a fetus. Such substances can also be administered, as with immunoglobulin shots.

passive transport   Transport of materials across the cell's plasma membrane that involves no expenditure of energy. Simple and facilitated diffusion are examples of passive transport.

pathogen   A disease-causing organism.

pedigree   A familial history. Pedigrees can be created, generally in the form of diagrams, that trace the history of inheritable diseases through a family.

pelagic zone   The vertical zone of the ocean, including all of the water from the surface to the floor.

perennial   A type of plant that lives for many years, such as trees, woody shrubs, and many grasses.

period   A division of geologic time that is a part of an era and in turn includes several epochs. The Cambrian and the Jurassic are periods.

peripheral nervous system (PNS)   The part of the nervous system that includes all of the neural tissue outside the central nervous system (brain and spinal cord). The PNS brings information to and carries it from the central nervous system; it also provides voluntary control of the skeletal muscles, and involuntary control of the smooth muscles, cardiac muscles, and glands.

peripheral protein   A protein of the plasma membrane that lies on the inside or outside of the membrane but is not attached to the membrane's hydrophobic interior.

peristalsis   A wave of muscle contractions that pushes materials through a tubular organ, such as the digestive tract.

permafrost   The permanently frozen ground that begins about a meter below the ground in the tundra biome of the far north. Neither roots nor water can penetrate this layer.

Permian Extinction   The greatest mass-extinction event in Earth's history. This event, in which up to 96 percent of all species on Earth were wiped out, occurred about 245 million years ago.

petal   The colorful, leaflike structure of a flower; petals attract pollinators.

petiole   The stem of a leaf, attaching it to a branch or trunk.

phagocyte   A type of white blood cell that can ingest other cells, parts of cells, or other particles. These cells remove both pathogens and parts of the body's own cells that have become cellular debris.

phagocytosis   A process of bringing relatively large materials into a cell by means of wrapping extensions of the plasma membrane around the materials to be brought in and fusing the extensions together.

pharyngeal slits   In animals, openings to the pharyngeal cavity. All chordates possess pharyngeal slits at some point in their development.

pharynx   In humans, the region at the rear of the mouth that forms the passageway between the mouth and both the digestive tract and the respiratory system.

phenotype   A physical function, bodily characteristic, or action of an organism that is the observable outcome of its genotype.

phloem   The tissue through which the food produced in photosynthesis, along with some hormones and other compounds, is conducted in vascular plants.

phloem sap   The material conducted by phloem tissue in plants, consisting mostly of sucrose and water but also including some hormones, amino acids, and other compounds.

phosphate group   A phosphorus atom surrounded by four oxygen atoms.

phospholipid   A type of lipid consisting of a polar "head," composed of glycerol and a phosphate group, and two nonpolar fatty-acid "tails." The cell's plasma membrane has a phospholipid bilayer as one of its chief components.

phospholipid bilayer   One of the chief components of the cell's plasma membrane, consisting of two layers of phospholipids, aligned with their hydrophobic "tails" pointing inward toward each other, and their hydrophilic "heads" pointing outward, toward the watery environment lying on either side of the plasma membrane.

photic zone   The vertical layer of an ocean or lake into which the Sun's rays can penetrate strongly enough to drive photosynthesis.

photoautotrophy   The nutritional mode in which an organism gets its carbon from carbon dioxide and its energy to make food from the Sun's rays. Plants, algae, and certain bacteria are photoautotrophs.

photoheterotrophy   The nutritional mode in which an organism gets its carbon from organic material and its energy to make food from the Sun's rays. A few bacteria and archaea are photoheterotrophs.

photoperiodism   The ability of a plant to respond to changes it experiences in the daily duration of darkness, relative to light.

photorespiration   The process in which the enzyme rubisco binds oxygen instead of carbon dioxide, thus reducing the amount of carbohydrate produced in photosynthesis.

photosynthesis   The process by which certain groups of organisms capture energy from sunlight and convert it into chemical energy, with this energy initially being stored in a carbohydrate.

photosystem   A collection of molecules that function as a unit in the light-dependent reactions of photosynthesis. Photosystems II and I receive electrons derived originally from water and facilitate the process by which these electrons are transferred and boosted to higher energy states through use of the Sun's energy.

phototropism   The bending of a plant's shoots in response to light. Generally, this capability helps a plant grow toward the Sun to get the most available sunlight.

pH scale   The scale, ranging from 0 to 14, that quantifies the concentration of hydrogen ions in solution. The lower the pH number, the more acidic the solution; the higher the pH, the more basic the solution. The scale is logarithmic, so a pH difference of 1 means a tenfold difference in acidity.

phylogeny   A hypothesis about the evolutionary relationships of a group of organisms.

phylum   A category of living things, directly subordinate to the category of kingdom, whose members share traits as a result of shared ancestry.

physical science   The natural sciences not concerned with life.

physiology   The study of the physical functioning of animals and plants.

phytoplankton   Microscopic photosynthesizing organisms that drift in the upper layers of oceans or bodies of freshwater, often forming the base of aquatic food webs.

pinocytosis   A form of endocytosis that brings into the cell a small volume of extracellular fluid and the solutes suspended in it.

pith   The area of ground tissue in the stem of dicotyledonous plants that is inside the ring of vascular bundles. The pith cells are specialized for storage.

placenta   A complex network of maternal and embryonic blood vessels and membranes that develops in mammals in pregnancy. The placenta allows nutrients and oxygen to flow to the embryo from the mother, while allowing carbon dioxide and waste to flow from the embryo to the mother.

placental mammal   A type of mammal that is nurtured before birth by the placenta, a network of maternal and embryonic blood vessels and membranes. Embryonic placental mammals derive their nutrition not from food stored in an egg, but directly from the mother's circulation.

plankton   Microscopic organisms, including both phytoplankton and zooplankton, that live in the upper layers of bodies of water and often form the base of aquatic food webs.

plasma   The fluid portion of blood, consisting mostly of water but also containing proteins and other molecules. Contrast with formed elements—the cells and cell fragments that form the nonfluid portion of blood.

plasma cell   A type of B cell produced in antibody-mediated immunity that exists in the lymph nodes and produces antibodies specific to a given infection.

plasma membrane   A membrane forming the outer boundary of many cells, composed of a phospholipid bilayer interspersed with proteins and cholesterol molecules and coated, on its exterior face, with short carbohydrate chains associated with proteins and lipids.

plasmid   A ring of DNA that lies outside the chromosome in bacteria. Plasmids can move into bacterial cells in the process called transformation, thus making them a valuable tool in biotechnology.

plasmodesmata (singular, *plasmodesma*) Tiny channels in a plant's cell wall that enable communication between cells by making the cytoplasm between cells continuous.

platelet A small fragment that has pinched off from a larger, complete blood cell. Platelets contain enzymes and clotting factors.

pleiotropy The phenomenon by which one gene has many effects.

pluripotent cell An animal cell that is capable of giving rise to almost all of the different kinds of cells or tissues in the body.

point mutation Mutation at a single location in the genome, rather than a whole chromosome aberration.

polar body A small haploid cell produced during meiosis in a female. During meiosis, the cells divide unevenly to produce one richly endowed egg cell and two or three small polar bodies. The polar bodies will eventually disintegrate into their constituent parts.

polar covalent bond A type of covalent bond in which electrons are shared unequally between atoms, so that one end of the molecule has a slight negative charge and the other end a slight positive charge.

polarity A difference in electrical charge at one end of a molecule, as compared to the other.

pollen tube A tube-like structure that sprouts from a pollen grain that has landed on the stigma of a plant. The pollen tube grows down toward the egg, and sperm cells then move down through the pollen tube.

pollination The transfer of pollen—by wind, animal, or other means—to a plant's female reproductive structure.

polygenic Having multiple genes affecting a given character, such as height in humans.

polygenic inheritance Inheritance of a genetic character that is determined by the interaction of multiple genes, with each gene having a small additive effect on the character.

polymer Large compound made up of many repeating molecular units called monomers.

polymerase chain reaction (PCR) A technique for generating many copies of a DNA sequence.

polypeptide A chain of 10 or more amino acids. When the chain is folded into its three-dimensional form, it is a protein.

polyploidy A process by which one or more sets of chromosomes are added to the genome of an organism. Human beings cannot survive in a polyploid state, but many plants flourish in it. Polyploidy is a means by which speciation can occur (most often in plants) in a single generation.

polysaccharide A carbohydrate built of many monosaccharides. Also known as complex carbohydrates, these compounds include starch, glycogen, cellulose, and chitin.

polyunsaturated fatty acid A type of fatty acid that has two or more double bonds between the carbon atoms of its hydrocarbon chain.

pons A part of the vertebrate brain stem that contains relay centers connecting several other brain structures. The pons also contains cells involved in involuntary control of breathing.

population All the members of a species living in a single geographic region.

positive feedback A system of control in which the product of a process stimulates the activity that produced the product. Uncommon in living things because it promotes instability.

post-anal tail A tail, existing at some point in the development of all chordates, that is located posterior to the anus.

posterior pituitary An endocrine gland that receives its hormones directly from the hypothalamus, then stores these hormones and later releases them.

postsynaptic neuron The neuron that receives the action potential's signal from neurotransmitters released by the presynaptic neuron at a synapse.

potential energy Stored energy. A rock perched at the top of a hill has potential energy.

predation The act of one free-standing organism feeding on parts or all of another organism. Predation includes animals eating other animals, and animals eating plants.

pressure-flow model The hypothesis that the force behind the movement of phloem sap comes from water that moves into the phloem by means of osmosis, following the energetic pumping of solutes into the phloem.

presynaptic neuron The neuron that transmits an action potential to a synapse, via neurotransmitters it releases into the synaptic cleft. These neurotransmitters then prompt propagation of the action potential in the postsynaptic neuron.

primary consumer Any organism that eats producers (organisms that make their own food).

primary growth The type of growth in plants that occurs at the tips of their roots and shoots and mainly increases their length.

primary oocyte A diploid cell produced in the female embryo that may mature into an egg, initially by giving rise to haploid secondary oocytes. After the female reaches puberty, an average of one oocyte per month is selected to continue the process of maturation in the ovary.

primary spermatocyte A diploid cell in a male that will undergo meiosis to produce haploid secondary spermatocytes, which ultimately give rise to mature sperm cells.

primary structure The sequence of amino acids in a protein. This sequence dictates the final shape of the protein because electrochemical bonding and repulsion forces act on the structure to create the folded-up protein.

primary succession In ecology, succession in which the starting state is one of little or no life and a soil that lacks nutrients.

primary visual cortex An area at the back of the brain that is the primary site of visual processing in human beings.

producer Any organism that manufactures its own food. Plants, algae, and certain bacteria are producers. By converting the Sun's energy into biomass, producers capture energy that is then passed along in food webs. Producers occupy the first trophic level.

product A substance formed in a chemical reaction. The products are written on the right side of a chemical equation.

profundal zone The vertical zone in a lake that occurs beneath the limnetic zone and in which not enough light penetrates for photosynthesis to occur.

progestin A class of hormones, produced by the ovaries, that prepares the body for possible pregnancy.

prokaryote An organism whose cells lack nuclei. Bacteria and archaea are prokaryotes.

prokaryotic cell A cell whose DNA is not located in the membrane-bound organelle known as a nucleus. Prokaryotes are microscopic forms of life, and all are either bacteria or archaea.

prolactin A hormone, secreted by the anterior pituitary, that stimulates the development of the mammary glands and their production of milk in female mammals.

promoter sequence A sequence of DNA at which transcription of a gene begins. More specifically, the binding site for the RNA polymerase enzyme that brings about genetic transcription by pairing messenger RNA nucleotides with their DNA counterparts.

prostate gland In human males, a gland surrounding the urethra near the urinary bladder that contributes fluids to the semen.

protein A class of biological molecule composed of one or more polypeptide chains folded into a specific three-dimensional form. Polypeptide chains are, in turn, composed of sequences of individual amino acids.

protein conformation The three-dimensional shape of a protein, determined by its order of amino acids and their resulting interactions. This shape is critical in allowing a protein to interact properly with other molecules to perform its function.

proteomics The study of sets of proteins, with a focus on both their functions and their interactions.

proton A basic constituent of an atom, found in the nucleus of the atom and having positive electrical charge. Elements are defined by the number of protons in their nucleus.

proximal tubule In the nephron tubule of the kidney, the convoluted portion that is closest to Bowman's capsule. Some reabsorption of salt and water occurs here.

pseudopodium A "false foot," or cell extension formed when a portion of the cell is thrust forward, anchored, and then cytoplasm streams into the extension. Amoebas move by using pseudopodia.

pulmonary circulation The system that circulates blood between the heart and the lungs. This system brings oxygen into and takes carbon dioxide away from the body.

pulmonary ventilation The physical movement of air into and out of the lungs through the respiratory passages; also known as breathing.

pupil In the eye, a central opening in the iris, which grows smaller or larger in response to contractions or relaxations of the iris muscles.

pyloric sphincter The circular muscle that regulates the flow of chyme (food mixed with gastric juices) between the stomach and small intestine.

quaternary structure The way in which two or more polypeptide chains come together to form a protein.

radial symmetry A type of animal symmetry in which body parts are distributed evenly about a central axis. Sea stars are radially symmetrical.

radiometric dating A technique for determining the age of objects by measuring the decay of the radioactive elements within them. The age of fossils can be determined with this technique.

reabsorption In the kidney, the process of drawing water and nutrients back from the fluid in the nephron tubules into the blood supply, thus conserving these valuable materials.

reactant A substance (atoms, molecules, or compounds) that goes into a chemical reaction. The reactants interact to form the product(s). The reactants are written on the left side of a chemical equation.

reaction center A molecular complex in a chloroplast that, in photosynthesis, transforms solar energy into chemical energy.

reaction wood Extra wood formed in a plant in reaction to pressure from one side, thus enabling the plant to resist being pushed over.

receptor-mediated endocytosis (RME) A form of endocytosis in which receptors on the surface of a cell bind to a substance and then move laterally through the plasma membrane to join other receptors of their type that have bound to the same substance. These bound receptors congregate in a cell pit that deepens and pinches off, creating a vesicle that moves into the cell.

receptor protein A protein with a binding site, located at the surface of the plasma membrane of a cell. The binding of a receptor protein with a specific molecule causes a change in the cell's activity.

recessive Not expressed in the heterozygous condition. When an organism possesses two different alleles of the same gene, but one allele of the two is not expressed, that allele is said to be recessive.

recessive condition A condition that occurs when only recessive, nonfunctional alleles are present. Red-green color blindness is an example of a recessive condition, because only one set of functional alleles need be present for normal color vision.

reciprocal altruism An exchange of altruistic acts by individual animals over time.

recognition sequence A series of DNA bases recognized by a particular restriction enzyme as the location at which to cut the DNA strand.

recombinant DNA Two or more segments of DNA that have been combined by humans into a sequence that does not exist in nature.

recombination The exchange of reciprocal pieces of chromosomes that occurs when homologous chromosome pairs intertwine during prophase I of meiosis. Also called crossing over.

rectum The terminal portion of the large intestine, and therefore the end of the digestive tract. In defecation, feces pass from the colon through the rectum and out the anus.

red blood cell (RBC) The blood cells, also known as erythrocytes, that transport oxygen to and carry carbon dioxide from every part of the body.

redox reaction A combination of a reduction and an oxidation reaction in which the electrons lost from one substance in oxidation are gained by another in reduction.

reduction The state of having gained one or more electrons in a chemical reaction. Such a substance has reduced its positive charge by gaining negative charge from electrons.

reflex A rapid, automatic nervous system response, triggered by a specific stimulus, that helps the organism avoid danger or preserve a stable physical state. The knee-jerk response is a well-known reflex.

regeneration   A process by which an animal can reproduce itself from a part of itself. An arm attached to part of the central disk from a sea star can regenerate into a whole sea star. The term is also used in referring to the repair of damaged tissue.

releaser   The critical element in an action or object that triggers an action pattern in animals. A releaser can be visual, such as a color; but odors, touch sensations, tastes, and sounds can also trigger behavior.

renal pelvis   The chamber inside the kidney of birds and mammals that receives urine from the collecting ducts and channels it into the ureter.

replication   The process by which DNA is duplicated.

reproductive cloning   The cloning of whole, complex living things.

reproductive isolating mechanism   Any factor that, in nature, prevents interbreeding between individuals of the same or closely related species. These factors keep species separate.

reproductive system   The organ system that develops gametes and delivers them to a location where they can fuse with other gametes to produce a new individual.

resource partitioning   The dividing of scarce resources among species that have similar requirements. Such partitioning allows species to coexist in the same habitat.

respiratory system   The organ system that brings oxygen into the body and expels carbon dioxide from the body. In humans, this system includes the lungs and passageways that carry air to the lungs.

restriction enzyme   A type of enzyme, occurring naturally in bacteria, that recognizes a specific series of DNA bases and cuts the DNA strand at that site. Restriction enzymes are used in biotechnology to cut DNA in specific places.

retina   An inner layer of tissue in the eye, containing cells that convert light signals into neural signals.

reversible reaction   A chemical reaction that can go in both directions.

ribonucleic acid (RNA)   A class of nucleic acids that takes several forms in living things, among them messenger RNA (mRNA), transfer RNA (tRNA) and ribosomal RNA (rRNA). RNA also is the genetic material of many viruses.

ribosomal RNA (rRNA)   A type of RNA that, along with proteins, forms ribosomes.

ribosome   An organelle, located in the cell's cytoplasm, that is the site of protein synthesis. Ribosomes can be embedded in the rough endoplasmic reticulum (rough ER), but may exist as "free" ribosomes, unattached to the rough ER. The translation phase of protein synthesis takes place within ribosomes.

ribozyme   Molecule composed of RNA that can both encode information and act as an enzyme. Molecules similar to these would have been necessary early in the evolution of life.

RNA polymerase   The enzyme that unwinds the DNA double helix and puts together a chain of RNA nucleotides complementary to the exposed DNA nucleotides.

rods   Photoreceptors that function in low-light situations, but that do not provide color vision.

root cap   A collection of cells at the very tip of a root that protects the adjacent meristematic cells; the root cap secretes a lubricant to help the root move through its growing medium.

root hair   In plants, a threadlike extension of a root cell. Root hairs greatly increase the surface area of roots, thus allowing greater absorption of water and nutrients.

rough endoplasmic reticulum (rough ER)   A cell organelle that is a folded continuation of the cell membrane, studded with ribosomes and functioning in protein synthesis and processing.

r-selected species   A species that tends to be relatively short lived, that tends to produce relatively many offspring for whom it provides little or no care, and whose population size tends to fluctuate widely in reaction to an environment that it experiences as highly variable. Also known as an opportunist species.

rubisco   The large enzyme that brings CO_2 and the sugar RuBP together in the first step of the Calvin cycle of photosynthesis.

rule of addition   In probability theory, the principle that, when an outcome can occur in two or more

different ways, the probability of that outcome is the sum of the respective probabilities.

rule of multiplication   In probability theory, the principle that the probability of any two events happening is the product of their respective probabilities.

sapwood   The lighter-colored, outer wood in a tree trunk where xylem tubes conduct xylem sap.

sarcomere   The functional unit of a striated muscle, which contracts when thin filaments slide past thick filaments. The sarcomeres shorten, thus contracting the whole muscle.

saturated fatty acid   A type of fatty acid that has no double bonds between carbon atoms of its hydrocarbon chain. At room temperature, these fats are solid. Saturated fatty acids have been linked with heart disease and are found primarily in animal products and tropical oils.

science   A process of learning about nature by observation and experiment, as well as a collection of knowledge and insights about nature.

scientific method   The testing of hypotheses through observation and experiment. The scientific method advances our understanding of science.

sclerenchyma cell   A type of plant cell that is dead at maturity, has thick cell walls containing cellulose, and helps the plant return to its original shape after it has been deformed. These cells provide strong support for the plant.

sebaceous gland   A type of gland in the skin that produces a waxy, oily secretion (sebum) that lubricates the hair shaft and inhibits bacterial growth in the surrounding area.

secondary consumer   Any organism that eats a primary consumer.

secondary growth   The type of lateral growth in plants that thickens the plant. Secondary growth occurs in woody plants such as trees, and thickens their trunks and branches.

secondary phloem   Phloem tissue produced to the outside of the vascular cambium in woody plants as part of secondary growth, or thickening.

secondary spermatocyte   A haploid cell formed when a primary spermatocyte undergoes meiosis I. After meiosis II, this cell becomes a spermatid and then a mature sperm cell.

secondary structure   The small-scale twisting or bending of a protein. The alpha helix and beta pleated sheet are two examples of secondary structure.

secondary succession   In ecology, succession in which the final state of a habitat has been disturbed by some force, but life remains and the soil has nutrients. A farmer's field that has been abandoned is a site of secondary succession.

secondary xylem   Xylem tissue produced to the inside of the vascular cambium in woody plants. This tissue, also called wood, is responsible for much of the thickening of woody plants.

second law of thermodynamics   Energy transfer always results in a greater amount of disorder (entropy) in the universe.

secretion   In the kidney, the process of moving hydrogen and potassium ions, as well as other compounds, from the capillaries into the distal tubules of the nephron.

secretory protein   A protein that is exported out of a cell. These proteins are generally synthesized on the ribosomes of the rough ER.

seed   A reproductive structure in plants that includes a plant embryo, its food supply, and a tough protective casing.

seed coat   The tough outer layer of a seed, derived from the integuments of the ovule, which protect the embryo from mechanical damage and water loss.

seedless vascular plant   A plant that has a vascular (fluid transport) structure but does not reproduce through use of seeds. One of the four principal varieties of plants. Ferns are the most familiar example.

semen   The mixture of sperm and glandular materials that is ejaculated from the human male through the urethra.

seminal vesicle   A gland that, in human males, contributes fluids to the semen as the semen moves down the urethra in the process of ejaculation.

seminiferous tubule   A convoluted tubule inside the testes where sperm development begins. Immature

sperm are eventually released into the interior cavity and travel to the epididymis, where sperm maturation is completed.

sensitive period   The time interval during which an animal can learn to respond to a given stimulus.

sensory neuron   A neuron that senses conditions both inside and outside the body and conveys that information to neurons in the central nervous system.

sepal   The leaflike structure that, with other sepals, protects the flower before it opens.

sessile   Fixed in location; organisms such as mushrooms and sponges are sessile.

sex chromosome   The chromosomes that determine the sex of an organism. The X or Y chromosomes in humans.

sexual selection   Differential reproductive success based on differential success in obtaining mating partners.

shoot apical meristem   The growing tip of plant shoots that produces the cells responsible for vertical growth of the plant.

sieve element   In plants, a type of cell in phloem tissue that conducts nutrients. These cells are stacked on each other and have ends studded with small holes, forming tubes for nutrient flow. The cells are alive at maturity but lack nuclei, making room for rapid conduction of nutrients.

sieve tube   A tube, formed by a stack of sieve elements, that conducts nutrients in phloem.

simple diffusion   Diffusion through the plasma membrane that requires only concentration gradients, as opposed to concentration gradients and special protein channels. Water, oxygen, carbon dioxide, and steroid hormones can all cross the plasma membrane through simple diffusion.

simple sugar   The building block of carbohydrates. Also known as a monosaccharide; glucose and fructose are familiar examples.

single bond   A type of covalent bond in which a single pair of electrons is shared between two atoms.

sink   In plants, an unloading site for phloem sap in the phloem transport system.

skeletal muscle tissue   The muscle tissue responsible for movement of the body. This tissue is made of long cells, each typically containing hundreds of nuclei.

skeletal system   The animal organ system that forms an internal supporting framework for the body and protects delicate tissues and organs. This system consists of all the bones and cartilages in the body, and the connective tissues and ligaments that connect the bones at the joints.

skin   An organ consisting of two layers, epidermis and dermis, and covering the outside of an animal. The skin protects the body and receives signals from the environment.

smooth endoplasmic reticulum   The part of the endoplasmic reticulum membrane, farther out from the cell nucleus, that has no ribosomes. Depending on the cell type, different lipids are synthesized there and potentially harmful substances may be detoxified.

smooth muscle tissue   Made of nonstriated muscle tissue, this tissue is responsible for contraction of blood vessels, air passages, and hollow organs such as the uterus, and is not under voluntary control.

solute   The ingredient being dissolved in a solvent to form a solution. For example, sugar is the solute in the sugar-water nectar you put in your hummingbird feeder.

solution   A homogeneous mixture of two or more substances in the same phase (gas, liquid, or solid). Frequently, solutions consist of a solute dissolved in water, and these are called aqueous solutions.

solvent   The substance in which a solute is dissolved to form a solution. In an aqueous solution, the solvent is water.

somatic cell   Any cell that is not and will not become an egg or sperm cell.

somatic nervous system   The subdivision of the efferent division of the peripheral nervous system that provides voluntary control over the skeletal muscles.

somite   One of the blocks of mesodermal tissue in the vertebrate embryo that lies on both sides of the notochord and gives rise to muscles and the vertebrae that enclose the spinal cord.

source   In plants, the loading site for phloem sap in the phloem transport system.

space-filling model A picture of a molecule showing the three-dimensional structure, particularly showing how big atoms are in relation to each other and the overall shape of the molecule.
speciation The development of new species through evolution.
species A group of actually or potentially interbreeding natural populations which are reproductively isolated from other such populations.
specific heat The amount of energy needed to raise the temperature of 1 gram of a substance by 1 degree Celsius. Water serves as an insulator because it has a high specific heat compared to other substances.
spermatid An immature sperm cell. This haploid cell will become a mature sperm cell after developing a tail-like flagellum and ridding itself of nearly all cellular organelles.
spermatocyte An immature sperm cell. Spermatocytes develop from spermatogonia; they develop into spermatids and eventually into mature sperm.
spermatogonia (singular, *spermatagonium*) Di-ploid cells that are the starting cell in sperm production in males. Some spermatogonia will produce more spermatogonia, while others will continue to develop into sperm.
sperm cell In flowering plants, either of two cells in a pollen grain, one of which fertilizes an egg, the other of which fertilizes the central cell in an embryo sac. In animals, the male gamete, which fertilizes the female gamete (the egg).
spinal nerve Paired nerves that connect the spinal cord to different parts of the body, including the head, neck, diaphragm, arms, chest, abdominal muscles, legs, bladder, bowels, sexual organs, and feet. In humans, there are 31 pairs of spinal nerves.
spirochetes Spiral-shaped bacteria.
spongy bone Type of bone that is porous and less dense than compact bone. Spongy bone fills the expanded ends of long bones.
spore A single, asexual reproductive cell that, without fusing with anything, has the ability to grow into an adult plant or fungus.
sporophyte generation The spore-producing plant generation. This generation is the dominant, visible generation in flowering plants. Contrast with **gametophyte generation**.
S-shaped growth In ecology, population growth that begins exponentially, but eventually slows and then stops, with the population stabilizing at a certain level. When plotted on a graph, this type of growth looks like the letter S. Also known as logistic growth.
stabilizing selection The type of selection in which intermediate forms of a given character are favored over either extreme. This process tends to maintain the average for the character. Compare to **directional** and **disruptive selection.**
stamen The male reproductive structure in a flower; consists of a filament (stalk) with an anther, bearing pollen, at the top.
starch A complex carbohydrate that serves as the major form of carbohydrate storage in plants. Starches—found in such forms as potatoes, rice, carrots, and corn—are important sources of food for animals.
stasis A situation in which there is no change. In geographical time, stasis refers to long periods in the fossil record when there seems to have been little or no change in a species.
stem cell Any cell with the capacity for prolonged self-renewal that can produce at least one type of differentiated daughter cell.
stereotyped Performed precisely the same way each time by individual animals of the same species.
steroid A type of lipid that has a linked set of four carbon rings, to which various side chains can be attached. Cholesterol, testosterone, and estrogen are steroids.
steroid hormone A type of hormone built from cholesterol molecules. These hormones have important roles in reproductive and behavioral functions in the body. Testosterone and estrogen are examples.
stigma The tip end of the carpel of a flower, where pollen grains are deposited before fertilization occurs.
stimulus Any signal that influences the activity of an organism or a part of the organism.
stomata (singular, *stoma*) Microscopic pores in the epidermal cells of plants, found mostly on the underside of leaves. Carbon dioxide enters and water vapor leaves through these openings. Stomata can be closed to conserve a plant's water.
stratosphere The layer of the Earth's atmosphere situated above the troposphere, at about 20 to 35 kilometers (13 to 21 miles) above sea level. The ozone layer lies in this level.
stroma The liquid material of chloroplasts.
structural formula A two-dimensional representation of a molecule that shows how the atoms are arranged and the types of bonds between them.
style The slender tube structure in the carpel of a flower, connecting the stigma and the ovary.
substrate The substance that is worked on by an enzyme.
succession In ecology, a series of replacements of community members at a given location until a stable final state is reached.
suppressor T cell The type of T cell that inhibits the production of cytotoxic (killer) T cells once the immediate infection has passed.
surface tension The cohesive strength of water at its interface with air, provided by the number and direction of hydrogen bonds.
sweat gland In the skin of animals, a type of duct-containing (exocrine) gland that produces perspiration.
swim bladder An inflatable organ in a fish that the fish can fill with gas to maintain neutral buoyancy—to float in place without expending energy.
symmetry An equivalence of size, shape, and relative position of parts across a dividing line or around a central point.
sympathetic division The division of the autonomic nervous system that generally has stimulatory effects on the body.
sympatric speciation A type of speciation that occurs without geographic separation of populations. Polyploidy is one form of sympatric speciation.
synapse A site where a neuron communicates with another cell. A synapse consists of the presynaptic neuron, a space (synaptic cleft), and a postsynaptic cell, which can be a neuron or effector cells.
synaptic cleft The gap between a presynaptic neuron and the postsynaptic cell. Neurotransmitter is released into this cleft by the presynaptic neuron.
synaptic terminal The structure at the end of an axon or branch of the axon that helps form a synapse with another neuron or effector cells. Neurotransmitter is released from the synaptic terminal and binds to receptors on the postsynaptic cell, bringing about communication between the cells.
synthetic work The work done in building up complex molecules from their simpler components.
systematics The field of biology dealing with the diversity and relatedness of organisms. Systematists study the evolutionary history of groups of organisms.
systemic circulation The system that circulates blood between the heart and the body, exclusive of the lungs. This system allows exchange of gases, nutrients, and wastes with the tissues.
taiga The biome consisting of boreal (northern) forest, characterized by cold, dry conditions, a relatively short growing season, and large expanses of coniferous trees.
taproot system The type of plant root system that consists of a large central root and many smaller lateral roots.
target cell A cell, bearing specific receptor molecules, that will be affected by a specific hormone.
taxis In animals, a genetically predisposed movement toward or away from a stimulus. A positive taxis is a movement toward the stimulus; a negative taxis is a movement away from the stimulus.
taxon A group of organisms in a category of the taxonomic system. In order of increasing inclusiveness, these categories are species, genus, family, order, class, phylum, kingdom, and domain.
taxonomic system A set of groupings used to classify every species of living thing on Earth.
tendon The connective tissue that attaches a skeletal muscle to a bone.
tendril A thin, modified leaflet that wraps around another object to help a plant climb upward toward light.
terminal bud Another term for the shoot apex, or the main apical meristem of a plant; used especially when it is dormant.
territoriality A phenomenon in which animals try to keep other animals out of a given area. In general, territoriality refers to efforts to keep members of an animal's own species from entering an area.
tertiary consumer Any organism that eats secondary consumers.
tertiary structure The large-scale twists and turns in a protein conformation.
testis The organ in the male reproductive system in which sperm begin development and testosterone is produced.
tetrad The grouping formed by two homologous chromosomes pairing up in recombination in prophase I of meiosis. The four chromatids in this grouping give it the name tetrad.
tetrapod A four-limbed vertebrate.
thalamus A part of the vertebrate brain stem that contains relay and processing centers for incoming sensory information.
theory A general set of principles, supported by evidence, that explains some aspect of nature.
theory of punctuated equilibrium The idea that evolution takes place in rapid bursts after long periods of stasis.
thermodynamics The study of energy.
thigmotropism The growth of a plant in response to touch. This capability allows tendrils wrap around other objects, thus helping a plant climb upward toward light.
thylakoid A flattened, membrane-bound sac in the interior of a chloroplast. The light-dependent reactions of photosynthesis occur in the thylakoids.
thylakoid compartment In plants, the interior watery space of thylakoids.
thyroid-stimulating hormone (TSH) A hormone, secreted by the anterior pituitary, that acts on the thyroid gland to trigger the release of thyroid hormones.
tissue An organized assemblage of similar cells that serves a common function. Nervous, epithelial, and muscle tissue are some familiar examples.
T-lymphocyte (T cell) A class of lymphocytes that plays a central role in cell-mediated immunity. T cells come in several varieties that play specific roles in recognizing and killing infected cells in the body.
top predator A species that preys on other species but is not preyed upon by any species.
totipotent cell A cell capable of giving rise to every kind of cell in an organism, and hence to a whole new organism.
tracheid A slender, tapered cell of the xylem that has perforations (bordered pits) that match up with adjacent cells to allow water to move between them. Tracheids are dead and empty of cellular contents at maturity, allowing conduction of xylem sap.
trait A variation in a character, such as green or yellow seed color in Mendel's pea plants.
transcript A length of messenger RNA (mRNA) produced by complementary base pairing with a strand of DNA. The transcript encodes the amino-acid sequence of one or more proteins.
transcription The process in which DNA's information is copied onto mRNA. This process, the first stage in protein synthesis, occurs in the cell nucleus.
transfer RNA (tRNA) A form of RNA that, in protein synthesis, bonds with amino acids, transfers them to ribosomes, and then bonds with messenger RNA.
transformation A cell's incorporation of genetic material from outside its boundary. Some bacteria readily undergo this process, and others can be induced to for uses in biotechnology.
translation The process in which a polypeptide chain is produced within a ribosome, based on the information encoded in messenger RNA. This process, the second major stage in protein synthesis (after transcription), occurs in the cytoplasm.
translocation The swapping of fragments by non-homologous chromosomes, resulting in gene sequences that are out of order on both chromosomes. Translocation can have phenotypic effects.
transpiration The process by which plants lose water when water vapor leaves the plant through open stomata. More than 90 percent of the water that enters a plant evaporates into the atmosphere via transpiration.
transport protein In a cell membrane, transport proteins form hydrophilic channels through the hydrophobic interior of the plasma membrane, allowing hydrophilic materials to pass down their concentration

gradient through the membrane. In blood, transport proteins carry compounds that would otherwise be filtered out or would not dissolve in the blood.

transport vesicle A sphere of membrane, carrying proteins or other molecules, that moves through the cytosol of a cell. Transport vesicles fuse with and bud off from other membranes.

transport work The work done in transporting molecules against their concentration gradient. The sodium-potassium pump does transport work.

triglyceride Form of glyceride in which three fatty acids link up with a single molecule of glycerol. Most substances we call fats are triglycerides.

trimester A period lasting about 3 months during a human pregnancy. There are three trimesters during the 9-month pregnancy.

triple bond A covalent bond in which three pairs of electrons are shared between atoms.

triplet code The phenomenon in genetics by which each three DNA base pairs specify three mRNA bases, but each mRNA triplet specifies a single amino acid or a start or stop command.

triploblastic In animals, the state of developing from three germ layers of cells.

trophic level A position in an ecosystem's food chain or web, with each level defined by a transfer of energy from one kind of organism to another. Plants and other photosynthesizers are producers of food and thus occupy the first trophic level. Organisms that consume producers are primary consumers, and occupy the second trophic level, and so on.

trophoblast The cells at the periphery of the developing mammalian embryo that establish physical links with the mother's uterine wall and eventually develop into the fetal portion of the placenta.

tropical savanna A type of tropical grassland that is characterized by seasonal drought, small seasonal changes in the generally warm temperatures, and stands of trees that punctuate the grassland. Tropical savanna exists in Africa, South America, and Australia.

troposphere The lowest layer of the Earth's atmosphere, extending from sea level to about 12 kilometers (7.4 miles) above sea level. This layer contains most of the gases in the atmosphere.

tube cell A cell in a pollen grain that, following arrival of the pollen grain on a stigma, germinates and forms a pollen tube that conducts sperm cells to the embryo sac.

umbilical cord The tissue linking the fetus with the placenta in mammals and containing the fetal blood vessels that carry blood between the fetus and the placenta.

urea An organic waste product produced in the liver of vertebrates and excreted in the urine.

ureter One of a pair of tubes that carries urine from the kidneys to the urinary bladder in reptiles, birds, and mammals.

urethra The tube that carries urine from the urinary bladder to the outside of the body of mammals. The urethra in males also carries semen in the process of ejaculation.

urinary bladder The muscular organ in vertebrates that stores urine prior to excretion.

urinary system The organ system that eliminates waste products from the blood through formation of urine. In mammals, this system consists of the kidneys where the urine is formed; and the ureters, urinary bladder, and urethra, which transport the urine outside the body.

uterine tube The tube that transports an ovulated egg from the ovary to the uterus in mammals. Fertilization occurs inside this tube. Also called the fallopian tube.

variable An element of an experiment that is changed compared to an initial condition.

vascular bundle Collections of xylem and phloem tubes that run in parallel in the stems of plants, forming a transport system.

vascular cambium The thin layer of tissue in woody plants between the primary xylem and phloem cylinders that generates the secondary xylem and phloem tissues. This meristematic tissue gives rise to secondary growth in woody plants.

vas deferens In human males, the tube that carries the sperm from the epididymis on top of the testis to the urethra for ejaculation.

vasoconstriction The decrease in diameter of a blood vessel caused by contraction of the smooth muscles in the wall of the vessel. Together with vasodilation, this process helps regulate blood pressure.

vasodilation The increase in diameter of a blood vessel caused by relaxation of the smooth muscles in the wall of the vessel. Together with vasoconstriction, this process helps regulate blood pressure.

vegetal pole The end of the zygote with relatively more yolk and lying farther from the cell's nucleus. The location of the egg's poles defines the orientation in which the embryo develops.

vegetative reproduction In plants, a form of asexual reproduction in which a portion of one plant gives rise to a whole new plant, the second plant being a genetic replica of the original plant.

vein A blood vessel that carries blood to the heart.

venule A small-diameter vein that receives blood from the capillaries and sends it on to larger-diameter veins on its way to the heart.

vertebral column A flexible column of bones extending from the anterior to posterior end of an animal. Also known as a backbone, the vertebral column distinguishes vertebrates from other chordates.

vertebrate An animal that has a flexible series of linked bones called a vertebral column (or backbone).

vessel element One of two types of cells that make up xylem tissue, which transports water and dissolved minerals through plants. At maturity, vessel elements are dead and empty of cellular contents, thus facilitating the rapid conduction of water.

villi (singular, *villus*) In vertebrates, the many fingerlike projections of the lining of the small intestine that greatly increase the surface area for absorption.

viroid A virus-like entity consisting of only a small strand of infectious RNA, lacking even a protein coat.

vitamin An organic compound required in a very small quantity for critical chemical reactions in the body.

viviparous A condition in animals in which fertilized eggs develop inside a mother's body.

wetlands Lands that are wet for at least part of the year. Wetlands are sites of great biological productivity, and they provide vital habitat for migrating birds.

white blood cell (WBC) A class of blood cells that are critical to immune function. Types of white blood cells include T cells, B cells, neutrophils, eosinophils, and macrophages, each playing a specific role in immune responses.

white matter The tissue within the central nervous system that has a white appearance from the myelin sheaths surrounding the axons that run through it.

wood Secondary xylem in a plant. This material in woody plants is responsible for most of their thickening.

woody plant A plant that has wood, or secondary xylem.

xenotransplantation The transplanting of organs from one species to another.

xylem The tissue through which water and dissolved minerals flow in vascular plants.

xylem sap The water and dissolved minerals that are conducted through xylem in plants.

zero population growth State of a population in which births exactly equal deaths in a given period.

zone of cell division In plants, the area just behind the apical meristem where more cells are produced. Cells then move on to the zones of elongation and differentiation.

zone of differentiation In plants, the area just behind the zone of elongation, near the apical meristem, where primary tissue types finish taking shape.

zone of elongation In plants, the area between the zones of cell division and differentiation, near the apical meristem, where most primary plant growth occurs through the elongation of cells.

zoologist A biologist who studies animals.

zooplankton Tiny animals and other heterotrophs that drift or move weakly in the upper layers of bodies of water.

zygote A fertilized egg. In humans, the developing organism from the time of fertilization through the time of the first division of the zygote (about 30 hours after fertilization).

Photo Credits

Index